Edible Medicinal and Non-Medicinal Plants

T.K. Lim

Edible Medicinal and Non-Medicinal Plants

Volume 1, Fruits

ISBN 978-90-481-8660-0   e-ISBN 978-90-481-8661-7
DOI 10.1007/978-90-481-8661-7
Springer Dordrecht Heidelberg London New York

Library of Congress Control Number: 2011932982

© Springer Science+Business Media B.V. 2012
No part of this work may be reproduced, stored in a retrieval system, or transmitted in any form or by any means, electronic, mechanical, photocopying, microfilming, recording or otherwise, without written permission from the Publisher, with the exception of any material supplied specifically for the purpose of being entered and executed on a computer system, for exclusive use by the purchaser of the work.

Printed on acid-free paper

Springer is part of Springer Science+Business Media (www.springer.com)

*To my wife, Chiew-Yee and children Maesie, Julian
and Meiyan, for their inspiration, encouragement and support.*

Acknowledgements

I would like to acknowledge Peter Johnson, Department of Agriculture and Food Western Australia; Lily Eng and G.F. Chung for several photo credits.

Disclaimer

The author and publisher of this work have checked with sources believed to be reliable in their efforts to confirm the accuracy and completeness of the information presented herein and that the information is in accordance with the standard practices accepted at the time of publication. However, neither the author nor publishers warrant that information is in every aspect accurate and complete and they are not responsible for errors or omissions or for consequences from the application of the information in this work. This book is a work of reference and is not intended to supply nutritive or medical advice to any individual. The information contained in the notes on edibility, uses, nutritive values, medicinal attributes and medicinal uses and suchlike included here are recorded information and do not constitute recommendations. No responsibility will be taken for readers' own actions.

Contents

Introduction ... 1

Part I Actinidiaceae

Actinidia arguta ... 5

Actinidia chinensis ... 12

Actinidia deliciosa ... 20

Part II Adoxaceae

Sambucus nigra ... 30

Part III Anacardiaceae

Anacardium occidentale ... 45

Bouea macrophylla ... 69

Bouea oppositifolia ... 72

Dracontomelon dao ... 75

Mangifera caesia ... 79

Mangifera foetida ... 82

Mangifera indica ... 87

Mangifera kemanga ... 121

Mangifera laurina ... 124

Mangifera odorata ... 127

Mangifera pajang ... 131

Mangifera quadrifida ... 135

Mangifera similis ... 138

Pentaspadon motleyi ... 140

Pistacia vera ... 142

Schinus molle 153

Spondias cytherea 160

Spondias purpurea 166

Part IV Annonaceae

Annona atemoya 171

Annona diversifolia 176

Annona glabra 180

Annona montana 186

Annona muricata 190

Annona reticulata 201

Annona squamosa 207

Rollinia mucosa 221

Stelechocarpus burahol 227

Uvaria chamae 231

Part V Apocynaceae

Carissa macrocarpa 237

Carissa spinarum 240

Willughbeia angustifolia 247

Willughbeia coriacea 249

Part VI Araceae

Monstera deliciosa 252

Part VII Arecaceae

Adonidia merrillii 257

Areca catechu 260

Areca triandra 277

Arenga pinnata 280

Bactris gasipaes 285

Borassus flabellifer 293

Cocos nucifera 301

Elaeis guineensis 335

Elaeis guineensis *var.* pisifera 393

Eleiodoxa conferta 396

Lodoicea maldivica 399

Nypa fruticans 402

Phoenix dactylifera 407

Phoenix reclinata 419

Salacca affinis 423

Salacca glabrescens 425

Salacca magnifica 427

Salacca wallichiana 429

Salacca zalacca 432

Serenoa repens 438

Part VIII Averrhoaceae

Averrhoa bilimbi 448

Averrhoa carambola 454

Averrhoa dolichocarpa 465

Averrhoa leucopetala 468

Part IX Betulaceae

Corylus avellana 471

Part X Bignoniaceae

Crescentia cujete 480

Kiegelia africana 486

Oroxylum indicum 497

Parmentiera aculeata 508

Parmentiera cereifera 512

Part XI Bixaceae

Bixa orellana 515

Part XII Bombacaceae

Adansonia digitata 527

Adansonia gregorii 536

Ceiba pentandra	540
Durio dulcis	550
Durio graveolens	552
Durio kinabaluensis	556
Durio kutejensis	559
Durio oxleyanus	563
Durio testudinarum	566
Durio zibethinus	569
Pachira aquatica	584
Pachira insignis	588
Quararibea cordata	590

Part XIII   Bromeliaceae

Ananas comosus	593

Part XIV   Burseraceae

Canarium decumanum	616
Canarium indicum	619
Canarium odontophyllum	624
Canarium vulgare	630
Dacryodes rostrata	633

Part XV   Cactaceae

Cereus repandus	636
Hylocereus megalanthus	640
Hylocereus polyrhizus	643
Hylocereus undatus	650
Nopalea cochenillifera	656
Opuntia ficus-indica	660
Opuntia monacantha	683
Opuntia stricta	687

Part XVI   Caricaceae

Carica papaya 693

Vasconcellea × heilbornii 718

Part XVII   Cuppressaceae

Juniperus communis 723

Part XVIII   Cycadaceae

Cycas revoluta 732

Medical Glossary 739

Scientific Glossary 794

Common Name Index 817

Scientific Name Index 824

Introduction

The world is endowed with a rich, natural biodiversity of flora and fauna species. Over 7,000 plant species have been recorded to be edible today. Unfortunately only a meager number of these are cultivated widely or used on a large, extensive scale. Those cultivated commercially on a wide scale include crops like maize, rice, wheat, oats and potatoes. Others also cultivated widely and commercially but on smaller scales includes both vegetables like cabbages, broccoli, garlic, onions, capsicums and fruit crops like apples, oranges, grapes, mango, banana and pineapple. The predominance of a dozen or two of such commodities and widely grown crops have captured the attention of global research and development resources and efforts because of their significance in food production, food supply, food security and international food trade. These have resulted in the voluminous information generated and published on all aspects of these crops. Not so for many of the lesser known edible crops especially those indigenous vegetables and fruit that are harvested from the wild in the bush, forest, jungle or from a wide array of anthropogenic environments commensal with farming. These include wastelands, fallow fields, boundaries, roadsides, waysides, irrigation canal edges, river banks, stream sides, ponds, swamps, trenches and gullies as well as those grown as backyard crops. Many of these crops are perishable, low yielding and their value as commercial crops has not been explored. Comparatively, little research has been conducted because of their lack of commercial importance and whatever little published information available are found scattered in many cases in obscure, local publication sources.

Compounding this is the fact that many of these lesser know edible indigenous crops are on the verge of disappearance in many countries owing to:
- rapid rate of urbanization,
- changes in the agroecology due to the increasing frequency of prolonged drought, salinity, flooding, etc. brought about by climate change,
- agroecological changes caused by overgrazing, bush fires, shifting cultivation, over-cultivation, deforestation, etc.,
- over exploitation and harvesting from the wild,
- disappearance and loss of indigenous knowledge from older generations of the nutritive benefits, medicinal attributes and application, preservation and utilization of traditional vegetables and fruits due to the rural urban migration of younger generations of indigenous population.

Sadly, many of these lesser known, underutilized, indigenous edible crops also suffer notable disregard and erosion. Many countries have their own unique assemblage of edible indigenous plant species as well as a wide assortment of introduced edible plant species that has become naturalized over the centuries. Many of the edible indigenous and naturalized crops are also considered as traditional crops in the sense that they are interwoven with the socio-cultural and customary fabric, food habits, indigenous knowledge, rustic lifestyle and traditional agricultural practices of the local communities. The almost total disregard for traditional farming system knowledge has been one of the glaring mistakes of modern agricultural research. Local knowledge which is often perceived as non-scientific knowledge, primitive and intuitive, has been looked down on and largely ignored.

Many lesser known edible, traditional crops have potential for food security and possess advantageous traits. Food security means having sufficient physical and economic access to safe and nutritious food to meet dietary needs and food preferences for an active and healthy lifestyle. For more than three decades, food security has remained a major development concern in the developing world and an important area for research and conceptual development. While wild food resources are given some meager recognition, especially as famine foods and use in crisis situations, their role and contribution have largely remained undervalued in the food security debate.

Many of these lesser known traditional edible crops especially the indigenous vegetables and fruits are rich in micronutrients and vitamins and are often consumed for health reasons. The staple foods provide calories needed for body energy but are very low in other nutrients while the traditional vegetables and fruits have very high nutritive value. Recent researches indicate the presence of bio-active substances or phytochemicals with potential therapeutic, disease-preventing or health-promoting mechanisms in many traditional indigenous food plants. Unfortunately, the consumers have not been exposed to appreciate the role of these crops in fulfilling the above human needs. These crops could be harnessed and be developed as a strategy to overcome malnutrition besides food security. Additionally, they are hardy, easy to grow, well adapted to local agro-ecological habitats, easy to harvest and preserve, and are free from the risk of pesticide residues as they are naturally and organically grown without the use of chemicals. Importantly too, they can contribute appreciably to household food and income, livelihood security and in many cases to the national economy.

Food plants including both wild and cultivated have medicinal, nutraceutical, pharmaceutical and health benefits. They are excellent sources of phenolic phytochemicals, especially as antioxidants. Antioxidants play an important role in processes that stimulate the immune system, or have antibacterial, antimutagenic or antiviral activities, for example flavonoids, tannins, pectins and saponins. While phenolic antioxidants from dietary sources have a history of use in food preservation, many increasingly have therapeutic and disease prevention applications. This role is becoming very significant at a time when the importance of phytochemicals in the prevention of oxidation-linked chronic diseases is gaining rapid recognition globally. Disease prevention and management through the diet can be considered an effective tool to improve health and/or reduce the increasing health-care costs for these oxidation-linked chronic diseases, especially in low-income countries.

Nutraceutical encompasses nutrition and pharmaceutical and is more popularly known as "functional food". Functional foods are foods or dietary components that may provide a health benefit beyond basic nutrition. They are an important part of a holistic healthful lifestyle that includes a balanced diet and physical activity. Functional foods include fresh foods (e.g. vegetables and fruits) that have specific health-related claims attached and processed food made from functional food ingredients, or fortified with health-promoting additives, like "vitamin-enriched" or "calcium fortified" products. Fermented foods with live cultures are often also considered to be functional foods with probiotic benefits. Probiotics are dietary supplements and live microorganisms containing potentially beneficial bacteria or yeasts that are taken into the alimentary system for healthy intestinal functions. Functional food also include prebiotics, defined as non-digestible food ingredients that beneficially affect the body by selectively stimulating the growth and/or activity of one or a limited number of bacteria in the colon, and thus improve body health.

Many medicinal plants may also have direct nutritional benefits of which we know little. The practice of using herbal infusions may add small daily quantities of essential micronutrients and vitamins to the intake of sick or malnourished individuals that may be sufficient to alter the metabolic uptake or restore the balance between nutrients and thereby improve body functioning.

This work aims to create awareness and interest in the rich and diverse array of edible medicinal and non-medicinal plants in a multi-compendium format. This work does not claim to be comprehensive to cover the heterogeneous assemblage of crop diversity to include all the edible crop species (both cultivated and wild), varieties, ecotypes, landraces and genetically modified (GMO) plants. Neither will this work cover cultivation aspects as the agronomy of many of the traditional and wild, indigenous species has been little studied or published and for those that are commonly or extensively cultivated there is a plethora of information that have been published in many text and reference books.

The edible species dealt with in the book include commonly and widely grown crops, wild and underutilized crops that are consumed as fruits, vegetables, culinary herbs, pulses, cereals, spices and condiments, root crops, beverages, mushrooms, edible oils and stimulants. Vegetables covered include species cultivated for their fruits, seeds, flowers, leaves, petioles, shoots and roots or tubers. Each volume will cover about a hundred edible species in alphabetical order according to botanical families. Under each species, information is arranged in a logical and easy-to-follow format that covers the botanical name and synonyms; common English and vernacular names; origin and distribution; agroecological requirements; edible plant parts and uses; plant botany; nutritive and medicinal properties, uses and relevant research; other uses and selected/cited references for further reading. The first few volumes will focus on plants whose fruits and or seeds are edible and consumed as fruits, nuts, vegetables, cereals, beverages, liquor, spices and flavouring agents, colouring agents, sweeteners, cooking oils and fats, masticatory stimulants and as sources of nutraceuticals. The first volume focuses on edible medicinal and non-medicinal plants in the following botanical families: Actinidiaceae, Adoxaceae, Anacardiaceae, Annonaceae, Apocynaceae, Arecaceae, Averrhoaceae, Betulaceae, Bignoniaceae, Bixaceae, Bombacaceae, Bromeliaceae, Burseraceae, Cactaceae, Caricaceae, Cupressaceae and Cycadaceae. This series of book are written with a broad spectrum of readers in mind – scientists (medical practitioners, pharmacologists, food nutritionists, herbologists, herbalists, horticulturist, botanist, ethnobotanist, conservationist, agriculturist), hobbyist, cooks, gardeners, teachers, students and also the layman and consumers. Many people are curious about what plants can be eaten, how to eat them and what nutritive and health benefits they provide. It is hope that this book will help readers to identify edible medicinal and non-medicinal plants and their edible plant parts and to learn about their nutritive and medicinal values and application. To help achieve this, copious color plates are provided together with relevant and up-to-date information (including scientific names, synonyms, vernacular names, nutritive and medicinal attributes and research findings) from multifarious sources including old and new books, general and scientific papers in various languages, unpublished local knowledge, databases and the web.

All appropriate and available synonyms are included as many publications or papers are often published under the synonyms. For scientific names the following publications and databases have been used:
– USDA, ARS Germplasm Resources Information Network – (GRIN)
– Mansfeld's Encyclopedia of Agricultural and Horticultural Crops
– Efloras
– Royal Botanic Gardens, Kew (2002). Electronic Plant Information Centre
– The International Plant Names Index (IPNI)
– Australian Plant Name Index (APNI)
– Tropicos.org. Missouri Botanical Garden.

Common, local and vernacular names are included whenever available as they are important to help in identification of the edible plants. Also

it is important to note that sometimes local names may refer to different or allied crops. Publications and databases used include:
- University of Melbourne's Searchable World Wide Web Multilingual Multiscript Plant Name Database
- Mansfeld's Encyclopedia of Agricultural and Horticultural Crops database
- Rehm's Multilingual Dictionary of Agronomic Plants
- Liberty Hyde Bailey Hortorium's Hortus Third
- USDA Tropicos, ARS Germplasm Resources Information Network – (GRIN)
- Pacific Island Ecosystems at Risk (PIER) database
- Foundation for Revitalisation of Local Health Traditions ENVIS database (Indian vernacular names)
- Stuart Philippine Alternative Medicine. Manual of Some Philippine Medicinal Plants (Filipino vernacular names).
- Publications by Ochse, J.J; Ochse, J.J. and Bakhuizen van den Brink R.C. (Indonesian); Burkill, H.M. (African); Burkill, I.H. (Malay); Ho, P.H. *Illustrated Flora of Vietnam*; Do Huy Bich et al. *Cay Thuoc va Dong Vat Lam Thuoc o Viet Nam (Medicinal Herbs and Animals in Viet Nam)* 2 Vols. (Vietnamese); Smitinand, T. (Thai); PROSEA publications and others listed under selected references under the appropriate crop.

Information on the nutritive food composition and values, has been collated from various books listed under the appropriate crop and from the following databases:
- USDA National Nutrient Database for Standard Reference, Release 22.
- FAO Food Composition Table for Use in Africa, Near East, East Asia.
- Food Standards Australia New Zealand Database
- The Danish Food Composition Databank.
- Nutrient Composition of Malaysian Foods

Books and scientific publications from where other information has been gleaned are listed under the appropriate crop.

Additionally, two glossaries are provided, one for medical terms and another for other scientific terms where it was not possible to avoid because of brevity and accuracy. Also two comprehensive indices are provided, one for scientific names and another for common/English names.

It is hope that this work will also serve as a useful reference guide for further research and investigation into edible medicinal and non-medicinal plants.

Actinidia arguta

Scientific Name

Actinidia arguta (Siebold & Zuccarini) Planchon ex Miquel

Synonyms

Actinidia arguta var. *curta* Skvortsov, *Actinidia arguta* var. *dunnii* H. Léveillé, *Actinidia arguta* var. *megalocarpa* (Nakai) Kitagawa, *Actinidia arguta* var. *purpurea* (Rehder) C. F. Liang, *Actinidia arguta* var. *rufa* (Siebold & Zucc.) Maxim., *Actinidia arguta* var. *cordifolia* (Miquel) Bean, *Actinidia callosa* var. *rufa* (Siebold & Zucc.) Makino, *Actinidia chartacea* Hu, *Actinidia cordifolia* Miq., *Actinidia giraldii* Diels, *Actinidia megalocarpa* Nakai, *Actinidia melanandra* Franchet var. *latifolia* E. Pritzel, *Actinidia platyphylla* A. Gray ex Miq., *Actinidia purpurea* Rehder, *Trochostigma argutum* Sieb. & Zucc., *Trochostigma rufa* Siebold & Zucc.

Family

Actinidiaceae

Common/English Names

Arctic Kiwi, Baby Kiwi, Bower Actinidia, Bower Vine, Cocktail Kiwi, Dessert Kiwi, Hardy Kiwi, Japanese Actinidia, Kiwiberry, Mini Kiwi, Siberian Gooseberry, Taravine, Vinepear.

Vernacular Names

Chinese: Ruan Zao Mi Hou Tao, Juan-Tsao-Mi-Huo-T'ao, Ruan-Zao-Zi, Juan-Tsao-Tzu, Temg-Gua, T'eng-Kua, Yuan Li, Hong Mao Mi Hou Tao;
Danish: Stikkelsbærkiwi, Stikkelsbær-Kiwi;
Dutch: Mini-Kiwi;
Eastonian: Teravahambaline Aktiniidia;
Finnish: Japaninlaikkuköynnös;
French: Actinidier À Chair Verte, Actinidier De Sibérie, Kiwai, Kiwi D'été, Kiwi De Sibérie, Souris Végétale;
German: Taravini, Japanische Stachelbeere, Scharfzahniger Strahlengriftel;
Hungarian: Kopasz Kivi;
Japanese: Saru Nashi, Kokuwa, Shima Saru Nashi;
Korean: Ta Rae Ta Rae Na Mu, Seom Da Rae Na Mu;
Polish: Aktinidia Ostrolistna, Mini Kiwi;
Russian: Aktinidiia Arguta, Aktinidiia Krupnaia, Aktinidia Ostraja, Kišmiš Krupnyi.;
Swedish: Krusbärsaktinidia.

Origin/Distribution

The species is native to eastern-most Siberia, Eastern and north eastern China, Korea, Japan, Sakhalin and Kuril Islands. It is cultivated in the native area, in different regions of the former Soviet Union, occasionally in Europe and North America, New Zealand and Australia.

Agroecology

A temperate species, most frost tolerant of the commercial *Actinidia* species. When fully dormant it can withstand temperatures down to −32°C. However, kiwiberry needs to acclimatise to cold slowly and any sudden drop in temperature may cause trunk splitting and subsequent damage to the vine. Late winter freezing temperatures will kill any exposed buds. To date, all cultivars that have been grown in both high chill and low chill areas have produced equally well. All cultivars require a certain period of winter chilling and their needs vary, according to cultivar.

In its native range, kiwiberry occurs in the moist areas in mountain forest as a climbing vine (liana), scaling more than 30 m up trees, in thickets and sides of streams from 700 to 3,600 m altitude. Under cultivation the crop is grown in a trellising system. Kiwiberry is intolerant of saline soils but is tolerant of other soil types including infertile soil. It thrives best in well-drained, moist, acidic soils from pH 5–6.5. Neutral soils are tolerable but the leaves may show nitrogen deficiency when the soils become too basic.

Edible Plant Parts and Uses

Ripe kiwiberry fruit is very rich in vitamin C and sweet (14–29% Brix). It is eaten fresh, whole with the smooth skin intact like grapes. Sliced kiwiberry can be made into pastries, pies, muffins, sweet omelette, salads such as kiwiberry salad and kiwiberry shrimp salad. It is also processed into puree for making a range of culinary delights such as jams, salsas, salad dressing, vinegars, appetizers, sauces for entrees especially baked and grill fish, cheese-cake, ice-cream, sherbet, yoghurt and sorbet. Kiwiberry sauce is excellent over a skinless breast of chicken. Kiwiberry is an excellent meat tenderiser and marinade. It goes well with orange, honey and chocolate. It can be made into non-alcoholic spritzers (consisting of kiwiberry puree, orange juice and soda), kiwiberry margarita (kiwiberry puree, melon liquer, tequila, splash of lime juice), health food drinks to wines, meads, and beers.

Botany

A robust, large, climbing, deciduous, dioecious liana with branchlets villose when young becoming glabrous with lamellate, white to brown pith. Leaves are borne on glabrous sometimes rusty, tomentose pinkish-brown petioles, 3–10 cm long. Leaf lamina is ovate to broadly ovate to oblong ovate, 6–12 cm by 5–10 cm, membranous, apex acuminate, base rounded to cordate, symmetric or oblique, margin sharply serrate, abaxially green, glabrous to rusty tomentose with tufts of brownish hairs in the nerve angles, adaxially dark green and glabrous. Inflorescences are cymose, axillary or lateral, 1-6-flowered, brown to pale brown tomentose; peduncles 7–10 mm; pedicels 8–14 mm; bracts linear, 1–4 mm. Flowers are creamy white or white, 1.2–2 cm across when fully open. Sepals 4–6, ovate to oblong, margin ciliate, both surfaces glandular-tomentose. Petals 4–6, cuneate-obovate to orbicular-obovate, 7–9 mm. Stamens with silky 1.5–3 mm filaments; black or dark purple, oblong, 1.5–2 mm anthers sagittate at base. Ovary is bottle-shaped, 6–7 mm, glabrous; styles many, filiform, 3.5–4 mm long. Fruit is green yellowish-green, or purple-red when mature, subglobose to oblong (Plate 1), 1.5–3 cm, smooth, weighing 5–20 g. Pulp is emerald green or tinged with purple depending

Plate 1 Kiwiberry fruits

on cultivar, juicy, sweet with good flavour. Seeds numerous tiny, 2.5 mm and black.

Nutritive/Medicinal Properties

Kiwiberry has been promoted as a nutritional powerhouse as they are the most nutrient dense of all the minor fruits, and is an extremely healthy food source. Containing almost 20 vital nutrients, Kiwiberry is deemed a super fruit (Anonymous 2009). They are rich in antioxidants which are known to reduce the risk of cardiovascular disease, cancer, and help slow the aging process.

The nutrient composition of kiwiberry was reported by Rothgerber (1993) as follows: protein <3% of dry matter, glucose 32.6% of dry matter or 8.47% of fresh fruit, starch 25% of dry matter, ash traces, vitamin C 80 mg/100 g (400 mg% of dry matter), vitamin PP 55 mg/100 g, pro-vitamin A (β-carotene) 0.28 mg/100 g, cellulose 15% of dry matter, pectin 5% of dry matter, acidity 5% dry matter or 1.29% of fresh fruit. Based on a serving of 170 g, kiwiberry would provide 130 cal, total fat 1 g, saturated fat 0 g, cholesterol 0 g, sodium 0 g, total carbohydrate 30 g, dietary fibre 8 g, sugars 16 g, protein 2 g, vitamin A 6%, vitamin C 120%, calcium 10% and iron 4% (Anonymous 2009).

Nishiyama et al. (2004) found wide variation among cultivars in vitamin C content, with concentrations ranging from 37 to 185 mg/100 g FW in *Actinidia arguta* fruit. In cultivar Gassan, Issai, and Mitsuko, the vitamin C content of the fruit was much higher than that of *Actinidia deliciosa* cv. Hayward. In *Actinidia arguta* fruit, the ratio of L-ascorbic acid to total ascorbic acid tended to be higher than that of other *Actinidia* species.

The concentrations of chlorophyll, lutein, and β-carotene in the fruit of *A. deliciosa* cv Hayward were reported as 1.65, 0.418, and 0.088 mg/100 g fresh weight, respectively (Nishiyama et al. 2005). In *Actinidia rufa*, *Actinidia arguta*, and their interspecific hybrids, the contents of chlorophyll, lutein, and β-carotene were much higher than in cv Hayward. In particular, these fruits were found to be the richest dietary source of lutein among commonly consumed fruits.

The compounds extracted from *Actinidia arguta* flowers largely comprised linalool derivatives including the lilac aldehydes (12a-d) and alcohols (13a-d), 2,6-dimethyl-6-hydroxyocta-2,7-dienal, 8-hydroxylinalool, sesquiterpenes, and benzene compounds that were presumed metabolites of phenylalanine and tyrosine (Matich et al. 2003). Extracts of fruit samples contained some monoterpenes, but were dominated by esters such as ethyl butanoate, hexanoate, 2-methylbutanoate and 2-methylpropanoate, and by the aldehydes hexanal and hex-E2-enal. A number of unidentified compounds were also detected, including 8 from flowers. A new flavonol triglycoside, quercetin 3-O-β-D-[2G-O-β-D-xylopyranosyl-6G-O-α-L-rhamnopyranosyl] glucopyranoside was identified amongst the 15 flavonols found in *Actinidia arguta* var. *Giraldii* (Webby 1991). This compound was characterized, along with quercetin 3-O-β-D-xylopyranosyl-(1-2)-O-β-D-glucopyranoside. The kaempferol analogues were also isolated.

Recent studies have shown that *Actinidia arguta* possesses anti-tumour, anti-inflammatory, immunomodulatory, anti-allergic diarrhoea, antidiabetic, anti-dermatitis, bone marrow cell promoting and pancrease lipase inhibition properties.

Anticancer Activity

The n-butyl alcohol extracts from *Actinidia arguta* were found to display potent inhibitory effect (87.2%) on Eca-109 cells (human oesophageal carcinoma cells) in a concentration time manner. The extract was found to significantly induce apoptosis of Eca-109 cells (Zhang et al. 2007). In another study, the chloroform and methanol extracts from *Actinidia arguta*, were found to possess antitumour activity against human leukemias, namely Jurkat T and U937 cells by inducing apoptosis (Park et al. 2000). The substance in the solvent extract was found to be a chlorophyll derivative (Cp-D). Its IC_{50} value was found to be 15 μg/ml for Jurkat T, 10 μg/ml for U937, and 11.4 μg/ml for HL-60 (human promyelocytic leukemia). Cp-D was more toxic to these leukemias than to solid tumours or normal fibroblast. The results suggested that inactivation of cyclin-dependent kinases during cell cycle progression in Jurkat T cells following a treatment with the chlorophyll derivative would lead to induction of apoptotic cell mortality.

Bone Marrow Cell Proliferation Activity

Methanol extract of *Actinidia arguta* stems was found to promote proliferation of cultured bone marrow cells (Takano et al. 2003). The extract also stimulated formation of myeloid colonies from bone marrow cells. (+)-catechin (1) and (−)-epicatechin (2) were isolated as bioactive compounds from the methanol extract. Both compounds stimulated cell proliferation in a dose-dependent fashion in the range of 1–100 mg/ml. Both compounds also promoted formation of myeloid colonies and augmented the effect of interleukin-3 (IL-3), increasing the number of colony forming-units in culture (CFU-c). In an ex-vivo experiment using a model mouse of decreasing bone marrow functions, orally administrated (+)-catechin (100 mg/kg/day) was found to promote IL-3-induced CFU-c formation of the bone marrow cells. When intraperitoneally injected (+)-catechin (1 and 10 mg/kg per day) accelerated the recovery of the number of leukocytes and platelets but did not affect the number of circulating erythrocytes (Takano et al. 2004). (+)-catechin also enhanced the number of myelocytes and splenocytes. It reversed the reduction of the population of different leukocytes in whole blood, spleen and bone marrow caused by 5-fluorouracil in mice. (+)-catechin also demonstrated appreciable recovery of Gr-1+ cells in all three types of tissues and of CD11b+ cells in the bone marrow cells. The findings suggested that (+)-catechin selectively enhanced the recovery of the population of granulocytes reduced by 5-fluorouracil in mice.

Immunomodulatory Activity

Actinidia arguata was found to have immunomodulatory effect (Park et al. 2005). Various allergic responses are thought to result from the unbalanced development of T(H)1 (T-Helper 1) and T(H)2 (T-Helper 2) pathways and, subsequently, the overproduction of IgE. Therefore the modulation of T(H)1 and T(H)2 responses is considered a rational approach for the treatment of allergic diseases. Oral administration of a water-soluble extract prepared from *Actinidia arguta*, (PG102), was found to modulate the level of T(H)1 and T(H)2 cytokines and to suppress the production of immunoglobulin E (IgE) in the ovalbumin-immunized murine model as well as in the in-vitro cell culture system. Further, it significantly improved dermatitis conditions in the NC/Nga murine model. Oral administration of PG102T and PG102E significantly decreased the level of selective T(H)2 cytokines, but increased the level of T(H)1 cytokines. This was accompanied by a rise in the plasma level of IgG2a and by a fall in the plasma levels of IgE and IgG1. The percentages of both IL-4-producing T cells and IgE-producing B cells were reduced. The concentration of IgE produced within B cells also appeared to be decreased. These data suggested that *Actinidia arguta* extract may have great potential as orally active immune modulators for the therapy of various allergic diseases.

Antiallergic Diarrhoea Activity

Recent studies showed that *Actinidia arguta* aqueous extract might have anti-allergic effects on allergic diarrhoea. Oral administration of the extract was found to suppress the incidence of diarrhoea in a murine allergic diarrheal model (Kim et al. 2009b). The amelioration of allergic diarrhoea by the extract was accompanied by inhibition of mast cell infiltration into the large intestine. The serum levels of IgE (immunoglobin E), interleukin-6 (IL-6) and MCP-1 (monocyte chemotactic protein-1) were reduced in the extract-treated mice. When splenocytes were isolated from the respective groups and cultured in the presence of ovalbumin, cells from PG102-treated animals were found to produce smaller quantities of IL-6 and MCP-1. The results indicated that *Actinidia arguta* extract had the potential to be used as a preventive for food allergic diseases.

Anti-inflammatory Activity

The suppression of dermatitis by *Actinidia arguta* aqueous extract (PG102) was accompanied by a decrease in the plasma level of IgE, IgG1, and IL-4 and also by an increase in that of IgG2a and IL-12 (Park et al. 2007). The splenic level of IL-4, IL-5, and IL-10 was reduced, whereas that of IFN-gamma and IL-12 was increased. The number of eosinophils and the expression of eotaxin and thymus and activation-regulated chemokine were decreased by PG102T or PG102E. Histological findings also indicated that the thickening of epidermis/dermis and the dermal infiltration of inflammatory cells including mast cells were greatly inhibited. These data suggested that PG102 may be effective and safe therapeutic agents for the treatment of atopic dermatitis (AD), a chronic inflammatory skin disease.

Kim et al. (2009a) found that *Actinidia arguta* aqueous extract (PG102) could decrease the level of IgE, interleukin IL-5, and IL-13 in in-vitro cell culture systems with IC_{50} being 1.12–1.43 mg/ml. PG102 ameliorated asthmatic symptoms, including AHR and eosinophilia in the lungs. Such improvement of asthmatic symptoms by PG102 was accompanied by the suppression of IL-5 and IgE. In PG102-treated mice, high level expression of heme oxygenase-1, a potent anti-inflammatory enzyme, was observed in alveolar inflammatory cells, while the mRNA levels of foxp3, TGF-beta1, and IL-10, important markers for regulatory T cells, were also augmented in the lung tissue. The data suggested that PG102 may have potential as a safe and effective reagent for the prevention or treatment of asthma.

DA-9102, a coded compound isolated from *Actinidia arguta* was tested as a candidate of natural medicine currently under Phase II clinical trial for atopic dermatitis in Korea (Choi et al. 2008). Studies showed that oral administration of DA-9102 at a dose of 100 mg/kg for 16 days substantially reduced the occurrence of spontaneous dermatitis. Eczematous skin lesions, water loss and scratching behavior were significantly reduced by DA-9102 in a dose-dependent fashion. Results from flow cytometry analysis of peripheral blood mononuclear cells indicated that DA-9102 downregulated activation of leukocytes. The reduction in the number of CD45RA+ cells was accompanied by a lower level of IgE in DA-9102 treated rats, and the decrease in the number of CD11b+ cells by DA-9102 in both periphery and skin was significant. CD11b+ cells are the major source of oxidative stress in UV radiation-irradiated skin and have possible role in photoaging and photocarcinogenesis. Further, DA-9102 not only inhibited the mRNA expression of T(H)2 cytokines including IL-4 and IL-10 in the lymph node but it also decreased the levels of inflammatory mediators such as nitric oxide and leukotriene B(4) (LTB(4)) in the serum. These results revealed DA-9102 to be an orally applicable potent immune modulator, capable of controlling the occurrence of atopic dermatitis-like skin disease. Another earlier study reported a water soluble polysaccharide from the stem stimulated significantly the immune functions in mice and thus could be used as an effective immunological modulator (Hou et al. 1995).

Antidermatitis Activity

Orally administered *A. arguta* extract was found to significantly reduce clinical dermatitis severity, epidermal thickness, mast cell dermal infiltration and degranulation in atopic dermatitis, 2-chloro-1,3,5-trinitrobenzene-treated NC/Nga mice (Kim et al. 2009c). Total levels of serum IgE and IgG(1) were also significantly downregulated. The suppression of total serum IgE and IgG(1) levels was accompanied by a reduction in interleukin IL-4 and an increase in interferon IFN-gamma expression in skin and splenocytes. Additionally, Toll-like receptor TLR-9 expression was augmented by oral administration of *A. arguta* extract. This study confirmed that *A. arguta* extract had potential as a dietary therapeutic agent for the treatment of atopic dermatitis. The findings also suggested that the clinical efficacy and mode of action of *A. arguta* extract against atopic dermatitis were associated with the modulation of biphasic T-helper (Th) 1/Th2 cytokines and the inhibition of Th2-mediated IgE overproduction, and possibly with the augmentation of Toll-like receptor-9.

Pancreatic Lipase Inhibition Activity

A new coumaroyl triterpene, 3-O-trans-p-coumaroyl actinidic acid (1), as well as five known triterpenes, ursolic acid (2), 23-hydroxyursolic acid (3), corosolic acid (4), asiatic acid (5) and betulinic acid (6) were isolated from an ethanol-acetate-soluble extract of *Actinidia arguta* roots (Jang et al. 2008). Amongst the isolates, 3-O-trans-p-coumaroyl actinidic acid was found to possess the highest inhibitory activity on pancreatic lipase, with an IC_{50} of 14.95 μM, followed by ursolic acid IC_{50} = 15.83 μM. The other four triterpenes (3-6) also exhibited significant pancreatic lipase inhibitory activity, with IC_{50} values ranging from 20.42 to 76.45 μM.

Antidiabetic Activity

Two flavan-3-ols, 6-(2-pyrrolidinone-5-yl)-(-)-epicatechin (1) and 8-(2-pyrrolidinone-5-yl)-(-)-epicatechin (2), as well as five known compounds, (-)-epicatechin (3), (+)-catechin (4), proanthocyanidin B-4 (5), (+)-pinoresinol, and p-hydroxybenzoic acid, were isolated from an ethyl acetate-soluble extract of *Actinidia arguta* roots (Jang et al. 2009). Of these, compounds 1-5 exhibited significant inhibitory activity against advanced glycation end products (AGEs) formation with IC_{50} values ranging from 10.1 to 125.2 μM. Advanced glycation end products (AGEs) may play an important adverse role and as proinflammatory mediators in gestational diabetes and in the process of atherosclerosis, aging and chronic renal failure.

Allergy Problem

On the downside, hardy kiwifruit has the potential to cause allergy. Allergy to green kiwifruit has become common since the fruit was introduced in North America and Europe three decades ago (Chen et al. 2006). Kiwiberry is a third species that is now cultivated in North America with potential application as a fresh fruit and in processed foods. Studies suggested that some kiwifruit-allergic individuals may suffer allergic cross-reactions if they consume raw kiwiberry. However, heat processing of the kiwiberry altered allergenic protein structure, dramatically reducing in-vitro IgE binding. They found that processing likely reduced the risk of eliciting an allergic response in those with allergies to raw kiwifruit.

Traditional Medicinal Uses

In folkloric medicine, fruits and cortex of the kiwiberry vines are used as laxans and anthelmintics.

Other Uses

Selection and breeding studies to combine excellent fruit quality of kiwi fruit (*Actinidia deliciosa*) and frost resistance and smooth fruit peel of *A. arguta* crossings were made in California and in New Zealand resulting in Fairchildii hybrids and cultivar 'AADA' respectively.

Comments

There are three recognised botanical varieties: *Actinidia arguta* var. *arguta*, *Actinidia arguta* var. *giraldii* and *Actinidia arguta* var. *purpurea*.

Selected References

Anonymous (2009) A nutritional powerhouse. http://www.kiwiberries.com/nutrition.htm

Chen L, Lucas JS, Hourihane JO, Lindemann J, Taylor SL, Goodman RE (2006) Evaluation of IgE binding to proteins of hardy (*Actinidia arguta*), gold (*Actinidia chinensis*) and green (*Actinidia deliciosa*) kiwifruits and processed hardy kiwifruit concentrate, using sera of individuals with food allergies to green kiwifruit. Food Chem Toxicol 44(7):1100–1107

Choi JJ, Park B, Kim DH, Pyo MY, Choi S, Son M, Jin M (2008) Blockade of atopic dermatitis-like skin lesions by DA-9102, a natural medicine isolated from *Actinidia arguta*, in the Mg-deficiency induced dermatitis model of hairless rats. Exp Biol Med 233(8): 1026–1034

Hou F, Sun Y, Chen F, Li X (1995) Studies on immunopharmacological effects of polysaccharide from the stem of *Actinidia arguta* (Sieb. et Zucc.) ex Miq. Zhongguo Zhongyao Zazhi 20:42–44 (In Chinese)

Hu SY (2005) Food plants of China. The Chinese University Press, Hong Kong, 844 pp

Jang DS, Lee GY, Kim J, Lee YM, Kim JM, Kim YS, Kim JS (2008) A new pancreatic lipase inhibitor isolated from the roots of *Actinidia arguta*. Arch Pharm Res 31(5):666–670

Jang DS, Lee GY, Lee YM, Kim YS, Sun H, Kim DH, Kim JS (2009) Flavan-3-ols having a gamma-lactam from the roots of *Actinidia arguta* inhibit the formation of advanced glycation end products in vitro. Chem Pharm Bull (Tokyo) 57(4):397–400

Kim D, Kim SH, Park EJ, Kang CY, Cho SH, Kim S (2009a) Anti-allergic effects of PG102, a water-soluble extract prepared from *Actinidia arguta*, in a murine ovalbumin-induced asthma model. Clin Exp Allergy 39(2):280–289

Kim D, Kim SH, Park EJ, Kim J, Cho SH, Kagawa J, Arai N, Jun K, Kiyono H, Kim S (2009b) Suppression of allergic diarrhea in murine ovalbumin-induced allergic diarrhea model by PG102, a water-soluble extract prepared from *Actinidia arguta*. Int Arch Allergy Immunol 150(2):164–171

Kim JY, Lee IK, Son MW, Kim KH (2009c) Effects of orally administered *Actinidia arguta* (Hardy Kiwi) fruit extract on 2-chloro-1,3,5-trinitrobenzene-induced atopic dermatitis-like skin lesions in NC/Nga mice. J Med Food 12(5):1004–1015

Li JQ, Li XW, Soejarto DD (2007) Actinidiaceae. In: Wu ZY, Raven PH, Hong DY (eds.) Flora of China, vol 12, Hippocastanaceae through Theaceae. Science Press/Beijing, and Missouri Botanical Garden Press, St. Louis

Matich AJ, Young H, Allen JM, Wang MY, Fielder S, McNeilage MA, MacRae EA (2003) *Actinidia arguta*: volatile compounds in fruit and flowers. Phytochem 63(3):285–301

Nishiyama I, Yamashita Y, Yamanaka M, Shimohashi A, Fukuda T, Oota T (2004) Varietal difference in vitamin C content in the fruit of kiwifruit and other *Actinidia* species. J Agric Food Chem 52(17):5472–5475

Nishiyama I, Fukuda T, Oota T (2005) Genotypic differences in chlorophyll, lutein, and beta-carotene contents in the fruits of *Actinidia* species. J Agric Food Chem 53(16):6403–6407

Park YH, Chun EM, Bae MA, Seu YB, Song KS, Kim YH (2000) Induction of apoptotic cell death in human Jurkat T cells by a chlorophyll derivative (Cp-D) isolated from *Actinidia arguta* Planchon. J Microbiol Biotechnol 10:27–34

Park EJ, Kim B, Eo H, Park K, Kim Y, Lee HJ, Son M, Chang Y-S, Cho S-H, Kim S, Jin M (2005) Control of IgE and selective TH1 and TH2 cytokines by PG102 isolated from *Actinidia arguta*. J Allergy Clin Immunol 116:1151–1157

Park EJ, Park KC, Eo H, Seo J, Son M, Kim KH, Chang YS, Cho SH, Min KU, Jin M, Kim S (2007) Suppression of spontaneous dermatitis in NC/Nga murine model by PG102 isolated from *Actinidia arguta*. J Invest Dermatol 127(5):1154–1160

Rothgerber P (1993) The overall composition of *A. arguta*. Kiwifruit Enthusiasts J 6:192 pp. http://www.kiwiberry.com/index.html

Takano F, Tanaka T, Tsukamoto E, Yahagi N, Fushiya S (2003) Isolation of (+)-catechin and (-)-epicatechin from *Actinidia arguta* as bone marrow cell proliferation promoting compounds. Planta Med 69:321–326

Takano F, Tanaka T, Aoi J, Yahagi N, Fushiya S (2004) Protective effect of (+)-catechin against 5-fluorouracil-induced myelosuppression in mice. Toxicology 201(1–3):133–142

Webby RF (1991) A flavonol triglycoside from *Actinidia arguta* var. Giraldii. Phytochemistry 30:2443–2444

Zhang L, Guo HL, Tian L, Cao SF, Du CH (2007) Study of inhibitory effect of extracts from *Actinidia arguta* on human carcinoma of esophagus cells. Zhong Yao Cai 30(5):564–566 (In Chinese)

Actinidia chinensis

Scientific Name

Actinidia chinensis Planchon

Synonyms

Actinidia chinensis Planchon var. *chinensis* C.F. Liang, *Actinidia chinensis* f. *jinggangshanensis* C.F. Liang, *Actinidia chinensis* var. *jinggangshanensis* (C.F. Liang) C.F. Liang & A.R. Ferguson, *Actinidia multipetaloides* H.Z. Jiang

Family

Actinidiaceae

Common/English Names

Chinese Actinidia, Gold Kiwi Fruit, Kiwifruit, Kiwi Gold, Yellow-Fleshed Actinidia.

Vernacular Names

Chinese: Zhong Hua Mi Hou Tao, Mi Hou Tao, Jīwéiguǒ, (Yuan Bian Zhong);
Czech: Aktinídie čínská;
Danish: Almindelig Kiwi, Kiwi-Slægten;
Dutch: Gele Kiwi;
Eastonian: Hiina aktiniidia;
French: Actinidier De Chine, Actinidier À Gros Fruits;
German: Chinesische Stachelbeere, Grossfrüchtige Aktinide;
Hungarian: Kínai Egres, Szőrös Kivi;
Italian: Kiwi;
Japanese: Oni Mata Tabi, Kiiui, Kiui Furuutsu, Shina Saru Nashi;
Polish: Aktinidia Chinska;
Portuguese: Actinídia Da China;
Russian: Aktinidiia Kitaiskaia, Aktinidija Kitajskaja;
Spanish: Grosella China, Kiwi;
Turkish: Aktinidya, Kivi.

Origin/Distribution

The species is indigenous to China – the central and eastern provinces.

Agroecology

The plant is a warm temperate species and is frost intolerant. In its native area, it occurs in sparse secondary forests, tall grassy thickets on low mountains and mountain slopes from 200 to 600 m elevation, and is also widely cultivated in Anhui, Fujian, Guangdong, Guangxi, Henan, Hubei, Hunan, Jiangsu, Jiangxi, S Shaanxi,

Yunnan and Zhejiang. It does well in full sun in well-drained, moist, fertile soils. Climatic and soil requirements are quite similar to that for *Actinidia deliciosa*.

Plate 1 Ripe gold kiwi fruit

Edible Plant Parts and Uses

Fruit pulp is sweet, juicy and nutritious. It is used and processed the same way as for *Actinidia deliciosa* and *Actinida arguata*. The pulp is more amenable to processing because of its yellow coloured flesh whereas the chlorophyll-based green colour pulp of *A. deliciosa* tends to change to brown on processing.

Botany

A vigorous, woody, climbing, deciduous, dioecious vine (liana) with white-brown, lamellate pith. Young branchlets and petioles white pubescent to roughly tomentose, glabrous when mature or not. Leaves broadly ovate to broadly obovate or suborbicular, 6–17×7–15 cm, papery, base rounded to truncate to cordate, margin setose-serrulate, apex truncate to emarginate to abruptly cuspidate or shortly acuminate adaxially usually glabrous, occasionally puberulent, especially more densely so on midvein and lateral veins, dark-green above (Plate 4). Inflorescences cymose, 1–3-flowered, white silky-tomentose or yellowish brown velutinous; peduncles 0.7–1.5 cm; pedicels 0.9–1.5 cm; bracts linear,. 1 mm. Flowers orange-yellow. Sepals usually 5, broadly ovate to oblong-ovate, both surfaces densely yellowish tomentose, persistent. Petals usually 5, broadly obovate, 1–2 cm, shortly clawed at base, rounded at apex. Filaments 5–10 mm with yellow oblong anthers. Ovary globose, 5 mm across, densely golden villous. Fruit subglobose to cylindric to obovoid or ellipsoidal, 4–6 cm, densely tomentose when young becoming glabrous (Plate 1). Pulp fleshy, juicy, firm, green, greenish yellow to golden yellow, sweet to subacid with many tiny, purplish black seeds around a white succulent center (Plate 2–3).

Plate 2 Gold kiwi fruit with greenish-yellow flesh

Plate 3 Gold kiwi fruit with golden-yellow flesh

Plate 4 Foliage of gold kiwi fruit

Nutritive/Medicinal Properties

The overall nutrient composition of the gold kiwi fruit (*Actinidia chinensis*) is similar to that for *Actinidia deliciosa* but differs in the levels of vitamin C, chlorophyll, lutein, β-carotene and different types and levels of other carotenoids and anthocyanins.

Fruit of *Actinidia deliciosa* cv. Hayward, the most common commercially available cultivar, was found to contain 65.5 mg/100 g fresh weight (FW) vitamin C. Vitamin C content in *A. deliciosa* fruit varied from 29 mg/100 g FW to 80 mg/100 g FW. In most cultivars of *A. chinensis*, vitamin C content in the fruit was higher than that of Hayward (Nishiyama et al. 2004). In particular, vitamin C content in *A. chinensis* cv. Sanuki Gold fruit reached more than 3-times that of Hayward on a weight for weight basis. Levels of whole fruit mean ascorbic acid in the different genotypes ranged from 98 to 163 mg/100 g of fresh weight (FW), whereas mean oxalate varied between 18 and 45 mg/100 g of FW (Rassam and Laing 2005). Ascorbic acid was highest in the inner and outer pericarp, whereas oxalate was concentrated in the skin, inner pericarp, and seed. Essentially no ascorbic acid was found in the seed.

The concentrations of chlorophyll, lutein, and β-carotene in the fruit of *A. deliciosa* cv. Hayward were 1.65, 0.418, and 0.088 mg/100 g fresh weight, respectively. In most cultivars of *A. chinensis*, although both chlorophyll and lutein contents were significantly lower than in Hayward, the β-carotene content tended to be slightly higher (Nishiyama et al. 2005).

Ripe fruit of *A. deliciosa* was found to contain chlorophylls a and b and the carotenoids normally associated with photosynthesis, β-carotene, lutein, violaxanthin, and 9'-cis-neoxanthin (McGhie and Ainge 2002). The carotenoids in *A. chinensis* were similar to those in *A. deliciosa* but also included esterified xanthophylls. Only trace amounts of chlorophyll were present in *A. chnensis*. The yellow colour of *A. chinensis* was mostly due to the reduction of chlorophyll rather than an increase in carotenoid concentration. In contrast to the three yellow/orange species, the green fruit of *A. deliciosa* retained chlorophyll during maturation and ripening, and esterified xanthophylls were not produced.

The yellow and green colours of the outer fruit pericarp were due to different concentrations and proportions of carotenoids and chlorophylls. The red color found mainly in the inner pericarp was attributed to anthocyanins. In the *A. chinensis* genotypes tested, the major anthocyanin was cyanidin 3-O-xylo(1–2)-galactoside, with smaller amounts of cyanidin 3-O-galactoside (Montefiori et al. 2005). In the *A. deliciosa* genotypes analyzed, cyanidin 3-O-xylo(1–2)-galactoside was not detected; instead, the major anthocyanins identified were cyanidin 3-O-galactoside and cyanidin 3-O-glucoside. However, the two species (*A. chinensis* and *A. deliciosa*) did not differ consistently in anthocyanin composition.

A new vitamin E, δ-tocomonoenol, with the structure, 2,8-dimethyl-2-(4,8,12-trimethyltridec-11-enyl)chroman-6-ol, was isolated from *Actinidia chinensis* fruit (Fiorentino et al. 2009). GC–MS analysis of peels and pulps of kiwi showed that the new compound, together with δ-tocopherol, was mainly present in the fruit peel, whilst α-tocopherol was present in a similar amount in both matrices. The compound was tested for its radical-scavenging and antioxidant capabilities, by measuring its ability to scavenge DPPH (2,2'-diphenyl-1-picrylhydrazyl radical) and anion superoxide radical, and inhibited the formation of methyl linoleate conjugated diene hydroperoxides and TBARS (thiobarbituric acid reactive species). Four 3-(methylsulfanyl)-propionate esters, ethyl 3-(methylsulfanyl)prop-2-enoate, two 2-(methylsulfanyl)acetate esters and their possible precursors 2-(methylsulfanyl) ethanol, 3-(methylsulfanyl)propanol and 3-(methylsulfanyl)propanal were quantified in ripe *Actinidia chinensis* 'Hort 16A' kiwifruit pulp (Günther et al. 2010). The majority of these compounds were specific for eating-ripe fruit and their levels increased in parallel with the climacteric rise in ethylene, accumulating towards the very soft end of the eating firmness.

Seven compounds were isolated from the root of *A. chinensis*, and the structures were identified as 2α-hydroxyoleanolic acid, 2α-hydroxyursolic acid, euscaphic acid, 23-hydroxyursolic acid, 3β-O-acetylursolic acid, ergosta 4, 6, 8, (14), 22-tetraen-3-one and β-sterol (Cui et al. 2007).

Scientific studies on gold kiwifruit reveal that it shares many pharmacological properties as the common green kiwifruit.

Anticancer and Antiviral Activities

Among the 12 phenolic compounds, including 4 novel skeleton phenolic compounds, planchols A-D (1–4), together with 4 pairs of isomeric flavanoids (5–12) isolated from the roots of *Actinidia chinensis*, compounds 1 and 2 showed remarkable cytotoxic activity against P-388 (leukemia) and A-549 (carcinomic human alveolar basal epithelial cells) cell lines (Chang and Case 2005). Seven compounds were isolated from the root of *A. chinensis*, and the structures were identified as 2α-hydroxyoleanolic acid (1), 2αa-hydroxyursolic acid (2), euscaphic acid (3), 23-hydroxyursolic acid (4), 3β-O-acetylursolic acid (5), ergosta4, 6, 8, (14), 22-tetraen-3-one (6), β-steriol (7). Compound 1 and 4 were also isolated from *Actinidia deliciosa roots* and exhibited antihepatotoxic activity. In separate studies, a new polysaccharide compound (ACPS-R) isolated from the root of *Actinidia chinensis* was found to have anti-tumorous activity (Lin 1988). When administered intraperitoneally to the transplantable tumour bearing mice at a dose of 75–125 mg/kg, the tumour inhibition rate was higher than 88.8% in Ehrlich ascitic cancer (EAC) or ascitic form of hepatoma (HepA) and higher than 49.6% in solid hepatoma (HepS). ACPS-R could also extend the life of Ehrlich ascitic cancer -or P388-bearing mice, and increased the percentage of Ehrlich ascitic cancer-free mice. A marked increase in cAMP (adenosine 3', 5'-cyclic monophosphate) levels and cAMP/cGMP (cyclic guanosine monophosphate) ratio of the spleen of EAC-bearing mice were observed after treatment of ACPS-R. The increases of cAMP, cAMP/cGMP and tumour remission had statistical significance. The results indicated that ACPS-R acted as a new antitumour polysaccharide, and the treatment effect of *Actinidia* root in folk medicine was probably related to ACPS-R. Studies also showed that *Actinidia chinensis* was among five Chinese herbal plants tested that contained antimutagenic factors against both picrolonic acid- and benzo[a]pyrene-induced mutations (Lee and Lin 1988).

Five fractions H1, H2 (hexane extract), A1, A2 (acetone extract) and M2 (methanol extract) of gold kiwi fruit showed selective cytotoxic activity against human oral tumour cell lines, which was more sensitive than human gingival fibroblasts (Motohashi et al. 2002). The more hydrophilic fractions [70M3, 70M4, 70M5] of 70% methanol extract displayed higher anti-HIV (human immunodeficiency virus) activity, radical generation and O^{2-} scavenging activity. The antibacterial activity of 70% methanol extracts [70M0, 70M1, 70M2, 70M3, 70M4] was generally lower than that of more lipophilic fractions (hexane, acetone, methanol extracts), although each fraction did not show any specific antimicrobial action. All fractions were inactive against *Helicobacter pylori*.

Two new triterpenoids (1, 2), together with one flavonoid glycoside and 13 known triterpenoids were isolated from the roots of *Actinidia chinensis* (Xu et al. 2010). The structures of the new constituents were elucidated as 12α-chloro-2α, 3β, 13β, 23-tetrahydroxyolean-28-oic acid-13-lactone (1), 2α, 3α, 19α, 23, 24-pentahydroxyurs-12-en-28-oic acid (2). Additionally, two known triterpenoids showed positive cytotoxic activity against LOVO (human adenocarcinoma) and human liver carcinoma (HepG2) cell lines.

Lee et al. (2010) found that both gold kiwifruit (GOK) and green kiwifruit (GRK) protected WB-F344 rat liver epithelial cells from H_2O_2-induced inhibition of gap-junction intercellular communication (GJIC). Both kiwi fruit types were found to block the H_2O_2-induced phosphorylation of connexin 43 (Cx43) and extracellular signal-regulated protein kinase 1/2 (ERK1/2) signalling pathways in WB-F344 cells. Quercetin alone attenuated the H_2O_2-mediated ERK1/2-Cx43 signalling pathway and consequently reversed H_2O_2-mediated inhibition of GJIC in WB-F344 cells. A free radical-scavenging assay using 1,1-diphenyl-2-picrylhydrazyl showed that the scavenging activity of quercetin was higher than that of a synthetic antioxidant, butylated hydroxytoluene, suggesting that the chemopreventive effect of quercetin on H_2O_2-mediated inhibition of ERK1/2-Cx43 signalling and GJIC may be mediated through its free radical-scavenging activity. Since the carcinogenicity of reactive oxygen species such as H_2O_2 was attributable to the inhibition of GJIC, both kiwi fruit types and quercetin may have chemopreventive potential by preventing the inhibition of GJIC. Gap-junction intercellular communication had been reported to propagate cell death in cancerous cells (Krutovshikh et al. 2002).

Keratinocyte Proliferation Stimulating Activity

Ten μg/ml of raw polysaccharide, neutral type-II-arabinogalactans, and acidic arabinorhamnogalacturonans of *Actinidia chinensis* fruits were found to stimulate cell proliferation of human keratinocytes (NHK, HaCaT) up to 30% significantly while mitochondrial activity was stimulated for about 25% in comparison to control cells (Deters et al. 2005). Fibroblasts were not as sensitive as keratinocytes but a dose of >130 μg/ml kiwifruit polysaccharides increased also proliferation and ATP-synthesis significantly. Results indicated that the polysaccharides led to a doubled collagen synthesis of fibroblasts compared to the normally strongly reduced biosynthetic activity.

Immunomodulatory Activity

Gold kiwifruit puree was found to enhance the response to ovalbumin by significantly increasing the levels of total immunoglobulins and immunoglobulin G specific for ovalbumin (Hunter et al. 2008). It also augmented the antigen-specific proliferation of cells from the draining mesenteric lymph nodes compared with mice fed a 20% sugar control. These results indicated that gold kiwifruit could modulate an antigen-specific immune response and provide a new type of functional food ingredient.

Across a range of concentrations, the gold kiwifruit extract was found to protect cultured cells from oxidative stress using neuroblastoma cells exposed to hydrogen peroxide at a level sufficient to induce cell mortality (Skinner et al. 2008). The gold kiwifruit-treated muscles displayed a marked increase in maximum force and a significant delay in fatigue commencement compared with the untreated control muscles in mice. Consumption of gold kiwifruit puree led to significant increases in ovalbumin-specific antibodies (IgG1, IgG2b and IgG2c) in the sera following sub-optimal immunisation conditions. Consumption also elevated IL-5 production from mesenteric lymph node cells. The results provided evidence that gold kiwifruit could protect from oxidative stress, ameliorate muscle performance, delay time to muscle fatigue and augment antigen-specific immunity. The researchers asserted that gold kiwifruit would be useful as a new type of functional food ingredient for sports drinks and others foods targeted at enhancing muscle performance and immune function.

Kiwifruit extracts (*Actinidia chinensis* and *Actinidia deliciosa*) were found to significantly enhance specific intestinal mucosal and serum antibody responses to vaccines and to stimulate interferon-γ and natural killer cell activities in BALB/c mice (Shu et al. 2008). No significant improvement was noted in proliferative response, phagocytic activity and interleukin-4 production. The overall results demonstrated the ability of kiwifruit extract to augment markers of innate and acquired immunity in the tested murine model.

Anti-inflammatory Activity

Aqueous extract of kiwifruit was demonstrated in-vitro to reduce lipopolysaccharide (LPS)-induced tumour necrosis factor (TNF)α and interleukin (IL)-1β production indicating its potential anti-inflammatory attribute (Farr et al. 2008). Aqueous extracts from ZESPRI™ GOLD (*Actinidia chinensis* 'Hort16A') and ZESPRI™ GREEN (*A. deliciosa* 'Hayward') Kiwifruit cultivars exhibited promising activity with human THP-1 monocytes and peripheral whole blood, although differential effects were seen in individual donors.

Antimicrobial and Prebiotic Activity

The extracts from various parts of *Actinidia chinensis* plant (leaves, fruits and stems) were found to exhibit antibacterial activity against eight Gram positive and Gram negative bacterial strains (Basile et al. 1997). The antibiotic activity of the fruit was found to reside essentially in the seeds. The antibacterial activity of extracts from vegetative plant parts was generally less active that from fruit extracts. Kiwifruit, and in particular the water extract of Gold Kiwifruit demonstrated an ability to positively influence intestinal bacterial enzymes, by inhibiting ß-glucuronidase activity and promoting the activity of ß-glucosidase (Molan et al. 2008). Further, extracts prepared from gold kiwifruit and green kiwifruit were able to promote the growth of faecal lactic acid bacteria (LAB) especially at high concentrations and to reduce the growth of *Escherichia coli*. This was explained by the reduction of the activity of ß-glucuronidase generated mainly by *E. coli*. The increase in activity of ß-glucosidase was postulated to be due to the stimulation of the growth of LAB, which had high levels of ß-glucosidase activity in comparison with the members of the gut microflora. It was concluded that both the edible flesh and, particularly, water extracts of gold and green kiwifruit, exhibited antimicrobial and prebiotic activities when tested under in-vitro conditions. Kiwifruits were also shown to beneficially modulate the intestinal bacterial enzymes, ß-glucosidase and ß-glucuronidase, in a manner considered beneficial for gastrointestinal health.

Hepatoprotective Activity

Two new triterpenoids, 1 and 2, were isolated from the hepatoprotective ethyl acetate fraction of the roots of *Actinidia chinensis*, together with eight known 12-en-28-oic acids of oleanane or ursane type, 3–10 (Zhou et al. 2009). The two new compounds were elucidated as 2α,3β-dihydroxyurs-12-en-28,30-olide and 2α, 3β, 24-trihydroxyurs-12-en-28,30-olide.

Allergy Problem

On the down side, kiwifruit can cause allergy. Allergy to green kiwifruit, *Actinidia deliciosa*, has become common since the fruit was introduced into North America and Europe three decades ago. *Actinidia chinensis* (gold kiwifruit) more recently introduced commercially, was shown to have in-vitro immunoglobulin E (IgE) cross-reactivity with green kiwi (Chen et al. 2006; Lucas et al. 2005). Gold kiwi fruit is allergenic and patients allergic to green kiwi are at risk of reacting to the gold kiwi fruit. Despite having different protein profiles and IgE-binding patterns, the two species have proteins that extensively cross-inhibit the binding to IgE. Allergic reactions to fruits and vegetables are among the most frequent food allergies in adults. Kiwi fruit (*Actinidia deliciosa*) is commonly involved, causing local mucosal, systemic, or both types of symptoms by an IgE-mediated mechanism (Pastorello et al. 1998). Studies demonstrated that the major allergen of kiwi fruit, Act c 1, was actinidin, a proteolytic enzyme belonging to the class of thiol-proteases. Two other allergens of 24 and 28 kDa were also identified. Gold kiwi fruit is also rich in kiwellin, an allergenic protein formerly isolated from green kiwi fruit (Tuppo et al. 2008). Kiwellin is a modular protein with two domains. It may undergo in-vivo proteolytic processing by actinidin, thus producing KiTH and kissper. When probed with sera recognizing kiwellin from green kiwi fruit, KiTH showed IgE binding, with reactivity levels sometimes different from those of kiwellin. The IgE-binding capacity of kiwellin from gold kiwi fruit appeared to be similar to that of the green species.

Other Uses

A. chinensis will become more important in breeding of new cultivars.

Comments

Refer also to chapters on *Actinidia arguata* and *A. deliciosa*.

Selected References

Bai XP, Qiu AY (2006) The liver-protecting effect of extract from the root of *Actinidia deliciosa* in mice. J Chin Inst Food Sci Technol 6:247–249 (In Chinese)

Basile A, Vuotto ML, Violante U, Sorbo S, Martone G, Castaldo-Cobianchi R (1997) Antibacterial activity in *Actinidia chinensis*, Feijoa sellowiana and Aberia caffra. Int J Antimicrob Agents 8(3):199–203

Chang J, Case R (2005) Cytotoxic phenolic constituents from the root of *Actinidia chinensis*. Planta Med 71(10):955–9

Chen L, Lucas JS, Hourihane JO, Lindemann J, Taylor SL, Goodman RE (2006) Evaluation of IgE binding to proteins of hardy (*Actinidia arguta*), gold (*Actinidia chinensis*) and green (*Actinidia deliciosa*) kiwifruits and processed hardy kiwifruit concentrate, using sera of individuals with food allergies to green kiwifruit. Food Chem Toxicol 44(7):1100–1107

Cui Y, Zhang XM, Chen JJ, Zhang Y, Lin XK, Zhou L (2007) Chemical constituents from root of *Actinidia chinensis*. Zhongguo Zhong Yao Za Zhi 32(16):1663–5 (In Chinese)

Deters AM, Schröder KR, Hensel A (2005) Kiwi fruit (*Actinidia chinensis* L.) polysaccharides exert stimulating effects on cell proliferation via enhanced growth factor receptors, energy production, and collagen synthesis of human keratinocytes, fibroblasts, and skin equivalents. J Cell Physiol 202(3):717–722

Farr JM, Hurst SM, Skinner MA (2008) Anti-inflammatory effects of kiwifruit. In: Proceedings of the Nutrition Society of New Zealand, vol 32, pp 20–25. Forty second annual conference held in combination with the Australian Nutrition Society at Massey University, Albany Campus, Auckland (Dec 2007)

Fiorentino A, Mastellone C, D'Abrosca B, Pacifico S, Scognamiglio M, Cefarelli G, Caputo R, Monaco P (2009) δ-Tocomonoenol: a new vitamin E from kiwi (*Actinidia chinensis*) fruits. Food Chem 115(1):87–192

Günther CS, Matich AJ, Marsh KB, Nicolau L (2010) (Methylsulfanyl)alkanoate ester biosynthesis in *Actinidia chinensis* kiwifruit and changes during cold storage. Phytochem 71(7):742–50

Hu SY (2005) Food plants of China. The Chinese University Press, Hong Kong, 844 pp

Hunter DC, Denis M, Parlane NA, Buddle BM, Stevenson LM, Skinner MA (2008) Feeding ZESPRI GOLD Kiwifruit puree to mice enhances serum immunoglobulins specific for ovalbumin and stimulates ovalbumin-specific mesenteric lymph node cell proliferation in response to orally administered ovalbumin. Nutr Res 28(4):251–7

Krutovshikh VA, Piccoli C, Ymasaki H (2002) Gap junction intercellular communication propagates cell death in cancerous cells. Oncogene 21(13):1989–9

Lee H, Lin JY (1988) Activity of extracts from anticancer drugs in Chinese medicine. Mutat Res 204(2):229–34

Lee DE, Shin BJ, Hur HJ, Kim JH, Kim J, Kang NJ, Kim DO, Lee CY, Lee KW, Lee HJ (2010) Quercetin, the active phenolic component in kiwifruit, prevents hydrogen peroxide-induced inhibition of gap-junction intercellular communication. Br J Nutr 104(2):164–70

Li JQ, Li XW, Soejarto DD (2007) Actinidiaceae. In: Wu ZY, Raven PH, Hong DY (eds.) Flora of China, vol 12, Hippocastanaceae through Theaceae. Science Press/Missouri Botanical Garden Press, Beijing/St. Louis

Liang CF, Ferguson AR (1986) The botanical nomenclature of the kiwifruit and related tax. NZ J Bot 24:183–184

Lin PF (1988) Antitumor effect of *Actinidia chinensis* polysaccharide on murine tumor. Zhonghua Zhong Liu Za Zhi 10(6):441–4 (In Chinese)

Lucas JS, Lewis SA, Trewin JB, Grimshaw KE, Warner JO, Hourihane JO (2005) Comparison of the allergenicity of *Actinidia deliciosa* (kiwi fruit) and *Actinidia chinensis* (gold kiwi). Pediatr Allergy Immunol 16(8):647–54

McGhie TK, Ainge GD (2002) Color in fruit of the genus *Actinidia*: carotenoid and chlorophyll compositions. J Agric Food Chem 50(1):117–21

Molan AL, Kruger MC, Drummond LN (2008) Kiwifruit: the ability to positively modulate key markers of gastrointestinal function. In: Proceedings of the Nutrition Society of New Zealand, 2008, vol 32, pp 66–71. Forty second annual conference held in combination with the Australian Nutrition Society at Massey University, Albany Campus, Auckland, Dec 2007

Montefiori M, McGhie TK, Costa G, Ferguson AR (2005) Pigments in the fruit of red-fleshed kiwifruit (*Actinidia chinensis* and *Actinidia deliciosa*). J Agric Food Chem 53(24):9526–30

Motohashi N, Shirataki Y, Kawase M, Tani S, Sakagami H, Satoh K, Kurihara T, Nakashima H, Mucsi I, Varga A, Molnár J (2002) Cancer prevention and therapy with kiwifruit in Chinese folklore medicine: a study of kiwifruit extracts. J Ethnopharmacol 81(3):357–64

Nishiyama I, Yamashita Y, Yamanaka M, Shimohashi A, Fukuda T, Oota T (2004) Varietal difference in vitamin C content in the fruit of kiwifruit and other *Actinidia* species. J Agric Food Chem 52(17):5472–5

Nishiyama I, Fukuda T, Oota T (2005) Genotypic differences in chlorophyll, lutein, and β-carotene contents in the fruits of *Actinidia* species. J Agric Food Chem 53(16):6403–7

Pastorello EA, Conti A, Pravettoni V, Farioli L, Rivolta F, Ansaloni R, Ispano M, Incorvaia C, Giuffrida MG, Ortolani C (1998) Identification of actinidin as the major allergen of kiwi fruit. J Allergy Clin Immunol 101(4 Pt 1):531–7

Rassam M, Laing W (2005) Variation in ascorbic acid and oxalate levels in the fruit of *Actinidia chinensis* tissues and genotypes. J Agric Food Chem 53(6):2322–6

Shu Q, Mendis De Silva U, Chen S, Peng W, Ahmed M, Lu G, Yin Y, Liu A, Drummond L (2008) Kiwifruit extract enhances markers of innate and acquired immunity in a murine model. Food Agric Immunol 19(2):149–161

Skinner MA, Hunter DC, Denis M, Parlane N, Zhang J, Stevenson LM, Hurst R (2008) Health benefits of ZESPRI GOLD Kiwifruit: effects on muscle performance, muscle fatigue and immune responses. In: Proceedings of the Nutrition Society of New Zealand, 2008, vol 32, pp 49–59. Forty second annual conference held in combination with the Australian Nutrition Society at Massey University, Albany Campus, Auckland, Dec 2007

Tuppo L, Giangrieco I, Palazzo P, Bernardi ML, Scala E, Carratore V, Tamburrini M, Mari A, Ciardiello MA (2008) Kiwellin, a modular protein from green and gold kiwi fruits: evidence of in vivo and in vitro processing and IgE binding. J Agric Food Chem 56(10):3812–7

Xu YX, Xiang ZB, Jin YS, Shen Y, Chen HS (2010) Two new triterpenoids from the roots of *Actinidia chinensis*. Fitoterapia 81(7):920–4

Zhou X-F, Zhang P, Pi H-F, Zhang YH, Ruan HL, Wang H, Wu J-Z (2009) Triterpenoids from the roots of *Actinidia chinensis*. Chem Biodivers 6(8):1202–7

Actinidia deliciosa

Scientific Name

Actinidia deliciosa (A. Chev.) C.F. Liang & A.R. Ferguson

Synonyms

Actinidia chinensis var. *deliciosa* (A. Chev) A. Chev.; *Actinidia chinensis* var. *hispida* C.F. Ling; *Actinidia latifolia* var. *deliciosa* A. Chev., *Actinidia chinensis* var. *setosa* Li

Family

Actinidiaceae

Common/English Names

Chinese Gooseberry, Chinese-Gooseberry, Green-Fleshed Actinidia, Green Kiwifruit, Kiwifruit, Strawberry Peach, Strawberry-Peach

Vernacular Names

Arabic: Kiwi;
Bohemian: Kiwi Voæe;
Brazil: Kiwi;
Chinese: Mei Wei Mi Hou Tao, Mi Yang Dao, Yang Dao, Yang Dao Deng, Yang Tao, Ji Wei Guo, Jwéigu, Mao Li, Mi Hou Tao, Qí Yì Gu, Teng Li, Yang Tao, Ying Mao Mi Hou Tao;
Czech: Aktinídie čínská;
Danish: Ægte Kiwi;
Dutch: Kiwi;
Eastonian: Kiivi-aktiniidia;
Finnish: Kiivi;
French: Actinidier Commun, Actinidier Cultivé, Groseille Chinoise, Groseille De Chine, Kiwi, Kiwi De Chine;
German: Chinesische Stachelbeere, Die Kiwi, Grossfrüchtige Aktinide, Kiwifrucht;
Hebrew: Kiuui;
Hungarian: Kínai Egres, Kivi, Szrös Kivi;
Indonesia: Buah Kiwi;
Japanese: Jiiueiguo, Kiui Furuutsu;
Korean: Cham Da Rae, Yang Da Rae;
Malaysia: Buah Kiwi;
New Zealand: Green Kiwi Fruit, kiwi fruit, Kuihipere (Maori);
Persian: Kiwi;
Polish: Kiwi;
Portuguese: Groselha-Chinesa, Kiwi;
Russian: Aktinidija Kitajskaja; Aktinidiia Kitaiskaia, Kivi;
Spanish: Groseille De Chine, Grosellas Chinas, Kiwi;
Swedish: Kiwi;
Taiwan: Mei Wei Mi Hou Tao;
Turkish: Kivi;

Origin/Distribution

The species is native to China in the temperate provinces of Chongqing, Gansu, Guangxi, Guizhou, Henan, Hubei, Hunan, Jiangxi, Shaanxi, Sichuan, Yunnan and Fukien and Zhejiang Province on the coast of eastern China. It is cultivated extensively in New Zealand, USA and southern Europe.

Agroecology

Actinidia deliciosa is a temperate species. In its native area, it occurs in the mountain forests at 800–1,400 m altitudes in Chongqing, Gansu, Guangxi, Guizhou, Henan, Hubei, Hunan, Jiangxi, Shaanxi, Sichuan, Yunnan. Elsewhere it is widely cultivated at altitudes of 600–2,000 m between latitudes 20° and 45° north and south. Kiwifruit is frost sensitive. Kiwifruit vines are killed by plunge in temperature below −1.67°C. In France, 1-year-old plants have been killed to the ground by frosts. In California, autumn frosts retard new growth and kill developing flower buds, or, if they occur after the flowers have opened, will prevent the setting of fruits. However, dormant vines have been reported to survive temperatures down to −12°C. A frost free growing period of 7–8 months is ideal. In the Bay of Plenty in New Zealand where kiwifruit is widely cultivated, the mean minimum daily winter temperatures are 4.5–5.5°C, mean maximum (14–16°C) and in summer the mean minimum is 13–14°C and the mean maximum 24–25°C; annual precipitation, 1,300–1,630 mm and relative humidity 76–78%. The vine does best in full sun although young vines require partial shade. Kiwifruit performs best in deep, fertile, moist but well-drained soil, preferably a friable, sandy loam. Heavy soils subject to water logging are completely unsuitable. In Kiangsi Province, China, the plants flourish in a shallow layer of 'black wood earth' on top of stony, red subsoil.

Edible Plant Parts and Uses

Ripe Kiwifruit is best eaten fresh by scooping out the pulp with a spoon or served as appetizers, as slices or chunks in fruit salads, salsas, with croissant, yogurt, ice-cream, in fish, chicken, lamb and meat dishes, shrimp cocktails, in pies, tarts, Danish pastries, bread, cake fillings or as chunks in kebab skewers. Slightly under-ripe fruits, which are high in pectin, must be chosen for making jelly, jam, chutney, pickles in vinegar and for drying. The peeled whole fruits may be pickled with vinegar, brown sugar and spices.

Slightly unripe fruits are used for jelly, jam, chutney. Optimally ripe ones are canned whole, halves or slices in syrup, and preserved by freezing. Freeze-dried kiwifruit slices are shipped to health food outlets in Sweden and Japan where they are sometimes coated with chocolate. Ripe kiwifruit are blended to make smoothies, sherbets slush, shaved ice and various beverages. Ripe and over ripe fruits are processed into purees, sauces, ice-cream, drinks, mixed juices, soft drinks, fizzy drinks and to wine and spirits. Ice cream may be topped with kiwifruit sauce or slices, and the fruit is used in breads and various beverages. Kiwifruit cannot be blended with yogurt because an enzyme conflicts with the yogurt process. Meat can be tenderized by placing slices of kiwifruit over it or by rubbing the meat with the flesh. The meat should then be cooked immediately.

The aqueous extract of kiwifruit puree contains health-beneficial constituents and can be considered a functional ingredient for gluten-free bread formulation (Sun-Waterhouse et al. 2009). Studies demonstrated that the aqueous extract had good stability and high phenolic and vitamin C contents and suitable for gluten-free breadmaking. The resultant kiwifruit extract-enhanced bread was acceptable to a taste panel, possessing softer and smoother texture than plain gluten-free bread.

Plate 1 (a, b) Whole and halved kiwi fruit (cross-section left, longitudinal section right) showing the green flesh and numerous tiny, black seeds

Plate 2 Leaf habit of kiwi fruit

Botany

A vigorous, climbing, woody, deciduous, dioecious vine with whitish to brown, large, lamellate pith. Young branchlets and petioles are brownish and strigose. Leaves are alternate, suborbicular or obovate orbicular, 6–17 × 7–15 cm, base subcordate, margin setose-serrulate, apex truncate to emarginate to abruptly cuspidate, usually glabrous adaxially, occasionally puberulent, especially more densely so on midvein and lateral veins (Plate 2). Inflorescences are cymose, 1-3-flowered, white silky-tomentose or yellowish brown, velutinous. Flowers are unisexual, orange-yellow on brown-villous pedicels. Sepals usually 5, persistent, broadly ovate to oblong-ovate, both surfaces densely yellowish tomentose. Petals usually 5, broadly obovate, shortly clawed at base, rounded at apex; stamens numerous with yellow, oblong anthers, ovary superior, globose, golden-strigose, styles free. Fruit is subglobose to cylindric or ovoid, 5–6 cm, densely hispid even when mature. The flesh is firm until fully ripe, juicy, sweet to subacid, bright-green with white, succulent center surrounded by numerous tiny, purplish-black seeds (Plates 1a, b).

Nutritive/Medicinal Properties

The nutrient composition of fresh kiwifruit (*Actinidia deliciosa*) per 100 g edible portion was reported by Wills (1987) as follows: water 82.8 g, energy 205 kJ, protein 1.4 g, total lipid (fat) 0.2 g, fibre (total dietary) 3.3 g, glucose 4.1 g, fructose 4.2 g, sucrose 1.4 g; minerals Ca 24 mg, Fe 0.4 mg, Mg 17 mg, K 280 g, Na 5 mg, Zn 0.2 mg; vitamins – vitamin C (ascorbic acid) 72 mg, thiamine 0.01 mg, riboflavin 0.04 mg, niacin 0.4 mg, β-carotene equivalent 0.07 mg, organic acids, malic acid 0.14 g, citric acid 1.04 g, quinic acid 0.58 g.

Food value of raw, kiwifruit per 100 g edible portion excluding 14% skin was reported as follows (USDA 2010): water 83.05 g, energy 61 kcal (255 kJ), protein 0.99 g, total lipid (fat) 0.44 g, ash 0.64 g, carbohydrate 14.88 g; total dietary fibre 3.4 g, minerals – calcium 26 mg, iron, 0.41 mg, magnesium 30 mg, phosphorus 40 mg, potassium 332 mg, sodium 5 mg; vitamins – vitamin C (total ascorbic acid) 75 mg, thiamine 0.020 mg, riboflavin 0.050 mg, niacin 0.50 mg, vitamin A 175 IU.

Kiwi fruit is outstanding because of its high vitamin C content, its dietary fibre and its antioxidant phytochemicals. It is also rich in potassium and magnesium.

Fruit of *A. deliciosa* cv. Hayward, the most common commercially available cultivar, was found to contain 65.5 mg/100 g fresh weight (FW) vitamin C (Nishiyama et al. 2004). Vitamin C content in *A. deliciosa* fruit varied from 29 mg/100 g FW to 80 mg/100 g FW. The total soluble solid content, acidity, vitamin C, ash and total nitrogen content of kiwifruit cv. Hayward were 7.32%, 1.64%, 108 mg/100 g, 0.71 g/100 g and 0.84%, respectively. The fresh fruits have 1.09 mg/100 g total chlorophylls and flesh colour data represented as L, a and b were 57.18, 17.25 and 37.46, respectively (Celik et al. 2007).

Ripe fruit of *Actinidia deliciosa* was found to contain chlorophylls a and b and the carotenoids normally associated with photosynthesis, β-carotene, lutein, violaxanthin, and 9'-cis-neoxanthin (Nishiyama et al. 2005). The green fruit of *A. deliciosa* was found to retain chlorophyll during maturation and ripening. This suggested that in fruit of *A. deliciosa* chloroplasts were not converted to chromoplasts as is typical for ripening fruit. The concentrations of chlorophyll, lutein, and β-carotene in the fruit of *A. deliciosa* Hayward were 1.65, 0.418, and 0.088 mg/100 g fresh weight, respectively.

Kiwifruit was found to contain many phytochemicals like phenols and flavonoid which contributed to its antioxidative and pharmacological activities. Vitamin E, 2,8-dimethyl-2-(4,8,12-trimethyltridec-11-enyl)chroman-6-ol, as well as α- and δ-tocopherol, 7 sterols, the triterpene ursolic acid, chlorogenic acid, and 11 flavonoids (Fiorentino et al. 2009) were found in its peels. Two caffeic acid glucosyl derivatives and two coumarin glucosides were found in the pulp, besides the three vitamin E, β-sitosterol, stigmasterol, and its Delta(7) isomer, campesterol, chlorogenic acid, and some flavone and flavanol molecules.

In the *Actinidia chinensis* genotypes tested the major anthocyanin was cyanidin 3-O-xylo (1-2)-galactoside, with smaller amounts of cyanidin 3-O-galactoside (Montefiori et al. 2005). In the *A. deliciosa* genotypes analyzed, cyanidin 3-O-xylo (1-2)-galactoside was not detected; instead, the major anthocyanins identified were cyanidin 3-O-galactoside and cyanidin 3-O-glucoside. However, the two species did not differ consistently in anthocyanin composition.

Antioxidant Activity

Kiwifruit was found to be a good example of a food with putative antioxidant properties, and was effective at decreasing oxidative DNA damage (Collins et al. 2001). Ex-vivo, consumption of kiwifruit was found to lead to an increased resistance of DNA to oxidative damage induced by H_2O_2 in isolated lymphocytes, in comparison with lymphocytes collected after a control drink of water. No effect was seen on endogenous DNA damage. In-vitro, a simple extract of kiwifruit, buffered to pH 7, was more effective than a solution of vitamin C (of equivalent concentration) at protecting DNA from damage, whereas at the highest concentrations tested, neither kiwi extract nor vitamin C had a protective effect. The results demonstrated that the significant antioxidant activity of kiwifruit ex-vivo and in-vitro, was not attributable entirely to the vitamin C content of the fruit.

Ethylene treatment of kiwifruits was found to lead to positive changes in most of the studied kiwifruit compounds and to an increase in the fruit antioxidant potential (Park et al. 2006a). Ethylene shortened the ripening time and improved fruit quality by decreasing its flesh firmness and acidity. Significant increases were found to occur in the contents of free sugars, soluble solids, endogenous ethylene production, sensory value, 1-aminocyclopropane-l-carboxylic acid (ACC) content, ACC synthase and ACC oxidase activities, total polyphenols and related antioxidant potential that was significantly higher than in untreated samples. During ethylene treatment, the bioactivity of kiwifruit was found to increase and to reach its peak at the 6th day and therefore represented the optimum time for kiwifruit consumption. On the 6th day of the ethylene treatment kiwifruit samples had the highest contents of polyphenols and flavonoids and the highest antioxidant activity. Total polyphenols were found to be the main contributor to the overall antioxidant activity of kiwifruit.

Prior et al. (2007) demonstrated that consumption of certain berries and fruits such as blueberries, mixed grape and kiwifruit, was associated with increased plasma antioxidant capacity in the postprandial state. It was also found that consumption of an energy source of macronutrients containing no antioxidants was associated with a decline in plasma antioxidant capacity. However, further long term clinical studies are needed to translate increased plasma antioxidant capacity into a potential decreased risk of chronic degenerative disease. Consumption of high antioxidant foods with each meal was recommended in order to prevent periods of postprandial oxidative stress.

Protease Activity

Kiwi fruit was found to be rich in actinidin, a meat-tenderizing cysteine protease (Lewis and Luh 1988). They found that the tenderness of broiled bovine semitendinosus (ST) steaks exhibiting an actinidin activity of 400 U/mL resulted in Kramer shear values and sensory tenderness scores equivalent to that produced by Adolph's papain-based meat tenderizer (18 U/mL) (Lewis and Luh 1988). Both were significantly more tender than steaks having no tenderizing treatment. Actinidin did not over-tenderize the steak surface as did Adolph's papain-based meat tenderizer. In separate studies, Mostafaie et al. (2008) demonstrated the collagenolytic effect of actinidin. Actinidin was found to hydrolyze collagen types I and II at neutral and alkaline buffers. Additionally, actinidin compared with type II or IV collagenase was found to be able to isolate intact human umbilical vein endothelial cells, hepatocytes, and thymic epithelial cells with viability more than 90%.

The results showed actinidin to be a novel and valuable collagenase, which could be used efficiently for hydrolysis of collagen and isolation of different cell populations from various solid tissues.

Laxative Activity

Increasing dietary fibre intake was found to be effective in relieving chronic constipation in Chinese patients (Chan et al. 2007). The mean CSBM (complete spontaneous bowl motion) increased after a 4 week treatment of kiwifruit twice daily. There was also improvement in the scores for bothersomeness of constipation, and satisfaction of bowel habit, and a decrease in days of laxative used. There was also improvement in transit time and rectal sensation. However, there was no change in the bowel symptoms or anorectal physiology in the healthy subjects. The regular use of kiwifruit appeared to lead to a bulkier and softer stool, as well as more frequent stool production. Rush et al. (2002) also found kiwifruit to be a natural remedy to improve laxation for elderly individuals who are otherwise healthy and to be palatable to most of the population. Their study suggested that a number of factors in the whole fruit were involved, but the nature of the stools suggested fibre was important. This study provided evidence of the potential for improvement in bowel function, health and well-being through changes in diet.

Cardiovascular Activity

Kiwi fruit which contains high amounts of vitamin C, vitamin E and polyphenols may be beneficial in cardiovascular disease. Studies reported Duttaroy and Jørgensen (2004) found that consuming two or three kiwifruits per day for 28 days reduced platelet aggregation response to collagen and ADP (adenosine diphosphate) by 18% compared with the controls. Platelets are known to be involved in atherosclerotic disease development and the reduction of platelet activity could reduce the incidence and severity of disease. Additionally, consumption of kiwifruit lowered blood triglycerides levels by 15% compared with control, whereas no such effects were observed in the case of cholesterol levels. The data indicated that consuming kiwifruit may be beneficial in cardiovascular disease.

Antimicrobial/Prebiotic Activity

Kiwi fruit peels also has medicinal efficacy. In studies conducted, both cytotoxic activity and multi-drug resistance reversal activity were found in the less polar fractions of kiwifruit peel extracts (Motohashi et al. 2001). Antibacterial activity was distributed into various fractions and all fractions were inactive against *Candida albicans* and *Helicobacter pylori*. Only 70% methanol extracts showed anti-human immunodeficiency virus activity. These fractions also effectively scavenged O^{2-} produced by the xanthine-xanthine oxidase reaction, suggesting a bimodal (pro-oxidant and antioxidant) action.

Kiwifruit, and in particular the water extract of gold kiwifruit demonstrated an ability to positively influence intestinal bacterial enzymes, by inhibiting ß-glucuronidase activity and promoting the activity of ß-glucosidase (Molan et al. 2008). In addition, extracts prepared from gold kiwifruit and green kiwifruit were able to promote the growth of faecal lactic acid bacteria (LAB) especially at high concentrations and reduce the growth of *Escherichia coli*. This was explained by the reduction of the activity of ß-glucuronidase generated mainly by *E. coli*. The increase in activity of ß-glucosidase was postulated to be due to the stimulation of the growth of LAB, which have high levels of ß-glucosidase activity in comparison with the members of the gut microflora. It was concluded that both the edible flesh and, particularly, water extracts of gold and green kiwifruit, exhibited antimicrobial and prebiotic activities when tested under in-vitro conditions. Kiwifruits were also shown to beneficially modulate the intestinal bacterial enzymes, ß-glucosidase and ß-glucuronidase, in a manner that was considered beneficial for gastrointestinal health.

Anti-inflammatory Activity

Aqueous extracts of kiwifruit was demonstrated in-vitro to reduce lipopolysaccharide (LPS)-induced tumour necrosis factor (TNF)α and interleukin (IL)-1β production indicating its potential anti-inflammatory activity (Farr et al. 2008). Aqueous extracts from ZESPRI™ GOLD (*Actinidia chinensis* 'Hort16A') and ZESPRI™ GREEN (*A. deliciosa* 'Hayward') kiwifruit cultivars showed promising activity with human THP-1 (Human acute monocytic leukemia cell line) monocytes and peripheral whole blood, although differential effects were seen in individual donors.

Anticancer Activity

Ethanol extracts of *Actinidia deliciosa* roots were found to possess anticancer properties in-vitro and in-vivo (Liang et al. 2007). Both ethanol and butanol extract of *A. deliciosa* roots exhibited anti-tumour effects to hepatoma S180-mice and prolonged hepatoma H22-mice life. Both extract were found to improve the immune function of bearing tumour mice. In a subsequent study, Lee et al. (2010) found that both gold kiwifruit (GOK) and green kiwifruit (GRK) protected WB-F344 rat liver epithelial cells from H_2O_2-induced inhibition of gap-junction intercellular communication (GJIC). The extracellular signal-regulated protein kinase 1/2 (ERK1/2)-connexin 43 (Cx43) signalling pathway was found to be crucial for the regulation of GJIC, and both kiwifruit types blocked the H_2O_2-induced phosphorylation of Cx43 and ERK1/2 in WB-F344 cells. Quercetin alone attenuated the H_2O_2-mediated ERK1/2-Cx43 signalling pathway and consequently reversed H_2O_2-mediated inhibition of GJIC in WB-F344 cells. A free radical-scavenging assay using 1,1-diphenyl-2-picrylhydrazyl showed that the scavenging activity of quercetin was higher than that of a synthetic antioxidant, butylated hydroxytoluene, suggesting that the chemopreventive effect of quercetin on H_2O_2-mediated inhibition of ERK1/2-Cx43 signalling and GJIC may be mediated through its free radical-scavenging activity. Since the carcinogenicity of reactive oxygen species such as H_2O_2 was attributable to the inhibition of GJIC, both kiwifruit types and quercetin may have chemopreventive potential by preventing the inhibition of GJIC. GJIC was reported to propagate cell death in cancerous cells (Krutovshikh et al. 2002).

Hepatoprotective Activity

Studies reported that ethanol extracts of *A. deliciosa* roots had anti-hepatotoxic effects (Bai and Qiu 2006). The main antihepatotoxic chemical constituents identified were two triterpenoids (TTP), ursolic acid (3b-hydroxy-urs-12-en-28-oic acid) and oleanolic acid (3b-hydroxy- olea-12-en-28-oic acid) (Bai et al. 2007). The ethanol-water extract of *A. deliciosa* root (EEAD) was fractionated into n-hexane (EEAD-He), ethyl acetate (EEAD-Ea), n-butanol (EEAD-Bu) and aqueous (EEAD-Aq) fractions according to their different polarity and solubility. Amongst the four fractions, it was found that EEAD-Bu was enriched with oleanolic acid (OLA). The studies showed that the EEAD-Bu had higher in-vitro antioxidant and in-vivo hepatoprotective activities than those of the other types of extracts in rats. When the carbon tetrachloride-induced rats were treated with 120 mg/kg EEAD-Bu, the activities of alanine transaminase (ALT) and aspartate transaminase (AST) in rat serum declined 90% and 81%, respectively, as compared with those of the carbon tetrachloride control rats. Further, the lipid peroxidation MDA (malondialdehyde) decreased 42% and glutathione (GSH) increased 114% in the rats liver homogenate, as compared with those of the control. The results also indicated that the hepatoprotective activity of the EEAD-Bu (at the dose of 120 mg/kg) was higher than that of the reference drug silymarin (at the dose of 60 mg/kg), and that OLA played an important role in dose-dependent protection against carbon tetrachloride hepatotoxicity.

Antihyperglycemic/Antidiabetic Activity

The methanol extract of kiwifruit leaf was found to suppress postprandial blood glucose level after an oral administration of soluble starch or sucrose in mice (Shirosaki et al. 2008). The mechanism of action was proposed to be due to the α-amylase-inhibiting activity in the 90% aqueous methanol fraction, and to α-glucosidase-inhibiting activity in the n-butanol fraction.

A 100% methanol fraction of methanolic extract from unripe kiwifruit (*Actinidia deliciosa*), designated KMF (kiwifruit methanol fraction) when applied to 3 T3-L1 preadipocyte cells, stimulated adipocyte differentiation, increased glycerol-3-phosphate dehydrogenase (GPDH) activity, and enhanced triglyceride (TG) content (Abe et al. 2010). KMF was found to markedly enhance mRNA expression of peroxisome proliferator-activated receptor gamma (PPARgamma)- the master adipogenic transcription factor and its target genes. Moreover, KMF increased mRNA expression and protein secretion of adiponectin, whereas mRNA expression and secretion of monocyte chemoattractant protein-1 (MCP-1) and interleukin-6 (IL-6) were downregulated. Compared with troglitazone, KMF downregulated the production of reactive oxygen species (ROS) and nuclear factor-kappaB (NFkappaB) activation. Glucose uptake was promoted by KMF in differentiated 3 T3-L1 adipocytes. The results indicated that KMF may exert beneficial effects against diabetes via its ability to regulate adipocyte differentiation and function.

Antihyperlipidemic Activity

In a study involving 43 subjects (13 males and 30 females) with hyperlipidemia, the HDL-C concentration was significantly augmented and the LDL cholesterol/HDL-C ratio and total cholesterol/HDL-C ratio were significantly reduced after 8 weeks of consumption of kiwifruit (Chang and Liu 2009). Vitamin C and vitamin E contents also increased significantly. In addition, the lag time of LDL oxidation and malondialdehyde + 4-hydroxy-2(E)-nonenal were significantly altered at 4 and 8 weeks during kiwifruit intervention. The study suggested that regular consumption of kiwifruit might exert beneficial effects on the antioxidative status and the risk factors for cardiovascular disease in hyperlipidemic subjects.

Immunomodulatory Activity

Kiwifruit extracts (*Actinidia chinensis* and *Actinidia deliciosa*) significantly increased specific intestinal mucosal and serum antibody responses to the vaccines and promoted interferon-γ and natural killer cell activities in BALB/c mice (Shu et al. 2008). No significant improvement was observed in proliferative response, phagocytic activity and interleukin-4 production. The study demonstrated the ability of kiwifruit extract to enhance markers of innate and acquired immunity in the tested murine model.

Wound Healing Activity

In recent studies, debridement and scar contraction were found to occur significantly faster in the kiwi fruit-treated group than in the untreated group of rats (Hafezi et al. 2010). After 20 days, all eschars had detached and fallen off in the kiwifruit intervention group. Following rapid enzymatic debridement, healing appeared to progress normally, with no evidence of damage to adjacent healthy tissue.

Allergy Problem

On the down side, kiwi fruit can cause allergy and its oxalate content is an antinutrient. Allergy to kiwi fruit was first described in 1981, and there have since been reports of the allergy presenting a wide range of symptoms from localized oral allergy syndrome (OAS) to life-threatening anaphylaxis (Lucas et al. 2003). *Actinidia chinensis* (gold kiwi) more recently introduced commercially has been shown to have in-vitro immunoglobulin E (IgE) cross-reactivity with green kiwi (Lucas et al. 2005). Gold kiwifruit is allergenic and patients allergic to green kiwi are at risk of reacting to the gold kiwi fruit. Despite having different protein profiles and IgE-binding patterns, the two species have proteins that extensively cross-inhibit the binding to IgE. Allergic reactions to fruits and vegetables are among the most frequent food allergies in adults. Kiwi fruit (*Actinidia deliciosa*) is commonly involved, causing local mucosal, systemic, or both types of symptoms by an IgE-mediated mechanism (Pastorello et al. 1998). Studies demonstrated that the major allergen of kiwifruit, Act c 1, was actinidin, a proteolytic enzyme belonging to the class of thiol-proteases. Two other allergens of 24 and 28 kd were also identified. Green fleshed Kiwi fruit was also found to be rich in kiwellin, an allergenic protein (Tuppo et al. 2008). Kiwellin is a modular protein with two domains. It may undergo in-vivo proteolytic processing by actinidin, thus producing KiTH and kissper. When probed with sera recognizing kiwellin, KiTH showed IgE binding, with reactivity levels sometimes different from those of kiwellin. The IgE-binding capacity of kiwellin from gold kiwifruit and green kiwifruit appeared to be similar.

Traditional Medicinal Uses

In Chinese folk medicine the fruits, stems and roots are regarded as diuretic, febrifuge and sedative. The vines and leaves are used against kidney stones, rheumatoid arthralgia, cancers of the liver and oesophagus. The root of *A. deliciosa* has been used as a traditional drug in China for a long time and has recently acquired interest due to its attractive potential application in indigenous drugs. It has been employed in folk medicine remedy for adult diseases, such as potent antihepatotoxic, anti-pyorrhoea and gingival inflammation (Bai and Qiu 2006; Bai et al.

2007). The branches and leaves are boiled in water and the liquid used for treating mange in dogs. In China, the fruit and the juice of the stalk are esteemed for expelling "gravel". The Chinese use kiwifruit extract as a tonic for growing children and for women after childbirth.

Other Uses

The bast fibres of the vines are used for ropes. From the leaves and the bark paper can be made. If the bark is removed in one piece from near the root and placed in hot ashes, it becomes very hard and can be used as a tube for pencils. California kiwifruit growers have found the vine trimmings unsuitable for mulch or disposal by burning but excellents for floral arrangements and are shipping them to florists.

Comments

This species has two botanical varieties: *Actinidia deliciosa* var. *chlorocarpa* and *Actinidia deliciosa* var. *deliciosa*.

In China there are four main cultivars: Zhong Hua (Chinese gooseberry), Jing Li (northern pear gooseberry), Ruan Zao (Soft date gooseberry) and Mao Hua (may be tight- or loose-haired). In New Zealand the predominant cultivars are: Abbott, Allison, Bruno, Hayward, Monty (Montgomery) and Greensill.

Selected References

Abe D, Saito T, Kubo Y, Nakamura Y, Sekiya K (2010) A fraction of unripe kiwi fruit extract regulates adipocyte differentiation and function in 3 T3-L1 cells. Biofactors 36(1):52–59

Bai XP, Qiu AY (2006) The liver-protecting effect of extract from the root of *Actinidia deliciosa* in mice. J Chin Inst Food Sci Technol 6:247–249 (In Chinese)

Bai X, Qiu A, Guan J, Shi Z (2007) Antioxidant and protective effect of an oleanolic acid-enriched extract of *A. deliciosa* root on carbon tetrachloride induced rat liver injury. Asia Pac J Clin Nutr 16(Suppl 1):169–173

Celik A, Ercisli S, Turgut N (2007) Some physical, pomological and nutritional properties of kiwifruit cv. Hayward. Int J Food Sci Nutr 58(6):411–418

Chan AO, Leung G, Tong T, Wong NY (2007) Increasing dietary fiber intake in terms of kiwifruit improves constipation in Chinese patients. World J Gastroenterol 13(35):4771–4775

Chang WH, Liu JF (2009) Effects of kiwifruit consumption on serum lipid profiles and antioxidative status in hyperlipidemic subjects. Int J Food Sci Nutr 60(8):709–716

Chen L, Lucas JS, Hourihane JO, Lindemann J, Taylor SL, Goodman RE (2006) Evaluation of IgE binding to proteins of hardy (*Actinidia arguta*), gold (*Actinidia chinensis*) and green (*Actinidia deliciosa*) kiwifruits and processed hardy kiwifruit concentrate, using sera of individuals with food allergies to green kiwifruit. Food Chem Toxicol 44(7):1100–1107

Collins BH, Horská A, Hotten PM, Riddoch C, Collins AR (2001) Kiwifruit protects against oxidative DNA damage in human cells and in vitro. Nutr Cancer 39(1):148–153

Duttaroy AK, Jørgensen A (2004) Effects of kiwi fruit consumption on platelet aggregation and plasma lipids in healthy human volunteers. Platelets 15(5):287–292

Farr JM, Hurst SM, Skinner MA (2008) Anti-inflammatory effects of kiwifruit. In: Proceedings of the Nutrition Society of New Zealand, 2008. vol 32, pp 20–25. Forty Second Annual Conference held in combination with the Australian Nutrition Society at Massey University, Albany Campus, Auckland, Dec 2007

Fiorentino A, D'Abrosca B, Pacifico S, Mastellone C, Scognamiglio M, Monaco P (2009) Identification and assessment of antioxidant capacity of phytochemicals from kiwi fruits. J Agric Food Chem 57(10): 4148–4155

Hafezi F, Rad HE, Naghibzadeh B, Nouhi AH, Naghibzadeh G (2010) *Actinidia deliciosa* (kiwifruit), a new drug for enzymatic debridement of acute burn wounds. Burns 36(3):352–355

Hu SY (2005) Food plants of China. The Chinese University Press, Hong Kong, 844 pp

Krutovshikh VA, Piccoli C, Ymasaki H (2002) Gap junction intercellular communication propagates cell death in cancerous cells. Oncogene 21(13):1989–1999

Lee DE, Shin BJ, Hur HJ, Kim JH, Kim J, Kang NJ, Kim DO, Lee CY, Lee KW, Lee HJ (2010) Quercetin, the active phenolic component in kiwifruit, prevents hydrogen peroxide-induced inhibition of gap-junction intercellular communication. Br J Nutr 104(2): 164–170

Lewis DA, Luh BS (1988) Application of actinidin from kiwifruit to meat tenderization and characterization of beef muscle protein hydrolysis. J Food Biochem 12(3):147–158

Li HL (1952) A taxonomic review of the genus *Actinidia*. J Arnold Arbor 33:31

Li JQ, Li XW, Soejarto DD (2007) Actinidiaceae. In: Wu ZY, Raven PH, Hong DY (eds) Flora of China, Vol. 12. (Hippocastanaceae through Theaceae). Science Press/Missouri Botanical Garden Press, Beijing/St. Louis

Liang CF, Ferguson AR (1986) The botanical nomenclature of the kiwifruit and related taxa. NZ J Bot 24:183–184

Liang J, Wang XS, Zhen HS, Zhong ZG, Zhang WY, Zhang WW, Li WL (2007) Study on anti-tumor effect of extractions from roots of *Actinidia deliciosa*. Zhong Yao Cai 30(10):1279–1282 (In Chinese)

Lucas JS, Lewis SA, Hourihane JO (2003) Kiwi fruit allergy: a review. Pediatr Allergy Immunol 14:420–428

Lucas JS, Lewis SA, Trewin JB, Grimshaw KE, Warner JO, Hourihane JO (2005) Comparison of the allergenicity of *Actinidia deliciosa* (kiwi fruit) and *Actinidia chinensis* (gold kiwi). Pediatr Allergy Immunol 16(8):647–654

McGhie TK, Ainge GD (2002) Color in fruit of the genus *Actinidia*: carotenoid and chlorophyll compositions. J Agric Food Chem 50(1):117–121

Molan AL, Kruger MC, Drummond LN (2008) Kiwifruit: the ability to positively modulate key markers of gastrointestinal function. In: Proceedings of the Nutrition Society of New Zealand, 2008, vol 32, pp 66–71. Forty Second Annual Conference held in combination with the Australian Nutrition Society at Massey University, Albany Campus, Auckland, Dec 2007

Montefiori M, McGhie TK, Costa G, Ferguson AR (2005) Pigments in the fruit of red-fleshed kiwifruit (*Actinidia chinensis* and *Actinidia deliciosa*). J Agric Food Chem 53(24):9526–9530

Morton JF (1987) Kiwifruit. In: Fruits of warm climates. Julia F. Morton, Miami, pp 293–300

Mostafaie A, Bidmeshkipour A, Shirvani Z, Mansouri K, Chalabi M (2008) Kiwifruit actinidin: a proper new collagenase for isolation of cells from different tissues. Appl Biochem Biotechnol 144(2):123–131

Motohashi N, Shirataki Y, Kawase M, Tani S, Sakagami H, Satoh K, Kurihara T, Nakashima H, Wolfard K, Miskolci C, Molnár J (2001) Biological activity of kiwifruit peel extracts. Phytother Res 15(4):337–343

Nishiyama I, Yamashita Y, Yamanaka M, Shimohashi A, Fukuda T, Oota T (2004) Varietal difference in vitamin C content in the fruit of kiwifruit and other *Actinidia* species. J Agric Food Chem 52(17):5472–5475

Nishiyama I, Fukuda T, Oota T (2005) Genotypic differences in chlorophyll, lutein, and beta-carotene contents in the fruits of *Actinidia* species. J Agric Food Chem 53(16):6403–6407

Park YS, Jung ST, Kang SG, Delgado-Licon E, Katrich E, Tashma Z, Trakhtenberg S, Gorinstein S (2006a) Effect of ethylene treatment on kiwifruit bioactivity. Plant Foods Hum Nutr 61(3):151–156

Park YS, Jung ST, Kang SG, Drzewiecki J, Namiesnik J, Haruenkit R, Barasch D, Trakhtenberg S, Gorinstein S (2006b) In vitro studies of polyphenols, antioxidants and other dietary indices in kiwifruit (*Actinidia deliciosa*). Int J Food Sci Nutr 57(1–2):107–122

Pastorello EA, Conti A, Pravettoni V, Farioli L, Rivolta F, Ansaloni R, Ispano M, Incorvaia C, Giuffrida MG, Ortolani C (1998) Identification of actinidin as the major allergen of kiwi fruit. J Allergy Clin Immunol 101(4 Pt 1):531–537

Prior RL, Gu L, Wu X, Jacob RA, Sotoudeh G, Kader AA, Cook RA (2007) Plasma antioxidant capacity changes following a meal as a measure of the ability of a food to alter in vivo antioxidant status. J Am Coll Nutr 26(2):170–181

Rush EC, Patel M, Plank LD, Ferguson LR (2002) Kiwifruit promotes laxation in the elderly. Asia Pac J Clin Nutr 11(2):164–168

Shirosaki M, Koyama T, Yazawa K (2008) Antihyperglycemic activity of kiwifruit leaf (*Actinidia deliciosa*) in mice. Biosci Biotechnol Biochem 72:1099–1102

Shu Q, Mendis De Silva U, Chen S, Peng W, Ahmed M, Lu G, Yin Y, Liu A, Drummond L (2008) Kiwifruit extract enhances markers of innate and acquired immunity in a murine model. Food Agric Immunol 19(2):149–161

Stuart RGA (1979) Chinese Materia Medica: vegetable kingdom. Southern Materials Centre, Taipei

Sun-Waterhouse D, Chen J, Chuah C, Wibisono R, Melton LD, Laing W, Ferguson LR, Skinner MA (2009) Kiwifruit-based polyphenols and related antioxidants for functional foods: kiwifruit extract-enhanced gluten-free bread. Int J Food Sci Nutr 60(S7):251–264

Tuppo L, Giangrieco I, Palazzo P, Bernardi ML, Scala E, Carratore V, Tamburrini M, Mari A, Ciardiello MA (2008) Kiwellin, a modular protein from green and gold kiwi fruits: evidence of in vivo and in vitro processing and IgE binding. J Agric Food Chem 56(10):3812–3817

U.S. Department of Agriculture, Agricultural Research Service (USDA) (2010) USDA National Nutrient Database for Standard Reference, Release 23. Nutrient Data Laboratory Home Page. http://www.ars.usda.gov/ba/bhnrc/ndl

Wills RBH (1987) Composition of Australian fresh fruit and vegetables. Food Technol Aust 39(11):523–526

Sambucus nigra

Scientific Name

Sambucus nigra L.

Synonyms

Sambucus alba Raf., *Sambucus arborescens* Gilib., *Sambucus florida* Salisb., *Sambucus laciniata* Mill., *Sambucus medullosa* Gilib., *Sambucus pyramidata* Lebas, *Sambucus virescens* Desf., *Sambucus vulgaris* Neck.

Family

Adoxaceae, also placed in Caprifoliaceae and Sambucaceae

Common/English Names

Black Elder, Common Elder, Pipe Tree, Bore Tree, Bour Tree, Danewort, Elder Bush, Elder, Elderberry, European Alder, European Black Elder, European Black Elderberry, European Elder, European Elderberry, Pipe Tree, Sambu, Tree Of Medicine, Tree Of Music

Vernacular Names

Brazil: Sabugueiro;
Czech: Bez Černý;
Danish: Almindelig Hyld, Hyld, Hyldebær;
Dutch: Gewone Viler;
Eastonian: Must Leeder;
Finnish: Mustaselja;
French: Grand Sureau, Seu, Sus, Sureau, Sureau Noir;
German: Flieder, Fliederbeerbusch, Hollerbusch, Schwarzer Holunder;
Hungarian: Fekete Bodza;
Icelandic: Svartyllir;
Italian: Sambuco, Sambuco, Commune, Sambuco Negro, Sambuco Nero, Zambuco;
Latvia: Melnais Plūškoks;
Lithuanian: Juoduogis Šeivamedis;
Norwegian: Hyll, Hærsbutre, Svarthyll;
Papiamento: Sauku;
Polish: Bez Czarny, Czarny Bez, Dziki Bez Czarny, Dziki Czarny Bez;
Portuguese: Sabugueiro-Negro;
Russian: Buzina Černaja;
Slovaščina: Črni Bezeg, Bezeg, Bezeg Črni;
Slovencina: Baza Čierna;
Spanish: Cañiler, Canillero, Caúco Negro, Sabuco, Sauch, Saúco;
Swedish: Äkta Fläder, Fläder, Hyll, Sommarfläder, Vanlig Fläder.

Origin/Distribution

The species is common to most of Europe including Scandinavia and Great Britain, northwest Africa, Asia Minor, Caucasus to Western Siberia. Its naturalised distribution area reaches 63°N

Sambucus nigra

latitude in western Norway (with scattered naturalised shrubs up to at least 68°N) and approximately 55°N in Lithuania. The eastern limit of its distribution is approximately 55°E.

Agroecology

Sambucus nigra is a European species with an oceanic to suboceanic, cool-temperate and west-mediterranean range. It is predominantly a shrub of open areas and associated with moderately to highly eutrophic and disturbed soils, for example in floodplains, coastal scrub or along forest margins and in forest gaps, or anthropogenically in hedgerows, abandoned fields and gardens, around farm houses, road margins, near railways and on post-industrial wasteland. It grows in a variety of conditions including both wet and dry fertile soils, primarily in sunny locations and abhors deep shade.

Edible Plant Parts and Uses

The ripe purple berries can be eaten raw or cooked, but the flavour of the raw berries are not pleasantly acceptable though when cooked they make delicious pies, jellies, jams, preserves, juice, wine, etc. In Hungary an elderberry brandy is produced from the fruits. In Beerse, Belgium, a variety of Jenever called *Beers Vlierke* is made from the berries. The fruit is commonly used to impart flavour and colour (anthocyanin) to preserves, jams, pies, sauces, soups and chutney. Elderberries when cooked go well with blackberries and with apples in pies. In Scandinavia and Germany, soup made from the elderberry is a traditional meal. They can also be dried for later use.

The flowers are also edible raw or cooked or dried for later use. They are fragrant and aromatic, and commonly used in flower infusions as a refreshing drink in Northern Europe and the Balkans. Commercially, these are sold as elderflower cordials. The flowers are also made into a syrup (in Romanian: *Socată*, in Swedish: *fläder(blom)saft*), which is diluted before drinking. The popularity of this traditional drink has recently encouraged some commercial soft drink producers to introduce elderflower-flavoured drinks (*Fanta Shokata, Freaky Fläder*). Wine is also made from the flowers. In south-western Sweden, it is traditional to make a snaps liqueur flavored with elderflower. It is also made and sold commercially, under the name Hallands Fläder. Elderflowers are also used in liqueurs such as St. Germain and a mildly alcoholic sparkling elderflower 'champagne'. The flowers are also used to flavour stewed fruits, jellies and jams. A sweet tea is made from dried flowers. The flowers can also be dipped into a light batter and then fried to make elderflower fritters.

The leaves are use to impart a green colouring to oils and fats.

Botany

Deciduous banching shrub to small tree to 10 m with erect shoots from the base and arching branches. Bark is brownish-gray, corky and deeply furrowed. Leaves are imparipinnate with 3- (5–7)-9 leaflets. Leaflets are 3–9 cm, ovate, ovate-lanceolate or ovate-elliptic, acuminate tip with serrate margins (Plate 1). Petioles are 3–4 cm long and deeply grooved on the adaxial side. Stalk-like extrafloral nectaries are present

Plate 1 Pinnate leaves and flower buds of elderberry

Plate 2 Opened flowers of elderberry

Plate 3 Close-up of elderberry flower

at the base of leaves and leaflets. Inflorescence is flat-topped, 10–20 cm across, corymbose with five primary rays. Flowers are pentamerous, bisexual, actinomorphic, calyx lobes small, corolla rotate, with a short tube and spreading lobes, 5 mm across, creamy-white and fragrant (Plates 2–3). Anthers are extrose, cream; style is short with 3–5 stigmas. Fruit is a drupe, 6–8 mm, globose, purplish-black, containing 3–5 small, compressed seeds.

Nutritive/Medicinal Properties

The nutrient composition of raw elderberries per 100 g edible portion was reported as: water 79.80 g, energy 73 kcal (305 kJ), protein 0.66 g, total lipid 0.50 g, ash 0.64 g, carbohydrate 18.40 g, total dietary fibre 7.0 g, Ca 38 mg, Fe 1.60 mg, Mg 5 mg, P 39 mg, K 280 mg, Na 6 mg, Zn 0.11 mg, Cu 0.061 mg, Se 0.6μg, vitamin C 36 mg, thaimin 0.070 mg, riboflavin 0.060 mg, niacin 0.5 mg, pantothenic acid 0.140 mg, vitamin B-6 0.230 mg, total folate 6μg, vitamin A 30μg RAE, vitamin A 600 IU, total saturated fatty acids 0.023 g, 16:0 (palmitic acid) 0.018 g, 18:0 (stearic acid) 0.005 g, total monounsaturated fatty acids 0.080 g, 18:1 undifferentiated 0.080 g, total polyunsaturated fatty acids 0.247 g, 18:2 undifferentiated 0.162 g, 18:3 undifferentiated 0.085 g, tryptophan 0.013 g, threonine 0.027 g, isoleucine 0.027 g, leucine 0.060 g, lysine 0.026 g, mehtionine 0.014 g, cystine 0.015 g, phenylalanine 0.040 g, tyrosine 0.051 g, valine 0.033 g, arginine 0.047 g, histidine 0.015 g, alanine 0.030 g, aspartic acid 0.058 g, glutamic acid 0.096 g, glycine 0.036 g, proline 0.025 g and serine 0.032 g (USDA 2010). Wild European elderberry plants in Turkey were found to be very rich in terms of health components with high protein content of 2.68–2.91%, high total antioxidant capacity of 6.37 mmol/100 g FW; total phenolic content of 6,432 mg GAE/100 g FW and total anthocyanin content of 283 mg cyaniding-3-glucoside/100 g FW (Akbulut et al. 2009).

Among the elderberry cultivars/selections tested the 'Haschberg' cultivar of *Sambucus nigra* was found to be the richest in organic acids (6.38 g/kg FW), and it contained the least sugar (68.5 g/kg FW) (Veberic et al. 2009a, b). The following major cyanidin based anthocyanins were identified in the fruit of black elderberry: cyanidin 3-sambubioside-5-glucoside, cyanidin 3,5-diglucoside, cyanidin 3-sambubioside, cyanidin 3-glucoside and cyanidin 3-rutinoside. The most abundant anthocyanin in elderberry fruit was cyanidin 3-sambubioside, which accounted for more than half of all anthocyanins identified in the berries. The 'Rubini' cultivar had the highest amount of the anthocyanins identified (1,265 mg/100 g FW) and the lowest amount was found in berries of the 'Selection 14' (603 mg/100 g FW). The 'Haschberg' cultivar contained a relatively low amount of anthocyanins in ripe berries (737 mg/100 g FW). From the quercetin group, quercetin, quercetin 3-rutinoside and quercetin 3-glucoside were identified. The cultivar with the highest amount of total quercetins was 'Selection 25' (73.4 mg/100 g FW), while the 'Haschberg' cultivar contained average amounts of quercetins (61.3 mg/100 g FW). The chemical composition of the 'Haschberg' cultivar, the most commonly planted cultivar, conformed to the standards for sugars, anthocyanins and quercetins and exceeded them in the content levels of organic acids, the most important parameter in fruit processing.

Both *Sambucus canadensis* and *S. nigra* were found to have cyanidin-based anthocyanins as major pigments in the fruits (Lee and Finn 2007). Trace levels of delphinidin 3-rutinoside were present in all elderberry samples except cv. 'Korsør'. Also, petunidin 3-rutinoside was detected in cvs 'Adams 2', 'Johns', 'Scotia', 'York', and 'Netzer' (*S. canadensis*). The identified polyphenolics of both species were mainly composed of cinnamic acids and flavonol glycosides. The major polyphenolic compounds present in *S. nigra* were chlorogenic acid and rutin, while neochlorogenic acid, chlorogenic acid, rutin, and isorhamnetin 3-rutinoside, were found to be major polyphenolic compounds in *S. canadensis*.

Elderberry wine was found to have a moderate ethanol concentration, intense red coloration, and higher pH value compared to most red wines (Schmitzer et al. 2010). Total phenolic content of elderberry must and wine ranged up to 2004.13 GAE/L. Antioxidative potential of elderberry wine was in the range of red wine, and a close correlation was detected between total phenolic content and antioxidative potential of elderberry wine.

The following triterpenoids and sterols were identified in the flowers of *Sambucus nigra* (Willuhn and Richter 1977): triterpenoids: α- and β-amyrin (major), lupeol, cycloartenol, 24-methylenecycloartanol, cycloeucalenol; sterols: cholesterol, campesterol, stigmasterol and sitosterol (major). These triterpenoids and sterols were present in the free form and in the esterified form, primarily as ester of palmitic acid. The four sterols were also present in the glycosidically bound form. Besides this, traces of α- and β-amyrinacetate and oleanolic acid were isolated, and ursolic acid was found to be a major component of the flowers. Additionally, the flowers contained primary alcohols of the homologous series from C17 to C30 in the free form and in the esterified form, the even-members being predominant. The major acids of the esters were found to be 16:0, 18:0, 18:1 and 18:2 acids. The secondary alcohols were unsaturated, ranging in the carbon number from C14 to C25 with the odd-members (C21, C23, C25) being predominant. They occurred only in the esterified form with 16:0, 18:0, 18:1, 20:0 and 20:1 acids as major components.

Forty eight volatile compounds were found in fresh elder flowers of five cultivars (Jørgensen et al. 2000). The odor of the volatiles was evaluated by the GC-sniffing technique. Cis-rose oxide, nerol oxide, hotrienol, and nonanal contributed to the characteristic elder flower odor, whereas linalool, alpha-terpineol, 4-methyl-3-penten-2-one, and (Z)-beta-ocimene provided floral notes. Fruity odors were associated with pentanal, heptanal, and beta-damascenone. Fresh and grassy odors were primarily associated with hexanal, hexanol, and (Z)-3-hexenol.

Nonsaponifiable lipid components of the pollen of elder extracted with chloroform-methanol afforded the following groups of compounds: hydrocarbons (8.7%), polycyclic aromatic hydrocarbons (0.2%), complex esters (5.2%), triglycerides (18.7%), hydroxy esters (27.9%), free fatty acids and alcohols (16.8%), free sterols (6.8%), and triterpenic alcohols (4.0%) (Stránsky et al. 2001).

The non-saponifiable fractions from petroleum and ether extracts of the bark of *Sambucus nigra* were found to contain α-amyrenone, α-amyrin, betulin, oleanolic acid and β-sitosterol (Lawrie et al. 1964). Inoue and Sato (1975) reported the following compounds from *S. nigra*: the triterpenes α- and β-amyrin, ursolic acid, oleanic acid, betulin, betulic acid and other components that include sitosterol, stigmasterol, campesterol, quercetin, rutin and n-alkanes.

The following phenolic acids: caffeic acid, p-coumaric acid, ferulic acid, gallic acid, syringic acid, 3, 4, 5-trimethoxybenzoic acid and chlorogenic acid were detected and identified in the bark of *Sambucus nigra* (Turek and Cisowski 2007).

Twenty-four aromatic metabolites belonging to cyanogenins, lignans, flavonoids, and phenolic glycosides were obtained from *Sambucus nigra* (D'Abrosca et al. 2001). Two compounds were isolated and identified as (2 S)-2-O-beta-D-glucopyranosyl-2-hydroxyphenylacetic acid and benzyl 2-O-beta-D-glucopyranosyl-2,6-dihydroxybenzoate. Cyanogenins had a mainly inhibiting effect on seed germination of *Lactuca sativa* (lettuce) and *Raphanus sativus* (radish) and *Allium cepa* (onion) while lignans stimulated radicle growth.

Antioxidant Activity

Spray-dried elderberry juice, containing high amounts of anthocyanin glucosides exhibited significant antioxidant and proxidant activities (Abuja et al. 1998). A strong, concentration-dependent prolongation of the lag phase of copper-induced oxidation of human low-density lipoprotein (LDL) was found, but the maximum oxidation rate was unaltered. Peroxyl-radical-driven LDL oxidation exhibited both extension of lag time and decrease of maximum oxidation rate. In the case of copper-mediated oxidation, low level concentrations (4μg/m) of elderberry anthocyanins were able to reduce α-tocopheroxyl radical to α-tocopherol. Clear prooxidant activity in copper-mediated oxidation was observed, depending on the time of addition of extract. No such effect was found in peroxyl-radical-mediated LDL oxidation. The ability of elderberry extract to provide antioxidant protection via inhibition of LDL-oxidation and free radical scavenging makes it a potentially valuable tool in the treatment of disease resulting from oxidative stress including cardiovascular disease, cancer, neurodegenerative disease, peripheral vascular disease, autoimmune diseases, and multiple sclerosis.

Several studies had shown that anthocyanin glycosides were indeed absorbed in humans, thus supplementing with elderberry extracts containing anthocyanins could provide significant antioxidant benefit. Wu et al. (2002) reported that two major anthocyanins in elderberry extract – 3-glucoside and cyanidin-3-sambubioside, as well as four metabolites: peonidin 3-glucoside, peonidin 3-sambubioside, peonidin monoglucuronide, and cyanidin-3-glucoside monoglucuronide were identified in the urine of elderly women within 4 hours of consumption. Their study demonstrated that in-vivo methylation of cyanidin to peonidin and glucuronide conjugate formation occurred after consumption of the anthocyanins and the low absorption and excretion of the anthocyanin compared with other flavonoids. Bitsch et al. (2004) found that within 7 hours, the urinary excretion of total anthocyanins was 0.04% and 0.37% of the administered dose following blackcurrant juice and elderberry extract consumption, respectively. Anthocyanin absorption was found to be significantly greater following the intake of elderberry extract than after the intake of blackcurrant juice as shown by the 5.3- and 6.2-fold higher estimates of dose-normalized Cmax and AUC(0-tZ) of total anthocyanins, respectively. Cao and Prior (1999) also reported that elderberry extract containing cyanidin 3-glucoside and cyanidin 3-sambubioside were absorbed into the human plasma. Milbury et al. (2002) also examined the bioavailability and pharmacokinetics of elderberry anthocyanins in humans. They reported that anthocyanins were detected as glycosides in both plasma and urine samples and that the elimination of plasma anthocyanins appeared to follow first-order kinetics and most anthocyanin compounds were excreted in urine within 4 hours after feeding. The current findings appeared to refute assumptions that anthocyanins were not absorbed in their unchanged glycosylated forms in humans. Pharmacokinetics studies by Frank et al. (2007) found that the fraction of orally administered anthocyanins from elderberry extract recovered unchanged in urine of the healthy volunteers indicating a low bioavailability of these compounds. The peak and average systemic exposure to the major elderberry anthocyanidin glycosides in plasma as well as their renal excretion exhibited approximate dose-dependent characteristics within the administered range.

Studies showed that the enrichment of endothelial cells with elderberry anthocyanins conferred significant protective effects in endothelial cells against the following oxidative stressors: hydrogen peroxide (H_2O_2); 2,

2'-azobis(2-amidinopropane) dihydrochloride (AAPH); and $FeSO_4$, ascorbic acid (Youdim et al. 2000). Elderberry extract containing four anthocyanins, were found to be incorporated into the plasma membrane and cytosol of human endothelial cells following 4 hours incubation at 1 mg/ml. However, incorporation within the cytosol was considerably less than that in the membrane. Uptake within membrane and cytosol seemed to be structure dependent, with monoglycoside concentrations greater than that of the diglucosides in both compartments. These results demonstrated that vascular endothelial cells could incorporate anthocyanins into the membrane and cytosol, imparting significant protective effects against damage from reactive oxygen species. The most pronounced affect was observed with protection against H_2O_2-induced loss in cell viability.

Antihyperlipidemic Activity

In a randomised placebo-controlled study of 34 healthy subjects, there was only a small, statistically non-significant alteration in cholesterol levels in the elderberry group (from 199 to 190 mg/daily) compared to the placebo group (from 192 to 196 mg/daily) at the end of a 2 weeks period (Murkovic et al. 2004). Small reductions were also reported in triglycerides, and HDL- and LDL-cholesterol. The resistance to copper-induced oxidation of LDL did not change within 3 weeks. Although improvements in lipid values were statistically insignificant, the dosage of elderberry extract was low. The researchers asserted that it was possible higher, but nutritionally relevant doses might significantly reduce postprandial serum lipids, however, further study on patients with elevated lipid levels was warranted.

Antiviral Activity

Elderberry has been used in folk medicine for centuries to treat influenza, colds and sinusitis (Zakay-rones et al. 1995). Numerous studies using Sambucol, a standardized elderberry preparation had shown it to neutralize and reduce the infectivity of influenza viruses A and B (Zakay-Rones et al. 1995, 2004), HIV strains and clinical isolates (Sahpira-Nahor et al. 1995), and *Herpes simplex* virus type 1 (HSV-1) strains and clinical isolates (Morag et al. 1997).

A standardized elderberry extract, Sambucol (SAM), was found to be effective in-vitro against ten strains of influenza virus (Zakay-Rones et al. 1995). SAM was found to decrease hemagglutination and to inhibit replication of human influenza viruses type A/Shangdong 9/93 (H3N2), A/Beijing 32/92 (H3N2), A/Texas 36/91 (H1N1), A/Singapore 6/86 (H1N1), type B/Panama 45/90, B/Yamagata 16/88, B/Ann Arbor 1/86, and of animal strains from Northern European swine and turkeys, A/Sw/Ger 2/81, A/Tur/Ger 3/91, and A/Sw/Ger 8533/91 in Madin-Darby canine kidney cells in a placebo-controlled, double blind study. A significant improvement of the symptoms, including fever, was seen in 93.3% of the cases in the SAM-treated group within 2 days, whereas in the control group 91.7% of the patients showed an improvement within 6 days. Complete healing was achieved within 2–3 days in nearly 90% of the SAM-treated group and within at least 6 days in the placebo group. Convalescent phase serum exhibited a higher antibody level to influenza virus in the Sambucol group, than in the control group. In another randomized, double-blind, placebo-controlled study, the efficacy and safety of oral elderberry syrup for treating influenza A and B infections was investigated in 60 influenza patients aged 18–54 years (Zakay-rones et al. 1995). Elderberry extract appeared to offer an efficient, safe and cost-effective treatment for influenza.

The elderberry extract was found to inhibit Human Influenza A (H1N1) infection in-vitro with an IC_{50} value of 252 µg/ml (Roschek et al. 2009). The Direct Binding Assay confirmed that flavonoids from the elderberry extract bound to H1N1 virions and obstructed the ability of the viruses to infect host cells. Two compounds were identified, 5,7,3',4'-tetra-O-methylquercetin (1) and 5,7-dihydroxy-4-oxo-2-(3,4,5-trihydroxyphenyl)chroman-3-yl-3,4,5-trihydroxycyclohexanecarboxylate (2), as H1N1-bound chemical species. Compound 1 and dihydromyricetin (3), the corresponding 3-hydroxyflavonone of 2, were synthesized and shown to inhibit H1N1 infection in-vitro by binding to H1N1 virions, obstructing host cell entry and/or recognition. Compound 1 gave an IC_{50} of 0.13 µg/mL (0.36 µM) for H1N1 infection inhibition, while dihydromyricetin (3) gave an IC_{50} of 2.8 µg/ml (8.7 µM). The H1N1 inhibitory activities of the elderberry flavonoids compared favorably to the known anti-influenza activities of Oseltamivir (Tamiflu; 0.32 µM) and Amantadine (27 µM).

Shapira-Nahor et al. (1995) found a significant reduction in the infectivity of HIV strains (ELI, LAI, HIV IIIb) in CD4+ cell lines (CEM and Molt 4) and human peripheral blood lymphocytes in the presence of Sambucol, by measuring the level of HIV core antigen p24 in supernatants of the infected cultures, as compared with controls without Sambucol. HIV− antigen was not detected 5 and 9 days post infection in cultures infected with patient isolates which were previously treated with Sambucol.

Morag et al., (1997) tested Sambucol (SAM) against four strains of HSV-1 (*Herpes simplex* Virus) including two acyclovir-resistant strains in MS 9 (human diploid fibroblasts) and buffalo green monkey cells. The replication of HSV-1 strains was completely prevented: (a) by preincubation with SAM before infection of the cells; (b) when SAM and the virus were added together to the cells; and (c) when SAM was added 30 min post viral adsorption on the cells.

A lyophilized infusion from flowers of *Sambucus nigra*, aerial parts of *Hypericum perforatum*, and roots of *Saponaria officinalis* (100 g; 70 g; 40 g) exhibited an antiviral effect (Serkedjieva et al. 1990). It inhibited the reproduction of different strains of influenza virus types A and B, both in-vitro and in-vivo, and herpes simplex virus type 1, in-vitro. The preparation was found to contain flavonoids, triterpene saponins, phenolic acids, tannins and polysaccharides which could be responsible for its antiviral properties.

A review on complementary medicine for treating or preventing influenza or influenza-like illness indicated that 2 or more trials out of 14 randomized controlled trials testing 7 preparations including *Sambucus nigra* extract showed some encouraging data (Guo et al. 2007). Another review of the common cold and influenza viruses, also highlighted select botanicals including *Sambucus nigra*, and nutritional considerations (vitamins A and C, zinc, high lactoferrin whey protein, N-acetylcysteine, and DHEA) that may help in the prevention and treatment of these conditions (Roxas and Jurenka 2007).

Immunomodulatory Activity

In addition to its antiviral properties, Sambucol Elderberry Extract and its formulations were found to stimulate the healthy immune system by increasing inflammatory cytokine production (Barak et al. 2001). Production of inflammatory cytokines (interleukin IL-1 beta, IL-6, IL-8, tumour necrosis alpha TNF-α) were significantly increased, mostly by the Sambucol Black Elderberry Extract (2–45 fold), as compared to LPS, a known monocyte activator (3.6–10.7 fold). The most notable increase was observed in TNF-α production (44.9 fold). In a follow-on study, the Sambucol preparations increased the production of five cytokines that included four inflammatory cytokines (interleukin-1 beta, tumour necrosis factor alpha, and IL-6 and IL-8) and one anti-inflammatory cytokine (IL-10) by 1.3–6.2 fold compared to the control (Barak et al. 2002). The three Sambucol formulations activated the healthy immune system by increasing inflammatory and anti-inflammatory cytokines production. Sambucol may therefore be beneficial to the immune system activation and in the inflammatory process in healthy individuals or in patients with various diseases. Sambucol could also have an immunoprotective or immunostimulatory effect when administered to cancer or AIDS patients, in conjunction with chemotherapeutic or other treatments.

Antidiabetic Activity

The results of studies by Gray et al. (2000) demonstrated the presence of insulin-releasing and insulin-like activity in the traditional antidiabetic plant, *Sambucus nigra*. In their study (2 × 2 factorial design), an aqueous extract of elder (AEE, 1 g/L) significantly enhanced 2-deoxy-glucose transport, glucose oxidation and glycogenesis of mouse abdominal muscle. In acute 20-minute tests, 0.25–1 g/L AEE incubated with rat pancreatic cells exhibited a dose-dependent stimulatory effect on insulin secretion. The insulin releasing effect of AEE (0.5 g/L) was significantly potentiated by 16.7 mmol/L of glucose and significantly decreased by 0.5 mmol/L of diazoxide. Sequential extraction with solvents exhibited activity in both methanol and water fractions, indicating a cumulative effect of more than one extract constituent. Known constituents of elder, including lectin, rutin and the lipophilic triterpenoid (lupeol) and sterol (ß-sitosterol), did not stimulate insulin secretion. In a separate study, extracts of elderflowers were found to activate peroxisome proliferator-activated receptor (PPAR) gamma and to stimulate insulin-dependent glucose uptake suggesting that they may have a potential use in the prevention and/or treatment of insulin resistance (Christensen et al. 2010). Type 2 diabetes (T2D) is caused by a combination of insulin resistance and beta-cell failure and can be treated with insulin sensitizing drugs that target the nuclear receptor PPAR gamma. Fractionation of a methanol extract of elderflowers resulted in the identification of two well-known PPARgamma agonists; α-linolenic acid and linolenic acid as well as the flavanone naringenin. Naringenin was found to activate PPARgamma without inducing adipocyte differentiation. Elderflower metabolites such as quercetin-3-O-rutinoside, quercetin-3-O-glucoside, kaempferol-3-O-rutinoside, isorhamnetin-3-O-rutinoside, isorhamnetin-3-O-glucoside, and 5-O-caffeoylquinic acid failed to activate PPARgamma. The findings suggested that flavonoid glycosides could not activate PPARgamma, whereas some of their aglycones were potential agonists of PPARgamma.

Recent studies disclosed that the glycosylated hemoglobin levels were much higher in the diabetic group but were significantly lower in the group protected by natural polyphenolic extracts from common elder fruit (Ciocoiu et al. 2009). The natural polyphenol compounds decreased the lipids peroxides, neutralized the lipid peroxil radicals and prevented LDL oxidation. It was found that due to the polyphenolic protection of the rats from the diabetic group treated with natural polyphenolic extracts, the atherogenic risk was maintained at normal limits. In contrast, following the perturbation of the lipid metabolism in the diabetic rats, atherogen risk exhibited significantly elevated values. The serum activity of glutathione-peroxidase and superoxide-dismutase had significantly lower values in the diabetic group as compared to the group protected by polyphenols. Through the hypoglycemic, hypolipidemic and antioxidant effects, *Sambucus nigra* represented a possible dietary adjunct for the treatment of diabetes and a potential source for the discovery of new orally active agent(s) for future diabetes therapy. Patients with gall-bladder inflammation that underwent laparoscopy were found to have higher percentage of specifically bound insulin-like growth factor binding protein 3(IGFBP-3) to *Sambucus nigra* agglutinin compared with healthy subjects due to the increased content of sialic acid (Baricević et al. 2006).

Anti-Inflammatory Activity

The elder flower extract was found to display useful anti-inflammatory properties that could be exploited therapeutically for the control of inflammation in human periodontitis (Harokopakis et al. 2006). The elder flower extract was found to potently inhibit all proinflammatory activities of major virulence factors from the periodontal pathogens *Porphyromonas gingivalis* and *Actinobacillus actinomycetemcomitans*. Investigation of the underlying mechanisms revealed that the anti-inflammatory extract inhibited activation of the nuclear transcription factor kappaB and of phosphatidylinositol 3-kinase.

Anticancer Activity

In the study to gauge anticarcinogenic potential, the acetone aqueous extracts of cultivated *S. nigra* and wild *S. canadensis* fruits demonstrated significant chemopreventive potential through strong induction of quinone reductase and inhibition of cyclooxygenase-2, which was indicative of antiinitiation and antipromotion properties, respectively (Thole et al. 2006). In addition to flavonoids, the presence of more lipophilic compounds such as sesquiterpenes, iridoid monoterpene glycosides, and phytosterols were confirmed.

A study on linkage-specific sialylation changes in oral carcinogenesis showed that linkage-specific lectins, *Sambucus nigra* (SNA) and *Maackia amurensis* (MAM) were able to detect α 2–6- and α 2–3-linked sialic acid, which were used to analyze linkage-specific sialyltransferases activity (TSA) and sialoproteins in oral carcinogenesis (Shah et al. 2008). Malignant tissues exhibited significantly higher concentrations of TSA, reactivity of SNA and MAM, and α 2,3-ST activity compared to the adjacent normal tissues. The data suggested potential utility of sialylation markers in early detection, prognostication and treatment monitoring of oral cancer. Colonic mucin sialylation in ulcerative colitis of Asians was found to be different from Europeans in that there was less binding of *Sambucus nigra* agglutinins (McMahon et al. 1997). This might have a role in lower colorectal carcinoma rates in Asians. In human colon carcinoma, elevated amounts of sialic acids were found that correlated with tumour progression. Murayama et al. (1997) found that colon carcinoma glycoproteins carrying α 2,6-linked sialic acid reactive with *Sambucus nigra* agglutinin (SNA) were not constitutively manifested in normal human colon mucosa and were different from sialyl-Tn antigen. The data suggested that SNA binding in human colon carcinoma was due to de novo expression of a specific sialic acid present on selected glycoproteins. Data from studies indicated that the expression of α 2,6-sialylated sugar chains was appreciably elevated in the majority of colon cancer specimens examined (Dall'Olio and Trerè 1993). SNA (*S. nigra* agglutinin) was unreactive with epithelial cells of all the 13 normal colon specimens, weakly reactive with 3 out of 8 benign lesions and strongly reactive for 23 out of 26 carcinomas. The quantitative binding pattern determination of *Sambucus nigra* agglutinins could be used to in routine diagnosis to identify moderate/severe dysplasia in colorectal adenomas (Bronckart et al. 1999).

The expressive levels of galectin-3(gal-3) and *Sambucus nigra* agglutinin (SNA) were found to have clinicopathological significance in the diagnosis of benign and malignant lesions of stomach (Zhou et al. 2009). The positive rates of gal-3 and SNA were significantly higher in gastric cancer tissues than those in peritumoral tissues and different types of benign lesions. The positive cases of gal-3 and/or SNA in peritumoral tissues and benign lesions exhibited mild- to severe-atypical hyperplasia of mucous epithelial cells. The positive rates of gal-3 and SNA were higher in lymphnode metastatic site N1. The results revealed that expressive levels of gal-3 and SNA may be important molecular markers of lectins for studying carcinogenesis, progression and biological behaviors in gastric cancer. Similar studies showed that the expressive level of gal-3 and SNA lectins might have clinicopathological significance on the carcinogenesis, progression and biologic behaviors of breast cancer (Chen et al. 2010). The positive rates and scoring means of gal-3 and SNA were significantly greater in breast cancer than those in benign lesions.

Sambucus species were found contain a number of two-chain ribosome inactivating proteins (RIPs) structurally and enzymatically related to ricin from *Ricinus communis*, and to have an enzymatic activity on ribosomes, leading to the inhibition of protein synthesis, higher than ricin, but lacked the tremendous non-specific toxicity of ricin (Girbes et al. 2003). Therefore, they were designated as non-toxic type 2 RIPs. The most representative and studied members were nigrin b present in the bark of the common (black) elder *Sambucus nigra* and ebulin 1 present in the leaves of the dwarf elder *Sambucus ebulus*. Conjugation of nigrin b or ebulin 1 to either transferrin or monoclonal antibodies provided highly active conjugates targeting cancer. Thus, these non-toxic type 2 RIPs could be promising tools for cancer therapy.

Protein Synthesis Inhibition

Sambucus nigra seed proteins were found to strongly inhibit protein synthesis and to display 28 S rRNA N-glycosidase activity characteristic of all types of ribosome-inactivating proteins (RIPs) (Citores et al. 1994). Western blot analysis showed several proteins that reacted with antibodies raised against the novel non-toxic type 2 ribosome-inactivating protein nigrin b isolated from elder bark, thus indicating the presence of a new type-2 RIP.

A strongly basic, two-chain ribosome-inactivating protein (RIP) named basic nigrin b, was found in the bark of *Sambucus nigra* (de Benito et al. 1997). The new protein did not agglutinate red blood cells, even at high concentrations and displayed an unusually and extremely high activity towards animal ribosomes (IC_{50} of 18 μg/ml for translation by rabbit reticulocyte lysates).

Detailed studies demonstrated that virtually all tissues of the elderberry tree contained multiple type-2 RIPs/lectins. Ribosome-inactivating proteins (RIPs) are a family of enzymes that trigger the catalytic inactivation of ribosomes. All elderberry type-2 RIPs/lectins could be classified into four groups (Chen et al. 2002). A first group comprised the tetrameric Neu5Ac(a2,6)Gal/GalNAc-specific type-2 RIPs similar to the bark type-2 RIP SNA-I (Van Damme et al. 1996b). The NeuAc(alpha-2,6)Gal/GalNAc binding lectin from elderberry bark (SNAI) was found to strongly inhibit cell-free protein synthesis in a rabbit reticulocyte lysate and to be a type-2 ribosome-inactivating protein. However, SNAI differed from all previously described type-2 ribosome-inactivating proteins by its specificity towards NeuAc(alpha-2,6)Gal/GalNAc and its unusual molecular structure. Dimeric galactose-specific type-2 RIP resembling SNA-V from the bark (Van Damme et al. 1996a) formed a second group, whereas the third group comprised the monomeric type-2 RIPs with 2an inactive B-chain, similar to SNLRP from the bark (Van Damme et al. 1997a). The occurrence of a type 2 ribosome-inactivating protein with an inactive B chain. A fourth group comprising the Gal/GalNAc-specific lectins similar to SNA-IVf and related to the dimeric galactosespecific type-2 RIPs, but are not RIPs because they are encoded by genes from which the complete A-chain is deleted (Van Damme et al. 1997c). Type-2 RIPs were found to compose of two structurally and functionally different polypeptides, called the A chain and the B chain. The A chain depurinated rRNA by means of an N-glycosidase activity and was a potent inhibitor of eukaryotic (and sometimes prokaryotic) protein synthesis at the ribosomal level. The B chain was catalytically inactive but exhibited a carbohydrate-binding activity comparable to that of lectins.

Sambucus nigra agglutinin 1 (SNA-I) was originally isolated from elderberry bark where it represented about 5% of the total soluble protein (Broekaert et al. 1984). The lectin was found to be a glycoprotein especially rich in asparagine/aspartic acid, glutamine/glutamic acid, valine and leucine. A second lectin (SNA-II) was isolated from elderberry (*Sambucus nigra* L.) bark (Kaku et al. 1990). This lectin was found to be a blood group nonspecific glycoprotein containing 7.8% carbohydrate and rich in asparagine/aspartic acid, glutamine/glutamic acid, glycine, valine, and leucine. The lectin's binding site appeared to be most complementary to GalNAc linked α to the C-2, C-3, or C-6 hydroxyl group of galactose. These disaccharide units were approximately 100 times more potent than melibiose, 60 times more potent than N-acetyllactosamine, and 30 times more potent than lactose. Interestingly, the blood group A-active trisaccharide containing an L-fucosyl group linked α1–2 to galactose was ten-fold poorer as an inhibitor than the parent oligosaccharide (GalNAcα1–3Gal), suggesting steric hindrance to binding by the α-L-fucosyl group and the failure of the lectin to exhibit blood group A specificity. A third elderberry (*Sambucus nigra*) lectin (SNA-III) was isolated from dry seeds (Peumans et al. 1991) This lectin was found to be a blood-group, nonspecific glycoprotein containing 21% of carbohydrate, and rich in asparagine (or aspartic acid), serine, glutamine (or glutamic acid), and glycine. A comparison of SNA-III to the previously described elderberry-bark lectins, SNA-I and SNA-II, indicated that the seed lectin was markedly different from them. A second NeuAc(alpha2,6)Gal/GalNAc binding type 2 ribosome-inactivating protein (RIP), called SNAI' was isolated from elderberry bark (Barre et al. 1997; Van Damme et al. 1997b). SNAI', a minor bark protein was found to closely resemble the previously described major Neu5Ac(alpha2,6)Gal/GalNAc binding type 2 RIP called SNAI with respect to its carbohydrate-binding specificity and ribosome-inactivating activity but had a dissimilar molecular structure.

A previously unknown haemagglutinin, named *Sambucus nigra* agglutinin-III (SNAIII), was purified from the fruit of the elder (*Sambucus nigra*) (Mach et al. 1991). SNA-III displayed a high affinity for oligosaccharides containing exposed Nacetylgalactosamine and galactose residues whereas elder bark agglutinin I (SNA-I) was highly specific for terminal alpha 2,6-linked sialic acid residues. Different N-terminal sequences and the amino acid composition distinguished the fruit lectin from elder bark agglutinin II (SNA-II), which exhibited a similar carbohydrate specificity.

A very similar lectin called *Sambucus nigra* fruit specific agglutinin I (SNA-If) was identified as a minor protein in ripe elderberry fruits (Peumans et al. 1998). Bioassays further showed that the transgenic plants were as sensitive as control plants towards infection with tobacco mosaic virus (TMV), indicating that SNA-If did not act as an antiviral protein in plants (Chen et al. 2002). The RIPs/lectins are believed to play a role in the plant's defence against bacterial, fungal or insect attack. Plant lectins have been widely used for the detection, isolation, and characterization of glycoconjugates using their characteristic carbohydrate binding properties and has importance in biological phenomena.

Purified SNA lectin from elderberry bark was found to precipitate highly sialylated glycoproteins such as fetuin, orosomucoid, and ovine submaxillary mucin (Shibuya et al. 1987b). Colon cancer tissues was found to display an increased activity of beta-galactoside alpha2,6 sialyltransferase (ST6Gal.I) and an increased reactivity with the lectin from *Sambucus nigra* (SNA), specific for alpha2,6-sialyl-linkages (Dall'Olio et al. 2000).

The *Sambucus nigra* type-2 ribosome-inactivating protein SNA-I' was found to exhibit in plants an antiviral activity in transgenic tobacco (Chen et al. 2002b). Expression of SNA-I' under the control of the 35 S cauliflower mosaic virus promoter enhanced the plant's resistance against infection with tobacco mosaic virus. Although the type-2 ribosome-inactivating proteins (SNA-I, SNA-V, SNLRP) from *Sambucus nigra* were all devoid of rRNA N-glycosylase activity towards plant ribosomes, some of them vividly showed polynucleotide-adenosine glycosylase activity towards tobacco mosaic virus RNA (Vandenbussche et al. 2004). This particular substrate specificity was exploited to further unravel the mechanism underlying the in planta antiviral activity of ribosome-inactivating proteins. This lectin was shown by immunochemical techniques to bind specifically to terminal Neu5Ac (α2–6)Gal/GalNAc residues of glycoconjugates (Shibuya et al. 1987a, b).

Recently, a major lectin called called SNAflu-I was found in whole inflorescence of *Sambucus nigra* and was GalNAc specific (Karpova et al. 2007). Two other lectins were found in the pollen. A major positively charged lectin called SNApol-I was Glc/Man specific, while the other pollen component (SNApol-II) was Gal specific. Only SNApol-I demonstrated the antagonistic activity against the phenazine inhibitors on *Bacillus subtilis* cells in-vivo. This lectin but not the SNAflu-I also inhibited transcription in-vitro.

Weight Reduction Activity

A supplement with *Sambucus nigra* and *Asparagus officinalis* was found to be good for weight reduction (Chrubasik et al. 2008). Eighty participants completed a diet regime of *Sambucus nigra* berry juice enriched with flower extract and tablets of berry powder containing a total of 1 mg anthocyanins, 370 mg flavonol glycosides and 150 mg hydroxycinnamates per day; the *Asparagus officinalis* powder tablets provided

19 mg saponins per day. After the diet, the mean weight, blood pressure, physical and emotional well-being and the quality of life were significantly improved.

Traditional Medicinal Uses

Sambucus nigra has a very long history of household use as a medicinal herb and is also much used by herbalists. The flowers, fruits, leaves, bark and roots of *S. nigra* are used for a diverse variety of ailments. Although *S. nigra* is not generally considered poisonous, isolated cases of poisoning in animals and man have been reported after eating the bark, leaves, berries, roots and stems (Atkinson and Atkinson 2002).

The flowers are the main part used in modern herbalism. The fresh flowers are employed in the distillation of 'Elder Flower Water'. The water is mildly astringent and a gentle stimulant. It is mainly used as a vehicle for eye and skin lotions. The dried flowers are diaphoretic, diuretic, expectorant, galactogogue and pectoral. An infusion is used for the treatment of chest complaints and as a wash for inflamed eyes. The flower infusion is regarded as tonic and blood cleanser. Tea from the flowers is used against cold, high temperature and scarlatina. Externally, the flowers are used in poultices to ease pain and alleviate inflammation; used as an ointment, it treats chilblains, burns, wounds, scalds.

The fruit is deemed to be depurative, weakly diaphoretic and gently laxative A tea made from the dried berries is believed to be effective for colic and diarrhoea. The leaves are purgative, diaphoretic, diuretic, expectorant and haemostatic. The juice is claimed to be a good remedy for inflamed eyes. An ointment made from the leaves is emollient and is used in the treatment of bruises, sprains, chilblains, wounds.

The inner bark is diuretic, purgative and emetic. It is used in the treatment of constipation and arthritic conditions. An emollient ointment is prepared from the green inner bark. A homeopathic remedy is made from the fresh inner bark of young branches. It relieves asthmatic symptoms and spurious croup in children. The pith of young stems is used in treating burns and scalds. The root is emetic and purgative and is effective against dropsy but is no longer used in herbal medicine.

Other Uses

Extracts from *S. nigra* are used in horticulture as a repellent against insects. Its shoots are put into the soil to scare off mice and moles. *S. nigra* has also been planted for erosion control.

It is not valued as a timber due to it small dimensions, but the wood is suitable for making pegs and other small wooden items because of its whiteness, close grain, good cutting and polishing properties. The pith from 1-year-old branches is used for making plant sections in microscopy.

Comments

Several closely related species native to Asia and North America have been subsumed as subspecies under *Sambucus nigra*.

Selected References

Abuja PM, Murkovic M, Pfannhauser W (1998) Antioxidant and prooxidant activities of elderberry (*Sambucus nigra*) extract in low-density lipoprotein oxidation. J Agric Food Chem 46:4091–4096

Akbulut M, Ercisli S, Tosun M (2009) Physico-chemical characteristics of some wild grown European elderberry (*Sambucus nigra* L.) genotypes. Pharmacog Mag 5(20):320–323

Atkinson MD, Atkinson E (2002) *Sambucus nigra* L. – biological flora of the British Isles, No. 225. J Ecol 90:895–923

Barak V, Halperin T, Kalickman I (2001) The effect of Sambucol, a black elderberry-based, natural product, on the production of human cytokines: I. inflammatory cytokines. Eur Cytokine Netw 12(2):290–296

Barak V, Birkenfeld S, Halperin T, Kalickman I (2002) The effect of herbal remedies on the production of human inflammatory and anti-inflammatory cytokines. Isr Med Assoc J 4:S919–S922

Baricević I, Malenković V, Jones DR, Nedić O (2006) The influence of laparoscopic and open surgery on the concentration and structural modifications of insulin-like growth factor binding protein 3 in the human circulation. Acta Physiol Hung 93(4):361–369

Barre A, Citores L, Mostafapous K, Rougé P, Girbés T, Goldstein IJ, Peumans WJ (1997) Elderberry (*Sambucus nigra*) bark contains two structurally different Neu5Ac(alpha2,6)Gal/GalNAc-binding type 2 ribosome-inactivating proteins. Eur J Biochem 245(3):648–655

Bitsch I, Janssen M, Netzel M, Strass G, Frank T (2004) Bioavailability of anthocyanidin-3-glycosides following consumption of elderberry extract and blackcurrant juice. Int J Clin Pharmacol Ther 42(5):293–300

Bolli R (1994) Revision of the genus *Sambucus*. Diss Bot 223:161

Bown D (1995) Encyclopaedia of herbs and their uses. Dorling Kindersley, London, 424 pp

Broekaert WF, Nsimba-Lubaki M, Peeters B, Peumans WJ (1984) A lectin from elder (*Sambucus nigra* L.) bark. Biochem J 221:163–169

Bronckart Y, Nagy N, Decaestecker C, Bouckaert Y, Remmelink M, Gielen I, Hittelet A, Darro F, Pector JC, Yeaton P, Danguy A, Kiss R, Salmon I (1999) Grading dysplasia in colorectal adenomas by means of the quantitative binding pattern determination of *Arachis hypogaea, Dolichos biflorus, Amaranthus caudatus, Maackia amurensis,* and *Sambucus nigra* agglutinins. Hum Pathol 30(10):1178–1191

Burbidge NT, Gray M (1970) Flora of the Australian capital territory. Australian National University Press, Canberra, 447 pp

Cao G, Prior RL (1999) Anthocyanins are detected in human plasma after oral administration of an elderberry extract. Clin Chem 45:574–576

Chen Y, Peumans WJ, Van Damme EJ (2002a) The *Sambucus nigra* type-2 ribosome-inactivating protein SNA-I' exhibits in planta antiviral activity in transgenic tobacco. FEBS Lett 516(1–3):27–30

Chen Y, Vandenbussche F, Rougé P, Proost P, Peumans WJ, Van Damme EJM (2002b) A complex fruit-specific type-2 ribosome-inactivating protein from elderberry (*Sambucus nigra*) is correctly processed and assembled in transgenic tobacco plants. Eur J Biochem 269:2897–2906

Chen G, Zou Q, Yang Z (2010) Expression of galectin-3 and *Sambucus nigra* agglutinin and its clinicopathological significance in benign and malignant lesions of breast. Zhong Nan Da Xue Xue Bao Yi Xue Ban 35(6):584–589 (In Chinese)

Chiej R (1984) The Macdonald encyclopaedia of medicinal plants. Macdonald & Co, London, 447 pp

Christensen KB, Petersen RK, Kristiansen K, Christensen LP (2010) Identification of bioactive compounds from flowers of black elder (*Sambucus nigra* L.) that activate the human peroxisome proliferator-activated receptor (PPAR) gamma. Phytother Res 24(Suppl 2):S129–S132

Chrubasik C, Maier T, Dawid C, Torda T, Schieber A, Hofmann T, Chrubasik S (2008) An observational study and quantification of the actives in a supplement with *Sambucus nigra* and *Asparagus officinalis* used for weight reduction. Phytother Res 22(7):913–918

Ciocoiu M, Mirón A, Mares L, Tutunaru D, Pohaci C, Groza M, Badescu M (2009) The effects of *Sambucus nigra* polyphenols on oxidative stress and metabolic disorders in experimental diabetes mellitus. J Physiol Biochem 65(3):297–304

Citores L, Iglesias R, Muñoz R, Ferreras JM, Jimenez P, Girbes T (1994) Elderberry (*Sambucus nigra* L.) seed proteins inhibit protein synthesis and display strong immunoreactivity with rabbit polyclonal antibodies raised against the type 2 ribosome-inactivating protein nigrin b. J Exp Bot 45(4):513–516

D'Abrosca B, DellaGreca M, Fiorentino A, Monaco P, Previtera L, Simonet AM, Zarrelli A (2001) Potential allelochemicals from *Sambucus nigra*. Phytochem 58(7):1073–1081

Dall'Olio F, Trerè D (1993) Expression of alpha 2,6-sialylated sugar chains in normal and neoplastic colon tissues. Detection by digoxigenin-conjugated *Sambucus nigra* agglutinin. Eur J Histochem 37(3):257–265

Dall'Olio F, Chiricolo M, Ceccarelli C, Minni F, Marrano D, Santini D (2000) Beta-galactoside alpha2,6 sialyltransferase in human colon cancer: contribution of multiple transcripts to regulation of enzyme activity and reactivity with *Sambucus nigra* agglutinin. Int J Cancer 88(1):58–65

de Benito FM, Citores L, Iglesias R, Ferreras JM, Camafeita E, Méndez E, Girbés T (1997) Isolation and partial characterization of a novel and uncommon two-chain 64-kDa ribosome-inactivating protein from the bark of elder (*Sambucus nigra* L.). FEBS Lett 413(1):85–91

Frank T, Janssen M, Netzel G, Christian B, Bitsch I, Netzel M (2007) Absorption and excretion of elderberry (*Sambucus nigra* L.) anthocyanins in healthy humans. Methods Find Exp Clin Pharmacol 29(8):525–533

Girbes T, Ferreras JM, Arias FJ, Muñoz R, Iglesias R, Jimenez P, Rojo MA, Arias Y, Perez Y, Benitez J, Sanchez D, Gayoso MJ (2003) Non-toxic type 2 ribosome-inactivating proteins (RIPs) from *Sambucus*: occurrence, cellular and molecular activities and potential uses. Cell Mol Biol (Noisy-le-grand) 49(4):537–545

Gray AM, Abdel-Wahab YHA, Flatt PR (2000) The traditional plant treatment, *Sambucus nigra* (elder), exhibits insulin-like and insulin-releasing actions in vitro. J Nutr 130:15–20

Grieve M (1971) A modern herbal, vol 2, Penguin. Dover publications, New York, 919 pp

Guo R, Pittler MH, Ernst E (2007) Complementary medicine for treating or preventing influenza or influenza-like illness. Am J Med 120(11):923–929

Harokopakis E, Albzreh MH, Haase EM, Scannapieco FA, Hajishengallis G (2006) Inhibition of proinflammatory activities of major periodontal pathogens by aqueous extracts from elder flower (*Sambucus nigra*). J Periodontol 77(2):271–279

Inoue T, Sato K (1975) Triterpenoids of *Sambucus nigra* and *S. canadensis*. Phytochem 14(8):1871–1872

Jørgensen U, Hansen M, Christensen LP, Jensen K, Kaack K (2000) Olfactory and quantitative analysis of aroma compounds in elder flower (*Sambucus nigra* L.) drink processed from five cultivars. J Agric Food Chem 48(6):2376–2383

Kabuce N (2006) NOBANIS – invasive alien species fact sheet – *Sambucus nigra*. Online Database of the North European and Baltic Network on Invasive Alien Species – NOBANIS www.nobanis.org. Accessed 10 Dec 2009

Kaku H, Peumans WJ, Goldstein IJ (1990) Isolation and characterization of a second lectin (SNA-II) present in elderberry (*Sambucus nigra* L.) bark. Arch Biochem Biophys 277(2):255–262

Karpova IS, Korets'ka NV, Pal'chykovs'ka LH, Nehruts'ka VV (2007) Lectins from *Sambucus nigra* L inflorescences: isolation and investigation of biological activity using procaryotic test-systems. Ukr Biokhim Zh 79(5):145–152, In Ukrainian

Lawrie W, Mclean J, Paton AC (1964) Triterpenoids in the bark of elder (*Sambucus nigra*). Phytochem 3(2):267–268

Lee J, Finn CE (2007) Anthocyanins and other polyphenolics in American elderberry (*Sambucus canadensis*) and European elderberry (*S. nigra*) cultivars. J Sci Food Agric 87:2665–2675

Lust JB (1974) The herb book. Bantam Books, New York, 174 pp

Mach L, Scherf W, Ammann M, Poetsch J, Bertsch W, Marz L, Glossl J (1991) Purification and partial characterization of a novel lectin from elder (*Sambucus nigra* L.) fruit. Biochem J 278(Pt 3):667–671

McMahon RF, Warren BF, Jones CJ, Mayberry JF, Probert CS, Corfield AP, Stoddart RW (1997) South Asians with ulcerative colitis exhibit altered lectin binding compared with matched European cases. Histochem J 29(6):469–477

Milbury PE, Cao G, Prior RL, Blumberg J (2002) Bioavailablility of elderberry anthocyanins. Mech Ageing Dev 123(8):997–1006

Morag AM, Mumcuoglu M, Baybikov T, Schelsinger M, Zakay-Rones Z (1997) Inhibition of sensitive and acyclovir-resistant HSV-1 strains by an elderberry extract *in vitro*. Z Phytother 25:97–98

Murayama T, Zuber C, Seelentag WK, Li WP, Kemmner W, Heitz PU, Roth J (1997) Colon carcinoma glycoproteins carrying alpha 2,6-linked sialic acid reactive with *Sambucus nigra* agglutinin are not constitutively expressed in normal human colon mucosa and are distinct from sialyl-Tn antigen. Int J Cancer 70(5):575–581

Murkovic M, Abuja PM, Bergmann AR, Zirngast A, Adam U, Winklhofer-Roob BM, Toplak H (2004) Effects of elderberry juice on fasting and postprandial serum lipids and low-density lipoprotein oxidation in healthy volunteers: a randomized, double-blind, placebo-controlled study. Eur J Clin Nutr 58(2):244–249

Peumans WJ, Kellens JT, Allen AK, Van Damme EJ (1991) Isolation and characterization of a seed lectin from elderberry (*Sambucus nigra* L) and its relationship to the bark lectins. Carbohydr Res 213:7–17

Peumans WJ, Roy S, Barre A, Rougé P, Van Leuven F, Van Damme EJM (1998) Elderberry (*Sambucus nigra*) contains truncated Neu5Ac (α-2,6)Gal/GalNac-binding type 2 ribosome-inactivating proteins. FEBS Lett 425:35–39

Roschek B Jr, Fink RC, McMichael MD, Li D, Alberte RS (2009) Elderberry flavonoids bind to and prevent H1N1 infection in vitro. Phytochem 70(10):1255–1261

Roxas M, Jurenka J (2007) Colds and influenza: a review of diagnosis and conventional, botanical, and nutritional considerations. Altern Med Rev 12(1):25–48

Sahpira-Nahor O, Zakay-Rones Z, Mumcuoglu M (1995) The effects of Sambucol® on HIV infection *in vitro*. Ann Israel Congress Microbiol, 6–7 Feb 1995

Schmitzer V, Veberic R, Slatner A, Stampar F (2010) Elderberry (*Sambucus nigra* L) wine: a product rich in health promoting compounds. J Agric Food Chem 58(18):10143–10146

Serkedjieva J, Manolova N, Zgorniak-Wowosielska I, Zawilińska B, Grzybek J (1990) Antiviral activity of the infusion (SHS-174) from flowers of *Sambucus nigra* L., aerial parts of *Hypericum perforatum* L., and roots of *Saponaria officinalis* L. against influenza and herpes simplex viruses. Phytother Res 4(3):97–100

Shah MH, Telang SD, Shah PM, Patel PS (2008) Tissue and serum alpha 2-3- and alpha 2-6-linkage specific sialylation changes in oral carcinogenesis. Glycoconj J 25(3):279–290

Shibuya N, Goldstein IJ, Broekaert WF, Nsimba-Lubaki M, Peeters B, Peumans WJ (1987a) Fractionation of sialylated oligosaccharides, glycopeptides, and glycoproteins on immobilized elderberry (*Sambucus nigra*) bark lectin. Arch Biochem Biophys 254(1):1–8

Shibuya N, Goldstein IJ, Broekaert WF, Nsimba-Lubaki M, Peeters B, Peumans WJ (1987b) The elderberry (*Sambucus nigra* L.) bark lectin recognizes the Neu5Ac(alpha 2–6)Gal/GalNAc sequence. J Biol Chem 262(4):1596–1601

Stránsky K, Valterová I, Fiedler P (2001) Nonsaponifiable lipid components of the pollen of elder (*Sambucus nigra* L.). J Chromatogr A 936(1–2):173–181

Thole JM, Kraft TFB, Sueiro LA, Kang YH, Gills JJ, Cuendet M, Pezzuto JM, Seigler DS, Lila MA (2006) A comparative evaluation of the anticancer properties of European and American elderberry fruits. J Med Food 9(4):498–504

Turek S, Cisowski W (2007) Free and chemically bonded phenolic acids in barks of *Viburnum opulus* L. and *Sambucus nigra* L. Acta Pol Pharm 64(4):377–383

Uphof JCT (1968) Dictionary of economic plants, 2nd edn. Cramer, Lehre, (1st ed. 1959)591 pp

U.S. Department of Agriculture, Agricultural Research Service (USDA) (2010) *USDA National Nutrient Database for Standard Reference, Release 23*. Nutrient Data Laboratory Home Page, http://www.ars.usda.gov/ba/bhnrc/ndl

Van Damme EJM, Barre A, Rougé P, Van Leuven F, Peumans WJ (1996a) Characterization and molecular cloning of *Sambucus nigra* agglutinin V (nigrin b), a GalNAc-specific type 2 ribosome-inactivating protein from the bark of elderberry (*Sambucus nigra*). Eur J Biochem 237:505–513

Van Damme EJM, Barre A, Rougé P, Van Leuven F, Peumans WJ (1996b) The NeuAc (a-2,6)-Gal/GalNAc binding lectin from elderberry (*Sambucus nigra*) bark, a type 2 ribosome inactivating protein with an unusual specificity and structure. Eur J Biochem 235:128–137

Van Damme EJM, Barre A, Rougé P, Van Leuven F, Peumans WJ (1997a) Isolation and molecular cloning of a novel type 2 ribosome-inactivating protein with an inactive B chain from elderberry (*Sambucus nigra*) bark. J Biol Chem 272:8353–8836

Van Damme EJM, Roy S, Barre A, Citores L, Mostafapous K, Rougé P, Van Leuven F, Girbés T, Goldstein IJ, Peumans WJ (1997b) Elderberry (*Sambucus nigra*) bark contains two structurally different Neu5Ac(alpha2,6)Gal /GalNAc-binding type 2 ribosome-inactivating proteins. Eur J Biochem 245(3):648–655

Van Damme EJM, Roy S, Barre A, Rougé P, Van Leuven F, Peumans WJ (1997c) The major elderberry (*Sambucus nigra*) fruit protein is a lectin derived from a truncated type 2 ribosome inactivating protein. Plant J 12:1251–1260

Vandenbussche F, Desmyter S, Ciani M, Proost P, Peumans WJ, Van Damme EJ (2004) Analysis of the in planta antiviral activity of elderberry ribosome-inactivating proteins. Eur J Biochem 271(8):1508–1515

Veberic R, Jakopic J, Stampar F, Schmitzer V (2009a) European elderberry (*Sambucus nigra* L.) rich in sugars, organic acids, anthocyanins and selected polyphenols. Food Chem 114(2):511–515

Veberic R, Jakopic J, Stampar F (2009b) Flavonols and anthocyanins of elderberry fruits (*Sambucus nigra* L.). Acta Hort (ISHS) 841:611–614

Willuhn G, Richter W (1977) On the constituents of *Sambucus nigra*. II. The lipophilic components of the flowers. Planta Med 31(4):328–343

Wu X, Cao G, Prior RL (2002) Absorption and metabolism of anthocyanins in elderly women after consumption of elderberry or blueberry. J Nutr 132:1865–1871

Youdim KA, Martin A, Joseph JA (2000) Incorporation of the elderberry anthocyanins by endothelial cells increases protection against oxidative stress. Free Radic Biol Med 29(1):51–60

Zakay-Rones Z, Varsano N, Zlotnik M, Manor O, Regev L, Schlesinger M, Mumcuoglu M (1995) Inhibition of several strains of influenza virus in vitro and reduction of symptoms by an elderberry extract (*Sambucus nigra* L.) during an outbreak of influenza B Panama. J Altern Complement Med 1(4):361–369

Zakay-Rones Z, Thom E, Wollan T, Wadstein J (2004) Randomized study of the efficacy and safety of oral elderberry extract in the treatment of influenza A and B virus infections. J Int Med Res 32(2):132–140

Zhou JP, Yang ZL, Liu DC, Zhou JP (2009) Expression of galectin 3 and *Sambucus nigra* agglutinin and their clinicopathological significance in benign and malignant lesions of stomach. Zhonghua Wei Chang Wai Ke Za Zhi 12(3):297–300 (In Chinese)

Anacardium occidentale

Scientific Name

Anacardium occidentale Linnaeus

Synonyms

Acajuba occidentalis (L.) Gaertner, *Anacardium amilcarianum* Machado, *Anacardium curatellifolium* A. St. Hilaire, *Anacardium kuhlmannianum* Machado, *Anacardium microcarpum* Ducke, *Anacardium occidentale* Linnaeus var. *americanum* de Candolle, *Anacardium mediterraneum* Vellozo, *Anacardium occidentale* Linnaeus var. *indicum* de Candolle, *Anacardium occidentale* Linnaeus var. *gardneri* Engler, *Anacardium occidentale* Linnaeus var. *longifolium* Presl, *Anacardium othonianum* Rizzini, *Anacardium rondonianum* Machado, *Anacardium subcordatum* Presl, *Cassuvium pomiferum* Lamarck, *Cassuvium reniforme* Blanco

Family

Anacardiaceae

Common/English Names

Cashew, Cashew nut, Kidney-nut

Vernacular Names

Angola: Cajueiro (Portuguese);
Arabic: Habb Al-Biladhir;
Belize: Cashew;
Benin: Yovotchan (Adja), Youbourou Somba, Yibo Somba (Bariba), Akaju, Casu, Kandju, Kadjoutin, Lakazu (Fon), Anacardier, Pomme Acajou (*French*), Darkassou, Kandju (Goum), Akaju (Mina), Ékadjou (Nagot), Yorubaakadiya (Yom), Kadjou, Kandju (Yoruba);
Bosnian: Beli Mahagoni, Indijanski Kašu-Orah, Pipak;
Bolivia: Marañón;
Brazil: Acajaiba, Acajou, Acajé (Tupi Indians), Rabuno-Eté (Carajá Indians), Cajú, Cajú, Amarelo, Cajú Banana, Cajú Da Praia, Cajú Do Campo, Cajú Manga, Cajú Vermelho, Cajueiro, Cajueiro Azedo, Cajueiro Doce, Cajuhy, Cajuhy Azedo, Cajuhy Doce, Oa-Caju (Portuguese);
Brazzaville: Pomme Cajou (*French*);
Burmese: Sihosayesi, Thayet Si, Thiho, Thiho-Thayet, Tihotiya-Si;
Chamorro: Casue, Kasoe, Kasoy;
Chinese: Yao Guo, Yao Guo Li, Guo Li, Yao Guo Shu;
Columbia: Marañón, Merey;
Comoros: Pomme Cajou (*French*), Mani Ya Mbibo (Great Comoros, Anjouan Island), Mbotza (Mohéli Island);
Cook Islands: Aratita Popaa, Kātū (Maori);
Costa Rica: Marañón;

Croatian: Beli Mahagoni, Indijanski Kašu-Orah, Pipak, Pipci;
Cuba: Cajuil, Marañón;
Czech: Ledvinovník Západní;
Danish: Acajounød, Cashew, Akajoutræ, Akajutræ;
Domincan Republic: Cajuil;
Dutch: Acajoeboom, Apennotenboom, Cashewnoot, Cashewsoort, Kasjoe, Kasjoeboom, Nierenboom, Westindische;
East Africa: Korosho (Kiswahili);
Finnish: Cashew-Paehkinae;
French: Acajou À Pommes, Cajou, Noix-Cajou, Pomme D'acajou; Noix D'acajou, Noix De Cajou; Acajou, Anacadier, Anacarde, Anacardes, Noix D' Cajou, Pomme Acajou, Pomme D' Cajou, Pommier Cajou;
French Guiana: Acajou, Acajou À Pomme, Anacadier, Cassoun;
German: Acajubaum, Akajoubaum, Cachunuss-Baum, Elefantenlaus-Baum, Kaschubaum, Kaschunuß, Kaschunuss, Kaschunußbaum, Nierenbaum, Westindische Elefantenlaus;
Ghana: Àntírìnyà (Dagomba);
Greek: Anakardia, Kasious;
Guatemala: Jocote Marañón;
Guyana: Cashew, Cashew Nut, Merche, Merehi; Wak-Roik-Yik, You-Ro-Yik, Youw-Rouii-Yik (Patamona),
Honduras: Jocote Marañón;
Hungarian: Akazsu, Kesu(Fa);
India: Kajubadam (Assamese), Hijli Badam, Hijuli, Kaju (Bengali), Kaju (Gujarati), Duk, Hijli-Badam Kaajuu, Kaju, Kaju-Ki-Gutli, Kajubadam, Khajoor (Hindi), Gaeru, Gaeru Beeja, Gaeru Kaayi, Gaeru Pappu, Gaerumara, Gerapoppu, Gerbija, Gerligai, Geru, Geru Pappu, Gerubija, Gerumara, Gerupoppu, Godambe, Godambi, Godambi Mara, Godambimara, Godambee, Godamber, Godami, Gokuda, Gonkuda, Gori, Govamba, Gove, Govambe, Gove Gaeru, Gove Hannu, Kempu Gaeru, Kempugeru, Kempukerubija, Kerubija, Mandiri Pappu, Tarukageru, Thuruka Geru, Turakageru, Turukagerujidi (Kannada), Kapa-Mava, Kapamava, Kappa-Mavakuru, Kappa-Mavu, Kappal-Cheru-Kuru, Kappamavu, Kappalcher, Kappaimavu, Kappalmavu, Kappalsera, Kappamavakum, Karmavu, Kashukavu, Kashumavu, Kasumavu, Parangimavu, Parankimava, Parankimavu, Parangi Mavu, Paringi Maavu, Paringimavu, Patirimavu, Portugimavu, Portukimavu, Pritikannavu (Malayalam), Bi, Kaju, Kajucha, Kajoo, Kaajoocha (Marathi), Kaju (Manipuri), Lanka Badam, Lonkabhalya (Oriya), Kaju (Punjabi), Agnikrita, Arushkara, Batada, Bhallataka, Guchhapushpa, Hijli Badam, Kajutah, Kajutaka, Kajutakah, Parvati, Prithagabija, Shoephahara, Sophahara, Sophara, Srigdhahapitaphala, Upapushpika, Venamrah, Vrittapatra, Vrkkabijah, Vrkkaphalah, Vrttarushkarah (Sanskrit), Andima, Andimangottai, Kaju, Kallaarmaa, Kallarma, Kolamavu, Kola Maavu, Kollamma, Kottai Munitirima, Kottai-Mundiri, Kottaimundiri, Matumancam, Mindiri Andima, Mindiri-Appazham, Mindiriparuppu, Mundhiri Paruppu, Mundiri-Kai, Mundiri-Kottai, Mundiri Kottae, Mundthri, Mundhiri, Mundiri, Munthri, Munthri-Kottei, Muntiri, Muntiri- P-Palam, Munthiri, Saram, Sigidima, Tammuntiri, Tirigai, Tirikai, Uttumabalam, Virai-Muntirikai (Tamil), Gidimavridi, Jaedima-Midi, Jeedimaamidi, Jidi Chettu, Jidi Kaaya, Jidi Mamidi, Jidi-Mamidi-Vittu, Jidimamidi, Jiidi Ma'midi, Mokka Mamidi, Mokkamamidi, Moonthamamidivittu, Muntamamidi, Muntamamidi-Vittu, Munta Mamidi, Munthamamidi, Mokkamaamidi, Munthamaamidi (Telugu), Kaajju (Urdu);
Indonesian: Buwa Jaki, Buwah Monyet, Jambu Dipa, Jambu Dipang, Jambu Gajus, Jambu Golok, Jambu Mede, Jambu Mete, Jambu Monyet, Jambu Seran, Jambu Terong, Jumbuk Jebet, Kacang Mede, Kacang Mete, Kanoke, Masapana, Wojakis;
Italian: Acajiú, Anacardio;
Jamaica: Cashew;
Japanese: Anakarudiumu Okushidentare, Kashuunatto No Ki;
Khmer: Svaay Chanti;
Madagascar: Diab, Mabibo, Mahabibo, Voambarika,;
Malaysia: Gajus, Jambu Golok, Jambu Mede, Jambu Monyet, Janggus;
Mali: Jibarani, Darakase (Bambara), Mali Sow (Bobo-Fing), Jibarani (Malinke), Komi Gason (Senoufo);

Mandinka: Kasuowo, Kasuwu;
Mexico: Marañón;
Nepalese: Kaajuu, Kaju;
Nigeria: Ikashu (Esan), Cashew, Kadinnia, Kanju (Hausa), Cashew (Ibibio), Kanshuu (Igbo), Ikashu (Igede), Kanju (Kanuri), Mashichada (Koma), Kasu (Okeigbo), Kaju (Ondo), Kaju (Yor) Kasu (Yoruba);
Northern Marianas: Apu Initia;
Papiamento: Kashu Kashipete;
Peru: Casho, Casú, Marañón
Philippines: Sasoi (Ibanag), Kosing (Igorot), Balogo, Balugo, Sambalduke (Iloko), Balubad (Pampangan), Marañón (Spanish), Kasul (Sulu),Balubad, Balubag, Baluban, Balubat, Batuban, Kachui, Sasoi (Tagalog);
Polish: Nanercz Wschodni;
Portuguese: Cajú, Cajú Do Camop, Cajueiro, Cajueiro-Do-Campo;
Puerto Rico: Pajuil;
Russian: Anakardium Zaladnyi, Derevo Kesh'iu, Indiiskii Orekh, Kesh'iu, Orekh Kesh'iu;
Samoan: 'Apu 'Initia, Apu Initia;
Senegal: Finzâ (Bambara), Anacardier, Pomme-Cajou, Pommier Cajou (Local French), Daf Duruba (Serere), Darkasu, Darkassou, Darcassou (Wolof);
Serbian: Beli Mahagoni, Indijanski Kašu-Orah, Pipak;
Seychelles: Bois Cachou;
Sierra Leone: Kushu (Krio), Kusui (Mende), E - Lil- E - Potho (Temne)'
Slovaščina: Epatca, Epatka, Indijski Oreščak;
South Africa: Kasjoe, Kasjoeboom (Afrikaans), Mkatshu (Zulu);
Spanish: Acaya, Anacardo, Cacajuil, Cajuil, Casho, Cashú, Casú, Marañon, Marañón, Pajuil;
Sri Lanka: Caju, Kaju, Montin-Kai (Sinhalese);
Surinam: Merehe, Mereke (Arawak) Olvi, Oroi, Orvi (Carib.), Djam,Boe Monjet (Malayan), Kadjoe, Sabana Kadjoe (Saramaccan), O-Roy (Tirio & Wayana), Boschkasjoe, Kasjoe, Kasjoen, Kasyu;
Swedish: Akajouäpple;
Tanzania: Korosho, Mbibo, Mkanju, Mkorosho (Swahili);
Taiwan: Gang-Ru-Shu, Kang-Ju-Shu;

Thai: Himmaphan, Mamuang, Mamuang Letlor, Yaruang, Mamuang Him, Maphan, Supardi, Pajuil;
Tongan: 'Apu, Kesiu;
Trinidad And Tobago: Cashew;
Venezuela: Marañón, Merey;
Vietnamese: Cây Diêù (South), Dào Lôn Hôt (North).

Origin/Distribution

The cultivated form of cashew is held to have originated in the *restinga* (low vegetation found along the sandy coast along eastern Brazil) of north-eastern Brazil. *Anacardium occidentale* is also probably indigenous to the savannas of Columbia, Venezuela and the Guianas (Mitchell and Mori 1987). It is a also a dominant native in the *cerrados* (savanna-like vegetation) of central and Amazonian Brazil. The plant was introduced into South America west of the Andes, Central America, West Indies and then to the East by early travellers. It is now found widely in the old and new tropics around the world between latitudes 27°N and 28°S. Cashew is cultivated especially in coastal sandy areas. The main cashews producers and exporters are India, Tanzania, Mozambique, Angola and Vietnam.

Agroecology

Cashew grows from 0–1,000 m above sea level. It prefers high temperatures with mean annual temperatures of 17–38°C and is extremely frost sensitive. Low temperatures have been reported to delay flowering. It is rather drought tolerant because of its deep and extensive root system and can grow in areas receiving only 762–1,275 mm of rain per year. It grows well in areas of evenly distributed rainfall of 500–3,500 mm, with a dry spell during the flowering and fruiting period and with 65–80% relative humidity. High rainfall and humidity favour diseases that destroy the flowers, fruits and reduce fruit set. It is adaptable to a wide range of soils with pH of 4.3–6.5 from poor sandy soils as in Bris soils in Terengganu, Malaysia, and somewhat saline soils to sterile, very shallow and impervious savanna soils, on which few other trees or crops will grow and lateritic soils as in Cochin and Kerala, India. But cashew is intolerant of poor soil drainage and calcareous or highly saline soils, pure clays and water-logged soils should be avoided. It thrives best in deep, fertile, friable sandy soils with soil pH 4.5–6.5. Cultivation should be done on nearly level areas of red-yellow podzols, quartziferous sands, and red-yellow latosols. The flowers are pollinated by bees, wasps, ants, flies and also humming birds. The fruits are dispersed primarily by frugivorous bats and also secondarily by water.

Plate 1 Yellow fruited variety

Plate 2 Orange fruited variety

Edible Plant Parts And Uses

Roasted cashew is relished as a snack, eaten with or without salt or sugar. Roasted nuts are used whole or in crushed pieces for snacks, cake, chocolate, candies, biscuits, ice-cream and dessert. The nut is also made into cashew flour and cashew butter. The raw nut is used in cooking vegetarian and meat dishes. Cashew nut is popular in Indian, Thai and Chinese cuisine. In India, the nut is ground into sauces such as '*shahi korma*' and used as garnish in Indian sweets and desserts. Cashew is also used in cheese alternatives for vegans. In the Philippines cashew is eaten with '*suman*' (glutinous rice-coconut milk dessert) and in another sweet dessert '*turrones de casuy*' (cashew marzipan wrapped in white wafer) in the city of Pampanga.

Young leaves and shoots are eaten raw or cooked as vegetable. In Malaysia, the young leaves and shoots (Plate 6) are eaten raw as '*ulam*' and usually with '*sambal belacan*' (shrimp paste mixed with chilli and lime juice). The fleshy, juicy hypocarp or cashew-apple is also eaten raw when ripe, used as vegetable and is used in fruit salads or used for making juice, beverage, wine, vinegar, syrups, clarified and cloudy cashew apple juice, cashew apple juice blended with other fruit juices and pulps like lime, pineapple, orange, mango, and papaya, cashew apple juice concentrate, cashew apple preserve and candy, cashew apple jam, cashew apple mixed fruit jam, cashew apple pickle and chutney, jelly, candies, preserves, or used as an ice-cream flavour. In Brazil, Mozambique and some parts of Indonesia, cashew wine (slightly fermented juice) is consumed at harvest time and can be distilled to produce stronger alcoholic drinks. In Goa, India, the fermented juice from the apple is processed into Fenni cashew apple wine and fenni cashew apple brandy and in Tanzania, a gin-like product called '*konyagi*' or a stronger product called '*gongo*'. In Mozambique, a strong liquor called '*agua ardente*' is made from the pseudo-apple.

Arabic and cashew gum microencapsulation of coffee extracts were observed to have similar aroma protection, external morphology and size distribution (Rodrigues and Grosso 2008). Biochemical, structural data and sensory analysis to examine flavour protection and consumer preference suggested that low cost cashew gum is a well suited alternative for odour microencapsulation to the more costly Arabic gum currently used in Brazil.

Anacardium occidentale 49

Plate 3 Red fruited variety

Plate 4 Bronze coloured juvenile leaves

Plate 5 Cashew nuts

Plate 6 Young cashew leaves sold as vegetables in a local market

Botany

A small-sized, much branched, evergreen perennial tree with low, spreading crown 6–12 m; grows to a height of 2–10(–15) m with a brown or grey bark, smooth to rough with longitudinal fissures and a deep root system. Leaves simple, alternate, coriaceous, glabrous, narrowly to broadly obovate, sometimes broadly oblong or elliptic, 10–18 × 8–15 cm, apically rounded or shallowly emarginate or shorly acuminate, entire, cuneate or obtuse at base at time attenuate or auriculate, prominently veined with 9–14 pairs of widely spreading lateral veins, short petiolate, reddish bronze when young and dark green when mature (Plate 4). The inflorescence is a sub-corymbose, terminal panicle,10–25 cm long, branched, sparsely puberulous to densely puberulous toward the apices of rachises, commonly bearing male, female and hermaphroditic flowers which are first white or pale green with red or pink lines at anthesis and truning dark red after fertilization. The male flowers are the most numerous, 5-merous and usually bear 1 exserted stamen and 9 small inserted ones and with a small pistillode, 0.3–1 mm long. Bisexual flowers have

5 lanceolate to narrowly ovate sepals; cylindric corolla of 5 petals; stamens 6–10 (–12) with usually one long exserted stamen, all with normal anthers, glabrous, one-locular, one-ovulate ovary with central awl-shaped style and punctiform stigma. Female flowers have staminodes instead. Fruit is a reniform achene (nut), about 3 cm long, 2.5 cm wide, attached to the distal end of an enlarged receptacle and hypocarp, called the cashew-apple which is shiny, orangey, red or yellowish, pear-shaped or rhomboid-to-ovate, soft, juicy, 10–20 cm long, 4–8 cm broad (Plates 1–3). The nut is green turning dark brown when mature and has an edible yellowish white reniform, hard kernel with two large white cotyledons and a small embryo (Plate 5), surrounded by a hard pericarp which consist of a double wall separated by honey-combed cellular fibrous tissues filled with a caustic resinous oil.

Nutritive/Medicinal Properties

Analyses carried out in the United States (USDA 2010) reported raw, cashew nut (minus the pericarp) to have the following proximate composition (per 100 g edible portion): water 5.20 g, energy 553 kcal (2,314 kJ), protein 18.22 g, total lipid 43.85 g, ash 2.54 g, carbohydrates 30.19 g, total dietary fibre 3.3 g, total sugars 5.91 g, Ca 37 mg, Fe 6.68 mg, Mg 292 mg, P 593 mg, K 660 mg, Na 12 mg, Zn 5.78 mg, Cu 2.195 mg, Mn 1.655 mg, Se 19.9 µg, vitamin C 0.5 mg, thiamine 0.423 mg, riboflavin 0.058 mg, niacin 1.062 mg, pantothenic acid 0.864 mg, vitamin B-6 0.417 mg, total folate 25 µg, vitamin E (α-tocopherol) 0.090 mg, β-tocopherol 0.03 mg, γ-tocopherol 5.31 mg, δ-tocopherol 0.36 mg, vitamin K (phylloquinone) 34.1 µg, total saturated fatty acids 7.783 g, 8:0 (caprylic acid) 0.015 g, 10:0 (capric acid) 0.015 g, 12:0 (lauric acid) 0.015 g, 14:0 (myristic acid) 0.015 g, 16:0 (palmitic acid) 3.916 g, 17:0 (margaric acid) 0.046 g, 18:0 (stearic acid) 3.223 g, 20:0 (arachidic acid) 0.266 g, 22:0 (behenic acid) 0.173 g, 24:0 (lignoceric acid) 0.101 g; total monounsaturated fatty acids 23.797 g, 16:1 undifferentiated (palmitoleic acid) 0.136 g, 18:1 undifferentiated (oleic acid) 23.523 g, 20:1 (gadoleic acid) 0.138 g; total polyunsaturated fatty acids 7.845 g, 18:2 undifferentiated (linoleic acid) 7.782 g, 18:3 undifferentiated (linolenic acid) 0.062 g; tryptophan 0.287 g, threonine 0.688 g, isoleucine 0.789 g, leucine 1.427 g, lysine 0.928 g, methionine 0.362 g, cystine 0.393 g, phenylalanine 0.951 g, tyrosine 0.508 g, valine 1.094 g, arginine 2.213 g, histidine 0.456 g, alanine 0.837 g, aspartic acid 1.795 g, glutamic acid 4.506 g, glycine 0.937 g, proline 0.812 g, serine 1.079 g, lutein + zeaxanthin 22 µg.

From the above, it is evident that cashew nut is nutritious, rich in calories, lipids (mainly unsaturated fatty acids) – oleic and linoleic acids, vitamin E (tocopherols), protein, amino acids, vitamin K and minerals like Fe, Mg, Zn, Cu and Selenium. It also has vitamin Bs, Vitamin B6, pantothenic acid, lutein and zeaxanthin.

Nuts like cashew, brazilnut, pecan, pinenut and pistachio contain bioactive constituents that elicit cardio-protective effects including phytosterols, tocopherols and squalene. Studies in Ireland reported that the total oil content of cashew nuts ranged from 40.4% to 60.8% (w/w) while the peroxide values ranged from 0.14 to 0.22 mEq O_2/kg oil. The most abundant monounsaturated fatty acid was oleic acid (C18:1), while linoleic acid (C18:2) was the most prevalent polyunsaturated fatty acid (Ryan et al. 2006). The levels of total tocopherols ranged from 60.8 to 291.0 mg/g. Squalene ranged from 39.5 mg/g oil in the pine nut to 1377.8 mg/g oil in the brazil nut. β-sitosterol was the most prevalent phytosterol, ranging in concentration from 1325.4 to 4685.9 mg/g oil. The data indicated cashew nuts to be a good dietary source of unsaturated fatty acids, tocopherols, squalene and phytosterols.

Raw cashew nut kernels were found to possess appreciable levels of certain bioactive compounds such as β-carotene (9.57 µg/100 g of DM), lutein (30.29 µg/100 g of DM), zeaxanthin (0.56 µg/100 g of DM), α-tocopherol (0.29 mg/100 g of DM), γ-tocopherol (1.10 mg/100 g of DM), thiamin (1.08 mg/100 g of DM), stearic acid (4.96 g/100 g of DM), oleic acid (21.87 g/100 g of DM), and linoleic acid (5.55 g/100 g of DM) (Trox et al. 2010). All of the conventional shelling methods including oil-bath roasting, steam roasting, drying, and open pan roasting were found to significantly reduce these bioactive compounds, whereas the Flores hand-cracking method exhibited similar levels of carotenoids, thiamin, and unsaturated fatty acids in cashew nuts when compared to raw unprocessed samples.

Cashew apple was found to contain 87.9% water, 0.2% protein, 0.1% fat, 11.6% carbohydrate, 0.2% ash, 0.01% Ca, 0.01% P, 0.002% Fe, 0.26% vitamin C, and 0.09% β-carotene (Nair et al. 1979; Duke 1983). The testa was found to have α-catechin, α-sitosterol, 1-epicatechin and proanthocyanadins, leucocyanadine, and leucopelargodonidine. The dark color of the nut was attributable to an iron-polyphenol complex. The shell oil contained about 90% anacardic acid ($C_{22}H_{32}O_3$) and 10% cardol ($C_{32}H_{27}O_4$). It yielded glycerides, linoleic, palmitic, stearic, and lignoceric acids, and sitosterol. Gum exudates were reported to contain arabinose, galactose, rhamnose, and xylose (Morton 1987).

The carotenoid composition were different in the red and yellow cashew-apples and two commercial brands of pasteurized cashew juice (Cecchi and Rodriguez-Amaya 1981). α-carotene, β-carotene, ζ-carotene, cis-β-carotene, cryptoxanthin, aurochrome, cryptochrome, and auroxanthin were found in all samples, with β-carotene as the major pigment. Colour differences reflected the total amount, not the type of carotenoid present. Processed juice, had lower total carotenoid content and vitamin A value but higher epoxide. No quantitative differences were observed among samples of the same brand or between brands of pasteurized juice. Cashew apple also contained flavonoids (Sousa de Brito et al. 2007; Michodjehoun-Mestres et al. 2009a) and tannins (Michodjehoun-Mestres et al. 2009b). One anthocyanin, 3-O-hexoside of methyl-cyanidin and thirteen glycosylated flavonols among them, the 3-O-galactoside, 3-O-glucoside, 3-O-rhamnoside, 3-O-xylopyranoside, 3-O-arabinopyranoside and 3-O-arabinofuranoside of quercetin, kaempferol 3-O-glucoside and myricetin were identified in the methanol–water extract of cashew apple (Sousa de Brito et al. 2007). Trace amounts of delphinidin and rhamnetin were found in the hydrolyzed extract. Monomeric phenols were extracted by acetone/water (60:40) from the skin and flesh of four cashew apple genotypes from Brazil and Bénin (West Africa) (Michodjehoun-Mestres et al. 2009a). Skins were found much richer than flesh in simple phenolics. Flavonol glycosides were preponderant with myricetin and quercetin hexosides (2 of each), pentosides (3 of each), and rhamnosides as dominant compounds. Anthocyanidin glycosides peonidin, petunidin and cyanidin 3-O-hexosides were found in skins from the two scarlet and orange pigmented genotypes but were absent from the flesh. Tannins were also extracted by acetone/water 60:40 from skin and flesh of four cashew apple genotypes from Brazil and Bénin (West Africa), and separated from monomeric phenols (Michodjehoun-Mestres et al. 2009b). Both skin and flesh tannins had high levels of (−)-epigallocatechin and (−)-epigallocatechin-O-gallate, followed by minor concentrations of (−)-epicatechin and (−)-epicatechin-3-O-gallate; 100% of the compounds were of the 2,3-cis configuration. Skin tannins were half as galloylated (20%) than flesh tannins (40%).

Cashew apple and its products were found to contain carotenoids and ascorbic acid (Assunção and Mercadante 2003a, b). In all the fruits from different varieties from different regions in Brazil, β-carotene (16.6–67.9 μg/100 g), β-cryptoxanthin (7.7–64.4 μg/100 g), α-carotene (5.9–51.9 μg/100 g) and 9-cis- + 13-cis-β-carotene (3.3–15.6 μg/100 g) were detected (Assunção and Mercadante 2003b). In general, the levels of carotenoids were greater in the red than in the yellow cashew apples, from both regions; for example, the levels of α- and β-carotene were about 1.8 and 1.3 fold more in the red than in the yellow fruits from the Southeast and Northeast, respectively. In contrast, ascorbic acid level was slightly higher in the yellow variety. Elongated red and yellow fruits also presented slightly higher ascorbic acid levels than the globose red fruits. The total carotenoid contents of the globose red fruits were 1.5 and 1.7 times lesser than those found in the yellow and red varieties, respectively, all coming from the Northeast region. Yellow fruits from the Northeast had

1.7 times more provitamin A contents than those from the Southeast whereas, for the red variety, the values were similar. The yellow and red varieties from the Northeast exhibited non-statistically higher ascorbic acid levels than those from the Southeast. The following products, concentrated juice, frozen pulp, nectar, ready-to-drink, and sweetened concentrated juice, had ascorbic acid contents varying from 13.7 to 121.7 mg/100 g and total carotenoid levels ranging from 8.2 to 197.8 µg/100 g (Assunção and Mercadante 2003a). β-Carotene was the dominant carotenoid in the majority of the products, followed by α-carotene, β-cryptoxanthin and 9-cis- +13-cis-β-carotene in similar proportions. However, in 10 of the 60 samples analyzed, another carotenoid profile was found with the presence of auroxanthin, 5,8-epoxy-cryptoxanthin, 5,8-epoxy-lutein, ζ-carotene and two unidentified carotenoids. Cashew apple products were proven to be excellent sources of vitamin C, but not very good sources of carotenoids for the human diet.

Cashew apple juice was found to contain the highest amount of vitamin C (203.5 mg/100 ml) of edible portion compared to other tropical fruits like pineapple, orange, grapes, mango and lemon (Akinwale 2000). Orange, grape, pineapple, mango and lemon contained average values of 54.7 mg, 45.0 mg, 14.70 mg, 30.9 mg and 33.7 mg vitamin C per 100 ml of juice respectively. It was found to contain almost four times the amount of vitamin C in the popular citrus fruits and more than four times as much vitamin C as in other fruits. Hence, when cashew apple was blended with other tropical fruits it boosted their nutritional quality. Conversely, these fruits enhanced the acceptability of cashew juice in terms of taste and flavour. The blends, even though significantly dissimilar in taste, colour and mouth-feel, were all acceptable to consumers with no significant difference in overall acceptability.

Cashew apple nectar, a secondary product from the production of cashew nuts was found to possess an exotic tropical aroma. Aroma volatiles in pasteurized and reconstituted (from concentrate) Brazilian cashew apple nectars were identified as methional, (Z)-1,5-octadien-3-one, (Z)-2-nonenal, (E,Z)-2,4-decadienal, (E,E)-2,4-decadienal, β-damascenone, and δ-decalactone (Valim et al. 2003). These compounds together with butyric acid, ethyl 3-methylbutyrate, 2-methylbutyric acid, acetic acid, benzaldehyde, homofuraneol, (E)-2-nonenal, γ-dodecalactone, and an unknown were the most intense aroma volatiles. Thirty-six aroma volatiles were detected in the reconstituted sample and 41 in the pasteurized sample. Thirty-four aroma active components were common to both samples. Ethyl 3-methylbutyrate and 2-methylbutyric acid were the character impact compounds of cashew apple (warm, fruity, tropical, sweaty). 2-methyl-3-furanthiol and bis(2-methyl-3-furyl) disulfide were identified for the first time in cashew apple. Both were aroma active (meaty).

Young tender cashew leafy shoots were found to have the following proximate composition per 100 g edible portion (Tee et al. 1997): moisture 89.1%, energy 35 kcal, carbohydrates 4.5 g, fat 0.2 g, protein 3.8 g, fibre 1.5 g, Ash 0.9 g, Ca 53 mg, P 29 mg, Fe 3.5 mg, carotene 2,076 µg, vitamin A 653 µg RE, vitamin B2 1.19 mg, niacin 0.5 mg, vitamin C 91 mg. Cashew leaves also contained lutein 773 µg and β-carotene 1,324 µg (Tee et al. 1997) The tender leafy shoots of cashew were also rich in protein, vitamin A and vitamin C.

Anacardic acids were found to be the major alkylphenols contained in both cold pressed raw and roasted cashew nut oils followed by cardol, cardanol and 2-methylcardol compounds (Gómez-Caravaca et al. 2010). Raw and roasted oils did not show different compositions except for cardanols. The oil produced from roasted cashew nut yielded a higher concentration of cardanols. Tocopherol content varied in a range of 171.48–29.56 mg/100 g from raw to roasted cashew nut oil, with β-tocopherol having a higher decrease (93.68%). Also minor polar compounds in cashew oil decreased after roasting from 346.52 to 262.83 mg/kg.

The gum exudate from the cashew-nut tree was found to contain traces of the reducing sugars, rhamnose (0.005%), arabinose (0.03%), mannose (0.007%), galactose (0.03%), glucose (0.02%), β-D-Galp-(1→6)-α β-D-Gal (0.05%), α-L-Rhap-(1→4)-α β-D-GlcA (0.008%) and α-L-Rhap-(1→4)-β-D-GlcpA-(1→6)-β-D-Galp-(1→6)-α-β-D-Gal (0.008%) (Menestrina et al. 1998). Rhamnose, arabinose, glucose and the three oligosaccharides represented components of the side-chains of the gum polysaccharide, which had a main chain of (1→3)-linked β-D-Galp units. Other new side-chain structures were characterized, particularly -α-D-Galp-(1→6)-D-Galp- and α-L-Araf-(1→6)-D-Galp-.

Various parts of the cashew nut tree contain many bioactive chemicals which impart many pharmacological and medicinal attributes that include:

Antioxidant Activity

With the exception of the ethyl acetate and hexane extracts of *Anacardium occidentale*, the methanol plant extract showed significant radical scavenging and reducing properties (Mokhtar et al. 2008). The methanol extract of *A. occidentale* was potent at scavenging the ABTS·+ and nitric oxide radicals. Total phenolic content of the plant extract showed high correlation with the antioxidant activities, suggesting that phenolic compounds present in the extract may be a major contributor of the observed antioxidant activities. The scientists concluded that the methanol extract of *A. occidentale* could be an alternative source of polyphenolics with potent antioxidant activities.

Anacardic acids, 6-pentadec(en)ylsalicylic acids isolated from the cashew nut and apple, were found to possess preventive antioxidant activity while salicylic acid did not display this activity (Kubo et al. 2006). These anacardic acids restricted superoxide radicals generation by inhibiting xanthine oxidase without radical-scavenging activity. Anacardic acid ($C_{15:1}$) inhibited the soybean lipoxygenase-1 catalyzed oxidation of linoleic acid with an IC_{50} of 6.8 μM. Anacardic acids acted as antioxidants in a variety of modes, including inhibition of various prooxidant enzymes involved in the production of the reactive oxygen species and chelate divalent metal ions such as Fe^{2+} or Cu^{2+}, but did not quench reactive oxygen species. The C_{15}-alkenyl side chain was found to be largely associated with the activity.

In another study, the content of anacardic acids, cardanols and cardols in cashew apple, nut (raw and roasted) and cashew nut shell liquid (CNSL) were found to be different (Trevisan et al. 2006). Higher levels (353.6 g/kg) of the major alkyl phenols, anacardic acids were found in CNSL followed by cashew fibre (6.1 g/kg) while the lowest (0.65 g/kg) quantities were found in roasted cashew nut. Cashew apple and fibre were found to contain anacardic acids exclusively, whereas CNSL had an abundance of cardanols and cardols. Cashew nut (raw and roasted) also contained low quantities of hydroxy alkyl phenols. The hexane extracts (10 mg/ml) of all cashew products tested plus CNSL, displayed significant antioxidant activity. Cashew nut shell liquid was more efficacious (inhibition = 100%) followed by the hexane extract of cashew fibre (94%) and cashewapple (53%). The antioxidant capacity correlated well with the concentration of alkyl phenols in the extracts. A mixture of anacardic acids (10.0 mg/ml) showed the higher antioxidant capacity ($IC_{50}=0.60$ mM) compared to cardols and cardanols ($IC_{50}>4.0$ mM). Of these substances, anacardic-1 was comparatively the more potent antioxidant ($IC_{50}=0.27$ mM) compared to cardol-1 ($IC_{50}=1.71$ mM) and cardanol-1 ($IC_{50}>4.0$ mM). The antioxidant capacity of anacardic acid-1 was related to inhibition of superoxide generation ($IC_{50}=0.04$ mM) and xanthine oxidase ($IC_{50}=0.30$ mM) than to scavenging of hydroxyl radicals Both cashew fibre and CNSL contained high amounts of anacardic acids and could be better utilized in functional food formulations and may represent cheap sources of cancer chemopreventive agents. Presently, cashew fibre is a waste product that is mostly used in formulations of animal or poultry feeds.

Immature cashew nut-shell liquid (iCNSL) exhibited excellent protective activities against oxidative damage induced by hydrogen peroxide and inhibited acetylcholinesterase activity in strains of *Saccharomyces cerevisiae* (De Lima et al. 2008). iCNSL was found to contain anacardic acid, cardanol, cardol, and 2-methyl cardol. Immature cashew nut oil also contained triacylglycerols, fatty acids, alkyl-substituted phenols, and cholesterol. The main constituents of the free fatty acids were found to be palmitic (C16:0) and oleic acid (C18:1). The scientists concluded that iCNSL may have an important role in protecting DNA against damage caused by reactive oxygen species, as well as hydrogen peroxide, generated by intra-cellular and extra-cellular mechanisms.

Cashew shell nut liquid CNSL was found to be a mixture of meta-alkylphenols with variable degree of unsaturation attached to the benzene ring (Rodrigues et al. 2006). Based on kinetic parameters, the sequence of antioxidant activity was: CNSL >> cardanol ≡ hydrogenated and alkylated cardanol > hydrogenated cardanol. The effect of CNSL could be attributed to the extra contribution of the other components besides cardanol and to the unsaturation on the long side chain. CNSL was found to be much more cost effective than its derivatives.

Cashew nut kernel oil was found to enhance the activities of SOD (sodium oxide dismutase), catalase, glutathione S-transferase (GST), methylglyoxalase I and levels of glutathione (GSH) in liver of Swiss albino mice treated orally with two doses (50 and 100 μl/animal/day) of kernel oil of cashew nut for 10 days. (Singh et al. 2004). The data suggested that cashew nut kernel oil possessed an ability to increase the antioxidant status of animals. The decreased level of lipid peroxidation supported this possibility. The tumour promoting property of the kernel oil was also examined and it was found that cashew nut kernel oil did not exhibit any solitary carcinogenic activity.

Antidiabetic, Hypoglycaemic Activity

Leaves, stem-bark and roots of *A. occidentale* were found to have hypoglycaemic activity. In one study, 3 days after streptozotocin (STZ) administration, there was a 48% elevation in blood glucose level in aqueous cashew extract pre-treated rats, compared with a 208% increase in diabetic control rats treated with STZ alone (Kamtchouing et al. 1998). Further, these pre-treated animals presented no glycosuria, had normal weight gain and a non-significant increase in food and fluid intake at the end of the treatment compared with the normal control. Diabetic control animals showed a positive glycosuria, body weight loss, a real polyphagia and polydypsia. These results indicated the protective role of *Anacardium occidentale* extract against the diabetogenic action of STZ.

Sokeng et al. (2001) showed that a single oral administration of the aqueous extract of cashew leaves at doses of 35, 175 and 250 mg/kg exhibited notable hypoglycemic effect at 1 hour 30 minutes after feeding. The maximum pronounced hypoglycemic effect was observed 3 hours after treatment with doses of 175 and 250 mg/kg eliciting 25.30% and 24.30% decrease, respectively. The most significant effect was observed with the dose of 175 mg/kg, which remained significantly active 8 hours after administration. A repeated administration of the extract (175 mg/kg) twice daily significantly decreased the blood glucose level by 43% in diabetic rats after 3 days. The lowest dose (35 mg/kg) showed a non-significant decrease of 4%. Conversely, a decrease in urine glucose levels was observed in diabetic treated rats. When administered before glucose load to normal rats, the extract at doses of 175 and 250 mg/kg attenuated the rise of blood glucose level 1 hour after administration. In another study, *Anacardium occidentale* leaves at the dose of 300 mg/kg/day, exhibited significant attenuation of blood glucose level, total protein excreted, glycosuria and urea in streptozotocin-induced diabetic rats (Tedong et al. 2006). *Anacardium* treatment, commenced 3 days after diabetes induction, attenuated destruction of renal structure and other metabolic disturbances more than when treatment was initiated 2 weeks after. Histopathological study showed that *A. occidentale* significantly attenuated accumulation of mucopolysaccharides in the kidneys of diabetic animals. The cashew leaf extract at the dose of 300 mg/kg had no nephrotoxic potential in normal rats. The study demonstrated the efficacy of *Anacardium occidentale* (hexane extract) in reducing diabetes-induced functional and histological alterations in the kidneys. In separate studies, after acute oral administration of the leaf hexane extract less than 6 g/kg were found not be toxic (Tédong et al. 2007). The leaf hexane extract of *A. occidentale* is used in Cameroon traditional medicine for the treatment of diabetes and hypertension. Signs of toxicity at high doses were asthenia, anorexia, diarrhoea, and syncope. The LD_{50} of the extract, determined in mice of both sexes after oral administration was 16 g/kg. At doses of 2, 6 and 10 g/kg of extract, repeated oral administration to mice reduced food intake, weight gain, and behavioural effects.

Tedong et al. (2010) found that the hydroethanolic extract of cashew seed (CSE) and its active component, anacardic acid, stimulated glucose transport into C2C12 myotubes in a concentration-dependent fashion. Extracts of other parts (leaves, bark and apple) were inactive. Significant synergistic effect on glucose uptake with insulin was oberved at 100 μg/ml CSE. CSE and anacardic acid caused stimulation of adenosine monophosphate-activated protein kinase in C2C12 myotubes resulting in enhanced glucose uptake. Further, the dysfunction of mitochondrial oxidative phosphorylation may increase glycolysis and contribute to elevated glucose uptake. No significant effect was noticed on Akt (serine/threonine protein kinase) and insulin receptor phosphorylation. Both CSE and anacardic acid exerted significant uncoupling of succinate-stimulated respiration in rat liver mitochondria. These results suggested that CSE may be a potential anti-diabetic nutraceutical.

Studies showed that relatively moderate-to-high doses of cashew stem-bark extracts (100–800 mg/kg per os) produced dose-dependent, significant reductions in the blood glucose concentrations in both fasted normal and fasted diabetic rats (Ojewole 2003). On their own, both insulin (5 μg/kg subcutaneously applied) and glibenclamide (0.2 mg/kg per os) produced major reductions in the blood glucose concentrations of both fasted normal and fasted diabetic rats. At single doses of 800 mg/kg per os, cashew stem-bark aqueous and methanolic extracts significantly lowered the mean basal blood glucose concentrations of fasted normal and fasted diabetic rats. The hypoglycemic effect of the methanolic plant extract was found to be slightly more pronounced than that of the aqueous plant extract in both the normal and diabetic rats examined. Although cashew stem-bark aqueous or methanolic extract was less potent than insulin as an antidiabetic agent, the results of this experimental animal study indicated that it possessed hypoglycemic activity, and thus supported the folkloric use of the plant in the management and/or control of adult-onset of type-2 diabetes mellitus among the Yoruba-speaking people of Western Nigeria.

Intravenous administration of the hexane extract of the bark of cashew in normal, healthy dogs produced a significant lowering of the blood glucose levels (Alexander-Lindo et al. 2004). The hypoglycaemic principle(s) in the hexane extract were found to be two compounds, stigmast-4-en-3-ol and stigmast-4-en-3-one. Both compounds produced significant hypoglycaemic activity after intravenous administration at a dose of 1.3 mg/kg body weight. The hypoglycaemic effect of the cashew bark was postulated to be due to the presence of these compounds.

Studies by Olatunji et al. (2005) showed that administration of the cashew stem-bark methanol extract significantly prevented changes in plasma glucose, triglyceride, total cholesterol/HDL-cholesterol ratio, malonyldialdehyde, urea, and creatinine induced by enriched fructose diet. The enriched fructose diet resulted in significant increases in plasma glucose, total cholesterol, triglyceride, total cholesterol/HDL-cholesterol ratio, malonyldialdehyde, total protein, urea, and creatinine in rats. Treatment with enriched fructose diet and/or extract did not have any significant effect on plasma alkaline phosphatase level. These results showed that chronic oral administration of methanol extract of cashew stem-bark at a dose of 200 mg/kg body weight may be a safe alternative antihyperglycemic agent with beneficial effect by improving plasma glucose and lipids in fructose-induced diabetic rats associated with a reduced lipid peroxidation.

The crude ethanolic extract of cashew root and pawpaw fruit showed hypoglycemic potencies, however, the crude ethanolic extract of cashew roots exhibited greater hypoglycemic potency than the crude extract of pawpaw fruit in both guinea pigs and rats (Egwim 2005). The ethanolic extracts decreased both plasma glucose as well as cholesterol and total lipids but exhibited no effect on plasma protein. The work concluded that ethanolic extract of cashew root and pawpaw fruit contained natural substances that could be harnessed for the treatment and management of diabetes mellitus.

Anti-inflammatory Activity

A mixture of tannins (hydrolysable and non-hydrolysable) obtained from the bark of *Anacardium occidentale*, on intraperitoneal injection, demonstrated apparent anti-inflammatory activity in carrageenan- and dextran-induced rat paw oedemas, cotton pellet granuloma test and adjuvant-induced polyarthritis in rats (Mota et al. 1985). At higher doses orally administered tannins also exhibited activity in carrageenan paw oedema and adjuvant arthritis experiments. The tannins also inhibited acetic acid-induced "writhing responses" in mice and were found to obstruct the permeability-increasing effects in rats of certain mediators of inflammation and to inhibit the migration of leucocytes to an inflammatory site.

Aqueous extract of *A. occidentale* stem-bark (800 mg/kg per os) like diclofenac (100 mg/kg per os), produced time-dependent, continous and significant reduction of the fresh egg albumin-induced acute inflammation of the rat hind paw (Ojewole 2004). However, the anti-inflammatory effect of the plant extract was found to be approximately 8–15-fold lower than that of diclofenac. Co-administration per os of grapefruit juice (5 ml/kg) with *A. occidentale* stem-bark aqueous extract 800 mg/kg or diclofenac 100 mg/kg significantly potentiated the anti-inflammatory effects of the crude plant extract and diclofenac on fresh egg albumin-induced rat paw edema. Although *A. occidentale* stem-bark aqueous extract was less potent than diclofenac, the results indicated that the plant extract possessed anti-inflammatory activity, and thus lend pharmacological support to the folkloric use of cashew plant in the management and/or control of arthritis and other inflammatory conditions among the Yoruba-speaking people of western Nigeria.

In another study, pre-treatment with *Anacardium occidentale* stem bark extract (25–200 mg/kg) caused a dose-dependent and significant reduction in the elevated levels of alanine and aspartate aminotransferases in the sera of D-galactosamine-primed mice injected with lipopolysaccharide (LPS) (Olajide et al. 2004). The highest dose of the extract (200 mg/kg) gave 100% protection against mortality from sepsis. Pentoxifylline (100 mg/kg) and L-NAME (L-nitro- arginine methyl ester) (5 mg/kg) offered 100% protection against LPS-induced septic shock, and produced marked reduction in elevated levels of transferases. A dose-dependent inhibition of LPS-induced microvascular permeability in mice was also produced by pentoxifylline, L-NAME and the extract.

Anticancer Activity

Anacardic acids, 6[80(Z), 110(Z),140-pentadecatrienyl]salicylic acid (C15:3) (1), 6[80(Z),110(Z)-pentadecadienyl] salicylic acid (C15:2) (2), and 6[80(Z)-pentadecenyl] salicylic acid (C15:1) (3), from cashew apple juice were found to be cytotoxic against BT-20 breast carcinoma cells (Kubo et al. 1993). Anacardic acids (1–4), cardols (58) and methylcardols (9–12) from cashew nut and nut shell oil also exhibited moderate cytotoxic activity against Bt-20 breast and HeLa epitheloid cervix carcinoma cells.

Substituted isobenzofuranones (Ia and Ib) designed from anacardic acids, the major natural cashew nut-shell phenolic lipid, exhibited anticancer activity (Logrado et al. 2010). Isobenfuranone Ia showed significant antiproliferative effect against HL-60 leukemia cells and moderate activity against SF295 SF295 glioblastoma and MDA-MB435 melanoma cell lines. Mechanisms involved in the cytotoxic activity were shown to be through DNA damage, triggering apoptosis or necrosis induction.

Konan et al. (2011) found the crude ethanolic extract of cashew leaves to be cytotoxic and to induce apoptosis in Jurkat (acute lymphoblastic leukemia) cells. In a sub-acute toxicity assay they found that the leaf ethanolic to elicite lymphopenia in rats. One of the four fractions obtained from the ethanol leaf extract led to the isolation of the biflavonoid agasthisflavone, which exhibited high anti-proliferative effect in Jurkat cells with an IC_{50} of 2.4 µg/ml (4.45 µM). Agathisflavone was shown to induce apoptosis in Jurkat cells. The effect of agathisflavone on the acute promyelocytic leukemia cell line HL60, Burkitt lymphoma Raji cells and Hep-2 laryngeal carcinoma cells was also determined. Agathisflavone exhibited a slight effect on Burkitt lymphoma Raji cells and Hep-2 laryngeal carcinoma cells.

Antiulcerogenic Activity

Cashew leaves exhibited anti-ulcerogenic effect (Konan and Bacchi 2007). The hydroethanolic extract of cashew leaves inhibited gastric lesions induced by HCL/ethanol in female rats, with an ED_{50} value of 150 mg/kg body weight. Extract doses higher than 100 mg/kg body weight were more effective than 30 mg/kg of lansoprazol in preventing gastric lesions. A methanolic fraction (257.12 mg/kg) reduced gastric lesion at 88.20% and was postulated to contain the active principle of the antiulcer effect. No signs of acute toxicity were observed when mice were treated with extract dose up to 2,000 mg/kg body weight. A chemical analysis of the extract identified phenolic compounds as the main active principles. Glycosylated quercetin, amentoflavone derivate and a tetramer of proanthocyanidin were identified. The level of total phenolics in the extract was evaluated at 35.5% and flavonoid content was 2.58%. Morais et al. (2010) found that anacardic acids from cashew nut shell liquid afforded a dose-associated gastroprotection against the ethanol-induced gastric damage in mice and further prevented the ethanol-induced changes in the concentrations of gluthathione, malondehyde, catalase, sodium oxide dismutase and nitrate/nitrite. However, they failed to modify the gastric secretion or the total acidity in 4-hour pylorus-ligated rat. It was observed that the gastroprotection by anacardic acids was greatly decreased in animals pretreated with capsazepine, indomethacin, L-NAME or glibenclamide. These results suggested that anacardic acids afforded gastroprotection principally through an antioxidant mechanism. Other complementary mechanisms included the stimulation of capsaicin-sensitive gastric afferents, activation of endogenous prostaglandins and nitric oxide, and opening of K(+) (ATP) channels. These combined effects were postulated to be accompanied by an enhancement in gastric microcirculation.

Antibacterial Activity

An in-vitro study showed that all of the studied plant extracts including that of cashew were able to inhibit and kill *Porphyromonas gingivalis* (Srisawat et al. 2005). Among the studied plants, *Anacardium occidentale* bark and leaves were most potent in inhibiting *Porphyromonas gingivalis* with minimal inhibitory concentration (MIC) of 1.068 and 2.652 mg/ml respectively and the minimal bactericidal concentration (MBC) of 1.754 (cashew bark) and 5.625 mg/ml (leaves). *Terminalia bellerica* seed exhibited the best effect in killing *Porphyromonas gingivalis*. This data further implicated that cashew plant extract may have potential for periodontal therapy use.

Ten flavour compounds of cashew apple exhibited antimicrobial activity (Muroi et al. 1993). Most of them exhibited some activity against one or more of 14 microorganisms. Notably, (E)-2-hexenal exhibited antimicrobial activity against all the tested microorganisms, including Gram-negative bacteria. Further, the antibacterial activity of indole against *Escherichia coli* was enhanced fourfold by combining it with a sub-lethal amount of (E)-2-hexenal.

Anacardium occidentale nut shell oil exhibited potent antibacterial activity against only Gram-positive bacteria, among which *Streptococcus mutans*, one of several bacteria responsible for tooth decay, and *Propionibacterium acnes*, one of the bacteria responsible for acne, were the most sensitive (Himejima and Kubo 1991). Anacardic acids also showed weak activity against molds (*Saccharomyces cerevisiae*, a yeast, and *Penicillium chrysogenum*). Anacardic acids having the C_{10} alkyl side chain were found to be the most potent against *Staphylococcus aureus* while those with C_{12} alkyl side chain was most effective against *Propionibacterium acnes*, *Streptococcus mutans* and *Brevibacterium ammoniagenes* (Kubo et al. 1993b). In separate studies, cashew nut shell liquid (CNSL) exhibited minimal inhibitory concentrations (MIC) against *Propionibacterium acnes* (1.56 µg/ml), *Corynebacterium xerosis* (6.25 µg/ml), and various strains of *Staphylococcus aureus* (25 µg/ml), but was inactive against the fungus *Pityrosporum ovale* (Bouttier et al. 2002). The -lactamase inhibition was evaluated by the agar diffusion method with penicillin G, against two penicillin-resistant strains of *S. aureus* but one strain being insensitive to all -lactamines antibiotics due to an alteration of protein linking penicillin (PLP 2a). The results showed clearly the -lactamase inhibitory activity of CNSL and anacardic acids on *Streptococcus aureus* enzyme, but without activity on PLP 2a. In another study, anacardic acids isolated from the cashew apple and a series of their synthetic analogs, 6-alkylsalicylic acids, were found to be bactericidal against *Streptococcus mutans* (Muroi and Kubo 1993). Synergistic bactericidal activity was found in the combination of anacardic acid [6-[8(Z),11(Z),14-pentadecatrienyl] salicylic acid] and anethole or linalool. Anacardic acids and (E)-2-hexenal cashew apple were also found to exhibit antibacterial activity against the Gram-negative bacterium *Helicobacter pylori*, causal agent of acute gastritis (Kubo et al. 1999). The same antibacterial compounds were also found to inhibit urease.

Antihypertensive Activity

Tannins in cashew apple (hypocarp) were found to impart anti-hypertensive activity, lending support to the use of the plant in folkloric medicine for the treatment of diabetes mellitus, diarrhoea and hypertension in south of Cameroon and other African countries (Paris et al. 1977).

Antileishmanial Activity

The hydroalcoholic extract of *A. occidentale* bark showed high in-vitro activity against the promastigotes of the species *Leishmania (Viannia) braziliensis* (Franca et al. 1993). However, in the in-vivo model no therapeutic activity was observed.

Molluscicidal Activity

The components of anacardic acids exhibited varying degree of toxicity to fresh water snails, *Biomphalaria glabrata* (Sullivan et al 1982). The triene component was the most toxic form (LC_{50} 0.35 ppm), the diene and monoene components were less toxic (LC_{50} 0.9 and 1.4 ppm), and the saturated component was relatively nontoxic ($LC_{50} > 5$ ppm). Since decarboxylated anacardic acid (cardanol) and salicylic acid did not kill snails at concentrations up to 5 ppm, it appeared that both, carboxyl group and unsaturated side chain were essential for molluscicidal activity. Twelve components of cashew nut shell oil including derivatives of anacardic acids, cardol, 2-methyl-cardol, and cardonol exhibited molluscicidal activity against *Biomphalaria glabrata*, snail vector of schistosomiasis (Kubo et al. 1986).

In another study, hexanolic extracts of cashew nut shells were found to exhibit molluscicidal activity (De Souza et al. 1992). The lethal concentration, CL_{90}, of sample 1 was from 2.0 to 2.2 ppm for adult snails of the three species *Biomphalaria glabrata*, *Biomphalaria tenagophila* and *Biomphalaria straminea*. With sample 2, the CL_{90} was 2.0, 0.5 and 30.0 ppm for *B. glabrata* adults, newly hatched snails and egg mass respectively. Non-extracted cashew shells caused 40–80% death of adult snails, 22–35% mortality of embryos and 40–55% reduction of egg production. The hexanolic extract, sample 2, were harmless to tadpoles and fish at 2 ppm. In the field, in pools of still water treatment with 20 ppm of extract, sample 1, caused a 97.1% mortality of *B. straminea* and 100% mortality of *B. glabrata* and *B. tenagophila*. In separate studies, sodium salts of cashew nut shell extracts (CNSL) and anacardic acids isolated from *Anacardium occidentale* demonstrated a strong anti-vectorial activity against *Biomphalaria glabrata* snails (Laurens et al. 1997).

Larvicidal Activity

Sodium salts of cashew nut shell extracts (CNSL) and anacardic acids isolated from *Anacardium occidentale* also demonstrated potent anti-vectorial activity against *Aedes aegypti* larvae (Laurens et al. 1997). *Aedes aegypti* larvae and pupae were found to be highly susceptible to 12 ppm CNSL, while *Anopheles subpictus* larvae and pupae, were found susceptible to 38 ppm of Cardanol/CNSL solution in laboratory and in field conditions in Andhra Pradesh (Mukhopadhyay et al. 2010).

Antiophidian Activity

Anacardium occidentale bark extract was found to neutralize *Vipera russelii* venom hydrolytic enzymes such as phospholipase, protease, and hyaluronidase in a dose dependent manner (Ushanandini et al. 2009). These enzymes are responsible for both local effects of envenomation such as local tissue damage, inflammation and myonecrosis, and systemic effects including dysfunction of vital organs and alteration in the coagulation components. Further, the extract neutralized pharmacological effects such as edema, hemorrhage, and myotoxic effects including lethality, caused by snake venom. Since, the bark extract inhibited both hydrolytic enzymes and pharmacological effects; the researchers concluded that it may be used as an alternative treatment to serum therapy and, further, as a rich source of potential inhibitors of hydrolytic enzymes involved in several physiopathological diseases.

Tyrosinase Inhibition Activity

Anacardic acids, 2-methylcardols, and cardols isolated from various parts of the cashew fruit were found to exhibit tyrosinase inhibitory activity (Kubo et al. 1994). Kinetic studies with the two principal active compounds, 6-[8(Z),11(Z),14-pentadecatrienyl]salicylic acid and 5-[8(Z),11(Z),14-pentadecatrienyl]resorcinol, indicated that both of these phenolic compounds exhibited characteristic competitive inhibition of the oxidation of L-3,4-dihydroxyphenylalanine (L-DOPA) by mushroom tyrosinase. In separate studies, cashew leaf extract (CLE) demonstrated potential as tyrosinase inhibitor that can be used as therapeutic in pigmentation problem (Abdul Gaffar et al. 2008). The extract inhibited tyrosinase and reduced melanin in human epidermal melanocytes exposed to the leaf extract for more than 24 h. The CLE extract also exhibited high antioxidant activity. Kubo et al. (2011) found that although anacardic acids and cardols inhibited tyrosinase, a key enzyme in melanin synthesis, melanogenesis in melanocytes was not suppressed in cultured murine B16-F10 melanoma cells but rather enhanced. Both anacardic acids and cardols exhibited moderate cytotoxicity.

Cardol triene, purified from cashew nut shell liquid was found to be potent irreversible competitive tyrosinase inhibitor exerting a complex type of tyrosinase inactivation (Zhuang et al. 2010). Two molecules of cardol triene could bind to one molecule of tyrosinase leading to the complete loss of its catalytic activity.

Antiviral Activity

Extracts from *Anacardium occidentale* leaves (4 µg/ml) and *Psidium guajava* leaves (8 µg/ml) showed activity only against simian rotavirus (82.2% and 93.8% inhibition, respectively) (Gonçalves et al. 2005). Rotaviruses have been recognized as the major agents of diarrhoea in infants and young children in developed as well as developing countries. In Brazil, diarrhoea is one of the principal causes of death, mainly in the infant population.

Anthelmintic Activity

Cashew nut shell oil was found to have anthelmintic activity against ascaridiasis (Varghese et al. 1971). It was as effective as piperazine citrate against *Ascaris galli* in the domestic fowl and produced no adverse effects.

Lipoxygenase and Prostaglandin Endoperoxide Synthase Inhibition Activity

Anacardic acid 8'Z-monoene exhibited the highest potency ($IC_{50} = 50$ uM) in inhibiting 15-lipoxygenase (soyabean lipoxygenase-1) followed by cardol 8'Z,11'Z,14'-triene ($IC_{50} = 60$ uM) (Shobha et al. 1994). Cardol and cardonol were ineffective. In another study, C22:1 omega 5-anacardic acid was found to be a good inhibitor of both potato lipoxygenase and ovine prostaglandin endoperoxide synthase with approximate IC_{50}'s of 6 and 27 µM, respectively (Grazzini et al. 1991). 6[8'(2)-Pentadecenyl]salicylic acid, otherwise known as anacardic acid (C15:1) was found to inhibit the linoleic acid peroxidation catalyzed by soybean lipoxygenase-1 with an IC_{50} of 6.8 µM (Ha and Kubo 2005). The inhibition of the enzyme by anacardic acid (C15:1) was a slow and reversible reaction without residual activity. The inhibition kinetics analyzed by Dixon plots indicated that anacardic acid (C15:1) was a competitive inhibitor. Although anacardic acid (C15:1) inhibited the linoleic acid peroxidation without being oxidized, 6[8'(Z),11'(Z)-pentadecadienyl]salicylic acid, otherwise known as anacardic acid (C15:2), was dioxygenated at low concentrations as a substrate. Further, anacardic acid (C15:2) was also found to exhibit time-dependent inhibition of lipoxygenase-1.

Toxicity Studies

Swiss albino mice administered ethanolic extract of *Anacardium occidentale* orally through orogastric tube for 25 days, exhibited no toxic effects (Ofusori et al. 2008). The histological findings indicated that the treated sections of the brain and kidney showed no cyto-architectural distortions and no evidence of apoptotic bodies. These findings suggested a non-toxic effect of ethanolic extract of *Anacardium occidentale* in Swiss albino mice. The results also suggested that the functions of the brain and kidney could be enhanced and in some cases, *Anacardium occidentale* could be recommended for the management of certain ailments associated with brain and kidney.

Studies using the Comet assay on Chinese hamster lung fibroblasts (V79 cells) revealed that the two lowest concentrations of cashew stem bark methanolic extract (500 µg/ml, 1,000 µg/ml) tested presented no genotoxic activity, whereas the highest dose of 2,000 µ/ml produced genotoxicity (Barcelos et al. 2007a). All of the concentrations showed protective activity in simultaneous and post-treatment in relation to methyl methanesulfonate (MMS) as a positive control. Cashew stem bark methanolic extract (CSBME) was also found to be antimutagenic (Barcelos et al. 2007b) The data obtained in the chromosome aberrations test on cell cultures of Chinese hamster lung fibroblasts (V79) showed a significant reduction in chromosome aberrations frequency in the cultures treated with doxorubicin and cashew stem bark methanolic extract in comparison with those that received only doxorubicin during the cell cycle phases G1 and S and throughout the entire cycle. The antimutagenic effect observed in this study reinforced the presence of the previously described therapeutic properties of cashew and indicated the safe use of this extract.

Separate studies showed that a crude cashew extract did not produce subacute toxicity and genotoxicity symptoms in rats in doses up to 2,000 mg/kg (Konan et al. 2007). The extract was shown to induce frame-shift, base pair substitution and damage to the chromosomes. However, this effect was less deleterious than the clastogenic effect of ciclophosphamide. Based on biochemical analyses of renal and hepato-biliary functions, such as the level of urea, creatinine, transaminases and alkaline phosphatase, the scientists determined that the extract was generally tolerated by rats. This was also confirmed by hematological and histopathological examinations. Genotoxicity was assessed by the Ames test in *Salmonella typhimurium* strains TA97, TA98, TA100, TA102 and by the bone marrow micronucleus test in mice.

The chemical constituents of both fresh cashew apple juice (CAJ) and the processed (cajuina) were analyzed and characterized as complex mixtures containing high concentrations of vitamin C, various carotenoids, phenolic compounds, and metals (Melo-Cavalcante et al. 2003). In the study, these beverages exhibited direct and rat liver S9-mediated mutagenicity in the *Salmonella*/microsome assay with strains TA97a, TA98, and TA100, which detected frame-shifts and base pair substitution. No mutagenicity was observed with strain TA102, which detected oxidative and alkylating mutagens and active forms of oxygen. Both CAJ and cajuina exhibited antioxidant activity as evaluated by a total radical-trapping potential assay. CAJ and cajuina protected strain TA102 against mutation by oxidative damage in co- and post-treatments. The antimutagenic effects during co-treatment with hydrogen peroxide were postulated to be due to scavenging free radicals and complexing extracellular mutagenic compounds. The protective effects in post-treatment were attributable to stimulation of repair and/or reversion of DNA damage. The results indicated that CAJ and cajuina had mutagenic, radical-trapping, antimutagenic, and comutagenic activity and that these properties could be related to the chemical constituents of the juices.

In a test for its potency in promoting the DMBA (7,12-dimethylbenz[α]anthracene)-initiated cells into papillomas in a murine two-stage skin tumorigenesis model system in Male Swiss albino mice, cashew nut shell oil was found to have a weak tumour enhancing effect (Banerjee and Rao 1992). The mean tumour incidences were found to be 1.1 and 2.5 in 1% and 2% oil treatment groups, respectively, while the corresponding figure vas 6.6 in the positive control group (DMBA and 1% croton oil in acetone).

Toxicity profiles of the methanol extract of cashew inner stem bark on some liver function parameters were evaluated following a sub-chronic oral administration at doses of 1.44 and 2.87 g/kg (Okonkwo et al. 2010). Sub-chronic administration of the extract significantly increased the serum levels of alanine aminotransaminase and aspartate aminotransaminase, which were indicative of liver damage but serum levels of alkaline phosphatase and total protein of the treated animals were not significantly elevated. Sub-chronic administration of the extract did not significantly depressed the function of hepatocytes in Wistar rats. The effects of sub-chronically administered extract on hepatocytes were minimal as the serum alkaline phosphatase, total bilirubin and total protein levels in treated animals were not significant. The extract was found to contain high levels of tannins, moderate saponins and traces of free reducing sugars. The anti-nutrient concentrations were 5.75% (tannins), 2.50% (oxalates), 2.00% (saponins), 0.25% (phytate) and 0.03% (cyanide). The amount of iron detected from dried crude was 8.92 mg/100 g, while lead and cadmium were non-detectable. The extract had LD_{50} of 2.154 g/kg per os in mice.

Allergy Problem

There have been a number of reports of allergic contact dermatitis following ingestion of cashew nut butter (Rosen and Fordice 1994), a cashew nut pesto (Hamilton and Zug 1998), cashew nuts (Marks et al. 1984) and cashew nuts contaminated with cashew nut shell oil among cashew nut workers (Pasricha et al. 1988; Diogenes et al. 1996). The most common reaction was the appearance of dermatitis which required treatment with high doses of oral steroids. Systemic contact dermatitis from the cashew nut shell oil resorcinol allergens, cardol and anacardic acid, was recognized clinically as a dermatitis (Hamilton and Zug 1998). This was characterised by flexural accentuation, typically on the extremities, groin, and buttocks which appeared 1–3 days after ingestion of raw cashew nuts contaminated with allergenic oil. They also reported a case of systemic contact dermatitis to raw cashew nuts, an atypical and unexpected ingredient flavouring in an imported pesto sauce. In another report, 44 individuals developed dermatitis after eating imported cashew nuts (Marks et al. 1984). The patients had very itchy, erythematous, maculopapular eruption that was intensified in the flexural areas of the body. Three had blistering of the mouth and four had rectal itching. A case of cashew nut urushiol dermatitis due to ingestion of homemade cashew nut butter contaminated by cashew nut shell oil was also reported (Rosen and Fordice 1994). Cashew nut shell liquid (CNSL) was found toxic to human skin on contact. The dermatitis caused in humans by CSNL was attributed to the presence of urushiol, a homologue of catechol or similar materials (Nair et al. 1979). However, the decarboxylated and refined CNSL was not toxic.

Three cases of anaphylaxis to cashew nuts were observed (García et al. 2000). It was found that in the skin prick tests and specific IgE to cashew and pistachio nuts were positive in the three patients. Both nuts showed several protein bands in SDS-PAGE. The strongest IgE-binding bands had similar molecular weights (15, 30 and 60 kDa) in cashew and pistachio nuts. These main bands were found to be sensitive to reducing agents. It was concluded that these three patients suffered immediate reactions to cashew nut due to an IgE-mediated mechanism.

Cashew nut allergy is the second most commonly reported tree nut allergy in the United States. The following allergen were identified in cashew nuts: Ana o-1 a vicilin-like protein, a major food allergen in cashews; anao2 legumin-like protein; and anao 3 a 2 S albumin protein (Wang et al. 2002; Wang et al. 2003; Robotham et al. 2005).

Traditional Medicinal Uses

Cashew apple, kernel, leaves, bark, wood and roots have been employed in traditional folk medicine. The leaves, bark and roots are used in the management and/or control of arthritis, diabetes and other inflammatory conditions among the Yoruba-speaking people of western Nigeria and for the treatment of diabetes mellitus, diarrhoea and hypertension in south of Cameroon and other African countries.

Buds and young leaves are used for skin diseases. In Java, old leaves are used as a paste applied to skin affections and burns. Pulped leaves are used for skin applications of infants for pemphigus neonatorum. In Peruvian herbal medicine today, cashew leaf tea (called *casho*) is employed as a common diarrhea remedy. Cashew leaf teas are prepared as a mouthwash and gargle for mouth ulcers, tonsillitis, and throat problems and are used for washing wounds. The Tikuna tribe in northwest Amazonia brew a tea of cashew leaves to treat diarrhoea. In India, leaves and branches are used by tribals as an antiseptic post-parturition bath additive.

Cashew apple is anti-scorbutic, astringent and diuretic, and is used for cholera and kidney troubles. In Java, cashew apple juice is used to arrest vomiting, as mouth wash for thrush, and as gargle for catarrh. Cashew syrup is considered a good remedy for coughs and colds. Cashew apple juice is said to be effective for the treatment of syphilis and is prescribed as a remedy for sore throat and chronic dysentery in Cuba and Brazil. In Brazilian herbal medicine, the cashew apple is taken for syphilis and as a diuretic, stimulant, and aphrodisiac. Fresh or distilled, it is a potent diuretic and is said to possess sudorific properties. The brandy is applied as a liniment to relieve the pain of rheumatism and neuralgia. Old cashew liquor in small doses cures stomach-ache. The kernel is a demulcent, an emollient and is used for diarrhoea. The resinous juice of seeds is used for mental derangement, heart palpitation, rheumatism; it is used to cure the loss of memory, a sequel to smallpox. Powdered seeds are used as anti-venom for snake bites.

Cashew nut shell oil is anti-hypertensive and purgative; it is used for blood sugar problems, kidney troubles, hookworms and cholera. In India, the oil is used for, cracks on soles of feet and for corns, leprous sores and warts in Johore, Malaysia. Oil from the kernel is a chemical antidote for irritant poisons and a good vehicle for liniments and other external applications.

Cashew bark is astringent, counterirritant, rubefacient, vesicant, and is used for ulcer. In Peninsular Malaysia, a decoction of the bark is used for severe diarrhoea and for constipation by the native Sakais. Root infusion is an excellent purgative. The Wayãpi tribe in Guyana uses a cashew bark tea as a diarrhea remedy and colic remedy for infants. In Brazil, a cashew bark tea is used as a douche for vaginal discharge and as an astringent to stop bleeding after a tooth extraction and for diarrhoeal remedies. The cashew bark is employed to treat eczema, psoriasis, scrofula, dyspepsia, genital problems, and venereal diseases, as well as for impotence, bronchitis, cough, intestinal colic, leishmaniasis, and syphilis-related cashew skin disorders. An infusion and/or maceration of the cashew bark are used to treat diabetes, weakness, muscular debility, urinary disorders, and asthma. An infusion of bark and leaves is used as a gargle to relieve toothache, sore gums, or sore throat.

Other Uses

The cashew nut shell oil or liquid (CNSL) contains mainly cardanol (decarboxylated anacardic acid), anacardic acids, cardol, and 2-methyl cardol (Kumar et al. 2002; De Lima et al. 2008). It is traditionally obtained as a byproduct during extraction of the kernel. CNSL has wide industrial applications and is a valuable raw material for a number of polymer based industries like paints and varnishes, resins, industrial and decorative laminates, brake linings, lubricants, rubber compounding resins and other synthetic resins (epoxy resins), adhesives, gums, impregnations, sealants for cracks and joints in buildings, floor coverings preservative to treat, for instance, wooden structures and fishing nets and other technical products (Nair et al. 1979; Ohler 1979; Murthy and Sivasamban 1985). CNSL and its derivatives are also use as bactericides, fungicides, herbicides and insecticides, and the preparation of antioxidants and drugs. Rice husks boards prepared from CNSL reside based resin as binder are used for false ceilings, as insulating panels and for acoustic purposes. A number of derivatives of 3-pentadecyl phenol were synthesised and have been used to produce good pesticidal and pharmacological compounds. Cardanol, the major constituent of CNSL, is pale coloured and is used in lacquers, coatings, insulating varnishes, furniture coatings and as media for quick drying enamels. The dark colour of CNSL prohibits its use in light coloured finishes. The CNSL distillation residue and its blends with alky are used in the preparation of air drying varnishes, cycle enamels and primers. The films of cardanol based resins and vinyl resins have excellent properties. Pale coloured adhesives have been obtained by the reaction of cardanol-formaldehye resin with epichlorohydrin. Chemical modification of the alkyl side-chain of cardanol can lead to the formation of polyfunctional compounds which were subsequently used for preparing water-soluble binders and modified alkyd resins (Madhusudhan and Murthy 1992). The properties of the coatings so formed from these binders such as flexibility, hardness and resistance to water and chemicals were found to be superior to those of coatings based on conventional cardanol media.

In Tanzania, CNSL has been used for making tribal marks and scars on the face and body and elsewhere employed in preserving fishing nets, preserving books and in softening chicken eggs and is effective against termites. The cake remaining after oil has been extracted from the kernel serves as animal food. The residue of the shell is often used as fuel in cashew nut shell liquid extraction plants. Seed coats are used as poultry feed. Anacardic acids have been used effectively against tooth abscesses due to their lethality to Gram-positive bacteria.

Recent studies by Oddoye et al. (2009) found that dried cashew pulp, the sun-dried residue after juice has been extracted from the cashew apple, could be used in growing pig diets up to a level of 200 g/kg without any adverse effects. Studies by Pacheco et al. (2010) found cashew apple bagasse to be an efficient support for *Saccharomyces cerevisiae* cell immobilization aiming at ethanol production. Ten consecutive fermentations of cashew apple juice for ethanol production were carried out using immobilized *Saccharomyces cerevisiae*. High ethanol productivity was attained from the third fermentation assay until the tenth fermentation. Ethanol concentrations (about 19.82–37.83 g/l, mean value) and ethanol productivities (about 3.30–6.31 g/l) were high and stable. Residual sugar concentrations were low in almost all fermentations (around 3.00 g/l) with conversions ranging from 44.80% to 96.50%, showing efficiency of 85.30–98.52% and operational stability of the biocatalyst for ethanol fermentation. Araújo et al. (2011) utilised two different strains of yeast *Saccharomyces cerevisiae* viz. SCP and SCT, for fermentation of cashew apple juice to produce an alcoholic product. The fermented products obtained from both yeast strains contained ethanol concentration above 6% v/v with total acidity below 90 mEq/l, representing a pH of 3.8–3.9. Methanol was not detected. The volatile compounds were characterized by the occurence of aldehyde (butyl aldehyde diethyl acetal, 2,4-dimethyl-hepta-2,4-dienal, and 2-methyl-2-pentenal) and ester (ethyl α-methylbutyrate) representing fruity aroma. The SCT strain was found to be more efficient, producing 10% more alcohol over that of the SCP strain.

Pulp from the wood is used to fabricate corrugated and hardboard boxes. The wood known as 'white mahogany' in Latin America is fairly hard and is use in various industrial applications – in wheel hubs, yoke, fishing boats, furniture, false ceilings and interior decoration. Boxes made from the wood are collapsible but are strong enough to compete with conventional wooden packing cases. Pulp from the wood is used to fabricate corrugated and hard-board boxes. The wood is popular for firewood and charcoal. Wood ash is rich in potassium and so used for applying as fertiliser to crop plants.

The bark contains an acrid sap of thick brown resin, which is used as indelible ink in marking and printing linens and cottons. The resin is also used as a varnish, a preservative for fishnets and a flux for solder metals. The stem also yields an amber-coloured gum, which is partly soluble in water, the main portion swelling into a jellylike mass. This gum is used as an adhesive (for woodwork panels, plywood, bookbinding), partly because it has insecticidal properties. The acrid sap of the bark contains 3–5% tannin and is employed in the tanning industry.

Comments

Other edible *Anacardium* species reported (Mitchell and Mori 1987) are: *Anacardium excelsum* (Bertero & Balbies ex Kunth) Skeels, *Anacardium humile* St.Hilaire, *Anacardium corymbosum* Barbosa Rodrigues, *Anacardium giganteum* Hancock ex Engler, *Anacardium microsepalum* Loesener.

Selected References

Abdul Gaffar R, Abdul Majid FA, Sarmidi MR (2008) Tyrisonase inhibition and melanin reduction of human melanocytes (HEMn-MP) using *Anacardium occidentale* L extract. Med J Malaysia 63(Suppl A):100–101

Akinwale TO (2000) Cashew apple juice: its use in fortifying the nutritional quality of some tropical fruits. Eur Food Res Technol 211(3):205–207

Alexander-Lindo RL, Morrison EY, Nair MG (2004) Hypoglycaemic effect of stigmast-4-en-3-one and its corresponding alcohol from the bark of *Anacardium occidentale* (cashew). Phytother Res 18(5):403–407

Araújo SM, Silva CF, Moreira JJ, Narain N, Souza RR (2011) Biotechnological process for obtaining new fermented products from cashew apple fruit by Saccharomyces cerevisiae strains. J Ind Microbiol Biotechnol [Epub ahead of print]

Assunção RB, Mercadante AZ (2003a) Carotenoids and ascorbic acid composition from commercial products of cashew apple (*Anacardium occidentale* L.). J Food Comp Anal 16(6):647–657

Assunção RB, Mercadante AZ (2003b) Carotenoids and ascorbic acid from cashew apple (*Anacardium occidentale* L.): variety and geographic effects. Food Chem 81(4):495–502

Banerjee S, Rao AR (1992) Promoting action of cashew nut shell oil in DMBA-initiated mouse skin tumour model system. Cancer Lett 62(2):149–152

Barcelos GR, Shimabukuro F, Maciel MA, Cólus IM (2007a) Genotoxicity and antigenotoxicity of cashew (*Anacardium occidentale* L.) in V79 cells. Toxicol In Vitro 21(8):1468–1475

Barcelos GR, Shimabukuro F, Mori MP, Maciel MA, Cólus IM (2007b) Evaluation of mutagenicity and antimutagenicity of cashew stem bark methanolic extract in vitro. J Ethnopharmacol 114(2):268–273

Bouttier S, Fourniat J, Garofalo C, Gleye C, Laurens A, Hocquemiller R (2002) β-lactamase inhibitors from *Anacardium occidentale*. Pharm Biol 40(3):231–234

Burkill IH (1966) A dictionary of the economic products of the Malay Peninsula. Revised reprint. 2 vols, vol 1 (A–H), vol 2 (I–Z). Ministry of Agriculture and Co-operatives, Kuala Lumpur, pp 1–1240, pp 1241–2444

Cecchi HM, Rodriguez-Amaya DB (1981) Carotenoid composition and vitamin a value of fresh and pasteurized cashew-apple (*Anacardium occidentale* L.) juice. J Food Sci 46(1):147–149

De Lima SG, Feitosa CM, Citó AM, Moita Neto JM, Lopes JA, Leite AS, Brito MC, Dantas SM, Cavalcante AA (2008) Effects of immature cashew nut-shell liquid (Anacardium occidentale) against oxidative damage in *Saccharomyces cerevisiae* and inhibition of acetylcholinesterase activity. Genet Mol Res 7(3):806–818

De Souza CP, Mendes NM, Janotti-Passos LK, Pereira JP (1992) The used of the shell of cashew nut, *Anacardium occidentale*, as an alternative molluscacide. Rev Inst Med Trop São Paulo 34(5):459–466 (In Portuguese)

Diogenes MJN, De Morais SM, Carvalho FF (1996) Contact dermatitis among cashew nut workers. Contact Dermat 35(2):115

Duke JA (1983) *Anacardium occidentale* L. In: Handbook of energy crops. http://www.hort.purdue.edu/newcrop/duke_energy/Anacardium_occidentale.html

Egwim E (2005) Hypoglycemic potencies of crude ethanolic extracts of cashew roots and unripe pawpaw fruits in guinea pigs and rats. J Herb Pharmacother 5(1):27–34

Foundation for Revitalisation of Local Health Traditions (2008) FRLHT database. htttp://envis.frlht.org

Franca F, Cuba CA, Moreira EA, Miguel O, Almeida M, Das Virgens M, Marsden PD (1993) An evaluation of the effect of a bark extract from the cashew (*Anacardium occidentale* L.) on infection by *Leishmania (Viannia) braziliensis*. Rev Soc Bras Med Trop 26(3):151–155 (In Portuguese)

García F, Moneo I, Fernández B, García-Menaya JM, Blanco J, Juste S, Gonzalo (2000) Allergy to Anacardiaceae: description of cashew and pistachio nut allergens. J Investig Allergo Clin Immunol 10(3): 173–177

Gómez-Caravaca AM, Verardo V, Caboni MF (2010) Chromatographic techniques for the determination of alkyl-phenols, tocopherols and other minor polar compounds in raw and roasted cold pressed cashew nut oils. J Chromatogr A 1217(47):7411–7417

Gonçalves JL, Lopes RC, Oliveira DB, Costa SS, Miranda MM, Romanos MT, Santos NS, Wigg MD (2005) In vitro anti-rotavirus activity of some medicinal plants used in Brazil against diarrhea. J Ethnopharmacol 99(3):403–407

Grazzini R, Hesk D, Heininger E, Hildenbrandt G, Reddy CC, Cox-Foster D, Medford J, Craig R, Mumma RO (1991) Inhibition of lipoxygenase and prostaglandin endoperoxide synthase by anacardic acids. Biochem Biophys Res Commun 176(2):775–780

Ha TJ, Kubo I (2005) Lipoxygenase inhibitory activity of anacardic acids. J Agric Food Chem 53:4350–4354

Hamilton TK, Zug KA (1998) Systemic contact dermatitis to raw cashew nuts in a pesto sauce. Am J Contact Dermat 9(1):51–54

Himejima M, Kubo I (1991) Antibacterial agents from the cashew *Anacardium occidentale* (Anacardiaceae) nut shell oil. J Agric Food Chem 39(2):418–421

Johnson D (1973) The botany, origin, and spread of cashew, Anacardium occidentale L. J Plantation Crops 1:1–7

Kamtchouing P, Sokeng DS, Moundipa PF, Pierre W, Jatsa BH, Lontsi D (1998) Protective role of *Anacardium occidentale* extract against streptozotocin induced diabetes in rats. J Ethnopharmacol 62:95–99

Kirtikar KR, Basu BD (1975) Indian medicinal plants, 4 vols, 2nd edn. Jayyed Press, New Delhi

Konan NA, Bacchi EM (2007) Antiulcerogenic effect and acute toxicity of a hydroethanolic extract from the cashew (*Anacardium occidentale* L.) leaves. J Ethnopharmacol 112(2):237–242

Konan NA, Bacchi EM, Lincopan N, Varela SD, Varanda EA (2007) Acute, subacute toxicity and genotoxic effect of a hydroethanolic extract of the cashew (*Anacardium occidentale* L.). J Ethnopharmacol 110(1):30–38

Konan NA, Lincopan N, Collantes Díaz IE, de Fátima Jacysyn J, Tanae Tiba MM, Pessini Amarante Mendes JG, Bacchi EM, Spira B (2011) Cytotoxicity of cashew flavonoids towards malignant cell lines. Exp Toxicol Pathol [Epub ahead of print]

Kubo I, Komatsu S, Ochi M (1986) Molluscicides from the cashew *Anacardium occidentale* and their large-scale isolation. J Agric Food Chem 34:970–973

Kubo I, Muroi H, Himejima M (1993a) Structure-antibacterial activity relations of anacardic acids. J Agric Food Chem 41:1016–1019

Kubo I, Ochi M, Vieira PC, Komatsu S (1993b) Antitumor agents from the cashew (*Anacardium occidentale*) apple juice. J Agric Food Chem 41:1012–1015

Kubo I, Kinst-Hori I, Yokokawa Y (1994) Tyrosinase inhibitors from *Anacardium occidentale* fruits. J Nat Prod 57(4):545–551

Kubo J, Lee JR, Kubo I (1999) Anti-*Helicobacter pylori* agents from the cashew apple. J Agric Food Chem 47(2):533–537

Kubo I, Masuoka N, Ha TJ, Tsujimoto K (2006) Antioxidant activity of anacardic acids. Food Chem 99(3):555–562

Kubo I, Nitoda T, Tocoli FE, Green IR (2011) Multifunctional cytotoxic agents from *Anacardium occidentale*. Phytother Res 25(1):38–45

Kumar PP, Paramashivappa R, Vithayathil PJ, Subba Rao PV, Srinivasa Rao A (2002) Process for isolation of cardanol from technical cashew (*Anacardium occidentale* L.) nut shell liquid. J Agric Food Chem 50(16):4705–4708

Laurens A, Fourneau C, Hocquemiller R, Cavé A, Bories C, Loiseau PM (1997) Antivectorial activities of cashew nut shell extracts from *Anacardium occidentale* L. Phytother Res 11(2):145–146

Logrado LP, Santos CO, Romeiro LA, Costa AM, Ferreira JR, Cavalcanti BC, Manoel de Moraes O, Costa-Lotufo LV, Pessoa C, Dos Santos ML (2010) Synthesis and cytotoxicity screening of substituted isobenzofuranones designed from anacardic acids. Eur J Med Chem 45(8):3480–3489

Madhusudhan V, Murthy BGK (1992) Polyfunctional compounds from cardanol. Prog Org Coat 20(1):63–71

Marks JG Jr, DeMelfi T, McCarthy MA, Witte EJ, Castagnoli N, Epstein WL, Aber RC (1984) Dermatitis from cashew nuts. J Am Acad Dermatol. 10(4): 627–31

Martin FW (1984) Handbook of tropical food crops. CRC Press, Boca Raton, 296 pp

Melo-Cavalcante AA, Rubensam G, Picada JN, Gomes da Silva E, Fonseca Moreira JC, Henriques JA (2003) Mutagenicity, antioxidant potential, and antimutagenic activity against hydrogen peroxide of cashew (*Anacardium occidentale*) apple juice and cajuina. Environ Mol Mutagen 41:360–369

Menestrina JM, Iacomini M, Jones C, Gorin PA (1998) Similarity of monosaccharide, oligosaccharide and polysaccharide structures in gum exudate of *Anacardium occidentale*. Phytochemistry 47(5):715–721

Michodjehoun-Mestres L, Souquet J-M, Fulcrand H, Meudec E, Reynes M, Brillouet J-M (2009a) Characterisation of highly polymerised prodelphinidins from skin and flesh of four cashew apple (*Anacardium occidentale* L.) genotypes. Food Chem 114(3):989–995

Michodjehoun-Mestres L, Souquet J-M, Fulcrand H, Bouchut C, Reynes M, Brillouet J-M (2009b) Monomeric phenols of cashew apple (*Anacardium occidentale* L.). Food Chem 112(4):851–857

Mitchell JD, Mori SA (1987) The cashews and its relatives (*Anardium: Anacardiaceae*). Mem NY Bot Gard.42:76 pp

Mohd Noor O (1972) Cashew nut in Trengganu. Agricultural leaflet no. 51. Ministry of Agriculture and Fisheries, Kaula Lumpur, 20 pp

Mokhtar NM, Kanthimathi MS, Aziz AA (2008) Comparisons between the antioxidant activities of the extracts of *Anacardium occidentale* and *Piper betle*. Malays J Biochem Mol Biol 16(1):16–21

Morais TC, Pinto NB, Carvalho KM, Rios JB, Ricardo NM, Trevisan MT, Rao VS, Santos FA (2010) Protective effect of anacardic acids from cashew (*Anacardium occidentale*) on ethanol-induced gastric damage in mice. Chem Biol Interact 183(1):264–269

Morton J (1987) Cashew apple. In: Fruits of warm climates. Julia F. Morton, Miami, pp 239–240

Mota ML, Thomas G, Barbosa Filho JM (1985) Antiinflammatory actions of tannins isolated from the bark of *Anacardium occidentale* L. J Ethnopharmacol 13(3):289–300

Mukhopadhyay AK, Hati AK, Tamizharasu W, Babu PS (2010) Larvicidal properties of cashew nut shell liquid (*Anacardium occidentale* L) on immature stages of two mosquito species. J Vector Borne Dis 47(4):257–260

Muroi H, Kubo I (1993) Bactericidal activity of anacardic acids against *Streptococcus mutans* and their potentiation. J Agric Food Chem 41(10):1780–1783

Muroi H, Kubo A, Kubo I (1993) Antimicrobial activity of cashew apple flavor compounds. J Agric Food Chem 41(7):1106–1109

Murthy BGK, Sivasamban MA (1985) Recent trends in CNSL utilization. Acta Hort (ISHS) 108:200–206

Nair MK, Bhaskara Rao EVV, Nambiar KKN, Nambiar MC (eds) (1979) Cashew (*Anacardium occidentale* L.) monograph on plantation crops –1. Central Plantation Crops Research Institute, Kasaragod, 169 pp

Nanjundaswamy AM, Radhakrishniah Setty G, Patwardhan MV (1985) Utilization of cashew apples for the development of processed products. Acta Hort (ISHS) 108:150–158

Oddoye EO, Takrama JF, Anchirina V, Agyente-Badu K (2009) Effects on performance of growing pigs fed diets containing different levels of dried cashew pulp. Trop Anim Health Prod 41(7):1577–1581

Ofusori D, Enaibe B, Adelakun A, Adesanya O, Ude R, Oluyemi K, Okwuonu C, Apantak O (2008) Microstructural study of the effect of ethanolic extract of cashew stem bark *Anacardium occidentale* on the brain and kidney of Swiss albino mice. Internet J Alternat Med 5(2)

Ohler JG (1979) Cashew (Department of Agricultural Research Communication no. 71). Royal Tropical Institute, Amsterdam, 260 pp

Ojewole JA (2003) Laboratory evaluation of the hypoglycemic effect of *Anacardium occidentale* Linn. (Anacardiaceae) stem-bark extracts in rats. Methods Find Exp Clin Pharmacol 25(3):199–204

Ojewole JA (2004) Potentiation of the antiinflammatory effect of *Anacardium occidentale* (Linn.) stem-bark aqueous extract by grapefruit juice. Methods Find Exp Clin Pharmacol 26(3):183–188

Okonkwo TJ, Okorie O, Okonta JM, Okonkwo CJ (2010) Sub-chronic hepatotoxicity of *Anacardium occidentale* (Anacardiaceae) inner stem bark extract in rats. Indian J Pharm Sci 72(3):353–357

Olajide OA, Aderogba MA, Adedapo AD, Makinde JM (2004) Effects of *Anacardium occidentale* stem bark extract on in vivo inflammatory models. J Ethnopharmacol 95(2–3):139–142

Olatunji L, Okwusidi J, Soladoye A (2005) Antidiabetic effect of *Anacardium occidentale* stem-bark in fructose-diabetic rats. Pharm Biol 43(7):589–593

Pacheco AM, Gondim DR, Gonçalves LR (2010) Ethanol production by fermentation using immobilized cells of *Saccharomyces cerevisiae* in cashew apple bagasse. Appl Biochem Biotechnol 161(1–8):209–217

Pacific Island Ecosystems at Risk (PIER) (1999) *Anacardium occidentale* L. Anacardiaceae. http://www.hear.org/Pier/scientificnames/..%5Cspecies%5Canacardium_occidentale.htm

Paris R, Plat M, Giono-Barber, Linhard J, Laurens A (1977a) Recherche chimique et pharmacologique sur les feuilles d'*Anacardium occidentale* L. (Anacardiacées). Bull Soc Med Afr Noire 22:275–281

Paris R, Plat M, Giono-Barber P, Linhard J, Laurens A (1977b) Chemical and pharmacological study of leaves of *Anacardium occidentale* L. (Anacardiaceae). Bull Soc Méd Afr Noire Lang Fr 22(3):275–281 (In French)

Pasricha JS, Srinivas CR, Shanker Krupa DS, Shanker DS, Kalpana Shenoy K (1988) Contact dermatitis due to cashew nut (*Anacardium occidentale*) shell oil, pericarp and kernel. Indian J Dermatol Venereol Leprol 54(1):36–37

Perry LM, Metzger J (1980) Medicinal plants of east and Southeast Asia. The MIT Press, Cambridge, 632 pp

Porcher M H et al (1995–2020) Searchable world wide web multilingual multiscript plant name database. Published by The University of Melbourne, Melbourne.

http://www.plantnames.unimelb.edu.au/Sorting/Frontpage.html

Robotham JM, Wang F, Seamon V, Teuber SS, Sathe SK, Sampson HA, Beyer K, Seavy M, Roux KH (2005) Ana o 3, an important cashew nut (*Anacardium occidentale* L.) allergen of the 2S albumin family. J Allergy Clin Immunol 115(6):1284–1290

Rodrigues RA, Grosso CR (2008) Cashew gum microencapsulation protects the aroma of coffee extracts. J Microencapsul 25(1):13–20

Rodrigues FHA, Feitosa JPA, Ricardo NMPS, de Franca FCF, Carioca JOB (2006) Antioxidant activity of cashew shell nut liquid (CNSL) derivatives on the thermal oxidation of synthetic cis-1,4-polyisoprene. J Braz Chem Soc 17(2):265–271

Rosen T, Fordice DB (1994) Cashew nut dermatitis. South Med J 87(4):543–546

Ryan E, Galvin K, O'Connor TP, Maguire AR, O'Brien NM (2006) Fatty acid profile, tocopherol, squalene and phytosterol content of brazil, pecan, pine, pistachio and cashew nuts. Int J Food Sci Nutr 57(3–4):219–228

Shobha SV, Ramadoss CS, Ravindranath B (1994) Inhibition of soybean lipoxygenase-1 by anacardic acids, cardols, and cardanols. J Nat Prod 57(12):1755–1757

Singh B, Kale RK, Rao AR (2004) Modulation of antioxidant potential in liver of mice by kernel oil of cashew nut (*Anacardium occidentale*) and its lack of tumour promoting ability in DMBA induced skin papillomagenesis. Indian J Exp Biol 42(4):373–377

Sokeng SD, Kamtchouing P, Watcho P, Jatsa HB, Moundipa PF, Lontsi D (2001) Hypoglycemic activity of *Anacardium occidentale* L. aqueous extract in normal and streptozotocin-induced diabetic rats. Diabetes Res 36:1–9

Sokeng SD, Lontsi D, Moundipa PF, Jatsa HB, Watcho P, Kamtchouing P (2007) Hypoglycemic effect of *Anacardium occidentale* L. methanol extract and fractions on streptozotocin-induced diabetic rats. Res J Med Med Sci 2(2):133–137

Sousa de Brito E, Pessanha de Araújo MC, Lin L-Z, Harnly J (2007) Determination of the flavonoid components of cashew apple (*Anacardium occidentale*) by LC-DAD-ESI/MS. Food Chem 105(3):1112–1118

Srisawat S, Teanpaisan R, Wattanapiromsakul C, Worapamorn W (2005) Antibacterial activity of some Thai medicinal plants against *Porphyromonas gingivalis*. In: International Association for Dental Research – 20th Annual Scientific Meeting of the Southeast Asia Division and Southeast Asia Association for Dental Education – 16th Annual Scientific Meeting, Malacca, 1–4 Sept 2005

Stuart GU (2010) Philippine alternative medicine. Manual of some Philippine medicinal plants. http://www.stuartxchange.org/OtherHerbals.html

Sullivan JT, Richards CS, Lloyd HA, Krishna G (1982) Anacardic acid: molluscicide in cashew nut shell liquid. Planta Med 44(3):175–177

Swarnalakshmi TK, Gomathi NS, Baskar EA, Parmar NS (1981) Anti-inflammatory activity of (−)-epicatechin, a biflavonoid isolated from *Anacardium occidentale*-Linn. Indian J Pharm Sci 43:205–208

Tedong L, Dimo T, Dzeufiet PDD, Asongalem AE, Sokeng DS, Callard P, Flejou JF, Kamtchouing P (2006) Antihyperglycemic and renal protective activities of *Anacardium occidentale* (Anacardiaceae) leaves in streptozotocin induced diabetic rats. Afr J Tradit Complement Altern Med 3(1):23–35

Tedong L, Madiraju P, Martineau LC, Vallerand D, Arnason JT, Desire DDP, Lavoie L, Kamtchouing P, Haddad PS (2010) Hydro-ethanolic extract of cashew tree (*Anacardium occidentale*) nut and its principal compound, anacardic acid, stimulate glucose uptake in C2C12 muscle cells. Mol Nutr Food Res 54(12):1753–1762

Tédong L, Dzeufiet PDD, Dimo T, Asongalem EA, Sokeng SN, Flejou J-F, Callard P, Kamtchouing P (2007) Acute and subchronic toxicity of *Anacardium occidentale* Linn (Anacardiaceae) leaves hexane extract in mice. Afr J Tradit Complement Altern Med 4(2):140–147

Tee ES, Noor MI, Azudin MN, Idris K (1997) Nutrient composition of Malaysian foods, 4th edn. Institute for Medical Research, Kuala Lumpur, p 299

Trevisan, M.T., Pfundstein, B., Haubner, R., Würtele, G., Spiegelhalder, B., Bartsch, H. and Owen, R.W (2006) Characterization of alkyl phenols in cashew (*Anacardium occidentale*) products and assay of their antioxidant capacity. Food Chem Toxicol 44(2):188–197

Trox J, Vadivel V, Vetter W, Stuetz W, Scherbaum V, Gola U, Nohr D, Biesalski HK (2010) Bioactive compounds in cashew nut (*Anacardium occidentale* L.) kernels: effect of different shelling methods. J Agric Food Chem 58(9):5341–5346

U.S. Department of Agriculture, Agricultural Research Service (2010) USDA National Nutrient Database for Standard Reference, Release 23. Nutrient Data Laboratory Home Page. http://www.ars.usda.gov/ba/bhnrc/ndl

Ushanandini S, Nagaraju S, Nayaka SC, Kumar KH, Kemparaju K, Girish KS (2009) The anti-ophidian properties of *Anacardium occidentale* bark extract. Immunopharmacol Immunotoxicol 31(4):607–615

Valim MF, Rouseff RL, Lin J (2003) Gas chromatographic-olfactometric characterization of aroma compounds in two types of cashew apple nectar. J Agric Food Chem 51(4):1010–1015

Varghese GC, Jacob DP, Georgekutty PT, Peter CT (1971) Use of cashew (*Anacardium occidentale*) nut shell oil as an anthelmintic against ascaridiasis in the domestic fowl. Kerala J Vet Sci 21(1):5–7

Wang F, Robotham JM, Teuber SS, Tawde P, Sathe SK, Roux KH (2002) Ana o 1, a cashew (*Anacardium occidentale* L.) allergen of the vicilin seed storage protein family. J Allergy Clin Immunol 110(1):160–166

Wang F, Robotham JM, Teuber SS, Sathe SK, Roux KH (2003) Ana o 2, a major cashew (*Anacardium occidentale* L.) nut allergen of the legumin family. Int Arch Allergy Immunol 132(1):27–39

Watt JM, Breyer-Brandwijk MG (1962) The medicinal and poisonous plants of Southern and Eastern Africa, 2nd edn. E. and S. Livingstone, Edinburgh, 1457 pp

Woodruff JG (1979) Tree nuts: production, processing, products, 2nd edn. Avi Publishing Co., Westport, 731 pp

Zhuang JX, Hu YH, Yang MH, Liu FJ, Qiu L, Zhou XW, Chen QX (2010) Irreversible competitive inhibitory kinetics of cardol triene on mushroom tyrosinase. J Agric Food Chem 58(24):12993–12998

Bouea macrophylla

Scientific Name

Bouea macrophylla Griffith

Synonyms

Bouea gandaria Blume ex Miq., *Tropidopetalum javanicum* Turcz

Family

Anacardiaceae

Common/English Names

Gandaria, Marian Plum, Plum Mango

Vernacular Names

Dutch: Gandaria;
German: Pflaumenmango;
Indonesia: Wetes (Alfoersch, North Sulawesi), Buwa Melawe (Buginese, Sulawesi), Barania (Dyak, Kalimantan), Lluber (Flores), Remie (Gajo), Gandaria (Java), Ramen (Lampung, Sumatra), Pao Gandaria (Madurese), Kalawasa, Rapo-Rapo Kebo (Makassar, Sulawesi), Kundangan (Malay, Sumatra), Gandaria (Malay, Manado), Gandoriah (Minangkabau, Sumatra), Gunarjah, Jatake, Jantake, Kendarah, (Sundanese),
Malaysia: Asam Suku, Kondongan, Kedungan Hutan, Kundang, Kundangan Medang Asam, Pako Kundangan, Rembunia, Remenya, Rumenia, Rumia, Serapoh, Serapok, Setar;
Philippines: Gandaria;
Spanish: Gandaria;
Thailand: Mapraang, Mayong, Somprang.

Origin/Distribution

Gandaria is native to north Sumatra, Peninsular Malaysia and west Java, and is cultivated widely as a fruit tree in Sumatra, the wetter parts of Java, Borneo and Ambon, Mauritius, Philippines, as well as in Thailand.

Agroecology

Gandaria is a species of the hot, humid tropics and thrives in well-drained, light and fertile soil. It occurs naturally in lowland rainforests below 300 m altitude, but has been successfully cultivated up to elevations of about 850 m.

Edible Plant Parts and Uses

Both the leaves and fruit can be eaten. Ripe sweet to sour fruit is eaten fresh, used in drinks or cooked in syrup or stew. The fruit are also used in pickles (*asinan*), compote, *sambal* or made into jam in Indonesia. The brownish-violet young leaves can be eaten raw or used in salads in Java. While the seed is edible, the cotyledon is generally bitter. Unripened fruit can be used to make *rojak* and *asinan*. In Thailand, young fruits are used as an ingredient of a special kind of dish, a chilli-based condiment, and in pickles; the young leaves are used as vegetables and consumed with chilli and shrimp paste.

Botany

A robust, evergreen, perennial tree, up to 27 m tall, with a trunk diameter of 55 cm and light brown to grey, lightly fissured bark (Plate 4). Leaves are elliptic, ovate-oblong to lanceolate, coriaceous 11.5–30 cm by 4–8 cm wide, glabrous, with acute to acuminate apex and cuneate to acute base, 15–25 pairs of nerves and reticulate venation, decussate, violet brown when young turning dark green when old, petiole 1–2 mm (Plate 3). Inflorescences are axillary panicles, 4–13 cm long. Flowers are mostly tetramerous, small; calyx lobes broadly ovate; petals oblong to obovate, 1.5–2.5 mm × 1 mm, yellowish or light yellowish-green; stamens 0.5–1 mm with apiculate anthers. Fruit is a subglobose drupe, 3.5–5 cm by 3.4–5 cm, green (Plate 2) turning to yellow or orange when ripe (Plate 1), with edible skin and juicy, sweet or sour, orangey yellow flesh surrounding a single seed with blue violet cotyledons.

Plate 2 Unripe green fruit

Plate 3 Large, glossy dark green with prominent lateral veins

Plate 1 Ripe subglobose fruit of *Bouea macrophylla*

Plate 4 Grey brown fissured bark

Nutritive/Medicinal Properties

Tee et al. (1997) reported the following nutrient composition of the fruit per 100 g edible portion: energy 49 kcal, water 87.5 g, protein 0.6 g, fat 1.2 g, carbohydrate 9.0 g, fibre 1.4 g, ash 0.3 g, Ca 5 mg, P 10 mg, Fe 0.2 mg, Na 2 mg, K 84 mg, carotene 331 μg, vitamin A 55 μg RE, vitamin B1 0.03 mg, vitamin B2 0.03 mg, niacin 0.1 mg, vitamin C 20 mg, lutein 457 μg, cryptoxanthin 155 μg, γ-carotene 52 μg, β-carotene 301 μg, other carotenoids 514 μg, total carotenoids 67 μg RE. The composition of a fruit sample from a single tree in Honduras per 100 g edible portion was reported as: water 85 g, protein 112 mg, fat 40 mg, fibre 600 mg, ash 230 mg, calcium 6 mg, phosphorus 10.8 mg, iron 0.31 mg, carotene 0.043 mg, thiamine 0.031 mg, riboflavin 0.025 mg, niacin 0.286 mg and vitamin C 75 mg (Morton 1987). Thai Department of Health reported the composition of a fruit sample per 100 g edible portion as: water 86.6 g, protein 40 mg, fat 20 mg, carbohydrates 11.3 g, dietary fibre 150 mg, ash 20 mg, calcium 9 mg, phosphorous 4 mg, iron 0.3 mg, α-carotene 23 mg, thiamine 0.11 mg, riboflavin 0.05 mg, niacin 0.5 mg and vitamin C 100 mg (Subhadrabandhu 2001).

The leaves have been used as poultice for headaches and their juice as a gargle for thrush in traditional medicine.

Other Uses

Timber is of minor importance and is used for light construction purposes and to make scabbards for kris (Malay dagger), arts and craft.

Comments

Another closely related edible species is *Bouea oppositifolia*.

Selected References

Backer CA, van den Bakhuizen Brink RC Jr (1965) Flora of Java (spermatophytes only), vol 2. Wolters-Noordhoff, Groningen, 641 pp

Burkill IH (1966) A dictionary of the economic products of the Malay Peninsula. Revised reprint, 2 vols. Ministry of Agriculture and Co-operatives, Kuala Lumpur, vol 1(A–H), pp 1–1240, vol 2(I–Z), pp 1241–2444

Ding Hou (1978) Anacardiaceae. In: van Steenis CGGJ (ed) Flora Malesiana, Ser. I, vol 8. Sijthoff & Noordhoff, Alphen aan den Rijn, pp 395–548, 577 pp

Morton JF (1987) Anacardiaceae. In: Fruits of warm climates. Julia F. Morton, Miami, pp 221–248

Ochse JJ, Bakhuizen van den Brink RC (1931) Fruits and fruitculture in the Dutch East Indies. G. Kolff & Co, Batavia, 180 pp

Rifai MA (1992) *Bouea macrophylla* Griffith. In: Coronel RE, Verheij EWM (eds.) Plant resources of South-East Asia. No. 2: Edible fruits and nuts. Prosea Foundation, Bogor, pp 104–105

Slik JWF (2006) Trees of Sungai Wain. Nationaal Herbarium Nederland. http://www.nationaalherbarium.nl/sungaiwain/

Subhadrabandhu S (2001) Under utilized tropical fruits of Thailand. FAO Rap publication 2001/26, Rome, 70 pp

Tee ES, Noor MI, Azudin MN, Idris K (1997) Nutrient composition of Malaysian foods, 4th edn. Institute for Medical Research, Kuala Lumpur, 299 pp

Bouea oppositifolia

Scientific Name

Bouea oppositifolia (Roxb.) Meisn.

Synonyms

Bouea angustifolia Bl., *Bouea burmanica* Griff., *Bouea burmanica* var. *kurzii* Pierre, *Bouea burmanica* var. *microphylla* (Griff) Engl., *Bouea burmanica* var. *roxburghii* Pierre, *Bouea diversifolia* Miq., *Bouea microphylla* Griff., *Bouea myrsinoides* Bl., *Mangifera oppositifolia* Roxb. (basionym), *Mangifera oppositifolia* var. *microphylla* (Griff.) Merr., *Mangifera oppositifolia* var. *roxburghii* (Pierre) Tard., *Matania laotica* Gagnep.

Family

Anacardiaceae

Common/English Names

Burmese Plum, Marian Plum, Marian Tree, Plum-Mango

Vernacular Names

Burmese: Meriam;
Borneo: Kedjauw Lepang, Tampusu, (Dayak), Asam Djanar, Bandjar, Kundang Rumania, Ramania Hutan, Ramania Pipit, Rengas, Tolok Burung, Umpas;
Indonesia: Kaju-Rusun, Kunangan, Raman Burung, Raman Padi, Raman Utan, Reiden Daun (Sumatra); Gandaria, Raman, Uris, Urisan (Banka), Ramania Pipit (Samarinda), Umpas (Kalimantan);
Malaysia: Perus (Sakai), Gemia, Gemis, Kemunia, Kundang, Kundang Daun Kecil, Kundang Rumenia, Kundang Siam, Merapoh Rumenia, Poko Rummiyah, Rambainyia, Ramunia, Rembunia, Remnia, Romaniah, Rumboi-Nigor, Rumenia, Rumenia Betul, Rumia, Rumiah, Rumiang (Peninsular);
Thailand: Ma Prang, Ma Prang Wan.

Origin/Distribution

The species is indigenous to the Andaman Islands, Myanmar, Laos, Cambodia, Thailand to Vietnam and South China (Yunnan, Hainan), Indonesia (Sumatra, Java, Kalimantan) and Malaysia.

Agroecology

Marian plum is a warm tropical species that occurs in undisturbed mixed dipterocarp lowland forest, peat forest, kerangas, and sometimes on limestone. In secondary forest it is usually present

Bouea oppositifolia

as a pre-disturbance remnant. The species is found from near sea-level up to 600 m altitude.

Edible Plant Parts and Uses

Fruits are sourish and edible fresh or cooked and are sometimes made into preserves and jams when in a near ripe state.

Botany

An erect, evergreen, perennial tree up to 32 m high and with a trunk diameter of 75 cm and grey-green light brown to purple-brown fissured bark (Plate 5). Leaves are opposite, coriaceous, elliptic to elliptic-oblong, lanceolate to oblanceolate, 2–15 cm by 1–5 cm, glabrous, acute to cuneate base and acuminate apex, nerves usually 8–14 (–26) pairs, light violet brown when young and dark green older, (Plate 4) petiole 0.5–2 cm. Inflorescence in axillary panicles, 2.5–6 cm long. Flowers white to pale yellow, with broadly ovate calyx lobes, oblong-obovate-oblong petals, stamens 0.6–1 mm with apiculate anthers. Fruit broadly ellipsoid, glabrous drupe, 2.5 by 1.5 cm, yellow or orange-red when ripe (Plates 1–2) with sourish juicy orangey flesh and one seed with violet purple cotyledons (Plate 3).

Plate 2 Ripe ellipsoid fruit resembling a small mango

Plate 3 Fruit sliced to show the violet cotyledon

Plate 4 Young and old leaves

Plate 1 Bunch of ellipsoid near-ripe fruit

Plate 5 Brown to purple-brown fissured bark

Nutritive/Medicinal Properties

The nutrient composition of raw Maprang fruit per 100 g edible portion analysed in Thailand was reported as: moisture 86.6 g, energy 53 kcal, protein 0.4 g, fat 0 g, carbohydrate 12.8 g, fibre 1.5 g, ash 0.2 g, Ca 9 mg, Fe 0.3 mg, thiamin 0.11 mg, and vitamin C 100 mg (Puwastien et al. 2000).

Other Uses

The timber is dark brown in colour, hard and durable and use for house posts and various purposes.

Comments

See also *Bouea macrophylla* the other edible species.

Selected References

Burkill IH (1966) A dictionary of the economic products of the Malay Peninsula. Revised reprint, 2 vols. Ministry of Agriculture and Co-operatives, Kuala Lumpur, vol 1(A–H), pp 1–1240, vol 2(I–Z), pp 1241–2444

Ding Hou (1978) Anacardiaceae. In: van Steenis CGGJ (ed) Flora Malesiana, ser. I, vol 8. Sijthoff & Noordhoff, Alphen aan den Rijn, pp 395–548, 577 pp

Molesworth AB (1967) Malayan fruits: an introduction to the cultivated species. Moore, Singapore, 245 pp

Puwastien P, Burlingame B, Raroengwichit M, Sungpuag P (2000) ASEAN food composition tables 2000. Institute of Nutrition, Mahidol University, Salaya, 157 pp

Rifai MA (1992) *Bouea macrophylla* Griffith. In: Coronel RE, Verheij EWM (eds) Plant resources of South-East Asia. No. 2: Edible fruits and nuts. Prosea Foundation, Bogor, pp 104–105

Slik JWF (2006) Trees of Sungai Wain. Nationaal Herbarium Nederland http://www.nationaalherbarium.nl/sungaiwain/

Dracontomelon dao

Scientific Name

***Dracontomelon dao* (Blanco) Merrill & Rolfe**

Synonyms

Comeurya cumingiana Baill., *Dracontomelon cumingiana* (Baill.) Baill., *Dracontomelon brachyphyllum* Ridl., *Dracontomelon celebicum* Koord., *Dracontomelon edule* (Blanco) Skeels, *Dracontomelon edule* Merr., *Dracontomelon lamiyo* Merr., *Dracontomelon laxum* K. Sch., *Dracontomelum mangiferum* Blume, *Dracontomelum mangiferum* (Blume) Blume nom. illeg., *Dracontomelon mangiferum* var. *puberulum* (Miq.) Engl., *Dracontomelon mangiferum* var. *pubescens* K. & V., *Dracontomelum puberulum* Miq, *Dracontomelon sylvestre* Blume, *Paliurus dao* Blanco, *Paliurus edulis* Blanco, *Paliurus lamiyo* Blanco, *Pomum draconum* Rumph., *Pomum draconum silvestre* Rumph., *Poupartia mangifera* Bl. nom. illeg.

Family

Anacardiaceae

Common/English Names

Argus Pheasant-Tree, Mon Fruit, New Guinea Walnut, Pacific Walnut, Papua New Guinea Walnut

Vernacular Names

Burmese: Nga-Bauk;
Chinese: Ren Mien Zi;
Dutch: Drakeboom;
Indonesia: Tarpati (Banda), Rau (Flores), Ngame, Tabulate (Halmahere), Dahu, Dau, Langsep Alas, Theuoh, Rahan, Rahu, Rao, Rau (Madurese), Gijubuk, Rahan, Rahu, Rao, Rau (Javanese), Jakan, Sangkuwang, Urui, Sengkuang, Singkuang, Talanjap (Kalimantan), Arouwsauw (Kwesten, Papua Barat), Basuong (Papau Barat), Rao, Dewu, Lolomao, Mabiru (Manado, Sulawesi), Senai (Manikiong, Papua Barat), Koili (Minah, Sulawesi), Rago (Muna Island, Sulawesi), Wuarau Takau (Tobela, Sulawesi), Bemangiohik, Biohiki, Nganin (Moratai, Sulawesi), Anglip Etem, Dau-Pajo (Simalur, Sumatra), Beka, Landur, Surian Keli (Palembang, Sumatra), Touuw (Tokuni, Papua Barat), Kiking (Malay, Sumatra), Kasuang (Sumbawa), Dahu (Sundanese), Leombawi (Talaud Island), Ngame, Ngawe (Ternate);
German: Drachenapfel;
Malaysia: Bengkuang, Cengkuang, Mati Ayak, Sakal, Sekuan, Sekuang, Sepul, Surgan, (Peninsular), Sangkuang, Unkawang (Sarawak), Sankuang, Sarunsab, Sorosob, Suronsub, Tarosoup, Tehrengzeb (Sabah);
Papua New Guinea: Daa (Amberbaken), Fa, Faila (Amele), Taa (Andai), Mon (Bembi), Onomba (Binendele), Wehm (Bogia), Jaap (Hattam), Gain (Jal), Kumbui (Karoon), Alaisoi, Mon, Rou (Madang), Ufaka (Minufia), Los (Mooi), Damoni (Motu), Imbur (Onjob), Mon

(Rawa), Touv (Sko), Aua (Vailala), Dores (West Evara);
Philippines: Mamakau (Bagobo), Alauihau, Dao, Kalauihau (Bikol), Batuan (Bisaya), Habas (Cebu-Bisaya), Lupigi (Ibanag), Hamarak, Kamarak, Makadaeg (Iloko), Ulandang, Ulandug (Kuyonon), Anduong, Makau, Mamakau (Manabo), Makau (Maguindanao) Bili-Bili, Dao, (Panay Bisaya), Bio (Pangasingan), Alauihau, Anduong, Dao, Kiakia Makau (Samar-Leyte-Bisaya), Aduas, Anangging-Puti, Dao, Lamio, Malaiyo, Maliyan, Olandag (Tagalog), Dao, Maliyan, Sengkuang;
Thai: Prachao Ha Pra-Ong, Dao, Ka-Kho, Sang-Kuan;
Vietnam: Sau.

Plate 1 Dense crown of a young Mon tree

Origin/Distribution

The species is indigenous to eastern India, Andaman Islands, South China, Myanmar, Indochina, Thailand, Peninsular Malaysia, Indonesia, Philippines, New Guinea and the Solomon Islands.

Agroecology

Dracontomelon dao occurs in undisturbed primary or secondary, evergreen or semi-deciduous (monsoon) forest in tropical areas with mean annual rainfall of 1,800–2,900 mm and elevations of 500–1,000 m. It is found scattered on well-drained to poorly-drained, clayey to stony soils, organosols, gley humus soils or red-yellow podsolic soil, on alluvial flats and in swampy areas and along river banks.

Edible Plant Parts and Uses

Young fruits are used to flavour curry, and the edible flesh covering the seed is sweet-sourish and eaten in Thailand. In Malaysia, the fruit is used as a sour relish with fish. The seed kernel is also edible. The flowers, fruits and young leaves are eaten cooked as vegetable e.g. in Vietnam, Thailand, East Malaysia and Papua New Guinea.

Plate 2 Large imparipinnate leaves with 4–9 pairs of leaflets

In Maluku, Indonesia, the flowers and leaves are used as spice for flavouring food.

Botany

A large, tall, erect, deciduous tree up to 36 m high with a large trunk of 70–100 cm diameter, greyish-brown, scaly bark, thick buttresses up to 5 m high and a dense large crown (Plate 1). Leaves are arranged spirally, crowded towards the ends of branchlets, large, imparipinnate with 6–44 cm long leaf rachis and 4–9 pairs of leaflets (Plate 2). Leaflets are alternate to opposite, chartaceous, subcoriaceous, elliptic-oblong, oblong, ovate-oblong to lanceolate, 4.5–27 cm × 2–10.5 cm, glabrous or sometimes pubescent below, lower

Plate 3 Young leaves eaten as vegetable and sold in the local markets

Plate 4 Immature green, hard mon fruit

surface pale green, upper surface dark green with 3–10 pairs of nerves. Young leaves are brownish-yellow in colour (Plate 3). Flowers are bisexual, stalked, actinomorphic, 5-merous, slightly fragrant, white to greenish-white, 7–10 mm long, in panicles of up to 50 cm long; petals are oblanceolate and valvate but imbricate at the apical part, puberulous outside or on both surfaces, or glabrous; stamens 10, in 2 whorls, filaments glabrous, anthers dorsifixed, disk intrastaminal, puberulous or glabrous; pistil composed of 5 carpels which are free but connate at base and apically, ovary superior, 5-celled with a single ovule in each cell, styles 5, stigma capitate. Fruit is a drupe, globose, 5-celled, or seemingly 1-celled by abortion, each cell with a distinct operculum, endocarp lentiform, woody and hard and with 5 oval markings on the upper side of the fruit (Plate 4). Seed conical, 1–20 mm diameter.

Nutritive/Medicinal Properties

Nutrient composition of the leaves and fruits have not been published.

The crude methanolic extracts of the leaves, stem and root barks of *Dracontomelon dao* and their subsequent partitioning (petrol, dichloromethane, ethyl acetate, butanol) gave fractions which demonstrated a very good level of broad spectrum antibacterial activity (Khan and Omoloso 2007). The dichloromethane and butanol fractions of the leaf were the most active. Only the leaf fractions had antifungal activity, particularly the dichloromethane and butanol. The bark is used against dysentery in Peninsular Malaysia.

Other Uses

The tree is use for firewood and provides a source of commercial dao timber which is soft and light and not very durable. The timber is used for veneers, panelling, furniture, plywood, flooring, interior trim and light frames, matches and boxes. The wood is also appreciated for joinery and turnery. The timber has the following trade names: sengkuang (Malaysia), paldao (Philippines), dar (West New Guinea), New Guinea Walnut (Papua New Guinea).

The tree is planted as an ornamental in roadside plantings.

Comments

The tree is usually propagated from seeds.

Selected References

Burkill IH (1966) A dictionary of the economic products of the Malay Peninsula. Revised reprint, 2 vols. Ministry of Agriculture and Co-operatives, Kuala

Lumpur, vol 1 (A–H), pp 1–1240, vol 2 (I–Z), pp 1241–2444
Conn BJ, Damas KQ (2006+) Guide to trees of Papua New Guinea (http://www.pngplants.org/PNGtrees)
Ding Hou (1978) Anacardiaceae. In: van Steenis CGGJ (ed) Flora Malesiana, ser I, vol 8. Sijthoff & Noordhoff, Alphen aan den Rijn, pp 395–548
French BR (1986) Food plants of Papua New Guinea: a compendium. Australia and Pacific Science Foundation, Ashgrove, 408 pp
Khan MR, Omoloso AD (2007) Antibacterial and antifungal activities of *Dracontomelon dao*. Fitoterapia 73(4):327–330
Kochummen KM (1989) Anacardiaceae. In: Ng FSP (ed.) Tree flora of Malaysia, vol 4. Longman Malaysia, Kuala Lumpur, pp 9–57
Lemmens RHMJ, Soerianegara I, Wong WC (eds) (1995) Plant resources of South-East Asia. No 5(2). Timber trees: minor commercial timbers. Backhuys Publishers, Leiden

Mangifera caesia

Scientific Name

Mangifera caesia Jack

Synonyms

Mangifera caesia Jack var. *verticillata* (C.B. Robinson) Mukherji, *Mangifera caesia* Jack var. *wanji* Kostermans, *Mangifera verticillata* C.B. Robinson.

Family

Anacardiaceae

Common/English Name

Binjai

Vernacular Names

Brunei: Belunu, Binjai (Malay);
Danish: Binjai;
French: Binjai;
Indonesia: Wani (Balinese), Binjai, Belu (Lampong, Sumatra), Binjai (Malay, Sumatra), Binglu (Sundanese, West Java);
Malaysia: Binjai, Binnu (Malay, Peninsular Malaysia, Sarawak), Beluno (Malay, Sabah), Bundo (Dusun, Sabah);
Philippines: Bauno. Bayuno (Cebu Bisaya), Baluno (Manobo);
Thailand: Bin-Yaa, Lam-Yaa (Malay, Pattani).

Origin/Distribution

The natural distribution of the species is in Peninsular Malaysia, Sumatra and the island of Borneo. Its cultivation has been extended to Peninsular Thailand, Bali, Java and to the Philippines in the Sulu archipelago and Mindanao.

Agroecology

The species is generally restricted to the wet tropical lowlands below 450 m elevation, rarely in forests and more frequently in periodically inundated areas, along riverbanks and marshes. It thrives in areas with high rainfall which is evenly distributed throughout the year.

Edible Plant Parts and Uses

The juicy, acid-sweet fruit flesh can be eaten fresh when ripe, or dipped in chilli with sugar and dark soy sauce. It is often made into *sambal*, *jeruk* and eaten with fish especially grilled fish. The flesh is also pickled and preserved with salt in jars, to be able to make this sambal when there is no fresh fruit. The flesh is excellent for making creamy juices. The wani variety which is mainly found in Bali but also in East Kalimantan and Sabah, is much relished fresh as the fruit is juicy and sweet, almost fibreless. This variety commands a high price in local markets.

and juicy, fibrous pulp, with a peculiar sourish taste and strong smell at maturity. The 'wani' form has ellipsoid, ellipsoid-oblong or rounded fruit, 9–11 cm × 6.5–7 cm, glossy pale green at

Botany

A large, perennial, erect tree reaching heights of 30–45 m with a bole of 50–12 cm, a dome-shaped crown with massive branches and greyish-brown, superficially fissured bark containing irritant sap. Leaves are alternate, often crowded at the end of stout branchlets, simple, elliptic to lanceolate to obovate, 7–30 cm long by 3–10 cm wide, glossy, medium green above, paler below, coriaceous, blunt or bluntly acuminate tip, penniveined with midrib thick, flattened, raised above, leaf base gradually decurrent and borne on stout, flattened 1.5–2.5 cm long petioles. Flowers are pale pink or pale lilac, bisexual, fragrant and borne in densely flowered, much-branched terminal panicles, 15–40 cm long. Flower is pentamerous, with linear petals up to 10 mm long, not strongly reflexed; one fertile stamen, filament 5 mm long, white at base, purple towards the apex; four teeth-like staminodes; narrow, stipe-like disc, pale green and obliquely globose; reddish brown ovary and excentric style, 8 mm long, white, becoming violet after anthesis. Fruit is an obovate-oblong drupe, necked at base, 12–20 cm × 6–12 cm with yellowish, greenish or pale brownish, smooth, thin (1 mm) skin (Plates 1–4) and whitish, soft

Plate 2 Yellowish binjai fruit with smooth, glossy skin

Plate 3 Pale brownish–pinkish smooth-skinned binjai variety

Plate 1 Yellowish variety of binjai

Plate 4 Pale green wani variety sold in Sabah markets

maturity, with milky white almost fibreless, sweet flesh with ellipsoid-lanceolate, non-woody endocarp.

Nutritive/Medicinal Properties

The nutrient composition of raw binjai fruit per 100 g edible portion was reported as: water 86.5 g, energy 47.8 kcal, protein 1.0 g, fat 0.2 g, carbohydrate (including fibre) 11.9 g, vitamin A 8.3 IU, thiamin 0.08 mg, ascorbic acid 58 mg (Pareek et al. 1998). Another analysis reported that raw binjai fruit per 100 g edible portion contained energy 64 kcal, water 81.2 g, protein 1.0 g, fat 0.2 g, carbohydrate 14.6 g, fibre 2.4 g, ash 0.6 g, Ca 7 mg, P 17 mg, Fe 0.3 mg, Na 1 mg, K 120 mg, vitamin B1 0.05 mg, vitamin B2 0.16 mg, niacin 1.2 mg and vitamin C 74.1 mg (Tee et al. 1997). The fruit is rich in vitamin C.

Sixty-three volatile compounds were identified in binjai fruit (Wong and Siew 1994). Terpenoids were absent in binjai fruit volatiles which contained esters (71.3%) and alcohols (23.2%) as the two major chemical classes and ethyl 3-methylbutanoate (40.0%) as the most abundant constituent.

Alkenylphenol and alkenylsalicylic acid were isolated from the bark of *Mangifera caesia* and found to have antibacterial activity (Masuda et al. 2002).

Other Uses

The wood is light red marbled with yellow, and is used for light construction and boards.

Comments

Mangifera caesia Jack closely resembles *Mangifera kemanga* Blume, but differs in having longer leaf petiole (subsessile in *M. kemanga*), shorter panicles (up to 75 cm long in *M. kemanga*) and the yellowish, whitish-green, or pale brownish smooth fruit (brown to deep brown and scurfy fruit in *M. kemanga*).

The sap from unripe fruit is extremely irritant to the skin and when ingested.

Selected References

Bompard JM (1992) *Mangifera caesia* Jack, *Mangifera kemanga* Blume. In: Coronel RE, Verheij EWM (eds.) Plant resources of South-East Asia. No. 2: Edible fruits and nuts. Prosea Foundation, Bogor, pp 207–209

Bompard JM (1993) The genus *Mangifera* re-discovered: the potential contribution of wild species to mango cultivation. Acta Hortic 341:69–77

Burkill IH (1966) A dictionary of the economic products of the Malay Peninsula. Revised reprint, 2 vols. Ministry of Agriculture and Co-operatives, Kuala Lumpur, vol 1 (A–H), pp 1–1240, vol 2 (I–Z), pp 1241–2444

Ding Hou (1978) Anacardiaceae. In: van Steenis CGGJ (ed) Flora Malesiana, ser I, vol 8. Sijthoff & Noordhoff, Alphen aan den Rijn, pp 395–548, 577 pp

Kostermans AJGH (1965) New and critical Malaysian plants VII *Mangifera caesia* Jack. Reinwardtia 7(1):19–20

Masuda D, Koyano T, Fujimoto H, Okuyama E, Hayashi M, Komiyama K, Ishibashi M (2002) Alkenylphenol and alkenylsalicylic acid from *Mangifera caesia*. Biochem Syst Ecol 30(5):475–478

Mukherji S (1949) A monograph on the genus *Mangifera* L. Lloydia 12:73–136

Ochse JJ, van den Brink RCB (1931) Fruits and fruit-culture in the Dutch East Indies. G. Kolff & Co., Batavia, 180 pp

Ochse JJ, van den Brink RCB (1980) Vegetables of the Dutch Indies, 3rd edn. Ascher & Co., Amsterdam, 1016 pp

Pareek OP, Sharma S, Arora RK (1998) Underutilized edible fruits and nuts: an inventory of genetic resources in their regions of diversity. IPGRI Office for South Asia, New Delhi, pp 191–206

Saidin I (2000) Sayuran Tradisional Ulam dan Penyedap Rasa. Penerbit Universiti Kebangsaan Malaysia, Bangi, pp 228 (In Malay)

Slik JWF (2006) Trees of Sungai Wain. Nationaal Herbarium Nederland. http://www.nationaalherbarium.nl/sungaiwain/

Tee ES, Noor MI, Azudin MN, Idris K (1997) Nutrient composition of Malaysian foods, 4th edn. Institute for Medical Research, Kuala Lumpur, 299 pp

Wong KC, Siew SS (1994) Volatile components of the fruits of bambangan (*Mangifera panjang* Kostermans) and binjai (*Mangifera caesia* Jack). Flav Fragr J 9(4):173–178

Mangifera foetida

Scientific Name

Mangifera foetida Lour.

Synonyms

Mangifera foetida var. *sphaeroidea* Blume, *Mangifera foetida* var. *leschenaultil* (March.) Engl., *Mangifera horsfieldii* Miq., *Mangifera leschenaultil* March., *Mangifera rubicunda* Jack ex Gage

Family

Anacardiaceae

Common/English Names

Bachang, Gray bacang, Grey bacang, Horse Bacang, Machang

Vernacular Names

Brunei: Bangbangan;
Borneo (Kalimantan, Sabah, Sarawak): Hambawang, Mangga Batjan, Tempajang (Balikpapan), Asampajang, Kedjan Lemah, Thulik Kaki (Dayak), Asam Pamas, Buah Assam (Iban), Ata, Baya, Pelam (Kayan), Asam Hambawang, Asam Mas (Kutai), Bachang, Machang (Kuching), Pauh Hutan, Pudan, Talantang (Malay), Hambawang Kambat (Samrinda);
Burmese: La-Mot, Thayetpoh;
Dutch: Batjan-Boom;
French: Mangue Fetide, Bachang;
German: Stinkender Bacangbaum;
Indonesia: Mancang (Aceh, Sumatra), Pata, Pate (Ambon, Maluku), Pau Kasi (Alor, Timor), Ambawa (Bari, Sulawesi), Batjang, Batjang Maros, Lemus (Batak), Fu Wau (Bima, Timor), Pao Daeko Cani, Pao Macang (Boegineesch, Sulawesi), Idamo (Boeol, Sulawesi), Batin Laka (Boeroe, Maluku), Bata (E. Ceram, Maluku), Pate, Hau Pa'a, Hatu Malaka (W. Ceram, Maluku), Aune, Hau Paolo, Losa (S. Ceram, Maluku), Ambacang (Dyak), Pau Kate (Flores), Berhul Mancang (Gajo, Sumatra), Dulamaya (Gorontalo, Sulawesi), Pakel, Plem Bawang, Poh (Javanese), Rawa (Karimon, Sumatra), Ambachang, Sitorngom (Kesarin, Sumatra), Limus, Leko (Lampong, Sumatra), Pao Bhapang (Madurese), Taipa Bacang (Makassar, Sulawesi), Dedko, Mangga Hutan, Umbawa (Malili, Sulawesi), Asam Bawang, Bacang, Bawang, Bedara Mangga Bachang, Mangga Padi (Malay), Limus (Manado, Sulawesi), Agbanan (Mentawai, Sumatra), Ambacang (Minangkabau), Hetapate, Patol, (Oelias, Maluku), Ambatjang (Pajakumbuh, Sumatra), Medang Pergam Pau Puti (Palembang, Sumatra), Abawang Dotan (Simalur, Sumatra), Pau Karane (Soemba, Timor), Limus Bacang, Limus (Sundanese), Mangga Papai (Timor);

Khmer: Svaay Saa;
Malaysia (Peninsular): Bachang, Machang Utan, Kurau, Machai, Mempening, Machang, Embachang, Kembachang, Membachang, Batel, Empelam, Sepam, Sepopn (Sakai), Pudu, Pelam (Kayan, Sarawak), Machang, Bacang (Sarawak), Asam (Sabah);
New Guinea: Boeja;
Philippines: Horse Bacang;
Thailand: Maa-Chang, Malamut (South) Ma Mud, Som Mud;
Vietnam: Xoai Hoi.

Origin/Distribution

M. foetida is endemic to the dipterocarp forests of Peninsular Malaysia, Peninsular Thailand, Sumatra, and Borneo. It was also collected apparently wild in Java. It was introduced to south Tenasserim (Burma), where it is popular. It is widely cultivated in its area of origin. In Vietnam, Cambodia and the Philippines, it is hardly cultivated and virtually unknown.

Agroecology

Mangifera foetida occurs mainly in primary lowland forest in the hot wet tropics from near sea level to 1,000 m altitude. The species is adapted to areas with abundant rainfall, uniformly distributed over the year. In Peninsular Malaysia, it is the most important representative of the machang trade group found scattered in natural forests. *M. foetida* occurs wild in dipterocarp forests of Peninsular Malaysia, Peninsular Thailand, Sumatra, and Borneo. It is widely cultivated in Malesia, sometimes as village backyard trees. It was introduced to south Tenasserim (Myanmar), where it is popular. It is very common in west Koetai in central east Borneo. Escapes are naturalized or indigenous in dryland, lowland primary forests and also on hill ridges.

Edible Plant Parts and Uses

The pulp of ripe fruit is rather savoury and has a strong pleasant turpentine odour and is eaten fresh. It is also excellent as component of fruit cocktail. The fruit is also used in curries or as pickles. Unripe fruit, washed in salted water and sliced is used in vegetable salads (*rujak*) and in a sour pickle (*asinan*). In Borneo, especially in East Kalimantan, the fruit commonly replaces tamarind as an acid ingredient in the preparation of *sambal*. In Malaysia, it is used to make chutneys, *sambal, acar, rojak* and *jeru*k (green pepper with lemon) as well as pickles and in curries.

Botany

Evergreen, perennial tree up to 30–40 m high, straight trunk with light brown to dark greyish-brown, shallowly fissured bark (Plate 4), containing a caustic whitish sap, massive branches and a dense crown of shiny dark green foliage. Leaves elliptic-oblong to broadly elliptic, sometimes oblanceolate, 15–40 cm × 9–15 cm, rigidly coriaceous, apex obtuse or slightly emarginate, base cuneate or attenuate, more or less bullate between the 15 and 33 pairs of nerves (Plate 3); petiole 1.5–8 cm, stout, swollen at the base. Panicles terminal to subterminal, upright, pyramidal,

Plate 1 Pale yellow flesh of immature bachang fruit

Plate 2 Ripe bachang fruit with yellow, fibrous flesh

Plate 3 Large, elliptic-oblong leaves with distinctive puckering between the main lateral veins

Plate 4 Straight trunk with dark greyish-brown, shallowly fissured bark

10–40 cm long, sparsely branched, rather densely flowered, deep reddish-pink, inflorescence axes stout, deeply red. Flowers pentamerous, faintly fragrant; sepals obovate-lanceolate, 4–5 mm long; petals narrowly lanceolate, 6–9 mm × 1.5–2.5 mm, pale reddish-pink at the base, pale yellow towards the apex, reflexed; stamens 5, 1(–2) fertile, filament 8 mm long, pinkish-purple, anthers dark violet, other ones smaller, filaments connate at the base; ovary subglobose, yellow, style excentric, white, 6–7 mm long. Fruit obliquely ovoid-oblong or almost globose drupe, 9–16 cm × 7–12 cm, dirty dark olive-green or yellowish-green, dull, with brown lenticels, skin 5 mm thick; flesh pale yellowish-white when immature (Plate 1) turning to yellow or golden-yellow when ripe, fibrous (Plate 2), juicy, savoury, with strong smell of turpentine. Stone plump, 6 cm × 5 cm × 3 cm, coarsely fibrous; seed monoembryonic.

Nutritive/Medicinal Properties

The nutrient composition of raw bachang fruit per 100 g edible portion was reported as: water 72.5 g, protein 1.4 g, fat 0.2 g, carbohydrate 25.4 g, Ca 21 mg, P 15 mg, vitamin A 363 IU, ascorbic acid 56 mg (Pareek et al. 1998). Another analysis reported that raw bachang fruit per 100 g edible portion contained: energy 76 kcal, water 78.5 g, protein 0.8 g, fat 0.2 g, carbohydrate 17.9 g, fibre 1.8 g, ash 0.8 g, Ca 16 mg, P 19 mg, Fe 0.2 mg, Na 2 mg, K 361 mg, carotenes 255 μg, vitamin A 43 μg, vitamin B1 0.09 mg, vitamin B2 0.04 mg, niacin 0.6 mg and vitamin C 47.4 mg (Tee et al. 1997).

In a study conducted in Malaysia, the total carotene content (mg/100 g) of selected underutilized tropical fruits in decreasing order (Khoo et al. 2008) was jentik-jentik (*Baccaurea polyneura*) 19.83 mg > cerapu 2 (*Garcinia prainiana*) 15.81 mg > durian nyekak 2 (*Durio kutejensis*) 14.97 mg > tampoi kuning (*Baccaurea reticulata*) 13.71 mg > durian nyekak 1 (*Durio kutejensis*) 11.16 mg > cerapu 1(*Garcinia prainiana*) 6.89 mg > bacang 1 (*Mangifera foetida*) 4.81 > kuini (*Mangifera odorata*) 3.95 mg > jambu susu

(*Syzygium malaccense*) 3.35 mg>bacang (2) 3.25 mg>durian daun (*Durio lowianus*) 3.04 mg>bacang (3) 2.58 mg>tampoi putih (*Baccaurea macrocarpa*) 1.47 mg>jambu mawar (*Syzygium jambos*) 1.41 mg. β-carotene content was found to be the highest in jentik-jentik 17,46 mg followed by cerapu (2) 14.59 mg, durian nyekak (2) 10.99 mg, tampoi kuning 10.72 mg, durian nyekak (1) 7.57 mg and cerapu (1) 5.58 mg. These underutilized fruits were found to have acceptable amounts of carotenoids and are potential antioxidant fruits.

Eighty-four volatile components were identified in the bachang fruit (*M. foetida*) and 73 in the kuini (*M. odorata*) (Wong and Ong 1993). Among the bachang fruit volatiles, esters (55.7%) and oxygenated monoterpenes (20.3%) were dominant, with ethyl butanoate (33.4%) the most abundant. Oxygenated monoterpenes (45.4%) and esters (33.0%) constituted the main classes of kuini fruit volatiles, and α-terpineol (31.9%) was the major component. The volatile composition of both fruits differed significantly from results reported for the fruits of some of the commercially important bacang (*M. indica*) cultivars.

Mangifera foetida, *Mangifera odorata* and *Mangifera pajang* fruits were found to contain isoflavones, a major group of phytoestrogen that may serve as health promoting compounds in the diet. All 3 species were found to contain the isoflavones diadzein and genistein (Khoo and Ismail 2008). Studies showed that at optimised condition, daidzein content of *M. foetida* (2.8–8.0 mg/100 g) was lower than in *M. pajang* (8.3–8.7 mg/100 g), but the genistein content of *M. foetida* (0.4–0.8 mg/100 g) was similar to that of *M. pajang* (0.4–0.6 mg/100 g). *M. odorata* had comparatively higher daidzein (9.4–10.5 mg/100 g) and genistein (1.6–1.7 mg/100 g) contents.

Antioxidant Activity

The range of antioxidant capacity as determined by FRAP, TEAC and β-carotene bleaching assay in *Mangifera foetida* was 852.8–2000.8 μmol Fe(II)/100 g edible portion (EP), 558.1–960.8 μg Trolox/100 g EP and 59.7–92.2% respectively (Tan and Amin 2008). On the other hand, total reducing activity, flavonoid, carotenoid and ascorbic acid content in the bachang extracts were in the range of 122.8–199.8 mg GAE/100 g EP, 493.5–2360.7 mg catechin/100 g EP, 96.5–153.0 μg β-carotene/100 g EP and 29.4–59.4 mg ascorbic acid/100 g EP respectively. The FRAP and TEAC values were in the order of fresh bachang>bacang fibre>bacang powder whereas for β-carotene bleaching assay, the order was bacang fibre>bacang powder>fresh bacang. Total reducing activity, flavonoid and carotenoid content in bacang extracts were in the order of bacang fibre>fresh bacang>bacang powder while the ascorbic acid content in bacang extracts was in the order of bacang fibre>bacang powder>fresh bacang. This study indicated that bacang fibre possessed the highest antioxidant properties compared with fresh bacang and bacang powder.

Traditional Medicinal Uses

In folk medicine, the leaves are said to be antipyretic and the seeds used against trichophytosis, scabies and eczema. The Orang Asli in Peninsular Malaysia reportedly used the sap to deepen tattoo scars. Sap from bark is used in lotion for treating ulcers.

Other Uses

The wood is not durable, but is suitable for light indoor constructions, temporary constructions and plywood. Streaked heartwood is suitable for the manufacture of furniture.

Comments

In Indonesia, different forms are recognized. Small, almost globose fruits (e.g. *'limus piit'* in West Java) are consistently distinguished from large and more oblong ones which are commonly sold in local markets. Another form called *'limus tipung'* with large, oblong fruits, is known for being hardly fibrous, and finer textured. A similar kind (*'asem linggau'*) is found in East Kalimantan, with a large proportion of fruit having abortive seeds.

Selected References

Boer E, Lemmens RHMJ, Keating WG, den Outer RW (1995) *Mangifera* L. In: Lemmens RHMJ, Soerianegara I, Wong WC (eds.) Plant resources of South-East Asia. No. 5(2): Timber tree: minor commercial timber. Prosea Foundation, Bogor, pp 325–328, 330–331

Bompard JM (1992) *Mangifera foetida* Lour. *Mangifera pajang* Kostermans. In: Verheij EWM, Coronel RE (eds.) Plant resources of South-East Asia. No. 2: Edible fruits and nuts. Prosea Foundation, Bogor, pp 209–211

Burkill IH (1966) A dictionary of the economic products of the Malay Peninsula. Revised reprint, 2 vols. Ministry of Agriculture and Co-operatives, Kuala Lumpur, vol 1 (A–H), pp 1–1240, vol 2 (I–Z), pp 1241–2444

Ding Hou (1978) Anacardiaceae. In: van Steenis CGGJ (ed.) Flora Malesiana, ser I, vol 8. Sijthoff & Noordhoff, Alphen aan den Rijn, pp 395–548, 577 pp

Khoo HE, Ismail A (2008) Determination of daidzein and genistein contents in *Mangifera* fruit. Malays J Nutr 14(2):189–198

Khoo HE, Ismail A, Mohd-Esa N, Idris S (2008) Carotenoid content of underutilized tropical fruits. Plant Foods Hum Nutr 63:170–175

Kostermans AJGH, Bompard JM (1993) The Mangoes: Their botany, nomenclature, horticulture and utilization. Academic, London, 352 pp

Mukherji S (1949) A monograph on the genus *Mangifera* L. Lloydia 12:73–136

Ochse JJ, van den Brink RCB (1931) Fruits and fruitculture in the Dutch East Indies. G. Kolff & Co., Batavia, 180 pp

Pareek OP, Sharma S, Arora RK (1998) Underutilized edible fruits and nuts: an inventory of genetic resources in their regions of diversity. IPGRI Office for South Asia, New Delhi, pp 191–206

Saidin I (2000) Sayuran Tradisional Ulam dan Penyedap Rasa. Penerbit Universiti Kebangsaan Malaysia, Bangi, p 228 (In Malay)

Slik JWF (2006) Trees of Sungai Wain. Nationaal Herbarium Nederland. http://www.nationaalherbarium.nl/sungaiwain/

Tan ST, Amin I (2008) Antioxidant properties (components and capacity) in fresh, powder and fibre products prepared from bacang (*Mangifera foetida*) fruits. Malays J Nutr 14(2(supplement)):S8–S9

Tee ES, Noor MI, Azudin MN, Idris K (1997) Nutrient composition of Malaysian foods, 4th edn. Institute for Medical Research, Kuala Lumpur, 299 pp

Wong KC, Ong CH (1993) Volatile components of the fruits of Bachang (*Mangifera foetida* Lour.) and Kuini (*Mangifera odorata* Griff.). Flav Fragr J 8(3):147–151

Mangifera indica

Scientific Name

Mangifera indica L.

Synonyms

Mangifera amba Forssk., *Mangifera anisodora* Blanco, *Mangifera arbor* Bonti, *Mangifera austroindica* Kosterm., *Mangifera austro-yunnanensis* Hu, *Mangifera balba* Genibrel, *Mangifera bompardii* Kosterm., *Mangifera domestica* Gaertn., *Mangifera equina* Genibrel, *Mangifera fragrans* Maingay, *Mangifera gladiata* Bojer, *Mangifera kukula* Blume, *Mangifera integrifolia* Genibrel, *Mangifera linnaei* Korth., *Mangifera maritima* Lechaume, *Mangifera mekongensis* anon., *Mangifera montana* Heybe, *Mangifera orophila* Kosterm., *Mangifera oryza* Genibrel, *Mangifera racemosa* Bojer, *Mangifera rostrata* Blanco, *Mangifera rubra* Bojer, *Mangifera rubropetala* Kosterm., *Mangifera sativa* Roem. & Schult., *Mangifera siamensis* Warb., *Mangifera sugenda* Genibrel, *Mangifera sylvatica* Roxb., *Mangifera viridis* Boj.

Family

Anacardiaceae

Common/English Names

Indian Mango, Mango, Mango Tree

Vernacular Names

Angola: Mangueira (Portuguese), Manga (Umbundu);
Arabic: Abanj, Abnig, Shajratul-Anbaj, Mangô, Manja, Manjo;
Armenian: Mang, Mank, Մանգ;
Banaban: Te Mangko;
Bangladesh: Am, Ambra;
Basque: Mango, Mangondo;
Benin: Mango (Adja), Mangokorou, Mangodanrou (Bariba), Dapotaga (Berba), Boubatouri (Bourma), Mango-Touri (Dendi), Yapéétaa (Ditammari), Anangatin Manga, Mangania, Yovoslotin, (Fon), Manga, Mangania (Gon), Manguier (French),Pétamou (Natimba), Mago (Peuhl), Mangoti (Sahoué), Égui-Mango (Savé), Mou-Piétamou, Boubatouri (Somba), Iporé, Tambo (Woaba), Mangounou (Yom);
Brazzaville: Onmangourou (Akwa), Loumangou (Beembe), Moumago (Laali), Manga (Laadi), Moumanga (*Sound*i), Moumanga (Yoombe);
Brazil: Manga, Mango,Mangueira;
Burkina Faso: Mangoro-Gô (Bissa), Manguier (Local French);
Burundi: Umwembe, Umwembe W'abungere (Kirundi);
Cameroon: Mangrou;
Central African Republic: Mango (Gbaya):
Chamorro: Mangga, Manggah Saipan;
Chinese: Mang Guo, Mang Guo;
Chuukese: Kangit, Mangko, Manko;
Cook Islands: Vī, Vī Varani (Maori);
Czech: Mangovník Indický;

Danish: Mango, Mangotræ;
Democratic Republic Of Congo: Omuhembe (Hunde), n'Manga (Kiyaka) Umwembe (Kinyarwanda), Manga (Lingala), Manga (Lokele), Mwembe (Mashi), Embe, Hembe, Manga, Mwembe (Swahili), Mwembe (Nyanga), Manga (Tapoke), Kimania (Tetela), Muti-A-Nsafu (Yanzi), Mueba Phutu;
Dutch: Bobbie Manja, Manga, Mangga, Manja, Mangoestanboom, Mangostanboom;
Eastonian: India Mangopuu;
Esperanto: Mango;
Fijian: Am, Mango, Maqo;
Finnish: Mango, Mangopuu;
French: Arbre De Mango, Mangier, Mangot, Mangue, Manguier;
Gabon: Wéba-Montanga (Apindji), Mundjku-A-Mutangani (Baduma), Humanga (Bahumu), Mwiba-Mutangani (Eshira), Andoc-Ntangha (Fang), Manguier (French), Oba W'atanga (Galoa), Oba Oba W'atanga (Nkomi), W'atanga (Mpongwé), Oba W'atanga (Orungu);
German: Indischer Mangobaum, Mango, Mangga, Mangofrucht, Mangobaum, Mangopalme, Muembo;
Ghana: Mango (Asante-Twi), Mango (Lobi), Mango (Wale);
Greek: Magko, Mangko;
Guinea Conakry: Mango;
Hawaiian: Manakō, Manakō Meneke, Meneke;
Hungarian: Mangó;
I-Kiribati: Te Mangko, Te Manko;
Icelandic: Mangó;
India: Am (Assamese), Aam, Am, Ul, (Bengali), Am Thekachu, (Garo), Aam, Am, Am-Ka-Per, Ama, Amb, Amba, Ambe, Ambo, Amra, Amuva, Anb-Ka-Per, Kairi, Keri (Hindu), Amba, Ambaro, Balli Maavu, Ballimavu, Candramavu, Chandra Maavu, Chandramavu, Jeerige Maavu, Jiragemavu, Kaadugenasina Gadde, Kaligaala Gida, Kukku, Kulike, Maavina Mara, Maavu, Manmatha, Marveen, Mava, Mavan, Mavena, Mavi, Mavina, Mavina-Hannu, Mavina-Mara, Mavinamara, Mavinamra, Mavu, Mawa, Rasala, Rasala Mara, Shimavu, Simavu, Suka, Toremavu (Kannada), Dieng Thlai Nar (Khasi), Ambo (Konkani), Amram, Cutam, Gomanna, Gomanne, Mangas, Mangga, Mango, Manna, Mao, Mau, Mav, Mava, Mavina, Mavu, Mavva, Moochi, Mucci, Much-Chi-Maram, Muchi, Nattumavu, Tenmavu, (Malayalam), Heinou (Manipuri), Aamba, Am, Amba, Ambo (Marathi), Theihai (Mizoram), Amba (Oriya), Alipriya, Amra, Amrah, Amravrikshaha, Amva, Atisairrabha, Bhramarapriya, Bhringabhishta, Chukralatamra, Chuta, Chutaka, Chutu, Cuta, Cutah, Gandhabandhu, Joota, Kamanga, Kamaphala, Kamarasa, Kamashara, Kamavallabha, Kamayudha, Kameshta, Keshavayudha, Kireshta, Kokilananda, Kokilavasa, Kokilotsava, Koshi, Madadhya, Madha-Dut, Madhavadruma, Madhuduta, Madhukara, Madhuli, Madhvavasa, Madirasakha, Makanda, Manjari, Manmathalaya, Manmathavasa, Manodna, Modakhya, Mrishalaka, Nilakapittha, Nriyapapriya, Parapushtamahotsava, Phalashreshtha, Phalotpatti, Pikapriya, Pikaraga, Pikavallabha, Priyambu, Rasala, Sahakara, Shareshta, Shatapadatilhi, Shukrapriya, Sidhurasa, Sripriya, Sumadana, Vanotswa, Vasantadru, Vasantaduta (Sanskrit), Adishelarayam, Akkapiram, Akkappiram, Akkaram, Akki, Ambiram, Amiram, Amiratam, Ampiram, Ampiram, Aticaurapam, Cakaram, Cakiyam, Cakkaram, Cakkaramaram, Cakkirakaram, Cakkiram, Cakkirattitam, Calam, Cankiraki, Catpatati, Catpatatiti, Cavakay, Cavakayamaram, Cavatavam, Cavatayilam, Cekaracam, Cekaram, Cekarajamaram, Cetaram, Cetika, Cincam, Cipivallapam, Cirima, Citaram, Civamatitam, Civamatitamaram, Cukapiccakam, Culli, Curan, Cutam, Ekin, Ilanci, Inar, Inarmamaram, Iracalam, Iracam, Iradam, Irai, Iraima, Iraimamaram, Iratam, Ivuli, Kachakkar, Kacakkarmaram, Kacakkar, Kaccukam, Kakkar, Kakkaram, Kakkoram, Kamacaram, Kamam, Kamankam, Kamaracam, Kamavallapam, Kanakkutirai, Kantamaturam, Kantapantu, Kapitanam, Karaccuvatam, Kentapattira, Kilimukkuma, Kocini, Kogilorsavam, Kokilanantam, Kokilanantamaram, Kokilavacam, Kokilorcavam, Kokku, Kokkukkali, Kokkukkalimaram, Kokkumaram, Kosam, Kosamaram, Kosi, Kotam, Kottulatam, Kutirai, Kutiraimaram, Ma, Ma-Maram, Maa, Maa Maram, Madi, Makakalam, Makantamaram, Makaracal, Makarakalam, Makarakantam, Makarantam,

Mangifera indica

Makarantika, Makarantikamaram, Mam, Mamaram, Mampalam, Mankai, Manga-Maram, Malai, Mankaputpam, Manmatalayam, Manmatankanai, Mantalati, Manti, Mantirakam, Mantirakamaram, Mati, Matimi, Matimimaram, Mattipakantam, Mattipakantamaram, Matiyakantam, Mattiyakantam, Matuli, Matulika, Matulikamaram, Matuma, Matututam, Nivimaram, Nuvanai, Omai, Palacirestamaram, Palacirettam, Palalatokatam, Palalatotakam, Palalatotakamaram, Palaltokatam, Palatokam, Palatokatam, Palorpati, Palorpatimaram, Palorpatti, Paramar, Paraputtamatorcavam, Picapantu, Pikapantu, Pikapentu, Pikarakam, Pikavallapam, Pikavalli, Pikkakam, Pikupantu, Piru, Tecalam, Tecamamaram, Tecaravam, Tecavam, Tevam, Tevamamaram, Tevamaram, Titalaka, Titalakamaram, Titalam, Titanam, Tumaram, Mirutakam, Mirutalakam, Mirutalakamaram, Niccanatari, Nilakkapittam, Uyirttavam, Vacantatiru, Vacantatutam, Vacattiru, Vauvaviyam, Vauvaviyamaram, Vauvutavi, Vauvutavimaram (Tamil), Amramu, Amu, Cutamu, Elamaavi, Elamavi, Guddzumamidi, Gujjumaamidi, Gujjumamidi, Maakandamu, Maamidi, Makandamu, Mamadi, Mamedi, Mamid, Mamidi, Mamide, Maumedee, Mamidi-Chettu, Maavi, Maavidi, Maavudi Mavi, Mavidi, Raacha Maamidi, Raasamaamidi, Racamamidi, Rasamamidi, Ratsamamidi, Sutamu, Suthamu, Tiyyamamidi, Thiyyammamidi (Telugu), Aam, Abanj, Amba, Anba, Murbiyat, (Urdu);
Indonesia: Lempelam (Alas, Sumatra), Kalimbong, Kwidi, Kawilei, Kombilei, Komboloi, Uai, Oii (Alforesch, Sulawesi), Pau (Alor), Mamplam (Aceh, Sumatra), Taipe (Bajo), Amplem, Getas, Poh (Bali), Mangga, Morpolom, Pau (Batak), Wowo (Bacan), Fu (Bima), Mangga (Boeol, Sulawesi), Maplane, Paplange, Pawen (Boero), Pau (Bugis), Balamo, Hau (North Ceram), Auaer, (East Ceram), Au Huwane, Apalane, Mabelang, Mapilane, Pota-Pota (West Ceram), Apalam, Mabelang (South Ceram), Ampalam, Hampalam, Mangga, Tekorang (Dyak, Kalimantan), Eokio (Enggano), Mo, Pau (Flores), Lempelam (Gajo), Ayile, Oile, Ombili (Gorontalo), Guawe, Wale (North Halmaheira), Lelit (South Halmaheira), Wei, (Jaurtefa), Mempelam, Pelem (Java), Pao (Madurese), Faw (Kai, Maluku), Kandora (Kambang), Mampalana (Kisar), Isem, Hampelom, Kapelam, Kapelom, Mangga, Pelem, Pelom (Lampong), Olai (Leti), Pao, Taipa (Makassar), Ampelam (Malay, Banjermasin), Mangga (Malay, Java), Empelam, Mampalang (Malay, Maluku), Mangga, Memperlam, Pelam (Malay), Pao (Mandar), Pegun, Peigu (Mentawai), Ampelam, Mangga, Marapalm (Minangkabau), Maga (Nias), Babulangan, Mablang, Mabulang, Mangga, Nambalana (Oeilias), Awa, Manilja, Pager, Pibereekari (West Papua), Mbao, Mapo, Pao (Roti), Pao (North Salajar), Taipa (South Salajar), Uai (Sangir), Paok, Taipah (Sasak), Pau (Sawoe), Balem, Manggi, Memplam (Simaloer), Ruwe, Weu (Soela), Upo, Pau, Porgo (Soemba), Pau (Solor), Buah, Manggah (Sundanese), Fa (Tanimbar), Guwae (Ternate), Guwae (Tidore), Hasor Ho, Heum, Upun, Soh (Timor), Taepang (Tolitoli), Taipa, Taripa, Po (Toraaja). Mamplam (Wetar);
Italian: Mango;
Ivory Coast: Manguier (Local French);
Japanese: Mangoo, Mangou;
Kenya: Mwembe (Embu), Mûembe, Mwlembe (Kikuyu);
Khmer: Svaay;
Kwara'Ae: Asai;
Korean: Mangko;
Laos: Mak Mouang, Mwàngx;
Lithuanian: Indinis Mangas;
Madagascar: Manga, Manguier;
Malaysia: Empelam, Hempelam, Manga, Mempelam;
Mali: Badulafen, Mangoro (Bambara), Manguier (Local French), Julasungalani, Mangora, Mununna, (Malinke), Jula Sungalani (Senoufo);
Marshallese: Mañko;
Mauritius: Manguier;
Myanmar: Ampelam, Mempalam, Thayat, Thayt-Hypu, Tharyetthi;
Nauruan: Damangko;
Nepalese: Aanpa, Amacura;
Niger: Manguaro (Hausa), Mangu (Zarma);
Nigeria: Mango (Ibibio), Mangoro (Huasa), Mangoro (Igede), Mangoro (Igbo), Mangora (Koma), Mangoro (Okeigbo), Mango (Ondo), Mangoro (Yoruba);

Niuean: Mago;
Norwegian: Mango;
Pakistan: Aam;
Palauan: Edel, Iedel;
Papiamento: Mango;
Papua New Guinea: Mango;
Persian: Amba, Ambeh, Anbeh, Anbh, Darakhte-Anbah, Darakhte-Naghzak Naghyak;
Philippines: Paho (Bisaya), Mango (Ilokano), Mangang Kalabau, Mangga (Tagalog);
Pohnpei: Kangit, Kanit, Kehngid;
Polish: Drzewo Mangowe, Mango Indyjskie;
Portuguese: Manga, Mangueira;
Republic Of Guinea: Mankoron (Manika), Mango (Pular), Mange (Soso);
Reunion: Manguier;
Rodrigues Island: Mnaguier;
Rotuman: Magko;
Russian: Манго, Mango;
Rwanda: Umwembe;
Samoan: Mago;
Senegal: Magoro (Bambara), Bumango (Diola), Manguier (Local French), Mago, Manngoyi (Peul), Mangara (Serere), Mago (Tocolor), Mangara (Wolof);
Sierra Leone: Mangoi (Kpaa Mende), Mangro (Krio), Mangoi (Mende), Imangourou (Niominka), A - -Mangkro (Temne), Matigo;
Slovakian: Mangovník Indický
Slovenian: Mango;
Spanish: Manga, Mango, Manguey;
Sri Lanka: Mangass, Etamba (Sinhalese);
Sudan: Mangora (Bambara), Mangora (Malinke), Mangora (Kasombe), Mangora (Sominke);
Suriname: Manja;
Swedish: Mango, Mango-Arter;
Tahitian: Tumi Vi, Tumu Vi, Vi, Vipopa'A;
Tanzania: Omunembe (Haya), Mwembe (Swahili);
Thailand: Mamong, Mamuang;
Tibetan: A Ma I Bras Bu, A Mai Bras Bu, Sin Lcan Go;
Togo: Mangoti (Ewé), Amagoti (Mina);
Tonga: Mango;
Tuamotuan: Mako;
Tuvalu: Mago;
Uganda: Emiyembe (Banyankole), Muyembe (Lusoga), Omuyembe (Sango Bay Area), Aeme;
Vietnam: Xoài;
Yap: Manga, Mangueira.

Origin/Distribution

Mangifera indica has its origin in India and Myanmar. From the specimens at Kew, it was also found wild in China (Fukien, Yunan, Hainan), Peninsular Malaysia (Kuantan in Pahang and Perak), and Sri Lanka. It is widely cultivated in the tropics and subtropics.

Agroecology

In its native range, it occurs scattered in primary and secondary lowland forests on hill slopes. Mango grows and produces fruit over a wide range of altitude from sea level up to 1,200 m, but above 600 m, production can be affected. Mango is adapted to a tropical or subtropical climatic regime. It thrives in a warm, frost-free climate with mean annual rainfall of 400–3,600 mm with a well defined winter dry season. Rain and high humidity during flowering and fruit development reduces fruit set and fruit yields and increases the incidence of diseases affecting the blossoms and fruit. Fruit yield is optimal when the dry periods lasts 1–3 months before flowering to after harvest but the tree needs to be adequately irrigated during fruit development. Mango is drought tolerant and can tolerate drought up to 8 months in certain situations. In the tropical areas close to the equator drought and dry periods have a strong influence on flowering and fruiting as moisture stress was found to stimulate flowering. In areas away from the equator and the subtropics, low temperature stress is necessary for floral induction. Mango has a lower temperature limit of 1–2°C and an upper limit of 42°C. Mango's optimum growing temperature is 24–27°C. Mango will grow outside this range, however, but frost will kill juvenile trees and severely defoliate mature trees. Mango performs best in full sun because its flowers and fruit are produced at the edge of the canopy (the outside of the tree) in full sun.

Mango is not fastidious of soil types, but is generally sensitive to saline soils and is intolerant of continuous salt sprays. It can be found growing on a wide range of soil: from light-textured, sandy soils (tin-tailings and bris soils in Peninsular Malaysia), sedentary inland soils such as sandy loams, loams, sandy clay loams, clays, clay loams, and sandy clays, calcareous soils to heavy-textured clayey soils and fertile alluvial and volcanic soils. It does best in well-drained, moist, fertile, light soils. Its optimal pH range is from 5.5 to 7.5. Low pH below 4.5 is detrimental to growth. Mango is tolerant of waterlogged and flooding for brief periods (1–2 months) as occurs in Thailand, Vietnam and in Florida.

Edible Plant Parts and Uses

Mango fruit can be eaten immature green, mature green or ripe. Immature green mangoes are usually eaten as pickled mangoes or used in curries. Mature green mangoes can be peeled, sliced and eaten fresh as it is, with salt or soya sauce or chilli powder, or sliced or julienned and used in fruit salad. There are many different recipes for green mango fruit salad such as in Malaysian or Indonesian *rujak* or petje which is eaten with *sambal*, Indian green mango salad, Fijian green mango salad, Thai green mango salad or Vietnamese green mango salad. One Vietnamese recipe of green mango salad comprises sliced green mangoes, ram-ram (Vietnamese mint, *Persicaria odorata*) coriander, mint, sliced red onion, fried shallots, grated palm sugar or dark brown sugar, crushed garlic, lime juice, fish sauce or vinegar. One Thai recipe has the following ingredients: julienned green mango, dry shredded unsweetened coconut, bean sprouts chopped fresh coriander, sliced spring onions, chopped peanuts or cashews, fresh basil, fish sauce or soy sauce, chilli, lime juice, brown sugar, Thai chilli sauce. Optionally sliced cooked chicken pieces or cooked shrimp or fried tofu can be added. One common recipe of green mango salad in Fiji consists of sliced green mangoes, coconut cream lemon juice or vinegar, chopped onion, salt and pepper. In the Philippines, green mango is eaten with *bagoong* (shrimp paste or sauce). In Central America (Guatemala, El Salvador, Honduras, Nicaragua and Costa Rica), mango is either eaten green with salt, pepper and hot sauce, or ripe in various forms. In Guatemala, green mangoes are eaten accompanied with toasted and ground pumpkin seed (called *Pepita*) with lime and salt. In Colombia, mango is also eaten either green with salt and/or lime. In Maharashtra, during the hot summer months, a cooling, sweet-sour-spicy drink summer drink called *panha* (in Marathi) and *panna* (across north India) is made with fresh green mango.

Unripe green mango is also consumed cooked or processed into various products. Green mangos are peeled, sliced, parboiled, and then combined with sugar, salt, and various spices and cooked, sometimes with raisins or other fruits, to make chutney; or they may be salted, sun-dried and kept for use in chutney and pickles. Thin slices, seasoned with turmeric, are dried, and sometimes powdered, and used to impart an acid flavour to chutneys, vegetables and soup. Green or ripe mangos may be used to make relish. In India, green mango is used as pickle (*aachar*), *amawat* (a gummy, chewy mango candy), *murraba* (savoury pickle), *amchur* (dried unripe mango powder), *sukhawata* & *chatni* or chutney. In Goa, *miscut* is a spicy mustard-oil pickle made from raw mangoes. *Fhodd* is a water-pickle where raw mangoes are preserved in a brine solution (with dried red chillies). Dried and powdered unripe mango is known as *amchur* in India and *ambi* in Urdu. Green mango pickle is a hot, spicy pickle with a sour taste that is eaten as a condiment. It is made from unripe green mangoes that are fermented with lactic acid bacteria. Preservation is through a combination of salt, increased acidity (lactic acid) and to a small extent the added spices. One recipe consists of green mango, salt, turmeric and spices like roasted mustard seed, roasted fenugreek seeds, roasted asafoetida, chilli powder and oil. In Malaysia, green mango is pickled in salt and/or vinegar; sliced, salted and dried as preserves and snacks and eaten raw in rujuk or cooked in curries. In Thailand, green mango is pickled called *ma mung dong*, soaked in water and sugar (*ma mung chaien*), salted and dried (*ma mung khem*). In Indonesia, green unripe mango is made into *manisan* (candied fruit) and *asinan* (pickle). The unripe flesh of the green mango is popular in Thailand, India and Malaysia. Grated green, its flesh contains enzymes useful for tenderizing meats. In a sun-dried, fine-powder form, green mango makes up a gray-brown seasoning (called *Aamchur*) used in Indian and Asian cooking for its tart flavour and to tenderize meats. The green mango is often used in combination with vegetable and lentil dishes. Green mature mangos are peeled and sliced as filling for pies, used for jelly, or made into sauce which, with added milk and egg whites, can be converted into mango sherbet. In the Philippines, green mango is made into dried green mangoes, green mango jam, puree and green mango wine, together with ripe, yellow mango wine. Green mango wine is a very good white meat tenderizer while the yellow mango wine is good for the red meat. Green mangoes can be dried and even spiced and will keep for several months when stored in a tightly sealed jar, tin or plastic bag. Green mango can be cooked in curries and various dishes such as mango and papaw relish, sweet and sour vegetables. One recipe for green mango curry comprises: cumin seeds, jaggery, red chillies, coriander seeds, grated coconut, sliced onion, salt, oil, green mangoes, garlic flakes, turmeric powder, ginger with optional meat added. A spicy, sweet and sour semi-liquid side-dish called *meth-amba* is made from unripe mango slices called *kairi*, jaggery and fenugreek seeds. They can be enjoyed with *poories* and *polies* (Indian bread), like jam.

Ripe mango is best eaten fresh or stored in the refrigerator for a few days or eaten fresh in mixed fruit salad or with ice cream They are good for breakfast, for dessert, or as a snack. They may be flavoured with lemon or lime juice or mixed with other fruits. Mango is also used in a variety of cereal products, such as in muesli and oat granola. Because it is soft and full of vitamins, ripe mango is an excellent food for babies. Mangoes may also be cooked in various dishes like mango chicken in coconut milk and mango-papaya relish. In Kerala, ripe mangoes are used in a dish called *mambazha kaalan* (ripe mango pulp cooked in yogurt coconut gravy). Ripe mango pulp can be used in almost any recipe requiring a sweet fruit, in pies, mango salsa, puddings, cakes, pastries, bread with and without other nuts, and sweets. Fresh ripe mangos are processed and preserved into a wide range of products including juices, frozen slices, ice-cream, sherbet, smoothies (mango *lassi*), yoghurt, mango ice, mango mousse, aseptic packed mango beverage, nectar, sauces, mango juice, dried mango slices, scoops, mango leather, mango bars, drinks, juices, dried slices, pulp (fruit leather), chutneys, jams, pickles, canned in syrup, and sliced in brine. Mango purees and essences are used to flavour many food products such as drinks, mixed fruit juices, ice creams, wines, teas, breakfast cereals, muesli bars, and biscuits. Mango puree is fantastic on grilled fish, chicken or pork tenderloin. In the Philippines, mango pie, mango salsa, dried mango strips of sweet, ripe mango are also popular, with those from Cebu being exported worldwide. In Mexico, mango puree is used to make juices, smoothies, ice cream, fruit bars, *raspada, aguas frescas* (fresh fruit drinks), pies and sweet chilli sauce, or mixed with *chamoy*, a sweet and spicy chilli paste. *Raspado* or *raspao*, is an ice dessert usually made from crushed ice and fruit syrup topped with condensed milk. When these ingredients are combined with fresh fruit it is called *cholado*. It is popular on a stick dipped in hot chilli powder and salt or also as a main ingredient in fresh fruit combinations. Pieces of mango can be mashed and used as a topping on ice cream or blended with milk and ice as milkshakes. In Thailand and other South East Asian countries, sweet glutinous rice is flavoured with coconut then served with sliced mango as a dessert. In Taiwan, mango puree is a topping that can be added to shaved ice along with condensed milk. The pulpy flesh is served as dessert, in fruit salad, on dry cereal, or in gelatin or custards, or on ice cream. The ripe flesh may be spiced and preserved in jars. Surplus ripe mangos are peeled, sliced and canned in syrup, or made into jam, marmalade, jelly or nectar. The extracted pulpy juice of fibrous types is used for making mango halva and mango leather. Sometimes corn flour and tamarind seed jellose are mixed in. Mango juice may be spray-dried and powdered and used in infant and invalid foods, or reconstituted and drunk as a beverage. The dried juice, blended with wheat flour has been made into "cereal" flakes. A dehydrated mango custard powder has also been developed in India, especially for use in baby foods.

Ripe mangos may be frozen whole or peeled, sliced and packed in sugar (one part sugar to ten parts mango by weight) and quick-frozen in moisture-proof containers. The diced flesh of ripe mangos, bathed in sweetened or unsweetened lime juice, to prevent discoloration, can be quick-frozen, as can sweetened ripe or green mango puree.

Alcoholic beverages made from mangos include wines and liquors as are made in Australia and India.

Mango peel extract exhibited good antioxidant activity in different systems and thus may be used in nutraceutical and functional foods. Studies reported that at a concentration of 0.05%, the antioxidant activity of mango peel dietary fibre was 0.75 times as effective as that of 2-tert-butyl-4-hydroxyanisol (BHA) and 1.4 and 3.4 times higher, respectively, than that of French Parad'Ox (a commercial polyphenols concentrate) and of DL-α-tocopherol (Larrauri et al. 1997). All Bran, Quaker Oats, lemon and apple fibre did not exhibit any antioxidant capacity.

In parts of India the seed is eaten as a boiled or baked vegetable or ground into starchy flour. In Java, the seeds are made into flour. This mango flour is made into a kind of *dodol* or pudding called *jenang pelok* with coconut milk and sugar. The seeds are also used for sambal. Specialty teas are occasionally flavoured with fragrant mango flowers. Young leaves, still rose or bronze-coloured, can be boiled to render them edible. Although the cooked leaves hold their shape and are attractive, their resinous flavour is an acquired taste. Some varieties are more suitable for eating in this manner. In Indonesia, the young tender leaves of certain varieties e.g. mangga gurih, manga wangi, manga kidang, pelem sengir are eaten fresh and uncooked as *lalab* with the rice table.

Botany

An evergreen tree, 10–30 m tall with symmetrical, rounded canopy ranging from low and dense to upright and open, trunk girth up to 3 m and a dense tap root. Cultivated trees are usually 3–10 m high with much reduced girth, depending on canopy management. Branchlets are brown and glabrous. Leaves are alternate and borne on petioles 2–6 cm long, shallowly grooved apically and inflated basally. Leaf lamina is simple, variable in shape, oval-lanceolate, lanceolate, ovate, oblong to oblong lanceolate, 12–45 cm long by 3–12 cm wide, apex acute to long acuminate, base cuneate to obtuse, margin entire, undulating, with 18–30 pairs of lateral veins, midrib distinct on both sides, leathery, glossy deep green adaxially and paler green and glabrous abaxially. Leaves are bronze-tan, purplish-brown when young turning light green and then deep green when mature. Inflorescences are paniculate (narrowly to broadly conical) 20–45 cm long, pseudo-terminal, glabrous to tomentose-pilose with lanceolate, pubescent bracts and bearing hundreds of fragrant, subsessile to shortly pedicellate male and hermaphrodite flowers. Hermaphrodite flowers are small 5–10 mm, with four to five ovate-lanceolate, pubescent sepals and four to five oblong, lanceolate, thinly pubescent, yellowish petals with prominent reddish pattern adaxially, becoming pink and recurved after anthesis. Stamens 4–5 but with only 1–2 fertile with ovate anther; single ovary borne centrally on the disc with the style arising from one side; disc is fleshy with 4–5 lobes bearing the nectaries. Male flowers are similar but without pistil. Fruit is a drupe with a thin gland-dotted epicarp, fleshy, yellow to orange, smooth to fibrous mesocarp surrounding a leathery laterally compressed endocarp containing a single seed. Fruit shape and colour is very variable depending on variety, globose to ovate, obovate to oblong to elongate and with variable lateral compression. The colour turns from green to pale green and yellow when ripe with blushes of pink, orange, red and purple depending on variety (Plates 1–15). The seed is monoembryonic or polyembryonic.

Plate 1 Mango cv. Kensington Pride in Australia

Plate 4 Thai mango cv Chokanan mature fruits

Plate 2 Harvested mature Kensington Pride mangoes

Plate 5 Ripe Chokanan fruits

Plate 3 (**a**, **b**) Mango cv. Celebration (sliced left and whole right) developed in Darwin, Australia

Nutritive/Medicinal Properties

The nutrient composition of raw mango per 100 g edible portion (exclude refuse of 31% seeds and skin) was reported as follows: water 81.71, energy 65 kcal (272 kJ), protein 0.51 g, fat 0.27 g, ash 0.5 g, carbohydrate 17 g, total dietary fibre 1.8 g, total sugars 14.80 g, Ca 10 mg, Fe 0.13 mg, Mg 9 mg, P 11 mg, K 156 mg, Na 2 mg, Zn 0.04 mg, Mn 0.027 mg, Se 0.6 μg, vitamin C 27.7 mg, thiamin 0.058 mg, riboflavin 0.057 mg, niacin 0.584 mg, pantothenic acid 0.160 mg, vitamin

Plate 6 Thai mango cv. Nam Dok Mai

Plate 7 Thai mango cv. Sala

Plate 8 Mango cv. Haden grown in Australia

Plate 9 Mango cv. Irwin grown in Australia

Plate 10 Vietnamese mango cv. Choiu

B6 0.134 mg, total folate 14 µg, total choline 7.6 mg, vitamin A 765 IU (38 µg RAE), vitamin E (α-tocopherol) 1.12 mg, vitamin K (phylloquinone) 4.2 µg, total saturated fatty acids 0.066 g, 12:0 (lauric acid) 0.001 g, 14:0 (myristic acid) 0.009 g, 16:0 (palmitic acid) 0.052 g, 18:0 (stearic acid) 0.003 g; total monounsaturated fatty acids 0.101 g, 16:1 undifferentiated

Plate 11 Indonesian mango cv. Harum manis

Plate 12 Indonesian mango cv. Gedung Gincu

Plate 13 Indonesian mango cv. Golek

Plate 14 Indonesian mango cv. Podang

aspartic acid 0.042 g, glutamic acid 0.060 g, gly-
(palmitoleic) 0.048 g, 18:1 undifferentiated (oleic acid) 0.054 g; total polyunsaturated fatty acids 0.051 g, 18:2 undifferentiated (linoleic acid) 0.014 g, 18:3 undifferentiated (linolenic acid) 0.037 g; amino acids - tryptophan 0.008 g, threonine 0.019 g, isoleucine 0.018 g, leucine 0.031 g, lysine 0.041 g, methionine 0.005 g, phenylalanine 0.017 g, tyrosine 0.010 g, valine 0.026 g, arginine 0.019 g, histidine 0.012 g, alanine 0.051 g, cine 0.021 g, proline 0.018 g, serine 0.022 g; β–carotene 445 µg, α-carotene 17 µg and β-cryptoxanthin 11 µg (USDA 2010).

Fifty four volatile acids were identified in mango (*Mangifera indica* var. Alphonso) and 5-hydroxy-(Z)-7-decenoic acid (2 mg/kg) and 3-hydroxyoctanoic acid (1.1 mg/kg) were confirmed

Plate 15 Ripe cv. Keitt mango grown in Australia

as the major constituents (Idstein et al. 1985). An unusual fatty acid, cis-9, cis-15-octadecadienoic acid, with the suggested trivial name, mangiferic acid was identified in the pulp lipids of mango grown in the Philippines (Shibahara et al. 1993). This butylene-interrupted dienoic fatty acid was concentrated in the pulp part of mango fruit and occupied 5.4% of total acyl groups in the pulp lipids; whereas the common octadecadienoic acid, linoleic acid, comprised a minor component (1.1%) in the same lipids.

Seven phenolic constituents: gallic acid, 3,4-dihydroxy benzoic acid, gallic acid methyl ester, gallic acid propyl ester, mangiferin, (+)-catechin, (−)-epicatechin, and benzoic acid and benzoic acid propyl ester were isolated from mango stem bark and used in a decoction (Núñez Sellés et al. 2002). This aqueous decoction of mango stem bark had been developed in Cuba on an industrial scale to be used as a nutritional supplement, cosmetic, and phytomedicine. The following alkyl gallates were identified from mango flowers: methyl gallate, n-propyl gallate, n-pentyl gallate, n-octyl gallate, 4- phenyl-n-butyl gallate, 6-phenyl-n-hexyl gallate and dihydrogallic acid (Khan and Khan 1989).

Two new terpenoidal saponins, indicoside A and indicoside B; and manghopanal, mangoleanone, friedelin, cycloartan-3β-30-diol and derivatives, mangsterol, manglupenone, mangocoumarin, n-tetacosane, n-heneicosane, n-triacontane and mangiferolic acid methyl ester and others were isolated from mango stem bark (Khan et al. 1993). 5-Alkyl- and 5-alkenylresorcinols, as well as their hydroxylated derivatives, were extracted from mango (*Mangifera indica*) peels (Knɸdler et al. 2007). Among the 15 compounds analyzed, 3 major and 12 minor C(15)-, C(17)-, and C(19)-substituted resorcinols and related analogues were identified. The occurrence of mono-, di-, and triunsaturated C(15)-homologues, saturated and triunsaturated C(17)-homologues, and mono- and diunsaturated C(19)-homologues were found for the first time in mango peels. Additionally, several hydroxylated C(15)- and C(17)-homologues, also not yet described in mango, and a C(14)-monoene, unique because of its even-numbered side chain, were tentatively identified.

In addition to its much relished nutritive fruit, the fruit and other parts of the mango tree contain phytochemicals with numerous medicinal and pharmacological attributes and are elaborated below.

Antioxidant Activity

Mango is a nutritious fruit packed with nutrients and phytochemicals qualifying it as a "super fruit". It is rich in prebiotic dietary fibre, antioxidant vitamins (vitamin A, C and E), antioxdant carotenoids like β–carotene, α-carotene, β-cryptoxanthin and others; antioxidant polyphenols (in the peels). It also has other vitamins (vitamin B6, K and B vitamins), essential minerals, omega-3 fatty acid (linolenic acid) and 17 essential minerals at good levels. Polyphenols that included five quercetin (Q) glycosides and one kaempferol glycoside were identified in mango puree (Schieber et al. 2000). The predominant flavonol glycosides were Q 3-galactoside (22.1 mg/kg fresh weight), Q 3-glucoside (16 mg/kg), and Q 3-arabinoside (5 mg/kg). Among the phenolic acids, gallic acid was predominant (6.9 mg/kg). Quantification of the C-glycoside mangiferin (4.4 mg/kg) was also achieved. A gallotannin consisting of glucose and four gallic acid units was detected. Due to the presence of both carotenoids and polyphenols, mangos can be considered as an especially rich source of antioxidants.

Nutritional and bioactive values of mango were found comparable with these indices in durian and avocado (Gorinstein et al. 2010). These fruits were found to contain high, comparable quantities of basic nutritional and antioxidant compounds, and to possess high antioxidant potentials. All fruits showed a high level of correlation between the contents of phenolic compounds (flavonoids and phenolic acids) and the antioxidant potential. The methods used (three-dimensional fluorescence, FTIR spectroscopy, radical scavenging assays) were suitable for bioactivity determination of these fruits. In order to receive best results, a combination of these fruits was recommended to be included in the diet. The contents of total fibre, proteins and fats were significantly higher in avocado, and carbohydrates were significantly lower in avocado than in the two other fruits. Similarity was found between durian, mango and avocado in polyphenols (9.88, 12.06 and 10.69, mg gallic acid equivalents/g dry weight,(d.w.)), and in antioxidant assays such as CUPRAC (27.46, 40.45 and 36.29, μM Trolox equivalent (TE)/g d.w.) and FRAP (23.22, 34.62 and 18.47, μoM TE/g d.w.), respectively.

Twenty–five carotenoids were identified in the pulp which imparted the yellow-orange colour to the flesh. All-trans-β-carotene was present in largest amount (29.34 μg/g), followed by cis isomers of β-carotene (9.86 μg/g), violaxanthin and its cis isomers (6.40 μg/g), neochrome (5.03 μg/g), luteoxanthin (3.6 μg/g), neoxanthin and its cis isomers (1.88 μg/g), zeaxanthin (1.16 μg/g) and 9- or 9′-cis-lutein (0.78 μg/g) (Chen et al. 2004). Another study reported that all-trans-β-carotene, all-trans-violaxanthin, and 9-cis-violaxanthin were the most abundant carotenoids in mango grown in Mexico (de Ornelas-Paz et al. 2007). The content of all-trans-β-carotene ranged between 0.4 and 2.8 mg/100 g, and cultivars 'Haden' and 'Ataulfo' mangoes had the highest amount. The amounts of all-trans-violaxanthin and 9-cis-violaxanthin (as dibutyrates) ranged between 0.5 and 2.8 mg/100 g and between 0.4 and 2.0 mg/100 g, respectively. The main xanthophylls from 'Ataulfo' mango were all-trans-violaxanthin dibutyrate and 9-cis-violaxanthin dibutyrate while all-trans-β-carotene was the more abundant carotene (Yahia et al. 2006). Lutteoxanthin was also identified in the pulp (Pott et al. 2003).

Mangoes are a good source of antioxidants in human diet. Differences were found in the fruit pulp among the four mango cultivars in all the antioxidant components carotenoids, total phenolic compounds and vitamin C analysed (Rocha Ribeiro et al. 2007). The content of phenolic compounds ranged from 48.40 (Haden) to 208.70 mg/100 g (Ubá); total carotenoid from 1.91 (Haden) to 2.63 mg/100 g (Palmer); β-carotene from 661.27 (Palmer) to 2,220 μg/100 g (Ubá) and total ascorbic acid ranged from 9.79 (Tommy Atkins) to 77.71 mg/100 g (Ubá). The pulp also contained very small amounts of gallotannins. Twenty-one and eight gallotannins were found in the kernels and pulp, respectively, no benzophenone derivatives was obtained. Gallotannins amounted to 1.4 mg/g dm (dry matter) in the peels (expressed as gallic acid), while only small amounts (0.2 mg/g dm) were found in the pulp. In contrast, mango kernels contained 15.5 mg/g dm and thus proved to be a rich source of gallotannins (Berardini et al. 2004). Mango pulp also contained the triterpene, lupeol with potent antioxidant and anti-cancerous properties (Prasad et al. 2007).

Mango peels are also rich in antioxidant polyphenols and carotenoids. The peel was found to have up to 4,860 mg/kg dry matter of total phenolic compounds while only traces could be detected in the flesh. In the peel of cv 'Tommy Atkins' 18 gallotannins and five benzophenone derivatives were detected and identified as galloylated maclurin and iriflophenone glucosides (Berardini et al. 2004). The peel also contained flavonol glycosides such as rhamnetin 3-O-β-galactopyranoside and rhamnetin 3-O-β-glucopyranoside. In the peels of red-coloured cultivars, cyanidin 3-O-galactoside and an anthocyanidin hexoside were identified. The contents and degrees of esterification of pectins extracted from the lyophilized peels ranged from 12.2 to 21.2% and from 56.3 to 65.6%, respectively, suggesting mango peels to be a promising source of high-quality pectin (Berardini et al. 2005). Flavonol O- and xanthone C-glycosides were extracted from mango (*Mangifera indica*

L. cv. "Tommy Atkins") peels and characterized (Schieber et al. 2003). Among the fourteen compounds analyzed, seven quercetin O-glycosides, one kaempferol O-glycoside, and four xanthone C-glycosides were found. On the basis of their fragmentation pattern, the latter were identified as mangiferin and isomangiferin and their respective galloyl derivatives. A flavonol hexoside with m/z 477 was tentatively identified as a rhamnetin glycoside, which had not yet been reported in mango peels. The results obtained in the study confirmed that peels originating from mango fruit processing to be a promising source of phenolic compounds that might be recovered and used as natural antioxidants or functional food ingredients.

Ripe mango peels were found to contain higher amounts of anthocyanins and carotenoids compared to raw peels while raw mango peel had high polyphenol content (Ajila et al. 2007). The study found that mango peel extract exhibited good antioxidant activity as evaluated by different systems antioxidant systems such as reducing power activity, DPPH free radical scavenging activity, iron induced lipid peroxidation of liver microsomes and soybean lipoxygenase inhibition. The IC_{50} values were found to be in the range of 1.39–5.24 µg of gallic acid equivalents. Thus, indicating mango peels would be beneficial in nutraceutical and functional foods.

Mango peels with high dietary was found to have potent antioxidant activity using the ferric thiocyanate colorimetric method (Larrauri et al. 1997). At a concentration of 0.05%, the antioxidant activity of mango peel dietary fibre was 0.75 times as effective as that of 2-*tert*-butyl-4-hydroxyanisol (BHA) and 1.4 and 3.4 times higher, respectively, than that of French PARAD'OX (a commercial polyphenols concentrate) and of DL-α-tocopherol. All Bran, Quaker Oats, lemon and apple fibre did not exhibit any antioxidant capacity.

An extract of *Mangifera indica* (Vimang) showed potent scavenging activity of hydroxyl radicals and hypochlorous acid and acted as an iron chelator (Martinez et al. 2000). The extract also showed a significant inhibitory effect on the peroxidation of rat-brain phospholipid and hindered DNA damage by bleomycin or copper-phenanthroline systems. Mango stem extract, vimang with formulations used in Cuba as food supplements, was found to exhibit in-vitro antioxidant and free radical scavenging activities (Garrido et al. 2008). Luminol-enhanced chemiluminescence was used to elucidate the effect of this extract on the generation of reactive oxygen species in PMA- or zymosan-stimulated human polymorphonuclear leukocytes and on superoxide radicals generated in the hypoxanthine–xanthine oxidase reaction. Chemiluminescence was reduced in a dose-dependent fashion at extract concentrations from 5 to 100 µg/ml, most probably by inhibiting the superoxide generation reaction. Part of this *M. indica* extract antioxidant activity was ascribed to the presence of mangiferin as its main component. Mangiferin exhibited antioxidant with an IC_{50} value of 7.5 µg/ml using the free radical 2,2-diphenyl-1-picrylhydrazyl (DPPH) assay (Stoilova et al. 2005).

The profiles of secondary plant substances such as xanthone C-glycosides, gallotannins, and benzophenones in different parts of mango tree (barks, kernels, peels, and old and young leaves) varied greatly but were fairly consistent across Brazilian mango cultivars (Barreto et al. 2008). Four of the major phenolic compounds, namely mangiferin, penta- O-galloyl-glucoside gallic acid, and methyl gallate were also evaluated in additional in-vitro bioassay systems such as oxygen radical absorbance capacity, 2, 2-diphenyl-1-picrylhydrazyl, and ferric reducing ability of plasma. Mangiferin in particular, was detected at high concentrations in young leaves (cv Coite = 172 g/kg), in bark (cv Momika = 107 g/kg), and in old leaves (cv Itamaraka = 94 g/kg), and exhibited an remarkably potent antioxidant capacity.

Mango fruit also contain many volatile compounds. Three hundred and seventy-two compounds were identified in the volatile component of mango fruit (Pino et al. 2005). The total concentration of volatiles hovered from 18 to 123 mg/kg of fresh fruit. Terpene hydrocarbons were the major volatiles of all cultivars, the dominant terpenes being δ-3-carene (in cvs. Haden, Manga amarilla, Macho, Manga blanca, San Diego, Manzano, Smith, Florida, Keitt, and Kent), limonene (in cvs. Delicioso, Super Haden, Ordoñez, Filipino, and La Paz), both terpenes (in cv. Delicia), terpinolene (in cvs. Obispo, Corazón, and Huevo de toro), and α-phellandrene (in cv. Minin).

Anticancer Activity

The methanol extract from the stem bark of *M. indica* elicited differential cellular sensitivity on the MDA-MB-231, MDA-MB-435, and MDA-N human breast cancer cell lines (Muanza et al. 1995). Mango juice and mango juice factions were found to inhibit chemically induced neoplastic transformation of mammalian cell lines (Percival et al. 2006). Incubation of HL-60 cells with whole mango juice and mango juice fractions resulted in inhibition of the cell cycle in the G(0)/G(1) phase. A fraction of the eluted mango juice with low peroxyl radical scavenging ability was most effective in arresting cells in the G(0)/G(1) phase. Whole mango juice was also effective in decreasing the number of transformed foci in the neoplastic transformation assay in a dose-dependent fashion.

Lupeol, a triterpene present in mango and other fruits, was shown to possess anticancer properties in in-vivo and in-vitro assays (Prasad et al. 2008a,b). Lupeol and mango pulp extract (MPE) exhibited apoptogenic activity in mouse prostate. Lupeol/MPE supplementation resulted in arrest of prostate enlargement in testosterone-treated animals. Lupeol and MPE supplementation resulted in a significantly high percentage of apoptotic cells in the hypodiploid region of mouse prostate. The apoptogenic response of lupeol-induced changes in LNCaP cells was summarized as early increase of reactive oxygen species followed by induction of mitochondrial pathway leading to cell death. Another paper reported that both pre- and post-treatment of lupeol showed significant protective effects in 7,12-dimethylbenz(a)anthracene (DMBA) induced DNA strand breaks in dose and time dependent fashion (Nigam et al. 2007). The pre-treatment of lupeol at the dose of 200 μg/mouse showed 56.05% prevention, and post-treatment at the same dose showed 43.26% prevention, at 96 hour time interval, against DMBA induced DNA strand breakage. The results suggested protective effects of lupeol on DMBA induced DNA alkylation damage in Swiss albino mice.

The mango cultivars, Ataulfo and Haden, selected on the basis of their superior antioxidant capacity compared to the other varieties were found to exhibit anticarcinogenic activity against several cancer cell lines that included Molt-4 leukemia, A-549 lung, MDA-MB-231 breast, LnCap prostate, and SW-480 colon cancer cells and the non-cancer colon cell line CCD-18Co (Noratto et al. 2010). SW-480 and MOLT-4 were statistically equally most sensitive to both cultivars followed by MDA-MB-231, A-549, and LnCap in order of decreasing efficacy as determined by cell counting. The efficacy of extracts from all mango varieties in the inhibition of cell growth was tested in SW-480 colon carcinoma cells, where Ataulfo and Haden demonstrated superior efficacy, followed by cvs. Kent, Francis, and Tommy Atkins. At 5 mg of GAE/L, Ataulfo retarded the growth of colon SW-480 cancer cells by approximately 72% while the growth of non-cancer colonic myofibroblast CCD-18Co cells was not inhibited. The growth inhibition exerted by Ataulfo and Haden polyphenolics in SW-480 was associated with an increased mRNA expression of pro-apoptotic biomarkers and cell cycle regulators, cell cycle arrest, and a decrease in the generation of reactive oxygen species. The data suggested that polyphenolics from several mango varieties exerted anticancer effects, with compounds from Haden and Ataulfo mango varieties exhibiting superior chemopreventive activity.

Lupeol, triterpene from mango, also had anti-skin tumour promoting effect (Saleem et al. 2004). Animals pre-treated with lupeol showed significantly decreased tumour incidence, lower tumour body burden and a significant lag in the latency period for tumour appearance. At the completion of the experiment at 28 weeks, 100% of the animals in 12-O-tetradecanoyl-phorbol-13-acetate (TPA)-treated group exhibited seven to eight tumours/mouse, whereas only 53% of the mice receiving lupeol prior to TPA treatment exhibited one to three tumours/mouse. These results indicated that Lupeol possessed anti-skin tumour-promoting effects in CD-1 mouse and inhibited conventional as well as novel biomarkers of tumour promotion.

Mangiferin, a C-glucosylxanthone (1,3,6,7-tetrahydroxyxanthone-C2-β-D-glucoside) was shown to have in-vivo growth-inhibitory activity against ascitic fibrosarcoma in Swiss mice (Guha et al. 1996). Following in-vivo or in-vitro treatment, mangiferin also enhanced tumour cell cytotoxicity of the splenic cells and peritoneal macrophages of normal and tumour-bearing mice. In-vitro treatment of the splenic cells of tumour-bearing mice with mangiferin resulted in augmented death of tumour cells, both resistant and sensitive to natural killer cells. Mangiferin was also found to antagonize in-vitro the cytopathic effect of HIV. The drug appeared to perform as a potent biological response modifier with antitumour and antiviral effect.

Yoshimi et al. (2001) reported that in the short-term assay, 0.1% mangiferin in a diet significantly inhibited the aberrant crypt foci development (tumorigensis) in rats treated with azoxymethane compared to rats treated with azoxymethane alone. In the long-term assay, the group treated with 0.1% mangiferin at the beginning of the experimental protocol had significantly lesser incidence and multiplicity of intestinal neoplasms induced by azoxymethane (47.3 and 41.8% reductions of the group treated with azoxymethane alone for incidence and multiplicity, respectively). Further, the cell proliferation in colonic mucosa was lowered in rats treated with mangiferin (65–85% reductions of the group treated with, azoxymethane alone). These results suggested that mangiferin had potential as a naturally-occurring chemopreventive agent. Peng et al. (2004) showed that five different concentrations of mangiferin (25–200 μmol/l) dose-dependently and time-dependently inhibited the proliferation of K562 leukaemia cells, and induced apoptosis in K562 cell line. Reverse transcriptase-PCR revealed that bcr/abl gene expression was down-modulated when K562 cells had been treated with different concentrations of mangiferin.

Studies reported that mangiferin increased susceptibility of rat liver mitochondria to calcium-induced mitochondria permeability transition (Pardo-Andreu et al. 2005). Mitochondrial permeability transition was found to be a Ca^{2+}-dependent, cyclosporine A-sensitive, non-selective inner membrane permeabilization induced by a wide range of agents or conditions, which had often been associated with necrotic or apoptotic cell death. The scientists postulated that Ca^{2+} increased levels of mitochondria-generated ROS (reactive oxygen species), which reacted with mangiferin producing quinoid derivatives, which in turn reacted with the most accessible mitochondrial thiol groups, thus triggering mitochondria permeability transition. They further proposed that accumulation of mangiferin quinoid products would occur in cells exposed to an overproduction of ROS, such as cancer cells, where the occurrence of mitochondria permeability transition-mediated apoptosis may be a cellular defence mechanism against excessive ROS formation.

The mango xanthone, mangiferin was also found to have a modulatory effect on mitochondrial lipid peroxidation (LPO), tricarboxylic acid (TCA) cycle key enzymes and electron transport chain complexes in Swiss albino mice with lung cancer induced by benzo(a)pyrene (Rajendran et al. 2008). Pre- and post-treatment with mangiferin prevented and restored decreased activities of electron transport chain complexes and TCA cycle key enzymes such as isocitrate dehydrogenase (ICDH), succinate dehydrogenase (SDH), malate dehydrogenase (MDH) and α-ketoglutarate dehydrogenase (α-KGDH), in lung cancer bearing animals. The study indicated that mangiferin had chemopreventive and chemotherapeutic effect.

Hepatoprotective Activity

Lupeol/mango pulp extract (MPE) supplementation was found to be effective in combating oxidative stress induced cellular injury of mouse liver by modulating cell-growth regulators (Prasad et al. 2007). Lupeol/MPE supplementation effectively influenced the 7,12-dimethylbenz(a) anthracene DMBA induced oxidative stress, reflected by restored antioxidant enzyme activities and decrease in lipid peroxidation. A reduction of apoptotic cell population in the hypodiploid region was observed in lupeol and MPE supplemented animals. The inhibition of apoptosis was preceded by reduction in reactive oxygen species (ROS) level and restoration of mitochondrial transmembrane potential followed by decreased DNA fragmentation. In DMBA treated animals, down-regulation of antiapoptotic Bcl-2 and up-modulation of proapoptotic Bax and Caspase 3 in mouse liver was observed. These changes were restored by lupeol/MPE.

Vimang® was found to protect liver from oxidative damage induced by oxygen-based free radicals in several in-vitro test systems conducted (Sánchez et al. 2003). The ischaemia/reperfusion model led to an increase of transaminase (ALT and AST), membrane lipid peroxidation, tissue neutrophil infiltration, DNA fragmentation, loss of protein -SH groups, cytosolic Ca2+ overload and a decrease of catalase activity in rats. Oral administration of Vimang® (50, 110 and 250 mg/kg, b.w.) 7 days before reperfusion, reduced transaminase levels and DNA fragmentation in a dose dependent way. Vimang® also restored the cytosolic Ca^{2+} levels and impeded polymorphonuclear migration at a dose of 250 mg/kg b.w., improved the oxidation of total and non-protein sulfhydryl groups and prevented modification in catalase activity, uric acid and lipid peroxidation markers. These data suggested that Vimang® could be a useful new natural drug for preventing oxidative damage during hepatic injury associated with free radical generation.

Further studies reported that the Vimang-Fe(III) mixture was more effective at scavenging 2,2-diphenyl-1-picrylhydrazyl (DPPH) and superoxide radicals, as well as in protecting against t-butyl hydroperoxide-induced mitochondrial lipid peroxidation and hypoxia/reoxygenation-induced hepatocytes injury, compared to Vimang alone (Pardo-Andreu et al. 2006b). Besides acting as antioxidant, Vimang extract or its mangiferin component decreased liver iron by increasing its excretion. Based on earlier in-vitro results, the scientists further postulated that Vimang and mangiferin provided therapeutically useful effects in iron overload related diseases (Pardo-Andreu et al. 2008a). Both Vimang or mangiferin treatment prevented iron overload in serum as well as liver oxidative stress, reduced serum and liver lipid peroxidation, serum GPx activity, and augmented serum and liver GSH, serum SOD and the animals overall antioxidant condition. Vimang® was shown to be a potent inhibitor of 2-deoxyribose degradation mediated by Fe (III)-EDTA plus ascorbate or superoxide (O-2). The results strongly indicated that Vimang® did not obstruct 2-deoxyribose degradation by simply trapping OH radicals. Rather, Vimang® appeared to act as an antioxidant by complexing iron ions, rendering them inactive or poorly active in the Fenton reaction.

Ethanol extract of seed kernels of Thai mango was found to exert ant-inflammatory effect against CCl_4 induced liver injury in rats (Nithitanakool et al. 2009a,b). Three phenolic compounds were isolated from the extract: 1,2,3,4,6-penta- O-galloyl-β-D-glucopyranose (PGG), methyl gallate (MG), and gallic acid (GA). The hepaprotective effect was attributed to the main principle, PGG, which exhibited potent antioxidant and anti-inflammatory activities. In recent studies, Agarwala et al. (2011) demonstrated the cytoprotective potential of mangiferin (MGN) a glucosylxanthone found in *Mangifera indica*, which may be attributed to quenching of the reactive oxygen species (ROS) generated in the cells due to oxidative stress induced by $HgCl_2$, restoration of mitochondrial membrane potential and normalization of cellular antioxidant levels. Pre-treatment of Hep G2 cells with mangiferin significantly decreased the percentage of $HgCl_2$ induced apoptotic cells. Similarly, the levels of ROS generated by the $HgCl_2$ treatment were inhibited significantly by MGN. MGN also significantly inhibited the $HgCl_2$-induced reduction in glutathione (GSH), glutathione-S-transferase (GST), superoxide dismutase (SOD), and catalase (CAT) levels at all post incubation intervals.

Pourahmad et al. (2010) reported that aqueous mango fruit extracts and gallic acid (100 μM) protected hepatocyte against all oxidative stress markers including cell lysis, ROS generation, lipid peroxidation, glutathione depletion, mitochondrial membrane potential decrease, lysosomal membrane oxidative damage and cellular proteolysis. Mango extracts (20, 50 and 100 μg/ml) were more effective than gallic acid (100 μM) in protecting hepatocytes against cumene hydroperoxide induced lipid peroxidation. In contrast gallic acid (100 μM) was more effective than mango extracts (20, 50 and 100 μg/ml) at preventing lysosomal membrane damage. There were no significance differences between all plant extracts and gallic acid (100 μM) in H_2O_2 scavenging activity. These results suggested a hepatoprotective role for mango extract against liver injury associated with oxidative stress.

Mangifera indica aqueous extract (Vimang) was found to modulate the activity of P450 isozymes system (Rodeiro et al. 2007). Cytochrome P450 isozymes are a superfamily of haemoprotein enzymes found on the membrane of endoplasmic reticulum predominantly in the liver but are also found in intestine, lungs, kidneys, brain etc. They are responsible for catalyzing the metabolism of large number of endogenous and exogenous compounds. Exposure of rat hepatocytes to Vimang at concentrations of up to 100 μg/ml produced a significant reduction (60%) in 7-methoxyresorufin-O-demethylase (MROD; CYP1A2) activity, an increase (50%) in 7-penthoxyresorufin-O-depentylase (PROD; CYP2B1) activity, while no significant effect was observed with other isozymes. The antioxidant properties of Vimang were also evaluated in t-butyl-hydroperoxide-treated hepatocytes. A 36-hours pre-treatment of cells with Vimang (25–200 μg/ml) strongly inhibited the decrease of GSH levels and lipid peroxidation induced by t-butyl-hydroperoxide in a dose- and time-dependent manner.

Antigenotoxic Activity

Pre-treatment with mangiferin was found to mitigate the cytotoxic and genotoxic effects of cadmium chloride ($CdCl_2$) treatment in HepG2 cells growing in-vitro (Satish Rao et al. 2009). Mangiferin enhanced cell survival and prevented cadmium chloride induced apoptotic cell death. A significant reduction in the micronuclei frequency and comet parameters after mangiferin pre-treatment to $CdCl_2$ clearly indicated its antigenotoxic potential. Similarly, the reactive oxygen species generated by the $CdCl_2$ treatment were inhibited significantly by mangiferin. Thus, the study revealed that the potent cytoprotective and antigenotoxic effect of mangiferin against $CdCl_2$ induced toxicity in HepG2 cell line may be attributed to decrease in $CdCl_2$ induced reactive oxygen species levels and resultant oxidative stress.

Viswanadh et al. (2010) reported on the protective role of mangiferin (MGN) against cadmium chloride ($CdCl_2$)-induced genotoxicity in Swiss albino mice. Treatment of mice to various doses of cadmium chloride, resulted in a dose-dependent increase in the frequency of micronucleated polychromatic (MnPCE) and normochromatic erythrocytes (MnNCE), with corresponding decrease in the polychromatic/normochromatic erythrocyte ratio (PCE/NCE ratio) at various post-treatment times. MGN (2.5 mg/kg body weight) pre-treatment significantly reduced the frequency of MnPCE, MnNCE and increased PCE/NCE ratio when compared with the DDW (doubled deionised water)+$CdCl_2$ group at all post-treatment times indicating its antigenotoxic effect. Additionally, pre-treatment of MGN decreased the lipid peroxidation (LPx) content in liver, while significant increase was observed in hepatic Glutathione (GSH), glutathione-S-transferase (GST), superoxide dismutase (SOD) and catalase (CAT) activity. The study revealed that MGN had potent antigenotoxic effect against $CdCl_2$-induced toxicity in mice, which may be due to the scavenging of free radicals and increased antioxidant status.

Antihyperglycaemic, Antidiabetic, Antihyperlipidemic Activities

Mango leaves and stem bark were found to have antihyperglycaemic, antidiabetic, antihyperlipidemic and antiatherogenic activities.

Studies showed that ethanol and water extracts of leaves and stem-barks of *M. indica* had significant antihyperglycemic effect in type 2 model rats when fed simultaneously with glucose load (Bhowmik et al. 2009). The ethanol extract of stem-barks exhibited significant antihyperglycemic effect when the extract was fed 30 minutes prior to the glucose load. The ethanol extracts of stem-barks reduced glucose absorption gradually during the whole perfusion period in type 2 rats. In another study, chronic intraperitoneal (i.p.) administration of mangiferin (10 and 20 mg/kg) once daily for 28 days exhibited antidiabetic activity by significantly lowering fasting plasma glucose level at different time intervals in STZ-diabetic rats (Muruganandan et al. 2005). Further, mangiferin (10 and 20 mg/kg, i.p.) showed significant antihyperlipidemic and antiatherogenic activities as evidenced by significant decrease in plasma total cholesterol, triglycerides, low-density lipoprotein cholesterol (LDL-C) levels coupled together with elevation of high-density lipoprotein cholesterol (HDL-C) level and diminution of atherogenic index in diabetic rats. Additionally, the chronic administration of mangiferin (10 and 20 mg/kg, i.p.) for 14 days significantly improved oral glucose tolerance in glucose-loaded normal rats indicating its potent antihyperglycemic activity. The study demonstrated that mangiferin possessed significant antidiabetic, antihyperlipidemic and antiatherogenic properties and could have beneficial effect in the treatment of diabetes mellitus associated with hyperlipidemia and related cardiovascular complications.

Muruganandan et al. (2002) found that repeated intraperitoneal injections of mangiferin (10 and 20 mg/kg) and insulin (6 U/kg) controlled streptozotocin (STZ)-induced lipid peroxidation and significantly protected Wistar rats against cardiac as well as renal damage. Diabetic rats exhibited significant increase in serum creatine phosphokinase (CPK) and total glycosylated haemoglobin. Kidney had tubular degeneration and decreased levels of superoxide dismutase (SOD) and catalase (CAT) with an elevation of malonaldehyde (MDA). Cardiac SOD, CAT and lipid peroxidation were significantly enhanced. Histopathological findings revealed cardiac hypertrophy with haemorrhages. Analysis of erythrocyte revealed significantly elevated levels of MDA with insignificant decrease in CAT and SOD. Results showed that oxidative stress appeared to play a major role in STZ-induced cardiac and renal toxicity as was evident from significant inhibition of antioxidant defence mechanism in renal tissue or a compensatory increase in antioxidant defence mechanism in cardiac tissue. Intraperitoneal administration of mangiferin exhibited significant decrease in glycosylated haemoglobin and CPK levels along with the amelioration of oxidative stress that was comparable to insulin treatment.

An earlier study reported that an aqueous mango leaf extract given orally (1 g/kg) did not change the blood glucose levels in either normoglycaemic or streptozotocin STZ-induced diabetic rats. In glucose-induced hyperglycaemia, however, antidiabetic activity was seen when the leaf extract and glucose were administered simultaneously and also when the extract was given to the rats 60 minutes before the glucose load (Aderibigbe et al. 1999). Perpétuo and Salgado (2003) found that diabetic rats administered diets containing 5, 10 and 15% mango flour for 30 days showed a significant decrease in blood glucose level in comparison to the diabetic controls with 0% mango diet. In a second study, rats consuming 5% mango diet had 66% lower blood glucose level than the controls. Further they found that the hepatic glycogen level of rats fed mangoes was higher than in the controls. The results suggested that mango flour could assist in the treatment of diabetes.

Flavonoids from *Emblica officinalis* and *Mangifera indica* were found to effectively reduce lipid levels in serum and tissues of rats induced hyperlipidemia. Lecithin-cholesterol acyltransferase (LCAT) exhibited elevated levels in rats fed flavonoids from both sources (Anila and Vijayalakshmi 2002). The degradation and elimination of cholesterol was highly enhanced in both the groups. In *E. officinalis*, the mechanism of hypolipidemic action was by the combined action of inhibition of synthesis and enhancement of degradation. In the *M. indica* group, inhibition of cholesterogenesis was not encountered but highly significant degradation of cholesterol was observed, which may be the pivotal factor for hypolipidemic activity. Though the mechanisms differed in the two cases, the net effect was to lower lipid levels.

Mango bark ethanolic extract was found to be the most potent in α-glucosidase inhibitory activity, with an IC_{50} value of 314 µg/ml when compared with the ethanolic extracts of *Lawsonia inermis* leaves, *Holarrhena antidysenterica* bark and *Swertia chirata* whole plant (Prashanth et al. 2001). In another study, simultaneous oral administration of Hawthorn, *Crataegus oxyacantha* berry extract and mango stem extract (0.5 ml/100 g body weight) to rats fed an atherogenic (4% cholesterol, 1% cholic acid, 0.5% thiouracil) diet prevented the elevation of lipids in the serum and heart and also caused a significant decrease in lipid accumulation in the liver and aorta reverting the hyperlipidaemic condition of these rats (Akila and Devaraj 2008). These extracts significantly normalised the activity of antioxidant enzymes such as superoxide dismutase, catalase, glutathione peroxidase, and glutathione, thereby restoring the antioxidant status of the organism to almost normal levels. The effect was attributed to the synergistic activity of flavonoids in Hawthorn berries and polyphenols of mangiferin.

The methanolic extract of *Mangifera indica* leaves exhibited potent in-vitro dipeptidyl peptidase IV (DPP-IV) inhibitory activity with an IC_{50} value of 182.7 µg/ml (Yogisha and Raveesha 2010). Inhibitors of DPP IV were reported to augment the level of glucagon-like peptide 1 (GLP-1), which improved glucose tolerance and increased insulin secretion. The results confirmed the inhibitory effect of *M. indica* on DPPIV, and the potential to be a novel, efficient and tolerable approach for treatment of diabetes. Morsi et al. (2010) studied the effect of aqueous mango leaves extract at doses of 30, 50 and 70 mg on alloxan-induced Sprauge-Dawley albino diabetic rats. They found that in streptozotocin-induced diabetic rats, blood glucose, triglycerides, total cholesterol, LDL, VLDL, urea, uric acid, creatinine and liver enzymes activities (AST and ALT) were significantly enhanced, while HDL, serum total protein, albumin and globulin were significantly decreased compared with the negative control rats. Treating diabetic rats with 30, 50 and 70 mg aqueous mango leaves extract caused a significant improvement in these biochemical measures and the best results were achieved by using 70 mg mango leaves extract followed by 50 and 30 mg aqueous extract, respectively. They suggested that aqueous mango leaf extract beings rich in total phenols and total flavonoids acted as powerful antioxidants and should be used in manufacture processes of the natural products as functional foods or as a dietary supplement with anti-diabetic activity.

Li et al. (2010) found that serum-advanced glycation end-products level, malonaldehyde level, sorbitol concentration of red blood cell, 24 hour albuminuria excretion were significantly decreased, whereas activity of serum superoxide dismutase and glutathione peroxidase and creatinine clearance rate were increased by mangiferin in a diabetic nephropathy rat model. Blood glucose level remained unaffected. Mangiferin significantly inhibited glomerular extracellular matrix expansion and accumulation and transforming growth factor-β 1 over-expression in glomeruli of diabetic nephropathy rats. Further, mangiferin was observed to inhibit proliferation of mesangial cells induced by high glucose and the over-expression of collagen type IV of mesangial cells induced by advanced glycation end products. The results indicated that mangiferin could significantly prevent progression of diabetic nephropathy and improve renal function.

Antiatherosclerotic Activity

Studies demonstrated that liver mitochondria isolated from the atherosclerosis-prone hypercholesterolemic LDL receptor knockout (LDLr(−/−)) mice had lower content of NADP(H)-linked substrates than the controls and, as consequence, higher sensitivity to oxidative stress and mitochondrial membrane permeability transition (MPT) (Pardo-Andreu et al. 2008b). Oral supplementation with Vimang extract or its main polyphenol mangiferin shifted the sensitivity of mice liver mitochondria to MPT to control levels. These in-vivo treatments with Vimang and mangiferin also significantly reduced ROS generation by both isolated mice liver mitochondria and spleen lymphocytes. Thus, Vimang and mangiferin spared the endogenous reducing equivalents (NADPH) in LDLr(−/−) mice mitochondria by rectifying their lower antioxidant capacity and restoring the organelle redox homeostasis. The effective bioavailability of these compounds could make them suitable antioxidants with potential use in atherosclerosis susceptible conditions.

Antioxidant Activity

Under prescribed experimental conditions, concentrations of MDA (malondialdehyde) and adenosine-5'-triphosphate (ATP) catabolites were affected in a dose-dependent way by H_2O_2 in a red blood cell system. Incubation with Vimang (0.1, 1, 10, 50 and 100 μg/mL), mangiferin (1, 10, 100 μg/mL) and epigallocatechin gallate (EGCG) (0.01, 0.1, 1, 10 μM) significantly enhanced erythrocyte resistance to H_2O_2-induced reactive oxygen species production (Rodríguez et al. 2006). In particular, the study demonstrated the protective activity of these mango compounds on ATP, GTP (guanosine triphosphate) and total nucleotides (NT) depletion after H_2O_2-induced damage and a reduction of NAD (nicotinamide adenine dinucleotide) and ADP (adenosine triphosphate), which both increased because of the energy consumption following H_2O_2 addition. Energy charge potential, reduced in H_2O_2-treated erythrocytes, was also restored in a dose-dependent way by these compounds. Their protective effects may be related to the strong free radical scavenging ability described for polyphenols.

The protective effect of peel extracts of unripe and ripe mango fruits of two varieties namely, Raspuri and Badami on hydrogen peroxide induced haemolysis, lipid peroxidation, degradation of membrane proteins and its morphological changes were reported (Ajila and Prasada Rao 2008). The oxidative haemolysis of rat erythrocytes by hydrogen peroxide was inhibited by mango peel extract in a dose dependent way. The IC_{50} value for lipid peroxidation inhibition on erythrocyte ghost membrane was found to be in the range of 4.5–19.3 μg gallic acid equivalents. The mango peel extract showed protection against membrane protein degradation caused by hydrogen peroxide. Morphological changes to erythrocyte membrane caused by hydrogen peroxide were protected by mango peel extract. The results demonstrated that mango peel extracts protected erythrocytes against oxidative stress and may impart health benefits and could be used as a valuable food ingredient or as nutraceutical product.

Mangifera indica stem bark extract (Vimang®) 50–250 mg/kg, mangiferin 50 mg/kg, vitamin C 100 mg/kg, vitamin E 100 mg/kg and β-carotene 50 mg/kg exerted protective effects against the 12-O-tetradecanoylphorbol-13-acetate (TPA)-induced oxidative damage in serum, liver, brain as well as in the hyper-production of reactive oxygen species (ROS) by peritoneal macrophages (Sánchez et al. 2000). The treatment of mice with Vimang®, vitamin E and mangiferin reduced the TPA-induced production of ROS by the peritoneal macrophages by 70, 17 and 44%, respectively. Similarly, the H_2O_2 levels were reduced by 55–73, 37 and 40%, respectively, when compared to the control group. The TPA-induced sulfhydryl group loss in liver homogenates was attenuated by all the tested antioxidants. Vimang®, mangiferin, vitamin C plus E and β-carotene decreased TPA-induced DNA fragmentation by 46–52, 35, 42 and 17%, respectively, in hepatic tissues, and by 29–34, 22, 41 and 17%, in brain tissues. Similar results were observed in respect to lipid peroxidation in serum, in hepatic mitochondria and microsomes, and in brain homogenate supernatants. Vimang® exhibited a dose-dependent inhibition of TPA-induced biomolecule oxidation and of H_2O_2 production by peritoneal macrophages. The studies demonstrated that Vimang® could be useful to prevent the production of ROS and the oxidative tissue damages in-vivo.

Vimang tablet supplementation in the elderly (>65 years) group increased extracellular superoxide dismutase activity (EC-SOD), activity and serum total antioxidant status (TAS) (Pardo-Andreu et al. 2006b). It also decreased significantly serum thiobarbituric reactive substances and GSH (glutathione) levels. The scientists found that the antioxidant components of the extract could have been utilized by the cells (especially blood and endothelial cells), sparing the intra- and extracellular antioxidant system and increasing serum peroxil scavenging capacity, thus obviating age-associated augmentation in GSH oxidation and lipid peroxidation. The studies demonstrated that Vimang tablets obviated age-associated oxidative stress in elderly humans, which could hinder the onset of age-associated disease, improving the quality of life for elderly persons.

Cardioprotective Activity

Mangiferin was found to have cardioprotective activity (Prabhu et al. 2006). Pre-treatment with mangiferin (100 mg/kg b.w. i.p.) for 28 days prevented these mitochondrial changes, oxidation with energy metabolism and normalised the TCA cycle enzyme activities following isoproterenol ISPH (20 mg/kg b.w.) sub-cutaneous administration twice at an interval of 24 h. The structural integrity of the heart was protected by mangiferin in ISPH administered rats when compared to the untreated controls. The present findings suggested that the prophylactic effect of mangiferin could be attributed to its reducing effect on oxidative damage and activation of mitochondrial energy metabolism.

Recent studies by Daud et al. (2010) showed that mangiferin, and mango extracts rich in mangiferin, increased endothelial cell migration. The dose-effect relationship for various extracts further suggested that this action of mangiferin was regulated by other components present in the extracts. The promigratory effect of mango extracts or mangiferin was unrelated to an effect on cell proliferation, and did not involve alteration in the production of matrix metalloprotease-2 or -9 by the endothelial cells. The results suggested that mangiferin present in mango extracts may have health promoting effects in diseases related to the impaired formation of new blood vessels. Results confirmed that mango extracts contained bioactive molecules capable of regulating endothelial cell migration, an essential step in angiogenesis. The formation of new blood vessels is an important therapeutic target for diseases such as limb ischemia, coronary infarction or stroke.

Gastroprotective/Antiulcerogenic Activity

Studies reported that mangiferin protected against gastric injury induced by ethanol and indomethacin most possibly through the antisecretory and antioxidant mechanisms (Carvalho et al. 2007). Mangiferin (3, 10 and 30 mg/kg, p. o.) significantly attenuated gastric damage induced by ethanol by 30, 35, and 63%, and of indomethacin by 22, 23 and 57%, respectively. Another study reported that mango flowers had gastroprotective and ulcer healing properties (Lima et al. 2006). Oral pre-treatment with mango flower decoction (250, 500 and 1,000 mg/kg) in rats with gastric lesions induced by ethanol, decreased the gastric lesions from 89.0 (control group) to 9.25, 4.50 and 0, respectively. Pre-treatment with the flower decoction (250, 500 and 1,000 mg/kg) to mice with HCl/ethanol- or stress-induced gastric lesions resulted in a dose-dependent significant reduction of lesion index. In the piroxicam-induced gastric lesions, the gastroprotective effect of the flower decoction was reducing with the increasing dose. In the pylorusligature, the flower decoction (p.o.) significantly decreased the acid output indicating the antisecretory property involved in the gastroprotective effect of *M. indica*. Treatment with the flower decoction during 14 consecutive days significantly hastened the healing process in subacute gastric ulcer induced by acetic acid in rats. Phytochemical screening showed the presence of steroids, triterpenes, phenolic compounds and flavonoids. Estimation of the global polyphenol content in the flower decoction was performed by Folin-Ciocalteu method and showed approximately

53% of total phenolic on this extract. These findings indicated the potential gastroprotective and ulcer-healing properties of aqueous decoction of *M. indica* flowers and further supported its popular use in gastrointestinal disorders in Caribbean.

Mango leaves and stem bark have anti-ulcerogenic activities. The administration of leaf decoction up to a dose of 5 g/kg (p.o.) did not produce any signs or symptoms of toxicity in the treated animals, while significantly decreasing the severity of gastric damage induced by several gastroprotective models (Severi et al. 2009). Oral pre-treatment with the mango leaf decoction (250, 500 or 1,000 mg/kg) in mice and rats with gastric lesions induced by HCl/ethanol, absolute ethanol, non-steroidal anti-inflammatory drug (NSAID) or stress-induced gastric lesions resulted in a significant decrease of gastric lesions. Phytochemical analyses of leaf decoction demonstrated the presence of bioactive phenolic compounds amounting to 57.3% of total phenolic content in the extract. Two main phenolic compounds were isolated, mangiferin (C-glucopyranoside of 1,3,6,7-tetrahydroxyxanthone) and C-glucosyl-benzophenone (3-C-β-D-glucopyranosyl-4′,2,4,6-tetrahydroxybenzophenone).

In another study, the ulcer-protective effects of the polyphenols were found to be primarily due to its anti-radical effects (Priya et al. 2009). The polyphenols from the stem bark exhibited good capacity for scavenging nitric oxide (NO) radical and ferric-reducing antioxidant power (FRAP) assay in-vitro. The severity of lesions was found to significantly decrease by the administration of polyphenols in both models. The increased gastric nitrite level of ethanol administered rats were decreased by polyphenol treatment. Administration of polyphenols reduced thiobarbituric acid reactive substance (TBARS) level and enhanced the GSH (glutathione) level significantly in a dose-dependent fashion in both models.

Anti-inflammatory and Antinociceptive Activities

The standard extract of *M. indica* was found to have analgesic and anti-inflammatory activities. Vimang (50–1,000 mg/kg, p.o.) exhibited a potent and dose-dependent antinociceptive effect against acetic acid test in mice (Garrido et al. 2001). The mean potency (ED_{50}) was 54.5 mg/kg and the maximal inhibition attained was 94.4%. Vimang (20–1,000 mg/kg, p.o.) dose-dependently inhibited the second phase of formalin-induced pain but not the first phase. The ED_50 of the second phase was 8.4 mg/kg and the maximal inhibition was 99.5%, being more potent than indomethacin at doses of 20 mg/kg. Vimang (20–1,000 mg/kg, p.o.) significantly inhibited oedema formation of both carrageenan- and formalin-induced oedema in rat, guinea-pigs and mice (maximal inhibitions: 39.5, 45.0 and 48.6, respectively). In subsequent studies, Vimang was confirmed to have anti-inflammatory and antinociceptive (Garrido et al. 2004a; 2006). Vimang administered orally (50–200 mg/kg body weight), reduced ear edema induced by arachidonic acid (AA) and phorbol myristate acetate (PMA) in mice. In the PMA model, the extract also decreased myeloperoxidase (MPO) activity. The extract inhibited the induction of prostaglandin PGE2 and LTB4 with IC_{50} values of 21.7 and 26.0 µg/ml, respectively. This extract p.o. administered also inhibited tumour necrosis factor α (TNFα) serum levels in both models of inflammation. Mangiferin (a glucosylxanthone isolated from the extract) also inhibited these arachidonic acid metabolites (PGE2, IC_{50} value = 17.2 µg/ml and leukotriene B4 (LTB4), IC_{50} value = 2.1 µg/ml). The anti-inflammatory response observed during treatment with the extract was suggested to be associated with inhibition of tumour necrosis factor α (TNFα) and nitric oxide (NO) production. Mangiferin, a main component in the extract, was involved in these effects. In another study, the anti-inflammatory action of Vimang and mangiferin was found to be related with the inhibition of iNOS and cyclooxygenase-2 expression, but not with its effect on vasoconstrictor responses (Beltrán et al. 2004). Vimang (0.5–0.1 mg/ml) and mangiferin (0.025 mg/ml) inhibited the interleukin-1β (1 µg/ml)-induced iNOS expression more in spontaneously hypertensive (SHR) rats than in Wistar Kyoto (WKY) rats, and cyclooxygenase-2 expression more in WKY rats than in SHR rats. Vimang (0.25–1 mg/ml) reduced noradrenaline - and U46619 – but not KCl-induced contractions. Mangiferin (0.05 mg/ml) did not affect noradrenaline-induced contraction.

M. indica stem-bark aqueous extract (MIE, 50–800 mg/kg i.p.) produced dose-dependent and significant analgesic effects against thermally and chemically induced nociceptive pain stimuli in mice (Ojewole 2005). MIE (50–800 mg/kg i.p.) also significantly inhibited fresh egg albumin-induced paw edema, and caused significant hypoglycemic effects in rats. It was suggested that the analgesic effects of MIE (50–800 mg/kg i.p.) may be peripherally and centrally mediated. The different chemical constituents of the plant, especially the polyphenolics, flavonoids, triterpenoids, mangiferin, and other chemical compounds present in the plant may be involved in the observed antiinflammatory, analgesic, and hypoglycemic effects of the plant's extract.

Mango peels contained alky-resourcinols such as 5-(11'Z-heptadecenyl)-resorcinol (1) and 5-(8'Z,11'Z-heptadecadienyl)-resorcinol that were found to exhibit potent cyclooxygenase (COX)-1 and COX-2 inhibitory activity with IC_{50} values ranging from 1.9 (2) to 3.5 μM (1) and from 3.5 (2) to 4.4 (1) μM, respectively, approximating the IC_{50} values of reference drugs (Knödler et al. 2008). 5-Lipoxygenase (5-LOX) catalysed leukotriene formation was only slightly inhibited.

Tawaha et al. (2010) isolated a new trimeric proanthocyanidin, epigallocatechin-3-O-gallat-(4β→8)-epigallocatechin-(4β→8)-catechin (1), together with three known flavan-3-ols, catechin (2), epicatechin (3), and epigallocatechin (4), and three dimeric proanthocyanidins, 5–7, from the air-dried leaves of *Mangifera indica*. Compound 1 was found to have a potent inhibitory effect on COX-2, while compounds 1 and 5–7 exhibited moderate inhibition against COX-1. Pharmacological inhibition of COX could provide relief from the symptoms of inflammation and pain.

Radioprotective Activity

Mangiferin was found to possess radioprotective activity (Jagetia and Baliga 2005). Treatment of mice with different doses of mangiferin, one hour before irradiation reduced the symptoms of radiation sickness and delayed the onset of mortality when compared with the non-drug treated irradiated controls. The radioprotective action of mangiferin increased in a dose dependent manner up to 2 mg/kg and declined thereafter. The highest radioprotective effect was observed at 2 mg/kg mangiferin, with highest number of animals surviving against the radiation-induced mortality. The study showed that mangiferin, protected mice against the radiation-induced sickness and mortality and the optimum protective dose of 2 mg/kg was 1/200 of LD50 dose (400 mg/kg) of mangiferin. Additional studies (Jagetia and Baliga 2005) showed that irradiation of cultured human peripheral blood lymphocytes (HPBLs) with 0, 1, 2, 3, or 4 Gy of γ-radiation evoked a dose-dependent increase in the MNBNC (micronucleated binucleate cells) frequency, while treatment of HPBLs with 50 μg/ml mangiferin 30 minutes before radiation resulted in significant falls in these frequencies. Mangiferin alone did not modify the proliferation index. Irradiation caused a dose-dependent decline in the proliferation index, while treatment of HPBLs with 50 μl/ml mangiferin significantly enhanced the proliferation index in irradiated cells. Mangiferin treatment reduced hydrogen peroxide-induced lipid peroxidation in HPBLs in a concentration-dependent manner.

Transcription Modulatory Activity

Mango components, quercetin and mangiferin and the aglycone derivative of mangiferin, norathyriol, were found to modulate the transactivation of peroxisome proliferator-activated receptor isoforms (PPARs) (Wilkinson et al. 2008). PPARs are nuclear transcription factors involved in lipid metabolism, differentiation, proliferation, cell death, angiogensis, thrombosis, coagulation and inflammation. Through the use of a gene reporter assay it was shown that quercetin inhibited the activation of all three isoforms of PPARs (PPARγ $IC_{50}=56.3$ μM; PPARα $IC_{50}=59.6$ μM; PPAR/β $IC_{50}=76.9$ μM) as did norathyriol (PPARγ $IC_{50}=153.5$ μM; PPARα $IC_{50}=92.8$ μM; PPARfβ

$IC_{50} = 102.4$ µM). In contrast, mangiferin did not inhibit the transactivation of any isoform. These findings suggested that mango components and metabolites may alter transcription and could contribute to positive health benefits via this or similar mechanisms.

Thyroid Stimulatory Activity

Studies showed that administration of fruit peel extracts of *Mangifera indica, Citrullus vulgaris* and *Cucumis melo* to Wistar albino male rats for ten consecutive days significantly increased both the thyroid hormones, serum triiodothyronine and thyroxin with a concomitant decrease in tissue lipid peroxidation (LPO), suggesting their thyroid stimulatory and antiperoxidative role. This thyroid stimulatory nature was also exhibited in propylthiouracil (PTU) induced hypothyroid animals where the three test peel extracts appeared to be stimulatory to thyroid functions and inhibitory to tissue LPO but only when treated individually (Parmar and Kar 2009).

P-Glycoprotein (P-gp) Inhibition Activity

Recent studies showed that the mango stem bark aqueous extract (MSBE) and related phenols inhibited P-glycoprotein (P-gp) activity in the HK-2 proximal tubule cell line. P-glycoprotein is a cell membrane-associated protein that transports a variety of drug substrates. P-gp is one of the main reasons cancer cells are resistant to chemotherapy drugs. All investigated compounds except for catechin and gallic acid inhibited P-gp activity in HK-2 cells, in the order of mangiferin < norathyriol < quercetin < MSBE. MSBE, quercetin and norathyriol also inhibited significantly esterase activity. Similar effects were obtained in resistant Caco-2/VCR cells. Studies demonstrated that *M. indica* and derived polyphenols may affect the activity of the multidrug transporter P-gp ABCB1.

Neuroprotective and Neurological Activity

Mango extracts were found to have neuroprotective activity (Lemus-Molina et al. 2009). Mango extracts (MiE) was recently reported to prevent glutamate-induced excitotoxicity in primary cultured neurons of the rat cerebral cortex. Maximal protection (56%) was obtained with 2.5 µg/ml of MiE. Mangiferin, an antioxidant polyphenol and a major component of MiE, was also effective in preventing neuronal death, oxidative stress and mitochondrial depolarization. Maximal protection (64%) was obtained at 12.5 µg/ml of mangiferin which also lessened oxidative stress and mitochondrial depolarization at the neuroprotective concentrations. The results indicated (i) MiE to be an efficient neuroprotector of excitotoxic neuronal death, (ii) mangiferin contributed substantially to the antioxidant and neuroprotective activity of the mango extract, and (iii) this natural extract had therapeutic potential to treat neurodegenerative disorders.

Studies showed that orally administered *Mangifera indica* extract, QF808 in the gerbil hippocampal CA1 sector was absorbed across the blood-brain barrier and reduced neuronal death of the hippocampal CA1 area after ischaemia-reperfusion (Martínez Sánchez et al. 2001). These protective effects were attributed to the antioxidant activity of QF808.

Systemic administration of Vimang was found to have effects on behavioural outcomes of neurological function in rats (Preissler et al. 2009). A single oral administration of Vimang produced an impairment of short- and long-term retention of memory for aversive training when given either 1 hour pre-training or immediately post-training, but not 8 hour post-training. Vimang did not affect open field behaviour or habituation. The results indicated that Vimang might induce deficits of emotionally motivated memory without affecting non-associative memory, locomotion, exploratory behaviour or anxiety.

Immunomodulatory Activity

The aqueous stem bark extract of *Mangifera indica* (Vimang) was found to protect T cells from activation-induced cell death (AICD) (Hernández et al. 2006). The extract reduced anti-CD3-induced accumulation of reactive oxygen species (ROS) and intracellular free Ca2+ and consequently, down-modulated CD95L mRNA expression and CD95-mediated AICD. *M. indica* extract enhanced T-cell survival by inhibiting AICD, a finding associated with a decrease in oxidative stress generated through the TCR (T cell receptor) signalling pathway in activated T cells.

Administration of alcoholic extract of mango stem bark containing 2.6% mangiferin augmented humoral antibody (HA) titre and delayed type hypersensitivity (DTH) in mice (Makare et al. 2001). The study indicated the mango bark extract to be a promising drug with immunostimulant properties.

Separate studies showed that components of Vimang, including the polyphenol mangiferin, had depressor effects on the phagocytic and ROS production activities of rat macrophages and, thus, may have potential in the treatment of immunopathological diseases characterized by the hyperactivation of phagocytic cells such as certain autoimmune disorders (García et al. 2002). Another study showed that orally administered Vimang and mangiferin (the major polyphenol present in Vimang) regulated mouse antibody responses induced by inoculation of microsporidian parasites spores (García et al. 2003b). Inoculation with spores caused splenomegaly, which was significantly attenuated by Vimang and significantly augmented by mangiferin. Vimang significantly inhibited IgG production, but had no effect on IgM. Mangiferin did not affect either IgM or IgG2a, but significantly augmented production of IgG1 and IgG2b. These results suggested that components of *Mangifera indica* extracts may be of potential value for modulating the humoral response in different immunopathological disorders.

Antiviral Activity

Zheng and Lu (1990) found that mangiferin and isomangiferin exhibited antiviralactivity against typeIherpessimplex virus (HSV-I). Isomangiferin performed better than such control drugs as acyclovir, idoxuridine, and cyclocytidine in HSV-1 inhibition in log determination by 0.27–0.50, but mangiferin was lower than isomangiferin in log by 0.53. The average plaque reduction rates of mangiferin and isomangiferin were 56.8% and 69.5% respectively. The antiviral effect of mangiferin and isomangiferin was postulated to be due to their capability in inhibiting viral replication within cells.

Mangiferin, extracted from mango leaves exhibited antiviral activity against herpes simplex virus type 2 (HSV-2) in-vitro with EC_{50} of 111 μg/ml, against HSV-2 plaque formation in HeLa cells (Zhu et al. 1993). At concentrations of 33 and 80 μg/ml, it reduced the virus replicative yields by 90% (EC_{90}) and 99% (EC_{99}), respectively. The therapeutic index (IC_{50}/EC_{50}) was 8.1. Mangiferin did not directly inactivate HSV-2 but the results of the drug addition and removal tests suggested that mangiferin inhibited the late event in HSV-2 replication.

Antimicrobial Activity

Mango seed kernel, leaves and stem exhibited antimicrobial activity. Leaf extracts of *M. indica* possessed some antibacterial activity against *Staphylococcus aureus, Esherichia coli* and *Pseudomonas aeruginosa* (Bbosa et al. 2007a,b). The extracts displayed weak antibacterial activity against the tested organisms compared with the positive control (gentamycin). The ethanolic extract was most active with minimum inhibitory concentration ranging from 5481.0 to 43750.0 μg/ml. Chemical tests showed the presence of saponins, steroids and triterpenoids in the ether fraction, alkaloids, anthracenocides, coumarins, flavonones, reducing sugars, catechol and gallic tannins, saponins, steroids and triterpenoids in the ethanolic portion and anthracenocides, flavonones, reducing sugars in the aqueous fraction of the leaf extract. Mangiferin in polyethylene glycol-400 solution exhibited antimicrobial activity against seven bacteria: *Bacillus pumilus, Bacillus cereus, Staphylococcus aureus, Staphylococcus citreus, Escherichia coli, Salmonella agona, Klebsiella pneumoniae* and 5 fungal species: *Saccharomyces cerevisiae, Thermoascus aurantiacus, Trichoderma reesei, Aspergillus flavus* and *Aspergillus fumigatus* (Stoilova et al. 2005). Both the ethereal and ethanolic fractions of mango leaf extract showed antibacterial activity against *Clostridium tetani* activity with an MIC of 6.25 and 12.5 mg/ml, respectively (Bbosa et al. 2007a, b).

Mango extract prepared from sun-dried and ground mango stick, at 50% concentration, showed maximum zone of inhibition on *Streptococcus mitis* (Prashant et al. 2007). Neem extract similarly prepared from neem stick, produced the maximum zone of inhibition on *Streptococcus mutans* at 50% concentration. Even at 5% concentration neem extract showed some inhibition of growth for all the four species of organisms. The study showed that a combination of neem and mango chewing sticks may provide the maximum benefit in preventing the development of dental caries.

Crude mango stem extract at a concentration of 50 mg/ml inhibited the growth of all the respiratory tract pathogens, *Escherichia coli, Staphylococcus aureus, Pseudomonas aeruginosa, Klebsiella pneumonia* and *Streptococcus pneumonia*, though with varying degrees of susceptibility depending on the bacterium and the extracting solvent. The MIC values ranged from 12.5 to 200 mg/ml, while the MBC values ranged from 50 to 400 mg/ml (El-Mahmood 2009).

Three pure hydrolyzable gallotannins, penta-, hexa-, and hepta-O-galloylglucose isolated from mango kernel exhibited antibacterial activity (Engels et al. 2009). Although the growth of lactic acid bacteria was not inhibited, the proliferation of Gram-positive food spoilage bacteria was prevented and the growth of Gram-negative *Escherichia coli* was reduced. As bacterial growth could be restored by the addition of iron to the medium, this study strongly supported the view that the inhibitory effects of hydrolyzable tannins were due to their iron-complexing properties. In subsequent studies, gallotannins (tetra-O-galloylglucose to deca-O-galloylglucose) fractionated from mango kernels by high-speed counter-current chromatography exhibited minimum inhibitory concentration against *Bacillus subtilis* in the range of 0.05–0.1 g/l in Luria-Bertani broth but up to 20 times higher in media containing more iron and divalent cations (Engels et al. 2010).

Antiprotozoal Activity

The mango stem extract exhibited a schizontocidal effect during early infection, on *Plasmodium yoelii nigeriensis* and also demonstrated repository activity (Awe et al. 1998). A reduction in yeast-induced hyperpyrexia was also produced by the extract. Bidla et al. (2004) reported that mango leaf extract at a concentration of 40 μg/ml exhibited inhibitory effect on the on the growth of *Plasmodium falciparum* with 50.4% inhibition. Mangiferin at 100 mg/kg/die had a significant anticryptosporidial activity and that its activity was similar to that produced by the same dose (100 mg/kg/die) of paromomycin. However, both mangiferin and paromomycin were not able to completely inhibit intestinal colonization of *Cryptosporidium parvum* but only to reduce it (Perrucci et al. 2006).

Antidiarrhoeal Activity

Mango seed exhibited anti-diarrhoeal activity (Sairam et al. 2003). Methanolic (MMI) and aqueous (AMI) extracts of seeds administered orally in the dose of 250 mg/kg to mice showed significant anti-diarrhoeal activity comparable with that of the standard drug loperamide. However, only the methanol extract markedly decreased intestinal transit in charcoal meal test in comparison to atropine sulphate (5 mg/kg; im). Both inhibited growth of *Streptococcus aureus* and *Proteus vulgaris*, and did not show any significant effect on growth of *Escherichia coli* and *Klebsiella*. The results illustrated that the extracts of *M. indica* had significant antidiarrhoeal activity and part of the activity of MMI may be attributed to its effect on intestinal transit.

Antityrosinase Activity

The alcoholic extract from seed kernels of Thai mango (*Mangifera indica*. cv. 'Fahlun') and its major phenolic principle (pentagalloylglucopyranose) exhibited potent, dose-dependent inhibitory effects on tyrosinase with respect to L-DOPA (L-3,4-dihydroxyphenylalanine) (Nithitanakool et al. 2009a,b). The scientists postulated a possible mechanism for their anti-tyrosinase activity which may involve an ability to chelate the copper atoms which were required for the catalytic activity of tyrosinase.

Anthelmintic, Antiallergic Activities

Treatment of mice infected experimentally with the nematode, *Trichinella spiralis* with Vimang or mangiferin (500 or 50 mg per kg body weight per day, respectively) throughout the parasite life cycle led to a significant fall in the number of parasite larvae encysted in the musculature; however, neither treatment was effective against adults in the gut (García et al. 2003a). Treatment with Vimang or mangiferin likewise evoked a significant decrease in serum levels of specific anti-*Trichinell*a IgE, throughout the parasite life cycle. Oral treatment of rats with Vimang or mangiferin, daily for 50 days, inhibited mast cell degranulation as evaluated by the passive cutaneous anaphylaxis test (sensitization with infected mouse serum with a high IgE titre, then stimulation with the cytosolic fraction of *Trichenella spiralis* muscle larvae). Since IgE played a key role in the pathogenesis of allergic diseases, these results suggested that Vimang and mangiferin may be useful in the treatment of such diseases.

Antivenin Activity

Mango seed kernel was found to have potent antivenin activity(Pithayanukul et al. 2009). Ethanol seed kernel extract and its major phenolic principle (pentagalloylglucopyranose) potently exhibited dose-dependent inhibitory effects on the caseinolytic and fibrinogenolytic activities of Malayan pit viper and Thai cobra venoms in in-vitro tests. Molecular docking studies revealed the phenolic principles could form hydrogen bonds with the three histidine residues and could chelate the Zn^{2+} atom of the snake venom metalloproteinases (SVMPs), which could potentially result in inhibition of the venom enzymatic activities and thereby inhibit tissue necrosis. The ethanolic extract and PGE exhibited dose-dependent inhibitory effects on enzymatic activities of phospholipase A(2), hyaluronidase and L-amino acid oxidase of *Calloselasma rhodostoma* and *Naja naja kaouthia* (NK) venoms in in-vitro tests (Leanpolchareanchai et al. 2009). The anti-hemorrhagic and anti-dermonecrotic activities of mango seed kernel extract against both venoms were clearly supported by in-vivo tests.

Toxicology, Genotoxicity and Cytoxicity Studies

Mango stem bark aqueous extract (MSBE) known by the brand name of Vimang® is sold in various formulations (e.g., tablets, capsules, syrup, vaginal oval, and suppositories) in Cuba. In view of the ethnomedical, preclinical, and clinical uses of this extract and the necessity to assess its possible toxicological effect on man, a toxicological analysis of a standard extract was conducted in Cuba (Garrido et al. 2009). Acute toxicity was evaluated in mice and rats by oral, dermal, and intraperitoneal (i.p.) administration. The extract, by oral or dermal administration, showed no lethality at the limit doses of 2,000 mg/kg body weight and no adverse effects were found. Deaths occurred with the i.p. administration at 200, but not 20 mg/kg in mice. MSBE was also studied on irritant tests in rabbits, and the results showed that it was non-irritating on skin, ocular, or rectal mucosa. The extract had minimal irritancy following vaginal application. In another study, Vimang extract showed evidences of slight cytotoxic activity but did not induce a mutagenic or genotoxic effects in a range of assays (Ames, Comet and micronucleus) employed (Rodeiro et al. 2006). The observed results affirmed that Vimang (200–5,000 μg/plate) did not increase the frequency of reverse mutations in the Ames test. Results of Comet assay showed that the extract did not induce single strand breaks or alkali-labile sites on blood peripheral lymphocytes of treated animals compared with controls. In contrast, the results of the micronucleus studies (in-vitro and in-vivo) showed Vimang induced cytotoxic activity, determined as cell viability or PCE/NCE ratio, but neither increased the frequency of micronucleated binucleate cells in culture of human lymphocytes nor in mice bone marrow cells under the experimental conditions.

Mango Dermatitis

Mango has been reported to contain allergens, believed to be alk(en)yl catechols and/or alk(en)yl resorcinols which cause dermatitis on sensitive people. Three resorcinol derivatives: heptadecadienylresorcinol, heptadecenylresorcinol and pentadecylresorcinol; collectively named 'mangol' were shown to elicit positive patch test reactions in mango-sensitive patients (Oka et al. 2004). Depending on cultivar, alk(en)ylresorcinol contents varied from 79.3 to 1850.5 mg/kg of dry matter (DM) in mango peels and from 4.9 to 187.3 mg/kg of DM in samples of mango pulp (Knödler et al. 2009). The profile of alk(en)ylresorcinols was found to be highly distinct, with an average homologue composition of C15:0 (6.1%), C15:1 (1.7%), C17:0 (1.1%), C17:1 (52.5%), C17:2 (33.4%), C17:3 (2.4%), C19:1 (2.1%), and C19:2 (0.8%). Mango puree samples prepared from peeled and unpeeled fruits revealed contents of 3.8 and 12.3 mg/kg of fresh weight, respectively. Content and homologue composition were not significantly affected during puree processing and thermal preservation. In nectar samples prepared from peeled and unpeeled fruits, contents of 1.4 and 4.6 mg/l, respectively, were found.

Traditional Medicinal Uses

As many as there are scientific studies on the pharmacological attributes of mango fruit and plant parts, likewise there have been many medicinal uses of mango plant parts in traditional folkloric medicine.

In the Philippines, a decoction of the root is considered diuretic, the bark and seeds are astringent and resin from the tree is used as a cure for aphthae. In Kampuchea, the bark is astringent and used in hot lotions for rheumatism and leucorrhoea. A solution of the gum obtained from the bark is swallowed for dysentery both in Kampuchea and in India. This gum resin mixed with coconut oil is applied to scabies and other parasitic diseases of the skin. A fluid extract, or an infusion, is used in menorrhagia, leucorrhoea, bleeding piles and in cases of haemorrhage from the lungs and for nasal catarrh. The gum resin of the bark and of the fruit is used as sudorific, and is an effective antisyphilitic. The root bark is a bitter aromatic and is used in India for diarrhoea and leucorrhoea. In Samoa, a bark infusion has been a traditional remedy for mouth infections in children (*pala gutu*).

The tender leaves dried and made into a powder are said to have medicinal uses such as aiding early stages of diabetes. The juice of the leaf is useful in bleeding dysentery. A decoction of the leaves with a little honey added is given for loss of voice. An infusion of the young leaves is prescribed for chronic diseases of the lungs, and for asthma and coughs. The young leaves are used as a pectoral in Sind. The ashes of the leaves are a popular remedy for burns and scalds; infusions of leaves of mango, the orange (*Citrus sinensis*), and other species are used to make a potion to treat relapse sickness (kita). In the Philippines, the leaves are prepared as tea.

The dried mango flowers, in decoction or powder, are used in diarrhoea, chronic dysentery and gleet. In India, a drink made from unripe mango fruit is used as a remedy for exhaustion and heat stroke. Half-ripe fruit eaten with salt and honey is used for a treatment of gastro-intestinal disorders, bilious disorders, blood disorders, and scurvy. Pickled mango is much used on account of its stomachic and appetizing qualities. Peeled and dried mango slices are used as antiscorbutic. Conversely, unripe fruit contains a lot of the toxic sap that when eaten in excess can cause throat irritation, indigestion, dysentery, and colic. Ripe mangos and juices are a rich source of vitamin A, and are used to treat vitamin A deficiencies such as night blindness and useful in nervous and atomic dyspepsia and constipation. Resin from the fruit is commonly regarded in India as a cure for scabies and other cutaneous affections.

The seed is medicinal in India, China, and Malaysia. It is bitter and acts as a vermifuge. It is also astringent, and in powdered form is given in obstinate diarrhoea, chronic dysentery, asthma and for bleeding piles. The kernel or stone from green mango is considered as an anthelmintic. The seed has been used in Johore, Malaysia for round worms, menorrhagia, and piles. In Pakistan, mango seeds are used traditionally for their antispasmodic, antiscorbutic, astringent and anthelmintic properties, and the flowers provide a fumigant against mosquitoes (Khan and Khan 1989).

Other Uses

Mangos trees are often planted as shelter and shade trees in villages, parks and home gardens. They are sometimes used in mixed-species windbreaks and as living fence posts. Mango blossoms are a rich source of nectar for honey bees. Livestock will graze on mango leaves and eat fallen fruit. Leaves have been fed to rabbits. Seeds and by-products of processed fruit have been used to feed cattle, poultry, and pigs. The bark yields a resinous gum consisting of 75% resin and 15% gum and provides a yellowish-brown dye used for silk. Properly seasoned mango timber has been used in furniture, for carving, wall and floor panelling, and utensil manufacture. Mango timber decays rapidly if exposed to the elements without preservation treatment. In French Oceania and the Cook Islands, mango wood is used for canoe and boat construction. Mango wood makes excellent charcoal.

Comments

Hundreds to thousands of mango varieties have been developed in the world. Each mango growing country has their unique range of cultivars and also cultivates common commercial cultivars from other countries. Some new and common popular mango varieties in Australia and southeast Asia are shown in Plates 1–15.

Selected References

Aderibigbe AO, Emudianughe TS, Lawal BA (1999) Antihyperglycaemic effect of *Mangifera indica* in rat. Phytother Res 13(6):504–507

Agarwala S, Nageshwar Rao B, Mudholkar K, Bhuwania R, Satish Rao BS (2011) Mangiferin, a dietary xanthone protects against mercury-induced toxicity in HepG2 cells. Environ Toxicol 26:n/a, 10.1002/tox–20620

Ajila CM, Prasada Rao UJ (2008) Protection against hydrogen peroxide induced oxidative damage in rat erythrocytes by *Mangifera indica* L. peel extract. Food Chem Toxicol 46(1):303–309

Ajila CM, Naidu KA, Bhat SG, Prasada Rao UJS (2007) Bioactive compounds and antioxidant potential of mango peel extract. Food Chem 105(3):982–988

Akila M, Devaraj H (2008) Synergistic effect of tincture of Crataegus and *Mangifera indica* L. extract on hyperlipidemic and antioxidant status in atherogenic rats. Vasc Pharmacol 49(4–6):173–177

Anila L, Vijayalakshmi NR (2002) Flavonoids from *Emblica officinalis* and *Mangifera indica*-effectiveness for dyslipidemia. J Ethnopharmacol 79(1):81–87

Awe SO, Olajide OA, Oladiran OO, Makinde JM (1998) Antiplasmodial and antipyretic screening of *Mangifera indica* extract. Phytother Res 12:437–438

Backer CA, van den Brink RCB Jr (1965) Flora of Java, (spermatophytes only), vol 2. Wolters-Noordhoff, Groningen, 641 pp

Bally ISE (2006) *Mangifera indica* (mango), ver. 3.1. In: Elevitch CR (ed) Species profiles for Pacific Island Agroforestry. Permanent Agriculture Resources (PAR), Holualoa. http://www.traditionaltree.org

Barreto JC, Trevisan MT, Hull WE, Erben G, de Brito ES, Pfundstein B, Würtele G, Spiegelhalder B, Owen RW (2008) Characterization and quantitation of polyphenolic compounds in bark, kernel, leaves, and peel of mango (*Mangifera indica* L.). J Agric Food Chem 56(14):5599–5610

Bbosa GS, Kyegombe DB, Ogwal-Okeng J, Bukenya-Ziraba R, Odyek O, Waako P (2007a) Antibacterial activity of *Mangifera indica* (L.). Afr J Ecol 45(Supplement 1):13–16

Bbosa GS, Lubega A, Musisi N, Kyegombe DB, Waako P, Ogwal-Okeng J, Odyek O (2007b) The activity of *Mangifera indica* leaf extracts against the tetanus causing bacterium, *Clostridium tetani*. Afr J Ecol 45:54–58

Beltrán AE, Alvarez Y, Xavier FE, Hernanz R, Rodriguez J, Núñez AJ, Alonso MJ, Salaices M (2004) Vascular effects of the *Mangifera indica* L. extract (Vimang). Eur J Pharmacol 499(3):297–305

Berardini N, Carle R, Schieber A (2004) Characterization of gallotannins and benzophenone derivatives from mango (*Mangifera indica* L. cv. 'Tommy Atkins') peels, pulp and kernels by high-performance liquid chromatography/electrospray ionization mass spectrometry. Rapid Commun Mass Spectrom 18(19):2208–2216

Berardini N, Fezer R, Conrad J, Beifuss U, Carle R, Schieber A (2005) Screening of mango (*Mangifera indica* L.) cultivars for their contents of flavonol O- and xanthone C-glycosides, anthocyanins and pectin. J Agric Food Chem 53(5):1563–1570

Bhowmik A, Khan LA, Akhter M, Rokeya B (2009) Studies on the antidiabetic effects of *Mangifera indica* stem-barks and leaves on nondiabetic, type 1 and type 2 diabetic model rats. Bangladesh J Pharmacol 4:110–114

Bidla G, Titanji VP, Jako B, Bolad A, Berzins K (2004) Antiplasmodial activity of seven plants used in African folk medicine. Indian J Pharmacol 36:245–246

Bompard JM (1993) The genus *Mangifera* re-discovered: the potential contribution of wild species to mango cultivation. Acta Hor 341:69–77

Burkill IH (1966) A dictionary of the economic products of the Malay Peninsula. Revised reprint, 2 vols. Ministry of Agriculture and Co-operatives, Kuala Lumpur, vol 1 (A–H), pp 1–1240, vol 2 (I–Z), pp 1241–2444

Campbell RJ, Campbell CW (1994) A guide to mangos in Florida. Fairchild Tropical Garden, Miami, 227 pp

Carvalho AC, Guedes MM, de Souza AL, Trevisan MT, Lima AF, Santos FA, Rao VS (2007) Gastroprotective effect of mangiferin, a xanthonoid from *Mangifera indica*, against gastric injury induced by ethanol and indomethacin in rodents. Planta Med 73(13): 1372–1376

Chen JP, Tai CY, Chen BH (2004) Improved liquid chromatographic method for determination of carotenoids in Taiwanese mango (*Mangifera indica* L.). J Chromatogr A 1054(1–2):261–268

Chieli E, Romiti N, Rodeiro I, Garrido G (2009) In vitro effects of *Mangifera indica* and polyphenols derived on ABCB1/P-glycoprotein activity. Food Chem Toxicol 47(11):2703–2710

Daud NH, Aung CS, Hewavitharana AK, Wilkinson AS, Pierson JT, Roberts-Thomson SJ, Shaw PN, Monteith GR, Gidley MJ, Parat MO (2010) Mango extracts and the mango component mangiferin promote endothelial cell migration. J Agric Food Chem 58(8):5181–5186

de Ornelas-Paz J, Yahia EM, Gardea-Bejar A (2007) Identification and quantification of xanthophyll esters, carotenes, and tocopherols in the fruit of seven Mexican mango cultivars by liquid chromatography-atmospheric pressure chemical ionization-time-of-flight mass spectrometry [LC-(APcI(+))-MS]. J Agric Food Chem 55(16):6628–6635

Ding Hou (1978) Anacardiaceae. In: van Steenis CGGJ (ed) Flora Malesiana, Series I, vol 8. Sijthoff & Noordhoff, Alphen aan den Rijn, pp 395–548

El-Mahmood MA (2009) Antibacterial efficacy of stem bark extracts of *Mangifera indica* against some bacteria associated with respiratory tract infections. Sci Res Essays 4(10):1031–1037

Engels C, Knödler M, Zhao YY, Carle R, Gänzle MG, Schieber A (2009) Antimicrobial activity of gallotannins isolated from mango (*Mangifera indica* L.) kernels. J Agric Food Chem 57(17):7712–7718

Engels C, Gänzle MG, Schieber A (2010) Fractionation of Gallotannins from mango (*Mangifera indica* L.) kernels by high-speed counter-current chromatography and determination of their antibacterial activity. J Agric Food Chem 58(2):775–780

Foundation for Revitalisation of Local Health Traditions (2008) FRLHT Database. htttp://envis.frlht.org

García D, Delgado R, Ubeira FM, Leiro J (2002) Modulation of rat macrophage function by the *Mangifera indica* L. extracts Vimang and mangiferin. Int Immunopharmacol 2(6):797–806

García D, Escalante M, Delgado R, Ubeira FM, Leiro J (2003a) Anthelminthic and antiallergic activities of *Mangifera indica* L. stem bark components Vimang and mangiferin. Phytother Res 17(10):1203–1208

García D, Leiro J, Delgado R, Sanmartín ML, Ubeira FM (2003b) *Mangifera indica* L. extract (Vimang) and mangiferin modulate mouse humoral immune responses. Phytother Res 17(10):1182–1187

Garrido G, González D, Delporte C, Backhouse N, Quintero G, Núñez-Sellés AJ, Morales MA (2001) Analgesic and anti-inflammatory effects of *Mangifera indica* L. extract (Vimang). Phytother Res 15(1):18–21

Garrido G, Delgado R, Lemus Y, Rodríguez J, García D, Núñez-Sellés AJ (2004a) Protection against septic shock and suppression of tumor necrosis factor α and nitric oxide production on macrophages and microglia by a standard aqueous extract of *Mangifera indica* L. (VIMANG). Role of mangiferin isolated from the extract. Pharmacol Res 50(2):165–172

Garrido G, González D, Lemus Y, García D, Lodeiro L, Quintero G, Delporte C, Núñez-Sellés AJ, Delgado R (2004b) In vivo and in vitro anti-inflammatory activity of *Mangifera indica* L. extract (VIMANG). Pharmacol Res 50(2):143–149

Garrido G, González D, Lemus Y, Delporte C, Delgado R (2006) Protective effects of a standard extract of

Mangifera indica L. (VIMANG) against mouse ear edemas and its inhibition of eicosanoid production in J774 murine macrophages. Phytomed 13(6):412–418

Garrido G, González D, Romay C, Núñez-Sellés AJ, Delgado R (2008) Scavenger effect of a mango (*Mangifera indica* L.) food supplement's active ingredient on free radicals produced by human polymorphonuclear cells and hypoxanthine–xanthine oxidase chemiluminescence systems. Food Chem 107(3): 1008–1014

Garrido G, Rodeiro I, Hernández I, García G, Pérez G, Merino N, Núñez-Sellés A, Delgado R (2009) In vivo acute toxicological studies of an antioxidant extract from *Mangifera indica* L. (Vimang). Drug Chem Toxicol 31(1):53–58

Gorinstein S, Haruenkit R, Poovarodom S, Vearasilp S, Ruamsuke P, Namiesnik J, Leontowicz M, Leontowicz H, Suhaj M, Sheng GP (2010) Some analytical assays for the determination of bioactivity of exotic fruits. Phytochem Anal 21(4):355–362

Guha S, Ghosal S, Chattopadyay U (1996) Antitumor, immunomodulatory and anti-HIV effect of mangiferin: a naturally occurring glucosylxanthone. Chemother 42(6):443–451

Hernández P, Delgado R, Walczak H (2006) Mangifera indica L. extract protects T cells from activation-induced cell death. Int Immunopharmacol 6(9):1496–1505

Idstein H, Bauer C, Schreier P (1985) Volatile acids in tropical fruits: cherimoya (*Annona cherimolia*, Mill.), guava (*Psidium guajava*, L.), mango (*Mangifera indica*, L., var. Alphonso), papaya (*Carica papaya*, L.). Z Lebensm Unters Forsch 180(5):394–397

Jagetia GC, Baliga MS (2005) Radioprotection by mangiferin in DBAxC57BL mice: a preliminary study. Phytomed 12(3):209–215

Khan MA, Khan MN (1989) Alkyl gallates of flowers of *Mangifera Indica*. Fitoterapia 60(3):284

Khan MN, Nizami SS, Khan MA, Ahmed Z (1993) New saponins from *Mangifera indica*. J Nat Prod 56:767–770

Knödler M, Conrad J, Wenzig EM, Bauer R, Lacorn M, Beifuss U, Carle R, Schieber A (2008) Anti-inflammatory 5-(11'Z-heptadecenyl)- and 5-(8'Z,11'Z-heptadecadienyl)-resorcinols from mango (*Mangifera indica* L.) peels. Phytochem 69(4):988–993

Knödler M, Reisenhauer K, Schieber A, Carle R (2009) Quantitative determination of allergenic 5-alk(en)ylresorcinols in mango (*Mangifera indica* L.) peel, pulp, and fruit products by high-performance liquid chromatography. J Agric Food Chem 57(9):3639–3644

Knφdler M, Berardini N, Kammerer DR, Carle R, Schieber A (2007) Characterization of major and minor alk(en)ylresorcinols from mango (*Mangifera Indica* L.) peels by high-performance liquid chromatography/atmospheric pressure chemical ionization mass spectrometry. Rapid Commun Mass Spectrom 21:945–951

Koestermans AJGH, Bompard JM (eds) (1993) The mangoes: their botany, nomenclature, horticulture and utilization. Academic, London, 352 pp

Larrauri JA, Rupérez P, Saura-Calixto F (1997) Mango peel fibres with antioxidant activity. Z Lebensm Unters Forsch A 205(1):39–72

Leanpolchareanchai J, Pithayanukul P, Bavovada R, Saparpakorn P (2009) Molecular docking studies and anti-enzymatic activities of Thai mango seed kernel extract against snake venoms. Molecules 14(4):1404–1422

Lemus-Molina Y, Sánchez-Gómez MV, Delgado-Hernández R, Matute C (2009) *Mangifera indica* L. extract attenuates glutamate-induced neurotoxicity on rat cortical neurons. Neurotoxicol 30(6):1053–1058

Li X, Cui X, Sun X, Li X, Zhu Q, Li W (2010) Mangiferin prevents diabetic nephropathy progression in streptozotocin-induced diabetic rats. Phytother Res 24(6):893–899

Lim TK, Khoo KC (1985) Diseases and disorders of mango in Malaysia. Tropical Press, Kuala Lumpur, 101 pp

Lima ZP, Severi JA, Pellizzon CH, Brito AR, Solis PN, Cáceres A, Girón LM, Vilegas W, Hiruma-Lima CA (2006) Can the aqueous decoction of mango flowers be used as an antiulcer agent? J Ethnopharmacol 106(1):29–37

Litz RE (ed) (1997) The mango: botany, production and uses, 1st edn. CAB International, Wallingford, 587 pp

Makare N, Bodhankar S, Rangari V (2001) Immunomodulatory activity of alcoholic extract of *Mangifera indica* L. in mice. J Ethnopharmacol 78(2–3): 133–137

Martin FW, Ruberté RM (1975) Edible leaves of the tropics. Agency for International Development Department of State, and the Agricultural Research Service, U.S. Department of Agriculture, Washington, DC, 235 pp

Martínez Sánchez G, Candelario-Jalil E, Giuliani A, León OS, Sam S, Delgado R, Núñez Sellés AJ (2001) *Mangifera indica* L. extract (QF808) reduces ischaemia-induced neuronal loss and oxidative damage in the gerbil brain. Free Radic Res 35(5):465–473

Martinez G, Delgado R, Perez G, Garrido G, Nunez Selles AJ, Leon OS (2000) Evaluation of the *in-vitro* antioxidant activity of *Mangifera indica* L: extract (Vimang). Phytother Res 14:424–427

Min T, Barfod A (2008) Anacardicaceae. In: Wu ZY, Raven PH, Hong DY (eds) Flora of China, vol 11, Oxalidaceae through Aceraceae. Science Press/Missouri Botanical Garden Press, Beijing/St. Louis

Morsi RMY, El-Tahan NR, El-Hadad AMA (2010) Effect of aqueous extract *Mangifera indica* leaves, as functional foods. J Appl Sci Res 6(6):712–721

Morton JF (1987) Mango. In: Fruits of warm climates. Julia Morton, Miami, pp 221–237

Muanza DN, Euler KL, Williams L, Newman DJ (1995) Screening for antitumor and anti-HIV activities of nine medicinal plants from Zaire. Int J Pharmacol 33(2):98–106

Mukherji S (1949) A monograph on the genus *Mangifera* L. Lloydia 12:73–136

Muruganandan S, Gupta S, Kataria M, Lal J, Gupta PK (2002) Mangiferin protects the streptozotocin-induced oxidative damage to cardiac and renal tissues in rats. Toxicol 176(3):165–173

Muruganandan S, Srinivasan K, Gupta S, Gupta PK, Lal J (2005) Effect of mangiferin on hyperglycemia and atherogenicity in streptozotocin diabetic rats. J Ethnopharmacol 97(3):497–501

Nigam N, Prasad S, Shukla Y (2007) Preventive effects of lupeol on DMBA induced DNA alkylation damage in mouse skin. Food Chem Toxicol 45(11):2331–2335

Nithitanakool S, Pithayanukul P, Bavovada R (2009a) Antioxidant and hepatoprotective activities of Thai mango seed kernel extract. Planta Med 75(10):1118–1123

Nithitanakool S, Pithayanukul P, Bavovada R, Saparpakorn P (2009b) Molecular docking studies and anti-tyrosinase activity of Thai mango seed kernel extract. Molecules 14(1):257–265

Noratto GD, Bertoldi MC, Krenek K, Talcott ST, Stringheta PC, Mertens-Talcott SU (2010) Anticarcinogenic effects of polyphenolics from mango (*Mangifera indica*) varieties. J Agric Food Chem 58(7):4104–4112

Núñez Sellés AJ, Vélez Castro HT, Agüero-Agüero J, González-González J, Naddeo F, De Simone F, Rastrelli L (2002) Isolation and quantitative analysis of phenolic antioxidants, free sugars, and polyols from Mango (*Mangifera Indica* L.) stem bark aqueous decoction used in Cuba as a nutritional supplement. J Agric Food Chem 50(4):762–766

Ochse JJ, van den Brink RCB (1931) Fruits and fruitculture in the Dutch East Indies. G. Kolff & Co., Batavia, 180 pp

Ochse JJ, van den Brink RCB (1980) Vegetables of the Dutch Indies, 3rd edn. Ascher & Co., Amsterdam, 1016 pp

Ojewole JA (2005) Antiinflammatory, analgesic and hypoglycemic effects of *Mangifera indica* Linn. (Anacardiaceae) stem-bark aqueous extract. Methods Find Exp Clin Pharmacol 27(8):547–554

Oka K, Saito F, Yasuhara T, Sugimoto A (2004) A study of cross-reactions between mango contact allergens and urushiol. Contact Dermat 51(5–6):292–296

Pacific Island Ecosystems at Risk (PIER) (2004) *Mangifera indica* L., Anacardiaceae. http://www.hear.org/pier/species/mangifera_indica.htm

Pardo-Andreu GL, Delgado R, Velho JA, Curti C, Vercesi AE (2005) Mangiferin, a natural occurring glucosyl xanthone, increases susceptibility of rat liver mitochondria to calcium-induced permeability transition. Arch Biochem Biophys 439(2):184–193

Pardo-Andreu GL, Delgado R, Núñez-Sellés AJ, Vercesi AE (2006a) *Mangifera indica* L. extract (Vimang) inhibits 2-deoxyribose damage induced by Fe (III) plus ascorbate. Phytother Res 20(2):120–124

Pardo-Andreu GL, Philip SJ, Riaño A, Sánchez C, Viada C, Núñez-Sellés AJ, Delgado R (2006b) *Mangifera indica* L. (Vimang) protection against serum oxidative stress in elderly humans. Arch Med Res 37(1):158–164

Pardo-Andreu GL, Sánchez-Baldoquín C, Avila-González R, Yamamoto ET, Revilla A, Uyemura SA, Naal Z, Delgado R, Curti C (2006c) Interaction of Vimang (*Mangifera indica* L. extract) with Fe(III) improves its antioxidant and cytoprotecting activity. Pharmacol Res 54(5):389–395

Pardo-Andreu GL, Barrios MF, Curti C, Hernández I, Merino N, Lemus Y, Martínez I, Riaño A, Delgado R (2008a) Protective effects of *Mangifera indica* L extract (Vimang), and its major component mangiferin, on iron-induced oxidative damage to rat serum and liver. Pharmacol Res 57(1):79–86

Pardo-Andreu GL, Paim BA, Castilho RF, Velho JA, Delgado R, Vercesi AE, Oliveira HC (2008b) *Mangifera indica* L. extract (Vimang) and its main polyphenol mangiferin prevent mitochondrial oxidative stress in atherosclerosis-prone hypercholesterolemic mouse. Pharmacol Res 57(5):332–338

Parmar HS, Kar A (2009) Protective role of *Mangifera indica*, *Cucumis melo* and *Citrullus vulgaris* peel extracts in chemically induced hypothyroidism. Chem Biol Interact 177(3):254–258

Peng ZG, Luo J, Xia LH, Chen Y, Song S (2004) CML cell line K562 cell apoptosis induced by mangiferin. Zhongguo Shi Yan Xue Ye Xue Za Zhi 12(5):590–594

Percival SS, Talcott ST, Chin ST, Mallak AC, Lounds-Singleton A, Pettit-Moore J (2006) Neoplastic transformation of BALB/3 T3 cells and cell cycle of HL-60 cells are inhibited by mango (*Mangifera indica* L.) juice and mango juice extracts. J Nutr 136(5):1300–1304

Perpétuo GF, Salgado JM (2003) Effect of mango (*Mangifera indica*, L.) ingestion on blood glucose levels of normal and diabetic rats. Plant Foods Hum Nutr 58:1–12

Perrucci S, Fichi G, Buggiani C, Rossi G, Flamini G (2006) Efficacy of mangiferin against *Cryptosporidium parvum* in a neonatal mouse model. Parasitol Res 99(2):184–188

Pino JA, Mesa J, Muñoz Y, Martí MP, Marbot R (2005) Volatile components from mango (*Mangifera indica* L.) cultivars. J Agric Food Chem 53(6):2213–2223

Pithayanukul P, Leanpolchareanchai J, Saparpakorn P (2009) Molecular docking studies and anti-snake venom metalloproteinase activity of Thai mango seed kernel extract. Molecules 14(9):3198–3213

Porcher MH et al (1995–2020) Searchable world wide web multilingual multiscript plant name database. Published by The University of Melbourne, Melbourne. http://www.plantnames.unimelb.edu.au/Sorting/Frontpage.html

Pott I, Breithaupt DE, Carle R (2003) Detection of unusual carotenoid esters in fresh mango. Phytochemistry 64(4):825–829

Pourahmad J, Eskandari MR, Shakibaei R, Kamalinejad M (2010) A search for hepatoprotective activity of fruit extract of *Mangifera indica* L. against oxidative stress cytotoxicity. Plant Foods Hum Nutr 65(1):83–89

Prabhu S, Jainu M, Sabitha KE, Shyamala Devi CS (2006) Effect of mangiferin on mitochondrial energy production in experimentally induced myocardial infarcted rats. Vasc Pharmacol 44(6):519–525

Prasad S, Kalra N, Shukla Y (2007) Hepatoprotective effects of lupeol and mango pulp extract of carcinogen induced alteration in Swiss albino mice. Mol Nutr Food Res 51(3):352–359

Prasad S, Kalra N, Shukla Y (2008a) Induction of apoptosis by lupeol and mango extract in mouse prostate and LNCaP cells. Nutr Cancer 60(1):120–130

Prasad S, Kalra N, Singh M, Shukla Y (2008b) Protective effects of lupeol and mango extract against androgen induced oxidative stress in Swiss albino mice. Asian J Androl 10(2):313–318

Prashant GM, Chandu GN, Murulikrishna KS, Shafiulla MD (2007) The effect of mango and neem extract on four organisms causing dental caries: *Streptococcus mutans, Streptococcus salivavius, Streptococcus mitis*, and *Streptococcus sanguis*: an in vitro study. Indian J Dent Res 18(4):148–151

Prashanth D, Amit A, Samiulla DS, Asha MK, Padmaja R (2001) α-Glucosidase inhibitory activity of *Mangifera indica* bark. Fitoterapia 72(6):686–688

Preissler T, Martins MR, Pardo-Andreu GL, Pêgas-Henriques JA, Quevedo J, Delgado R, Roesler R (2009) *Mangifera indica* extract (Vimang) impairs aversive memory without affecting open field behaviour or habituation in rats. Phytother Res 23(6):859–862

Prinsley RT, Tucker G (1987) Mangoes – a review. Commonwealth Science Council, London, 159 pp

Priya TT, Sabu MC, Jolly CI (2009) Role of *Mangifera indica* bark polyphenols on rat gastric mucosa against ethanol and cold-restraint stress. Nat Prod Res 14:1–12

Rajendran P, Ekambaram G, Sakthisekaran D (2008) Effect of mangiferin on benzo(a)pyrene induced lung carcinogenesis in experimental Swiss albino mice. Nat Prod Res 22(8):672–680

Rocha Ribeiro SM, Queiroz JH, de Queiroz LRME, Campos FM, Pinheiro Sant'ana HM (2007) Antioxidant in mango (*Mangifera indica* L.) pulp. Plant Foods Hum Nutr 62(1):13–17

Rodeiro I, Cancino L, González JE, Morffi J, Garrido G, González RM, Nuñez A, Delgado R (2006) Evaluation of the genotoxic potential of *Mangifera indica* L. extract (Vimang), a new natural product with antioxidant activity. Food Chem Toxicol 44(10):1707–1713

Rodeiro I, Donato MT, Jimnenez N, Garrido G, Delgado R, Gomez-Lechon MJ (2007) Effects of *Mangifera indica* L. aqueous extract (Vimang) on primary culture of rat hepatocytes. Food Chem Toxicol 45(12):2506–2512

Rodríguez J, Di Pierro D, Gioia M, Monaco S, Delgado R, Coletta M, Marini S (2006) Effects of a natural extract from *Mangifera indica* L, and its active compound, mangiferin, on energy state and lipid peroxidation of red blood cells. Biochim Biophys Acta 1760(9):1333–1342

Sairam K, Hemalatha S, Kumar A, Srinivasan T, Ganesh J, Shankar M, Venkataraman S (2003) Evaluation of anti-diarrhoeal activity in seed extracts of *Mangifera indica*. J Ethnopharmacol 84(1):11–15

Saleem M, Afaq F, Adhami VM, Mukhtar H (2004) Lupeol modulates NF-kappaB and PI3K/Akt pathways and inhibits skin cancer in CD-1 mice. Oncogene 23(30):5203–5214

Sánchez GM, Re L, Giuliani A, Núñez-Sellés AJ, Davison GP, León-Fernández OS (2000) Protective effects of *Mangifera indica* L. extract, mangiferin and selected antioxidants against TPA-induced biomolecules oxidation and peritoneal macrophage activation in mice. Pharmacol Res 42(6):565–573

Sánchez GM, Rodríguez HMA, Giuliani A, Núñez Sellés AJ, Rodríguez NP, Fernández OSL, Re L (2003) Protective effect of *Mangifera indica* L. extract (Vimang) on the injury associated with hepatic ischaemia reperfusion. Phytother Res 17(3):197–201

Satish Rao BS, Sreedevi MV, Nageshwar Rao B (2009) Cytoprotective and antigenotoxic potential of Mangiferin, a glucosylxanthone against cadmium chloride induced toxicity in HepG2 cells. Food Chem Toxicol 47(3):592–600

Schieber A, Ullrich W, Carle R (2000) Characterization of polyphenols in mango puree concentrate by HPLC with diode array and mass spectrometric detection. Int J Food Sci Nutr 1:161–166

Schieber A, Berardini N, Carle R (2003) Identification of flavonol and xanthone glycosides from mango (*Mangifera indica* L. Cv. 'Tommy Atkins') peels by high-performance liquid chromatography-electrospray ionization mass spectrometry. J Agric Food Chem 51(17):5006–5011

Severi JA, Lima ZP, Kushima H, Brito AR, Santos LC, Vilegas W, Hiruma-Lima CA (2009) Polyphenols with antiulcerogenic action from aqueous decoction of mango leaves (*Mangifera indica* L.). Molecules 14(3):1098–1110

Shibahara A, Yamamoto K, Shinkai K, Nakayama T, Kajimoto G (1993) Cis-9, cis-15-octadecadienoic acid:a novel fatty acid found in higher plants. Biochim Biophys Acta 1170:245–252

Stoilova I, Gargova S, Stoyanova A, Ho L (2005) Antimicrobial and antioxidant activity of the polyphenol mangiferin. Herb Polonica 51:37–44

Sukonthasing S, Wongrakpanich M, Verheij EWM (1992) *Mangifera indica* L. In: Verheij EWM, Coronel RE (eds) Plant resources of South-East Asia, No. 2. Edible fruits and nuts. PROSEA Foundation, Bogor, pp 211–216

Tawaha K, Sadi R, Qa'dan F, Matalka KZ, Nahrstedt A (2010) A bioactive prodelphinidin from *Mangifera indica* leaf extract. Z Naturforsch C 65(5–6):322–326

U.S. Department of Agriculture, Agricultural Research Service (2010) USDA National Nutrient Database for Standard Reference, Release 23. Nutrient Data Laboratory Home Page. http://www.ars.usda.gov/ba/bhnrc/ndl

Viswanadh EK, Rao BN, Rao BS (2010) Antigenotoxic effect of mangiferin and changes in antioxidant enzyme levels of Swiss albino mice treated with cadmium chloride. Hum Exp Toxicol 29(5):409–418

Whiley AW (1994) Mango. In: Page PE (ed) Tropical tree fruits for Australia. Queensland Department of Primary Industries (QDPI), Brisbane, pp 25–31

Wilkinson AS, Monteith GR, Shaw PN, Lin CN, Gidley MJ, Roberts-Thomson SJ (2008) Effects of the mango components mangiferin and quercetin and the putative mangiferin metabolite norathyriol on the transactivation of peroxisome proliferator-activated receptor isoforms. J Agric Food Chem 56(9):3037–3042

Yahia EM, Omelas-PAZ JJ, Gardea A (2006) Extraction, separation and partial identification of 'Ataulfo' mango fruit carotenoids. Acta Hort (ISHS) 712:333–338

Yogisha S, Raveesha KA (2010) Dipeptidyl peptidase IV inhibitory activity of *Mangifera indica*. J Nat Prod 3:76–79

Yoshimi N, Matsunaga K, Katayama M, Yamada Y, Kuno T, Qiao Z, Hara A, Yamahara J, Mori H (2001) The inhibitory effects of mangiferin: a naturally occuring glucosylxanthone, in bowel carcinogenesis of male F344 rats. Cancer Lett 163:163–170

Zheng MS, Lu ZY (1990) Antiviral effect of mangiferin and isomangiferin on herpes simplex virus. Chin Med J 103(2):160–165

Zhu XM, Song JX, Huang ZZ, Wu YM, Yu MJ (1993) Antiviral activity of mangiferin against herpes simplex virus type 2 in vitro. Zhongguo Yao Li Xue Bao 14(5):452–454 (In Chinese)

Mangifera kemanga

Scientific Name

Mangifera kemanga Blume

Synonyms

Mangifera caesia Jack var. *kemanga* (Blume) Kostermans, *Mangifera polycarpa* Griffith

Family

Anacardiaceae

Common/English Names

Kemang

Vernacular Names

Danish: Kemang;
French: Kemang;
Indonesia: Kemang (Sundanese, West Java), Kemang (Malay, Sumatra), Palong (Kutai, East Kalimantan);
Malaysia: Kemang, Kemanga (Peninsular), Beluno (Sabah);

Origin/Distribution

Kemang is indigenous to Peninsular Malaysia, Sumatra and Borneo. It is commonly cultivated in western Java, especially near Bogor, but less frequently in Borneo where binjai dominates.

Agroecology

Its agroecological requirements are similar to that of binjai, *Mangifera caesia*.

Edible Plant Parts and Uses

Ripe Kemang fruit is eaten fresh, pickled or made into a juice. Near ripe Kemang fruit is also relish in rojak and ripe fruit are also used in curries for its sour flavour. Occasionally a dish is made from fresh, grated seeds, with fermented soya beans and spices. In west Java, the young tender leaves are eaten in lalap.

Botany

Mangifera kemanga is closely similar to *M. caesia* in tree habit, leaf, flower and fruit characteristics but differ from *M. caesia* in having subsessile with narrowly decurrent margins; its panicles which are longer (up to 75 cm long), more open and contain fewer flowers; its fruits which are dull greenish-brown to yellowish-brown at maturity, gibbous at the base and scurfy (Plates 1–4) with pale creamy-white flesh (Plate 4).

Nutritive/Medicinal Properties

Its nutritive value would be similar to that reported for *Mangifera caesia*.

Plate 3 Pale brown, gibbous and scurfy kemang fruit (Indonesia)

Plate 1 Brown and scurfy kemang fruit (Indonesia)

Plate 4 Ripe beluno fruit Sabah

Plate 2 Greenish-brown, gibbous and scurfy kemang fruit (Indonesia)

Plate 4 Pale creamy-white flesh of kemang

Other Uses

Its timber is pinkish and better than binjai, being more compact and firmer.

Comments

Some botanists have united both species under *M. caesia* but Bompard considered that it should be treated as a distinct species because of some distinct morphological differences as listed above. Hybrids of this species with *Mangifera caesia* have been found in Kalimantan.

Selected References

Bompard JM (1992) *Mangifera caesia* Jack, *Mangifera kemanga* Blume. In: Coronel RE, Verheij EWM (eds) Plant resources of South-East Asia, No. 2. Edible fruits and nuts. Prosea Foundation, Bogor, pp 207–209

Bompard JM (1993) The genus *Mangifera* re-discovered: the potential contribution of wild species to mango cultivation. Acta Hort 341:69–77

Burkill IH (1966) A dictionary of the economic products of the Malay Peninsula. Revised reprint, 2 vols. Ministry of Agriculture and Co-operatives, Kuala Lumpur, vol 1 (A–H), pp 1–1240, vol 2 (I–Z), pp 1241–2444

Ding Hou (1978) Anacardiaceae. In: van Steenis CGGJ (ed) Flora Malesiana, Series I, vol 8. Sijthoff & Noordhoff, Alphen aan den Rijn, pp 395–548, 577 pp

Kostermans AJGH (1965) New and critical Malaysian plants VII *Mangifera caesia* Jack. Reinwardtia 7(1): 19–20

Mukherji S (1949) A monograph on the genus *Mangifera* L. Lloydia 12:73–136

Ochse JJ, Bakhuizen van den Brink RC (1931) Fruits and fruitculture in the Dutch East Indies. G. Kolff & Co., Batavia-C, 180 pp

Ochse JJ, Bakhuizen van den Brink RC (1980) Vegetables of the Dutch Indies, 3rd edn. Ascher & Co., Amsterdam, 1016 pp

Mangifera laurina

Scientific Name

Mangifera laurina Blume

Synonyms

Mangifera longipes Griffith, *Mangifera parih* Miq., *Mangifera sumatrana* Miq.

Family

Anacardiaceae

Common/English Names

Egg Mango, Padi Mango, Water Mango

Vernacular Names

Burma: Thayet The Nee, Thayet;
China: Chang Geng Mang Guo;
Indonesia: Mangga Pari, Parih, (Sundanese, West Java), Pelem Kecik (Javanese, East Java), Empelem (Malay, Kalimantan), Asam Pun, Ampelan Dotan, Empelem, Mangga Tiakar, Pauh Gadang (Sumatra), Pauh Pong (Flores Islands);
Malaysia: Mempelam, Emplam (General, Peninsular Malaysia, Sabah, Sarawak), Tanh Chai (Cantonese, Peninsular Malaysia), Mangga Telur, Mempelam Melur, Mempelam Padi, Pauh Telur, Pauh Padi (Malay, Peninsular Malaysia) Mangga Ayer, Manga Hutan, Pauh Kijang (Malay, Sabah);
Philippines: Apali;
Thailand: Mamuang Kaleng, Mamuang Khee Kwaang (Peninsular);
Vietnam: Cây Nui, Xoái Nui, Queo.

Origin/Distribution

Wild distribution of this species is found in Myanmar, Indochina (Kampuchea, Vietnam) and Malesia, from peninsular Thailand to New Guinea, in lowland tropical rain forest. It was probably brought into cultivation long before the introduction of *Mangifera indica* L. in the region. In most parts of Borneo it is still widely cultivated, but less so in Peninsular Malaysia, Sumatra and Java. Probably introduced into the Philippines.

Agroecology

This species has the same agroecological requirements as the common mango, *Mangifera indica*. It is found in the lowland rainforests up to 300 m altitude.

Edible Plant Parts and Uses

Both mature, ripe and unripe fruit are edible (Plates 1–3). The ripe fruit is very acid and has been processed into drinks with sugar or honey.

Plate 1 Mature, unripe padi mango on sale in local market in Malaysia

Plate 2 Ripe padi mango fruit

Plate 3 Padi mango fruit whole and sliced showing the deep yellow flesh

Plate 4 Habit of young padi mango tree

More often the fruit is harvested immature, sliced and served in fruit salads with a spicy sauce ('*ruja*k'). Mature, unripe fruits are halved or sliced for use in pickles and chutney. The slices are also dried and salted as preserves and consumed as snacks.

Botany

A medium-sized, evergreen, erect tree 20–30 m high, with 40–190 cm diameter, rough, greyish trunk (Plates 4 and 5). Leaf is borne on slender, 2.5–5.5 cm long petiole that is grooved above and inflated at the base. Lamina is oblong-lanceolate, elliptic-lanceolate to lanceolate, 15–26 cm by 4–6 cm, leathery, glabrous, base cuneate to broadly cuneate, margin entire and slightly undulate, apex acuminate or caudate-acuminate, lateral veins prominent on both sides (Plate 5). Inflorescence is paniculate, sub-terminal to terminal, 40 cm long, with slender primary branches; pedicels are about 1.5 mm long,

Plate 5 Trunk and leaves of padi mango

sparsely pubescent to subglabrous, lax and loosely flowered. Flowers are whitish-green to pale yellow, pentamerous, small and fragrant on long pedicel; calyx 5-lobed with acute sepals; petals 5, linear-oblong; stamens 5 only 1 fertile, rest are staminodes; ovary is ovoid to subglobose and style subterminal. Fruit is a drupe-like small mango, obliquely oblong, 5–7 cm by 4 cm, not beaked, medium green (Plate 1) turning greenish-yellow to yellow at maturity (Plates 2 and 3); flesh yellow, juicy, very acid and fibrous (Plate 3). Stone is large, 4–5 cm by 2–3 cm. Seed is polyembryonic.

Nutritive/Medicinal Properties

No information has been published on its nutritive and medicinal values.

Other Uses

The seeds being polyembryonic are used often as rootstocks for the common mango, *Mangifera indica* in Malaysia.

Comments

The tree is usually grown from seeds.

Selected References

Bompard JM (1991) *Mangifera laurina* Blume *Mangifera pentandra* Hooker f. In: Verheij EWM, Coronel RE (eds) Plant resources of South-East Asia, No. 2. Edible fruits and nuts. Pudoc, Wageningen, pp 216–218

Bompard JM, Schnell RJ (1997) Taxonomy and systematics. In: Litz RE (ed) The mango botany, production and uses. CAB International, Wallingford, pp 21–47

Ding Hou (1978) Anacardiaceae. In: van Steenis CGGJ (ed) Flora Malesiana, Series I, vol 8. Sijthoff & Noordhoff, Alphen aan den Rijn, pp 395–548, 577 pp

Min T, Barfod A (2008) Anacardicaceae. In: Wu ZY, Raven PH, Hong DY (eds) Flora of China, vol 11. Oxalidaceae through Aceraceae. Science Press/ Missouri Botanical Garden Press, Beijing/St. Louis

Mukherjee SK (1949) A monograph of the genus *Mangifera* L. Lloydia 12:73–136

Mangifera odorata

Scientific Name

Mangifera odorata Griffith

Synonyms

Mangifera foetida Lour. var. *bakkill* (Miq.); *Mangifera foetida* Lour. var. *bombom* Blume; *Mangifera foetida* Lour. var. *kawini* Blume; *Mangifera foetida* Lour. var. *mollis* Blume; *Mangifera foetida* Lour. var. *odorata* (Griffith) Pierre; *Mangifera oblongifolia* Hook.f.; *Mangifera odorata* Griffith var. *pubescens* Engl.

Family

Anacardiaceae

Common/English Names

Fragrant Mango, Kuini, Kurwini Mango, Kwini, Kwini Mango, Saipan Mango

Vernacular Names

Dutch: Kweni;
French: Mangue Odorante, Manguier À Fruit Odorant, Manguier Odorant, Kuweni;
Indonesia: Kalimbang Kuini, Kawidei Kepaya, Kawilei Kapaya, Kawilei Kuini, Kombilei Chapya, Kmbilo Insam, Kmbiloi Koini, Oi Kapaya, Oi Kuini, Uwai Kuini (Alfoersch, N. Sulawesi), Mancang (Acheh, Sumatra), Kweni, Weni (Bali), Ancami (Bari, Sulawesi), Ambasang, Embasang, Gorat, Kuweni (Batak), Fu Huni (Bima, Timor), Pao Daeko (Boegineesch, Sulawesi), Mangga Kuini (Boeol, Sulawesi), Ruone (Boeroe, Maluku), Ori Asula Kowini (E. Ceram, Maluku), Kuin, Kuine (W. Ceram, Maluku), Guin, Kuini (S. Ceram, Maluku), Ambacang, Pangi (Dyak), Pao Jawa (Flores), Lukup (Gajo, Sumatra), Mangga Kuini, Oile Koini (Gorontalo, Sulawesi), Kuweni, Pakel, Kebembem (Java), Guawe Sitingki, Hitinki, Sitingki (N. Halmaheira), Lelit Salo (S. Halmaheira), Kuini (Lampong, Sumatra), Taipa Macang (Makassar), Asam Membacnag, Kweni (Jambi, Sumatra), Kuini (Lingga, Sumatra), Bine, Bemberm, Beni, Kaeni, Kaweni (Madurese), Gandarasa, Kebembem, Keweni, Macang, Membacang (Malay), Lekup, Mangga Kuini (Manado, Sulawesi), Ambacang, Embacang, Kuwini, Lakuik (Minangkabau), Hambawa (Nias, Sumatra), Babkang, Koine, Mamblang, Koine, Mangga Koim (Oelias, Maluku), Bembem, Kaweni (Sundanese), Guwae Stinki (Ternate), Guwa Stinki (Tidore, Ternate);
Malaysia: Mangga Wani, Wani (Sabah), Bachang Beto (In Semang), Kwini Boli, Kwini, Kuini, Kohini;
Philippines: Kuwini, Uani (Cebu, Bisaya), Huani, Kandupeh; Thailand: Mamuang Chingreet, Mamuang Paa (Central);
Vietnamese: Xoai Huong, Cay Muong.

Origin/Distribution

The exact origin of the species is unknown. It is only known in cultivation as wild distribution has not been encountered. The species is primarily cultivated in Malesia – Peninsular Malaysia – (Perak, Kelantan, Malacca, Pahang); Singapore; Sumatra – (Lampong, Sibolangit); Java (Pekalongan, Bantam, Batavia); and Sabah – (Sandakan, Sipitang, Bukit Penampang). It is also cultivated to a lesser extent in Southern Thailand, south Vietnam, Philippines, Christmas Island and Guam.

Agroecology

M. odorata is strictly a tropical species. It grows below 1,000 m altitude in areas with a fairly heavy rainfall that is uniformly distributed throughout the year, although it grows even with a moderate rainfall (1,200 mm) provided there are no prolonged dry periods. It can be found in lowland mixed forests but is mainly cultivated.

Edible Plant Parts and Uses

The pulp of the ripe fruit is eaten raw. Ripe and unripe fruit are also eaten as rujak. They must be peeled thick because of the presence of an acrid juice in the skin, which can also be reduced by steeping in diluted lime-water before eating. The unripe fruit is also used for making chutney and for pickles with salt and also used in curries. In Java the seeds are used for making a kind of flour for preparing the traditional delicacy *jenang pelok* (a thick pappy preparation from *Curcuma* rhizomes) or dodol (with gelatinous rice).

Botany

A medium-sized, evergreen tree, 10–30 m, rarely more than 30 m high with a round or broadly ovoid canopy, straight trunk with grey, smooth or fissured bark, containing caustic sap. Leaves are coriaceous-chartaceous, oblong-lan-

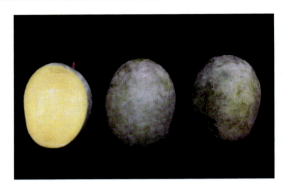

Plate 1 Oblongish kwini fruit with yellow flesh

ceolate, 12–35 cm × 4–10 cm, glabrous, with entire margin, shortly acuminate apex, cuneate or obtuse base with 15–25 pairs of lateral nerves, prominent reticulated veins especially on the lower surface, petioles are 3–7 cm and swollen at base. Panicles are terminal, divaricate, pyramidal, 15–50 cm long, rather densely flowered, rachis is yellowish-green tinged with reddish-brown. Flowers are 5(–6)-merous, about 6 mm across, fragrant; sepals ovate, 3–4 mm long, brown-red or partly green; petals lanceolate, 5–6 mm × 1.2–2 mm, yellowish at the base, pale pinkish towards the apex, reflexed, with 3–5 ridges on 2/3 of the length of the petals, joined at the base, pale yellow becoming dark red; stamens 5(–6), only 1 fertile, with 5 mm long filament and oblong anthers, staminodes 1.5–2 mm long; ovary subglobose, yellowish, style excentric, 3–5 mm long, dark red. Fruit an obliquely ellipsoid-oblong or oblong drupe, 10–13 cm × 6–9 cm, green to yellowish-green, sparingly spotted with dark brown lenticels; fairly thick (3–4 mm) rind; flesh yellow to orange-yellow, firm, fibrous (Plates 1 and 2), sweet to acid-sweet, juicy, with a strong smell and flavour of turpentine. Stone (endocarp) is 8–10 cm × 4.5–5 cm × 2.5–3 cm. Seed is usually polyembryonic.

Nutritive/Medicinal Properties

The nutrient composition of the flesh of *M. odorata* per 100 g edible portion was reported as: water 80 g, energy 69.3 kcal., protein 0.9 g,

Plate 2 Obliquely elliptic-oblong kwini fruit with golden yellow fibrous flesh

fat 0.1 g, carbohydrate (including fibre), vitamin A 600 IU, thiamin 0.04 mg, riboflavin 0.06 mg, niacin 0.7 mg, ascorbic acid 13 mg. (Pareek et al. 1998). Analyses conducted in Malaysia by Tee et al. (1997) reported the following nutrient composition per 100 g edible portion: energy 83 kcal, water 79.5 g, protein 1.0 g, fat 1.8 g, carbohydrate 15.6 g, fibre 1.4 g, ash 0.7 g, Ca 9 mg, P 13 mg, Fe 0.4 mg, Na 2 mg, K 187 mg, carotenes 1888ug, vitamin A 315ug RE, vitamin B1 0.05 mg, vitamin B2 0.07 mg, niacin 0.5 mg, vitamin C 23 mg.

Mangifera odorata, *Mangifera foetida* and *Mangifera pajang* fruit contain isoflavones, a major group of phytoestrogen that may serve as health promoting compounds in our diet. All three species were found to contain the isoflavones diadzein and genistein (Khoo and Ismail 2008). Studies showed that at optimised condition, *M. odorata* had comparatively higher daidzein (9.4–10.5 mg/100 g) and genistein (1.6–1.7 mg/100 g) contents. Daidzein content of *M. pajang* (8.3–8.7 mg/100 g) was higher than *M. foetida* (2.8–8.0 mg/100 g), but the genistein content of *M. pajang* (0.4–0.6 mg/100 g) was similar to that of *M. foetida* (0.4–0.8 mg/100 g).

In a study conducted in Malaysia, the total carotene content (mg/100 g) of selected underutilized tropical fruit in decreasing order (Khoo et al. 2008) was jentik-jentik (*Baccaurea polyneura*) 19.83 mg > Cerapu 2 (*Garcinia prainiana*) 15.81 mg > durian nyekak 2 (*Durio kutejensis*) 14.97 mg > tampoi kuning (*Baccaurea reticulata*) 13.71 mg > durian nyekak (1) 11.16 mg > cerapu 1 6.89 mg > bacang 1 (*Mangifera foetida*) 4.81 > kuini (*Mangifera odorata*) 3.95 mg > jambu susu (*Syzygium malaccense*) 3.35 mg > bacang (2) 3.25 mg > durian daun (*Durio lowianus*) 3.04 mg > bacang (3) 2.58 mg > tampoi putih (*Baccaurea macrocarpa*) 1.47 mg > jambu mawar (*Syzygium jambos*) 1.41 mg. β-carotene content was found to be the highest in jentik-jentik 17,46 mg followed by cerapu (2) 14.59 mg, durian nyekak (2) 10.99 mg, tampoi kuning 10.72 mg, durian nyekak (1) 7.57 mg and cerapu (1) 5.58 mg. These underutilized fruits were found to have acceptable amounts of carotenoids and to be potential antioxidant fruits.

Seventy-three volatile components were identified in the kuini (*M. odorata*) and 84 volatiles in bachang fruit (*M. foetida*). Oxygenated monoterpenes (45.4%) and esters (33.0%) constituted the main classes of kuini fruit volatiles, and α-terpineol (31.9%) was the major component (Wong and Ong 1993). Among the bachang fruit volatiles, esters (55.7%) and oxgenated monoterpenes (20.3%) were dominant, with ethyl butanoate (33.4%) the most abundant. The volatile composition of both fruits differs significantly from results reported for the fruits of some of the commercially important mango (*M. indica*) cultivars.

In folk medicine, the bark is recommended for external application in hystero-epilepsy, in the form of a compound like a cosmetic mixture. The sap from the tree and unripe fruit is caustic and has been reported to cause dermatitis.

Other Uses

The wood is used locally as bachang *(M. foetida)*, but is reportedly of poor quality.

Comments

Mangifera odorata is a highly polymorphic species and apparently includes many forms of a hybrid swarm. In West Java, several forms are distinguished: – '*kebembem or bembem*', the fruit has a strong smell and taste of turpentine reminiscent of the fruit of *M. foetida*, the leaves are coriaceous, restricted to the Soenda provinces. *Bembem* is identical with *Pelem lengis* from Solo; – '*keweni*', with less fibrous flesh and a mild taste of turpentine; the most common form and the best forms are very palatable; – '*gandarassa*' of the Bantem area in West Java, a rare and poorly known form which is said to be superior to '*keweni*', less sweet but more juicy and with a mild odour. Another form worthy of mention is *Keweni loorik*. In the Philippines '*sangay*', known from Jolo, is distinguished by its yellow colour at maturity from the greenish '*huani*' fruit. Ding Hou (1978) suggested *M. odorata* to be a hybrid between *M. indica* L. and *M. foetida* Lour. due to morphological intermediacy. Recent studies using amplified fragment length polymorphism (AFLP) analysis confirmed the hybrid status of *M. odorata* (Teo et al. 2002). All the *M. odorata* samples additively inherit bands specific to *M. indica* and *M. foetida*, which strongly suggested the hybrid origin. *Mangifera odorata* was closer to *M. foetida* than to *M. indica*, indicating that backcrossing with *M. foetida* might have taken place.

Selected References

Bompard JM (1992) *Mangifera odorata* Griffith. In: Verheij EWM, Coronel RE (eds) Plant resources of South-East Asia, No. 2. Edible fruits and nuts. Prosea Foundation, Bogor, pp 218–220

Burkill IH (1966) A dictionary of the economic products of the Malay Peninsula. Revised reprint, 2 vols. Ministry of Agriculture and Co-operatives, Kuala Lumpur, vol 1 (A–H), pp 1–1240, vol 2 (I–Z), pp 1241–2444

Ding Hou (1978) Anacardiaceae. In: van Steenis CGGJ (ed) Flora Malesiana, Series I, vol 8. Sijthoff & Noordhoff, Alphen aan den Rijn, pp 395–548, 577 pp

Khoo HE, Ismail A (2008) Determination of daidzein and genistein contents in *Mangifera* fruit. Malays J Nutr 14(2):189–198

Khoo HE, Ismail A, Mohd.-Esa N, Idris S (2008) Carotenoid content of underutilized tropical fruits. Plant Foods Hum Nutr 63:170–175

Kostermans AJGH, Bompard JM (1993) The mangoes: their botany, nomenclature, horticulture and utilization. Academic, London, 233 pp

Ochse JJ, Bakhuizen van den Brink RC (1931) Fruits and fruitculture in the Dutch East Indies. G. Kolff & Co., Batavia, 180 pp

Pareek OP, Sharma S, Arora RK (1998) Underutilized edible fruits and nuts: an inventory of genetic resources in their regions of diversity. IPGRI Office for South Asia, New Delhi, pp 191–206

Slik JWF (2006) Trees of Sungai Wain. Nationaal Herbarium Nederland. http://www.nationaalherbarium.nl/sungaiwain/

Tee ES, Noor MI, Azudin MN, Idris K (1997) Nutrient composition of Malaysian foods, 4th edn. Institute for Medical Research, Kuala Lumpur, 299 pp

Teo LL, Kiew R, Set O, Lee SK, Gan YY (2002) Hybrid status of kwuini, *Mangifera odorata* Griff. (Anacardiaceae) verified by amplified fragment length polymorphism. Mol Ecol 11(8):1465–1469

Wong KC, Ong CH (1993) Volatile components of the fruits of Bachang (*Mangifera foetida* Lour.) and Kuini (*Mangifera odorata* Griff.). Flav Fragr J 8(3): 147–151

Mangifera pajang

Scientific Name

Mangifera pajang Kostermans

Synonyms

None

Family

Anacardiaceae

Common/English Names

Bambangan, Pajang (common names)

Vernacular Names

Brunei: Membangan (Kedayan), Bambangan;
Kalimantan (Indonesia): Limum (Sangkulirang), Asem Pajang (Kutai), Alim (Kenyah, Dayak), Hambawang (Malay), Pangin (Punam, Malinau), Lampayang (Lun Daye, Mentarang), Pangaeng (Merap, Malinau), Alieng (Kenyah Uma' Lung);
Sabah (Malaysia): Embang (Kayan, Malay), Bambangan, Embawang;
Sarawak (Malaysia): Mawang (Iban), Embang, Buah Pangin (Malay).

Origin/Distribution

Its origin is in Borneo – Sarawak, Sabah, Brunei, Kalimantan. The species is found mostly wild in the forests or in scattered backyard plantings in villages. Areas of wild distribution are found mostly around Kapit, Ulu Dapoi, Long Silat in Sarawak; Sipitang, Sandakan, Beaufort in Sabah; Sangkulirang and west Kutei in Kalimantan.

Agroecology

Most of the trees grow wild in the forests where the tropical agroclimatic regime is humid and shaded. For instance in the lowland dipterocarp forest and in forest on well-drained alluvial soils. Areas with heavy rainfall and strong wind are not suited as conditions are detrimental to flower induction and fruit set. It can withstand prolong dry weather and flooding. Bambangan can be grown on a wide range of soils including upland soils. It thrives best on, well-drained alluvial clay to sandy loam with pH ranging from 5 to 7. It will not tolerate acid peat swamps or the leached coastal sandy beaches.

Edible Plant Parts and Uses

The thick, acid sweet and mildly banana flavoured, fibrous pulp is eaten fresh. The thick skin is edible when pickled or eaten fresh with sambal belacan.

Plate 1 Large, globose, brownish and rough pajang fruit

Plate 4 Jeruk bambangan packaged in plastic containers

Plate 2 Large, broad-ovoid pajang fruits

Plate 3 Deep yellow, fibrous flesh and thick rind

One food item that is widely consumed in the Kadazan-dusun community in Sabah is the '*nonsom Bambangan*' or '*jeruk Bambangan*' (Plate 3) The *nonsom Bambangan* or the Dickled bambangan is relatively easy to make. The skin is removed and the flesh cut into bite sizes. It is then sprinkled with salt and then the seeds are cut open and grated. The grated seeds, in powder form are then mixed thoroughly with the concoction. It is then left to keep for weeks before being eaten. Some pickles are known to last for a year. These pickles are commonly available at all local markets. In Sarawak, the acid fruits are eaten with sambal belacan or cooked in a curry. Fresh young leaf flushes are sold in local markets and are cooked and consumed with sambal belacan or tempoyak.

Botany

A tall, perennial tree, 15–50 m high with huge, inaccessible columnar straight trunk with branches raised high above the ground and a bole of 30–70 cm with grey, smooth or broadly fissured bark and an umbrella-shaped canopy high up the tree. Mature leaves are rigidly coriaceous, elliptic-oblong or occasionally obovate-oblong, 17.5–45 cm long by 7–15 cm wide, cuneate to attenuated base, acute to mucronate apex, entire margin, with 14–30 pairs of prominent nerves, usually dark green in colour and borne on petioles up to 12 cm long (Plates 4–6). Young flushes are bronze to pink in colour. Panicles are terminal and also in upper leaf axils, pyramidal, to 30 cm long with lateral branches up to 1 cm long and densely flowered. Flower is red with ovate or oblong-ovate bracts 2 mm long; calyx pentamerous, lobes ovate and glabrous; petals 5, purplish-red on the inner surface and pinkish white on the outside, elliptic-oblong or lanceolate and with ridges; stamens 5–2

Plate 5 Leafy shoot of bambangan

Plate 6 Close-up of lower surface of bambangan leaf

fertile with 7 mm long filaments and broad-ovoid anthers, staminodes 5 mm long; ovary ellipsoid with excentric style. Fruit is a drupe, brownish, globose to broad-ovoid, 15–20 cm across, roughish (Plates 1 and 2), fruit can weigh up to 2–3 kg. Flesh yellow, fibrous, acid to acid sweet, mildly fragrant with a thick rind 5–10 mm thick (Plate 3) enclosing one large stone.

Nutritive/Medicinal Properties

Fifty components were identified in bambangan fruit volatiles (Wong and Siew 1994); monoterpene hydrocarbons (91.3%) and esters (7.6%) predominated, with α-pinene (67.2%) and α-phellandrene (11.0%) constituting the two most abundant components.

Mangifera pajang, *Mangifera foetida* and *Mangifera odorata* fruits contain isoflavones, a major group of phytoestrogen that may serve as health promoting compounds in our diet. All three species were found to contain the isoflavones diadzein and genistein. Studies showed that at optimised condition, *M. odorata* had comparatively higher daidzein (9.4–10.5 mg/100 g) and genistein (1.6–1.7 mg/100 g) contents (Khoo and Ismail 2008). Daidzein content of *M. pajang* (8.3–8.7 mg/100 g) was higher than *M. foetida* (2.8–8.0 mg/100 g), but the genistein content of *M. pajang* (0.4–0.6 mg/100 g) was similar to that of *M. foetida* (0.4–0.8 mg/100 g).

Antioxidant Activity

Mangifera pajang kernel extract was reported to display the highest free radical scavenging and ferric reducing activities compared to *Artocarpus odoratissimus* (Abu Bakar et al. 2009). Total phenolic content of the samples were in the range of 5.96–103.3 mg gallic acid equivalent/g. The kernel and flesh of *Mangifera pajang* contained the highest and lowest total flavonoid content with the values of 10.98 and 0.07 mg rutin equivalent/g, respectively. The antioxidant activities of extracts were significantly correlated with the total phenolic and flavonoid content (but not the anthocyanins content). The phytochemicals and antioxidant properties of *M. pajang* especially their by-products (kernel/seed), indicated that they may impart health benefits when consumed and should be regarded as a valuable source of antioxidant-rich nutraceuticals.

Anticancer Activity

Abu Bakar et al. (2010) reported that *Mangifera pajang* crude kernel extract induced cytotoxicity in MCF-7 (hormone-dependent breast cancer) cells and MDA-MB-231 (non-hormone dependent breast cancer) cells with IC_{50} values of 23 and 30.5 µg/ml, respectively. The kernel extract induced cell cycle arrest in MCF-7 cells at the sub-G1 (apoptosis) phase of the cell cycle in a time-dependent manner. For MDA-MB-231 cells, the kernel extract induced strong G2-M arrest in cell cycle progression at 24 hours, resulting in substantial sub-G1 (apoptosis) arrest after 48 and

72 h of incubation. Apoptosis occurred early in both cell types, 36 h for MCF-7 cells and 24 hours for MDA-MB-231 cells, with 14.0% and 16.5% of the cells respectively undergoing apoptosis at these times. This apoptosis appeared to be dependent on caspase-2 and −3 in MCF-7 cells, and on caspase-2, -3 and −9 in MDA-MB-231 cells. These findings suggested that *M. pajang* kernel extract has potential as a potent cytotoxic agent against breast cancer cell lines.

Other Uses

The species is also grown for its timber.

Comments

The species is listed as vulnerable by the World Conservation Monitoring Centre (1998).

Selected References

Abu Bakar MF, Mohamed M, Rahmat A, Fry J (2009) Phytochemicals and antioxidant activity of different parts of bambangan (*Mangifera pajang*) and tarap (*Artocarpus odoratissimus*). Food Chem 113(2):479–483

Abu Bakar MF, Mohamad M, Rahmat A, Burr SA, Fry JR (2010) Cytotoxicity, cell cycle arrest, and apoptosis in breast cancer cell lines exposed to an extract of the seed kernel of *Mangifera pajang* (bambangan). Food Chem Toxicol 48(6):1688–1697

Bompard JM (1992) *Mangifera foetida* Lour. *Mangifera pajang* Kostermans. In: Verheij EWM, Coronel RE (eds) Plant resources of South-East Asia, No. 2. Edible fruits and nuts. Prosea Foundation, Bogor, pp 209–211

Ding Hou (1978) Anacardiaceae. In: van Steenis CGGJ (ed) Flora Malesiana, Series I, vol 8. Sijthoff & Noordhoff, Alphen aan den Rijn, pp 395–548

Khoo HE, Ismail A (2008) Determination of daidzein and genistein contents in *Mangifera* fruit. Malay J Nutr 14(2):189–198

Kostermans AJGH, Bompard JM (1993) The mangoes: their botany, nomenclature, horticulture and utilization. Academic, London, 233 pp

Munawaroh E, Purwanto Y (2009) Studi hasil hutan non kayu di Kabupaten Malinau, Kalimantan Timur. (In Indonesian). Paper presented at the 6th basic science national seminar, Universitas Brawijaya, Malang, 21 Feb 2009. http://fisika.brawijaya.ac.id/bss-ub/proceeding/PDF%20FILES/BSS_146_2.pdf

Voon BH, Chin TH, Sim CY, Sabariah P (1988) Wild fruits and vegetables in Sarawak. Sarawak Department of Agriculture, Sarawak, 114 pp

Wong KC, Siew SS (1994) Volatile components of the fruits of bambangan (*Mangifera panjang* Kostermans) and binjai (*Mangifera caesia* Jack). Flav Fragr J 9(4):173–178

World Conservation Monitoring Centre (1998) *Mangifera quadrifida var. quadrifida*. In: IUCN 2010. IUCN Red List of Threatened Species. Version 2010.4. http://www.iucnredlist.org

Mangifera quadrifida

Scientific Name

Mangifera quadrifida Jack

Synonyms

Mangifera maingayi Hook.f., *Mangifera langong* Miq., *Mangifera longipetiolata* King, *Mangifera quadrifida* var. *spathulaefolia* (Bl.) Engl., *Mangifera rigida* Bl., *Mangifera spathulaefolia* Bl.

Family

Anacardiaceae

Common/English Names

Asam Kumbang (common name)

Vernacular Names

Brunei: Rancha-Rancha;
Indonesia: Ambatjang Rawanghalus (Balai Selasa, Sumatra), Batjang Utan, Boanu, Fuluh, Uding, Langong (Simalur, Sumatra), Putian (Palembang, Sumatra), Sempiat, Masam Koembang, Asam Kumbang, Asem Rawa, Asam Rarawa (Sumatra), Asam Putarum, Rarawa, Rawa-Rawa Biasa; Rawa-Rawa Pipit, Ubab (Kalimantan);
Malaysia: Asam Kumbang, Sepam, Lekub, Pauh (Peninsular), Mangga Hutan, Damaran (Bajan), Asam (Sabah), Asam Kumbang (Sarawak).

Origin/Distribution

The species is endemic to Malesia – Brunei (Tutong); Malaysia: Sabah (Lahad Datu, Sandakan, Beaufort, Sepilok, Tawau), Sarawak, Peninsular Malaysia (Perak, Pahang, Johore Kelantan); Singapore (Upper Mandai); Indonesia: Sumatra (Riau, Palembang, Beneden, Bangka Island, Ketambe, Acheh, Simalur Island, Balai Selasa, Rau and Pelem Bay), Kalimantan (Balikpapan, Bangko), Java (Bogor).

Agroecology

This species is strictly a tropical species. It is found in the hot, wet and humid undisturbed lowland forest, often on inundated land or along riversides, up to 1,000 m altitude but usually below 500 m. It is found often in villages or forest gardens as a cultivated backyard tree.

Edible Plant Parts and Uses

Fruit is edible raw. In Sarawak, both green and ripe acid-sweet fruit is consumed with salt, *sambal belacan* or pickled.

Botany

Tree medium-sized, evergreen, 10–32 m high, with 25–90 cm bole, mildly scaly bark and dense crown (Plate 2). Leaves are alternate or in whorls, simple, coriaceous, elliptic to elliptic lanceolate, ovate-oblong, 6.5–30 by 3–9 cm, base cuneate or rounded, apex acute to obtuse, nerves 7–22 pairs, reticulate venation, petiole 1–7 cm long. Panicles are terminal, pyramidal and glabrous. Juvenile leaves are bronze in colour (Plate 3) and soft turning dark, glossy green with age (Plate 4). Flowers are white to greenish–white, fragrant, 4-merous. Sepals – 4 lobed, elliptic to ovate-oblong, 2–3.5 long, glabrous; petals – 4, ovate to broadly elliptic, 3.5–4.5 × 1.5–2.5 mm, with 3–5 ridges half length of petals, stamens 4, 1 fertile, 2–5 mm long, anthers oblong, 1 mm, staminodes 0.75 mm , ovary subglobose, 1.5–2 mm, style excentric, 1.5–2 mm. Drupe is subglobose to broadly ellipsoid, reddish-purple (Plate 1), 8–10 cm by 5.5–7 cm, with pale brown or greenish lenticels, flesh pale yellow to white, acid sweet and fibrous covering the stone.

Plate 2 Tree habit of *Mangifera quadrifida*

Nutritive/Medicinal Properties

No published information is available on its nutritive or medicinal values.

Plate 3 Bronze coloured juvenile foliage

Plate 1 Subglobose reddish-purple fruit

Other Uses

A timber species but is mainly cultivated for its edible fruits.

Plate 4 Mature leaf flush

Comments

The species has been listed as a threatened species by the World Conservation Monitoring Centre (1998).

Selected References

Ding Hou (1978) Anacardiaceae. In: van Steenis CGGJ (ed.) Flora Malesiana, ser I, vol 8. Sijthoff & Noordhoff, Alphen aan den Rijn, pp 395–548, 577 pp

Kochummen KM (1989) Anacardiaceae. In: Ng FSP (ed.) Tree flora of Malaysia, vol 4. Longman, Kuala Lumpur, pp 9–57

Kostermans AJGH, Bompard JM (1993) The mangoes: their botany, nomenclature, horticulture and utilization. Academic, London, 233 pp

Mukherji S (1949) A monograph on the genus *Mangifera* L. Lloydia 12:73–136

Slik JWF (2006) Trees of Sungai Wain. Nationaal Herbarium Nederland. http://www.nationaalherbarium.nl/sungaiwain/

Voon BH, Chin TH, Sim CY, Sabariah P (1988) Wild fruits and vegetables in Sarawak. Sarawak Department of Agriculture, Sarawak, 114 pp

World Conservation Monitoring Centre (1998) *Mangifera quadrifida var. quadrifida*. In: IUCN 2010. IUCN Red List of Threatened Species. Version 2010.4. http://www.iucnredlist.org

Mangifera similis

Scientific Name

Mangifera similis Blume

Synonyms

Mangifera torquenda Kosterm.

Family

Anacardiaceae

Common/English Name

Lamantan (common name)

Vernacular Names

East Malaysia: Asam (Sandakan, Sabah), Lamantan (Sarawak);
Indonesia: Fais, Fajas, Masam Humbang, Membaljang Bubuk, Paias, Tajas (Palembang, Sumatra), Pelem Kera, Penkajang Utan (Bengkalis, Sumatra), Asem Rawa, Asem Telor (Bangka, Sumatra), Asam Pipit, Putaram (Kutei, Kalimantan), Embang Duan Kecil (Kalimantan).

Origin/Distribution

The species is scattered, mainly in Borneo – Sarawak, Sabah (Sandakan, Tawau) and Kalimantan (Kutei, Samarinda, Balikpapan). It is also found in Sumatra (Bangka, Palembang and Bengkalis).

Agroecology

Lamantan is strictly a tropical species, thriving in the hot, wet and humid lowland forests up to 150 m elevation.

Edible Plant Parts and Uses

The fruit is edible. Both green and ripe fruit are used mainly in the preparation of *ulam* with *sambal belacan* and pickles. The ripe fruit is acid sweet and has a pleasant flavour.

Botany

The tree is medium-sized, evergreen, up to 32 m high, bole 53–100 cm diameter and smooth, brown bark. Leaves are coriaceous, elliptic oblong to lanceolate, 7–21 cm by 3–9 cm, base cuneate or

Plate 1 Mature Lamantan fruit

attenuate, apex acute to shortly acuminate, nerves 14–20 pairs, reticulate venation, petiole 4–8 cm long. Panicles are terminal and pyramidal in outline. Flowers are greenish–white, sweetly fragrant, 4-merous; sepals with triangular or ovate lobes, petals ovate to broadly elliptic with ridges half length of petal, stamens 4, 1 fertile, 2–5 mm long, anthers are ovoid, staminodes 0.5 mm, ovary is subglobose with a lateral style. Fruit is a globose drupe, smooth, yellowish-green, 10 cm diameter (Plate 1), with pale yellow, acid-sweet, fibrous flesh covering the stone.

Nutritive/Medicinal Properties

No published information is available on its nutritive and medicinal values.

Other Uses

It is also listed as a timber species (Oey 1951).

Comments

The species has been listed as threatened by the World Conservation Monitoring Centre (1998).

Selected References

Ding Hou (1978) Anacardiaceae. In: van Steenis CGGJ (ed.) Flora Malesiana, ser I, vol 8. Sijthoff & Noordhoff, Alphen aan den Rijn, pp 395–548, 577 pp

Kostermans AJGH, Bompard JM (1993) The mangoes. Their botany, nomenclature, horticulture and utilization. Academic, London, 233 pp

Mukherji S (1949) A monograph on the genus *Mangifera* L. Lloydia 12:73–136

Oey DS (1951) Specific gravity of Indonesian woods and its significance for practical use. Rapport No 46 Bosbouw-proefstation, Bogor, 183 pp

Voon BH, Chin TH, Sim CY, Sabariah P (1988) Wild fruits and vegetables in Sarawak. Sarawak Department of Agriculture, Sarawak, 114 pp

World Conservation Monitoring Centre (1998) *Mangifera similis*. In: IUCN 2010. IUCN Red List of Threatened Species. Version 2010.4. http://www.iucnredlist.org

Pentaspadon motleyi

Scientific Name

Pentaspadon motleyi Hook. f.

Synonyms

Nothoprotium sumatranum Miq., *Pentaspadon moszkowskii* Lauterb., *Pentaspadon minutiflora* B. L. Burtt, *Pentaspadon officinalis* Holmes ex King, *Rhus novoguineensis* Laut.

Family

Anacardiaceae

Common/English Names

Pelajau, Pelong Tree, Plajau, White Pelon Tree

Vernacular Names

Indonesia: Plajau (Sumatra), Djuping, Empit, Empelanjau, Kedondong, Letjut, Panjau, Pelajau, Peladjau, Pelasit, Pilajau, Plajau, Planjau, Polajo, Praju, Tampison, Umpit, Vpie. (Kalimantan);
Malaysia: Pelong Licin, Kedondong (Peninsular), Lakacho, Plajau, Plasin Uping (Sarawak), Empelanjau, Pelajau (Sabah);
New Guinea: Ailala, Laleua, Laleva.

Origin/Distribution

Pelajau is native to Malaysia (peninsular, Sarawak and Sabah); Indonesia (Kalimantan, Irian Jaya, Sumatra), Papua New Guinea (Gulf and Madang Provinces), Bougainville and the Solomon Islands.

Agroecology

In its native tropical range, Pelajau is found in lowland undisturbed primary forests up to 200 m altitude. It occurs mostly in swamps, periodically inundated areas and along rivers and streams on sandy to clayey soils. In secondary forests it is usually present as a pre-disturbance remnant.

Edible Plant Parts and Uses

The fruits are edible fresh, fried, cooked or roasted. The edible kernel is shaped like an almond, but more flattened, and has a pleasant taste when fried or roasted. It is also cooked in a delicious soup. There is a mildly irritating property in the sponge-like shells, so care should be taken when opening the fruit.

Botany

An emergent, tall tree up to 51 m tall and 70 cm dbh with thin spreading buttresses and scaly, grey-brown bark (Plate 2). Leaves are alternate,

Plate 1 Imparipinnate leaves of Pelajau

Plate 2 Large trunk and rough, scaly bark of Pelajau

compound, imparipinnate, 10–20 cm long with 7–9 dark green leaflets which are pink when young (Plate 1). Leaflets are opposite, 5–13 cm × 2–6 cm, pointed apex, rounded base, penni-veined with 6–10 pairs of secondary nerves and with hairy domatia in axils of secondary nerves. Flowers are small, 4 mm in diameter, cream, placed in much–branched panicles, Flowers are bisexual with 5-lobed calyx, 5 cream coloured, imbricate petals, 5 fertile stamens alternating with 5 sterile staminodes, 1-celled ovary with short style. Fruits are, fleshy ovoid, pointed drupe, 3–5 cm by 2–2.75 cm, greenish-brown containing one large, ovoid, oblong compressed seed.

Nutritive/Medicinal Properties

No information on its nutritive value has been published.

The sap is used as an oil as a remedy for itch, scabies and other skin diseases in Sarawak.

Other Uses

The timber is not very durable and only used for cheap flooring.

Comments

The species has been listed as a threatened species by the World Conservation Monitoring Centre (1998).

Selected References

Burkill IH (1966) A dictionary of the economic products of the Malay Peninsula. Revised reprint. 2 vols. Ministry of Agriculture and Co-operatives, Kuala Lumpur, vol 1 (A–H), pp 1–1240, vol 2 (I–Z), pp 1241–2444

Chai PPK (2006) Medicinal plants of Sarawak. Lee Ming Press, Kuching, 212 pp

Corner EJH (1988) Wayside trees of Malaya, vol 1. Malayan Nature Society, Kuala Lumpur, 476 pp

Eswani N, Abd. Kudu K, Nazre M, Awang Noor AG, Ali M (2010) Medicinal plant diversity and vegetation analysis of logged over hill forest of Tekai Tembeling Forest Reserve, Jerantut Pahang. J Agr Sci 2(3): 189–210

Kochummen KM (1989) Anacardiaceae. In: Ng FSP (ed.) Tree flora of Malaya, vol 4. Longman Malaysia, Petaling Jaya, Malaysia, pp 9–57

Lemmens RHMJ, Soerianegara I, Wong WC (eds.) (1995) Plant resources of South-East Asia. No 5(2). Timber trees: minor commercial timbers. Backhuys Publishers, Leiden

Slik JWF (2006) Trees of Sungai Wain. Nationaal Herbarium Nederland http://www.nationaalherbarium.nl/sungaiwain/

World Conservation Monitoring Centre (1998) *Mangifera quadrifida var. quadrifida*. In: IUCN 2010. IUCN Red List of Threatened Species. Version 2010.4. http://www.iucnredlist.org

Pistacia vera

Scientific Name

Pistacia vera L.

Synonyms

Pistacia nigricans Crantz, *Pistacia officinarum* Aiton, *Pistacia reticulata* Willd., *Pistacia terebinthus* Mill., *Pistacia trifolia* L., *Pistacia variifolia* Salisb.

Family

Anacardiaceae, also placed in Pistaciaceae

Common/English Names

Common Pistache, Common Pistachio, Green Almond, Pistachio, Pistachio Nut, Pistacia Nut, Terebinth Nut

Vernacular Names

Afrikaans: Pistachio;
Arabic: Fustuq, Post Berune Pista;
Brazil: Pistache;
Catalan: Pistatxo;
Chinese: A Yue Hun Zi;
Croatian: Zelenožut;
Czech: Pistácie Pravá;
Danish: Ægte Pistacie, Pistacie, Pistacienød;
Dutch: Pistache, Pimpernoot;
Eastonian: Harilik Pistaatsia;
Finnish: Mantelipistaasi, Pistaasimanteli;
French: Pistache, Pistache Comestible, Pistachier, Pistachier Cultivé, Pistachier Vrai;
Galician: Pistache;
German: Echte Pistazie, Grüne Mandel, Italienich Pimpernuss, Pistakinuss, Pistazie, Pistazienbaum, Pistazienkern;
Hungarian: Pisztácia;
Icelandic: Pistasíu;
India: Buzaganja, Guli Pistah, Pistaa, (Hindi), Abhisuka, Mukulakam, Nikocakam (Sanskrit), Maghz Pista, Pista, Pistaa Ilak-Ul-Anbat, Gul Pista, Poast Bairun, Poast Pista, Roghan Pista (Urdu);
Indonesia: Badam Hijau;
Iran: Pista;
Irish: Pistéise
Italian: Pistacchio;
Japanese: Pisutakia, Pisutachio, Pisutachio Nattsu;
Latvian: Pistācija;
Lithuanian: Pistacija;
Malaysia: Badm Hijau;
Maltese: Pistaċċi;
Morocco: Pistâsh;
Norwegian: Pistaj;
Pakistan: Pista;
Persian: Pistaa, Pistah, Pistaka;
Philippines: Pistasiyo (Tagalog);
Polish: Pistacja Właściwa;
Portuguese: Alfóncico, Pistácio, Pistácia, Pistacheira;

Romanian: Fistic;
Russian: Fistashka;
Slovak: Pistácie;
Slovaščina: Pistacija;
Spanish: Alfóncigo, Pistacho, Pistachero, Alfonigo, Alfónsigo;
Swedish: Pistasch, Pistaschmandel;
Turkish: Fistik;
Vietnamese: Cây Hồ Trăn, Màu Hồ Trăn, Quả Hồ Trăn;
Welsh: Pistasio.

Origin/Distribution

Pistachio is indigenous to central Asia (Northeast Iran, North Afghanistan, Southeast Uzbekistan, West Tajikistan, Turkmenistan and West Kyrgyzstan). Pistachio is widely cultivated in the Mediterranean, Middle East, Central Asia including parts of Russia and USA. Iran, Turkey, USA, Syria and Greece are the major producers in descending order. Smaller commercial areas are found in Italy, Cyprus, Tunisia and Jordan. Pistachios are also cultivated in Pakistan, India, New Zealand and Australia.

Agroecology

Pistachios thrive in areas with long, hot and dry summers and moderately cold winters. It can withstand summer temperature in excess of 38°C as in Iran and central Asia, pistachios grows in deserts and semi-deserts at high altitudes. Cool winter chilling estimated at 600–1,500 hours, is required to break dormancy for bud burst. It is frost hardy, and will withstand sub-zero temperatures down to −18°C to −24°C. It is also drought tolerant and prefers full sun, it cannot grow in the shade. The tree abhors high humidity or dampness as it is unfavourable for nut development. A rainfall regime of 400–750 mm has been reported to be required for regular cropping. Pistachio does best in deep, well-drained, moist, sandy to loamy soil. However, it can survive in poor stony, calcareous, slightly acidic or alkaline soils and even saline soils.

Edible Plant Parts and Uses

Pistachio nuts (kernels) are eaten whole, fresh or roasted and salted as dessert nut. The nuts are used in ice cream, sweets, meat dishes, pies, cakes, confectionery such as *baklava* and cold cuts like *mortadella*. Mortadella is a type of Italian sausage flavored with spices, including whole or ground black pepper, myrtle berries, nutmeg, coriander and pistachios. In the Midwest of United States, pistachios is eaten fresh, in salads, pistachio pudding, cool whip, canned fruit and sometimes cottage cheese or marshmallows. The fruits can be made into a flavourful marmalade. On Chios, a Greek island, the flesh of the pistachio fruit surrounding the nut is cooked and preserved in syrup, and is traditionally offered as a sweet delicacy to guests. In Greece, the unripe whole fruits are also used for the preparation of traditional spoon sweets. An edible oil with a pleasant mild flavour is extracted from the seed but is not produced commercially due to the high price of the seed.

Botany

A small dioecious, deciduous, branched tree growing to 6–10 m high with separate male and female trees. Leaves are imparipinnate with 3–10 dark-green leaflets. Leaflets entire, ovate to suborbiculate, 5.5–10 cm by 3–6.5 cm, with prominent veins. Rachis barely winged, pubescent. Inflorescence an axillary panicle. Flowers small, apetalous. Male flowers with 3-5-(−7) stamens with short filaments adnate to disk and large ovoid anthers, pistillode reduced or absent. Female flowers bracteates, staminode absent; disk minute or absent; ovary superior, 1-locule with 1 ovule, style short with 3 spreading stigmas. Drupe ovoid-oblong, 12–30 × 10–12 mm, apiculate, red at maturity enclosing a hard, ivory-coloured, dehiscent endocarp (shell) containing a non-endospermic seed (Plates 1 and 2). Seed (nut) has a mauve coloured testa and light green kernel (flesh) (Plate 3).

Plate 1 Pistachio fruit

Plate 2 Developing pistachio fruit

Plate 3 Pistachio nuts

Nutritive/Medicinal Properties

Food value of raw, unroasted pistachio nuts per 100 g edible portion excluding 47% shell was reported as follows (USDA 2010): water 3.97 g, energy 557 kcal (2,332 kJ), protein 20.61 g, total lipid (fat) 44.44 g, ash 3.02 g, carbohydrate 27.97 g; fibre (total dietary) 10.3 g, total sugars 7.64 g, sucrose 7.00 g, glucose (dextrose) 0.27 g, fructose 0.17 g, maltose 0.20 g, starch 1.67 g; minerals – calcium 107 mg, iron, 4.15 mg, magnesium 121 mg, phosphorus 490 mg, potassium 1,025 mg, sodium 1 mg, zinc 2.20 mg, copper 1.300 mg, manganese 1.200 mg, Se 7.0 μg; vitamins – vitamin C (total ascorbic acid) 5.0 mg, thiamine 0.870 mg, riboflavin 0.160 mg, niacin 1.30 mg, pantothenic acid 0.520 mg, vitamin B-6 1.700 mg, folate (total) 51 μg, vitamin A 553 IU, β-carotene 332 μg, vitamin E (α tocopherol) 2.30 mg, γ-tocopherol 22.60 mg, δ-tocopherol 0.80 mg; lipids – fatty acids (total saturated) 5.440 g, 16:0 (palmitic acid) 4.889 g, 18:0 (stearic acid) 0.466 g, 20:0 (arachidic acid) 0.043 g, 22:0 (behenic acid) 0.043 g; fatty acids (total monounsaturated) 23.319 g, 16:1 undifferentiated (palmitoleic acid) 0.463 g, 18:1 undifferentiated (oleic acid) 22.686 g, 20:1 (gadoleic acid) 0.170 g; fatty acids (total polyunsaturated) 13.455 g, 18:2 undifferentiated (linoleic acid) 13.201 g, 18:3 undifferentiated (linolenic acid) 0.254 g; phytosterols 214 mg, stigmasterol 5 mg, campesterol 10 mg, β-sitosterol 198 mg; amino acids – tryptophan 0.273 g, threonine 0.673 g, isoleucine 0.900 g, leucine 1.554 g, lysine 1.151 g, methionine 0.338 g, cystine 0.357 g, phenylalanine 1.062 g, tyrosine 0.415 g, valine 1. g, arginine 2.028 g, histidine 0.507 g, alanine 0.921 g, aspartic acid 1.817 g, glutamic acid 3.819 g, glycine 0.953 g, proline 0.812 g and serine 1.225 g.

Psitachio nuts are rich in protein, amino acids (17), carbohydrates, dietary fibre, vitamin A, iron, calcium, potassium, phosphorus, magnesium, selenium, copper and monounsaturated (oleic acid) and polyunsaturated fat (linoleic acid). Moreover, they contain significant amounts of thiamin, vitamin Es and vitamin B6, as well as smaller amounts of other B vitamins, phytosterols, stigmasterol, campesterol and β-sitosterol.

Pistachios, sesame seeds, and peanuts are the richest sources of ubiquinone or co-enzyme 10 (UQ10, Q10 or CoQ10) amongst nuts and seeds, with levels over 20 mg/kg (Pravst et al. 2010).

UQ10 is an essential nutrient in the human body; a component of the electron transport chain and participates in aerobic cellular respiration, generating energy in the form of ATP (adenosine triphosphate).

Dried pistachio kernels were found to contain about 55% oil (Clarke et al. 1976). The unsaturated fatty acids oleic and linoleic constituted about 80% of this oil. Carbohydrate analysis of a hydrolysed sample showed the predominant sugar to be glucose followed by galactose and mannose with traces of fucose, arabinose and xylose. The total amino acid content was about 25% by weight. Nine essential amino acids were present with only cystine in a limiting amount and with a high level of lysine.

Sixteen different 3-alkylphenols (cardanols) with a saturated, monounsaturated and diunsaturated chain were detected in pistachio kernels (Saitta et al. 2009). The most abundant cardanols were 3-(8-pentadecenyl)-phenol, 3-(10-pentadecenyl)-phenol, 3-pentadecyl-phenol and 3-(10-heptadecenyl)-phenol. Total amount of cardanols in the oils (mean of five samples) was determined as 440 mg/kg.

Fresh unripe pistachio fruits were found to be richer in essential oil (0.5%, w/w) than the leaves (0.1%, w/w) (Tsokou et al. 2007). Twenty one compounds were identified in the essential oil of the fruits and the major components were (+)-α-pinene (54.6%) and terpinolene (31.2%). The enantiomeric ratio of the major constituents of the essential oil of the fruits was determined and it was found that the ±-α-pinene ratio was 99.5:0.5, ±-limonene 80:20, ±-β-pinene 96:4, and ±-α-terpineol 0:100. Thirty three compounds were identified in the essential oil of the leaves and the major components were found to be α-pinene (30.0%), terpinolene (17.6%) and bornyl acetate (11.3%). The major components in all essential oil (EO) from the fruits, peduncles, and leaves of pistachio were monoterpenes (Dragull et al. 2010). Limonene was the primary component with 80–85% in fruit EO and 78–81% in leaf EO. The lowest limonene levels (50–52%) were found in peduncle EO. The closely related isomer, α-terpinolene, was generally the second highest compound with 6–10% in fruit EO and 9–10% in leaf EO. Terpinolene was also the second highest compound in peduncle EO, in most instances with an average of 14% at higher level relative to limonene. Peduncles were unique in containing relatively high levels of α-thujene. Fruits and leaves were virtually devoid of α-thujene (<1%). All other components were relatively minor. Minor compounds that were above the 1% threshold include five related monoterpenes: bornyl acetate, camphene, Δ3Carene, myrcene, and α-pinene.

The total polyphenol levels in pistachios, measured as mg of gallic acid equivalent (GAE), were: 201, 349 and 184.7 mg GAE/100 g DM in unripe, ripe and dried ripe samples, respectively (Ballistreri et al. 2009). Most phenolics in ripe pistachios were found to be anthocyanins. They increased with ripening, while the sun drying process caused a substantial loss. Flavonoids found in all pistachio samples were daidzein, genistein, daidzin, quercetin, eriodictyol, luteolin, genistin and naringenin, which decreased both with ripening and drying. Before the drying process both unripe and ripe pistachios showed a higher content of trans-resveratrol than dried ripe samples. γ-tocopherol was the major vitamin E isomer found in pistachios. The total content (of α- and γ-tocopherols) decreased, both during ripening and during the drying process. These results suggested that unpeeled pistachios can be considered an important source of phenolics, particularly of anthocyanins.

Of the products typically consumed as snack foods, pistachio and sunflower kernel were found to be richest in phytosterols (270–289 mg/100 g) (Phillips et al. 2005). β-sitosterol, δ-5-avenasterol, and campesterol were predominant. Campestanol ranged from 1.0–12.7 mg/100 g. Only 13 mg/100 g β-sitosterol was found in pistachio seed kernel, although total sterol content was high (265 mg/100 g). Phytosterol concentrations were greater than reported in existing food composition databases, probably due to the inclusion of steryl glycosides, which represent a significant portion of total sterols in nuts and seeds. Phytosterols and phytostanols have received much attention in the last 5 years because of their cholesterol-lowering properties (Moreau et al.

2002). Over the years two spreads had been commercialized, one containing phytostanyl fatty-acid esters and the other phytosteryl fatty-acid esters, and were shown to significantly lower serum cholesterol at dosages of 1–3 g per day. The popularity of these products caused the medical and biochemical community to focus more attention on phytosterols and consequently research activity on phytosterols had increased dramatically.

Studies in Ireland presented data that indicated nuts like brazil nut, pecan, pine-nut, pistachio and cashew nuts are a good dietary source of unsaturated fatty acids, tocopherols, squalene and phytosterols (Ryan et al. 2006). Nuts were found to contain bioactive constituents including phytosterols, tocopherols and squalene that elicited cardio-protective effects. The total oil content of the nuts was found to range from 40.4–60.8% (w/w) while the peroxide values ranged from 0.14–0.22 mEq O_2/kg oil. The most abundant monounsaturated fatty acid was oleic acid (C18:1), while linoleic acid (C18:2) was the most prevalent polyunsaturated fatty acid. The levels of total tocopherols ranged from 60.8–291.0 mg/g. Squalene ranged from 39.5 mg/g oil in the pine nut to 1377.8 mg/g oil in the brazil nut. β-sitosterol was the most prevalent phytosterol, ranging in concentration from 1325.4–4685.9 mg/g oil.

Pistachios possess many pharmacological properties which are elaborated below.

Antioxidant Activity

Pistachio nuts are a rich source of phenolic compounds, and have recently been ranked among the first 50 food products highest in antioxidant potential (Tomaino et al. 2010). The total content of phenolic compounds in pistachios was shown to be significantly higher in skins than in seeds of Bronte pistachio. By HPLC analysis, gallic acid, catechin, eriodictyol-7-O-glucoside, naringenin-7-O-neohesperidoside, quercetin-3-O-rutinoside and eriodictyol were found both in pistachio seeds and skins. Additionally, genistein-7-O-glucoside, genistein, daidzein and apigenin appeared to be present only in pistachio seeds, while epicatechin, quercetin, naringenin, luteolin, kaempferol, cyanidin-3-O-galactoside and cyanidin-3-O-glucoside were found only in pistachio skins. Pistachio skins were shown to possess a better activity with respect to seeds in all different antioxidant assays (DPPH assay, Folin-Ciocalteau colorimetric method and TEAC assay, SOD-mimetic). The excellent antioxidant activity of pistachio skins could be explained by its higher content of antioxidant phenolic compounds. By HPLC-TLC analysis, gallic acid, catechin, cyanidin-3-O-galactoside, eriodictyol-7-O-glucoside and epicatechin appeared to be responsible for the antioxidant activity of pistachio skin, together with other unidentified compounds. The scientists asserted that introduction of pistachios in the daily diet may be beneficial in protecting human health and well-being against cancer, inflammatory diseases, cardiovascular pathologies and, more generally, pathological conditions related to free radical overproduction. On the other hand, pistachio skins could be successfully employed in food, cosmetic and pharmaceutical industry.

The total antioxidant activity was found to be much higher (50-fold) in the hydrophilic than in the lipophilic extract of a Sicilian variety of pistachio nut and contained substantial amounts of total phenols (Gentile et al. 2007). The hydrophilic extract inhibited dose-dependently both the metal-dependent and -independent lipid oxidation of bovine liver microsomes, and the Cu^{2+}-induced oxidation of human low-density lipoprotein (LDL). Peroxyl radical-scavenging as well as chelating activity of nut components were suggested to explain the observed inhibition patterns. Among tocopherols, γ-tocopherol was the only vitamin E isomer found in the lipophilic extract that did not contain any carotenoid. Vitamin C was found only in a modest amount. The hydrophilic extract was a source of polyphenol compounds which contained trans-resveratrol, proanthocyanidins, and an appreciable amount of the isoflavones daidzein and genistein, 3.68 and 3.40 mg per 100 g of edible nut, respectively. With the exception of isoflavones that appeared unmodified, the amounts of other bioactive molecules were considerably reduced in the pistachio nut after roasting, and the total antioxidant activity decreased by about 60%. The findings provided evidence that the Sicilian pistachio nut may be considered for its bioactive components and could effectively contribute to a healthy status.

Studies in Turkey showed that consumption of pistachio as 20% of daily caloric intake enhanced serum paraoxonase activity by 35% and arylesterase activity by 60%, and inhibited LDL (low density lipoproteins) cholesterol oxidation, compared with the control group (Aksoy et al. 2007). However, increased antioxidant activity was subdued when pistachio intake was increased to 40% of daily caloric intake. The results showed that consumption of pistachio as 20% of daily caloric intake led to significant improvement in HDL (high density lipoproteins) and TC (total cholesterol)/HDL ratio and inhibited LDL cholesterol oxidation. The results suggested that pistachios may be beneficial for both prevention and treatment of coronary artery disease. Separate studies in Turkey (Kocyigit et al. 2006) demonstrated that after 3 weeks of pistachio diet, the mean plasma total cholesterol, MDA (malondialdehyde) levels and, total cholesterol/high-density lipoprotein HDL and LDL/HDL ratios were found to be significantly decreased but HDL and AOP (antioxidant potential) levels and AOP/MDA ratios were significantly enhanced. Triglyceride and LDL (low-density lipoprotein) levels also decreased but this was not statistically significant. These results indicated that consumption of pistachio nuts decreased oxidative stress, and improved total cholesterol and HDL levels in healthy volunteers.

Antihyperlipidemic/ Antihypercholesterolemic Activities

Studies in Virginia showed that a diet consisting of 15% of calories as pistachio nuts (about 2–3 oz per day) over a 4 week period could favourably improve some lipid profiles in subjects with moderate hypercholesterolemia and may reduce risk of coronary disease (Sheridan et al. 2007). On the pistachio nut diet, a statistically significant decrease was seen for percent energy from saturated fat in TC/HDL-C and in LDL-C/HDL-C. On the pistachio nut diet, statistically significant increases were seen for percent energy from polyunsaturated fat, fibre intake and in HDL-C. No statistically significant differences were seen for total cholesterol, triglycerides, LDL-C, VLDL-C, apolipoprotein A-1 or apolipoprotein B-100. No statistically significant differences were found for total energy or percent of energy from protein, carbohydrate or fat and no changes were observed in BMI (Body Mass Index) or blood pressure.

Extensive studies conducted over the years in Pennsylvania State confirmed that nut consumption, in particular pistachio nuts, lowered cardiovascular disease (CVD) risk (Kay et al. 2007; Gebauer et al. 2007, 2008; West et al. 2007). Pistachios in particular significantly reduced levels of low-density lipoprotein (LDL cholesterol) while increasing antioxidant levels in the serum of volunteers. Pistachios have high levels of lutein, β-carotene and γ-tocopherol relative to other nuts (Kay et al. 2007).

Kay et al. (2007), in a randomized controlled crossover feeding study in 28 hypercholesterolemic adults found that pistachio diets of 1.5 oz/day pistachios (1.5 oz PD; 30% total fat, 8% saturated fatty acids (SFA)) and with 3.0 oz/day pistachios (3.0 oz PD; 34% total fat, 8% SFA) significantly reduced oxidized LDL from baseline compared to the Step-1 diet without pistachios (25% total fat, 8% SFA). Furthermore, participants receiving the pistachio enriched diets had significant increases in lutein compared to baseline and significantly higher lutein than the Step-1 diet. The 3.0 oz PD had greater increases in β-carotene and γ-tocopherol compared to the Step-1 diet. Regression analysis showed that increases in lutein following the 3.0 oz pistachio diet were correlated with a decrease in oxidized LDL, indicating that lutein in the pistachio-rich diets may improve CVD risk through reducing oxidized LDL. In a parallel paper, West et al. (2007) in a randomized, controlled, crossover feeding study of 28 hypercholesterolemic adults with normal blood pressure found that both pistachio diets decreased mean systolic BP, although the magnitude of this effect was larger on the 1.5 oz diet. The 3 oz diet had larger effects on systemic hemodynamics and significantly reduced total peripheral vascular resistance (TPR) and heart rate. These changes were offset by increases in cardiac output. The results suggested a novel dose-dependent mechanism for the cardioprotective effects of pistachios. In another a randomized crossover controlled-feeding study, involving 28 individuals with LDL cholesterol ≥2.86 mmol/l, the two pistachio diets comprising 1 serving/day of a pistachio diet (1 PD; 10% of energy from pistachios; 30% total fat; 8% SFAs, 12% MUFAs (monounsaturated fatty acids), and 6% PUFAs (polyunsaturated fatty acids)), and 2 servings/day of a pistachio diet (2 PD; 20% of energy from pistachios; 34% total fat; 8% SFAs, 15% MUFAs, and 8% PUFAs) decreased total cholesterol (−8%), LDLcholesterol (−11.6%), non-HDL cholesterol (−11%), apo B (−4%), apo B/apo A-I (−4%), and plasma SCD activity (−1%) compared with the control diet with no pistachios [25% total fat; 8% saturated fatty acids (SFAs), 9% MUFAs, and 5% PUFAs] (Gebauer et al. 2007; 2008). The two pistachio diets, respectively, educed a dose-dependent lowering of total cholesterol/HDL cholesterol (−1% and −8%), LDL cholesterol/HDL cholesterol (−3% and −11%), and non-HDL cholesterol/HDL cholesterol (−2% and −10%). The data showed that inclusion of pistachios in a healthy diet beneficially affected CVD risk factors in a dose-dependent manner, which may reflect effects on plasma stearoyl-CoA desaturase activity (SCD). Kay et al. (2010) also reported that when participants consumed the pistachio-enriched diets, they had higher plasma lutein, α-carotene, and β-carotene concentrations than after the baseline diet. Participants also had greater plasma lutein and γ-tocopherol relative to the lower-fat control diet. After the 2 PD diet period, participants also had lower serum oxidized-LDL concentrations than following the baseline diet period and after the control diet period. The change in oxidized-LDL from baseline correlated positively with the change in LDL-cholesterol across all treatments.

Studies in California confirmed that eating pistachio nuts instead of other dietary fat calories could improve lipid profiles, thereby reducing coronary risk (Edwards et al. 1999). After 3 weeks, there was a decline in total cholesterol, an increase in HDL, a decrease in the total cholesterol/HDL ratio and a decrease in the LDL/HDL ratio in patients with moderate hypercholesterolemia. Triglycerides and LDL levels decreased, but not significantly. Body weight and blood pressure remained constant throughout the study.

Marinou et al. (2010) found that during short-term administration concurrently with atherogenic diet, both the methanolic (ME) and cyclohexane (CHE) extracts of *Pistachio vera* were found to be beneficial on HDL-, LDL-cholesterol and aortic intimal thickness. The ME additionally exhibited an antioxidant effect and significant decrease of aortic surface lesions. ME group of animals had lower MDA (malondialdehyde) and lower γ-GT (γ glutamyl transpeptidase). The results indicated that *P. vera* dietary inclusion, in particular its ME, could be potentially beneficial in atherosclerosis management.

In recent studies, Sari et al. (2010) found that pistachio nut consumption besides lipid lowering, elicited other favourable effects that included improved blood glucose level, endothelial function, and some indices of inflammation and oxidative status. Compared with the Mediterranean diet, the pistachio diet reduced glucose, low-density lipoprotein, total cholesterol and triacylglycerol significantly and high-density lipoprotein non-significantly. Total cholesterol/high-density lipoprotein and low-density lipoprotein/high-density lipoprotein ratios declined significantly. The pistachio diet significantly improved endothelium-dependent vasodilation, reduced serum interleukin-6, total oxidant status, lipid hydroperoxide, and malondialdehyde and increased superoxide dismutase. However, there was no significant change in C-reactive protein and tumour necrosis factor-α levels.

Antinociceptive/Anti-inflammatory Activities

Hydroalcoholic gum extract of *Pistacia vera* exhibited antinociceptive, anti-inflammatory effect (Parvardeh et al. 2002b) The intraperitoneal (i.p.)

injection of pistachio extract (0.25–2 g/kg) in mice showed significant and dose-dependent antinociceptive activity in hot plate assay. The extract (0.25 and 0.5 g/kg, i.p.) exhibited antinociceptive activity against acetic acid-induced writhing, which was not blocked by naloxone. The extract (0.25–1 g/kg, i.p.) also showed antinociceptive activities in the formalin test. In the xylene ear edema assay, P. vera extract (0.25–1 g/kg, i.p.) showed significant anti-inflammatory activity in mice and exerted appreciable activity against chronic inflammation induced by the cotton pellet in rats as well. It was concluded that ethanolic extract of P. vera gum had antinociceptive effect that may be mediated by opioid receptors, as well as inhibition of inflammatory mediators. The extract showed also anti-inflammatory effect against acute and chronic inflammation.

Oleoresin from P. vera was shown to possess a marked anti-inflammatory activity against carrageenan-induced hind paw edema model in mice without inducing any gastric damage at both 250 and 500 mg/kg doses whereas the rest of the extracts were totally inactive (Orhan et al. 2006b). While the oleoresin was found to display significant antinociceptive activity at 500 mg/kg dose, the ethanolic and aqueous extracts belonging to fruit, leaf, branch and peduncle of *Pistacia* vera did not exhibit any noticeable antinociception in p-benzoquinone-induced abdominal contractions in mice. Fractionation of the oleoresin indicated the n-hexane fraction to be active, which further led to recognition of some monoterpenes, mainly α-pinene (77.5%).

Hepatoprotective Activity

Pre-treatment of rats with P. vera aqueous ethanol gum extracts (0.5 and 1 g/kg, i.p.) reduced SGPT (serum glutamic pyruvic transaminase) value in CCl_4 treated animals but had no effect on SGOT (serum glutamic oxaloacetic transaminase) level (Parvardeh et al. 2002a). LD_{50} value and the maximum non fatal dose of the gum were 3.77 and 1 g/kg (i.p.) in mice, respectively. The extract was found to contain flavonoids, saponins and hydrotannins. The results suggested that the ethanolic extract of P. vera gum had a protective activity against liver damage induced by CCl_4 in rats.

Neuroprotective Activity

Treatment with P. vera gum extract significantly and in a non-dose-dependent manner reduced brain malondialdehyde (MDA) level by 63% and increased antioxidant power of brain by 235% in comparison to the controls (Mansouri et al. 2005). The gum was found to contain saponins, tannins, and flavonoids. The results suggested that P. vera gum may exhibit neuroprotective effects against ischemia.

Antimutagenic Activity

The results of antimutagenicity studies showed that phenolic compounds of pistachio green hull exhibited antimutagenic activity against direct mutagen of 2-nitrofluorene (Rajaei et al. 2010).

Antimicrobial/Antiviral Activities

Aqueous, chloroformic and ethanolic extracts of P. vera hull were found to inhibit the growth and acid production of tested bacteria – *Streptococcus mutans, Streptococcus salivarious, Streptococcus sobrinus* and *Streptococcus sanguis* (Kamrani et al. 2007). Ethanolic extract displayed the most potent antimicrobial activity. In-vitro studies showed that the ethanolic extract at concentrations of 2, 6 and 10% w/v, could inhibit the growth as well as the acid-producing ability of *Streptococcus mutans*. In-vivo study showed that mouth rinse prepared by suspending 10% of ethanolic extract in PEG (polyethylene glycol) 300 (w/v) could reduced salivary bacterial count by more than 55%. It also elicited about 75% inhibition of adherence and inhibition of salivary glycolysis up to 3 hours post rinsing.

The essential oil obtained by hydrodistillation of *Pistacia vera* gum exhibited bacteriostatic activity against 12 clinical isolates of *Helicobacter pylori* (Ramezani et al. 2004). All isolates were sensitive to the essential oil, and the minimum inhibition concentration (MIC) was 1.55 mg/ml for all isolates.

P. vera essential oil was found to contain about 89.67% monoterpenes, 8.1% oxygenated monoterpenes and 1.2% diterpenes (Alma et al. 2004). α-Pinene (75.6%), β-pinene (9.5%), trans-verbenol (3.0%), camphene (1.4%), trans-pinocarveol (about 1.20%), and limonene (1.0%) were the major components. The antimicrobial results showed that the essential oil inhibited nine bacteria and all the yeasts studied, and the activities were considerably dependent upon concentration and its bioactive compounds such as carvacrol, camphene, and limonene (Alma et al. 2004). The essential oil of the gum was found to be a more effective yeasticide than Nystatin, a synthetic yeasticide, but the antibacterial activities of the essential oil were lower than those of standard antibiotics, ampicillin sodium, and streptomycin sulfate.

Fifteen lipophilic extracts obtained from different pistachio plant parts (leaf, branch, stem, kernel, shell skins, seeds) were screened against both standard and the isolated strains of *Escherichia coli*, *Pseudomonas aeruginosa*, *Enterococcus faecalis*, *Staphylococcus aureus*, *Candida albicans* and *Candida parapsilosis* and against Herpes simplex (DNA) and Parainfluenza viruses (RNA) by using Vero cell line (Ozçelik et al. 2005). Ampicillin, ofloxocin, ketoconazole, fluconazole, acyclovir and oseltamivir were used as the control agents. The extracts showed little antibacterial activity between the range of 128–256 μg/ml concentrations whereas they had noticeable antifungal activity at the same concentrations. Pistachio kernel and seed extracts showed significant antiviral activity against Herpes simplex (DNA) and Parainfluenza viruses (RNA) compared to the rest of the extracts and the controls.

Studies showed that essential oil resin of *P. vera* L. exhibited antimicrobial activity against Gram negative bacteria (*Escherichia coli* and *Proteus* spp.) as well as Gram positive bacteria (*Staphylococcus aureus*) with *Proteus* spp. the most inhibited (Ghalem and Mohamed 2009). Aqueous and purified pistachio green hull extracts inhibited the growth of Gram positive bacteria with *Bacillus cereus* being the most susceptible with MIC of 1 mg/ml and 0.5 mg/ml for the crude and purified extracts, respectively (Rajaei et al. 2010).

Antiprotozoal Activity

At 4.8 μg/ml concentration, the n-hexane extract of *Pistacia vera* branch (PV-BR) significantly inhibited (77.3%) the growth of *Leishmania donovani*, whereas the dry leaf extract (PV-DL) was active against *Plasmodium falciparum* (60.6% inhibition) (Orhan et al. 2006a). The IC_{50} values of these extracts were determined as 2.3 μg/ml (PV-BR, *Leishmania donovani*) and 3.65 μg/ml (PV-DL, *Plasmodium falciparum*). None of the extracts induced cytotoxicity on mammalian cells.

Immunoglobulin E (IgE)-binding Activity

In recent studies, Noorbakhsh et al. (2010) found that the IgE-binding capacity was significantly lower for the protein extract prepared from steam-roasted than from raw and dry-roasted pistachio nuts. The results of sensory evaluation analysis and hedonic rating found no significant differences in color, taste, flavor, and overall quality of raw, roasted, and steam-roasted pistachio nut treatments. The most significant finding of the present study was the successful reduction of IgE-binding by pistachio extracts using steam-roast processing without any significant changes in sensory quality of product.

Like other members of the Anacardiaceae family (which includes poison ivy, sumac, mango, and cashew), pistachios were found to contain urushiol, an irritant that could cause allergic reactions (Aguilar-Ortigoza et al. 2003).

Traditional Medicinal Uses

In China, the plant is used in traditional folk medicine for the treatment of abdominal ailments, abscesses, amenorrhoea, bruises, chest ailments, circulation, dysentery, gynecopathy, pruritus, rheumatism, sclerosis of the liver, sores and trauma. The seed is said to be sedative and tonic.

Other Uses

Pistachio seed yields up to 40% of a non-drying oil but is not used commercially due to the high value of the seed for food. Male trees yield a small quantity of a high grade resin which is used in paints, lacquers etc. Research has shown that the bio-oil obtained from soft shell of pistachio can be used as a renewable fuel and chemical feedstock (Demiral et al. 2009). The maximum bio-oil yield was 33.18%.

Activated carbons, a type of lignocellulosic can be processed from pistachio-nut shells (Yang and Lua 2003). Although commercial activated carbon is a preferred adsorbent for toxic pollutants (minerals, phenols) from waste water removal, its widespread use is restricted due to the high cost. As such, alternative non-conventional adsorbents have been investigated. The natural materials, waste materials from industry and agriculture and bioadsorbents such as activated carbons from pistachio shells can be employed as inexpensive adsorbents.

Comments

Selected References

Aguilar-Ortigoza CJ, Sosa V, Aguilar-Ortigoza M (2003) Toxic phenols in various Anacardiaceae species. Econ Bot 57(3):354–364

Aksoy N, Aksoy M, Bagci C, Gergerlioglu HS, Celik H, Herken E, Yaman A, Tarakcioglu M, Soydinc S, Sari I, Davutoglu V (2007) Pistachio intake increases high density lipoprotein levels and inhibits low-density lipoprotein oxidation in rats. Tohoku J Exp Med 212(1):43–48

Alma MH, Nitz S, Kollmannsberger H, Digrak M, Efe FT, Yilmaz N (2004) Chemical composition and antimicrobial activity of the essential oils from the gum of Turkish pistachio (*Pistacia vera* L.). J Agric Food Chem 52(12):3911–3914

Ballistreri G, Arena E, Fallico B (2009) Influence of ripeness and drying process on the polyphenols and tocopherols of *Pistacia vera* L. Molecules 14(11):4358–4369

Clarke JA, Brar GS, Procopiou J (1976) Fatty acid, carbohydrate and amino acid composition of pistachio (*Pistacia vera*) Kernels. Plant Foods Hum Nutr 25(3–4):219–225

Demiral I, Atilgan NG, Sensoz S (2009) Production of biofuel from soft shell of pistachio (*Pistacia vera* L.). Chem Eng Commun 196(1–2):104–115

Dragull K, Beck JJ, Merrill GB (2010) Essential oil yield and composition of *Pistacia vera* 'Kerman' fruits, peduncles and leaves grown in California. J Sci Food Agric 90(4):664–668

Duke JA, Ayensu ES (1985) Medicinal plants of China, vols 1 &2. Reference Publications, Algonac, 705 pp

Edwards K, Kwaw I, Matud J, Kurtz I (1999) Effect of pistachio nuts on serum lipid levels in patients with moderate hypercholesterolemia. J Am Coll Nutr 18(3):229–232

Facciola S (1990) Cornucopia. A source book of edible plants. Kampong, Vista, 677 pp

Gebauer SK, Kay CD, West SG, Alaupovic P, Psota TL, Kris-Etherton PM (2007) Pistachios beneficially affect multiple lipid and apolipoprotein CVD risk factors. FASEB J 21:682.3

Gebauer SK, West SG, Kay CD, Alaupovic P, Bagshaw D, Kris-Etherton PM (2008) Effects of pistachios on cardiovascular disease risk factors and potential mechanisms of action: a dose-response study. Am J Clin Nutr 88(3):651–659

Gentile C, Tesoriere L, Butera D, Fazzari M, Monastero M, Allegra M, Livrea MA (2007) Antioxidant activity of Sicilian pistachio (*Pistacia vera* L. var. Bronte) nut extract and its bioactive components. J Agric Food Chem 55(3):643–648

Ghalem BR, Mohamed B (2009) Antimicrobial activity evaluation of the oleoresin oil of *Pistacia vera* L. Afr J Pharm Pharmacol 3(3):92–96

Kamrani YY, Amanlou M, Esmaeelian B, Bidhendi SM, Jamei MS (2007) Inhibitory effects of a flavonoid-rich extract of *Pistacia vera* hull on growth and acid production of bacteria involved in dental plaque. Int J Pharmacol 3(3):219–226

Kay CD, GebauerSK, West SG, Kris-Etherton PM (2007) Pistachios reduce serum oxidized LDL and increase serum antioxidant levels. FASEB J 21:847.19

Kay CD, Gebauer SK, West SG, Kris-Etherton PM (2010) Pistachios increase serum antioxidants and lower serum oxidized-LDL in hypercholesterolemic adults. J Nutr 140(6):1093–1098

Kocyigit A, Koylu AA, Keles H (2006) Effects of pistachio nuts consumption on plasma lipid profile and oxidative status in healthy volunteers. Nutr Metab Cardiovasc Dis 16(3):202–209

Kroon AHJ (1969) The pistachio nut (*Pistacia vera* L.). Trop Abstr 24:73–75

Mansouri SMT, Naghizadeh B, Hosseinzadeh H (2005) The effect of *Pistacia vera* L. gum extract on oxidative damage during experimental cerebral ischemia-reperfusion in rats. Iran Biomed J 9(4):181–185

Marinou KA, Georgopoulou K, Agrogiannis G, Karatzas T, Iliopoulos D, Papalois A, Chatziioannou A, Magiatis P, Halabalaki M, Tsantila N, Skaltsounis LA, Patsouris E, Dontas IA (2010) Differential effect of *Pistacia vera* extracts on experimental atherosclerosis in the rabbit animal model: an experimental study. Lipids Health Dis 9:73

Moreau RA, Whitaker BD, Hicks KB (2002) Phytosterols, phytostanols, and their conjugates in foods: structural diversity, quantitative analysis, and health-promoting uses. Prog Lipid Res 41(6):457–500

Noorbakhsh R, Mortazavi SA, Sankian M, Shahidi F, Jam SM, Nasiraii LR, Falak R, Sima HR, Varasteh A (2010) Influence of processing on the allergenic properties of pistachio nut assessed in vitro. J Agric Food Chem 58(18):10231–10235

Orhan I, Aslan M, Sener B, Kaiser M, Tasdemir D (2006a) In vitro antiprotozoal activity of the lipophilic extracts of different parts of Turkish *Pistacia vera* L. Phytomed 13(9–10):735–739

Orhan I, Küpeli E, Aslan M, Kartal M, Yesilada E (2006b) Bioassay-guided evaluation of anti-inflammatory and antinociceptive activities of pistachio, Pistacia vera L. J Ethnopharmacol 105(1–2):235–240

Ozçelik B, Aslan M, Orhan I, Karaoglu T (2005) Antibacterial, antifungal, and antiviral activities of the lipophilic extracts of *Pistacia vera*. Microbiol Res 160(2):159–164

Padulosi S, Caruso T, Barone E (1996) Taxonomy, distribution, conservation and uses of *Pistacia* genetic resources. IPGRI, Rome, 69 pp

Parvardeh S, Niapour M, Hosseinzadeh H (2002a) Hepatoprotective activity of *Pistacia vera* L. gum extract in mice. J Med Plants 1:27–34

Parvardeh S, Niapour M, Nassiri Asl M, Hosseinzadeh H (2002b) Antinociceptive, anti-inflammatory and toxicity effects of *Pistacia vera* extract in mice and rats. J Med Plants 1:59–68

Phillips KM, Ruggio DM, Ashraf-Khorassani M (2005) Phytosterol composition of nuts and seeds commonly consumed in the United States. J Agric Food Chem 53(24):9436–9445

Pravst I, Mitek K, Mitek J (2010) Coenzyme Q10 contents in foods and fortification strategies. Crit Rev Food Sci Nutr 50(4):269–280

Rajaei A, Barzegar M, Mobarez AM, Sahari MA, Esfahani ZH (2010) Antioxidant, anti-microbial and antimutagenicity activities of pistachio (*Pistachia vera*) green hull extract. Food Chem Toxicol 48(1):107–112

Ramezani M, Khaje-karamoddin M, Karimi-fard V (2004) Chemical composition and anti-Helicobacter pylori activity of the essential oil of *Pistacia vera*. Pharm Biol 42(7):488–490

Ryan E, Galvin K, O'Connor TP, Maguire AR, O'Brien NM (2006) Fatty acid profile, tocopherol, squalene and phytosterol content of Brazil, pecan, pine, pistachio and cashew nuts. Int J Food Sci Nutr 57(3–4):219–228

Saitta M, Giuffrida D, La Torre GL, Potortì AG, Dugo G (2009) Characterisation of alkylphenols in pistachio (*Pistacia vera* L.) kernels. Food Chem 117(3):451–455

Sari I, Baltaci Y, Bagci C, Davutoglu V, Erel O, Celik H, Ozer O, Aksoy N, Aksoy M (2010) Effect of pistachio diet on lipid parameters, endothelial function, inflammation, and oxidative status: a prospective study. Nutrition 26(4):399–404

Sheridan MJ, Cooper JN, Erario M, Cheifetz CE (2007) Pistachio nut consumption and serum lipid levels. J Am Coll Nutr 26(2):141–148

Tomaino A, Martorana M, Arcoraci T, Monteleone D, Giovinazzo C, Saija A (2010) Antioxidant activity and phenolic profile of pistachio (*Pistacia vera* L., variety Bronte) seeds and skins. Biochimie 92(9):1115–1122

Tsokou A, Georgopoulou K, Melliou E, Magiatis P, Tsitsa E (2007) Composition and enantiomeric analysis of the essential oil of the fruits and the leaves of *Pistacia vera* from Greece. Molecules 12:1233–1239

U.S. Department of Agriculture, Agricultural Research Service (USDA) (2010) USDA National Nutrient Database for Standard Reference, Release 23. Nutrient Data Laboratory Home Page. http://www.ars.usda.gov/ba/bhnrc/ndl

West SG, Kay CD, Gebauer SK, Savastano DM, Diefenbach CM, Kris-Etherton PM (2007) Pistachios reduce blood pressure and vascular responses to acute stress in healthy adults. FASEB J 21:682.6

Whitehouse WE (1957) The pistachio nut - a new crop for the western United States. Econ Bot 11:281–321

Yang T, Lua AC (2003) Characteristics of activated carbons prepared from pistachio-nut shells by physical activation. J Colloid Interface Sci 267(2):408–417

Schinus molle

Scientific Name

Schinus molle L.

Synonyms

Schinus angustifolia Sessé & Moc. nom. illeg., *Schinus angustifolia* Sessé & Moc., *Schinus areira* L, *Schinus bituminosus* Salisb., *Schinus huigan* Molina, *Schinus molle* var. *areira* (L.) DC., *Schinus molle* var. *argentifolius* Marchand, *Schinus molle* var. *huigan* (Molina) Marchand, *Schinus molle* var. *huyngan* (Molina) March., *Schinus occidentalis* Sessé & Moc.

Family

Anacardiaceae

Common/English Names

American Pepper, Australian Pepper, Brazil Pepper-Tree, Brazilian Peppertree, California Pepper Tree, California Pepper-Tree, California Peppertree, Chilean Pepper Tree, Escobilla, False Pepper, Mastic Tree, Molle, Molle Del Peru, Pepper Berry Tree, Pepper Rose (Madagascar), Pepper Tree, Pepperina, Peppertree, Peruvian Mastic Tree, Peruvian Pepper, Peruvian Pepper-Tree, Peruvian Peppertree, Pink Pepper, Pink Peppercorns (Fruits), Weeping Pepper.

Vernacular Names

Amharic: Qundo Berbere;
Arabic: Felfel-Kazib, Filfilrafie;
Argentina: Aguaribay, Aguaraiba, Pimienta De America, Pimientillo Escobilla, Chichita, Gualeguay, Falso Pimentero, Molle, Molle Del Perú, Molle Del Incienso (Spanish);
Catalan: Pebre Del Perú;
Chinese: Jia Zhou Hu Jiao;
Columbia: Muelle (Spanish);
Danish: Peruansk Pebertrae, Peruviansk Pebertrae;
Dutch: Braziliaanse Peperboom, Californische Peperboom, Mastixbaum, Peruanischer Pfefferbaum, Peruaanse Peperboom, Peperstruik;
Eastonian: Pehme Skiinus;
Ethiopia: Qundo Berbere (Amargna);
Finnish: Brasilianpippuripuu;
French: Faux Poivrier, Molée Des Jardins, Poivrier d'Amérique, Poivre Rosé;
German: Brasilianischer Pfeffer, Peruanischer Pfeffer, Peruanischer Pfefferbaum, Rosa Pfeffer, Rosé-Pfeffer;
Guatamala: Pirú;
Hebrew: Pilpelon Damui-Aley;
Hungarian: Perui Bors;
Italian: Falso Pepe Peruviano;
Lithuanian: Švelnusis Pirulis;
Mexico: Pirú;
Peru: Cullash, Huinan, Mulli, Molle;
Portuguese: Anacauíta, Araguaraíba, Aroeira, Aroeira-Do-Amazonas, Aroeira Folha De Salso, Aroeira-Mansa, Aroeira-Mole, Aroeira-Periquita,

Aroeira-Salso, Aroeira-Vermelha, Corneiba Dos Tupis, Pimenteira-Do-Peru, Fruto-De-Sabiá, Pimenteira Bastarda, Pimenteiro, Terebinto;
Russian: Perets Peruanskii, Peruanskii Perets, Shinus Miagkii;
South Africa: Umpelempele (Zulu);
Spanish: Arveira, Pimienta, Pimientero Falso, Lentisco Del Perú, Pirul;
Swahili: Mpilipili,
Swedish: Peruanskt Pepparträd, Rosépeppa;
Tigrina: Berbere-Tselim, Berebere-Tselim;
Turkish: Yalancı Karabiber Ağacı;
Uruguay: Aguaribay, Anacahuita, Molle Del Perú, Pimentero, Pirul.

Origin/Distribution

Schinus molle is found in the subtropical regions of South America – Argentina, Bolivia, Brazil, Chile, Columbia, Ecuador, Peru and Uruguay. It was introduced elsewhere and became naturalized in South Europe as an ornamental plant.

Agroecology

Schinus molle grows in full sun in the dry temperate to subtropical regions. It prefers a mean annual temperature of 15–20°C with mean annual rainfall of 300–60 mm. It tolerates warmer temperatures and once established is extremely drought tolerant. It is also resistant to frost and to temperatures as low as −10°C. In the wild, it occurs in abandoned fields, dry forests, along the banks of waterways and rivers, and slopes up to 2,400 m altitude. Sandy and well-drained soils are preferable but it is tolerant of waterlogged, poorly drained and infertile soils. It is also tolerant of alkalinity and salinity. It is shallow rooted and can be brittle; therefore, it is likely to be blown over or have its branches broken off in strong wind and needs wind protection.

Edible Plant Parts And Uses

The fruit is relished by children in Ethiopia who consumed the fruit during normal times. The dried and roasted berries are used as a pepper spice substitute. The fruits are pulverised and used in cooling drinks called '*horchatas*' which is relished in South America. The natives of the Andes also make a light alcoholic drink, the "*chicha de molle*" from the fruits. In Mexico, the fruit is ground and mixed with other substances to make beverages. In Mexico, it is used as ingredient for the famous "*pulque*" to produce an intoxicating liquor known as '*copalocle*' or '*copalote*'. The fruits are boiled and used to prepare vinegar or syrup or mixed with maize to make a nourishing light porridge. The seeds are sometimes used to adulterate pepper. An edible oil distilled from the fruit is used as a spice in baked products and candy. A gum that exudes from the bark is used for chewing.

Botany

A small to medium sized, branched, evergreen, dioecious trees, 5–15 m high, with slender pendant (drooping) branches forming a spreading crown. Leaves are alternate, imparipinnate on a winged rachis, with 9–20 pairs of opposite or alternate, sessile leaflets which are narrowly lanceolate to linear-lanceolate, 1.5–6 cm long by 0.2–0.8 cm wide, terminal leaflet smaller than lateral ones, grayish green, glabrous to sparsely puberulent with generally entire margins (Plate 1). Flowers yellowish white borne profusely in pendant, axillary clusters. Sepals deltoid, 0.4 mm long; petals yellowish-white, narrowly ovate, 1.5–2 mm long; styles and stigmas 3. Fruit small, globose, 5–8 mm in diameter, pink to red to black (Plates 1 and 2). The bark, leaves and fruit are all very aromatic.

Nutritive/Medicinal Properties

The fruit and leaves of *Schinus molle* contained many bioactive compounds with pharmacological properties. Forty-six compounds were identified in the essential oil obtained by steam distillation of the fruit of *Schinus molle* which included 9 monoterpene hydrocarbon, 1 aromatic compound, 1 aliphatic acid ester, 2 monoterpene esters, 16 sesquiterpene hydrocarbons and 17

Plate 1 Pendant fruit bunches and leaves of *Schinus molle*

Plate 2 Close-up of globose berries of *Schinus molle*

other sesquiterpenoids (Bernhard et al. 1983). Major components of the fruit oil were myrcene, α-phellandrene, δ-cadinene, limonene, α-cadinol and γ-phellandrene. Other compounds identified included sabine, α-terpinene, δ-terpinene, terpinolene, methyl n-octanoate, bourbonene, n-transbergamotene, caryophyllene, α-terpineaol, germacene D, δ-cadinene, transterpine ferrojone, β-spathulene (Jennings and Bernhard 1975). Of the terpene hydrocarbon fraction the main components were α-pinene, β-pinene, α-phellandrene, β-phellandrene, myrcene, D-limonene, camphene, p-cymene and three unidentified constituents (Bernhard and Wrolstad 1963). The main components in the leaf essential oil were β-phellandrene (13.6%) and limonene (13.4%) (Baser et al. 1997).

Among 57 and 62 compounds (% [mg/100 g dry matter]) identified in the essential oils from berries of *Schinus molle* and *Schinus terebinthifolius*, the main compounds were α-phellandrene (46.52% and 34.38%), β-phellandrene (20.81% and 10.61%), α-terpineol (8.38% and 5.60%), α-pinene (4.34% and 6.49%), β-pinene (4.96% and 3.09%) and p-cymene (2.49% and 7.34%), respectively (Bendaoud et al. 2010). A marked quantity of γ-cadinene (18.04%) was also identified in the *S. terebinthifolius* essential oil whereas only traces (0.07%) were detected in the essential oil of *S. molle*.

Some of the reported pharmacological activities of the leaves and fruit extract of *Schinus molle* are elaborated below.

Antioxidant Activity

The methanol extract of *Schinus molle* leaves yielded two new acylated quercetin glycosides, named isoquercitrin 6″-O-p-hydroxybenzoate (12) and 2″-O-α-L-rhamnopyranosyl-hyperin 6″-O-gallate (13), together with 12 known polyphenolic metabolites namely gallic acid (1), methyl gallate (2), chlorogenic acid (3), 2″-α-L-rhamnopyranosyl-hyperin (4), quercetin 3-O-β-D-neohesperidoside (5), miquelianin (6), quercetin 3-O- β-D-galacturonopyranoside (7), isoquercitrin (8), hyperin (9), isoquercitrin 6″-gallate (10), hyperin 6″-O-gallate (11) and (+)-catechin (14) (Marzouk et al. 2006). Compounds 4–9 and 11 exhibited moderate to strong radical scavenging properties on lipid peroxidation, hydroxyl radical and superoxide anion generations with the highest activities shown by 6 and 7 in comparison with that of quercetin as a positive control in-vitro. Twelve sesquiterpenoids, six tirucallane-type triterpenoids, and four flavonoids isolated from the fruits of *Schinus molle* exhibited antioxidant activity (Ono et al. 2008). Among them, three flavonoids exhibited almost identical antioxidative activity as that of α-tocopherol by the ferric thiocyanate method. In addition, one flavonoid showed a stronger radical-scavenging effect on 1,1-diphenyl-2-picrylhydrazyl than that of α-tocopherol. Essential oil of *S. terebinthifolius* berries expressed stronger antioxidant activity in the 2,2′-Azinobis(3-ethylbenzothiazoline-6-sulfonic acid) (ABTS)assay, with an IC_{50} of 24 mg/l, compared to *S. molle* (IC_{50} = 257 mg/l) berries (Bendaoud et al. 2010).

Anticancer Activity

Of 4 plants tested that showed cytotoxic activity against a human hepatocellular carcinoma cell line (Hep G2), *Schinus molle* was the most active with an IC_{50} of 50 µg/ml (Ruffa et al. 2002). Another study found 42 constituents in the leaf oil, accounting for 97.2% of the total oil (Diaz et al. 2008). The antioxidant activity of the leaf essential oil showed an IC_{50} of 36.3 µg/ml. The major constituents of the oil were β-pinene and α-pinene. The essential oil was cytotoxic in several cell lines, showing that it was more effective on breast carcinoma and leukemic cell lines. The LD_{50} for cytotoxicity at 48 hours in K562 (human erythromyeloblastoid leukemia cell line) corresponded to 78.7 µg/ml, which was very similar to the LD_{50} obtained when apoptosis was measured. The essential oil did not induce significant necrosis up to 200 µg /ml, The results indicated that apoptosis was the main mechanism of toxicity induced by *S. molle* essential oil in this cell line. The report concluded that some of the oil components could have potential anti-tumoural effects, either alone or in combination. *S. terebinthifolius* berries essential oil was more effective against human breast cancer cell line (MCF-7) (IC_{50} = 47 mg/l) than that from *S.molle* (IC_{50} = 54 mg/l) (Bendaoud et al. 2010).

Anti-inflammatory Activity

Three compounds with anti-inflammatory activity were isolated from *Schinus molle* fruits (Zeng et al. 2003). Two of the compounds were identified as triterpenes: 3-epi-isomasticadienolalic acid, isomasticadienonalic acid and biflavone chamaejasmin. On the phospholipase A (2)-induced mouse paw oedema, only isomasticadienonalic acid was active (30 mg/kg, 66% inhibition at 60 minutes), whereas all compounds reduced the chronic model of inflammation (48–26% of swelling reduction), but only triterpenes reduced leukocyte infiltration, measured as tissue peroxidase activity. In the case of the carrageenan-induced mouse paw oedema, only chamaejasmin resulted in a swelling reduction 3 hours after challenge (50 mg/kg, 46% oedema inhibition). In addition, chamaejasmin inhibited the LTB (lymphotoxin β) (4) production in rat peritoneal polymorphonuclear leukocytes with an IC_{50} value of 29.8 µM, while triterpenes showed toxicity against cells at 100 µM.

Antidepressant Activity

Hexanic extract of *Schinus molle* leaves was found to produce antidepressant-like effects in the tail suspension test in mice that appeared to be dependent on its interaction with the serotonergic, noradrenergic and dopaminergic systems (Machado et al. 2007, 2008). Further investigations indicated the antidepressant-like effect of the ethanolic extract of *S. molle* in the tail suspension test may be dependent on the presence of rutin (a flavonoid) that was postulated to exert its antidepressant-like effect by increasing the availability of serotonin and noradrenaline in the synaptic cleft.

Angiotensin Converting Enzyme (ACE) and Acetylcholine Inhibition Activities

The triterpenes from fractionated extracts of *S. molle* leaves showed moderate angiotensin converting enzyme(ACE)-inhibitory activity with IC_{50} about 250 µM (Olafsson et al. 1997). The pharmacological activity of the dichloromethanol extract of *Schinus molle* was analysed in in-vitro models. Studies showed that pre-incubation of the isolated guinea-pig ileum or rat uterus preparations with the dichloromethanol extract of *Schinus molle* (100 µg/ml) erased the contractile effects of histamine and serotonin respectively

(Bello et al. 1998). At the same dose, the extract partially reduced the contractile effects of acetylcholine on the isolated rat duodenum. A 10 μg/ml dose showed an inhibitory effect on histamine and serotonin, but not on acetylcholine-induced contractions. No significant effect was found with a 1 μg/ml dose.

Antimicrobial Activity

Leaf oil of *Schinus molle* exhibited maximum fungitoxic activity during the screening of some essential oils against some common storage and animal pathogenic fungi (Dikshit et al. 1986). It showed absolute toxicity against animal pathogens and mild activity against storage fungi. The minimum fungistatic concentration of *S. molle* oil were 300, 200 and 200 ppm against *Microsporum gypseum*, *Trichophyton mentagrophytes* and *Trichophyton rubrum* respectively. The minimum fungicidal concentrations were 900 ppm against *Trichophyton mentagrophytes* and 400 ppm against *Trichophyton rubrum*. The oil fungistatically checked the growth of other animal pathogens including *Epidermophyton floccosum*, *Histoplasma capsulatum*, *Microsporum canis*, *Microsporum ferrugineum*, *Trichophyton equinum* and *Trichophyton tonsurans* from a group of 19 fungi tested. The minimum fungistatic concentration of *S. molle* oil were 60, 75 and 55 times more active against *Microsporum gypseum*, *Trichophyton mentagrophytes* and *Trichophyton rubrum* respectively when compared with Multifungin (a synthetic ringworm lotion). In terms of minimum fungicidal concentrations, *S. molle* oil was 125 times more effective against *Trichophyton rubrum* and 55.5 times more effective against *Trichophyton mentagrophytes*. Of 50 compounds in the oil, 10 were identified that included α-pinene 1.96%, myrcene 12.59%, α-phellandrene 12.59%12.59%, limonene 15.46%, β-phellandrene 15.46%, ρ-cymene 8.41%, β-caryophyllene 11.43%, cryptone 2.93%, α-terpineol 5.96% and cavacrol 1.52%.

Volatile oil from *S. molle* leaves also exhibited significant activity against the following bacterial species (Gundidza 1993): *Klebsiella pneumoniae*, *Alcaligenes faecalis*, *Pseudomonas aeruginosa*, *Leuconostoc cremoris*, *Enterobacter aerogenes*, *Proteus vulgaris*, *Clostridium sporogenes*, *Acinetobacter calcoacetica*, *Escherichia coli*, *Beneckea natriegens*, *Citrobacter freundii*, *Serratia marcescens*, *Bacillus subtilis* and *Brochothrix thermosphacata*. The storage fungal species: *Aspergillus ochraceus*, *Aspergillus parasiticus*, *Fusarium culmorum* and *Alternaria alternata* exhibited significant sensitivity to the volatile oil.

Oleoresin from Aguaribay (*S. molle*) fruit exhibited antibacterial activity against both Gram positive and Gram negative bacteria (Padin et al. 2007). The minimal bacterial concentration was determined at 2 mg/ml for *Listeria monocytogenes*, at 15 mg/ml for *Staphylococcus aureus*, *Bacillus cereus* and *Escherichia coli*, and at 14 mg/ml for *E. coli* O157:H7, *Salmonella enteridis*, *Pseudomonas aeruginosa* and *Klebsiella pneumonia*. These concentrations produced 100% inhibition. The oleoresin maintained its inhibitory activity of the bacteria studied even after storage. The results indicated the promising potential for the oleoresin to be used as a natural preservative in the food industry due to its bactericidal effect on food spoilage microorganisms and those which cause food-toxin-infection.

Traditional Medicinal Uses

In traditional medicine, various plant parts of *Schinus molle* have been used in treating a variety of ailments such as amenorrhoea, bronchitis, gingivitis, gonorrhoea, gout, tuberculosis, tumour, ulcer, urethritis, wart, wounds, and urogenital and venereal diseases. It also has been used for wounds and infections due to its antibacterial and antiseptic properties. It has also been used as an antidepressant, a diuretic, and for toothache, rheumatism and menstrual disorders. Leaf juice has been used to treat ophthalmia and rheumatism. The bark extract infusion has been employed for diarrhoea. The resin from the bark known as 'American Mastic' or 'Molle resin' is a potent purgative and has been used to treat digestive disorders. Other known medicinal uses of the plant include as an astringent, a balsamic, diuretic, expectorant, masticatory, stomachic, tonic and vulnerary.

Other Uses

Schinus molle is often cultivated as an ornamental plant or as shade tree in subtropical and temperate regions. The wide, multi-branched crown provides good shade and acts as a suitable windbreak. The tree is also planted for soil conservation, as a live fence, barrier or support. *S. molle* also has been grown as indoor bonsai plants.

The tree provides a durable termite resistant wood that is used for posts and for both charcoal and firewood. The aromatic resin is used as mastic and tannin from the bark is used for tanning skins. Latex is produced from many parts of the tree. The cortex and leaves also yield a yellow dye. Ripe berries and leaves are often cut and used fresh or dried in floral displays.

Ethanolic and hexanic extracts from fruits and leaves of *Schinus molle* showed ability to control several insect pests and could have potential use as natural alternatives to synthetic insecticides. Potential vertebrate toxicity associated with insecticidal plants requires toxicity investigation before institutional promotion. Acute and subacute toxicity tests of ethanolic extracts from fruits of *Schinus molle* in rats suggest that ethanolic extracts from fruits and leaves of *Schinus molle* should be relatively safe to use as insecticide (Ferrero et al. 2007a). Fruit and leaf essential oils of *Schinus molle* showed insect repellent and insecticidal activity against food storage pests *Trogoderma granarium* and *Tribolium castaneum* (Abdell-Sattar et al. 2009). Hexanic extracts from leaves and fruits of *Schinus molle* were found to display repellent and insecticidal activities against first instar nymphs and eggs of *Triatoma infestans*, the vector of Chagas' disease (Ferrero et al. 2006). Leaf and fruit extracts were highly repellent for first nymphs. Fruit extracts had also ovicidal activity. Ethanol and petroleum ether extracts from leaves and fruits of *Schinus molle* produced significant repellent effect and mortality against adults of *Blattella germanica*, the German cockroach (Ferrero et al. 2007b).

Comments

Schinus molle can become a serious invasive weed in some dry temperate to subtropical areas.

Selected References

Abdel-Sattar E, Zaitoun AA, Farag MA, El Gayed SH, Harraz FM (2009) Chemical composition, insecticidal and insect repellent activity of *Schinus molle* L. leaf and fruit essential oils against *Trogoderma granarium* and *Tribolium castaneum*. Nat Prod Res 25:1–10

Baser KHC, Kürkçüoğlu M, Demirçakmak B, Ülker N, Beis SH (1997) Composition of the essential oil *Schinus molle* L. grown in Turkey. J Essent Oil Res 9:693–696

Bello R, Beltrán B, Moreno L, Calatayud S, Primo-Yúfera E, Esplugues J (1998) In vitro pharmacological evaluation of the dichloromethanol extract from *Schinus molle* L. Phytother Res 12(7):523–525

Bendaoud H, Romdhane M, Souchard JP, Cazaux S, Bouajila J (2010) Chemical composition and anticancer and antioxidant activities of *Schinus molle* L. and *Schinus terebinthifolius* Raddi berries essential oils. J Food Sci 75:C466–C472

Bernhard RA, Wrolstad R (1963) The essential oil of *Schinus molle*, the terpene hydrocarbon fraction. J Food Sci 28:59–63

Bernhard RA, Schibamoto T, Yamaguchi K, White E (1983) The volatile constituents of *Schinus molle* L. J Agric Food Chem 31:463–466

Díaz C, Quesada S, Brenes O, Aguilar G, Cicció JF (2008) Chemical composition of *Schinus molle* essential oil and its cytotoxic activity on tumour cell lines. Nat Prod Res 22(17):1521–1534

Dikshit A, Naqvi AA, Husain A (1986) *Schinus molle*: a new source of natural fungitoxicant. Appl Environ Microbiol 51(5):1085–1088

Ferrero AA, González JOW, Chopa CS (2006) Biological activity of *Schinus molle* on *Triatoma infestans*. Fitoterapia 77(5):381–383

Ferrero A, Minetti A, Bras C, Zanetti N (2007a) Acute and subacute toxicity evaluation of ethanolic extract from fruits of *Schinus molle* in rats. J Ethnopharmacol 113(3):441–447

Ferrero AA, Chopa CS, González JOW, Alzogaray RA (2007b) Repellence and toxicity of *Schinus molle* extracts on *Blattella germanica*. Fitoterapia 78(4):311–314

Goldstein DJ, Coleman RC (2004) *Schinus molle* L. (Anacardiaceae) chicha production in the central Andes. Econ Bot 58(4):523–529

Gundidza M (1993) Antimicrobial activity of essential oil from *Schinus molle* Linn. Cent Afr J Med 39(11): 231–234

Jennings WG, Bernhard RA (1975) Identification of components of the essential oil from the California pepper tree (*Schinus molle* L.). Chem Mikrobiol Technol Lebensm 4:95–96

Kramer FL (1957) The pepper tree, *Schinus molle* L. Econ Bot 11:322–326

LHB Hortorium (1976) Hortus third. A concise dictionary of plants cultivated in the United States and Canada. Liberty Hyde Bailey Hortorium, Cornell University, Wiley, New York, 1312 pp

Machado DG, Kaster MP, Binfaré RW, Dias M, Santos AR, Pizzolatti MG, Brighente IM, Rodrigues AL (2007) Antidepressant-like effect of the extract from leaves of *Schinus molle* L. in mice: evidence for the involvement of the monoaminergic system. Prog Neuropsychopharmacol Biol Psychiatry 31(2):421–428

Machado DG, Bettio LE, Cunha MP, Santos AR, Pizzolatti MG, Brighente IM, Rodrigues AL (2008) Antidepressant-like effect of rutin isolated from the ethanolic extract from *Schinus molle* L. in mice: evidence for the involvement of the serotonergic and noradrenergic systems. Eur J Pharmacol 587(1–3):163–168

Marzouk MS, Moharram FA, Haggag EG, Ibrahim MT, Badary OA (2006) Antioxidant flavonol glycosides from *Schinus molle*. Phytother Res 20(3):200–205

Olafsson K, Jaroszewski JW, Smitt UW, Nyman U (1997) Isolation of angiotensin converting enzyme (ACE) inhibiting triterpenes from *Schinus molle*. Planta Med 63(4):352–355

Ono M, Yamashita M, Mori K, Masuoka C, Eto M, Kinjo J, Ikeda T, Yoshimitsu H, Nohara T (2008) Sesquiterpenoids, triterpenoids, and flavonoids from the fruits of *Schinus molle*. Food Sci Technol Res 14(5):499–508

Padin EV, Pose GN, Pollio ML (2007) Antibacterial activity of oleoresin from Aguaribay (*Schinus molle* L.). J Food Technol 5(1):5–8

Ruffa MJ, Ferraro G, Wagner ML, Calcagno ML, Campos RH, Cavallaro L (2002) Cytotoxic effect of Argentine medicinal plant extracts on human hepatocellular carcinoma cell line. J Ethnopharmacol 79(3):335–339

Uphof JCTh (1968) Dictionary of economic plants, 2nd edn. Cramer, Lehre, 591 pp, (1st edn, 1959)

Zeng YQ, Recio MC, Máñez S, Giner RM, Cerdá-Nicolás M, Ríos JL (2003) Isolation of two triterpenoids and a biflavanone with anti-Inflammatory activity from *Schinus molle* fruits. Planta Med 69(10):893–898

Spondias cytherea

Scientific Name

Spondias cytherea Sonnerat

Synonyms

Condonum malaccense Rumphius, *Poupartia dulcis* (Forst. f.) Blume, *Evia amara* var. *tuberculosa* Blume, *Evia dulcis* (Forst. f.) Commelin ex Blume, *Spondias dulcis* Sol. ex Parkinson, *Spondias mangifera* Willd., *Spondias mangifera* var. *tuberculosa* (Blume) Engler

Family

Anacardiaceae

Common/English Names

Ambarella, Golden Apple, Hog Plum, Indian Mombin, Jew Plum, June Plum, Otaheite Apple, Polynesian Plum, Tahitian Quince, Vi Apple, Wi Apple, Yellow Plum

Vernacular Names

Antigua: Golden Apple;
Barbados: Golden Apple;
Belize: Golden Plum;
Brazil: Cajá-Assú, Caja-Manga, Caja-Mangueira, Cajarana, Imbuseirro, Tapiriba Do Sertao (Portuguese);
Burmese: Gway;
Colombia: Hobo De Racimos;
Costa Rica: Juplón, Yuplon;
Cuba: Cireula Dulce;
Czech: Mombín Sladký;
Danish: Evi, Sød Mombinblomme, Soed Mombinblomme, Stor Mombin;
Dutch: Ambarella;
Ecuador: Manzana De Oro;
French: Arbre De Cythère, Casamangue, Mombin Espagnol, Pomme Cythere, Pommier De Cythère, Prune Cythère Pomme De Cythère, Prune Cythère, Prune De Cythère, Prunier De Cythère, Prunier d'Amérique;
Eastonian: Magus Mombinipuu, Vili: Mombin;
Fiji: Polynesian Plum, Polynesian Vi Apple, Wi;
Finnish: Otaheite-Omena;
German: Goldpflaume, Süsse Mombinpflaume, Suesse Mombinpflaume, Tahiti-Apfel;
Grenada: Golden Apple;
Guadeloupe: Prune De Cythere;
Guinea-Bissau: Cajamanga (Crioulo);
Hawaii: Tahitian Quince, Vi, Vi-Apple, Wi, Wi-Apple;
India: Mirrey Manga (Tamil), Ambra, Amra Jouru, Mamedi (Telugu);
Indonesia: Kedongdong (Batak), Inji (Bima), Kadondo (Gorontalo), Kedongdong (Lampong), Dongdong, Kedongdong, Kedongdong Sem (Java), Kedongdong, Kedongdong Manis (Malay),

Kadungdung (Madurese), Kedungdung (Lingga), Desa (Sangir), Kedongdong (Singkep), Woa, Injung Maradda (Soemba), Ahang, Ehe, Li, Leheng, Lesem (Solor), Kadongdong (Sundanese), Jolo (Ternate), Ojo (Tidore);
Italian: Ambarella;
Jamaica: Jew Plum, June Plum;
Japanese: Tamagonoki;
Khmer: Mokak;
Laotian: Kook Hvaan;
Malaysia: Kedongdong;
Martinque: Prune De Cythere;
Mauritius: Prune De Cythere;
New Guinea: Aimemiek, Awiminik, Arama, Baramijan, Bermoi, Bikato, Dren, Gi, Gungkia, Huneg, Hunek, Iopeia, Juwut, Kanureskara, Karisi, Kedongdong utan, Maar, Mur, Ona, Sutiek, Pehjet, Warea, Wutiel, Unimi, Vain, Witosu;
Peru: Manzana De Oro;
Philippines: Hevi (Tagalog);
Polynesia: Evi, Polynesian Plum, Vee, Vi;
Puerto Rico: Pomarosa;
Reunion Island: Evi;
Rotuman: Vi;
Sierra Leone: Iŋglish-Plɔm (Krio);
Solomon Islands: Air;
Spanish: Ambareaal, Cireula, Ciruela Judia, Ciruelo Judío, Citara, Jobo De La India, Manzana De Otahiti, Spondias Dorata;
Sri Lanka: Ambarella (Sinhalese);
St Kitts and Nevis: Golden Apple;
Suriname: Fransi Mape, Pomme De Cythere
Swedish: Cytheraaepple, Guldplommon, Soett Balsamplommon, Tahitiaepple;
Tahiti: Evi;
Thai: Makok, Ma-Kok-Farang, Má Kok Wan;
Tongan: Vi;
Trinidad & Tobago: Pommecythere;
Venezuela: Jobo De La India, Jobo De Indio, Mango Jobo;
Vietnam: Cóc.

Origin/Distribution

The species is indigenous to south and southeast Asia. The species is extensively distributed throughout Indo-Malesia and from Melanesia to Polynesia and is widely cultivated in these areas, making it impossible to differentiate between indigenous and naturalised occurrences. It is now cultivated pantropically in south America, central America, the Caribbean, Florida, Africa and Australia.

Agroecology

The tree flourishes in humid and wet tropical areas. In new Guinea, it is rather common in lowland primary forests, sometimes in secondary forests up to 1,000 m usually occurring on well-drained soils, sometimes in flood plains rarely on limestone with a thin clay layer. It grows on all types of soil, including acidic soils and oolitic limestone in Florida, as long as they are well-drained. The tree is intolerant of strong winds and need protection from wind breaks and is moderately drought tolerant.

Edible Plant Parts and Uses

The fruit, young leaves and tender inflorescence are edible. Fully ripe fruit are best eaten raw and fresh. Ripe fruits are relish in fresh juice drinks, cordials and made into nectar which can be used for yogurt. The fruit is juiced, and goes under the popular name umbra juice in Malaysia. Studies conducted in the West Indies showed that yoghurts with 15% and 20% golden apple nectar were more liked than the control (0% nectar) yoghurt in all sensory attributes (Bartoo and Badrie 2005). A 226 g yoghurt serving provided an excellent source of phosphorus and was good in protein. Ripe fruit are also eaten in stews and prepared as jams and sauces. The fruit is also much appreciate when it is mature, unripe and green. Unripe fruit is excellent for making chutney or preserved as pickles in vinegar and sugar and made into *manisan*. Ambarellas are sometimes simmered in curries, but are more commonly cooked with sugar and a stick of cinnamon. The fruit is eaten as compote or in Indonesia the unripe fruit is often mixed with other fruits in *rujak*, and eaten in *sambal goring* as the unripe fruit is excellent as a substitute for tamarind *asam*. In Indonesia and Malaysia, unripe fruit is peeled and sliced and eaten with a thick black salty-sweet, fermented shrimp paste called *hayko* or dipped in chilli powder and salt and eaten. The fruits are also used in stews, jellies, jams and sauce. Being high in pectin, it is sometimes added to other low-pectin fruit to obtain a good set when making jams and jellies. In the French West Indies (Martinique and Guadeloupe) and Grenada, unripe, mature, green fruit is used to make a refreshing drink that is much appreciated by the consumers of these regions. The slight acidity and astringency, and its olive green colour caused by the green pigments from the outer layers of the fruit, make this drink a much sought craze.

Young leaves and sprouts are used as vegetable. In Indonesia, young leaves are eaten steamed as *lalab* or use in *sayur* (salad). Young leaves are added to salted fish and eaten as side dish with rice. Young leaves are often used as a substitute for *asam* or lemon to impart the sour, savory flavor to dishes. Young leaves are also cooked with tough meat to tenderize them. Young, tender panicles can be eaten steamed or dressed as salad. In India, leaves are eaten as greens; fruit cooked into curries, and pickled.

Pectin from ambarella fruit peel extracted with oxalic acid/ammonium oxalate solution gives the highest pectin yield, with high molar mass and degree of methylation, making the extract suitable for use as additives in food.

Plate 1 Leaves and fruits of ambarella

Botany

A medium sized tree, 10–25 m high may grow to 45 m high, with a straight trunk, greyish to light brown bark, bole usually 45 cm and sparse, irregular crown, branchlets thick, grey with fallen leaf scars. Leaves alternate, imparipinnate with 4–10 pairs of leaflets and a terminal one, glabrous, rachis 11–20 cm, petiole 9–15 cm (Plate 1). Leaflets opposite to sub-opposite, sub-coriaceous petiolulate (7.5 mm) ovate–oblong to lanceolate, 5–20 cm long by 1.5–8 cm wide, dark green, glossy above paler green beneath, base acute, oblique, apex shortly acuminate to acumi-

Plate 2 Inflorescence of ambarella

nate, margin serrate or crenulated with 14–24 pairs of nerves. Inflorescence appearing before leaves or accompanied by very small ones, paniculate, terminal 25–40 cm long, glabrous (Plate 2). Flowers white or creamy-white, calyx glabrous, 5-partite, segments deltoid, petals 5–6, reflexed, glabrous, ovate-oblong with 3 longitu-

Plate 3 Cluster of mature ambarella fruits

Plate 4 Ripe ambarella fruits

dinal nerves, stamens 10 longer than petals, glabrous wit oblong anthers. Style 5 free, ovary subglobose, 4–5 loculed. Fruit an indehiscent drupe, ellipsoid or oblong, glabrous, green turning greenish-yellow, yellow or golden yellow when ripe (Plates 3–5), 4–10 cm by 3–8 cm, with thick, yellow, sub acid to acid, juicy pulp enclosing a woody, fibrous endocarp with 5 flanges, the flanges often indirectly connected with a peripheral layers of meshes formed by numerous spinose and fibrous processes. The endocarp contains several flat seeds.

Nutritive/Medicinal Properties

The fruit flesh is a good source of vitamin C and iron. When unripe it was found to contain about 10% pectin. One hundred g of fresh fruit was found to contain 0.8 g protein, 11.1 g carbohydrate, 1.2 g or crude fibre and 0.6 g ash. The nutritional value was reported as: 20 mg calcium, 2 mg phosphorous and 1.2 g of iron as well as 1,382 IU of vitamin A, 70 mg vitamin C, 0.4 mg niacin, 0.02 mg riboflavin and 0.06 mg thiamine (Subhadrabanhdu 2001).

Analyses made in Malaysia (Tee et al. 1997) reported the following nutrient composition of the fruit in g per 100 g edible portion: energy 32 kcal, water 89.4 g, protein 0.7 g, fat 0.1 g, carbohydrate 7.0 g, fibre 2,3 g, ash 0.5 g, Ca 17 mg, P 50 mg, Fe 3.4 mg, Na 2 mg, K 191 mg, carotenes 145 µg, vitamin A 24 µg RE, vitamin B1 0.05 mg, vitamin B2 0.18 mg, niacin 0.3 mg, vitamin C 22 mg.

Analyses made in Hawaii and the Philippines (Morton 1987) reported the following food value for ambarella per 100 g edible portion, energy 157.30 cal, total solids 14.53–40.35%, moisture 59.65–85.47%, protein 0.50–0.80%, fat 0.28–1.79%, sugar (sucrose) 8.05–10.54%, acid 0.47%, crude fibre 0.85–3.60%, ash 0.44–0.65%.

Mature-green ambarella with a low pH (2.6) and a high titratable acidity (1.3 g citric acid Eq./100 g fresh material) was found comparable to the lemon. The fruit was found to be rich in vitamin C (52.0 mg/100 g), phenols (349.5 mg gallic acid equivalent/100 g) and starch (7.1 g/100 g). The pale green colour of the pulp was attributed to pheophytins A and B (Franquin et al. 2005).

The fruits were found to contain fat (0.34–0.53%), protein (1.76–2.33%) and ash (6.23–6.78%)

Plate 5 (**a, b**) Ripe ambarella fruits Golden yellow (*left*) and greenish yellow fruits (*right*)

(Siti Aishah et al. 2005). The pH, soluble solids and insoluble matter increased with the maturity of the fruit. Ripe fruits were significantly less acidic than green and half-ripe ones. The total phenolic content decreased with maturity. Vitamin C content ranged from 4.65 to 5.86 mg/100 g. Water-soluble pectins (WSP) in ripe fruits contributed to the softening of the fruit, which caused fruit firmness to decrease with advanced maturity. The amount of P (3.93–4.52 g/100 g) was highest compared with the other minerals (Na, Mg, Ca and Zn), while total chlorophyll content decreased in ripe fruits, contributing to changes in fruit colour. Total dietary fibre content was lowest (17.20, dry basis) in ripe fruits.

The main aroma compounds from extracts of ripe and unripe fruits were found to be minor alcohols and esters, monoterpenes, hexane derivatives and fatty acids. The major compounds among more than 50 constituents identified were 1,8-cineole, α-pinene, β-pinene, terpinolene, limonene, α-terpineol, butyl acetate, γ-terpinene and terpinen-4-ol (Jirovetz et al. 1999). Another analysis reported that the volatile constituents of the green (unripe) and ripe fruits were very similar, being dominated by C6 compounds and monoterpene hydrocarbons, with (E)-hex-2-enal and α-pinene constituting the two major components. The ripe fruit, however, gave a higher yield of total volatiles and a greater number of minor constituents which were mostly esters (Wong and Lai 1995).

The total carbohydrate content of the intact pulp of *Spondias cytherea* was 41%. Polysaccharides were mainly a type I rhamnogalacturonan with arabinogalactan branches. Cell eliciting activity in a dose-depending pattern was observed in-vitro on peritoneal macrophages treated with soluble fractions containing the polysaccharide (Iacomini et al. 2005). The polysaccharides isolated from *Spondias purpurea var. lutea* and *Spondias cytherea* contained galactose, as main component, arabinose, mannose, xylose and rhamnose residues (Koubala et al. 2008). *S. cytherea* gum had the highest intrinsic viscosity (63 ml/g). The high solubility of the *Spondias* gums studied and the absence of tannin may have potential industrial application.

The antioxidant activity of the ambarella fruit was determined by free radical 1,1 diphenyl-2-picrylhydrazyl (DPPH) method to be $IC_{50} = 3.0$ mg/ml and phenolic compound content of 165 mg/100 g, which was much higher than that reported for orange (*Citrus sinensis*) $IC_{50} = 18.3$ mg/ml and phenolic compound content of 64.2 mg/ml (Franco, 2006).

Traditional Medicinal Uses

There are diverse uses of fruits, leaves and bark in traditional medicine in different parts of the world. The treatment of wounds, sores and burns is reported from several countries. In Cambodia, the astringent bark is used with various species of *Terminalia* as a remedy for diarrhoea. The leaves have been used for naso-pharygneal affections.

Other Uses

The wood is light-brown and buoyant and in the Society Islands has been used for canoes. The bar yields a resinous gum. Extract of *Spondias cyntherea* has nematicidal activity as it exhibited strong anti-nematodal activity (MED: 1·25–2·5 mg per ball) against *Bursaphelenchus xylophilus*, the pine wood nematode (Mackeen et al. 1997). The fruit is fed to the pigs and the leaves given to cattle.

Comments

Selected References

Backer CA, Bakhuizen van den Brink RC Jr (1965) Flora of Java (spermatophytes only), vol 2. Wolters-Noordhoff, Groningen, 641 pp

Bartoo AS, Badrie N (2005) Physicochemical, nutritional and sensory quality of stirred 'dwarf' golden apple (*Spondias cytherea* Sonn) yoghurts. Int J Food Sci Nutr 56(6):445–454

Burkill IH (1966) A dictionary of the economic products of the Malay Peninsula. Revised reprint, 2 vols. Ministry of Agriculture and Co-operatives, Kuala Lumpur, vol 1 (A–H), pp 1–1240, vol 2 (I–Z), pp 1241–2444

Burkill HM (1985) The useful plants of west tropical Africa, vol 1. Families A-D. Royal Botanical Gardens Kew, Richmond, 960 pp

de Pinto GL, Martnez M, Sanabria L, Rincon F, Vera A, Beltran O, Clamens C (2000) The composition of two *Spondias* gum exudates. Food Hydrocoll 14(3):259–263

Ding Hou (1978) Anacardiacea. In: van Steenis CGGJ (ed) Flora Malesiana, ser I, vol 8. Sijthoff & Noordhoff, Alphen aan den Rijn, pp 395–548, 577 pp

Franco EM (2006) Actividad antioxidante in vitro de las bebidas de frutas. Bebidas-Alfa Editores Técnicos, Junio/Julio:20–27. (In Spanish)

Franquin S, Marcelin O, Aurore G, Reynes M, Brillouet J-M (2005) Physicochemical characterisation of the mature-green Golden apple (*Spondias cytherea* Sonnerat). Fruits 60:203–210

Iacomini M, Serrato RV, Sassaki GL, Lopes L, Buchi DF, Gorin PA (2005) Isolation and partial characterization of a pectic polysaccharide from the fruit pulp of *Spondias cytherea* and its effect on peritoneal macrophage activation. Fitoterapia 76(7–8):676–683

Jirovetz L, Buchbauer G, Ngassoum MB (1999) Analysis of the aroma compounds of the fruit extracts of *Spondias cytherea* (ambarella) from Cameroon. Z Lebensm Unters Forsch 208(1):74–76

Kennard WC, Winters HF (1960) Some fruits and nuts for the tropics. USDA Agric Res Serv Misc Publ 801: 1–135

Koubala BB, Mbome LI, Kansci G, Tchouanguep Mbiapo F, Crepeau M-J, Thibault J-F, Ralet M-C (2008) Physicochemical properties of pectins from ambarella peels (*Spondias cytherea*) obtained using different extraction conditions. Food Chem 106(3):1202–1207

Ledin RB (1957) Tropical and subtropical fruits in Florida (other than *Citrus*). Econ Bot 11:349–376

Mackeen MM, Ali AM, Abdullah MA, Nasir RM, Mat NB, Razak AR, Kawazu K (1997) Antinematodal activity of some Malaysian plant extracts against the pine wood nematode, *Bursaphelenchus xylophilus*. Pest Manag Sci 51(2):165–170

Morton JF (1961) Why not use and improve the fruitful ambarella. Hort Adv 5:3–16

Morton J (1987) Ambarella. In: Fruits of warm climates. Julia F. Morton, Miami, pp 240–242

Ochse JJ, Bakhuizen van den Brink RC (1931) Fruits and fruitculture in the Dutch East Indies. G. Kolff & Co, Batavia, 180 pp

Ochse JJ, Bakhuizen van den Brink RC (1980) Vegetables of the Dutch Indies, 3rd edn. Ascher & Co, Amsterdam, 1016 pp

Siti Aishah I, Noryati I, Mohd Azemi MN, Hanisah A (2005) Some physical and chemical properties of ambarella (*Spondias cytherea* Sonn.) at three different stages of maturity. J Food Compos Anal 18(8): 819–827

Subhadrabanhdu S (2001) Under utilized tropical fruits of Thailand, FAO Rap publication 2001/26. FAO, Rome, 70 pp

Tee ES, Noor MI, Azudin MN, Idris K (1997) Nutrient composition of Malaysian foods, 4th edn. Institute for Medical Research, Kuala Lumpur, 299 pp

Verheij EWM (1992) *Spondias cytherea* Sonnerat. In: Verheij EWM, Coronel RE (eds.) Plant resources of South-East Asia. No. 2: Edible fruits and nuts. Prosea Foundation, Bogor, pp 287–288

Wong KC, Lai FY (1995) Volatile constituents of ambarella (*Spondias cytherea* Sonnerat) fruit. Flav Fragr J 10(6):375–378

Spondias purpurea

Scientific Name

Spondias purpurea L.

Synonyms

Spondias cirouella Tussac, *Spondias crispula* Beurl., *Spondias dulcis* Blanco, *Spondias mexicana* S. Watson, *Spondias mombin* L., nom. illeg. *Spondias myrobalanus* L., *Spondias myrobalanus* Jacq. nom. illeg., *Spondias nigrescens* Pittier, *Spondias purpurea* var. *munita* I.M. Johnst., *Spondias radlkoferi* Donn. Sm., *Warmingia pauciflora* Engl.

Family

Anacardiaceae

Common/English Names

Brazilian Plum, Chile Plum, Ciruela, Hog Plum, Jamaica Plum, Jocote, Ovo, Purple Jobo, Purple Mombin, Purple Plum, Red Mombin, Scarlet Plum, Spanish Plum, Wild Plum

Vernacular Names

Bolivia: Jocote;
Brazil: Ambu, Cajá, Ciriguela, Ciruela, Cirouela Serigüela, Imbu, Imbuzeiro, Serigüela, Ombuzeiro, Umbuzeiro (Portuguese);
Columbia: Canajo, Cirgüelo, Cirgüelo, Ciruela, Ciruela Colorada, Ciruela Común, Ciruela Común, Ciruela De Hueso, Ciruela De Monte, Ciruela Del Fraile, Ciruela Española, Ciruela Jobo, Ciruela Mexicana, Ciruela Morada, Ciruela Roja, Ciruelo, Ciruelo De Hueso, Ciruelo Mexicano, Hobo, Hobo Blanco Hobo Colorado;
Costa Rica: Jobillo, Jobito, Jobito;
Czech: Mombín červený;
Danish: Mombinblomme, Rød Ciruela, Rød Mombin, Rød Mombinblomme, Roed Mombinblomme;
Dutch: Makka Pruim, Rode Mombinpruim;
Eastonian: Purpur-mombinipuu;
Finnish: Punamombin
French: Cirouelle, Mombin Rouge, Mombin Rouge (fruit), Prune Café (Fruit), Prune Café, Prune D'espagne, Prune D'espagne (Fruit), Prune D'espagne, Prune Du Chili, Prunier D'espagne (Plant), Prunier D'espagne, Prunier Des Antilles, Prunier Des Antilles (Plant), Prune Jaune (Yellow Form), Prune Rouge;
Eastonian: Purpur-mombinipuu;
Ecuador: Hobo Colorada;
El Savador: Jobo, Jobo Colorado, Jobo Gusanero, Jocose, Jocote, Jocote ácido, Jocote Común, Jocote Común, Jocote De Azucarón, Jocote De Iguana, Jocote De Pitarrío, Jocote De Verano, Jocote Tronador;
German: Mombinpflaume, Rote Mombinpflaume, Spanische Pflaume
Italian: Susina Mombin Rossa;
Netherlands Antilles: Prunier Des Antilles, Prunier D'espagne Noba, Makka Pruim;

Philippines: Sineguelas, Si Sireguelas (Bikol), Sereguelas (Cebu Bisaya), Saraguelas (Ibanag), Saguelas, Sarguelas (Iloko), Sineguelas, Sirhuelas (Tagalog);
Portuguese: Ambu, Ambuzeiro, Ameixa Da Espanha, Cajá Vermelha (Yellow Form), Ciriguela, Ciroela, Cirigüela, Cirouela, Ciruela, Imbu, Imbuzeiro, Serigüela, Umbu, Umbuzeiro;
Spanish: Ajuela Ciruela, Chiabal, Cirgüelo, Ciruela, Ciruela Agria, Ciruela Calentana, Ciruela Campechana, Ciruela Colorada, Ciruela Común, Ciruela De Coyote, Ciruela Del Fraile, Ciruela De Hueso, Ciruela De Mexico Ciruela De Monte, Ciruela Del Fraile, Ciruela Del País, Ciruela Española, Ciruela Jobo, Ciruela Morada, Ciruela Roja, Ciruela Sanjuanera, Ciruelo, Ciruelo De Hueso, Gusanero, Hobo, Hobo Colorado, Hobo Blanco, Ismoyo, Jobillo, Jobito, Jobo, Jobo Colorado, Jobo Francés, Jocote, Jocote Agrio, Jocote Amarillo (Yellow Form), Jocote Común, Jocote De Corona, Jocote De Iguana, Jocote Iguanero, Jocote Tronador, Jocotillo, Ovo, Pitarillo, Sua, Taperibá, Ubo, Yocote;
Swedish: Mombinplommon, Roett Balsamplommon;
Thai: Makok Farang.

Origin/Distribution

Spondias purpurea is indigenous to tropical America from southern Mexico through to northern Peru and Brazil and the Antilles. It is an ancient crop of the Mayas in Yucatán and is widely cultivated in central America. It is now cultivated pantropically for its edible fruit. It has also naturalised in some countries, including the Philippines, Galapagos and Nigeria.

Agroecology

S. *purpurea* is tropical, its natural populations is found from sea level to an altitude of 1,200 m in areas with alternating wet and dry seasons from Sinaloa and Jalisco in Mexico to Colombia. It grows in regions of high and low humidity and remains leafless during the dry season just before flowering. It is a hardy tree, thriving in both wet and dry sites but is cold sensitive. It is adaptable to a wide range of soils – from sand, gravel, heavy clay loam, loams to calcareous soils.

Edible Plant Parts and Uses

The savoury pulp of ripe fruits are commonly eaten fresh, out-of-hand. They can be stewed whole, with sugar, and consumed as dessert. The fruit can be preserved by boiling in brine and drying in the sun for future use or eaten as dried fruit snacks. Another way to preserve the fruit is to heat it in unsalted water and dry it in the sun, while a third process used in Mexico is to obtain *ciruelo negro*, this consists of pricking the skin of the fruit, placing it in syrup and letting it simmer until the sugar burns or becomes concentrated. The strained juice of cooked fruits yields an excellent jelly and is also used for making jam, marmalade, syrups, soft drinks, wine *chicha* (maize liquor) and vinegar. The juice makes pleasant fresh beverage on its own or mixed with other fruit juices. Other uses of the pulp include as an *atole*, mixed with maize flour and sugar. In Mexico, unripe fruits are made into a tart, green sauce, or are pickled in vinegar and eaten with salt and chilli peppers. The new shoots and leaves are acid and eaten raw or cooked as vegetables in northern Central America.

Botany

Spondias purpurea is a small, deciduous tree, growing 4–8 m, with a spreading crown, irregular, stout trunk and thick brittle, spreading branches. Leaves are alternate, pinnately compound, bronze-red or purplish when young, with 5–12 pairs of elliptical-ovate, sub-sessile leaflets (Plates 1–2, 4), 2–4 cm long, oblique base, and shallowly toothed towards the acute apex, green, and fall before the flowering period. Flowers are small, male, female or bisexual with deltoid calyx lobes, 4–5 red or purplish, ovate oblong petals, 8–10 stamens and 4–5 styles. Flowers are borne in short up to 4 cm long, pubescent panicles (few

Plate 1 Fruiting branches of red mombin

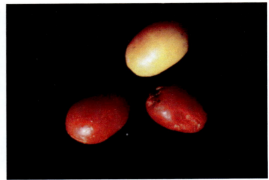

Plate 3 Close-up of red mombin fruits

Plate 2 Compound leaves and fruits of red mombin

Plate 4 Reddish-purple fruited variety

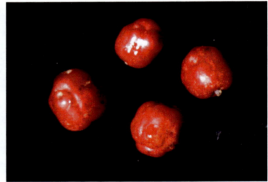

Plate 5 Close-up of reddish-purple fruited variety

flowered) in the axils of fallen leaves along the branches. Fruit is a drupe, oval or oblong, 2.5–5 cm long, purple, dark- or bright-red, orange, yellow, or red-and-yellow, glabrous, glossy (Plates 1–4). Mesocarp is yellow, fibrous, juicy, acid-sweet, aromatic adhering to a fibrous, hard, oblong, knobbly pale stone which contains up to 5 small, usually abortive seeds.

Nutritive/Medicinal Properties

Food value of jocote fruit per 100 g of edible portion based on analyses made in Ecuador, Central America and Brazil (amino acids) (Morton 1987) was reported as: moisture 65.9–86.6 g, protein 0.096–0.261 g, fat 0.03–0.17 g, fibre 0.2–0.6 g, ash 0.47–1.13 g, calcium 6.1–23.9 mg, phosphorus

31.5–55.7 mg, iron 0.09–1.22 mg, carotene 0.004–0.089 mg, thiamine 0.033–0.103 mg, riboflavin 0.014–0.049 mg, niacin 0.540–1.770 mg, ascorbic acid 26.4–73.0 mg; amino acids (mg per g nitrogen [N=6.25]) – lysine 316 mg, methionine 178 mg, threonine 219 mg, tryptophan 57 mg.

Analyses of the fresh fruit showed that the percentage of moisture in the flesh ranged from 76% to 86%, protein and fat were very low and calcium, phosphorus, iron and ascorbic acid were found in appreciable quantities (Kozio and Macía 1998). Ovo had the highest caloric density of the fruits studied, 74 kcal/100 g edible portion versus 39–58 kcal/100 g for peach, apricot, plum, mango, and cherry. This higher caloric density was attributed principally to its higher concentration of total carbohydrates (19.1%); fructose, glucose, and sucrose together account for 65% of the soluble matter. Unlike the other fruits, ovo was found to retain a fair amount of starch in the mesocarp. It had a moderate level of potassium (250 mgl 100 g edible portion) and an excellent quantity of vitamin C (49 mg/100 g edible portion). Analysis of volatile flavour compounds showed 2-hexenal to be the main flavour compound present.

The volatile composition of the siriguela fruit was found to consist of ketones, alcohols, aldehydes, esters and terpenic hydrocarbons (Ceva-Antunes et al. 2006). The major compounds identified were: hexanal, trans-2-hexenal, 3-hexen-1-ol, 2-hexen-1-ol, ethyl acetate and hexyl acetate.

Spondias purpurea was found to produce a polysaccharide gum constituted by galactose (59%), arabinose (9%), mannose (2%), xylose (2%), rhamnose (2%), uronic acids (26%) and a proteinaceous material (2%) (Martinez et al. 2008). Some plants parts have a few pharmacological properties such as antioxidant and antimicrobial activities.

Antioxidant Activity

The antioxidant activity of red mombin fruit was determined by free radical 1,1-diphenyl-2-picrylhydrazyl (DPPH) method to be $IC_{50}=5.3$ mg/ml and phenolic compound content of 140 mg/100 g, which was better than that reported for orange (*Citrus sinensis*) $IC_{50}=18.3$ mg/ml and phenolic compound content of 64.2 mg/ml (Franco 2006).

Antimicrobial Activity

Spondias purpurea is one of 10 plants among 84 plants screened for antibacterial activity in Guatemala and was found to exhibit good antibacterial activity against five enterobacteria pathogenic to man (enteropathogenic *Escherichia coli*, *Salmonella enteritidis*, *Salmonella typhi*, *Shigella dysenteriae* and *Shigella flexneri*) to varying degrees (Caceres et al. 1990, 1993).

Traditional Medicinal Uses

In Mexico, the fruits are regarded as diuretic and antispasmodic. In folk medicine, the fruit decoction is used to bathe wounds and heal sores in the mouth. A syrup prepared from the fruit is administered to treat chronic diarrhoea. Cubans used the fruit as emetic; Haitians use the fruit syrup for angina and the Dominicans use it as laxative. In French Guiana, the fruit has been used as ingredient in laxative marmalade and shoots are considered astringent.

In Mexico, the astringent bark decoction is used as a remedy for mange, ulcers, dysentery and for bloating caused by intestinal gas in infants. In the Philippines, decoction of the fruit used for diarrhoea, dysentery and gonorrhoea. The sap of the bark is used to treat stomatitis in infants and a decoction of the bark used for dysentery and infantile tympanites. In the Amazon, Tikunas Indians used the decoction of bark for pain and excessive menstrual bleeding, for stomach pains, diarrhoea, and for washing wounds.

The bark has been used for minor skin ulcers The juice of the fresh leaves has been reported as a therapy for thrush. A decoction of the leaves and bark is used as a febrifuge. The gum-resin of the tree is blended with pineapple or soursop juice for treating jaundice. Amazonian Indians use a daily cup of decoction for permanent sterility.

In south-western Nigeria, an infusion of shredded leaves is employed for washing cuts, sores and burns. Researchers at the University of Ife have found that an aqueous extract of the leaves has antibacterial action, and an alcoholic extract is even more effective (Morton 1987).

Other Uses

The fruits, leaves and bark are fairly rich in tannin. Fruit has been used to remove stains from clothing and for washing hands. Cattle eat the leaves. Large stumps are used as fence posts. The soft and light wood is suitable for pulp.

Comments

The plant can be easily propagated by cuttings as it grows very slowly from seed.

Selected References

Caceres A, Cano O, Samayoa B, Aguilar L (1990) Plants used in Guatemala for the treatment of gastrointestinal disorders. 1. Screening of 84 plants against enterobacteria. J Ethnopharmacol 30(1):55–73

Cáceres A, Fletes L, Aguilar L, Ramirez O, Figueroa L, Taracena AM, Samayoa B (1993) Plants used in Guatemala for the treatment of gastrointestinal disorders. 3. Confirmation of activity against enterobacteria of 16 plants. J Ethnopharmacol 38(1):31–38

Ceva-Antunes PMN, Ribeiro Bizzo H, Silva AS, Carvalho CPS, Antunes OAC (2006) Analysis of volatile composition of siriguela (*Spondias purpurea* L.) by solid phase microextraction (SPME). Lebensm Wiss Technol 39(4):437–443

Coronel RE (1992) *Spondias purpurea* L. In: Verheij EWM, Coronel RE (eds) Plant resources of South-East Asia, vol 2, Edible fruits and nuts. Prosea Foundation, Bogor, pp 288–290

Fouqué A (1973) Espèces fruitières d'Amérique tropicale. Famille des Anacardiacées. Fruits 28:144–149

Franco EM (2006) Actividad antioxidante in vitro de las bebidas de frutas. Bebidas-Alfa Editores Técnicos, Junio/Julio:20–27 (In Spanish)

Hernández Bermejo JE, León J (eds) (1994) Neglected crops 1492 from a different perspective, vol 26, FAO plant production and protection series. Food and Agriculture Organization of the United Nations, Rome

Kennard WC, Winters HF (1960) Some fruits and nuts for the tropics. U S D A Agric Res Serv Misc Publ 801: 1–135

Kozio MJ, Macía MJ (1998) Chemical composition, nutritional evaluation, and economic prospects of *Spondias purpurea* (Anacardiaceae). Econ Bot 52(4): 373–380

León L, Shaw PE (1990) Spondias the red mombin and related fruits. In: Nagy S, Shaw PE, Wardowsky WF (eds) Fruits of tropical and subtropical origin. FSS, Lake Alfred, pp 116–126

Martínez M, de Pinto GL, de González MB, Herrera J, Oulyadi H, Guilhaudis L (2008) New structural features of *Spondias purpurea* gum exudates. Food Hydrocoll 22(7):1310–1314

Morton JF (1987) Purple mombin. In: Fruits of warm climates. Julia F. Morton, Miami, pp 242–245

Stuart GU (2010) Philippine alternative medicine. Manual of some Philippine medicinal plants. http://www.stuartxchange.org/OtherHerbals.html

Tropicos.Org. Missouri Botanical Garden. <http://www.tropicos.org/Name/3200271>

Annona atemoya

Scientific Name

Annona × *atemoya* **Mabb.**

Synonyms

Annona squamosa × *Annona cherimola*, *Annona atemoya* Hort.

Family

Annonaceae

Common/English Names

Atemoya, Custard Apple, Pineapple Sweetsop

Vernacular Names

Cuba: Mammon;
Lebanon: Achta;
Russia: Atemoyia,
Spanish: Atemoya;
Venezuela: Chirimorinon;
Vietnam: Mãng Cầu Dai.

Origin/Distribution

Atemoya originated from a cross of *Annona squamosa* and *Annona cherimola* in Florida. The first cross was made by the horticulturist, P.J. Wester, at the United States Department of Agriculture's subtropical laboratory, Miami, in 1908. Seedlings were planted out in 1910. Other crosses made in 1910 fruited in 1911 and seeds were taken by Wester to the Philippines and elsewhere.

Agroecology

Atemoya grows best in the lowlands in the warm tropical to subtropical areas. Optimum growth development is reported to occur between a mean minimum temperature of 13–20°C and a mean maximum of 22–32°C. For fruit maturation, optimum temperature ranges from 22°C to 26°C. Temperatures cooler than this often results in fruits failing to mature properly. Atemoya is intolerant of frost, young trees are damaged by low temperatures of 4°C and subzero temperatures will cause fruit splitting and discoloration. It thrivers best in areas with 70–80% relative humidity and with adequate rainfall occurring during the growing period. Heavy rain during fruit development will cause fruit splitting and irrigation is required in drier areas. It is also susceptible to strong winds because of its shallow rooting system and a wind break is essential. It grows best in sandy and loamy soils but will grow on red basalt or well-structured, well drained clay loams. Optimum soil pH is from 6 to 6.5. Waterlogging is detrimental to the tree.

Edible Plant Parts and Uses

Atemoya, preferably chilled, is one of the most delicious of fruits. The pulp is eaten fresh or made into juice, sherbets, shakes, ice cream and yoghurt. Slices or cubes of the pulp may be added to fruit cups or salads or various dessert recipes. It can be blended with orange juice, lime juice and cream and frozen as ice cream.

Botany

A small tree growing to 8 m high with a short trunk and low , drooping branches. The leaves are deciduous, alternate, elliptical, leathery, slightly pubescent and up to 15 cm long. Flowers are axillary, solitary or in clusters of 2–3, bisexual, long pedicellate, triangular, yellow, 6 cm long by 4–5 cm wide. Flower has a convex or cone-shaped receptacle containing numerous carpels (gynoecium) at the top half and the androecium in the lower circumference of the receptacle. The receptacle is encased by 3–4 sepals and 6–8 fleshy yellow petals in two whorls, nectaries are present near the petal bases. The carpels are sessile with club-shaped stigma and ovary with 1 ovule. The fruit is a fleshy syncarp formed by fusion of the carpels and the receptacle (Plates 1–5). The fruit is conical or heart-shaped, 8–12 cm long by 9–10 cm wide, some weighing as much as 2–2.5 kg, pale bluish-green to pea-green, and

Plate 2 Developing fruit of cv African Pride

Plate 3 Close-up of fruit of cv African Pride

Plate 1 An immature cv. Pink's Mammoth fruit

Plate 4 Custard apple fruit – *left* cherimola, *centre and right* cv African Pride, *bottom centre* cv Gefner

Plate 5 Custard apple mainly cv African Pride

slightly yellowish between the areoles. The rind is 3–4 mm thick, composed of fused, prominent and angular areoles with rounded or slightly upturned tips, firm, pliable, and indehiscent. The flesh is white, finely texture, fragrant, almost solid, not conspicuously divided into segments, with fewer seeds than the sugar apple and sweet. The seeds are cylindrical, 2 cm by 1 cm, dark brown to black, hard and smooth.

Nutritive/Medicinal Properties

The proximate food value of atemoya (African Pride) per 100 g edible portion (72%) was reported by Wills et al. (1984, 1986) as follows: water 78.7 g, fat 0.6 g, protein 1.4 g, energy 326 kJ, fibre 2.5 g, available carbohydrates 15.8 g, total dietary fibre 2.5 g, starch 1.1 g, sugars 14.7 g, (fructose 5.4 g, glucose 5.6 g, sucrose 3.7 g), ash 0.4 g, Ca 17 mg, K 250 mg, Mg 32 mg, Fe 0.3 mg, Zn 0.2 mg, Na 4 mg, Cu 2.4 mg, ascorbic acid (vitamin C) 43 mg, thiamine 0.05 mg, riboflavin 0.08 mg, niacin 0.8 mg, α-carotene 10 μg, β-carotene equivalent 5 μg, total saturated fatty acids 0.2 g, total monounsaturated fatty acids 0.1 g, total polyunsaturated fatty acids 0.2 g, citric acid 0.3., total soluble solids (TSS) or brix 22.3.

The fruit is rich in vitamin C, sugars and potassium and also has citric and malic acid.

Forty compounds were identified in the volatile component in custard apple (*Annona atemoya*) predominated by esters (94%) (Bartley 1987). The major components were methyl and ethyl butanoate and methyl hexanoate. Among the terpenes, major components were α- and β-pinene, E-ocimene and germacrene-D.

Acetogenins and Anticancer Activity

As in many of the Annonaceae family, the seeds of *Annona atemoya* were found to contain many acetogenins which were isolated and identified besides N-fatty acyl tryptamines (Wu et al. 2005). A series of N-fatty acyl tryptamines, including a mixture of N-nonadecanoyltryptamine, N-behenoyltryptamine, N-lignoceroyltryptamine, N-cerotoyltryptamine, and N-octacosanoyl tryptamine, and a mixture of N-tricosanoyl-4,5-dihydroxytryptamine, N-lignoceroyl-4,5-dihydroxytryptamine, N-pentacosanoyl-4,5-dihydroxytryptamine, and N-heptacosanoyl-4,5-dihydroxytryptamine, along with two alkaloids, atemoine and cleistopholine, were isolated from the ethyl acetate extract of seeds of *Annona atemoya*.

The following acetogenins were isolated and identified from Atemoya seeds: almunequin, annonacin, annonisin, artemoin-A, artemoin-B, artemoin-C, artemoin-D, artemoyin, asimicin, atemotetrolin, atemoyacin B, Atemoyacin E, bulladecin, bullatacin, cherimolin-1, cherimolin-2, cleistopholine, desacetyluvaricin, isodesacetyluvaricin, molvizarin, motrilin, neoannonin, parviflorin, rolliniastatin-1, rolliniastatin-2, rollinicin, squamocin, 12,15-cis-squamostatin-D, 12,15-cis-squamostatin-A (Chang et al. 1999; Chen et al. 1995; Chih et al. 2001; Chiu et al. 2003; Duret et al. 1995, 1997,1998; Wu et al. 2001). Many of the bioactive acetogenins exhibited anticancerous activities. Many of the acetogenins were cytotoxic and exhibited potent cytotoxicity against Hep G2, Hep 2,2,15, KB, CCM2 and CEM cancer cell lines (Chang et al. 1999).

The acetogenin, bullatacin, isolated from the fruit of *Annona atemoya* was shown to inhibit proliferation of cell line 2.2.15 (human hepatoma HepG2 cells transfected with hepatitis B virus DNA plasmid) by apoptosis induction (Chih et al. 2001). Bullatacin induced cytotoxicity of 2.2.15

cells in a time- and dose-dependent manner. Fifty percent effective dose (ED_{50}) on day 1 of exposure to bullatacin were 7.8 µM for 2.2.15 cells. Bullatacin-treatment also reduced concentrations of hepatitis B surface antigen (HBsAg) in the cultured medium released from 2.2.15 cells, coincident with the decrease in cell proliferation. Different concentrations (10^{-3}–1.0 µM) of bullatacin induced apoptosis in a concentration-dependent manner at 16 h. The determination of intracellular cyclic AMP (cAMP) and cyclic GMP (cGMP) levels in 2.2.15 cells after exposure to bullatacin demonstrated that bullatacin caused both to decrease in a time- and concentration-dependent manner (Chiu et al. 2003). The concentration-dependent decrease of both cAMP and cGMP induced by bullatacin was parallel with the magnitude of apoptosis induced by various concentrations (10^{-3}–1.0 nM) of bullatacin. Additionally, the bullatacin-induced apoptosis was inhibited by the addition of cAMP and cGMP elevating agents (forskolin and S-nitrosoglutathione). The results suggested that a decrease of both cAMP and cGMP levels may play a crucial role in bullatacin-induced apoptosis in 2.2.15 cells.

In separate studies, atemoyacin B and bullatacin exhibited cytotoxicity on cell culture of human MDR (multi drug resistance) breast adenocarcinoma cells, MCF-7/Dox and human KBv200 cells, and their parental sensitive cell lines MCF-7 and KB (Fu et al. 1999). IC_{50} of atemoyacin B for MCF-7/Dox, MCF-7, KBv200, and KB cells were 122, 120, 1.34, and 1.27 nmol.l^{-1}, respectively. IC_{50} of bullatacin for MCF-7/Dox, MCF-7, KBv200, and KB cells were 0.60, 0.59, 0.04, and 0.04 nmol/l, respectively. The cytotoxicities of bullatacin and atemoyacin B to MDR cells were similar to those to parental sensitive cells. Bullatacin and atemoyacin B markedly increased cellular Fura-2 and Dox accumulation in MCF-7/Dox cells, but not in MCF-7 cells. The rates of apoptosis in MDR cells were similar to those in sensitive cells induced by atemoyacin B. There was no cross-resistance of P-glycoprotein positive MCF-7/Dox and KBv200 cell lines to bullatacin and atemoyacin B as compared with their sensitive P-glycoprotein negative MCF-7 and KB cell lines. The mechanism of the circumvention of multi drug resistance was associated with the decrease of P-glycoprotein function and the increase of cellular drug accumulation in multi drug resistance cells.

The compound acetogenin 89–2 from *Atemoya* showed potent cytotoxicity in KBv200 and KB cells in nude mice, and the mean IC_{50} of acetogenin 89–2 to KBv200 and KB cells was 48.7 and 64.6 nmol/l, respectively (Fu et al. 2003). The IC_{50} of acetogenin 89–2 to multidrug resistant (MDR) cells was similar to that to the parental drug-sensitive cells. In the models of KBv200 and KB cell xenografts in nude mice, acetogenin 89–2 (0.90 mg/kg,) exhibited 52.3% and 56.5% inhibition of the growth of xenografts, respectively. Both MDR KBv200 cells and parental drug-sensitive KB cells were sensitive to the treatment of acetogenin 89–2 in-vitro and in-vivo. The mechanism of overcoming MDR was found to be associated with the decrease of P-glycoprotein function MDR cells.

(Refer also to notes under *Annona muricata* and other *Annona* species).

Other Uses

The tree is cultivated primarily for its edible fruits.

Comments

Three popular, commonly cultivated cultivars with good commercial value are Gefner, African Pride and Pink's Mammoth (Plates 1–5). The male and female parts of the flower mature at different times as such fruit set by natural pollination is poor. The pollens are discharged later after the stigma loses its receptivity and dries up. Nitidulid beetles have been found to enhance pollination, and for commercial crops artificial hand pollination is essential.

Selected References

Bartley JP (1987) Volatile constituents of custard apple. Chromatographia 23:129–131

Chang FR, Chen JL, Lin CY, Chiu HF, Wu MJ, Wu YC (1999) Bioactive acetogenins from the seeds of *Annona atemoya*. Phytochem 51(7):883–889

Chen WS, Yao ZJ, Wu YL (1995) Study on the chemical constituents of *Annona atemoya* Hort and the isolation and structure of atemoyacin B. Acta Chim Sinica 53(5):16–520

Chih HW, Chiu HF, Tang KS, Chang FR, Wu YC (2001) Bullatacin, a potent antitumor annonaceous acetogenin, inhibits proliferation of human hepatocarcinoma cell line 2.2.15 by apoptosis induction. Life Sci 69(11):1321–1331

Chiu HF, Chih TT, Hsian YM, Tseng CH, Wu MJ, Wu YC (2003) Bullatacin, a potent antitumor Annonaceous acetogenin, induces apoptosis through a reduction of intracellular cAMP and cGMP levels in human hepatoma 2.2.15 cells. Biochem Pharmacol 65(3):319–327

Duret P, Hocquemiller R, Laurens A, Cave A (1995) Atemoyin, a new bis-tetrahydrofuran acetogenin from the seeds of *Annona atemoya*. Nat Prod Res 5(4):295–302

Duret P, Hocquemiller R, Cave A (1997) Annonisin, a bis-tetrahydrofuran acetogenin from *Annona atemoya* seeds. Phytochem 45(7):1423–1426

Duret P, Hocquemiller R, Cave A (1998) Bulladecin and atemotetrolin, two bis-tetrahydrofuran acetogenins from *Annona atemoya* seeds. Phytochem 48(3):499–506

Fu LW, Pan QC, Liang YJ, Huang HB (1999) Circumvention of tumor multidrug resistance by a new annonaceous acetogenin: atemoyacin-B. Zhongguo Yao Li Xue Bao 20(5):435–439

Fu LW, He LR, Liang YJ, Chen LM, Xiong HY, Yang XP, Pan QC (2003) Experimental chemotherapy against xenografts derived from multidrug resistant KBv200 cells and parental drug-sensitive KB cells in nude mice by annonaceous acetogenin 89–2. Yao Xue Xue Bao 38(8):565–570

Morton J (1987) Atemoya. In: Fruits of warm climates. Julia F. Morton, Miami, pp 72–75

Sanewski GM (1988) Growing custard apples. Department of Primary Industries, Queensland Government Information Series Q187014, Brisbane, 86 pp

Wills RBH, Poi A, Greenfield H, Rigney C (1984) Post-harvest changes in fruit composition of Atemoya during ripening and effects of storage temperature on ripening. Hort Sci 19(1):96–97

Wills RBH, Lim JSK, Greenfield H (1986) Composition of Australian foods. 31. Tropical and sub-tropical fruit. Food Technol Aust 38(3):118–123

Wu P, Chen WS, Hu TS, Yao ZJ, Wu YL (2001) Atemoyacin E, a bis-tetrahydrofuran annonaceous acetogenin from *Annona atemoya* seeds. J Asian Nat Prod Res 3(3):177–182

Wu YC, Chang FR, Chen CY (2005) Tryptamine-derived amides and alkaloids from the seeds of *Annona atemoya*. J Nat Prod 68(3):406–408

Annona diversifolia

Scientific Name

Annona diversifolia Saff.

Synonyms

None

Family

Annonaceae

Common/English Names

Cherimoya of the Lowlands, Ilama

Vernacular Names

Antilles: Anona Blanca;
Bolivia: Anona Blanca;
Costa Rica: Papauce;
El Salvador: Anona Blanca, Anona Rosada;
French: Cherimole Des Terres Basses;
Guatemala: Anona, Anona Blanca, Papauce;
Mexico: Chirimoya De Las Tierras Bajas, Hilama, Ilama, Ilama-Zapote, Izlama, Papuas (Spanish), Ilamazapotl (Nahuatl);
Spanish: Annona Blanca Ilama, Ilama, Perpauce, Papauce.

Origin /Distribution

The origin of the species is in Central America. It is indigenous in the foothills of south-western coast of Mexico, the Pacific coast of Guatemala, and El Salvador.

Agroecology

Ilama is a tropical plant. The climatic requirements of the ilama are similar to those of the sugar-apple and the custard-apple. It thrives best in warm climates where there is a long dry season followed by heavy and plentiful rainfall. Ilama is found in its native habitats at relatively low elevations in the foothills of the southwest coast of Mexico and of the Pacific coast of Guatemala and El Salvador. It is grown up to 610 m in Mexico, although in El Salvador it is cultivated up to 1,500 m and in Guatemala up to 1,800 m. Ilama does best in deep, rich, friable, loamy soils.

Edible Plant Parts and Uses

The fruit pulp is eaten fresh, scooped out with a spoon or made into drinks and sherbets. The ilama is usually chilled when served. It is sometimes served with a little cream and sugar to intensify the flavour, or with a drop of lime or lemon juice to impart a tart and bitter tinge.

Botany

A small, perennial tree, growing to 7.5 m high with a slender trunk, 23 cm diameter across which tends to branch off from the ground level to form 3–6 main stems. It has aromatic, pale brownish-gray, furrowed bark. Leaves are alternate, glossy, simple, thin, broadly elliptic to oblanceolate, 5–15 cm long with a rounded apex and cuneate base and entire margin, coppery when young turning green (Plate 1). A distinguishing characteristic of this species is the presence of orbicular leaf-like bracts, 2.5–3.5 cm long at the bases of the flowering branchlets. The flowers are solitary, regular, long-pedicellate, maroon-coloured, 2.3 cm long with three small rusty hairy sepals, three linear-oblong, pubescent outer petals and three inner minute petals. Stamens numerous in many rows on the torus around the gynoecium. Ovaries numerous, crowded into a pyramidal mass. Fruit formed by the confluence of the numerous ovaries after fertilisation. The fruit is conical, heart-shaped, or ovoid globose, about 15 cm long and may weigh up to 1 kg. Generally, the fruit is studded with more or less pronounced, triangular protuberances, though fruits on the same tree may vary from rough to fairly smooth. The rind, pale-green to deep-pink or purplish, is coated with a dense, velvety, gray-white bloom (Plate 1). The green type has a flesh that is white and sweet, while in the pink type, the flesh is a rose colour and has a tart taste. The center of both types are somewhat fibrous but smooth and custardy near the rind. The flesh varies from being dry to being fairly juicy, and contains 25–80 small, hard, smooth, brown, cylindrical seeds, 2 cm × 1 cm.

Nutritive/Medicinal Properties

The food composition of the fruit based on analyses made in El Salvador (Morton 1987) per 100 g of edible portion was reported as follows: water, 71.5 g, protein 0.447 g, fat 0.16 g, fibre 1.3 g, ash 1.37 g, Ca 31.6 mg, P 51.7 mg, Fe 0.70 mg, carotene 0.011 mg, thiamin 0.235 mg, riboflavin 0.297 mg, niacin 2.177 mg and ascorbic acid 13.6 mg.

Some plant parts have been reported to have various pharmacological properties.

Anti-epileptic Activity

In Mexico, the leaves of ilama have been used in traditional folkloric medicine for epilepsy. Studies conducted by González-Trujano et al. (2001) reinforced the traditional use of this plant in epilepsy. Palmitone (16-hentriacontanone) was found as the only anticonvulsant active compound in the leaves of *A. diversifolia*. This aliphatic ketone was highly effective in attenuating pentylenetetrazole (PTZ)-induced clonic-tonic seizures and toxicity. Also, it prolonged the latency for onset of seizures and a decrease in the death rate produced by 4-aminopyridine (4-AP) and bicuculline (BIC). However, it failed to inhibit the kainic acid (KA)- and strychnine (STC)-induced seizures. Palmitone did not produce motor incoordination and loss of righting reflex which were employed as signs of neurological impairment. Palmitone ($ED_{50}=1.85$ mg/kg) was found to be a more potent antiepileptic drug against the PTZ-induced seizures than etosuximide ($ED_{50}=59.6$ mg/kg), sodium valproate ($ED_{50}=63$ mg/kg), and carbamazepine ($ED_{50}>300$ mg/kg) and it was only four-fold less potent than diazepam ($ED_{50}=0.48$ mg/kg). In subsequent study, oral administration of hexane (100 mg/kg IP or PO) and ethanol (100 mg/kg, PO) *A. diversifolia* leaf extracts significantly decreased spike frequency, whereas intraperitone-

Plate 1 Fruits and leaves of Ilama

ally administered hexane extract and palmitone (10 mg/kg, IP) only reduced after-discharge duration in rat amygdala kindling (González-Trujano et al. 2009). Hexane extract and palmitone exhibited anticonvulsant properties and delayed establishment of a kindling state as observed with diazepam (0.3 mg/kg IP). These results reinforced the anticonvulsant properties of this plant, palmitone and other constituents that were responsible for the anti-epileptic effects. Palmitone extracted from the leaves was also found to possess anticonvulsant properties against penicillin-, 4-AP-, and pentylenetetrazole (PTZ)-caused seizure in adult animals (Cano-Europa et al. 2010). Palmitone did not prevent the PTZ-caused seizure but palmitone prevented the PTZ-caused neuronal damage in the CA3 hippocampal region.

Depressant Activity

Ethanol extract of *Annona diversifolia* leaves produced a depressant activity on the central nervous system. The extract and palmitone isolated from its leaves did not produce a sedative-hypnotic effect although both of them were effective in reducing the severity of behavioral and EEG (electroencephalographic) seizures induced by penicillin in rats (González-Trujano et al. 2006b). The results suggested that the reduction in the paroxystic activity by *A. diversifolia* was likely produced by palmitone through GABAergic (γ aminobutyric acid) neurotransmission. Intraperitoneal administration of the leaf extract delayed the onset of chronic seizures induced by pentylenetetrazole and delayed the time in the 'rota-rod' and 'swimming' test (González-Trujano et al. 1998). In addition, the extract increased the duration of sleeping time induced by sodium pentobarbital.

Anxiolytic Activity

Palmitone produced anxiolytic effect on test animals (González-Trujano et al. 2006a). In the elevated plus-maze test, palmitone (0.3, 1, 3,10 and 30 mg/kg, i.p.) lengthened, from 50% to 199%, the time spent in the open arm region of the maze at all doses tested, as compared to the vehicle group. In relation to the rearing activity in the exploratory cylinder, palmitone significantly altered in a dose-dependent manner, this activity by reducing the number of rearings with an effective dose value (ED_{50}) and 95% confidence limits (CL_{50}) of 0.79 (0.23–2.68) mg/kg. In addition, in the hole-board test, nose-poking was also significantly decreased in a dose-dependent manner [ED_{50} (CL_{50}) = 9.07 (4.51–18.26) mg/kg]. Furthermore, palmitone at any dose caused no change in motor activity nor disruption in traction performance. In contrast, diazepam, used as reference drug, produced an anxiolytic effect with a significant and dose-dependent decrease in motor coordination accompanied by disruption of the traction performance. Behavioral studies suggested an anti-anxiety effect produced by palmitone, but its neuropharmacological profile differed from that observed for benzodiazepines such as diazepam.

Antinociceptive Activity

Studies by Carballo et al. (2010) showed that *Annona diversifolia* possessed antinociceptive activity, giving support to their traditional use for treatment of spasmodic and arthritic pain. The ethanol extract caused a 25% significant and dose-dependent recovery of limb function in rats. Additionally, this extract produced a similar antinociceptive response (ED_{50} = 15.35 mg/kg) to that of the reference drug tramadol (ED_{50} = 12.42 mg/kg) when evaluated in the writhing test in mice. Bio-guided fractionation yielded hexane and acetone active fractions from which the presence of palmitone and flavonoids was respectively detected. Palmitone produced an antinociceptive response with an ED_{50} = 19.57 mg/kg in the writhing test. The results suggested the participation of endogenous opioids and 5-HT(1A) receptors in this antinociceptive response.

Anticancer Activity

Annona diversifolia was found to contain antitumorous acetogenins. Laherradurin and cherimolin-2, acetogenins isolated from *Annona diversifolia* seeds, exhibited inhibitory potential on HeLa and SW-480 cells when injected once daily into athymic mice bearing these cancer cell lines (Schlie-Guzmán et al. 2009). Laherradurin was more active than cherimolin-2, and it produced a similar IC_{50} in both neoplasic lines in-vitro proliferation assays. In athymic mice, laherradurin administered in 10×, 100× and 500× doses, reduced the size of HeLa tumours, and with 100× and 500× doses, affected the SW-480 tumour development. These doses were similar to results found with the control drug doxorubicin. In contrast, cherimolin-2 had an effect on HeLa tumour cells at 100× and 500× doses.

Antibacterial Activity

Rolliniastatin-2, an acetogenin with elevated cytotoxic activity isolated from the hexanic extract of *Annona diversifolia* seeds was found to have antibacterial activity (Luna-Cazares and Gonzalez-Esquinca 2010). *Escherichia coli, Pseudomonas aeruginosa* and *Salmonella typhi* were susceptible to rolliniastatin-2 (28–40% at 2 mg/ml), while the bacterial spheroplasts showed an increased susceptibility (55–58%, at 1 ng/ml). The most susceptible bacterium was *P. aeruginosa*, while the spheroplasts of *E. coli* and *S. typhi* were most sensitive to rolliniastatin-2.

Other Uses

It is mainly cultivated for its delicious fruits.

Comments

See also other edible Annonaceous species.

Selected References

Cano-Europa E, González-Trujano ME, Reyes-Ramírez A, Hernández-García A, Blas-Valdivia V, Ortiz-Butrón R (2010) Palmitone prevents pentylenetetrazole-caused neuronal damage in the CA3 hippocampal region of prepubertal rats. Neurosci Lett 470(2):111–114

Carballo AI, Martínez AL, González-Trujano ME, Pellicer F, Ventura-Martínez R, Díaz-Reval MI, López-Muñoz FJ (2010) Antinociceptive activity of *Annona diversifolia* Saff. leaf extracts and palmitone as a bioactive compound. Pharmacol Biochem Behav 95(1):6–12

Davidson A (1999) Ilama. In: Oxford companion to food. Oxford University Press, New York, USA, pp 395–396, 912 pp

González-Trujano ME, Navarrete A, Reyes B, Hong E (1998) Some pharmacological effects of the ethanol extract of leaves of *Annona diversifolia* on the central nervous system in mice. Phytother Res 12(8):600–602

González-Trujano ME, Navarrete A, Reyes B, Cedillo-Portugal E, Hong E (2001) Anticonvulsant properties and bio-guided isolation of palmitone from leaves of *Annona diversifolia*. Planta Med 67(2):136–141

González-Trujano ME, Martínez AL, Reyes-Ramírez A, Reyes-Trejo B, Navarrete A (2006a) Palmitone isolated from *Annona diversifolia* induces an anxiolytic-like effect in mice. Planta Med 72(8):703–707

González-Trujano ME, Tapia E, López-Meraz L, Navarrete A, Reyes-Ramírez A, Martínez AL (2006b) Anticonvulsant effect of *Annona diversifolia* Saff. and palmitone on penicillin-induced convulsive activity. A behavioral and EEG study in rats. Epilepsia 47(11):1810–1817

González-Trujano ME, López-Meraz L, Reyes-Ramírez A, Aguillón M, Martínez A (2009) Effect of repeated administration of *Annona diversifolia* Saff. (ilama) extracts and palmitone on rat amygdala kindling. Epilepsy Behav 16(4):590–595

Luna-Cazares LM, Gonzalez-Esquinca AR (2010) Susceptibility of complete bacteria and spheroplasts of *Escherichia coli, Pseudomonas aeruginosa* and *Salmonella typhi* to rolliniastatin-2. Nat Prod Res 24(12):1139–1145

Morton J (1987) Ilama. In: Fruits of warm climates. Julia F. Morton, Miami, pp 83–85

Popenoe W (1974) Manual of tropical and subtropical fruits. Facsimile of the 1920 edition. Hafner Press, New York, USA.

Safford WE (1912) *Annona diversifolia*, a custard-apple of the Aztecas. J Wash Agric Sci 2:118–125

Schlie-Guzmán MA, García-Carrancá A, González-Esquinca AR (2009) In vitro and in vivo antiproliferative activity of laherradurin and cherimolin-2 of *Annona diversifolia* Saff. Phytother Res 23(8):1128–1133

Annona glabra

Scientific Name

Annona glabra L.

Synonyms

Annona australis St.-Hil., *Annona chrysocarpa* Lepr. ex Guillemin & Perr., *Annona humboldtiana* Kunth, *Annona humboldtii* Dunal, *Annona klainii* Pierre ex Engl. & Diels, *Annona klainii* Pierre Ex Engler & Diels var. *moandensis* De Wild., *Annona laurifolia* Dunal, *Annona palustris* L. nom. illeg., *Annona peruviana* Humb. & Bonpl. ex Dunal, *Annona pisonis* St.-Hil. & Tul., *Annona uliginosa* H.B.K., *Guanabanus palustris* Gomez de la Maza.

Family

Annonaceae

Common/English Names

Alligator Apple, Bob Wood, Corkwood, Cow Apple, Mangrove Annona, Monkey Apple, Pond Apple

Vernacular Names

Brazil: Araticum-Bravo, Araticum Do Bréjo, Araticum-Caca, Araticum-Cortiça, Araticupana, Araticum-Do-Mangue, Araticum De Mangue, Araticum-D'água, Araticum-Da-Lagoa, Araticum-Da-Praia, Araticum-De-Boi, Araticum-De-Jangada, Araticum-Do-Bréjo, Araticum Do Rio, Araticunzeiro-Do-Brejo, Caroáo, Cortisso, Maçã-De-Cobra, Mulato, Cortiça, Birba, Jaca-de-pobre (Portuguese);
Chinese: Niu Xin Guo, Yuan Xi Fan Li Zhi;
Czech: Láhevník Lysý;
Danish: Mangroveannona;
Eastonian: Sile Annoona;
Fijian: Kaitambo, Kaitambu, Uto Ni Bulumakau, Uto Ni Mbulumakau;
French: Anone Des Marais, Bois Flot, Cachiman Cochon, Corossol Des Marais, Guanamin, Mamain;
German: Alligatorapfel, Wasserapfel, Mangroven-Annone;
Guatemala: Guanaba;
Martinique: Mamain;
Nicaragua: Anona De Río;
Japanese: Pondo Appuru;
Portuguese: Araticum Do Bréjo, Araticum De Rio, Araticurana;

Spanish: Anona Lisa, Anón Liso, Anon De Puerco, Anonillo Cabuye, Bagá, Cayur, Cayuda, Corcho, Cortisso, Chirimoya De Los Pantanos, Chirimoya Cimarrona, Corazón cimarrón, Palo Bobo; ***Vietnamese***: Bình Bát.

Origin/Distribution

Wild distribution of the species is found in the mangroves of tropical South America (Venezuela), West Indies and West Africa. It has been introduced elsewhere in the tropics as rootstock for other *Annona* species when grown in wet soils.

Agroecology

In its natural habitat in the tropics, pond apple occurs in fresh and brackish wetlands in tropical North, Central and South America and coastal West Africa from sea level to 50 m elevation. In Australia, it occurs on creek and river banks and farm drainage systems, wetlands and mangrove swamps, and the high tide area of the littoral zone on beaches. *A. glabra* can behave as a 'freshwater or brackish water mangrove' as it can survive root immersion at high tide and prolonged freshwater flooding. Both the fruit and the seed float (an adaptation which facilitates dispersal in flowing water). The hard seeds can remain viable for considerable periods in either fresh, brackish or sea water. Pond apple is also tolerant of salinity.

Edible Plant Parts and Uses

The pulp of ripe fruit is edible fresh although scarcely desirable.

Botany

A small, semi-deciduous, woody tree usually 3–6 m high but can reach a height of 12 m, with a trunk buttressed at the base. Leaves are alternate, oblong-elliptical, with acute or shortly acuminate apex and cuneate to rounded base, 7–15 cm long, up to 6 cm broad, glabrous and glossy green with a prominent mid-rib (Plate 1). Flowers are solitary and axillary and borne on curved, stout peduncles. Sepals are reniform-cordate, 5–6 mm, glabrous. Outer petals are creamy-white, ovate-cordate, adaxially concave, 2.5–3 cm, apex acute and inner petals are creamy-white and deep purple at the inside base, oblong-ovate, 2–2.5 cm. Stamens are linear, 3–4 mm with connective thickened above anther tip. Carpels are conically massed and connate. Fruit is an ovoid syncarp up to 15 cm long, 8 cm broad, green turning yellow when ripe, pulp pinkish-orange, rather dry, pungent-aromatic (Plates 1–3). Seeds numerous, about 140, ellipsoid to obovoid, 1–1.5 cm and light brown.

Plate 1 Axillary fruit and alternate leaves of pond apple

Plate 2 Developing pond apple fruit with pronounced flat areoles

Plate 3 Close up of immature pond apple fruit

Nutritive/Medicinal Properties

Pond Apple fruit was reported to have 887.9 mg/kg of total volatile compounds (Pino et al. 2002). Major volatile components were myrcene (47.1%), (Z)-β-ocimene (16.3%), limonene (11.2%) and α-pinene (9.5%).

The following compounds were isolated from fresh fruit of *A. glabra*: a kaurane diterpenoid, annoglabasin G (16α-hydro-19-acetoxy-ent-kauran-17-al), along with 27 compounds including 18 kaurane diterpenoids (Hsieh et al. 2004), 16β-hydro-ent-kauran-17-oic acid, 16α-hydro-ent-kauran-17-oic acid, 19-nor-ent-kauran-4α-ol-17-oic acid, 16α-hydro-19-ol-ent-kauran-17-oic acid, ent-kaur-16-en-19-oic acid, 16α-hydroxy-ent-kauran-19-oic acid, 16α,17-dihydroxy-ent-kauran-19-oic acid, 16β,17-dihydroxy-ent-kauran-19-oic acid, 16α-hydro-ent-kauran-17, 19-dioic acid, 16β-hydroxy-17-acetoxy-ent-kauran-19-oic acid, 16β-hydro-17-hydroxy-ent-kauran-19-al, 16α-hydro-17-hydroxy-ent-kauran-19-al, 16β,17-dihydroxy-ent-kauran-19-al, 16α-hydro-19-al-ent-kauran-17-oic acid, 16α-hydro-17-acetoxy-ent-kauran-19-al, 16α-hydro-19-acetoxy-ent-kauran-17-oic acid, ent-kaur-15-en-19-oic acid and ent-kaur-15-en-17-ol-19-oic acid; four acetogenins, annomontacin, annonacin, isoannonacinone and squamocin; four steroids, β-sitosterol, stigmasterol, β-sitosteryl-D-glucoside and stigmasteryl-D-glucoside and one oxoaporphine, liriodenine.

Twenty compounds including a dioxoaporphine, annobraine; two oxoaporphines, liriodenine and lysicamine; five aporphines, (−)-nornuciferine, (−)-anonaine, (−)-N-formylanonaine, (−)-asimilobine and (+)-nordomesticine; one proaporphine, (+)-stepharine; two protoberberines, (−)-kikemanine and dehydrocorydalmine; one azaanthraquinone, 1-aza-4-methyl-2-oxo-1,2-dihydro-9,10-anthracenedione; two amides, N-trans-feruloyltyramine and N-p-coumaroyltyramine; one ionone, blumenol A; and five steroids, β-sitosterol, stigmasterol, β-sitosteryl-D-glucoside, stigmasteryl-D-glucoside and 6-O-palmitoyl-β-sitosteryl-D-glucoside, were isolated from the fruits and stems of *Annona glabra* (Chang et al. 2000). Three new kaurane diterpenoids, annoglabasin C (16α-acetoxy-ent-kauran-19-oic acid-17-methyl ester), annoglabasin D (16α-acetoxy-ent-kauran-19-al-17-methyl ester) and annoglabasin E (16α-hydro-19-ol-ent-kauran-17-oic acid), and a new norkaurane diterpenoid, annoglabasin F (16α-acetoxy-19-nor-ent-kauran-4α-ol-17-methyl ester) (4), along with 13 known kaurane derivatives were isolated from the stems of *Annona glabra* (Chen et al. 2000). The seeds of *Annona glabra* were found to have cyclopeptides, glabrin C [cyclo-(prolyl- glycyl -tyrosyl-valyl leucyl alanyl leucyl-valyl)] and glabrin D [cyclo-(prolyl-prolyl-valyl tyrosyl glycyl prolyl- glutamyl)] (Li et al. 1999).

Pharmacological properties of various plant parts include:

Anticancer Activity

The leaves of *Annona glabra* were found to contain acetogenins which showed varying degrees of cytotoxic and anti-cancerous activities against several human cancer cell lines. Studies conducted in Purdue University reported that glacins A and B showed potent and selective in-vitro cytotoxicities against several human solid tumour cell lines (Liu et al. 1998b). Another study reported that two acetogenins from fractionated ethanolic leaf extract exhibited anti-cancerous activities. Annoglaxin while less potent than 27-hydroxybullatacin exhibited interesting selectivity for the human breast carcinoma (MCF-7) cell line (Liu et al. 1999b). Acetogenin 27-hydroxybullatacin was at least 100,000 times more potent than adriamycin against the human kidney carcinoma (A-498), prostate carcinoma (PC-3), and pancreatic carcinoma (PACA-2) cell lines. The scientists isolated two bioactive mono-tetrahydrofuran Annonaceous acetogenins, annoglacins A and B, from the fractionated ethanolic extracts of the leaves of *Annona glabra*. Annoglacins A and B were selectively 1,000 and 10,000 times, respectively, more potent than adriamycin against the human breast carcinoma (MCF-7) and pancreatic carcinoma (PACA-2) cell lines (Liu et al. 1999a). They also isolated two bioactive bis-tetrahydrofuran Annonaceous acetogenins, glabracins A and B, and two previously known acetogenins, javoricin and bullatanocin, from the leaves of *Annona glabra* by activity-directed fractionation using the brine shrimp lethality test (Liu et al. 1998a). Glabracins A and B exhibited selective cytotoxicities to certain human tumour cell lines, and glabracin A was significantly more potent. Three bis-tetrahydrofuran acetogenins; squamocin-C, squamocin-D, and annonin I were isolated from the methanolic extract of *Annona glabra* seeds (Abdel-lateff et al. 2009). All compounds were assayed for their cytotoxicity towards brine shrimp and five in-vitro cancer cell lines (A549, HT29, MCF 7, RPMI, and U251), and exhibited significant activity.

Studies reported that the alcoholic extracts prepared from pond apple leaves, pulp and seed, were not cytotoxic to normal human lymphocytes. However, extracts were highly cytotoxic to drug sensitive (CEM) and multidrug-resistant leukemia (CEM/VLB) cell lines (Cochrane et al. 2008). The seed extract was more potent than leaf and pulp extracts, but the cytotoxicity values were significantly lower than that for adriamycin. Treatment of CEM and CEM/VLB cells with the seed extract induced apoptosis and necrosis in both sensitive and resistant leukemia cells in a concentration-dependent fashion. The findings supported the traditional use of *A. glabra* and the alcoholic seed extract provided a potent source of anticancer compounds that could be utilized pharmaceutically.

Chinese scientists reported that cunabic acid (ent-kaur-16-en-19-oic acid) and ent-kauran-19-al-17-oic acid had strong inhibitory effect on the proliferation of human liver cancer (HLC) cell lines (Zhang et al. 2004). The proliferation of hepatoma cell line SMMC-7721 cells was inhibited after being treated with cunabic acid at the concentration >5 μmol/L and ent-kauran-19-al-17-oic acid >10 μmol/L. The biggest inhibitory effect was 81.05% when treated with cunabic acid at the concentration of 25 μmol/L. The effect had a linear relationship with concentration. The apoptotic rates of the cells treated by cunabic acid and ent-kauran-19-al-17-oic acid were 43.31% and 24.95%, respectively. Another compound, annoglabayin from the fruit produced apoptotic events in Hep G2 cells, by inducing changes in mitochondria (Chen et al. 2004). From the cytotoxic ethanol extract of *Annona glabra* seeds, a new mono-tetrahydrofuranic (mono-THF) acetogenin, glabranin, as well as a pair of 22-epimer *bis*-THF acetogenins, were isolated together with four known mono-THF acetogenins, annonacin, annonacinone, corossolin and corossolone which were found to be potent inhibitors of complex I of the mitochondrial respiratory chain (Gallardo et al. 1998).

Antiviral Activity

Phytochemical analysis of the fruits of *Annona glabra* yielded two new kaurane diterpenoids, annoglabasin A (methyl-16 β-acetoxy-19-al-ent-kauran-17-oate) and annoglabasin B (16 α-hydro-19-acetoxy-ent-kauran-17-oic acid), along with 11 known kaurane derivatives (Chang et al. 1998). Among these methyl-16-hydro-19-al-ent-kauran-17-oate showed mild activity against HIV replication in H9 lymphocyte cells, and 16α-17-dihydroxy-ent-kauran-19-oic acid exhibited significant inhibition of HIV-reverse transcriptase.

Anti-acetylcholinesterase Activity

The active fraction of the ethanol extract of the stem of *Annona glabra* exhibited activity against acetylcholinesterase yielded 20 compounds (Tsai and Lee 2010). Four of these were new (7 S,14 S)-(−)-N-methyl-10-O-demethylxylopinine salt (3), S-(−)-7,8-didehydro-10-O-demethylxylopininium salt (10), S-(−)-7,8-didehydrocorydalminium salt (11), and 5-O-methylmarcanine D (17). In addition, compounds 10 and 11 represent the first natural occurrence of 7,8-didehydroprotoberberines. Compound 3, pseudocolumbamine (12), palmatine (15), and pseudopalmatine (16) showed anti- acetylcholinesterase IC_{50} values of 8.4, 5.0, 0.4, and 1.8 µM, respectively.

Neuroprotective Activity

FLZ a novel cyclic derivative of squamosamide from *Annona glabra*, exhibited neuroprotective effects on neuronal apoptosis by increasing the Bcl-2/Bax ratio and decreasing the active caspase-3 fragment/caspase-3 ratio in the hippocampus of APP/PS1 mice (Li and Liu 2010). Compared with vehicle-treated APP/PS1 mice, FLZ (150 mg/kg) significantly increased brain-derived neurotrophic factor (BDNF) and NT3 (neurotrophin 3) expression in the hippocampus of APP/PS1 mice. Furthermore, FLZ promoted BDNF high-affinity receptor TrkB phosphorylation and activated its downstream ERK (extracellular signal-regulated kinases), thus increasing phosphorylation of the protein CREB (cAMP response element-binding) at Ser133 in the hippocampus of APP/PS1 mice. The results suggested that FLZ could be explored as a potential therapeutic agent in long-term Alzheimer's disease therapy.

Antimicrobial, Insecticidal, Sporicidal Activities

Preliminary screening studies reported that appreciable antimicrobial, antifungal and moderate insecticidal, sporicidal and cytotoxic activities were observed for the hexane extract of *Annona glabra* stem bark (Padmaja et al. 1995). Chromatographic fractionation of this extract led to the isolation of kaur-16-en-19-oic acid in a large amount as the main constituent, which was found to be largely responsible for the biological activities possessed by the crude extract.

Other Uses

The cultivation and usefulness of pond apple is more for its medicinal attributes than as a fruit tree. *Annona glabra* is used as a root stock for other edible *Annona* species for its tolerance and adaptability to water-logged conditions. The seeds can serve as insecticide. In Japan, three acetogenins isolated from seeds of pond apple: squamocin, asimicin and desacetyluraricin had varying insecticidal activity against *Mamestra brassica, Plutella xylostella, Crocidolomia binotalis, Nephotettix cincticeps, Tetranychus urticae* and *Henosepilachna vigintiocto-punctata* (Ohsawa et al. 1991). The light and soft wood is used to substitute cork in fishing nets. The fruit can be used as forage for cattle.

Comments

Pond apple is listed as a restricted invasive weed of national significance in Australia.

Selected References

Abdel-Lateff A, El-Menshawi BS, Haggag MY, Nawwar MA (2009) Cytotoxic acetogenins from *Annona glabra* cultivated in Egypt. Phcog Res 1:130–135

Agriculture and Resource Management Council of Australia and New Zealand, Australian and New Zealand Environment and Conservation Council and Forestry Ministers (2000) Weeds of national significance pond apple (*Annona glabra*) strategic plan. National Weeds Strategy Executive Committee, Launceston

Chang FR, Yang PY, Lin JY, Lee KH, Wu YC (1998) Bioactive kaurane diterpenoids from *Annona glabra*. J Nat Prod 61(4):437–439

Chang FR, Hsieh TJ, Cho CP, Wu YC, Chen CY (2000) Chemical constituents from *Annona glabra* III. J Chin Chem Soc 47:913–920

Chen CY, Chang FR, Cho CP, Wu YC (2000) ent-Kaurane diterpenoids from *Annona glabra*. J Nat Prod 63(7):1000–1003

Chen CH, Hsieh TJ, Liu TZ, Chern CL, Hsieh PY, Chen CY (2004) Annoglabayin, a novel dimeric kaurane diterpenoid, and apoptosis in Hep G2 cells of annomontacin from the fruits of *Annona glabra*. J Nat Prod 67(11):1942–1946

Cochrane CB, Nair PK, Melnick SJ, Resek AP, Ramachandran C (2008) Anticancer effects of *Annona glabra* plant extracts in human leukemia cell lines. Anticancer Res 28(2A):965–971

Fouqué A (1973) Espèces fruitières d'Amérique tropicale. XV Fruits 28(7–8):548–558

Gallardo T, Aragon R, Tormo JR, Amparo Blazquez M, Zafra-Polo MC, Cortes D (1998) Acetogenins from *Annona glabra* seeds. Phytochem 47(5):811–816

Hsieh TJ, Wu YC, Chen SC, Huang CS, Chen CY (2004) Chemical constituents from *Annona glabra*. J Chin Chem Soc 51(4):869–876

Li N, Liu GT (2010) The novel squamosamide derivative FLZ enhances BDNF/TrkB/CREB signaling and inhibits neuronal apoptosis in APP/PS1 mice. Acta Pharmacol Sin 31(3):265–272

Li C-M, Tan N-H, Zheng H-L, Qing M, Hao X-J, He Y-N, Zhou J (1999) Cyclopeptides from the seeds of *Annona glabra*. Phytochem 50(6):1047–1052

Liu XX, Alali FQ, Hopp DC, Rogers LL, Pilarinou E, McLaughlin JL (1998a) Glabracins A and B, two new acetogenins from *Annona glabra*. Bioorg Med Chem 6(7):959–965

Liu XX, Alali FQ, Pilarinou E, McLaughlin JL (1998b) Glacins A and B: two novel bioactive mono-tetrahydrofuran acetogenins from *Annona glabra*. J Nat Prod 61(5):620–624

Liu XX, Alali FQ, Pilarinou E, McLaughlin JL (1999a) Two bioactive mono-tetrahydrofuran acetogenins, annoglacins A and B, from *Annona glabra*. Phytochem 50(5):815–821

Liu XX, Pilarinou E, McLaughlin JLJ (1999b) Two novel acetogenins, annoglaxin and 27-hydroxybullatacin, from *Annona glabra*. J Nat Prod 62(6):848–852

Ohsawa K, Atsuzawa S, Mitsui T, Yamamoto I (1991) Isolation and insecticidal activity of three acetogenins from seeds of pond apple, *Annona glabra* L. J Pestic Sci 16:93–96

Padmaja V, Thankamany V, Hara N, Fujimoto Y, Hisham A (1995) Biological activities of *Annona glabra*. J Ethnopharmacol 48(1):21–24

Pino J, Marbot R, Agüero J (2002) Volatile components of baga (*Annona glabra* L.) fruit. J Essent Oil Res 14(4):252–253

Tsai SF, Lee SS (2010) Characterization of acetylcholinesterase inhibitory constituents from *Annona glabra* assisted by HPLC microfractionation. J Nat Prod 73(10):1623–1625

Zhang YH, Peng HY, Xia GH, Wang MY, Han Y (2004) Anticancer effect of two diterpenoid compounds isolated from *Annona glabra* Linn. Acta Pharmacol Sin 25(7):937–942

Annona montana

Scientific Name

Annona montana Macfady

Synonyms

Annona marcgravii Mart., *Annona muricata* Vell. nom. illeg., *Annona pisonis* Mart., *Annona sphaerocarpa* Splitg.

Family

Annonaceae

Common/English Names

Mountain Soursop, Wild Custard Apple, Wild Soursop

Vernacular Names

Brazil: Araticú, Araticunzeiro;
Chinese: Shan Ci Fan Li Zhi;
Cuba: Guanábana Cimarrona, Guanábana De Lama;
Czech: Láhevník Horský;
Dominican Republic: Guanábana Cimarrona;
Dutch: Boszuurzak;
French: Corossolier Bâtard, Cachiman Morveux, Cachiman Montage, Corossol Zombie, Corossolier Bâtard, Kachiman Montan;
French Guiana: Busi Atuku, Corossolier Sauvage, Manigl;
German: Schleimapfel;
Guayana: Busi Atuku, Corossolier Sauvage, Manigl;
Haiti: Corossol Zombie (French), Kowosol Zombie (Creole);
Honduras: Anona, Anona Cimarrona, Anone;
Hungarian: hegyi annóna;
Japanese: Yama Toge Banreishi;
Martinique: Kachiman Montan (Créole);
Peru: Chirimoya, Guanabana, Guanábana, Guanábano Sirimbo, Huanabana;
Philippines: Ponhe;
Portuguese: Araticum, Araticum Açú, Araticum Apé;
Spanish: Guanábana De Monte, Cimarrón, Guanábana Cimarrona, Guanábana De Perro, Guanábana De Lama, Guanábana De Las Montañas, Taragus, Turagua;
Suriname: Boszuurzak (Dutch), Busi Atuku, Manigl;
Venezuela: Guanábana, Guanábana Cimarrona, Guanábana De Perro, Guanábana Brasileiro, Guanobano Cimarrón, Catuche Cimarron, Turagua.

Origin/Distribution

This species is indigenous to South America, Central America and West Indies - Bolivia, Brazil, Colombia, Costa Rica, Cuba, Dominican Republic, Ecuador, Guadeloupe, Jamaica,

Panama, Paraguay, Peru, Puerto Rico, Suriname and Venezuela.

Agroecology

The climatic requirement for this species is tropical – dry tropical to the wet tropical forest zones but it also grows in the dry to moist subtropics. It is found from near sea level to 650 m altitude in its native areas. Optimum mean annual temperature range is from 21°C to 26°C and annual precipitation from 60 to 400 mm. It is drought tolerant and will grow well in dry conditions but cannot withstand prolonged water-logging. It thrives in full sun on many soil types in the pH range of 5.8–8 but optimum range is from pH 6 to 6.5. It is much more cold hardy than the soursop and will withstand sub-zero temperatures for short periods.

Plate 1 Developing fruit and mature leaves of mountain soursop

Edible Plant Parts and Uses

Yellow, aromatic pulp is eaten fresh in desserts but is used more for juice. Its eating quality is inferior to soursop.

Botany

A small deciduous tree to 10 m high with a spreading crown and dark, grey or brown bark. Leaves are alternate, distichous, short petiolate, oblong or elliptic 7–18 cm long by 2.5–8 cm wide, with tapering apex and rounded base, leathery, dark green above and pale green beneath, glabrous and glossy (Plate 1). Flowers are solitary or in pairs in older twigs, with stout peduncle. Sepals three, broad and pubescent; petals 6 in two whorls, inner three rounded; stamens numerous and crowded in rounded mass. Fruit is a syncarp, formed by the fusion of numerous carpels. It is subglobose to oval, up to 15 cm long by 7–13 cm across, green turning yellow when ripe and covered with soft, 4 mm long spines (Plates 1 and 2). The pulp is yellow and subacid, containing numerous light brown, oblong seeds, 18 mm long.

Plate 2 Close-up of immature fruit of mountain soursop

Nutritive/Medicinal Properties

Nutritive composition of the edible fruits has not been published.

Phytochemicals and Associated Pharmacological Activities

The seeds, leaves and stem of *Annona montana* have yielded many acetogenins which showed potent cytotoxicity and anti-tumorous activities against a range of human cancer cell lines. Acetogenins are a series of apparently polyketide-derived fatty acid derivatives that possess tetrahydrofuran rings and a methylated γ-lactone (sometimes rearranged to a methyl ketolactone) with various hydroxyl, acetoxyl, and/or ketoxyl groups along the hydrocarbon chain. They exhibit a broad range of potent biological activities which include antitumor, cytotoxicity, antimalarial, anti-inflammatory, antimicrobial, immunosuppressant, antifeedant, and pesticidal (Rupprecht et al. 1990) (see also notes under *Annona muricata*).

From Seeds

Studies reported that nine monotetrahydrofuranic Annonaceous acetogenins, montalicins A-E, cis-annoreticuin, montalicins F, I and J, montalicins G and H, monlicins A and B, (+)-monhexocin, monhexocin along with eight known acetogenins 10–17, isolated from the seeds of *Annona montana* showed selectively potent cytotoxicity against two human cancer cell lines, 1A9 and Hep G2 (Liaw et al. 2004a). Many other acetogenins from seeds such as annomonysvin, annomontacin, annomontacin, annonacinone and annonacin were cytotoxic against murine leukemia L1210, human breast adenocarcinoma MDA-MB231, and human breast carcinoma MCF7 cell lines (Jossang et al. 1991). Four mono-tetrahydrofuranic Annonaceous acetogenins, montalicins G and H, and monlicins A and B, and two new linear acetogenins, (+)-monhexocin and (−)-monhexocin, as well as three known compounds, murisolin, 4-deoxyannomontacin, and muricatacin, isolated from the seeds of *Annona montana* showed selective cytotoxicity against human hepatoma cells, Hep G2 (Liaw et al. 2005). Two new monotetrahydrofuran Annonaceous acetogenins, montacin (1), and cis-montacin (2), along with four known acetogenins, annonacin, cis-annonacin, annomontacin, and cis-annomontacin, were isolated from the seeds of *Annona montana* (Liaw et al. 2004b). The two new acetogenins exhibited moderate in-vitro cytotoxic activity against the 1A9 human ovarian cancer cell line. When Ca^{2+} was present, compounds 1 and 2 became three to tenfold more active against 1A9 ovarian carcinoma cells and the PTX10 ovarian carcinoma subline. Other cytotoxic mono-THF acetogenins isolated form the seeds included murisolin and annonacin. Two known monotetrahydrofuran acetogenins, murisolin and annonacin, were also isolated from the seeds.

Four new cyclopeptides, cyclomontanins A-D (1–4), annomuricatin C (5), and (+)-corytuberine were isolated from a methanol extract of *Annona montana* seeds (Chuang et al. 2008). Compounds 1, 3, and 4 exhibited anti-inflammatory activity in-vitro using the J774.1 macrophage model.

From Leaves

Three monotetrahydrofuran (mono-THF) acetogenins, anmontanins A−C (1–3) were isolated from the leaves (Mootoo et al. 2000). A novel acetogenin, montanacin F, with a new type of terminal lactone unit, was isolated from leaves of *Annona montana*. Its in-vitro cytotoxic activity against Lewis lung carcinoma (LLC) and in-vivo antitumour activity of annonacin with LLC tumour cells in mice were determined (Wang et al. 2002). Further studies on leaves of *Annona montana* led to isolation of one iso-acetogenin, montanacin G, three pairs of acetogenins, montanacin H-J and 34-epi-montanacin H-J, together with four known acetogenins, gigantetrocins A and B, annonacin and cis-annonacin (Wang et al. 2001). The cytotoxic activities of these compounds, together with previously reported acetogenins, montanacins B and C, were examined against Meth-A and LLC tumour cell lines in-vitro. Four novel mono-tetrahydrofuran acetogenins, montanacins B-E (1–4), were isolated from the ethanolic extract of the leaves of *Annona montana* (Wang et al. 2000). The leaves of *Annona montana* afforded one novel cytotoxic phenanthrene alkaloid, annoretine [3-hydroxy-4-methoxy-N-methyl-tetrahydropyrido(4,3-a)phenanthrene], together with two cytotoxic alkaloids, argentinine and liriodenine, as well as a new oxoaporphine alkaloid, annolatine, and one steroid, sitosterol-β-D-glucoside, which were inactive (Wu et al. 1993). Four annonaceous acetogenins, montanacins B-E(1–4) were isolated from *Annona montana* (Wang et al. 1999). Montanacins D and E contain a tetrahydropyran ring along with a nonadjacent tetrahydrofuran ring, belonging to the most unusual type of acetogenins.

Tucupentol, a mono-tetrahydrofuran-pentahydroxy-acetogenin isolated from leaves and twigs together with known acetogenins, annonacin-A, cis-annonacin-10-one, aromin, and giganteronenin were found to act as selective inhibitors of mitochondrial complex I in the 0.8–5.4-nM range (Colom et al. 2009). Mitochondrial complex I contributes to oxidative stress and perhaps ageing.

From Stem

Related acid amides, N-trans-feruloyltyramine, N-p-coumaroyltyramine, and N-trans-caffeoyltyramine, one lignan, (−)-syringaresinol, one aromatic aldehyde, syringaldehyde, and two steroids, β-sitosterol and β-sitosterol-β-D-glucoside were isolated from the stem parts of *Annona montana* (Wu et al. 1995). All these compounds and their acetate derivatives were subjected to the antiplatelet aggregation and cytotoxicity bioassay where some of them showed significant activities. A naturally occurring phenanthrene-1,4-quinone, annoquinone-A, along with parietin (physcion) and β-sitostenone were isolated from the stem bark of *Annona montana* (Wu et al. 1987). Annoquinone-A demonstrated potent antimicrobial activity against *Bacillus subtilis* and *Micrococcus luteus* as well as cytotoxicity in the KB human tumour cells (ED_{50}=0.16 μg/ml) tissue culture assay. β-Sitostenone also showed significant cytotoxicity.

Other Uses

The seeds are pounded and macerated in water to make an insecticide. This species is also used as rootstock for other edible *Annona* spp. The wood is used as fire wood.

Comments

The species is more appreciated as a rootstock for other *Annona* species and medicinal attributes than as an edible fruit crop.

Selected References

Chuang PH, Hsieh PW, Yang YL, Hua KF, Chang FR, Shiea J, Wu SH, Wu YC (2008) Cyclopeptides with anti-inflammatory activity from seeds of *Annona montana*. J Nat Prod 71(8):1365–1370

Colom OA, Neske A, Chahboune N, Zafra-Polo MC, Bardón A (2009) Tucupentol, a novel monotetrahydrofuranic acetogenin from *Annona montana*, as a potent inhibitor of mitochondrial complex I. Chem Biodivers 6(3):335–340

Fouqué A (1973) Espèces fruitières d'Amérique tropicale. XV Fruits 28(7–8):548–558

Jossang A, Dubaele B, Cavé A, Bartoli MH, Bériel H (1991) Annomonysvin: a new cytotoxic γ-lactone-monotetrahydrofuranyl acetogenin from *Annona montana*. J Nat Prod 54(4):967–971

Liaw CC, Chang FR, Wu CC, Chen SL, Bastow KF, Hayashi K, Nozaki H, Lee KH, Wu YC (2004a) Nine new cytotoxic monotetrahydrofuranic Annonaceous acetogenins from *Annona montana*. Planta Med 70(10):948–959

Liaw CC, Chang FR, Wu YC, Wang HK, Nakanishi Y, Bastow K, Lee KH (2004b) Montacin and *cis*-Montacin, two new cytotoxic monotetrahydrofuran Annonaceous acetogenins from *Annona montana*. J Nat Prod 67:1804–1808

Liaw CC, Chang FR, Chen SL, Wu CC, Lee KH, Wu YC (2005) Novel cytotoxic monotetrahydrofuranic Annonaceous acetogenins from *Annona montana*. Bioorg Med Chem 13:4767–4776

Mootoo BS, Ali A, Khan A, Reynolds WF, McLean S (2000) Three novel monotetrahydrofuran Annonaceous acetogenins from *Annona montana*. J Nat Prod 63(6):807–811

Morton J (1987) Related species. In: Fruits of warm climates. Julia F. Morton, Miami, pp 87–88

Rupprecht JK, Hui YH, McLaughlin JL (1990) Annonaceous acetogenins: a review. J Nat Prod 53(2):237–278

Wang LQ, Zhao WM, Qin GW, Cheng KF, Yang RZ (1999) Four novel Annonaceous acetogenins from *Annona montana*. Nat Prod Res 14(2):83–90

Wang L-Q, Nakamura N, Meselhy R, Hattori M, Zhao W-M, Cheng K-F, Yang R-Z, Qin G-W (2000) Four mono-tetrahydrofuran ring acetogenins, montanacins B-E, from *Annona montana*. Chem Pharm Bull Tokyo 48(8):1109–1113

Wang LQ, Li Y, Min BS, Nakamura N, Qin GW, Li CJ, Hattori M (2001) Cytotoxic mono-tetrahydrofuran ring acetogenins from leaves of *Annona montana*. Planta Med 67(9):847–852

Wang L-Q, Min B-S, Li Y, Nakamura N, Qin G-W, Li C-J, Hattori M (2002) Annonaceous acetogenins from the leaves of *Annona montana*. Bioorg Med Chem 10(3):561–565

Wu TS, Jong TT, Tien HJ, Kuoh CS, Furukawa H, Lee KH (1987) Annoquinone-A, an antimicrobial and cytotoxic principle from *Annona montana*. Phytochem 26(6):1623–1625

Wu YC, Chang GY, Duh CY, Wang SK (1993) Cytotoxic alkaloids of *Annona montana*. Phytochem 33(2):497–500

Wu YC, Chang GY, Ko FN, Teng CM (1995) Bioactive constituents from the stems of *Annona montana*. Planta Med 61(2):146–149

Annona muricata

Scientific Name

Annona muricata L.

Synonyms

Annona bonplandiana H.B.K., *Annona cearaensis* Barb. Rodr., *Annona macrocarpa* auct., *Annona macrocarpa* Werckle, *Annona muricata* var. *borinquensis* Morales, *Annona sericea* Dunal, *Guanabanus muricatus* Gómez de la Maza.

Family

Annonaceae

Common/English Names

Brazilian Pawpaw, Guanabana, Prickly Custard Apple, Soursop, Sugar Apple.

Vernacular Names

Angola: Sap-Sap (Portuguese);
Antilles: Cachiman Épineux, Corossol;
Banaban: Te Tiotabu;
Benin: Gniglou, Niglou (Adja), Sap-Sap (Bariba), Okoungoro, Otrigbogbo, Gnibo (Dassa), Chapou Chapou (Dendi), Yovo Wouinglé (Fon), Annone Muriculée, Cachiman Épineux, Cachimantier, Corossolier Épineux, Grand Corossol, Grand Corossolier (French), Cha-Cham, Chap-Chap (Goum), Doukoumé Porto (Peuhl), Wouinglo (Sahoué), Agniglo (Watchi), Sapi Sapi (Yoruba);
Brazil: Araticum-Grande, Araticum-Manso, Guanabano, Condessa, Coração-De-Rainha, Jaca-De-Pobre, Jaca-Do-Pará, Cabeça-De-Negro, Fruta-Do-Conde (Portuguese);
Brazzaville: Corossolier (French);
Bolivia: Sinini;
Brunei: Durian Salat, Durian Belanda;
Cameroon: Saba-Saba (Eséka), Ebom Beti (Mbalmayo);
Chamorro: Laguaná, Laguana, Laguanaha, Syasyap;
Chinese: Ci Guo Fan Li Zhi;
Chuukese: Saasaf, Saasaf, Saasap, Sasaf;
Comoros: Konokono (Anjouan Island);
Cook Islands: Kātara'Apa, Kātara'Apa Papa'Ā, Naponapo Taratara (Maori);
Cuba: Guanábana;
Danish: Pigget Annona, Sukkeræble;
Democratic Republic Of Congo: Mustafeli (Shi), Mustaferi;
Dominican Republic: Guanábana;
Dutch: Zuurzak;
Eastonian: Oga-Annoona, Vili: Oga-Annoona;
Fijian: Sarifa, Seremai;
French: Anone, Anone Muriquee, Cacheimantier Èpineux, Cachiman Èpineux, Corasol, Corossol, Corossol, Corossol Èpineux, Corossolier, Sapotille;

German: Sauersack, Stachelannone, Stachel-Annone;
Guatemala: Guanaba;
Guinea Conakry: Doukoume Porto (Foula Du Fouta-Djallon), Mete (Soussou)
Hungarian: Savanyúalma, Tüskés Annóna;
India: Mullu Raamana Phala, Mullu Ramaphala, Mullu Raama Phala (Kannada), Mullancakka, Mullanchukka, Vilattinuna (Malayalam), Mamaphal, Mamphal (Marathi), Jeuto (Oriya), Mullu Citta, Mulluccitta, Mulcita, Mulluccita, Mutcita, Mutcitamaram, Pulippala (Tamil), Lakshmana Phalamu (Telugu);
Indonesia: Nangka Seberang, Nangka Landa, Nangka Landi, Nangka Manila, Sirsak (Java), Mandalika, Nangka Belanda, Nangka Walanda (Sundanese), Nangka Englan, Nangka Mooris (Madurese);
Japanese: Toge Banreishi;
Khmer: Tiep Banla, Tiep Barang;
Kosraean: Sosap;
Laotian: Khan Thalot, Khiềp Thét;
Madagascar: Corossol, Karasôly, Kaorasaly, Senasena;
Malaysia: Datu Alo (Kelabit), Rian Balanda (Iban), Buah Sirsak, Sirsak, Sirkaya Belanda, Nangka Belanda (Malay);
Marquesan: Koroso;
Marshallese: Jojaa;
Mauritius: Corossol;
Mexico: Catuche;
Mokilese: Truka Shai;
Niuean: Talapo Fotofoto;
Palauan: Sausab;
Papiamento: Sòrsaka;
Papua New Guinea: Saua Sap, Sapsap;
Philippines: Ilabanos, Labanos (Bikol), Atti, Gayubano (Ibanag), Bayubana, Gayubano (Iloko), Babana (Panay Bisaya), Guiabano (Sambali), Labanus (Sulu), Guyabana, Guayabano, Guabana, Guanabano, Guiabano (Tagalog);
Rotuman: Ai' Pen Mamami;
Pohnpeian: Sae, Sei;
Portuguese: Graviola, Pinha, Pinha Azeda;
Puerto Rico: Guanábana;
Reunion: Coprossol, Corrosol, Sapoty;
Rodrigues Island: Coronsol, Corossol, Soursop;
Russian: Annona Murikata;
Samoan: Sanalapa, Sasalapa, Sasalapa;
Senegal: Corossolier (Diola);
Seychelles: Corosso (French);
Sierra Leone: Soursap (Krio), Soursapi (Mende);
Spanish: Anona, Guanabana, Guanábano, Zapote Agrio, *Catche, Catoche, Catuche*;
Tahitian: Tapotapo Papa'A, Tapotapo Urupe;
Thai: Rian-Nam, Thurien-Thet, Thurian-Khaak;
Togo: Anyigli, Anyiklé (Ewé), Corossolier (French);
Tongan: 'Apele 'Initia;
Tuamotuan: Korosor, Tapotaporatara;
Venezuela: Catuche;
Vietnamese: Mang Cân Xiem, Mang Câù Xiê;
Yapese: Sausau.

Origin/Distribution

The species probably originated in the Antilles in the Caribbean, Central America and in northern South America. It is found wild and cultivated from sea level to 1,000 m elevation in the Antilles and from southern Mexico to Peru and northern Argentina. Now it is widely distributed in Asia from south India to northern Australia and Polynesia, and in Africa in the hot lowlands in the west and east.

Agroecology

Soursop grows in the humid tropical and subtropical lowlands with relatively warm winters below 1,000 m altitude. Optimum mean annual temperature range is 25–30°C and mean annual rainfall of above 1,000 mm. It is frost and low temperature (below 5°C) sensitive and abhors water-logging. It is common on the coast and is found on slopes and become wild or naturalized in thickets, pastures and along roads. The species is commonly cultivated in home gardens and is found in rural garden areas on many soil types including volcanic and calcareous soils, where it is occasionally naturalized. It thrives best in friable, fairly rich, deep loams with a pH range of 5.5–6.5.

Edible Plant Parts and Uses

The fruit pulp is eaten fresh when ripe or made into refreshing drinks, beverages, shakes, sherbets, sorbets, yogurt and ice-cream. Bottled and canned soursop beverages are popular in southeast Asia and are much sought after as health drinks. The pulp is also processed into desserts, sauce, jellies, candies and tarts. In Mexico it is a common fruit often used for dessert as the only ingredient, or as an *agua fresca* beverage. Ice-cream, and fruit bars made of soursop are also very popular. Immature, young fruits and young sprouts are cooked as vegetables in Indonesia and Papua New Guinea and the tender young leaves are similarly used in the latter country.

Botany

A small, much branched, perennial tree, 3–8 m tall, with terete, reddish-brown glabrous branchlets. Leaves are biseriate, shortly petiolate, pale green to bright green, oblong-obovate or oblong-elliptic, 6–18 cm by 3–7 cm wide, glossy leaves with acuminate apex and cuneate or acute base. Flowers are regular, green to yellowish green (to yellow on a yellow fruited variety), on short axillary, 1–2 flowered branchlets (Plate 3). Flower has three dark green, ovate deltoid, coriaceous sepals which are pubescent and persistent; six broadly ovate, coriaceous, usually green to subsequently yellowish-green, tomentose, cordate base petals; numerous stamens in rows around the gynoecium; numerous ovary, pubescent with styles shaped like soft prickles. Fruit is oblong or somewhat curved, 15–35 cm by 10–15 cm wide, beset with soft spines with a thin, green or yellow, coriaceous rind (Plates 1, 2, and 4). Seeds are numerous, obovoid and flattened, dark brown to black, glabrous and glossy, embedded in firm, white, fleshy, acid-sweet, juicy pulp.

Plate 1 The common green variety of soursop grown in Malaysia, Indonesia and Northern Australia

Plate 2 An unusual yellow-fruited variety of soursop grown in Darwin, Australia

Plate 3 Flower of the yellow-fruited soursop (*left*) and green soursop (*right*)

Plate 4 Green soursop variety grown in Vietnam

Nutritive/Medicinal Properties

Analyses carried out in the United States reported raw, soursop fruit (exclude 33% seeds and skin) to have the following nutrient composition (per 100 g edible portion): water 81.16, energy 66 kcal (276 kJ), protein 1.00 g, total lipid 0.30 g, ash 0.70 g, carbohydrates 16.84 g, total dietary fibre 3.3 g, total sugars 13.54 g, Ca 14 mg, Fe 0.60 mg, Mg 21 mg, P 27 mg, K 278 mg, Na 14 mg, Zn 0.10 mg, Cu 0.086 mg, Se 0.6 µg, vitamin C 20.6 mg, thiamin 0.070 mg, riboflavin 0.050 mg, niacin 0.900 mg, pantothenic acid 0.253 mg, vitamin B-6 0.059 mg, total folate 14 µg, choline 7.6 mg, vitamin A 2 IU, β-carotene 1 µg, vitamin E (α-tocopherol) 0.08 mg, vitamin K (phylloquinone) 0.4 µg, total saturated fatty acids 0.051 g, 16:0 (palmitic) 0.040 g, 18:0 (stearic) 0.011 g; total monounsaturated fatty acids 0.090 g, 16:1 undifferentiated (palmitoleic) 0.004 g, 18:1 undifferentiated (oleic) 0.085 g; total polyunsaturated fatty acids 0.069 g, 18:2 undifferentiated (linoleic) 0.069 g; tryptophan 0.011 g, lysine 0.060 g, and methionine 0.007 g (USDA 2010).

The flesh of the fruit consist of a white edible pulp that is high in carbohydrates and sugars and fair amounts of vitamin C, vitamin B1, vitamin B2, potassium and dietary fibre. It is poor in vitamin A.

The aroma volatile components of soursop were found to consists mainly of esters (80%). Methyl benzanoate (31%) and methyl 2-hexenoate, (27%) were the most abundant (Macleod and Pieris 1981). Another analysis reported forty-one compounds in the volatile component of the fruit (Pino and Marbot 2001). The major volatiles were methyl 3-phenyl-2-propenoate, hexadecanoic acid, emthyl(E)-2-hexenoate and methyl 2-hydroxy-4-methyl valerate.

The seeds of *A. muricata* were found to contain (per 100 g D.M.) 7.7% moisture, 40% crude oil, 8.5% crude proteins, 34.1% carbohydrate, 5.2% crude fiber, 9.7% ash, K 357.14 mg, Ca 149.1 mg, P 136 mg, Na 17.35 mg, Mg 12.57 mg (Kimbonguila et al. 2010). The seed oil was found to contain high levels of unsaturated fatty acids, especially oleic (up to 41.41%) and linoleic (up to 30.60%). The dominant saturated acid was palmitic (up to 20.33%). The oil extracts exhibited good physicochemical properties and could be useful as edible oils and for industrial applications. High unsaponifiable matters content (1.34%) of the oil assured its potential application in cosmetics industry.

Essential oil from leaves of *Annona muricata* was found to consist mainly of sesquiterpene derivatives and alcohols (Kossough et al. 2007). Large amounts of sesquiterpene derivatives: β-elemene, isocaryophyllene, β-caryophyllene, αhumulene, γ-muurolene, OC-selinene, α-muurolene, germacrene A, β-cadinene and the alcohols: elemol, (E)-nerolidol, spathulenol, globulol, epi-globulol, 1-epi-cubenol, d-epi-cadinol, á-epi-muurolol, and á –cadinol were obtained. The most abundant constituents were β-caryophyllene (13.6%), δ-cadinene (9.1%), epi-α-cadinol (8.4%), α-cadinol (8.3%). Another analysis identified 59 compounds from the leaf oil principally ß-caryophyllene (31.4%), Ô-cadinene (6.7%), á-muurolene (5.5%) T- and á-cadinols (4.3%) (Pélissier et al. 1994). Analysis conducted in Cameroon reported the following predponderant components in the oil of soursop leaves: ß-caryophyllene (40%) and in seeds: oc-phellandrene (25%) (Boyom et al. 1996).

Recent studies have supported many of soursop's traditional medicinal uses and also showed that various parts of the tree contains acetogenins, which have been shown to be responsible for its myriad array of its medicinal attributes. To date approximately 82 acetogenins from 10 different groups have been reported from A. muricata. The Annonaceous acetogenins are a series of apparently polyketide-derived fatty acid derivatives that possess tetrahydrofuran rings and a methylated γ-lactone (sometimes rearranged to a methyl ketolactone) with various hydroxyl, acetoxyl, and/or ketoxyl groups along the hydrocarbon chain. Acetogenins exhibit a broad range of potent biological activities (cytotoxicity, antitumour, antimalarial, antiprotozoal, antispasmodic, antiviral, antimicrobial, immunosuppressant, antifeedant, anthelmitic, and pesticidal) (Rupprecht et al. 1990).

Antioxidant Activity

Soursop fruit was found to have high antioxidant activity with IC_{50} of 2.0 mg/ml as determined by free radical 1,1- diphenyl-2-picrylhydrazyl (DPPH) method and high phenolic compound content of 368 mg/100 g, which was much higher than that reported for orange (Citrus sinensis) $IC_{50} = 18.3$ mg/ml and phenolic compound content of 64.2 mg/ml (Franco 2006).

The extracts of A. muricata possessed potent in-vitro antioxidant activity as compared to leaves of A. squamosa and A. reticulata suggesting its role as an effective free radical scavenger, augmenting its therapeutic application. The ethanolic extract of A. muricata leaves at 500 μg/ml showed maximum scavenging activity (90.05%) of ABTS (2,2-azinobis-(3-ethylbenzothizoline-6-sulphonate)) radical cation followed by the scavenging of hydroxyl radical (85.88%) and nitric oxide (72.60%) at the same concentration (Baskar et al. 2007). However, the extract showed only moderate lipid peroxidation inhibition activity.

Anticancer Activity

In the studies conducted by separate research groups all confirmed significant anti-tumour, anti-cancer and selective toxicity against several different types of cancer cells which they studied. One study showed that cis-annonacin one of the five acetogenins namely cis-annonacin, cis-annonacin-10-one, cis-goniothalamicin, arianacin and javoricin isolated from seeds, was selectively cytotoxic to colon adenocarcinoma cells, with a potency 10,000 times that of adriamycin (Rieser et al. 1996). Other acetogenins isolated from the seeds muricins A-G, muricatetrocin A, muricatetrocin B, longifolicin, corossolin and corossolone showed significantly selective in-vitro cytotoxicities toward the human hepatoma cell lines Hep G(2) and 2,2,15 (Chang and Wu 2001). Solamin, a cytotoxic mono-tetrahydrofuranic γ-lactone acetogenin from Annona muricata seeds was cytotoxic in KB and VERO cell lines in in-vitro tests (Saw et al. 1991). Other bioactive acetogenins found in the seeds include annocacin, muricatacin (Rieser et al. 1991) and two γ-lactone acetogenins, epomuricenins A and B (Roblot et al. 1993); murihexol, donhexocin, annonacin A and annonacin (Yu et al. 1998). A cyclopeptide, annomuricatin B was also isolated from the seeds (Li et al. 1998). Three new monotetrahydrofuran annonaceous acetogenins, muricin H, muricin I and cis-annomontacin, along with five known acetogenins, annonacin, annonacinone, annomontacin, murisolin, and xylomaticin, were isolated from the seeds of Annona muricata (Liaw et al. 2002). Muricin H, muricin I and cis-annomontacin exhibited significant activity in in-vitro cytotoxic assays against two human hepatoma cell lines, Hep G(2) and 2,2,15. The acetogenin murisolin, was reported to show cytotoxic activity against human tumour cell lines with potency from 105 to 106 times that of adriamycin (Kojima 2004).

In separate studies, monotetrahydrofuran Annonaceous acetogenins isolated from leaves of Annona muricata, namely annomutacin, (2,4-trans)-10R-annonacin-A-one, and (2,4-cis)-10R- annonacin-A-one showed selective cytotoxicities against the human A-549 lung tumour cell line (Wu et al. 1995d). A known bioactive amide, N-p-coumaroyl tyramine, was also found. Five monotetrahydrofuran acetogenins, annonacin A, (2,4-trans)-isoannonacin, and (2,4-cis)-isoannonacin, and muricatocins A and B were found in the leaves and the last two acetogenins: muricatocins A and B exhibited enhanced cytotoxicity against the A-549 human lung tumour cell line (Wu et al. 1995a). Other monotetrahydrofuran acetogenins from the leaves, annomuricin C and muricatocin C significantly augmented the cytotoxicities against the A-549 human lung and the MCF-7 human breast solid tumour cell lines (Wu et al. 1995b). One known onotetrahydrofuran acetogenin, gigantetronenin was also found. The leaves also contained other bioactive acetogenins, annohexocin (Zeng et al. 1995a), murihexocins A and B which showed significant inhibitory effects among six human tumour cell lines with interesting selectivity to the prostate (PC-3) and pancreatic (PACA-2) cell lines (Zeng et al. 1995b); muricoreacin and murihexocin C which also showed significant cytotoxicities among six human tumour cell lines with selectivity to the prostate adenocarinoma (PC-3) and pancreatic carcinoma (PACA-2) cell lines (Kim et al. 1998a, b). Acetogenins from the leaves, annopentocins A (1), B (2), and C (3), and cis- and trans-annomuricin-D-ones (4, 5) exhibited cytoxicity to several cancer cell lines (Zeng et al. 1996). Compound 1 was selectively cytotoxic to pancreatic carcinoma cells (PACA-2), and compounds 2 and 3 were selectively cytotoxic to lung carcinoma cells (A-549). The mixture of 4 and 5 was selectively cytotoxic for the lung (A-549), colon (HT-29), and pancreatic (PACA-2) cell lines with potencies equal to or exceeding those of adriamycin. Additionally, eight monotetrahydrofuran acetogenins were found in the leaves: annomuricins A and B, gigantetrocin A, annonacin-10-one, muricatetrocins A and B, annonacin, and goniothalamicin (Wu et al. 1995c). Two new monotetrahydrofuran annonaceous acetogenins, cis-corossolone and annocatalin, together with four known ones, annonacin, annonacinone, solamin, and corossolone, were isolated from the leaves (Liaw et al. 2002). These new acetogenins exhibited significant activity in in-vitro cytotoxic assays against two human hepatoma cell lines, Hep G(2) and 2,2,15. Annocatalin displayed a high selectivity towards the Hep 2,2,15 cell line.

It was reported that the Annonaceous acetogenins irrespective of their various structural types classes of bis-adjacent, bis-non-adjacent, and single tetrahydrofuran (THF) ring compounds and their respective ketolactone rearrangement products selectively inhibited the growth of cancerous cells (Oberlies et al. 1995; 1997a, b). They also inhibited the growth of adriamycin resistant tumour cells and non-resistant tumour cells at the same levels of potency while exhibiting only minimal toxicity to 'normal' non-cancerous cells. Amongst a series of bis-adjacent THF ring acetogenins, those with the stereochemistry of threo-trans-threo-trans-erythro (from C-15 to C-24) were found to be the most potent against the growth of adriamycin resistant human mammary adenocarcinoma (MCF-7/Adr) cells with as much as 250 times the potency of adriamycin (Oberlies et al. 1997a). This cell line was found to be resistant to treatment with adriamycin, vincristine, and vinblastine. Several single-THF ring compounds were also quite potent with gigantetrocin A being the most potent compound tested. The acetogenins may, thus, have chemotherapeutic potential, especially with regard to multidrug resistant tumours. The Annonaceous acetogenin, bullatacin, was found to be effectively cytotoxic to the human mammary adenocarcinoma (MCF-7/Adr) cells while it was more cytostatic to the parental non-resistant wild type (MCF-7/wt) cells (Oberlies et al. 1997b). ATP depletion was found to be the mode of action of the Annonaceous acetogenins. Biologically, Annonaceous acetogenins exhibited their potent bioactivities through depletion of ATP levels via inhibiting complex I of mitochondria and inhibiting the NADH oxidase of plasma membranes of tumour cells (Alali et al. 1999). They thwarted ATP-driven resistance mechanisms.

Antiviral Activity

Studies by Padma et al. (1998) confirmed the antiviral activity of *Annona muricata* against the Herpes simplex virus. The stem bark extract inhibited the cytopathic effect of HSV-1 on vero cells which was indicative of anti-HSV-1 potential. The minimum inhibitory concentration of ethanolic extract of *A. muricata* was found to be 1 mg/ml. Bioactive acetogenins isolated from the roots included cis-solamin, cis-panatellin, cis-uvariamicin IV, cis-uvariamicin I, cis-reticulatacin, and cis-reticulatacin-10-one and solamin (Gleye et al. 1998), cohibins A and B (Gleye et al. 1997), and sabadelin (Gleye et al. 1999). Annonaceous acetogenin precursor compounds, epoxymurins A and B, were also isolated from the hexane extract of the stem bark (Hisham et al. 1993).

Antidiabetic Activity

In Nigeria, *Annona muricata* leaf extract was found to protect and to preserve pancreatic β- cell integrity of test diabetic rats subjected to streptozotocin (STZ)-induced oxidative stress by directly quenching lipid peroxides and indirectly augmenting production of endogenous antioxidants (Adewole and Caxton-Martins 2006). *A. muricata*-treated rats showed a significant decrease in elevated blood glucose, malondialdehyde (MDA) and nitric oxide. Further, *A. muricata* treatment significantly increased antioxidant enzymes' activities, as well as pancreatic/serum insulin contents. Adewole and Ojewole (2009) suggested that *A. muricata* extract had a protective, beneficial effect on hepatic tissues subjected to streptozotocin-induced oxidative stress, possibly by reducing lipid peroxidation and indirectly increasing production of insulin and endogenous antioxidants. Treatment of streptozotocin (STZ)-induced diabetic rats with aqueous leaf extract(AME) resulted in significant reduction in elevated blood glucose, reactive oxygen species (ROS), thiobarbituric acid reactive substances (TBARS), triglycerides (TG), total cholesterol (TC), and low density lipoprotein (LDL). Additionally, AME treatment significantly increased antioxidant enzymes' activities, as well as serum insulin levels. Treatment of rats with STZ (70 mg/ kg i.p.) resulted in hyperglycaemia, hypoinsulinaemia, and increased TBARS, ROS, TC, TG and LDL levels. STZ treatment also significantly decreased catalase (CAT), glutathione (GSH), superoxide dismutase (SOD), glutathione peroxidase (GSH-Px) activities, and high density lipoprotein (HDL) levels.

Studies by Adeyemi et al. (2009) reported that methanolic extracts of *Annona muricata* exhibited anti-hyperglycemic activities. A significant difference was found between the low blood glucose concentrations of treated (4.22 mmol/l) and the high blood glucose level of untreated (21.64 mmol/l) hyperglycemic groups of rats.

Antinociceptive and Anti-inflammatory Activities

Studies by de Sousa et al. (2010) found that *A. muricata* contained bioactive compounds with antinociceptive and anti-inflammatory activities. The ethanol leaf extract decreased the number of abdominal contortions by 14.42% (at a dose of 200 mg/kg) and 41.41% (400 mg/kg). Doses of 200 and 400 mg/kg (p.o.) were observed to inhibit both phases of the time paw licking: first phase (23.67% and 45.02%) and the second phase (30.09% and 50.02%), respectively. The extract (p.o.) enhanced the reaction time on a hot plate at doses of 200 (30.77% and 37.04%) and 400 mg/kg (82.61% and 96.30%) after 60 and 90 minutes of treatment, respectively. The paw edema was decreased by the ethanol extract (p.o.) at doses of 200 (23.16% and 29.33%) and 400 mg/kg (29.50% and 37.33%) after 3–4 hours of application of carrageenan, respectively. Doses of 200 and 400 mg/kg (p.o.), administered 4 hours before the carrageenan injection, reduced the exudate volume (29.25% and 45.74%) and leukocyte migration (18.19% and 27.95%) significantly.

Antidepressant Activity

Scientific research have validated through animal studies the use of various parts of the graviola tree for hypertension, as a vasodilator, as an antispasmodic (smooth muscle relaxer), and as cardiodepressant (slowing of heart rate). The fruit and the leaves of *Annona muricata* are used in traditional medicine for their tranquillizing and sedative properties. Acetogenin compounds annonain, nomuciferine and asimilobine isolated from the fruit were found to exhibit anti-depressive effects on test animals (Hasrat et al. 1997). Extracts of the plant were observed to inhibit binding of [3H]rauwolscine to 5-HTergic 5-HT1A receptors in calf hippocampus, and three alkaloids, annonaine (1), nornuciferine (2) and asimilobine (3), isolated from the fruit were found to have IC_{50} values of 3 μM, 9 μM and 5 μM, respectively. Studies in Nigeria demonstrated that graviola leaf extract significantly reduced the incidence of tonic pentylenetetrazol (PTZ) seizures and mortality, and lengthened the onset to clonic PTZ seizures (N'gouemo et al. 1997). The study indicated that extract of the leaves of *A. muricata* contained active substance(s) that preferentially suppressed the tonic phase of PTZ-induced clonic-tonic seizures. Another study found that administration of an ethanol extract of the stem bark of *Annona muricata* significantly inhibited cold immobilization stress-induced increase in lipid peroxidation in the liver and brain of albino rats (Padma et al. 1997).

Antileishmanial Activity

Acetogenins annonacin, annonacin A and annomuricin A, isolated from the fruit pericarp showed cytotoxicity and antileishmanial activity (Jaramillo et al. 2000). Hexane, ethyl acetate and methanol extracts of *Annona muricata* pericarp were inhibitory in-vitro against *Leishmania braziliensis* and *Leishmania panamensis* promastigotes, and cytotoxic against cell line U-937.

Extracts from *Annona muricata*, *Rollinia exsucca*, *Rollinia pittieri* and *Xylopia aromatica* were found to be active against promastigotes of three *Leishmania* species, epimastigotes of *Trypanosoma cruzi* exhibiting IC_{50} values lower than 25 μg/ml. These plants and/or plant parts could be useful in the treatment of leishmaniasis and Chagas' disease.

Antiprotozoal and Molluscicidal Activities

Acetoginin was found also to have pesticidal activity. Methanolic extracts of *Annona muricata* and *Annona cherimola* seeds exhibited antiparasitic activity against *Entamoeba histolytica*, *Nocardia brasiliensis*, *Molinema desetae* and *Artemia salina* (Bories et al. 1991). The acetogenins isolated from these extracts were found to be responsible for the important activity on infective larvae of *Molinema desetae*. Annonacin, isoannonacin and goniothalamicin isolated from leaves exhibited high molluscicidal activity, the extracts were highly toxic to adult forms of the snail *Biomphalaria glabrata* and to larvae of the brine shrimp *Artemia salina* (Luna et al. 2002). The leaves of *Annona muricata* also exhibited antimalarial action against *Plasmodium falciparum* cultured in-vitro (Gbeassor et al. 1990). *Annona muricata* was also reported to have antiprotozoal activity (Osorio et al. 2007).

Adverse Effects

On the down side, atypical parkinsonism in Guadeloupe (French West Indies) was found to be associated with the consumption of fruit and infusions or decoctions prepared from leaves of *Annona muricata*, which contained high concentration of annonacin (Champy et al. 2005). Recent experimental studies performed in midbrain cell cultures identified annonacin, a selective mitochondrial complex I inhibitor contained in the fruit and leaves of soursop, as a probable etiological factor (Lannuzel et al. 2003, 2008). Studies demonstrated that annonacin promoted dopaminergic neuronal death by impairment of energy production. Consistent with this view, chronic administration of annonacin to rats was found able to reproduce the brain lesions characteristic of the human disease.

Traditional Medicinal Uses

All parts of the tree have been used in traditional folkloric medicine in various cultures for various ailments and complaints. The leaves and seeds of the tree have long been used by native peoples of various cultures for an astounding variety of ailments, ranging from parasites (the seeds), to high blood pressure and cancer. Seeds are emetic,

Pulverized seeds and seed oil effective for head lice. Flowers are regarded to be antispasmodic and pectoral and used to alleviate catarrh. Infusion of leaves have been used as sudorific, antispasmodic and emetic. In the Caribbean, it is believed that laying the leaves of the soursop on a bed below a sleeping person with a fever will break the fever by the next morning. Also, boiling the leaves and drinking was believed to help induce sleep. Decoctions of leaves have been used as compresses for inflammation and swollen feet and similar decoction used for treating head lice and bedbugs. Poultice of mashed leaves and sap of young leaves have been employed for eczema and skin eruptions. Ripe fruit is reported antiscorbutic and is also used as an anthelmintic. Pulp of soursop has been used as poultice to draw out chiggers; and the juice of ripe fruit used as diuretic and for hematuria and urethritis. The unripe and dried fruit, also astringent, is used in diarrhoea, among the Amerindians. Unripe fruit and seeds are astringent and used for dysentery. The tea, fruit, and juice are used medicinally to treat illness ranging from stomach ailments to worms. The bark is used in powdered form for diarrhoea and dysentery and is used by the Chinese and Malays in Malaysia as tonic.

Other Uses

The pulverized seeds have been used as insecticide. Six acetogenin compounds, representing the mono-tetrahydrofuran (THF) (gigantetrocin A, annomontacin), adjacent bis-THF (asimicin, parviflorin), and nonadjacent bis-THF (sylvaticin, bullatalicin) caused high percentages of mortality and delays in development of the fifth instars of both insecticide-resistant and insecticide-susceptible strains of the German cockroach, *Blattella germanica* (Alali et al. 1998). Insecticide-susceptible nymphal development was mainly affected by gigantetrocin A and annomontacin, whereas insecticide-resistant nymphal development was mainly affected by gigantetrocin A and bullatalicin. Most tested acetogenins performed better than the conventional insecticides against both stages of both strains.

In India and Sri Lanka, the plant is also used as rootstock for *Annona squamosa*. The fruit is used as a bait in fish traps. The bark has been used in tanning and the wood is a potential source of paper pulp.

Comments

The common soursop has a green skin that turns pale green when ripe with white, mucilaginous, fleshy, acid-sweet pulp. There is a yellow-fruited variety in Darwin that remains yellow when ripe with white, mucilaginous, fleshy, acid-sweet pulp.

Selected References

Adewole SO, Caxton-Martins EA (2006) Morphological changes and hypoglycemic effects of *Annona muricata* Linn. (Annonaceae) leaf aqueous extract on pancreatic B-cells of Streptozotocin-treated diabetic rats. Afr J Biomed Res 9:173–187

Adewole SO, Ojewole JAO (2009) Protective effects of *Annona Muricata* Linn. (Annonaceae) leaf aqueous extract on serum lipid profiles and oxidative stress in hepatocytes of streptozotocin-treated diabetic rats. Afr J Tradit Complement Altern Med 6(1):30–41

Adeyemi DO, Komolafe OA, Adewole OS, Obuotor EM, Adenowo TK (2009) Anti hyperglycemic activities of *Annona muricata* (Linn). Afr J Tradit Complement Altern Med 6(1):62–69

Alali FQ, Kaakeh W, Bennett GW, McLaughlin JL (1998) Annonaceous acetogenins as natural pesticides; potent toxicity against insecticide-susceptible and resistant German cockroaches (Dictyoptera: Blattellidae). J Econ Entomol 91(3):641–649

Alali FQ, Liu XX, McLaughlin JL (1999) Annonaceous acetogenins: recent progress. J Nat Prod 62(3):504–540

Backer CA, van den Brink RCB jr (1963) Flora of java (Spermatophytes only), vol 1. Noordhoff, Groningen, 648

Baskar R, Rajeswari V, Kumar TS (2007) In vitro antioxidant studies in leaves of *Annona* species. Indian J Exp Biol 45(5):480–485

Bories C, Loiseau P, Cortes D, Myint SH, Hocquemiller R, Gayral P, Cavé A, Laurens A (1991) Antiparasitic activity of *Annona muricata* and *Annona cherimolia* seeds. Planta Med 57(5):434–436

Boyom FF, ZoIIo RHA, Menut C, Lamaty G, Besslere JM (1996) Aromatic plants of Tropical Central Africa. XXVII comp study volatile constituents five

Annonaceae species growing Cameroon. Flav Fragr J 11:333–338

Burkill IH (1966) A dictionary of the economic products of the Malay Peninsula. Revised reprint, 2 vols. Ministry of Agriculture and Co-operatives, Kuala Lumpur, vol 1 (A–H), pp 1–1240, vol 2 (I–Z) pp. 1241–2444

Champy P, Alice Melot A, Guérineau V, Christophe Gleye C, Fall D, Höglinger GU, Ruberg M, Lannuzel A, Laprévote O, Alain Laurens A, Hocquemiller R (2005) Quantification of acetogenins in *Annona muricata* linked to atypical parkinsonism in Guadeloupe. Mov Disord 20(12):1629–1633

Chang FR, Wu YC (2001) Novel cytotoxic annonaceous acetogenins from *Annona muricata*. J Nat Prod 64(7):925–931

Council of Scientific and Industrial Research (1948) The wealth of India. A dictionary of Indian raw materials and industrial products. (Raw materials 1). Publications and Information Directorate, New Delhi

de Sousa OV, Vieira GD-V, de Pinho JJRG, Yamamoto CH, Alves MS (2010) Antinociceptive and anti-inflammatory activities of the ethanol extract of *Annona muricata* L. leaves in animal models. Int J Mol Sci 11(5):2067–2078

De Q Pinto AC, Cordeiro MCR, de Andrade SRM, Ferreira FR, de C Filgueiras HA, Alves RE, Kinpara DI (2005) *Annona* species. International Centre for Underutilised Crops, University of Southampton, Southampton

Foundation for Revitalisation of Local Health Traditions (2008) FRLHT Database. htttp://envis.frlht.org

Fouqué A (1973) Espèces fruitières d'Amérique tropicale. XV Fruits 28(7–8):548–558

Franco EM (2006). Actividad antioxidante in vitro de las bebidas de frutas. Bebidas -Alfa Editores Técnicos, Junio/Julio:20–27. (In Spanish)

French BR (1986) Food plants of Papua New Guinea – a compendium. Australia and Pacific Science Foundation, Sheffield, Tasmania, 408 pp

Gbeassor M, Kedjagni AY, Koumaglo K, de Souza C, Agbo K, Aklikokou K, Amegbo KA (1990) In vitro antimalarial activity of six medicinal plants. Phytother Res 4(3):115–117

Gleye C, Laurens A, Hocquemiller R, Laprevote O, Serani L, Cavé A (1997) Cohibins A and B, acetogenins from roots of *Annona muricata*. Phytochemistry 44(8):1541–1545

Gleye C, Duret P, Laurens A, Hocquemiller R, Cavé A (1998) Cis-monotetrahydrofuran acetogenins from the roots of *Annona muricata*. J Nat Prod 61(5):576–579

Gleye C, Laurens A, Laprevote O, Serani L, Hocquemiller R (1999) Isolation and structure elucidation of sabadelin, an acetogenin from roots of *Annona muricata*. Phytochem 52(8):1403–1408

Hasrat JA, De Bruyne T, De Backer J-P, Vauquelin G, Vlietinc KA (1997) Isoquinoline derivatives isolated from the fruit of *Annona muricata* a 5-Htergic 5HT1A receptor agonists in rats: unexploited antidepressive (lead) products. J Pharm Pharmacol 49(11):1145–1149

Hisham A, Sreekala U, Pieters L, De Bruyne T, Van den Heuvel H, Claeys M (1993) Epoxymurins A and B, two biogenetic precursors of annonaceous acetogenins from *Annona muricata*. Tetrahedron 49(31):6913–6920

Jaramillo MC, Arango GJ, González MC, Robledo SM, Velez ID (2000) Cytotoxicity and antileishmanial activity of *Annona muricata* pericarp. Fitoterapia 71(2):183–186

Kim GS, Zeng L, Alali F, Rogers LL, Wu FE, McLaughlin JL, Sastrodihardjo S (1998a) Two new mono-tetrahydrofuran ring acetogenins, annomuricin E and muricapentocin, from the leaves of *Annona muricata*. J Nat Prod 61(4):432–436

Kim GS, Lu F, Alali F, Rogers LL, Wu F-E, Sastrodihardjo S, McLaughlin JL (1998b) Muricoreacin and murihexocin C, mono-tetrahydrofuran acetogenins, from the leaves of *Annona muricata* in honour of professor G. H. Neil Towers 75th birthday. Phytochem 49(2):565–571

Kimbonguila A, Nzikou JM, Matos L, Loumouamou B, Ndangui CB, Pambou-Tobi NPG, Abena AA, Silou T, Scher J, Desobry S (2010) Proximate composition and physicochemical properties on the seeds and oil of *Annona muricata* grown in Congo-Brazzaville. Res J Environ Earth Sci 2(1):13–18

Koesriharti (1992) *Annona muricata* L. In: Verheij EWM, Coronel RE (eds) Plant resources of South-East Asia 2. Edible fruits and nuts. Prosea Foundation, Bogor, pp 75–78

Kojima N (2004) Systemic synthesis of antitumor annonaceous acetogenins. Yakugaku Zasshi 124(10):673–681 (In Japanese)

Kossouoh C, Moudachirou M, Adjakidje V, Chalchat J-C, Figueredo G (2007) Essential oil chemical composition of *Annona muricata* L. leaves from Benin. J Essent Oil Res 19(4):307–309

Lannuzel A, Michel PP, Höglinger GU, Champy P, Jousset A, Medja F, Lombès A, Darios F, Gleye C, Laurens A, Hocquemiller R, Hirsch EC, Ruberg M (2003) The mitochondrial complex I inhibitor annonacin is toxic to mesencephalic dopaminergic neurons by impairment of energy metabolism. Neuroscience 121(2):287–296

Lannuzel A, Ruberg M, Michel PP (2008) Atypical parkinsonism in the Caribbean island of Guadeloupe: etiological role of the mitochondrial complex I inhibitor annonacin. Mov Disord 23(15):2122–2128

Li CM, Tan NH, Zheng HL, Qing M, Hao XJ, He YN, Zhou J (1998) Cyclopeptide from the seeds of *Annona muricata*. Phytochem 48(3):555–556

Liaw CC, Chang FR, Lin CY, Chou CJ, Chiu HF, Wu MJ, Wu YC (2002) New cytotoxic monotetrahydrofuran annonaceous acetogenins from *Annona muricata*. J Nat Prod 65(4):470–475

Luna JS, De Carvalho JM, De Lima MR, Bieber LW, Bento ES, Franck X, Sant'ana AE (2002) Acetogenins in *Annona muricata* L. (Annonaceae) leaves are potent molluscicides. Nat Prod Res 20(3):253–257

MacLeod AJ, Pieris NM (1981) Volatile flavor components of soursop (*Annona muricata*). J Agric Food Chem 29(3):488–490

Morton JF (1966) The soursop or guanábana (*Annona muricata* Linn.). Proc Fla State Hort Soc 79:355–366

Morton JF (1987) Soursop. In: Fruits of warm climates. Julia F. Morton, Miami, pp 75–80

N'gouemo P, Koudogbo B, Tchivounda HP, Akono-Nguema C, Etoua MM (1997) Effects of ethanol extract of *Annona muricata* on pentylenetetrazol-induced convulsive seizures in mice. Phytother Res 11(3):243–245

Oberlies NH, Jones JL, Corbett TH, Fotopoulos SS, McLaughlin JL (1995) Tumor cell growth inhibition by several Annonaceous acetogenins in an in vitro disk diffusion assay. Cancer Lett 96(1):55–62

Oberlies NH, Chang CJ, McLaughlin JL (1997a) Structure-activity relationships of diverse Annonaceous acetogenins against multidrug resistant human mammary adenocarcinoma (MCF-7/Adr) cells. J Med Chem 40(13):2102–2106

Oberlies NH, Croy VL, Harrison ML, McLaughlin JL (1997b) The Annonaceous acetogenin bullatacin is cytotoxic against multidrug-resistant human mammary adenocarcinoma cells. Cancer Lett 115(1):73–79

Ochse JJ, Bakhuizen van den Brink RC (1931) Fruits and fruitculture in the Dutch East Indies. G. Kolff & Co, Batavia-Ct, 180 pp

Osorio E, Arango GJ, Jiménez N, Alzate F, Ruiz G, Gutiérrez D, Paco MA, Giménez A, Robledo S (2007) Antiprotozoal and cytotoxic activities in vitro of Colombian Annonaceae. J Ethnopharmacol 111(3):630–635

Pacific Island Ecosystems at Risk (PIER) (1999) *Annona muricata*. L., Annonaceae. http://www.hear.org/pier/species/annona_muricata.htm

Padma P, Chansouria JPN, Khosa RL (1997) Effect of alcohol extract of *Annona muricata* on cold immobilization stress induced tissue lipid peroxidation. Phytother Res 11(4):326–327

Padma P, Pramod NP, Thyagarajan SP, Khosa RL (1998) Effect of the extract of *Annona muricata* and *Petunia nyctaginiflora* on Herpes simplex virus. J Ethnopharmacol 61(1):81–83

Pélissier Y, Marion C, Kone D, Lamaty G, Menut C, Besslere JM (1994) Volatile components of *Annona muricata* L. J Essent Oil Res 6:411–414

Pino JA, Marbot R (2001) Volatile components of soursop (*Annona muricata* L.). J Essent Oil Res 13(2):140–141

Rieser MJ, Kozlowski JF, Wood KV, McLaughlin JL (1991) Muricatacin: A simple biologically active acetogenin derivative from the seeds of *Annona muricata* (Annonaceae). Tetrahedron Lett 32(9):1137–1140

Rieser MJ, Gu ZM, Fang XP, Zeng L, Wood KV, McLaughlin JL (1996) Five novel mono-tetrahydrofuran ring acetogenins from the seeds of *Annona muricata*. J Nat Prod 59(2):100–108

Roblot F, Laugel T, Lebœuf M, Cavé A, Laprévote O (1993) Two acetogenins from *Annona muricata* seeds. Phytochem 34(1):281–285

Rupprecht JK, Hui YH, McLaughlin JL (1990) Annonaceous acetogenins: a review. J Nat Prod 53(2):237–278

Saw HM, Cortes D, Laurens A, Hocquemiller R, Lebœuf M, Cave A, Cotte J, Quero A-M (1991) Solamin, a cytotoxic mono-tetrahydrofuranic γ-lactone acetogenin from *Annona muricata* seeds. Phytochem 30(10):3335–3338

U.S. Department of Agriculture, Agricultural Research Service (USDA) (2010) USDA National nutrient database for standard reference, release 23. Nutrient Data Laboratory Home page. http://www.ars.usda.gov/ba/bhnrc/ndl

Wu FE, Zeng L, Gu ZM, Zhao GX, Zhang Y, Schwedler JT, McLaughlin JL, Sastrodihardjo S (1995a) Muricatocins A and B, two new bioactive monotetrahydrofuran Annonaceous acetogenins from the leaves of *Annona muricata*. J Nat Prod 58(6):902–908

Wu FE, Zeng L, Gu ZM, Zhao GX, Zhang Y, Schwedler JT, McLaughlin JL, Sastrodihardjo S (1995b) New bioactive monotetrahydrofuran Annonaceous acetogenins, annomuricin C and muricatocin C, from the leaves of *Annona muricata*. J Nat Prod 58(6):909–915

Wu FE, Zeng L, Gu ZM, Zhao GX, Zhang Y, Schwedler JT, McLaughlin JL, Sastrodihardjo S (1995c) Two new cytotoxic monotetrahydrofuran Annonaceous acetogenins, annomuricins A and B, from the leaves of *Annona muricata*. J Nat Prod 58(6):830–836

Wu FE, Zhao GX, Zeng L, Zhang Y, Schwedler JT, McLaughlin JL, Sastrodihardjo S (1995d) Additional bioactive acetogenins, annomutacin and (2, 4-trans and cis)-10R-annonacin-A-ones, from the leaves of *Annona muricata*. J Nat Prod 58(9):1430–1437

Yu J, Gui H, Luo X, Sun L (1998) Murihexol, a linear acetogenin from *Annona muricata*. Phytochem 49(6):1689–1692

Zeng L, Wu FE, Gu ZM, McLaughlin JL (1995a) Annohexocin, a novel mono-THF acetogenin with six hydroxyls, from *Annona muricata* (Annonaceae). Bioorg Med Chem Lett 5(16):1865–1868

Zeng L, Wu FE, Gu ZM, McLaughlin JL (1995b) Murihexocins A and B, two novel mono-THF acetogenins with six hydroxyls, from *Annona muricata* (Annonaceae). Tetrahedron Lett 36(30):5291–5294

Zeng L, Wu FE, Oberlies NH, McLaughlin JL, Sastrodihadjo S (1996) Five new monotetrahydrofuran ring acetogenins from the leaves of *Annona muricata*. J Nat Prod 59(11):1035–1042

Annona reticulata

Scientific Name

Annona reticulata L.

Synonyms

Annona lutescens Saff., *Annona humboldtiana* H.B.K., *Annona humboldtii* Dunal, *Annona laevis* H.B.K., *Annona longifolia* Sessé & Moç.

Family

Annonaceae

Common/English Names

Bullock's Heart, Bull's Heart, Common Custard Apple, Custard Apple, Jamaican Apple, Netted Custard Apple, Ox-Heart, Sweetsop

Vernacular Names

Aztec: Quaultzapotl;
Bolivia: Anonillo, Chirimoya Roia;
Brazil: Araticum Ape, Araticum Do Mato, Coração De Boi, Fruta-De-Condessa, Fruta-Do-Conde;
Carolinian: Anoonas;
Chamorro: Annonas, Annonas, Anonas, Anonas;
Chinese: Niu Xin Fan Li Zhi, Niu Xin Guo, Niu Xin Li;
Colombia: Anon Pelon;
Cook Islands: Naponapo Papa'Lā, Tapotapo Kirimoko, Tapotapo Papa Ā (Maori);
Cuba: Anón Injerto, Anón Manteca, Cherimoya, Mamón;
Czech: Láhevník ostnitý;
Danish: Atta, Guanbana, Netannona, Tornet Corosol;
Dominican Republic: Anón Injerto, Anón Manteca, Mammon;
Dutch: Buah Nona;
Eastonian: Vili: võrkannoona, Võrkannoona;
El Salvador: Anona Rosada;
Fijian: Chotka Sarifa, Uto Ni Mbulamakau, Uto Ni Mbulumakau;
French: Cachiman, Coeur De Boeuf, Annone Reticule;
German: Netzannone, Ochsenherz, Rahmapfeln, Süsssack, Zuckersack;
Guatemala: Anona Roja, Anona Colorada, Anona Pelon, Anona Corazón, Anona De Corazón Rojo, Anona De Seso;
Honduras: Anona De Cuba, Anona De Redecilla;
India: Gom, Nona, Nona Ata (Bengali), Ramphal (Gujarati), Kapri, Lavni, Lona, Luvuni, Nona, Ramphal (Hindu), Ram-Phal, Ramaphala, Raamaphala, Ramapala, Ramphal, Ramphala (Kannada), Anon (Konkani), Manilanilam (Malayalam), Sitaphal (Manipuri), Ram-Phal, Ramphal, Raamphal (Marathi), Ramphalo Ramopholo (Oriya), Krishnabija, Lavali, Lavani,

Mriduphala, Raktatvatch, Ramawhaya, Ramphala, Vasanta (Sanskrit), Aninuna, Atta, Iramacitta, Iramapalai, Manilatta, Manilayatta, Parankiyaninuna, Ram Citta, Ram-Sitaphalam, Ramachita, Rama Sitha, Ramaseethapazham, Ramacitta, Ramapalam (Tamil);
Indonesia: Buah Nona, (Malay), Manowa (Medan), Kanowa, Kemluwa, Kluwa, Manowa, Menuwa, Menowan, Mulwa (Java), Buah Nona (Sundanese), Binoa, Binowa (Madurese), Serba Rabsa (Aceh), Buah Nonang, Manunang (North Sulawesi), Buah Njonja (Ambon), Buah Nona (Batak), Sirikaya (Bima), Sirikaya Lasetedong, Sirikaya Uso, Sirikaya Susu (Buginese), Amanona, Anona, Aunona (Flores), Buah Nona (Gorontalo), Boi Non (Kai Islands), Buah Jus, Buah Nona Jambu Ala, Jambuluna (Lampong), Sirikaya Doke, Sirikaya Susu (Makassar), Nona, Nonas (Oelias, Ambon), Nona Daelok (Roti), Tariwang Jawa (Soemba, Timor), Ata Kase (Timor);
Japanese: Gyuushinri;
Khmer: Mean Bat, Mo Bat;
Kwara'Ae: Beretetutu;
Laotian: Khan Tua Lot;
Malaysia: Lonang, Nona, Nona Kapri;
Maya: Pox, Oop, Op, Tsulimay, Tsulilpox, Tsulipox;
Mexico: Manzana De Ilán, Saramuyo;
Nepalese: Anti, Raamaphal, Ramphal, Srii Raamaphal;
Netherlands Antilles: Kasjoema;
Nicaragua: Anona De Cuba, Anona De Redecilla;
Niuean: Talapo;
Philippines: Sarikaya, Anonas;
Portuguese: Fruto Da Condessa, Coracao De Boi, Graviola, Milolo;
Rotuman: Fat Manao, Fat Manaova;
Spanish: Anona, Anona Colorado, Anona De Manteca, Corazon, Anon Injerto, Anon, Mamon, Anonillo, Anona Corazón, Corazón De Buey, Mamán;
Surinam: Boeah Nona, Kasjoema;
Thailand: Noi Nong (Central), Noi Nang (South), Manong (North);
Tuamotuan: Mafatu Puakatoro, Taptapu;
Vietnam: Binh Bat (South), Qua Na (North);
West Indies: Cachiman, Cachiman Coeur De Boeuf, Corossol Sauvage (French);

Origin/Distribution

Annona reticulata is a native of the West Indies, probably in the Antilles but it was carried in early times through Central America to southern Mexico. It has long been cultivated and naturalized as far south as Peru and Brazil. It is commonly grown in the Bahamas and occasionally in Bermuda and southern Florida and was introduced into African and Asian countries many centuries ago.

Agroecology

Annona reticulata is a warm, tropical climate species but will survive in sub-tropical conditions but it is not cold hardy. It is too tender for California and trees introduced into Palestine succumbed to the cold. It is less drought-tolerant than the sugar apple and prefers a humid atmosphere. With the advent of winter or a prolonged dry season the leaves are shed the tree will remain dormant in winter as experienced in Florida. It flourishes in the lowlands but can be grown up to an elevation of 1,500 m. It grows on many soil types from pH 5 to 8, but thrives best in low-lying, deep, rich soil with ample moisture and good drainage. It is intolerant of water-logging.

Edible Plant Parts and Uses

Pulp of ripe fruit is eaten fresh with a spoon and also used for sauce, juice, ice cream and puddings.

Botany

A small, erect tree, growing to 3–10 m high, widely branched near the base with trunk diameter of 25–35 cm, irregular canopy and pubescent terete branchlets (Plate 3). All parts ill-smelling

Plate 1 Red fruited variety of bullock's heart

Plate 2 Brownish-pink variety of bullock's heart

Plate 3 Opened, straggly, irregular canopy of bullock's heart

when crushed. Leaves distichous, alternate, shortly petiolate, coriaceous, simple, oblong-lanceolate, 10–20 cm long by 2.5–3.5 cm wide, acuminate apex, entire margin with conscious veins, dark green above and dull green beneath. Flowers sweet smelling, greenish, in branched two to ten flowered axillary inflorescence, on 1.5–3 cm long pedicel, bracts small, ovate deltoid and densely pubescent. Sepals 3, small, connate at the base, triangular or cordate, pilose with shortly adpressed rusty hairs, deciduous. Petals 6, biseriate, with three outer fleshy, oblanceolate, petals, 17–19 × 6–7 mm, thick, triquetrous, obtuse, basally concave within, pilose externally, puberulent within; inner petals minute, frequently absent. Receptacle convex, glabrous. Stamens numerous in many whorls on the raised receptacle round the gynoecium, linear, 1.5 mm long, filament less than half the length of anther, connective-tip rounded. Ovaries numerous, crowded into a pyramidal mass, pilous, small with subsessile, glabrous stigmas. Fruit ovoid-cordate, 10–15 × 7.5–12.5 cm, subglobose to ovoid-cordate, more or less smooth, rhomboidal or hexagonal areoles separated by a marked reticulation but not tuberculate, reddish brown or red when ripe, pulp white or cream, aromatic, granular, adhering closely to the seeds (Plates 1 and 2). Seeds numerous 50–75, oblong, laterally compressed, smooth, glossy dark brown.

Nutritive/Medicinal Properties

Analyses carried out in the United States reported raw, bullock's heart fruit (exclude 42% seeds and skin) to have the following nutrient composition (per 100 g edible portion): water 71.50 g, energy 101 kcal (423 kJ), protein 1.70 g, total lipid 0.60 g, ash 1.00 g, carbohydrates 25.20 g, total dietary fibre 2.4 g, Ca 30 mg, Fe 0.71 mg, Mg 18 mg, P 21 mg, K 382 mg, Na 4 mg, vitamin C 19.2 mg, thiamin 0.080 mg, riboflavin 0.100 mg, niacin 0.500 mg, pantothenic acid 0.135 mg, vitamin B-6 0.221 mg, vitamin A 33U, total saturated fatty acids 0.231 g, tryptophan 0.007 g, lysine 0.037 g and methionine 0.004 g (USDA 2010).

One hundred and eighty compounds were identified in the aroma extracts of *Annona muricata* fruit, of which α-pinene, β-pinene, myrcene, limonene, terpinen-4-ol, and germacrene D were found to be the major constituents (Pino et al. 2003). Another analysis reported 52 volatile components in bullock heart fruit, predominated by terpenoids (98.3%) mainly terpinen-4-ol and d-terpineol as the major constituents (Wong and Khoo 1993). Thirty-nine components were characterized in the leaf essential oil of *Annona muricata*. These consisted of 18 monoterpenes amounting to 29.0%, 20 sesquiterpenes totaling 52.9% and one aromatic esters making up 10.9%. The oil contained (E,E)-farnesyl acetate (19.0%), ar-turmerone (12.0%), benzyl benzoate (10.9%) and γ-terpinene (7.4%) as the major constituents (Ogunwande et al. 2006).

Pharmacological properties of various plant parts include:

Anticancer Activity

The root bark yielded three bioactive alkaloids: anonaine, liriodenine and reticuline (muricinine) (Chang et al. 1995). The acetogenins present in the bark, fruit, seeds and leaves were reported to have anticancer properties.

Ethyl acetate extract of seeds of *Annona reticulata* yielded a new cytotoxic γ-lactone acetogenin, cis-/trans-isomurisolenin, along with six known cytotoxic acetogenins, annoreticuin, annoreticuin-9-one, bullatacin, squamocin, cis-/trans-bullatacinone and cis-/trans-murisolinone (Chang et al. 1998). Some of the compounds isolated, showed potent cytotoxicities against Hep. 2,2,15, Hep. G2, KB and CCM2 cancer cell-lines. Squamocin, a cytotoxic bis-tetrahydrofuran acetogenin, from the seeds was found to be cytotoxic to all the cancer lines tested (Yuan et al. 2006). Additionally, squamocin arrested T24 bladder cancer cells at the G1 phase and caused a selective cytotoxicity on S-phase-enriched T24 cells. It induced the expression of Bax and Bad pro-apoptotic genes, augmented caspase-3 activity, cleaved the functional protein of PARP (Poly (ADP-ribose) polymerase) and triggered cell apoptosis. Annonacin, a cytotoxic mono-tetrahydrofuran acetogenin, from the seeds caused significant cell death in various cancer cell lines (Yuan et al. 2003). T24 bladder cancer cells at the S phase were more vulnerable to the cytotoxicity of annonacin. Moreover, annonacin activated p21 in a p53-independent manner and arrested T24 cells at the G1 phase. It also induced Bax expression, increased caspase-3 activity, and produced apoptotic cell death in T24 cells. Both squamocin and annonacin represented potentially promising anticancer compounds. Squamocin and rolliniastatin I were isolated from the seeds (Vu et al. 1993). The seeds also contain a series of N-fatty acyl tryptamines, in which the acyl portion ranged from hexadecanoyl to hexacosanoyl (Maeda et al. 1993).

A methanol extract of *Annona reticulata* leaves showed good dose-dependant cytotoxicity against both Caco-2 and Hep G2 cell lines but did not show cytotoxicity against HEK (human embryonic kidney) cells up to a concentration of 20 μg/ml (Mondal et al. 2007). In the human recombinant caspase inhibition assay, the extract failed to show promising inhibition against caspase-3, while at 5 and 10 μg/ml the extract exhibited 56.02 and 66.64% inhibition against caspase-6 and 76.35 and 87.03% inhibition against caspase-9, respectively. The extract was cytotoxic to colon and liver cancer cells. Caspases (cysteine-dependant aspartate specific proteases) are involved in apoptosis (programmed cell death), inflammatory responses and cellular proliferation and differentiation. The initiator and executioner caspases have gained considerable importance in drug research as caspase induction is known to have role in cancer cell death, while inhibition of caspases can be used in the therapy in acute cellular degenerative diseases such as ischemic damage and Alzheimer's disease.

In separate studies using cytotoxicity as a guide to fractionation, one novel acetogenin, annoreticuin-9-one, and four known cytotoxic acetogenins, squamone, solamin, annomonicin and rolliniastatin 2, were isolated from active extracts of the leaves (Chang et al. 1993).

The acetogenin, reticulatacin, was isolated from a 95% ethanol extract of the bark together with bullatacin, and diterpenes, (−)-kau-16-en-19-oic acid and methyl 16β,17-dihydro-(−)-kauran-19-oate and a known alkaloid, liriodenine (Saad et al. 1991). Some of these compounds showed selective cytotoxic activities for certain human tumour cell lines. Reticulatacin showed cytotoxicities on three human tumour cell lines including human lung carcinoma, human breast carcinoma, and human colon adenocarcinoma. A new acetogenin, reticulacinone, was isolated together with rolliniastatin-2 (= bullatacin = annonin-VI) and molvizarin from a hexane extract of the stem bark (Hisham et al. 1994).

Antioxidant Activity

Annona reticulata leaves was reported to have higher antioxidant activity than the leaves of *A. squamosa* (Baskar et al. 2007). The ethanol leaf extract showed better activity in quenching DPPH1 (1-diphenyl-2-picryl hydrazyl) (89.37%) and superoxide radical (80.88%) respectively.

Anthelmintic Activity

Leaves of *A. reticulata* was reported to have anthelmintic activity (Nirmal et al. 2010). Studies on adult Indian earthworms, *Pherentima posthuma* showed that ethanol leaf extract took less time to cause paralysis of the earthworms.

Traditional Medicinal Uses

Ethnomedically, the leaves are used internally as a vermifuge against worms, and externally as poultice to treat boils, abscesses and ulcers. The leaves are anthelmintic and antiphlogistic. Unripe fruits and the astringent bark are rich in tannins and are used to treat diarrhea and dysentery. In severe cases, the leaves, bark and green fruits are all boiled together for 5 minutes in water to make an exceedingly potent decoction. Malays and Chinese in Malaysia have used the astringent bark for tonic effect and the leaves are used to maturate abscesses. Unripe fruit is also anthelmintic. Fragments of the root bark are packed around the gums to relieve toothache. The root decoction is taken as a febrifuge.

Other Uses

The pulverized seeds and leaves serve as insecticide. The leaf juice kills lice. The young shoots and the leaves are used in tanning and dyeing black leather. Custard apple wood is yellow, rather soft, fibrous but durable and moderately close-grained; and has been used to make yokes for oxen and utensils. It is reported to be a good rootstock for cherimoya and soursop.

Comments

See also notes on other edible Annonaceous fruits.

Selected References

Backer CA, van den Brink RCB jr (1963) Flora of Java (spermatophytes only), vol 1. Noordhoff, Groningen, 648 pp

Baskar R, Rajeswari V, Kumar TS (2007) In vitro antioxidant studies in leaves of *Annona* species. Indian J Exp Biol 45(5):480–485

Burkill IH (1966) A dictionary of the economic products of the Malay Peninsula. Revised reprint, 2 vols. Ministry of Agriculture and Co-operatives, Kuala Lumpur, vol 1 (A–H), pp 1–1240, vol 2 (I–Z), pp 1241–2444

Chang FR, Wu YC, Duh CY, Wang SK (1993) Studies on the acetogenins of Formosan annonaceous plants. II. Cytotoxic acetogenins from *Annona reticulata*. J Nat Prod 56(10):1688–1694

Chang FR, Chen KS, Ko F-N, Teng C-M, Wu Y-C (1995) Bioactive alkaloids from *Annona reticulata*. Chin Pharm J 47(5):483–491

Chang F-R, Chen J-L, Chiu H-F, Wu M-J, Wu Y-C (1998) Acetogenins from seeds of *Annona reticulata*. Phytochem 47(6):1057–1061

Foundation for Revitalisation of Local Health Traditions (2008) FRLHT Database. http://envis.frlht.org

Hisham A, Sunitha C, Sreekala U, Pieters L, De Bruyne T, Van den Heuvel H, Claeys M (1994) Reticulacinone, an acetogenin from *Annona reticulata*. Phytochem 35(5):1325–1329

Jansen PCM, Jukema J, Oyen LPA, van Lingen RG (1992) Minor edible fruits and nuts. In: Verheij EWM, Coronel RE (eds) Plant resources of South-East Asia. No.2: Edible fruits and nuts. Prosea Foundation, Bogor, pp 313–370

Ledin RB (1957) Tropical and subtropical fruits in Florida (other than *Citrus*). Econ Bot 11:349–376

Maeda U, Hara N, Fujimoto Y, Srivastava A, Gupta YK, Sahai M (1993) N-fatty acyl tryptamines from *Annona reticulata*. Phytochem 34(6):1633–1635

Molesworth Allen B (1967) Malayan fruits. An Introduction to the cultivated species. Moore, Singapore, 245 pp

Mondal SK, Mondal NB, Mazumder UK (2007) In vitro cytotoxic and human recombinant caspase inhibitory effect of *Annona reticulata* leaves. Indian J Pharmacol 39(5):253–254

Morton JF (1987) Custard apple. In: Fruits of warm climates. Julia F. Morton, Miami, pp 80–83

Nirmal SA, Gaikwad SB, Dhasade VV, Dhikale RS, Kotkar PV, Dighe SS (2010) Anthelmintic activity of *Annona reticulata* leaves. Res J Pharm Biol Sci 1(1):115–118

Ochse JJ, Bakhuizen van den Brink RC (1931) Fruits and fruitculture in the Dutch East Indies. G. Kolff & Co, Batavi, 180 pp

Ogunwande IA, Ekundayo O, Olawore NO, Kasali AA (2006) Essential oil of *Annona reticulata* L. leaves from Nigeria. J Essent Oil Res 18(4):374–376

Pino JA, Marbots R, Fuentes V (2003) Characterization of volatiles in bullock's heart (*Annona reticulata* L.) fruit cultivars from Cuba. J Agric Food Chem 51:3826–3839

Saad JM, Hui YH, Rupprecht JK, Anderson JE, Kozlowski JF, Zhao GX, Wood KV, McLaughlin JL (1991) Reticulatacin: a new bioactive acetogenin from *Annona reticulata* (Annonaceae). Tetrahedron 47(16–17):2751–2756

Subhadrabandhu S (2001) Under utilized tropical fruits of Thailand, FAO Rap publication 2001/26. FAO, Rome, 70 pp

U.S. Department of Agriculture, Agricultural Research Service (USDA) (2010) USDA national nutrient database for standard reference, Release 23. Nutrient Data Laboratory Home Page. http://www.ars.usda.gov/ba/bhnrc/ndl

Vu TT, Bui CH, Chappe B (1993) Squamocin and rolliniastatin I from the seeds of *Annona reticulata*. Planta Med 59(6):576

Watt JM, Breyer-Brandwijk MG (1962) The medicinal and poisonous plants of Southern and Eastern Africa, 2nd edn. E. & S. Livingstone, Edinburgh, 1457 pp

Wong KC, Khoo KH (1993) Volatile components of Malaysian Annona fruits. Flav Fragr J 8(1):5–10

Yuan SSF, Chang HL, Chen HW, Yeh YT, Kao YH, Lin KH, Wu YC, Su JH (2003) Annonacin, a mono-tetrahydrofuran acetogenin, arrests cancer cells at the G1 phase and causes cytotoxicity in a Bax- and caspase-3-related pathway. Life Sci 72(25):2853–2861

Yuan SSF, Chang HL, Chen HW, Kuo FC, Liaw CC, Su JH, Wu YC (2006) Selective cytotoxicity of squamocin on T24 bladder cancer cells at the S-phase via a Bax-, Bad-, and caspase-3-related pathways. Life Sci 78(8):869–874

Annona squamosa

Scientific Name

Annona squamosa L.

Synonyms

Annona asiatica L., *Annona biflora* Moç & Sessé, *Annona cinerea* Dunal, *Annona forskahlii* DC., *Annona glabra* Forssk., *Guanabanus squamosus* M. Gómez, *Xylopia frutescens* Sieb. ex Presl.

Family

Annonaceae

Common/English Names

Custard Apple, Scaly Custard Apple, Sugar-Apple, Sweetsop

Vernacular Names

Antilles: Pomme Cannelle;
Arabic: Gishta, Ahta;
Argentina: Cachiman;
Bahamas: Sweetsop;
Brazil: Atá, Cabeça-De-Negro, Condessa, Coração-De-Boi, Fruta Do Conde, Fruta De Condessa, Frutiera Deconde, Pinha, Araticutitaia, Ati;
Bolivia: Anon;
Burmese: Awza Thee;
Chamorro: Ates, Atis;
Chinese: Shijia, Fan Li Zhi;
Cook Islands: Katara'Āpa Māori, Kātara'Apa Māori, Naponapo, Naponapo, Tapotapo, Tapotapo Māori, Tapotapo Māori, Tapotapo Māori (Maori);
Columbia: Chirimoya Verrugosa, Chirimoyo, Fruta Del Conde, Mocuyo, Anon De Azucar, Anon Domestico, Hanon, Mocuyo;
Comoros: Mkonokono (Anjouan Island);
Costa Rica: Anon;
Cuba: Anón;
Czech: Láhevník Šupinatý;
Danish: Sød Annona, Flasketræ, Sukkeræbletræ;
Dominican Republic: Anón Candonga, Anona Blanca;
Dutch: Kannelappel;
Eastonian: Soomusannoona, Vili: Kaneelannoona;
Ecuador: Chirimoya;
El Salvador: Anona De Castilla;
Fijian: Noi-Na, Rinon; Apeli;
French: Anone Écailleuse, Pomme Canelle, Corossol Écailleux, Attier, Cachiman Cannelle;
French Guiana: Pomme Cannelle;
Gabon: Atte;
German: Rahmapfel, Süßsack, Schuppenannone, Schuppen-Annone, Zimtapfel, Zuckerapfel;
Ghana: Apre (Asante-Twi);
Grenadines: Applebush;
Guadeloupe: Pomme Cannelle;
Guatemala: Anona, Anona Blanca, Chirimoya;
Haiti: Cachiman Cannelle;
Honduras: Anona Blanca;

Hungarian: Gyömbéralma;
India: Atakathal, Katal (Assamese), Ata, Mandargom, Madar (Bengali), At, Atasitaphal, Sharilfah, Shariphal, Sitaphal, Sherifa, Sharifa, Sarifa (Hindi), Amritaphala, Duranji, Sitaphal, Sitaphala, Amrutaphala, Duranji Mara, Seethaaphala, Seethaphala, Amrita, Amritakayida, Amrytakayi, Amrytakayida, Amrytaphala, Amucikayi, Seethaphal, Sita, Sitahala, Sitahannu, Amrithaphala, Amuchi Kaayi, Duranji Hannu (Kannada), Antacheecha, Atamaram, Attaccakka, Attachchakka, Sirpa, Sirpha, Sitapalam, Sitappalam, Sutakanni, Ata, Atha, Atta, Suthakanni (Malayalam), Sitaphal (Manipuri), Sitaphal, At, Ath, Seethaphal (Marathi), Sitaphalo (Oriya), Agrimakhya, Atripya, Bahubijaka, Ganda, Gandagatra, Gandhagatra, Gutea, Krishnabija, Krisnabija, Shubha, Sitapalam, Sitaphala, Subha, Suda, Vaidehivallabha (Sanskrit), Annila, Atta, Citta, Cintamaram, Cita, Cittamaram, Cupakalavam, Cupakarajamaram, Letcumi, Letcumimaram, Nilampu, Nilampuccitta, Pilavai-Kolli, Sitapalam, Sith-Tha, Seetha Pazham, Sitappalam, Sitha, Sithapalam, Sita, Tirilokeyam, Tirilokeyamaram (Tamil), Gandagatramu, Seetapandu, Sitapandu, Sitaphalamu, Ganda Gathram, Ganda Gaathramu, Seethapandu, Seethaphalamu (Telugu);
Indonesia: Atu Su Walanda (Alfoersch, Northern Sulawesi), Delima Bintang, Serba Bintang (Acheh, Sumatra), Sirkaya, Srikaya, (Bali), Sarikaya (Batak, Sumatra), Sirikaya (Buginese, Sulawesi), Garoso, Geroso M'bojo (Bima, Timor), Sirikaya (Boeol, Sulawesi), Sarikaya (Dyak, Kalimantan), Ata (Flores), Atis, Danoya, Sirikaya (Gorontalo, Sulawesi), Sirkaya, Srikaya, Srikawis, Surikaya (Javanese), Sarikaya, Sarkaja, Sarkajeh (Madurese), Sirikaya (Makassar, Sulawesi), Srikaya (Malay), Ata (Malay, Timor), Atis, Serikaya (Malay, Manado), Buah Serihkaya (Malay, Lingga), Sarikaya (Minangkabau, Sumatra), Atisi, Hirikaya (N. Halmaheira, Ternate), Perse (Kisar, Timor), Jambu Muluku, Seraikaya, Serikaya (Lampong, Sumatra), Wo Ajai Jawa (Sawoe, Timor), Wo Tari Wang (Soemba, Timor), Ata (Solor, Timor), Atis (Ternate), Atis (Tidore, Ternate), Hau Fua (Timor);
Italian: Mela Canella, Pomo Canella;
Jamaica: Sweetsop;
Japanese: Banreishi;
Khmer: Tiep Baay, Tiep Srôk;
Laotian: Khièb;
Madagascar: Konikony, Kônokônovazaha;
Malaysia: Buah Nona, Nona Sri Kaya, Sri Kaya;
Mali: Tubabu Sunsun (Bambara);
Marquesan: Noni Haoe, Nonihao;
Mauritius: Atta, Pomme Cannelle (French), Zatte;
Mexico: Ahate, Saramulla, Saramuya;
Nepal: Shariiphaa, Sitaaphal, Sariphal;
Netherlands Antilles: Scopappel;
Nicaragua: Anona De Guatemala;
Niuean: Talapo;
Pakistan: Sharifa (Urdu);
Palauan: Ngel Ra Ngebard;
Panama: Anon;
Papiamento: Skopapel;
Persian: Kaj, Sharifah;
Philippines: Ates, Atis (Tagalog);
Polish: Cukrowe Jabłko;
Portuguese: Ateira, Ata, Ata Pinha, Fruta Da Condessa, Fruta De Condessa, Fruta Do Conde, Fruteira De Conde, Pinha, Pinha Da Bahia, Pinheira;
Puerto Rico: Anón;
Pukapukan: Kātalāpa;
Reunion Island: Pomme Cannelle (French);
Rotuman: Fat Manaova, Ai Piol Moumon;
Spanish: Anón, Anona, Anona Blanca, Ata Del Brazil, Chirimoya, Rinón, Saramuyo, Anona, Anonillo Candongo, Fructo Do Conde, Pinha, Saramuya;
Surinam: Kaneelappel;
Swahili: Mastaferi, Mtomoko, Mtopetope;
Tahitian: Tapo Tapo, Tapotapo;
Taiwan: Sakya, Sekkhia, Sek-Kia;
Tanzania: Mastaferi (Swahili);
Thai: Lanang, Makkhiap, Manonah, Noi-Na;
Tongan: 'Apele Papalangi;
Tuvaluan: Nameana;
Venezuela: Rinón;
Vietnamese: Mang Câu Ta, Na;
West Africa: Pomme Cannelle (French).

Origin/Distribution

Only cultivated, its origin is uncertain, probably indigenous to the Antilles or South Mexico. Now distributed worldwide and naturalized in the tropics and warm subtropics. In Jamaica, Puerto Rico, Barbados, and in dry regions of North Queensland, Australia, it has escaped from cultivation and is found wild in pastures, forests and along roadsides.

Agroecology

Sweetsop requires a tropical or sub-tropical climate with summer temperatures from 25–41°C, and mean winter temperatures above 15°C. It is sensitive to cold and frost, being defoliated below 10°C and generally killed by sub-zero temperatures. It is only moderately drought-tolerant, requiring rainfall above 700 mm, and not producing fruit well during droughts. Trees will grow in hot and relatively dry climates such as those of the low-lying interior plains of many tropical countries but adequate moisture during the growing season, responding well to supplementary irrigation. During the flowering season, drought interferes with pollination and it is, therefore, concluded that the sugar apple should have high atmospheric humidity but no rain when flowering. In severe droughts, the tree sheds its leaves and the fruit rind hardens and will split with the advent of rain. Its altitudinal range is from 0 to 2,000 m above sea level. Sweetsop tolerates a wide variety of soils. It has performed well on sand, oolitic limestone, heavy loam with good drainage and in rich, well-drained, deep rocky soils. But it prefers friable, sandy loams with pH from 5.5 to 6.5.

Edible Plant Parts and Uses

Sweet sop has a fragrant, juicy, sweet, delicious pulp with a high quantity (35–42 mq/100 g) of vitamin C, slightly higher than in grapefruit. The pulp of the ripe fruit is eaten fresh or utilized as a flavouring for ice-cream, sherbets, shakes or made into juice and refreshing drinks. The ripe sugar apple is usually broken open and the fleshy carpel segments enjoyed while the hard seeds are separated in the mouth and spat out. It is so luscious that it is well worth the trouble. In Malaysia, the flesh is pressed through a sieve to eliminate the seeds and is then added to ice cream or blended with milk to make a cool beverage. In the West Indies, the fruit is fermented to prepare a kind of cider.

Botany

A small, semi-deciduous tree, 3–8 m in height, with a short trunk, broad, open crown and irregularly spreading branches. Leaves are alternate, simple, petiolate (4–22 mm long). Lamina narrowly elliptic to oblong or lanceolate, 5–17 × 2.5–6 cm, base broadly cuneate to rounded, apex acute to obtuse; surfaces glaucous and pale green, abaxially variably pubescent, adaxially glabrous. Inflorescences are supra- axillary, consist of solitary flowers or fascicles of 2–4 flowers; peduncle slender, to 2 cm, becoming enlarged in fruit. Flowers are greenish-yellow, fragrant with hairy, green, triangular sepals about 16 mm long. Corolla with three outer petals oblong, thick and rounded at the tips, fleshy, yellow-green, slightly hairy, light yellow inside and keeled with a purplish spot at the base and three inner petals reduced to minute, ovate, pointed scales. Stamens very numerous, crowded in whorls around the gynoecium, white, club-shaped, curved, 1–3 mm with dilated, flattened and truncate connective. Carpels with pale green ovaries, white styles, conically massed on the raised torus. The fruit is sub-globose, heart shaped, ovate or conical, 5–11 cm in diameter, muricate with many round protuberances; greenish-yellow (or pinkish-purple in a purplish cultivar) when ripe, with a white, powdery bloom. The fruit is a pseudo-syncarp formed of loosely coherent or almost free carpels with the rounded ends projecting to render the surface tuberculate (Plates 1–7). The fruit pulp is white, edible and sweetly aromatic; in each carpel is embedded a blackish or dark brown seed, ellipsoid to obovoid, 1–1.4 cm, shiny and smooth.

Plate 1 Common green variety of Sweetsop

Plate 4 A ripe green fruited sweetsop variety

Plate 2 Purple-fruited sweet sop variety

Plate 5 Green fruited sweetsop in a Vietnamese market

Plate 6 Sweetsop in bamboo baskets in Thailand

Plate 3 Green fruited sweetsop (*left*) and red fruited variety (*right*)

Nutritive/Medicinal Properties

Analyses carried out in the United States reported raw, sweetsop fruit (exclude 45% seeds and skin) to have the following nutrient composition (per 100 g edible portion): water 73.23, energy 94 kcal

Plate 7 Ripe sweetsop being sold in a Thai market

(393 kJ), protein 2.06 g, total lipid 0.29 g, ash 0.78 g, carbohydrates 23.64 g, total dietary fibre 4.4 g, total sugars 19.24 g, Ca 24 mg, Fe 0.60 mg, Mg 21 mg, P 32 mg, K 247 mg, Na 9 mg, Zn 0.10 mg, Cu 0.086 mg, Se 0.6 μg, vitamin C 36.3 mg, thiamine 0.110 mg, riboflavin 0.113 mg, niacin 0.883 mg, pantothenic acid 0.226 mg, vitamin B-6 0.20 mg, total folate 14 μg, vitamin A 6 IU, vitamin A 9 RAE, total saturated fatty acids 0.048 g, 16:0 (palmitic acid) 0.028 g, 18:0 (stearic acid) 0.020 g, total monounsaturated fatty acids 0.114 g, 18:1 undifferentiated (oleic acid) 0.114 g, total polyunsaturated fatty acids 0.040 g, 18:2 undifferentiated (lioleic acid) 0.040 g, tryptophan 0.010 g lysine 0.055 g and methionine 0.007 g (USDA 2010).

The flesh of the fruit consist of a white edible pulp that is high in carbohydrates and sugars, Vitamin C, and fair amounts of Vitamin B1, Vitamin B2, Potassium and dietary fibre. It is low in vitamin A.

The amount of sugars in the fruit pulp were found to be quite high (58% of dry mass) (Andrade et al. 2001). The triglyceride concentration was found to be very low. Considerable amount of the diterpenoid compound kaur-16-en-18-oic acid (0.25% of dry mass) was detected in the lipid fraction. The major compounds of the essential oil of the fruit were α-pinene (25.3%), sabinene (22.7%) and limonene (10.1%). The occurrence of the isoricinoleic acid previously reported in the seed oil could not be confirmed.

Recent scientific studies supported many of the traditional medicinal uses of extracts of the various parts of the *Annona squamosa* tree. *Annona squamosa* has abundant acetogenins especially in the seeds and bark, many of which have antitumorous activities and other bioactive compounds in various plant parts. In seeds, some of the acetogenins identified included: annonacin, anonacin A, dieposabadelin, squamocenin, lepirenin and dotistenin, corepoxylone, diepomuricanins A and B, diepoloreticenine, tripoxyrollin, bullatencin, glabrencin B, reticulatain-1, reticulatain-2, uvariamicin I, uvariamicin II, uvariamicin III, erythrosolamin, annotemoyin-1, annotemoyin-2, solamin, squamostolide, squamocin-O1, squamocin-O2, cyclosquamosin B, cyclosquamosin D, squafosacin B, squadiolin A, squadiolin B, squamostanin-A, squamostanin-B, squamotacin (Araya et al. 2002; ba Ndob et al. 2009; Liaw et al. 2008; Lieb et al. 1990; Morita et al. 2006; Mukhlesur Rahman et al. 2005; Souza et al. 2008; Xie et al. 2003; Yang et al. 2008; Yu et al. 2005). Four non-adjacent bis-tetrahydrofuranic acetogenins, designated squamostatins-B (2), -C (3), -D (4) and -E (5), were isolated from the petroleum ether extract of Annona squamosa seeds (Fujimoto et al. 1994). Two new annonaceous acetogenins designated as squamostanin-C and squamostanin-D were isolated from 95% ethanolic extract of *Annona squamosa* seeds and their cytotoxicities were evaluated by the MTT method (Yang et al. 2009a, b). Li et al. (2010) isolated the following chemicals from *A. squamosa* seeds: uvarigrandin A, squamocin, motrilin, bullatacin, neo-expoxyrolin, neo-desacetyluvaricin, annonareticin, epoxyrolin B, murisolin, squamostatin C, annoglaxin, squamin-A, erythritol, D,L-threitol, D-mannitol, stigmasterol, heptadecanoic acid, nonadecanoic acid, and stearic acid.

From the stem bark, some acetogenins isolated included: bullatacin, bullatacinone, squamone, tetrahydrosquamone, squamotacin, bullacin B, 4-deoxyannoreticuin, cis-4-deoxyannoreticuin, 2,4-cis-squamoxinone and 2,4-trans-squamoxinone (Hopp et al. 1996; Hopp et al. 1998a, b; Li et al. 1990; Yang et al. 2004; Yeh et al. 2005). The seed also contained cholesteryl glucopyranoside, samoquasine A (a benzoquinazoline alkaloid) quercetin (flavonoid) and annosquamosin A (cyclopetide) (Li et al. 1997; Morita et al. 2000; Mukhlesur Rahman et al. 2005).

The kaurane diterpenoids annosquamosin A and annosquamosin B and 10 other kaurane compounds were isolated from the fruit (Wu et al. 1996). Podophyllotoxin (a non alkaloid toxin lignan compound), and its demethyl derivative, 4'-demethylpodophyllotoxin, liriodenine and (−)-kaur-16-en-19-oic acid were isolated from the branches (Hatano et al. 2002). The ent-kaurane diterpenoids: annomosin A, annosquamosin C, annosquamosin D, annosquamosin E, annosquamosin F annosquamosin G were found in the stems (Yang et al. 2002). The leaves contained quercetin-3-O-glucoside (a flavonoid), palmitone, alkaloids ((−)-xylopine; (+)-O-methylarmepavine; lanuginosine), terpenes and sesquiterpenes mainly B-caryophyllene. Squamolone (a diazepine) (Yang and Chen 1972) and aporphine alkaloids anonaine, roemerine, norcorydine, corydine, norisocorydine, isocorydine and glaucine were isolated from *A. squamosa* (Bhakuni et al. 1972). Also cytotoxic bis-tetrahydrofuran squamocin and squamostatin-A were isolated from *A. squamosa* (Fujimoto et al. 1988, 1990).

Annonaceous acetogenins are a new class of compounds that have been reported to have potent cytotoxic, anticancerous, pesticidal, antimalarial, parasiticidal, antimicrobial, cell growth inhibitory activities.

Anticancer Activities

Studies in India found that organic and aqueous extracts of the defatted seeds extracts induced apoptosis in human tumour cell lines, MCF-7 and K-562 cells but not in COLO-205 cells (Pardhasaradhi et al. 2005). Treatment of MCF-7 and K-562 cells with aqueous and organic extracts produced nuclear condensation, DNA fragmentation, induction of reactive oxygen species (ROS) generation and decreased intracellular glutathione levels. These observations suggested that the induction of apoptosis by *A. squamosa* seed extracts could be selective for certain types of cancerous cells.

An acetogenin, squamostolide isolated from the seeds of *Annona squamosa* exhibited cytotoxic activity in-vitro against bel-7402 and CNE2 human tumour cell lines (Xie et al. 2003). Squadiolins A and B, acetogenins from the seeds showed high potency against human Hep G2 hepatoma cells and significant cytotoxic activity against human MDA-MB-231 breast cancer cells (Liaw et al. 2008). Squafosacin B also exhibited significant cytotoxic activity against human Hep G2 and 3B hepatoma and MCF-7 breast cancer cells. Cyclosquamosin D from the seeds showed an inhibitory effect on the production of pro-inflammatory cytokines within lipopolysaccharide and Pam3Cys-stimulated J774A.1 macrophages (Yang et al. 2008). Another two acetogenins, squamostanin-A and squamostanin-B isolated from 95% ethanol extract of the seeds, exhibited cytotoxic activity (Yang et al. 2009a, b).

The acetogenin squamotacin from the bark of sugar apple showed selective cytotoxicity for the human prostate tumour cell line (PC-3), with a potency greater than 100 million times that of adriamycin (Hopp et al. 1996). Another acetogenin from the bark bullacin B was selectively cytotoxic in a panel of six human tumour cell lines with a potency of a million times that of adriamycin against the MCF-7 (human breast adenocarcinoma) cell line (Hopp et al. 1998b). Additionally, 4-deoxyannoreticuin, cis-4-deoxyannoreticuin, and (2,4-cis and trans)-squamoxinone isolated from the bark showed moderate, but significant, cytotoxicities against a panel of six human tumour cell lines with (2,4 cis and trans)-squamoxinone exhibiting promising selectivity against the pancreatic cell line (PACA-2) (Hopp et al. 1998a).

Highly bioactive acetogenins bullatacin and bullatacinone, squamone, tetrahydrosquamone and liriodenine and (−)-kaur-16-en-19-oic acid were isolated from the stem bark (Li et al. 1990). The cytotoxicities of squamone were enhanced significantly by reduction of the two keto groups to hydroxyls. Both bullatacinone and tetrahydrosquamone displayed selective cytotoxicities to MCF-7 human breast carcinoma.

Podophyllotoxin (a non alkaloid toxin lignan compound), and its demethyl derivative, 4'-demethylpodophyllotoxin from branches of *A. squamosa* were found to exhibit high cytotoxic activity against human lung and colon cancer cells in-vitro

(Hatano et al. 2002). Both the aqueous and organic extracts from defatted seeds of *A. squamosa* caused significant apoptotic tumour cell death in the AK-5 cell line with increased caspase-3 activity, decrease of antiapoptotic genes Bcl-2 and BclXL, and increase in the generation of intracellular reactive oxygen species, which correlated well with the decreased levels of intracellular GSH (Pardhasaradhi et al. 2004). Furthermore, the extracts was confirmed to induce apoptosis in tumour cells through the oxidative stress.

Studies by Suresh et al. (2010) suggested that *A. squamosa* bark extracts protected the cell surface glycoconjugates during 7,12-dimethyl benz(a)anthracene (DMBA) induced hamster buccal pouch carcinogenesis. Oral administration of aqueous and ethanolic extracts of *A. squamosa* bark at a dose of 500 mg/kg b.w. and 300 mg/kg b.w. reduced the total number of tumours, tumour burden, tumour volume as well as the incidence of tumour and normalized levels of glycoconjugates in tumour-bearing animals.

Antiviral Activity

The fruits of *Annona squamosa* yielded 12 known kaurane derivatives and two new kaurane diterpenoids, annosquamosin A (16β-hydroxy-17-acetoxy-ent-kauran-19-al) and annosquamosin B (19-nor-ent-kaurane-4α,16β,17-triol) (Wu et al. 1996). Among these 14 compounds, 16β, 17-dihydroxy-ent-kauran-19-oic acid showed significant activity against HIV replication in H9 lymphocyte cells with an EC_{50} value of 0.8 μg/ml and a therapeutic index >5.

Antidiabetic and Antioxidant Activities

Studies in India using streptozotocin–nicotinamide induced type 2 diabetic rats provided supporting evidence to the claim by some tribal populations in parts of Northern India that the young leaves of *Annona squamosa* had antidiabetic properties (Shirwaikar et al. 2004). Separate studies in India reported that administration of the leaf extract also improved the lipid profile of the treated groups indicating thereby that the high levels of triglyceride and total cholesterol associated with diabetes could also be significantly managed with the extract (Gupta et al. 2008). They found that *Annona squamosa* leaves possessed antioxidant activity as shown by increased activities of scavenging enzymes, catalase (CAT), superoxide dismutase (SOD), reduced glutathione (GSH), glutathione reductase (GR) and glutathione-s-transferase (GST) and decrease in malondialdehyde levels present in various tissues. The same group of scientists reported that the fruit pulp could reduce the total cholesterol level by 45–46% in normal and 32.4% in diabetic animals with increased HDL-cholesterol (Gupta et al. 2005b). Feeding fruit pulp was found to improve the liver function in normal as well as diabetic rabbit as shown by reduction in the serum SGOT (serum glutamate oxaloacetate transaminase), SGPT (serum glutamate pyruvate transaminase), ALKP (alkaline phosphatase) and serum bilirubin levels. Further, fruit pulp decreased urine sugar, urine protein and glycohemoglobin in diabetic rabbits. Feeding fruit pulp also augmented utilization of dietary protein, body weight as well as the ratio of gain in body weight per gram of protein consumed. It exhibited a protective effect on liver and heart as indicated by reduction in the SGOT, SGPT, ALKP and serum bilirubin levels. Treatment of severely-diabetic rabbits for 15 days with a dose of 350 mg/kg of extract decreased fasting blood glucose FBG by 52.7% and urine sugar by 75%. It decreased the level of total cholesterol (TC) by 49.3% and low-density lipoprotein (LDL) by 71.9%, triglycerides (TG) 28.7% and increased high-density lipoprotein (HDL) by 30.3%.

Studies showed that ethanolic leaf extract reduced fasting blood sugar (FBG) by 26.8% and enhanced glucose tolerance by 38.5 and 40.6% at 1 and 2 h, respectively, during glucose tolerance test (GTT) in alloxan-induced diabetic rabbits (Gupta et al. 2005a). In streptozotocin (STZ)-induced diabetic rats a decline of 13.0% in FBG and an enhancement in glucose tolerance by 37.2 and 60.6% at 1 and 2 hours, respectively, was observed during GTT. It reduced the level of total cholesterol (TC) by 49.3% and produced an increase of 30.3% in high-density lipoprotein (HDL) and decrease of 71.9 and 28.7% in low-density lipoprotein (LDL) and triglycerides (TG) levels, respectively.

Quercetin-3-O-glucoside isolated from *Annona squamosa* leaves was found to regulate alloxan-induced hyperglycemia and lipid peroxidation (LPO) in rats (Panda and Kar 2007b). In alloxan treated animals, an increase in the concentration of serum glucose with a parallel decrease in insulin level was observed. However, administration of 15 mg/kg/day of isolated quercetin-3-O-glucoside for 10 consecutive days to the hyperglycemic animals reversed these effects and simultaneously inhibited the activity of hepatic glucose-6-phosphatase. It further decreased the hepatic and renal LPO with a concomitant increase in the activities of antioxidative enzymes, such as catalase (CAT) and superoxide dismutase (SOD) and in glutathione (GSH) content, indicating its safe and antiperoxidative effects. These findings suggested the potential of quercetin-3-O-glucoside in the amelioration of diabetes mellitus and tissue lipid peroxidation.

A. squamosa leaf aqueous extract supplementation was found useful in controlling the blood glucose level, improving the plasma insulin, lipid metabolism and was beneficial in preventing diabetic complications from lipid peroxidation and antioxidant systems in experimental diabetic rats; implicating its potential for prevention or early treatment of diabetes mellitus (Kaleem et al. 2006, 2008). Oral administration of *A. squamosa* (300 mg/kg) leaf aqueous extract to diabetic rats for 30 days significantly reduced blood glucose, urea, uric acid and creatinine, but increased the activities of insulin, C-peptide, albumin, albumin/globulin ratio and restored all marker enzymes to near control levels. The results showed that *A. squamosa* leaf extract had an antihyperglycaemic effect and consequently might alleviate liver and renal damage associated with streptozotocin-induced diabetes mellitus in rats.

Ethanolic extract of *Annona squamosa* fruit showed antioxidative activity by its ability to scavenge 1,1-diphenyl-2-picrylhydrazyl (DPPH) free radical with an IC_{50} of 250 µg/ml (Soubir 2007).

Lipoxygenase Inhibitory Activity

The seeds of *Annona squamosa* yielded a novel lipoxygenase inhibitor fatty acid ester, (+)-annonlipoxy (Sultana 2008). Annonlipoxy exhibited inhibitory activity against lipoxygenase with IC_{50} of 69.05 µm. The crude ethanolic extract of fruit pulp and seeds of *Annona squamosa* also exhibited lipoxygenase inhibitory activity with 22.2% and 26.7% inhibition; while the petroleum ether extract of seeds of *A. squamosa* exhibited 52.7% inhibition at a concentration of 40 µg/200 ml. Lipoxygenase is a family of iron-containing enzymes that catalyse the dioxygenation of polyunsaturated fatty acids in lipids.

Antithyriodal Activity

Recent studies reported that simultaneous administration of the *Annona squamosa* seed extract (200 mg/kg) or quercetin (10 mg/kg) to T(4)-induced hyperthyroid animals for 10 days, reversed all these effects indicating their potential in the regulation of hyperthyroidism (Panda and Kar 2003; 2007a). Further, the seed extract did not enhance, but reduce the hepatic lipid peroxidation (LPO) suggesting its safe and antiperoxidative nature. Quercetin also decreased hepatic LPO. The presence of quercetin in the seed extract was confirmed and the results on the effects of quercetin suggested its involvement in the mediation of anti-thyroidal activity of *Annona squamosa* seed extract. Further research revealed that *Annona squamosa* leaf extract exhibited thyroid inhibitory effects in mice, but altered hepatic LPO in a dose-dependent manner. At low concentration it appeared to be anti-thyroidic as well as antiperoxidative, whereas, at higher concentration it was found to be thyroid inhibitory but hepatotoxic as indicated by enhanced LPO, thereby suggesting the unsafe nature of the highest dose.

Immunomodulatory and Antiinflammatory Activities

Eleven ent-kauranes isolated from the stems of *Annona squamosa* exhibited immunomodulating effects in leukocytes. All ent-kauranes showed significant inhibitory effect on O^{2-} generation by human neutrophils in response to formyl-L-methionyl- L-leucyl- L-phenylalanine (Yang et al. 2004). In contrast, phorbol myristate acetate (PMA)-induced O^{2-} generation was not suppressed by any ent-kauranes. None of the compounds showed inhibitory effect on nitric oxide generation. Human neutrophils are important in the pathogenesis of rheumatoid arthritis, ischemia-reperfusion injury, chronic obstructive pulmonary disease, asthma and other inflammatory diseases. Further studies showed that the ent-kaurane compound 16β,17-dihydroxy-ent-kauran-19-oic acid inhibited the generation of superoxide anion, the formation of ROS, and the release of elastase in formyl-L-methionyl-L-leucyl-L-phenylalanine (FMLP)-activated human neutrophils in a concentration-dependent manner with IC_{50} values of 3.95, 12.20, and 12.52 μM, respectively (Yeh et al. 2005). The anti-inflammatory actions were not attributable to cytotoxicity because incubation of the neutrophils with the compound did not result in lactate dehydrogenase release. The compound did not display antioxidant or superoxide anion-scavenging activity. The inhibitory effects of this compound on respiratory burst and degranulation of human neutrophils were through the inhibition of cytosolic calcium mobilization, but not via the cAMP-dependent pathways.

Analgesic and Anti-inflammatory Activities

Caryophyllene oxide was isolated from an unsaponified petroleum ether extract of *Annona squamosa* bark (Chavan et al. 2010). The unsaponified petroleum ether extract at a dose of 50 mg/kg body weight and caryophyllene oxide at the doses of 12.5 and 25 mg/kg body weight exhibited significant central as well as peripheral analgesic activity, along with anti-inflammatory activity. 18-Acetoxy-ent-kaur-16-ene was isolated from petroleum ether extract of *A. squamosa* bark and exhibited analgesic and anti-inflammatory activities (Chavan et al. 2011). The compound at the doses of 12.5 and 25 mg/kg, and the petroleum extract at a dose of 50 mg/kg exhibited significant analgesic and anti-inflammatory activity.

Antiplatelet Aggregation Activity

Six new ent-kaurane diterpenoids (acetogenins), annomosin A (16beta-hydroxy-19-al-ent-kauran-17-yl 16beta-hydro-19-al-ent-kauran-17-oate) (1), annosquamosin C (16alpha-hydro-17-hydroxy-19-nor-ent-kauran-4alpha-ol) (2), annosquamosin D (16beta-acetoxy-17-hydroxy-19-nor-ent-kauran-4alpha-ol) (3), annosquamosin E (16beta-hydroxy-17-acetoxy-19-nor-ent-kauran-4alpha-formate) (4), annosquamosin F (16beta-hydroxy-17-acetoxy-18-nor-ent-kauran-4beta-hydroperoxide) (5), and annosquamosin G (16beta,17-dihydroxy-18-nor-ent-kauran-4beta-hydroperoxide) (6), along with 14 known ent-kaurane diterpenoids (Yang et al. 2002). The two ent-kaurane diterpenoids, ent-Kaur-16-en-19-oic acid (9) and 16alpha-hydro-19-al-ent-kauran-17-oic acid (17) showed complete inhibitory effects on rabbit platelet aggregation at 200 μM.

Vasorelaxant Activity

Research in Japan reported that cyclosquamosin B isolated from the seeds of *Annona squamosa*, showed inhibition effect on vasocontraction of depolarized rat aorta with high concentration potassium, but moderately inhibitory effect on norepinephrine-induced contraction in the presence of nicardipine (Morita et al. 2006). These results showed that the vasorelaxant effect by cyclosquamosin B may be attributed mainly to inhibition of calcium influx from extracellular space through voltage-dependent calcium channels.

Antifertility and Abortifacient Effect

Annona squamosa has been reported to show varied medicinal effects, including insecticide, antiovulatory and abortifacient. Recent studies showed that aqueous *A. squamosa* seed extract did not interfere with the reproductive performance of pregnant rats (Damasceno et al. 2002). Treatment with aqueous extract of *A. squamosa* seeds caused no morphological change in the endometrium. The absence of morphological alterations in uterine epithelial cells in treated groups permitted a viable embryonic implantation, as verified by the number of embryos in development at day 10 of pregnancy.

Anti-head Lice Activity

In Thailand, research demonstrated the anti-head lice activity of *A. squamosa* seed extracts (Intaranongpai et al. 2006). Extract of custard apple seeds in coconut oil at the ratio of 1:2 killed 98% of head lice within two hours, while the leaf extract had less potency. The petroleum ether extract of the leaves and seeds dissolved in coconut oil at a ratio of 1:1, killed 90% of head lice in-vitro within 53 and 26 minutes, respectively. A 20% cream (oil/water) preparation of petroleum ether extract of custard apple seeds killed 93% of head lice within 3 h. Scientists found that 20 g of 20% freshly prepared cream could kill 94.5% of head lice within 3 h of application to school girls' hair. The cream preparation of custard apple seed was also found to be biologically stable for at least 12 months.

Antimicrobial Activity

Research in Bangladesh reported that annotemoyin-1, annotemoyin-2, squamocin and cholesteryl glucopyranoside isolated from the seeds of *Annona squamosa* showed remarkable antimicrobial and cytotoxic activities (Mukhlesur Rahman et al. 2005). Studies found that palmitone (11-hydroxy-16-hentriacontanone) which constituted up to 48% of the total leaf cuticular wax of sweetsop exhibited significantly higher antibacterial activity against selected Gram-positive and Gram-negative bacterial strains than that of the isomeric hydroxy ketones, but their antifungal activities were comparable (Shanker et al. 2007).

Six major volatile components from the bark oil were identified: 1 H-Cycloprop(e)azulene (3.46%), germacrene D (11.44%), bisabolene (4.48%), caryophyllene oxide (29.38%), bisabolene epoxide (3.64%) and kaur-16-ene (19.13%). The oil exhibited a significant antimicrobial activity against *Bacillus subtilis* and *Staphylococcus aureus* (Chavan et al. 2006). Of 4 different solvent extracts of *A. squamosa* leaves, the highest zone of inhibition was observed in methanol extract against *Pseudomonas aeruginosa* (MIC: 130 µg/ml) followed by petroleum ether extract against *Pseudomonas aeruginosa* (MIC: 165 µg/ml) and methanol extract against *Escherichia coli* (MIC: 180 µg/ml) (Patel and Kumar 2008). The presence of some phytochemicals (linalool, borneol, eugenol, farnesol, and geraniol) in the extracts were found responsible for the antibacterial activity.

Antigenotoxic Activity

Annona squamosa bark extracts was found to exhibit antigenotoxic effect in 7,12 dimethylbenz(a)anthracene (DMBA)-induced genotoxicity in golden Syrian hamsters (Suresh et al. 2008). Oral administration of aqueous and ethanolic bark extracts significantly decreased the frequency of micronucleated polychromatic erythrocytes and chromosomal aberration in DMBA-treated hamsters. The effect of the ethanolic extract was found to be more significant than the aqueous extract.

Traditional Medicinal Uses

Various parts of *A. squamosa* tree have been used in traditional folkloric medicine.

In India, the crushed leaves are sniffed to overcome hysteria and fainting spells; they are also applied on ulcers and wounds and a leaf decoction is taken in cases of dysentery. Throughout tropical America, a decoction of the leaves alone or with those of other plants is imbibed either as an emmenagogue, febrifuge, tonic, cold remedy, digestive, or to clarify the urine. The leaf decoction is also employed in baths to alleviate rheumatic pain. In the Philippines, the leaves are applied as a poultice to children with dyspepsia. A boiled decoction of the leaves is used to induce or hastens menstrual flow and used to treat dysentery, colds and fever. Crushed leaves are inhaled for dizziness and fainting. The leaf decoction is also used for bathing to alleviate rheumatic pain. The bruised leaves, with salt, make a good cataplasm to induce suppuration. The leaves are used as an anthelmintic.

The green, very astringent fruit, is employed against diarrhoea in El Salvador. In the Philippines, infected insect bites are cured by applying the juice from an unripe fruit. In India, the crushed ripe fruit, mixed with salt, is applied as a maturant to malignant tumours to hasten suppuration.

The bark and roots are both highly astringent. The bark is considered a powerful astringent and a decoction is given as a tonic and to arrest diarrhoea. The root, because of its strong purgative action, is administered as a drastic treatment for dysentery and other ailments.

A paste of the seed powder or crushed seeds with coconut oil are applied on the scalp to rid it of lice in Thailand but must be kept away from the eyes as it is a powerful irritant to the conjunctiva and can cause blindness. A decoction of the seeds is used as an enema for the children with dyspepsia. Crushed seeds in a paste with water, is used as an abortifacient and applied to the uteri in pregnant women.

Other Uses

The unripe fruits, seeds, roots and leaves have vermicidal and insecticidal properties. Indonesian scientists reported that aqueous seed extracts and an aqueous emulsion of ethanolic seed extracts were toxic to the diamondback moth, *Plutella xylostella* L. (Lepidoptera: Plutellidae) and the cabbage looper, *Trichoplusia ni* (Hübner) (Lepidoptera: Noctuidae) and were less so to natural enemies in particular, *Chrysoperla carnea* (Stephens) (Neuroptera: Chrysopidae) (Leatemia and Isman 2004).

Powdered seeds and also pounded dried fruits served as fish poison and insecticides in India. Heat-extracted oil from the seeds has been employed against agricultural pests. In Mexico, the leaves are rubbed on floors and put in hen's nests to repel lice. In Brazil, researchers isolated a C37 trihydroxy adjacent bistetrahydrofuran acetogenin from *Annona squamosa* seeds extracts that showed anthelmintic activity against *Haemonchus contortus*, the main nematode of sheep and goat in north-eastern Brazil (Souza et al. 2008). All plant extracts (of leaf, bark, and seed) showed moderate parasitic effects after 48 h exposure for egg hatching and larval development assay of the sheep parasite, *Haemonchus contortus*, respectively; however, 100% egg hatching and larvicidal inhibition were found in the methanol extracts of *A. squamosa,* at 25 mg/ml (Kamaraj et al. 2011). The effect was similar to positive control of Albendazole (0.075 mg/ml) and Ivermectin (0.025 mg/ml) against *H. contortus*, respectively. The leaf and seed ethyl acetate, acetone and methanol of *Annona squamosa,* showed complete inhibition (100%) at the maximum concentration tested (50 mg/ml) of invitro ovicidal and larvicidal activities on *Haemonchus contortus (*Kamaraj and Rahuman 2011).

The leaves yield an excellent oil rich in terpenes and sesquiterpenes, mainly β-caryophyllene, which finds limited use in perfumes, giving a woody spicy accent. Fibre extracted from the bark has been employed for cordage. The tree serves as host for lac-excreting insects. The ethanolic extract of *Annona squamosa* leaves exhibited potent activity against *Sitophilus oryzae* pest (Ashok Kumar et al. 2010) Hundred percent mortality was achieved at 39.6 minutes and 14.5 minutes for 1% w/v and 5% w/v concentration respectively. No mortality of the insects was found in any of the controls up to 100 hours. The ethyl acetate extract of *A. squamosa* bark showed larvicidal activity against fourth instar larvae of the malaria vector, *Anopheles stephensi* and *Culex quinquefasciatus*, lymphatic filariasis vector (Kamaraj et al. 2010).

An alkaloid extract from *Annona squamosa* plant was found to inhibit corrosion of c38 steel in normal hydrochloric acid medium (Lebrini et al. 2010). The corrosion inhibition efficiency increased on increasing plant extract concentration and was also temperature dependent.

Comments

See also notes on other edible Annonaceous species.

Selected References

Andrade EHA, Zoghbi MdGB, Maia JGS, Fabricius H, Marx F (2001) Chemical characterization of the fruit of *Annona squamosa* L. occurring in the Amazon. J Food Comp Anal 14(2):227–232

Araya H, Sahai M, Singh S, Singh AK, Yoshida M, Hara N, Fujimoto Y (2002) Squamocin-O1 and squamocin-O2, new adjacent bis-tetrahydrofuran acetogenins from the seeds of *Annona squamosa*. Phytochem 61(8):999–1004

Ashok Kumar J, Rekha T, Devi SS, Kannan M, Jaswanth A, Gopal V (2010) Insecticidal activity of ethanolic extract of leaves of *Annona squamos*. J Chem Pharm Res 2(5):177–180

ba Ndob IB, Champy P, Gleye C, Lewin G, Akendengué B (2009) Annonaceous acetogenins: precursors from the seeds of *Annona squamosa*. Phytochem Lett 2(2):72–76

Backer CA, van den Brink RCB Jr (1963) Flora of Java (spermatophytes only), vol 1. Noordhoff, Groningen, 648 pp

Bhakuni DS, Tewari S, Dhar MM (1972) Aporphine alkaloids of *Annona squamosa*. Phytochem 11(5):1819–1822

Bhaumik PK, Mukherjee B, Juneau JP, Bhacca NS, Mukherjee R (1979) Alkaloids from leaves of *Annona squamosa*. Phytochem 18(9):1584–1586

Burkill IH (1966) A dictionary of the economic products of the Malay Peninsula. Revised reprint, 2 vols. Ministry of Agriculture and Co-operatives, Kuala Lumpur, vol 1 (A–H), pp 1–1240, vol 2 (I–Z), pp 1241–2444

Chavan MJ, Shinde DB, Nirmal SA (2006) Major volatile constituents of *Annona squamosa* L. bark. Nat Prod Res 120(8):754–757

Chavan MJ, Wakte PS, Shinde DB (2010) Analgesic and anti-inflammatory activity of Caryophyllene oxide from *Annona squamosa* L. bark. Phytomed 17(2):149–151

Chavan MJ, Wakte PS, Shinde DB (2011) Analgesic and anti-inflammatory activities of 18-acetoxy-ent-kaur-16-ene from *Annona squamosa* L. bark. Inflammopharmacol 19(2):111–115

Damasceno DC, Volpato GT, Sartori TCF, Rodrigues PF, Perin EA, Calderon IMP, Rudge MVC (2002) Effects of *Annona squamosa* extract on early pregnancy in rats. Phytomed 9(7):667–672

Foundation for Revitalisation of Local Health Traditions (2008) FRLHT database. htttp://envis.frlht.org

Fujimoto Y, Eguchi T, Kakinuma K, Ikekawa N, Sahai M, Gupta YK (1988) Squamocin, a new cytotoxic bis-tetrahydrofuran containing acetogenin from *Annona squamosa*. Chem Pharm Bull Tokyo 36(12):4802–4806

Fujimoto Y, Murasaki C, Kakinuma K, Eguchi T, Ikekawa N, Furuya M, Hirayama K, Ikekawa T, Sahai M, Gupta YK, Ray AB (1990) Squamostatin-A: unprecedented bis-tetrahydrofuran acetogenin from *Annona squamosa*. Tetrahedron Lett 31(4):535–538

Fujimoto Y, Murasaki C, Shimada H, Nishioka S, Kakinuma K, Singh S, Singh M, Gupta YK, Sahai M (1994) Annonaceous acetogenins from the seeds of *Annona squamosa* non- adjacent bis-tetrahydrofuranic acetogenins. Chem Pharm Bull 42(6):1175–1184

George AP, Nissen RJ (1992) *Annona cherimola* Miller, *Annona squamosa* L., *A. cherimola* x *A. squamosa*. In: Verheij EWM, Coronel RE (eds) Plant resources of South-East Asia. No. 2: Edible fruits and nuts. PROSEA, Bogor, pp 71–75

Gupta RK, Kesari AN, Murthy PS, Chandra R, Tandon V, Watal G (2005a) Hypoglycemic and antidiabetic effect of ethanolic extract of leaves of *Annona squamosa* L. in experimental animals. J Ethnopharmacol 99(1):75–81

Gupta RK, Kesari AN, Watal G, Murthy PS, Chandra R, Tandon V (2005b) Nutritional and hypoglycemic effect of fruit pulp of *Annona squamosa* in normal healthy and alloxan-induced diabetic rabbits. Ann Nutr Metab 49:407–413

Gupta RK, Kesari AN, Diwakar S, Tyagi A, Tandon V, Chandra R, Watal G (2008) In vivo evaluation of anti-oxidant and anti-lipidimic potential of *Annona squamosa* aqueous extract in Type 2 diabetic models. J Ethnopharmacol 118(1):21–25

Hatano H, Aiyama R, Matsumoto S, Nishisaka F, Nagaoka M, Kimura K, Makino T, Shishido Y, Hashimoto S (2002) Cytotoxic constituents in the branches of *Annona squamosa* grown in Philippines. Ann Rep Yakult Inst Microbiol Res 22:5–9 (In Japanese)

Hopp DC, Zeng L, Gu ZM, McLaughlin JL (1996) Squamotacin: an annonaceous acetogenin with cytotoxic selectivity for the human prostate tumor cell line (PC-3). J Nat Prod 59(2):97–99

Hopp DC, Alali FQ, Gu ZM, McLaughlin JL (1998a) Mono-THF ring annonaceous acetogenins from *Annona squamosa*. Phytochem 47(5):803–809

Hopp DC, Alali FQ, Gu ZM, McLaughlin JL (1998b) Three new bioactive bis-adjacent THF-ring acetogenins from the bark of *Annona squamosa*. Bioorg Med Chem 6(5):569–575

Intaranongpai J, Chavasiri W, Gritsanapan W (2006) Anti-head lice effect of *Annona squamosa* seeds. Southeast Asian J Trop Med Public Health 37(3): 532–535

Kaleem M, Asif M, Ahmed QU, Bano B (2006) Antidiabetic and antioxidant activity of *Annona squamosa* extract in streptozotocin-induced diabetic rats. Singapore Med J 47(8):670–675

Kaleem M, Medha P, Ahmed QU, Asif M, Bano B (2008) Beneficial effects of *Annona squamosa* extract in streptozotocin-induced diabetic rats. Singapore Med J 49(10):800–804

Kamaraj C, Rahuman AA (2011) Efficacy of anthelmintic properties of medicinal plant extracts against *Haemonchus contortus*. Res Vet Sci (In press)

Kamaraj C, Abdul Rahman A, Bagavan A, Abduz Zahir A, Elango G, Kandan P, Rajakumar G, Marimuthu S, Santhoshkumar T (2010) Larvicidal efficacy of medicinal plant extracts against *Anopheles stephensi* and *Culex quinquefasciatus* (Diptera: culicidae). Trop Biomed 27(2):211–219

Kamaraj C, Rahuman AA, Elango G, Bagavan A, Zahir AA (2011) Anthelmintic activity of botanical extracts against sheep gastrointestinal nematodes, *Haemonchus contortus*. Parasitol Res (In press)

Leatemia JA, Isman MB (2004) Toxicity and antifeedant activity of crude seed extracts of *Annona squamosa* (Annonaceae) against lepidopteran pests and natural enemies. Int J Trop Insect Sci 24:150–158

Lebrini M, Robert F, Roos C (2010) Inhibition effect of alkaloids extract from *Annona Squamosa* plant on the corrosion of c38 steel in normal hydrochloric acid medium. Int J Electorchem Sci 5:1698–1712

Li X-H, Hui Y-H, Rupprecht JK, Liu Y-M, Wood KV, Smith DL, Chang C-J, Mclaughlin JL (1990) Bullatacin, bullatacinone, and squamone, a new bioactive acetogenin, from the bark of *Annona squamosa*. J Nat Prod 53(1):81–86

Li CM, Tan NH, Mu Q, Zheng HL, Hao XJ, Wu Y, Zhou J (1997) Cyclopeptide from the seeds of *Annona squamosa*. Phytochem 45(3):521–523

Li X, Chen XL, Chen JW, Sun DD (2010) Annonaceous acetogenins from the seeds of *Annona squamosa*. Chem Nat Comp 46(1):101–105

Liaw CC, Yang YL, Chen M, Chang FR, Chen SL, Wu SH, Wu YC (2008) Mono-tetrahydrofuran Annonaceous acetogenins from *Annona squamosa* as cytotoxic agents and calcium ion chelators. J Nat Prod 71(5):764–771

Lieb F, Nonfon M, Wachendorff-Neumann U, Wendisch D (1990) Annonacins and annonastatin from *Annona squamosa*. Planta Med 56(3):317–319

Morita H, Sato Y, Chan KL, Choo CY, Itokawa H, Takeya K, Kobayashi J (2000) Samoquasine A, a benzoquinazoline alkaloid from the seeds of *Annona squamosa*. J Nat Prod 63(12):1707–1708

Morita H, Iizuka T, Choo CY, Chan KL, Takeya K, Kobayashi J (2006) Vasorelaxant activity of cyclic peptide, cyclosquamosin B, from *Annona squamosa*. Bioorg Med Chem Lett 16(1):4609–4611

Morton J (1987) Sugar apple. In: Fruits of warm climates. Julia F. Morton, Miami, pp 69–72

Mukhlesur Rahman M, Parvin S, Ekramul Haque M, Ekramul Islam M, Mosaddik MA (2005) Antimicrobial and cytotoxic constituents from the seeds of *Annona squamosa*. Fitoterapia 76(5):484–489

Ochse JJ, Bakhuizen van den Brink RC (1931) Fruits and fruitculture in the Dutch East Indies. G. Kolff & Co, Batavia, 180 pp

Panda S, Kar A (2003) Possible amelioration of hyperthyroidism by the leaf extract of *Annona squamosa*. Curr Sci 84(11):1402–1404

Panda S, Kar A (2007a) *Annona squamosa* seed extract in the regulation of hyperthyroidism and lipid-peroxidation in mice: possible involvement of quercetin. Phytomed 14(12):799–805

Panda S, Kar A (2007b) Antidiabetic and antioxidative effects of *Annona squamosa* leaves are possibly mediated through quercetin-3-O-glucoside. Biofactors 31(3–4):201–210

Pardhasaradhi BVV, Reddy M, Ali AM, Kumari AL, Khar A (2004) Antitumour activity of *Annona squamosa* seed extracts is through the generation of free radicals and induction of apoptosis. Indian J Biochem Biophys 41(4):167–172

Pardhasaradhi BVV, Reddy M, Ali AM, Kumari AL, Khar A (2005) Differential cytotoxic effects of *Annona squamosa* seed extracts on human tumour cell lines: role of reactive oxygen species and glutathione. J Biosci 30:237–244

Patel JD, Kumar V (2008) Annona squamosa L.: phytochemical analysis and antimicrobial screening. J Pharm Res 1(1):35–38

Popenoe W (1974) Manual of tropical and subtropical fruits. Facsimile of the 1920 edition. Hafner Press, New York, USA

Prest RL, Ross AA (1955) The custard apple. Qd Agric J 80:17–21

Rehm S (1994) Multilingual dictionary of agronomic plants. Kluwer, Dordrecht/Boston/London, 286 pp

Shanker KS, Kanjilal S, Rao BV, Kishore KH, Misra S, Prasad RB (2007) Isolation and antimicrobial evaluation of isomeric hydroxy ketones in leaf cuticular waxes of *Annona squamosa*. Phytochem Anal 18(1): 7–12

Shirwaikar A, Rajendran K, Kumar CD, Bodla R (2004) Antidiabetic activity of aqueous leaf extract of *Annona squamosa* in streptozotocin–nicotinamide type 2 diabetic rats. J Ethnopharmacol 91(1):171–175

Soubir T (2007) Antioxidant activities of some local Bangladeshi fruits (*Artocarpus heterophyllus, Annona squamosa, Terminalia bellirica, Syzygium samarangense, Averrhoa carambola* and *Olea europa*). Sheng Wu Gong Cheng Xue Bao 23(2):257–261

Souza MM, Bevilaqua CM, Morais SM, Costa CT, Silva AR, Braz-Filho R (2008) Anthelmintic acetogenin from *Annona squamosa* L. seeds. An Acad Bras Ciênc 80(2):271–277

Sultana N (2008) Lipoxygenase inhibition by novel fatty acid ester from *Annona squamosa* seeds. J Enzym Inhib Med Chem 23(6):877–881

Suresh K, Manoharn S, Blessy D (2008) Protective role of *Annona squamosa* Linn bark extracts in DMBA induced genotoxicity. Kathmandu Univ Med J 6(23): 364–369

Suresh K, Manoharan S, Vijayaanand MA, Sugunadevi G (2010) Modifying effects of *Annona squamosa* Linn on glycoconjugates levels in 7, 12-dimethylbenz (A) anthracene (DMBA) induced hamster buccal pouch carcinogenesis. J Appl Sci Res 6(8):973–979

U.S. Department of Agriculture, Agricultural Research Service (USDA) (2010) USDA national nutrient database for standard reference, release 23. Nutrient Data Laboratory Home Page. http://www.ars.usda.gov/ba/bhnrc/ndl

Wu Y-C, Hung Y-C, Chang F-R, Cosentino M, Wang H-K, Lee K-H (1996) Identification of ent-16β,17-dihydroxykauran-19-oic acid as an anti-HIV principle and isolation of the new diterpenoids annosquamosins A and B from *Annona Squamosa*. J Nat Prod 59(6): 635–637

Xie HH, Wei XY, Wang JD, Liu MF, Yang RZ (2003) A new cytotoxic acetogenin from the seeds of *Annona squamosa*. Chin Chem Lett 14(6):588–590

Yang H, Zhang N, Li X, He L, Chen J (2009a) New non-adjacent bis-THF ring acetogenins from the seeds of *Annona squamosa*. Fitoterapia 80(3):177–181

Yang HJ, Li X, Zhang N, Chen JW, Wang MY (2009b) Two new cytotoxic acetogenins from *Annona squamosa*. J Asian Nat Prod Res 11(3):250–256

Yang TH, Chen CM (1972) Structure of squamolone, a novel diazepine from *Annona squamosa* L. J Chin Chem Soc Taipei 19:149–151

Yang YL, Chang FR, Wu CC, Wang WY, Wu YC (2002) New ent-kaurane diterpenoids with anti-platelet aggregation activity from *Annona squamosa*. J Nat Prod 65(10):1462–1467

Yang YL, Chang FR, Hwang TL, Wu YC (2004) Inhibitory Effects of *ent*-Kauranes from the stems of *Annona squamosa* on the superoxide anion generation by human neutrophils. Planta Med 70:256–258

Yang YL, Hua KF, Chuang PH, Wu SH, Wu KY, Chang FR, Wu YC (2008) New cyclic peptides from the seeds of *Annona squamosa* L. and their anti-inflammatory activities. J Agric Food Chem 56(2):386–392

Yeh SH, Chang FR, Wu YC, Yang YL, Zhuo SK, Hwang TL (2005) An anti-inflammatory *ent*-Kaurane from the stems of *Annona squamosa* that inhibits various human neutrophil functions. Planta Med 71:904–909

Yu JG, Luo XZ, Sun L, Li DY, Huang WH, Liu CY (2005) Chemical constituents from the seeds of *Annona squamosa*. Yao Xue Xue Bao 40(2):153–158 (In Chinese)

Rollinia mucosa

Scientific Name

Rollinia mucosa (Jacq.) Baillon

Synonyms

Annona biflora Ruiz & Pav. ex G. Don nom illeg., *Annona biflora* Sessé & Moc., *Annona microcarpa* R. & P. ex G. Don nom illeg., *Annona mucosa* Jacq., *Annona obtusiflora* Tussac, *Annona obtusifolia* DC., *Annona pterocarpa* Ruiz & Pavon ex G. Don nom illeg., *Annona pteropetala* Ruiz & Pav. ex E.A. López nom illeg., *Annona pteropetala* R. & P. ex R.E. Fries, *Annona reticulata* var. *mucosa* (Jacq.) Willd., *Annona sieberi* A. DC., *Rollinia curvipetala* R. E. Fries, *Rollinia deliciosa* Safford, *Rollinia jimenezii* Saff., *Rollinia jimenezii* var. *nelsonii* R.E. Fr., *Rollinia mucosa* subsp. *aequatorialis* R. E. Fr., *Rollinia mucosa* subsp. *portoricensis* R.E. Fr., *Rollinia mucosa* var. *macropoda* R. E. Fr., *Rollinia mucosa* var. *neglecta* (R.E. Fr.) R.E. Fr., *Rollinia neglecta* R. E. Fr., *Rollinia obtusiflora* Tussac, *Rollinia orthopetala* A. DC., *Rollinia orthopetala* Correa, *Rollinia permensis* Standl., *Rollinia pterocarpa* G. Don, *Rollinia pulchrinervia* A.DC., *Rollinnia sieberi* A. DC.

Family

Annonaceae

Common/English Names

Biriba, Biribah, Rollinia, Lancewood, Wild Cachiman, Wild Custard Apple, Wild Sugar Apple

Vernacular Names

Belize: Anonilla, Wild Custard Apple;
Brazil: Araticu, Araticum, Araticum Pitaya, Biriba De Pernambuco, Fruta Da Condessa, Graviola Brave, Jaca De Pobre, (Portuguese);
Colombia: Anón, Anón De Enorme, Anón De Montana, Anón De Silvestre, Anona, Mulato, Chirimoya, Cuero Negro;
Dominican Republic: Anón, Anona Candongo, Candón, Candóngo Candongo;
Ecuador: Ananas Panga, Anuna, Aparina Cara, Cabeza De Negro, Chirimoya, Chirimoya Arisco;
French: Cachiman Crème, Cachiman Morveux;
German: Schleimapfel, Unechtes Lunzenholz;
Guadeloupe: Cachiman Cochon, Cachiman Montagne, Cachiman Morveux; Guatemala: Anona, Anonilla, Saramullo Silveste;
Guyana: Casimo;
Honduras: Anona, Anono;
Martinique: Cachiman Montagne, Cachiman Morveux;
Mexico: Anonilla, Annóna, Anona Babosa, Anonita, Anonita De Monte, Zambo;
Panama: Toreta;

Peru: Anón, Anóna, Anonilla, Anuna;
Puerto Rico: Cachiman, Anon Cimarrón;
Portuguese: Biribarana;
Trinidad: Cachiman, Slimy Rollinia, Wild Cashima, Wild Sugar Apple;
Venezuela: Anón De Monte, Anonilla, Biriba, Chirimoya, Condessa, Rinon, Rinon De Monte.

Origin/Distribution

This species has an extensive natural range, from Peru and northern Argentina, Paraguay and Brazil and northward to Guyana, Venezuela, Colombia and southern Mexico; Trinidad, the Lesser Antilles including Guadeloupe, Martinique and St. Vincent; and Puerto Rico and Hispaniola. It is much cultivated around Iquitos, Peru, and Rio de Janeiro, Brazil and the fruits are marketed in abundance. It is the favorite fruit in western Amazonia. The species was also introduced into Northern Australia and southeast Asia for trial purposes.

Agroecology

Biriba occurs in hot, humid, tropical areas from 20° north to 30° south latitudes in tropical America in areas with >1,250 mm of annual rainfall. It is usually found in the lowland from 0 to 500 m elevation, but in Puerto Rico, it occurs at elevations between 150 and 600 m. In Brazil, the tree grows naturally in low areas along the Amazon subject to periodic flooding. It grows in southern Florida but is susceptible to frost and subzero temperatures. In the Philippines it flourishes where the rainfall is equally distributed throughout the year. It grows in tropical northern Australia around Darwin in the Northern Territory and in northern Queensland and in Taiwan. Calcareous soils do not seem to be unsuitable in Florida or Puerto Rico as long as they are moist. The tree and fruit develop best in clay soils, deep, well-drained and rich in organic matter It can grow in full sun to partial shade.

Edible Plant Parts and Uses

Pulp of ripe fruit is eaten fresh and in desserts, processed into refreshing drinks, ice-creams and is fermented to make wine.

Botany

A small- to medium-sized, deciduous, vigorous tree, growing to 4–15 m high with brown pubescent branchlets. Leaves are alternate, medium green, oblong-elliptic or ovate-oblong, 10–25 cm long by 4–12 cm wide, pointed apex and obtuse to rounded base, entire margin somewhat coriaceous and pubescent on the underside (Plate 1). Flowers occur in 1–3 flowered cluster at the axils of leaves, 2–3.5 cm across, deltoid, with three pubescent, pale green sepals, three large, fleshy outer petals with upturned or horizontal wings, and three rudimentary inner petals (Plate 1). Fruit is variable shape, oblong-cylindrical (5–8 cm by 10–15 cm long), cordate, obovate to oblate(6.5–12.5 diameter), indehiscent with a leathery, tough, rough rind, turning yellow when ripe and bearing hexagonal or conical segments from which arises conical or triangular, curved, stiff, 3–5 mm thick, 7–12 mm long wart-like protuberances (Plate 2a–d). The pulp is white, mucilaginous, translucent, juicy, subacid to sweet. There is a slender, opaque-white core with numerous dark-brown, elliptic or obovate seeds 1.6–2 cm long.

Plate 1 Leaves and flowers of biriba

Plate 2 (a–d) Different varieties of biriba with different shapes and protuberances

Nutritive/Medicinal Properties

The food value per 100 g of edible portion of raw Biriba fruit based on analyses made in Brazil (Morton 1987) was reported as: energy 80 Cal, moisture 77.2 g, protein 2.8 g, lipids 0.2 g, glycerides 19.1 g, fibre 1.3 g, ash 0.7 g, calcium 24 mg, phosphorus 26 mg, iron 1.2 mg, vitamin b1 0.04 mg, vitamin b2 0.04 mg, niacin 0.5 mg, ascorbic acid 33.0 mg, amino acids (mg per g of nitrogen (N=6.25)): lysine 316 mg, methionine 178 mg, threonine 219 mg and tryptophan 57 mg.

Rollinia mucosa was found to contain more than 50 different acetogenins of diverse structure and other compounds that were isolated mainly from the leaves but also from the stem, fruit and seeds. Many of the bioactive acetogenins and other bioactive compounds were reported to have anti-tumorous and anti-cancerous properties. Some bioactive acetogenins and alkaloids also exhibited antileishmanial and anti-platelet aggregation activities respectively.

Anticancer Activity

The leaves of *Rollinia mucosa* yielded the most acetogenins. Of these, mucoxin, an acetogenin containing a hydroxylated trisubstituted tetrahydrofuran (THF) ring was found to be a highly potent and specific antitumour agent against MCF-7 (breast carcinoma) cell lines ($ED_{50} = 3.7 \times 10^{-3}$ μg/ml compared to adriamycin, $ED_{50} = 1.0 \times 10^{-2}$ μg/ml) (Narayan and Borhan 2006). (+)-Muconin, a sequential tetrahydrofuran or tetrahydropyran (THF/THP)-possessing acetogenin, exhibited potent and selective in-vitro cytotoxicity toward pancreatic and breast tumour cell lines (Yoshimitsu et al. 2004). Two mono-THF ring acetogenins, rollinecins A and B showed equivalent and selective in-vitro activities against several human solid tumour cell lines (Shi et al. 1996c). Acetogenins, rollitacin, rollinacin and javoricin were isolated from the ethanol extract of the leaves and both rollitacin and rollinacin exhibited selective inhibitory effects among six human solid tumour cell lines (Shi et al. 1997). The acetogenins, rollidecin C isolated from leaves exhibited selective cytotoxicity towards the colon tumour cell line (HT-29), while rollidecin D showed only borderline cytotoxicity in a panel of six human tumour cell lines (Gu et al. 1997a, b). Muricatetrocin C, rollidecin A, and rollidecin B, three bioactive annonaceous acetogenins bearing vicinal diols, isolated from leaves, showed potent and selective inhibitory effects against several human cancer cell lines (Shi et al. 1996a). Mucocin, a tetrahydropyran (THP) ring-bearing acetogenin, and another closely related acetogenin, muconin were identified in the bioactive leaf extracts of *Rollinia mucosa* (Shi et al. 1996b). A new furofuranic lignan named (+)-epimembrine together with known (+)-epieudesmine and (+)-epimagnoline were also isolated from leaves of *R. mucosa* (Estrada-Reyes et al. 2002). Rollipyrrole 1, a novel pyrromethenone derivative was isolated from the leaves of *Rollinia mucosa* in Taiwan (Kuo et al. 2001a, b).

From the seeds of *Rollinia mucosa* a mixture of eight novel tryptamine amides: N-palmitoyltryptamine, N-stearoyltryptamine, N-arachidoyltryptamine, N-behenoyltryptamine, N-tricosanoyltryptamine, N-pentacosanoyltryptamine and N-cerotoyltryptamine were isolated (Chávez et al. 1999). Two lignans (pinoresinol dimethyl ether and magnolin) and six acetogenins: membranacin, desacetyluvaricin, rolliniastatin 1, bullatacin, squamocin, and motrilin were also isolated. The cytotoxicity of membranacin and desacetyluvaricin was determined against six human solid tumour cell lines.

From the unripe fruits of *Rollinia mucosa*, rollicosin was reported as a potential antitumour agent (Liaw et al. 2003). Four new compounds, a mixture of 20,23-cis-2,4-trans-bullatalicinone (1) and 20,23-cis-2,4-cis-bullatalicinone (2), rollimusin (3), and rolliacocin (4), along with eight known acetogenins, were isolated from an ethyl acetate extract of the unripe fruits of *Rollinia mucosa* (Liaw et al. 1999). Also isolated from fruits were two γ-lactone acetogenins, epomusenins A and B, and acetogenins, rolliniastatin-1, rolliniastatin-2 and squamocin (Chen et al. 1996b). Romucosine a novel aporphinoid alkaloid possessing a *N*-(methoxycarbonyl) group, along with 12 known compounds including eight isoquinoline alkaloids, anonaine, glaucine, purpureine, liriodenine, oxoglaucine, oxopurpureine, berberine, and tetrahydroberberine, and four lignans, yangambin, magnolin, eudesmin, and membrin, were also isolated and characterized from the fresh unripe fruits of *Rollinia mucosa* (Chen et al. 1996a). The lignans (+)-yangambine, (+)-magnoline, (+)-eudesmin, (+)-epieudesmin and (+)-membrine were identified from mature fruits (Paulo et al. 1991). *Rollinia mucosa* produces furofuranic lignans (magnolin, epiyangambin, yangambin) that are antagonists of platelet-activating factor (PAF) (Figueiredo et al. 1999). The major accumulation of lignans was observed in leaves. In the mature leaves of in-vivo grown seedlings magnolin and yangambin predominated, in contrast to leaves from in-vitro propagated plants that presented epiyangambin as the major lignin.

Antiplatelet Aggregation Activity

From the stems of *Rollinia mucosa,* N-methoxycarbonyl aporphine alkaloids, romucosine A (1), romucosine B (2), romucosine C (3), and romucosine D (4), along with the known alkaloid, N-methoxylcarbonyl-nornuciferine (5) were isolated (Kuo et al. 2001a, b). Alkaloids 1 and 4 exhibited significant inhibition of collagen, arachidonic acid, and platelet activating factor-induced platelet aggregation, and alkaloid 3 also showed an inhibitory effect on arachidonic acid induced platelet aggregation. Other alkaloids isolated from *R. mucosa* included a new N-(methoxycarbonyl) phenanthrene alkaloid, romucosine I, along with three known N-(methoxycarbonyl) alkaloids, romucosine C, tuduranine and promucosine (Kuo et al. 2004).

CNS Depressant Activity

The hexane extract of leaves from *Rollinia mucosa* was found to induce anxiolytic-like actions similar to those induced by diazepam in the avoidance exploratory behavior model (Estrada-Reyes et al. 2010). Its significant activity was shown at doses from 1.62 to 6.25 mg/kg.

It also enhanced pentobarbital-induced hypnosis time, and at high doses produced motor coordination impairment. The plant extract reduced benzodiazepine (BDZ) receptor binding in the hippocampus (29%), amygdala (26%), and temporal cortex of mice (36%). The findings supported the hypothesis that *R. mucosa* may induce central nervous system (CNS) depressant effects, presumably through an interaction with the GABA/benzodiazepine receptor complex.

Antileishmanial Activity

Acetogenins also exhibited interesting in-vitro antileishmanial activity (Raynaud-Le Grandic et al. 2004). The IC_{50} of the 12 acetogenins having anti-leishmanial activity against promastigote forms of *Leishmania donovani* was in the range of 4.7–47.3 μM. The most active compound was rolliniastatin 1 (IC_{50} at 4.7 μM). On the intramacrophage amastigote in-vitro model, only seven compounds exhibited measurable anti-leishmanial activity with IC_{50} values in a range of 2.5–29.7 μM. Rollinistatin 1 was the most interesting compound with IC_{50} of 2.5 μM and it appeared as the most promising one on the basis of therapeutic index (18.08). Isoannonacin, which was active against intramacrophagic amastigotes (IC_{50} of 6.2 μM) with a therapeutic index of 2.05, exhibited a strong action on drug-resistant strains (IC_{50} from 5.1 to 9.8 μM).

Traditional Medicinal Uses

In traditional Brazilian folk medicine, the fruit is regarded as refrigerant, analeptic and antiscorbutic. The powdered seeds are said to be a remedy for enterocolic disorder. Leaves are used against rheumatism.

Other Uses

The wood of the tree is yellow, hard, heavy and strong, and is used for ribs for canoes, boat masts, boards, boxes and construction. Seeds are used as necklaces and serve as insecticides.

Comments

See also notes on other edible Annonaceous species.

Selected References

Chávez D, Acevedo LA, Mata R (1999) Tryptamine derived amides and acetogenins from the seeds of *Rollinia mucosa*. J Nat Prod 62(8):1119–1122

Chen YY, Chang FR, Wu YC (1996a) Isoquinoline alkaloids and lignans from *Rollinia mucosa*. J Nat Prod 59:904–906

Chen YY, Chang FR, Yen HF, Wu YC (1996b) Epomusenins A and B, two acetogenins from fruits of *Rollinia mucosa*. Phytochem 42(4):1081–1083

Estrada-Reyes R, Alvarez ALC, Lopez-Rubalcava C, Rocha L, Heinze G, Moreno J, Martinez-Vazquez M (2002) Lignans from leaves of *Rollinia mucosa*. Z Naturforsch C 57:29–32

Estrada-Reyes R, López-Rubalcava C, Rocha L, Heinze G, González Esquinca AR, Martínez-Vázquez M (2010) Anxiolytic-like and sedative actions of *Rollinia mucosa*: possible involvement of the GABA/benzodiazepine receptor complex. Pharm Biol 48(1):70–75

Figueiredo SFL, Viana VRC, Simões C, Albarello N, Trugo LC, Kaplan MAC, Krul WR (1999) Lignans from leaves, seedlings and micropropagated plants of *Rollinia mucosa* (Jacq.) Baill. – annonaceae. Plant Cell Tissue Organ Cult 56(2):121–124

Gu ZM, Zhou D, Wu J, Shi G, Zeng L, McLaughlin JL (1997a) Screening for Annonaceous acetogenins in bioactive plant extracts by liquid chromatography/mass spectrometry. J Nat Prod 60(3):242–248

Gu ZM, Zhou D, Lewis NJ, Wu J, Shi G, McLaughlin JL (1997b) Isolation of new bioactive annonaceous acetogenins from *Rollinia mucosa* guided by liquid chromatography/mass spectrometry. Bioorg Med Chem 5(10):1911–1916

Kuo RY, Chang FR, Wu YC (2001a) A new propentdyopent derivative, rollipyrrole, from *Rollinia mucosa* Baill. Tetrahedron Lett 42(44):7907–7909

Kuo RY, Chang FR, Chen CY, Teng CM, Yen HF, Wu YC (2001b) Antiplatelet activity of N-methoxycarbonyl aporphines from *Rollinia mucosa*. Phytochem 57(3):421–425

Kuo RY, Chen CY, Lin AS, Chang FR, Wu YC (2004) A new phenanthrene alkaloid, romucosine I. from *Rollinia mucosa* Baill. Z Naturforsch 59B:334–336

Liaw CC, Chang FR, Chen YY, Chiu HF, Wu MJ, Wu YC (1999) New annonaceous acetogenins from *Rollinia mucosa*. J Nat Prod 62(12):1613–1617

Liaw CC, Chang FR, Wu MJ, Wu YC (2003) A novel constituent from *Rollinia mucosa*, rollicosin, and a new approach to develop annonaceous acetogenins as potential antitumor agents. J Nat Prod 66(2):279–281

Maas PJM, Lubbert YThW, Collaborators (1992) Rollinia (Flora Neotropica Monograph No. 57). Organization for Flora Neotropica, The New York Botanical Garden Press, New York, 188 pp

Morton JF (1987) Biriba. In: Fruits of warm climates. Julia F. Morton, Miami, pp 88–90

Narayan RS, Borhan B (2006) Synthesis of the proposed structure of mucoxin via regio- and stereoselective tetrahydrofuran ring-forming strategies. J Org Chem 71(4):1416–1429

Paulo MQ, Kaplan MAC, Laprevote O, Rublot F, Hocquemiller R, Cave A (1991) Lignans and other non-alkaloid constituents from *Rollinia mucosa*. Fitoterapia 62:150–152

Popenoe W (1974) Manual of tropical and subtropical fruits Facsimile of the 1920 edition. Hafner Press, New York, USA

Raynaud-Le Grandic S, Fourneau C, Laurens A, Bories C, Hocquemiller R, Loiseau PM (2004) In vitro antileishmanial activity of acetogenins from Annonaceae. Biomed Pharmacother 58(6–7):388–392

Shi G, Gu ZM, He K, Wood KV, Zeng L, Ye Q, MacDougal JM, McLaughlin JK (1996a) Applying Mosher's method to acetogenins bearing vicinal diols. The absolute configurations of muricatetrocin C and rollidecins A and B, new bioactive acetogenins from *Rollinia mucosa*. Bioorg Med Chem 4(8):1281–1286

Shi G, Kozlowski JF, Schwedler JT, Wood KV, MacDougal JM, McLaughlin JL (1996b) Muconin and mucoxin: additional nonclassical bioactive acetogenins from *Rollinia mucosa*. J Org Chem 61(23):7988–7989

Shi G, Ye Q, He K, McLaughlin JL, MacDougal JM (1996c) Rollinecins A and B: two new bioactive annonaceous acetogenins from *Rollinia mucosa*. J Nat Prod 59(5):548–551

Shi G, MacDougal JM, McLaughlin JL (1997) Bioactive annonaceous acetogenins from *Rollinia mucosa*. Phytochem 45(4):719–723

Tropicos.Org. Missouri Botanical Garden. http://www.tropicos.org/Name/1600784

Yoshimitsu T, Makino T, Nagaoka H (2004) Total synthesis of (+)-muconin. J Org Chem 69(6):1993–1998

Stelechocarpus burahol

Scientific Name

Stelechocarpus burahol (Blume) Hook. f. & Thompson

Synonyms

Uvaria burahol Blume

Family

Annonaceae

Common/English Names

Burahol, Kepel, Kepel Apple

Vernacular Names

Dutch: Kepel;
French: Kepel;
Indonesia: Kecindul (Javanese), Burahol (Sundanese).

Origin/Distribution

It is indigenous to Java, Indonesia, and is grown also in Southeast Asia throughout Malesia as far as the Solomons. However, in the Philippines and Australia, it is a recent introduction. Its cultivation as a fruit tree seems to be limited to Java.

Agroecology

Kepel is a tropical species, grows in its native habitat in hot, humid secondary forest on deep, moist clay soils in Java. It is found from sea level to an elevation of 600 m.

Edible Plant Parts and Uses

Fruit pulp is eaten and the fruit is sold in local markets in Java.

Botany

Large, erect evergreen tree, up to 25 m tall. Trunk up to 40 cm in diameter with dark grey-brown to black, bark characteristically covered with numerous thick tubercles. Leaves are elliptic-oblong to ovate-lanceolate, 12–27 cm × 5–9 cm, with a distinct mid-rib, soft pink to burgundy-red when young, dark green, glabrous and thin leathery when mature, borne on 1.5 cm long petioles (Plates 2a, b). Flowers are unisexual, green turning whitish, fascicled on tubercles; male flowers are ramiflorous on older branches, 8–16 together, up to 1 cm in diameter with 3- ovate-deltoid, obtuse sepals and 7–8 mm long imbricate petals in 2 whorls of three. Female flowers are

Plate 1 (**a, b**) Ramiflorous female flowers arising from the main trunk and knobby tubercles on the trunk

Plate 2 (**a, b**) Young, reddish leaves (*left*) and dark green, glossy, old leaves (*right*)

cauliflorous on the lower part of the trunk as well as ramiflorous on the main branches, up to 3 cm in diameter with 3 oval-ovate, obtuse sepals and imbricate petals in 2 whorls of three; ovaries many, 6–8 ovuled with hairy sessile stigma (Plates 1a, b). Fruit sub-globose to obovoid, brownish, 5–6 cm in diameter, juicy, edible, yellow pulp enclosing 4–6 ellipsoid seeds, 3–3.5 cm long (Plates 3–5).

Nutritive/Medicinal Properties

Fruit has a fresh weight of 62–105 g, the edible flesh amounting to 49%, and the seed to 27% of fresh weight (Sunarto 1992).

Anticancer Activity

Two phenanthrene lactams were isolated from *S. burahol* stem bark, aristolactam BI and aristolactam BII (Sunardi et al. 2003). Both compounds exhibited cytotoxic activities on leukemia cell L1210 with IC_{50} values of 0.87 and 0.66 μg/ml, respectively.

Antioxidant Activity

Flavonoids were isolated from the ethanolic fraction of *S. burahol* leaves which exhibited free radical DPPH (1,1-diphenyl-2-picrylhydrazyl) scavenging antioxidant activity (Sunarni

Plate 3 Ramiflorous fruit arising from the main branches

Plate 4 Cauliflorous fruit arising from the trunk

Plate 5 Ripe fruit with yellow pulp

et al. 2007). One of them, B4b exhibited a strong free radical scavenging with an EC_{50} value of 6.43 μg/ml.

Traditional Medicinal Uses

In traditional medicine in Indonesia, the fruit pulp is a diuretic and is use to prevent kidney inflammation and also as a deodorant to make sweat fragrant. The fruit pulp is commonly used in Jamu "*awet ayu*", by the local cosmetic industry. The pulp has been reported to cause temporary sterility in women.

Other Uses

Kepel has been used traditionally as a perfume in Indonesia by aristocratic ladies in Java. Its use was traditionally restricted to the consorts of the sultans of Jogjakarta. The timber is suitable for household articles; the straight trunk, after immersion in water for several months, is used in house building and is durable for more than 50 years.

Kepel is also planted as an ornamental tree for its beautiful foliage that changes from light pink into a burgundy red colour before turning a brilliant green.

Comments

To ensure postharvest quality, the fruit is bagged 1–2 months before harvesting, using plaited sleeves of bamboo or coconut leaflets, or polythene bags. The fruit is packed in baskets or bags and handles well; it can be kept for 2–3 weeks at room temperature.

Selected References

Backer CA, Bakhuizen van den Brink Jr RC (1963) Flora of Java, (spermatophytes only), vol 1. Noordhoff, Groningen, 648 pp

Ochse JJ (1927) Indische Vruchten. Volkslectuur Weltevreden. 330 pp (In Dutch)

Sangat HM, Larashati I (2002) Some ethnophytomedical aspects and conservation strategy of several medicinal plants in Java, Indonesia. Biodiversitas 3(2):231–235

Sunardi C, Padmawinata K, Kardono LBS, Hanafi M, Usuki Y, Iio H (2003) Identification of cytotoxic alkaloid phenanthrene lactams from *Stelechocarpus burahol*. ITE Lett Batter New Technol Med 4(3):328–331

Sunarni T, Pramono S, Asmah R (2007) Flavonoid antioksidan penangkap radikal dari daun kepel (*Stelechocarpus burahol* (Bl.)) Hook f. & Th. (Antioxidant–free radical scavenging of flavonoid from the leaves of *Stelechocarpus burahol* (Bl.) Hook f. & Th.). Majalah Farmasi Indonesia, 18(3):111–116 (In Indonesian)

Sunarto AT (1992) Stelechocarpus burahol (Blume) Hook. f. & Thomson. In: Verheij EWM, Coronel RE (eds.) Plant resources of South-East Asia. No. 2: Edible fruits and nuts. Prosea Foundation, Bogor, pp 290–291

Van Heusden ECH (1995) Revision of the Southeast Asian genus *Stelechocarpus* (Annonaceae). Blumea 40:429–438

Uvaria chamae

Scientific Name

Uvaria chamae P. Beauv.

Synonyms

Unona macrocarpa DC., *Uvaria cristata* R.Br., *Uvaria cylindrica* Schumach. & Thonn., *Uvaria echinata* A.Chev., *Uvaria nigrescens* Engl. & Diels, *Xylopiastrum macrocarpum* (DC.) Roberty

Family

Annonaceae

Common/English Names

Bush Banana, Finger Root

Vernacular Names

Ghana: Akotompo, Akotompotsen (Akan-Fante), Aura (Dagaari), Aŋmɛdãa, Aŋweda (Ga), Agbana-Asile (Gbe-Vhe), Atore, Darigaza (Huasa), Worsalla (Lobi), Sai (Songhai-Zarma), Akotumpetsin, Anweda (Twi), Agbana (Vhe, Awalan), Gbanagbana (Vhe, Kpando);
Guinea: Boélémimbo, Boïlé (Fula-Pulaar);
Guinea-Bissau: Banana Santcho (Crioulo), Guélè-Bálè (Fula-Pulaar), Sambafim (Manding-Mandinka), Begundja, Bugunha (Mandyak), Furigna (Maninka), Begundja, Bugunha (Mankanya), Gundje (Pepel);
Ivory Coast: Ado Massa (Anyi);
Nigeria: Mmimi Ohia (Igbo), Kas Kaifi (Huasa), Akisan (Yoruba), Okó-Ajá (Lagos);
Senegal: Sézei (Balanta), Sikaral (Banyun), Buhal Bare, Buléo, Bu Lèv, Bu Riay (Diola, Fogny), Bananaru, Bananiaroli (Diola, Pointe), Boléo (Diola, Tentouck), Buleo (Diola-Flup), Boélénimbo, Kélen Baley (Fula-Pulaar), Sâbafim, Sâbéfin, Sâbifiri (Manding-Mandinka), Boguna (Mankanya), Mbélam, Yidi (Serer), Hasao, Sédada (Wolof);
Sierra Leone: Kɛmbɔŋyundɔŋ, Koi-Yondoe (Kissi), Finga, Finger, Fingers, Finger-Root (Krio), Hondowa (Loko), Bilui, Ndɔgbɔ-Jɛlɛ, Ndɔgbɔ-Jɛlɛ-J Gbɔu, Jɛlɛ:, Gbɔu:, Nɛgbɔta:, Njɔpɔ-Jɛgbɔu Njɔpɔ, Jɛ: (Mende), Nɛgbɔta (Mende-Kpa), Kan-Yaduŋga-Na (Susu-Dyalonke), An-Lanɛ (Temne);
The Gambia: Bu Lev, Bu Riay, Fuléafo (Diola);
Togo: Liasa, Padiwin (Yoruba-Ife), Pereng (Difale, Kabre).

Origin/Distribution

The species is native to tropical West and Central Africa. It has been introduced to other parts of Africa and elsewhere in the tropics as a curiosity plant because of its finger-like, ornamental fruits.

Agroecology

The species is found in the lowland wet and dry tropical forests, savanna and coastal scrublands in West and Central Africa extending eastwards and southwards into Zaïre.

Edible Plant Parts and Uses

The aril around the seed is sweet and edible. The fruit is widely eaten in its native range.

Botany

A scandent or scrambling shrub to about 3–4 m with dark brown stem, twining branchlets rusty-pubescent becoming glabrous. Leaves, alternate, shortly petiolate, coriaceous, elliptical oblong or oval, apex obtusely pointed, base obtuse to slightly cordate, with minute stellate hairs becoming glabrous with age and midrib impressed above with minute hairs on midrib in lower surface, 9–13 pairs of laterals, entire with slightly undulating margin, shiny green above dull green below (Plates 1 and 2). Flower, 2.20–2.5 cm across, greenish-brown, bisexual, in 2–5 flowered cluster usually axillary with a more or less hemispherical receptacle. Calyx brownish green, tomentose, three sepals valvate, connate near the base, Petals greenish-beige, usually six, distinct, the outer three often larger and differentiated from the inner, slightly imbricate. Stamens usually numerous, packed into a ball-like or disk-like configuration, distinct, carpels rusty- tomentose, numerous. Fruit-carpels 20 or fewer, rusty-pubescent, oblong, terete, on stipes, sometimes rough, with irregularly spaced, blunt projecting points. Fruits occur in finger-like clusters and are yellow when ripe with sweet aril enclosing the seeds. (Plates 1 and 2). Seeds few, more or less compressed, shining and pale brown.

Plate 2 Ripe fruits whole and halved

Plate 1 Finger-like clusters of fruit and alternate leaves on slender branchlets

Nutritive/Medicinal Properties

No published data is available on the nutritive value of the fruits.

Sesquiterpenes and aromatic compounds were reported as the predominant constituents of the leaf and root bark oils, respectively (Oguntimein et al. 1989). The major constituent of *Uvaria chamae* leaf oil was δ-cadinene while thymoquinoldimethyl ether and benzyl benzoate were the major components of the root oil.

Two benzyldihydrochalcones and the known benzyl benzoate were identified in the roots of *Uvaria chamae* (Okorie 1977). A monobenzylated monoterpene, chamanen, and the dimethyl ether of thymoquinol were isolated from the root bark of *Uvaria chamae* (Lasswell and Hufford 1977a). From root bark of *U. chamae*, 5-metyl ether derivatives of chamanetin and dichamanentin were isolated (El-Sohly et al. 1979).

A novel acetogenin, cis-bullatencin together with eight known mono-THF acetogenins – bullatencin, annotemoyin-1, solamin, uvariamicin-I, -II, -III, cis-reticulatacin and cis-uvariamicin-I were isolated from a cyclohexane extract of *Uvaria chamae* roots (Fall et al. 2002).

Anti-inflammatory and Oxytocic Activities

The extracts of *Uvaria chamae*, *Clerodendron splendens* and aspirin were found to inhibit carrageenan-induced paw oedema on albino rats and mice with a strong activity in aspirin having (80.43%) inhibition while *U. chamae* and *Clerodendron splendens* have 69.57% and 47.83% inhibition respectively (Okwu and Iroabuchi 2009). The ethanolic extracts of the plants showed uterine contraction activity. *U. chamae* had activity comparable to that of oxytocins. This may be due to its high flavonoids and phenolic content. The plant extract appeared to protect the vascular system and strengthen the tiny capillaries that carry oxygen and essential nutrients to all cells. *U. chamae* was found to contain bioactive components comprising flavonoids 5.70 mg/100 g, alkaloids 0.81 mg/100 g, tannins 0.40 mg/100 g, saponins 0.38 mg/100 g and phenols 0.10 mg/100 g.

Antimicrobial Activity

The antimicrobial activities of a number of cytotoxic C-benzylated flavonoids from *Uvaria chamae* were found to have antimicrobial activity (Hufford and Lasswell 1978). The minimum inhibitory concentration values of these flavonoids and some of their derivatives against *Straphylococcus aureus, Bacillus subtilis,* and *Mycobacterium smegmatis* compared favorably with those of streptomycin sulphate. Uvarinol, a tribenzylated flavanone from stem bark of *U. chamae*, showed significant antibacterial activity against *Staphyloccocus aureus, Bacillus subtilis* and *Mycobaterium smegmatis* (Hufford et al. 1979).

The ethyl acetate-soluble fraction of the methanol extract of *Uvaria chamae* stem bark was found to have antimicrobial activity (Ebi et al. 1999). Several sub-fractions, containing glycosides (8, 11–15) and tannins (18), exhibited activity against a number of microorganisms, being in some cases more active than penicillin G and chloramphenicol. Various extracts of *U. Chamae* were found to exhibit in-vitro antibacterial activity (Ogbulie et al. 2007). Cold and hot water extracts moderately inhibited the growth of *Staphylococcus aureus* and *Streptococcus pyogenes* with zone of inhibition of between 9 and 15 mm. Cold or hot ethanol extracts strongly inhibited the growth of *Staphylococcus aureus, Streptococcus pyogenes, Escherichia coli* and *Salmonella typhi* with inhibition zones of 13–21 mm. While *Pseudomonas aeruginosa* was slightly inhibited by the ethanol extract. Soxhlet extracts of *U. chamae,* strongly inhibited the growth of all four bacteria with zones of inhibition ranging from 13 to 22 mm. *Vibrio* spp. were not inhibited by the cold and hot extract as well as the soxhlet extracts tested.

Root bark and leaves of *Uvaria chamae* used traditionally in the treatment of diarrhea, cough and urinary tract infections (UTI), were found to have in-vitro antibacterial activity (Oguekea et al. 2007). The ethanol extracts of *U. chamae* root bark and leaves were more inhibitory on the bacterial isolates than the water extracts. The minimum inhibitory concentration (MIC) of the ethanol root bark extracts on *Escherichia coli, Staphylococcus aureus, Pseudomonas aeruginosa* and *Salmonella typhi* were 72.44 mg/ml, 50.19 mg/ml, 45.71 mg/ml and 131.83 mg/ml respectively while the MIC of the ethanol leaves extracts were 61.66 mg/ml, 79.43 mg/ml, 125.89 mg/ml and 105.93 mg/ml, respectively. The MIC of the water extracts ranged from 120.23 mg/ml, observed on *Staphylococcus aureus* to 131.83 mg/ml, observed on *E. coli*. The observed antibacterial effects were believed to be due to the presence of alkaloids, tannins and flavonoids identified in the extracts. The results supported their use in traditional medicine especially in the treatment of UTI and diarrhea, and are of significance in the health care delivery system.

Anticancer Activity

Two cytotoxic C-benzylflavanones uvaretin and isouvaretin were isolated from the stem bark of *Uvaria chamae* (Hufford and Lasswell 1976). Ethanolic extract of *Uvaria chamae* demonstrated in-vivo activity against P-388 lymphocytic leukemia (PS) in mouse and against cells derived from human carcinoma of the nasopharynx (KB) (Lasswell and Hufford 1977b). Fractionation of the extracts yielded the known flavanones pinocembrin and pinostrobin, the C-benzylated flavanones chamanetin, isochamenatin and dichasmanetin and the C-benzylated dihydrochalcones uvaretin, isouvaretin and diuvaretin. Uvarinol, a tribenzylated flavanone from stem bark of *U. chamae*, showed cytoxicity against KB cells at ED50 5.9 ug/ml and against PS cell cultures at 9.7 ug/ml (Hufford et al. 1979). Benzylisoquinoline alkaloids (+)-armepavine (1) and racem. O,O-dimethylcoclaurine (2) together with aporphines nornantenine (3), nantenine (4) and corydine (7) were isolated from *U. chamae* leaves (Philipov et al. 2000). The alkaloids exhibited cytotoxic activity against L 929 (murine aneuploid fibrosarcoma) transformed cells. The highest activity was shown by 1, 3, and 5. At a concentration corresponding to their IC_{50} against L929 cells, they were nontoxic against mouse thymocytes.

A cytotoxic acetogenin, chamuvarinin, containing a tetrahydropyran ring with an adjacent bis-tetrahydrofuran ring, was isolated from the roots of *Uvaria chamae*, together with the previously reported acetogenins squamocin, desacetyluvaricin, and neoannonin (Fall et al. 2004). Chamuvarinin showed significant cytotoxicity toward the epidermoid carcinoma KB3-1 cell line ($IC_{50} = 8 \times 10^{-10}$ M). A novel acetogenin, joolanin, along with eight known acetogenins, squamocin, desacetyluvaricin, chamuvarinin, tripoxyrollin, diepoxyrollin, dieporeticanin-1, dieporeticanin-2 and dieporeticenin were isolated from the seeds of *Uvaria chamae* (Fall et al. 2006). Joolanin, with the first adjacent bis-tetrahydrofuranic ring acetogenin bearing a ketone group at C-5, showed significant cytotoxicity towards the epidermoid carcinoma KB3-1 cell line ($IC_{50} = 0.4$ nM).

Mutagenic Effect

Chamuvaritin, a dihydrobenzylchalcone isolated from *Uvaria chamae* was found to be mutagenic in *Salmonella typhimurium* tester strains TA94-98 and TA100-1535 and required activation by the hepatic S-9 microsomal enzyme preparation (Uwaifo et al. 1979).

Antihepatotoxic Activity

The root bark extract of *U. chamae* was found to have antihepatotoxic and trypanocidal activities (Madubunyi et al. 1996). The LD 50 of the ethanol extract of in mice at 24 h was 166 mg/kg (i.p.). Intraperitoneal injection of the ethanol extract in mice showed no significant effect on pentobarbitone-induced hypnosis. Oral administration of the extract (60 mg/kg) significantly reduced pentobarbitone-induced sleep in CCl_4-poisoned rats. The augmentation of serum glutamate oxaloacetate transaminase, glutamate pyruvate transaminase, alkaline phosphatase and urea induced by CCl_4 intoxication in rats was also significantly reduced by the ethanol extract. In a subsequent paper, oral admi-nistration of the methanol extract of *U. chamae* roots (60 mg/kg) significantly reduced pentobarbitone-induced sleep in rats poisoned with acetaminophen (Madubunyi 2010). Intraperitoneal injection of the methanol extract into rats elicited no significant effect on pentobarbitone-induced hypnosis. The increase of serum glutamate oxaloacetate transaminase, glutamate pyruvate transaminase, alkaline phosphatase, and urea induced by paracetamol intoxication in rats was also significantly attenuated by the methanol extract. The methanol extract did not influence the concentration of microsomal proteins in the serum. This in-vivo efficacy was substantiated by significant hepatoprotection on acetaminophen (AA)-induced hepatotoxicity in isolated rat hepatocytes. The methanol extract, at a dose of 1 mg/ml, reduced appreciably the leakage of lactate dehydrogenase in primary cultured rat hepatocytes and had a significant effect on lipid peroxidation. The AA-induced elevation of the lipid peroxidation in rats was significantly attenuated in the presence of the root bark methanol extract. A protection of 56% against the tert-butyl hydroperoxide-induced lipid peroxidation in rats was obtained by pretreatment with the methanol extract. The methanol extract also showed a significant antioxidant effect.

Antimalarial Activity

The extract of *Uvaria chamae*, when given orally at 300–900 mg/kg/day to mice, exhibited significant antimalarial activity against both early and established infections by *Plasmodium berghei berghei* (Okokon et al. 2006). When established infections were treated, the mean survival time of the mice observed with this extract at 900 mg/kg/day was similar to that seen with the positive control, chloroquine at 5 mg/kg/day. *U. chamae* is used traditionally as malarial remedy in southern Nigeria.

Antiparasitic Activity

Uvaria chamae root bark ethanol extract exhibited a significant trypanocidal effect which was comparable to that of diminazine aceturate (Madubunyi et al. 1996). Reduction of existing parasitaemia in mice experimentally infected with *Trypanasoma brucei brucei* was dose-dependent. Addae- Kyereme et al. (2001) found *Uvaria chamae* extract to have significant antiplasmodial activity against *Plasmodium falciparum* (strain K1). The extract of *U. chamae* was toxic to brine shrimps and could be a cause of concern. Extracts of stem barks and roots of *Uvaria chamae* showed antiparasitic activity against *Trypanosomia brucei* (Fall et al. 2003). Thirteen acetogenins, from the roots of *Uvaria chamae* and *Annona senegalensis*, were identified.

Traditional Medicinal Uses

In folk medicine, extracts of the roots, barks and leaves are used to treat gastroenteritis, vomiting, diarrhoea, dysentery, wounds, sore throats, inflamed gums and a number of other ailments (Dalziel 1955; Irvine 1961; Bouquet and Debray 1974; Kerharo and Adam 1974; Burkill 1985; Iwu 1993; Okwu and Iroabuchi 2009). The juice from the roots, stems or leaves is commonly applied to wounds and sores and is said to promote rapid healing. A leaf-infusion is used as an eyewash and a leaf-decoction as a febrifuge. The plant is used to make a pomade in Ghana. In Senegal, leaves and roots are macerated for internal use as a cough mixture, and mixed with those of *Annona senegalensis*, dried and pulverized are considered strong medicine for renal and costal pain.

In Nigeria the root is used as a purgative; the root bark is used for respiratory catarrh and the root extract is used in phytomedicine for the treatment of piles, menorrhagia, epistaxis, haematuria and haemalysis. A root-decoction is also used as stomachic and vermifuge, and is used as a lotion; sap from the root and stem is applied to wounds; and the root is made into a drink and a body-wash for oedematous conditions. In Ghana, the root is used in a stock hospital prescription for dysentery. Its attributes astringent and styptic, and it is used in native medicine as a specific for piles, useful also for menorrhagia, epistaxis, haematuria, haematemesis and haemoptysis. Severe abdominal pain is treated by a root-infusion with native pepper in gin, and the root with Guinea grains is used in application to the fontanelle for cerebral diseases. In Sierra Leone the root is credited with having purgative and febrifugal properties; the roots are also used for healing sores and a concoction called *n'taba* is reputed to cure infantile rickets. The root or the root-bark is boiled with spices and the decoction drunk for fevers. Amongst the Fula peoples of Senegal, the root has a reputation as the 'Medicine of Riches', and is taken for conditions of lassitude and senescence. It is also considered to be a 'woman's medicine' used for amenorrhoea and to prevent miscarriage. In the Ivory Coast the root is added in a treatment for a form of jaundice. In Togo a root-decoction is given for the pains of childbirth.

Other Uses

The plant is regarded more for its medicinal attributes and uses than as an edible fruit tree.

Comments

Selected References

Addae-Kyereme J, Croft SL, Kendrick H, Wright CW (2001) Antiplasmodial activities of some Ghanaian plants traditionally used for fever/malaria treatment and of some alkaloids isolated from *Pleiocarpa mutica*; in vivo antimalarial activity of pleiocarpine. J Ethnopharmacol 76(1):99–103

Arbonnier M (2004) Trees, shrubs and lianas of West African Dry Zones. CIRAD, Paris; Margraf, Weikersheim; MNHN, Paris, 574 pp

Bongers F, Parren MPE, Traore D (eds) (2005) Forest climbing plants of West Africa: diversity, ecology and management. CAB International, Wallingford/Cambridge, 271 pp

Bouquet A, Debray M (1974) Plantes Médicinales de la Côte d'Ivoire, vol 32. Mémoires O.R.S.T.O.M, Paris, 231 pp

Burkill HM (1985) The useful plants of West Tropical Africa, vol 1. Families A-D. Royal Bot, Gardens Kew, 960 pp

Dalziel JM (1955) The useful plants of West Tropical Africa (Reprint of 1937 ed.) Crown Agents for Overseas Governments and Administrations, London, 612 pp

Ebi GC, Ifeanacho CJ, Kamalu TN (1999) Antimicrobial properties of *Uvaria chamae* stem bark. Fitoterapia 70(6):621–624

El-Sohly HN, Lasswell WL Jr, Hufford CD (1979) Two new C-benzylated flavanones from *Uvaria chamae* and 13 C NMR analysis of flavanone methyl ethers. J Nat Prod 42(3):264–270

Fall D, Gleye C, Franck X, Laurens A, Hocquemiller R (2002) Cis-bullatencin, a linear acetogenin from roots of *Uvaria chamae*. Nat Prod Lett 16(5):315–321

Fall D, Badiane M, Ba D, Loiseau P, Bories C, Gleye C, Laurens A, Hocquemiller R (2003) Antiparasitic effect of Senegalese Annonaceae used in traditional medicine. Dakar Méd 48(2):112–116 (In French)

Fall D, Duval RA, Gleye C, Laurens A, Hocquemiller R (2004) Chamuvarinin, an acetogenin bearing a tetrahydropyran ring from the roots of *Uvaria chamae*. J Nat Prod 67(6):1041–1043

Fall D, Pimentel L, Champy P, Gleye C, Laurens A, Hocquemiller R (2006) A new adjacent bis-tetrahydrofuran annonaceous acetogenin from the seeds of *Uvaria chamae*. Planta Med 72(10):938–940

Hufford CD, Lasswell WL Jr (1976) Uvaretin and isouvaretin. Two novel cytotoxic c-benzylflavanones from *Uvaria chamae* L. J Org Chem 41(7):1297–1298

Hufford CD, Lasswell WL Jr (1978) Antimicrobial activities of constituents of *Uvaria chamae*. Lloydia 41(2):156–160

Hufford CD, Lasswell WL Jr, Hirotsu K, Clardy J (1979) Uvarinol: a novel cytotoxic tribenzylated flavanone from *Uvaria chamae*. J Org Chem 44(25):4709–4710

Irvine FR (1961) Woody plants of Ghana: with special reference to their uses. Oxford University Press, London, 868 pp

Iwu MM (1993) Handbook of African medicinal plants. CRC Press, Boca Raton, 464 pp

Kerharo J, Adam JG (1974) La Pharmacopée Sénégalaise Traditionnelle. Plantes Médicinales et Toxiques. Editions Vigot Frères, Paris, 1011 pp (In French)

Lasswell WL Jr, Hufford CD (1977a) Aromatic constituents from *Uvaria chamae*. Phytochem 16(9):1439–1441

Lasswell WL Jr, Hufford CD (1977b) Cytotoxic C-benzylated flavonoids from *Uvaria chamae*. J Org Chem 42(8):1295–1302

Madubunyi II (2010) Hepatoprotective activity of *Uvaria chamae* root bark methanol extract against acetaminophen-induced liver lesions in rats. Comp Clin Pathol, In press

Madubunyi II, Njoko CJ, Ibeh EO, Chime AB (1996) Antihepatotoxic and trypanocidal effects of the root bark extract of *Uvaria chamae*. Pharm Biol 34(1):34–40

Ogbulie JN, Ogueke CC, Nwanebu FC (2007) Antibacterial properties of *Uvaria chamae*, *Congronema latifolium*, *Garcinia kola*, *Vemonia amygdalina* and *Aframomium meleguetae*. Afr J Biotechnol 6(13):1549–1553

Oguekea CC, Ogbulie JN, Anyanwu BN (2007) The effects of ethanolic and boiling water extracts of root barks and leaves of *Uvaria chamae* on some hospital isolates. J Am Sci 3(3):68–73

Oguntimein B, Ekundayo O, Laakso I, Hiltunen R (1989) Volatile constituents of *Uvaria chamae* leaves and root bark. Planta Med 55(3):312–313

Okokon JE, Ita BN, Udokpoh AE (2006) The in-vivo antimalarial activities of *Uvaria chamae* and *Hippocratea africana*. Ann Trop Med Parasitol 100(7):585–590

Okorie DA (1977) New benzyldihydrochalcones from *Uvaria chamae*. Phytochemistry 16(10):1591–1594

Okwu DE, Iroabuchi F (2009) Phytochemical composition and biological activities of *Uvaria chamae* and *Clerodendoron splendens*. E J Chem 6(2):553–560

Oliver-Bever B (1986) Medicinal plants in Tropical West Africa. Cambridge University Press, Cambridge, 375 pp

Philipov S, Ivanovska N, Istatkova R, Velikova M, Tuleva P (2000) Phytochemical study and cytotoxic activity of alkaloids from *Uvaria chamae* P Beauv. Pharmazie 55(9):688–689

Uwaifo AO, Okorie DA, Bababunmi EA (1979) Mutagenicity of chamuvaritin: a benzyldihydrochalcone isolated from a medicinal plant. Cancer Lett 8(1):87–92

Carissa macrocarpa

Scientific Name

Carissa macrocarpa (Ecklon) A. DC.

Synonyms

Arduina grandiflora E. Mey. nom. illeg., *Arduina macrocarpa* Eckl., *Carissa africana* A.DC., *Carissa carandas* Lour. sensu auct., *Carissa grandiflora* (E. Mey.) A.DC., *Carissa praetermissa* Kupicha, *Jasminonerium africanum* (A.DC.) Kuntze, *Jasminonerium grandiflorum* (E. Mey.) Kuntze nom. illeg., *Jasminonerium macrocarpum* (Eckl.) Kuntze.

Family

Apocynaceae

Common/English Names

Carissa, Big Num Num, Large Num Num, Natal Plum

Vernacular Names

Chinese: Da Hua Jia Hu Ci
French: Carisse;
German: Karanda Wachsbaum, Karandang, Natal-Pflaume;
India: Karaunda (Hindu), Avighna, Karamarda (Sanskrit);
Portuguese: Carandeira;
South Africa: Amatungula (Zulu), Grootnoem-Noem (Afrikaans), Big Num Num, Large Num Num;
Spanish: Ameixa-Do-Cabo Caranda, Cereza De Natal.

Origin/Distribution

The carissa is native to the coastal region of Natal, South Africa, and is cultivated far inland in the Transvaal. Cultivated in tropics and subtropics world-wide.

Agroecology

A subtropical to near-tropical species that is tolerant to light frost, enduring temperatures as low as −3.89°C when well-established. It grows best in full sun. The plant has moderate drought tolerance and high tolerance to soil salinity and salt spray. It cannot stand water-logging. It is adaptable to a wide range of soil types red clay, sandy loam, and light sand – thrives in dry, rocky terrain in Hawaii; in red clay or sandy loam in California, and in sandy or alkaline soils in Florida. It is somewhat drought-resistant.

Edible Plant Parts and Uses

The fruit flavour is subacid and agreeable and the fruit is much prized in Natal for preserving. Ripe fruit is eaten raw out of hand, enjoyed in fruit salads, adding to gelatins and using as topping for cakes, puddings and ice cream; or stew with sugar. Carissa can be cooked to a sauce or used in pies and tarts. Fruit is also suitable for sweet pickles, jams, other preserves and syrup. Slightly under-ripe fruits, or a combination of ripe and unripe to enhance the colour are made into jellies. Fruity vinegars are produced from overripe fruit.

Botany

A vigorous, woody, much-branched, evergreen shrubs or small trees to 5 m tall with profuse white latex. Branches are armed with stout, single or double-pronged spines 2–4 cm long (Plates 1–3).

Plate 1 Mature fruit, flower, leaves (*under surface*) and forked spines

Plate 2 Close up of Ripe fruit, flower, leaves (*upper surface*) and forked spines

Plate 3 Leaves with bi-forked spines

Leaves are opposite, broadly ovate, 2.5–7.5 × 2–5 cm, thick leathery, dark green, glossy, glabrous, base rounded to obtuse, apex mucronate, lateral veins obscure (Plates 1–3). Flowers are fragrant, pentamerous, borne on short, 2–3 mm pedicels in 1–3 flowered, terminal cymes. Sepals are very narrowly ovate, 3–6 mm. Corolla is white, 1.1–1.8 cm tubular, pubescent inside with oblong lobes, 0.9–2.4 cm, overlapping to left (Plates 1 and 2). Stamens included in top of tube, filaments short, anthers free from stigma, Ovary glabrous 2 celled, each cell with 4-ovules. Berries, ovoid, 2–5 cm, bright red (Plates 1 and 2) to violet when ripe, containing 8–16 small, thin, flat, brown seeds.

Nutritive/Medicinal Properties

Analyses made in the Philippines showed the following values: calories, 270/lb (594/kg); moisture, 78.45%; protein, 0.56%; fat, 1.03%; sugar,

12.00%; fibre, 0.91%; ash, 0.43% (Morton 1987). According to another analysis made in Hawaii, the chief chemical constituents of the fruit were: total solids 21.55%, ash 0.43%% acids 1.19%, protein 0.56%, total sugars 12.00%, fat 1.03%, and fiber 0.91% (Popenoe 1974) .

Natal plum is an excellent source of vitamin C, containing somewhat more than in the average orange; one analysis found 53 mg per 100 g of whole fruit (Wehmeyer 1986).

Other Uses

The plant makes a good protective hedging and an ornamental shrub used in landscaping.

Comments

The white latex of the leaf and stem is poisonous.

Selected References

Backer CA, van den Brink RCB Jr (1965) Flora of Java (spermatophytes only), vol 2. Wolters-Noordhoff, Groningen, 641 pp

Brown WH (1951–1957). Useful plants of the Philippines. Reprint of the 1941–1943 edition, 3 vols. Technical Bulletin 10. Department of Agriculture and Natural Resources, Bureau of Printing, Manila, The Philippines. Vol 1 (1951), 590 pp, vol 2 (1954), 513 pp, vol 3 (1957), 507 pp

Facciola S (1990) Cornucopia. A source book of edible plants. Kampong Publication, Vista, 677 pp

Govaerts R (2008) World Checklist of Apocyanaceae. The Board of Trustees of the Royal Botanic Gardens, Kew. Published on the Internet. http://www.kew.org/wcsp/

Jansen PCM, Jukema J, Oyen LPA, van Lingen TG (1992) Carissa macrocarpa (Ecklon) A. DC. In: Verheij EWM, Coronel RE (eds) Plant resources of South-East Asia. No. 2: Edible fruits and nuts. Prosea Foundation, Bogor, p 324

Li B, Leeuwenberg AJM, Middleton DJ (1995) Apocynaceae A. L. Jussieu. In: Wu ZY, Raven PH (eds.) Flora of China, vol 16, Gentianaceae through Boraginaceae. Science Press/Missouri Botanical Garden Press, Beijing/St. Louis, 479 pp

Morton JF (1987) Carissa. In: Fruits of warm climates. Julia F. Morton, Miami, pp 420–422

National Research Council (2008) Lost crops of Africa, vol 111. Fruits. The National Academies Press, Washington, DC, 380 pp

Popenoe W (1974) Manual of tropical and subtropical fruits. Facsimile of the 1920 edition. Hafner Press, New York, USA

Purseglove JW (1968) Tropical crops: dicotyledons, vol 1 & 2. Longman, London, 719 pp

Uphof JCT (1968) Dictionary of economic plants, 2nd edn. Cramer, Lehre, 591 pp

Wehmeyer AS (1986) Edible wild plants of Southern Africa: data on nutrient contents of over 300 species. Council for Scientific and Industrial Research, Pretoria

Carissa spinarum

Scientific Name

Carissa spinarum L.

Synonyms

Arduina brownii K. Schum., *Arduina campenonii* Drake, *Arduina edulis* (Forssk.) Spreng., *Arduina inermis* (Vahl) K. Schum., *Arduina laxiflora* (Benth.) K.Schum., *Arduina xylopicron* (Thouars) Baill., *Antura edulis* Forssk., *Antura hadiensis* J.F. Gmel. nom. illeg., *Cabucala brachyantha* Pichon, *Carandas edulis* (Forssk.) Hiern, *Carissa axillaris* Roxb., *Carissa brownii* F.Muell. nom. illeg., *Carissa brownii* var. *angustifolia* Kempe, *Carissa brownii* var. *ovata* (R.Br.) Maiden & Betche nom. inval., *Carissa campenonii* (Drake) Palacky, *Carissa candolleana* Jaub. & Spach, *Carissa carandas* Lodd. sensu auct., *Carissa carandas* var. *paucinervia* (A.DC.) Bedd., *Carissa cochinchinensis* Pierre ex Pit., *Carissa comorensis* (Pichon) Markgr., *Carissa congesta* Wight, *Carissa carandas* var. *congesta* (Wight) Bedd., *Carissa coriacea* Wall. ex G. Don, *Carissa cornifolia* Jaub. & Spach, *Carissa dalzellii* Bedd., *Carissa densiflora* Baker, *Carissa densiflora* var. *microphylla* Danguy ex Lecomte, *Carissa diffusa* Roxb., *Carissa dulcis* Schumach. & Thonn., *Carissa edulis* (Forssk.) Vahl, *Carissa edulis* f. *nummularis* (Pichon) Markgr., *Carissa edulis* f. *pubescens* (A.DC.) Pichon, *Carissa edulis* subsp. *madagascariensis* (Thouars) Pichon, *Carissa edulis* var. *ambungana* Pichon, *Carissa edulis* var. *comorensis* Pichon, *Carissa edulis* subsp. *continentalis* Pichon, *Carissa edulis* var. *densiflora* (Baker) Pichon, *Carissa edulis* var. *horrida* (Pichon) Markgr., *Carissa edulis* var. *lucubea* Pichon, *Carissa edulis* var. *major* Stapf, *Carissa edulis* var. *microphylla* (Danguy ex Lecomte) Pichon, *Carissa edulis* var. *nummularis* Pichon, *Carissa edulis* var. *revoluta* (Scott-Elliot) Markgr., *Carissa edulis* var. *sechellensis* (Baker) Pichon, *Carissa edulis* var. *septentrionalis* Pichon, *Carissa edulis* var. *subtrinervia* Pichon, *Carissa edulis* var. *tomentosa* (A.Rich.) Stapf nom. illeg., *Carissa gangetica* Stapf ex Gamble, *Carissa hirsuta* Roth, *Carissa horrida* Pichon, *Carissa inermis* Vahl, *Carissa lanceolata* Dalzell nom. illeg., *Carissa lanceolata* R.Br., *Carissa laotica* Pit., *Carissa laotica* var. *ferruginea* Kerr, *Carissa laxiflora* Benth., *Carissa macrophylla* Wall. ex G. Don, *Carissa madagascariensis* Thouars, *Carissa mitis* Heynh. ex A.DC., *Carissa obovata* Markgr., *Carissa oleoides* Markgr., *Carissa opaca* Stapf ex Haines, *Carissa ovata* R.Br., *Carissa ovata* var. *pubescens* F.M. Bailey, *Carissa ovata* var. *stolonifera* F.M. Bailey, *Carissa papuana* Markgr., *Carissa paucinervia* A.DC., *Carissa pilosa* Schinz nom. illeg., *Carissa pubescens* A.DC., *Carissa revoluta* Scott-Elliot, *Carissa richardiana* Jaub. & Spach, *Carissa scabra* R.Br., *Carissa sechellensis* Baker, *Carissa septentrionalis* (Pichon) Markgr., *Carissa spinarum* Lodd. ex A. DC. nom. illeg., *Carissa stolonifera* (F.M. Bailey) F.M. Bailey ex Perrot & Vogt nom. inval., *Carissa suavissima* Bedd. ex Hook.f., *Carissa tomentosa* A. Rich. nom. illeg.,

Carissa velutina Domin, *Carissa villosa* Roxb., *Carissa xylopicron* Thouars, *Carissa yunnanensis* Tsiang & P.T. Li, *Damnacanthus esquirolii* H. Lév., *Jasminonerium densiflorum* (Baker) Kuntze, *Jasminonerium dulce* (Schumach. & Thonn.) Kuntze, *Jasminonerium edule* (Forssk.) Kuntze, *Jasminonerium inerme* (Vahl) Kuntze, *Jasminonerium laxiflorum* (Benth.) Kuntze, *Jasminonerium madagascariense* (Thouars) Kuntze, *Jasminonerium ovatum* (R.Br.) Kuntze, *Jasminonerium pubescens* (A.DC.) Kuntze, *Jasminonerium sechellense* (Baker) Kuntze, *Jasminonerium suavissimum* (Bedd. ex Hook.f.) Kuntze, *Jasminonerium tomentosum* (A. Rich.) Kuntze, *Jasminonerium xylopicron* (Thouars) Kuntze.

Family

Apocynaceae

Common/English Names

Bengal Currant, Carandas Plum, Carissa, Christ' Thorns, Karanda

Vernacular Names

Burmese: Hkan Ping;
Chinese: Ci Huang Guo;
German: Karanda Wachsbaum, Karandang;
India: Bainchi, Karamcha, Karenja, Kurumia (Bengali), Garinga, Gotho, Kantakregi, Karaunda, Karaunta, Karonda, Karondi, Karonti, Karrona, Karumcha, Karunda, Korada, Timukhia, Timukha (Hindi), Dodda Kalaa, Dodda Kavali, Doddakavale, Doddakavali, Garacha, Garchinakai, Garaja, Garji, Harikalivi, Heggarichige, Heggaricige, Heggarjige, Hirikalavi, Hirikavali, Kalaagida, Kalavige, Kalivi, Kalla, Kamarika, Kamrdepuli, Karanda, Karande, Karande Pli, Karavadi, Karekai, Karekayi, Kareki, Karevati, Kari, Karice, Kariche, Karichi, Karichina Kaayi, Karicinakayi, Karinda, Karndepuli, Kauligida, Kavale, Kavali, Kavali Gida, Kawliballi, Korindamalekalaavu (Kannada), Kalavu, Karakka, Karanta, Karekai, Keelay, Kilai, Klavu, Kulay, Perumklavu (Malayalam), Boranda, Boronda, Haradundi, Karanda, Karandi, Karaunda, Karavanda, Karavanad Karavandi, Karwand-Karanja (Marathi), Sushena (Oriya), Avighna, Avighnah, Avinga, Bahudala, Bolekarambuka, Dimdima, Dridhakantaka, Guchhi, Jalipushpa, Kanachuka, Kantaki, Karamarda, Karamardaka, Karamla, Karamlaka, Karinkara, Krishna-Pakphula, Krishnapakaphala, Krishnaphala, Krisnapakaphala, Kshiraphala, Kshiri, Ksiraphala, Pakakrishna, Pakaphala, Panimarda, Phalakrishna, Supushpa, Susena, Vanalaya, Vanekshudra, Vasha (Sanskrit), Aintarikam, Aintarikamaram, Alarukam, Alarukamaram, Cenkala, Cirapalam, Cirukala, Kala, Kalakka, Kalakkay, Perumkla, Perungala, Kalaaha, Kalaka, Kalar, Kalarva, Karavintai, Kila, Kilakkai, Kilakki, Kilamaram,Kilatti, Kilay, Kirusnapakapalam, Kirusnapalai, Kiruttinapakapalam, Periyakala, Perukala, Perunikila,Perunkila, Perungila-Maram, Perungkala, Perunkala, Perunkala Ver, Yokatumacceti, Yokatumam (Tamil), Kalay, Kali, Kali-Kai, Kalikai, Kaliva, Kalive, Kalivi, Kalli, Kallia, Kalumi, Kaluva, Kavila, Oka, Okalive, Pedda Kalive, Peddakalavi, Peddakalive, Peddakalivi, Peddavaka, Peddavaaka Kaaya, Peddavakakaya, Vaka, Vakalive,Vakalivi, Vakalvi, Vakudu, Waaka, Waka, Wakay, Wyaka, Vaaka Chettu, Vaaklive, Yaakudu (Telugu);
Indonesia: Karandan, Karendang;
Malaysia: Berenda, Kerenda, Kerandang;
Philippines: Caramba, Caranda, Caraunda, Perunkila;
Pakistan: Gerna, Karanda, Kakranda;
Portuguese: Carandeira;
Spanish: Caranda;
Thai: Nam Phrom, Namdaeng (Bangkok), Manaao Ho (Southern Peninsula), Naam Khee Haet (Chiang Mai);
Timor: Senggaritan;
Vietnam: Cay Siro.

Origin/Distribution

The karanda is native and common throughout much of India, Burma and dry areas of Ceylon. The species was introduced into other countries in south east Asia – Malaysia, Indonesia, Cambodia, South Vietnam and the Philippines and to East Africa and the new world. In Java, it has naturalised and run wild.

Agroecology

Carissa spinarum is more cold-tolerant than *Carissa macrocarpa*. It grows from sea-level to 600 m but can be found up to an altitude of 1,800 m e.g. in the Himalayas. It occurs on a wide range of soil types. In India, it grows wild on the poorest and rockiest soils and is grown as a hedge plant in dry, sandy or rocky soils and in Florida on sand or limestone. It thrives best on deep, well-drained soil in full sun.

Plate 1 Oblong leaves with sharp forked thorns

Edible Plant Parts and Uses

The sweeter types may be eaten raw out-of-hand but the more acid ones are best stewed with plenty of sugar, they make a pleasant conserve. They are used also as seasoning with curry and in puddings, pies and tarts. Ripe fruit makes an excellent acid jelly for serving with meats or fish. Unripe fruit makes a good pickle, preserves and chutney. The fruit exudes much gummy latex when being cooked but the rich-red juice becomes clear and is much used in cold beverages and syrups. The syrup has been successfully utilized on a small scale by at least one soda-fountain operator in Florida. The ripe fruits can be processed also into wine.

Plate 2 Unripe broadly ovoid fruit

Botany

A much-branched, straggly, woody, climbing evergreen shrub or small tree growing to 5 m tall. The branches are armed with sharp thorns, simple or forked, up to 5 cm long, in pairs in the leaf axils (Plate 1). Leaves are opposite, broadly ovate to oblong, 3–7 × 1.5–4 cm, with broadly cuneate to rounded base, short apiculate apex and 8 pairs of lateral veins. Leaves are dark-green, leathery, glossy on the upper surface, lighter green and dull on the underside (Plate 1). Inflorescences are terminal cymes usually 3-flowered, with fragrant flowers on 1.5–2.5 cm long peduncles with minute bracteoles. Calyx is synsepalous, 5-partite, with very slender, pointed, and hairy segments with many basal glands inside. Corolla is synpetalous, 5-lobed, salverform, the lobes are oblong lanceolate, overlapping to the right, pubescent, and the corolla tube is cylindrical to 2 cm, dilated at the throat, pubescent, white or pale rose. Stamens have basifixed, introrse anthers with longitudinal dehiscence. Pistil comprise an ellipsoid 2-carpelled, syncarpous ovary with 2 locules with axile placentation and 2 ovules in each locule; filiform style and minutely bifid stigma. The fruit is a drupe, broadly ovoid or ellipsoid, 1.5–2.5 cm long, bluntly pointed, white to pinky-white and pink turning blackish or reddish-purple when ripe (Plates 2–5). It contains 2–4 small, flat, oblongoid, brown seeds embedded in a sour reddish-purple pulp.

Plate 3 Fruit at various stages of development

Plate 4 Ripe, purplish-black fruit

Plate 5 Black, glossy ripe fruit

Nutritive/Medicinal Properties

Analyses made in India and the Philippines showed the following values for the ripe karanda: calories, 338–342/lb (745–753/kg); moisture, 83.17–83.24%; protein, 0.39–0.66%; fat, 2.57–4.63%; carbohydrate, 0.51–0.94%; sugar, 7.35–11.58%; fiber, 0.62–1.81%; ash, 0.66–0.78%. Ascorbic acid content had been reported as 9–11 mg per 100 g (Morton 1987).

One hundred and fifty compounds were identified in the volatile aroma concentrate of karanda fruit, of which isoamyl alcohol, isobutanol and [β]-caryophyllene were found to be the major constituents (Pino et al. 2004).

Karanda fruit was also found to contain anthocyanin, identified as cyanidin 3 –rhamnoglucoside that has been reported to have potential as a natural colouring agent for products requiring milder processing treatment and low temperature storage (Iyer and Dubash 1993).

Four pentacyclic triterpenoids including one new constituent carissin (3 β-hydroxy-27- E -feruloyloxyurs-12-en-28-oic acid) and two hitherto unreported compounds 2 and 3 were isolated from *C. carandas* leaves (Siddiqui et al. 2003). Two new compounds, the sesquiterpene glucoside carandoside and (6 S,7R,8R)-7a-[(β-glucopyranosyl)oxy] lyoniresinol, were isolated from the stem of *Carissa carandas*, together with three known lignans (Wangteeraprasert and Likhitwitayawuid 2009). Pal et al. (1975) isolated a lignan, carinol from *C. congesta* roots. The following were also isolated from the roots: a mixture of sesquiterpenes, namely carissone and carindone as a novel type of C31 terpenoid (Joshi and Boyce 1957; Rastogi et al. 1966, 1967; Singh and Rastogi 1972). Ursolic acid and carissic acid 1b were isolated from *Carissa* fruit and leaves (Naim et al. 1988).

Antioxidant Activity

Carissa carandas fruit was found to have an antioxidant IC_{50} value of 62.97 μg/ml as evaluated using 1,1 diphenyl-2-picrylhydrazyl (DPPH) radical scavenging but with little potential use in hyperuricemia due to poor xanthine oxidase inhibition (Ahmad et al. 2006). Studies conducted in Malaysia reported that the chloroform and methanol extracts of the unripe Karanda fruits showed strong antioxidant activities which was much higher than BHT, a commercial antioxidant (Sulaiman et al. 2008a, c). Flavonoid compounds from the half-ripe fruit juice were isolated and characterized. The major flavonoid compound in the extract was identified as apigenin 6-C-rhamnosil-7-O-rhamnoside. The minor components found in the extract were pelargonidin 3-O-glucoside, chrysoeriol 7-O-glycoside and quercetin 3-O-methy-7-O-glucoside. The roots of *C. carandas* was also found to possess antioxidant property and could serve as free radical inhibitors or scavengers (Bhaskar and Balakrishnan 2009b). In the DPPH radical scavenging assay the IC_{50} value for the ethanol root extract of *C. carandas* was 178.84 µg/ml. The aqueous extract exhibited 50% scavenging activity at 208.07 µg/ml on H_2O_2 radical. The reducing power of the extracts increased dose dependently. All the extracts exhibited significant antioxidant activity.

Anticancer and Antiviral Activities

Sulaiman et al. (2008b) reported that *Carissa carandas* extracts exhibited anticancer effects against the human ovarian carcinoma, Caov-3 and the lung cancer cells, NCI. Chloroform extract from leaves showed good anticancer activity against the Caov-3 with the EC_{50} value of 7.702 ug/ml while the n-hexane extract of the unripe fruit exhibit appreciable activity towards the lung cancer cells, NCI with the EC_{50} value of 2.942 ug/ml. In Nepal, scientists reported that *Carrisa carandas* extracts exhibited the most impressive antiviral activity when tested against three mammalian viruses: herpes simplex virus, Sindbis virus and poliovirus (Taylor et al. 1996).

Hepatoprotective Activity

Studies conducted in India showed that oral pretreatment with ethanolic extract of the roots of *C. carandas* (at 100, 200 and 400 mg/kg, p.o.) exhibited significant hepatoprotective activity against CCl_4 and paracetamol induced hepatotoxicity(Hegde and Joshi 2009). It decreased the activities of serum marker enzymes, bilirubin and lipid peroxidation, and significantly increased the levels of uric acid, glutathione, super oxide dismutase, catalase and protein in a dose dependent manner. This was also was confirmed by the decrease in the total weight of the liver and histopathological examination. Data also showed that ethanolic root extract possessed strong antioxidant activity, which could have contributed to the promising hepatoprotective activities the root extract. These findings supported the traditional belief on hepatoprotective effect of the roots of *C. carandas*. Ethanolic root extract of C. *carandas* was also found to exhibit anticonvulsant effects via non-specific mechanisms since it reduced the duration of seizures produced by maximal electroshock as well as delayed the latency of seizures produced by pentylenetetrazole and picrotoxin in mice (Hegde et al. 2009). However, only 200 and 400 mg/kg of the extract conferred protection (25% and 50%, respectively).

Analgesic, Anti-inflammatory and Antipyretic Activities

The ethanol and aqueous extracts from roots of *C. carandas* were found to exhibit significant analgesic, anti-inflammatory and antipyretic activities in rodents at the doses of 100 and 200 mg/kg body weight (Bhaskar and Balakrishnan 2009a). In analgesic activity, the highest percentage of inhibition of abdominal constriction (72.67%) was observed for ethanol extracts of *C.carandas* at a dose of 100 mg/kg body weight. The ethanol and aqueous extracts of *C. carandas* were found to reduce significantly the formation of edema induced by carrageenan after 2 hours. The extracts showed significantly antipyretic activity in Brewer's yeast induced hyperpyrexia in rats after 2 hours.

Cardiotonic Activity

The alcoholic extract of roots of *C. congesta* was found to exhibit cardiotonic activity and prolonged blood pressure lowering effect (Rastogi et al. 1966, 1967; Vohra and De 1963). An amorphous water-soluble polyglycoside possessing significant cardiac activity was isolated. The cardiac activity of water-soluble fraction was attributed to the presence of the glucosides of odoroside.

Cardiac Depressant Activity

Other studies reported also that the alcoholic plant extract of *Carissa carandas* exhibited cardiac depression activity. It showed significant negative inotropic activity with negative chronotropic effects (Sajid et al. 1996).

Miscellaneous Activities

Separate studies reported that the roots contained compounds which possessed histamine liberating activity supporting its traditional use in the treatment of helminthiasis and as antidote for snake poisoning (Joglekar and Gaitonde 1970).

Traditional Medicinal Uses

The astringent, sour unripe fruit is used in traditional medicine as an astringent, antipyretic, aphrodisiac, thermogenic, appetiser and constipating agent. Unripe fruit is taken for diarrhoea, anorexia, hyperdipsia, haematemesis and intermittent fever. The ripe fruit is taken as an antiscorbutic, carminative, expectorant and is useful for biliousness as a stomachic. It also has a cooling effect, and used for haematemesis, antidote for poisons and as appetiser. Ripe fruits are also used to treat fever, mouth ulcer and sore-throats, otalgia, diarrhoea, burning sensation, skin disorders including scabies. The pulverised fruit, mixed with camphor and lime juice is used to remove pruritus.

The leaf decoction is employed for intermittent fever, diarrhoea, oral inflammation, soreness of the mouth and throat, syphilitic pains and earache. The root is employed as a bitter stomachic and vermifuge and it is an ingredient in a remedy for pruritis (itches) and scabies. The root is also used as remedy for gonorrhoea, pyrexia, indigestion and chronic ulcer.

Other Uses

This species is used frequently as a hedge plant. The fruits have been utilised as agents in tanning and dyeing. The leaves have been used as fodder for the tussar silkworm and the leaves used for tanning. A paste of the pounded roots served as a fly repellent. The white or yellow wood is hard, smooth and useful for fashioning spoons, combs, household utensils, furniture and miscellaneous products of turnery. It is sometimes burned as fuel.

Comments

The white latex of the plant is caustic.

Selected References

Ahmad NS, Farman M, Najmi MH, Mian KB, Hasan A (2006) Activity of polyphenolic plant extracts as scavengers of free radicals and inhibitors of xanthine oxidase. J Appl Basic Sci 2(1)

Backer CA, Bakhuizen van den Brink Jr RC (1965) Flora of Java (spermatophytes only), vol 2. Wolters-Noordhoff, Groningen, 641 pp

Bhaskar VH, Balakrishnan N (2009a) Analgesic, anti-inflammatory and antipyretic activities of *Pergularia daemia* and *Carissa carandas*. DARU 17(3):168–174

Bhaskar VH, Balakrishnan N (2009b) In vitro antioxidant property of lactiferous plant species from Western Ghats Tamilnadu. India Int J Health Res 2(2):163–170

Brown WH (1951–1957) Useful plants of the Philippines. Reprint of the 1941–1943 edition, 3 vols. Technical Bulletin 10. Department of Agriculture and Natural Resources. Bureau of Printing, Manila, vol 1 (1951), 590 pp, vol 2 (1954), 513 pp, vol 3 (1957), 507 pp

Burkill IH (1966) A dictionary of the economic products of the Malay Peninsula. Revised reprint, 2 vols. Ministry of Agriculture and Co-operatives, Kuala Lumpur, vol 1 (A–H), pp 1–1240, vol 2 (I–Z), pp 1241–2444

Chopra RN, Nayar SL, Chopra IC (1986) Glossary of Indian medicinal plants (including the supplement). Council Scientific Industrial Research, New Delhi, 330 pp

Council of Scientific and Industrial Research (1950) The wealth of India. A dictionary of Indian raw materials and industrial products (Raw materials 2). Publications and Information Directorate, New Delhi

Foundation for Revitalisation of Local Health Traditions (2008). FRLHT database. htttp://envis.frlht.org

Govaerts R (2008) World Checklist of Apocyanaceae. The Board of Trustees of the Royal Botanic Gardens, Kew. Published on the Internet. http://www.kew.org/wcsp/

Hegde K, Joshi AB (2009) Hepatoprotective effect of *Carissa carandas* Linn root extract against CCl_4 and paracetamol induced hepatic oxidative stress. Indian J Exp Biol 47(8):660–667

Hegde K, Thakker SP, Joshi AB, Shastry CS, Chandrashekhar KS (2009) Anticonvulsant activity of *Carissa carandas* Linn. root extract in experimental mice. Trop J Pharm Res 8(2):117–125

Iyer CM, Dubash PJ (1993) Anthocyanin of Karwand (*Carissa carandas*) and studies on its stability in model systems. J Food Sci Technol 30(4):246–248

Joglekar SN, Gaitonde BB (1970) Histamine releasing activity of *Carissa carandas* roots (Apocyanaceae). Jpn J Pharm 20(3):367–372

Joshi DV, Boyce SF (1957) Chemical investigation of root of *Carissa congesta*. Santapau J Org Chem 22:95–97

Li B, Leeuwenberg AJM, Middleton DJ (1995) Apocynaceae A. L. Jussieu. In: Wu ZY, Raven PH (eds) Flora of China, vol 16, Gentianaceae through Boraginaceae. Science Press/Missouri Botanical Garden Press, Beijing/St. Louis, 479 pp

Morton JF (1987) Karanda. In: Fruits of warm climates. Julia F. Morton, Miami, pp 422–424

Naim Z, Khan MA, Nizami SS (1988) Isolation of a new isomer of ursolic acid from fruits and leaves of *Carissa carandas*. Pak J Sci Ind Res 31(11):753–755

Orwa C, Mutua A, Kindt R, Jamnadass R, Simons A (2009) Carissa congesta. Agroforestree database: a tree reference and selection guide version 4.0. (http://www.worldagroforestry.org/af/treedb/)

Pal R, Kulshreshtha DK, Rastogi RP (1975) A new lignan from *Carissa carandas*. Phytochem 14(10):2302–3

Pino JA, Marbot R, Vázquez C (2004) Volatile flavor constituents of karanda (*Carissa carandas* L.) fruit. J Essent Oil Res 16(5):432–434

Quisumbing E (1978) Medicinal plants of the Philippines. Katha Publishing Co., Quezon, 1262 pp

Rastogi RC, Vohara MM, Rastogi RP, Dhar ML (1966) Studies on *Carissa carandas* Linn. Part I. Isolation of the cardiac active principles. Indian J Chem 4(3):132–133

Rastogi RC, Rastogi RP, Dhar ML (1967) Studies on *Carissa carandas* Linn. II. Polar glycosides. Indian J Chem 5(5):215–21

Sajid TM, Rashid S, Ahmad M, Khan U (1996) Estimation of cardiac depressant activity of ten medicinal plant extracts from Pakistan. Phytother Res 10(2):178–80

Siddiqui BS, Ghani U, Ali ST, Usmanin SB, Begum S (2003) Triterpenoidal constituents of the leaves of *Carissa carandas*. Nat Prod Res 17(3):153–8

Singh B, Rastogi RP (1972) The structure of carindone. Phytochemistry 11(5):1797–80

Subhadrabandhu S (2001) Under utilized tropical fruits of Thailand, FAO Rap Publication 2001/26. FAO, Rome, 70 pp

Sulaiman SF, Khoo HS, Yusof SR (2008a) The antioxidant effect of Carissa carandas unripe fruit extracts and fractions. Project Report, Monograph Universiti Sains Malaysia

Sulaiman SF, Wong ST, Ooi KL, Yusof SR, Tengku Muhammad TS (2008b) Anticancer study of Carissa carandas extracts. Project Report, Monograph Universiti Sains Malaysia

Sulaiman SF, Yusof SR, Mohamed Daud MZ (2008c) The correlation between the antioxidant properties of the Carissa carandas extracts with their flavonoids occurrences. Project Report, Monograph Universiti Sains Malaysia

Taylor RSL, Hudson JB, Manandhar NP, Towers GHN (1996) Antiviral activities of medicinal plants of southern Nepal. J Ethnopharmacol 53(2):97–104

Tun UK, Than UP, staff of TIL (2006) Myanmar medicinal plant database. http://www.tuninst.net/MyanMedPlants/indx-DB.htm

Vohra MM, De NN (1963) Comparative cardiotonic activity of *Carissa carandas* {L}. and *Carissa spinarum* {A}. Ind J Med Res 51(5):937–940

Wangteeraprasert R, Likhitwitayawuid K (2009) Lignans and a sesquiterpene glucoside from *Carissa carandas* stem. Helv Chim Acta 92:1217–1223

Willughbeia angustifolia

Scientific Name

Willughbeia angustifolia (Miq.) Markgr.

Synonyms

Chilocarpus brachyanthus Pierre, *Urnularia rufescens* (Dyer ex Hook.f.) Pichon, *Vahea angustifolia* Miq., *Willughbeia angustifolia* var. *gracilior* Markgr., *Willughbeia apiculata* Miq., *Willughbeia elmeri* Merr., *Willughbeia rufescens* Dyer ex Hook.f., *Willughbeiopsis rufescens* (Dyer ex Hook.f.) Rauschert.

Family

Apocynaceae

Common/English Names

Kubal Madu, Pitabu, Serapit

Vernacular Names

Kalimantan: Gitak Madu, Pitabu;
Sabah: Serapit, Kabal Madu;
Sarawak: Akar Kubal, Kubal Madu.

Origin/Distribution

The species is indigenous to Nicobar Island to Malesia. It is commonly found wild in Borneo (Brunei, Sarawak, Sabah and Indonesian Kalimantan).

Agroecology

Strictly a tropical species. It occurs wild in mixed dipterocarp forests and riverine forests in Borneo.

Edible Plant Parts and Uses

The mucilaginous pulp of the ripe fruit is eaten. The orange pulp of kubal madu variety is sweet and has excellent flavour liken to that of orange sherbet.

Botany

A perennial, evergreen woody climber with branches bearing tendrils formed from modified inflorescences. Leaves are opposite, distichous on short petioles, lamina ovate-lanceolate, apex acute to acuminate, base obtuse ovate, coriaceous and penninerve with a distinct midrib (Plate 1). Inflorescence of axillary cymes, rarely terminal,

Plate 1 Ripe fruits and leaves of Kubal madu

Plate 2 Hard rind and orange pulp of the fruit

inflorescence axis is shorter than or as long as subtending petiole. Flowers are pentamerous, yellowish-white and actinimorphic; calyx deeply divided with ovate-oblong lobes; corolla lobes in bud overlapping to the left forming a cone or cylinder of erect lobes; corolla tube cylindrical, inflated and short, stamens inserted in tube with ovate anthers; ovary single, glabrous, unilocular, with 2 parietal placentas. Fruit is a fleshy berry, globose, golden yellow to orangey yellow, indehiscent, hard-walled which exude white latex when cut (Plate 1 and 2). Fruit has a few to many seeds. Seeds are compressed ovoid smooth, with a very thin endosperm and thick horny cotyledons and embedded in mucilaginous orangey-yellow pulp (Plate 2).

Nutritive/Medicinal Properties

No published information is available on the nutritive value of the fruit.

Latex is reported to be used as treatment for shingles in Sarawak.

Other Uses

The plant yields a rubber-like latex which soon becomes resinous and hard.

Comments

The vine contains a sticky white latex.

Selected References

Chai PPK (2006) Medicinal plants of Sarawak. Lee Ming Press, Kuching, 212 pp

Govaerts R (2008) World Checklist of Apocyanaceae. The Board of Trustees of the Royal Botanic Gardens, Kew. Published on the Internet; http://www.kew.org/wcsp/

Lee YF, Gibot A (1986) Indigenous edible plants of Sabah. FRC Publication No. 25. Sabah Forestry Department, Sandakan, Sabah, 9 pp

Meijer W (1969) Fruit trees in Sabah (North Borneo). Malay Forester 32(3):252–265

Middleton DJ (1993) A taxonomic revision of *Willughbeia* Roxb (Apocyanaceae). Blumea 38:1–24

Middleton DJ (2001) The apocynaceae of the Crocker Range National Park. In: Ghazaly IL, Lamri A (eds.) Crocker Range National Park Sabah, vol 1, Natural Ecosystem and Species Components. Asean Academic Press, London, pp 65–87

Middleton DJ (2002) The Apocynaceae of the Crocker Range National Park Sabah. ASEAN Rev Biodivers Environ Conserv ARBEC July–Sept 2002:1–15

Middleton DJ (2004) Apocynaceae. In: Soepadmo E, Saw LG, Chung RCK (eds) Tree flora of Sabah and Sarawak, vol 5. Kuala Lumpur Forest Research Institute, Ampang Press, Kuala Lumpur, pp 1–61

Voon BH, Chin TH, Sim CY, Sabariah P (1988) Wild fruits and vegetables in Sarawak. Department of Agriculture, Sarawak, 114 pp

Willughbeia coriacea

Scientific Name

Willughbeia coriacea Wall.

Synonyms

Ancylocladus coriaceus (Wall.) Kuntze, *Ancylocladus firmus* (Blume) Kuntze, *Ancylocladus minutiflorus* Pierre, *Ancylocladus nodosus* Pierre, *Ancylocladus vriesianus* Pierre, *Tabernaemontana macrocarpa* Korth. ex Blume, *Willughbeia burbidgei* Dyer, *Willughbeia firma* Blume, *Willughbeia firma* var. *oblongifolia* Blume, *Willughbeia firma* var. *macrophylla* Boerl., *Willughbeia firma* var. *obtusifolia* Boerl., *Willughbeia firma* var. *platyphylla* Boerl., *Willughbeia minutiflora* (Pierre) K.Schum., *Willughbeia nodosa* (Pierre) K.Schum., *Willughbeia vrieseana* (Pierre) K.Schum.

Family

Apocyanaceae

Common/English Names

Kubal Tusu, Borneo Rubber

Vernacular Names

Indonesia: Lumbu (General), Gitan Sussu, Karet Akar Grutuk (Sumatra), Tjungkangkang (Sunda Islands);
Malaysia: Akar Jela, Akar Getah Gaharu, Akar Getah Menjawa, Akar Gerit Hitam, Akar Gĕrit-Gĕrit Bĕsi, Akar Getah Gegerip Jantan, Akar Garok, Akar Getah Jela, Akar Getah Ujul, Akar Senggerip Merah, Puchong Kapur, Akar Jauloh, Akar Jolok, Lumbu Jawa (Peninsular), Kelang (Lundayeh, Sabah), Kubal Tusu, Kubal Arang (Iban, Sarawak).

Origin/Distribution

The species is native to Peninsular Thailand to West Malesia. It is cultivated in Java and Sumatra.

Agroecology

A strictly tropical species, common in mixed dipterocarp or riverine forests in its native range.

Edible Plant Parts and Uses

The flesh of the ripe fruit is sweet and eaten fresh.

Botany

The plant is a large, evergreen, perennial, woody climber with tendrils modified from the peduncle. The leaves are leathery, opposite, distichous, large, oblong-elliptic, 8–14 cm long with tertiary venation faint or obscure. Inflorescences are sessile in axillary, glabrous, puberulent cymes. Flowers are white, pentamerous and actinomorphic, calyx lobes ovate-oblong, >1 × as long as wide, corolla tube cylindric, inflated only around stamens, >3 mm long inserted near base of corolla tube; ovary, unilocular with parietal placentation and numerous ovules, style glabrous, and stigma not collared. Fruit is an indehiscent, pear-shaped, ellipsoid, glabrous hard-walled, berry (Plates 1 and 2), golden-yellow, orange to dull red when ripe, up to 10 cm in diameter. Seeds a few to many, compressed, embedded in white or orange-coloured, sweet, mucilaginous pulp (Plate 2).

Plate 1 Ripe fruit of Kubal tusu

Plate 2 Sliced fruit with thick, hard wall, orange flesh and seeds

Nutritive/Medicinal Properties

No published information is available on the nutritive value of the edible fruit.

The extract of the bark of aerial roots of *Willughbeia coriacea* showed 1,1-diphenyl-2-picrylhydrazyl (DPPH) radical-scavenging activity of more than 70% at 100 µg/ml; tyrosinase inhibitory activity of more than 40% using L-tyrosine as a substrate at 500 µg/ml and tyrosinase inhibitory activity of more than 40% using L-DOPA as a substrate at 500 µg/ml (Arung et al. 2009). The root bark extract at 200 µg/ml, strongly inhibited the melanin production of B16 melanoma cells without significant cytotoxicity. The data indicated that the root bark extract of *W. coriaceae* has potential as ingredients for skin-whitening cosmetics.

Traditional Medicinal Uses

In Peninsular Malaysia, the roots have been prescribed for jaundice: in infants externally for icterus neonatorum; and internally for malaria complicated by jaundice. It has also been prescribed as a draught for heartburn. The root has also been used as a poultice for puru or yaws but more commonly the latex was used. The roots have also been pulped and used in a preparation for applying over a man's body to cure hysteria.

Other Uses

The latex provides birdlime not rubber. The plant furnishes caoutchouc (getah susu, Borneo-caoutchouc), which is collected from wild and cultivated plants. Towards the end of the nineteenth century, the plant was cultivated experimentally in southeast Asia but the trials were abandoned as the latex it produced proved to be inferior to *Hevea*.

The hardened latex from the plant was used for wrapping the heads of sticks used for beating gongs and drums.

Comments

Selected References

Arung ET, Kusuma IW, Christy EO, Shimizu K, Kondo R (2009) Evaluation of medicinal plants from Central Kalimantan for antimelanogenesis. J Nat Med 63(4):473–480

Boer E, Ella AB (eds) (2000) Plants producing exudates. Plant resources of South-East Asia no. 18. Plants producing exudates. Backhuys Publishers, Leiden, pp 189

Burkill IH (1966) A dictionary of the economic products of the Malay Peninsula. Revised reprint, 2 vols. Ministry of Agriculture and Co-operatives, Kuala Lumpur, vol 1 (A–H), pp 1–1240, vol 2 (I–Z), pp 1241–2444

Govaerts R (2008) World Checklist of Apocyanaceae. The Board of Trustees of the Royal Botanic Gardens, Kew. Published on the Internet; http://www.kew.org/wcsp/

Hoare A (2003) Food resources and changing patterns of resource use among the Lundayeh of the Ulu Padas. Sabah Borneo Res Bull 34:94–119

Keng H (1990) The concise flora of Singapore, vol 1, Gymnosperms and dicotyledons. NUS Press, Singapore, 364 pp

Middleton DJ (1993) A taxonomic revision of *Willughbeia* Roxb (Apocynaceae). Blumea 38:1–24

Middleton DJ (2001) The apocynaceae of the Crocker Range National Park. In: Ghazaly IL, Lamri A (eds) Crocker range National Park Sabah, vol 1, Natural ecosystem and species components. Asean Academic Press, London, pp 65–87

Middleton DJ (2002) The apocynaceae of the Crocker Range National Park Sabah. ASEAN Rev Biodivers Environ Conserv ARBEC July–Sept 2002:1–15

Middleton DJ (2004) Apocynaceae. In: Soepadmo E, Saw LG, Chung RCK (eds) Tree flora of Sabah and Sarawak, vol 5. Forest Research Institute/Ampang Press, Kuala Lumpur, pp 1–61

Voon BH, Chin TH, Sim CY, Sabariah P (1988) Wild fruits and vegetables in Sarawak. Department of Agriculture, Sarawak, 114 pp

Monstera deliciosa

Scientific Name

Monstera deliciosa Liebm.

Synonyms

Monstera borsigiana K.Koch, *Monstera deliciosa* var. *borsigiana* Engl., *Monstera deliciosa* var. *sierrana* G.S.Bunting, *Monstera lennea* K.Koch, *Monstera tacanaensis* Matuda, *Philodendron anatomicum* Kunth, *Tornelia fragrans* Gut. ex Schott nom. illeg.

Family

Araceae

Common/English Names

Breadfruit-Vine, Ceriman, Cut-Leaf-Philodendron, Delicious Monster, Fruit Salad Plant, hurricane-plant, Mexican Breadfruit, Monster Fruit, Monstera, Locust and Wild Honey, Pine Fruit Tree, Split-Leaf Philodendron, Swiss Cheese Plant, Window-Leaf.

Vernacular Names

Brazil: Banana-Do-Brejo, Banana-Do-Mato (Portuguese);
Chinese: Gui Bei Zhu;
Columbia: Hojadillo;
Danish: Fingerfilodendron;
Dutch: Gatenplant, Vensterblad, Vingerplant;
Finnish: Peikonlehti;
French: Ananas De Mexico, Anana Du Pauvre, Cériman, Monstère Délicieux, Monstera, Philodendron Monstéra, Philodendron À Feuilles Incisées;
French Guiana: Arum Du Pays, Arum Troud;
German: Philodendron, Fensterblatt, Köstlicher Kolbenriese, Zimmer-Philodendron;
Guadeloupe: Caroal, Liane Percee, Liane Franche;
Guatemala: Harpon, Arpon Comun;
Hungarian: Könnyezöpálma;
India: Amar-Phal;
Italian: Monstera;
Japanese: Monsutera Derishioosa;
Martinique: Siguine Couleurr;
Mexico: Piñanona;
Nepalese: Lahare Karkalo;
Norwegian: Vindusblad;
Polish: Monstera Dziurawa;
Portuguese: Banana De Brejo, Banana De Macao, Banana Do Mato, Deliciosa, Fruta De México, Costela De Adão, Balaço, Tornélia;
Russian: Monstera Delikatesnaia, Monstera Lakomaia, Monstera Prelestnaia, Monstera Privlekatel'naia;
Spanish: Balazo, Pinanona, Costilla de Adán, Piñanona monster, Cerimán De México, Hojadillo, Huracán, Ojal, Piña Anona, Piñanona Monstera;

Swedish: Monstera;
Venezuela: Ojul, Huracan.

Origin/Distribution

The ceriman is indigenous to the wet tropical forests of southern Mexico, Guatemala and parts of Costa Rica and Panama. Now it is pantropical and has naturalised in many areas for instance in Florida and coastal areas of North Coast and Central Coast, also scattered throughout the lower to mid Blue Mountains (Central Coast) of New South Wales in Australia.

Agroecology

Being a tropical plant, ceriman grows best at temperatures of 20–30°C, requires high humidity, and needs full or partial shade. It is found all over tropical America in the hot humid rainforest from 1 to 600 m elevation and where the annual rainfall is above 1,000 mm. It can withstand cold conditions provided it is sheltered from frost and cold winds. Growth ceases below 10°C and frost will kill it. The plant grows vigorously in almost any soil, including calcareous but flourishes best in well-drained, loamy soils rich in organic matter. It is intolerant of saline conditions. It is especially suited for use as an ornamental on fences and tree stumps. It flowers around 3 years after it is planted in ideal conditions, and takes more than a year for the fruit to ripen.

Edible Plant Parts and Uses

The elongated fruit is ready to harvest when the tile-like caps (scales) of the fruitlets at the base start to separate and show creamy colour between them, usually, about 12 months after flowering (Plate 5). When mature, the fruit can be detached, leaving the flower stem. Ripe sectioned of the fruit can be recognised by the separation of the scales. Do not eat from the section where the scales have not been shed, for severe irritation of the throat can be caused by oxalate and also by the small black scales between the edible segments within the fruit. To facilitate ripening after harvest, the fruit is placed in a paper bag in a warm position and in a day or so the green scales will fall from the ripened section at the base of the fruit and expose the edible portion beneath. The unripe section is left in the paper bag until the next portion is ready to eat. Alternatively, the whole fruit can be ripened for eating at one time by standing the base in water and keeping it in the dark for a few days.

The pulp of ripe fruit is eaten fresh, made into jellies and jams. Ripe pulp has a flavour liken to a blend of pineapple and banana. It may be served as dessert with a little light cream, or may be added to fruit cups, salads or ice cream. To make a preserve, rinsed segments can be stewed for 10 min in water, too which is added sugar or honey and lime juice and the mixture is simmered again for 20 min and then preserved in sterilized jars.

Botany

A robust, fast growing, stout, herbaceous or woody, scrambling or climbing vine growing to 20 m high (Plate 1). The stem is cylindrical, heavy, 6.0–7.5 cm diameter, rough with leaf scars, and producing numerous, long, tough, fibrous aerial roots. Leaves are large, leathery, glossy green, cordate, 25–90 cm long by 25–75 cm broad (Plate 1). Juvenile plants start out with smaller and entire leaves with no lobes or holes, but older leaves and plant soon produce leaves with deeply cut strips or lobes around the margins and perforated on each side of the midrib with elliptic or oblong holes of various sizes. Several inflorescences arise in a group from the leaf axils on tough, cylindrical stalks. Inflorescence is an erect, elongated, cylindrical (corn-like) spadix, creamy-white turning to bluish green and usually enclosed in a white, boat-shaped spathe (Plate 2). The spadix is fleshy upright spike growing to 20 cm long with tiny flowers on it covered by hexagonal scales or tiles. The spadix develops into the compound fruit (Plates 2–5). The hexagonal

Plate 1 Ceriman plant habit

Plate 2 Ceriman fruit and inflorescence spathe (*white*)

Plate 3 Ceriman fruit at various stages of development

Plate 4 Close-up of ceriman fruit

scales dry out and separate as the fruit ripens from the base upwards, revealing the white or pale yellow, sweet pulp and the thin, black particles (floral remnants) between the scales, usually there are no seed (Plate 5).

Nutritive/Medicinal Properties

Analysis conducted in the Philippines recorded the ceriman fruit to have the following nutrient composition per 100 g edible portion (Morton 1987): calories, 737/kg; moisture, 77.88%; protein, 1.81%; fat, 0.2%; sugar, 16.19%; fibre 0.57%; ash, 0.85%. The pulp had also been reported to be rich in potassium and vitamin C.

Plate 5 Ripe ceriman fruit with tile-like scales removed

During ripening *Monstera deliciosa* fruit undergo a pronounced climacteric with the concomitant large increase in ethylene production. The ripe fruit was found to contain 19.1% soluble solids with 7–8 meq/100 g titratable acidity and 0.41% oxalic acid as the major component (Peters and Lee 1977). Acridity was attributed to the oxalic acid raphides and associated proteins in the developing fruit and floral remnants and all other parts of the plant. The ripe fruit juice is considered safe for human consumption with respect to the levels of saponin, hydrocyanic acid and oxalic acid by the researchers.

About 400 volatile components of the fruit were detected and the components most responsible for the characteristics flavour consisted mainly of esters and lactones (Peppard 1992) that included: 6-Dimethylocta-2,7-dien-6-ol; Limonene; β-myrcene; 1,6-Cyclodecadiene, 1-methyl-5-methylene-8-(1-methylethyl)-,[s-(e,e)]-; Benzeneethanol; β-phellandrene;1-Butanol, 3-methyl-; Butanoic acid, ethyl ester; Hexanoic acid, ethyl ester; 2-Heptanone; 2-Furanmethanol, 5-ethenyltetrahydro-α,α,5-trimethyl-, cis-; Octanoic acid, ethyl ester; Acetic acid, butyl ester; 1-Butanol, 3-methyl-, acetate; 1H-cyclopropa[a]naphthalene, 1a,2,3,5,6,7,7a,7b-octahydro-1,1,7,7a-tetramethyl-, [1ar-(1aα,7α,7aα,7bα)]-; Butanoic acid, 2-methyl-, ethyl ester; Benzaldehyde, 4-hydroxy-3-methoxy-; 1-Butanol, 2-methyl-; Acetic acid, 2-phenylethyl ester; Benzoic acid, methyl ester; Propanoic acid, 2-methyl-, ethyl ester; Decanoic acid, ethyl ester; Benzoic acid, ethyl ester; Propanoic acid, ethyl ester; Hexadecanoic acid, methyl ester; Hexanoic acid, methyl ester ; 2-Pentadecanone, 6,10,14-trimethyl-; Butanoic acid, 3-methyl-, ethyl ester; Isobutyl acetate; Butanoic acid, methyl ester; γ-Butyrolactone; n-Propyl acetate; Octanoic acid, methyl ester; Hexadecanoic acid, ethyl ester; 2(3H)-Furanone, 5-hexyldihydro-; 5-Hydroxy decanoic acid, lactone; Dodecanoic acid, ethyl ester; Pentanoic acid, ethyl ester; Butanoic acid, 2-methyl-, methyl ester; Butyl butyrate; 2(3H)-Furanone, 5-ethyldihydro-; 2(3H)-furanone, 5-butyldihydro-; Benzeneacetic acid, ethyl ester; Heptanoic acid, ethyl ester; Tetradecanoic acid, ethyl ester; 2-Pentadecanone; Butanoic acid, 3-hydroxy-, ethyl ester; Butanoic acid, propyl ester; Hexanoic acid, butyl ester; 2,3-Butanediol; 2-Hexenal; 1-Butanol, 2-methyl-, acetate; 2H-pyran-2-one, tetrahydro-6-propyl-; Nonanoic acid, ethyl ester; Butanoic acid, 3-methyl-, methyl ester; Propyl caproate; Benzeneacetic acid, methyl ester; Hexanoic acid, 3-hydroxy-, ethyl ester; 3-(Methylthio)propanoic acid ethyl ester; 2-Butenoic acid, ethyl ester, (E)-; Pentadecanoic acid, ethyl ester; Propyl octanoate; Butanoic acid, 3-oxo-, ethyl ester; Ethyl 2-hexenoate; Undecanoic acid, ethyl ester; Undecanolactone; Naphthalene, decahydro-; Ethyl tridecanoate; 2H-pyran-2-one, tetrahydro-6-methyl-; 2,4-Hexadienoic acid, ethyl ester; Ethyl tiglate; 1H-2-benzopyran-1-one, 3,4-dihydro-8-hydroxy-3-methyl-; Ethyl 9-hexadecenoate; 1,2-Ethanediol, 1-phenyl-; Butanoic acid, 2-hydroxy-2-methyl-, ethyl ester; 3-Hexenoic acid, ethyl ester; 1H-indole-3-ethanol; Ethyl 4-octenoate; 3-Octenoic acid, ethyl ester, (z)-; ethyl 4,7-octadienoate; 2H-pyran-2-one, 4-methoxy-6-methyl-; Benzenemehanol, 4-hydroxy-3-methoxy, acetate.

Antidiabetic Activity

Monstera deliciosa exhibited insulin secretagogue activity at 1 μg/mL and was found to be a potential natural resource for antidiabetic compounds (Hussain et al. 2004). Insulin secretagogue therapy is a logical part of therapy of type 2 diabetes when diet and lifestyle modifications fail (Schmitz et al. 2002). Type 2 diabetes is characterized by impaired insulin secretion, usually with concomitant impaired insulin sensitivity.

Traditional Medicinal Uses

In Mexican traditional medicine, a leaf or root infusion is taken daily to relieve arthritis. A preparation of the root is employed in Martinique as a cure for snakebite.

Other Uses

The plant is widely cultivated as an ornamental for home and tropical gardens. The aerial roots have been used as ropes in Peru. In Mexico, they are fashioned into coarse, strong baskets.

Comments

This plant is capable of causing intense injury to the mouth, tongue and alimentary tract if bitten, chewed or eaten. Unripe fruit contains so much oxalic acid that it is poisonous, causing immediate and painful blistering and irritation, swelling, itching, of mucous membranes and loss of voice and may lead to death.

Selected References

Govaerts R (2008) World Checklist of Araceae. The Board of Trustees of the Royal Botanic Gardens, Kew. Published on the Internet; http://www.kew.org/wcsp/

Hussain Z, Waheed A, Qureshi RA, Burdi DK, Verspohl EJ, Khan N, Hasan M (2004) The effect of medicinal plants of Islamabad and Murree region of Pakistan on insulin secretion from INS-1 cells. Phytother Res 18(1):73–77

Kennard WC, Winters HF (1960) Some fruits and nuts for the tropics. U S D A Agric Res Serv Misc Publ 801:1–135

Morton J (1987) Ceriman. In: Fruits of warm climates. Julia F. Morton, Miami, pp 15–17

Peppard TL (1992) Volatile flavor constituents of *Monstera deliciosa*. J Agric Food Chem 40(2):257–262

Peters RE, Lee TH (1977) Composition and physiology of *Monstera deliciosa* fruit and juice. J Food Sci 42:1132–1133

Purseglove JW (1972) Tropical crops, vol 1 & 2, Monocotyledons. Longman, London, 607 pp

Schmitz O, Lund S, Andersen PH, Jonler M, Porksen N (2002) Optimizing insulin secretagogue therapy in patients with type 2 diabetes: a randomized double-blind study with repaglinide. Diabetes Care 25: 342–346

Adonidia merrillii

Scientific Name

Adonidia merrillii (Becc.) Becc.

Synonyms

Normanbya merrillii Becc., *Veitchia merrillii* (Becc.) H. E. Moore.

Family

Arecaceae

Common/English Names

Adonidia Palm, Christmas Palm, Chinese Betel-Nut, Dwarf Royal Palm, Kerpis Palm, Manila Palm, Merrill Palm, Veitchia Palm.

Vernacular Names

Chamorro: Pugua Chena;
Chinese: Ma Ni La Ye Zi;
Dutch: Kerstpalm;
French: Palmier De Manille, Palmier De Noël, Palmier Nain Royal, Palmier Des Philippines;
German: Manilapalme, Weihnachtspalm;
Italian: Palma Di Manila;
Japanese: Manira Yashi;
Malay: Palma Manila;
Philippines: Lugos (Sulu), Bunga De Jolo, Bunga De China, Oring-Oring (Tagalog)
Spanish: Chaguaramo Enano, Palmade Manila, Palma De Navidad;
Thai: Maak Nuan, Mak-Nual, Paam Nuan;
Vietnamese: Cau Trắng.

Origin/Distribution

Manila palm is native to Palawan and neighbouring islands in the Philippines and east Coast of Sabah in East Malaysia (Dransfield et al. 2008).

Agroecology

Manila palm thrives in a tropical climate. In its native range mean annual temperature ranges from 22°C to 32°C, with a mean annual temperature of 27.5°C, and average annual relative humidity of 80–90% and mean annual rainfall of 2,000–2,500 mm. It can be grown in the subtropics but is rather cold sensitive. In its native Palawan it occurs on karst limestone cliffs (Dransfield et al. 2008). It thrives in a sunny position on well-drained, fertile soil. It is moderately salt tolerant and has good wind resistance.

Edible Plant Parts and Uses

The fruit is used as a masticatory for chewing when ripe but is an inferior substitute for betel-nut (Whitmore 1973; Haynes and McLaughlin 2000).

Botany

An erect, solitary, unarmed, monoecious palm, 5–10 m high. The stem is slender, 15–20 cm diameter, ringed with close leaf scars, becoming longitudinally striate, grey. Leaves are pinnate, strongly arched, sheaths forming a prominent crownshaft, covered with deciduous grey tomentum and scattered dark scales. Petiole is short and tomentose with two short triangular auricles at the base. Leaflets crowded, regularly arranged, but diverging in slightly different planes, flat, intense-green, broad, tapering to base and apex, strongly curved and rather irregularly bent near the tip, apex oblique, truncate, acute, or acuminate or toothed, with midrib and marginal veins prominent. Inflorescences occur below the leaves, branched to 3–4 orders basally, fewer orders distally, branches tomentose when newly emerged. Flowers inconspicuous, creamy-white, unisexual, occurs in clusters of 2–3 or solitary along the rachillae in twisted spikes. Rachis and rachillae white. Staminate flowers bullet-shaped, with three imbricate sepals, three valvate petals longer than sepals, 45–50 stamens with long slender filaments and dorsifixed anthers and with a trifid or bifid pistillode. Pistillate flowers, ovoid, with three imbricate sepals and three imbracite petals, slightly longer than sepals, six connate staminodes, gynoecium ovoid, unilocular, uniovulate, style thick, stigmas three sessile. Fruit ovoid, 3–4 cm long, beaked, pale green becoming bright red at maturity (Plates 1 and 2), perianth whorls enlarged, persistent, stigmatic remains apical. The fruit has a thin epicarp and dry, yellowish, thin-fleshy mesocarp, and thin, fragile endocarp. Seed is ovoid, truncate basally, pointed apically, with a ruminate endosperm and embryo basally.

Plate 1 Clusters of unripe and rip fruit occurring way below the crown

Plate 2 Close-up of ripe fruits

Nutritive/Medicinal Properties

No published information is available on its nutritive and medicinal properties and uses.

Other Uses

A popular ornamental palm grown in roadside plantings, parks and gardens. It can also be grown as a container plant and grown in well-lit offices and reception area. The palm is also popularly used in interior landscape in shopping malls and atria.

Comments

This palm is very susceptible to the dreaded lethal yellowing disease.

Selected References

Dransfield J, Uhl N, Asmussen C, Baker WJ, Harley M, Lewis C (2008) *Genera Palmarum*. The evolution and classification of palms. Kew Publishing, Royal Botanic Gardens, Kew

Haynes J, McLaughlin J (2000) Edible palms and their uses. University Florida/Miami-Dade County Extension Office Fact Sheet MDCE-00-50

Huxley AJ, Griffiths M, Levy M (eds) (1992) The new RHS dictionary of gardening, 4 vols. MacMillan, London

Moore HE Jr (1957) Veitchia Gentes Herb 8:501–506

Whitmore TC (1973) Palms of Malaya. Oxford University Press, Singapore, 129 pp

Zona S, Fuller D (1999) A revision of *Veitchia* (Arecaceae-Arecoideae). Harv Pap Bot 4:543–557

Areca catechu

Scientific Name

Areca catechu L.

Synonyms

Areca cathecu Burman f., *Areca faufel* Gaertner, *Areca himalayana* H. Wendland, *Areca hortensis* Loureiro, *Areca nigra* H. Wendland, *Sublimia areca* Comm. ex Mart. nom. inval.

Family

Arecaceae

Common/English Names

Areca nut, Areca nut palm, Betel palm, Betelnut, Betel nut, Betelnut, Betelnut palm, Indian nut, Pinang palm

Vernacular Names

Arabic: Fûfal, Fofal, Foufal, Faufil, Fufal, Fofal, Fouzal, Fofal, Foufal, Kawthal, Tânbûl;
Armenian: Arygni;
Burmese: Kunthi Pin Kun, Kunya;
Chamorro: Pugua, Mama'on;
Chinese: Bing Lang, Bing Lang, Bing Lang Zi, Guo Ma, Da Fu Bi, Da Fu Pi;
Chuuk: Pu;
Czech: Palma Katechová;
Danish: Betelpalme, Betelnød, Pisangpalme;
Dutch: Arecapalm, Arekanoot, Betelnoot Betelnootpalm, Betelpalm, Pinangnoot;
Eastonian: Beetli-Areekapalm;
Fijian: Areca, Niu Solomon;
Finnish: Arekapalmu;
French: Arec, Aréquier, Noix De Betel Pugua, Noix D'arec, Arec Cachou, Arec De L'inde;
German: Arekanuss, Arekapalme, Betel-Palme, Betelnüsse, Betelnuß Palme Betelnuss-Palme, Betelpalme, Catechu-Palme, Pinang;
Guam: Pugua;
Hungarian: Arékapálma, Beteldió, Bételpálma;
India: Tamol, Guwa (Assamese), Supari (Bengali), Ayrike, Sopari (Gujerati), Chalia, Chamarpushpa, Guvak, Guwa, Khapur, Pug, Pugi, Supari, Supiari, Suppari, Supyari, Udveg (Hindu), Adake, Adaki, Adike, Adike Kaayi, Adike Mara, Betta, Bettadake, Bettadike, Bette, Chautaki, Chikaniyadike, Cikaniyadike,Chikke, Cikke, Gotadike, Kangu, Kaungu, Khapura, Khhapura, Kowngu, Puga, Pugiphala, Tambula, Thaamboolagotu (Kannada), Kwai (Khasi), Pophala, Supari (Konkani), Adaka, Adakai, Adakka, Atakka, Ataykkamaram, Atekka, Atekkai, Caunga, Cavooghoo, Chempalukka, Cuanga, Ghhonta, Ghonta, Kalunnu, Kamugu, Kamuk, Kamuka, Kamuku, Kamunnu, Kavung, Kavungu, Kavunnu, Kazhangu, Kazhunnu, Khhapuram, Kramukam, Pakavakka, Pakka, Pakku, Pugam (Malayalam), Kwa, Kwa Pambi (Manipuri), Madi, Pophal, Pophali, Pophala, Pug,

Pugaphal, Pung, Supari, Supaaree, (Marathi), Kuva, Kuvathing (Mizoram), Trynodrumo (Oriya), Akota, Akoth, Akotaja, Chamarpushpa, Chhataphala, Chikkana, Dirghapadapa, Dridhavalkala, Ghonta, Gopadala, Gubak, Guvaka, Kabukah, Kapitana, Karamattam, Khapur, Khapura, Khipura, Kramuka, Kramukah, Kramukam, Kuvara, Phalam, Pooga, Puga, Puga-Phalam, Pugah, Pugaphal, Pugh, Pugi, Rajatala, Suranjana, Tambula, Tantusara, Udveg, Udvegam, Valkataruh (Sanskrit), Ataikkay, Akotam, Cakuntam, Cakuntikai, Cakuntikaimaram, Cattamarkkam, Chakuntam, Chamara-Putpam, Ciram, Curancanam, Inippilapatitam, Inippilatitamaram, Iracatalam, Kaiccikam, Kaiccikamaram, Kalacattiram, Kalaymaram, Kamugu, Kamuku, Kandi, Kanti, Kapuram, Katti, Kiramugam, Kiramamuki, Kiramamukimaram, Kiramukam, Kirantimukam, Kirumukam, Kontai, Kottai Paakku, Kottai-Pakku, Kottai Pakku, Kugagam, Kukacamaram, Kukakam, Kuntal, Kuntar-Kamaku, Kuvakam, Maturapakam, Maturapakamaram, Nattukkamuku, Paak, Pakku, Pakkumaram, Pakkuppanai, Paku-Kotai, Palacankiyam, Palacinikiyamam, Piramataru, Piramatarucam, Pirumaniyam, Pugam, Pukam, Pukaram, Putakam, Talattiram, Tampulavatanimaram, Tantucaram, Tantucaramaram, Tarpati, Taru, Tiritavalkam, Tuvarkkay, Tuvarkkaymaram, Vimpu (Tamil), Chikinamu, Chikini, Gautupoka, Ghonta, Kazhangu, Khapuramu, Kolapoka, Kramukamu, Oppulu, Oppuvakkulu, Pakavakka, Pogamu, Poka, Poka Chettu, Poka-Vakka, Prakka, Pugamu, Vakka, Vakkalu (Telugu), Chalia, Chalia Sokhta, Chalia Purani Jalai Hai, Chalia Jalai Hui, Chikni Chhalia, Chhalia, Chhalia Kohna Sokhta, Chikni Pisi Hui, Gond Supari, Gul Supari, Supari, Supari Katha (Urdu);
Indonesia: Boa (Bali), Jambé, Wohan (Java), Jambe (Sundanese), Pinang (Sumatra);
Italian: Areca, Avellana D'india, Noce Del Betel, Noce Di Arec, Noce Di Betel, Palma Arec;
Japanese: Areka Yashi, Binrou, Binrouju;
Kwar'Ae: Angiro;
Laos: Kok Hma:k;
Malaysia: Bluk, Blokn, Blukn, Blockn (Sakai), Kavugu, Pinang, Pinang Sirih, Pokok Pinang;
Nepalese: Supaarii;

New Ireland: Buai (Kuanua), Vua (Lamekot), Buei (Pala);
New Guinea: Hwatiy;
Palau: Bua;
Papua New Guinea: Buai;
Persian: Gird-Chob, Girdchob, Popal, Pupal;
Philippines: Takobtob (Bikol), Búñga (Bisaya), Boá, Buá (Ibanag), Boá (Iloko), Boá, Huá, Va (Itagon), Dapiau (Ivatan), Lúyos (Pampangan), Búñga, Lúgos (Sulu), Búñga, Tempak Siri, Nga Nga (Tagalog), Pasa (Yakan);
Pohnpei: Poc;
Polish: Palma Arekowa;
Portuguese: Areca Catecú, Arequeira, Arequiera, Noz De Areca, Noz De Bétele;
Russian: Areka Katekhu, Arekovaia Pal'ma, Betel'naia Pal'ma, Pal'ma Katekhu;
Santa Cruz Islands: Kalwa, Matapa;
Spanish: Areca, Bonga, Nuez De Areca, Nuez De Betel, Palma Catecu, Palma De Areca;
Sri Lanka: Puak, Puwak;
Swedish: Betel N T, Betelpalm;
Taiwan: Binlang;
Thailand: Luukmaak, Luuktan, Maak, Mak, Maak Khiaw, Maaklang; Maak Mia, Maak Muu (Shan); Phla, Pi Nae (Malay), Sae, Si Sa (Karen), Sa La (Khmer), Maak Sonk (Seeds) Maaksuan, Phonmaak, Siat;
Tibetan: Go Yu, Zu (M) Khan, Gla Gor Zo Sa, Kra Ma Ka (P), Kra Mra Sa, Sla Bor Se Sa (D);
Vietnam: Cau, Binh Lang, Tan Lang, May Lang (Tay), Po Lang (K'ho);
Yap: Bu.

Origin/Distribution

Areca nut is found originally in tropical southeast Asia in the Philippines and or the Malay Archipelago, truly wild populations presumably do not exist.

Agroecology

Being indigenous to the tropics, areca nut requires high temperature and high relative humidity. Its optimum temperature range hovers between 21°C and 32°C and annual average rainfall of 1,500–5,000 mm evenly distributed throughout the year. It is cold sensitive, intolerant of strong winds and drought. It grows at altitudes from near sea-level to 900 m in the humid tropical lowland, maritime tropical, subtropical wet and tropical wet forest. It thrives on a wide range of soil types from pH 5–8 including lateritic, calcareous, volcanic, loamy clays and loams that has good drainage and good water holding capacity. Areca nut prefers friable, moist and fertile deep clay loams rich in organic matter and does best in full sun but requires partial (50%) shading when young. Light sandy soils are unsuitable unless adequately irrigated and well manured.

Edible Plant Parts and Uses

In Asia and the Pacific, betel nuts are harvested at various stages of maturity and ripeness and sold in the market (Plates 2–6). Areca nuts minus the husk are also available in local markets (Plate 7).

In India and throughout much of Asia betel nuts are peeled, dried and often undergo a pickling process. They are then distributed in the markets and consumed as desired. The dry nuts are extremely hard and are processed by cracking, slicing or shredding. Cracked nuts ('supari') are often ground to a fine powder just before consumption in a chewing quid. Sliced nuts are chewed as they are. Shredded nuts are added to tobacco to make '*gutkha*', or to fennel seeds to make '*paan masala*'.

In Asia and the south Pacific, kernel (seed) of green and ripe fruit are sliced into thin small pieces (Plate 8) and chewed as an astringent and stimulant masticatory, often with the leaves or fruit of betel-pepper (*Piper betle*) and lime.

In Papau New Guinea and the Solomon Islands it is eaten with dakka (*Piper betle*) inflorescence or leaf. Betelnut chewing has recently spread to Vanuatu and Guam.

The betel leaves and areca juices are used ceremonially in Vietnamese weddings. The betel and areca represents important symbols of love and marriage and the Vietnamese phrase "matters of betel and areca" (*chuyện trầu cau*) is synonymous with marriage.

In India along with lime, betel nut is chewed together with clove or cardamom for extra flavouring, or in many cases chewed with a small quantity of tobacco. In Assam,

Plate 1 Clusters of areca fruits arising from the axils of leaf scars

Plate 2 Immature areca fruits

Plates 3 & 4 Semir-ripe areca fruit (*left*) and immature fruits (*right*)

Plate 5 Ripe detached areca fruits

Plate 7 Areca nut with husk removed

Plate 6 Bunches of ripe areca fruits

Plate 8 Slices of areca nut on sale in a Vietnamese market

India it is a tradition to offer *Pan-tamul* (Betel leaves and raw betel nut) to guests after tea or meals. In Assam betel nuts also have a variety of uses during religious and marriage ceremonies, where it takes on fertility symbolisms.

The terminal bud or palm heart or palm cabbage, is edible, and is either eaten raw as a salad or cooked. In Java, it is eaten as lalab or made into pickles. In the Philippines, the cabbage is eaten raw as salad, or cooked. The tender shoots are eaten after cooking in syrup. In the Philippines, the flowers are sometimes added to salads.

Botany

The palm has an erect, slender, cylindrical and solitary trunk, 15–25 m high, and marked with annular scars (vestiges of fallen fronds). The leaves are up to 2.5 m long, pinnate with numerous linear, dark green, leaflets, 60–90 cm long, with the upper ones confluent and petioles expanded at the base. Inflorescence consists of a much branched and compressed spadix encased in a caduceous spathe. Spadix has filiform branches above, bearing very numerous, somewhat distichous fragrant, male flowers, with three greenish-yellow sepals 5 mm long, three white petals, six stamens. Female flowers occur at the bases of the branches and in axils, about 1 cm long with identical perianth and a 3-celled ovary. The fruit is ovoid to ellipsoid, smooth, green or greenish-yellow turning orange or reddish-yellow when ripe, 4–6 cm long; with a thin somewhat fleshy pericarp and a hard, fibrous mesocarp (Plates 1–6) enclosing a conical, brown, hard kernel (seed) (Plate 7).

Nutritive/Medicinal Properties

Constituents of Areca nut expressed as percentage values (range) calculated on a dry basis (except moisture) were reported by Bavappa et al. (1982) follows with values for ripe nuts unbracketed and values for green nut in brackets: moisture: 38.9–56.7% (69.4–74.1%), total water extractives 23.3–29.9% (32.9–56.5%), polyphenols 11.1–17.8% (17.2–29.8%), arecoline (extraction method) 0.12–0.24% (0.11–0.14%), fat 0.12–0.24% (8.1–12.0%), crude fiber 11.4–15.4% (8.2–9.8%), total polysaccharides 17.8–25.7% (17.3–23.0%), crude protein 6.2–7.5% (6.7–9.4%), ash 1.1–1.5% (1.2–2.5%).

The areca seed was found to contain 50–60% sugars, 15% lipid (glyceride of lauric, myristic and oleic acid), 15% condensed tannins (phlobatannin, catechin), polyphenolics (NPF-86IA, NPF-86IB, NPF-86A and NPF-86B) (Reijiro et al. 1988) and 0.2–0.5% alkaloids (arecoline, arecaidine, guvacine and guvacoline) (Wetwitayaklung et al. 2006).

The nut is official in the British, Chinese, Danish, French, German, Swedish, Swiss and Vietnamese Pharmacopoeias.

Antioxidant Activity

Areca seed was found to have strong radical scavenger activities that could be considered as natural antioxidant for medicinal uses. Studies in Korea and Thailand reported that extracts of *Areca catechu* var. *dulcissima* showed higher antioxidant activity than resveratrol in all experiments (Lee et al. 2003). Most of the plant extracts used in this study inhibited the H_2O_2-induced apoptosis of Chinese hamster lung fibroblast (V79-4) cells. The extracts of *Areca catechu* var. *dulcissima, Paeonia suffruticosa, Alpinia officinarum, Glycyrrhiza uralensis*, and *Cinnamomum cassia* strongly enhanced viability against H_2O_2-induced oxidative damage in V79-4 cells. Relatively high levels of 1,1-diphenyl-2-picrylhydrazyl (DPPH) radical scavenging activity were detected in extracts of *Areca catechu* var. *dulcissima, Paeonia suffruticosa,* and *Cinnamomum cassia* ($IC_{50} < 6.0$ µg/ml). The activities of superoxide dismutase (SOD), catalase (CAT) and glutathione peroxidase (GPX) were dose-dependently enhanced in V79-4 cells treated with most of the plant extracts. These results suggested that the plant extracts prevented oxidative damage in normal cells probably because of their antioxidant characteristics.

Thai scientists reported that the aqueous and methanol extracts of the betel nut at various developmental stages presented higher concentrations of tannin and total phenols than extracts of the other parts of areca tree (Wetwitayaklung et al. 2006). The methanol extracts of the seeds at various ages gave higher antioxidant activities than the extracts of other aerial parts (leaves, crown-shafts, fruit shells (4 and 8 months), root and adventitious root). The sequence of antioxidant activities of methanol betel nut extracts from high to low were 4, 8, 6, 3, 2 and 1-month nut.

Antiaging, Antiinflammatory and Whitening Activities

Research suggested that the phenolic substance purified from *A. catechu* had an anti-ageing effect by protecting connective tissue proteins (Lee et al. 2001). Extract of *Areca catechu* was found to have significant inhibitory effects on the ageing and inflammation of skin tissues. The bioactive principle in the extract was found to be a phenolic compound with inhibitory activity against elastase. IC_{50} values of this phenolic substance were 26.9 µg/ml for porcine pancreatic elastase (PPE) and 60.8 µg/ml for human neutrophil elastase (HNE). This phenolic substance showed more potent activity than that of reference compounds, oleanolic acid (76.5 µg/ml for PPE, 219.2 µ/ml for HNE) and ursolic acid (31.0 µg/ml for PPE, 118.6 µg/ml for HNE). The phenolic substance from *A. catechu* effectively inhibited hyaluronidase activity (IC_{50}=210 µg/ml). The results suggested that the phenolic substance purified from *A. catechu* had an anti-ageing effect by protecting connective tissue proteins. Lee and Choi (1999a) found that *Areca catechu* extract (CC-516) containing a high proportion of proline (13%) of free amino acid content, exhibited 37 to 90% inhibition on elastase at 10 to 250 µg/ml concentration. CC-516 showed protection of elastic fibre against degradation by the enzyme in an ex-vivo assay. It increased proliferation of human fibroblast cell by 85% and collagen synthesis by 40%. The treatment with CC-516 improved skin hydration, the skin elasticity, and skin wrinkles suggesting that it could be as a new anti-aging component for cosmetics.

Studies indicated that *Areca catechu* ethanol extract had anti-inflammatory activity/anti-melanogenesis activity, and could be used as a new anti-aging component for cosmetics (Lee and Choi 1998). It enhanced collagen synthesis, improved skin hydration, skin elasticity and skin wrinkles. In another study Lee and Choi (199b) reported that, ethanolic extract, (CC-516) from *Areca catechu* showed potent anti-oxidative, free radical scavenging, and anti-hyaluronidase activity. Anti-oxidative effect of the extract (IC_{50}=45.4 µg/ml) was lower than butylated hydroxytoluene (IC_{50}=5 µg/ml), but similar to tocopherol and higher than ascorbic acid. The extract exhibited relatively high free radical scavenging activity (IC_50=10.2 µg/ml) compared to control. The extract inhibited effectively hyaluronidase activity (IC_{50}=416 µg/ml), exhibited inhibition in-vivo on delayed hypersensitivity as well as croton-oil induced ear edema in mice when it was topically applied. These results strongly suggested that the ethanolic extract may reduce immunoregulatory/inflammatory skin trouble. Also, from the results, the scientists elucidated that the ethanolic extract showed anti-allergic and anti-cytotoxicity activity. The whitening effect of the ethanolic extract was shown by the inhibition of mushroom tyrosinase activity with IC_{50} of 0.48 mg/ml and of melanin synthesis in B16 melanoma cells.

Anticancer Activity

Areca catechu, was found to be rich in tannins. A study found that condensed tannins from catechu potently inhibited animal fatty acid synthase (FAS) (Zhang et al. 2008). Among them, trimeric condensed tannin showed the most potent inhibition with IC_{50} of 0.47 µg/ml and it also exhibited strong time-dependent inhibition. Additionally, condensed tannins were found to suppress the growth of MCF-7 breast cancer cells, and the effect was related to their activity of FAS inhibition. FAS was augmented in breast cancers. The inhibition of both FAS activity and MCF-7 growth was exhibited by low concentrations of condensed tannins without FAS being precipitated. These results suggested that tannins in areca nut would be a valuable resource of bioactive substances.

An autophagy-inducing areca nut ingredient (AIAI) was identified in the partially purified

30–100 kDa fraction of areca nut extract (ANE), designated as ANE 30–100 K (Lin et al. 2008). Before disintegration, most ANE 30–100 K-treated cells exhibited rounding morphology, cytoplasmic clearance, and nuclear shrinkage, distinct from arecoline- and cisplatin-induced cellular apoptosis. The cytotoxicity of ANE 30–100 K was further shown to be sensitive to cellulase and proteinase K digestion suggesting AIAI in ANE 30–100 K to be a proteoglycan (or glycoprotein). Thus, although ANE contained apoptosis-inducing ingredients such as arecoline, it predominantly triggered autophagic cell death by this natural autophagy-inducing areca nut ingredient.

Wound Healing Activity

Seeds of areca were found to contain catechin, tannins (15%), gallic acid fat, gum and alkaloids like arecoline (0.07%) and arecaine (1%) (Azeez et al. 2007). Arecaidine and guvacoline, guvacine and choline were present in trace amounts. Among these, arecoline, a parasympathomimetic (cholinomimetic), was the most important alkaloid and exhibited potent cholinergic activity. It increased salivary, sweat and gastric secretions and contracted the pupils. When used in low dosage it stimulated the contractility of smooth muscles and in high doses it induced muscular paralysis. Research showed that the arecoline alkaloid, polyphenol of betelnut and the combined formulation enhanced the breaking strength in incision wound model in wistar rats (Azeez et al. 2007). All the extracts increased wound contraction on the 4th and 16th day and the period of epithelization. In the dead space model, only the areca alkaloid fraction enhanced the tensile strength of granulation tissue. The results showed that the alkaloid and polyphenols of betel nut could be used to enhance the healing of burn wounds, leg ulcers and skin graft surgery.

Parasympathomimetic/Laxative Activity

Betel nut extract was found to contain cholinomimetic and acetylcholinesterase (AChE) inhibitory constituent which acted as parasympathomimetic on muscarinic receptor (Anwar et al. 2004). At high dose, it acted on nicotinic receptor, increased smooth muscle tone, dilated blood vessel, decreased blood pressure and increased secretion (saliva and sweat). The crude extract of betel nuts or *Areca catechu* caused a dose-dependent (0.3–300 μg/ml) spasmogenic effect in the isolated rabbit jejunum. The spasmogenic effect was blocked by atropine, similar to that of acetylcholine (ACh), suggestive of muscarinic receptor mediated effect. Both the extract (0.3–10 μg/ml) and physostigmine (0.1–3.0 μM) potentiated the effect of a fixed dose of ACh (10 μM) in a dose-dependent fashion, suggesting acetylcholinesterase (AChE) inhibitory effect. In the in-vivo model of gastrointestinal transit, the crude extract (10–30 mg/kg) increased the travel of charcoal meal and also had a laxative effect in mice.

Antivenin Activity

Studies showed polyphenols from the extracts of *Areca catechu* had therapeutic potential against necrosis in snakebite victims (Leanpolchareanchai et al. 2009). The polyphenols inhibited phospholipase A(2), proteases, hyaluronidase and L-amino acid oxidase of *Naja naja kaouthia* Lesson and *Calloselasma rhodostoma* Kuhl venoms in in-vitro tests. In in-vivo assays, the extract inhibited the hemorrhagic activity of *Calloselasma rhodostoma* venom and the dermonecrotic activity of *Naja naja kaouthia* venom. The inhibitory activity of plant polyphenols against local tissue necrosis induced by snake venoms may be caused by inhibition of inflammatory reactions, hemorrhage, and necrosis.

Psychostimulant Activity

In Taiwan, the common reasons to chew betel-nut among the recruits after military services were curiosity (33.3%) and as a stimulant (29.8%) (Lin et al. 2004), but the mechanism of psychostimulant effect of areca quid was unknown. However, beneficial effects of areca nut were outweighed by the fact that long-term chewing of the nut was found to be carcinogenic.

Areca Nut Consumption and Cancer

Constituents of betel nut are potentially carcinogenic and this was confirmed by many recent studies. Long-term use had been associated with oral submucous fibrosis (OSF), pre-cancerous oral lesions (mouth wounds), and squamous cell carcinoma (cancer). Adverse, acute effects of betel chewing included worsening of asthma, low blood pressure, and rapid heart-beat (Lin et al. 2004). Areca quid caused strong adverse dental health, hastened teeth decay and discolored teeth and mouth but it was also implicated as a carcinogen that caused oral mucosa lesions and oral cancer. The International Agency for Research on Cancer (IARC) deemed betel nut to be a known human carcinogen. In Asian countries where it is commonly and frequently consumed, oral cancer was reported to comprise up to 50% of malignant cancers. Betel nut chewers in Taiwan were found to have a 28 times higher risk of acquiring oral cancer. Although a substantial proportion of the cancers were found to be caused by the tobacco rather than the betel nut and leaves in the quid, according to WHO, betel chewing without tobacco also led to cancer of the mouth. Chewing of betel nut with lime had been reported to cause mouth, cheek and tongue cancers. The awareness regarding submucous fibrosis, a premalignant condition characterized by excessive collagen production by mucosal fibroblasts had dealt a major blow to the masticatory use of areca nut. Researchers reported that polyphenols of areca stabilized collagen production by human and bacterial collagenases in a concentration dependent manner thus promoting the development of submucosal fibrosis following damage to the oral epithelium. Betel-nut chewing had also been reported to be one of several factors that affected asthma control and severity of attacks.

Betel Nut and Oral Cancer

Oral cavity squamous cell carcinoma (OSCC), a part of head and neck cancer (HNC) had been reported to be increasing in the betel quid chewing population in recent years (Chen et al. 2008). Head and neck cancer is one of the 10 most frequent cancers worldwide, with an estimated over 500,000 new cases being diagnosed annually. The overall 5-year survival rate in patients with head and neck cancer was one of the lowest among common malignant neoplasms and had not significantly changed during the last two decades. During 2006, OSCC had become the sixth most common type of cancer in Taiwan, and it was also the fourth most common type of cancer among men; wreaking a high social and personal cost. Environmental carcinogens such as betel quid chewing, tobacco smoking and alcohol drinking were identified as major risk factors for head and neck cancer. There is growing interest in understanding the relationship between genetic susceptibility and the prevalent environmental carcinogens for head and neck cancer prevention.

A variety of oral mucosal lesions and conditions had been reported in association with quid and tobacco use, and the association of these conditions with the development of oral cancer emphasized the importance of education to limit the use of quid (Avon 2004). In most cases, cessation of the habit was found to ameliorate incidence of mucosal lesions as well as in clinical symptoms. Quid is a mixture of substances that is placed in the mouth or actively chewed over an extended period, thus remaining in contact with the mucosa. It usually contains one or both of two basic ingredients, tobacco and areca nut. Betel quid or paan is a mixture of areca nut and slaked lime, to which tobacco can be added, all wrapped in a betel leaf. The specific components of this product vary between communities and individuals. Cultural and dietary risk factors had been reported to play an important role in oral precancer and cancer (Zain 2001). The quid habit had played a major social and cultural role in communities throughout the Indian subcontinent, Southeast Asia and locations in the western Pacific. Following migration from these countries to North America, predominantly to inner city areas, the habit had remained prevalent among its practitioners.

Studies showed that the markedly enhancing effects of areca nut extracts on peripheral blood mononuclear cells -released inflammatory cytokines, interleukin (IL)-6 and IL-8 might cause a sustained cytokine-rich inflammatory milieu in oral cavity of area quid chewers (Chang et al. 2006). Both ripe and tender areca nut extracts also modified mRNA expression of IL-6. These excessive cytokines from areca nut-treated immune cells may impair periodontal health. Chang et al. (2008) found that areca nut extracts augmented expression of inflammatory cytokines, tumour necrosis factor-α, interleukin-1beta, interleukin-6 and interleukin-8, in peripheral blood mononuclear cells. Both extracts of ripe areca nut (< or = 40 µg/ml) and extracts of tender areca nut significantly elevated the production of tumour necrosis factor-α and interleukin-1β in peripheral blood mononuclear cells in a dose-dependent and time-dependent fashion. The stimulatory effects of areca nut extracts on the secretion of tumour necrosis factor-α, interleukin-1β, interleukin-6 and interleukin-8 and on the mRNA expression of tumour necrosis factor-α, interleukin-1β and interleukin-6 at 4 hours of incubation were depressed by curcumin (20–50 µM). However, the level of interleukin-8 transcripts was not affected by curcumin. The complex cytokine profile induced by areca nut extracts-treated peripheral blood mononuclear cells implied the possibility of enhanced local inflammation and altered immune functions by the areca chewing habit. The inhibitory effects of curcumin on cytokine expression suggested that oxidative stress might be involved in areca nut extract-associated immune alteration.

Feeding Swiss albino mice of either sex (4 weeks old) with diets containing 0.25%, 0.5%, or 1% arecanut (w/w) for 5 weeks, caused a significant enhancement in hepatic levels of cytochrome b5, cytochrome P-450, malondialdehyde (MDA), and glutathione S-transferase (GST). The hepatic acid soluble sulfhydryl content was depressed by 0.5% and 1% arecanut diets (Singh and Rao 1995b). Further, long-term (36 weeks) feeding of a 1% arecanut diet in a group of mice evoked changes similar to those seen following treatment for 5 weeks. The modulatory influence of areca nut on biotransformation of hepatic enzymes and antioxidant levels were suggestive of its influence in the process of carcinogenesis caused by bioactivated electrophilic species of potential chemical carcinogens among habitual areca nut chewers. Additionally, animals placed on a 1% arecanut diet and treated with a standard two-stage protocol for tumour induction with 7,12-dimethylbenz[a]anthracene (DMBA) developed a 5.41 tumour burden (control value: 5.76) along with 100% incidence of mice bearing papillomas (control value: 94.4%), thus signifying that dietary intake of 1% areca nut for 18 weeks could not induce or alter the mouse skin tumorigenesis pattern. The researchers (Singh and Rao 1995a) also found that supplementation of mice diet with 0.5% or 1% BHA (2(3)-tert-butyl-4-hydroxy anisole) during the last 10 days of a 45 days 1% areca nut diet treatment significantly modulated the hepatic detoxification system enzymes, -SH content and malondialdehyde (MDA) levels in the liver of the mice. BHA-induced alterations in hepatic GST and -SH content were depressed while cytochrome b5, cytochrome P-450 and MDA levels were further enhanced by the arecanut treatment.

A case of oral lichen planus, a chronic mucocutaneous disease on the skin, tongue and oral mucosa, induced by betel quid use in a 79-year-old Cambodian woman was also reported (Stoopler et al. 2003). Recent studies in mice confirmed that areca nut extract may impair oral fibroblasts and then modulate the progression of oral epithelial oncogenesis via their secreted molecules (Lu et al. 2008). Studies showed that areca nut chewing was associated with an increase in the incidence of oral neoplastic or inflammatory diseases. Aberrations in matrix metalloprotease (MMP) expression were associated with the pathogenesis of oral diseases. Additional evidence was provided whereby matrix metalloprotease −2(MMP)-2 secreted from areca nut extract-induced senescent gingival fibroblasts facilitated the invasiveness of polymorphonuclear leukocytes, which could be associated with the oral inflammatory process in areca chewers (Lu et al. 2009).

Over-expression of epidermal growth factor receptor (EGFR) was found in 27.9% (48 of 172) of the areca quid associated oral cavity squamous cell carcinomas (OSCCs) and this was associated with lymph node metastasis and extracapsular spread (Huang et al. 2009). Only 1 (0.58%) OSCC displayed somatic EGFR mutation but in a silent form. The data indicated that EGFR-targeted therapy might have some potential in areca quid -associated OSCCs.

Lin et al. (2009) found that arecanut extract inhibited the growth of Chinese hamster ovary cells (CHO-K1) in a dose- and time-dependent fashion. Intracellular reactive oxygen species (ROS) levels and micronuclei (MN) frequency were significantly elevated following arecanut treatment in CHO-K1 cells. Arecanuta was found to cause G2/M arrest, cytokinesis failure in CHO-K1 cells, and an accumulation of hyperploid/aneuploid cells. These events were associated with an elevation in intracellular H_2O_2 level and actin filament disorganization.

Areca Nut and Oral Submucous Fibrosis (OSF)

Areca nut chewing was found to be the prime etiological factor in the pathogenesis of oral submucous fibrosis (OSF). A recent 2009 study found positive connective tissue growth factor (CTGF) staining in fibroblasts and endothelial cells in all 20 OSF cases caused by areca nut chewing (Deng et al. 2009). Connective tissue growth factor (CTGF) was associated with the onset and progression of fibrosis in many human tissues. The results indicated that arecoline-induced CTGF synthesis was mediated by ROS, NF-kappaB, JNK, P38 MAPK pathways and curcumin could be a useful agent in controlling OSF.

Areca quid chewing was found to be a major risk factor for oral submucous fibrosis and oral cancer. Clinical evidence suggested that the pathophysiology of such oral diseases was closely associated with immune deterioration. Exposure of naïve splenic lymphocytes to areca nut extract (ANE) significantly enhanced apoptosis in a time- and concentration-dependent manner (Wang et al. 2009). Further, an elevated level in the intracellular reactive oxygen species was detected in ANE-treated lymphocytes undergoing apoptosis. The results demonstrated the pro-apoptotic effect of ANE in primary lymphocytes, which was partially mediated by the activation of the mitochondrion-dependent pathway and oxidative stress.

Oral submucous fibrosis (OSF) was found to be associated with the habitual areca nut chewing; it was characterized by excessive collagen production by mucosal fibroblasts stimulated by areca nut extract (Harvey et al. 1986). Studies revealed that [3H]-arecoline was metabolized predominantly to [3H]-arecaidine concomitant with a concentration-dependent stimulation of collagen synthesis and cell proliferation. Arecaidine was potently more stimulating than arecoline. The rate of hydrolysis of a series of synthetic arecaidine esters (methyl, ethyl, butyl, propyl and pentyl) by fibroblasts was closely correlated with the degree of stimulation of collagen synthesis. Exposure of buccal mucosa fibroblasts to these alkaloids in-vivo may contribute to the accumulation of collagen in OSF. Another study reported that treatment of reconstituted collagen fibrils and pieces of rat dermis with the crude betel nut extract, purified tannins or (+)-catechin from betel nut increased their resistance to both human and bacterial collagenases in a concentration-dependent manner (Scutt et al. 1987). These tanning agents may stabilise collagen in-vivo following damage to the oral epithelium and may promote the occurrence of sub-epithelial fibrosis in betel nut chewers.

In a recent study spanning 2004–2008 in India involving 239 patients, 197 (82.4%) patients chewed areca nut/dohra, 14 (5.8%) were smokers and 2 (0.8%) patients were habitual alcoholics, 89(37.2%) patients reported difficulty in opening of the mouth (trismus), 51 (57.4%) patients were found to have stage II (2–3 cm) trismus while the remainder had stage I (>3 cm) and III (<2 cm) trismus (Pandya et al. 2009). The buccal mucosa was found to be the most commonly involved site. On the basis of histopathological examination, 52(21.7%) were classified as oral submucous fibrosis OSF grade I, 75(31.3%) patients as grade II and 112(46.8%) had grade III disease. The study indicated the widespread habit of chewing dohra/paan masala to be a major risk factor of OSF, especially in the younger age group. In this study, an increase in histopathological grading was found with severity and duration of addiction habit. However no significant correlation was found between clinical staging and histopathological grading.

In oral submucous fibrosis specimens, heme oxygenase-1 expression was significantly higher and expressed mainly by fibroblasts, endothelial cells, and inflammatory cells (Tsai et al. 2009). Heme oxygenase-1 (HO-1) is known as a stress-inducible protein and functions as an antioxidant enzyme. OSF demonstrated significantly higher HO-1 mRNA expression than buccal mucosa fibroblasts. Arecoline was also found to increase HO-1 mRNA and protein expression in a dose-dependent manner. Overall, the results demonstrated that HO-1 expression was significantly upregulated in OSF from areca quid chewers, and arecoline may be responsible for the enhanced HO-1 expression in-vivo.

A 2008 study indicated that hypermethylation may be involved in the pathogenesis of oral precancerous lesions associated with betel-quid chewing in Sri Lanka (Takeshima et al. 2008). A high frequency of hypermethylation of p14, p15 and p16 was detected in the pre-cancerous lesions including epithelial dysplasia and submucous fibrosis, although no hypermethylation was found in normal epithelium. The frequency of hypermethylation was higher than that of positive staining for p53 mutation except in the case of p16 in mild dysplasia. No significant correlation was observed between p53-positive reactions and hypermethylation in any lesions. The hypermethylation was highly detectable even in p53-negative lesions, suggesting that hypermethylation of p14, p15 and p16 occurred regardless of whether the lesions had p53 mutations or not.

Histopathological features of oral submucous fibrosis (OSMF) based on biopsy of 35 specimens were obtained from patients presenting with oral submucous fibrosis (Isaac et al. 2008). Diffuse nonspecific chronic inflammation with fibrosis was found in all specimens (100%). Classical picture of OSMF was present in 20 (57.4%) of specimens while 16 (45.7%) exhibited Lichenoid reaction as well. OSMF with ulceration was detected in 14 (40%) specimens. Pseudoepitheliomatous hyperplasia was found in 9 (25.7%) specimens. Dysplastic changes were detected in 3 (8.6%) specimens.

Betel Nut and Leukoplakia

Studies conducted on the population in Taiwan on the linkage of betel nut chewing, smoking and alcohol on the occurrence of leukoplakia and its malignant transformation to oral carcinoma reported that cessation of smoking may reduce by 36% leukoplakia cases, while elimination of betel nuts may prevent 62% of leukoplakia and 26% of malignant transformation to oral carcinoma (Shui et al. 2000). The annual incidence rate (per year) of leukoplakia was estimated as 0.35% in male betel quid chewers (Yen et al. 2008). The average dwelling times (ADTs) were 24 years for leukoplakia and 7 years for erythroleukoplakia. Annual incidence rate of leukoplakia with high consumption and long duration of betel quid and smoking was higher. Both quantity and duration of smoking and betel quid chewing were found to play minor roles in the influence of ADT. The risks of developing oral cancer after 20 years of follow-up were 42.2% for leukoplakia and 95.0% for erythroleukoplakia. The study elucidated effects of betel quid chewing and smoking on multistate progressions between oral pre-malignancies. These results could be applied to predict long-term risk of malignant transformation varying with different duration and quantity of betel quid and cigarette.

Betel Nut and Periodontitis

In a study of 64 patients with the habits of chewing betel nut, the occlusal surfaces of all patients were found to have mild to severe abrasion (Yin et al. 2003). The occlusal abrasion became more severe with the increase in chewing time and frequency. The severe abrasion was accompanied by periapical periodontitis and the resorption of alveolar bone. The study indicated that long-term chewing betel nut could result in the abrasion of occlusal surface.

Betel Nut and Oesophageal Cancer

A study was conducted in Taiwan to investigate the impact of different habits of betel nut chewing on esophageal squamous cell carcinoma (SCC) and its interaction with cigarette use and alcohol consumption (Wu et al. 2006). Oesophageal cancer was the sixth leading cause of death among men in Taiwan, but it was the fastest increasing (70%) alimentary tract cancer. The results revealed that smoking, alcoholic beverage drinking and low education level were independent risk factors for esophageal cancer. Although betel nut chewers only had a borderline significant higher risk than non-chewers, those who chewed with a piece of betel inflorescence and swallow betel-quid juice had a significant higher risk. Betel nut chewing was found to play a relevant role in the development of esophageal SCC and compounded the carcinogenetic effect of smoking and alcohol drinking. Direct mucosal contact of betel juice may contribute to its carcinogenesis.

A 2004 paper in India reported that in 90 patients with esophagus cancer studied, the risk for esophageal cancer was 3.5 times higher with alcohol consumption, 2.5 times higher for tobacco users, and 2.8 times higher each for betel nut chewers and smokers (Chitra et al. 2004). The calculated odds ratio for the social habits and diet factors was significant amongst cases of cancer esophagus.

Betel Nut and Genomic Damage (Genotoxicity)

Cytogenetic studies revealed the genomic damage caused by areca nut consumption (without tobacco and not as a component of betel quid), among areca nut chewers, which included normal people who chewed areca nuts, patients with oral submucous fibrosis, and patients with oral cancer (Dave et al. 1992). The analysis revealed statistically significant increases in the frequencies of sister chromatid exchanges and chromosome aberrations in peripheral blood lymphocytes and the percentage of micronucleated cells in exfoliated cells of buccal mucosa among all three groups of chewers when compared with those of the controls. Data highlighted that this popular masticatory was erroneously considered "safe" and that it enhanced the genomic damage further when chewed without tobacco.

In a separate study, no detectable levels of clastogenic activity were observed in the saliva of non-chewing individuals; however, after 5 minutes of chewing betel quid, betel nut, betel leaves, quid ingredients and Indian tobacco, the saliva samples showed relatively potent clastogenic activities (chromosome aberrations such as chromatid breaks and chromatid exchanges) in Chinese hamster ovary (CHO) cells (Stitch and Stitch 1982). The addition of transition metals Mn^{2+} and Cu^{2+} to the saliva samples of betel nut and Indian tobacco chewers enhanced their clastogenic activities, whereas Fe^{3+} increased the clastogenicity of the betel nut saliva but decreased the genotoxic effect of the saliva of Indian tobacco chewers. After removal of the betel quid or its components from the mouth, the clastogenic activity disappeared within 5 minutes. The western-type chewing tobacco did not produce a genotoxic activity in the saliva of chewers. A possible association between the genotoxicity in the saliva of betel quid chewers and the development of oral, pharyngeal and esophageal carcinomas was also discussed.

Betel Nut and Liver Cirrhosis/ Hepatocellular Carcinoma

A recent population based study in Taiwan showed that betel chewing enhanced liver cirrhosis (LC) and hepatocellular carcinoma (HCC) risk 4.25-fold in current chewers and 1.89-fold in ex-chewers versus never-chewers. (Wu et al. 2008). Subjects without hepatitis B/C infections had 5.0-fold increased risk of LC/HCC versus never-chewers, and betel chewing had an additive synergistic effect on hepatitis B/C-related risks. Risk reduction with betel habit cessation could exceed that expected from immunization programmes for hepatitis B and C. The study showed that increased risks of cirrhosis and hepatocellular cancer were found in betel chewers free of hepatitis B/C infection, and these risks were synergistically additive to those of hepatitis B/C infections. Estimated risk reduction from effective anti-betel chewing programmes would be considerable. Another aetiological study on the role of betel quid and hepatocellular carcinoma (HCC) showed that risk on HCC increased as duration and quantity of betel quid chewing increased (Tsai et al. 2001). Multivariate analysis indicated that betel quid chewing, serum hepatitis B surface antigen HBsAg, antibodies to hepatitis C virus and educational duration of less than 10 years were independent risk factors of HCC. Further, an additive interaction was found between betel quid chewing and chronic infection with either hepatitis B virus (synergy index, 5.37) or hepatitis C virus (synergy index, 1.66).

Betel Nut and Diabetes

A population-based cross-sectional survey and a multiple-disease-screening programme showed that areca nut chewers had higher age-adjusted prevalence rates for hyperglycaemia (11.4% versus 8.7%) and type 2 diabetes (10.3% versus 7.8%) compared with non-chewers (Tung et al. 2004). Areca nut chewing independently enhanced the risk of hyperglycaemia and type 2 diabetes. The independent effects of duration of chewing were dose-dependent for type 2 diabetes as were the effects of increased rates of areca nut chewing. Similar findings were observed for hyperglycaemia. The findings indicated that the habit of chewing areca nut independently contributed to the risk of both hyperglycaemia and type 2 diabetes in Taiwanese men and that this association was dose-dependent with respect to the duration of areca nut use and the quantity of areca nut chewed per day.

Mannan et al. (2000) found that of CD1 mice fed with betel-nut or associated nitrosamines 8.5% developed glucose intolerance with marked obesity. Many nitrosamines were diabetogenic, causing both type 2 and type 1 diabetes. Of 993 'healthy' Bangladeshis examined for glycaemia and anthropometric risk markers for type 2 diabetes in relation to betel usage, 12% had known diabetes. The findings revealed that waist size was strongly related to betel usage and independent of other factors such as age. Betel use interacted with sex, relating to increasing glycaemia only in females. Since waist and age were the prominent markers of increasing glycaemia the researchers suggested that betel chewing, a habit common to about 10% of the world population (more than 200 million people) may contribute to the risk of developing type 2 diabetes mellitus.

Recent studies showed that in subjects with type 2 diabetes, the severity of these variables was related to glycemic levels rather than gutka consumption (Javed et al. 2008). Gutka is a preparation of crushed betel nut, tobacco, catechu, lime and sweet or savory flavourings.

Studies using a cross-sectional telephone survey conducted in Taiwan involving a total of 81,226 (37,226 men and 44,000 women) patients with type 2 diabetes mellitus (T2DM) revealed that betel nut chewing was significantly associated with hypertension in Taiwanese patients with T2DM and the association was stronger in women (Tseng 2008). Hypertension was defined by a positive history or reported systolic blood pressure > or =140 mm Hg and/or diastolic blood pressure > or =90 mm Hg. Betel nut chewing was significantly associated with blood pressure, with regression coefficients of 0.958 for systolic and 0.441for diastolic blood pressure in men; and the respective values for women were 1.805 and 1.198.

Betel Nut and Asthma

In the UK, the rate of hospital admission for acute asthma was found to be higher among Asians than among other groups in the population; betel-nut chewing was postulated to be one of several contributing factors. (Taylor et al. 1992). Two Asian patients admitted to hospital with acute severe asthma had been chewing betel nut immediately before the attacks. In-vitro, arecoline caused dose-related contraction of human bronchial smooth-muscle strips, with one-tenth the potency of methacholine. In a double-blind challenge study, inhalation of arecoline caused bronchoconstriction in six of seven asthmatic patients and one of six healthy subjects; methacholine caused bronchoconstriction in all the asthmatic patients and in five controls.

Betel Nut and Pregnancy

A survey of 1264 Aboriginal women in Taiwan revealed that maternal areca nut chewing during pregnancy was found to be significantly associated with both birth weight loss (-89.54 g) and birth length reduction (-0.43 cm) (Yang et al. 2008). A significantly lower male newborn rate was observed among aboriginal women with a habit of betel quid chewing during pregnancy. The use of this substance conveyed a 2.40- and 3.67-fold independent risk of low birth weight and full-term low birth weight, respectively. An enhanced risk (aOR = 3.26–5.99) of low birth weight was observed among women concomitantly using betel quid, cigarette and alcohol during gestation. The findings suggested that betel quid chewing during pregnancy had a significant effect on a number of birth outcomes, including sex ratio at birth, lower birth weight and reduced birth length.

Betel Nut and Schizophrenia

In a preliminary study conducted in the North Colombo Teaching Hospital, Sri Lanka, it was observed that a higher proportion of patients with schizophrenia chewed betel compared with control subjects. The frequency of chewing betel was also higher among the patients with schizophrenia (Kuruppuarachchi and Williams 2003). In contrast, an earlier study from Micronesia reported that betel chewing may in fact have a beneficial effect on patients with schizophrenia in terms of reducing both positive and negative symptoms (Sullivan et al. 2000). They postulated that the muscarinic agonist action of the betel nut alkaloid, arecoline, may provide an explanation.

Betel Nut and Heart Disease

Non-pregnant adults aged 20–64 years (n = 1932, 52% women) from the nationally representative Nutrition and Health Survey in Taiwan (1993–1996) were studied for independent associations between betel-quid use and heart disease after adjustment for lifestyle factors, age, obesity, diabetes mellitus, hypertension, and levels of serum total cholesterol and HDL cholesterol (Guh et al. 2007). Findings showed that the prevalence of betel-quid use was higher in men than in women (31% compared with 2.4%). The prevalence of heart disease was not significantly different between men and women (3.3% compared with 2.3%). The prevalence of betel-quid use decreased, whereas the prevalence of heart disease increased, with age. Betel-quid users were younger, drank more, had a lower dietary fruit intake, had a higher Framingham risk score, and had higher serum triacylglycerol concentrations than did the nonusers. Betel-quid use was found to be independently associated with heart disease in women.

Traditional Medicinal Uses

The seed has anthelmintic, antifungal, antibacterial, anti-inflammatory stimulant, astringent, vermifuge and antioxidant activities. It is reported that some time back women in Malaya use the young green shoots as an abortifacient in early pregnancy. Large doses of areca nut cause vomiting and diarrhoea. In China the bark is used for choleric affections, and for flatulence, dropsical, and obstructive diseases.

Its seed is also claimed as taeniacide and is used to remove tapeworms and other intestinal worms (vermifuge). Its therapeutic uses include treatment for oliguria, oedema, dysuria, indigestion, gingivitis, bleeding from the gastrointestinal tract, menorrhagia, menstrual disorders, hyperemesis of pregnancy and skin infections. Ash of the fruit is externally used for skin infection with coconut oil. Areca nut in decoction is considered an abortifacient and the nut as an emmenagogue. Seeds are said to be purgative. It is also a powerful sialogogue.

Other Uses

Areca nut is often planted as an ornamental. It provides timber that is used for walling, flooring, battens, elaborate crematory and temporary structures. Tannins are a by-product of boiling the nuts during processing the commercial product. An extract of betel nut provides black and red dyes. Lipid from the betel nut is used as an extender for cocoa butter. Powdered betel nut is used as a constituent in some tooth powders. Extracts of the seeds are also insecticidal and larvicidal. Dry seeds are used as taeniacide especially in horses. The nuts, husks, young shoots, buds, leaves, and roots are used in various medicinal preparations. Fallen fronds, bracts, inflorescences and husks removed from the fruits during processing could be used for fuel; culled trees could also be used as firewood. The tough leaf bases are used in hats, inner soles for slippers, and is an excellent paper pulp source. Husks are used for insulating wool, boards, and for manufacturing furfural (a solvent). In the Philippines, the husk is used to make toothbrushes. The leaf sheaths and spathes are used as wrapping and as a substitute for cardboard packing material. In the Philippines, the leaf sheaths are used to make book covers. In Sri Lanka, the leaf sheaths are used as plates, bags, and for wrapping.

Betel nut chewing is socio- culturally important in many Asian and Pacific societies. It is used as a masticatory stimulant in the Pacific nations of Papua New Guinea, the Solomon Islands, Fiji, Vanuatu, Palau, Guam, Yap, much of Micronesia, Taiwan, the Philippines, as well as in Malaysia, Indonesia, Thailand, Laos, Cambodia, and Vietnam. The fragrant flowers are used in weddings and funerals in some southeast Asian countries. Furthermore, the whole inflorescences are used in religious rituals in Sri Lanka and are displayed on the front of vehicles during pilgrimages, to bring good luck. The trunks are used to construct crematory and temporary ceremonial structures in several Asian countries.

Comments

Betel nut is a major commercial economic crop in Papua New Guinea, the Solomon Islands, and the Caroline Islands of Micronesia.

Selected References

Anwar H, Gilani M, Ghayur N, Saify ZS, Ahmed SP, Choudhary MI, Khalid A (2004) Presence of cholinomimetic and acetylcholinesterase inhibitory constituents in betel nut. Life Sci 75:2377–2389

Avon SL (2004) Oral mucosal lesions associated with use of quid. J Can Dent Assoc 70:244–248

Azeez S, Amudhan S, Adiga S, Rao N, Rao N, Udupa LA (2007) Wound healing profile of *Areca catechu* extracts on different wound models in wistar rats. Kuwait Med J 39(1):48–52

Backer CA, van den Brink RCB Jr (1968) Flora of Java (spermatophytes only), vol 3. Wolters-Noordhoff, Groningen, 761 pp

Bavappa KVA, Nair MK, Kumar TP (eds) (1982) The arecanut palm (Areca catechu Linn.). Central Plantation Crops Research Institute, Kasaragod, 340 pp

Brontonegoro S, Wessel M, Brink M (2000) *Areca catechu* L. In: van der Vossen HAM, Wessel M (eds) Plant resources of South-East Asia no. 16. Stimulants. Backhuys Publishers, Leiden, pp 51–55

Burkill IH (1966) A dictionary of the economic products of the Malay Peninsula. Revised reprint, 2 vols. Ministry of Agriculture and Co-operatives, Kuala Lumpur, vol 1 (A—H), pp 1–1240, vol 2 (I—Z), pp 1241–2444

Chang LY, Wan HC, Lai YL, Liu TY, Hung SL (2006) Enhancing effects of areca nut extracts on the production of interleukin-6 and interleukin-8 by peripheral blood mononuclear cells. J Periodontol 77(12): 1969–1977

Chang LY, Wan HC, Lai YL, Kuo YF, Liu TY, Chen YT, Hung SL (2008) Areca nut extracts increased expression of inflammatory cytokines, tumor necrosis factor-alpha, interleukin-1beta, interleukin-6 and interleukin-8, in peripheral blood mononuclear cells. J Periodont Res 44(2):175–183

Chen YJ, Chang JT, Liao CT, Wang HM, Yen TC, Chiu CC, Lu YC, Li HF, Cheng AJ (2008) Head and neck cancer in the betel quid chewing area: recent advances in molecular carcinogenesis. Cancer Sci 99(8):1507–1514

Chitra S, Ashok L, Anand L, Srinivasan V, Jayanthi V (2004) Risk factors for esophageal cancer in Coimbatore, Southern India: a hospital-based case-control study. Indian J Gastroenterol 23(1):19–21

Dave BJ, Trivedi AH, Adhvaryu SG (1992) Role of areca nut consumption in the cause of oral cancers. A cytogenetic assessment. Cancer 70(5):1017–1023

Deng YT, Chen HM, Cheng SJ, Chiang CP, Kuo MY (2009) Arecoline-stimulated connective tissue growth factor production in human buccal mucosal fibroblasts: modulation by curcumin. Oral Oncol 45(9):e99–e105

Facciola S (1990) Cornucopia. A source book of edible plants. Kampong Publications, Vista, 677 pp

Guh JY, Chen HC, Tsai JF, Chuang LY (2007) Betel-quid use is associated with heart disease in women. Am J Clin Nutr 85(5):1229–1235

Harvey W, Scutt AM, Meghji S, Caniff JP (1986) Stimulation of human buccal mucosal fibroblast in vitro by betel nut alkaloid. Arch Oral Biol 31:45–49

Huang Z, Xiao B, Wang X, Li Y, Deng H (2003) Betel nut indulgence as a cause of epilepsy. Seizure 12(6):406–408

Huang SF, Chuang WY, Chen IH, Liao CT, Wang HM, Hsieh LL (2009) EGFR protein overexpression and mutation in areca quid-associated oral cavity squamous cell carcinoma in Taiwan. Head Neck 31(8):1068–1077

Isaac U, Issac JS, Ahmed Khoso N (2008) Histopathologic features of oral submucous fibrosis: a study of 35 biopsy specimens. Oral Surg Oral Med Oral Pathol Oral Radiol Endod 106(4):556–560

Javed F, Altamash M, Klinge B, Engström PE (2008) Periodontal conditions and oral symptoms in gutka-chewers with and without type 2 diabetes. Acta Odontol Scand 66(5):268–273

Kuruppuarachchi KALA, Williams SS (2003) Betel use and schizophrenia. Br J Psychiatry 182:455

Leanpolchareanchai J, Pithayanukul P, Bavovada R (2009) Anti-necrosis potential of polyphenols against snake venoms. Immunopharmacol Immunotoxicol 31(4):556–562

Lee KK, Choi JD (1998) *Areca catechu* L. extract. II. Effects on inflammation and melanogenesis. J Soc Cosmet Chemists 49(6):351–352

Lee KK, Choi JD (1999a) The effects of *Areca catechu* L extract on anti-inflammation and anit-melanogensis. Int J Cosmet Sci 21(4):275–84.

Lee KK, Cho J, Park E, Choi JD (2001) Anti-elastase and anti-hyaluronidase of phenolic substance from *Areca catechu* as a new anti-ageing agent. Int J Cosmet Sci 23(6):341–346

Lee SE, Hwang HJ, Ha JS, Jeong HS, Kim JH (2003) Screening of medicinal plant extracts for antioxidant activity. Life Sci 73:167–179

Lin YS, Chu NF, Wu DM, Shen MH (2004) Prevalence and factors associated with the consumption of betel-nut among military conscripts in Taiwan. Eur J Epidemiol 19:343–351

Lin MH, Liu SY, Liu YC (2008) Autophagy induction by a natural ingredient of areca nut. Autophagy 4(7):967–968

Lin CC, Chang MC, Chang HH, Wang TM, Tseng WY, Tai TF, Yeh HW, Yang TT, Hahn LJ, Jeng JH (2009) Areca nut-induced micronuclei and cytokinesis failure in Chinese hamster ovary cells is related to reactive oxygen species production and actin filament deregulation. Environ Mol Mutagen 50(5):367–374

Lu HH, Liu CJ, Liu TY, Kao SY, Lin SC, Chang KW (2008) Areca-treated fibroblasts enhance tumorigenesis of oral epithelial cells. J Dent Res 87(11):1069–1074

Lu HH, Chen LK, Cheng CY, Hung SL, Lin SC, Chang KW (2009) Areca nut extract-treated gingival fibroblasts modulate the invasiveness of polymorphonuclear leukocytes via the production of MMP-2. J Oral Pathol Med 38(1):79–86

Mannan N, Boucher BJ, Evans SJ (2000) Increased waist size and weight in relation to consumption of *Areca catechu* (betel-nut); a risk factor for increased glycaemia in Asians in East London. Br J Nutr 83(3):267–273

National Institute of Materia Medica (1999) Selected medicinal plants in Vietnam, vol 1. Science and Technology Publishing House, Hanoi, 439 pp

Padmaja PN, Bairy KL, Kulkarni DR (1994) Effects of polyphenols of areca in wound healing. Fitoterapia 45:298–300

Pandya S, Chaudhary AK, Singh M, Singh M, Mehrotra R (2009) Correlation of histopathological diagnosis with habits and clinical findings in oral submucous fibrosis. Head Neck Oncol 1(1):10

Reijiro U, Toshiharu M, Masaya I, Yasuhiro T, Akira F (1988) New 5 –nucleotidase inhibitors NPF-86IA, NPF-86IB, NPF-86 A and NPF-86 B from *Areca catechu*. Part: Isolation and biological properties. Planta Med 54:419–422

Scutt AM, Meghji S, Caniff JP, Harvey W (1987) Stabilization of collagen by betel nut polyphenol as a mechanism in oral submucous fibrosis. Experimentia 43:391–393

Shiu MN, Chen TH, Chang SH, Hahn LJ (2000) Risk factors for leukoplakia and malignant transformation to oral carcinoma: a leukoplakia cohort in Taiwan. Br J Cancer 82(11):1871–1874

Singh A, Rao AR (1995a) Modulatory influence of areca nut on antioxidant 2(3)-tert-butyl-4-hydroxy anisole-induced hepatic detoxification system and antioxidant defence mechanism in mice. Cancer Lett 91:107–114

Singh A, Rao AR (1995b) Modulatory influence of areca-nut on the mouse hepatic xenobiotic detoxication system and skin papillomagenesis. Teratog Carcinog Mutagen 15(3):135–146

Staples GW, Bevacqua RF (2006) *Areca catechu* (betel nut palm), ver. 1.3. In: Elevitch CR (ed) Species profiles for Pacific Island Agroforestry. Permanent Agriculture Resources (PAR), Holualoa. http://www.traditionaltreeorg

Stich HF, Stich W (1982) Chromosome-damaging activity of saliva of betel nut and tobacco chewers. Cancer Lett 15:193–202

Stoopler ET, Parisi E, Sollecito TP (2003) Betel quid-induced oral lichen planus: a case report. Cutis 71(4):307–311

Sullivan RJ, Allen JS, Otto C, Tiobech JA, Nero K (2000) Effects of chewing betel nut (*Areca catechu*) on the symptoms of people with schizophrenia in Palau. Micronesia Br J Psychiatry 177:174–178

Takeshima M, Saitoh M, Kusano K, Nagayasu H, Kurashige Y, Malsantha M, Arakawa T, Takuma T, Chiba I, Kaku T, Shibata T, Abiko Y (2008) High frequency of hypermethylation of p14, p15 and p16 in oral pre-cancerous lesions associated with betel-quid chewing in Sri Lanka. J Oral Pathol Med 37(8):475–479

Taylor RF, al-Jarad N, John LM, Conroy DM, Barnes NC (1992) Betel-nut chewing and asthma. Lancet 339(8802):1134–1136

Tsai JF, Chuang LY, Jeng JE, Ho MS, Hsieh MY, Lin ZY, Wang LY (2001) Betel quid chewing as a risk factor for hepatocellular carcinoma: a case-control study. Br J Cancer 84(5):709–713

Tsai CH, Yang SF, Lee SS, Chang YC (2009) Augmented heme oxygenase-1 expression in areca quid chewing-associated oral submucous fibrosis. Oral Dis 15(4):281–286

Tseng CH (2008) Betel nut chewing is associated with hypertension in Taiwanese type 2 diabetic patients. Hypertens Res 31(3):417–423

Tung TH, Chiu YH, Chen LS, Wu H-M, Boucher BJ, Chen TH-H (2004) A population-based study of the association between areca nut chewing and type 2 diabetes mellitus in men (Keelung Community-based Integrated Screening programme no. 2). Diabetologia 47(10):1776–1781

Wang CC, Liu TY, Cheng CH, Jan TR (2009) Involvement of the mitochondrion-dependent pathway and oxidative stress in the apoptosis of murine splenocytes induced by areca nut extract. Toxicol In Vitro 23(5):840–847

Wetwitayaklung P, Phaechamud T, Limmatvapirat C, Keokitichai S (2006) The study of antioxidant capacity in various parts of *Areca catechu* L. Naresuan Univ J 14(1):1–14

Wu IC, Lu CY, Kuo FC, Tsai SM, Lee KW, Kuo WR, Cheng YJ, Kao EL, Yang MS, Ko YC (2006) Interaction between cigarette, alcohol and betel nut use on oesophageal cancer risk in Taiwan. Eur J Clin Invest 36(4):236–241

Wu GH-M, Boucher BJ, Chiu Y-H, Liao C-S, Chen TH-H (2008) Impact of chewing betel-nut (*Areca catechu*) on liver cirrhosis and hepatocellular carcinoma-: a population-based study from an area with a high prevalence of hepatitis B and C infections. Public Health Nutr 12(1):129–135

Yang MS, Lee CH, Chang SJ, Chung TC, Tsai EM, Ko AM, Ko YC (2008) The effect of maternal betel quid exposure during pregnancy on adverse birth outcomes among aborigines in Taiwan. Drug Alcohol Depend 95(1–2):134–139

Yen AM, Chen SC, Chang SH, Chen TH (2008) The effect of betel quid and cigarette on multistate progression of oral pre-malignancy. J Oral Pathol Med 37(7):417–422

Yin XM, Peng JY, Gao YJ (2003) Clinical study on the relationship between tooth abrasion and the habits of chewing betel nut. Hunan Yi Ke Da Xue Xue Bao 28(2):171–173

Zain RB (2001) Cultural and dietary risk factors of oral cancer and precancer – a brief overview. Oral Oncol 37:205–210

Zhang SY, Zheng CG, Yan XY, Tian WX (2008) Low concentration of condensed tannins from catechu significantly inhibits fatty acid synthase and growth of MCF-7 cells. Biochem Biophys Res Commun 371(4):654–658

Areca triandra

Scientific Name

Areca triandra Roxb. ex Buch.-Ham.

Synonyms

Areca alicae W. Hill. ex F. Muell., *Areca borneensis* Becc., *Areca humilis* Blanco ex H. Wendl., *Areca laxa* Buch.-Ham., *Areca nagensis* Griff., *Areca polystachya* (Miq.) H. Wendl., *Areca triandra* var. *bancana* Scheff., *Nenga nagensis* (Griff.) Scheff., *Ptychosperma polystachyum* Miq.

Family

Arecaceae

Common/English Names

Australian Areca Palm, Bungua, Clumping Betel Nut, Triandra Palm, Wild Areca Palm

Vernacular Names

Brazil: Areca Bangua, Areca-Vermelha (Portuguese);
Burmese: Taw-Kun-Thi;
Chinese: San Yao Bing Lang;
French: Aréquier Bangua;
India: Uvai (Mizoram);
Indonesia: Buring Utan;
Japanese: Kabu-Dachi-Binro;
Malaysia: Pinang Rumpun;
Portuguese: Areca Bangua;
Thailand: Mak Cha Waek , Mak Ling (Chanthaburi), Mak Nang Ling (Trat), Mak No, Mak Iak (Northern), Mak Hom (Bangkok), Mak Khiao (Narathiwat), Krue-Do (Malay-Narathiwat).

Origin/Distribution

Triandra palm is indigenous to tropical and subtropical Asia – India, Andaman Islands, Bangladesh, and southeast Asia – Myanmar, Thailand, Peninsular Malaysia.

Agroecology

Triandra palm is a tropical to subtropical species. The species is found in mixed hill and montane dipterocarp forest and kerangas forest, from 600 to 1,300 m above sea level. An easily grown palm for both the tropics and sub-tropics, it needs shade when young, as well as a moist, well-drained rich soil. It is drought and frost tender. *Areca triandra* is much smaller but hardier than *Areca catechu*. Propagation is by seed which has a short viability of 2 weeks.

Edible Plant Parts and Uses

The nut is edible and is used as a masticatory stimulant but is deemed an inferior substitute for *Areca catechu*. The ubod, or palm heart or palm cabbage, is edible, and is either eaten raw as a salad or cooked.

Botany

The monoecious, small, clump-forming, evergreen palm has clustered, slender, erect, grey smooth and ringed stems (5–10 cm diameter) and grows to about 7 m tall, stilt roots may be present and the crown shaft is slightly swollen and green (Plate 1). Leaves are dark green and pinnate to 2 m long on petioles up to 3 m long; pinnae are broad, 3.5–5 cm wide, truncate, toothed at the apex. Inflorescence 15–30 cm by 5–15 cm, prophyll 30 cm long with cream-coloured, lemon–scented flowers on green stalks, and formed on the branching spadix below the crownshaft. Male flowers are restricted to the apical half of the spike with both male and female (one large female with a small male either side) together towards the base of the spike either in a spiral or two-ranked arrangement. Fruit, ellipsoid narrowing to a beaked tip, up to 2.5 cm long and turning from green to brown or orange-red to scarlet when ripe (Plates 2–3).

Nutritive/Medicinal Properties

The active alkaloid in *Areca catechu* and the other *Areca* species including *A. triandra* contains arecoline a very potent alkaloid that should only be consumed in small quantities at a time. Arecoline, a major areca nut alkaloid, is considered to be the most important etiologic factor in areca quid chewing which has been linked to oral submucous fibrosis and oral cancer the areca nut (Chang et al. 2001). 2 mg of the pure alkaloid is a strongly

Plate 1 Clumping habit of Triandra palm

Plate 2 Cluster of triandra fruits below the crown shaft

Plate 3 Ellipsoid triandra fruits at various stages of maturity

stimulating dose and it is recommended not to exceed 5 mg at once. The maximum dosage of the nuts should in any case be less than 4 g, while 8 g can already be fatal. The alkaloid content ranges between 0.3% and 0.6%. Nicotine has a synergistic effect with arecoline, which explains the popular combination of betel nuts with tobacco but this can enhance the risk of oral cancers. Studies in Taiwan reported that arecoline may render human buccal mucosal fibroblasts more vulnerable to other reactive agents in cigarettes (Chang et al. 2001). The compounds of tobacco products may act synergistically in the pathogenesis of oral mucosal lesions in areca quid chewers. The data presented partly explain why patients who combined the habits of areca quid chewing and cigarette smoking are at greater risk of contracting oral cancer.

Other Uses

Triandra palm is commonly cultivated as an ornamental palm. The leaves are used for thatching and the trunk used as posts.

Comments

It is propagated from seeds which takes 6–10 months to germinate.

Selected References

Backer CA, van den Brink RCB Jr (1968) Flora of Java, vol 3. Wolters-Noordhoff, Groningen, 761 pp

Burkill IH (1966) A dictionary of the economic products of the Malay Peninsula. Revised reprint. 2 volumes. Ministry of Agriculture and Co-operatives, Kuala Lumpur, Malaysia. Vol 1 (A—H) pp 1–1240, vol 2 (I—Z). pp 1241–2444

Chang YC, Hu CC, Tseng TH, Tai KW, Lii CK, Chou MY (2001) Synergistic effects of nicotine on arecoline-induced cytotoxicity in human buccal mucosal fibroblasts. J Oral Pathol Med 30(8):458–464

Johnson D (1991) Palms for human needs in Asia: palm utilization and conservation in India, Indonesia, Malaysia and the Philippines. A.A. Balkema, Rotterdam, 258 pp

Johnson DV (1998) Tropical palms. FAO Regional Office for Asia and the Pacific, Food and Agriculture Organization of the United Nations, Rome, 166 pp

Jones D (1984) Palms in Australia. Reed Books Pty Ltd, Balgowlah, 278 pp

Whitmore TC (1973) Palms of Malaya. Oxford University Press, Singapore, 129 pp

Arenga pinnata

Scientific Name

Arenga pinnata (Wurmb) Merr.

Synonyms

Arenga gamuto Merr., *Arenga saccarifera* Labill., *Borassus gomutus* Lour., *Caryota onusta* Blanco, *Gomutus saccharifer* Spr., *Sagus gomutus* Perr., *Saguerus australasicus* Drude & Wendl., *Saguerus pinnatus* Wurmb, *Saguerus pinnatus saccharifer* Wurmb., *Saguerus rumphii* Roxb., *Saguerus saccharifer* Blume.

Family

Arecaceae

Common/English Names

Areng Palm, Arenga Palm, Black-Fiber Palm, Black Sugar Palm, Gomuti Palm, Sagwire-Palme, Sugar Palm, Toddy Palm.

Vernacular Names

Arabic: Nakhlet Es Sukkar;
Chinese: Guang Lang, Suo Mu, Sha Tang Ye Zi, Tang Shu;
Burmese: Taung, Taung-Ong;
French: Palmier À Sucre, Palmier Aren, Palmier Areng, Palmier Condiar;
Dutch: Arengpalm, Arenpalm, Gomoetoepalm, Sagoeweerpalm, Suikerpalm;
German: Zuckerpalme, Echte Zuckerpalme, Gomuti-Palme;
India: Thangtung, Thanglung (Mizoram), Alam Panai (Tamil);
Indonesia: Inau, Anau, Enau, Nau, Hanau, Peluluk, Biluluk, Kabung, Kabung Enau, Juk, Ijuk, Bergat, Mergat (Sumatra), Aren, Lirang, Nanggung (Java), Kawung, Taren (Sundanese), Akol, Akel, Akere, Inru, Indu (Sulawesi), Moka, Moke, Tuwa, Tuwak (Nusa Tenggara), Ejow, Gomuti, Kaong;
Italian: Palma Da Zucchero;
Japanese: Satou Yashi;
Khmer: Chuek, Chrae;
Malaysia: Berkat, Bakeh (Semang), Enau, Habong, Henau, Inau, Kabong, Nau;
Philippines: Hidiók (Bikol), Bagátbat, Bagóbat, Bat-Bát, Iliók, Idióg, Idiók, Onay, Unau (Cebu Bisaya), Rapitan (Iloko), Hiliók (Manabo), Hidiók, Igók (Panay Bisaya), Irók (Sambali), Kábo-Négro, Káong, Káuing (Tagalog);
Portuguese: Gomuteira;
Russian: Sakharnia Pal'ma;
Spanish: Barú, Bary, Palma De Azúcar, Palmera Del Azúcar;
Thai: Aren, Chok, Kaong Tao, Luk Chid;
Vietnamese: Bung Bang, Doac Dot.

Origin/Distribution

Tropical south Asia to South-east Asia – from India, Bangladesh to Malaysia, Indonesia and Philippines to the east. Widely cultivated in the tropics and the Pacific.

Agroecology

Arenga pinnata in its native range is found growing in some forested areas but never far from settled areas, in ravines, along streams, on slopes and areas under semi-cultivation. It is also occasionally found in virgin forest since its fruits are scattered by fruit bats, wild pigs and civet cat. It does best in a hot wet and humid climate from sea level to 1,400 m elevation. At higher elevation over 1,000 m, it takes more than 15 years to flower at lower elevation (<700 m), it takes 7–10 years.

Edible Plant Parts and Uses

To obtain the sugary sap from the male and female inflorescence, the peduncle is bruised by beating over several days and then an incision is made in the peduncle for the sap to flow out which is then collected in a vessel (Plate 4). The fresh sap is collected from the vessel twice a day using fresh containers. Yield of sap from male palms are higher and easier to extract. The sugary sap from the cut inflorescence makes a fresh drink called *saguir, nira* or *lahang*. On standing for several days it becomes fermented and provides *toddy, arrack* or *tuak* containing about 30% alcohol. On fermentation for several weeks it is converted to vinegar (acetic acid) and also with

Plate 2 Unripe Arenga fruit harvested for various edible uses

Plate 1 Arenga fruit split open to show the yellow, fleshy pulp

Plate 3 Harvested Arenga fruit bunches are covered with weed stubble to protect from the sun

Plate 5 Arenga palm sugar cylinders are wrapped in banana leaves

Plate 4 Harvesting of sap from a tall, old Arenga palm

Plate 6 Cylinders of Arenga palm sugar

further distillation into wine. The juice can be boiled down to make arenga syrup or allow to set in moulds to form arenga sugar (Plate 5–6), called *gur* in India. Arenga sugar is dark brown in colour and contains much impurities which causes it to have a short shelf life. It has a distinctive characteristic taste. Arenga syrup or sugar is used in all kind of dishes, sweets, beverages and preserves. It acts as the perfect sweetener and flavour stimulant for cakes, pastries, pancakes, toast, sandwiches, cereals, beverages, fruit juices, dessert toppings and many other uses. To obtain vinegar, the sap is placed in vitrified earthen jars for 3–4 weeks. The vinegar is then pasteurised and bottled.

The young crown top or terminal bud called 'palm cabbage' is also eaten as a salad or cooked.

The endospermum of immature seeds are also eaten as vegetable or dessert. Immature seeds are much eaten by the Filipinos, being usually boiled with sugar to form a kind of sweetmeat. The endosperms of immature seeds are widely consumed in the Philippines and are made into canned fruits after boiling them in sugar syrup. In Indonesian, the white endosperms of immature seeds are extracted and undergo 10–20 days of water immersion and then boiled with sugar to yield *buah atap* or *kolang-kaling* which is used as a sweetmeat popularly used in *es campur* (shave ice delicacy) or in *kolak,* a local refreshment. Both dishes are specially served during the fasting month of Ramadan.

Sago (East Indian sago, Java sago, East Indian tapioca) a starch is also made from the tender pith of trunk of old palms. In Indonesia, arenga sago is used asw an ingredient of *bakso* (noodles), cakes and other dishes. In Java, a syrup called *chendol* is make from it.

In India, the Manipuris has been reported to eat the very young, tender white leaves as a pickle.

Botany

Tall, unbranched, solitary palm from 12 to 20 m high with a stout trunk of 30–60 cm diameter. The trunk is marked with rather distinct bases of broken leaves and long black fibres top by a dense crown of leaves. The leaves are 6–10 m long, ascending, pinnate, the basal part of the petiole covered with a sheath of stout, black fibres. The leaflets are up to 160 on each side, linear, 1.5–1.8 m long, the tip lobed and variously toothed, the base 2-auricled, the lower surface white or pale. The inflorescence is axillary, usually unisexual, pendulous with a stout, peduncle bearing female flowering spikes at the top and male flowering spikes lower down the peduncle and appearing later. Flowers are trimerous with a 3-lobed tubular corolla. There are up to 11,500 male flowers per inflorescence, greenish –yellow with many stamens; female flowers up to 1,500 per inflorescence with a globose, trilocular ovary. The fruit is a rounded or ellipsoid drupe, about 5–7 cm in diameter, green when immature (Plates 1–2) turning yellow and gradually black and contain from two to three black seeds.

Nutritive/Medicinal Properties

The young roots of *Arenga pinnata* have been reported to be used in a medicine for treating kidney stones and old roots for toothache in Malaysia. In southeast Asia, consumption of arenga sugar is used as a traditional and homeopathic remedy, believed to revitalise the body. In the Philippines, the stems and petioles are reported to be diuretic and antithermic. The root has been deemed stomachic and pectoral. Root decoction is held to be beneficial to the lungs; assists digestion and improves appetite. The petiole fuzz (fibres) are styptic, used as a haemostatic and cicatrizant for applying to wounds.

Other Uses

This crop has potential to be developed into a major resource of biofuel (bioethanol). Boiled starch extracted as sago from the palm is used as hogs feed. The flowers are a good source of nectar for honey production. A toy beetle has been made from the seed.

The leaves are used for thatching houses. Leaves are used for rough brooms and woven into coarse baskets. The youngest leaves are sometimes used as cigarette paper. Old woody leaf bases as well as the long leaves can be used for fuel. The petioles when split are used for basketry and a form of marquetry work on tables, stands, screen, boxes and other light pieces for furniture. Leaf petioles and the wood of the stem are also made into walking sticks and the Pagan tribes in Peninsular Malaysia used the wood for the butts of blow-pipes. The very hard outer part of the trunk is used for barrels, flooring and furniture, posts for pepper vines, boards, tool handles and musical instruments like drums are all made from the wood of the trunk. The pith of the leaf rachis is an ideal shape for use as a drinking cup.

Fibres (gomuti, ijok or yonot fibre) especially from that surrounding the trunk are used for ship cordage as they are durable in water. The black horsehair-like fibres of the leaf bases serve the production of rigging and brushes as well as for covers of underwater telegraph cables and for caulking of boats as the fibres can stand long exposure to either fresh or salt water and is also fire resistant. The same fibres are used for stuffing mattresses, for thatching materials, as sieves, brooms and brushes and other minor products. The hairs found on the base of the leaf sheaths are very good tinder for igniting fire. Fibres are also obtained from the trunk pith, leave sheaths and roots. Fibres are also used for making hats in Indo China. Recent studies showed that *Arenga pinnata* fibres can be used as a natural filament and epoxy resin as a matrix (Sastra et al. 2006).

The *Arenga* fibers were mixed with epoxy resin at the various fibre weight percentages of 10%, 15%, and 20% *Arenga pinnata* fibre and with different fibre orientations such as long random, chopped random, and woven roving. The results indicated that the woven roving *Arenga pinnata fibre* has a better bonding between its fibre and matrix compared to long random *Arenga pinnata* fibre and chopped random *Arenga pinnata* fibre.

Roots are used for matting and the root fibres have been used for making capes in Sulawesi.

Comments

The husk of the fruit contains very numerous, microscopic, needle-like, stinging crystals (raphides), and this part of the fruit is exceedingly irritating and poisonous.

Selected References

Backer CA, van den Brink RCB Jr (1968) Flora of Java, vol 3, Spermatophytes only. Wolters-Noordhoff, Groningen, 761 pp

Burkill IH (1966) A dictionary of the economic products of the Malay Peninsula. Revised reprint. 2 volumes. Ministry of Agriculture and Co-operatives, Kuala Lumpur, Malaysia. Vol 1 (A—H) pp 1–1240, vol 2 (I—Z). pp 1241–2444

Council of Scientific and Industrial Research (1948) The wealth of India. A dictionary of Indian raw materials and industrial products. (Raw materials 1). Publications and Information Directorate, New Delhi

Florido HB, de Mesa PB (2003) Sugar palm [*Arenga pinnata* (Wurmb.) Merr.]. Research Information Series on Ecosystems, Bureau Plant Industry, Government of the Philippines. Vol 15. No. 2

Miller RH (1964) The versatile sugar palm. Principes 8:115–147

Mogea J, Seibert B, Smits W (1991) Multipurpose palms: the sugar palm (*Arenga pinnata* (Wurmb) Merr.). Agroforest Syst 13:111–129

Perry LM (1980) Medicinal plants of East and Southeast Asia. Attributed properties and uses. MIT Press, Cambridge – USA/London – UK, 620 pp

Sastra HY, Siregar JP, Sapuan SM, Hamdan MM (2006) Tensile properties of *Arenga pinnata* fiber-reinforced epoxy composites. Polym Plastics Technol Eng 45(1–3):149–155

Smits WTM (1996) *Arenga pinnata* (Wurmb) Merr. In: Flach M, Rumawas F (eds) Plant resources of South East Asia no. 9. Plants yielding non-seed carbohydrates. Prosea Foundation, Bogor, pp 53–58

Stuart GU (2010) Philippine alternative medicine. Manual of some Philippine medicinal plants. http://www.stuartxchange.org/OtherHerbals.html

Bactris gasipaes

Scientific Name

Bactris gasipaes Kunth

Family

Arecaceae

Synonyms

Bactris ciliata (Ruiz & Pav.) Mart. nom. illeg., *Bactris insignis* Drude, *Bactris insignis* (Mart.) Baill., *Bactris macana* (Mart.) Pittier, *Bactris speciosa* (Martius) H. Karsten, *Bactris speciosa* var. *chichagui* H. Karst., *Bactris utilis* (Oerst.) Benth. & Hook.f. ex Hemsl., *Guilelma chontaduro* H. Karst. & Triana, *Guilielma ciliata* (Ruiz & Pav.) H. Wendland ex Kerchove, *Guilielma gasipaes* (Kunth) L.H. Bailey, *Guilielma gasipaes* (Martius) Bailey, *Guilielma gasipaes* var. *chichagui* (H. Karst.) Dahlgren, *Guilielma gasipaes* var. *chontaduro* (H. Karst. & Triana) Dugand, *Guilielma gasipaes* var. *coccinea* (Barb. Rodr.) L.H. Bailey, *Guilielma gasipaes* var. *flava* (Barb. Rodr.) L.H. Bailey, *Guilielma gasipaes* var. *ochracea* (Barb. Rodr.) L.H. Bailey, *Guilielma insignis* Mart., *Guilielma macana* Mart., *Guilielma microcarpa* Huber, *Guilielma speciosa* Martius, *Guilielma speciosa* var. *coccinea* Barb. Rodr., *Guilielma speciosa* var. *flava* Barb. Rodr., *Guilielma speciosa* var. *mitis* Drude, *Guilielma speciosa* var. *ochracea* Barb. Rodr., *Guilielma utilis* Oersted, *Martinezia ciliata* Ruiz & Pav. nom. rej.

Common/English Names

Peach Palm, Palm Chestnut, Pejibaye, Pewa Nut

Vernacular Names

There are more than 200 vernacular names for *Bactris asipaes* Kunth. The most common vernacular names are:
Brazil: Papunha, Pupunha, Pupunheira;
Bolivia: Chonta, Comer, Palma De Castilla, Tembe;
Colombia: Cachipay, Chantaduro, Chenga Chonta, Chontaduro, Chichagai, Chichaguai, Choritadura, Contaruro, Jijirre, Macanilla, Pijiguay, Pipire, Pirijao, Pupunha, Tenga;
Costa Rica: Pejibaye, Pejivalle;
Ecuador: Chantaduro, Chonta, Chontaduro;
French: Palmier-Pêche;
French Guiana: Parepon;
German: Pfirsich-Palme, Pfirsichpalme;
Guatamala: Pejibaye;
Guyana: Paripie;
Indonesia: Pewa (Malay);
Nicaragua: Pejibaye;
Panama: Pixbae;

Peru: Chonta Ruru, Chonta Dura, Pejwao, Pifuayo, Pijuanyo, Pisho-Guayo, Sara-Pifuayo;
Spanish: Pejivalle, Peyibaye;
Surinam: Amana, Paripe, Paripoe;
Trinidad And Tobago: Peach Palm, Pewa Nut;
Venezuela: Bobi, Cachipaes, Gachipaes, Macana, Melocoton, Pichiguao, Pihiguao, Pijiguao, Piriguao, Pixabay, Rancanilla.

Origin/Distribution

Peach palm is native to the tropical forests of the South and Central America but the exact centre(s) of origin and domestication is still unresolved. Some authors hypothesized a single origin in the western Amazon Basin, others proposed a multiple origin including areas in the western Amazon Basin, western and north-western side of the Andes, and lower Central America. The cultivation of peach palm covers a wide geographical distribution since pre-Columbian times, extending from central Bolivia to north-eastern Honduras and from the mouth of the Amazon River and Guayanas to the Pacific coast of Ecuador and Colombia into Central America, extending north to Mexico and eastwards to some Caribbean Islands. It has also been introduced into other countries including Australia, Indonesia, Malaysia (Sabah), Papua New Guinea and the Hawaiian Islands.

Agroecology

In its native range in the hot, humid tropics, peach palm occurs scattered not in extensive stands in disturbed natural ecosystems, principally along river beds and in primary forest gap, also in secondary forest fallows that develop after slash-and-burn agriculture. It grows best in areas with profuse and well distributed rainfall of 2,000–5,000 mm annually and with average temperatures above 24°C at low to mid altitudes of <800 m. It is adaptable to a wide range of edaphic conditions. It produces relatively well on low fertility soils including acid and highly eroded laterites with 50% aluminium-saturated acid soils. It thrives best and is most productive on deep, fertile, well-drained, loamy soils. It cannot tolerate heavy clays or water-logged soils. Symbiotic associations with vesicular-arbuscular mycorrhizae have been reported to improve growth. Seedlings require partial shade but adult palms need full sunlight for optimal production of flowers, fruits and offshoots. Peach palm is perennial for both fruit and heart-of-palm production. Cutting the main stem and offshoots (suckers) for palm-heart does not kill the plant, but instead allows preformed buds to develop into new offshoots.

Two systems of planting are currently adapted for commercial production of peach palm. For palmetto or palm-heart production a high density planting system of 3,000–20,000 plants is employed and for fruit production a lower planting density of 400–500 plants/ha is used.

Edible Plant Parts and Uses

Peach palm has been used as staple food by the indigenous pre-Columbian Amerindian communities in Central and South America for centuries. The native Amerindian utilised both the fruit in various ways and also consumed the palm heart. The fruit is usually boiled in water or salted water or roasted and the cooked mesocarp is eaten. Another mode of preparation was to mash the cooked mesocarp, ferment for 1–2 days and then add water to the fermented mash to make a drink for breakfast and other times. Leaving the mash to ferment for a week or more would result in an alcoholic beverage, *chicha* which is popularly consumed to celebrate festive occasions. The cooked mash is also preserved as a silage by compressing in a hole (silo) in the ground lined with leaves of *Musa* spp. The silage is consumed after a month and is diluted with water to make beverages. Whole fruits can be dried or smoked to preserve them. Nowadays the boiled mesocarp, with various seasonings, is a popular *hors d'œuvre* in many regions. The cooked pulp is eaten as a local delicacy in many local dishes and is an important source of vitamins, particularly vitamin A, niacin, vitamin C, protein, essential amino acids and iron.

The pulp is both starchy and oily, not at all sweet, with an acceptable flavour that is likened to that of chestnut. The cooked mesocarp is also blended with sugar and water and fermented for 1–2 days to make beverages. In Columbia, the cooked mesocarp is mixed with milk, sugar and several condiments to prepare a commercial beverage. The cooked fruit is also eaten with salt or honey and used for compotes and jellies. Fruits must be boiled and preferably peeled before consuming or processing to dissolve the calcium oxalate crystals and to eliminate the adverse effect of trypsin inhibitor. Uncooked fruit contains anti-nutrients such as calcium oxalate and the trypsin inhibitor which interferes with protein digestion. The fruits are amenable to canning. Canned fruits are being marketed in Costa Rica: these include whole or half fruits, either peeled or unpeeled, with or without the seed. Chips are also made from the fruit.

Flour derived from the pulp can be used for making soups, tortillas, pastas, noodles, bread, cakes, infant formula, other confectionery and preparations. Studies showed that pejibaye flour in a mixture containing 15% pejibaye flour (PF) and 85% wheat flour can be used for making food pasta, spaghetti and twist noodles and it did not significantly alter its characteristics of quality and texture (De Oliveira et al. 2006). Pejibaye flour does not contain gluten and would be good for those with gluten allergy. It also has good potential as animal feed as a substitute for grain or supplement for grain, in the manufacture of concentrates and fermented ensilage. The fruit can be processed into other derived products such as oil, β-carotene and starch. The fruit also yields edible cooking oil which is rich in unsaturated fatty acids, notably oleic acid.

Heart-of-palm, a good source of dietary fibre, magnesium and iron, is already an important commercial crop, particularly for the gourmet market. The Brazilian domestic market for heart-of-palm is about five times bigger than the external one. Palm heart is increasingly used in international cookery and is being exported to Europe and USA. Over 2,000 ha are grown in Costa Rica for marketing as fresh, dried or canned products and processing of the heart-of-palm for the international market could become a major agro-industry for other producing countries. The palm heart is differentiated into two product categories for the international market: true or 'quality heart-of-palm' is a cylinder composed of a tender petiole-sheath enveloping the developing leaves above the apical meristem, and the 'caulinar heart' or tender tubular stem tissue below the apical meristem. In Brazil, palm heart is further differentiated into three categories: thin (1.5–2.5 cm) hearts to be canned for the internal and export markets; medium (2–4 cm) hearts for the fresh market in Brazil; thick (3–6 cm) hearts for the Brazilian *churrascaria* market (restaurants that specialize in barbecued meat with thick hearts-of-palm as garnish). For hearts-of-palm with 2–4 cm diameter, offshoots (suckers) are harvested when they attain stem diameters of >9 cm, measured at 20–30 cm above the ground. As for the fruit, the palm hearts have to be cooked to detoxify the anti-nutrients it contains. Cooked palm heart is used in salads and as vegetables in various recipes.

The sap from the trunk, either unfermented or fermented to varying degrees, is used to prepare nutritional beverages and intoxicating drinks. The young inflorescences are also eaten roasted "the herdsman's way", without opening the protective spathe. The cooked male flowers are used as condiment. Flowers of excess racemes are sometimes cut when very young, chopped and eaten cooked with eggs.

Botany

Bactris gasipaes is a caespitose (multi-stemmed, 1–13 stems), armed (spiny) (Plates 1–3), monoecious palm with an extensive but fairly superficial adventitious root system, reaching heights of 6–20 m. The stems are straight, cylindrical, unbranched (Plates 1–3), 12–26 cm in diameter, with internodes covered with circular rows of dark spines, alternating with nodes without spines. Suckers (off-shoots) arise from basal axillary buds, and usually vary in number from 1 to 12. The crown has 10–30 pinnate leaves. The petiole is spiny and the petiole-sheath is 50–180 cm long and the rachis is 180–400 cm

Plate 1 Clumping habit of Pejibaye palm

Plate 2 Multi-stemmed, spiny, straight trunk

long The pinnate leaves 2.5–3.6 m long, with many linear, pointed, bifurcated leaflets 60–115 cm long by 3–6 cm wide; dark green above, pale beneath, spiny on the veins on both surfaces and margins. The inflorescence arises from axils of leaves and is covered with two spathes, the outer deltoid, short, hard and thick, the rachis is branched with thousands of male flowers intermixed with a few hundred female flowers, which are slightly bigger than the males. Male flowers are cream-light yellow, 2–6 mm long and 2–6 mm wide, with six stamens arranged in pairs on the sides of the corolla. Female flowers are usually yellow, or rarely green, 3–13 mm long and 4–12 mm wide. The gynoecium is syncarpous, trilocular. The fruit is a drupe, usually shiny orange, red or yellow (Plates 4–7) and may have superficial striations and are produced in axillary clusters (Plate 6) in the crown at the top of the trunk (Plate 5). Fruit is ovoid, oblate, cylindrical or conical, 2–7 cm by 2–6 cm, cupped at the base by a green, leathery, three-pointed calyx (Plates 6–7). A single stem may bear five or six clusters at a time. The epicarp is thin, and the mesocarp is yellow to light-orange, dry and mealy. The dark endocarp, containing the seed, is usually located centrally in the fruit, but may occur at the distal end. The endocarp is ovoid, elliptic, round, oblong or cuneiform, 1–4 cm long by 1–2 cm wide and has three pores. Normally there is a single conical seed 2 cm long, with a hard, thin shell and a white, oily, coconut-flavoured kernel. Some fruits are seedless because of parthenocarpy.

Nutritive/Medicinal Properties

The mean nutrient composition of *Bactris gasipaes* fruit mesocarp complied from various sources by Mora-Urpí et al. (1997) was reported

Plate 3 Close-up of thorns on the trunk

Plate 5 A cluster-bunch of Pejibaye fruit

Plate 6 Ripe pejibaye fruit, note the persistent, leathery green calyx

Plate 4 Pejibaye fruit cluster formed near the crown apex

as follows: moisture 53.5% FW (fresh weight), protein 7.8% DW (dry weight), oil 15.8% DW, carbohydrate 67.8% DW, fibre 6.7% DW, ash 2.1% DW; minerals – mg/10 g FW – calcium 10.9 mg, Fe 6.1 mg, Mg 11.7 mg, K 162.8 mg, Na 2.7 mg, Zn 2.1 mg; essential amino acids (%

Plate 7 Ripe pejibaye fruit with growth cracks

of total Nitrogen) – arginine 6%, glycine 4.3%, histidine 2.2%, isoleucine 2.3% leucine 3.6%, lysine 4.3%, methionine 1.5%, threonine 3.05, tryptophan 0.9%, tyrosine 2.0%, valine 3.1%; non-essential amino acids – alanine 3.9%, aspartic acid 4.8%, glutamic acid 5.5%, phenylalanine 2.0%, proline 2.8%, serine 3.7%; saturated fatty acids (% of total oil) 38.8% – palmitic acid (16:0) 38.2%, stearic acid (18:0) 1.0%, unsaturated fatty acids 60.9% – palmitoleic acid (16:1) 7.4%, oleic acid (18:1) 46.3%, linoleic acid (18:2) 6.2%, linolenic acid (18:3) 1.4%.

Analysis of the fruit mesocarp conducted in Brazil (Yuyama et al. 2003) reported that the mean protein levels ranged from 1.8% to 2.7%, lipid levels ranged from 3.5% to 11.1%, the nitrogen free fraction ranged from 24.3% to 35%, food fibre ranged from 5.2% to 8.7%, and energy ranged from 179.1 to 207.4 kcal%. All essential, as well as non-essential, amino acids were present, with tryptophan and methionine presenting the lowest mean concentrations. The mono-unsaturated oleic acid predominated in the oil, ranging from 42.8% to 60.8%, and palmitic acid was the most abundant saturated fatty acid, ranging from 24.1% to 42.3%. Among the essential fatty acids, linoleic acid was the most abundant, with a maximum of 5.4% in cv. Pampa-8. The most important mineral elements were potassium, selenium and chromium, respectively corresponding to 12%, 9% and 9% of daily recommended allowances.

The full nutrient composition of the fruit mesocarp was reported by Yuyama et al. (2003) as follows percent composition (% fresh weight) of 100 g of the mesocarp of the peach palm fruit from three populations produced in Central Amazonia: moisture 53.0, protein 2.3, N-free extract 29.7, lipids 7.7, ashes 0.6, soluble fibre 1.1, insoluble fibre 5.5, total fibre 6.6, energy 197.7 cal. The amino acid composition (% g N) of the protein in the mesocarp of peach palm fruits from the Pará 85 population sample produced in Central Amazonia (Yuyama et al. 2003) was reported as: essential amino acids – leucine 3.14, phenylalanine 2.04, lysine 1.67, valine 2.83, isoleucine 1.70, threonine 2.71, methionine 0.80, tryptophan 0.45; non-essential amino acids – proline 2.57, áspartic acid 4.33, serine 2.72, glutamic acid 4.98, glycine 2.87, alanine 3.51 1. The composition of the major fatty acids in the oil extracted from the mesocarp from three population samples produced in Central Amazonia (as % oil) was found as follows: saturated fatty acids 39.2%, C16:0 (palmitic) 38.2, C18:0 (stearic) 1.0, mono-unsaturated fatty acids 53.7%, C16:1 (palmitoleic) 7.4, C18:1 (oleic) 46.3, polyunsaturated fatty acids 6.9%, C18:2 (linoleic) 6.2, C18:3 (linolenic) 1.4. Polyunsaturated/saturated ratio 1.8. Essential mineral macro- and microelement content in 100 g of the mesocarp of the fruits of three peach palm population samples (Pampa-8, Pampa-40 and Pará-85 respectively) produced in Central Amazonia was reported as: Ca 24.7, 10.2, 21.8 mg, K 289.3, 225.8, 206.4 mg, Na 0.2, 0.2, 12.6 mg, Mg 16.9, 16.9, 17.6 mg, Mn 84.3, 115.1,82.6 mg, Zn 277.7, 258.5, 278.3 µg, Se 3.5, 3.3, 11.4 µg, Fe 565.6, 470, 739,3 µg, and Cr 8.2, 12.2, 13.9 µg (Yuyama et al. 2003).

Jatunov et al. (2010) found significant differences in total carotenoid content (1.1–22.3 mg/100 g) among the six varieties studied. The carotenoids detected in the fruit mesocarp included Z-β-carotene, E-β-carotene, α-carotene, E-γ-carotene, Z-γ-carotene, E-lycopene, Z-lycopene and xanthophylls. Boiling the fruits for 30 minutes did not affect total carotenoid content, but did change the amount of some carotenoids, mainly by the production of Z-isomers. Peach palm varieties had the same carotenoids, but in different proportions, presenting mainly, all E-β–carotene (26.2–47.9%), Z-γ-carotene (18.2–34.3%)

and Z-lycopene (10.2–26.8%). When antioxidant activity was evaluated using DPPH, it was observed that the variety with higher percentages of β-carotene (54.1%) had higher activity. Carotenoids content in raw and cooked pejibaye was similar, 4.8–29.6% DB and 4.8–29.9% DB, respectively, giving a nutrient retention greater than 85% after cooking. Average edible portion was 68% (Blanco and Muñoz 1992).

No significant differences were found in fatty acid content between cooked and uncooked mesocarp samples, but there were differences among races in the content of four fatty acids (Fernández-Piedra et al. 1995). Pejibaye fat comprised mainly mono-unsaturated fatty acids (45.6%) and had a low poly-unsaturated to saturated fatty acid ratio (0.5). The fatty acid profile of uncooked pejibaye samples was: oleic acid, 32.6–47.8%; palmitic acid, 30.5–40.3%; linoleic acid, 11.2–21.1%; palmitoleic acid, 5.7–7.1%; linoleic acid, 1.5–5.5%; and stearic acid, 1.7–2.4%.

Yuyama et al. (1991; 1996) conducted feeding studies on Wistar rats with diets containing 0.0, 2.0, 4.0 and 8.0 IU retinyl palmitate/g ration, respectively (control groups), and diets containing 4.0 IU vitamin A from mango and 4.0 IU vitamin A from peach palm/g ration (test groups). Results obtained indicated that vitamin A in peach palm was highly bioavailable, with 171% efficiency, followed by mango (82% efficiency) when compared with the control groups. Mean basal vitamin A levels were 0.17 µg/g in liver and 3.50 µg% in plasma. At the end of the experiment, mean vitamin A levels in liver were 53.45 µg/g (peach palm group), 5.76 µg/g (mango group), and 9.52 µg/g (4 IU group), and vitamin A levels in plasma were 28.33 µg%, 29.80 µg% and 32.36 µg% for the same groups, respectively. When the control and test groups were compared for liver and plasma carotene concentrations, a significant difference between groups was detected only for liver carotene.

Nutrient composition of pejibaye flour (fresh basis per 100 g) was reported as: energy 413 cal, moisture 12 g, protein 3.8 g, fat 8.9 g, ash 1.3 g, crude fibre 2.1 g, carbohydrate 72.1 g, vitamin A 1.2ug, vitamin B1 0.1 mg, vitamin B2 0.3 mg, niacin 2.5 mg, vitamin C 62.2 mg, Ca 10.9 mg, Fe 6.1 mg, Na 2.7 mg, K 162.8 mg, Mg 11.7 mg and Zn 2.1 mg (Metzler et al. 1992).

The mean nutrient composition of *Bactris gasipaes* heart-of-palm complied from various sources by Mora-Urpí et al. (1997) was reported as follows:

% fresh weight – moisture 90.1%, protein 2.5%, oil 1.0%, carbohydrate 3.25, fibre 1.0%, ash 1.0%; minerals (mg/100 g dry weight) Ca 114 mg, Fe 4.3 mg, Mg 80 mg, P 94 mg, K 337.6 mg, Na 1.3 mg, Zn 0.8 mg.

Peach palm is used as a folk remedy for headache and stomach-ache and the roots are also used medicinally.

Other Uses

Peach palm has been utilised as shade tree in coffee and cacao plantations. Fruit provide meals that are valued as feed for livestock (pigs, chickens) and fish culture. Pejibaye flour was found to have a higher content of energy than corn and that it was not necessary to separate the seeds from the fruits in animal feeds (Zumbado and Murillo 1984). The level of indispensable aminoacids was considerably low, especially methionine, which was lower than in corn. Most of the free fatty acids were present in a ratio of 2:1 in unsaturated to saturated acids. The predominant fatty acids in whole pejibaye meal were oleic and palmitic acids with adequate levels of linoleic acid. Saturated fatty acids were predominant in the seed, with a very high content of lauric and myristic acids.

Excess flowers are chopped and fed to chickens. The wood from the trunk is used to manufacture building materials such as parquet, panels; luxury furniture; spears, bows, arrows, harpoons, fishing poles, bed-boards, hammer handles, siding for houses, beaters and spindles for weaving, carvings and other handicraft items, taking advantage of its beauty, great strength and elasticity. The long fibres on the inside of the trunk show promise for use in fibrocement products. Hollowed out trunks serve as water troughs or conduits, or as flower planters. The long spines are used as sewing needles. The leaves are utilized for thatching and basketry and as animal feed. A green dye can be extracted from the leaves for dyeing other fibres. The roots provide a vermicide. Unused leaf and stem parts can be used to manufacture paper, organic fertilisers and animal food supplement.

Comments

Spineless trunk cultivars have been developed.

Selected References

Arkcoll DB, Aguiar JPL (1984) Peach palm (*Bactris gasipaes* H.B.K.), a new source of vegetable oil from the wet tropics. J Sci Food Agric 35:520–526

Blanco A, Muñoz L (1992) Content and bioavailability of carotenoids from peach palm fruit (*Bactris gasipaes*) as a source of vitamin A. Arch Latinoam Nutr 42(2):146–154, Article in Spanish

Clement CR (1988) Domestication of the pejibaye palm (*Bactris gasipaes*): past and present. In: Ballick (ed) The palm – tree of life, vol 6, Advances in economic botany. Scientific Publications Office, New York Botanical Garden, New York, pp 155–174

Clement CR, Mora-Urpí J (1987) Pejibaye palm (*Bactris gasipaes*, Arecaceae): multi-use potential for the lowland humid tropics. Econ Bot 41(2):302–311

Clement CR, Manshardt RM, DeFrank J, Cavaletto CG, Nagai NY (1996) Introduction of pejibaye for heart-of-palm in Hawaii. Hortscience 31(5):765–768

De Oliveira MKS, Martinez-Flores HE 2, De Andrade JS, Garnica-Romo MG, Chang YK (2006) Use of pejibaye flour (*Bactris gasipaes* Kunth) in the production of food pastas. Int J Food Sci Technol 41(8):933–937

Duke JA (1983) Handbook of energy crops. http://www.hort.purdue.edu/newcrop/duke_energy/dukeindex.html

Fernández-Piedra M, Blanco-Metzler A, Mora-Urpí J (1995) Fatty acids contained in 4 pejibaye palm species, *Bactris gasipaes* (Palmae). Rev Biol Trop 43(1–3):61–6, In Spanish

Henderson A, Galeano G, Bernal R (1995) Field guide to the palms of the Americas. Princeton University Press, Princeton, p 352 pp

Jatunov S, Quesada S, Diaz C, Murillo E (2010) Carotenoid composition and antioxidant activity of the raw and boiled fruit mesocarp of six varieties of *Bactris gasipaes*. Arch Latinoam Nutr 60(1):99–104

Johannessen CL (1966) Pejibayes in commercial production. Turrialba 16:181–187

Johannessen CL (1967) Pejibaye palm: physical and chemical analysis of the fruit. Econ Bot 21:371–378

Metzler AB, Campos MM, Piedra MF, Mora-Urpí J (1992) Pejibaye palm fruit contribution to human nutrition. Principes 36(2):66–69

Mora-Urpí J, Weber JC, Clement CR (1997) Peach palm. *Bactris gasipaes* Kunth. Promoting the conservation and use of underutilized and neglected crops. 20. Institute of Plant Genetics and Crop Plant Research, Gatersleben, International Plant Genetic Resources Institute, Rome, Italy.

Morton JF (1987) Pejibaye. In: Morton JF (ed) Fruits of warm climates. Julia F. Morton, Miami, pp 12–14

Tropicos Org (Jan 2009) Missouri Botanical Garden. http://www.tropicos.org

Yuyama LK, Cozzolino SM (1996) Effect of supplementation with peach palm as source of vitamin A: study with rats. Rev Saúde Pública 30(1):61–6, (In Portuguese)

Yuyama LKO, Favaro RMD, Yuyama K, Vannucchi H (1991) Bioavailability of vitamin a from peach palm (*Bactris gasipaes* H.B.K.) and from mango (*Mangifera indica* L.) in rats. Nutr Res 11(10):1167–1175

Yuyama LK, Aguiar JP, Yuyama K, Clement CR, Macedo SH, Fávaro DI, Afonso C, Vasconcellos MB, Pimentel SA, Badolato ES, Vannucchi H (2003) Chemical composition of the fruit mesocarp of three peach palm (*Bactris gasipaes*) populations grown in central Amazonia. Braz Int J Food Sci Nutr 54(1):49–56

Zumbado ME, Murillo MG (1984) Composition and nutritive value of pejibaye (*Bactris gasipaes*) in animal feeds. Rev Biol Trop 32(1):51–6

Borassus flabellifer

Scientific Name

Borassus flabellifer L.

Synonyms

Borassus flabelliformis Roxb., *Borassus flabellifer* L. var. *aethiopicum* Warb.; *Borassus sundaicus* Becc., *Borassus tunicatus* Lour., *Pholidocarpus tunicatus* (Lour.) H. Wendl, *Thrinax tunicata* (Lour.) Rollisson.

Family

Arecaceae

Common/English Names

African Fan Palm, Asian Palmyra Palm, Borassus Palm, Brab Tree, Cambodian Palm, Doleib, Doub Palm, Great Fan Palm, Ice-Apple, Lontar Palm, Palmyra Palm, Ron Palm, Sea Apple, Tal-Palm, Tala Palm, Toddy Palm, Sugar Palm, Wine Palm

Vernacular Names

Arabic: Dôm, Tâl, Shag El Mûql;
Brazil: Boraço;
Burmese: Tan Bin;
Chinese: Shan Ye Shu Tou Zong, Shan Ye Tang Zong, Guo Dan;
Danish: Palmyrapalme;
Dutch: Lontar, Lontarpalm, Palmyrapalm, Jager-Boom, Weingeevende Palm-Boom;
East Timor: Akadiru;
Finnish: Palmyrapalmu;
French: Borasse, Palmier De Palmyre, Rondier, Rônier;
German: Borassuspalme, Lontaro, Palmyrapalme, Weinpalme;
Guinea-Bissau: Cicibes;
India: Taadi (Andra Pradesh), Tal (Assamese), Taala, Tal (Bengali), Tao (Divehi), Taad, Taada, Tadfali (Gujarati), Taad, Tad, Tad Mar, Tala, Tari, Tal, Taltar, Tar, Tarkajhar, Taduka, Tad, Tariya, Trinaraaj (Hindu), Karatale, Kari Thaale Karitale, Ole, Olegari, Oleya, Panai, Panal, Pane, Pani, Tala, Tale, Tali, Talimara, Tari, Thaale Mara, Thaatiningu Mara, Thaathinungu Mara, Thruna Raaja, Trinaraja, Trynaraja, Vole, (Kannada), Eroal, Targula (Konkani), Ampana, Carim-Pana, Carimpana, Eta, Karimapana, Karimpana, Karrumpana, Pana, Talam, Trinarajan, Trynarajan (Malayalam), Kona (Manipuri), Rotam, Taad, Tad, Tad-Mad, Tad Tamar, Tadh, Talatmad, Tamar, Tamas, Thad, Vet (Marathi), S-Iallu, Sial Lu (Mizoram), S-Iallu, Thaalo Gatcho (Oriya), Taalah, Tal, Tala, Taladrumah, Talah (Sanskrit), Acavattiru, Acavattirumaram, Acavatturu, Ailantal, Ailantalam, Ailantalmaram, Ailantar, Aintar, Aintaram, Anbanai, Carruppanai, Carupanai, Catapalam, Cattuppanai, Cirpaki, Civantikkiriyam, Civantikkiriyamaram, Edagam, Etakam, Etkai,

Etakamaram, Etkam, Kamam, Karadalam, Karambanai, Karampanai, Karakalam, Karpakam, Karatalam, Karatalamaram, Karimpanai, Karumpanai, Karumpul, Karumpuram, Karupuram, Karupuramaram, Kayinpanai, Kaympanai, Kayppanai, Kirusnakaya, Kirusnakentam, Kuliram, Makapattiram, Makonnatam, Mal, Maturacam, Maturaca, Maturacamaram, Narpanai, Netini, Netumi, Netuncevikam, Netuncevikamaram, Neyam, Nilam, Nilamani, Nilamanimaram, Nonkupanai, Nungu, Panai, Panai Maram, Panaimaram, Pandi, Panei, Pondai, Pondu, Pul, Purbadi, Puttrani, Talam, Tali, Pakarpali, Pana, Pennai, Pennaimaram, Pirancutirkkam, Ponantai, Pontukam, Pontukamaram, Pullutiyam, Purappi, Purpati, Purrali, Puttali, Talam, Talatalam, Talavilacam, Tali Z, Taruracam, Taruracan, Tarurajamaram, Taruviracan, Taruvirakam, Taruvirakamaram, Taruvirakan, Tatti, Tiranaracan, Tirunapati, Tirunaracan, Turapokam, Turarokam, Turumacirettam, Turumecuvar, Turumecuvaram, Turumekam, Turumekamaram, Ulokapattiram, Upatakam, Utupatakam, Varanikam, Varanikamaram (Tamil), Karatalamu, Karathaalamu, Naamathaadu, Namatadu, Namatody, Patootody, Penti Tadi-Female, Pentitadi, Pentitadu, Penthithaati Chettu, Pentithaadu, Pentytody, Pothuthaadu, Potutadi, Potutadi-Male, Potutadu, Taadi, Tadi, Tati, Tadu, Tatichettu, Thaati, Thaadi Chettu, Thaadu, Thaati Kullo, Thadi, Thati, Trinarajamu, Trynarajamu, (Telugu), Munjal, Taad (Urdu);
Indonesia: Ental, Lontar, Tal, Etal, Savalen, (Java) Pohon Siwalan (Sumatera), Tarebung, Ta'al (Madura), Tal (Lombok), Tua-Hu (Roti), Tua-Hua, Kali (Timor), Kepac-Duren, Duwe (Savu), Tala, Tola (Sulawesi);
Italian: Palma Del Ferro, Palma Del Sagù, Palma Di Palmira;
Ivory Coast: Keue;
Japanese: Parumira Yashi, Ougi Yashi, Ougi Yashi;
Khmer: Tnaot, Dom Thuot, Ta-Not;
Malaysia: Lontar, Tah, Tai;
Nepalese: Taadii, Taal;
Persian: Darakhte-Tari;
Portuguese: Broção, Palmira, Palmeira De Leque, Palmeira De Palmira, Panaguera, Palmieira Macha Brava, Cibes;
Russian: Lontarovaia Pal'ma, Pal'mira, Pal'mirova Pal'ma;
Senegali: Ronn;
Sri Lanka: Tal Gaha (Sinhalese);
Spanish: Boraso, Palma Palmira;
Swedish: Palmyrapalm;
Taiwan: Shan Ye Zi;
Thailand: Maktan (Don Daeng), Tan, Tanta Note (General), Ta-Not (Khmer), Tan Tanot, Tan Yai (Central), Than (Shan-Mae Hong Son), Tho-Thu (Karen-Mae Hong Son), Tha-Nao (Khmer-Phratabong), Thang (Karen-Tak, Chiang Mai), Not (Peninsular);
Tibetan: Sin Ta La;
Turkish: Tal;
Vietnamese: Thot Lot, Cay Thot, Lot.

Origin/Distribution

There is mixed view on its origin. Most botanists believed that it originated from Africa and then was introduced into India a long time ago. Another view is that it is native to South Asia, southeast Asia to New guinea Tropical Africa. Palmyra palm is extensively cultivated in south Asia and southeast Asia. It is cultivated or found in semi-wild stands in India, Sri Lanka, Burma, Thailand, Cambodia, China, West Malaysia, Indonesia, and New Guinea. In India, it is planted as a windbreak on the plains. It can also be found growing in Hawaii and southern Florida.

Edible Plant Parts and Uses

Economically, palmyra palm is very important. It is a source of sugar and toddy, the sweet sap is extracted from the cut made in the inflorescence from male or female trees. From the sap the following by-products are obtained:
– Sweet toddy – when care is taken to avoid fermentation the sap is consumed as a mild beverage 'sweet toddy'
– Toddy – fermented sap gives palm wine known as toddy and is a popular alcoholic drink in Asian countries

- Arrack/arak – toddy upon distillation a stronger alcoholic drink called arrack is obtained
- Vinegar – is made from fermented toddy
- Jaggery (Sri Lanka, *gur* (India)) – mixture of reducing and non-reducing sugars, solidified into moulds of wicker or oil paper – poor keeping quality. Palmyra palm sugar called *jaggery (gur)* is produced by allowing the filtered sap to boil until it thickens and is then poured into half shells of coconut to solidify, in order to keep the *jaggery* dry they are rolled in flour or rice flour is added to the boiling thickening sap This crude sugar is called *gula jawa* in Indonesia and is widely used in the Javanese cuisine
- Treacle – sap boiled to supersaturation and after clarification, viscous product is bottled
- Sugar – white crystalline sugar may be separated by centrifugation from supersaturated syrup
- Sugar candy when crystallisation is allowed to proceed undisturbed for a long period, large pieces of candy are obtained
- Molasses the mother liquor left over from crystallisation process can be used as a substrate for fermentation and the production of alcohol.

Small fruits are pickled in vinegar. The fibrous outer layer of the ripe palm fruits are also boiled or roasted and eaten. The gelatinous translucent pale-white, endosperm of young fruit is juicy, nutritious and edible and is a refreshing delicacy in India and Sri Lanka and southeast Asia. It is called *Nungu Fruit* in Tamil and *Thati Munjalu* in Telugu. The shell of the seed can be punctured with a finger and the sweetish liquid sucked out for refreshment like coconut water. The tender nutritious endosperm is also canned in Thailand and exported globally. It can also be roasted, sun-dried, conserved and boiled. Jams and cordials are also being tried in Sri Lanka. Palmyra fruit pulp (PFP) is widely used to manufacture many food products including dried PFP (*punatoo*), which has been consumed in North-East Sri Lanka for centuries.

Sprouts consisting of the first tender leaves and fleshy hypocotyl from fresh germinated nuts provide a rich source of vegetable starch and are boiled for immediate consumption or dried or conserved or roasted and pounded to make a meal. The sprouts are known as *Panai Kizhangu* or *Panamkizhangu* in Tamil. The germinated seed's hard shell is also cut open to take out the crunchy kernel which tastes like a sweeter water chestnut. It is called *dhavanai* in Tamil. Palm hearts or *palmito* or *Pananchoru* in Tamil, the tender apical bud is tasty and high prized and delicious. However, harvesting the palm heart/cabbage will destroy the palm. The fusiform roots are also eaten especially in times of famine. In Andhra Pradesh, the young fleshy roots are baked and eaten.

Studies reported that palmyra fruit pulp which is regarded as a waste product, appeared to be best used as an alcoholic fermentation base, with the possibility of using the carotenoids (by-product) as a food colorant (Ariyasena et al. 2001).

Botany

A solitary, large, erect palm (Plate 1) with rough and black stem, 20–30 m tall and the trunk may have a circumference of 1.7 m at the base. Petiole is 60–120 cm long, robust and grooved, semi-terete with edges armed with hard irregular spines. Leaf blade is 60–120 cm long, palmately fan-shaped with 60–80, linear-lanceolate, induplicate segments, 0.6–1.2 m long with marginally spiny segments. Male inflorescence (Plate 2) is

Plate 1 Large tall palmyra palm with a skirt of dried leaves hiding the trunk

Plate 2 Crown of palmyra palm with dried male inflorescences and palmately fan-shaped leaves

Plate 3 Tight cluster of palmrya fruit

90–150 cm long, much branched with primary and secondary branches; male flowers are subsessile, with narrowly cuneate sepals with truncate inflexed tips and obovate-spatulate and shorter petals, large anthers. Female inflorescence has a flowering portion to 30 cm long, with 8–16 flowers spirally arranged. Female flowers have fleshy, large, reniform sepals, smaller petals, subtrigonous ovary, recurved stigmas and sessile. Fruits are broadly ovoid, 15–20 cm across with a fibrous and fleshy, mesocarp and with a tightly adhering persistent large calyx at the base (Plates 3–4). Pyrenes usually three, are obcordate, 6–7 mm broad and black. Within the mature seed is a solid white kernel. When the fruit is very young, this kernel is hollow, soft and translucent like jelly, and is accompanied by a watery sweetish liquid.

Plate 4 Ripe palmyra fruit with persistent green calyx

Nutritive/Medicinal Properties

The fresh fruit pulp is reportedly rich in vitamins A and C. Nutritional composition (g/100 cm^3) of palmyra sweet sap (Davis and Johnson 1987) was found to comprise nitrogen 0.056 g, protein 0.35 g, total sugar 10.93 g, reduced sugar 0.96 g, ash 0.54 g, Ca traces, P 0.14 g, Fe 0.4 g, vitamin 13.25 mg, vitamin B 1 3.9 IU, vitamin B complex negligible, SG 1.07, ph 6.7–6.9. Nutritional composition (g/100 cm^3) of palmyra sugar (Jaggery) (Davis and Johnson 1987) was reported as: protein 0.24%, fat 0.37%, mineral matter 0.5%, carbohydrate (% by difference) 98.89%, (%, direct polarimetry) 98.4, Ca 0.08%, P 0.064%, Fe 30 mg/100 g, nicotinic acid 4.02μg/100 g, riboflavin 229μg/100 g, and energy 398 Cal. The fresh sap is reportedly a good source of vitamin B complex. Palmyra palm jaggery (gur) is much more nutritious than crude cane sugar, containing 1.04% protein, 0.19% fat, 76.86% sucrose, 1.66% glucose, 3.15% total minerals, 0.861% calcium, 0.052% phosphorus; also 11.01 mg iron per 100 g and 0.767 mg of copper per 100 g.

Bains et al. (1956) reported that the tender kernels of *B. flabellifer* contained free sugars, namely, sucrose (0.38%), fructose (1.46%), and glucose (3.21%). Besides the free sugars, the kernel contained polysaccharides, of which the major polysaccharide had been characterized as a galactomannan. The purified polysaccharide on hydrolysis yielded, on a moisture-free basis,

27.64% D-galactose and 78.32% D-mannose. The presence in the kernel of an insoluble fibrous material probably a type of mannose-cellulose was also indicated.

Four types of palmyra fruit were found with varying carotenoid profiles (Pathberita and Jansz 2005). Type 1 had the following carotenoid content (μg/100 g FW): phytoene 286 μg, phytofluene 30.6 μg, β-carotene 4.2 μg, ζ–carotene 64.6 μg, neurosperene 79.8 μg, and retinol equivalent 0.7. Type IIA had: phytoene 461 μg, phytofluene 48.8 μg, β-carotene 5 μg, ζ–carotene 176 μg, neurosperene 96 μg, and retinol equivalent 0.8. Type III had: phytoene 353 μg, phytofluene 9 μg, ζ–carotene 19.5 μg, neurosperene 183 μg, and retinol equivalent 0. Type IIB had: phytofluene 474 μg, β-carotene 43.8 μg, ζ–carotene 165 μg, neurosperene 690 μg, β-zeacarotene 30 μg, unidentified 35 μg, and retinol equivalent 9.8. Two carotenoids patterns were found. Type I, II-A and Type II fruit pulps were dominated by phytoene and neurosperene and were largely non-provitamin A. Type II-B fruit pulp were rich in phytofluene and had a retinol equivalent of 9.8/100 g FW from β-carotene and β-zeacarotene. The pulp contained both hydrophilic and lipophilic antioxidants with total antioxidant ranging from 2.4 to 21.8 /trolox equivalent as evaluated by ABTS.

Of four flabelliferins found in palmyra fruit pulp, one designated F-II corresponded to the bitter flabelliferin tetraglycoside, the other three were designated F_B, F_C, F_D (Nikawela et al. 1998). F_B was most effective in foam stabilisation while F_B and F_C had the highest haemolytic action. Ariyasena et al. (2002) found six types of flabelliferin in palmyra fruit pulp, F_F (flabelliferin monoglucoside M.W. 576), F_D (flabelliferin diglucoside, M.W. 722), F_E (flabelliferin diglucoside, M.W. 734), F_B (branched flabelliferin triglucoside, M.W. 868), F_C (linear flabelliferin monoglucoside, M.W. 868), FII – flabelliferin (tetraglycoside M.W. 1030).

Palmyra (*Borassus flabellifer* L.), wild date (*Phoenix sylvestris* Roxb.) and coconut (*Cocos nucifera* L.) sap were found to have high nutritive values before and after 24 hours partial natural fermentation (Barh and Mazumdar 2008). Total sugars, Ca Fe, Zn, Cu, P, niacin were highest in fresh wild date sap followed by male palmyra palm which was richest in vitamin-A (74.0 IU/100 ml) and coconut was richest in Na and K content. Palmyra male also contained highest amount of total proteins (19.70 mg/100 ml) followed by female palmyra and wild date (14.80 mg/100 ml). Mg content was higher in both male and female palmyra than that of wild date and coconut. Thiamin, riboflavin and niacin showed dramatic increase in fermented date and male palmyra sap with no significant change in micronutrients.

Palmyra palm flowers have a few pharmacological properties.

Antidiabetogenic Activity

Recent studies reported that the methanolic extracts from male flowers exhibited anti-diabetogenic activity. The methanolic extract from the male flowers of *Borassus flabellifer* was found to inhibit the increase of serum glucose levels in sucrose-loaded rats at a dose of 250 mg/kg, p.o. (Yoshikawa et al. 2007). From the methanolic extract, six new spirostane-type steroid saponins, borassosides A-F were isolated together with 23 known constituents. Additionally, the principal steroid saponin, dioscin inhibited the increase of serum glucose levels in sucrose-loaded rats at a dose of 50 mg/kg, p.o. Mice fed with 10% palmyra fruit pulp exhibited reduced weight gain due to the bitter steroidal saponin, flabelliferin II (Uluwaduge et al. 2006). At a dose of 10 mg per mouse, mixed falbellerins (containing 2.5 mg flabelliferin II) was found to reduce blood glucose after glucose challenge, increased faecal glucose and intestinal glucose. The results suggested that F-II had potential for use as an anti-obesity or antidiabetic food component. As the pulp was non-toxic it may find application as a functional food. A subsequent study reported that pinattu (dried Palmyrah fruit pulp) could be used as an anti-hyperglycemic agent (Uluwaduge et al. 2007). The active principle was found to be a steroidal saponin, flabelliferin-II which inhibited intestinal ATPase in mice. In a clinical trial, all mild diabetic patients treated with pinattu registered a significant reduction in serum glucose level.

Anti-inflammatory Activity

In a recent study using the nystatin-induced rat paw edema model, the ethanol extract of *B. flabellifer* male flowers at doses 200 mg/kg body weight (b.w.) and 400 mg/kg b.w. and diclofenac sodium (standard) at 100 mg/kg b.w. exhibited significant anti-inflammatory and antiarthritic activity, as compared to control (Paschapur et al. 2009). The result showed that the extract at both the doses gave a significant reduction of rat paw oedema from 2 to 72 hours when assessed at different time intervals. There was significant decreases in levels of serum glutamate pyruvate transaminase (SGPT), serum glutamate oxaloacetate transaminase (SGOT), lipid peroxidation and alkaline phosphatase (ALP) at either dose as compared to the control. Diclofenac sodium, a COX-inhibitor at the dose of 100 mg/kg, also significantly reduced the paw oedema. In FCA (Freund's Complete Adjuvant) induced arthritis, the elevated levels of hydroxy proline, hexosamine and total protein observed in edematous tissue during FCA injection were significantly inhibited by the extract at both the doses and diclofenac sodium treatment. The results confirmed the traditional use of *Borassus flabellifer* for the treatment of painful inflammatory conditions and in arthritic pain.

Adverse Effects

Animal models studies had reported on the adverse effects of palmyra flour which was consumed by humans in some Asian and African countries (Kangwanpong et al. 1981). Such adverse biological effects of palmyra flour included carcinogenic (induction of malignant lymphomas), immunosuppression, neurotoxicity, mutagencity and clastogenicity. The aqueous extracts of the flour were found to be clastogenic, mainly to group A chromosomes, producing chromatid and chromosome gaps, and chromatid and chromosome breaks with some formation of large and small acentric fragments. In another study, both boiled and raw flour showed dose related mutagenic responses in the base pair substitution sensitive *Salmonella typhimurium* strain TA100 (using Ames test) and *Escherichia coli* WP2uvrA, CM881 and CM891(Andersen and Poulsen 1985). The flour from boiled palmyra shoots exerted a somewhat stronger mutagenic effect on *Salmonella typhimurium* than flour from the raw shoots. Another study showed that oral feeding of mice with palmyra flour induced the generation of T suppressor cells which were able to suppress the delayed-type hypersensitivity response to sheep red blood cells (SRBC) (Devi et al. 1985). Inbred mice fed with a 40% mixture of palmyra flour pellets showed a significantly decreased capacity to mount a delayed-type hypersensitivity (DTH) response to sheep red blood cells (SRBC). Maximal suppression of DTH was observed after 6 days of feeding but was detectable even after 2 days. Maximal suppression occurred when flour feed was maintained into the sensitization period (with SRBC).

Recent studies found that the neurotoxic symptoms reported earlier could be eliminated by heat-detoxification of the palmyra flour at 80°C for 45 minutes (Sumudunie et al. 2004). The study showed that feeds containing 100% and 70% palmyra flour resulted in very little or no feed consumption, and that deaths reported in earlier studies could have been interpreted as being due to starvation. It was interpreted that the nutritional status of the diet influenced that manifestation of the neurotoxic effect; the effect being suppressed by a nutritious diet.

Traditional Medicinal Uses

There are innumerable medicinal uses for all parts of the palmyra palm in traditional local medicine. The young plant is said to relieve biliousness, dysentery, and gonorrhoea. Young roots are reported to be diuretic and anthelmintic, and a decoction is given in certain respiratory diseases. The ash of the spadix is taken to relieve heartburn and enlarged spleen and liver. The bark decoction, with salt, is used as a mouth wash, and charcoal made of the bark serves as a dentifrice. Sap from the flower stalk is prized as a tonic, diuretic, stimulant, laxative and anti phlegmatic and amoebicide. Sugar made from this sap is said to counteract poisoning, and it is prescribed in liver disorders. Candied, it is a remedy for coughs and various pulmonary complaints. Fresh toddy, heated to promote fermentation, is bandaged onto all kinds of ulcers. The cabbage, leaf petioles, and dried male flower spikes all have diuretic activity. The pulp of the mature fruit relieves dermatitis.

Other Uses

The tree yields a black timber with desirable unique properties – hard, heavy, and durable and is highly valued for house construction, particle board, and furniture or made into handles, sticks. The timber and leaves are also used as fuelwood. The chopped leaves are employed as green manure in rice fields. The leaves are seasoned, dried and used for thatching, mats, baskets, fans, hats, umbrellas, and as writing material. In Indonesia, the leaves of this plant are formerly used in the ancient culture as papers, known as *lontar*. The leaf petioles are used to make fences and also produce a strong, wiry fibre suitable for cordage, baskets and brushes. The palm yields five types of fibre a loose fibre from the leaf base, a long fibre from the leaf stalk, a fibre from the interior of the trunk, a fibre coir derived from the fruit pericarp and the fibrous material from leaf. The fibres, extracted from the leaf bases (palmyra fibre), are used in the manufacture of many types of brushes. The fibres from the midrib of the leaf are used to make brooms. Fibres for weaving are obtained from the adaxial part of the petiole and from leaves not yet unfolded. Fibres extracted from the frond bases exhibit valuable qualities of resistance to chemicals, termite and water and are exported from the provinces of Andra Pradesh, Tamil Nadu, Kerala. Tender fruits that fall prematurely are fed to cattle. The roots, the sap of the plant, and the pulp serve medicinal purposes. In India also planted in windbreaks. It is also used as a natural shelter by birds, bats and wild animals. Recent studies reported that palmyra palm fruit shell extract inhibited the corrosion of mild steel and could have industrial application (Vijayalakshmi et al. 2010).

Comments

This palm is propagated solely from seeds.

Selected References

Andersen PH, Poulsen E (1985) Mutagenicity of flour from the palmyrah palm (*Borassus flabellifer*) in *Salmonella typhimurium* and *Escherichia coli*. Cancer Lett 26(1):113–9

Ariyasena DD, Jansz ER, Abeysekera AM (2001) Some studies directed at increasing the potential use of palmyrah (*Borassus flabellifer* L) fruit pulp. J Sci Food Agric 81(14):1347–1352

Ariyasena DD, Jansz ER, Baeckstrom P (2002) Direct isolation of flabelliferins of palmyrah by MPLC. J Natn Sci Foundation Sri Lanka 30(1–2):55–60

Bains GS, Natarajan CP, Bhatia DS (1956) The carbohydrates of tender kernel of palmyra palm (*Borassus flabellifer*, L.). Arch Biochem Biophys 60(1):27–34

Balick MJ (1988) The palm – tree of life: biology, utilization and conservation. In: Proceedings of a symposium at the 1986 annual meeting of the society for economic botany held at The New York Botanical Garden, Bronx, New York, 13–14 June 1986. Advances Econ Bot, 6, 282 pp

Barh D, Mazumdar BC (2008) Comparative nutritive values of palm saps before and after their partial fermentation and effective use of wild date (*Phoenix sylvestris* Roxb.) sap in treatment of anemia. Res J Med Med Sci 3(2):173–176

Burkill IH (1966) A dictionary of the economic products of the Malay Peninsula. Revised reprint, 2 volumes, vol 1 (A–H) pp 1–1240, vol 2 (I–Z) pp 1241–2444. Ministry of Agriculture and Co-operatives, Kuala Lumpur

Chopra RN, Nayar SL, Chopra IC (1956) Glossary of Indian medicinal plants (Including the supplement). Council Scientific Industrial Research, New Delhi, p 330 pp

Council of Scientific and Industrial Research (1948) The wealth of India: A dictionary of Indian raw materials and industrial products (Raw materials 1). Publications and Information Directorate, New Delhi

Davis TA, Johnson DV (1987) Current utilization and further development of the palmyra palm (*Borassus flabel-*

lifer L., Arecaceae) in Tamil Nadu State, India. Econ Bot 41:247–266

Devi S, Arseculeratne SN, Pathmanathan R, McKenzie IF, Pang T (1985) Suppression of cell-mediated immunity following oral feeding of mice with palmyrah (*Borassus flabellifer* L.) flour. Aust J Exp Biol Med Sci 63(4):371–9

Flach M, Paisooksantivatana Y (1996) *Borassus flabellifer* L. In: Flach M, Rumawas F (eds) Plant resources of South-East Asia No. 9. Plants yielding non-seed carbohydrates. Prosea Foundation, Bogor, pp 59–63

Foundation for Revitalisation of Local Health Traditions (2008). FRLHT Database. htttp://envis.frlht.org.

Gupta RK, Kanodia KC (1968) Plants used during scarcity and famine periods in the dry regions of India. J Agric Trop Bot Appl 15:265–285

Johnson D (ed) (1991) Palms for human needs in Asia: palm utilization and conservation in India, Indonesia, Malaysia and the Philippines. A.A. Balkema, Rotterdam, p 258

Johnson DV (1998) Tropical palms. FAO Regional Office for Asia and the Pacific, Food and Agriculture Organization of the United Nations, Rome, p 166 pp

Kangwanpong D, Arseculeratne SN, Sirisinha S (1981) Clastogenic effect of aqueous extracts of palmyrah (*Borassus flabellifer*) flour on human blood lymphocytes. Mutat Res 89(1):63–8

Kovoor A (1983) The palmyrah palm: potential and perspectives. FAO plant production and protection paper 52. FAO, Rome, 77 pp

Morton JF (1988) Notes on distribution, propagation, and products of *Borassus* palms (Arecaceae). Econ Bot 42(3):420–441

Nikawela JK, Abeysekara AM, Jansz ER (1998) Flabelliferins – steroidal saponins from palmyrah (*Borassus flabellifer* L.) fruit pulp. J Natl Sci Coun Sri Lanka 26(1):9–18

Paschapur MS, Patil MB, Kumar R, Patil SR (2009) Influence of ethanolic extract of *Borassus flabellifer* L. male flowers (inflorescences) on chemically induced acute inflammation and poly arthritis in rats. Int J Pharm Technol Res 1(3):551–556

Pathberita LG, Jansz ER (2005) Studies on the carotenoids and in vitro antioxidant capacity of palmyrah fruit pulp from Mannar. J Natn Sci Foundation Sri Lanka 33(4):269–272

Reddy KN, Pattanaik C, Reddy CS, Raju VS (2007) Traditional knowledge on wild food plants in Andhra Pradesh, India. Indian J Tradit Knowl 6(1):223–229

Sumudunie KAV, Jansz ER, Jayasekera S, Wickramasinghe SM (2004) The neurotoxic effect of palmyrah (*Borassus flabellifer*) flour re-visited. Int J Food Sci Nutr 55(8):607–14

Uluwaduge I, Thabrew MI, Jansz ER (2006) The effect of flabelliferins of palmyrah fruit pulp on intestinal glucoses uptake in mice. J Natn Sci Foundation Sri Lanka 34(1):37–41

Uluwaduge I, Parera A, Jansz E, Thabrew I (2007) A pilot study on palmyrah pinattu (dried fruit pulp) as an antidiabetic food component. Int J Biol Chem Sci 1(3):250–254

Vijayalakshmi PR, Rajalakshmi R, Subhashini S (2010) Inhibitory action of *Borassus flabellifer* Linn. (palmyra palm) shell extract on corrosion of mild steel in acidic media. E-J Chem 7(3):1055–1065

Yoshikawa M, Xu F, Morikawa T, Pongpiriyadacha Y, Nakamura S, Asao Y, Kumahara A, Matsuda H (2007) Medicinal flowers. XII. (1)) New spirostane-type steroid saponins with antidiabetogenic activity from *Borassus flabellifer*. Chem Pharm Bull (Tokyo) 55(2):308–16

Cocos nucifera

Scientific Name

Cocos nucifera L.

Synonyms

Calappa nucifera (L.) Kuntze, *Cocos indica* Royle, *Cocos nana* Griff., *Cocos nucifera* var. *synphyllica* Becc., *Palma cocos* Mill. nom. illeg.

Family

Arecaceae

Common/English Names

Coconut, Coconut Palm

Vernacular Names

Albania: Arre kokosi;
Arabic: Gawz El Hind, Jadhirdah, Jauz Al Hind, Nârgîl;
Armenian: Hantkakan Engouz;
Australia: Kunga (Lockhart River, Aboriginal), Warraber, Warraper (Torres Strait, Aboriginal);
Banaban: Te Ni;
Brazil: Coco, Coco-Da-Bahia, Coco-Da-Baía, Coco Da India, Coco-Da-Praia, Coqueiro, Coqueiro-Da-Bahia, Coqueiro-Da-Praia;
Bulgarian: Kokos;
Burmese: Mak Un On, Ong, Ung, Ungbin;
Chinese: Ye Zi, Ye Shu;
Cook Islands: Nu;
Chuuk: Nu;
Croatian: Kokosov Orah;
Czech: Kokos, Kokosový Ořech, Kokosovník Ořechoplodý;
Danish: Kokos, Kokosnød, Kokosnoed, Kokospalme;
Dutch: Kokos, Kokosnoot, Kokosnootpalm, Kokospalm, Klapperboom;
Estonian: Kookospalm;
Ethiopia: Kokas, Kokonet;
Fijian: Niu;
Finnish: Kookospähkinä, Kookospalmu;
French: Coco, Cocotier, Cocoyer, Coq Au Lait, Nariyal, Noix De Coco;
German: Kokos, Kokosnuß, Kokosnuss, Kokospalme;
Greek: Kokkofoinika, Karida, Karides, Karyda;
Guam: Niyog;
Guyana: Coconut;
Hawaii: Ōʻlo (Unripe nut with jelly-like translucent flesh), Hao Hao (Maturing nut with shell still white and flesh soft and white), ʻIli Kole (Half-ripe nut), Niu Oʻo (Nut mature but husk not dried), Niu Maloʻo (Nut mature, husk dry), Niu Ō Kaʻa (Old nut with no water and flesh separated from shell);
Hebrew: Kokus, Kokosnus;
Hungarian: Kókusz, Kókuszdió, Kókuszpálma;
Icelandic: Kokoshenta;

India: Narikol (Assamese), Daab, Naarakel Naarakela, Narikel (Bengali), Nanivaara, Nariyal (Gujarati); Narel, Nariyal, Naariyal Kaa Per, Khopar (Hindi); Kobari, Naarikela, Thenginkai (Kannada), Nalikeram, Thenga (Malayalam); Naaral (Marathi), Naarial (Punjabi), Kalpavriksha, Naarikela (Sanskrit), Tengai, Tennai Marama, Thenga, Tengu, Thenkaii (Tamil), Kobbari, Kobbera, Narikelamu, Tenkaya (Telugu), Naariyal, Naariyal Kaa Per (Urdu);
Indonesia: Bhungkana, Krambil (Java), Kelapa, Nyior;
Isle of Man: Crobainery;
Italian: Cocco, Cocos Nucifera, Noce Di Cocco, Palma Del Cocco;
Japanese: Koko Yashi, Koko Yashi, Natsume Yashi;
Kenya: Dafu;
Khmer: Doung;
Kiribati: Te Ni;
Korean: K'o K'o Neos;
Kosrae: Lu;
Laotian: Kok Mak Phao, Phaawz;
Malaysia: Kelapa, Nyior;
Marshall Islands: Ni;
Nepalese: Narival, Nariwal;
North Mariana Islands: Nizok;
Palauan: Iru;
Papua New Guinea: Kokonas, Niu;
Persian: Nârgil;
Philippines: Lubi (Cebu-Bisaya), Iniug, Oñgot (Ibanag), I-Ing (Itogon), Laying (Manobo), Uñgut (Pampangan), Lubi (Panay Bisaya), Lobi (Samar-Leyte Bisaya), Gira-Gira (Sambali), Ponlaing (Subanum), Boko, Buko, Niyog, Ubod (Palm Heart) (Tagalog), Punlaing (Yakan);
Pohnpei: Ni;
Polish: Kokos;
Portuguese: Coco Da Bahia, Coco Da India, Coqueiro, Coqueiro De Bahia, Noz De Coco;
Polynesia: Niu;
Romanian: Cocotier;
Rotuman: Niu, Nariel;
Russian: Kokos, Kokos Orekhonosnmi, Kokosobaia Pal'ma;
Society Islands: Ha'Ari;
South Africa: Khokhonate;
Slovenian: Kokos, Kokosa;
Spanish: Coco, Coco De Agua, Cocotero, Nuez De Coco, Palma De Coco, Palmera De Coco, Palmera De Coco;
Sri Lanka: Polgaha (Sinhalese), Kobbai, Kobbera (Tamil);
Swahili: Mnazi;
Swedish: Kokos, Kokosnöt, Kokospalm;
Taiwan: Ke Ke Ye Zi;
Tanzania: Dafu;
Thai: Dung (Chong-Chan-Thaburi), Khosaa (Karen-Mae Hong Son), Maphrao, Má-Práao, Ma Phrao, Maprow, Ma Phrao On, Maak Muu;
Turkish: Hindistancevizi, Kafa;
Ukrainian: Kokos;
Vietnamese: Dừa;
Yap: Lu;
Zaire: Dafu.

Origin/Distribution

The coconut has been the subject of much interpretation and speculation in regard to its natural origin because of its wide distribution and cultivation in the tropics and subtropics. It is assumed that the coconut palm is native to coastal areas (the littoral zone) of southeast Asia (Malaysia, Indonesia, Philippines) and Melanesia. It is believed that in prehistoric times wild forms (niu kafa) eastwards into the central Pacific to the tropical Pacific islands (Polynesia, and Micronesia) and westward to coastal India, Sri Lanka, East Africa, and tropical islands (e.g., Seychelles, Andaman, Mauritius) in the Indian Ocean. Coconut is either an introduction in the pre-Columbian times or possibly native to the Pacific coast of Central America.

Agroecology

Coconut grows in climates with annual temperature of 21–30°C and mean of 25.7°C and annual sunlight hours above 2,000 h and relative humidity above 60%; with annual rainfall of 1,000–2,000 mm in summer, winter, bimodal, and uniform rainfall patterns. Coconut palm thrives best in full sun. It is cold sensitive, periods with mean daily temperatures below 21°C adversely affect the growth and yield of the palms. Cold temperatures below 7°C damages the palm and frost is fatal to seedlings and young palms. It is found mostly at low altitude from sea level to 500 m but can be found up to 1,000 m. It is adaptable to a to wide range of soil types light, medium and heavy, – peaty soil, sandy, calcareous, saline to clayey soils. It tolerates alkaline soils up to pH 8 (on coralline atolls) and acid soils with pH 4.5 or higher. The ideal pH range is 5.5–7. Although coarse sand is its natural habitat, best growth is obtained on free draining, deep soils with good physical and chemical properties. It is thus widely grown on loams as well as clays that are well drained. Coconut is intolerant of water-logging will not survive more than 2 weeks of surface water-logging. Coconut tolerates drought poorly. Symptoms include desiccation of older fronds, spears (emerging fronds) failing to open normally, and shedding of young nuts. Coconut is able to withstand salt spray very well. It is able to withstand cyclonic (hurricane) winds if roots are well anchored. Coconut is also cultivated in plantations as monocrop (Plate 4) or intercrop with other cash crops or fruit trees, cocoa or coffee.

Plate 1 "Red" exocarp coconut variety

Plate 2 Yellow-exocarp coconut variety

Edible Plant Parts And Uses

Several aerial plant parts of the coconut tree in particularly the fruit and its by-products have a wide array of edible uses (Philippine Coconut Authority undated; Ohler and Magat 2001; Chan and Elevitch 2006; van der Vossen and Chipungahelo 2007). Young and mature coconut are available on sale in various forms – whole fruit, dehusked, exocarp shaven and dehusked clean kernel (Plates 5–7).

At the apex of the tree is the apical meristem or pith which comprised of tightly packed, yellow-white, cabbage-like leaves, is the palm heart or palm cabbage or *ubod* in Tagalog, Palm heart (Plate 8) is esteemed as an expensive, delicious delicacy, once harvested the tree dies. It serves as a vegetable, used in salads, or in food dishes like *lumpia, achara*, spring roll and also made into pickles in coconut vinegar. Pith of the stem

Plate 3 Green-exocarp coconut variety

contains starch which may be extracted and used as flour. If a matured nut is allowed to germinate, its cavity fills with a spongy mass called 'bread' which is extracted from the germinated nut eaten

Plate 4 A coconut monocrop plantation

Plate 7 Dehusked coconut kernel on sale

Plate 5 Coconut dehusked for sale

Plate 8 Coconut 'heart' on sale in the market

Plate 6 Forms of young coconut on sale in the market

raw, baked or toasted in shell over fire. It makes a soft food for babies and elderly people. Sometimes it is scraped out, mixed with toddy, and eaten with fish. Sprouting germinating seeds may be eaten like celery.

The unopened inflorescence can be tapped for its sweet sap which is rich in nutrients and has a vitamin C content of 16–30/100 g. On fermentation the sap is converted into an alcoholic beverage with alcohol content up to 8%, called *'toddy'* in India, Sri Lanka and Malaysia; *'Nam-tan-mao'* in Thailand; *'tuba'* in Philippines and Mexico; and *'tuwak'* in Indonesia, or *'tuak'* in East Malaysia. Coconut toddy is consumed fresh as an alcoholic beverage. After standing a few weeks, it becomes a vinegar. On distillation, a liquor called *'lambanog'* in Tagalog, or *arrack,* is obtained from toddy. In Goa, India, commercial arrack obtained by distillation of coconut toddy is known as coconut fenny. The sap can be processed into coconut sugar or jaggery after collection and filtration. The filtrate is boiled down into a sticky brown, thick syrup called coconut molasses. The molasse is poured into tin cans or semi-spherical moulds and allow to crystallised into a rich dark sugar, almost exactly like maple sugar. A technique developed by the Thailand Horticultural Research Institute (Chomchalow 1999) produces granulated coconut sugar for use in instant coffee or tea. This process involves heating the collected sap down to 80% concentration, then removing the pan from the fire and continuously stirring until the syrup turns into granulated sugar which is passed through a sieve to obtain uniformed-sized granules. The granulated sugar is yellow in colour and has total sugar as 90% saccharose. It is packed in bottles or plastic bags. Sometimes the dark sugar is mixed with grated coconut for candy. Toddy was also used as a source of yeast. Coconut nectar is an extract from the young bud, a very rare type of nectar collected and used as morning breakfast drink in the islands of Maldives, and is reputed to have energetic power. A by-product, a sweet honey-like syrup called *dhiyaa hakuru* is used as a creamy sugar for desserts.

Young, green coconut and mature, old coconuts are harvested and eaten. Fermented ripe nuts have a tough flesh that is sour and oily and some people relished its special flavour. A variety of coconut found in some Pacific atolls has part of the coconut husk that can be eaten. The sweet and fairly soft husk is chewed. The small nuts formed after the flower has set on trees of this variety are also eaten. Children like this sweet-flavoured treat. The coconut water from young green coconuts makes a refreshing, nutritious drink chilled or otherwise and the young soft white meat is scooped out and mixed with the coconut water and eaten. In Thailand, coconut water is sold in plastic bags, canned, or frozen. The meat from young coconuts is also eaten separately as snack or in fruit salads or mixed with other ingredients in various dessert. The flesh is soft and spongy, making an excellent baby food. It is also used in confectioneries and in making pies (*buko* pie in the Philippines). In the Philippines, macapuno is a freak coconut with soft, tender and jelly-like meat which fills up the whole nut cavity. *Macapuno* is used in confectionery and bakery items, ice-cream and fruit salads. The kernel meat from mature coconut is made into chips, desiccated coconut and grated coconut for extraction of coconut milk. Coconut chips are prepared from thinly sliced, dried matured coconut meat (not copra meat). They are usually served as snack food in parties. Fresh, grated coconut meat is packed and sold in plastic bags in local markets for immediate use. Grated coconut is used in confectioneries and bakery – biscuits cakes and candies. In the Philippines *bukayo* is a popular dessert delicacy made from freshly grated coconut, molasses and corn syrup. Grated coconut are pressed and filtered through a cloth to obtain the white rich, coconut milk or *gata* in Tagalog, *kathi* in Thailand, *santan* in Malay and *katas ngungut* in Kapampangan. Coconut milk is packed and sold in tin cans with a shelf-life of 1 year. Coconut milk is widely used in Asian and Pacific cuisine; in vegetable, prawn, fish, chicken, beef curries; laksa noodles; soups (e.g. the famous Thai *Tom Yam Kung* soup); cooked with rice to make Panama's famous *arroz con coco*; also cooked with taro leaves, and used in coffee as cream. In the Philippines, coconut milk mixed with other ingredients makes a very refreshing drink such as *pina colada* which is also available in cans. Coconut jam is a mixture of coconut milk and brown sugar, popularly used as bread spread and dessert. Coconut honey is prepared from coconut milk and is a good substitute for honey and is used as bread spread. Coconut milk is used extensively in various types of desserts: with red beans or green gram porridge; black glutinous rice dessert called *pulut hitam* in Malaysia; with rice or tapioca starch desserts; fruits like jackfruit, bananas; root crops like taro, tapioca, sweet potato, and vegetables like pumpkin. Desiccated coconut is the white dehydrated, shredded kernel meat of coconut, used mainly in a variety of edible food processing, including the production of confectionery (in chocolate bars and as filler in nut-based chocolate products, candies and liquorice) and bakery products (viz, cakes, biscuits, macaroons, ice cream, chocolates), canned and frozen foods (ice-cream and other frozen products) and as a commercial and domestic cooking ingredient. Desiccated coconut is available in granular flakes or shredded form of varying sizes and shred lengths to match the need of confectionery, pastry or baking requirements. It is available in tins or plastic packets. Coconut flour is the pulverized, clear white powdery, food grade residue from pressed and or solvent extracted, dehydrated coconut meat. It has a mean protein content of 24%. A combination of 10 parts coconut flour to 90 parts wheat flour is used in the baking of bread and biscuits to increase the food's protein value.

Coconut water can be fermented with brown sugar and yeast to obtain coconut vinegar. Coconut vinegar is used in the preservation of fruits and vegetables and the preparation of pickles, sauces, chutneys and other manufactured products. *Nata de Coco*, a popular gelatinous dessert is produced by brewing a species of the vinegar group of bacteria in sugared coconut water and coconut skim milk media. It is also used as an ingredient in the preparation of other desserts and confectionery. *Nata de Coco* is also used as clinical dextran for isolation studies and an appreciable volume is being exported to the United States from the Philippines.

Coconut oil is the natural oil obtained from copra or fresh kernel meat. It is colourless to brownish yellow in appearance. Copra can be made by smoke drying, sun drying, or kiln drying. In India, edible grade of copra is consumed as a dry fruit and used for religious purposes. Edible grade copra is distinct from milling copra which is used for oil extraction. Copra is the dried kernel meat of a matured coconut and is eaten as emergency food in the Pacific islands. Coconut oil is commonly used in cooking, especially when frying. In communities where coconut oil is widely used in cooking, the unrefined oil is the one most commonly used. Coconut oil is commonly used to flavor many South Asian curries. It does not create any harmful byproducts when heated. Hydrogenated or partially-hydrogenated coconut oil is often used in non-dairy creamers, and snack foods. RBD (refined, bleached, and deodorized) coconut oil is usually made from copra and is used for home cooking and commercial food processing besides cosmetic, industrial, and pharmaceutical purposes.

The main edible products made from coconut oil are:

(a) Cooking/frying oil – coconut oil is widely used for popcorn popping because of it high temperature stability, long shelf life and bland flavour.
(b) Shortening and baking fats – coconut oil has been combined with a small amount of hydrogenated palm or cotton seed oil to produce shortening of high quality with wide application that include bread, cakes, cookies, crackers, pastries, etc.
(c) Vanaspati – coconut oil either hardened or unhardened, alone or in combination with other oils is used in some countries for making vanaspati (ghee). Coconut oil can be incorporated in existing vanaspati formulation up to 20% without any appreciable change in the latter's physical and performance characteristics.
(d) Margarine.
(e) Imitation dairy products – filled milk, filled cheese, fat component in ice cream, mellorines (filled ice milk or ice cream with 4–12% fat content), coffee whiteners (hydrogenated or partially-hydrogenated coconut oil is preferred for these non-dairy creamers), whipped dessert toppings, spray oils (coconut oil makes a superior spray oil for crackers, cookies and cereal – enhance flavour, shelf-life and glossy appearance).
(f) Confectionary – candy bars, chocolates, toffees, caramels etc.
(g) Baby food.
(h) Salad oil.

Virgin Coconut oil (VCO) is oil extracted from fresh coconut (not copra) meat by mechanical or natural means with or without low, prolonged heating or chemical processing (Fife 2006). To protect the oil's essential properties, the production of virgin coconut oil does not undergo chemical refining, bleaching, or deodorizing. VCO has been shown to be high in vitamins and minerals; it is fit for consumption without the need for further processing. Since it is a solid most of the time at room temperature or when refrigerated, it can be a butter or margarine substitute for spreads or for baking. Any recipe calling for butter, margarine, or any other oil can be substituted with Virgin Coconut Oil.

Botany

Tall, erect, single-stemmed, monoecious palm, up to 30 m tall, with a slender, more or less curved or inclined, stout trunk, 30–45 cm diameter marked with leaf scars, arising rising from a swollen base surrounded by mass of adventitious roots, Leaves frond-like, 2–6 m long, clustered at top of trunk, pinnatisect, leaflets 0.6–1 m long, narrow, linear-lanceolate, bright green, inserted on rachis in 2 ranks. Inflorescence called a spadix, axillary, with male and female flowers on the same inflorescence .Spadix consists of a main axis 1–1.5 m in length with 40–60 branches or spikelets bearing the flowers and enclosed by a woody sheathe called spathe. Each spike-let carries from zero to three female flowers ("buttons") at its base and several hundred male flowers above. Thus a spadix will have several thousand male flowers but only 40–60 buttons. Male flowers with six creamy-yellow perianth segments surrounding six stamens, pistillode topped by 3 nectar glands. Female flowers, sweet-scented, larger, globose, with 6 creamy-yellow perianth segment in two whorls, with 6 staminodes, 3-locule ovary, and a trifid stigma. Following pollination, only one carpel develops into the seed, the other two aborting. The perianth persists at the base of the mature fruit. The fruit is a fibrous drupe. It consists of, from the outside in, a thin hard skin (exocarp), a thicker layer of fibrous mesocarp (husk), the hard endocarp (shell), the white albuminous endosperm (kernel), and a large cavity filled with fluid ("coconut water"). When immature, the exocarp is usually green, yellow or sometimes bronze, becoming brown when mature. Wide variation in fruit shape exocarp colour, green, yellow, reddish-brown, etc. and size exist within types and populations (Plates 1–3). Fruit shapes vary from elongated, ovoid, ellipsoid to almost spherical and weigh between 0.85 and 3.7 kg when mature. The seed comprises the dark brown shell, kernel (Plate 7), cavity and embryo. The nut has three micropyles or "eyes," one of which is soft and indicates the position of the viable embryo embedded in the kernel.

Nutritive/Medicinal Properties

Coconut Kernel (Endosperm) Meat

Food value of raw, coconut meat from mature nut (refuse 48% brown shell, skin and water) per 100 g edible portion was reported as follows (USDA 2010): water 44.99 g, energy 354 kcal (1,481 kJ), protein 3.33 g, total lipid (fat) 33.49 g, ash 0.97 g, carbohydrate 15.23 g; fibre (total dietary) 9.0 g, total sugars 6.23 g, minerals – calcium 14 mg, iron 2.43 mg, magnesium 32 mg, phosphorus 113 mg, potassium 356 mg, sodium 20 mg, zinc 1.10 mg, copper 0.435 mg, manganese 1.5 mg, selenium 10.1 µg; vitamins – vitamin C (total ascorbic acid) 3.3 mg, thiamin 0.066 mg, riboflavin 0.020 mg, niacin 0.540 mg, pantothenic acid 0.300 mg, vitamin B-6 0.054 mg, folate (total) 26 µg, total choline 12.1 mg, vitamin E (α tocopherol) 0.24 mg, γ-tocopherol 0.53 mg, vitamin K (phylloquinone) 0.2ug, lipids – fatty acids (total saturated) 29.698 g, 6:0 (caproic acid) 0.191 g, 8:0 (caprylic acid) 2.346 g, 10:0 (capric acid) 1.864 g, 12:0 (lauric acid) 14.858 g, 14:0 (myristic acid) 5.866 g, 16:0 (palmitic acid) 2.839 g, 18:0 (stearic acid) 1.734 g; fatty acids (total monounsaturated) 1.425 g, 18:1 undifferentiated (oleic acid) 1.425 g; fatty acids (total polyunsaturated) 0.366 g, 18:2 undifferentiated (linoleic acid) 0.366 g; phytosterols 47 mg; amino acids – tryptophan 0.039 g, threonine 0.121 g, isoleucine 0.131 g, leucine 0.247 g, lysine 0.147 g, methionine 0.062 g, cystine 0.066 g, phenylalanine 0.169 g, tyrosine 0.103 g, valine 0.202 g, arginine 0.546 g, histidine 0.077 g, alanine 0.170 g, aspartic acid 0.325 g, glutamic acid 0.761 g, glycine 0.158 g, proline 0.138 g and serine 0.172 g.

The major protein in coconut endosperm was found to be cocosin (Balasundaresan et al. 2002; Angelia et al. 2008). Angelia et al. (2008) purified cocosin and determined its physicochemical and functional properties toward its commercial applications in food systems. The total globulins were composed of 73% 11S and 27% 7S. The physicochemical properties of cocosin were found to be comparable with the other 11S globulins like the soybean and pea globulins. Hence, cocosin may be used as a potential ingredient for processed foods. Moreover, the ability of cocosin to produce stable emulsions in the absence of salt was found to be unique and could be the basis for developing new processed foods.

Coconut Water

Food value of coconut water per 100 g edible portion was reported as follows (USDA 2010): water 94.99 g, energy 19 kcal (79 kJ), protein 0.72 g, total lipid (fat) 0.20 g, ash 0.39 g, carbohydrate 3.71 g; fibre (total dietary) 1.1 g, total sugars 2.61 g, minerals – calcium 24 mg, iron 0.29 mg, magnesium 25 mg, phosphorus 20 mg, potassium 250 mg, sodium 105 mg, zinc 0.10 mg, copper 0.040 mg, manganese 0.142 mg, selenium 1.0 µg; vitamins – vitamin C (total ascorbic acid) 2.4 mg, thiamin 0.030 mg, riboflavin 0.057 mg, niacin 0.080 mg, pantothenic acid 0.043 mg, vitamin B-6 0.032 mg, folate (total) 3 µg, total choline 1.1 mg, lipids – fatty acids (total saturated) 0.176 g, 6:0 (caproic acid) 0.001 g, 8:0 (caprylic acid) 0.014 g, 10:0 (capric acid) 0.011 g, 12:0 (lauric acid) 0.088 g, 14:0 (myristic acid) 0.035 g, 16:0 (palmitic acid) 0.017 g, 18:0 (stearic acid) 0.010 g; fatty acids (total monounsaturated) 0.008 g, 18:1 undifferentiated (oleic acid) 0.008 g; fatty acids (total polyunsaturated) 0.002 g, 18:2 undifferentiated (linoleic acid) 0.002 g; amino acids – tryptophan 0.008 g, threonine 0.026 g, isoleucine 0.028 g, leucine 0.053 g, lysine 0.032 g, methionine 0.013 g, cystine 0.014 g, phenylalanine 0.037 g, tyrosine 0.022 g, valine 0.044 g, arginine 0.118 g, histidine 0.017 g, alanine 0.037 g, aspartic acid 0.070 g, glutamic acid 0.165 g, glycine 0.034 g, proline 0.030 g, serine 0.037 g.

Coconut water (coconut liquid endosperm), with its many applications, is one of the world's most versatile natural product. This refreshing beverage is consumed worldwide as it is nutritious and beneficial for health (Yong et al. 2009). There is increasing scientific evidence that supports the role of coconut water in health and medicinal applications. Coconut water is traditionally used as a growth supplement in plant tissue culture/micropropagation. The wide applications of coconut water can be justified by its unique chemical composition of sugars, vitamins, minerals, amino acids and phytohormones (Yong et al. 2009; Ma et al. 2008). Various classes of phytohormones identified from young coconut water included indole-3-acetic acid (IAA), indole-3-butyric acid (IBA), abscisic acid (ABA), gibberellic acid (GA), zeatin (Z), N(6)-benzyladenine (BA), alpha-naphthaleneacetic acid (NAA) and 2,4-dichlorophenoxyacetic acid (2,4-D) (Ma et al. 2008). Coconut water, which contains many uncharacterized phytohormones is extensively used as a growth promoting supplement in plant tissue culture.

Coconut Milk

Food value of raw, coconut milk (liquid expressed from grated meat and water) from mature nut (refuse 0%) per 100 g edible portion was reported as follows (USDA 2010): water 67.62 g, energy 230 kcal (962 kJ), protein 2.29 g, total lipid (fat) 23.84 g, ash 0.72 g, carbohydrate 5.54 g; fibre (total dietary) 2.2 g, total sugars 3.34 g, minerals – calcium 16 mg, iron 1.64 mg, magnesium 37 mg, phosphorus 100 mg, potassium 263 mg, sodium 15 mg, zinc 0.67 mg, copper 0.266 mg, manganese 0.916 mg, selenium 6.2 µg; vitamins – vitamin C (total ascorbic acid) 2.8 mg, thiamin 0.026 mg, riboflavin 0.000 mg, niacin 0.760 mg, pantothenic acid 0.183 mg, vitamin B-6 0.033 mg, folate (total) 16 µg, total choline 8.5 mg, vitamin E (α tocopherol) 0.15 mg, vitamin K (phylloquinone) 0.1ug, lipids – fatty acids (total saturated) 21.140 g, 6:0 (caproic acid) 0.136 g, 8:0 (caprylic acid) 1.67 g, 10:0 (capric acid) 1.327 g, 12:0 (lauric acid) 10.576 g, 14:0 (myristic acid) 4.176 g, 16:0 (palmitic acid) 2.021 g, 18:0 (stearic acid) 1.234 g; fatty acids (total monounsaturated) 1.014 g, 18:1 undifferentiated (oleic acid) 1.014 g; fatty acids (total polyunsaturated) 0.261 g, 18:2 undifferentiated (linoleic acid) 0.261 g; phytosterols 1 mg; amino acids – tryptophan 0.027 g, threonine 0.083 g, isoleucine 0.090 g, leucine 0.170 g, lysine 0.101 g, methionine 0.043 g, cystine 0.045 g, phenylalanine 0.116 g, tyrosine 0.071 g, valine 0.139 g, arginine 0.376 g, histidine 0.053 g, alanine 0.117 g, aspartic acid 0.224 g, glutamic acid 0.524 g, glycine 0.108 g, proline 0.095 g and serine 0.118 g.

Coconut Oil

Food value of coconut oil (refuse 0%) per 100 g edible portion was reported as follows (USDA 2010): energy 862 kcal (3,607 kJ), total lipid (fat) 100 g, iron 0.04 mg, total choline 0.3 mg, vitamin E (α tocopherol) 0.09 mg, γ-tocopherol 0.20 mg, vitamin K (phylloquinone) 0.5 ug, lipids – fatty acids (total saturated) 86.5 g, 6:0 (caproic acid) 0.600 g, 8:0 (caprylic acid) 7.500 g, 10:0 (capric acid) 6.000 g, 12:0 (lauric acid) 44.600 g, 14:0 (myristic acid) 16.800 g,16:0 (palmitic acid) 8.200 g, 18:0 (stearic acid) 2.800 g; fatty acids (total monounsaturated) 5.800 g, 18:1 undifferentiated (oleic acid) 5.800 g; fatty acids (total polyunsaturated) 1.800 g, 18:2 undifferentiated (linoleic acid) 1.800 g and phytosterols 86 mg.

Among the fatty acids in coconut endosperm oil, lauric acid (50.26–50.50%) exhibited the least variation among hybrids and parents (Laureles et al. 2002). The HPLC chromatogram of triacylglycerols (TAG) showed eight major peaks which differed in carbon number (CN) by two: identified as TAG CN 30, 32, 34, 36, 38, 40, 42, and 44. Medium chain triacylglycerols (MCTs), ranged from 13.81% to 20.55%.

Virgin Coconut Oil

Quality characteristics of Virgin coconut oil (VCO) as reported by Abdurahman et al. (2009) were as follows: C6 (caproic acid) 0.54–0.63%, C8 (caprylic acid) 8.77–9.20%, C10 (capric acid) 6.96–7.19%, C12 (lauric acid) 47.16–47.64%, C14 (myristic acid) 16.7–16.85%, C16 (palmitic acid) 8.32–8.58%, C18:0 (stearic acid) 2.35–2.43%, C18:1 (oleic acid) C18:2 (linoleic acid)1.27–1.32%, C20 (arachidic acid) + C18:3 (a-linolenic acid) 0.21%, C20:1 (eicosanoic acid) 0.20%, rancimat@100 = 108.87–123.73, moisture 0.07–0.16%.

A study on the commercial virgin coconut oil (VCO) available in the Malaysian and Indonesian market revealed the following chemical characteristics and fatty acid composition of VCO (Marina et al. 2009b). There was no significant difference in C12 (lauric acid) content (46.64–48.03%) among VCO samples. The major triacylglycerols obtained for the oils were 22.78–25.84% of LaLaLa, 14.43–16.54% of CCLa, 19.2021.38% of CLaLa, 13.62–15.55% of LaLaM and 7.399.51% of LaMM (where La = lauric; C = capric; M = myristic). Iodine value ranged from 4.47 to 8.55, indicative of only few unsaturated bond presence. The low content of iodine value indicating the high degree of saturation in VCO. The low degree of unsaturation conferring high resistance to oxidative rancidity. Saponification value ranged from 250.07 to 260.67 mg KOH/g oil. As compared to other vegetable oils, VCO had very high saponification value, which indicated that VCO contained a higher amount of short chain fatty acids. The low peroxide value (0.21–0.57 mequiv oxygen/kg) signified its high oxidative stability, while anisidine value ranged from 0.16 to 0.19. Oils with high peroxide values are unstable and easily become rancid. Anisidine value test measures secondary oxidation. Oils with an anisidine value below 10 were considered as good quality. Free fatty acid content of 0.15–0.25 was fairly low, showing that VCO samples were of good quality. Free fatty acids are responsible for undesirable flavor and aromas in fats. All chemical compositions were within the limit of Codex Alimentarius (2003) standard for edible coconut oil. Total phenolic contents of VCO samples (7.78–29.18 mg GAE/100 g oil) were significantly higher than refined, bleached and deodorized (RBD) coconut oil (6.14 mg GAE/100 g oil). The antioxidant activity of VCO samples ranged from 52% to 80%. These results suggested VCO to be as good as RBD coconut oil in chemical properties with the added benefit of being higher in phenolic content and antioxidant activity.

Che Man et al. (1997) found that the most efficient coconut cream separation was obtained at the 1:1 ratio of grated coconut meat to water at

70 °C, followed by 6 h settling time. Fermentation was then conducted on coconut cream emulsion with the sample from 1:1 ratio, 70 °C, and 6-h settling time. Oil yield from the fermentation process with 5% inoculum of *Lactobacillus plantarum* C was 95.06% after 10 h at 40°. Quality characteristics of the extracted oil were as follows: moisture content, 0.04%; peroxide value, 5.8 meq oxygen/kg; anisidine value, 2.10; free fatty acid, 2.45%; iodine value, 4.9; and color, 0.6 (Y+5R). Extraction of coconut oil from coconut meat with *L. plantarum* was significantly improved in both oil yield and quality over the traditional wet process.

Phosphorus-31 nuclear magnetic resonance spectroscopy ((31)P NMR) was used to differentiate virgin coconut oil (VCO) from refined, bleached, deodorized coconut oil (RBDCO) (Dayrit et al. 2008). Principal components analysis showed that the 1,2- diglycerides, 1,3- diglycerides, and free fatty acids were the most important parameters for differentiating VCO from RBDCO. On the average, 1-monoglyceride was found to be higher in VCO (0.027%) than RBDCO (0.019%). Total diglycerides were lower in VCO (1.55%) than in RCO (4.10%). Total sterols were higher in VCO (0.096%) compared with RBDCO (0.032%), and the free fatty acid content was eight times higher in VCO than RBDCO (0.127% vs 0.015%).

Coconut Flour

Five protein fractions (albumins, globulins, prolamines, glutelins-1, and glutelins-2) from defatted coconut flour were fractionated and then characterized (Kwon et al. 1996). Amino acid analysis demonstrated that the coconut proteins had a relatively high level of glutamic acid, arginine, and aspartic acid.

Coconut Shell

Polyphenolic compounds were isolated from the aqueous extract of green coconut shell (Akhter et al. 2010). Benzoyl ester derivatives were prepared from these polyphenols. Monobenzoyl and dibenzoyl derivatives of a polyphenol were separated and characterized.

Leaf Waxes

The epicuticular wax of coconut pinnae contained skimmiwallinol derivatives: isoskimmiwallin, skimmiwallin, skimmiwallinin, isoskimmiwallinol acetate, skimmiwallinol acetate, isoskimmiwallinone and skimmiwallinone (Escalante-Erosa et al. 2007, 2009).

Various parts of the coconut palm were found to contain bioactive phytochemicals that endowed the tree with a diverse assortment of pharmacological properties and potential medicinal applications.

Antioxidant Activity

Fresh sample of coconut water was found to scavenge 1,1-diphenyl-2-picrylhydrazyl (DPPH) 2,2'-azino-bis(3-ethylbenz-thiazoline-6-sulfonic acid) (ABTS) and superoxide radicals but stimulated the production of hydroxyl radicals and increased lipid peroxidation (Mantena et al. 2003). The activity was most significant for fresh samples of coconut water and diminished significantly upon heat, acid or alkali treatment or dialysis. Maturity of coconut drastically reduced the scavenging ability of coconut water against DPPH, ABTS and superoxide radicals. Coconut water protected hemoglobin from nitrite-induced oxidation to methemoglobin when added before the autocatalytic stage of the oxidation. Acid, alkali or heat treated or dialyzed coconut water showed diminished ability in protecting hemoglobin from oxidation. The scavenging ability and protection of hemoglobin from oxidation may be partly attributed to the ascorbic acid, an important constituent of coconut water.

An aqueous extract from the husk fiber of *Cocos nucifera* exhibited in-vitro and in-vivo analgesic and free radical scavenging activities (Alviano et al. 2004). The orally administered aqueous extract (200 or 400 mg/kg) inhibited the acetic acid-induced writhing response in mice. Tail flick and hot plate assays demonstrated that treatment of animals with the extract at 200 mg/kg cause a decrease in the response to a heat stimulus. A LD_{50} of 2.30 g/kg was obtained in acute toxicity tests. Topic treatment of rabbits with the coconut extract indicated that it did not induce any significant dermic or ocular irritation. In-vitro experiments using the 2,2-diphenyl-1-picryl-hydrazyl-hydrate (DPPH) photometric assay demonstrated that this fiber extract also possessed free radical scavenging properties.

The antioxidant activity of the methanolic extract prepared from different stages of *Cocos nucifera* mesocarp was demonstrated by DPPH, FRAP and deoxyribose assays, and suggested the potential of the mesocarp extract to be used for therapeutic purposes (Chakraborty and Mitra 2008). Separate studies showed that polyphenol fraction from virgin coconut oil (VCO) was found to have greater inhibitory effect on microsomal lipid peroxidation compared to that from the groundnut and copra oils (Nevin and Rajamohan 2006). Groundnut oil, rich in polyunsaturated fatty acids, reduced the levels of antioxidant enzymes and increased lipid peroxidation, indicated by the very high malonaldehyde (MDA) and conjugate diene content in the tissues of male Sprague-Dawley rats. VCO with more unsaponifiable components namely vitamin E and polyphenols than copra oil exhibited increased levels of antioxidant enzymes and inhibited the peroxidation of lipids in both in-vitro and in-vivo conditions. These results showed that VCO was superior in antioxidant action than copra oil and groundnut oil and confirmed VCO to be beneficial as an antioxidant.

Studies demonstrated that blended and interesterified oils of coconut oil (CO) with groundnut oil (GNO) or olive oil (OLO) increased hepatic antioxidant enzymes, reduced lipid peroxidation in liver and decreased the susceptibility of LDL (low density lipoproteins) to oxidation (Nagaraju et al. 2008). The hepatic lipid peroxidation (LPO) levels in the rats fed CO:GNO blend and interesterified oils were enhanced by 31% and 21%, when compared to the rats given CO. The superoxide dismutase activity was enhanced by 31% and 28%, and catalase (CAT) activity was augmented by 37% and 39%, respectively, in rats administered blends and interesterified oils of CO:GNO, as compared to those given CO. The activities of glutathione peroxidase (GPx) were enhanced by 17% and 20% in rats given CO:GNO blend and interesterified oils. Glutathione transferase (GST) level was also found to be augmented by 26% and 31% after feeding blended and interesterified oils of CO:GNO, compared to those given CO. The LDL oxidation, which was elevated by feeding GNO, was found to be diminished by 10% and 14%, respectively, in the groups given blended and interesterified oils of CO:GNO. Similarly, CO:OLO blended oils increased SOD, CAT, GPx and GST activities by 34%, 43%, 27% and 23%, respectively, compared to the rats fed CO-containing diets. The corresponding increases in theses antioxidant enzyme activities when CO:OLO interesterified oils were given to rats were 38%, 50%, 28% and 26%, respectively, when compared to rats fed CO-containing diets. There was a significant reduction in hepatic LPO as well as oxidation of LDL, when blended and interesterified oils of CO:OLO were fed to rats.

The antioxidant properties of virgin coconut oil produced through chilling and fermentation were investigated and compared with refined, bleached and deodorized coconut oil. Virgin coconut oil showed better antioxidant capacity than refined, bleached and deodorized coconut oil (Marina et al. 2009a). The virgin coconut oil produced through the fermentation method exhibited the strongest scavenging effect on 1,1-diphenyl-2-picrylhydrazyl and the highest antioxidant activity based on the β-carotene-linoleate bleaching method. However, virgin coconut oil obtained through the chilling method exhibited the highest reducing power. The major phenolic acids detected were ferulic acid and *p*-coumaric acid. Very significant correlations were found between the total phenolic content and scavenging activity ($r=0.91$), and between the total phenolic content and reducing power ($r=0.96$). There was also a high correlation between total phenolic acids and β-carotene bleaching activity. The study indicated that the contribution of antioxidant capacity in virgin coconut oil could be due to phenolic compounds.

Anticancer Activity

The results of studies conducted by Pillai et al. (1999) suggested hat coconut kernel fiber could protect colon cells in rats fed a high-fat diet for 15 weeks from loss of oxidative capacity with the administration of the procarcinogen 1,2-dimethylhydrazine (DMH). The DMH-treated fiber group showed higher levels of lipid peroxides than the control group treated with DMH at the preneoplastic and neoplastic stages. Free fatty acid levels were found to decrease significantly in the DMH-treated control group, whereas it was near normal in the fiber groups. Superoxide dismutase and catalase activity also were found to be enhanced in the liver, intestine, proximal colon, and distal colon. Glutathione levels in all the tissues studied showed significant reductions in the fiber group. Preliminary study by Kirszberg et al. (2003) suggested that the efficacy of the antitumoral activity of the catechin rich extract of *C. nucifera*, typical A variety, could be extended to erythroleukaemia cells having a multidrug-resistant phenotype. A dose-dependent inhibitory effect was observed on tumour cells and on lymphocytes activated by phytohemaglutinin (PHA) or phorbol ester.

Studies showed that aqueous extracts of the husk fiber of the typical A and common varieties of *Cocos nucifera* showed almost similar antitumoral activity against the leukemia cell line K562 (60.1% and 47.5% for the typical A and common varieties, respectively) (Koschek et al. 2007). Fractions ranging in molecular weight from 1 to 10 kDa exhibited higher cytotoxicity. Moreover, *C. nucifera* extracts were also active against Lucena 1, a multidrug-resistant leukemia cell line. Their cytotoxicity against this cell line was about 50% (51.9 and 56.3 9 for varieties typical A and common, respectively). The results suggested that coconut husk extract could be an inexpensive source of new antineoplastic and anti-multidrug resistant drugs that warranted further investigations.

Coconut cake exhibited a protective effect against 1,2-dimethyl hydrazine (DMH) induced colon cancer by virtue of its ability to decrease the activities of the microbial enzymes β-glucuronidase and mucinase (Nalini et al. 2004). Average number of tumours in the colon as well as the incidence of cancer was significantly increased in 1,2-dimethyl hydrazine (DMH) -treated rats which was markedly reduced on supplementing with coconut cake. DMH injections significantly elevated both the activities of β-glucuronidase in distal colon, distal intestine, liver and colon contents and mucinase in colon and fecal contents as compared to the control rats. The increase in β-glucuronidase activity may augment the hydrolysis of glucuronide conjugates, liberating the toxins, while the increase in mucinase activity may enhance the hydrolysis of the protective mucins in the colon. Coconut cake supplementation to DMH-treated rats significantly decreased the incidence and number of tumours as well as the activity of β-glucuronidase and mucinase.

Antimicrobial/Antiviral Activities

Water extract obtained from coconut husk fiber and its fractions exhibited antimicrobial activity against *Staphylococcus aureus* (Esquenazi et al. 2002). The crude extract and one of the fractions rich in catechin also showed inhibitory activity against acyclovir-resistant herpes simplex virus type 1 (HSV-1-ACVr). All fractions were inactive against the fungi *Candida albicans, Fonsecaea pedrosoi* and *Cryptococcus neoformans*. Catechin and epicatechin together with condensed tannins (B-type procyanidins) were demonstrated to be the components of the water extract. In separate studies chloroform and ethyl acetate extracts of coconut husk showed significant broad-spectrum antibacterial activity against *Escherichia coli, Pseudomonas aeruginosa, Bacillus subtilis, Bacillus pumilus* and *Staphylococcus aureus* at 100 µg/disc (Srinivas et al. 2003).

Coconut oil exhibited in-vitro antifungal activity against some isolates of *Candida* (Ogbolu et al. 2007). *Candida albicans* had the highest susceptibility to coconut oil (100%), with a minimum inhibitory concentration (MIC) of 25% (1:4 dilution), while fluconazole had 100%

susceptibility at an MIC of 64 μg/mL (1:2 dilution). *Candida krusei* showed the highest resistance to coconut oil with an MIC of 100% (undiluted), while fluconazole had an MIC of >128 μg/mL. Antimicrobial activity of the crude coconut mesocarp extract exhibited a potent anti-staphylococcal activity against *Staphylococcus aureus*. The probable compounds responsible for the bioactivity were identified as 5-O-caffeoylquinic acid (chlorogenic acid), dicaffeoylquinic acid and three tentative isomers of caffeoylshikimic acid (Chakraborty and Mitra 2008). Peptides from green coconut were found to have antibacterial activity (Mandal et al. 2009). Three peptides lower than 3 kDa were purified and identified from green coconut showing molecular masses of 858, 1,249 and 950 Da. The first peptide, named Cn-AMP1, was extremely efficient against both Gram-positive and Gram-negative bacteria. Cn-AMPs also displayed outstanding potential for the development of novel antibiotics from natural sources.

Coconut oil is rich in lauric acid. In the presence of lauric acid (C12), the production of infectious vesicular stomatitis virus (VSV) was inhibited in a dose-dependent manner (Hornung et al. 1994). The inhibitory effect was reversible; after removal of lauric acid, the antiviral effect vanished. Moreover, the chain length of the monocarboxylic acids was ascertained to be crucial, as those with shorter or longer chains were less effective or had no antiviral activity. Concurrent with the lauric acid -induced inhibition was the stimulation of triacylglycerol synthesis, increasing the amount up to ninefold. Analysis of the antiviral mechanism of lauric acid revealed that the correct assembly of the viral components was disturbed, but viral RNA and protein synthesis remained unimpaired. The results suggested that the newly synthesized triacylglycerols might interact with the host cell plasma membrane and interfere with virus maturation. Bartolotta et al. (2001) found that the most active inhibitor of arenavirus was lauric acid (C12), which decreased virus yields of several attenuated and pathogenic strains of Junin virus (JUNV) in a dose dependent manner, without affecting cell viability. Fatty acids with shorter or longer chain length had a reduced or negligible anti-JUNV activity. Lauric acid did not inactivate virion infectivity neither interacted with the cell to induce a state refractory to virus infection. Lauric acid hindered a late maturation stage in the replicative cycle of JUNV. Viral protein synthesis was not affected by the compound, but the expression of glycoproteins in the plasma membrane was attenuated. A direct correlation between the inhibition of JUNV production and the stimulation of triacylglycerol cell content was demonstrated, and both lauric-acid induced effects were dependent on the continued presence of the fatty acid. Thus, the decreased insertion of viral glycoproteins into the plasma membrane, apparently due to the increased incorporation of triacylglycerols, appeared to cause an inhibition of JUNV maturation and release.

Cardiovascular Diseases and Coconut Oil/Virgin Coconut Oil

Earlier studies by Prior et al. (1981) reported that cardiovascular disease was uncommon in the Polynesian population of Pukapuka and Tokelau atolls whose habitual diets were high in saturated fat but low in dietary cholesterol and sucrose. In both populations, there was no evidence of the high saturated fat intake having a harmful effect in these populations. Coconut was reported as the chief source of energy for both groups. Tokelauans obtained a much higher percentage of energy from coconut than the Pukapukans, 63% compared with 34%, so their intake of saturated fat was higher. The serum cholesterol levels were 35–40 mg higher in Tokelauans than in Pukapukans. These major differences in serum cholesterol levels were demed to be due to the higher saturated fat intake of the Tokelauans. Analysis of a variety of food samples, and human fat biopsies showed a high lauric (12:0) and myristic (14:0) content. A case-control study involving 93 cases in the Case group and 189 individuals in the control group was conducted among the Minangkabau community in West Sumatra to ascertain whether high intakes of saturated fat that could be associated with elevated serum cholesterol concentrations and increased coronary heart disease (CHD) mortality (Lipoeto et al. 2004). Similar intakes of saturated and unsaturated fatty acids between the cases and controls indicated that the consumption of total fat or saturated fat, including that from coconut, was not a predictor for CHD in their food culture. However, the intakes of animal foods, total protein, dietary cholesterol and less plant derived carbohydrates were predictors of CHD.

In a study, the diet of group of 16 free living young adults, ages 16–21, was shifted from their usual diet to a similar one in which the coconut oil was replaced by whole milk powder and corn oil (Mendis et al. 1989). The results indicated that their blood cholesterol, cholesteryl ester, and several other related circulating blood lipid values, as well as the platelet factor 4 values, were elevated prior to the diet change. Many of these factors, considered as risk factors for atherogenesis, were substantially diminished at the end of the diet change after 8 weeks. The only plasma components which were changed substantially were the triglycerides and the HDL cholesterol. These results suggested that the special atherogenic effects of coconut oil that had been demonstrated in so many animal models may be similarly active in humans.

The effect of daily consumption of coconut fat and soya-bean fat on plasma lipids and lipoproteins of young normolipidaemic men was reported by Mendis and Kumarasunderam (1990). Twenty-five young normolipidaemic males were given isoenergetic diets containing 30% of energy as fat, with a polyunsaturated: saturated fat ratio of 4.00 or 0.25, for 8 weeks. Approximately 70% of the fat energy was provided by the test fats: soyabean fat and coconut fat. During the soya-bean-fat-eating period the total plasma cholesterol level decreased significantly compared with baseline values and during the coconut-fat-eating phase total plasma cholesterol level increased significantly compared with the soya-bean-eating period. On the soya-bean-fat diet, high-density-lipoprotein (HDL)-cholesterol fell by 15% (range 6–35%) and plasma triacylglycerols felled by 25% (range 13–37%). Results of the study showed that even when the proportion of total fat in the diet was low, a high intake of linoleic acid reduced both total plasma cholesterol and HDL-cholesterol, while a high intake of saturated fat (coconut fat) increased both these lipid fractions. Studies by Mendis et al. (2001) reported that the decrease of dietary saturated fat with partial replacement of unsaturated fat elicited changes in total cholesterol, HDL- and LDL-cholesterol that were associated with a lower cardiovascular risk. In the first phase study (8 weeks), total fat was lowered from 31% to 25% energy (polyunsaturated fatty acids (PUFA):saturated fatty acids (SFA) ratio increased from 0.2 to 0.4) by lowering the quantity of coconut fat (CF) in the diet from 17.8% to 9.3% energy intake. At the end of phase 1, there was a 7.7% decrease in cholesterol and 10.8% decrease in LDL and no significant change in HDL and triacylglycerol. In phase 2, subjects were randomised to groups A and B. In group A total fat was lowered from 25% to 20% energy (PUFA:SFA ratio increased from 0.4 to 0.7) by decreasing the quantity of CF in the diet from 9.3% to 4.7% total energy intake. In group B, the saturated fat content in the diet was similar to group A. In addition a test fat (a mixture of soyabean oil and sesame oil, PUFA:monosaturated fatty acids ratio 2) contributed 3.3% total energy intake and total fat contributed 24% energy intake (PUFA:SFA ratio increased from 0.7 to 1.1). At the end of phase 2, the reduction in cholesterol in both groups was only about 4% partly due the concomitant rise in HDL. The reduction in LDL at 52 weeks was significantly higher in group B (group A mean reduction 11%, and group B mean reduction 16.2%). In phase 2, triacylglycerol levels showed a mean reduction of 6.5% in group 2A and a mean increase of 8.2% in group 2B. The reduction of saturated fat in the diet was associated with a lipoprotein profile that would be expected to reduce cardiovascular risk.

Pehowich et al. (2000) showed that while the coconut water fraction was low in lipid content, coconut milk contained about 24% of the fat content of coconut oil and the cream and meat fractions about 34%. The other coconut constituents contained significant amounts of medium-chain triglycerides that were formed from fatty acids of chain length 8:0 to 14:0. It was these fatty acids, primarily 14:0, that were thought to be atherogenic. In contrast, medium-chain triglycerides may be advantageous under some circumstances in that they were absorbed intact and did not undergo degradation and re-esterification processes. As a result, medium-chain triglycerides provide a ready source of energy and may be useful in baby foods or in diet therapy. Nevertheless, the possible negative effects of the saturated fatty acids and the absence of the essential fatty acid linolenic acid from all coconut components suggested that the coconut milk, oil and cream should not be used on a regular basis in adults.

Studies conducted by Muller et al. (2003) indicated that a coconut oil-based diet (HSAFA-diet) [high fat diet − 38.4% of energy (E%) from fat, polyunsaturated /saturated ratio 0.14] lowered postprandial tissue plasminogen activator antigen (t-PA antigen) concentration, and this favorably affected the fibrinolytic system and the fasting lipoprotein (a) Lp(a) concentration compared with the HUFA-diet (PUFA, 38.2 E% from fat P/S ratio 1.9). The proportions of dietary saturated fatty acids rather than the percentage of saturated fat energy appeared to have a beneficial influence on Lp(a) levels. Elevated levels of haemostatic factors including tissue plasminogen activator (t-PA) antigen were found to be associated with coronary heart disease (Forouhi et al. 2003).

Nevin and Rajamohan (2008) reported that administration of virgin coconut oil (VCO) showed significant antithrombotic effect compared to copra oil and the effects were comparable with sunflower oil. The antioxidant vitamin levels were found to be higher in VCO fed animals than other groups. LDL isolated from VCO fed animals when subjected to oxidant (Cu^{2+}) in-vitro showed significant resistance to oxidation as compared to the LDL isolated from other two groups. Dietary administration of VCO reduced the cholesterol and triglyceride levels and maintained the levels of blood coagulation factors. Results also indicated that VCO feeding could prevent the oxidation of LDL from oxidants. These properties of VCO may be attributed to the presence of biologically active unsaponifiable components viz. vitamin E, provitamin A, polyphenols and phytosterols.

Hypolipidemic/Antihypercholesteromic Activities

Sindhurani and Rajamohan (1989) reported that hemicullose component of coconut fibre possessed hypolipidemic activity. The neutral detergent fibre (NDF) isolated from coconut kernel was digested with cellulase and hemicellulase and the residual fibre rich in hemicellulose (without cellulose) and cellulose (with out hemicellulose) were fed to rats and compared with a fiber free group. The results indicated that hemicellulose rich fibre showed decreased concentration of total cholesterol, LDL + VLDL cholesterol and increased HDL cholesterol, while cellulose rich fibre showed no significant alteration. There was increased HMG CoA reductase activity and increased incorporation of labeled acetate into free cholesterol. Rats fed hemicellulose rich coconut fibre produced lower concentration of triglycerides and phospholipids and lower release of lipoproteins into circulation. There was increased concentration of hepatic bile acids and increased excretion of faecal sterols and bile acids. These results indicated that the hemicellulose component of coconut fibre was responsible for the observed hypolipidemic effect. In subsequent studies, inclusion of coconut protein and L-arginine into ethanol (to induce hyperlipidemia) fed male albino rats produced lower levels of total cholesterol, LDL + VLDL cholesterol, triglycerides and atherogenic index in the serum (Mini and Rajamohan 2004). Concentration of tissue cholesterol and triglycerides was also lower in these groups. Administration of coconut protein and L-arginine in the ethanol fed rats caused decreased activity of HMG-CoA reductase in the liver and increased activity of lipoprotein lipase in the heart. The activities of malic enzyme and glucose-6-phosphate dehydrogenase were also lower in these groups. Feeding coconut protein and L-arginine in ethanol treated rats showed increased concentration of hepatic bile acids and fecal excretion of neutral sterols and bile acids. All these effects were comparable in rats fed coconut protein and those fed L-arginine. These observations indicated that the major factor responsible for the hypolipidemic effect of coconut protein was due to the high content of L-arginine.

In separate studies, feeding coconut kernel along with coconut oil in human volunteers was found to reduce serum total and LDL cholesterol when compared to feeding coconut oil alone (Padmakumaran et al. 1999). Similar effect of the kernel was also observed in rats. Feeding kernel protein resulted in lower levels of cholesterol, phospholipids and triglycerides in the serum and most tissues when compared to casein fed animals. Rats fed kernel protein had increased hepatic degradation of cholesterol to bile acids and hepatic cholesterol biosynthesis, and decreased esterification of free cholesterol. In the intestine, however, cholesterogenesis was decreased. The kernel protein also caused decreased lipogenesis in the liver and intestine. This beneficial effect of the kernel protein was attributed to its very low lysine/arginine ratio 2.13% lysine and 24.5% arginine.

Animal studies indicated that those fed coconut protein and those fed L-arginine showed significantly lower levels of total cholesterol, LDL+VLDL cholesterol, triglycerides and phospholipids in the serum and higher levels of serum HDL cholesterol (Salil and Rajamohan 2001). The concentration of total cholesterol, triglycerides and phospholipids in the tissues were lower in these groups. There was increased hepatic cholesterogenesis which was evident from the higher rate of incorporation of labeled acetate into free cholesterol. Increased conversion of cholesterol to bile acids and increased fecal excretion of bile acids were observed. Feeding coconut protein resulted in decreased levels of malondialdehyde in the heart and increased activity of superoxide dismutase and catalase. Supplementation of coconut protein caused increased excretion of urinary nitrate indicating higher rate of conversion of arginine into nitric oxide. In the study, the arginine supplemented group and the coconut protein fed group gave similar effects. These studies clearly demonstrated that coconut protein was able to reduce hyperlipidemia and peroxidative effect induced by high fat cholesterol containing diet and these effects were mainly mediated by the L-arginine present in it.

Separate studies in New Zealand reported by Hodson et al. (2001) showed that when replacing saturated fat with either n-6 polyunsaturated fat or monounsaturated fat in the diet, total fat intakes decreased by 2.9% energy and 5.1% energy, respectively. In trial I, replacing saturated fat with n-6 polyunsaturated fat lowered plasma total cholesterol by 19% [from 4.87 to 3.94 mmol/l], low density lipoprotein cholesterol by 22% [from 2.87 to 2.24 mmol/l], and high density lipoprotein cholesterol by 14% [from 1.39 to 1.19 mmol/l]. In trial II, replacing saturated fat with monounsaturated fat decreased total cholesterol by 12%, low density lipoprotein cholesterol by 15%, and high density lipoprotein cholesterol by 4%. The change in the ratio of total to high density lipoprotein cholesterol was similar during trial I and trial II.

The following foods were tested on 21 human subjects with moderately raised serum cholesterol using a double-blind randomized crossover design: corn flakes as the control food, oat bran flakes as the reference food, and corn flakes with 15% and 25% dietary fiber from coconut flakes (made from coconut flour production) (Trinidad et al. 2004). Results showed a significant percent reduction in serum total and low-density lipoprotein (LDL) cholesterol (in mg/dL) for all test foods, except for corn flakes, as follows: oat bran flakes, 8.4 and 8.8, respectively; 15% coconut flakes, 6.9 and 11.0, respectively; and 25% coconut flakes, 10.8 and 9.2, respectively. Serum triglycerides were significantly reduced for all test foods: corn flakes, 14.5%; oat bran flakes, 22.7%; 15% coconut flakes, 19.3%; and 25% coconut flakes, 21.8%. Only 60% of the subjects were considered for serum triglycerides reduction (serum triglycerides >170 mg/dL). The scientists concluded that both 15% and 25% coconut flakes reduced serum total and LDL cholesterol and serum triglycerides of humans with moderately raised serum cholesterol levels. Coconut flour was found to be a good source of both soluble and insoluble dietary fiber, and both types of fiber may have significant role in the reduction of the above lipid biomarker. Results from this study served as a good basis in the development of coconut flakes/flour as a functional food.

Sandhya and Rajamohan (2006) showed that cholesterol feeding caused a marked increase in total cholesterol, very low-density lipoprotein (VLDL)+low-density lipoprotein (LDL) cholesterol, and triglycerides in serum. Administration of tender coconut water and mature coconut water counteracted the increase in total cholesterol, VLDL+LDL cholesterol, and triglycerides, while high-density lipoprotein cholesterol was higher. Lipid levels in the tissues of the liver, heart, kidney, and aorta were markedly decreased in cholesterol-fed rats supplemented with coconut water. Feeding coconut water resulted in elevated activities of 3-hydroxy-3-methylglutaryl-CoA reductase in liver, lipoprotein lipase in heart and adipose tissue, and plasma lecithin:cholesterol acyl transferase, while lipogenic enzymes showed decreased activities. An increased rate of cholesterol conversion to bile acid and an increased excretion of bile acids and neutral sterols were observed in rats fed coconut water. Histopathological studies of liver and aorta revealed much less fatty accumulation in these tissues in cholesterol-fed rats supplemented with coconut water. Feeding coconut water resulted also in increased plasma L-arginine content, urinary nitrite level, and nitric oxide synthase activity. These results indicated that both tender and mature coconut water had beneficial effects on serum and tissue lipid parameters in rats fed cholesterol-containing diet. In a subsequent study, coconut water or lovastatin supplementation was found to lower the levels of serum total cholesterol, VLDL+LDL cholesterol, triglycerides and increased HDL cholesterol in experimental hyperlipidemic rats (Sandhya and Rajamohan 2008). Coconut water feeding lowered activities of hepatic lipogenic enzymes and elevated HMG CoA reductase and lipoprotein lipase activity. Coconut water supplementation increased hepatic bile acid and fecal bile acids and neutral sterols. Coconut water had lipid lowering effect similar to the drug lovastatin in rats fed fat-cholesterol enriched diet.

Virgin coconut oil (VCO) obtained by wet process produced a beneficial effect in lowering lipid components compared to copra oil (Nevin and Rajamohan 2004). It reduced total cholesterol, triglycerides, phospholipids, LDL, and VLDL cholesterol levels and increased HDL cholesterol in serum and tissues. The polyphenol fraction of virgin coconut oil was also found to be capable of preventing in-vitro LDL oxidation with reduced carbonyl formation. The results demonstrated the potential beneficiary effect of virgin coconut oil in lowering lipid levels in serum and tissues and LDL oxidation by physiological oxidants. This property of VCO may be attributed to the biologically active polyphenol components present in the oil.

Studies showed that neonatal chick provided a suitable model in which to study the role of very-low-density lipoproteins in atherogenesis and the rapid response to saturated fatty acids with 12–14 carbons (Castillo et al. 1996). Supplementation of 10% or 20% coconut oil in the diet for 1–2 weeks produced a significant hypercholesterolemia in neonatal chicks. Plasma triacylglycerol concentration significantly increased after the addition of 20% coconut oil for 2 weeks. These results showed that newborn chicks were more sensitive to saturated fatty acids from coconut oil than adult animals. Coconut oil supplementation in the diet (20%) for 2 weeks increased cholesterol concentration in all the lipoprotein fractions, while 10% coconut oil only increased cholesterol in low-density and very-low-density lipoproteins, an increase that was significant after 1 week of treatment. Similar results were obtained for triacylglycerol concentration after 2 weeks of treatment. Changes in phospholipid and total protein levels were less profound. Coconut oil decreased low-density and very-low-density lipoprotein fluidity, measured as total cholesterol/phospholipid ratio. Changes in esterified cholesterol/phospholipid and triacylglycerol/phospholipid ratios suggested that coconut oil affected the distribution of lipid components in the core of very-low-density particles. Likewise, the esterified cholesterol/triacylglycerol ratio was clearly increased in the low-density, and especially in the very-low-density, fraction after the first week of coconut oil feeding.

Antiinflammatory Activity

Bibby and Grimble (1990) compared the effects of corn oil and coconut oil diets on tumour necrosis factor-α and endotoxin induction of the inflammatory prostaglandin E2 (PGE2) production. Rats fed maize oil exhibited increased prostaglandin E2 production in the hypothalami. Rats fed coconut oil did not produce an increase in PGE2. The results indicated that coconut oil exerted a modulatory effect causing a reduction of membrane phospholipid arachidonic acid content. In separate studies, crude extract of coconut husk fibre was found to have antinociceptive and anti-inflammatory activities (Rinaldi et al. 2009). Crude extract (CE, 50, 100, and 150 mg/kg), fraction 1 (F1, molecular weight lesser than 1 kDa, 1, 10, and 50 mg/kg), fraction 2 (F2, molecular weight higher than 1 kDa, 1, 10, and 50 mg/kg), significantly developed peripheral and central antinociceptive activity but with less effect on supra-spinal regions of the brain. Administration of the opioid antagonist, naloxone (5 mg/kg) inhibited the antinociceptive effect indicating that *Cocos nucifera* crude extract and fractions may be acting in opioid receptors. CE and F1 also inhibited rat paw edema induced by histamine, and serotonin. The results demonstrated that *Cocos nucifera* and its fractions confirmed the popular use of this plant in several inflammatory disorders.

Recent studies suggested anti-inflammatory, analgesic, and antipyretic properties of virgin coconut oil (VCO) (Intahphuak et al. 2010). In acute inflammatory models, VCO showed moderate anti-inflammatory effects on ethyl phenylpropiolate-induced ear edema in rats, and carrageenin- and arachidonic acid-induced paw edema. VCO exhibited an inhibitory effect on chronic inflammation by reducing the transudative weight, granuloma formation, and serum alkaline phosphatase activity. VCO also showed a moderate analgesic effect on the acetic acid-induced writhing response as well as an antipyretic effect in yeast-induced hyperthermia.

Antixerosis Activity

A randomized double-blind controlled clinical trial was conducted on mild to moderate xerosis in 34 patients randomized to apply either coconut oil or mineral oil on the legs twice a day for 2 weeks (Agero and Verallo-Rowell 2004). Xerosis is a common skin condition (1) characterized by dry, rough, scaly, and itchy skin, (2) associated with a defect in skin barrier function, and (3) can be treated with moisturizers. Coconut oil and mineral oil exhibited comparable effects on xerosis. Both oils showed efficacy through significant improvement in skin hydration and increase in skin surface lipid levels. Safety for both was further demonstrated by negative patch-test results prior to the study and by the absence of adverse reactions during the study. The results showed that coconut oil was as effective and safe as mineral oil when used as a moisturizer. In another study, virgin coconut oil (VCO) and virgin olive oil (VOO) were compared for effectiveness in treating atopic dermatitis (AD) – dry skin prone to colonization by *Staphylococcus aureus* (Verallo-Rowell et al. 2008). VCO was found to be more effective than VOO, post intervention, only 1 (5%) VCO subject remained positive for *Staphylococcus aureus* compared to 6 (50%) for VOO. Relative risk for VCO was 0.10, significantly superior to that for VOO (10:1).

Antiobesity Activity

In animal studies, coconut-oil enriched diet was found effective in activating uncoupling protein (UCP1) expression during ad libitum feeding and in preventing its down-regulation during food restriction, and this was accompanied with a decrease of the white fat stores (Portillo et al. 1998). UCP1 is an uncoupling protein found in the mitochondria of brown adipose tissue used to generate heat by non-shivering thermogenesis. White fats are adipocytes important in energy metabolism, heat insulation and mechanical cushioning.

Recent studies suggested that dietetic supplementation with coconut oil did not cause dyslipidemia and appeared to promote a reduction in abdominal obesity (Assunção et al. 2009). In a randomised, double-blind, clinical trial groups of abdominal obese women with waist circumferences greater than 88 cm were administered daily dietary supplements comprising 30 mL of either soy bean oil (group S; n=20) or coconut oil (group C; n=20) over a 12-week period, during which all subjects were instructed to follow a balanced hypocaloric diet and to walk for 50 min per day. One week after dietary intervention, group C presented a higher level of HDL (48.7 vs. 45.00) and a lower LDL:HDL ratio (2.41 vs. 3.1). Reductions in BMI (Body Mass Index) were observed in both groups at week 2, but only group C exhibited a reduction in waist circumferences. Group S exhibited an increase in total cholesterol, LDL and LDL:HDL ratio, whilst HDL diminished. Such alterations were not observed in group C.

Conjugated linoleic acid (CLA) was found to induce a body fat loss that was enhanced in mice fed coconut oil (CO), which lacked essential fatty acids (EFA) (Hargrave et al. 2005). Mice fed CO during only the 2-week CLA-feeding period did not differ from control mice in their adipose EFA content but still tended to be leaner than mice fed soy oil (SO). Mice raised on CO or fat free diets and fed CLA were leaner than the SO+CLA-fed mice. Mice raised on CO and then replenished with linoleic, linolenic, or arachidonic acid were leaner when fed CLA than mice raised on SO. Body fat of CO+CLA-fed mice was not affected by EFA addition. In summary, CO-fed mice not lacking in tissue EFA responded more to CLA than SO-fed mice. Also, EFA addition to CO diets did not alter the enhanced response to CLA. Therefore, the increased response to CLA in mice raised on coconut oil or fat free diets appeared to be independent of a dietary EFA deficiency.

Antidiabetic/Hypoglycemic Activity

The effect of neutral detergent fibre (NDF) from coconut kernel (*Cocos nucifera*) in rats fed 5%, 15% and 30% level on the concentration of blood glucose, serum insulin and excretion of minerals was studied by Sindurani and Rajamohan (2000). Increase in the intake of fiber resulted in significant decrease in the level of blood glucose and serum insulin. Faecal excretion of Cu, Cr, Mn, Mg, Zn and Ca was found to increase in rats fed different levels of coconut fibre when compared to fibre free group. The results suggested that inclusion of coconut fibre in the diet resulted in significant hypoglycemic action.

The need for a better lipid system to satisfy the fuel requirements of patients while avoiding the adverse effects of current systems had led to suggestions that medium chain fatty acids (MCFs) be incorporated into TPN (total parenteral nutrition) -lipid emulsions (Garfinkel et al. 1992). Using an isolated perifused mouse islet model, various doses of medium chain fatty acids and the essential fatty acid, linoleic acid, were tested and compared. The possibility of an additive effect of an insulinotropic MCF and linoleate when both were provided together was also examined. It was found that the ability of 5 mM of a given MCF to stimulate insulin secretion was dependent upon its chain length. Thus, while adipic acid (C6) had no effect, caprylic acid (C8) had a minimal effect that was not statistically significant, but capric acid (C10) and lauric acid had very potent effects that were of the same magnitude to the effect of linoleate on insulin secretion.

Trinidad et al. (2003) examined the glycaemic index (GI) of commonly consumed bakery products supplemented with increasing levels of coconut flour in ten normal and ten diabetic subjects, using a randomized crossover design. The significantly low-GI (<60) foods investigated were: macaroons (GI 45.7) and carrot cake (GI 51.8), with 200–250 g coconut flour/kg. The test foods with 150 g coconut flour/kg had GI ranging from 61.3 to 71.4. Among the test foods, pan de sal (GI 87.2) and multigrain loaf (GI 85.2) gave significantly higher GI with 50 and 100 g coconut flour/kg respectively. In contrast, granola bar and cinnamon bread with 50 and 100 g coconut flour/kg respectively gave a GI ranging from 62.7 to 71.6 and did not differ significantly from the test foods with 150 g coconut flour/kg. A very strong negative correlation was observed between the GI and dietary fibre content of the test foods supplemented with coconut flour. The results indicated that the GI of coconut flour-supplemented foods decreased with increasing levels of coconut flour and this may be due to its high dietary fibre content. The results of the study may form a scientific basis for the development of coconut flour as a functional food for the control and management of diabetes mellitus.

Antihypertensive Activity

Hypertensive subjects that received coconut water, mauby (*Colubrina arborescens*) and the mixture had significant decreases in the mean systolic blood pressure of 71%, 40% and 43% respectively (Alleyne et al. 2005). For these groups, the respective proportions showing significant decreases in the mean diastolic pressure were 29%, 40% and 57%. For the group receiving the mixture, the largest decreases in mean systolic and mean diastolic pressure were 24 and 15 mmHg respectively; these were approximately double the largest values seen with the single interventions.

Periodontic Activity

Studies found that coconut water may be better alternative to Hank's balanced salt solution (HBSS) or milk in terms of maintaining periodontal ligament cell viability after avulsion and storage (Gopikrishna et al. 2008). Fifty freshly extracted human teeth were divided into 3 experimental groups and 2 control groups. The experimental teeth were stored dry for 30 min and then immersed in 1 of the 3 media: coconut water, Hank's balanced salt solution (HBSS), and milk. The teeth were then treated with Dispase grade II and Collagenase for 30 min. Statistical analysis demonstrated that coconut water kept significantly more periodontal ligament cells on avulsed teeth viable compared to either HBSS or milk.

Neuroprotective Activity

Young coconut juice exhibited neuroprotective activity attributable to its estrogen-like characteristics (Radenahmad et al. 2009). Some degenerative diseases of the nervous system had been linked to hormonal imbalance in postmenopausal women. It was argued that young coconut juice (YCJ) could have some estrogen-like characteristics. Studies showed that the rat group which received young coconut juice (YCJ) had its serum E2 level significantly higher than the overiectomized group which did not receive any treatment, and the sham-operated group. A similar trend was observed with the group which received estradiol benzoate (EB) injections, but no significant difference was noted when the latter was compared with the sham-operated group. Further, a significant reduction in neuronal cell death was observed in the YCJ-treated group, as compared to the overiectomized group which did not receive any treatment. This was reflected by the significantly higher number of neurons which were immunopositive for anti-neurofilament (NF200) and anti-parvalbumin (PV) antibodies. Additionally, the number of these neurons was also significantly higher in the YCJ group, as compared to the EB group. The study confirmed the argument that young coconut juice had estrogen-like characteristics, and it also added more support to the observation that hormonal imbalance could induce some brain pathologies in females.

Antiulcerogenic Activity

Studies conducted by Nneli and Woyike (2008) showed that coconut milk (2 ml daily oral feeding) produced a stronger percentage (54%) decrease in the mean ulcer area than coconut water (39%) in male Wistar albino rats with ulcers induced by subcutaneous administration of indomethacin (40 mg/kg). The effect of coconut milk was similar to the effect of sucralfate (a conventional cytoprotective agent) that reduced the mean ulcer area by 56% in this study. The results showed that coconut milk and water via macroscopic observation had protective effects on the ulcerated gastric mucosa. It is concluded that coconut milk offered stronger protection on indomethacin-induced ulceration than coconut water in rats.

Renal Protective Activity

Studies by Monserrat et al. (1995) reported that coconut oil at a 20% concentration with 1%

safflower showed a protective effect in weanling Wistar male rats with renal necrosis and renal failure induced by methyl-deficient diet. The renal protective effect was evidenced by less or no mortality and increased survival time in the methyl-deficient rats receiving coconut oil, as well as by a reduced incidence percentage and severity of the renal lesions as evaluated by renal weight, and type (tubular and cortical necrosis or repair) and extent (grade) of the renal damage. The lack of a protective outcome when coconut oil was fed at 14%, along with the fact that in those rats receiving coconut oil at 20% the protection was greater when the diet was supplemented with 1% safflower oil, indicated that the protective effect could be attributed to the type of fatty acids in coconut oil and not to their shortage of essential fatty acids.

Pneumonia Therapeutic Activity

In a single blinded randomized controlled trial conducted on 40 children admitted because of community acquired pneumonia, virgin coconut oil (VCO) was found to be an effective adjunct therapy for pediatric community acquired pneumonia in accelerating the normalization of respiratory rate and resolution of crackles (Erguiza et al. 2008). The respiratory rate of the VCO group was significantly normalized earlier than the control group (32.6 h vs 48.2 h). More patients under the control group were still observed to have crackles at 72 h compared to VCO group [12/20 (60%) vs 5/20 (25%)]. VCO supplementation resulted in earlier time to normalize temperature (18.8 h vs 24.6 h) and oxygen saturation (60.9 h vs 74.15 h) and shorter hospitalization (75.9 h vs 91.85 h) than the Control group but was not statistically significant. Forty percent developed soft stools, 5% had vomiting and 55% had no adverse effect with VCO supplementation.

β-Mannanase Activity

β-mannanase was extracted from coconut (*Cocos nucifera*) haustorium and purified (Soumya and Abraham 2010) .The enzyme was used for the preparation of neutraceutical dietary supplement from galactomannans of guar gum and tender coconut kernel having a β-(1,4)-linked d-mannose backbone. Depolymerized guar gum has 92% of oligosaccharides with a degree of polymerization of 3 and 7. Tender coconut kernel has a degree of polymerization of 9–39 oligosaccharides along with disaccharides and trisaccharides. Hence this mannanase will be useful to depolymerize β-(1,4)-linked d-mannose polysaccharides from most plant sources to produce prebiotics in a cost-effective technique.

Leishmanicidal Activity

The polyphenolic-rich extract from the husk fibre of *Cocos nucifera* exhibited in-vitro leishmanicidal effects on *Leishmania amazonensis* (Mendonca-Filho et al. 2004). The minimal inhibitory concentration of the polyphenolic-rich extract from *C. nucifera* to completely abolish parasite growth was 10 μg/ml. Pre-treatment of peritoneal mouse macrophages with 10 μg/ml of *C. nucifera* polyphenolic-rich extract reduced approximately 44% the association index between these macrophages and *L. amazonensis* promastigotes, with a concomitant increase of 182% in nitric oxide production by the infected macrophage in comparison to nontreated macrophages. These results provided new perspectives on drug development against leishmaniasis, since the extract of *C. nucifera* at 10 μg/ml was a strikingly potent leishmanicidal substance which inhibited the growth of both promastigote and amastigote developmental stages of *L. amazonensis* after 60 minutes, presenting no in-vivo allergic reactions or in-vitro cytotoxic effects in mammalian systems.

Anthelmintic Activity

Bark extracts of green coconut (LBGC) displayed anthelmintic activity (Costa et al. 2010). A dose of 1,000 mg/kg of butanol bark extract had 90.70% efficacy in reducing the mouse intestinal nematode burden. Phytochemical tests revealed the presence of triterpenes, saponnins and condensed tannins in the LBGC and butanol extracts. These results suggested that *Cocos nucifera* extracts may be useful in the control of intestinal nematodes.

Insecticidal (Head Lice) Activity

In the school head lice control trial, percentage cures at day 8 were 14% (permethrin, n = 7) and 61% (CDE (coconut-derived emulsion), n = 37). In the family trial where all family members were treated, cure rate was 96% (n = 28), and when the shampoo was subsequently used as a cosmetic shampoo, only 1 of 12 children became re-infested after 10 weeks. The study showed that coconut-derived emulsion (CDE) shampoo was a novel effective method of controlling head lice and used after treatment as a cosmetic shampoo can aid in the reduction of re-infestation.

Wound Healing Activity

Coconut oil has wound healing activity (Srivastava and Durgaprasad 2008). Animal studies demonstrated that there was significant improvement in burn wound contraction in group (IV) treated with the combination of coconut oil and silver sulphadiazine. The period of epithelialization also decreased significantly in groups III (pure coconut oil) and IV. The results indicated coconut oil to be an effective burn wound healing agent.

Adverse Acrylamide Problem

Thai scientists analyzed Thai curry samples and found acrylamide at very low concentrations in the range of less than 60–606 ng/g dry weight (Na Jom et al. 2008). Acrylamide, a probable human carcinogen, was detected in various heat-treated carbohydrate-rich foods. Acrylamide was formed in solely heated coconut milk at 121°C. Changes in 5-(hydroxymethyl)-2-furfuraldehyde, fructose, glucose and glutamic acid contents in coconut milk during heat treatment were observed as progress parameters for the Maillard reaction. Moreover, acrylamide was determined in equimolar model system of glutamic acid with glucose or fructose (1 mM), and yielded acrylamide, approximately 0.1% and 0.06% (w/w), respectively. After 11.3 years of follow-up, 327, 300, and 1,835 cases of endometrial, ovarian, and breast cancer, respectively, were documented. The researchers in the Netherlands (Hogervorst et al. 2007) reported that compared with the lowest quintile of acrylamide intake (mean intake, 8.9 μg/day), multivariable-adjusted hazard rate ratios (HR) for endometrial, ovarian, and breast cancer in the highest quintile (mean intake, 40.2 μg/day) were 1.29, 1.78, and 0.93, respectively. For never-smokers, the corresponding HRs were 1.99, 2.22 and 1.10. The studies indicated increased risks of postmenopausal endometrial and ovarian cancer with increasing dietary acrylamide intake, particularly among never-smokers. Risk of breast cancer was not associated with acrylamide intake.

Traditional Medicinal Uses

Very young, immature, aborted coconut has been used in dysentery as an astringent mixed with opium and some flavouring material in Java, and as external applications for haemorrhoids, ulceration of the bone and ulcerative colitis and internally for fever in Peninsular Malaysia. Coconut water is believed to be an antidote against choleric attack and a diuretic assisting in removal of irritant substances in the blood. It is regarded as antiscorbutic and somewhat anthelmintic. In India, coconut water has been reported to be a tonic for the old and sick, to stop urinary infections and to be effective in treatment of kidney and urethral stones. Water from the young coconut has been used as a substitute for dextrose infusion in emergent situations during World War II and drank for diarrhoea and vomiting. Santan (coconut milk) has been commonly used as a vehicle for Malay medicaments internally used and externally applied. It is also found to be good on the skin as moisturiser. In the Philippines, gata (coconut milk) is employed for constipation. Grated kernel meat with scraping of porcupine quills are burned together and the ashes administered as a remedy for cholera. Ash from burnt fruit stalks is used after confinement to protect against evil spirits. Kernel of young fruit mixed with other ingredients are sand-rubbed on stomach against diarrhoea. Roots are considered antipyretic and diuretic. Roots are also believed to be antiblenorrhagic, antibronchitis, febrifugal, and antigingivitic. In the Philippines, young roots are used astringent for sore throats. In Indonesia, the root infusion is used for dysentery and other intestinal complaints and for coughs. In Peninsular Malaysia a decoction of pulped roots has been employed for small pox and as poultice in syphilis, gonorrhoea and rheumatism. Immature coconut flowers are regarded as astringent and chewed for gonorrhoea. Pounded coconut leaves placed in water is sprinkled over fever patients. Coconut oil is used to treat dandruff and dry skin. Ash of the bark is used for scabies in the Philippines. The Indian system of medicine known as Ayurveda is replete with instances of the use of coconut oil in various medicinal preparations. The decoction of *Cocos nucifera* husk fibre is used in northeastern Brazil traditional medicine for treatment of diarrhea and arthritis (Esquenazi et al. 2002; Koschek et al. 2007) and tea from the husk fibre is widely used to treat several inflammatory disorders (Rinaldi et al. 2009).

Other Uses

The coconut tree has often been dubbed the 'tree of life', 'tree of plent', 'tree of abundance' or 'tree of a thousand uses'. Indeed an extremely useful tree where every part of the tree from the roots, trunk, growing shoot, leaves, inflorescence, flowers, fruit and fruit parts are ubiquitously and economically used for a myriad of edible and non-edible products and applications (Burkill 1966; Child 1974; Philippine Coconut Authority undated; Philippine Council for Agriculture and Resources Research and Development (PCAARD) and Philippine Coconut Authority 1985; Haas and Wilson 1985; Whistler 1991; Abbott 1992; Arancon 1997; Ohler and Magat 2001; Foale 2003; Chan and Elevitch 2006; van der Vossen and Chipungahelo 2007).

Coconut palm is an important intercrop or agroforestry species in smallholder mixed cropping systems on most small islands and in coastal areas. It provides shade for cocoa plantings and are often intercropped with other annual cash crops. Coconut plantings especially in monocropping situations are sometimes undergrazed by cattle or other animals. It is useful as an ornamental and landscape plants especially in beach resorts and hotels. Coconut palm is steeped in mythology, legends, songs, proverbs, and riddles throughout Asia Pacific. It is featured in socio-cultural events and ceremonies; its leaves reflects a sign of high rank in Polynesia and Micronesia. In Tuvalu, specific trees or two trees planted together serve as boundary markers.

The edible products, and the medicinal attributes and associated uses of the coconut tree have been elaborate in the preceding sections and the non-edible uses of the tree are covered here, starting with the most economically important part, the fruit.

Fruit

Coconut Copra and Crude Coconut Oil

Coconut oil is the natural oil derived from copra or fresh meat of the fruit. Copra is the dried meat or kernel of a matured coconut. Copra contains from 57% to 66% oil and copra meal, the residue after oil extraction has a meal recovery of 32–35%. Copra meal is mainly used as livestock feed for cattle as it contains about 20% protein and pigs; and when combined with other ingredients makes a very good poultry feed. Crude coconut oil is further refined by alkali-neutralisation treatment or acid pretreatment. The former treatment yield neutralised coconut oil which is further processed into neutralised bleached coconut oil and refined, bleached and deodorised coconut oil (RBDCO); and soap stock which undergo saponification to produce soap and acidulation to produce acid oil. The latter pretreatment is further processed into pretreated bleached coconut oil and refined, bleached and deodorised coconut oil and fatty acids. RBDCO is further hydrogenated to produce refined, bleached hydrogenated coconut oil. RBDCO is used for home cooking, commercial food processing, and cosmetic, industrial, and pharmaceutical purposes. Fractionated coconut oil is a fraction of the whole oil, in which the long-chain fatty acids are removed leaving only medium chain saturated fatty acids. Lauric acid, a 12 carbon chain fatty acid, is often removed as well because of its high value for industrial and medical purposes. Fractionated coconut oil may also be referred to as caprylic/capric triglyceride oil or medium chain triglyceride (MCT) oil because it is primarily the medium chain caprylic (8C) and capric (10C) acids that make up the bulk of the oil. MCT oil is most frequently used for medical applications and special diets.

Non-edible uses of coconut oil include crude oil, diesel oil substitute, biofuel, laundry soap, toilet soap, pomade, hair oil, shampoos/bubble baths/shower bath, cosmetic/toiletries, surfactants for enhanced oil recovery, foam boosters, industrial cleaning applications, emulsifiers for agriculture chemicals, emulsifiers for oil drilling, tin plating, glycerine, coco-chemicals, fuel extenders, lubricants for textile and leather, surfactants for synthetic detergent, viscosity modifiers for lube oil, plasticizers for plastic, synthetic resins, paper processing, paints and varnishes and inks, pharmaceuticals, anti-sucker agents, toothpaste, explosives, humectants, suppositories and chemical products. General types of chemical products derived from coconut oil and their end-uses are as follows:

- Esters – surfactants, plasticizers, cosmetics, lubricants, diesel oil substitute, anti sucker agent;
- Fatty acids and derivatives – surfactants, non-ionic detergents, solubiliser softeners, dispersant defoamers, cosmetics;
- Fatty acid salts – soaps;
- Short carbon chain fatty acid -plasticizers, lubricants, coatings, cosmetics, perfumes, flavours, adhesives;
- Amines – dyes, pharmaceuticals, plasticizers, emulsifiers, detergents, fabric softener, lubricants;
- Amides – adhesives, resins, inks, surfactants, foam stabilizer;
- Glycerine -toothpaste, suppositories, humectants, pharmaceuticals, explosives.

Virgin Coconut Oil

A recent and economically important product from coconut is virgin coconut oil (VCO). Virgin Coconut oil is oil extracted from fresh coconut (not copra) meat by mechanical or natural means with or without low, prolonged heating or chemical processing (Fife 2006). To protect the oil's essential properties, the production of virgin coconut oil does not undergo chemical refining, bleaching, or deodorizing. VCO has been shown to be high in vitamins and minerals, it is fit for consumption without the need for further processing. VCO products are often organically certified and commands a significant premium over industrial coconut oil. VCO is marketed for its medicinal properties, for processing into high quality soaps, shampoos and body lotions and cooking oil. VCO has potent antioxidant property and has a myriad of medicinal benefits and applications as discussed in the preceding section. VCO is currently produced using one of four methods: the direct micro expelling method (Etherington 2005), the cold press wet milling method, hot oil immersion drying and fermentation and gentle heating.

Coconut Oil as Fuel and Biodiesel

Coconut oil can be used as illuminant and is used in lamps and torches. Coconut oil can be used in diesel fuel engines in a number of ways: straight coconut oil can be used in modified or unmodified engines, used along with diesel in dual systems in modified engines or be processed into biodiesel. Coco-biodiesel or coconut methyl esters (CME) is a renewable and biodegradable diesel fuel extracted from coconut oil. Studies on diesel combustion by Myo et al. (2004) demonstrated that the thermal efficiency of coconut oil methyl ester (CME) was almost the same as other test fuels, rapeseed, palm oil methyl esters. Smoke emission from CME was the lowest among the test fuels because of the highest oxygen content. Nitrous oxide (NOx) emission from

CME was lower than gas oil. Their studies showed that CMF can be used as an alternative petroleum diesel fuel. Earlier in Maldives, Shaheed and Swain (1999) compared the combustion data of coconut oil, coconut oil methyl esters and diesel fuel using a naturally aspirated single-cylinder diesel engine. The engine performed well on the three fuels except for initial engine starting problems with coconut oil. In addition, there was a small reduction in power using coconut oil, consistent with the lower calorific value of the fuel. Diaz and Galindo (2006) maintained that coconut methyl ester (CME), derived from coconut oil, is close to that of an ideal diesel fuel. CME has the high level of saturation (91%) consisting 62% "easy-to-burn" medium saturated carbon chain which gives its chemical stability (i.e., not prone to oxidation or bacterial degradation), easy burnability, high cetane number which enhances combustion and acceleration response and low nitrous oxides (NOx) emissions.

Coconut Shell

Coconut shell is hard and fine-grained, and may be carved into all kinds of objects including cups, plates, small bowls, cooking vessels, ladle, dippers, scoops, spoons, bailers, funnels, smoking pipe bowls, and collecting cups for rubber latex, kitchen utensils, storage containers, fish hooks and lures, floats, knee drums, cymbals, lagging discs, quoits, combs, buttons, earrings, brooches, necklaces trays, bags, necklaces, interior décor pieces, picture frames, coco shell money bank, toys, backing for musical instruments, blowpipes, a wide range of handicrafts, activated carbon, shell charcoal, and shell flour. Completed dwelling system with unit area $25M^2$ has been developed using coconut shell sandwich panels for the construction of cross walls (Menon et al. 1989).

Coconut Shell Flour

Coconut shell flour is the clear, light brown, free flowing powder obtained by pulverising the dried shells of matured coconut. Coconut shell flour is used as compound filler for synthetic resin glues, thermosetting moulds and mosquito coils, filler and extender for phenolic moulding powders, filler in plastics (Schueler 1946); soft blast to clean piston engines, coating for electric arc-welding electrodes. It is also used for specialized surface finishes, liquid products, plastic adhesives, resin casting, mils abrasive products, hand clearance and bitumen products. It imparts a smooth and lustrous finish to moulded articles and improves their resistance to moisture and heat.

Coconut Shell Charcoal and Char

Coconut shell charcoal is derived from partial carbonisation of the shell of fully matured coconuts. Coconut shell charcoal is used for cooking fires, air filters, in gas masks, submarines, and cigarette tips. The shells are also burned as fuel for copra kilns or housefires. Coconut shell charcoal is used also for melting gold and silver, used in brewery to colour beer bottles. Well powdered coconut shell charcoal is used to a small extent as dentifrice and provides an important ingredient in the making of car batteries and calcium carbide. Coconut shell charcoal in the form of briquets is packed in polythene for export. The briquets are used as fuel for barbeques and grills. Coconut shell coke can be used for treating metallic effluents (Feroz et al. 2005). Coconut shell char are used in Al-alloy coconut shell char particulate composites (Murali et al. 1982) as they are readily available or are naturally renewable at affordable cost. The specific applications of these composites include engine blocks, pistons, brake-system components, seals, solid lubricants, wear- and abrasion-resistant structures, electromechanic contacts, and chassis components (Ejiofor and Reddy 1997).

Activated Carbon

Activated carbon is obtained from coconut shell charcoal through activation. Activated carbon is used for general water purification, for pure water purification in exchange resins, for alcoholic beverages (remove hogtrack odour and improves alcohol quality), for cane sugar (minimise foaming in evaporators and speed ups crystallisation), in purification of air, gas and liquids, gold recovery, chemicals, as catalyst carrier, for solvent recovery in dry cleaning, as decolouring and deodorising agent, in cigarette filters, control of formation odours, preservation of fruit and flower storage, pharmaceuticals, solvent recovery, as adsorbent for removal of toxic chemicals and metals from water, in kitchen hoods and odour tents in hospitals. Coconut shells are the most widely used raw material for the production of activated carbon suitable for gold production by cyanide leaching (Yalcin and Arol 2002). Activated coconut shell carbon can be used for removal of the insecticide cypermethrin from manufacturing process wastewater (Bhuvaneswari et al. 2007). Activated carbons prepared from coconut shells can be developed as potential adsorbents for natural gas storage applications (Azevedo et al. 2007).

Coconut Coir Fibre and Pith

Coir or coconut fibre is extracted by retting from fruit husk of mature and young, green coconut. Retting partially decomposes the husk's pulp, allowing it to be separated into coir fibers and a residue called coir pith. Freshwater retting is used for fully ripe coconut husks, and saltwater retting is used for green husks.

Fibres from mature husk – the unarranged bristle fibre or shorter fibres are used for the manufacture of yarn, rubberized coir for car seats, brushes, bed carpet underlay, gym mats, upholstery cushioning, potting medium; the arranged bristle fibres or longer fibres are used for brush mats, doormats, sacks, ropes, cordage (sennet), strings, coir-tufted carpets, mattress fibre, baskets or carry bags, needled pads, filtration pads, fly whisks, brooms, upholstery stuffing, boat and house lashings, fishnets and lines, measuring tapes for garden lands, hammocks, belts, reef-walking sandals, caulking materials for boats. Fibre from green husks are used for yarn material for carpets, rugs and mats. Coir fibres are also used to make figurines, flower holder, plant fibre pot or planting media, lye, hollow brick, insulator, compost and also burn as tinder and fuel. Fresh husks contains more tannin than old husks. In certain parts of South India, the shell and husk also are burned for smoke to repel mosquitoes. Coconut husk has been used as fertiliser when buried in the ground and for purpose of moisture conservation (Shanmugam 1970). Coir fiber is resistant to sea water and is used for cables and rigging on ships. Coir matting has been developed as a barricade material for staving in mines (Sinha 1974). Coconut husk fibres, an agricultural waste has been thoroughly investigated as sorbent for the removal of toxic cadmium, mercury and other elements like zinc selenium, strontium, iodine, chromium from aqueous medium (Hasany and Ahmad 2006). This cheap material has potential applications in analytical chemistry, water decontamination, industrial effluent treatment and in pollution abatement. Coconut coir pith can be used for removal of arsenic from aqueous solution (Anirudhan and Unnithan 2007).

Another important and popular byproduct is Coco Peat also known as coir pith, coir fibre pith, coir dust which is made from coconut fibre makes an excellent growing medium with good water holding capacity for hydroponics, soil mixes, and container plant growing (Yahya et al. 1997; Abad et al. 2002). Coir pith is used as a soil conditioner, surface mulch/ rooting medium and desiccant. Composted coir pith is an excellent organic manure for indoor plants as well as for horticulture crops. Being a good absorbent, dry coco peat can be used as an oil absorbent on slippery floors. Coco peat is also used as a bedding in animal farms and pet houses to absorb animal waste so the farm is kept clean and dry. Coco peat is usually shipped in the form of compressed bales, briquettes, slabs or discs, the end user usually expands and aerates the compressed coco peat by the addition of water. Coir geotextile/mat is another product. It is a type of fabric made from woven coconut coir fibre and used for soil ersoin prevention and control in watershade areas, bunds, pond banks, stream banks, slopes, wetlands and hillsides and fire mitigation applications (Anil and Roshini 2003; Vishnudas et al. 2005). Handmade paper has been produced from pulp obtained from the different parts of the coconut palm such as leaves, husk (young and mature), petiole and rachis and spathes (Prasad et al. 1986). Coconut husk dust is used for manuring in Sri Lanka.

Coconut coir and pith have wide applications in structural materials buildings and manufacturing industries. Insulation boards prepared from different fractions of husk and pith have been found to possess good thermal insulating properties (Iyengar et al. 1961). Particle boards have been made from coir dust (pith), >10% coir fibre and 87% urea resin (Ogawa 1977). Coconut pith (husk particles) has been used in three layered particle boards in the core with wood particles on the faces and tannin formaldehyde resin used as binder, making the board tough, highly fire-resistant and with smooth surfaces (George and Bist 1962.)

Coconut pith was found to have a thermal conductivity equivalent to granulated cork and when bound into blocks with rubber latex was found suitable as insulating material for fish boxes (Pillai and Varier 1952). Coconut pith can be used in fibre resistant building boards (Shirsalkar et al. 1964); thermal insulating concrete (Jain and George 1970); thermal insulation boards (Rao et al. 1971). Coonite, fibreboard was developed from the coir fibres from immature coconuts, mixed with waste paper, resin and alum to provide a strong, tough fibreboard that displayed good sound and insulating properties (Menon 1944). Coir fibre has been used in particle boards (George 1961); building boards (George and Joshi 1959); hard boards (George and Joshi 1960); insulating material for acoustic control (Anonymous 1974); corrugated roofing sheet and building materials (Singh 1975; Singh 1978); coir fibre reinforced cement as a low cost roofing material (Cook et al. 1978); coconut fibre reinforced corrugated slabs for use in low cost housing particularly for developing countries (Paramasivam et al., 1984); cement-bonded coir fibre boards (Aggarwal 1992); coconut-fibre reinforced slag mortar (John et al. 2005); and coir-based lightweight cement boards for used as building components for energy conservation (Asasutjarit et al. 2007). Coconut fibre can be used as additives for strengthening adobe blocks (Filho et al. 1990). Use of coir fibre in Portland cements mortars was found to increase the impact strength of the matrix up to 220% and the tensile strength up to 175% but reduced the compressive strength (Savastano 1990). Coir fibres have been used as reinforcement in various composites to improve textural strength – coir reinforced cement paste composites for house building construction (Das Gupta et al. 1978); organic matrix resin composite (Varma et al. 1988); coconut fiber-reinforced rubber composites (Arumugam et al. 1989); coconut fibre-polypropylene composites (Rozman et al. 2000); and low-density polyethylene composite (Brahmakumar, et al. 2005). The use of coir fibre reinforced composite panels provides better dimensional stability and behaviour towards fire when compared with traditional materials such as ply wood, wood particleboard and wood fibreboard (Aggarwal 1991). Radiation-treated coconut fibre provides an inexpensive filler composite material in thermoplastics with acceptable strength that could be produced with coir content as high as 30% (Owolabi and Czvikovsky 1988).

Coconut Water

Fresh coconut water has been used by Dyers in Peninsular Malaysia to obtain a green colour after using indigo (Burkill 1966). Ayer banyar, a fermentation byproduct of coconut water has been used by dyers in the east coast of Peninsular Malasyia to darken the colour of silk purple and black following immersion in indigo. The vinegar obtained from fermentation of coconut water has been used to bleach palm leaf hats in Indo-China and has been used as latex coagulant in rubber plantations in Peninsular Malaysia. Coconut water from immature nuts has been used to wash off residues of arsenic in colouring kris-blades. Coconut water also has medicinal uses in traditional medicine (see above).

Stem

Coconut stem is a valued base material in the production of sawn timber and other wood articles. Coconut wood is unlike other conventional timbers, has no distinct grain patter, dries uniformly and has no cross-sectional distortion. It is extremely amenable to machining and fabrication into various forms. Coconut wood has been successfully used for frames, floorings (parquet), floor joists, stairs and railings, roofing shingles, rafters, beams, trusses, posts and other load bearing surfaces in buildings of all kinds. It is also made into door panels, window jambs, sidings, ceiling, jalousies, studs, purlins, exterior walls, panels, interior walls, furniture, dining sets, coco chairs and sofa, office furniture – desks, conference tables, classroom chairs, food containers, tools, spears and weapons, drums, canoe planking, small canoe hulls and paddles, walking sticks, spear shaft and lances, fish clubs, kitchen utensils, axe handles, novelty products – cigar boxes, chest and jewel boxes, crating and packaging boxes, canes and sticks, household implements plates, bowls, cups handles, gavels, glass holders, candle holder, paper weight, ink stand, pencil holder, ash tray, lampshade stand, name plate, laminated baseball bat, flower vases, cloth hangers, fixtures show case, shelves, cabinet dividers, balusters, headboards, radio and television cabinets, boat side planking, street sign posts, road guard rails. Coconut stem needs treatment with chromate copper arsenate for prolonged durability, especially if used outdoors for fencing and animal pens. In Hawai'i, the base of the trunk has been used to make food containers and hula drums and the bark used for scenting body oil and for smoking skirts. In Malaysia and Indonesia, fallen coconut logs are used as temporary bridges across drains and streams in villages. In the Cook Islands, the hollowed-out trunk is used as a vessel in which "bush beer" is fermented. Sections of stem, after scooping out pith, are used as flumes or gutters for carrying water. Pith of stem contains starch which may be extracted and used as flour. The trunk when injured exudes a gum which swells in water. Trunk made into charcoal and charcoal briquettes afford higher heating value, are easily handled and produce less smoke compared to wood. Briquettes have good crushing strength and burning properties. Binders used for charcoal briquettes include lime, cement, molasses, sorghum grains and clay.

Coconut Fronds

The unfurled immature leaves are used in Malaysia and parts of Indonesia for *ketumpat* (wrapped leaf packet with uncooked rice for cooking and subsequent consumption) and elsewhere to make skirts, body ornamentation, hats, eyeshades, baskets, fans, and fishing lures. Similarly in the Philippines, rice is wrapped in coconut leaves for cooking and subsequent storage and these packets are called *puso*. In Peninsular Malaysia, the large frond are variously shaped by bending and plaiting and used as festooning materials on buildings on ceremonial occasions. Young leaves from germinating nut are used to make a coconut-tree climbing bandage or foot-harness that is tied between the feet. Mature green fronds are commonly used as thatch for village dwellings in the Pacific islands. Green fronds are woven into all sizes of baskets for carrying food and other goods. Mature and young leaves used for weaving baskets, food containers and parcels, mats, housing thatch, tables or table mats for feasts, trays, fans, balls, weirs/barricades for communal fish drives, and other plaited ware. Leaves are employed in magic, particularly garden magic, and tied around trees in plantations as boundary markers or to ward off evil spirits and as a sign of no trespassing, or tapu. Old dried leaves and husks are used as mulching materials for vegetables and other crops and as fencing. Midribs of leaflets or pinnules are used for making brooms, bottle corks when bound together, in weaving, toy windmills, for fishing lures and shrimp snares, to spear mudworms, small arrows, musical instruments, headdresses, combs, for stringing fish, fastening thatch segments, for strengthening bonito hooks, cooking skewers (satay skewers), toy canoes, and etc. Juice from the leaflets were use to darken the edge of kris. In the Pacific islands, woody leaf base and midribs of fronds are used for house flooring and rafters, for sandals, carrying poles, toy boats, rattles, sledges or clappers, clubs or mallets, and to beat water during fish drives and for pounding and stabilizing banks of taro beds. In Tuvalu, midribs of young fronds are used for fishermen's belts. The midrib was used to string oil-rich nuts, such as candlenut (*Aleurites moluccana*), for torches. Dried fronds are used as torches for night fishing and night social ceremonies. The burlap-like fibrous sheath at base of fronds is used as tinder, toilet paper, gauze and a filter or strainer. It is also utilised as a filter/strainer or to squeeze medicinal plants or coconut. It is also use to wrap bait for deep-water fishing and the earth ball on the roots of seedlings when transplanting.

Coconut Flowers

Unopened flowers are protected by sheath, which is often used to fashion shoes, caps, and even a kind of pressed helmet for soldiers. The inflorescence sheet and dried fronds are used to make torches for night fishing and for major night-time ceremonial occasions. Dried sheaths are used as splints to set broken bones; flower spathe sheath and frond midribs are used to splint broken bones in Tuvalu. Opened flowers provide a good honey for bees. In Hawai'i, the male flowers are heated in coconut oil to perfume tapa cloth. In India, the buttons are used as a source of tannin for the leather industry (Pillai and Pandalai 1959). Coconut flowers are used in connection with religious rituals in Tahiti.

Coconut Roots

Coconut roots are used as a source for dyestuff and medicines (mouthwash and medicine for dysentery). The roots are also used to make fish traps, toothbrushes, floating cages, sand screens, and other objects.

Comments

Coconut is grown in more than 80 countries worldwide, with the Philippines, Indonesia and India as the top three producers. Other leading producing countries include Brazil, Sri Lanka, Thailand, Mexico and Vietnam.

Selected References

Abad MP, Noguera P, Puchades R, Maquiera A, Noguera V (2002) Physico-chemical and chemical properties of some coconut dusts for use as a peat substitute for containerized ornamental plants. Bioresour Technol 82:241–245

Abbott IA (1992) Lā'au Hawai'i: traditional Hawaiian uses of plants. Bishop Museum Press, Honolulu

Abdurahman HN, Mohammed FS, Yunus RM, Arman A (2009) Demulsification of virgin coconut oil by centrifugation method: a feasibility study. Int J Chem Technol 1:59–64

Agero AL, Verallo-Rowell VM (2004) A randomized double-blind controlled trial comparing extra virgin coconut oil with mineral oil as a moisturizer for mild to moderate xerosis. Dermatitis 15(3):109–116

Aggarwal LK (1991) Development of coir fiber reinforced composite panels. Res Ind 36(4):213–274

Aggarwal LK (1992) Studies on cement-bonded coir fibre boards. Cem Concr Compos 1(1):63–69

Akhter A, Zaman S, Umar Ali M, Ali MY, Jalil Miah MA (2010) Isolation of polyphenolic compounds from the green coconut (*Cocos nucifera*) shell and characterization of their benzoyl ester derivatives. J Sci Res 2(1):186–190

Alleyne T, Roache S, Thomas C, Shirley A (2005) The control of hypertension by use of coconut water and mauby: two tropical food drinks. West Indian Med J 54(1):3–8

Alviano DS, Rodrigues KF, Leitão SG, Rodrigues ML, Matheus ME, Fernandes PD, Antoniolli AR, Alviano CS (2004) Antinociceptive and free radical scavenging activities of *Cocos nucifera* L. (Palmae) husk fiber aqueous extract. J Ethnopharmacol 92(2–3):269–273

Angelia MRN, Garcia RN, Prak K, Utsumi S, Tecson-Mendoza EM (2008) Physicochemical and functional characteristics of cocosin and its unique emulsifying behaviour. In: 38th annual CSSP scientific conference, Iloilo City, Philippines, 12–16 May 2008, *Phil. J. Crop Sci.*, pp 80–81

Anil KR, Roshini S (2003) Effect of coir geotextiles in soil conservation and slope land cultivation in varying land slopes with forest loam soil. In: Mathur GN, Rao GV, Chawla AS (eds) Case histories of geosynthetics in infrastructure. CBIP Publications, New Delhi, pp 73–77

Anirudhan TS, Unnithan MR (2007) Arsenic (V) removal from aqueous solutions using an anion exchanger derived from coconut coir pith and its recovery. Chemosphere 66(1):60–66

Anonymous (1974) Coir as an insulating material for acoustic control. Coir 19(1):18

Arancon Jr. RN (1997) Asia-Pacific forestry sector outlook study: focus on coconut wood. [Internet] FAO, Rome, Italy. http://www.fao.org/documents/show_cdr.asp?url_file=/DOCREP/W7731E/ w7731e07.htm

Arumugam N, Selvy K, Rao K, Rajalingam P (1989) Coconut fiber-reinforced rubber composites. J Appl Sci 37(9):2645–2649

Asasutjarit C, Hirunlabh J, Khedari J, Charoenvai S, Zeghmati B, Cheul Shin U (2007) Development of coconut coir-based lightweight cement board. Constr Build Mater 21(2):277–288

Assunção ML, Ferreira HS, dos Santos AF, Cabral CR Jr, Florêncio TM (2009) Effects of dietary coconut oil on the biochemical and anthropometric profiles of women presenting abdominal obesity. Lipids 44(7):593–601

Azevedo DCS, Araújo JCS, Bastos-Neto M, Torres AEB, Jaguaribe EF, Cavalcante CL (2007) Microporous activated carbon prepared from coconut shells using chemical activation with zinc chloride. Microporous Mesoporous Mater 100(1–3):361–364

Balasundaresan D, Sugadev R, Ponnuswamy MN (2002) Purification and crystallization of coconut globulin cocosin from *Cocos nucifera*. Biochim Biophys Acta 1601(1):121–122

Bartolotta S, García CC, Candurra NA, Damonte EB (2001) Effect of fatty acids on arenavirus replication: inhibition of virus production by lauric acid. Arch Virol 146(4):777–790

Bhuvaneswari K, Prasad RP, Sarma PN (2007) Adsorption studies on wastewaters from cypermethrin manufacturing process using activated coconut shell carbon. J Environ Sci Eng 49(4):265–272

Bibby DC, Grimble RF (1990) Tumour necrosis factor-alpha and endotoxin induce less prostaglandin E2 production from hypothalami of rats fed coconut oil than from hypothalami of rats fed maize oil. Clin Sci (London) 79(6):657–662

Brahmakumar M, Pavithran C, Pillai RM (2005) Coconut fibre reinforced polyethylene composites: effect of natural waxy surface layer of the fibre on fibre/matrix interfacial bonding and strength of composites. Compos Sci Technol 65(3–4):563–569

Burkill IH (1966) A dictionary of the economic products of the Malay Peninsula. Revised reprint, 2 vols, vol 1 (A–H) pp 1–1240, vol 2 (I–Z) pp 1241–2444. Ministry of Agriculture and Co-operatives, Kuala Lumpur

Castillo M, Hortal JH, García-Fuentes E, Zafra MF, García-Peregrín E (1996) Coconut oil affects lipoprotein composition and structure of neonatal chicks. J Biochem 119(4):610–616

Chakraborty M, Mitra A (2008) The antioxidant and antimicrobial properties of the methanolic extract from *Cocos nucifera* mesocarp. Food Chem 107(3): 994–999

Chan E, Elevitch CR (2006) *Cocos nucifera* (coconut), ver. 2.1. In: Elevitch CR (ed) Species profiles for Pacific Island agroforestry. Permanent Agriculture Resources (PAR), Hōlualoa, Hawai'i. <http://www.traditionaltree.org>

Che Man YB, Abdul Karim MIB, Teng CT (1997) Extraction of coconut oil with *Lactobacillus plantarum* 1041 IAM. J Am Oil Chem Soc 74(9):1115–1119

Child R (1974) Coconuts, 2nd edn. Longmans, London, p 335 pp

Chomchalow N (1999) Amazing Thai coconut. Horticultural Research Institute, Department of Agriculture, Bangkok, 28 pp

Codex Alimentarius (2003) Codex standard for named vegetable oils. Codex Stan 210, pp 5–13

Connolly M, Stafford KA, Coles GC, Kennedy CT, Downs AM (2009) Control of head lice with a coconut-derived emulsion shampoo. J Eur Acad Dermatol Venereol 23(1):67–69

Cook DJ, Pama RP, Weerasingle HLSD (1978) Coir fibre reinforced cement as a low cost roofing material. Build Environ 13(1):193–198

Costa CTC, Bevilaqua CML, Morais SM, Camurça-Vasconcelos ALF, Maciel MV, Braga RR, Oliveira LMB (2010) Anthelmintic activity of *Cocos nucifera* L. on intestinal nematodes of mice. Res Vet Sci 88(1):101–103

Council of Scientific and Industrial Research (1950) The wealth of India. A dictionary of Indian raw materials and industrial products (Raw materials 2). Publications and Information Directorate, New Delhi

Das Gupta NC, Paramasivam P, Lee SL (1978) Mechanical properties of coir reinforced cement paste composites. Housing Sci 2:391–406

Dayrit FM, Buenafe OE, Chainani ET, de Vera IM (2008) Analysis of monoglycerides, diglycerides, sterols, and free fatty acids in coconut (*Cocos nucifera* L.) oil by 31P NMR spectroscopy. J Agric Food Chem 56(14): 5765–5769

Diaz Jr. RS, Galindo FC (2006) Coco-biodiesel – a perfect natural diesel. Asian Institute of Petroleum Studies, Inc (AIPSI). http://www.biofuels.com.ph/resources/perfect_natural_diesel.pdf

Ejiofor JU, Reddy RG (1997) Developments in the processing and properties of particulate Al-Si composites. JOM 49(11):31–37

Erguiza GS, Jiao AG, Reley M, Ragaza S (2008) The effect of virgin coconut oil supplementation for community-acquired pneumonia in children aged 3 to 60 months admitted at the Philippine Children's Medical Center: a single blinded randomized controlled trial. Chest 134:p139001

Escalante-Erosa F, Arvízu-Méndez GE, Peña-Rodríguez LM (2007) The skimmiwallinols – minor components of the epicuticular wax of *Cocos nucifera*. Phytochem Anal 18(3):188–192

Escalante-Erosa F, Fernández-Concha GC, Peña-Rodríguez LM (2009) Skimmiwallinin, an additional skimmiwallinol derivative from the cuticular wax of *Cocos nucifera*. Nat Prod Res 23(10):948–952

Esquenazi D, Wigg MD, Miranda MMFS, Rodrigues HM, Tostes JBF, Rozental S, da Silva AJR, Alviano CS (2002) Antimicrobial and antiviral activities of polyphenolics from *Cocos nucifera* Linn. (Palmae) husk fiber extract. Res Microbiol 153(10):647–652

Etherington D (2005) Bringing hope to remote island communities with virgin coconut oil production. In: Adkins SW, Foale M, Samosir YMS (eds) Coconut revival – new possibilities for the 'Tree of Life'. Proceedings of the international coconut forum held in Cairns, Australia, 22–24 Nov 2005. ACIAR Proceedings No. 125, pp 57–65

Feroz S, King P, Prasad VS (2005) Treatment of metallic effluents using coconut shell coke. J Environ Sci Eng 47(2):109–114

Fife B (2006) Virgin coconut oil: natures miracle medicine. Piccadilly Books, 80 pp

Filho RDT, Barbosa NP, Ghavami K (1990) Application of sisal and coconut fibres in adobe block. In:

Vegetables plants and their fibres as building materials. Proceedings of the 2nd international symposium sponsored by the International Union of Testing and Research Laboratories for Materials and Structures (RILEM), , Salvador, Bahia, 17–21 Sept 1990, pp 139–249

Foale M (2003) The coconut Odyssey: the Bounteous possibilities of the tree of life. ACIAR Monograph 101, ACIAR, Kingston

Foale M, Lynch PW (eds) (1994) Coconut improvement in the South Pacific. In: Proceedings of a workshop in Taveuni, Fiji Islands, 10–12 Nov 1993. ACIAR Proceedings 53, ACIAR, Kingston

Forouhi NG, Rumley A, Lowe GD, McKeigue P, Sattar N (2003) Specific elevation in plasma tissue plasminogen activator antigen concentrations in South Asians relative to Europeans. Blood Coagul Fibrinolysis 14(8):755–760

Garfinkel M, Lee S, Opara EC, Akkwari OE (1992) Insulinotropic potency of lauric acid: a metabolic rational for medium chain fatty acids (MCF) in TPN formulation. J Surg Res 52:328–333

George J (1961) Complete utilization of coconut husk. 3. Particle board from by-products of the coir industry. Indian Pulp Pap 15:613–616

George J, Bist BS (1962) Complete utilization of coconut husk. Part IV. Three layer particle boards with coconut husk particle core. Indian Pulp Pap 16(7):437–438

George J, Joshi HC (1959) Complete utilization of coconut husk. 1. Building boards from coconut husk. Indian Coconut J 12:46–51

George J, Joshi HC (1960) Hard boards from coconut fibre. Res Ind 5:66–68

Gopikrishna V, Thomas T, Kandaswamy D (2008) A quantitative analysis of coconut water: a new storage media for avulsed teeth. Oral Surg Oral Med Oral Pathol Oral Radiol Endod 105(2):e61–e65

Haas A, Wilson L (eds) (1985) Coconut wood: processing and use. FAO, Rome, 58 pp

Handy ESC, Handy EG (1972) Native planters in Old Hawaii: their life, lore, and environment. Bishop Museum Press, Honolulu

Hargrave KM, Azain MJ, Miner JL (2005) Dietary coconut oil increases conjugated linoleic acid-induced body fat loss in mice independent of essential fatty acid deficiency. Biochim Biophys Acta 1737(1): 52–60

Harries HC (1990) Malesian origin for a domestic *Cocos nucifera*. In: Baas P, Kalkman K, Geesink R (eds) The plant diversity of Malesia. Proceedings of the flora Malesiana symposium, Leiden, Netherlands, Aug 1989. Kluwer, Dordrecht, pp 351–357, 420 pp

Hasany SM, Ahmad R (2006) The potential of cost-effective coconut husk for the removal of toxic metal ions for environmental protection. J Environ Manage 81(3):286–295

Hodson L, Skeaff CM, Chisholm WA (2001) The effect of replacing dietary saturated fat with polyunsaturated or monounsaturated fat on plasma lipids in free-living young adults. Eur J Clin Nutr 55(10):908–915

Hogervorst JG, Schouten LJ, Konings EJ, Goldbohm RA, van den Brandt PA (2007) A prospective study of dietary acrylamide intake and the risk of endometrial, ovarian, and breast cancer. Cancer Epidemiol Biomar Prev 16(11):2304–2313

Hornung B, Amtmann E, Sauer G (1994) Lauric acid inhibits the maturation of vesicular stomatitis virus. J Gen Virol 75:353–361

Intahphuak S, Khonsung P, Panthong A (2010) Anti-inflammatory, analgesic, and antipyretic activities of virgin coconut oil. Pharm Biol 48(2):151–157

Iyengar NVR, Anandaswamy B, Raju PV (1961) Thermal insulating materials from agricultural wastes: coconut (*Cocos nucifera* Linn.) husK and pith. J Sci Ind Res India 20D(7):276–279

Jain RK, George J (1970) Utilisation of coconut pith for thermal insulating concrete. Coir 14(3):25–32

John VM, Cincotto MA, Sjöström C, Agopyan V, Oliveira CTA (2005) Durability of slag mortar reinforced with coconut fibre. Cem Concr Compos 27(5): 565–574

Kirszberg C, Esquenazi D, Alviano CS, Rumjanek VM (2003) The effect of a catechin-rich extract of *Cocos nucifera* on lymphocytes proliferation. Phytother Res 17:1054–1058

Koschek PR, Alviano DS, Alviano CS, Gattass CR (2007) The husk fiber of *Cocos nucifera* L. (Palmae) is a source of anti-neoplastic activity. Braz J Med Biol Res 40:1339–1343

Kwon K, Park KH, Rhee KC (1996) Fractionation and characterization of proteins from coconut (*Cocos nucifera* L.). J Agric Food Chem 44(7):1741–1745

Laureles LR, Rodriguez FM, Reano CE, Santos GA, Laurena AC, Mendoza EMT (2002) Variability in fatty acid and triacylglycerol composition of the oil of coconut (*Cocos nucifera* L.) hybrids and their parentals. J Agric Food Chem 50(6):1581–1586

Lipoeto NI, Agus Z, Oenzil F, Wahlqvist M, Wattanapenpaiboon N (2004) Dietary intake and the risk of coronary heart disease among the coconut-consuming Minangkabau in West Sumatra, Indonesia. Asia Pac J Clin Nutr 13(4):377–384

Ma Z, Ge L, Lee AS, Yong JW, Tan SN, Ong ES (2008) Simultaneous analysis of different classes of phytohormones in coconut (*Cocos nucifera* L.) water using high-performance liquid chromatography and liquid chromatography-tandem mass spectrometry after solid-phase extraction. Anal Chim Acta 610(2): 274–281

Mandal SM, Dey S, Mandal M, Sarkar S, Maria-Neto S, Franco OL (2009) Identification and structural insights of three novel antimicrobial peptides isolated from green coconut water. Peptides 30(4):633–637

Mantena SK, Jagadish, Badduri SR, Siripurapu KB, Unnikrishnan MK (2003) In vitro evaluation of antioxidant properties of *Cocos nucifera* Linn. water. Mol Nutr Food Res 47(2):126–131

Marina AM, CheMan YB, Nazimah SAH, Amin I (2009a) Antioxidant capacity and phenolic acids of virgin coconut oil. Int J Food Sci Nutr 60(S2):114–123

Marina AM, Che Man YB, Nazimah SAH, Amin I (2009b) Chemical properties of virgin coconut oil. J Am Oil Chem Soc 86(4):301–307

Mendis S, Kumarasunderam R (1990) The effect of daily consumption of coconut fat and soya-bean fat on plasma lipids and lipoproteins of young normolipidaemic men. Br J Nutr 63:547–552

Mendis S, Wissler RW, Bridenstine RT, Podbielski FJ (1989) The effects of replacing coconut oil with corn oil on human serum lipid profiles and platelet derived factors active in atherogenesis. Nutr Rep Int 40(4):773–782

Mendis S, Samarajeewa U, Thattil RO (2001) Coconut fat and serum lipoproteins: effects of partial replacement with unsaturated fats. Br J Nutr 85(5):583–589

Mendonca-Filho RR, Rodrigues IA, Alviano DS, Santos ALS, Soares RMA, Alviano CS, Lopes AHCS, Rosa MDS (2004) Leishmanicidal activity of polyphenolic-rich extract from husk fiber of *Cocos nucifera* Linn. (Palmae). Res Microbiol 155(3):136–143

Menon SRK (1944) Coconite (fibreboard from immature coconuts). J Sci Ind Res 2:172–174

Menon D, Prabhu BTS, Pillai SU (1989) Construction of low-cost vault-shaped dwelling units using coconut shell composites. J Inst Eng India, Part AR Archit Eng Div 69:37–43

Mini S, Rajamohan T (2004) Influence of coconut kernel protein on lipid metabolism in alcohol fed rats. Indian J Exp Biol 42(1):53–57

Monserrat AJ, Romero M, Lago NA, Aristi C (1995) Protective effect of coconut oil on renal necrosis occurring in rats fed a methyl-deficient diet. Ren Fail 17:525–537

Muller H, Lindman AS, Blomfeldt A, Seljeflot I, Pedersen JL (2003) A diet rich in coconut oil reduces diurnal postprandial variations in circulating tissue plasminogen activator antigen and fasting lipoprotein (a) compared with a diet rich in unsaturated fat in women. J Nutr 133(11):3422–3427

Murali TP, Surappa MK, Rohatgi PK (1982) Preparation and properties of Al-alloy coconut shell char particulate composites. Metall Trans B (Process Metallurgy) 13B(3):485–494

Myo T, Hamasaki K, Kinoshita E, Sawaoka J (2004) Diesel combustion characteristics of coconut oil methyl ester. Nihon Kikai Gakkai Kanto Shibu Sokai Koen Ronbunshu 10:157–158 (In Japanese)

Na Jom K, Jamnong P, Lertsiri S (2008) Investigation of acrylamide in curries made from coconut milk. Food Chem Toxicol 46(1):119–124

Nagaraju A, Lokesh R, Belur LR (2008) Rats fed blended oils containing coconut oil with groundnut oil or olive oil showed an enhanced activity of hepatic antioxidant enzymes and a reduction in LDL oxidation. Food Chem 108(3):950–957

Nalini N, Manju V, Menon VP (2004) Effect of coconut cake on the bacterial enzyme activity in 1,2-dimethyl hydrazine induced colon cancer. Clin Chim Acta 342(1–2):203–210

Nevin KG, Rajamohan T (2004) Beneficial effects of virgin coconut oil on lipid parameters and in vitro LDL oxidation. Clin Biochem 37(9):830–835

Nevin KG, Rajamohan T (2006) Virgin coconut oil supplemented diet increases the antioxidant status in rats. Food Chem 99(2):260–266

Nevin KG, Rajamohan T (2008) Influence of virgin coconut oil on blood coagulation factors, lipid levels and LDL oxidation in cholesterol fed Sprague–Dawley rats. e-SPEN, Eur e-J Clin Nutr Metab 3(1):1–8

Nneli RO, Woyike OA (2008) Antiulcerogenic effects of coconut (*Cocos nucifera*) extract in rats. Phytother Res 22(7):970–972

Ogawa H (1977) Utilization of coir dust for particle-boards. Osaka Kogyo Gijutsu Shikensho Kiho 28(3):231–233 (In Japanese)

Ogbolu DO, Oni AA, Daini OA, Oloko AP (2007) In vitro antimicrobial properties of coconut oil on *Candida* species in Ibadan. Nig J Med Food 10(2):384–387

Ohler JG, Magat SS (2001) *Cocos nucifera* L. In: Van der Vossen HAM, Umali BE (eds) Plant resources of South East Asia no.14. Vegetable oils and fats. Prosea Foundation, Bogor, pp 76–84

Owolabi O, Czvikovsky T (1988) Composite materials of radiation-treated coconut fiber and thermoplastics. J Appl Polym Sci 35(3):573–582

Padmakumaran NKG, Rajamohan T, Kurup PA (1999) Coconut kernel protein modifies the effect of coconut oil on serum lipids. Plant Foods Hum Nutr 53(2):133–144

Paramasivam P, Nathan GK, Das Gupta NC (1984) Coconut fibre reinforced corrugated slabs. Int J Cement Compos. Lightweight Concrete 6(1):19–27

Pehowich DJ, Gomes AV, Barnes JA (2000) Fatty acid composition and possible health effects of coconut constituents. West Indian Med J 49(2):128–133

Perry LM (1980) Medicinal plants of East and Southeast Asia: attributed properties and uses. MIT Press, Cambridge/London, 620 pp

Philippine Coconut Authority, undated. Spectrum of coconut products, 25 pp

Philippine Council for Agriculture and Resources Research and Development (PCAARD) and Philippine Coconut Authority (1985) The Philippines recommends for coconut timber utilization. Philippine Council for Agriculture and Resources Research and Development. PCAARD Technical Bulletin Series No 60, 93 pp

Piggott CJ (1964) Coconut growing. Oxford University Press, London, 109 pp

Pillai NG, Pandalai KM (1959) Total tannins in shed coconut buttons. Sci Cult 25:265–267

Pillai KS, Varier NS (1952) Coconut pith as an insulating material. Indian Coconut J 5:159–161

Pillai MG, Thampi BS, Menon VP, Leelamma S (1999) Influence of dietary fiber from coconut kernel (*Cocos nucifera*) on the 1,2-dimethylhydrazine-induced lipid peroxidation in rats. J Nutr Biochem 10(9):555–560

Porcher MH et al. (1995–2020) Searchable world wide web multilingual multiscript plant name database. The University of Melbourne, Victoria.. http://www.plant-names.unimelb.edu.au/Sorting/Frontpage.html

Portillo MP, Serra F, Simon E, del Barrio AS, Palou A (1998) Energy restriction with high-fat diet enriched with coconut oil gives higher UCP1 and lower white fat in rats. Int J Obes Relat Metab Disord 22:974–979

Prasad SV, Pillai CKS, Satyanarayana KG (1986) Paper and pulp board from coconut leaves. Res India 31(2):93–96

Prior IA, Davidson F, Salmond CE, Czochanska Z (1981) Cholesterol, coconuts, and diet on Polynesian atolls: a natural experiment: the Pukapuka and Tokelau Island studies. Am J Clin Nutr 34:1552–1561

Purseglove JW (1968) The origin and distribution of the coconut. Trop Sci 10(4):191–199

Purseglove JW (1972) Tropical crops. Monocotyledons 2. Wiley, New York

Radenahmad N, Saleh F, Sawangjaroen K, Rundorn W, Withyachumnarnkul B, Connor JR (2009) Young coconut juice significantly reduces histopathological changes in the brain that are induced by hormonal imbalance: a possible implication to postmenopausal women. Histol Histopathol 24(6):667–674

Rao CVN, Prabhu PV, Venkataraman R (1971) Thermal insulation boards from coconut pith. Fish Technol 8(2):185–188

Rinaldi S, Silva DO, Bello F, Alviano CS, Alviano DS, Matheus ME, Fernandes PD (2009) Characterization of the antinociceptive and anti-inflammatory activities from *Cocos nucifera* L. (Palmae). J Ethnopharmacol 122(3):541–546

Rozman HD, Tan KW, Kumar RN, Abubakar A, Mohd. Ishak ZA, Ismail H (2000) The effect of lignin as a compatibilizer on the physical properties of coconut fiber–polypropylene composites. Eur Polym J 36(7):1483–1494

Salil G, Rajamohan T (2001) Hypolipidemic and antiperoxidative effect of coconut protein in hypercholesterolemic rats. Indian J Exp Biol 39(10):1028–1034

Sandhya VG, Rajamohan T (2006) Beneficial effects of coconut water feeding on lipid metabolism in cholesterol-fed rats. J Med Food 9(3):400–407

Sandhya VG, Rajamohan T (2008) Comparative evaluation of the hypolipidemic effects of coconut water and lovastatin in rats fed fat-cholesterol enriched diet. Food Chem Toxicol 46(12):3586–3592

Savastano H (1990) The use of coir fibres as reinforcements to Portland cements mortars. In: Vegetables plants and their fibres as building materials. Proceedings of the 2nd international symposium sponsored by the international union of testing and research laboratories for materials and structures (RILEM), 17–21 Sept 1990, Salvador, Bahia, Brazil, pp 150–158

Schueler GBE (1946) Uses of coconut products. Mod Plast 23(10):118–119

Shaheed A, Swain E (1999) Combustion analysis of coconut oil and its methyl esters in a diesel engine. Proceedings of the institution of mechanical engineers, part A: J Power Energy 213(5): 417–425

Shanmugam KS (1970) Coconut husk – best to bury. Coconut Bull 1(5):3–6

Shirsalkar MN, Jain RK, George J (1964) Fibre resistant building boards from coconut pith. Res Ind 9(12):359–360

Sindhurani JA, Rajamohan T (1989) Hypolipidemic effect of hemicellulose component of coconut fiber. Indian J Exp Biol 36(8):786–789

Sindurani JA, Rajamohan T (2000) Effects of different levels of coconut fiber on blood glucose, serum insulin and minerals in rats. Indian J Physiol Pharmacol 44(1):97–100

Singh SM (1975) Corrugated roofing sheet and building panel from coir wastes. Coir 19(4):3–6

Singh SM (1978) Coconut husk – a versatile building panels materials. J Ind Acad Wood Sci 9(2):80–87

Sinha KN (1974) Evaluation of coir matting as a barricade material for staving in mines -a feasibility investigation. Coir 19(4):5–9

Soumya RS, Abraham ET (2010) Isolation of β-mannanase from *Cocos nucifera* Linn haustorium and its application in the depolymerization of β-(1,4)-linked d-mannans. Int J Food Sci Nutr 61(3):271–281

Srinivas K, Vijayasrinivas S, Kiran HR, Prasad PM, Rao MEB (2003) Antibacterial activity of *Cocos nucifera* Linn. Indian J Pharm Sci 65(4):417–418

Srivastava P, Durgaprasad S (2008) Burn wound healing property of *Cocos nucifera*: an appraisal. Indian J Pharmacol 40:144–146

Stuart GU (2010) Philippine alternative medicine. Manual of some Philippine medicinal plants. http://www.stuartxchange.org/OtherHerbals.html

Trinidad TP, Valdez DH, Loyola AS, Mallillin AC, Askali FC, Castillo JC, Masa DB (2003) Glycaemic index of different coconut (*Cocos nucifera*)-flour products in normal and diabetic subjects. Br J Nutr 90(3):551–556

Trinidad TP, Loyola AS, Mallillin AC, Valdez DH, Askali FC, Castillo JC, Resaba RL, Masa DB (2004) The cholesterol-lowering effect of coconut flakes in humans with moderately raised serum cholesterol. J Med Food 7(2):136–140

U.S. Department of Agriculture, Agricultural Research Service (2009) USDA national nutrient database for standard reference, release 22. Nutrient Data Laboratory Home Page. http://www.ars.usda.gov/ba/bhnrc/ndl

van der Vossen HAM, Chipungahelo GSE (2007) *Cocos nucifera* L. [Internet] record from protabase. In: van der Vossen HAM, Mkamilo GS (eds) PROTA (Plant Resources of Tropical Africa/Ressources végétales de l'Afrique tropicale), Wageningen. <http://database.prota.org/search.htm>

Varma DS, Varma M, Ananthakar SR, Varma LK (1988) Evaluation of coir and jute fibers as reinforced in organic matrix resin. Integr Fundam Polym Sci Technol 1987:599–603

Verallo-Rowell VM, Dillague KM, Syah-Tjundawan BS (2008) Novel antibacterial and emollient effects of coconut and virgin olive oils in adult atopic dermatitis. Dermatitis 19(6):308–315

Vishnudas S, Savenije HHG, van Der Zaag P, Anil KR, Balan K (2005) Experimental study using coir geotextiles in watershed management. Hydrol Earth Syst Sci Discuss 2:2327–2348

Whistler WA (1991) Polynesian plant introductions. In: Cox PA, Banack SA (eds) Islands, plants, and

Polynesians – an introduction to Polynesian ethnobotany. Dioscorides Press, Portland

Yahya A, Safie H, Kahar S (1997) Properties of coco-peat-based growing media and their effects on two annual ornamentals. J Trop Agric Food Sci 25:151–157

Yalcin M, Arol AI (2002) Gold cyanide adsorption characteristics of activated carbon of non-coconut shell origin. Hydrometallurgy 63(2):201–206

Yong JW, Ge L, Ng YF, Tan SN (2009) The chemical composition and biological properties of coconut (*Cocos nucifera* L.) water. Molecules 14(12):5144–5164

Elaeis guineensis

Scientific Name

Elaeis guineensis Jacq.

Synonyms

Elaeis dybowskii Hua, *Elaeis guineensis* f. *androgyna* A.Chev., *Elaeis guineensis* f. *caryolitica* Becc., *Elaeis guineensis* f. *dioica* A.Chev., *Elaeis guineensis* f. *dura* Becc., *Elaeis guineensis* f. *fatua* Becc., *Elaeis guineensis* f. *ramosa* A.Chev., *Elaeis guineensis* f. *semidura* Becc., *Elaeis guineensis* f. *tenera* Becc., *Elaeis guineensis* subsp. *nigrescens* A.Chev., nom. inval., *Elaeis guineensis* subsp. *virescens* A.Chev., *Elaeis guineensis* var. *albescens* Becc., *Elaeis guineensis* var. *angulosa* Becc., *Elaeis guineensis* var. *ceredia* A.Chev., *Elaeis guineensis* var. *compressa* Becc., *Elaeis guineensis* var. *gracilinux* A.Chev., *Elaeis guineensis* var. *idolatrica* A.Chev., *Elaeis guineensis* var. *intermedia* A.Chev., *Elaeis guineensis* var. *leucocarpa* Becc., *Elaeis guineensis* var. *macrocarpa* A.Chev., *Elaeis guineensis* var. *macrocarya* Becc., *Elaeis guineensis* var. *macrophylla* A.Chev., *Elaeis guineensis* var. *macrosperma* Welw., *Elaeis guineensis* var. *madagascariensis* Jum. & H.Perrier, *Elaeis guineensis* var. *microsperma* Welw., *Elaeis guineensis* var. *pisifera* A.Chev., *Elaeis guineensis* var. *repanda* A.Chev., *Elaeis guineensis* var. *rostrata* Becc., *Elaeis guineensis* var. *sempernigra* A.Chev., *Elaeis guineensis* var. *spectabilis* A.Chev., *Elaeis macrophylla* A.Chev., nom. nud., *Elaeis madagascariensis* (Jum. & H.Perrier) Becc., *Elaeis melanococca* Gaertn., *Elaeis melanococca* var. *semicircularis* Oerst., *Elaeis nigrescens* (A.Chev.) Prain, nom. inval., *Elaeis virescens* (A.Chev.) Prain, *Palma oleosa* Mill.

Family

Arecaceae

Common/English Names

African Oil Palm, Macaw Fat, Oil Palm, Palm Oil

Vernacular Names

Arabic: Nakhlet Ez Zayt;
Brazil: Caiaué, Dendê (Portuguese);
Burmese: Si-Ohn, So-Htan;
Chinese: You Zong;
Cook Islands: Nū Tāmara (Maori);
Czech: Olejnice Guinejská;
Danish: Oliepalme;
Dutch: Afrikaansche Oliepalm;
Eastonian: Aafrika Õlipalm;
Finnish: Öljypalmu;
French: Palmier À Huile, Palmier À Huile d'Afrique;
German: Afrikanische Ölpalme, Ölpalme;
Indonesia: Kelapa Sawit, Kalapa Ciung, Lalpa, Omyak (Sudanese);

Italian: Palma Avoira, Palma Da Olio, Palma Oleaginosa Africana;
Japanese: Abura Yashi;
Khmer: Doong Preeng;
Malaysia: Kelapa Sawit;
Pohnpeian: Apwiraiasi;
Portuguese: Dendê, Dendezeiro, Dihoho, Palmera Dendém, Palmeira Andim, Palmeira Do Azeite, Palmeira Do Dendê;
Russian: Gvineiskaia, Maslichnaia Pal'ma, Pal'ma Maslichnaia;
Slovak: Oljna Palma;
Spanish: Corojo De Guinea, Palma Africana, Palmera De Aceite, Palma Oleaginosa Africana;
Swahili: Mchikichi, Miwesi, Mjenga;
Swedish: Oljepalm;
Thai: Ma Phraao Hua Ling, Maak Man, Paam Namman;
Vietnamese: Co Dau, Dua Dau.

Origin/Distribution

Oil Palm is indigenous to the tropical rain forest region of West and Central Africa in a region about 200–300 km wide along the coastal belt between Senegal and north Angola, 12°N and 10°S latitudes. It occurs wild in riverine forests or in freshwater swamps. The palm has now been introduced and cultivated throughout the tropics between 16°N and °S latitudes.

Agroecology

Oil palm is heliophilous, grows best under bright sunshine and with unrestricted water availability. It thrives in areas with well distributed rainfall of 1,780–2,280 mm rainfall per year and mean maximum temperature of 30–32°C and mean minimum of 21–24°C. Seedling growth is arrested below 15°C. Oil palm is best developed on lowlands, with 2–4 month dry period. It can tolerate temporary flooding or a fluctuating water table, as might be found along rivers. It thrives best on coastal marine alluvial clays, soils of volcanic origin, acid sands and other coastal alluviums with pH of 4–6 and also on latosols and peat soils. Good soils should have soil depth of >1.5 m, soil availability at field capacity 1–1.5 mm/cm depth, organic carbon >1.5% in the top soil, cation exchange capacity >100 mmol/kg and well drained. Waterlogged, highly lateritic, extremely sandy, stony or highly peaty soils should be avoided.

Edible Plant Parts and Uses

In Africa, palm wine is obtained by tapping the sap from the unopened male inflorescences or the stem just below the apex of standing or felled palms. The sap contains about 4.3 g/100 ml of sucrose and 3.4 g/100 ml of glucose. The sap ferments quickly, and is an important source of Vitamin B complex in diet of people of West Africa. The palm heart or palm cabbage consisting of the soft tissues of the undeveloped leaves around the apical bud is eaten as vegetables (Plate 4).

Two kinds of oil are obtained from the oil palm fruit – palm oil and palm kernel oil. The use of crude, unrefined palm oil dates back 5,000 years (Sarmidi et al. 2009). Today crude, unrefined palm oil is still being used in certain villages in Africa. Palm oil is extracted from the fleshy mesocarp of the fruit and has balanced ratio of unsaturated and saturated fatty acids. It contains 40% oleic acid (monounsaturated fatty acid), 10% linoleic acid (polyunsaturated fatty acid), 45% palmitic acid and 5% stearic acid (saturated fatty acid). This composition results in an edible oil that is suitable for use in a myriad of food applications – cooking oils, margarine, hydrogenated vegetable oil or ghee (vanaspati), shortenings, frying fats, bakery and biscuits, potato crisps, chips, instant noodles, pastry, confectionery, ice cream, creamers, chocolate, instant soup and other snack foods. It varies from light yellow to orange-red in colour, and melts from 25°C to 50°C. For edible fat manufacture, the oil is bleached. The red crude palm oil is an essential diet of the West Africans.

Palm Oil

The average Malaysian virgin palm oil does not contain more than 3.5% free fatty acids, and these are readily removed by physical refining to produce refined, bleached, and deodorized (RBD) palm oil (Ong and Goh 2002). Dry fractionation of this oil yields palm olein (liquid fraction) and palm stearin (solid fraction). From these the refining industry produces 14 processed palm oils for various applications based on physical processes without the use of solvents, in contrast to seed oils such as soybean oil. The multiple applications of these oil palm based products in the food industry have been reported by Berger (1996), Pantzaris (2000), Pantzaris and Ahmad (2001), Ong and Goh (2002), and Sarmidi et al. (2009). Palm oil is widely used in shortenings, frying fats, ice cream, biscuits, cookies, crackers, cake and dessert mix, rending or curry mixes, sardines, baked beans, break-fast cereals, shrimp-paste powder, bouillon, peanut butter, beverages, instant noodles; and moderately used in margarines, and icings and less so in non-dairy creamers. Palm olein is widely used in shortenings, margarines, frying fats, instant noodles; moderately in cooking oil (warm climate), and less so in biscuits, crackers. Palm stearin (soft) is widely used in shortenings, margarines, fats and coatings; moderately in frying fats, biscuits, cookies, crackers, cake mix, instant noodles and less so in icings and non-dairy creamers. Palm stearin (hard) is used moderately in shortenings, and less so in margarines, biscuits, cookies, crackers, cake mix. Hardened palm oil is widely used in frying fats and cooking oil in warm climate. Double-fractionated palm olein is widely used in frying fats and cooking oil. Palm mid fraction is moderately used in icings and less so in shortenings, margarines, frying fats, fats and coatings. Palm kernel oil is widely used in margarines, fats and coatings, crackers and non-dairy creamers, moderately in biscuits and less so in shortenings. In addition to those listed, palm oil-based applications can be extended to the following: *trans*-free margarine formulation, expanded and extruded snacks, reduced fat spreads, nuts (fried), bakery fat, doughnuts, cocoa butter replacers and equivalents, milk with palm fat replacing milk fat, manufacture of food emulsifiers, vanaspati (a vegetable ghee), palm-based processed cheese, pastry, yogurt with palm fat replacing milk fat, flour confectionery, soup mixes, sugar confectionery, salad dressing, peanut butter, cooking sauces, frying chips (french fries) and ready-made meals.

Palm Olein

Palm olein is the liquid fraction obtained by fractionation of palm oil after crystallization at controlled temperatures. Palm oil and palm olein (liquid fraction) have similar compositions as monounsaturated oil. The physical characteristics of palm olein differ from those of palm oil. The clear olein fraction is preferred over the red crude palm oil in Southeast Asia. Palm olein is widely used as frying oil. Its popularity is attributable to its good resistance to oxidation and formation of breakdown products at frying temperatures and longer shelf life of finished products. In fact, palm olein is deemed as the gold standard in frying and is perhaps, the most widely used frying oil in the world. Palm olein also blends well with other popular traditional vegetable oils used in many parts of the world; earning a nickname of 'blending partner'. For example, in Japan, refined palm olein is blended with rice bran. In Malaysia, it is blended with groundnut oil. Palm olein has been found very suitable for use in infant food formulae when blended with other vegetable oils (Nelson et al. 1996). The low melting olein contains 10–15% palmitic acid in the 2-position of the glycerol chain which impart high digestibility of the product. RBD (refined, bleached & deodorised) palm olein is used as salad oil. Another application of palm oil in food is the use of refined red palm oil in cooking (Sarmidi et al. 2009). Red palm oil or red palm oil olein not only supplies fatty acids essential for proper growth and development, but also it contains an assortment of vitamins particularly in vitamin E (tocopherols and tocotrienols), antioxidants – phenolics, carotenes and other phytonutrients such as sterols, squalenes, co-enzymes (coenzyme Q_{10}) important for good health (Choo et al. 1996). Red palm oil derives its name from its characteristic dark red color, which comes from carotenes such as α-carotene, β-carotene and lycopene. The deep red colour of red palm oil blends well with curries and chillies making the dishes more appealing and nutritious. Red palm oil provides a potential means of combating vitamin A deficiency which is prevalent in many Asian and African countries.

Palm Stearin

Palm stearin is the more solid fraction obtained by fractionation of palm oil after crystallization at controlled temperatures. It is thus a co-product of palm olein. It is always traded at a discount to palm oil and palm olein; making it a cost effective ingredient in several applications. The physical characteristics of palm stearin differ significantly from those of palm oil and it is available in a wider range of melting points and iodine values. Palm stearin is a very useful source of fully natural hard fat component for products such as shortening and pastry and bakery margarines.

Other Palm Oil Fractions

In addition to palm olein and stearin, there are a dozen other fractions, obtained from palm oil including various grades of double fractionated palm olein (aka superolein) and palm mid fractions. Where pourability and clarity can be issues for palm olein, especially in temperate countries, superolein finds uses as frying oil and cooking oil, usually in blends with seed oils. Palm mid fraction is commonly used as a highly versatile natural ingredient in the manufacture of tub margarine and in CBE (Cocoa Butter Equivalent) manufacture and confectioneries.

Palm Kernel Oil

Palm kernel oil (PKO) is extracted from the kernel of the endosperm, and contains about 50% oil. Similar to coconut oil, with high content of saturated acids, mainly lauric, it is solid at normal temperatures in temperate areas, and is nearly colourless, varying from white to slightly yellow. PKO is more unsaturated and so can be hydrogenated to a wider range of products for the food industry. PKO a non-drying oil and its hydrogenated and fractionated products are widely used either alone or in blends with other oils for the manufacture of cocoa butter substitutes and other confectionery fats, biscuit doughs and filling creams, cake icings, ice cream, mayonnaise, imitation whipping cream, filling creams, sharp-melting creaming, table margarines, coffee whiteners and many other food products (Pantzaris 2000; Pantzaris and Ahmad 2001). Palm kernel oil may be used as cooking oils, sometimes in blends with coconut oils. There is a growing trend to use palm kernel oil products as an ingredient in the production of non-hydrogenated *trans* fat free margarine. Special margarines which whip easily to produce light filling creams for use in layer cakes and other bakery products rely on high levels of lauric fat to confer good eating and handling properties. Other general purpose and table margarines often contain PKO or hydrogenated PKO to give them quick melting properties or compensate for the use of other higher melting fats or for the absence of *trans* fats which are also sharp melting. PKO and CNO are the best fats for non-dairy ice cream because of their combination of high SFC at about $0°C$, low melting point and perfectly bland taste PKO is also used in the manufacture of medium chain triglycerides which consists of $C8 - C10$ chain lengths, imparting low viscosity and higher oxidative stability. A certain amount of fat is essential in human diets to carry the fat soluble vitamins and synthesize the prostaglandins, etc., and medium chain triglycerides meet the first of these functions. Most substitute chocolates for biscuit coating and cakes are made from Hydrogenated PKO products with and without interesterification Most of the leading international brands of non-dairy creamers (coffee whiteners) are made with HPKO. Pure HPKO, PKS and their blends are the standard fats used for imitation whipping cream.

Likewise Palm Kernel Olein and Palm Kernel Stearin can be obtained by fractionation of palm kernel oil after crystallization at controlled temperatures. Palm kernel olein is the liquid component of palm kernel oil obtained from fractionation.

Palm Kernel Stearin

Palm kernel stearin is the more solid fraction of palm kernel oil obtained from fractionation. Palm kernel oil, palm kernel olein and palm kernel stearin find uses in margarine, confectioneries, coffee whitener, filled milk, biscuit cream and coating fats; with little or no further processing. Palm kernel stearin is widely used to substitute for the more expensive cocoa butter in many of its traditional applications. In some instances, particularly when hydrogenated, palm kernel stearin exhibits performance superior to that of cocoa butter. The physical properties of Palm kernel stearin resemble particularly closely those of cocoa butter and it is generally acknowledged that the best type of CBS (Cocoa butter substitute) as opposed to CBE, are made from this fat. Substantial quantities of PKO are therefore fractionated in Europe, USA and Malaysia for this purpose. Hydrogenated PKS of SMP (Slip Melting Point) about 35°C has even higher SFC (solid fat content) at 20°C than cocoa butter, it gives higher contraction on cooling and tolerates better the softening effect of full cream milk powder. These properties make it especially suitable for hollow moulded chocolate confectionery such as Easter eggs, moulded bars and the finest biscuit creams. PKS is used in blends with HPKO for manufacture of imitation whipping creams.

Palm Kernel Olein

Even lower cost coating fats and filling creams are made from hydrogenated PKOo (Palm kernel oil olein). The fats used for toffees and caramels are usually based on hydrogenated PKO (HPKO) or hydrogenated PKOo (HPKOo), because of their sharp melting properties and long shelf-life. Incorporating some PKS in them gives even better eating quality. These two products are very similar to each other and essentially consist of boiled sugar and glucose, but usually some fat is also incorporated in their formulations in order to adjust their texture, hardness, chewiness and richness of taste PKOo are used to make chocolate for dipping ice cream.

In Japan, the principal uses of palm oil are in making margarine and shortening, and for deep frying instant noodles, tempura, and snack foods (Mori and Kaneda 1994). Corn oil, rice-bran oil, and rapeseed oil have traditionally been used for frying snack foods and rice crackers. Palm oil is now increasingly used with blends of these oils to improve the heat and oxidation stability and the price competitiveness of the products. Palm oil now enjoys a 22% share of all oils and fats used in preparing these foods. In addition, palm oil fractions such as palm olein, palm stearin, and palm midfraction are used increasingly in a variety of other food products, ranging from vegetable ghee and hard butter to make chocolate and ice cream. Palm oil contains 2-oleoyl 1,3-dipalmitin (POP), a symmetric triglyceride that can be separated by solvent fractionation, which is used to make hard butters for chocolate production. Palm olein or a mixture of POP and other palm oil fractions has melting properties similar to those of butter fat and can be used in ice cream and milk-product analogues. Super olein has been used in mayonnaise. Over the recent years, China has emerged as a major importer of Malaysian palm oil in the form of its "liquid" fraction, palm olein. It is now widely used in the Chinese food industry, particularly in the manufacture of instant noodles, snack foods, milk powder, margarine, and shortening (Fan and Chen 1994). Although palm olein is widely used throughout the tropics as a cooking oil in household kitchens, its high melting point makes it unsuitable for this purpose in the temperate and northern regions of China. However, this problem could be overcome by blending appropriate proportions of palm olein with other locally available vegetable oils.

Other Edible Palm Parts

The sap obtained by the tapping of the male inflorescence of oil palm is sweet when fresh and contains sugars and on fermentation is converted into palm wine (Eze and Ogan 1988; Lasekan et al. 2007). The sap ferments quickly, and is an important source of vitamin B complex in diet of people of West Africa. Oil palm wine is called *ngasi* in the Democratic Republic of Congo. Sugar can be produced from the sap. Gin is distilled from the fermented wine. Palm cabbage or palm heart, the soft core tissues around the apical bud in the crown, can be harvested and consumed as a vegetable delicacy.

Palm Oil Frying and Cooking Effects

Studies conducted by Che Man and Liu (1999) showed that RBDPO (refined, bleached and deodorized palm olein) was superior to SBO (soybean oil) in frying performance in terms of resistance to oxidation, changes in iodine value, anisidine value, foaming tendency, % polar component and polymer content. However, SBO performed better than RBDPO with respect to color, viscosity, free fatty acid content and smoke point. Sensory evaluation showed that potato chips fried in both RBDPO and SBO were equally acceptable by the panellists. Matthäus (2007) reported that palm oil and its products have a similar frying performance compared to the new so-called "high-oleic" oils. Today, palm oil and its products, mainly palm olein, belong to the most important oils used for the preparation of fried food. The reason is that the oil is relatively cheap, it is available in huge amounts, it has a high oxidative stability and results in high-quality and tasty foods. Thus, the oil meets the demands of the consumer and the producer of fried food.

More recent studies on the frying performance of palm oil and palm olein was reported by Ismail (2005). In the first study, about 4 tones of potato chips were continuously fried in palm olein 8 h a day and 5 days a week. The palm olein used (with proper management) performed well and was still in excellent condition and usable at the end of the trial. This was reflected in its low free fatty acid (FFA) content of around 0.23%, peroxide value of 4 meq/kg, anisidine value of 16, low polar and polymer contents of 10% and 2%, respectively, induction period (OSI) of 21 h and high content of tocopherols and tocotrienols of 530 ppm even after >1,900 h. In the second study in which an average 12 tones pre-fried frozen French fries were continuously fried daily for 5 days a week, palm oil performed excellently as reflected by its low FFA of 0.34%, food oil sensor reading of 1.1, low polar and polymer contents of 17% and 2.8%, respectively, over the 12 days of trial. In the third study in which palm shortening, palm oil and palm olein were simultaneously used to intermittently fry chicken parts in the laboratory simulating the conditions in fast food outlets, the three frying oils also performed very satisfactorily as reflected by their reasonably low FFA of <1%, smoke points of >180°C, and polar and polymer contents of <25% and <6%, respectively, after 5 days of consecutive frying. All the quality indicators did not exceed the maximum discard points for frying oils/fats in the three applications, while the fried food product was well accepted by the in-house train sensory panel using a-nine point hedonic score.

Al-Saqer et al. (2004) reported on the antioxidant vitamin E (tocopherols and tocotrienols) contents of two functional foods (pan bread and sugar-snap cookies), prepared using red palm olein (RPOL) and red palm shortening (RPS). The total vitamin E contents in the palm oils used in this study were significantly higher (ranging from 717.8 to 817.5 mg/kg fat) than those in the palm shortenings (ranging from 451.0 to 479.9 mg/kg fat). Tocotrienols were found to be the predominant compounds (about 81–84% of the total vitamin E) in all of the palm oil and palm shortening samples. The total tocopherol and total tocotrienol contents were significantly higher in bread made with control shortening (325.0 and 468.0 mg/kg fat) than in that made with a 75% replacement level of RPOL (192 and 407 mg/kg fat) or RPS (101.4 and 300 mg/kg fat). Similar trends were observed in the white bread and the brown bread made with RPOL and RPS. The tocotrienol contents in cookies made with varying levels of replacement RPS ranged from 229.1 to 317.8 mg/kg fat, compared with 379.4–672.3 mg/kg fat in cookies made with RPOL. The tocotrienols were found to be the predominant fraction in cookies made either with RPS (58–68%) or with RPOL (about 69–83% of the total vitamin E). Cookies, being higher in fat contents, would be better providers of these desirable phytochemicals and antioxidant vitamins than breads.

Effect of different cooking methods of crude palm oil like baking, seasoning, deep frying and shallow frying on retention of β-carotene was studied and it was observed that 70–88% of it was retained in the cooked foods (Manorama and Rukmini 1991). Repeated deep frying, using the oil five times consecutively, resulted in a total loss of β-carotene by the fourth frying stage itself and alteration of its organoleptic, physical and chemical properties. Hence, CPO may be suitable for single frying operations only or for preparations which involve a short heating time and completely take up the oil into the cooked product, e.g. 'Cake', 'Upma', 'Kichidi', 'Suji halwa' (Manorama and Rukmini 1991). Probably 'Suji halwa' could be selected as an ideal choice for vitamin A supplementation to vulnerable children because it was well accepted, and to be cheaper per RDA serving and it retained 70% of its β-carotene after cooking.

Palm Xylose and Xylooligosaccharides as Functional Food Ingredients

Recent studies clearly indicated the potential utilization of oil palm frond, an abundantly available lignocellulosic biomass, for the production of xylose and xylooligosaccharides which can serve as functional food ingredients (Sabiha-Hanim et al. 2011). In the study, oil palm frond fibres were subjected to an autohydrolysis treatment using an autoclave, operated at 121 °C for 20–80 min, to facilitate the separation of hemicelluloses. The hemicellulose-rich solution (autohydrolysate) was subjected to further hydrolysis with 4-16U of mixed *Trichoderma viride* endo-(1,4)-β-xylanases per 100 mg of autohydrolysate. Autoclaving of palm fronds at 121 °C for 1 hour recovered 75% of the solid residue, containing 57.9% cellulose and 18% Klason lignin, and an autohydrolysate containing 14.94% hemicellulose, with a fractionation efficiency of 49.20%. Subsequent enzymatic hydrolysis of the autohydrolysate with 8U of endoxylanase at 40 °C for 24 h produced a solution containing 17.5% xylooligosaccharides and 25.6% xylose.

Botany

A tall perennial, armed solitary palm, 8.5–30 m tall, erect, stout, trunks ringed; monoecious, male and female flowers in separate clusters, but on same tree with dead leaf sheaths from base to the crown. Crown dark green crown with a skirt of dead leaves and 60 live leaves in 8 spirals; trunk 30 cm in diameter, adherent leaf-bases; petioles 1.3–2.3 m long, 12.5–20 cm wide, saw-toothed, broadened at base, fibrous, green. Leaves steeply arching blade pinnate, 3.3–5 m long, with 100–150 pairs of leaflets in 4 ranks; leaflets 60–120 cm long, 3.5–5 cm broad; central nerve very strong, especially at base, green on both surfaces. Inflorescence dense with numerous stout spikes tipped by short spines. Flowers are trimerous. Staminate flowers on short furry branches 10–15 cm long, set close to trunk on short pedicels, the perianth surrounds an androecium, usually consisting of two whorls of three stamens with filaments connate, enclosing a rudimentary gynoecium. Female flowers with the perianth surrounding a sterile androecium of usually six rudimentary stamens (staminodes) enclosing a syncarpous gynoecium composed of three carpels and consequently fruits in large clusters of 200–300, close to trunk on short heavy pedicels, About 5 female inflorescences are produced per year; each inflorescence weighing about 8 kg, the fruits weighing about 3.5 g each. Fruit ovoid-oblong to 3.5 cm long and about 2 cm wide, deep violet (Plate 2) ripening to orange red (Plates 1 & 3), oval broadly triangular, 1.5 in long tipped by persistent style with 3 recurved stigma. The fruit weighs from 6 to 20 g, is made up of an outer skin (the exocarp), a fibrous, oil-laden pulp (mesocarp), a central nut consisting of a hard stony shell (endocarp) with 3 pores, and the kernel (embryo) with oily endosperm.

Plate 1 Mature oil palm bunch

Plate 2 Immature oil palm bunch

Plate 3 Close-up of harvested mature oil palm bunch

Plate 4 Oil palm heart pieces on sale in a local market

Nutritive/Medicinal Properties

An ideal oil palm fruit bunch composition was described as having a bunch weight of 23–26 kg, fruit/bunch 60–65%, oil/bunch 21–23%, kernel/bunch 5–7%, mesocarp/bunch 44–46%, mesocarp/ fruit 71–76%, kernel/fruit 21–23%, shell per fruit 10–11% (Poku 2002).

Nutritional composition of African oil palm fruit, *Elaeis guineensis* (per 100 g) was reported as follows: energy 746 cal., protein 2.2%, fat 81.9%, carbohydrate 14.6%, fibre 3.8%, calcium 136.1 mg, phosphorus 61.1 mg, iron 5.6 mg, carotene 50,688.6 μg, thiamine 0.35 mg, niacin 1.81 mg, riboflavin 0.17 mg, ascorbic acid 12.5 mg (Atchley 1984). Tee et al. (1997) reported the following nutrient composition of crude palm oil (per 100 g): energy 900 kcal, fat 100 g, carotenes 57,100 μg, retinol equivalent 9,520 μg, vitamin B1-0.1 mg; and oil pal olein as: energy 900 kcal, fat 100 g, calcium 4 mg, sodium 5 mg and niacin 4 mg.

Analysis of proximate composition of the pulp and kernels of the *Eliaes guineensis* palm fruits grown in the Northeast region of Brazil revealed the lipid content of the dried pulp and kernels to be 73.2% and 32.6%, respectively (Bora et al. 2003). Hexane extracted oils from the pulp and kernels yielded similar refractive indices, specific gravity but different peroxide, acid, iodine and saponification values. Gas chromatographic analysis revealed the presence of 24 and 18 fatty acids in pulp and kernel oils, respectively. The principal saturated acid of the pulp oil was found to be palmitic acid (36.9% of the total), and lauric acid (53.3%) for kernel oil. Oleic acid was the predominant monounsaturated fatty acid in both the oils though its concentration in the pulp and kernel oils was 45.29% and 5.5%, respectively. In relation to the essential amino acids, pulp proteins presented a better profile than the kernel proteins. In comparison to the FAO reference protein, the pulp proteins were found to be deficient in methionine, lysine and threonine (16.8%, 51.6% and 93.5% of FAO reference protein) but contained leucine, valine, isoleucine and phenylalanine in optimal concentrations. With exception to phenylalanine and valine (102.2% and 111.4% of reference protein, respectively), the kernel proteins were deficient in all other essential amino acids.

Determination of the antioxidants as total carotenoids (vitamin A) and vitamin E showed that vitamin A content in *E. pisifera, E. dura* and *E. tenera* were 8677.17, 8927.65 and 880.80 ug/g respectively, while vitamin E obtained for the species were 476.88, 443.52 and 181.92 mg/100 g for Pisifera, Dura and Tenera species respectively (Njoku et al. 2010). The protein content in the three species were 2.21, 2.30 and 2.70% for Pisifera, Dura and Tenera respectively while the iodine values were 44.38, 46.93 and 44.05 for Pisifera, Dura and Tenera respectively. The acid value were 0.65, 0.67 and 0.57 mg kOH/g for Pisifera, Dura and Tenera respectively while the saponification values were 191.45, 185.77 and 188.48 mg KoH/g for Pisifera, Dura and Tenera respectively. Also the percentage moisture content were 0.05, 0.04 and 0.05 for the three species in same order above. The dietary magnesium for the Pisifera, Dura and Tenera were 0.95, 1.13 and 0.37 mg/dm3 respectively and 0.08, 0.24 and 0.05 mg/dm3 for dietary zinc. The dietary calcium level were 0.46, 0.41 and 0.34 mg/dm3 and the potassium 0.46, 0.39 and 0.48 mg/dm3 for Pisifera, Dura and Tenera respectively. The iron concentration levels were 38.30, 67.70 and 78.30 mg/dm3 for the three in the same order above.

Oil Palm Wine

The pure unfermented sap from oil palm was found to have the following sugar composition(w/v): 9.59–10.59% sucrose, 0.13–0.73% of glucose and fructose and 0.13–0.35% raffinose (Eze and Ogan 1988). During fermentation of the sap to produce wine sucrose steadily decreased, fructose attained a peak at 1.51% and thereafter declined, while glucose and raffinose remained continuously low; all sugars disappeared beyond the 33 rd hour. pH decreased from pH 6.60 at zero time and stabilized at pH 3.30 after 48 h, while titratable acidity increased continuously up until the 96th hour. A total of 41 compounds were identified, 32 of them previously unknown in palm wine (*Elaeis guineensis*) (Lasekan et al. 2007). From these, a total of 13 compounds were quantified and the 13 key odorants revealed that the earthy-smelling 3-isobutyl-2-methoxypyrazine, the buttery-smelling acetoin, the fruity compounds ethyl hexanoate, 3-methylbutyl acetate and the popcorn-like-smelling 2-acetyl-1-pyrroline were likely to be important odorants of palm wine, with 3-isobutyl-2-methoxypyrazine, acetoin, and 2-acetyl-1-pyrroline being reported here for the first time as aroma constituents of palm wine.

Palm Oil

Two types of oil are obtained from the fruit, crude palm oil (CPO) from the fibrous pulp mesocarp, and palm kernel oil (PKO) and palm kernel cake obtained from the seed (kernel). The mesocarp was found to contain about 49% palm oil and the kernel about 50% palm kernel oil (Chen et al. 2007). Both oils CPO and PKO were found to have different chemical and physical composition and different applications. Palm oil and palm kernel oil were found to compose of fatty acids, esterified with glycerol just like any ordinary fat. Both were high in saturated fatty acids, about 50% and 80%, respectively. The oil palm imparted its name to the 16-carbon saturated fatty acid palmitic acid found in palm oil; monounsaturated oleic acid was also found to be a constituent of palm oil while palm kernel oil contained mainly lauric acid (C12:0). Most of the fatty acids in CPO were C16 and higher, while in PKO, they were C14 and lower (Chen et al. 2007; Pantzaris and Ahmad 2001). Interesterification catalysis of CPO, resulted in more medium and long chain triacylglycerols (TAG), with MMM/OLL, MMP, OOO and PPP (M, myristic acid; O, oleic acid; L, linoleic acid; P, palmitic acid) increasing in concentration. In contrast, catalysis of PKO resulted in the formation of more short and medium chain TAG, with LaLaC and LaMC (La, lauric acid; M, myristic acid; C, capric acid) in significant increments (Chen et al. 2007). PKO was found to be similar to coconut oil in fatty acid composition and properties. CPO had an iodine value (IV) 50 minimum, while PKO had 21 maximum.

Palm oil was found to consist mainly of glycerides made up of a range of fatty acids (Choo 1994). Triglycerides constituted the major component, with small proportions of diglycerides and monoglycerides. Palm oil also contained other minor constituents, such as free fatty acids and non-glyceride components. This composition was found to determine the oil's chemical and physical characteristics. About 50% of the fatty acids were saturated, 40% mono-unsaturated, and 10% polyunsaturated. It contained adequate amounts of n-6, 18:2 essential fatty acid. In its content of monounsaturated 18:1 acid, palm oil was found to be similar to olive oil. Tan and Oh (1981) reported the following fatty acid composition of Malaysian palm oil: C 12:0 (lauric) 0.1–1.0% mean 0.2%; C14:0 (myristic) 0.9–1.5%, mean 1.1%, C16:0 (palmitic), 41.86–46.8%, mean 44%; C16:1 (palmitoleic) 0.1–0.3%, mean 0.1%; C 18:0 (stearic) 4.2–5.1%, mean 4.5%; C 18:1 (oleic acid) 37.3–40.8%, mean 39.2%; C18:2 (linoleic) 9.1–11.0%, mean 10.1%; C 18:3 (α-linolenic) 0–0.6%, mean 0.4%, C 20:0 (arachidic) 0.2–0.7%, mean 0.4%. Cottrell (1991) reported palm oil to contain several saturated and unsaturated fats in the forms of glyceryl laurate (0.1%, saturated), myristate (0.1%, saturated), palmitate (44%, saturated), stearate (5%, saturated), oleate (39%, monounsaturated), linoleate (10%, polyunsaturated), and linolenate (0.3%, polyunsaturated). Ong and Goh (2002) reported virgin palm oil (crude palm oil, CPO) to contain mainly triacylglycerols (more than 95%), diacylglycerols, and free fatty acids (approximately 3.5%) and minor components (approximately 1%). There are more than 33 triacylglycerols in CPO with the major triglyceride as the symmetrical POP, POO, POL, and PLP glycerides where P=palmitic, o=oleic, l=limoleic. CPO was also found to contain 10–80 ppm ubiquinone-10 (UQ_{10}) (Coenzyme Q_{10}) and about 5 ppm of ubiquinone-9(UQ_9) (Hamid et al. 1995). Coenzyme Q_{10} is a component of the electron transport chain and participates in aerobic cellular respiration, generating energy in the form of ATP. Ninety-five percent of the human body's energy is generated this way (Ernster and Dallner 1995). The importance of coenzyme Q_{10} was confirmed when it afforded promising results when administered to patients with cardio-vascular or heart diseases (Lenaz 1985; Yamamura 1985). In addition, it was shown to be effective in the prevention of lipid peroxidation and oxidative damage in haemoglobin (Lenaz 1985; Beyer 1989; Ernster and Dallner 1995).

Another analysis by (Pantzaris and Ahmad 2001) reported palm oil (PO) to have 0.2% C12 (lauric acid), 1.1% C14 (myristic), 44.1% C16 (palmitic), 4.4% C18 (stearic), 39.0% C18:1 (oleic acid), 10.6% C18:2 (linoleic) and 0.75% other fatty acids.

Goh et al. (1985) found crude palm oil to contain about 1% of minor components including carotenoids, tocopherols, sterols, triterpene alcohols, phospholipids, glycolipids and terpenic and paraffinic hydrocarbons. Crude palm oil was also reported to contain approximately 1% of minor components: carotenoids (500–700 ppm) (Jacobsberg 1974), vitamin E (tocopherols and tocotrienols) (600–1,000 ppm) (Tan and Oh 1981), sterols (326–527 ppm) (Rossell et al. 1983), phospholipids (5–130 ppm) (Goh et al. 1982, 1984), glycolipids, triterpene alcohol (40–80 ppm) (Itoh et al. 1973a, b), methyl sterols (40–80 ppm) (Itoh et al. 1973a), squalene 200–500 ppm (Goh and Gee 1984), aliphatic alcohols (100–200 ppm) (Jacobsberg 1974), aliphatic hydrocarbons (50 ppm) (Goh and Gee 1984) and co-enzyme Q10 (10–80 ppm) (Ng et al. 2006). Ooi et al. (1994) found the carotenoid fraction of palm oil contained more than 80,000 ppm carotenoids with α- and β-carotenes as the predominant components. The content of sterols in palm oil was found to be about 0.03% of its total composition (Sambanthamurthi et al. 2000). Cholesterol (2.2–6.7%), Δ5-avenasterol (0–2.8%) Δ7-stigmasterol (0–2.8%) and Δ7-avenasterol (0–4%) were also found in the sterol fraction (326–627 mg/kg) of palm oil. The major components of palm oil phytosterols found were β-sitosterol, campesterol, stigmasterol and cholesterol (Ab Gapor et al. 1985; Lau et al. 2005). These compounds play a significant role in stabilizing palm oil as well as in the refinability of the oil. Extraction of some of these high-value compounds has provided several nutraceutical products.

Palm oil can be physically refined and fractionated into various fractions, ranging from very hard palm stearin with iodine values below 10 to palm superolein with iodine values as high as 72 (Gee 2007). Palm mid fractions consisting of symmetrical triacylglycerols provide sharp-melting fats for niche applications. The wide range of palm oil fractions provides versatility for different food applications, with the additional advantage that natural palm oil is trans free and genetic modification free. Palm oil and its fractions are widely used for direct blending with other oils or are interesterified with other oils to meet the trans-free fat requirements of the food industry. The sn-2 position of palm oil triacylglycerols is mainly esterified with oleic and linoleic acid. This provides better bioavailability of oleic acid as monounsaturated fatty acid and linoleic acid as an essential fatty acid, as compared to oils or fats with similar composition but with randomized fatty acid distribution. Crude palm oil also contains highly valuable minor components, including carotenoids and tocotrienols, which are potent fat-soluble antioxidants. Recent research findings on potential chemo-preventive and chemotherapeutic roles of tocotrienols are extremely encouraging.

At least five groups of chiral compounds were found in palm oil, namely asymmetrical acylglycerols, phospholipids, tocols, carotenoids and sterols (Loh and Choo 2003). Chiral compound are asymmetric molecules i.e. molecule that lacks all symmetry and chirality has an effect on the biological activity of the compound. Asymmetrical acylglycerols comprised monoacylglycerols – fatty acid monoesters of glycerol, diacylglycerols – fatty acid diesters of glycerol and triacylglycerols – fatty acid triesters of glycerol. The content of monoacylglycerols in crude palm oil was less than 1%. They existed in two isomeric forms as α- and β-isomers. Diacylglycerols comprised 2–7% of the crude palm oil composition. The arrangement of two different fatty acid moieties to any two of the three carbon atoms (sn-1 and sn-2 or sn-1 and sn-3) resulted in two stereoisomers as α, β- and α, α'-isomers. Each stereoisomer existed as a pair of enantiomers. Triacylglycerols were found to form the major component in crude palm oil (more than 90%). The acylation of three or two different fatty acid moieties to the carbon atoms of glycerol resulted in the formation of several stereoisomers respectively. Considering only the major fatty acids in crude palm oil, i.e. palmitic acid (C16:0), stearic acid (C18:0), oleic acid (C18:1) and linoleic acid (C18:2), a total of 25 triglycerides (including 15 chiral triglycerides) were documented as the major constituents in palm oil. The major chiral triglycerides in palm oil were POO, POS, PPO, PLO, POL, SOO, SLP and PPS. The triacylation of glycerol by three dissimilar fatty acids was found to produce achiral and symmetrical triglycerides.

Phospholipids (lipids containing phosphorus) were found to be important metabolic intermediates. All natural representatives of this group were found to be diacyl derivatives of the sn-glycerol-3-phosphate. The major chiral phospholipids of palm oil included phosphatidylcholine (PtdCho), phosphatidylethanolamine (PtdEtn), phosphatidylinositol (PtdIns), phosphatidylglycerol (PtdGro), phosphatidic acid (PtdA) and phosphatidylserine (PtdSer) (Loh and Choo 2003).

Tocols in palm oil were found to consist of two series of compounds; tocopherols and tocotrienols. In palm oil, α-tocopherol, γ-tocopherol, α-tocotrienol, γ-tocotrienol and δ-tocotrienol were found (Rossell et al. 1983). However, only the α-tocopherol and the γ-tocopherol were chiral. Most of the commercial vitamin E capsules derived from δ-α-tocopheryl acetate were found to be 58% more biologically active than the l-form. Thus chirality factor played a vital role in the performance of vitamin E therapy. The characteristic reddish colour of crude palm oil was caused by carotenoids. α- and β-carotenes are the major components, with δ-carotene, γ-carotene, xanthophylls and lycopene present in minor quantities The chiral carotenoids in palm oil were α-carotene and δ-carotene. Although β-carotene did not possess chiral value, it was found to be the most important pro-vitamin A. The chirality features of these compounds were found to also act as natural hypocholesterolic agents as well as intermediates to steroids (Goh et al. 1985).

The major phospholipid components of the mesocarp oil of *Elaeis guineensis* were found to be phosphatidylcholine (PC), phosphatidylethanolamine (PE), phosphatidylinositol (PI) and phosphatidylglycerol (PG) (Goh et al. 1982). Minor components were phosphatidic acid (PA), diphosphatidylglycerol (DPG) and lysophosphatidylethanolamine (LPE), and traces of lysophosphatidylcholine (LPC) and phosphatidylserine (PS) are detectable. An artifact from enzymatic transphosphatidylation in methanolic solvents was isolated and characterized as phosphatidylmethanol (PM). Phospholipids were only present at low levels (20~80 ppm) in commercial crude palm oil and they usually accounted for a minor part of the total elemental phosphorus of the oil. It was deemed desirable to have low levels of phospholipids in the oil to obtain better oxidative stability and bleaching properties. The role played by phospholipids was frequently misunderstood because they could act in two ways, i.e. as an antioxidant synergist and a surface active agent to disperse impurities in oil (Goh et al. 1982, 1984). Among the sterols, cholesterol constituted too small a percentage to be of much concern. Sterols, triterpenoids and terpenoid hydrocarbons could be also potentially useful side products should recovery technology became available.

The nutritionally and health important components such as carotenes and tocopherols also improve stability of the oil. Crude Palm Oil is the richest natural source of tocotrienols, which is a member of the vitamin E family comprising of tocotrienols and tocopherols. Tocopherols, being natural antioxidants, need to be carefully preserved during milling, refining, fractionation and modification of palm oils. The accumulation of tocopherols in the palm fatty acid distillate promises to provide a new source for the recovery of this valuable substance.

Red palm oil or red palm olein (RPo) is an unconventional oil produced from crude palm oil through a new process involving deacidification and deodorization using molecular distillation under milder conditions (Ooi et al. 1966). Red palm olein is the richest dietary source of provitamin A carotenes. It was found to have the same 11 types of carotenoids found in CPO with α-carotene and β-carotene constituting about 80% of the total carotenoids (Tay and Choo 2000). It had a total carotenoid content of 665 µg/g made up of phytoene (0.61–0.68%), phytofluene (0.15–0.17%), β-carotene (40–42%), α-carotene (40.6–41.9%), cis-α-carotene (9–11.4%), ζ-carotene (0.5–0.72%), γ-carotene (0.45–1.07%), δ-carotene (0.72–0.83%), neurosporene (0.11–0.26%), β-zeacarotene (1.17–1.33%), α-zeacarotene (0.5–0.56%), lycopene (0.86–1.07%). It also possessed xanthophylls: dehydro-retinal, ζ-carotendione and β-caroten-5,6-epoxide and vitamin E content of 717–863 ppm comprising of α-tocopherol (19%), α-tocotrienol (29%), γ-tocotrienol (41%) and δ-tocotrienol (10%). The sterol content ranged from 717 to 863 ppm consisting of β-sitosterol (59%), campesterol (22%), stigmasterol (17%) and cholesterol (<2.6%). The fatty acid composition comprised 46.7% monounsaturated, 12.8% polyunsaturated and 40.5% saturated lipids. The major saturated fats found were mainly palmitic (C16:0) 36.6% and stearic (C18:0) 3.7%, with very low lauric acid C12:0 (0.2%), and myristic C14:0 content of 1.10%. The monounsaturated fatty acid was C18: 1 oleic (46.7%) and polyunsaturated fatty acid of C18:2, linoleic acid (12.8%) In addition, red palm olein was found to have ubiquinone (UQ10) 18–25 ppm.

RBDO (Refined, Bleached, and Deodorized Palm Oil) was found to have 25%, a-tocopherol, 29% a-tocotrienol, 36% lambda-tocotrienol, 10% δ-tocotrienol and 5,515–800 ppm of vitamin E (Ab Gapor 1990).

Palm Kernel Oil

Palm kernel oil (PKO) was found to have 0.3% C6 (caproic), 4.2% C8 (caprylic), 3.7% C10 (capric), 48.7% C12 (lauric), 15.6% C14 (myristic), 7.5% C16 (palmitic), 1.8% C18 (stearic), 14.8% C18:1 (oleic acid), 2.6% C18:2 (linoleic) and 0.1% other fatty acids (Pantzaris and Ahmad 2001). The major fatty acids in PKO were C12 (lauric acid) about

48%, C14 (myristic acid) about 16% and C18:1 (oleic acid) about 15% No other fatty acid was present at more than 10% and this heavy preponderance of lauric acid, endowed PKO and CNO (Coconut Nut Oil), their sharp melting properties, meaning hardness at room temperature combined with a low melting point (Pantzaris 2000). This outstanding property of lauric oils was found to determine their use in the edible field and justified their usually higher price compared with most other oils. Because of their low unsaturation, the lauric oils were also very stable to oxidation. Semisolid in temperate climates, PKO could be fractionated into solid and liquid fractions known as stearin and olein respectively. These were then physically refined, bleached and deodorized or chemically neutralized, bleached and deodorized to give the RBD (Refined, Bleached & Deodorised) and NBD (Neutralised, Bleached & Deodorised) grades used in the food industry. The fractionation could be done either before or after the refining, according to circumstances.

Palm kernel stearin was found to have the following fatty acid profile (Pantzaris and Ahmad 2001): 2% C8 (caprylic), 3% C10 (capric), 56.5% C12 (lauric), 22.5% C14 (myristic), 8% C16 (palmitic), 2% C18 (stearic), 6% C18:1 (oleic acid), 1% C18:2 (linoleic), and iodine value 5.8–8.0. Palm kernel olein had the following fatty acid composition (Pantzaris and Ahmad 2001): 0.5% C6 (caproic), 4.5% C8 (caprylic), 4% C10 (capric), 44.5% C12 (lauric), 14% C14 (myristic), 9% C16 (palmitic), 2.5% C18 (stearic), 18% C18:1 (oleic acid), 3% C18:2 (linoleic), and iodine value 20.6–26.0.

Palm-Pressed Fibre

Fresh palm-pressed mesocarp fibre (PPF) oil was found to be a rich natural source of neutraceuticals components with high value addition of bouruet of vitamin E (tocopherol and tocotrienols), carotenes, squalene and phytosterols (Choo et al. 1996; Lau et al. 2006; Lau et al. 2007). Residual oil (5–6% on dry basis) extracted from palm press fibers contains a significant quantity of carotenoids (4,000–6,000 ppm), vitamin E (2,400–3,500 ppm), and sterols (4,500–8,500 ppm). The major identified carotenoids were α-carotene (19.5%), β-carotene (31.0%), lycopene (14.1%), and phytoene (11.9%). In terms of vitamin E, α-tocopherol constituted about 61% of the total vitamin E present, the rest being tocotrienols (α-, γ-, and δ-) (Choo et al. 1996). PFF oil was found to have higher total carotene than CPO (800–1,000 mg/kg) (Lau et al. 2007). Vitamin E in PFF oil was dominated by 60–70% of a-tocopherol (Lau et al. 2006), N-hexane extract of PFF oil consisted of 62.9% a-tocopherol, 14.2% α-tocotrienol, 20.9% γ-tocotrienol, 2.0% δ–tocotrienol with 1,981 mg/Kg total vitamin E (Lau et al. 2007). Sterols and squalene were higher in PFF oil than in CPO (Choo et al. 1996; Goh et al. 1985; Lau et al. 2006). The major sterols found were 70% β-sitosterol, 19% stigmasterol, 8% campesterol, 3% cholesterol, and total sterol 10,877 mg/kg and squalene 9,690 mg/kg (Lau et al. 2007). Palm pressed fibres also contained extractable phytochemicals such as sterols, vitamin E, carotenoids, phospholipids, squalene and phenolics (Tan et al. 2007b).

The wax esters isolated from the exocarp extract of oil palm fruit *Elaeis guineensis* were found to compose of mainly even numbered homologous esters between C40 and C60 (Goh et al. 1988). Methanolysis revealed that the fatty acid and fatty alcohols of these wax esters had carbon chain length ranging from C16 to C28 and C16 to C34, respectively. Two pentacyclic triterpene methyl ethers, viz. cylindrin and its $\Delta 12$-isomer, were isolated, the former being the dominant triterpenoid component. Traces of wax esters (5 ppm) in crude palm oil were detected.

Palm Vitamin E Content and Activity

Crude palm oil is rich in vitamin E (tocols) containing 600–1,000 ppm of tocols in the form of tocopherols and tocotrienols (Ng et al. 2004a). The vitamin E in palm oil was found to consist of various isomers of tocopherols and tocotrienols viz. α-tocopherol (α-T), α-tocotrienol, γ-tocopherol, γ-tocotrienol, and δ-tocotrienol (Ng et al. 2004b). α-T constituted about 3–4% (40 ppm) of vitamin E in crude palm oil (CPO) and was found in the phytonutrient concentrate (350 ppm) from palm oil, whereas its concentration in palm fibre oil (PFO) was about 11% (430 ppm). α-T constituted about 36% of the total vitamin E in CPO, at a level of 10% in the phytonutrient concentrate. In contrast, the composition of γ-tocotrienol increased from 31% in CPO to 60% in the phytonutrient concentrate. Vitamin E was present at 1,160 ppm, and its concentrations in PFO and the phytonutrient concentrate were 4,040 and 13,780 ppm, respectively. The minor components (carotenes, vitamin E, sterols, and squalene) in crude palm oil (CPO) and the residual oil from palm-pressed fibre were isolated using the quicker and more advanced supercritical fluid chromatography (Choo et al. 2005b). The carotenes, vitamin E, sterols, and squalene were isolated in less than 20 minutes. The individual vitamin E isomers present in palm oil were also isolated into their respective components, α-tocopherol, α-tocotrienol, γ-tocopherol, γ-tocotrienol, and δ-tocotrienol. A-tocotrienols was 40–60 times more potent than normal tocopherols making it one of the most powerful lipid soluble anti oxidants available.

Vitamin E in palm oil was reported to comprise various isomers of tocopherols and tocotrienols [α-tocopherol (α-T)], α-tocotrienol, γ-tocopherol, γ-tocotrienol, and δ-tocotrienol. In separation of vitamin E from palm oil, an α-tocomonoenol (α−T1) isomer was quantified and characterized in crude oil palm, (CPO) (Ng et al. 2004b). α−T1 constituted about 3–4% (40 ppm) of vitamin E in crude palm oil (CPO) and was found in the phytonutrient concentrate (350 ppm) from palm oil, and its concentration in palm fibre oil (PFO) was about 11% (430 ppm). The relative content of each individual vitamin E isomer before and after interesterification/transesterification of CPO to CPO methyl esters, followed by vacuum distillation of CPO methyl esters to yield the residue, remained the same except for α-tocopherol and γ-tocotrienol. Whereas α-tocopherol constituted about 36% of the total vitamin E in CPO, it was present at a concentration of 10% in the phytonutrient concentrate. In contrast, the composition of γ-tocotrienol increased from 31% in CPO to 60% in the phytonutrient concentrate. Vitamin E was present at 1160 ppm in CPO, and its concentrations in PFO and the phytonutrient concentrate were 4,040 and 13,780 ppm, respectively.

The biological activities of palm oil vitamin E (tocopherols and tocotrienols) are elaborated in various sections below.

Pro-vitamin A Activity

Crude palm oil with 30,000RE of retinol (provitamin A) equivalent was found to be the richest natural plant source of carotenoids (Choo 1994). It contained about 15–300 times more retinol equivalents as carrots (2,000RE), leafy green vegetables (685RE), and tomatoes (100RE), both considered to have significant quantities of provitamin A activity. These carotenoids imparted an orangey-red colour to crude palm oil. The total carotenoid content in *E. guineensis* var. *tenera* was found to be 673 ppm and in red palm oil is 545 ppm. The carotenoid composition of *E. guineensis* var *tenera* and red palm oil was reported by Choo (1994) as follow: phytoene, 1.27, 2%; phytofluene 0.06, 1.2%; cis-β-carotene 0.68, 0.8%; β-carotene 56.02, 47.4%; α-carotenes 35.16, 37%; cis-α-carotene 2.49, 6.9%; ζ-carotene 0.69, 1.3%; δ-carotene 0.83, 0.6%; γ-carotene 0.33, 0.5%; neurosporene 0.29, traces; β-zeacarotene 0.74, 0.5%; α-zeacarotene 0.23, 0.3%; lycopene 1.3, 1.5% respectively. A multitude of *cis*-isomers of α-, β- and γ-carotene were separated from the carotene extract from the fruits of the oil palm (*Elaeis guineensis*) (Mortensen 2005). α- and β-carotene are found to be the most abundant carotenoids comprising 12.3% and 17.9%, respectively, of a (roughly) 30% oil suspension of oil palm carotenes in vegetable oil. A large proportion (about 40%) of α- and β-carotene was in the form of *cis*-isomers. The γ-carotene content was 0.38% and other carotenes like phytoene, phytofluene, ζ-carotene, lycopene and possibly β-zeacarotene were present as well but were not quantified.

In the mesocarp oil of fresh palm fruits (MOFPF), the isomers of α-carotene were all trans, 13-cis, 13′-cis and 9-cis and the isomers of β-carotene all trans and 13-cis (Tay et al. 2002). The mesocarp oil of sterilized palm fruits (MOSPF) had similar geometrical isomers plus two unidentified cis α- and β-carotenes. Crude palm oil (CPO) had the geometrical isomers of MOSPF plus 9-cis-β-carotene. The isomer profiles of carotenoid extracts from red palm olein (RPOo) and carotene concentrates (CC1 and CC2) had similar patterns to that of CPO. The percentage compositions of total cis-isomers of α- and β-carotene in MOFPF, MOSPF, CPO, RPOo, CC1 and CC2 were 18.2%, 44.9%, 43.5%, 45.4%, 55.9%, and 38.8%, respectively.

Crude palm oil (CPO) and red palm olein (RPO) were found to be the richest dietary source of pro-vitamin. The α- and β-carotenes including their cis-isomers constituted more than 93% of the carotenes in CPO (Choo et al. 1996) and more than 80% in RPO (Tay and Choo 2000). The carotene profile of CPO showed higher content of β-carotene than α-carotene whereas they were almost equal in RPO. Both these carotenes possessed provitamin A activity. Other minor carotenes such as γ-carotene and β-zeacarotene also possessed vitamin A activity (Choo 1994). Crude palm oil (CPO) could serve as a promising source of β-carotene to treat vitamin A deficiency in developing countries (Manorama and Rukmini 1992). Indigenously produced CPO were found to be nutritionally adequate and toxicologically safe for human consumption. CPO was found to be well accepted in preparations where its yellow/orange colour blended well with the natural colour of certain Indian food items. 1:1 blend of CPO and ground nut oil (GNO) preparations were better accepted. Total carotene in CPO estimated was 540 μg/g and β-carotene was 370 μg/g, accounting for approximately 70% of total carotenes. Cooking losses of β-carotene in a range of products were observed to be 22–30% and total carotenes 24–32%. Repeated frying using the oil five times consecutively, resulted in a steep drop followed by a total loss of β-carotene accompanied by alteration of its physico-chemical and organoleptic properties.

In an incomplete balanced crossover study with 69 healthy adult volunteers, 4 days of supplementation with natural palm oil carotenoids (7.6 mg/day of α-carotene, 11.9 mg/day of all-trans-β-carotene, 7.5 mg/day of cis-β-carotene) or synthetic β-carotene (23.8 mg/day of all-trans-β-carotene, 4.4 mg/day of cis-β-carotene), added to a mixed meal, significantly elevated plasma levels of the supplied carotenoids as compared to consumption of a low-carotenoid meal: 7.2-fold increase in α-carotene and 3.5-fold increase in all-trans-β-carotene following palm oil carotenoids; 6.9-fold increase in all-trans β-carotene following synthetic β-carotene (Van Het Hof et al. 1999). The presence of α-carotene did not affect the bioavailability of β-carotene from palm oil. It was concluded that 4 days of supplementation with palm oil carotenoids or synthetic β-carotene enhanced the plasma β-carotene status substantially, whereas α-carotene was additionally delivered by the palm oil supplement. Palm oil carotenoids being a mixture of α- and β-carotenes, may also be applied as a functional food ingredient because of the provitamin A activity of α- and β-carotenes and their proposed beneficial roles in the prevention of chronic diseases.

Studies by Canfield et al. (2001) in lactating mothers in Colonia Los Pinos, Honduras showed that red palm oil (90-mg β-carotene) in the maternal diet for 10 days elevated provitamin A carotenoids in breastmilk and serum of the mother-infant dyad. Maternal serum α-carotene and β-carotene concentrations were increased 2 times by palm oil compared with 1.2-fold by β-carotene supplements. Increases in breastmilk β-carotene were higher for the palm oil group (2.5-fold) than for the β-carotene supplement group (1.6-fold) and increases in milk α-carotene concentrations (3.2-fold) were slightly higher than those of β-carotene. There were also small but significant changes among groups in breastmilk lutein and lycopene. Breastmilk retinol was not significantly different among the groups over the treatment period.

As part of a European multicentre placebo-controlled intervention investigation involving 400 healthy male and female volunteers, aged 25–45 years, several studies were undertaken to test whether the consumption of foods rich in carotenoids reduces oxidative damage to human tissue components. Elevated levels of oxidative stress have been implicated in tissue damage and the development of chronic diseases, and dietary antioxidants may reduce the risk of oxidative tissue damage. In one such study, supplementation with α-tocopherol was found to increase serum α-tocopherol levels, while producing a marked decrease in serum γ-tocopherol. Supplementation with α-carotene+β-carotene (carotene-rich palm oil) resulted in 14-fold and 5-fold increases respectively in serum levels of these carotenoids. In Spanish volunteers, additional data showed that the serum response to carotenoid supplementation reached a plateau after 4 weeks, and no significant side effects (except carotenodermia) or changes in biochemical or haematological indices were observed throughout the study.

Among the following vitamin A supplementation of preschool children, red palm oil (5 and 10 ml), afforded higher gain in retinol and β-carotene levels compared to ground nut oil fortified with 400 and 800 retinol equivalent retinol palmitate, and ground nut oil (5 and 10 ml), (Sivan et al. 2002). Administration of 10 ml did not offer any substantial improvement over the 5-ml daily dose.

Antioxidant Activity

Aside from pro-vitamin A activity, the carotenes such as β-carotenes, α-carotene and lycopene in oil palm were found to be effective antioxidants and singlet oxygen quenchers (Di Mascio et al. 1990). The singlet oxygen quenching ability decreased in the following sequence: lycopene, γ-carotene, astaxanthin, canthaxanthin, α-carotene, β-carotene, bixin, zeaxanthin, lutein, bilirubin, biliverdin, tocopherols and thiols. Singlet molecular oxygen had been shown to be generated in biological systems and was capable of damaging proteins, lipids and DNA. A study was performed to determine the tissue distribution of carotenoids in palm oil and to correlate their accumulation with protection against oxidative stress in rats by Serbinova et al. (1992a). After 2 weeks of feeding, β-carotene in the rat liver increased from 7.3 to 30 µg/g wet tissue; α-carotene and lycopene after 10 weeks of feeding were 74 and 49 µg/g wet tissue respectively; β-carotene content in heart and hind limb skeletal muscles increased after 10 weeks to 17 and 6 µg/g wet tissue respectively. No carotenoids were detected in the brain, adipose and skin during the period of feeding. After in-vitro induction of lipid peroxidation in liver homogenates by an azo-initiator of peroxyl radicals an inverse correlation between tissue carotenoid level and accumulation of lipid peroxidation products was observed. Using palm-based carotenes, they showed that the order of strength in in-vitro lipid oxidation inhibition was α-carotene>lycopene>β-carotene. The anti-oxidant activity elucidated the importance of the linkage of β-carotene, α-carotene and other carotenoids to decreased risk of some cancer as shown by scientific data from epidemiological and animal studies (Murakoshi et al. 1989; Ziegler 1989; Ziegler et al. 1996).

Studies in spontaneously hypertensive rats indicated that antioxidant supplement of γ-tocotrienol may prevent development of increased blood pressure, reduce lipid peroxides in plasma and blood vessels and enhance total antioxidant status including plasma superoxide dismutase activity (Newaz and Nawal 1999). After 3 months of antioxidant trial with γ-tocotrienol, it was found that all the treated groups (γ1, 15 mg/kg diet; γ2, 30 mg/kg diet and γ3, 150 mg/kg diet) exhibited reduced plasma lipid peroxides concentration but was only significant for group γ1.

The total phenolics content of the crude and ethanol extracts oil palm fruits as determined by the Folin-Ciocalteu method were found to be 40.3 and 49.6 mg GAE/g extract (dry basis), respectively (Balasundram et al. 2003; 2005). The radical scavenging activity of the extracts as determined using 2,2 diphenyl-1-picrylhydrazyl radical (DPPH) showed that both crude and ethanol extracts exhibited hydrogen-donating capacity, and have antiradical power (ARP) comparable to ascorbic acid. The reducing power of crude and ethanol extracts at 1 mM GAE were found to be comparable to that of 0.3 mM gallic acid. The extracts exhibited complete scavenging of hydrogen peroxide at concentrations above 0.4 mM GAE. These findings suggested that the crude and ethanol extracts were able to scavenge free radicals, by either hydrogen or electron donating mechanisms, and could therefore act as primary antioxidants. In another study, the phenolic compound of oil palm fruit was determined as follows: total phenolic content (TPC) 5.64 and 83.97 g/l gallic acid equivalent (GAE), total flavonoid content (TFC) 0.31–7.53 g/l catechin equivalent, o-diphenols index 4.90–93.20 g/l GAE, hydroxycinnamic acid index 23.74–77.46 g/l ferulic acid equivalent, flavonols index 3.62–95.33 g/l rutin equivalent and phenol index 15.90–247.22 g/l GAE (Neo et al. 2008). The antioxidant assay, 2,2-diphenyl-2-picrylhydrazyl radical scavenging assay, showed antioxidative activities in all the extracts with results ranging from 4.41 to 61.98 g/l trolox equivalent. The high antioxidant activities of the oil palm fruit phenolics were also found to increase with increasing total phenolic content and total flavonoid content.

The antioxidant effectiveness of palm oil α-carotene was compared with β-carotene in organic solution containing egg-yolk phosphatidycholine (EYPC) in the presence of lipid soluble 2,2^1-azobis (2,4-dimethyl valeronitrile) (AMVN)-generated peroxyl radicals by measuring the formation of phosphatidylcholine hydroperoxide (PCOOH) and thiobarbituric acid reacting substances (TBARS) (Farombi and Britton 1999.) The carotenes were found to be more speedily oxidised than the xanthophylls, lutein and zeaxanthin in the solution. α-Tocopherol, α-carotene, β-carotene, lutein and zeaxanthin reduced PCOOH accumulation by 78, 65, 40, 60 and 43%, respectively. AMVN incubated with EYPC for 2 h induced the formation of TBARS compared to the control. α-carotene significantly reduced the TBARS formation by 68% whilst β-carotene, lutein and zeaxanthin elicited 50, 64 and 53% suppressions, respectively. α-Tocopherol inhibited the TBARS formation by 80%. These results suggested that α-carotene, a carotenoid abundantly present in human diets, especially red palm oil, may better attenuate peroxyl radical-dependent lipid peroxidation than β-carotene in organic solution.

Oil palm leaves were found to contain water soluble antioxidative compounds with varying concentrations (Ng and Choo 2010). The dried leaves extracts showed higher antioxidative activity than the wet leaves extracts. Four extracts of the oil palm leaves, SFP (soluble free phenolics), ISBP (insoluble bound phenolics), EF (Esterified phenolics) and TPE (total phenolic extract) were subjected to different analyses to determine the total flavonol index (FI), hydroxycinnamic acid index (HCAI), O-diphenol indices (ODPI) and total flavonoid content. The flavonol indices (FI) of the ISBP extract from dried leaves was much higher than the SFP and EFP extracts of both dried and wet leaves. The FI of the wet leaves extracts ranged from 0.060 to 0.38 mg rutin equivalent (RE) per gram extract while the FI of dried leaves extracts varied from 0.07 to 1.00 mg RE per g extract. The hydroxycinnamic acid indices (HCAI) of the wet leaves extracts varied from 0.17 to 0.50 mg ferulic acid equivalents (FAE) per g extract while the HCAI of dried leaves ranged from 0.19 to 0.92 g FAE per g extract. The ODPI of wet leaves extracts ranged from 0.50 to 2.10 mg gallic acid equivalent (GAE) per g extract. ODPI of dried leaves extracts were found to be higher at 0.62–4.80 mg GAE per g extract Thus, the could be used as an indicator of the amount of total phenolic compounds present in the oil palm leaves. In the wet and dried oil palm leaves extracts, the TPC (total phenolic content) of wet and dried leaves extracts ranged from 5.1 to 10.2 mg GAE per g extract. All four extracts of the oil palm leaves, SFP, ISBP, EF and TPE displayed antioxidant activities when determined by the DPPH assay. The antioxidative activities of the wet leaves extract ranged from 65% to 88% while the dried leaves extracts showed activities from 56% to 93%.

Palm Oil and Cardiovascular Diseases

Numerous scientific studies (as discussed below) and reviews (Chong and Ng 1991; Ng 1994; Ebong et al. 1999; Edem 2002; Sundram et al. 2003; Ong and Goh 2002; Oguntibeju et al. 2009), in the past three decades have reported on the beneficial effects of dietary palm oil on cardiovascular diseases and health.

Antihypercholesterolemic Activity

Ample evidence established that the intramolecular fatty acid distribution in the triglycerides influenced the cholesterolaemic impact of dietary fats (Elson 1992; Kritchevsky 1988; McGandy et al. 1970). The position of the saturated and unsaturated fatty acid chains in a triacylglycerol backbone of the fat molecule determines the preferential absorption of fatty acids and their subsequent elevating or reducing effects on the cholesterol level in the blood (Goh 1999; Willis et al. 1998; Small 1991; Kritchevsky 2000; Kritchevsky et al. 1996). The highly structured triacylglycerols of palm oil with predominant unsaturation at the 2 position lead to these "unexpected" good properties. In palm oil, 87% of the fatty acids found in position 2 of the triglyceride molecule are unsaturated (Willis et al. 1998; Siew 2000; Small 1991), whereas the saturated acids mainly occupy positions 1 and 3. This was proposed as the likely explanation why palm oil was not cholesterol elevating and also possessed other associated beneficial properties.

In a study to compare the effects of dietary palmitic acid (16:0) vs oleic acid (18:1) on serum lipids, lipoproteins, and plasma eicosanoids, 33 normocholesterolemic subjects (20 males, 13 females; ages 22–41 years) were challenged with a coconut oil-rich diet for 4 weeks. Subsequently they were assigned to either a palm olein-rich or olive oil-rich diet followed by a dietary crossover during two consecutive 6-week periods. Dietary myristic acid (14:0) and lauric acid (12:0) from coconut oil were found to significantly enhance all the serum lipid and lipoprotein parameters measured (Ng et al. 1992). Subsequent one-to-one exchange of 7% energy between 16:0 (palm olein diet) and 18:1 (olive oil diet) resulted in identical serum total cholesterol (192, 193 mg/dl), low-density lipoprotein cholesterol (LDL-C) (130, 131 mg/dl), high-density lipoprotein cholesterol (HDL-C) (41, 42 mg/dl), and triglyceride (TG) (108, 106 mg/dl) concentrations. Effects attributed to gender included higher HDL in females and higher TG in males associated with the tendency for higher LDL and LDL/HDL ratios in men. However, both sexes were equally responsive to changes in dietary fat saturation. The results indicated that in healthy, normocholesterolemic humans, dietary 16:0 can be exchanged for 18:1 within the range of these fatty acids normally present in typical diets without affecting the serum lipoprotein cholesterol concentration or distribution. Further, replacement of 12:0+14:0 by 16:0+18:1, but especially 16:0 or some component of palm olein, appeared to have a beneficial impact on an important index of thrombogenesis, i.e., the thromboxane/prostacyclin ratio in plasma.

A comparative randomised crossover study in 21 young Australian adults (men and women) showed that the total blood cholesterol, triglycerides and HDL-cholesterol levels of those fed on palm oil (palm olein) and olive oil were lower than those fed on the usual Australian diet (Choudhury et al. 1995). They showed that young Australian adults fed on palm oil diets had the same total blood cholesterol, triglycerides and 'good' HDL-cholesterol levels as those fed on olive oil. The results indicated that 16:0, though saturated, was not always a plasma cholesterol-raising fatty acid. In a double-blind crossover study, in 17 normocholesterolemic male volunteers, the palmitic 16:0-rich diet produced a 9% lower serum cholesterol concentration than the lauric+myristic acids (12:0+14:0) -rich diet, reflected by a lower (11%) low-density-lipoprotein-cholesterol concentration and, to a lesser extent, high-density-lipoprotein cholesterol (Sundram et al. 1994). The data indicated that a dietary 12:0+14:0 combination produced a higher serum cholesterol concentration than does 16:0 in healthy normocholesterolemic young men fed a low-cholesterol diet.

Reanalysis of published data suggested that dietary myristic (14:0) and palmitic (16:0) acids exerted dissimilar effects on cholesterol metabolism in primates and humans (Hayes and Khosla 1992; Khosla and Hayes 1994). As established in an earlier human study, 14:0 appeared to be the principal saturated fatty acid that raised plasma cholesterol whereas 18:2 lowered it. Oleic acid (18:1) appeared neutral. The effect of palmitic may vary and appeared to have no effect on the plasma cholesterol in normocholesterolemic subjects when dietary cholesterol intake was below a certain critical level (400 mg/day). However, in hypercholesterolemic subjects (greater than 225 mg/dl) and especially those consuming diets providing cholesterol intakes of greater than or equal to 400 mg/day, dietary palmitic acid may expand the plasma cholesterol pool. These differential effects of palmitic acid on plasma cholesterol were attributable to differences in LDL-receptor status. Collectively these data implied that, for most of the world's population, palm oil would be an inexpensive and readily metabolized source of dietary energy with minimal impact on cholesterol metabolism.

Unlike most polyunsaturated vegetable oils, palm oil requires little or no hydrogenation before its incorporation into food products (Teah and Ahmad 1991; Ong and Berger 1992). The benefits of having a *trans*-free fat cannot be overemphasized, as sufficient data have documented the dangers of *trans* fatty acids from partially hydrogenated fats. Scientific papers have reported that dietary transfatty acids raised the atherogenic factors LDL (Mensink and Katan 1990; Judd et al. 1994; Khosla and Hayes 1996) and lipoprotein (a) (Mensink et al. 1992; Khosla and Hayes 1996; Sundram et al. 1997) and decreased the protective HDL (Mensink and Katan 1990; Judd et al. 1994; Khosla and Hayes 1996). Studies by Pedersen et al. (2005) compared the effects of three different margarines, one based on palm oil (PALM-margarine), one based on partially hydrogenated soybean oil (TRANS- margarine) and one with a high content of polyunsaturated fatty acids (PUFA-margarine), on serum lipids in 27 young women. They reported that dietary palm oil may have a more favourable effect on the fibrinolytic system compared to partially hydrogenated soybean oil. They concluded that from a nutritional point of view, palmitic acid from palm oil may be a reasonable alternative to trans fatty acids from partially hydrogenated soybean oil in margarine if the aim was to avoid trans fatty acids. A palm oil based margarine was, however, less favourable than one based on a more polyunsaturated vegetable oil.

Studies over the past two decades, reported that palm oil's impact on blood cholesterol and lipoprotein profiles were beneficial. In a double-blind cross-over trial involving 38 male Dutch volunteers, the palm-oil diet induced an 8% decrease in low-density-lipoprotein (LDL):HDL2 + HDL3-cholesterol ratio as well as a 9% reduction in triacylglycerols in the low-density-lipoprotein fractions (Sundram et al. 1992). Palm oil consumption produced a 4% increase in serum apolipoprotein AI and a 4% decrease in apolipoprotein B compared to the control diet. The B:AI apolipoprotein ratio was lowered by 8%. Although the observed differences were relatively modest, the present study, nonetheless, indicated that dietary palm oil, when replacing a major part of the normal fat content in a Dutch diet, may slightly reduce the lipoprotein- and apolipoprotein-associated cardiovascular risk profiles.

It was well established that palm oil being very rich in tocotrienols were not only potent antioxidants but were also natural inhibitors of cholesterol synthesis (Ng 1994; Ong and Goh 2002). Tocotrienols were reported to inhibit cholesterol synthesis in-vivo thereby exerting a hypocholesterolaemic effect in humans (Tan et al. 1991; Qureshi et al. 1991a; Zhang et al. 1997) and in animal models (Tan et al. 1991; Qureshi et al. 1991a, b). In a study conducted by Tan et al. (1991), all volunteers administered one palmvitee (capsulated palm-oil-vitamin E concentrate) capsule per day for 30 consecutive days exhibited lower serum total cholesterol (TC) and low-density-lipoprotein cholesterol (LDL-C) levels. Each palmvitee capsule contained approximately 18 mg tocopherols, 42 mg tocotrienols, and 240 mg palm olein. The results demonstrated that the palmvitee had a hypocholesterolemic effect. In separate double-blind, crossover, 8-week study, Qureshi et al. (1991b) compared the effects of the tocotrienol-enriched fraction of palm oil (200 mg palmvitee capsules/day) with those of 300 mg corn oil/day on serum lipids of hypercholesterolemic human subjects (serum cholesterol 6.21–8.02 mmol/l). Concentrations of serum total cholesterol (−15%), LDL cholesterol (−8%), Apo B (−10%), thromboxane (−25%), platelet factor 4 (−16%), and glucose (−12%) decreased significantly only in the 15 subjects given palmvitee during the initial 4 weeks. The crossover confirmed these actions of palmvitee. There was a carry over effect of palmvitee. Serum cholesterol concentrations of seven hypercholesterolemic subjects (greater than 7.84 mmol/L) decreased 31% during a 4-week period in which they were given 200 mg γ-tocotrienol per day. This indicated that γ-tocotrienol may be the most potent cholesterol inhibitor in palmvitee capsules.

Sundram et al. (1990) found no differences in the body and organ weights of rats fed various diets of corn oil, soybean oil, palm oil, and palm olein and palm stearin. Plasma cholesterol concentrations of rats fed soybean oil were significantly lower than those fed corn oil, palm oil, palm olein or palm stearin. Significant differences between the plasma cholesterol content of rats fed corn oil and rats fed the three palm oils were not evident. HDL cholesterol was raised in rats fed the three palm oil diets compared to the rats fed either corn oil or soybean oil. The cholesterol-phospholipid molar ratio of rat platelets was not influenced by the dietary fat type. The formation of 6-keto-PGF1 α was significantly greater in palm oil-fed rats compared to all other dietary treatments. Fatty acid compositional changes in the plasma cholesterol esters and plasma triglycerides were diet regulated with significant differences between rats fed the polyunsaturated corn and soybean oil compared to the three palm oils.

Ng et al. (1991) studied the effects on serum lipids of diets prepared with palm olein, corn oil, and coconut oil providing approximately 75% of the fat calories in three matched groups of healthy volunteers (61 males, 22 females, aged 20–34 years). Group I received a coconut-palm-coconut dietary sequence; group II, coconut-corn-coconut; and group III, coconut oil during all three 5-week dietary periods. Compared with entry-level values, coconut oil elevated the serum total cholesterol concentration higher than 10% in all three groups. Subsequent feeding of palm olein or corn oil significantly lowered the total cholesterol (−19%, −36%), the LDL cholesterol (−20%, −42%) and the HDL cholesterol (−20%, −26%) concentrations, respectively. Although the entry level of the ratio of LDL to HDL was not significantly altered by coconut oil, this ratio was reduced 8% by palm olein and 25% by corn oil. Serum triglycerides were unaffected during the palm-olein dietary period but were significantly reduced during the corn-oil dietary period. Marzuki et al. (1991) studied the effects of saturated (palm olein) and polyunsaturated (soybean oil) cooking oils on the lipid profiles of Malaysian male adolescents eating normal Malaysian diets for 5 weeks. Diets cooked with palm olein did not significantly change plasma total-cholesterol, LDL cholesterol, and HDL cholesterol concentrations or the ratio of total cholesterol to HDL cholesterol compared with diets cooked with soybean oil. However, the diet cooked with palm olein significantly raised apolipoprotein A-I (11%) and apolipoprotein B (9%) levels. Unexpectedly, soybean-oil-cooked diets induced a significant increase (47%) in plasma triglycerides compared with palm-olein-cooked diets. This suggested that palm olein, when used as cooking oil, had no detrimental effects on plasma lipid profiles in Malaysian adolescents. In a study on the effects on serum lipids, palm oil used in Chinese diets were compared with those of soybean oil, peanut oil and lard in 120 normocholesterolemic men (Zhang et al. 1997). The palm oil diet caused major reductions in serum TC, LDL-C and TC/HDL-C during the first 6 weeks and also a significant reduction in TC/HDL-C during the second 6 weeks. The palm oil diet had no significant influence on serum lipids in either experimental period.

Studies showed that hamsters palm oil refined, bleached and deodorized palm oil (RBPO) diet for 8 weeks had significantly lower serum total cholesterol concentrations than those palm oil 20% (w/w) of coconut oil (CNO) diet (Khor and Tan 1992b). No significant differences in the HDL cholesterol and serum triglyceride levels were detected in the CNO and RBPO groups. Hamsters fed the palm oil stripped of its nonsaponifiable lipids (POTG) diet had similar serum total cholesterol level as those given RBPO diet; however, the difference in serum total cholesterol level for the POTG group was intermediate between the CNO and RBPO groups. Since POTG was stripped of its nonsaponifiable lipids, these results indicated that the triglyceride fraction of RBPO was not hyper-cholesterolemic in hamsters. Results of further studies by Khor and Tan (1992a) showed that palm oil (PO) feeding significantly decreased serum total cholesterol (TC) and low density lipoprotein cholesterol (LDL-C) concentrations but did not produce any significant changes in high density lipoprotein cholesterol (HDL-C) and serum triglyceride (TG) levels in hamsters as compared to coconut oil (CNO) feeding. When hamsters were fed the palm oil triglyceride fraction (POTG), their serum TC and LDL-C level were not significantly increased compared to those fed on PO, indicating that POTG was not hypercholesterolemic. The results also showed that olive oil (OLO) feeding did not result in lower serum TC and LDL-C levels as compared to PO feeding. These results indicated that increasing the oleic acid and lowering the palmitic acid levels may not improve the lipidemic response of a diet. In contrast, analysis of the liver and heart lipids showed that dietary fats had a distinct effect on the cholesterol content of these tissues. PO, POTG and OLO feeding increased the liver TC and cholesterol ester (CE) content when compared to CNO feeding. On the contrary, CNO feeding increased the heart TC and CE contents while PO, POTG and OLO feeding did not. These results could be due to redistribution of cholesterol pools between plasma and tissues in response to the different dietary fats.

The results of studies in hamsters conducted by Khor and Chieng (1996) indicated that tocotrienols possessed hypo-cholesterolemic effect and tocopherols may have slight hypercholesterolemic effect in the hamster. Supplementation with high level of tocotrienols did not increase the hypocholesterolemic effect of the tocotrienols. Supplementation of diets containing palm olein triglycerides (POTG) with 72 ppm of tocopherols (POTG-T) raised the serum TC, LDL-C, HDL-C levels and lowered the serum TG level as compared to that of the POTG but the differences were not statistically significant. Supplementation of POTG with 162 ppm of tocotrienols (POTG-T3L) decreased the serum TC level significantly, but it did not significantly decrease the LDL-C, HDL-C and serum TG levels as compared with the POTG group. Supplementation of POTG with higher amounts of tocotrienols i.e. 1,000 ppm did not give further reduction in serum TC, LDL-C, HDL-C and TG levels as compared to the POTG-T3L group. In contrast, the POTG-T3H diet raised LDL-C level significantly when compared to the POTG-T3L diet. Analysis of the serum and liver vitamin E levels show that α-T was the main vitamin E in the serum and liver irrespective of whether tocopherols or tocotrienols were added to the oils. Small amounts of tocotrienols appeared in the serum and liver only when tocotrienols were supplemented at high levels of 1,000 ppm.

Animal experiments showed that tocotrienols inhibited the enzyme HMGCoA reductase and consequently the synthesis of cholesterol (Parker et al. 1993; Khor et al. 1995). Tocotrienols were found to influence the mevalonate pathway in mammalian cells by post-transcriptional suppression of HMG-CoA reductase, and appeared to specifically regulate the intracellular mechanism for controlled degradation of the reductase protein, an activity that mirrored the actions of the putative non-sterol isoprenoid regulators derived from mevalonate (Parker et al. 1993). Tocotrienols, natural farnesylated analogues of tocopherols were found to decrease hepatic cholesterol production and reduce plasma cholesterol levels in animals. Khor et al. (1995) showed that tocotrienols isolated from palm oil fatty acid distillate (PFAD) inhibited liver MMG CoA reductase activity in the guinea pig and the inhibitory effect of tocotrienols on HMG CoA reductase appeared to be dose-dependent.

Studies conducted by Wilson et al. (2005) concluded that hamsters fed the three palm oil preparations RPO (Red palm oil), RBD-PO (refined, bleached and deodorized palm oil) and RBD-PO plus red palm oil extract (reconstituted RBD-PO) had significantly lower plasma total cholesterol (TC), plasma triglyceride (TG) and non-HDL-C (non-high-density lipoprotein-cholesterol) and higher HDL-C levels while accumulating less aortic cholesterol concentrations compared to hamsters fed coconut oil. The plasma γ-tocopherol concentrations were higher in the coconut oil-fed hamsters compared to the hamsters fed the RPO (60%), RBD-PO (42%) and the reconstituted RBD-PO (49%), while for plasma α-tocopherol concentrations, the coconut oil-fed hamsters were significantly higher than only the RPO-fed hamsters (21%). The coconut oil-fed hamsters also exhibited significantly higher plasma lipid hydroperoxide concentrations compared to RBD-PO (112%) and the reconstituted RBD-PO (485%). The hamsters fed the coconut oil diet excreted appreciably more fecal total neutral sterols and cholesterol compared to the hamsters fed the RBD-PO (158% and 167%, respectively). The coconut oil-fed hamsters had significantly higher levels of aortic total, free and esterified cholesterol compared to the hamsters fed the RPO (74%, 50% and 225%, respectively), RBD-PO (57%, 48% and 92%, respectively) and the reconstituted RBD-PO (111%, 94% and 94%, respectively). Also, aortic free/ester cholesterol ratio in the aortas of hamsters fed RPO was significantly higher than in those fed the coconut oil (124%). In separate animal studies conducted by Matthan et al. (2009), the results suggested that an oil relatively high in palmitoleic acid (16:1) such as palm oil did not adversely affect plasma lipoprotein profiles or aortic cholesterol accumulation and was similar to other unsaturated fatty acid-rich oils. This could be linked to the lower susceptibility of palmitoleic acid (16:1) to oxidation compared to PUFA.

Ladeia et al. (2008) reported that boiled crude palm oil may have a mild, triacylglycerol-reducing effect in young, healthy individuals and may also show a mild LDL-C–increasing effect in males. In their studies, involving 34 medical students (18–26 year old), they observed a decrease in all lipid fractions, with a mild, statistically significant decrease in concentrations of very LDL-C (19.41 versus 17.18 mg/dL) and triacylglycerol (97. versus 85. mg/dL). Males (61.9%) also exhibited a mildly significant increase in LDL-C, whereas females experienced a mildly significant reduction in all lipid fractions, except for HDL-C.

Antiatherogenic Activity

A Netherlands study was conducted on rabbits to test the effect of palm oil on atherosclerosis (Hornstra 1988). After feeding the rabbits for 18 months, palm oil and sunflower oil diets was found to cause the lowest extent of atherosclerosis in comparison with fish oil, linseed oil and olive oil. Similarly Klurfeld et al. (1990) in the United States, also using the rabbit model, compared the effects of palm oil with hydrogenated coconut oil, cottonseed oil, hydrogenated cottonseed oil, and an American fat blend containing a mixture of butterfat, tallow, lard, shortening, salad oil, peanut oil and corn oil. At the end of the 14-month feeding period, coconut oil fed rabbits were found to have the most atherosclerotic lesions, while in palm oil-fed rabbits the number of lesions was no different from that with the other oils. Rand et al. (1988) reported that although palm oil contains about 50% saturated fatty acids, it did not increase arterial thrombosis tendency and tended to decrease platelet aggregation, as compared with highly polyunsaturated sunflower seed oil. Aggregation of platelets in whole blood activated with collagen was not altered by palm oil feeding in rats, but was enhanced in the sunflower seed oil group, compared with the control. The concomitant formation of thromboxane A2 was reduced by palm oil feeding, although formation of prostacyclin did not change; the ratio of thromboxane / prostacyclin formed was reduced significantly in the palm oil group. Qureshi et al. (1991a) reported that hypercholesterolemic pigs fed the tocotrienol-rich fraction isolated from palm oil (TRF) supplement showed a 44% decrease in total serum cholesterol, a 60% decrease in low-density-lipoprotein (LDL)-cholesterol, and significant decreases in levels of apolipoprotein B (26%), thromboxane-B2 (41%), and platelet factor 4 (PF4; 29%). The declines in thromboxane B2 and PF4 suggested that TRF exhibited a marked protective effect on the endothelium and platelet aggregation.

Studies conducted by Tomeo et al. (1995) suggested that antioxidants, such as tocotrienols found in Palm Vitee, (γ-tocotrienol-, and α-tocopherol enriched fraction of palm oil), may alter the course of carotid atherosclerosis. Bilateral duplex ultrasonography revealed apparent carotid atherosclerotic regression in seven and progression in 2 of the 25 tocotrienol patients, while none of the control group exhibited regression and 10 of 25 showed progression. Serum thiobarbituric acid reactive substances, an ex-vivo indicator of maximal platelet peroxidation, was found to decrease in the tocotrienol treatment group from 1.08 to 0.80 μM/L after 12 months, and in the placebo group, they increased non-significantly from 0.99 to 1.26 μM/L. Both tocotrienol and placebo groups displayed significantly attenuated collagen-induced platelet aggregation responses as compared with start values. Serum total cholesterol, low density lipoprotein cholesterol, and triglyceride levels remained unaltered in both groups, as did the plasma high density lipoprotein cholesterol values. Kooyenga et al. (1997) found in a 2 year study that the γ-tocotrienol and α-tocopherol-enriched fraction of palm oil, could reduce carotid artery restenosis in patients with carotid atherosclerosis.

Red palm oil was significantly less atherogenic than refined, bleached and deodorized (RBD) palm oil supporting the hypothesis that carotenoids and vitamin E may protect against atherosclerosis (Kritchevsky et al. 2000). Red palm oil, the oil initially obtained from the palm fruit contained carotenes and vitamin E that were removed during refining. The oils tested had similar effects on serum and liver lipids. Studies conducted by Ismail et al. (2000) found that there were no differences in the total cholesterol and triacylglycerol levels between the rabbits fed a 2% cholesterol diet (AT) group and (PV) group of rabbits fed a 2% cholesterol diet with oral palm vitamin E (60 mg/kg body weight). The PV group had a significantly higher concentrations of HDL-c and a lower TC/HDL-c ratio compared to the AT group. The aortic tissue content of cholesterol and atherosclerotic lesions were comparable in both the AT and PV groups. However, the PV group had a lower content of plasma and aortic tissue malondialdehyde. The findings suggest that despite a highly atherogenic diet, palm vitamin E improved some important plasma lipid parameters, reduced lipid peroxidation but did not have an effect on the atherosclerotic plaque formation.

Tocotrienols have a beneficial effect in reducing atherogenesis. The effects of vitamin E and β-carotene were evaluated on apoliprotein (apo)E +/− female mice, which developed atherosclerosis only when fed diets high in triglyceride and cholesterol (Black et al. 2000). Mice were fed a nonpurified control diet (5.3 g/100 g triglyceride, 0.2 g/100 g cholesterol), an atherogenic diet alone (15.8 g/100 g triglyceride, 1.25 g/100 g cholesterol, 0.5 g/100 g Na cholate) or the atherogenic diet supplemented with either 0.5 g/100 g (+)-α-tocopherol (mixed isomers); 0.5 g/100 g palm tocopherols (palm-E; 33% α-tocopherol, 16.1% α-tocotrienol, 2.3% β-tocotrienol, 32.2% γ-tocotrienol, 16.1% δ-tocotrienol); 1.5 g/100 g palm-E; or 0.01 g/100 g palm-carotenoids (58% β-carotene, 33% α-carotene, 9% other carotenoids). Compared with mice fed the nonpurified control diet, plasma cholesterol was four times higher in mice fed the atherogenic diet. Mice fed the 1.5 g/100 g palm-E supplement had 60% lower plasma cholesterol than groups fed the other atherogenic diets. Mice fed the atherogenic diet had appreciably higher VLDL, intermediate density lipoprotein (IDL) and LDL cholesterol and appreciably lower HDL cholesterol than the controls. Lipoprotein patterns in mice augmented with α-tocopherol or palm carotenoids were similar to those of the mice fed the atherogenic diet alone, but the pattern in mice augmented with 1.5 g/100 g palm-E was similar to that of mice fed the control diet. In mice fed the atherogenic diet, the hepatic cholesterol plus cholesterol ester concentration was 4.4-fold higher than in mice fed the control diet. Augmenting with 1.5 g/100 g palm-E decreased hepatic cholesterol plus cholesterol ester concentration 66% compared with the atherogenic diet alone. Mice fed the atherogenic diet had large atherosclerotic lesions at the level of the aortic valve. Augmentation with 0.5 g/100 g palm-E or 1.5 g/100 g palm-E, reduced the size of the lesions by 92 or 98%, respectively. The 0.5 g/100 g α-tocopherol and palm carotenoid supplements had no effect. Supplements did not alter mRNA abundance for apolipoproteins A1, E, and C3. The beneficial effect of tocotrienols on atherogenesis, the plasma lipoprotein profile and accumulation of hepatic cholesterol esters was not related to their antioxidant properties.

α-tocopherol and its esterified derivatives were shown to be effective in reducing monocytic-endothelial cell adhesion. Theriault et al. (2002) demonstrated that α-T3 (α-tocotrienol) significantly hindered the surface expression of vascular cell adhesion molecule-1 in TNF-α activated HUVEC in a dose- and time-dependent fashion. The optimal inhibition was observed at 25 µmol/l α-T3 within 24 h (77%) without cytotoxicity. Compared to α-tocopherol and α-tocopheryl succinate, α-T3 displayed a more marked inhibitory effect on adhesion molecule expression and monocytic cell adherence. The above results suggested α-T3 to be a potent and effective agent in the inhibition of cellular adhesion molecule expression and monocytic cell adherence. Similarly Naito et al. (2005) found that compared to α-tocopherol, tocotrienols displayed a more profound inhibitory effect on adhesion molecule expression and monocytic cell adherence. They observed that β-tocotrienol exerted a most significant inhibitory action on monocytic cell adherence when compared to α-tocopherol and α-, β-, and γ-tocotrienols. Tocotrienols were found to accumulate in human aortic endothelial cells (HAECs) to levels 25–95-fold greater than that of α-tocopherol. Additonally, the findings indicated that tocotrienols had a profound inhibitory effect on monocytic cell adherence to HAECs relative to α-tocopherol via inhibition of the vascular cell adhesion molecule-1 (VCAM-1) expression.

Studies in oestrogen-deficient postmenopausal rats by Adam et al. (2009) showed that fresh palm oil may have a protective effect on the aorta but such a protective action may be lost when the palm oil is repeatedly heated. The ultrastructural changes were minimal with once-heated palm oil, in which there was a focal disruption of the endothelial layer. The focal disruption was more pronounced with five-times-heated palm oil. The study may be clinically important for all postmenopausal women who are susceptible to atherosclerosis. Further studies demonstrated that five-times-heated palm oil evoked a significant increase in TBARS (thiobarbituric acid reactive substances) and total cholesterol in ovariectomized rats compared to control ovariectomized rats fed 2% cholesterol diet only throughout the study period (Adam et al. 2008). There was a significant elevation in serum homocysteine in the control as well as five-times heated palm oil group compared to ovariectomized rats fed the fresh and once-heated palm oil groups. The findings suggested that repeatedly heated palm oil increased lipid peroxidation and total cholesterol. Ovariectomy increased the development of atherosclerosis as seen in this study. Feeding with fresh and once-heated palm oil did not cause any detrimental effect but repeatedly heated oil may be harmful because it caused oxidative damage thereby predisposing to atherosclerosis. Similarly, Xin et al. (2008) found that fresh palm oil had no detrimental effects on blood pressure and cardiac tissue but portracted consumption of repeatedly heated palm oil may result in an increase in blood pressure level with necrosis of cardiac tissue. Rats fed basal diet fortified with 15% weight/weight (w/w) fresh palm oil did not exhibit any significant alterations in blood pressure and heart histology. However, there was a significant increase in blood pressure in rats fed the basal diet fortified with 15% w/w five times heated-palm oil and rats fed basal diet fortified with 15% w/w ten times heated-palm oil. However, blood pressure in the ten times heated-palm oil rats was higher than in the five times heated-palm oil group. Histological sections of the heart showed necrosis in cardiac tissue in both groups with the latter group showing greater damage.

Tocotrienol enriched palm oil was found to prevent atherosclerosis through modulating the activities of peroxisome proliferators-activated receptors α, γ, and δ (PPARα, PPARγ, and PPARδ) (Li et al. 2010). PPAR are ligand regulated transcription factors that play essential preventive roles in the development of atherosclerosis through regulating energy metabolism and inflammation.

Antithrombotic/Antiplatelet Activity

Palm oil consumption may enhance HDL levels and reduce platelet aggregation (Chong and Ng 1991). Hornstra (1988) in the Netherlands first demonstrated the palm oil had anti-clotting effect, and to be as anti-thrombotic as the highly unsaturated sunflower seed oil. Other supporting evidence showed that a palm oil diet either increased the production of a hormone that prevented blood-clotting (prostacyclin) or decreased the formation of a blood-clotting hormone (thromboxane) (Rand et al. 1988; Abeywardena et al. 1991; Sugano and Imaizumi 1991; Qureshi et al. 1991a, b, 1995). Rand et al. (1988) found that palm oil feeding decreased the formation of thromboxane A2, although formation of prostacyclin did not change; the ratio of thromboxane/prostacyclin formed was reduced significantly in the palm oil fed group. Compared with the control diet, platelet membrane fluidity, measured by fluorescence polarization, was not altered in the palm oil group and was significantly increased only by sunflower seed-oil feeding. Thus, although palm oil contained about 50% saturated fatty acids, it did not enhance arterial thrombosis tendency and tended to decrease platelet aggregation, as compared with highly polyunsaturated sunflower seed oil. Abeywardena et al. (1991) found that polyunsaturated fatty acids (PUFAs) of the myocardium showed significant changes in response to dietary lipid treatment. Prostacyclin production was unaffected whereas both fish oil and chemically refined palm oil caused a significant inhibition of myocardial thromboxane A2 (TXA2). They speculated that n-3 (omega-3) PUFAs may act as specific inhibitors of TXA2 synthetase whereas the chemically refined oil palm effect was unlikely to be mediated via fatty acids. Similarly, Sugano and Imaizumi (1991) found that palm oil, being very rich in palmitic acid, influenced the serum cholesterol concentration and tissue eicosanoid profile of rats. The ratio of prostacyclin produced by the aorta to thromboxane A2 in plasma was not simply predicted by the dietary level of saturated fatty acid. Palm oil tended to facilitate the utilization of arachidonate for prostacyclin in-vitro in peritoneal macrophages compared with safflower oil. Fatty acid profiles of palm oil rather than the glyceride structure or tocotrienol appeared to be the major determinant for the specific features of lipid metabolism observed in rats fed palm oil.

In a double-blind, crossover, 8-week study on hypercholesterolemic human subjects (serum cholesterol 6.21–8.02 mmol/L), the administration of a tocotrienol-enriched fraction of palm oil (200 mg palmvitee capsules/day) was found to decrease the concentrations of serum total cholesterol (−15%), LDL cholesterol (−8%), Apo B (−10%), thromboxane (−25%), platelet factor 4 (−16%), and glucose (−12%) (Qureshi et al. 1991b). Serum cholesterol concentrations of seven hypercholesterolemic subjects (greater than 7.84 mmol/L) decreased 31% during a 4-wk period in which they were given 200 mg γ-tocotrienol/day. This indicated that γ-tocotrienol may be the most potent cholesterol inhibitor in palmvitee capsules. In an animal study, Qureshi et al. (1991a) found that hypercholesterolemic pigs fed tocotrienol-rich fraction (TRF) from palm oil (50ug/g) supplement showed a 44% decrease in total serum cholesterol, a 60% decrease in low-density-lipoprotein (LDL)-cholesterol, and significant decreases in levels of apolipoprotein B (26%), thromboxane-B2 (41%), and platelet factor 4 (PF4; 29%). The declines in thromboxane B2 and PF4 suggested that TRF had a marked protective effect on the endothelium and platelet aggregation. The effect of the lipid-lowering diet persisted only in the hypercholesterolemic swine after 8 week feeding of the control diet. This was further confirmed in subsequent studies (Qureshi et al. 1995). Plasma apolipoprotein B and ex-vivo generation of thromboxane B2 were similarly responsive to the tocotrienol preparations. Studies on nine female human subjects conducted by Müller et al. (2001) indicated that dietary trans fatty acids from partially hydrogenated soybean oil had an adverse effect on postprandial t-PA (tissue plasminogen activator) activity and thus possibly on the fibrinolytic system compared with palm oil. The diurnal postprandial state level of tissue plasminogen activator (t-PA) activity was significantly decreased on the trans-diet compared with the palm-diet. t-PA activity was also decreased on the PUFA-diet compared with palm-diet but the difference was not statistically significant.

Cardioprotective Activity

Studies conducted by Charnock et al. (1991) reported that a dietary saturated animal fat (SF) supplement was found to increase rats' susceptibility to develop cardiac arrhythmia under ischemic stress whereas the polyunsaturated fatty acids of sunflower seed oil (SSO) reduced this susceptibility, and diets supplemented with either (PO-I) (chemically refined palm oil) or (PO-Il) (physically refined palm oil) gave results that were generally intermediate in value between the SF and the SSO groups. However, during reperfusion of a previously ischemic heart, both PO-I- and PO-Il- supplemented diets appeared to be as effective as sunflower seed oil (SS0) in reducing ventricular premature beats in rats. Further, the incidence of animals exhibiting severe ventricular fibrillation was much less after palm-oil consumption than it was after saturated animal fat (SF) feeding.

Serbinova et al. (1992b) demonstrated that palm oil vitamin E was more efficient in the protection of isolated Langendorff heart against ischemia/reperfusion injury than tocopherol as measured by its mechanical recovery when subjected to 40 min of global ischemia. Palm oil vitamin E completely curbed lactate dehydrogenase enzyme leakage from ischemic hearts, prevented the decrease in ATP and creatine phosphate levels and suppressed the formation of endogenous lipid peroxidation products. The data indicated that a palm oil vitamin E mixture containing both α-tocopherol and α-tocotrienol may be more efficient than α-tocopherol alone in the protection of the heart against oxidative stress induced by ischemia-reperfusion. Serbinova and Packer (1994) assessed the antioxidant properties of α-tocopherol, α-tocotrienol, and palm oil vitamin E, which contained 45% tocopherols and 55% tocotrienols. When vitamin E-deficient rats were fed either α-tocopherol- or α-tocotrienol-enriched diets, α-tocotrienol accumulated in the hearts and liver more slowly than α-tocopherol. The rate of lipid peroxidation induced in-vitro in heart homogenate from rats supplemented with α-tocotrienol was approximately two-thirds as high as that from rats with an equivalent concentration of α-tocopherol. Thus, palm oil vitamin E may be more efficient than α-tocopherol alone in protecting the heart against injury from ischaemia and reperfusion. Further, supplementation with α-tocopherol or α-tocotrienol protected skeletal muscles against exercise-induced increased in protein oxidation. Thus, palm oil vitamin E protected biological systems against both lipid and protein oxidation.

Studies conducted by Esterhuyse et al. (2005a) demonstrated in rats that Red Palm Oil (RPO)-supplementation protected hearts against the consequences of ischemia/reperfusion injury. RPO consists of saturated (SFAs), mono-unsaturated (MUFAs) and polyunsaturated (PUFAs) fatty acids and is an antioxidant rich in natural β-carotene and vitamin E (tocopherols and tocotrienols). The findings suggested that dietary RPO protected through the NO-cGMP (cyclic guanosine monophosphate) pathway and/or changes in PUFA composition during ischemia/reperfusion. Their studies suggested that that dietary red palm oil (RPO) containing fatty acids, carotonoids, tocopherols and tocotrienols may improve myocardial ischaemic tolerance by increasing bioavailability of NO and improving NO-cGMP signaling in the heart (Esterhuyse et al. 2005b, 2006). RPO-supplementation improved aortic output recovery and increased myocardial ischaemic cGMP levels. Simulated ischaemia (hypoxia) raised cardiomyocyte nitric oxide concentrations in the two RPO supplemented groups, but not in control non-supplemented groups. RPO supplementation also increased hypoxic nitric oxide concentrations in the control diet fed, but not the cholesterol fed rats. Further studies by Engelbrecht et al. (2006) suggested that dietary RPO supplementation caused differential phosphorylation of the MAPKs and PKB/Akt during ischemia/reperfusion-induced injury. These changes in phosphorylation were associated with improved functional recovery and reduced cleavage of an apoptotic marker, indicating that dietary RPO supplementation may confer protection through the MAPK and PKB/Akt signaling pathways during ischemia/reperfusion-induced injury. Further studies were conducted by the group in relation to the effects of RPO supplementation of a high-cholesterol diet on ischaemia-reperfusion injury (Kruger et al. 2007). Cholesterol supplementation caused a poor aortic output recovery compared with the control group, but when RPO was added, the percentage aortic output increased significantly. The cholesterol group's poor aortic output was accompanied by a significant increase in p38-MAPK phosphorylation, whereas the CRPO-supplemented group showed a significant reduction in p38-MAPK phosphorylation when compared with the cholesterol-supplemented group. This significant reduction in p38-MAPK was also associated with reduced apoptosis as indicated by significant reductions in caspase-3 and poly(ADP-ribose) polymerase (PARP) cleavage. They also demonstrated in recent studies that PI3-K inhibition induced PARP cleavage (marker of apoptosis) in the heart during ischaemia/reperfusion injury and that RPO supplementation counteracted this effect (Engelbrecht et al. 2009).

In an earlier study, Das et al. (2005), reported that the cardioprotective properties of tocotrienol-rich fraction (0.035%) of palm oil (TRF) appeared to be due to inhibition of c-Src activation and proteasome stabilization. As expected, ischemia-reperfusion caused ventricular dysfunction, electrical rhythm disturbances, and increased myocardial infarct size. TRF could reverse the ischemia-reperfusion-mediated cardiac dysfunction. Ischemia-reperfusion also upregulated c-Src expression and phosphorylation. In subsequent studies, Das et al. (2008) demonstrated that tocotrienol isoforms reduced c-Src but increased the phosphorylation of Akt, thus generating a survival signal. The γ-isoform of tocotrienol was the most cardioprotective of all the isomers followed by the – and δ-isoforms.

Abeywardena et al. (2002) found that oil palm frond extract (rich in polyphenols) resulted in considerable relaxation (>75%) in rings of rat thoracic aorta and isolated perfused mesenteric vascular beds and was found to be endothelium-dependent as removal of endothelium or inhibition of endogenous nitric oxide (NO) led to a total loss in relaxant activity. Of the extracts tested, palm fronds also demonstrated the highest antioxidant capacity, as determined by the ferric reducing activity/potential assay, and resulted in a significant delay in the oxidation of LDL. In toto, these preliminary findings lend further support to the potential cardiovascular actions of plant polyphenols and also identified oil palm fronds as containing constituents that promoted vascular relaxation via endothelium-dependent mechanisms.

Anticancer Activity

Carotenoids and tocopherols have been reported to exert a protective action against some types of cancer.

Studies showed that rats fed a high dietary intake of lard (20%) and beef tallow(20%), but not vegetable fat (5% (normal fat) corn oil; 20% (high fat) corn oil; 20% palm oil, from weaning until only 1 week after dimethylbenz(a)anthracene (DMBA) administration, exhibited significantly enhanced mammary tumorigenesis (Sylvester et al. 1986). Separate studies by Sundram et al. (1989), showed that high palm oil diets did not promote DMBA- induced mammary tumorigenesis in female rats when compared to high corn oil (CO) or soya bean oil (SOB) diets. CO and SBO differed greatly from the palm oils in their contents of tocopherols, tocotrienols, and carotenes. Rats fed 20% CO or SBO diet had higher tumour incidence than rats fed on palm oil (PO) diets; however differences of mean tumour latency periods among the groups were not statistically significant. At autopsy, rats fed on high CO or SBO diets had significantly more tumours than rats fed on the three palm oil diets viz. crude palm oil (CPO), refined, bleached, deodorized palm oil (RBDPO) and metabisulfite-treated palm oil (MCPO). In the DMBA induced rat mammary tumour model, palm oil appeared to be less tumourigenic than corn oil, soy bean oil, beef tallow and lard. DMBA rats administered the refined bleached and deodorized palm oil (RBDPO) diet after 5 months had the lowest tumour incidence, tumour count and tumours per rat compared to animals fed the corn oil (CO) diet (Nesaretnam et al. 1992). This suggested that corn oil had a greater tumour promoting effect than palm oil. Supplementation of TRF (tocotrienol rich fraction) at 500 ppm to corn oil did not provide protection. However, at the higher concentration of

1,000 ppm of TRF, the corn oil fed animals had significantly greater median latency period, lower tumour incidence and tumour count than the corn oil fed group as well as the group supplemented with either 500 ppm of TRF or 135 ppm α-tocopherol. The results appeared to indicate that supplementation of corn oil with adequate amounts of TRF may give some protection against the cancer promoting effect of corn oil. The results also suggested that the favourable effects observed on a palm oil diet may be attributed to the presence of naturally occurring tocotrienols in palm oil.

Tan and Chu (1991) reported on the in-vitro and in-vivo carcinogenenic effects of palm-oil carotenoid [β-carotene, α-carotene, or canthaxanthin] on benzo(a)pyrene (BaP) metabolism in the rat hepatic cytochrome P450-mediated monooxygenase system. They found that the order of anticarcinogenic strength was palm-oil carotenoid (β-carotene) > α-carotene > canthaxanthin. Increased formation of the detoxification intermediate, 3-hydroxy BaP, was noted with increased carotenoid concentration. The order of detoxification strength was β-carotene greater than α-carotene = canthaxanthin. The presence of carotenoids in-vivo inhibited BaP metabolism. The order of the antioxidative activity was palm oil (with carotenoids) greater than β-carotene greater than canthaxanthin greater than palm oil (without carotenoids). β-carotene and α-carotene may selectively modify the rat-liver microsomal enzymes, thus providing a biochemical mechanism for the inhibitory effect of palm carotenoids on chemical carcinogenesis. Manorama et al. (1993) showed that red palm oil (RPO) effectively prevented chemical carcinogenesis in rats more than refined, bleached deodorised olein (RBDO), and attributed the effectiveness of RPO to its carotenoids. RPO was found to enhance the activity of one of the detoxifying phase II enzymes, glutathione-S-transferase (GSH-T) but not the detoxifying phase I enzymes that were known to metabolize phenobarbitone and polycyclic aromatic hydrocarbons. This suggested that RPO afforded protection against chemical carcinogens, probably because of its carotenoid content. Significantly higher levels of GSH-T were observed in F_{2b} third generation rats and F_{3b} rats fourth generation rats given RPO than in those given groundnut oil or refined bleached deodorized palmolein oil (RBDPO).

Murakoshi et al. (1989) found that α-carotene inhibited the proliferation of the human neuroblastoma cell line GOTO in a dose- and time-dependent manner. In addition, it was about 10 times more inhibitory than β-carotene. α-carotene caused maximum suppression of the level of the N-myc messenger RNA of GOTO cells. When GOTO cells were exposed to α-carotene, they were arrested in the G0-G1 phase of their cell cycle. In subsequent studies, Murakoshi et al. (1992) found that the inhibitory effect of natural α-carotene, obtained from palm oil, was greater than that of β-carotene on spontaneous liver carcinogenesis in C3H/He male mice. The mean number of hepatomas per mouse was significantly decreased by α-carotene supplementation (per os administration in drinking water at a concentration of 0.05%, ad libitum) as compared with that in the control group. In contrast, β-carotene, at the same concentration as α-carotene, did not show any such significant difference from the control group. In the two-stage mouse lung carcinogenesis (initiator, 4-nitroquinoline 1-oxide; promoter, glycerol) activity, α-carotene, but not β-carotene, reduced the number of lung tumours per mouse to about 30% of that in the control group. The higher potency of the antitumour- promoting action of α-carotene compared to β-carotene was confirmed in other experimental systems; e.g., α-carotene was also found to have a stronger effect than β-carotene in suppressing the promoting activity of 12-O-tetradecanoylphorbol-13-acetate on skin carcinogenesis in 7,12-dimethylbenz[a]anthracene-initiated mice. These results suggested that not only β-carotene, but also other types of carotenoids, such as α-carotene, may play an important role in cancer prevention.

The tocotrienol-rich fraction (TRF) of palm oil consisting of tocotrienols and some α-tocopherol (α-T) was found to inhibit human breast cancer cell proliferation (Nesaretnam et al. 1995). TRF, α-, γ- and δ-tocotrienols inhibited proliferation of estrogen receptor-negative MDA-MB-435 human breast cancer cells with 50% inhibitory concentrations (IC_{50}) of 180, 90, 30 and 90 µg/ml, respectively, whereas α-tocopherol had no effect at concentrations up to 500 µg/ml. It was also found that TRF inhibited plating efficiency (PE) of the MDA-MB-435 estrogen-receptor-negative human breast cancer cells whereas α T had no effect. Further experiments with estrogen receptor-positive MCF-7 cells showed that tocotrienols also inhibited their proliferation, as measured by [3H] thymidine incorporation (Guthrie et al. 1997). The IC_{50}s for TRF, α-tocopherol, α-, γ- and δ-tocotrienols were 4, 125, 6, 2 and 2 µg/ml, respectively. Tamoxifen, a widely used synthetic antiestrogen inhibited the growth of MCF-7 cells with an IC_{50} of 0.04 µg/ml. Also, 1:1 combinations of TRF, α-tocopherol and the individual tocotrienols with tamoxifen in both cell lines were found to be synergistic in the MDA-MB-435 cells. In the MCF-7 cells, only 1:1 combinations of γ- tocotrienol or δ-tocotrienol with tamoxifen showed a synergistic inhibitory effect on the proliferative rate and growth of the cells. The inhibition by tocotrienols was not overcome by addition of excess estradiol to the medium. The results suggested that tocotrienols were effective inhibitors of both estrogen receptor-negative and estrogen receptor-positive cells and that combinations with tamoxifen should be considered as a possible improvement in breast cancer therapy.

In further studies, Nesaretnam et al. (1998) found that the tocotrienol-rich fraction (TRF) of palm oil inhibited growth of estrogen-responsive (ER+) MCF7 human breast cancer cells in both the presence and absence of estradiol with a nonlinear dose-response such that complete suppression of growth was achieved at 8 µg/mL. The estrogen-unresponsive (ER-) MDA-MB-231 human breast cancer cells were also inhibited by TRF but with a linear dose-response such that 20 µg/ml TRF was needed for complete growth suppression. Separation of the TRF into individual tocotrienols revealed that all fractions could inhibit growth of both ER + and ER- cells and of ER + cells in both the presence and absence of estradiol. However, the γ- and δ-fractions were the most inhibitory. Complete inhibition of MCF7 cell growth was achieved at 6 µg/ml of γ-tocotrienol/δ-tocotrienol (γT3/δT3) in the absence of estradiol and 10 µg/ml of δT3 in the presence of estradiol, whereas complete suppression of MDA-MB-231 cell growth was not attained even at doses of 10 µg/ml of δT3. By contrast to these inhibitory effects of tocotrienols, αT had no inhibitory effect on MCF7 cell growth in either the presence or the absence of estradiol, nor on MDA-MB-231 cell growth. These results confirmed studies using other sublines of human breast cancer cells and demonstrated that tocotrienols could exert direct inhibitory effects on the growth of breast cancer cells. Subsequent studies showed that the tocotrienol-rich fraction (TRF) of palm oil and individual fractions (α, γ and δ) could also inhibit the growth of another responsive human breast cancer cell line, ZR-75-1 (Nesaretnam et al. 2000). At low concentrations in the absence of oestrogen, tocotrienols promoted growth of the ZR-75-1 cells, but at higher concentrations in the presence or the absence of oestradiol, tocotrienols inhibited cell growth strongly. As for MCF7 cells, α-tocopherol had no effect on growth of the ZR-75-1 cells in either the absence or presence of oestradiol. In studying the effects of tocotrienols in combination with antioestrogens, it was found that TRF could further inhibit growth of ZR-75-1 cells in the presence of tamoxifen (10^{-7} M and 10^{-8} M). Individual tocotrienol fractions (α, γ and δ) could inhibit growth of ZR-75-1 cells in the presence of 10^{-8} M oestradiol and 10^{-8} M pure antioestrogen ICI 164,384. These results provided evidence of wider growth-inhibitory effects of tocotrienols beyond MCF7 and MDA-MB-231 cells, and with an oestrogen-independent mechanism of action, suggesting a possible clinical advantage in combining administration of tocotrienols with antioestrogen therapy.

McIntyre et al. (2000) conducted studies to determine the comparative effects of tocopherols and tocotrienols on preneoplastic (CL-S1), neoplastic (-SA), and highly malignant (+SA) mouse mammary epithelial cell growth and viability in-vitro. Over a 5-day culture period, treatment with 0–120 µM α- tocopherol and γ-tocopherol had no effect on cell proliferation, whereas growth was inhibited 50% (IC_{50}) as compared with controls by treatment with the following: 13, 7, and 6 µM tocotrienol-rich-fraction of palm oil (TRF); 55, 47, and 23 µM δ-tocopherol; 12, 7, and 5 µM α-tocotrienol; 8, 5, and 4 µM γ-tocotrienol; or 7, 4, and 3 µM δ-tocotrienol in CL-S1, -SA and +SA cells, respectively. Acute 24-h exposure to 0–250 µM α-tocopherol or γ-tocopherol (CL-S1, -SA, and +SA) or 0–250 µM δa-tocopherol (CL-S1) had no effect on cell viability, whereas cell viability was reduced 50% (LD_{50}) as compared with controls by treatment with 166 or 125 µM δ-tocopherol in -SA and +SA cells, respectively. Additional LD_{50} doses were determined as the following: 50, 43, and 38 µM TRF; 27, 28, and 23 µM α-tocotrienol; 19, 17, and 14 µM γ-tocotrienol; or 16, 15, or 12 µM δ-tocotrienol in CL-S1, -SA, and +SA cells, respectively. Treatment-induced cell death resulted from activation of apoptosis, as indicated by DNA fragmentation. Results also showed that CL-S1, -SA, and +SA cells preferentially accumulated tocotrienols as compared with tocopherols, and this may partially explain why tocotrienols displayed greater biopotency than tocopherols. These data also showed that highly malignant +SA cells were the most sensitive, whereas the preneoplastic CL-S1 cells were the least sensitive to the antiproliferative and apoptotic effects of tocotrienols, and suggested that tocotrienols may have potential health benefits in preventing and/or reducing the risk of breast cancer in women. Studies conducted by Shah et al. (2003) showed that treatment with tocotrienol-rich-fraction of palm oil (TRF) and g-tocotrienol, but not a-tocopherol, induced a dose-dependent decrease in +SA mouse mammary epithelial cell viability. TRF- and γ-tocotrienol-induced cell death resulted from apoptosis. Additional studies demonstrated that tocotrienol-induced apoptosis in +SA mammary cancer cells was mediated through activation of the caspase-8 signaling pathway and was independent of caspase-9 activation.

Data from studies conducted by Agarwal et al. (2004) suggested that tocotrienol-rich-fraction (TRF) of palm oil-induced apoptosis in human colon carcinoma RKO cells was mediated by p53 signaling network which appeared to be independent of cell cycle association. TRF treatment also resulted in alteration in Bax/Bcl2 ratio in favor of apoptosis, which was associated with the release of cytochrome c and induction of apoptotic protease-activating factor-1. Boateng et al. (2006) conducted a study to compare the inhibitory effects of red palm oil (RPO) (7% and 14% levels) and soybean oil (SBO) (7% and 14%) on azoxymethane (AOM) induced aberrant crypt foci. After 17 weeks, the numbers of aberrant crypt foci in the proximal and distal colon were: 39.9, 53.8, 26.0, 27.5 and 118.2, 125.6, 41, 52.3 in rats fed 7% SBO, 14% SBO, 7% RPO and 14% RPO, respectively. The results of this study showed that RPO reduced the incidence of AOM induced aberrant crypt foci and may therefore have a beneficial effect in reducing the incidence of colon cancer.

Tocotrienol-rich fraction of palm oil was found to induce cell cycle arrest and apoptosis selectively in human prostate cancer cells (Srivastava and Gupta 2006). Tocotrienol-rich fraction (TFR) resulted in significant reductions in cell viability and colony formation in all three prostate cancer cell lines. The IC_{50} values after 24 hours TRF treatment in LNCaP (androgen-sensitive human prostate adenocarcinoma cells), PC-3 prostate cancer cells, and DU145 prostate cancer cells were in the order 16.5, 17.5, and 22.0 µg/ml. Evidence from DNA fragmentation, fluorescence microscopy, and ELISA showed that TRF treatment resulted in significant apoptosis in all the cell lines, whereas the normal human prostate epithelial cells (PrEC) and normal human prostate epithelial cells (PZ-HPV-7 cells) did not undergo apoptosis, but showed modestly reduced cell viability only at a high dose of 80 µg/ml. TRF treatment (10–40 µg/ml) produced a dose-dependent G0/G1 phase arrest and sub G1 accumulation in all three cancer cell lines but not in PZ-HPV-7 cells. These results suggested that the palm oil derivative TRF was capable of selectively inhibiting cellular proliferation and accelerating apoptotic events in prostate cancer cells. TRF offers significant promise as a chemopreventive and/or therapeutic agent against prostate cancer.

Palm oil was found to possess anti-skin tumour promoting activities (Kausar et al. 2003). Topical application of palm oil 1 hour prior to application of TPA (12-O-Tetradecanoyl-phorbol-13-acetate) resulted in significant prophylaxis against skin tumour promotion. The animals pre-treated with palm oil showed a reduction in both tumour incidence and tumour yield as compared to the TPA (alone)-treated group. Palm oil application also reduced the development of malignant tumours. The results of the study thus suggested that palm oil possessed anti-skin tumour promoting effects, and that the mechanism of such activities may involve the inhibition of tumour promoter-induced epidermal ornithine decarboxylase (ODC) activity, [(3)H]thymidine incorporation and cutaneous oxidative stress.

Recent studies in China reported that γ-tocotrienol, a major component of the tocotrienol-rich fraction of palm oil, exhibited antitumour activity against human colon carcinoma HT-29 cells (Xu et al. 2009) and human gastric adenocarcinoma SGC-7901 cells (Sun et al. 2008; 2009). The results of studies of Xu et al. (2009) indicated that γ-tocotrienol inhibited cell proliferation and induced apoptosis in human colon carcinoma HT-29 cells in a time- and dose-dependent fashion, and that the process was accompanied by cell-cycle arrest at G(0)/G(1), an increased Bax/Bcl-2 ratio, and activation of caspase-3. The data also indicated that nuclear factor-kappaB p65 protein may be involved in those effects. (Sun et al. 2008, 2009) reported that γ-tocotrienol could induce the apoptosis on human gastric cancer SGC-7901 cells through mitochondria-dependent apoptosis pathway in a concentration- and time-dependent mode. The results suggested that key regulators in tocotrienol-induced apoptosis may be Bcl-2 families and caspase-3 in SGC-7901 cells via down regulation of the Raf-ERK signalling pathway. Further studies confirmed the inhibitory effects of γ-tocotrienol on invasion and metastasis of human gastric adenocarcinoma SGC-7901 cells (Liu et al. 2010). γ-tocotrienol significantly reduced the matrigel invasion capability through down-regulation of the mRNA expressions of MMP-2 and MMP-9, and up-regulation of tissue inhibitor of metalloproteinase-1 (TIMP-1) and TIMP-2 in SGC-7901 cells.

Recent studies showed that tocotrienols (T3) of palm oil inhibited growth and induced apoptosis in human HeLa cells (Wu and Ng 2010). Results showed that the antiproliferative effect of α-tocotrienol (IC_{50}: 3.19 μM) and γ-tocotrienol (IC_{50}: 2.85 μM) was more potent than δ-tocotrienol (IC_{50}: >100 μM) and α-tocotrienol (IC_{50}: 69.46 μM). Both α-tocotrienol and γ-tocotrienol also demonstrated a dose-dependent and time-dependent induction of cell death. They caused cell cycle arrest at G2/M phase and triggered apoptosis. The results suggest that α-tocotrienol and γ-tocotrienol were more effective than δ-tocotrienol and α-tocotrienol in inhibiting HeLa cell proliferation, and their mode of action could be through the upregulation of IL-6, and the downregulation of cyclin D3, p16, and CDK6 expression in the cell cycle signaling pathway.

Antiangiogenic Activity

Studies had revealed that tocotrienol-rich fractions (TRF) from palm oil inhibited the proliferation and the growth of solid tumours. For both prevention and therapy of various human ailments including cancer, natural compounds such as tocotrienols are preferred over synthetic ones due to their lower toxicity (Nesaretnam 2008). The anticancer activity of tocotrienol-rich fractions from palm oil (TRF) was suggested to be caused by several mechanisms, one of which was antiangiogenesis (Wong et al. 2009). In-vitro investigations of the antiangiogenic activities of TRF, δ-tocotrienol, and α-tocopherol (αToc) were carried out in human umbilical vein endothelial cells (HUVEC) (Wong et al. 2009). TRF and δ-tocotrienol significantly inhibited cell proliferation from 4 μg/ml onward. Cell migration was inhibited the most by δ-tocotrienol at 12 μg/ml. Anti-angiogenic properties of TRF were carried out further in-vivo using the chick embryo chorioallantoic membrane (CAM) assay and BALB/c mice model. TRF at 200 μg/ml reduced the vascular network on CAM. TRF treatment of 1 mg/mouse significantly reduced 4 T1 tumour volume and in vascular endothelial growth factor (VEGF) level BALB/c mice. The study showed that palm tocotrienols exhibited anti-angiogenic properties that may assist in tumour regression.

Studies reported that tocotrienol found in rice bran and palm oil, especially δ, β, and γ- tocotrienol had potent anti-angiogenic activity in-vitro and in-vivo (Inokuchi et al. 2003; Miyazawa et al. 2008). Angiogenesis is the formation of new blood vessels from preexisting vascular, is of fundamental importance in several pathological states such as tumour growth, rheumatoid arthritis, and diabetic retinopathy. Angiogenesis was reported to involve a set of steps, including activation and movement of endothelial cells and tube formation, inhibition of growth factor-induced proliferation, migration and tube formation in human umbilical vein endothelial cells. Tocotrienol showed inhibition of tumour cell-induced angiogenesis in mouse dorsal air sac (DOS) assay. These results indicated that tocotrienol was a potent anti-angiogenesis compound. Tocopherol did not inhibit angiogenesis. Anti-angiogenic mechanism of tocotrienol mediated inhibition of growth factor induced survival, migration and angiogenesis signals. These findings suggested that tocotrienol may have potential for preventing angiogenic disorders in humans and may have potential use as a therapeutic dietary supplement for minimizing tumour angiogenesis.

Tocotrienols were found to selectively inhibit the activity of mammalian DNA polymerase λ (pol λ) in-vitro (Mizushina et al. 2006). These compounds did not influence the activities of replicative polymerase such as α, δ, and ε, or even the activity of polymerase β which was thought to have a very similar three-dimensional structure to the polymerase β-like region of polymerase λ. Since δ-tocotrienol had the strongest inhibitory effect among the four (α- to δ-) tocotrienols, the isomer's structure might be an important factor in the inhibition of polymerase λ. The inhibitory effect of δ-tocotrienol on both intact pol X (residues 1–575) and a truncated pol λ lacking the N-terminal BRCA1 C-terminus (BRCT) domain (residues 133–575, del-1 pol λ) was dose-dependent, with 50% inhibition observed at a dose of 18.4 and 90.1 μM, respectively. However, del-2 pol X containing the C-terminal pol β-like region was unaffected. Tocotrienols also inhibited the proliferation of and formation of tubes by bovine aortic endothelial cells, with δ-tocotrienol having the greatest effect. These results indicated that tocotrienols targeted both pol X and angiogenesis as anti-cancer agents.

Amelioration of Chronic Pancreatitis Activity

Studies showed that tocotrienols caused apoptosis and autophagy of activated rat pancreatic stellate cells (PSCs) through the mitochondrial death pathway (Rickmann et al. 2007). Tocotrienol rich fraction from palm oil (TRF), but not α-tocopherol, reduced viability of activated pancreatic stellate cells by setting up a full death program, independent of cell cycle regulation. Activated pancreatic stellate cells died both through apoptosis, as evidenced by increased DNA fragmentation and caspase stimulation, and through autophagy, as indicated by the formation of autophagic vacuoles and LC3-II (membrane-bound microtubule-associated protein 1 light chain 3) accumulation. β-, γ-, and δ-tocotrienol, but not α-tocotrienol nor α-tocopherol, emulated TRF actions on activated PSCs. TRF death induction was confined to activated PSCs because it did not induce apoptosis either in quiescent PSCs or in acinar cells. The data showed that tocotrienols selectively elicit activated pancreatic stellate cell mortality by targeting the mitochondrial permeability transition pore. The findings disclosed a novel potential for tocotrienols to ameliorate the fibrogenesis associated with chronic pancreatitis.

Neuroprotective Activity

Palm oil derived from *Elaeis guineensis* represents the richest source of the lesser characterized vitamin E, α-tocotrienol (Sen et al. 2010). One of eight naturally occurring and chemically distinct vitamin E analogs, α-tocotrienol has unique biological activity that is independent of its potent antioxidant power. Developments in α-tocotrienol research demonstrated neuroprotective properties for the lipid-soluble vitamin in brain tissue rich in polyunsaturated fatty acids (PUFAs). Arachidonic acid (AA), one of the most abundant PUFAs of the central nervous system, was found to be highly susceptible to oxidative metabolism under pathologic conditions. A number of neurodegenerative conditions in the human brain had been reported to be associated with disturbed PUFA metabolism of arachidonic acid, including acute ischemic stroke. Palm oil-derived α-tocotrienol at nanomolar concentrations was shown to weaken both enzymatic and nonenzymatic mediators of arachidonic acid metabolism and neurodegeneration. On a concentration basis, this represented the most potent of all biological functions exhibited by any natural vitamin E molecule.

Numerous studies had reported on the neuroprotective function of toctrienols. The tocotrienol-rich-fraction (TRF) from palm oil, was found to significantly inhibit oxidative damage in-vitro to both lipids and proteins in rat brain mitochondria induced by ascorbate-Fe^{2+}, the free radical initiator azobis (2-amidopropane) dihydrochloride (AAPH) and photosensitisation (Kamat and Devasagayam 1995). The inhibitory effect was both time-and concentration-dependent. At a low concentration of 5 μM, TRF could significantly inhibit oxidative damage to both lipids and proteins. The inhibitory effect of TRF appeared to be primarily due to γ-tocotrienol and to a lesser extent α- and δ-tocotrienols. TRF was significantly more effective than α-tocopherol. This fraction from palm oil could be considered a natural antioxidant supplement capable of protecting the brain against oxidative damage and thereby from the ensuing adverse alterations. Six different isoforms of vitamin E were evaluated on the ischemic brain damage in the mice middle cerebral artery occlusion model (Mishima et al. 2003). α-tocopherol and its derivatives were demonstrated to be effective in attenuating cerebral ischemia-induced brain damage. α-tocopherol (2 mM), α-tocotrienol (0.2 and 2 mM) and γ-tocopherol (0.2 and 2 mM) were found to significantly decrease the size of the cerebral infarcts 1 day after the middle cerebral artery occlusion, while γ-tocotrienol, δ-tocopherol and δ-tocotrienol showed no effect on the cerebral infarcts. These results suggested α-tocotrienol and γ-tocopherol to be potent and effective agents for preventing cerebral infarction induced by middle cerebral artery occlusion.

Studies conducted by Osakada et al. (2004) found that a tocotrienol-rich fraction of edible oil derived from palm oil (Tocomin 50%), containing α-tocopherol, and α-, γ- and δ-tocotrienols, significantly inhibited hydrogen peroxide (H_2O_2)-induced neuronal death. Each of the tocotrienols, purified from Tocomin 50%, significantly attenuated H_2O_2-induced neurotoxicity, whereas α-tocopherol did not. α-, γ- and δ-tocotrienols also provided significant protection against the cytotoxicity of a superoxide donor, paraquat, and nitric oxide donors, S-nitrosocysteine and 3-morpholinosydnonimine. Moreover, tocotrienols obstructed oxidative stress-mediated cell mortality with apoptotic DNA fragmentation caused by an inhibitor of glutathione synthesis, L-buthionine-[S,R]-sulfoximine. Further, αa-tocotrienol, but not γ- or δ-tocotrienol, prevented oxidative stress-independent apoptotic cell death, DNA cleavage and nuclear morphological changes caused by a non-specific protein kinase inhibitor, staurosporine. These findings suggested that α-tocotrienol could exert anti-apoptotic neuroprotective action independently of its antioxidant property. Among the vitamin E analogs examined, α-tocotrienol exhibited the most potent neuroprotective actions in rat striatal cultures.

Tocotrienols were found to be more effective than α-tocopherol in preventing glutamate-induced death in HT4 hippocampal neuronal cells (Sen et al. 2000). Uptake of tocotrienols from the culture medium was more efficient compared with that of α-tocopherol. Results showed that at low concentrations, tocotrienols may have protected cells by an antioxidant-independent mechanism. Examination of signal transduction pathways revealed that protein tyrosine phosphorylation processes had a dominant role in the execution of death in neuronal cells. Nanomolar concentrations of α-tocotrienol, but not α-tocopherol, thwarted glutamate-induced death by suppressing early activation of c-Src kinase. At a concentration 4–10-fold lower than levels detected in plasma of supplemented humans, tocotrienol modulated unique signal transduction processes that were not sensitive to comparable concentrations of tocopherol. Further studies confirmed α-tocotrienol as a potent neuroprotective form of vitamin E (Khanna et al. 2003). Their study on HT4 and immature primary cortical neurons suggested that 12-lipoxygenase (12-LOX) had a dominant role in executing glutamate-induced neurodegeneration. The activity of 12-LOX, the key mediator of glutamate-induced neurodegeneration was found to be modulated by α-tocotrienol. The data indicated that in neuronal cells, nmol/L concentrations of α-tocotrienol (TCT), but not α-tocopherol (TCP), prevented glutamate-induced mortality by suppressing early activation of c-Src kinase and 12-lipoxygenase. Further studies (Khanna et al. 2005) showed that subattomole quantity of TCT, but not TCP, protected neurons from glutamate challenge. Pharmacological as well as genetic approaches revealed that 12-Lox was rapidly tyrosine phosphorylated in the glutamate-challenged neuron and that this phosphorylation was catalyzed by c-Src. 12-Lox-deficient mice were more resistant to stroke-induced brain injury than their wild-type controls. Oral supplementation of TCT to spontaneously hypertensive rats led to elevated TCT levels in the brain. TCT-supplemented rats exhibited more protection against stroke-induced injury compared with challenged controls. Such protection was associated with lower c-Src activation and 12-Lox phosphorylation at the stroke site. The findings showed that the natural vitamin E, TCT, acted on key molecular checkpoints to thwart glutamate- and stroke-induced neurodegeneration.

In a series of earlier studies Sen et al. (2000) and Khanna et al. (2003, 2005) found that nanomolar α-tocotrienol, not α-tocopherol, was potently neuroprotective. In subsequent studies, Khanna et al. (2006) found that the neuroprotective of oral α-tocotrienol may involve antioxidant-independent as well as antioxidant-dependent mechanisms, this was confirmed by separating the antioxidant-independent and antioxidant-dependent neuroprotective properties of α-tocotrienol using two different triggers of neurotoxicity, homocysteic acid (HCA) and linoleic acid. Both HCA and linoleic acid caused neurotoxicity with comparable features, such as increased ratio of oxidized to reduced glutathione GSSG/GSH, raised intracellular calcium concentration and compromised mitochondrial membrane potential. Mechanisms underlying HCA-induced neurodegeneration were comparable to those in the path implicated in glutamate-induced neurotoxicity. Micromolar, but not nanomolar, α-tocotrienol functioned as an antioxidant which was confirmed in a model involving linoleic acid-induced oxidative stress and cell death. Oral supplementation of α-tocotrienol to humans resulted in a peak plasma concentration of 3 μ.

Administration of tocotrienol to individuals with familial dysautonomia (FD) resulted in increased expression of both functional IKAP (I-KappaB-associated protein) and MAO A (monoamine oxidase) A transcripts in their peripheral blood cells (Anderson and Rubin 2005). Familial dysautonomia (FD), a recessive neurodegenerative disease, is caused by mutations in the IKBKAP (I-KappaB-Associated protein) gene that result in the production of nonfunctional IKAP protein. Monoamine oxidases are key isoenzymes responsible for degrading biogenic and dietary monoamines, in individuals with FD. Manifestations of FD include autonomic crises characterized by hypertension, tachycardia, diaphoresis, and vomiting. These findings provide new insight into the pathophysiology of FD and demonstrate the value of therapeutic approaches designed to elevate cellular levels of functional IKAP and monoamine oxidase A.

Palm Tocotrienols and Antiosteoporotic Activities

Studies conducted by Ima-Nirwana et al. (1998) found that, significant decline in bone mineral density (BMD) due to testosterone deficiency in male rats was best seen in the proximal femur, the site of predominantly cortical bone. Palm vitamin E was found to be effective in long-term maintenance of BMD in the proximal femur of testosterone-deficient male rats, while palm oil was not. Whole body BMD was higher in orchidectomised rat group supplemented with palm vitamin E compared to palm oil. BMD of the whole femur was higher in the SH control group compared to the ORX group supplemented with palm oil. In further studies, they found that palm vitamin E decreased bone resorption to a greater extent than bone formation in thyrotoxic rats, suggesting a net reduction in bone loss (Ima-Nirwana et al. 1999). The action of palm vitamin E was probably related to its antioxidant properties. Survival rates were also significantly increased in thyrotoxic rats given palm vitamin E, suggesting the role of free radicals in the overall morbidity and mortality in thyrotoxicosis. Ima-Nirwana and Suhaniza (2004) determined the effects of two isomers of vitamin E (i.e., palm oil-derived γ-tocotrienol and the commercially available α-tocopherol, 60 mg/kg of body weight/day) on body composition and bone calcium content in adrenalectomized rats replaced with two doses of dexamethasone, 120 and 240 µg/kg daily for 8 weeks. γ-tocotrienol (60 mg/kg of body weight/day) was found to reduce body fat mass and increase the fourth lumbar vertebra bone calcium content in these rats, while α-tocopherol (60 mg/kg of body weight/day) was ineffective. They concluded that palm oil-derived γ-tocotrienol had the potential to be utilized as a prophylactic agent in prevention of severe side effects such as obesity and osteoporosis brought about by long-term glucocorticoid use.

Norazlina et al. (2000) reported that vitamin E could impact on bone metabolism. Their findings in Sprague-Dawley female rats indicated that supplementation with both palm vitamin E and α-tocopherol maintained bone mineral density in ovariectomised rats but caused conflicting effects on bone calcium content. Alkaline phosphatase activity was elevated in ovariectomised rats supplemented with palm vitamin E 30 mg/kg rat weight and α-tocopherol 30 mg/kg rat weight compared to the intact rats. α-tocopherol also reduced the activity of tartrate-resistant acid phosphatase post-ovariectomy.

Studies conducted by Watkins et al. (2001) showed that rats demonstrated significant differences in tissue fatty acid composition that reflected the dietary lipid treatments. The concentrations of saturated fatty acids and monounsaturated fatty acids were higher but polyunsaturated fatty acids (PUFA) lower in tibia marrow of rats given the palm oil treatments compared with those administered soybean oil. The decrease in total n-6 PUFA concentration in bone of rats given palm oil was related to higher bone specific alkaline phosphatase activity and endosteal mineral apposition rate and bone formation rate in tibia. No difference was found in ex-vivo prostaglandin E2 production in femur from rats. The histomorphometric analyses indicated that red palm olein might support bone formation and bone modeling in growing rats. The studies suggested that palm oil or palm antioxidants may play a role in bone formation. In subsequent studies, Norazlina et al. (2002) found that rats on vitamin E deficient (VED) diet and 50% vitamin E deficient (50%VED) diets had lower bone calcium content in the left femur and L5 vertebra compared with rats fed on normal rat chow (RC, control group). Supplementing the VED group with PVE60 (60 mg/kg palm vitamin) improved bone calcification in the left femur and L5 vertebra while supplementation with PVE30 (30 mg/kg palm vitamin E) improved bone calcium content in the L5 vertebra. However, supplementation with ATF (30 mg/kg pure α-tocopherol) did not change the lumbar and femoral bone calcium content compared to the VED group. Supplementing the RC group with PVE30, PVE60 or ATF did not cause any significant changes in bone calcium content. They concluded that vitamin E deficiency impaired bone calcification. Supplementation with the higher dose of palm vitamin E improved bone calcium content, but supplementation with pure ATF alone did not. This effect may be attributed to the tocotrienol content of palm vitamin E. Therefore, tocotrienols play an important role in bone calcification.

Tocotrienol was found to confer better protection than tocopherol from free radical-induced damage of rat bones (Ahmad et al. 2005). Free radicals generated by ferric nitrilotriacetate (FeNTA) could activate osteoclastic activity and this was associated with elevation of the bone resorbing cytokines interleukin (IL)-1 and IL-6. In their study, ferric nitrilotriacetate (FeNTA) was found to elevate levels of interleukin (IL)-1 and IL-6 in Wistar rats. Only the palm oil tocotrienol mixture at doses of 60 and 100 mg/kg was able to thwart FeNTA-induced increases in IL-1. Both the palm oil tocotrienol mixture and α-tocopherol acetate, at doses of 30, 60 and 100 mg/kg, were able to reduce FeNTA -induced increases in interleukin IL-6. Therefore, the palm oil tocotrienol mixture was better than pure α-tocopherol acetate in protecting bone against FeNTA (free radical)-induced elevation of bone-resorbing cytokines. Supplementation with the palm oil tocotrienol mixture or α-tocopherol acetate at 100 mg/kg restored the reduction in serum osteocalcin levels due to ageing. All doses of the palm oil tocotrienol mixture lowered urine deoxypyridinoline cross-link (DPD) significantly compared with the control group, whereas a trend for decreased urine DPD was only observed for doses of 60 mg/kg onwards of α-tocopherol acetate. Bone histomorphometric analyses showed that FeNTA injections significantly reduced mean osteoblast number and the bone formation rate, but elevated osteoclast number and the ratio of eroded surface/bone surface compared with the saline (control) group. Supplementation with 100 mg/kg palm oil tocotrienol mixture was able to prevent all these FeNTA-induced changes, but a similar dose of α-tocopherol acetate was found to be effective only for mean osteoclast number. Injections of FeNTA were also shown to reduce trabecular bone volume and trabecular thickness, whereas only supplementation with 100 mg/kg palm oil tocotrienol mixture was able to prevent these FeNTA-induced changes.

Further studies found that palm tocotrienol mixture was better than α-tocopherol in reversing the effects of nicotine on serum IL-1 and IL-6 (Norazlina et al. 2007). Both forms of vitamin E were not able to restitute the nicotine-induced bone calcium loss, but the group supplemented with α-tocopherol suffered a greater loss and had lower urine deoxypyridinoline (DPD) after treatment. Palm tocotrienol at the dose of 100 mg/kg body weight significantly lowered the thiobarbituric acid-reactive substance (TBARS) level in the rat femur with a significant enhancement in glutathione peroxidase activity compared to the age-matched control group (Maniam et al. 2008). These were not observed in the α-tocopherol groups. Palm tocotrienol was more effective than pure α-tocopherol acetate in curbing lipid peroxidation in bone. Palm tocotrienol exhibited better protective effect against free radical damage in the femur compared to α-tocopherol. The study suggested that palm tocotrienol played an important role in preventing imbalance in bone metabolism due to free radicals.

Tocotrienol appeared to be superior to α-tocopherol in combating against the adverse effect of nicotine on bone. Hermizi et al. (2009) found that rats in the nicotine (N, nicotine for 2 months), nicotine cessation (NC), groups had lower trabecular bone volume, mineral appositional rate (MAR), and bone formation rate (BFR/BS) and higher single labeled surface and osteoclast surface compared to the control group. Vitamin E treatment reverted these nicotine effects. Both the tocotrienol-enhanced fraction (TEF), γ-tocotrienol (GTT) groups, but not the α-tocopherol (ATF) group, had a significantly higher trabecular thickness but lower eroded surface (ES/BS) than the control group. The tocotrienol-treated groups had lower ES/BS than the ATF group. The GTT group showed a significantly higher MAR and BFR/BS than the TEF and ATF groups. They concluded that nicotine induced significant bone loss, while vitamin E supplements not only reversed the effects but also stimulated bone formation significantly above baseline values. Tocotrienol was shown to be slightly superior compared to tocopherol. Thus, vitamin E, especially GTT, may have therapeutic potential to repair bone damage caused by chronic smoking.

Ingestion of both fresh and once-heated palm oils, mixed with rat chow given daily for 6 months were able to offer protection against the negative effects of ovariectomy, but these protections were lost when the oils were heated five times (Shuid et al. 2007). Ovariectomy had caused negative effects on the bone histomorphometric parameters. Soy oil that was heated five times actually worsened the histomorphometric parameters of ovariectomised rats. Therefore, it may be better for postmenopausal who are at risk of osteoporosis to use palm oil as frying oil especially if they practice recycling of frying oils.

Gastroprotective Activity

Jaarin et al. (2000) found that palm vitamin E exhibited healing of ethanol-induced gastric lesions in Sprague Dawley rats. Palm vitamin E only caused a significant reduction in gastric malondialdehyde (MDA) but exhibited no significant effects on prostaglandin E 2 and gastric acid concentration. The mean ulcer index of palm vitamin E supplemented group killed after 1 week of ethanol exposure was significantly lower compared to the respective control. The gastric acid concentration was significantly higher in the group treated with palm vitamin E killed 1 week after ethanol exposure compared to control. The gastric tissue MDA was significantly lower in the palm vitamin E supplemented group compared to control. There was no significant difference in gastric mucus content of the both groups. The ulcer healing which occurred in the presence of a high gastric acid suggests that the effect of palm vitamin E on the healing of gastric lesions was not mediated via a reduction in gastric acid nor was it mediated through increasing prostaglandin E 2 or mucus production. The most probable mechanism was via reducing lipid peroxidation as reflected by a significant decreased in gastric tissue MDA content. In further studies, Jaarin et al. (2002) found that palm vitamin E in doses of 60 mg, 100 mg and 150 mg/kg food as well as tocopherol in doses of 20 mg, 30 mg and 50 mg/kg food were equally effective in preventing aspirin-induced gastric lesions. The gastric lesions index was significantly lower in all the vitamin E groups compared to control. The lowest ulcer index was observed in the groups that received 100 mg of palm vitamin E and 30 mg tocopherol in the diet. However, there was no significant difference in ulcer indices between palm vitamin E and tocopherol-treated groups. The lower ulcer index was only accompanied by lower gastric malondialdehyde content. The most probable mechanism was through their ability in limiting the lipid peroxidation involved in aspirin-induced gastric lesions. Similarly, Nafeeza et al. (2002) found that tocotrienol-rich fraction (TRF) obtained from palm oil and tocopherol were equally effective in preventing aspirin-induced gastric lesions.

Separate studies showed that the gastric acid concentration and serum gastrin level in stressed rats were significantly reduced compared to the non-stressed rats in the control and tocopherol (TF) groups (Azlina et al. 2005). However, the gastric acidity and gastrin levels in the tocotrienol (TT) group were comparable in stressed and non-stressed rats. The findings suggested that tocotrienol was able to preserve the gastric acidity and serum gastrin level which were changed in stressed conditions. The PGE(2) content and the plasma GLP-1 level were, however, comparable in all stressed and non-stressed groups indicating that these parameters were not altered in stress and that supplementation with TF or TT had no effect on the gastric PGE2 content or the GLP-1 level. The malondialdehyde, an indicator of lipid peroxidation was higher from gastric tissues in the stressed groups compared to the non-stressed groups. These findings implicated that free radicals may play a role in the development of gastric injury in stress and supplementation with either TF or TT was able to reduce the lipid peroxidation levels compared to the control rats. It was concluded that both tocopherol and tocotrienol were comparable in their gastro-protective ability against damage by free radicals generated in stress conditions, but only tocotrienol had the ability to block the stress-induced changes in the gastric acidity and gastrin level.

Anti-inflammatory Activity

Studies indicated that γ-tocopherol and its major metabolite possessed anti-inflammatory activity and that γ-tocopherol at physiological concentrations may be important in human disease prevention (Jiang et al. 2000). γ-tocopherol (γT) was found to decrease PGE(2) synthesis in both lipopolysaccharide (LPS)-stimulated RAW264.7 macrophages and IL-β-treated A549 human epithelial cells with an apparent IC_{50} of 7.5 and 4 μM, respectively. The major metabolite of dietary γ-tocopherol 2,7,8-trimethyl-2-(β-carboxyethyl)-6-hydroxychroman (γ-CEHC), also exhibited an inhibitory effect, with an IC_{50} of approximately 30 μM in these cells. In contrast, α-tocopherol at 50 μM slightly reduced (25%) PGE(2) (prostaglandin E(2)) formation in macrophages, but had no effect in epithelial cells. The inhibitory effects of γ-tocopherol and γ-CEHC stemmed from their inhibition of COX-2 activity, rather than affecting protein expression or substrate availability, and appeared to be independent of antioxidant activity. γ-CEHC also inhibited PGE(2) synthesis when exposed for 1 h to Cyclooxygenase-2 (COX-2)-2-preinduced cells followed by the addition of arachidonic acid (AA), whereas under similar conditions, γ-tocopherol required an 8- to 24-hours incubation period to cause the inhibition. In a subsequent study, Jiang and Ames (2003) found that administration of γT (Tocotrienol) (33 or 100 mg/kg) and γ-CEHC (2 mg/pouch), to male Wistar rats but not αT (33 mg/kg), significantly reduced PGE2 synthesis at the site of inflammation. γT, but not αT, significantly inhibited the formation of leukotriene B4, a potent chemotactic agent synthesized by the 5-lipoxygenase of neutrophils. Although γT had no effect on neutrophil infiltration, it significantly attenuated the partial loss of food consumption caused by inflammation-associated discomfort. Administration of γT led consistently led to a significant reduction of inflammation-mediated increase in 8-isoprostane, a biomarker of lipid peroxidation. γT at 100 mg/kg reduced TNF-α (65%), total nitrate/nitrite (40%), and lactate dehydrogenase activity (30%). Collectively, γT inhibited proinflammatory PGE2 and LTB4, decreased TNF-α, and attenuated inflammation-mediated damage. These findings provided strong evidence that γT has anti-inflammatory activities in-vivo that may be important for human disease prevention and therapy.

In separate studies, Yam et al. (2009) found that tocotrienol-rich fraction (TRF) at 10 μg/ml and all tocotrienol isoforms namely δ-, γ-, and α-tocotrienol significantly inhibited the production of interleukin-6 and nitric oxide in lipopolysaccharide-induced RAW264.7 macrophages. However, only α-tocotrienol demonstrated a significant effect in lowering tumour necrosis factor-α production. Besides, TRF and all tocotrienol isoforms except γ-tocotrienol reduced prostaglandin E(2) release. It was accompanied by the downregulation of cyclooxygenase-2 gene expression by all vitamin E forms except α-tocopherol. Collectively, the data suggested that tocotrienols are better anti-inflammatory agents than α-tocopherol and the most effective form is δ-tocotrienol. Zainal et al. (2009) modified a palm olein fraction by transesterification with the n-3 polyunsaturated fatty acids, α-linolenate or eicosapentaenoic acid (EPA). On stimulation of chondrocyte cultures with interleukin-1α, all the oils showed increased expression of cyclooxygenase-2, the inflammatory cytokines tumour necrosis factor-α (TNF-α), IL-1α and IL-1β and the proteinase ADAMTS-4. This increased expression was not affected by challenge of the cultures with palm olein alone but showed concentration-dependent reduction by the modified oil in a manner similar to EPA. These results showed that it was possible to modify palm oil conveniently to produce a nutraceutical with effective anti-inflammatory properties.

Antidiabetic Activity

Palm oil tocotrienol-rich fractions exhibited beneficial effects on streptozotocin-induced diabetic rats (Budin et al. 2009). Tocotrienol-rich fractions treatment reduced serum glucose and glycated hemoglobin concentrations. The tocotrienol-rich fractions group also showed significantly lower levels of plasma total cholesterol, low-density lipoprotein cholesterol, and triglyceride, as compared to the untreated group. The tocotrienol-rich fractions group had higher levels of high-density lipoprotein cholesterol, as compared to the untreated group. Superoxide dismutase activity and levels of vitamin C in plasma were increased in tocotrienol-rich fractions-treated rats. The levels of plasma and aorta malondealdehyde+4-hydroxynonenal (MDA+4-HNE) and oxidative DNA damage were significant following tocotrienol-rich fractions treatment. Tocotrienol-rich fractions supplementation resulted in a protective effect on the vessel wall. The results showed that tocotrienol-rich fractions lowered the blood glucose level and improved dyslipidemia. Levels of oxidative stress markers were also reduced by administration of tocotrienol-rich fractions. Vessel wall integrity was maintained due to the positive effects mediated by tocotrienol-rich fractions.

Kuhad et al. (2009) evaluated the impact of tocotrienol mixture on cognitive function and neuroinflammatory cascade in streptozotocin-induced diabetic rats. After 10 weeks of streptozotocin injection, the rats produced significant increase in transfer latency which was coupled with enhanced acetylcholinestrease activity, increased oxidative-nitrosative stress, TNF-α, IL-1β, caspase-3 activity in cytoplasmic lysate and active p65 subunit of NFκβ in nuclear lysate of cerebral cortex and hippocampus regions of diabetic rat brain. Interestingly, co-administration of tocotrienol significantly and dose-dependently prevented behavioral, biochemical and molecular changes associated with diabetes. Diabetic rats treated with insulin-tocotrienol combination produced more pronounced effect on molecular parameters as compared to their per se groups. Collectively, the data revealed that activation of NFκβ signaling pathway was associated with diabetes induced cognitive impairment and point towards the therapeutic potential of tocotrienol in diabetic encephalopathy.

Tocotrienol-enriched palm oil were shown to help reduce blood glucose levels in patients and preclinical animal models. Fang et al. (2010) presented data indicating that tocotrienols within palm oil functioned as PPAR (peroxisome proliferator-activated receptors) modulators. Specifically, both α- and γ-tocotrienol activated PPARα, while δ-tocotrienol activated PPARα, PPARγ, and PPARδ in reporter-based assays. Tocotrienols enhanced the interaction between the purified ligand-binding domain of PPARα with the receptor-interacting motif of coactivator PPARγ coactivator-1α. In addition, the tocotrienol-rich fraction of palm oil improved whole body glucose utilization and insulin sensitivity of diabetic Db/Db mice by selectively regulating PPAR target genes. The data collectively suggested that PPARs represent a set of molecular targets of tocotrienols.

Skin Protective Activity

Ikeda et al. (2000) used a vitamin E mixture extracted from palm oil, to examine the tissue distribution of dietary tocotrienols and tocopherols in rats and mice. Wistar rats (4-week-old) were fed a diet containing 48.8 mg/kg α-tocopherol, 45.8 mg/kg α-tocotrienol and 71.4 mg/kg γa-tocotrienol for 8 weeks. Nude mice (BALB/c Slc-nu, 8-week-old) and hairless mice (SKH1, 8-week-old) were fed the same diet for 4 weeks. α-tocopherol was abundantly retained in the skin, liver, kidney and plasma of rats and mice. α-tocotrienol and γ-tocotrienol were detected slightly in the liver, kidney and plasma, while substantial amounts of these tocotrienols were detected in the skin of both rats and mice. The study suggested that the skin exhibited unique ability to discriminate between various vitamin E analogs. Further studies showed that dietary sesame seeds elevated α- and γ-tocotrienol concentrations in skin and adipose tissue of rats fed the tocotrienol-rich fraction extracted from palm oil (Ikeda et al. 2001). Dietary sesame seeds markedly elevated the tocopherol concentration in rats. α- and γ-tocotrienol accumulated in the adipose tissue and skin, but not in plasma or other tissues of the rats fed tocotrienols. Dietary sesame seeds elevated tocotrienol concentrations in the adipose tissue and skin, but did not affect their concentrations in other tissues or in plasma. The γ-tocopherol concentration in all tissues and plasma of rats fed γ-tocopherol was extremely low but was elevated in many tissues by feeding sesame seeds. These data suggested that the transport and tissue uptake of vitamin E isoforms were different. Dietary tocotrienol was found to reduce UVB-induced skin damage and sesamin enhances tocotrienol effects in hairless mice (Yamada et al. 2008). Tocotrienols were detected in the skin of mice fed T-mix (tocotrienols rich fraction extracted from palm oil), but their concentrations were significantly lower than for α-tocopherol. Sesamin lignans elevated tocotrienol contents in the skin. In spite of the high α-tocopherol contents, the effects of α-tocopherol on sunburn and incidence of tumour were low. T-mix fed groups reduced the extent of sunburn and incidence of tumour, and further reduction of sunburn and incidence of tumour were noted in the T-mix with sesamin group. The results suggested that dietary tocotrienols protected the skin more strongly than α-tocopherol against damage induced by UVB and sesamin enhanced tocotrienol effects.

Ocular Activity

Topical administration of tocotrienols was found to be an effective way to enhance ocular tissue vitamin E concentration (Tanito et al. 2004). Rats were administered 5 μL of pure tocopherol or tocotrienol to each eye daily for 4 days. Various tissues of the eyes were separated and analyzed for tocopherol and tocotrienol concentrations. The concentration of α-tocotrienol increased appreciably in every tissue to which it was administered; however, no significant increase was noted in the case of α-tocopherol. The intraocular penetration of γ-tocopherol and γ-tocotrienol did not differ significantly. Further, a significant elevation in total vitamin E level was observed in ocular tissues, including crystalline lens, neural retina, and eye cup, with topical administration using a relatively small amount (5 μL) of vitamin E, whereas no significant increase was observed when the same amount of vitamin E was administered orally.

Nephroprotective Activity

Gupta and Chopra (2009) reported that administration of a nephrotoxic agent, ferric nitrilotriacetate (Fe-NTA), appreciably increased blood urea nitrogen (BUN) and serum creatinine levels, which were coupled with a marked lipid peroxidation, reduced activity of glutathione levels, and morphological alterations in rat kidney. Pre-treatment with tocotrienol-rich fraction, a product from palm oil, (50 mg/kg/day) and tocopherol (50 mg/kg/day) for 7 days before Fe-NTA administration significantly reduced the serum creatinine and BUN levels, reduced lipid peroxidation in a significant manner, and restored levels of reduced glutathione and superoxide dismutase. Tocotrienol pre-treatment also attenuated the serum tumour necrosis factor-α levels, as compared to pre-treatment with tocopherol, and restored normal renal morphology. These findings suggested a strong correlation between iron-induced oxidative stress and renal dysfunction and indicated the protective effects of tocotrienol in Fe-NTA-induced renal injury.

Exercise Endurance Activity

Lee et al. (2009) showed that the tocotrienol-rich fraction (TRF) -treated animals (268.0 min for TRF-25 and 332.5 min for TRF-50) swam significantly longer than the control (135.5 min) and T-25 (25 mg/kg D-α-tocopherol)-treated (154.1 min) animals, whereas there was no difference in the performance between the T-25 and control groups. The TRF-treated rats also exhibited significantly greater concentrations of liver glycogen, superoxide dismutase (SOD), catalase (CAT), and glutathione peroxidase (GPx), as well as of muscle glycogen and SOD than the control and the T-25-treated animals, but lower levels in blood lactate, plasma and liver TBARS, and liver and muscle protein carbonyl. Collectively, these results suggested that TRF was able to improve the physiological condition and reduce the exercise-induced oxidative stress in forced swimming rats.

Wound Healing Activity

The results of studies by Sasidharan et al. (2010) showed that the *E. guineensis* methanol leaf extract had potent wound healing activity, as evident from better wound closure, improved tissue regeneration at the wound site, and supporting histopathological parameters pertaining to wound healing. Assessment of granulation tissue every fourth day showed a significant reduction in microbial count. The presence of tannins, alkaloids, steroids, saponins, terpenoids, and flavonoids were established in the extract. The extract exhibited significant activity against *Candida albicans* with an MIC value of 6.25 mg/mL. The results showed that *E. guineensis* leaf extract accelerated wound healing in rats, thus supporting this traditional use.

Traditional Medicinal Uses

According to Hartwell (1967–1971), palm oil has been reported to be used as a liniment for indolent tumours. It is also reported to be anodyne, antidotal, aphrodisiac, diuretic, and vulnerary. *E. guineensis* leaf extract has been used traditionally for wound healing in some parts of Africa (Sasidharan et al. 2010). *Elaeis guineensis* leaf extract prepared by boiling and administered orally has been used as a herbal remedy for malaria in the Dangme West District of Ghana (Asase et al. 2010).

Other Uses

Palm oil, and various parts of the tree have a diversified array of non-edible uses (Salmiah 1995; Rajanaidu et al. 2000; Pantzaris and Ahmad 2001; Poku 2002; Ghazali et al. 2006; Sarmidi et al. 2009; Obire and Putheti 2010).

Palm oil has a wide range of applications, about 80% are used for food industries and the rest are used as feedstock for a number of non-food applications. The non-food uses of palm oil and palm kernel oil can be classified into two categories; using the oils directly or by processing them to oleochemicals (chemicals derived from oils or fats) (Salmiah 2000) and direct product applications include the use of CPO as a diesel fuel substitute, soaps, drilling mud, and epoxidised palm oil products (EPOP), polyols, polyurethanes and polyacrylates, PVP plastics (Salmiah 2000).

Oleochemicals

Oleochemicals are produced from the hydrolysis or alcoholysis of oils and fats. oleochemicals or derivatives based on C12-C14 and C16-C18 chain lengths have wide variety of uses. The five basic oleochemicals are fatty acids, fatty methyl esters, fatty alcohol, fatty nitrogen compounds and glycerols. (Salmiah 2000).

(a) Fatty acids-high temperature and high pressure splitting of Palm Oil or Palm kernel Oil to produce crude fatty acids and glycerin as a by product.
 – Distillation of the crude fatty acids to produce distilled or fractionated fatty acids which is a high purity fatty acids. Fatty acids are the most important oleochemicals.
 – Medium chain triglycerides for use in the flavour and fragrance industries.
 – Processing aids for rubber products, for softening and plasticising effect. Stearic acid is an essential ingredient in the vulcanisation of natural rubber and it imparts plasticity to the finished products.
 – Production of candles. Incorporation of palm based fatty acids in candle formulation improves the shrinkage property of the product.
 – Manufacture of cosmetic products from myristic, palmitic and stearic acids.
 – Production of soaps via a neutralisation process.
 – Production of non-metallic or non-sodium soaps.
 – Fatty acids are also widely used in the production of food emulsifiers.

(b) Fatty esters – Transesterification of palm oil or palm kernel oil with methanol to produce crude methyl ester and glycerin as a by product. Distillation of the methyl ester to produce distilled or fractionated fatty methyl ester. These products are then used in various applications such as detergents, plasti-

cizers, lubricants, cosmetics, and pharmaceutical products.
- Production of pure soap – better quality than soaps from fatty acids.
- α-sulphonated methyl esters (α-MES) as active ingredients for washing and cleaning.
- Liquid and powder detergents (anionic surfactants) (Salmiah et al. 2002).
- α-sulphonated methyl esters (α-MES) palm based soap for deinking newspaper printed with oil palm based ink (Ooi and Salmiah 1997).
- Palm-based methyl esters as a substitute for diesel fuel for vehicles and engines.

(c) Fatty alcohols – Hydrogenation of distilled or fractionated methyl ester at high temperature and pressure in the presence of catalyst to produce crude fatty alcohol. Small amounts of fatty alcohols and fatty alcohols ethoxylates derived from palm and lauric oils are used in cosmetic creams and lotions. Other uses include emulsifiers and shampoos, lubricants, detergents and fabric softeners.
- Fatty alcohol sulphates (anionic surfactants) production of personal care products.
- Fatty alcohol ethoxylates (nonionic surfactants) washing.
- Fatty alcohol ether sulphates (anionic surfactants) cleaning products.

(d) Fatty nitrogen compounds The most common fatty nitrogen compounds are fatty amides, nitriles, amines and important quartenary ammonium compounds which are cationic surfactants used as fabric softeners, hair conditioners, anti-caking agents for fertilizers, antistatic agents for bitumen in road construction.
- Imidazolines with good surface active properties (rust prevention).
- Esterquats as softeners.

(e) Glycerol (monoglycerides and diglycerides).
- Wide range of applications such as a solvent for pharmaceutical products, personal care products, humectants in cosmetics and tobacco, stabilisers, lubricants, conditioners, surface coatings, antifreeze, etc. (Rubaah 1999)
- Palm and lauric oil are the major sources of natural glycerine. It is a by-product of fat splitting and it finds wide applications in pharmaceuticals, toiletries, industrial explosives, and as alkyd resins in the paint industry.

Palm-based oleochemical products applications include

(a) Cosmetics and personal cares: lotions, creams, foundations, compacts powders, eye make-up, lipsticks, hair dyes, hair shampoos and conditioners, shower gels, shower cream, shower foam; toners, cleansers, moisturizer, toothpaste, mouthwashes, deodorants, baby care, perfumes and fragrances.

(b) Soaps: sodium soap, toilet soap, laundry soap, potassium soap – liquid soap, metallic soap-animal feed. Palm-based soaps are high quality soaps with advantages including better perfume retention and being a vegetable oil fat it is therefore acceptable to all religions. The palm-based soaps are exported as bar soaps or soap chips/noodles.

(c) Candles: decorative candles, lighting purposes, warming purposes.

(d) Pharmaceuticals: ointment, emulsion, gel, creams, culture media, tabletting aids.

(e) Lubricants and grease: food grade lubricants, lubricants, greases, food grade purposes, multi-purpose greases.

(f) Surfactants: cleaning powder, hair conditioner, fabric softener.

(g) Industrial chemicals: industrial cleaners-hospitals, bottles cleaning; textiles processing aids; petroleum explorations-drilling fluids, drilling mud; polymer processing aids-plasticizer, stabilizer, additives.

(h) Agrochemicals: as solvents, emulsifiers, carriers.

(i) Coatings: for wood surfaces, metal surfaces, plastic surfaces, paper coatings.

(j) Paints and lacquers: metal and plastic surfaces.

(k) Electronics: insulation and special-purpose plastic components.

(l) Leather: softening, dressing, polishing, treating agents.

(m) Food: Emulsifier and specialty fat for cakes, pastries, margarine, ice-cream and other food products, cocoa butter substitutes, filler condensed milk, etc.

Recent studies reported that palm esters could be synthesized through enzymatic transesterification of various palm oil fractions with oleyl alcohol using Lipozyme RM IM as the catalyst (Keng et al. 2009). Palm oil esters was non-irritant, in dermal irritation assay palm oil esters registered a Human Irritancy Equivalent (HIE) score below 0.9. Furthermore, an increase in skin hydration of 40.7% after 90 min after application in an acute moisturizing test, has proven the suitably of palm oil esters to be used in the cosmetics formulation. Biocatalytic synthesis provided a promising environmentally friendly process for the production of biodiesel (Raita et al. 2010). Addition of *tert*-butanol was found to markedly increase the biocatalyst activity and stability resulting in improved product yield. Optimized reactions (20%, w/w PCMC-lipase to triacylglycerol and 1:4 fatty acid equivalence/ethanol molar ratio) led to the production of alkyl esters from palm olein at 89.9% yield on molar basis after incubation at 45 °C for 24 h in the presence of *tert*-butanol at a 1:1 molar ratio to triacylglycerol. Crude palm oil and palm fatty acid distillate were also efficiently converted to biodiesel with 82.1% and 75.5% yield, respectively, with continual dehydration by molecular sieving. This work thus shows a promising approach for biodiesel synthesis with microcrystalline lipase which could be further developed for cost-efficient industrial production of biodiesel.

Direct Applications of Crude Palm Oil

Polyols

Polyols are processed from epoxidised palm oil and palm oil products (EPOPs), (Ooi et al. 2006) A polyol is an alcohol with more than two reactive hydroxyl groups per molecule. Polyols can be used to produce various types of polyurethane products such as polyurethane sheets ceiling panels, flora foams, wall panels, sandwich panels, laminated boards, cushion, flexible foams, automotive parts, etc. Polyurethane sheets are used for in laying children's playground and sports tracks. Polyurethane (PU) foams can be prepared from mixtures of the palm oil polyol and polyethylene glycol (PEG) or diethylene glycol (DEG) and an isocyanate compound (Tanaka et al. 2008). Thermal and mechanical analyses showed that the chain motion of polyurethane becomes more flexible at the higher palm oil-based polyol content in the whole polymer, which indicates that the monoglyceride molecules work as soft segments. Flexible foam with lazy characteristics can also be produced from palm-based polyols. These are meant for products like furniture, cushions and mattresses. These foams can be used as packaging materials shock absorption materials, *etc*. Flexible polyurethane slabstock has been developed. A new polyol (PolyMo from palm glycerol and oleic acid) has been developed for use as adhesives and coatings. Two coatings were formulated, one for exterior and another for interior applications on different surfaces like wood, plastics and polyesters. Slow-release fertilizers coated with palm-based polyurethane coating materials has been developed that showed that the release of N, P and K were found to be comparable to two commercial coated fertilizers, of which one is soya-based and the other, petroleum polyurethane-based. Polyurethane products are widely used in the automotive industry to make car seats, head rests, liners, dashboards, carpet underlay, bumpers, energy adsorption parts and sound insulation. Rigid/semi-rigid palm oil-based polyurethane foams can be produced by incorporating palm fibre in the manufacturing process. The optimum amount of fibre for blending with palm oil-based polyols is 30% for frond and 35% for trunk and empty fruit bunch fibres. Blended polyols (palm-based: petroleum-based polyols, 50:50) showed optimal density with the inclusion of 20% frond, trunk and empty fruit bunches.

Two radiation curable acrylated polyester prepolymers, PEPP-1 (from refined, bleached, and deodorized palm oil) and PEPP-2 (from crude palm oil), prepared from palm oil-based polyol are found to be suitable as a main ingredient in UV radiation -curable coating materials for wood coating applications (Azam Ali et al. 2001). Various formulations have been developed with different reactive diluents, photoinitiators and curing agents. The production of UV curable acrylated polyol ester prepolymer from palm oil and its downstream products offer potential and promising materials for applications such as polymeric film preparation and coatings (Cheong et al. 2009). Palm olein polyol was reacted with acrylic acid in the presence of a catalyst and inhibitors via condensation esterification process.

Biofuel/Biodiesel
Palm oil can be used as a biofuel either directly (as CPO) or as palm oil methyl esters (PME). Research results have shown that crude palm oil can be used directly as a fuel for cars with suitably modified engines (Ahmad and Salmah 1998; Choo et al. 2005a). PME has very similar properties to petroleum diesel, and thus can be used directly in conventional unmodified diesel engines. Agreements were made with two bus companies for road tests on three fuels, Malaysian diesel, 50:50 blend of PME and diesel, and pure PME. The results showed that the buses with normal diesel engines could just as well run on PME or the blend (Schäfer 1998). PME can also be used as feedstock for many oleochemical derivatives. Further research has indicated that a blend of 5% refined, bleached and deodorized (RBD) palm oil in petroleum diesel is suitable as a biofuel (B5 biofuel) (Ghazali et al. 2006). The optimum conditions for biodiesel production from palm oil using KF/ZnO catalyst were found as follows: methanol/oil ratio of 11.43, reaction time of 9.72 hours and catalyst amount of 5.52 weight %. The optimum biodiesel yield was 89.23% (Hameed et al. 2009).

Mud Drilling and Industrial Applications
In drilling for oil, palm oil has been found to be a non-toxic alternative to diesel as a base for drilling mud. Palm oil-based greases (lithium grease) have been developed. A food-grade grease for palm oil mills, refineries and other food related industries has also been developed. This grease does not contain any heavy metals and is not carcinogenic. A palm oil-based printing ink was developed in 1992 (Ooi et al. 1992) and tested through several continuous commercial runs in a newspaper and the print quality found to be clearer and brighter. Therefore, the quality of palm-based printing inks is comparable, if not better, than that of petroleum-based inks. Palm oil is used extensively in tin plate industry and sheet steel manufacturing, steel cold rolling processes, metallic soaps for the manufacture of lubricating grease and metallic dryers, protecting cleaned iron surfaces before the tin is applied (Pantzaris and Ahmad 2001). Palm oil is also used as lubricant, in the textile and rubber industries.

Esters and Phytonutrients
Crude palm oil (CPO) with varying amounts of free fatty acids can be converted to esters in a continuous process by combining the esterification and transesterification processes (Choo et al. 2005b). The patented process can also be applied to other raw materials such as crude palm stearin, crude palm olein, crude palm kernel oil and palm kernel products. Alternatively, the process can also be applied to produce methyl esters of neutralized palm oil, refined, bleached and deodorized (RBDPO) palm oil and RBD palm olein. In this respect, only transesterification section is sufficient to convert these raw materials into their respective methyl esters. High-value phytonutrients such as carotenes (pro-vitamin A), vitamin E, sterols, squalene, phospholipids (better known as lecithin) and co-enzyme Q10 can be recovered from CPO. Several processes which incorporate technologies such as short path distillation, supercritical fluid technology, saponification, crystallation and solvent treatment have been developed in Malaysia to recover them. Many important palm oil fractions such as natural carotenes, vitamin E, sterols, squalines, co-enzymes and phenolic compounds are used for many pharmaceutical applications (Choo et al. 1996; Lau et al. 2008).

Palm Fatty Acid Distillate

Palm fatty acid distillate (PFAD) a by-product from the physical refining of palm oil. Typical PFAD is composed of free fatty acids (81.7%), glycerides (14.4%), squalene (0.8%), vitamin E (0.5%), sterols (0.4%) and other substances (2.2%) (Ab Gapor et al. 2000; Tan et al. 2007b). PFAD provides an important natural source of useful bioactive phytochemicals such as vitamin E (tocopherol and toctrienols), phytosterols (β-sitosterol, campesterol and stigmasterol) squalene for preparation of encapsulated health products and for fortification of foods wherever appropriate, falling into the category of nutraceuticals or functional foods (Ab Gapor et al. 2002).

A primary market for PFAD is the animal feed industry. PFAD is an attractive ingredient as a fat supplement for livestock; it is readily available, relatively stable to oxidant rancidity and relatively inexpensive (Palmquist 2004).

Other Palm-Based Byproducts

Paper coated with palm-based products such as RBD palm stearin and hydrogenated palm olein are quite similar to that of paraffin wax (also fatty acid and fatty alcohol) can replace paraffin wax film used in food wrapping (Abdul Rashid et al. 1997). Palm-based coating paper has the potential to be used as bread wrappers, fish, meat or vegetable wrappers, drinking cups, frozen food and paper milk cartons. Suppositories is another important pharmaceutical use of palm kernel oil products (Pantzaris and Ahmad 2001). Suppositories are designed to supply various medicines through the rectum, where supplying by mouth or by injection is for some reason, less acceptable. HPKS (Hydrogenated palm kernel stearin) or HPKO (Hydrogenated palm kernel olein), offer advantages in ease of production, less discolouration over time, have no need for tempering, their consistency is more easily adjusted and also have lower cost. Palm kernel oil is also used as an alternative to coconut oil in high quality soaps and detergents, as a source of short-chain and medium-chain fatty acids. Both types of oil have further industrial uses in the metal and leather industries. The use of refined red palm oil as an alternative means of combating vitamin A deficiency in many countries (Scrimshaw 2000). Palm-based solvents and surfactants are used as inert ingredients in non-flammable insecticides (due to higher flash points) that are environmental friendly (Ismail et al. 2004).

Palm Oil as Animal Feed

Palm oil can be used as a lipid feed for aquaculture (Ng 2004a). Studies showed that *Clarius gariepinus*, catfish responded significantly in a positive manner to palm oil additions of up to 8% (in isoenergetic diets) (Ng 2004a). Palm oil contains abundant saturates (48%) and monoenes (42%). Protein sparing effects resulting in higher protein retention in fish fed RBDPO supplemented diets were also observed. Palm oil is unique because tocotrienols represent about 70–80% of the vitamin E content. The deposition of palm vitamin E in the fillets of tilapia fed a tocotrienol-rich fraction extracted from palm oil imparted higher oxidative stability. Palm oil (0,1, 2 and 3%) could replace soybean oil in feeds for tilapia, *Oreochromis niloticus* fingerlings without any negative effect on the fish growth or body composition (Al-Owafeir and Belal 1996). Palm oil can be used successfully as a substitute for fish oil in the culture of Atlantic salmon in sea water (Bell et al. 2002). Studies showed that palm oil or palm olein in the form of calcium soap can be used as supplementation in dietary feed for lambs (Castro et al. 2005).

By-Products from Solid Wastes of Oil Palm Industry

Palm oil industry generates massive amount of solid wastes in form of lignocellulosic residue. These come from empty fruit bunches (EFB), palm press fibre (PPF), palm kernel cake (PKC), palm kernel shell (PKS), sludge cake or decanter cake and palm oil mill effluent (POME). Based on the percentage of fresh fruit bunch (FFB), the expected wastes are 20–28% EFB, 11–12% PPF and 5–8% PKS (Sarmidi et al. 2009).

Empty Fruit Bunches (EFB), Palm Press Fibre (PPF)

Palm oil mill solid wastes such as PPF and PKS are used as boiler fuel and significant fraction can be recycled inside the palm oil industry. Empty bunch stalks, mesocarp fibres and shells from the cracked nuts are used as fuel for the boilers at the oil palm mills. Refuse in the form of empty fruit bunches after stripping the bunches is used for mulching and manuring, recycling of nutrients back to the fields. About 25% of the harvested biomass may be returned to the field as a nutrient rich mulch, providing opportunities for growers to recycle nutrients and biomass from more fertile to less fertile parts of the estate. At the palm oil mill, the empty fruit bunches are mainly incinerated to produce bunch ash which contains a high percentage of potassium. The ash is distributed back to the field as fertilizer for the palms and sometimes used in soap-making. However, the incineration process results in severe air pollution. In some oil palm mills, the empty bunches are pressed to reduce their moisture content and burnt as fuel to generate steam to be sold to nearby factories. Palm pressed fibre is obtained during the pressing of sterilized fruits and contains about 60% w/w dry matter. Though produced in large quantities, the availability of the palm pressed fibre is limited because it is presently utilized in the mill as a source of energy. Empty fruit bunch is used to make – polyurethane (EFB–PU) composites; produced by reacting EFB and polyethylene glycol (PEG) with diphenylmethane diisocyanate (MDI). The empty fruit bunches is also converted into a light weight potting media for ornamental plants. Fibres from empty fruit bunches are used to produce pulp, paper, chipboard and plywood (see below). Recent research showed that bio oil can be obtained from EFB by means of fast pyrolysis process (Abdullah and Gerhauser 2008).

Recently, oil palm fibres were used successfully for the production of activated carbon for many environmental and industrial applications (Tan et al. 2007a; Hameed et al. 2008). EFB were also used successfully as an alternative raw material to obtain cellulosic pulp (Wan Rosli et al. 2007; Leh et al. 2008; Rodriguez et al. 2008). More recently, EFB and oil palm derived cellulose were used in polypropylene composite for the production of biodegradable lignocellulosic filler. This form short fibre reinforced polymer composite and used in automotive interior components and geotextile (Khalid et al. 2008). Oil palm empty fruit bunches (EFB), a lignocellulosic waste from palm oil mills was successfully treated by acid hydrolysis to produce xylose, glucose, furfural and acetic acid for industrial chemical usage (Rahman et al. 2006). Likewise acid hydrolysis of oil palm trunk yielded glucose (Lim et al. 1997). EFB can also be used as substrate for production of many industrially important enzymes such as cellulases, hemicellulases and lignin peroxidases (Umilalsom et al. 1997).

Palm Kernel Shell (PKS)

Shells from the cracked nuts are used as fuel for the boilers at the oil palm mills. In Africa, PKS is used as aggregates in foundation of local buildings and for the flooring of houses. Charcoal made with the inner kernel shells is very clean and hot burning, and is very popular with blacksmiths. Activated carbon can be obtained from PKS.

Palm Oil Mill Effluent (POME)

Based on the high carbon content and high nutritional value based on the traces of vitamins, carotenes, fatty acids and high potassium content of palm oil mill effluent (POME) it is successfully used as soil fertilizers (Esther 1997). However, based on the high nutritional value of palm oil mill wastes, they are good feed supplements. POME could be used as ruminant feed at 10–20% of the diet and its quality could be improved by adding 5–6% ammonium hydroxide or urea and fermenting for 2–3 weeks. Moreover, based on the its high carbon to nitrogen ratio, POME is used as carbon source in culture medium for the production of many microbial metabolites such as: alcohols, organic solvents, (Mun et al. 2004), organic acids (Jamal et al. 2005) and biopolymer(Hassan et al. 1997; Loo et al. 2005). Pretreated POME (after removal of suspension solids and reducing ash content) were supplemented with ammonium sulphate to obtain the C:N ratio of 20:1 and used thereafter as cultivation medium for penicillin production by *Penicillium chrysogenum* (Suwandi and mohamad 1984).

Beside the direct energy generation by means of direction combustion, POME is also used as potential low cost carbon source for biogas production using thermophilic anaerobic microorganisms.

Palm Kernel Cake (PKC)

Palm Kernel Cake (PKC) is the press cake, after extraction of oil from the kernels, is a useful source of protein and energy for livestock (Hishamudin 2001) and it is commonly used in animal feed especially for ruminants. Wan Zahari and Alimon 2004 reviewed the use of palm kernel cake in compound feed for animals. For feedlot cattle and buffalo, PKC can be used as the main ingredient (Wan Zahari and Alimon 2004). For dairy cattle rations, PKC can be used as a source of fibre and energy at the inclusion level of 30–50%. Feeding palm kernel-based diets to dairy cows has been shown to increase the milk-fat content, enhance the firmness of butter and produce good quality meat (Witt 1952). The milk of dairy cattle fed with PKC tends to produce a firm butter and a ration of 2–3 kg of PKC daily is satisfactory for adult cattle (Gohl 1981). Almost all exported Malaysian PKC is used in dairy cow feed (Osman 1986). PKC is used as a common ingredient in German and Dutch with dairy ration approximately 10% of the cake in the ration, whereas in Malaysia, dairy farmers use more than 50% (Osman and Hisamuddin 1999). The biological value of PKC is 61–80% for sheep (Devendra 1978). For sheep rations PKC recommended inclusion level is 30%, for goat PKC inclusion can be up to 50% and one such formulation is PKC 50%:grass/hay 30%, rice bran 10%: soybean 9% and mineral premix 1%. PKC is also suitable for swine at 20–25% diet inclusion for growers and finisher. Studies in Thailand found that up to 30% PKC can be used in diets for goats (Intarapichet et al. 1994) and palm oil with supplementation of conjugated linoleic acid was found suitable for pigs (Intarapichet et al. 2008). Poultry brolier can tolerate up to 20% PKC and up to 15% palm oil sludge in their diets without affecting their growth performance and feed efficiency (Yeong 1982). For layers, PKC can be included up to 25% without deleterious effects on egg production and quality (Radim et al. 2000). PKC has been accepted as one of the components in animal feeds, especially in the European Union.

Palm Kernel Cake (PKC) is also useful as aquaculture feed (Ng 2004a; Ng 2004b). The use of PKC in tilapia and catfish diets have generated encouraging results with fish growing well on dietary levels as high as 20%. PKC can be tolerated up to 30% in catfish (*Clarias gariepinus*) and 20% in tilapia (*Oreochromis niloticus*) with no deleterious effect on growth and performance (Sukkasame 2000; Saad et al. 1994). Studies have shown that tilapia fed PKC pre-treated with commercial feed enzymes consistently showed better growth and feed utilization efficiency compared to fish fed similar levels of raw PKC (Ng et al. 2002). Studies with grass carp were even more encouraging in terms of higher levels of raw PKC being used in their diets.

Oil Palm Biomass and Wood Based Products and Applications

Oil palm biomass can be used in various wood based products and application (Husin et al. 2002).

Trunk – Wood

Traditionally, oil palm trunk is used as good material for the furniture industry. Sawing and drying of palm wood are difficult because of density variation inside trunk and resulted in dimensional instability of the finished product and thus limited its application. In some countries, the trunk is sawn into timber and used in the construction of fences, roofing, as pillar beams and in reinforcing buildings. It is also used as fuel for cooking and heating. Oil palm trunk is also used in making various types of board (see below). Previous studies comparing oil-palm trunks as a roughage feed (Oshio et al. 1990) with rice straw supported the use of the oil-palm materials as a source of roughage for ruminants, as did a long-term feeding trial of oil-palm trunks for beef production (Abu Hassan et al. 1991).

Pulp and Paper

Untreated whole empty fruit bunch (EFB) is digested, and the pulp yield is between 26% and 32%, with a very high lignin content of over 6% (Husin et al. 2002). The pulp brightness is between 35% and 40%. However, the pulp yield improves significantly (to over 40%) when the EFB is pretreated with 3% soda for 1 h at 50–60°C. The addition of anthraquinone (AQ) (Kraft-AQ pulping) improves the yield only slightly but the lignin content is reduced to about 1% only. The properties of the pulp are superior to those of hardwood Kraft pulp (KP) such as beech and birch. The brightness is good (40%), compared to hardwood pulp (25–30%). For oil palm fronds (OPF) pretreated and pulped in a similar pretreated and pulped in a similar.

For oil palm fronds treated in a similar manner as the EFB, the optimum active alkali (AA) was 15% and AQ 0.1%, and the yield obtained was about 44%. However, pulp brightness is low at only 20%. The strength properties are excellent and are only slightly lower compared to softwood UKP. Oil palm frond pulp could used as a reinforcement component in newsprint production using softwood thermomechanical fibers (Wan Rosli et al. 2007). Morphologically, the frond fibers are comparable to those of hardwood. They contain high content of holocellulose but low in lignin. Chemical pulps of 45–50% yield produced either by soda-sulfite or soda process exhibit acceptable papermaking properties comparable to those of hardwood kraft pulps.

Fibre, Blockboard and Particle Boards

EFB was found to be suitable for the manufacture of medium density fibreboard (MDF). EFB can be an important source of particles for particleboard manufacture. Blockboard can be manufactured from oil palm trunk (OPT) as the centre core. It is lighter than blackboard manufactured from normal wood. The physical and mechanical properties indicate that blockboard from OPT is suitable for non-load bearing applications, such as paneling and furniture applications. OPT is also used in making particle boards with chemical binders. A study on homogenous particleboards and three-layered particleboards from oil palm trunk vascular bundles indicated that at 700 kg m^{-3} and 8% adhesive content, the boards passed the minimum standard stipulated in the BS 5669:1979 Type 1. For three-layered boards, it was necessary to have 10% resin content for surface particle and 8% for the core with an overall density of 700 kg m^{-3} in order to pass the minimum requirements as stipulated in the standard. OPT and fronds have also been used as raw materials in moulded particleboard manufacture.

The process of manufacturing moulded particle board is generally the same as normal particleboard, except that the particle mat is placed in a mould with the desired shape before pressing. Several items, such as table tops of various sizes and shapes and chair seats, have been successfully tried in industrial scale plant. Oil palm biomass can also be used as raw material for gypsum bonded particleboard and cement-bonded particleboard.

Other Palm Waste Products and By Products

Planting Medium

Oil palm biomass can also be utilized as a planting medium, as an alternative to rockwool. Initial study on planting medium from oil palm biomass that have been planted with tomatoes and brinjal showed that these plants were able to sustain their growth. The advantages of using planting medium from oil palm biomass were biodegradability, good water/fertilizer retention and availability of readily renewable resources.

Oil Palm Fronds

Fronds are used in plantation and left to rot in between the rows for many purposes such as: soil conservation, increase the soil fertility, increase water retention in the soil, erosion control and to provide a source of nutrient to the growing oil palm trees (Sumathi et al. 2008). In Africa, fronds are used for thatching; petioles and rachises for fencing and for protecting and reinforcing the tops of retted walls. The leaflets are used to make palm skirts used by local masquerades and dancers or it can be used to tie together mats or racks that are used to dry other farm produce copra, fish, pepper and many woven items, The midribs of leaflets can be used to make mats, baskets, brooms. Leaf fibers and are also employed in the manufacture of chipboard and plywood.

Of the commercial plantation crops, oil palm produces the most abundant biomass, and oil palm fronds have been shown to be a very promising source of roughage for ruminants (Abu Hassan et al. 1995, 1996). The ruminant sector in particular is well suited to maintaining competitiveness through the use of plantation and processing by-products. Oil-palm fronds belong to the category of fibrous crop residues, which also includes by-products such as rice straw. The large quantities of fronds produced by a plantation each year make these a very promising source of roughage feed for ruminants. Oil-palm fronds can be used as a substitute for grasses in cases where forage or fodder is a limiting factor. The recommended level of oil-palm fronds in the total mixed rations (on a dry matter basis) are 50% for beef cattle, and 30% for dairy cattle and goats (Abu Hassan et al. 1996). A digestibility study conducted using mature Kedah-Kelantan bulls (Abu Hassan and Ishida 1992) indicated a dry matter digestibility value of about 45% for oil-palm frond silage. This encouraging result was further tested for the suitability of oil-palm fronds in long-term feeding/production trials on beef cattle (growing and finishing), and also on lactating dairy cows (Abu Hassan et al. 1993; Ishida et al. 1994).

Comments

The commercial variety of *E. guineenis* grown in southeast Asia and Africa belong to the tenera variety (with a much thicker mesocarp and a thinner shell) which is a cross between the *Dura* (thick shell, thin mesocarp) and *Pisifera* (shell less) varieties.

Selected References

Ab Gapor MT (1990) Content of vitamin E in palm oil and its antioxidant activity. Palm Oil Dev 12:25–27

Ab Gapor MT, Murui T, Watanabe H, Kawada T (1985) Studies on minor components in palm fatty acid distillate. II. Occurrence of esters of fatty acids. J Jpn Oil Chem Soc 34(8):634–637

Ab Gapor MT, Wan Hasamudin WH, Sulon M (2000) Phytochemicals for nutraceuticals from the by-product of palm oil refining. Palm Oil Dev 32:13–18

Ab Gapor MT, Sulong M, Rosnah MS (2002) Production of phytosterols from palm fatty acid distillates. MPOB Information Series No. 173

Abdul Rashid AS, Ooi TL, Salmiah A (1997) Potential of paper coating with palm-based materials for food wrappers. Palm Oil Dev 26:7–12

Abdullah N, Gerhauser H (2008) Bio-oil derived palm empty-fruit-bunch fibre as substrate. Fuel 87: 2606–2613

Abeywardena MY, McLennan PL, Charnock JS (1991) Changes in myocardial eicosanoid production following long-term dietary lipid supplementation in rats. Am J Clin Nutr 53(4):1039S–1041S

Abeywardena M, Runnie I, Nizar M, Momamed S, Head R (2002) Polyphenol-enriched extract of oil palm fronds (*Elaeis guineensis*) promotes vascular relaxation via endothelium-dependent mechanisms. Asia Pac J Clin Nutr 11(1):S467–S472

Abu Hassan O, Ishida M (1992) Status of utilization of selected fibrous crop residues and animal performance with emphasis on processing of oil palm fronds (OPF) for ruminant feed in Malaysia, Tropical Agriculture Research Center TARS No. 25. Ministry of Agriculture, Forestry and Fisheries, Tsukuba, Japan, pp 134–143

Abu Hassan O, Oshio S, Ismail AR, Mohd. Jaafar D, Nakanishi N, Dahlan I, Ong SH (1991) Experiences and challenges in processing, treatment, storage and feeding of oil palm trunks based diets for beef production. In: Proceeding of seminar on oil palm trunks and other palmwood utilization, Oil Palm Tree Utilization Committee of Malaysia, Kuala Lumpur, Malaysia, 4–5 March 1991, pp 231–245

Abu Hassan O, Azizan AR, Ishida M, Abu Bakar C (1993) Oil palm fronds silage as a roughage source for milk production in Sahiwal-Friesian cows. In: Proceeding of 16th Malaysian Society of Animal Production, Pulau Langkawi, Malaysia, 8–9 June 1993, pp 34–35

Abu Hassan O, Ishida M, Ahmad Tajuddin Z (1995) Oil palm fronds (OPF) technology transfer and acceptance: a sustainable insitu utilization for animal feeding. In: Proceeding of the 17th Malaysian Society of Animal Production, Penang, Malaysia, 28–30 May 1995, pp 134–137

Abu Hassan O, Ishida M, Mohd Shukri I, Tajuddin ZA (1996) Oil palm fronds as a roughage feed source for ruminants in Malaysia. Extension Bulletin 420, Food & Fertiliser Technology Center for ASPAC Region, Taipei, Taiwan.

Adam SK, Soelaiman IN, Umar NA, Mokhtar N, Mohamed N, Jaarin K (2008) Effects of repeatedly heated palm oil on serum lipid profile, lipid peroxidation and homocysteine levels in a post-menopausal rat model. Mcgill J Med 11(2):145–151

Adam SK, Das S, Jaarin K (2009) A detailed microscopic study of the changes in the aorta of experimental model of postmenopausal rats fed with repeatedly heated palm oil. Int J Exp Pathol 90(3):321–327

Agarwal MK, Agarwal ML, Athar M, Gupta S (2004) Tocotrienol-rich fraction of palm oil activates p53, modulates Bax/Bcl2 ratio and induces apoptosis independent of cell cycle association. Cell Cycle 3(2):205–211

Ahmad H, Salmah J (1998) Palm oil as diesel fuel: field trial on cars with Elsbett engine. In: Proceeding of the 1998 PORIM international biofuel and lubricant conference, PORIM, Bangi, pp 165–174

Ahmad NS, Khalid BA, Luke DA, Ima-Nirwana S (2005) Tocotrienol offers better protection than tocopherol from free radical-induced damage of rat bone. Clin Exp Pharmacol Physiol 32(9):761–770

Al-Owafeir MA, Belal IEH (1996) Replacing palm oil for soybean oil in tilapia, *Oreochromis niloticus* (L.), feed. Aquacult Res 27(4):221–224

Al-Saqer JM, Sidhu JS, Al-Hooti SN, Al-Amiri HA, Al-Othman A, Al-Haji L, Ahmed N, Mansour IB, Minal J (2004) Developing functional foods using red palm olein. IV. Tocopherols and tocotrienols. Food Chem 85(4):579–583

Anderson SL, Rubin BY (2005) Tocotrienols reverse IKAP and monoamine oxidase deficiencies in familial dysautonomia. Biochem Biophys Res Commun 336(1):150–156

Asase A, Akwetey GA, Achel DG (2010) Ethnopharmacological use of herbal remedies for the treatment of malaria in the Dangme West District of Ghana. J Ethnopharmacol 129(3):367–376

Atchley AA (1984) Nutritional value of palms. Principes 28(3):138–143

Azam Ali M, Ooi TL, Salmiah A, Ishiaku US, Mohd. Ishak ZA (2001) New polyester acrylate resins from palm oil for wood coating application. J Appl Polym Sci 79(12):2156–2163

Azlina MF, Nafeeza MI, Khalid BA (2005) A comparison between tocopherol and tocotrienol effects on gastric parameters in rats exposed to stress. Asia Pac J Clin Nutr 14(4):358–365

Balasundram N, Bubb W, Sundram K, Samman S (2003) Antioxidants from palm (*Elaeis guineensis*) fruit extracts. Asia Pac J Clin Nutr 12(Suppl):S37

Balasundram N, Ai TY, Sambanthamurthi R, Sundram K, Samman S (2005) Antioxidant properties of palm fruit extracts. Asia Pac J Clin Nutr 14(4):319–324

Bell JG, Henderson RJ, Tocher DR, McGhee F, Dick JR, Porter A, Smullen RP, Sargent JR (2002) Substituting fish oil with crude palm oil in the diet of Atlantic salmon (*Salmo salar*) affects muscle fatty acid composition and hepatic fatty acid metabolism. J Nutr 132:222–230

Benjumea P, Agudelo J, Agudelo A (2008) Basic properties of palm oil biodiesel-diesel blends. Fuel 87:2069–2075

Berger KG (1996) Food uses of palm oil. MPOPC Palm Oil Information Series, Kuala Lumpur, 25 pp

Bergert DL (2000) Management strategies of *Elaeis guineensis* (oil palm) in response to localized markets in south eastern Ghana. Master of Science in Forestry, Michigan Technological University, West Africa

Beyer RE (1989) The role of coenzyme Q in free radical formation and antioxidation. In: Minguel J, Quinanilha AT, Weber H (eds) CRC handbook of free radicals and antioxidants in biomedicine, vol II. CRC, Boca Raton, pp 45–62

Black TM, Wang P, Maeda N, Coleman RA (2000) Palm tocotrienols protect ApoE +/− mice from diet-induced atheroma formation. J Nutr 130(10):2420–2426

Boateng J, Verghese M, Chawan CB, Shackelford L, Walker LT, Khatiwada J, Williams DS (2006) Red palm oil suppresses the formation of azoxymethane (AOM) induced aberrant crypt foci (ACF) in Fisher 344 male rats. Food Chem Toxicol 44(10):1667–1673

Bora PS, Rocha RVM, Narain N, Moreira-Monteiro AC, Moreira RA (2003) Characterization of principal nutritional components of Brazilian oil palm (*Eliaes guineensis*) fruits. Biores Technol 87(1):1–5

Budin SB, Othman F, Louis SR, Bakar MA, Das S, Mohamed J (2009) The effects of palm oil tocotrienol-rich fraction supplementation on biochemical parameters, oxidative stress and the vascular wall of streptozotocin-induced diabetic rats. Clin Sao Paulo 64(3):235–244

Canfield LM, Kaminsky RG, Taren DL, Shaw E, Sander JK (2001) Red palm oil in the maternal diet increases provitamin A carotenoids in breastmilk and serum of the mother-infant dyad. Eur J Nutr 40(1):30–38

Castro T, Manso T, Mantecón AR, Guirao J, Jimeno V (2005) Fatty acid composition and carcass characteristics of growing lambs fed diets containing palm oil supplements. Meat Sci 69(4):757–764

Charnock JS, Sundram K, Abeywardena MY, McLennan PL, Tan DT (1991) Dietary fats and oils in cardiac arrythmia in rats. Amer J Clin Nutri 53:1047S–1049S

Che Man YB, Liu J (1999) Quality changes of RBD palm olein, soybean oil and their blends during deep-fat frying. J Food Lipids 6(3):181–193

Chen CW, Chong CL, Ghazali HM, Lai OM (2007) Interpretation of triacylglycerol profiles of palm oil, palm kernel oil and their binary blends. Food Chem 100(1):178–191

Cheong MY, Ooi TL, Salmiah A, Wan Md Zin WY, Dzulkefly K (2009) Synthesis and characterization of palm-based resin for UV coating. J Appl Polym Sci 111(5):2353–2361

Chong YH, Ng TK (1991) Effects of palm oil on cardiovascular risk. Med J Malays 46(1):41–50

Choo YM (1994) Palm oil carotenoids. Food Nutr Bull 15(2):130–137

Choo YM, Yap SC, Ooi CK, Ma AN, Goh SH, Ong ASH (1996) Recovered oil from palm-pressed fiber: a good source of natural carotenoids, vitamin E and sterols. J Am Oil Chem Soc 73:599–602

Choo YM, Ma AN, Cheah KY, Abdul Majid R, Yap AKC, Lau HLN, Cheng SF, Yung CL, Puah CW, Ng MH, Yusof Basiron Y (2005a) Palm diesel: green and renewable fuel from palm oil. Palm Oil Dev 42:3–7

Choo YM, Ng MH, Ma AN, Chuah CH, Ali HM (2005b) Application of supercritical fluid chromatography in the quantitative analysis of minor components (carotenes, vitamin E, sterols, and squalene) from palm oil. Lipids 40(4):429–432

Choudhury N, Tan L, Truswell AS (1995) Comparison of palm olein and olive oil: effects on plasma lipids and Vitamin E in young adults. Am J Clin Nutr 61:1043–1051

Cottrell RC (1991) Introduction: nutritional aspects of palm oil. Am J Clin Nutr 53(4 Suppl):989S–1009S

Das S, Powell SR, Wang P, Divald A, Nesaretnam K, Tosaki A, Cordis GA, Maulik N, Das DK (2005) Cardioprotection with palm tocotrienol: antioxidant activity of tocotrienol is linked with its ability to stabilize proteasomes. Am J Physiol Heart Circ Physiol 289:H361–H367

Das S, Lekli I, Das M, Szabo G, Varadi J, Juhasz B, Bak I, Nesaretam K, Tosaki A, Powell SR, Das DK (2008) Cardioprotection with palm oil tocotrienols: comparison of different isomers. Am J Physiol Heart Circ Physiol 294:H970–H978

Devendra C (1978) Utilization of feedingstuffs from the oil palm. In: Devendra C, Hutagalung RI (eds) Feeding stuffs for livestock in South East Asia. Malaysia Society of Animal Production, Selongor, pp 113–139

Di Mascio P, Devasagayam TP, Kaiser S, Sies H (1990) Carotenoids, tocopherols and thiols as biological singlet molecular oxygen quenchers. Biochem Soc Trans 18(6):1054–1056

Ebong PE, Owu DU, Isong EU (1999) Influence of palm oil (*Elaeis guineensis*) on health. Plant Foods Hum Nutr 53(3):209–222

Edem DO (2002) Palm oil: biochemical, physiological, nutritional, hematological, and toxicological aspects: a review. Plant Foods Hum Nutr 57(3–4):319–341

Elson CE (1992) Tropical oils: nutritional and scientific issues. Crit Rev Food Sci Nutr 31(1/2):79–102

Engelbrecht AM, Esterhuyse J, du Toit EF, Lochner A, van Rooyen J (2006) p38-MAPK and PKB/Akt, possible role players in red palm oil-induced protection of the isolated perfused rat heart? J Nutr Biochem 17(4):265–271

Engelbrecht AM, Odendaal L, Du Toit EF, Kupai K, Csont T, Ferdinandy P, van Rooyen J (2009) The effect of dietary red palm oil on the functional recovery of the ischaemic/reperfused isolated rat heart: the involvement of the PI3-kinase signaling pathway. Lipids Health Dis 8:18

Ernster L, Dallner G (1995) Biochemical, physiological and medical aspects of ubiquinone function. Biochim Biophys Acta 1271:195–204

Esterhuyse AJ, du Toit E, Rooyen JV (2005a) Dietary red palm oil supplementation protects against the consequences of global ischemia in the isolated perfused rat heart. Asia Pac J Clin Nutr 14(4):340–347

Esterhuyse AJ, du Toit EF, Benadè AJS, van Rooyen J (2005b) Dietary red palm oil improves reperfusion cardiac function in the isolated perfused rat heart of animals fed a high cholesterol diet. Prostag Leukotr Ess Fatty Acids 72(3):153–161

Esterhuyse JS, van Rooyen J, Strijdom H, Bester D, du Toit EF (2006) Proposed mechanisms for red palm oil induced cardioprotection in a model of hyperlipidaemia in the rat. Prostag Leukotr Ess Fatty Acids 75(6):375–384

Esther U (1997) Anaerobic digestion of palm soil mill effluent and its utilization as fertilizer for environmental protection. Renew Energy 10:291–294

Eze MO, Ogan AU (1988) Sugars of the unfermented sap and the wine from the oil palm, *Elaeis guinensis*, tree. Plant Foods Hum Nutr 38(2):121–126

Fan WX, Chen XS (1994) Food uses of palm oil in China. Food Nutr Bull 15:147–148

Fang F, Kang Z, Wong C (2010) Vitamin E tocotrienols improve insulin sensitivity through activating peroxisome proliferator-activated receptors. Mol Nutr Food Res 54(3):345–352

Farombi EO, Britton G (1999) Antioxidant activity of palm oil carotenes in organic solution: effects of structure and chemical reactivity. Food Chem 64(3):315–321

Gee PT (2007) Analytical characteristics of crude and refined palm oil and fractions. Eur J Lipid Sci Technol 109(4):373–379

Ghazali R, Yusof M, Salmiah A (2006) Non-food applications of palm-based products – market opportunities and environmental benefits. Palm Oil Dev 44:8–14

Goh SH (1999) Cholesterol. Malays Oil Sci Technol 8(2):58–59

Goh SH, Gee PT (1984) Unreported constituents from *Elaeis guineensis*. In: Singh MM, Eng LS (eds) Proceedings of priochem Asia 1984 chemical conference. Malaysian Institute of Chemistry, Kuala Lumpur, Malaysia, pp 507–13

Goh SH, Khor HT, Gee PT (1982) Phospholipids of palm oil (*Elaeis guineensis*). J Am Oil Chem Soc 59:296–299

Goh SH, Tong SL, Gee PT (1984) Total phospholipids in crude palm oil: quantitative analysis and correlations with oil quality parameters. J Am Oil Chem Soc 61:1597–1600

Goh SH, Choo YM, Ong ASH (1985) Minor constituents of palm oil. J Am Oil Chem Soc 62:237–240

Goh SH, Lai FL, Gee PT (1988) Wax esters and triterpene methyl ethers from the exocarp of *Elaeis guineensis*. Phytochemistry 27(3):877–880

Gohl B (1981) Tropical feeds. Food and Agricultural Organization, Rome, pp 365

Gupta A, Chopra K (2009) Effect of tocotrienols on iron-induced renal dysfunction and oxidative stress in rats. Drug Chem Toxicol 32(4):319–325

Guthrie N, Gapor A, Chambers AF, Carroll KK (1997) Inhibition of proliferation of estrogen receptor-negative MDA-MB-435 and -positive MCF-7 human breast cancer cells by palm oil tocotrienols and tamoxifen, alone and in combination. J Nutr 127(3):544S–548S

Hameed BH, Tan IAW, Ahmed AL (2008) Optimization of basic dye removal by oil palm fibre- based activated carbon using response surface methodology. J Hazard Mater 158:324–332

Hameed BH, Lai LF, Chin LH (2009) Production of biodiesel from palm oil (*Elaeis guineensis*) using heterogeneous catalyst: an optimized process. Fuel Process Technol 90(4):606–610

Hamid HA, Choo YM, Goh SH, Khor HT (1995) The ubiquinones of palm oil. Nutr Lipids Health Dis, AOCS Press, USA, pp 122–128

Hardon JJ, Rajanaidu N, van der Vossen HAM (2002) *Elaeis guineensis* Jacq. In: van der Vossen HAM, Umali BE (eds) Plant resources of South-East Asia No 14. Vegetable oils and fats. Prosea Foundation, Bogor, pp 85–93

Hartley CWS (1988) The oil palm, 3rd edn. Longam, London, 761 pp

Hartwell JL (1967–1971) Plants used against cancer. A survey. Lloydia, vol. 30–34

Hassan MA, Shirai Y, Kusubayashi N, Abdul Karim MI, Nakanishi K, Hasimato K (1997) The production of polyhydroxyalkanoate from anaerobically treated palm oil mill effluent by *Rhodobacter sphaeroides*. J Ferm Bioeng 83:485–488

Hayes KC, Khosla P (1992) Dietary fatty acid thresholds and cholesterolemia. FASEB J 6(8):2600–2607

Hermizi H, Faizah O, Ima-Nirwana S, Nazrun AS, Norazlina M (2009) Beneficial effects of tocotrienol and tocopherol on bone histomorphometric parameters in sprague-dawley male rats after nicotine cessation. Calcif Tissue Int 84(1):65–74

Hishamudin MA (2001) Malaysian palm kernel cake as animal feed. Palm Oil Dev 34:4–6

Hornstra G (1988) Dietary lipids and cardiovascular disease. Effects of palm oil. Oleagineux 43:75–81

Husin M, Ramli R, Mokhtar A, Wan Hassan WH, Hassan K, Mamat R, Aziz AA (2002) Research and development of oil palm biomass utilization in wood-based industries. Palm Oil Dev 36:1–5

Ikeda S, Niwa T, Yamashita K (2000) Selective uptake of dietary tocotrienols into rat skin. J Nutr Sci Vitaminol (Tokyo) 46(3):141–143

Ikeda S, Toyoshima K, Yamashita K (2001) Dietary sesame seeds elevate alpha- and gamma-tocotrienol concentrations in skin and adipose tissue of rats fed the tocotrienol-rich fraction extracted from palm oil. J Nutr 131(11):2892–2897

Ima-Nirwana S, Suhaniza S (2004) Effects of tocopherols and tocotrienols on body composition and bone calcium content in adrenalectomized rats replaced with dexamethasone. J Med Food 7(1):45–51

Ima-Nirwana S, Kiftiah A, Zainal AG, Norazlina M, Gapor MTA, Khalid BAK (1998) Influence of palm oil and palm vitamin E on bone mineral density in orchidectomised male rats. Toxicol Lett 95(Suppl 1):214

Ima-Nirwana S, Kiftiah A, Sariza T, Gapor MTA, Khalid BAK (1999) Palm vitamin E improves bone metabolism and survival rate in thyrotoxic rats. Gen Pharmacol 32(5):621–626

Inokuchi H, Hirokane H, Tsuzuki T, Nakagawa K, Igarashi M, Miyazawa T (2003) Anti-angiogenic activity of tocotrienols. Biosci Biotech Biochem 67(7):1623–1627

Intarapichet K, Pralomkarn W, Chinajariyawong C (1994) Influence of genotypes and feeding on growth and sensory characteristics of goat meat. ASEAN Food J 9:151–155

Intarapichet K, Maikhunthod B, Thungmanee N (2008) Physicochemical characteristics of pork fed palm oil and conjugated linoleic acid supplements. Meat Sci 80(3):788–794

Ishida M, Abu Hassan O, Nakui T, Terada F (1994) Oil palm fronds as ruminant feed. Japan Int Res Agric Sci JIRCAS Newsl Int Collab 2(1):12–13

Ismail R (2005) Palm oil and palm olein frying applications. Asia Pac J Clin Nutr 14(4):414–419

Ismail NM, Abdul Ghafar N, Jaarin K, Khine JH, Top GM (2000) Vitamin E and factors affecting atherosclerosis in rabbits fed a cholesterol-rich diet. Int J Food Sci Nutr 51(Suppl):S79–S94

Ismail AR, Ooi TL, Salmiah A (2004) Environment friendly palm-based inert ingredient for EW-insecticide formulations. MPOB Information Series No. 243

Itoh T, Tamura T, Matsumoto T (1973a) Methylsterol compositions of 19 vegetable oils. J Am Oil Chem Soc 50:300–303

Itoh T, Tamura T, Matsumoto T (1973b) Sterol composition of 19 vegetable oils. J Am Oil Chem Soc 50:122–125

Jaarin K, Renuvathani M, Nafeeza MI, Gapor MT (2000) Effect of palm vitamin E on the healing of ethanol-induced gastric injury in rats. Int J Food Sci Nutr 51(1):S31–S41

Jaarin K, Gapor MT, Nafeeza MI, Fauzee AM (2002) Effect of various doses of palm vitamin E and tocopherol on aspirin-induced gastric lesions in rats. Int J Exp Pathol 83(6):295–302

Jacobsberg B (1974) Palm oil characteristics and quality. In: Chai OS, Awalludin A (eds) Proceedings of the 1st MARDI workshop on oil palm technology. Malaysia Agricultural Research and Development Institute (MARDI), Kuala Lumpur, Malaysia, pp 48–70

Jamal P, Alam MZ, Ramlan M, Salleh M, Nadzir MM (2005) Screening of *Aspergillus* for citric acid production from palm oil mill effluent. Biotechnol 4:275–278

Jiang Q, Ames BN (2003) Gamma-tocopherol, but not alpha-tocopherol, decreases proinflammatory eicosanoids and inflammation damage in rats. FASEB J 17(8):816–822

Jiang Q, Elson-Schwab I, Courtemanche C, Ames BN (2000) γ-Tocopherol and its major metabolite, in contrast to α-tocopherol, inhibit cyclooxygenase activity in macrophages and epithelial cells. Proc Natl Acad Sci U S A 97(21):11494–11499

Judd JT, Clevidence BA, Muesing RA, Wittes J, Sunkin ME, Podczasy JJ (1994) Dietary trans fatty acids. Effect on plasma lipids and lipoproptreins of healthy men and women. Am J Clin Nutr 59:861–868

Kamat JP, Devasagayam TPA (1995) Tocotrienols from palm oil as potent inhibitors of lipid peroxidation and protein oxidation in rat brain mitochondria. Neurosci Lett 195(3):179–182

Kausar H, Bhasin G, Zargar MA, Athar M (2003) Palm oil alleviates 12-O-tetradecanoyl-phorbol-13-acetate-induced tumor promotion response in murine skin. Cancer Lett 192(2):151–160

Keng PS, Basri M, Zakaria MRS, Abdul Rahman MB, Ariff AB, Abdul Rahman RNZ, Salleh AB (2009) Newly synthesized palm esters for cosmetics industry. Ind Crops Prod 29(1):37–44

Khalid M, Ratnam CT, Chuah TG, Ali S, Choong TSY (2008) Comparative study of polypropylene composites reinforced with oil palm empty fruit bunch fiber and oil palm derived cellulose. Mater Des 29:173–178

Khanna S, Roy S, Ryu H, Bahadduri P, Swaan PW, Ratan RR, Sen CK (2003) Molecular basis of vitamin E action: tocotrienol modulates 12-lipoxygenase, a key mediator of glutamate-induced neurodegeneration. J Biol Chem 278(44):43508–43515

Khanna S, Roy S, Slivka A, Craft TK, Chaki S, Rink C, Notestine MA, DeVries AC, Parinandi NL, Sen CK (2005) Neuroprotective properties of the natural vitamin E alpha-tocotrienol. Stroke 36(10):2258–2264

Khanna S, Roy S, Parinandi NL, Maurer M, Sen CK (2006) Characterization of the potent neuroprotective properties of the natural vitamin E alpha-tocotrienol. J Neurochem 98(5):1474–1486

Khor HT, Chieng DY (1996) Effect of dietary supplementation of tocotrienols and tocopherols on serum lipids in the hamster. Nutr Res 16(8):1393–1401

Khor HT, Tan DTS (1992a) Studies on the lipidemic property of dietary palm oil: comparison of the responses of serum, liver and heart lipids to dietary palm oil, palm oil trigylcerides, coconut oil and olive oil. Nutr Res 12(Suppl 1):S105–S115

Khor HT, Tan DTS (1992b) The triglycerides in palm oil are not hypercholesterolemic in the hamster. Nutr Res 12(4–5):621–628

Khor HT, Chieng DY, Ong KK (1995) Tocotrienols inhibit HMG-COA reductase activity in the guinea pig. Nutr Res 15:537–544

Khosla P, Hayes KC (1994) Cholesterolaemic effects of the saturated fatty acids of palm oil. Food Nutr Bull 15(2):119–125

Khosla P, Hayes KC (1996) Dietary trans monounsaturated fatty acids negatively impact plasma lipids in humans: critical review of the evidence. J Am Coll Nutr 15:325–329

Klurfeld D, Davidson LM, Lopez-Guisa JM (1990) Palm and other edible oils: atherosclerosis study in rabbits. FASEB J 4: A651.15

Kooyenga DK, Gerler M, Watkins TR, Gapor A, Diakoumakis E, Bierenbaum ML (1997) Palm oil antioxidant effects in patients with hyperlipidaemia and carotid stenosis-2 year experience. Asia Pac J Clin Nutr 6(1):72–75

Kritchevsky D (1988) Effects of triglyceride structure on lipid metabolism. Nutr Rev 46:177–181

Kritchevsky D (2000) Impact of red palm oil on human nutrition and health. Food Nutr Bull 21:182–188

Kritchevsky D, Tepper SA, Chen SC, Meijer GW (1996) Influence of triglyceride structure on experimental atherosclerosis in rabbits. FASEB J 10:A187

Kritchevsky D, Tepper SA, Kuksis A, Wright S, Czarnecki SK (2000) Cholesterol vehicle in experimental atherosclerosis. 22. Refined, bleached, deodorized (RBD) palm oil, randomized palm oil and red palm oil. Nutr Res 20(6):887–892

Kruger MJ, Engelbrecht AM, Esterhuyse J, du Toit EF, van Rooyen J (2007) Dietary red palm oil reduces ischaemia-reperfusion injury in rats fed a hypercholesterolaemic diet. Br J Nutr 97(4):653–660

Kuhad A, Bishnoi M, Tiwari V, Chopra K (2009) Tocotrienol attenuates diabetes-associated cognitive deficits in rats. Alzheimers Dement 5(4):482

Ladeia A, Costa-Matos E, Barata-Passos R, Guimarães AC (2008) A palm oil–rich diet may reduce serum lipids in healthy young individuals. Nutrition 24(1):11–15

Lasekan O, Buettner A, Christlbauer M (2007) Investigation of important odorants of palm wine (*Elaeis guineensis*). Food Chem 105(1):15–23

Lau HL, Puah CW, Choo YM, Ma AN, Chuah CH (2005) Simultaneous quantification of free fatty acids, free sterols, squalene, and acylglycerol molecular species in palm oil by high-temperature gas chromatography-flame ionization detection. Lipids 40(5):523–528

Lau HLN, Choo MY, Ma AN, Chuah CH (2006) Quality of residual oil from palm-pressed mesocarp fiber (*Elaeis guineensis*) using supercritical CO2 with and without ethanol. J Am Oil Chem Soc 83(10):893–898

Lau HLN, Choo MY, Ma AN, Chuah CH (2007) Selective extraction of palm carotene and vitamin E from fresh palm-pressed mesocarp fiber (*Elaeis guineensis*) using supercritical CO2. J Food Eng 84(2):289–296

Lau HLN, Choo YM, Ma AN, Chuah CH (2008) Selective extraction of palm carotene and vitamin E from fresh palm-pressed mesocarp fiber (*Elaeis guineensis*) using supercritical CO. J Food Eng 84:289–296

Lee SP, Mar GY, Ng LT (2009) Effects of tocotrienol-rich fraction on exercise endurance capacity and oxidative stress in forced swimming rats. Eur J Appl Physiol 107(5):587–595

Leh CP, Wan Rosli WD, Zeinuddin Z, Tanaka R (2008) Optimisation of oxygen delignification in production of totally chlorine-free., cellulose pulps from oil palm empty fruit bunch fibre. Ind Crops Prod 28:260–267

Lenaz G (1985) A biochemical rationale for the therapeutic effects of coenzyme Q. In: Lenaz G (ed) coenzyme Q. Wiley, New York, pp 474–475

Li FJ, Tan WJ, Kang ZF, Wong CW (2010) Tocotrienol enriched palm oil prevents atherosclerosis through modulating the activities of peroxisome proliferators-activated receptors. Atherosclerosis 211(1):278–282

Lim KH (1990) Soil erosion control under mature oil palms on slopes. In: Proceedings of the 1989 PORIM international palm oil development conference. Module II Agriculture, Palm Oil Research Institute of Malaysia, Kuala Lumpur

Lim KO, Faridah Hanum A, Vizhi SM (1997) A note on the conversion of oil-palm trunk to glucose via acid hydrolysis. Biores Technol 59:33–35

Liu HK, Wang Q, Li Y, Sun WG, Liu RJ, Ymag YM, Xu WL, Sun XR, Chen BQ (2010) Inhibitory effects of γ-tocotrienol on invasion and metastasis of human gastric adenocarcinoma SGC-7901 cells. J Nutr Biochem 21(3):206–213

Loh SK, Choo YM (2003) Palm-based chiral compounds. J Oil Palm Res 15(1):6–11

Loo CY, Lee WH, Tsuge T, Doi Y, Sudul K (2005) Biosynthesis and characterization of poly (3-hydroxybutyrate co-3-hydroxyhexanoate) from palm oil products in a *Wutersia eutropha* mutant. Biotechnol Lett 27:1405–1410

Maniam S, Mohamed N, Shuid AN, Soelaiman IN (2008) Palm tocotrienol exerted better antioxidant activities in bone than alpha-tocopherol. Basic Clin Pharmacol Toxicol 103(1):55–60

Manorama R, Rukmini C (1991) Effect of processing on β-carotene retention in crude palm oil and its products. Food Chem 42(3):253–264

Manorama R, Rukmini C (1992) Crude palm oil as a source of beta-carotene. Nutr Res 12(Suppl 1):S223–S232

Manorama R, Chinnasamy N, Rukmini C (1993) Effect of red palm oil on some hepatic drug-metabolizing enzymes in rats. Food Chem Toxico 31(8):583–588

Marzuki A, Arshad F, Tariq AR, Kamsiah J (1991) Influence of dietary fat on plasma lipid profiles of Malaysian adolescents. Am J Clin Nutr 53:1010S–1014S

Matthan NR, Dillard A, Lecker JL, Ip B, Lichtenstein AH (2009) Effects of dietary palmitoleic acid on plasma lipoprotein profile and aortic cholesterol accumulation are similar to those of other unsaturated fatty acids in the F1B golden Syrian hamster. J Nutr 139(2):215–221

Matthäus B (2007) Use of palm oil for frying in comparison with other high-stability oils. Eur J Lipid Sci Technol 109:400–409

McGandy RB, Hegsted DM, Myers ML (1970) Use of semi-synthetic fats in determining the effects of specific dietary fatty acids on serum lipids in man. Am J Clin Nutr 23(10):1288–1298

McIntyre BS, Briski KP, Gapor A, Sylvester PW (2000) Antiproliferative and apoptotic effects of tocopherols and tocotrienols on preneoplastic and neoplastic mouse mammary epithelial cells. Proc Soc Exp Biol Med 224(4):292–301

Mensink RP, Katan MB (1990) Effect of dietary trans fatty acids on high-density and low-density lipoprotein cholesterol levels in healthy subjects. N Engl J Med 323:439–445

Mensink RP, Zock PL, Katan MB, Hornstra G (1992) Effect of dietary cis and bans fatty acids on serum lipoprotein(a) levels in humans. J Lipid Res 33(10):1493–1501

Mishima K, Tanaka T, Pu F, Egashira N, Iwasaki K, Hidaka R, Matsunaga K, Takata J, Karube Y, Fujiwara M (2003) Vitamin E isoforms α-tocotrienol and γ-tocopherol prevent cerebral infarction in mice. Neurosci Lett 337(1):56–60

Miyazawa T, Shibata A, Nakagawa K, Tsuzuki T (2008) Anti-angiogenic function of tocotrienol. Asia Pac J Clin Nutr 17(Suppl 1):253–256

Mizushina Y, Nakagawa K, Shibata A, Awata Y, Kuriyama I, Shimazaki N, Koiwai O, Uchiyama Y, Sakaguchi K, Miyazawa T, Yoshida H (2006) Inhibitory effect of tocotrienols on eukaryotic DNA polymerase λ and angiogenesis. Biochem Biophys Res Comm 339:949–955

Mori H, Kaneda T (1994) Food uses of palm oil in Japan. Food Nutr Bull 15(2):144–146

Mortensen A (2005) Analysis of a complex mixture of carotenes from oil palm (*Elaeis guineensis*) fruit extract. Food Res Int 38(8–9):847–853

Müller H, Seljeflot I, Solvoll K, Pedersen JI (2001) Partially hydrogenated soybean oil reduces postprandial t-PA activity compared with palm oil. Atherosclerosis 155(2):467–476

Mun LT, Ishizaki A, Yoshino S, Furukawa K (2004) Production of acetone, butanol and ethanol from palm oil waste by *Clostridium saccharoperbutylacetonicum* N1-4. Biotechnol Lett 17:649–654

Murakoshi M, Takayasu J, Kimura O, Kohmura E, Nishino H, Iwashima A, Okuzumi J, Sakai T, Sugimoto T, Imanishi J, Iwasaki R (1989) Inhibitory effect of α-carotene on proliferation of the human neuroblastoma cell line GOTO. J Natl Cancer Inst 81:1649–1652

Murakoshi M, Nishino H, Satomi Y, Takayasu J, Hasegawa T, Tokuda H, Iwashima A, Okuzumi J, Hidetoshi Okabe H, Kitano H, Iwasaki R (1992) Potent preventive action of a-carotene against carcinogenesis: Spontaneous liver carcinogenesis and promoting stage of lung and skin carcinogenesis in mice are suppressed more effectively by α-carotene than by β-carotene. Cancer Res 52:6583–6587

Nafeeza MI, Fauzee AM, Kamsiah J, Gapor MT (2002) Comparative effects of a tocotrienol-rich fraction and tocopherol in aspirin-induced gastric lesions in rats. Asia Pac J Clin Nutr 11(4):309–313

Naito Y, Shimozawa M, Kuroda M, Nakabe N, Manabe H, Katada K, Kokura S, Ichikawa H, Yoshida N, Noguchi N, Yoshikawa T (2005) Tocotrienols reduce 25-hydroxycholesterol-induced monocytes-endothelial cell interaction by inhibiting the surface expression of adhesion molecules. Atherosclerosis 180(1):19–25

Nelson SE, Rogers RR, Frantz JA, Ziegler EE (1996) Palm olein in infant formula: absorption of fat and minerals by normal infants. Am J Clin Nutr 64:291–296

Neo YP, Ariffin A, Tan CP, Tan YA (2008) Determination of oil palm fruit phenolic compounds and their antioxidant activities using spectrophotometric methods. Int J Food Sci Technol 43(10):1832–1837

Nesaretnam K (2008) Multitargeted therapy of cancer by tocotrienols. Cancer Lett 269(2):388–395

Nesaretnam K, Khor HT, Ganeson J, Chong YH, Sundram K, Gapor A (1992) The effect of vitamin E tocotrienols from palm oil on chemically-induced mammary carcinogenesis in female rats. Nutr Res 12(1):63–75

Nesaretnam K, Guthrie N, Chambers AF, Carroll KK (1995) Effect of tocotrienols on the growth of a human breast cancer cell line in culture. Lipids 30(12):1139–1143

Nesaretnam K, Stephen R, Dils R, Darbre P (1998) Tocotrienols inhibit the growth of human breast cancer cells irrespective of estrogen receptor status. Lipids 33(5):461–469

Nesaretnam K, Dorasamy S, Darbre PD (2000) Tocotrienols inhibit growth of ZR-75-1 breast cancer cells. Int J Food Sci Nutr 51(Suppl):S95–S103

Nesaretnam K, Gomez PA, Selvaduray KR, Razak GA (2007) Tocotrienol levels in adipose tissue of benign and malignant breast lumps in patients in Malaysia. Asia Pac J Clin Nutr 16(3):498–504

Newaz MA, Nawal NN (1999) Effect of gamma-tocotrienol on blood pressure, lipid peroxidation and total antioxidant status in spontaneously hypertensive rats (SHR). Clin Exp Hypertens 21(8):1297–1313

Ng TKW (1994) A critical review of the cholesterolaemic effects of palm oil. Food Nutr Bull 15:112–118

Ng WK (2004a) Palm oil as a novel dietary lipid source in aquaculture feeds. Palm Oil Dev 41:14–18

Ng WK (2004b) Researching the use of palm kernel cake in aquaculture feeds. Palm Oil Dev 41:19–21

Ng WK, Lim HA, Lim SL, I CO (2002) Nutritive value of palm kernel meal pretreated with enzyme or fermented with *Trichoderma koningii* (Oudemans) as a dietary ingredient for red hybrid tilapia (*Oreochromis* sp.) Aquacult Res 33(15):1199–1207

Ng MH, Choo YM (2010) Determination of antioxidants in oil palm leaves (*Elaeis guineensis*). Am J Appl Sci 7:1243–1247

Ng TKW, Hassan K, Lim JB, Lye MS, Ishak R (1991) Non-hypercholesterolemic effects of a palm oil diet in Malaysian volunteers. Am J Clin Nutr 53:1015S–1020S

Ng TKW, Hayes KC, de Witt GF, Jegathesan M, Satgunasingham N, Ong ASH, Tan DTS (1992) Dietary palmitic and oleic acids exert similar effects on serum cholesterol and lipoprotein profiles in normocholesterolemic men and women. J Am Coll Nutr 11(4):383–390

Ng MH, Choo YM, Ma AN, Chuah CH, Ali HM (2004a) Isolation of palm tocols using supercritical fluid chromatography. J Chromatogr Sci 42(10):536–539

Ng MH, Choo YM, Ma AN, Chuah CH, Ali HM (2004b) Separation of vitamin E (tocopherol, tocotrienol, and tocomonoenol) in palm oil. Lipids 39(10):1031–1035

Ng MH, Choo YM, Ma AN, Chuah CH, Ali HM (2006) Separation of coenzyme [Q.sub.10] in palm oil by supercritical fluid chromatography. Am J Appl Sci 3(7):1929–1932

Njoku PC, Egbukole MO, Enenebeaku CK (2010) Physiochemical characteristics and dietary metal levels of oil from *Elaeis guineensis* species. Pak J Nutr 9(2):137–140

Norazlina M, Ima-Nirwana S, Gapor MT, Khalid BA (2000) Palm vitamin E is comparable to alpha-tocopherol in maintaining bone mineral density in ovariectomised female rats. Exp Clin Endocrinol Diab 108(4):305–310

Norazlina M, Ima-Nirwana S, Gapor MTA, Khalid AKB (2002) Tocotrienols are needed for normal bone calcification in growing female rats. Asia Pac J Clin Nutr 11(3):194–199

Norazlina M, Lee PL, Lukman HI, Nazrun AS, Ima-Nirwana S (2007) Effects of vitamin E supplementation on bone metabolism in nicotine-treated rats. Singapore Med J 48(3):195–199

Obire O, Putheti RR (2010) The oil palm tree: a renewable energy in poverty eradication in developing countries. Drug Invent Today 2(1):34–41

Oguntibeju OO, Esterhuyse AJ, Truter EJ (2009) Red palm oil: nutritional, physiological and therapeutic roles in improving human wellbeing and quality of life. Br J Biomed Sci 66(4):216–222

Olmedilla B, Granado F, Southon S, Wright AJ, Blanco I, Gil-Martinez E, van den Berg H, Thurnham D, Corridan B, Chopra M, Hininger I (2002) A European multicentre, placebo-controlled supplementation study with alpha-tocopherol, carotene-rich palm oil, lutein or lycopene: analysis of serum responses. Clin Sci (Lond) 102(4):447–456

Ong ASH, Berger KG (1992) Food uses of palm oil. Malays Oil Sci Tech (MOST) 1(1):12–19

Ong ASH, Goh SH (2002) Palm oil: a healthful and cost-effective dietary component. Food Nutr Bull 23(1):11–22

Ooi TL, Salmiah A (1997) Deinking of waste paper. Palm Oil Dev 26:4–6

Ooi TL, Ying NC, Kifli H (1992) Palm oil-based ink – a new downstream product. PORIM Information Series No. 7

Ooi CK, Choo YM, Yap SC, Basiron Y, Ong ASH (1994) Recovery of carotenoids from palm oil. J Amer Oil Chem Soc 71:423–426

Ooi CK, Choo YM, Yap SC, Ma AN (1996) Refining of red palm oil. Elaeis 8(1):20–28

Ooi TL, Ng CY, Hamirin K, Chacko M (1997) Development of palm oil-based printing ink. Palm Oil Dev 26:1–3

Ooi TL, Salmiah A, Hazimah AH, Chong YJ (2006) An overview of R&D in palm oil-based polyols and polyurethanes in MPOB. Palm Oil Dev 44:1–7

Osakada F, Hashino A, Kume T, Katsuki H, Kaneko S, Akaike A (2004) Alpha-tocotrienol provides the most potent neuroprotection among vitamin E analogs on cultured striatal neurons. Neuropharmacol 47(6):904–915

Oshio S, Abu Hassan O, Takigawa A, Mohd. Jaafar D, Abe A, Dahlan I, Nakanishi N (1990). In: Processing and utilization of oil palm by-products for ruminants. MARDI/TARC @ JIRCAS collaborative study report, pp 110

Osman A (1986) End uses and marketing of palm kernel cakes. Workshop on palm kernel products. Palm Oil Research Institute of Malaysia, Bangi, 31 pp

Osman A, Hisamuddin MA (1999) Oil palm and palm oil products as livestock feed. Palm oil familiarization programme. Palm Oil Research Institute of Malaysia, Bangi, 12 pp

Pacific Island Ecosystems at Risk (PIER) (1999) *Elaeis guineensis* N. Jacquin, Arecaceae. http://www.hear.org/pier/species/elaeis_guineensis.htm

Palmquist DL (2004) Palm fats for livestock feeding. Palm Oil Dev 40:10–16

Pantzaris TP (2000) Pocketbook of palm oil uses, 5th edn. MPOB, Bangi

Pantzaris TP, Ahmad MJ (2001) Properties and utilization of palm kernel oil. Palm Oil Dev 35:11–23

Parker RA, Pearce BC, Clark RW, Gordon DA, Wright JJK (1993) Tocotrienols regulate cholesterol production in mammalian cells by post-transcriptional suppression of 3-hydroxy-3-methylglutaryl-coenzyme A reductase. J Biol Chem 268:11230–11238

Pedersen JI, Muller H, Seljeflot I, Kirkhus B (2005) Palm oil versus hydrogenated soybean oil: effects on serum lipids and plasma haemostatic variables. Asia Pac J Clin Nutr 14(4):348–357

Poku K (2002) Small-scale palm oil processing in Africa. FAO Agr Serv Bull, 148. FAO Rome

Prasertsan P, Prasertsan S (2004) Palm oil industry residues. In: Pandey A (ed) Concise encyclopedia of bioresource technology. Haworth Press, New York, pp 460–465

Qureshi AA, Qureshi N, Hasler-Rapacz JO, Weber FE, Chaudhary V, Crenshaw TD, Gapor A, Ong AS, Chong YH, Peterson D (1991a) Dietary tocotrienols reduce concentrations of plasma cholesterol, apolipoprotein B. thromboxane B2, and platelet factor 4 in pigs with inherited hyperlipidemias. Am J Clin Nutr 53:1042S–1046S

Qureshi AA, Qureshi N, Wright JJ, Shen Z, Kramer G, Gapor A, Chong YH, DeWitt G, Ong A, Peterson DM (1991b) Lowering of serum cholesterol in hypercholesterolemic humans by tocotrienols (palmvitee). Am J Clin Nutr 53:1021S–1026S

Qureshi AA, Bradlow BA, Brace L, Manganello J, Peterson DM, Pearce BC, Wright JJ, Gapor A, Elson CE (1995) Response of hypercholesterolemic subjects to administration of tocotrienols. Lipids 30(12):1171–1177

Radim D, Alimon AR, Yunista Y (2000) Replacing corn with rice bran and palm kernel cake in layer ration. In: Proceeding of the 22nd Malaysian Society of Animal Production, Kota Kinabalu. pp 177–178

Rahman SHA, Choudhury JP, Ahmad AL (2006) Production of xylose from oil palm empty fruit bunch using sulfuric acid. Biochem Eng J 30: 97–103

Raita M, Champreda V, Laosiripojana N (2010) Biocatalytic ethanolysis of palm oil for biodiesel production using microcrystalline lipase in tert-butanol system. Process Biochem 45(6):829–834

Rajanaidu N, Kushari D, Raffii MY, Modh Din A, Maizura I, Isa ZA, Jalani BS (2000) Oil palm genetic resources and utilisation: a review. In: Proceedings international symposium on oil palm genetic resources and utilisation, Kuala Lumpur, 8–10 June 2000, Malaysian Oil Palm Board, Malaysia, pp A1–A55

Rand ML, Hennissen AA, Hornstra G (1988) Effects of dietary palm oil on arterial thrombosis, platelet responses and platelet membrane fluidity in rats. Lipids 23(11):1019–1023

Rickmann M, Vaquero EC, Malagelada JR, Molero X (2007) Tocotrienols induce apoptosis and autophagy in rat pancreatic stellate cells through the mitochondrial death pathway. Gastroenterol 132(7):2518–2532

Rodriguez A, Serrano L, Moral A, Perez A, Jimenez L (2008) Use of high-boiling point organic solvents for pulping oil palm empty fruit bunches. Biores Technol 99:1743–1749

Rossell JB, King B, Downes MJ (1983) Detection of adulteration. J Am Oil Chem Soc 60:333–339

Rubaah M (1999) Goat's milk lotion with palm vitamin E for dry skin (TT no. 63) Goat's milk cream with palm vitamin E for normal skin (TT no. 64) Goat's milk skin freshener (TT no. 65). MPOB Information Series No 88

Saad CR, Cheah CH, Kamaruddin MS (1994) Suitability of using palm kernel cake. (PKC) in catfish practical diet. Science and Technology Congress, 15–18 August 1994, Kuala Lumpur. pp 167–171

Sabiha-Hanim S, Noor MA, Rosma A (2011) Effect of autohydrolysis and enzymatic treatment on oil palm (*Elaeis guineensis* Jacq.) frond fibres for xylose and xylooligosaccharides production. Bioresour Technol 102(2):1234–1239

Salmiah A (2000) Non-food uses of palm oil and palm kernel oil. MPOPC Palm Oil Information Series, Kuala Lumpur, 24 pp

Salmiah A, Zahariah I, Mohtar Y (1998) Oleochemicals for soaps and detergents. Paper presented at the international seminar and exposition world scenario in oils, oleochemicals and surfactants industries, OTAI, Lucknow, India

Salmiah A, Zahariah I, Zulina M, Haliza AA, Yoo CK (2002) Palm-based methyl ester sulphonates. MPOB Information Series No. 160

Sambanthamurthi R, Sundram K, Tan YA (2000) Chemistry and biochemistry of palm oil. Prog Lipid Res 39(6):507–558

Sarmidi MR, Hesham AEE, Mariani AH (2009) Oil palm: the rich mine for pharma, food, feed and fuel industries. Am Eurasian J Agric Environ Sci 5(6):767–776

Sasidharan S, Nilawatyi R, Xavier R, Latha LY, Amala R (2010) Wound healing potential of *Elaeis guineensis* Jacq leaves in an infected albino rat model. Molecules 15(5):3186–3199

Schäfer A (1998) Palm oil fatty acid methyl ester (POME) as diesel fuel in Malaysia. In: Proceeding of the 1998 PORIM international biofuel and lubricant conference, PORIM, Bangi, pp 23–43

Scrimshaw NS (2000) Nutritional potential of red palm oil for combating vitamin A deficiency. Food Nutr Bull 21:195–201

Sen CK, Khanna S, Roy S, Packer L (2000) Molecular basis of vitamin E action. Tocotrienol potently inhibits glutamate-induced pp60(c-Src) kinase activation and death of HT4 neuronal cells. J Biol Chem 275(17):13049–13055

Sen CK, Rink C, Khanna S (2010) Palm oil-derived natural vitamin E alpha-tocotrienol in brain health and disease. J Am Coll Nutr 29(3 Suppl):314S–323S

Serbinova EA, Packer L (1994) Antioxidant and biological activities of palm oil vitamin E. Food Nutr Bull 15(2):138–143

Serbinova E, Choo YM, Packer L (1992a) Distribution and antioxidant activity of a palm oil carotene fraction in rats. Biochem Int 28:881–886

Serbinova E, Khwaja S, Catudioc J, Ericson J, Torres Z, Gapor A, Kagan V, Packer L (1992b) Palm oil vitamin E protects against ischemia/reperfusion injury in the isolated perfused Langendorff heart. Nutr Res 12(Suppl 1):S203–S215

Shah S, Gapor A, Sylvester PW (2003) Role of caspase-8 activation in mediating vitamin E-induced apoptosis in murine mammary cancer cells. Nutr Cancer 45(2):236–246

Shuid AN, Chuan LH, Mohamed N, Jaarin K, Fong YS, Soelaiman IN (2007) Recycled palm oil is better than soy oil in maintaining bone properties in a menopausal syndrome model of ovariectomized rat. Asia Pac J Clin Nutr 16(3):393–402

Siew WL (2000) Characteristics of palm olein from *Elaeis guineensis* palm oil. Malaysian Palm Oil Board, MPOB Technology 2000, Kuala Lumpur, Malaysia. Paper No. 23

Sivan YS, Jayakumar AY, Arumughan C, Sundaresan A, Jayalekshmy A, Suja KP, Soban Kumar DR, Deepa SS, Damodaran M, Soman CR, Kutty RV, Sarma SP (2002) Impact of vitamin A supplementation through different dosages of red palm oil and retinol palmitate on preschool children. J Trop Pediatr 48(1):24–28

Small DM (1991) The effects of glyceride structure on absorption and metabolism. Ann Rev Nutr 11:413–434

Srivastava JK, Gupta S (2006) Tocotrienol-rich fraction of palm oil induces cell cycle arrest and apoptosis selectively in human prostate cancer cells. Biochem Biophys Res Commun 346(2):447–453

Sugano M, Imaizumi K (1991) Effect of palm oil on lipid and lipoprotein metabolism and eicoanoid production in rats. Am J Clin Nutr 53:1034S–1038S

Sukkasame N (2000) Effect of palm kernel cake levels on growth performance of the Nile tilapia (*Oreochromis niloticus* Linn.). The 26th Cong. on Sci. and Tech. of Thailand, 18–20 October 2000, Bangkok

Sumathi S, Chai SP, Mohamad AR (2008) Utilisation of oil palm as a source of renewable energy in Malaysia. Renew Sustain Energ Rev 12:2404–2421

Sun WG, Wang Q, Chen BQ, Liu JR, Liu HK, Xu WL (2008) Gamma-tocotrienol-induced apoptosis in human gastric cancer SGC-7901 cells is associated with a suppression in mitogen-activated protein kinase signalling. Br J Nutr 99(6):1247–1254

Sun WG, Xu WL, Liu HK, Wang Q, Zhou J, Dong FL, Chen BQ (2009) Gamma-Tocotrienol induces mitochondria-mediated apoptosis in human gastric adenocarcinoma SGC-7901 cells. J Nutr Biochem 20(4):276–284

Sundram K, Khor HT, Ong ASH, Pathmarathan R (1989) Effect of dietary palm oils on mammary carcinogenesis in female rats induced by 7,12- dimethylbenz (a) anthracene. Cancer Res 49:1447–1451

Sundram K, Khor HT, Ong ASH (1990) Effect of dietary palm oil and its fractions on rat plasma and high density lipoprotein. Lipids 25(4):187–193

Sundram K, Hornstra G, Houwelingen AC (1992) Replacement of dietary fat with palm oil: Effect on human serum lipid, lipoprotein and apoprotein. Bri J Nutr 68:677–692

Sundram K, Hayes KC, Siru OH (1994) Dietary palmitic acid results in a lower serum cholesterol than a lauric-myristic acid combination in normolipemic humans. Am J Clin Nutr 59:841–846

Sundram K, Anisah I, Hayes KC, Jeyamalar R, Pathmanathan R (1997) Trans elaidic fatty acids adversely impact lipoprotein profile relative to specific saturated fatty acids in humans. J Nutr 127:514S–520S

Sundram K, Sambanthamurthi R, Tan YA (2003) Palm fruit chemistry and nutrition. Asia Pac J Clin Nutr 12(3):355–362

Suwandi MS, Mohamed AA (1984) Growth of Penicillium chrysogenum in palm oil mill effluent concentrate. In: Augustin MA, Ghazali HM (eds) Proceeding of the regional seminar-workshop on biotechnology in industrial development, Serdang, Selangor, Malaysia, pp 261–267

Sylvester PW, Russell N, Ip MM, Ip C (1986) Comparative effects of different animal and vegetable fats fed before and during carcinogen administration on mammary tumorigenesis, sexual maturation and endocrine function in rats. Cancer Res 46:757–762

Tan B, Chu FL (1991) Effects of palm carotenoids in rat hepatic cytochrome P450-mediated benzo(a)pyrene metabolism. Am J Clin Nutr 53(4 Suppl):1071S–1075S

Tan BK, Oh FCH (1981) Malaysian palm oil chemical and physical characteristics. PORIM Technol 3:1–5

Tan DTS, Khor HT, Low WHS, Ali A, Gapor A (1991) Effect of a palm-oil vitamin E concentrate on the serum and lipoprotein lipids in humans. Am J Clin Nutr 53:1027S–1030S

Tan IAW, Hameed BH, Ahmed AL (2007a) Equilibrium and kinetic studies on basic dye adsorption by oil palm fibre activated carbon. Chem Eng J 127:111–119

Tan YA, Sambanthamurthi R, Sundram K, Mohd Basri W (2007b) Valorisation of palm by-products as functional components. Eur J Lipid Sci Technol 109(4):380–393

Tanaka R, Hirose S, Hatakeyama H (2008) Preparation and characterization of polyurethane foams using a palm oil-based polyol. Biores Technol 99(9):3810–3816

Tanito M, Itoh N, Yoshida Y, Hayakawa M, Ohira A, Niki E (2004) Distribution of tocopherols and tocotrienols to rat ocular tissues after topical ophthalmic administration. Lipids 39(5):469–474

Tay YPB, Choo YM (2000) Valuable minor constituents of commercial red palm olein: carotenoids, vitamin E, ubiquinones and sterols. J Palm Oil Res 12(1):14–25

Tay YPB, Choo YM, Ee CLG, Goh SH (2002) Geometrical isomers of the major provitamin a palm carotenes, α- and β-carotenes in the mesocarp oil of fresh and sterilized palm fruits, crude palm oil and palm carotene-based products: red palm olein and carotene concentrates. J Oil Palm Res 13(2):23–32

Teah YK, Ahmad I (1991) Hydrogenation is often necessary with palm oil. In: Yusof B, Cheah SC, Chong YH (eds) Palm oil development, PORIM, Kuala Lumpur, Malaysia, pp 9–14

Tee ES, Noor MI, Azudin MN, Idris K (1997) Nutrient composition of Malaysian foods, 4th edn. Institute for Medical Research, Kuala Lumpur, p 299

Theriault A, Chao JT, Gapor A (2002) Tocotrienol is the most effective vitamin E for reducing endothelial expression of adhesion molecules and adhesion to monocytes. Atherosclerosis 160:21–30

Tomeo AC, Geller M, Watkins TR, Gapor A, Bierenbaum ML (1995) Antioxidant effects of tocotrienols in patients with hyperlipidemia and carotid stenosis. Lipids 30(12):1179–1183

Turner PD, Gillbanks RA (1974) Oil palm cultivation and management. Kuala Lumpur Society of Planters, Kuala Lumpur, 672 pp

Umerie SC (2000) Caramel production from saps of African oil palm (*Elaeis guineensis*) and wine palm (*Raphia hookeri*) trees. Biores Technol 75(2):167–169

Umilalsom MS, Ariff AB, Shamsuddin ZH, Tong CC, Hassan MA, Karim MIA (1997) Production of cellulase by wild strain of *Chaetomium globosum* using delignified oil palm empty fruit bunches as substrates. Appl Microbiol Biotechnol 47:590–595

Van Het Hof KH, Gärtner C, Wiersma A, Tijburg LBM, Weststrate JA (1999) Comparison of the bioavailability of natural palm oil carotenoids and synthetic β-carotene in humans. J Agr Food Chem 47(4):1582–1586

Wan Zahari M, Alimon AR (2004) Use of palm kernel cake and oil palm by-products in compound feed. Palm Oil Dev 40:5–9

Wanrosli WD, Zainuddin Z, Law KN, Asro R (2007) Pulp from oil palm fronds by chemical processes. Ind Crops Prod 25:89–94

Watkins BA, Li Y, Rogers LL, Hoffmann WE, Iwakiri Y, Allen KGD, Seifert MF (2001) Effect of red palm olein on bone tissue fatty acid composition and histomorphometric parameters. Nutr Res 21(1):199–213

Willis WM, Lencki RW, Marangoni AJ (1998) Lipid modification strategies in the production of nutritionally functional fats and oils. Crit Rev Food Sci Nutr 38:639–674

Wilson TA, Nicolosi RJ, Kotyla T, Sundram K, Kritchevsky D (2005) Different palm oil preparations reduce plasma cholesterol concentrations and aortic cholesterol accumulation compared to coconut oil in hypercholesterolemic hamsters. J Nutr Biochem 16(10):633–640

Witt M (1952) Studies of the effect of palm kernel and coconut cakes and meals of different fat contents on milk yield and the fat content of milk. Arch Tierernährung 3:80–101

Wong W-Y, Selvaduray KR, Ming CH, Nesaretnam K (2009) Suppression of tumor growth by palm tocotrienols via the attenuation of angiogenesis. Nutr Cancer 61(3):367–373

Wu SJ, Ng LT (2010) Tocotrienols inhibited growth and induced apoptosis in human HeLa cells through the cell cycle signaling pathway. Integr Cancer Ther 9(1):66–72

Xin FL, Aishah A, Nor Aini U, Das S, Jaarin K (2008) Heated palm oil causes rise in blood pressure and cardiac changes in heart muscle in experimental rats. Arch Med Res 39(6):567–572

Xu WL, Liu JR, Liu HK, Qi GY, Sun XR, Sun WG, Chen BQ (2009) Inhibition of proliferation and induction of apoptosis by gamma-tocotrienol in human colon carcinoma HT-29 cells. Nutrition 25(5):555–566

Yam ML, Abdul Hafid SR, Cheng HM, Nesaretnam K (2009) Tocotrienols suppress proinflammatory markers and cyclooxygenase-2 expression in RAW264.7 macrophages. Lipids 44(9):787–797

Yamada Y, Obayashi M, Ishikawa T, Kiso Y, Ono Y, Yamashita K (2008) Dietary tocotrienol reduces UVB-induced skin damage and sesamin enhances tocotrienol effects in hairless mice. J Nutr Sci Vitaminol (Tokyo) 54(2):117–123

Yamamura Y (1985) A survey of the therapeutic uses of coenzyme Q. In: Lenaz G (ed) coenzyme Q. Wiley, New York, pp 479–495

Yeong SW (1982) The nutritive value of palm oil by-products for poultry [chickens; Malaysia]. Asian-Australasian Animal Science Congress, Serdang, Selangor, Malaysia, 2–5 Sep 1980, pp 217–222

Zainal Z, Longman AJ, Hurst S, Duggan K, Hughes CE, Caterson B, Harwood JL (2009) Modification of palm oil for anti-inflammatory nutraceutical properties. Lipids 44(7):581–592

Zeven AC (1972) The partial and complete domestication of the oil palm (*Elais guineensis*). Econ Bot 26:274–279

Zhang J, Ping W, Chunrong W, Shou CX, Keyou G (1997) Nonhypercholesterolemic effects of a palm oil diet in Chinese adults. J Nutr 127(3):509S–513S

Ziegler RG (1989) A review of epidemiologic evidence that carotenoids reduce the risk of cancer. J Nutr 119:116–122

Ziegler RG, Colavito EA, Hartge P, McAdams MJ, Schoenberg JB, Mason TJ, Fraumeni JF Jr (1996) Importance of α-carotene, β-carotene and other phytochemicals in the etiology of lung cancer. J Natl Cancer Inst 88:612–618

Elaeis guineensis var. *pisifera*

Scientific Name

Elaeis guineensis Jacq. var. *pisifera* A.Chev.

Synonyms

None

Family

Arecaceae

Common/English Names

Pisifera Oil Palm, Pisifera Palm

Vernacular Names

Origin/Distribution

There are two native varieties of oil palm, *E. guineensis* in West Africa, *Dura* and *Pisifera*. Dura is the main variety found in natural groves, and has been the source of palm oil for decades – well before modern methods of oil palm cultivation were introduced to Africa in the second quarter of the twentieth century. Plant breeders created the hybrid variety *Tenera* by crossing the *Dura* and *Pisifera*.

Agroecology

Refer to notes in *E. guineensis*.

Edible Plant Parts and Uses

Refer to notes in *E. guineensis*.

Botany

Refer to notes under *Elaeis guineensis*. The *Pisifera* variety cannot be differentiated from the *Dura* or *Tenera* varieties using the vegetative morphological characteristics of the tree (Plate 1) or leaves. It can be differentiated by the size of its inflorescence and the fruit characters. Its inflorescence (Plate 2) and fruit bunch is smaller than the *Dura* or the commercial *Tenera* varieties. The *Tenera* fruit bunch is much larger than *Dura*. The *Dura* has a large nut with a thick shell (endocarp) 2–8 mm thick and medium mesocarp content (35–55% of fruit weight) (Latiff 2000). The *Tenera* has thinner shell 0.5–3 mm thick and high mesocarp content of 60–95%, its mesocarp containing much more oil and fat (chemically saturated oil) than either of its parents. The *Tenera* nut is small and is easily shelled to release the

Plate 1 Tree habit (G.F. Chung)

Plate 2 Inflorescence (G.F. Chung)

palm kernel. The *Pisifera* is a small fruit with no shell (endocarp). the *Pisifera* palms have no shell (endocarp) and about 95% mesocarp. Sometimes, the shell (endocarp) reduced, with small pea-like kernels in fertile fruits. Yields of *Pisifera* are low. As *Pisifera* palms are predominantly female sterile, they cannot be exploited for commercial planting. They are instead used as pollen parent for crossing with the *Dura* palm to produce the monofactorial *Tenera* (DxP) hybrid.

Genetically, the thick-shelled *Dura* is homozygous for the shell gene (Sh+, Sh+), the thin-shelled *Tenera*, heterozygous for the shell gene (Sh+, Sh-) and the homozygous ressessive shell-less *Pisifera* (Sh-, Sh-). In addition, the *Pisifera* is female sterile while the *Tenera* due to the thin shell, thicker oil bearing mesocarp gives more oil than the *Dura*. All three varieties can be easily differentiated using random amplified polymorphic DNA (RAPD) markers (Toruan-Mathius et al. 1997; Sathish and Mohankumar 2007). RAPD markers are useful in hybrid-oil palm-plant purity-control tests and oil palm types of *Dura, Pisifera*, and *Tenera* identification This approach is advantageous in its rapidity and simplicity; particularly as an alternative for *Dura, Pisifera*, and *Tenera* for which lengthy and costly phenotypic tests are currently used. RAPDs have high discriminatory power and can be successfully applied to reveal genetic diversity among the three varieties of oil palm. The RAPD markers provide a reliable method for identifying the varieties by the analysis of DNA polymorphism.

Sathish and Mohankumar (2007) used RAPD markers to identify the three oil palm varieties. Based on the specificity of amplification, the primers were categorized into three groups for identifying each variety separately. But, the two primers, viz. P12 and P15 showed an amplification pattern that could be used for identifying all the three varieties. In the case of primer P12, *Dura, Pisifera* and *Tenera* showed a common band of 750 bp, which acted as marker for *Dura* variety because other bands shared by *Pisifera* and *Tenera* were absent in *Dura*. The presence of 600 bp in *Pisifera* differentiated it from *Dura* and *Tenera*. Apart from the shared bands between *Tenera* and *Pisifera*, a specific band was present in *Pisifera* that was absent in *Tenera*, indicating the identity of *Tenera*. Using primer P15, a gradual reduction in band pattern was observed in the order that *Dura* had five bands with a unique band with a size of 1,000 bp; *Pisifera* with four bands, without the band of 1,000 bp; and *Tenera* with three bands, without the unique bands observed in parentals.

Using the primers P7 and P10, it was possible to identify the *Dura* variety specifically with a band size of 700 bp and 1,000 bp, respectively. The primer P10 was also useful in identifying pisifera with a band size of 700 bp. The parental variety *Pisifera* could be exclusively identified with a band size of 650 and 800 bp for the primers P6 and P19, respectively. The 10-mer primer P28 was found specific for identifying the *Tenera* variety from the parentals with a 1,100 bp band.

Nutritive/Medicinal Properties

Recently, Njoku et al. (2010) reported the following physio-chemical characteristics for *E. guineensis var. pisifera*, *E. guineensis var. dura* and *E. guineensis var. tenera*.

E. guineensis var. pisifera: vitamin A content (carotenoid) 8677.17 µg/g, vitamin E 476.88 mg/g, protein, 2.21%, moisture, 0.05%, calcium 0.46 mg/dm^3, potassium 0.46 mg/dm^3, magnesium 0.95 mg/dm^3, iron 38.30 mg/dm^3, zinc 0.08 mg/dm^3, acid value 0.65 mg KOH/g, saponification value 191.45 mg KOH/g and iodine value 44.38.

E. guineensis var. dura: vitamin A content (carotenoid) 8927.65 µg/g, vitamin E 443.52 mg/g, protein, 2.30%, moisture, 0.04%, calcium 0.41 mg/dm^3, magnesium 1.13 mg/dm^3, potassium 0.39 mg/dm^3, iron 67.70 mg/dm^3, zinc 0.24 mg/dm^3, acid value 0.67 mg KOH/g saponification value 185.77 mg KOH/g and iodine value 46.93.

E. guineensis var. tenera: vitamin A content (carotenoid) 880.80 µg/g, vitamin E 181.92 mg/g, protein, 2.70%, moisture, 0.05%, calcium 0.34 mg/dm^3, potassium 0.48 mg/dm^3, magnesium 0.37 mg/dm^3, iron 78.30 mg/dm^3, zinc 0.05 mg/dm^3, acid value 0.57 mg KOH/g, saponification value 188.48 mg KOH/g and iodine value 44.05.

The carotenoid composition of *E. guineensis* var *pisifera* was reported by Choo (1994) as follows: phytoene, 1.68%, phytofluene 0.90%, cis-β-carotene 0.10%, β-carotene 54.39%, α-carotenes 33.11%, cis-α-carotene 1.64%, ζ-carotene 1.12%, δ-carotene 0.27%, γ-carotene 0.48%, neurosporene 0.63%, β-zeacarotene 0.97%, α-zeacarotene 0.21% and lycopene 4.5%.

Other Uses

Pisifera is of the great importance in breeding commercial palms. They are instead used as pollen parent for crossing with the *Dura* palm to produce the monofactorial *Tenera* (DxP) hybrid.

Comments

Elaeis guineensis Jacq. var. *pisifera* A.Chev. is not synonymous with *Elaeis oleifera* (Kunth) Cortés from Central and South America.

Selected References

Choo YM (1994) Palm oil carotenoids. Food Nutr Bull 15(2):130–137

Latiff A (2000) The biology of the genus *Elaeis*. In: Basiron Y, Jalani BS, Chan KW (eds.) Advances in oil palm research, vol 1. Malaysian Palm Oil Board, Kuala Lumpur, pp 19–38

Njoku PC, Egbukole MO, Enenebeaku CK (2010) Physio-chemical characteristics and dietary metal levels of oil from *Elaeis guineensis* species. Pak J Nutr 9(2):137–140

Sathish DK, Mohankumar C (2007) RAPD markers for identifying oil palm (*Elaeis guineensis* Jacq.) parental varieties (Dura & Pisifera) and the hybrid Tenera. Indian J Biotechnol 6:354–358

Toruan-Mathius N, Hutabarat T, Djulaicha U, Purba AR, Utomo T (1997) Identification of oil palm (*Elaeis guineensis* Jacq.) Dura, Pisifera, and Tenera by RAPD markers. In: Jenie U et al (eds) Proceedings of the Indonesian biotechnology conference, vol II. IPB, Bogor, pp 237–248

Eleiodoxa conferta

Scientific Name

Eleiodoxa conferta (Griff.) Burret

Synonyms

Salacca conferta Griff., *Salacca scortechinii* Becc., *Eleiodoxa microcarpa* Burret, *Eleiodoxa orthoschista* Burret, *Eleiodoxa scortechinii* (Becc.) Burret, *Eleiodoxa xantholepis* Burret.

Family

Arecaceae

Common/English Names

Asam Paya, Edible-Fruited Salak Palm, Kelubi

Vernacular Names

French: Salacca De Borneo, Salacca De Sarawak, Coeur De Palmier (Vegetable);
Indonesia: Asam Payo, Kelumbi, Kuwai-Kuwai (Sumatra);
Malaysia: Asam Paya, Asam Kelubi, Buah Maram, Kelubi, Kelumi, Sumbalig, Salak Hutan, Bayam (Semang), Kubi (Besisi);
Thailand: Lumphi (Narathiwat, Pattani), Kra-Lu-Bi, Lu-Bi (Malay-Narathiwat).

Origin/Distribution

The crop is indigenous to Peninsular Thailand, Malaysia, Sumatra and Borneo.

Agroecology

It is found in peat swamp forests throughout tropical Thailand, Malaysia, Borneo and Sumatra where they form large, highly gregarious and dense colonies. A strictly tropical species, it thrives in warm, sheltered, shady and very moist swamp forests. It prefers organic rich, fertile, acidic soil.

Edible Plant Parts and Uses

The fruit is exceedingly sour and used in the same way as tamarind. The Chinese in Malaysia have processed the fruit into a candied sweetmeat. The fruit are often pickled or make into relishes. The palm heart is edible.

Botany

An acaulescant, dioecious, clustering palm which forms dense thickets. The palm is hapaxanthic i.e. individual trunks are determinate and die after flowering. The short clustering subterranean stem give rise to crowns of very large, erect, pinnate leaves to up to 8 m long with long, spiny 3 m long

Plate 1 Kelubi fruit red-skinned variety

Plate 3 Kelubi fruit – brown-skinned variety

Plate 2 Kelubi fruit yellow-skinned variety

petioles that are armed with whorls of 5–7 cm long black spines. The broad green leaflets, 1.5 m in length, are regularly arranged along the rachis, and toothed along the margins and silvery on the lower surface. All parts of the foliage are armed with black spines. The inflorescence emerges at ground level, bearing either male or female flowers, in the latter forming distinctively scaly fruits that are up to 3 cm long, pear-shaped, with one or occasionally two seeds. The fruit eventually ripens from green to a dull yellow, brownish or red-orange colour (Plates 1–3).

Nutritive/Medicinal Properties

The proximate nutrient composition of the asam paya (*Eleiodoxa conferta*) fruit per 100 g edible portion (pulp) based on analyses made in Sarawak (Voon and Kueh 1999) was reported as: water 82.8%, energy 78 kcal, protein 0.8%, fat 3.1%, carbohydrates 11.8%, crude fibre 0.8%, ash 0.7%, P 10 mg, K 227 mg, Ca 26 mg, Mg 22 mg, Fe 5.5 mg, Mn 5 ppm, Cu 2.9 ppm, Zn 8.9 ppm, vitamin C 0.6 mg.

The Semang natives in Malaysia used a decoction of the pounded stem for cough and hoarseness. A decoction of the scaly fruit husks has been used as a cough mixture by the Chinese in Peninsular Malaysia.

Other Uses

The leaves are used for thatching and for making mats.

Comments

Eleiodoxa is a monotypic genus, which means that *E. conferta* is the only species in the genus.

Selected References

Burkill IH (1966) A dictionary of the economic products of the Malay Peninsula. Revised reprint. 2 vols. Ministry of Agriculture and Co-operatives, Kuala Lumpur, vol 1 (A–H), pp 1–1240, vol 2 (I–Z), pp 1241–2444

Pearce KG (1991) Palm utilization and conservation in Sarawak (Malaysia). In: Johnson D (ed) Palms for human needs in Asia. Palm utilization and conservation in India, Indonesia, Malaysia and the Philippines. A.A. Balkema, Rotterdam, pp 131–173

Riffle RL, Craft P (2003) An encyclopedia of cultivated palms. Timber Press, Portland, p 528

Voon BH, Kueh HS (1999) The nutritional value of indigenous fruits and vegetables in Sarawak. Asia Pac J Clin Nutr 8(1):24–31

Whitmore TC (1973) Palms of Malaya. Oxford University Press, Singapore, 129 pp

Lodoicea maldivica

Scientific Name

Lodoicea maldivica (J.F. Gmel.) Pers.

Synonyms

Borassus sonneratii Giseke, *Cocos maldivica* J. F. Gmel., *Cocos maritima* Comm. ex H. Wendl., *Lodoicea callipyge* Comm. ex J. St.-Hil., *Lodoicea seychellarum* Labill., *Lodoicea sonneratii* (Giseke) Baill.

Family

Arecaceae

Common/English Names

Coco-De-Mer, Coco De Mer Palm, Double Coconut, Double-Coconut, Love Nuts, Maldive Coconut, Sea Coconut, Sea-Coconut, Seychelles Coconut, Seychelles Nut, Seychelles Nut Palm, Seychelles Palm Nut Tree

Vernacular Names

Arabic: Nârgîl Bahhrî;
Chinese: Hai Ye Zi;
Dutch: Dubbele Cocosnoot, Maldivische Noot, Seychellennoot;
French: Coco De Mer, Coco-De-Mer, Cocotier Des Séchelles, Coco-Fesse;
German: Seychellenpalme;
India: Daryanunariyal (Gujarati), Darya-Ka-Naryal (Hindu), Kataltenna (Malayalam), Jahari-Naral (Marathi), Ubdie-Narikaylum (Sanskrit), Kadalthengai (Tamil), Samudraputenkaya (Telugu);
Italian : Cocco Di Maldiva;
Turkish: Deniz Hindistan Cevizi.

Origin/Distribution

Seychelles nut is indigenous to the coastal rainforests on two Seychelles islands in West Indian Ocean, Valleé de Mai in Praslin and in Curieuse. Subsidiary populations have been established on Mahé and Silhouette Islands in the Seychelles to help conserve the species. The plant has been introduced to other tropical botanical gardens around the world.

Agroecology

In its natural habitat in Seychelles, it grows in the hot, humid tropical rainforest where temperature varies little throughout the year and annual rainfall is about 2,200 mm. Both islands lie outside the cyclone belt and is relatively free from strong winds and thunderstorms. There is a drier season during the south-east monsoon (April to September) and a wetter season during north-east monsoon (October to March).

The palm is very slow growing taking 200 years to reach its full size of 25–30 m. Trees generally take 30–60 years to begin flowering and may continue to do so for another 100–150 years. The fruit can take up to 10 years to ripen, but often ripen in just 5 years.

Edible Plant Parts and Uses

The soft flesh within the nut is edible. The palm heart also called "palm cabbage" (apical bud and underdeveloped leaf bases and leaves) is also edible.

Botany

The palm is perennial, robust, solitary, 25–30 m tall with an erect, spineless, stem which is ringed with leaf scars and a crown of dense stiff foliage. Leaves are large, fan-shaped, grey-green (Plate 1), 7–10 m long and 4.5 m wide with base splitting into two not spiny, blade on a 2–4 m greyish-green petiole (Plate 2). The plant is dioecious, with male and female plants, and the seeds take up to 7 years to mature on the plant. Inflorescence is unbranched, emerging through split leaf bases, stout spikes 1–2 m long. Female flowers are large, 5–13 to a spike. The pistillate flowers are solitary and borne at the angles of the rachis and are partially sunken in it in the form of a cup. They are ovoid with three petals and three sepals. Male flowers are much smaller and occur in groups, arranged spirally and are flanked by very tough leathery bracts. Each staminate flower has a small bracteole, three sepals forming a cylindrical tube, a three-lobed corolla and 17–22 stamens. Fruit is huge, ovoid, bilobed, and pointed, 40–50 cm in diameter and weighs 15–30 kg (Plates 1–3). It contains usually one large seed the largest seed in

Plate 2 Long petioles and immature fruit

Plate 1 Crown of fan-shaped leaves

Plate 3 Cluster of developing fruit and an emerging inflorescence

the plant kingdom. The epicarp is smooth and the mesocarp is fibrous. The endosperm is thick, relatively hard, hollow and homogenous. The embryo sits in the sinus between the two lobes.

Nutritive/Medicinal Properties

No information has been published on the nutritive value of the fruit or palm heart.

The endosperm is astringent and has been converted into medicinal copra and reported to be useful in colic, fever, diarrhoea, cholera, paralysis, epilepsy and other nervous diseases.

Other Uses

A great ornamental plant particularly for the botanical gardens and parks. Shells have been used as containers, water vessels and ornamental boxes. Endosperm provides vegetable ivory. Young leaves supply plaiting materials for hats, baskets, etc. Old leaves make useful thatch.

Comments

It was originally named only from floating seeds; erroneously thought to have originated in the Maldive Islands, and assigned the scientific species name 'maldivica'.

Selected References

Backer CA, Bakhuizen van den Brink RC Jr (1968) Flora of Java (spermatophytes only), vol 3. Wolters-Noordhoff, Groningen, 761 pp

Bailey LH (1949) Manual of cultivated plants most commonly grown in the continental United States and Canada, Revised edn. The Macmillan Co., New York, 1116 pp

Burkill IH (1966) A dictionary of the economic products of the Malay Peninsula. Revised reprint. 2 vols. Ministry of Agriculture and Co-operatives, Kuala Lumpur, vol 1 (A–H), pp 1–1240, vol 2 (I–Z), pp 1241–2444

Facciola S (1990) Cornucopia. A source book of edible plants. Kampong Publ, Vista, 677 pp

Whitmore TC (1973) Palms of Malaya. Oxford University Press, Singapore, 129 pp

Nypa fruticans

Scientific Name

Nypa fruticans Wurmb

Synonyms

Cocos nypa Lour., *Nypa arborescens* Wurmb ex H. Wendl., *Nipa fruticans* Thunb., *Nipa litoralis* Blanco, *Nypa fruticans* var. *neameana* F.M. Bailey.

Family

Arecaceae

Common/English Names

Mangrove Palm, Nipa, Nipa Palm, Nipah, Nypa Palm, Spelt Nipa, Water Coconut, Water Palm

Vernacular Names

Andaman Islands: Poothada;
Australia: Rola (Tiwi Islands), Ki-Bano (Cardwell), Tacannapoon (Pasco River);
Bangladesh: Golpata;
Brazil: Palmeira-Do-Mangue, Palmeira-Nipa (Portuguese);
Burmese: Dani;
Chinese: Shui Ye;
French: Palmier D'eau, Palmier Nipa, Palmier Des Marais, Palmier Des Marécages;
German: Attapalme, Mangrovenpalme, Nipapalme;
India: Gabna, Gulag (Bengali), Nipamu (Telugu);
Indonesian: Nipah (General), Buyuk (Javanese), Bobo (Moluccas);
Italian: Palma Delle Paludi;
Japanese: Nippa Yashi;
Khmer: Chak;
Malaysia: Nipah;
Nigeria: Ayamatangh, Ayangmbakara;
Papua New Guinea: Biri-Biri (Koriki);
Philippines: Anipa, Pinok, Tata (Ibanag), Pinóg (Itogon), Sasa (Pampangan), Saga (Sambali), Nipa (S-Filipino); Lasá, Pauid, Pawid, Sasa (Tagalog);
Portuguese: Palmeira Nipa;
Russian: Nipa;
Singapore: Attap Palm;
Sri Lanka: Gim-Pol;
Thai: Chaak, Lukchaak, Atta;
Vietnamese: Dừa Nước, Dùa Lá.

Origin/Distribution

This is the only species in the genus Nypa and is native to South Asia – Sri Lanka, Bangladesh and Southeast Asia, northern Australia and to the Pacific Islands – Solomon, Marianas, Caroline islands. This species has a fairly restricted distribution in Australia. In the Northern Territory, populations are found at Maxwell and Tjipripu Creeks, Melville Island and Trepang Bay on Cobourg Peninsula. In Queensland it is found on the north-east coast, with its eastern most location being Herbert River. It is now pan-tropical in distribution and is common in the coasts and estuarine rivers that flow into the Indian and Pacific oceans.

Agroecology

Nipa is strictly a tropical palm growing in areas with a warm humid climate – a minimum temperature of 20°C and maximum of 32–35°C and with more than 1,000 mm rainfall per month evenly distributed through the year.

Nipa palms thrives in soft, saline mud and slow moving tidal, brackish water and estuarine river waters that bring in nutrients in the silt. They can be found inland, in the upper tidal reaches of rivers, in brackish, waters as far as the tide can deposit the floating fruit and often form large stands or dense colonies as found in Borneo and Sumatra. In north Queensland it can be found upstream in rivers of the wet, tropical areas where there are low salinities and calm water. It can tolerate infrequent inundation, so long as the soil does not dry out for too long.

Edible Plant Parts and Uses

The young inflorescence before flower opening can be tapped to yield a copious sap for making sugar called *gula Malacca*, molasses or treacle, alcohol or vinegar, slightly fermented sap (*toddy*) or "*tuba*" in Philippines. Tuba is consumed as local beer in the Philippines. *Tuba* is also stored in tapayan (balloon vases) for several weeks to make vinegar in the Philippines, commonly known as *Sukang Paombong*. Sugar extracted from wounded young inflorescence is also used to make whisky. During the second World War in Malaysia, nip whisky and nip brandy were made from the sugary palm sap and also used as a source of biofuel ethanol. Young nipa shoots are also edible and the flower petals can be infused to make an aromatic tisane. Heart cabbage is eaten as vegetable. The white jelly-like endosperm of the immature seed is sweet, translucent, and is eaten raw, or preserves in syrup or made into gelatinous sweetmeat balls used as a dessert ingredient in Malaysia and Singapore and is called *Attap chee* ("chee" meaning "seed" in several Chinese dialects). In Papua New Guinea, a vinegar is made from the sap of the Nipa palm. From the ash of petioles and stems salt is extracted.

Botany

The Nipa palm is a clumping, trunkless palm growing to 10 m in height with a subterranean horizontal stem (rhizome) that grows beneath the ground and only the rosette leaves and flower stalk grow upwards above the ground surface (Plates 1–2). The leaves can extend up to 9 m long, are stiffly erect, pinnate, with two ranks of 60–130 cm long, alternating, stiff, lanceolate leaflets, pointing upwards from the overlapping stout leaf bases. Leaflets number up to 120 per leaf, and have a shiny green upper and powdery lower surface. The leaf bases, and the rhizomes are very light and spongy. Leaf bases are below ground, although they are sometimes exposed by erosion. Nipa is monoecious and its flowers dimorphic. The yellow inflorescences are subterminal on long, sturdy 1 m long stalks arising from the base of the plant. The stalk has big long sheathing spathes and few ascending spathed branches, spathes

Plate 1 Natural grove of Nipa palms in brackish water

Plate 2 Close-up of Nipa palm

Plate 3 Nipa fruit bunches

Plate 4 Close-up of a nipa fruit bunch

Plate 5 Excised edible nipa endosperm

orange tipped olive green. The female inflorescence is a densely packed, spherical head of flowers. The male inflorescence is a club-shaped spike of closely arranged flowers emerging from lateral stalks below the female inflorescence. Male flowers are cream-coloured, borne on branches and the central stalk; female flowers are lemon yellow and borne in a round head size of a golf ball terminating the central stalk. Male flower has three strap sepals with inflexed tips, petals are similar but smaller, with three stamens on a central column protruding from the perianth. Female flower has six small tepals and three woody angular carpel each with a slit below the narrow tip. Each flower develops into a fibrous, chestnut-brown, ovoid, 10–15 cm long, 5–8 cm wide, angular fruit and forms a large spherical infructescence, 30–45 cm in diameter (Plates 3–4). The heavy weight of the infructescence causes the inflorescence stalk to droop, but it is supported by seawater during tidal inundation. Each fruit holds one egg-shaped seed with homogenous, soft, edible endosperm (Plate 5).

Nutritive/Medicinal Properties

The nipa fruit was reported to have the following nutrient composition per 100 g edible portion (Brand Miller et al. 1993): 31 kJ, moisture 88.9 g, nitrogen 0.26 g, protein 1.6 g, fat 0.1 g, available

Plate 6 Dried Nipa leaves used as cigarette-tobacco wrappers

carbohydrate 0 g, Ca 37 mg, Fe 0.4 mg, Mg 69 mg, K 980 mg, Na 63 mg rich, niacin (derived form tryptophan or protein) 0.3 mg, niacin equivalents 0.3 mg and vitamin C 2 mg.

Traditional Medicinal Uses

There are a few folkloric uses of nipa in traditional medicine. In Malaysia, juice from the shoots is mixed with coconut milk for treating herpes and the residue pulp after the juice is expressed is applied. The ash from burnt leaves and roots gives an ash-salt called *garam nipah* which is used for headache and tooth-ache in Borneo. The ash is used with wood tar in blackening teeth. In the Philippines, the decoction of fresh leaves is used as a lotion for indolent ulcers. Fresh leaves are much used in the treatment of ulcers in the form of cataplasm or lotion. The alcohol from the toddy diluted with water is used as an eye-wash in inflammations of the eyelids and of the conjunctiva.

Other Uses

Nipah has a very high sugar-rich sap yield. It can be fermented and processed into ethanol and has been reported to yield much more litres of biofuel per hectare than sugar cane and corn – 15,000–20,000 l of the biofuel per hectare compare with sugarcane at 5,000–8,000 l, or corn, at 2,000 l.

Dried fronds are used as roofing material for thatched houses and dwellings and is called *attap* in Malay and *nipa* in the Philippines and for house walls. Leaflets and mid ribs are woven into mats, baskets, brooms, sunhats, raincoats, bags, umbrellas and other household and handicraft items. Young, dried unfurled leaflets are used to roll cigarettes and are called *rokok nipah* in Malaysia (Plate 6). Mature seeds has been used for buttons. The timber can be used for beams and poles for buildings. Its horizontal subterranean rhizomes stabilises river banks preventing soil erosion. On the islands of Roti and Savu, Indonesia, Nipah sap is fed to pigs during the dry season and is believed to impart a sweet flavour to the pork meat. Nipa palm is regarded as a dreaming or totemic plant for some Tiwi people. Large mud mussels (*Geloina coaxans*) that grow around the base of the stems in deep mud are eaten after roasting.

Comments

Nipa palm is one of the oldest angiosperm plants and perhaps the oldest palm species about 13–63 million years old.

Selected References

Brand Miller J, James KW, Maggiore P (1993) Tables of composition of Australian aboriginal foods. Aboriginal Studies Press, Canberra

Bureau of Plant Industry (2005) Medicinal plants of the Philippines. Department of Agriculture Republic of Philippines. http://www.bpi.da.gov.ph/Publications/mp/mplants.html

Burkill IH (1966) A dictionary of the economic products of the Malay Peninsula. Revised reprint. 2 vols, Ministry of Agriculture and Co-operatives, Kuala Lumpur, vol 1 (A–H), pp 1–1240, vol 2 (I–Z), pp 1241–2444

Council of Scientific and Industrial Research (1966) The wealth of India. A dictionary of Indian raw materials and industrial products (Raw materials 7). Publications and Information Directorate, New Delhi

Dowe J, Tucker R (1993) Notes on the mangrove palm *Nypa fruticans* Wurmb. in Queensland, Palm and Cycad Societies of Australia. Available at: http://www.pacsoa.org.au/palms/Nypa/fruticans.html

Duke NC (2006) Australia's mangroves: the authoritative guide to Australia's mangrove plants. University of Queensland, Brisbane, 200 pp

Hamilton LS, Murphy DH (1988) Use and management of nipa palm (*Nypa fructicans*, Arecaceae): a review. Econ Bot 42:206–213

Paivoke AEA (1996) *Nypa fruticans* Wurmb. In: Flach M, Rumawas F (eds) Plants yielding non-seed carbohydrates, vol 9, Plant resources of South East Asia. Prosea Foundation, Bogor, pp 133–137

Tan R (2001) Nipah palm. *Nypa fruticans*. Mangrove and wetland wildlife at Sungei Buloh Nature Park. http://www.naturia.per.sg/buloh/plants/palm_nipah.htm

Whitmore TC (1970) Palms of Malaya. Oxford University Press, Singapore, p 132

Wightman G (2006) Mangroves of the Northern Territory, Australia: identification and traditional use. Northern Territory. Department of Natural Resources, Environment and the Arts, Palmerston, 168 pp

Phoenix dactylifera

Scientific Name

Phoenix dactylifera L.

Synonyms

Palma dactylifera (L.) Mill., *Palma major* Garsault, *Phoenix atlantica* A. Chev., *Phoenix atlantica* var. *maroccana* A. Chev., *Phoenix atlantidis* A. Chev., *Phoenix chevalieri* D. Rivera, S. Ríos & Obón, *Phoenix dactylifera* var. *adunca* D.H. Christ ex Becc., *Phoenix dactylifera* var. *costata* Becc., *Phoenix dactylifera* var. *cylindrocarpa* Mart., *Phoenix dactylifera* var. *gonocarpa* Mart., *Phoenix dactylifera* var. *oocarpa* Mart., *Phoenix dactylifera* var. *oxysperma* Mart., *Phoenix dactylifera* var. *sphaerocarpa* Mart., *Phoenix dactylifera* var. *sphaerosperma* Mart., *Phoenix dactylifera* var. *sylvestris* Mart., *Phoenix excelsior* Cav. nom. illeg., *Phoenix iberica* D. Rivera, S. Ríos & Obón.

Family

Arecaceae

Common/English Names

Date, Date Palm, Dates, Wild Date Palm

Vernacular Names

Amharic: Yetemir Zaf;
Arabic: Balaha, Khurmae-Yabis, Nakhleh, Doqu, Tamar, Tamer Yabis, Temer;
Brazil: Tamareira (Portuguese);
Chinese: Hai Zao, Ye Zao, Zao Ye, Zao Ye Zi;
Creole: Date;
Czech: Datlovník Obecný;
Danish: Ægte Daddelpalme, Daddelpalme;
Dutch: Dadelpalm, Echte Dadelpalm, Gewone Dadelpalm;
Eastonian: Harilik Datlipalm;
Ethiopia: Tamar, Temri, Timer;
Finnish: Taatelipalmu;
French: Datte, Dattie, Dattier, Dattier Commun, Palmier Dattier;
German: Dattel, Dattelpalme, Echte Dattelpalme;
Hungarian: Datolyapálma, Főnixpálma;
India: Khajur (Bengali), Khaji, Khajur, Pindakhejur, Pindkhajur Salma, Sendhi, (Hindi), Gajjira, Gajjirahannu, Gijjira-Hanny, Kajjuri, Kajura, Karika, Karjora, Karjura, Khajjuri, Kharjoora, Kharjura, Kurjoora, Uttati, Uttatti (Kannada), Itta, Ittappalam, Ittappana, Tenicca, Tenich-Chan-Kaya, Tenitta (Malayalam), Khajur, Khajuri, Kharik, Kharjoor, Kharjur (Marathi), Pindakharjura (Oriya), Agraja, Dipya, Hayabhaksha, Kharjjuraha, Kharjura, Kharjurah, Kharjuram, Kharjuri, Kharjurika, Madhurasraoa, Mudarika, Phalapushpa, Pindakharjura,

Pindakharjurika, Pindi, Pindiphala, Rajajambu, Sapinda, Svadupinda, Swadumastaka (Sanskrit), Caki, Cakituru, Ciramakarivarukkam, Ciravani, Cirnaparnam, Cuvatumattam, Iccai, Iccamaram, Iccan Koluntu, Iccu, Iccucceti, Ichu, Incu, Inju, Inti, Kaccil, Kaccuram 1, Kamaracakikam, Kamirani, Kamiranicceti, Kantukapatikam, Kantukapatitamaram, Karacakamata, Karccur, Karchuram, Karcur, Karcuram 1, Karjjiram, Kasayava, Kokam, Kurampai, Kuranci, Kuravam, Kuravikam, Kuravikamaram, Makacarakkantam, Malatiyam, Malatiyamaram, Pantuyavanam, Paruciyam, Pereecham, Pereecham Pazham, Periccan, Periccankay, Periccu, Perich-Chankay, Perichchangayi, Perichchankay, Perichehu, Perindu, Perinju, Perincu, Periyaincu, Purusako, Titti, Tummitti, Tummuttikam, Tummuttikamaram, Mirutupalam, Nattuiccai, Vilalmurimaram, Yiccamaram (Tamil), Gajjooramu, Gajjuramu, Ita, Kharjooramu, Kharjoorapu Chettu, Karjurakaya, Kharjuramu, Karjuru-Kaya, Manciyita, Manjiyita, Muddakharjooram, Muddakarjuramu, Muddakharjuramu, Muddakharjurapu, Peridu, Perita, Seemakharjooramu, Simakharjuramu, Simakharajuramu, Simakajjura (Telugu), Chhuhara, Khurma, Pend Khajur (Urdu);
Italian: Datteri, Dattero, Palma Da Datteri, Palma Di Dattero, Palmizio;
Japanese: Natsume Yashi, Natsume Yashi;
Nepalese: Chohoraa;
Norwegian: Daddel, Daddelpalme;
Persian: Gour-E-Qurma, Khurma, Khurma Khushk, Khurmae-Khushk, Mung;
Polish: Daktylowiec;
Portuguese: Tâmara, Tamareira;
Russian: Finikovaia Pal'ma, Pal'ma Finikovaia;
Slovascina: Datelj;
Spanish: Dátil, Datilera, Palmera, Palmera Datilera, Palmera De Dátiles;
Swahili: Mtende;
Swedish: Dadelpalm;
Thai: Inthaphalam (Fruit), Ton Inthaphalam (Palm);
Turkish: Hurma A;
Vietnamese: Trái Chà Là.

Origin/Distribution

The date palm is native to North Africa or the Persian Gulf region, however its exact origin is uncertain. The date palm was one of the first plants to be cultivated by humans. People in north Africa and the Middle East have depended on dates as a food source for thousands of years. Dates are cultivated in Algeria, Egypt, Eritrea, Ethiopia, India, Iran, Iraq, Israel, Jordan, Kenya, Nigeria, Ethiopia, Syria, Israel, Lebanon, Libyan Arab Jamahiriya, Namibia, Pakistan, Saudi Arabia, Somalia, Spain, Sudan, Tunisia, Turkey and Zanzibar. It is also cultivated in Namibia (Naute /Keetmanshoop, Hardap/Mariental, Aussenkehr/Karasburg, Eersbegin/Kunene), Republic of South Africa (Henkries Fontein, Kakamas, Klein Pella), Tanzania (Tabora), Spain (Elche), Greece, Mexico, Brazil, Argentina and in the United States of America (California and Arizona). In Australia, dates are cultivated in the Alice Springs and Deep Well (Northern Territory), Pera Bore (New South Wales), Hergott Springs now Maree, Lake Harry (South Australia), Gascoyne Junction (Western Australia) and Cunnamulla-Elo (Queensland).

The world's top ten commercial producer and exporter of dates are Iraq, Egypt, Saudi Arabia, Tunisia, Algeria, United Arab Emirates, Oman, Libya Arab Jamahiriya, Pakistan and Sudan. Europe and USA are the important export markets for dates.

Agroecology

Date palm grows in arid and semi-arid regions which are characterised by long and hot summers, no (or at most low) rainfall, and very low relative humidity level during the ripening period. The extreme limits of date palm distribution are between 10°N (Somalia) and 39°N (Elche in Spain or Turkmenistan). Favourable areas are located between 24°N and 34°N (Morocco, Algeria, Tunisia, Libya, Israel, Egypt, Iraq, Iran, etc.). In USA, date palm is found between 33°N and 35°N. In the southern hemisphere, date palm distribution is found from 5°S in Tabora (Tanzania) to 33°51'S (Pera Bore in New South Wales, Australia). In date growing areas, the average maximum daily temperatures range from 28.2°C to 40.2°C in summer and average daily minimum from 1.1°C to 15.9°C in winter and in terms of mean annual temperature of 20.7–31.8°C. Date palm will endure exceptional high temperatures (56°C) for several days under irrigation. During winters, temperatures below 0°C are also endured. The zero vegetation point of a date palm is 7°C, above this level growth is active and reaches its optimum at about 32°C; the growth will continue at a stable rate until the temperature reaches 38°C/40°C when it will start decreasing. Inflorescences are heavily damaged by frost, from −9°C to −15°C, foliage are damaged and dry out. In date growing areas in the northern hemisphere, mean annual rainfall ranges from 0 to 40 mm and in the southern hemisphere from 3 to 457 mm while in the Sahel date growing areas mean annual rainfall ranges from 20 to 457 mm.

In altitudes, date palm is grown from 392 m below sea level to 1,500 m elevation. It thrives in full sun and is intolerant of shade. Date palm can withstand strong, hot and dusty summer wind. The date can withstand protracted periods of drought though, for heavy bearing, it has a high water requirement which is supplied by periodic flooding from the rivers in North Africa and by subsurface water rather than by rain. High temperatures and low humidity is preferable for flowering, fruit setting and ripening. Date palm is not fastidious of soil types and will grow on almost any type of soil, from almost pure sand, sandy loams, clays to heavy alluvial soils, provided they furnish the basic needs of anchorage to the palm, minerals, water penetration and good drainage. The palm is tolerant of alkali and will tolerate a moderate degree of salinity but excessive salt will stunt growth and reduces the quality of the fruit. It propagates by suckers or seeding in spring.

Edible Plant Parts and Uses

Date palm fruits are classified commercially on the basis of skin texture as soft, semi-dry or dry (Hussein et al. 1976). Soft dates have >30% moisture, low sucrose sugar but with >70% reducing sugars (glucose and fructose). Semi-dry dates are dates with 20–30% moisture, 18–30% sucrose and 45–54% reducing sugars. Dry dates contain <20% moisture with almost equal proportions of sucrose and reducing sugars (33–46%).

Dry, semi-dry or soft dates are best eaten out-of-hand, or may be pitted and stuffed with various types of fillings such as almonds, walnuts, candied lemon and orange peel, tahina (paste made of ground sesame seeds), marzipan (almond and sugar paste) or cream cheese. Pitted dates are also referred to as "stoned dates". Dates can be chopped and used in a great multitude of ways in sweet and savoury dishes: in Moroccan tagines (slow cooked stews), puddings, *ka'ak* (Arabian baked cakes), biscuits, pastries, candy bars, bread, date preserves and cardiments, caramel products and with cereal, ice cream, whipped cream and other dessert. Semi-finished date products include whole pitted date, macerated chips, date paste and date paste mixtures, extruded date pieces and diced dates, dehydrated dates, date flour (dietetic baby foods) and breakfast foods (dates with other dried fruits, cereals, almonds and nuts). Surplus dates are also processed into cubes, paste called "*ajwa*", spread, date syrup or "honey" called "*dibs*" or "*ru*b" in Libya, powder (date sugar), jam, jelly, syrup, wine or alcohol. Decoloured and filtered date juice yields a clear invert sugar solution. Liquid sugar (saccharin as a low calorie sweetener for soft drinks), protein yeast, alcohol and vinegar are also derived from dates. Libya is the leading producer of date syrup and alcohol. In many

Islamic countries, dates and yogurt or milk are a traditional first meal during *maghrib* when the sun sets during Ramadan. Some recent uses of dates include chocolate-coated dates, beverages and sparkling date juice, used in some Islamic countries as a non-alcoholic version of champagne, for special occasions and religious times such as Ramadan. In northern Nigeria, dates and peppers added to the native beer are believed to make it less intoxicating.

In India, North Africa, Ghana and Côte d'Ivoire, date palms especially those with poor fruiting are tapped for the sweet sap which is converted into palm sugar (known as *jaggery* or *gur*), molasses or alcoholic beverages. The terminal bud or palm heart is also eaten and when cut out for eating, the cavity fills with a thick, sweet fluid (called *lagbi* in India) that is drunk for refreshment but is slightly purgative. *Lagbi* ferments in a few hours and is highly intoxicating. Cutting the palm heart will kill the palm. Fresh spathes, by distillation, yield an aromatic fluid enjoyed by the Arabian people. Young date leaves are cooked and eaten as a vegetable. The finely ground seeds are mixed with flour to make bread in times of scarcity. Roasted whole date seeds are popular in Libya. The Bedouin in Saudi Arabia and Libya used the roasted seeds to prepare a special dish called *canua*. In India, date seeds are roasted, ground, and used to adulterate coffee. The seeds also yield an edible oil; which also can be used for cosmetics and pharmaceutical purposes. The flowers of the date palm are also edible. Traditionally the female flowers are the most available for sale. The flower buds are used in salad or ground with dried fish to make a condiment for bread. When the palm is cut down the tender terminal bud is eaten as vegetable or salad.

Botany

The date is a dioecious palm up to 30–36 m high, with a erect, cylindrical and columnar trunk of the same girth 1–1.10 m all the way up and covered from the ground up with upward-pointing, overlapping, persistent, woody leaf bases. Date palm has no tap root; its root system is fasciculated and roots are fibrous. Leaves are grouped in 13 nearly vertical columns, spiralling slightly to the left on some palms and to the right on others. Leaves of a date palm are large, 3–6 m long (usually 4 m) with a relatively short, 50 cm rachis base or petiole; pinnate, with 120–240 leaflets per frond. Leaflet is, entire, lanceolate, folded longitudinally and oblically attached to the petiole, 15 to over 100 cm long and 1–6.3 cm wide. Leaflets near the base are modified into 7.6–10.2 cm long spines. Being dioecious, male and female flowers are produced in clusters called spadix in the axils of leaves on separate palms. The spadix is enclosed in a hard covering/envelope known as spathe which splits open as the flowers mature exposing the entire inflorescence for pollination purposes. The spadix consists of a central stem called rachis and several strands or spikelets (usually 50–150 lateral branches), paniculate. The male spathes are shorter and wider than the female ones. Each spikelet carries a large number of tiny flowers as many as 8,000–10,000 in female and more in male inflorescence. The male flower is sweet-scented and normally has six stamens, surrounded by three, waxy scale-like, valvate petals and 3-lobed sepals. Each stamen is bears two little yellowish anthers. The female flower is about 3–4 mm across, The three sepals and three petals are united together so that only tips diverge with rudimentary stamens and three carpels closely pressed together and a superior ovary with a terminal stigma. Only one ovule per flower is fertilised, leading to the development of one carpel which in turn gives a fruit called a date; the other ovules aborted. The fruits hang down in pendant clusters (Plates 1–2). The date fruit is a single, oblong, 2.5–7.5 cm long, one-seeded berry, green turning bright red to bright yellow in colour when unripe, depending on variety becoming dark-brown, reddish, or yellowish-brown when ripe with thin or thickish skin, thick, sweet flesh (astringent until fully ripe) a fleshy pericarp and a membranous endocarp (between the seed and the flesh) (Plate 3). The seed is usually oblong, ventrally grooved, with a small embryo, and a hard endosperm.

Plate 1 Clusters of young dates

Plate 2 Pendant clusters of maturing dates

Plate 3 Ripe dates

Nutritive/Medicinal Properties

Nutrient composition of fresh dates (Medjool) per 100 g edible portion is: water 21.32 g, energy 277 kcal (1.160 kJ), protein 1.81 g, total lipid 0.15 g, ash 1.74 g, carbohydrate 74.97 g, total dietary fibre 6.7 g, total sugars 66.74 g, sucrose 0.53 g, glucose 33.68 g, fructose 31.95 g, maltose 0.30 g, Ca 64 mg, Fe 0.90 mg, Mg 54 mg, P 62 mg, K 696 mg, Na 1 mg, Zn 0.44 mg, Cu 0.362 mg, Mn 0.296 mg, vitamin C 0.0 mg, thiamine 0.050 mg, riboflavin 0.060 mg, niacin 1.610 mg, pantothenic acid 0.805 mg, vitamin B-6 0.249 mg, total folate 15 μg, choline 9.9 mg, betaine 0.4 mg, vitamin A 149 IU, vitamin K (phylloquinone) 2.7 mg, tryptophan 0.007 g, threonine 0.042 g, isoleucine 0.045 g, leucine 0.082 g, lysine 0.054 g, methionine 0.017 g, cystine 0.046 g, phenylalanine 0.048 g, tyrosine 0.016 g, valine 0.066 g, arginine 0.060 g, histidine 0.029 g, alanine 0.078 g, aspartic acid 0.220 g, aspartic acid 0.22 g, glutamic acid 0.265 g, glycine 0.090 g, proline 0.111 g, serine 0.062 g, β-carotene 89 μg, lutein + zeaxanthin 23 μg (USDA 2010).

Thirty-four date (*Phoenix dactylifera*) varieties, from start of *Tam*r stage of maturity, were analysed for moisture, protein, lipid and ash (Sahari et al. 2007). The mean percent of moisture, protein, lipid and ash were 29.35, 3.3, 0.42 and 2.25 g/100 g (fresh weight basis), respectively. Predominant sugars were fructose (12.62–43.31 g/100 g) and glucose (16.41–54.23 g/100 g, fresh weight basis). Sucrose was not practically detected in most varieties (except in Zark variety). Mineral elements such as Na, Mg, K and Ca were in the range of 4.46–47.74, 18.44–79.35, 203.61–982.97 and 23.24–73.85 mg/100 g (dry weight basis), respectively.

Chemical analysis of the fruit showed that glucose and fructose increased rapidly with maturation from *Kimri* through *Khalal* and *Rutab* to *Tamr* (Ahmed et al. 1995). Total sugars may represent over 50% of the fresh weight at *Tamr,* and these values, together with low moisture contents, encourage resistance to fungal spoilage after harvest. Minerals accumulated in the fruits as well, and the date could be an important source of potassium for regular consumers.

The major carotenoid pigment present in dates was found to be lutein followed by β-carotene, with an evident carotenoid disappearance during ripening from the *khallal* to the *tamr* stage (Boudries et al. 2007). The different date fruits presented a total carotenoid content in the range of 61.7–167, 32.6–672, and 37.3–773 μg/100 g fresh weight (FW) in Deglet-nour, Tantebouchte and Hamraya varieties, respectively. The *rutab* stage of Tantebouchte showed the lowest carotenoid content of 32.6 μg/100 g FW, whereas the *khallal* stage of Hamraya presented the highest value, 773 μg/100 g FW, followed by Tantebouchte with 672 μg/100 g FW. Provitamin A value (due exclusively to b-carotene) increased from 0.4 to 0.5 RE/100 g in Deglet-Nour fruits, but decreased from 11.7 to 1.6 RE/100 g and from 3.9 to 0.5 RE/100 g in Tantebouchte and Hamraya fruits, respectively, during ripening. The lowest value was found at the *tamr* stage of the Deglet-Nour variety (0.5 RE/100 g) whereas the highest provitamin A content was found at the *khallal* stage of the Tantebouchte variety (11.7 RE/100 g).

The chemical profiles of the extracted oil fatty acid (on a dry-weight basis) obtained from the date pulp and pits were found to be: total sugars 63.38% and 8.12%, reducing sugars 51.56% and 6.63%, sucrose 11.82% and 1.49%, protein 3.86% and 5.31%, oil 0.26% and 8.33% respectively (Saafi et al. 2008). The major unsaturated fatty acid was linoleic acid (32.77%) for the pulp and oleic acid (47.66%) for the pits, while the main saturated fatty acid was palmitic acid (20.55%) for the pulp and lauric acid (17.39%) for the pits. Myristic, stearic and linolenic acids were also found in both the pulp and seeds.

Date seeds of two cultivars Deglet Nour and Allig, were found to contain on a dry-weight basis: protein 5.56% and 5.17%, oil 10.19% and 12.67%, ash 1.15% and 1.12% and total carbohydrate 83.1% and 81.0% respectively (Besbes et al. 2004). The main saturated fatty acid was lauric acid (17.8%) for the Deglet Nour cultivar and palmitic acid for the Allig cultivar (15.0%). Capric, myristic, myristoleic, palmitoleic, stearic, linoleic and linolenic acids were also found.

The alkali-soluble polysaccharides in dates seeds were found to contain D-xylose and 4-O-methyl-D-glucuronic acid in a molar ratio of 5:1 (Ishurd et al. 2003). An aldobiouronic acid from hemicellulose was characterized, the hemicellulose consists of a polymer of $(1 \rightarrow 4)$-linked D-xylopyranosyl residues having branches of D-xylopyranosyl and 4-O-methyl-α-D-glucopyranosyluronic acid.

Some of the pharmacological attributes of date fruits and palm parts are as follows:

Antioxidant Activity

Using the ferric reducing antioxidant power (FRAP) assay, the highest total antioxidant activity was found at *biser* (unripe) stage, with a mean FRAP value of 5.71 mmol/100 g fresh weight (FW), followed by rutab (soft and ripped) with FRAP values of 1.2 mmol/100 g FW and tamer (dried fruit) 0.94 mmol/100 g FW (Allaith 2008). Total antioxidant activities of locally collected and imported *tamer* were similar. There was about 12- and 6-fold difference between cultivars possessing the highest and lowest antioxidant values at *biser* and *rutab* stages, respectively. A sharp decrease in antioxidant activity was found to be associated with fruit ripening. The average of phenolics at *biser* and *rutab* stage were 196.8 and 116.7 mg/100 g fresh weight, whereas the average of ascorbic acid was 6.6 and 3.3 mg/100 g fresh weight, respectively. Significant correlation was found between the antioxidant activity and total phenolic content and ascorbic acid at *biser* stage, but not at *rutab* stage. It was concluded that the phenolics were the major contributor for the antioxidant activity.

Edible parts of date palm (*Phoenix dactylifera*) fruits from Iran were analyzed for their antioxidant activities (AA) using Trolox equivalent antioxidant capacity (TEAC) method, 2,2′-azinobis (3-ethylbenzothiazoline-6-sulphonic acid) radical cation (ABTS+) assays and the ferric reducing/antioxidant power method (FRAP assay) (Biglari et al. 2008). The total phenolic content (TPC) and total flavonoid content (TFC) of the fruits were measured using Folin–Ciocalteau and aluminum chloride colorimetric methods, respectively. The samples used included four types of soft dates (SD) namely Honey date, Bam date, Jiroft date and Kabkab date; three types of semi-dry dates (SDD) namely Sahroon date, Piarom date and

Zahedi date and one type of dry date (DD) which was Kharak date. The AA (ABTS assay) of the fruits were 22.83–41.17, 47.6–54.61 and 500.33 µmol Trolox equivalents/100 g dry weights (dw) for SD, SDD and DD, respectively. The AA (FRAP assay) per 100 g dw sample were 11.65–20, 19.12–29.34 and 387.34 µmol FRAP for SD, SDD and DD, respectively. The TPC ranged from 2.89 to 4.82, 4.37 to 6.64 and 141.35 mg gallic acid equivalents (GAE)/100 g dw, while TFC ranged from 1.62 to 3.07, 1.65 to 4.71 and 81.79 mg catechin equivalents (CEQ)/100 g DW sample for soft dates, semi-dry dates and dry dates, respectively. Correlation analyses indicated that there was a linear relationship between antioxidant activities and the total phenolic content or total flavonoid content of date palm fruit.

Dates at the *rutab* and *tamar* maturity and ripening stages were found to contain a wide array of phenolic antioxidants, but little is known about the composition of phenolic compounds in dates at the *khalal* stage of ripening (Hong et al. 2006). Studies using liquid chromatography-electrospray ionization-tandem mass spectrometry (LC-ESI/MS/MS) identified 13 flavonoid glycosides of luteolin, quercetin, and apigenin, 19 when considering isomeric forms. Mass spectra indicated that both methylated and sulfated forms of luteolin and quercetin were present as mono-, di-, and triglycosylated conjugates whereas apigenin was present as only the diglycoside. LC-ESI/MS/MS spectra indicated that quercetin and luteolin formed primarily O-glycosidic linkages whereas apigenin was present as the C-glycoside.

Dates were found to have antioxidant and antimutagenic properties in-vitro (Praveen and Vayalil 2002). The aqueous extract of date fruit exhibited a dose-dependent inhibition of superoxide and hydroxyl radicals. The amount of fresh extract required to scavenge 50% of superoxide radicals was equivalent to 0.8 mg/mL of date fruit in the riboflavin photoreduction method. An extract of 2.2 mg/mL of date fruit was needed for 50% hydroxyl-radical-scavenging activity in the deoxyribose degradation method. Concentrations of 1.5 and 4.0 mg/mL completely inhibited superoxide and hydroxyl radicals, respectively. Aqueous date extract was also found to inhibit significantly the lipid peroxidation and protein oxidation in a dose-dependent manner. In the Fe^{2+}/ascorbate system, an extract of 1.9 mg/ml of date fruit was needed for 50% inhibition of lipid peroxides. In a time course inhibition study of lipid peroxide, at 2.0 mg/ml concentration of date extract, there was a complete inhibition of TBARS formation in the early stages of the incubation period that increased during later stages of the incubation. Similarly, in the high Fe^{2+}/ascorbate induction system a concentration of 2.3 mg/ml inhibited carbonyl formation by 50% as measured by DNPH reaction. Additionally, a concentration of 4.0 mg/mL completely inhibited lipid peroxide and protein carbonyl formation. Date fruit extract also produced a dose-dependent inhibition of benzo(a)pyrene-induced mutagenecity on *Salmonella* tester strains TA-98 and TA-100 with metabolic activation. Extract from 3.6 and 4.3 mg/plate was required for 50% inhibition of His + revertant formation in TA-98 and TA-100, respectively. These results indicated that antioxidant and antimutagenic activity in date fruit was quite potent and implicated the presence of compounds with potent free-radical-scavenging activity.

Antioxidant compounds (total phenolic content (TPC) and total flavonoid content (TFC)) of both soft (SD, Bam) and dry (DD, Kharak) date varieties increased following storage at 4°C for 6 months followed by an additional 1 week storage at 18°C (Biglari et al. 2009). Cluster analysis applied on TPC and TFC data before and after storage obtained two statistically significant clusters of SD and DD indicating that TPC and TFC had different behaviors according to the types of dates. As antioxidant concentration in dates was dependent on type of dates, selection of variety with high antioxidant compounds may be recommended. Further cold storage of up to 6 months followed by 1 week storage at 18°C may further improve the level of antioxidant compounds.

The simple phenolic content in selected common date cultivars ranged from 15.73 (Mermella variety) to 54.66 mg/100 g fresh weight (Korkobbi variety) (Chaira et al. 2009). Korkobbi had also the highest content of total flavonoids (54.46 quercetin equivalents/100 g fresh weight). The percentage of inhibition can reach 83% in the lipoperoxyl radical of the Korkobbi variety, while it was about 95% with the Rotbi variety for OH. The Trolox equivalent antioxidant capacity value provided a ranking in decreasing order at 6 minutes: Korkobbi > Bouhattam > Baht = Smiti > Bekreri = Garn ghzal > Mermilla = Kenta > Nef zaoui = Rotbi. It appeared that the highest level of flavonoids in the Korkobbi variety was principally responsible for the highest antiradical efficiency of this cultivar.

Phoenix dactylifera, Citrus aurantifolia had significantly (p<0.05) higher total phenol, and iron chelation ability, DPPH scavenging activity, and (%AA) than *Loranthus europeas* and *Zingiber officinalis* (Ahmed and Rocha 2009). Meanwhile the water extracts of *Phoenix dactylifera* and *Citrus aurantifolia* had the highest protective ability and this probably was attributed to its higher antioxidant activity (AA%), total phenol content, iron chelation and DPPH scavenging activity.

Hepatoprotective Activity

Dates have hepatoprotective activity. Treatment with aqueous extract of date flesh or pits significantly reduced CCl4-induced elevation in plasma enzyme and bilirubin concentration and ameliorated morphological and histological liver damage in rats (Al-Qarawi et al. 2004). Recent studies showed that date palm fruit may be useful for the prevention of oxidative stress induced hepatotoxicity by subchronic exposure to dimethoate (Saafi et al. 2010). Oral administration of dimethoate caused hepatotoxicity as monitored by the increase in the levels of hepatic markers enzymes (transaminases, alkaline phosphatase, γ-glutamyl transferase and lactate dehydrogenase), as well as in hepatic malondialdehyde thus causing drastic alteration in antioxidant defence system. Particularly, the activities of superoxide dismutase (SOD) and glutathione peroxidase (GPx) were increased by dimethoate while catalase (CAT) activity was reduced significantly. These biochemical alterations were accompanied by histological changes marked by appearance of vacuolization, necrosis, congestion, inflammation, and enlargement of sinusoids in liver section. Pre-treatment with date palm fruit extract restored the liver damage induced by dimethoate, as revealed by inhibition of hepatic lipid peroxidation, amelioration of SOD, GPx and CAT activities and improvement of histopathology changes.

Antiatherogenic Activity

Total phenolics concentration in the Hallawi versus Medjool dates was greater by 20–31% (Rock et al. 2009). The major proportion of the soluble phenolics in both date varieties consisted of phenolic acids, mainly ferulic acid and coumaric acid derivatives, and also chlorogenic and caffeic acid derivatives. Unlike the Medjool dates, Hallawi dates contained a significant proportion of catechins as well. In addition, both varieties contained a quercetin derivative. Both date varieties possessed antioxidative properties in-vitro, but the ferric ion reducing antioxidant power of Hallawi versus Medjool dates was higher by 24%. Studies showed that consumption of Hallawi or Medjool dates did not significantly affect the subjects' body mass index (BMI), their serum total cholesterol, or their cholesterol levels in the VLDL, LDL, or HDL fractions (Rock et al. 2009). Most important, fasting serum glucose and triacylglycerol levels were not increased after consumption of either date variety, and serum triacylglycerol levels even significantly decreased, by 8 or 15% after Medjool or Hallawi date consumption, respectively. Basal serum oxidative status was significantly decreased by 33%, as compared to the levels observed before consumption, after Hallawi (but not Medjool) date consumption. Similarly, the susceptibility of serum to AAPH (amidinopropane hydrochloride) -induced lipid peroxidation decreased by 12%, but only after Hallawi date consumption. In agreement with the above results, serum activity of the HDL-associated antioxidant enzyme paraoxonase 1 (PON1) significantly increased, by 8%, after Hallawi date consumption. It was concluded that date consumption (especially the Hallawi variety) by healthy subjects, despite their high sugar content, demonstrated beneficial effects on serum triacylglycerol and oxidative stress and did not aggravate serum glucose and lipid/lipoprotein patterns, and thus could be considered an anti-atherogenic food.

Nephroprotective Activity

Gentamicin treatment significantly increased the plasma concentrations of creatinine and urea and induced a marked necrosis of the renal proximal tubules in rats (Al-Qarawi et al. 2008). The date flesh and pits were effective in significantly reducing the increases in plasma creatinine and urea concentrations induced by gentamicin nephrotoxicity and ameliorating the proximal tubular damage. Antioxidant components in the date (e.g., melatonin, vitamin E, and ascorbic acid) were suggested to be the basis of the nephroprotection.

Gastroprotective Activity

Studies indicated that the aqueous and ethanolic extracts of the date fruit and, to a lesser extent, date pits, were effective in ameliorating the severity of gastric ulceration and mitigating the ethanol-induced increase in histamine and gastrin concentrations, and the decrease in mucin gastric levels in rats (Al-Qarawi et al. 2005). The ethanolic, undialyzed extract was more effective than the rest of the other extracts used. It was postulated that the basis of the gastroprotective action of date extracts may be multi-factorial, and may include an anti-oxidant action. The results supported a local folk medicinal claim that dates are beneficial in gastric ulcers in humans.

Anticancer Activity

An antitumour glucan compound, was isolated from Libyan dates (Ishurd and Kennedy 2005). The anti-tumourous activity was postulated to be associated to their $(1 \rightarrow 3)$-β-d-glucan linkages.

Hypocholesterolemic, Hypolipidemic Activity

Five diets were investigated which included basal diet, cellulose diet and three defatted date seed fibre diets containing 1.5%, 2.5% and 5.2% date seed fibre in Wistar rats (Al-Maiman 2005) The results revealed that diet based on date seed fibre had good potential as a source of dietary fibre. Diet containing 1.5% date seed fibre was the most appropriate because of its action in reducing LDL cholesterol, total cholesterol and triglyceride. However, such level (1.5%) of date seeds fibre had no effect on HDL cholesterol level.

Antidiarrhoeal Activity

Date palm spathe aqueous extract was found to have anti-spasmodic activity (Al-Taher 2008). Castor oil-induced diarrhoea, enteropooling and gastrointestinal transit test were elucidated in rats. The results showed that aqueous extract significantly reduced both castor-oil induced intestinal transit and frequency of diarrhoea effects. However, the extract did not affect castor oil induced intestinal fluid accumulation.

Fertility, Antifertility Effects

Experimentally, date extracts have been shown to have a gonadotrophic effect. The extract was found to increase sperm count in guinea pigs and to enhance spermatogenesis and increase the concentration of testosterone, follicle stimulating hormone, and luteinizing hormone in rats (El-Mougy et al. 1991). The pollen grains of date palm had been used by Egyptians to improve fertility in women (Amin et al. 1969).

A study found that treatment with normal date pits (14%) significantly increased the body weight of rats (Ali et al. 1999). The other treatments (7% and 14% acid-treated powdered date pits) did not significantly affect body weight. At concentrations of 7% and 14%, the normal date pits significantly increased the concentration of testosterone in plasma, while the acid-treated date pits (14%) slightly but significantly increased the concentration of luteinizing hormone. Plasma oestradiol concentration and the other measured variables were not affected by any of the treatments. Treatment of female rats for 10 days with a lyophilized extract of date pits, or with a polar or non-polar fraction prepared from it, at oral and intraperitoneal doses of 500 mg/kg, did not significantly affect body or uterine weights, nor did it affect the degree of vaginal orifice opening. Oestradiol (2 mg/kg, subcutaneously) increased uterine weight and degree of vaginal orifice opening. Compared to the control, the lyophilized extract and the polar fraction significantly reduced plasma oestradiol concentration by about 25% and 36%, respectively. The non-polar fraction, however, increased the hormone level by about 12%, but this was not significant. In another study, diet of date pits was found to have an antifertility effect on female rats, proximate analysis of date pits indicated that the nitrogen-free extract (NFE) was 71.5%, in which 3% was starch (Aldhaheri et al. 2004). When chemical analysis was based on dietary fibre, the NFE was found to be 26.7% only, in which 20.9% was mannose. The dietary treatments comprising five isonitrogenous (23% Crude Protein) and isocaloric (2.8 Mcal/kg) diets with or without date pits had no effect on testosterone level in male rats. While, oestradiol concentration in the serum of female rats decreased significantly as the percentage of date pits increased, this may be due to the estrogenic effect of date pits, which may cause reduction in the fertility of the female rats.

Traditional Medicinal Uses

In folkloric traditional medicine, dates are deemed as a demulcent, an expectorant and a laxative, and are used to treat respiratory diseases and fever. As an infusion, decoction, syrup, or paste, dates may be administered for sore throat, colds, bronchial catarrh, and taken to relieve fever and number of other complaints. The tree yields a gum used in treating diarrhoea and gastro-urinary ailments in India. One traditional belief is that it can counteract alcohol intoxication. The seed powder is also used in some traditional medicines. The roots are used against toothache.

Other Uses

Date palm is a majestic and spectacular palm for landscaping large areas. It is also ideally suitable for planting in streets, avenues and driveways. Date palm has been used for decades for the revegetation of salt affected lands in the Mediterranean region. The trunk can be used for aqueducts, bridges, various kinds of construction, and also parts of dhow. Date logs are used as crossbeams for ceilings, posts and rafters for huts, and sweeps for wells. All left over parts of the trunk are burnt for fuel.

Dates can also be dehydrated, ground and mixed with grain to form a nutritious stockfeed. Dried dates are fed to camels, horses and dogs in the Sahara. In Pakistan, a viscous, thick syrup made from the ripe fruits is used as a coating for leather bags and pipes to prevent leaking. Stripped fruit clusters are used as brooms. Fruit stalks are used as fuel.

Date seeds are soaked and ground up and fed to horses, cattle, camels, sheep and goats. Dried and ground up, they are now included in chicken feed. Date seeds can also be processed chemically as a source of oxalic acid. The seeds are also burned to make charcoal for silversmiths, and can be strung in necklaces. Studies showed that date seed oil could be used in cosmetics, pharmaceutical and food products (Besbes et al. 2004).

Date palm leaves are used for Palm Sunday by Christians. The leaves are also used as a *lulav* in the Jewish holiday of Sukkot. In North Africa, whole leaves are commonly used for thatching huts and coverings. Mature leaves are also made into mats, screens, baskets, fans, hats, crates and furniture. Fibres encircling the trunk are made into coarse cloth, ropes, cordage and mattresses. Ropes are also manufactured from the fruit stalks. Dried date leaves with their stiff, woody rachis are used for fencing. The leaves are also applied in sand dune stabilization and leaf prunings used as green manure. Processed leaves can be used for insulating board. Dried leaf petioles are a source of cellulose pulp, used for walking sticks, brooms, fishing floats and fuel. Bases of the leaves are used as fuel.

Comments

Date fruit has five (5) distinct stages of fruit development, known as *Hababouk, Kimri, Khalal, Rutab* and *Tamar* (Samarawira 1983; Zaid and Arias-Jimenez 2002).

(a) *Hababouk (Habbabok, Hababauk)* stage

This stage commences soon after fertilisation and continues until the beginning of the kimri stage. Fruit at this stage is immature with very slow growth rate and is completely covered by the calyx and only the sharp end of the ovary is visible.

(b) *Kimri (Khimri, Jimri,* green stage) stage

At this stage the fruit is quite hard, the colour is apple green and it is not suitable for eating. This stage can be further divided it into two different phases. The first phase is characterized by a rapid increase in weight and volume, high acidity, high moisture content, slight increase in total reducing sugars. The second phase is associated with generally reduced total sugar accumulation, acidity and a high moisture content.

(c) *Khalal (Khalaa)* stage

Also referred to as the colour stage. The fruit is physiologically mature, hard ripe and the colour changes completely from green to greenish yellow, yellow, pink, red or scarlet depending on the variety. At the end of this stage, date fruit reaches its maximum weight and size, but sugar concentration (saccharose), total sugar and active acidity have a rapid increase associated with a decrease in water content (around 50–85% moisture content). It is to be noted that date fruit accumulate most of their sugar, both the sucrose type and the reducing sugar type, as sucrose during the Khalal stage. At this stage colour of the seed changes at the end from white to brown.

(d) *Rutab (Routab)* stage

This is the ripening stage when the fruit changes in colour to brown or black and becomes soft. With softening, the last of the tannin under the skin is precipitated in an insoluble form, so that the fruit loses any astringency. It is a very good stage for consumption as a hard ripe date. With the exception of a few varieties, fruit at this stage is very sweet.

(e) *Tamar (Tamer, Tamr)* stage

This is also called full ripe stage or final stage in the ripening when the dates are fully ripe, and they completely change the colour from yellow to dull brown or almost black. The texture of the flesh is soft. The skin in most varieties adheres to the flesh, and wrinkles as the flesh shrinks. At this stage, the fruit contains the maximum total solids and has lost most of its water to such an extent (below 25% down to 10% and less) that it makes the sugar water proportion sufficiently high to prevent fermentation.

Selected References

Ahmed SH, Rocha JB (2009) Antioxidant properties of water extracts for the Iraqi plants *Phoenix dactylifera, Loranthus europeas, Zingiber officinalis* and *Citrus aurantifolia.* Mod Appl Sci 3(3):161–165

Ahmed IA, Ahmed AWK, Robinson RK (1995) Chemical composition of date varieties as influenced by the stage of ripening. Food Chem 54(3):305–309

Aldhaheri A, Alhadrami G, Aboalnaga N, Wasfi I, Elridi M (2004) Chemical composition of date pits and reproductive hormonal status of rats fed date pits. Food Chem 86(1):93–97

Ali BH, Bashir AK, Al Hadrami G (1999) Reproductive hormonal status of rats treated with date pits. Food Chem 66:437–441

Allaith AAA (2008) Antioxidant activity of Bahraini date palm (*Phoenix dactylifera* L.) fruit of various cultivars. Int J Food Sci Technol 43(6):1033–1040

Al-Maiman SA (2005) Effect of date palm (*Phoenix dactylifera*) seed fibers on plasma lipids in rats. J King Saud Univ Agric Sci 17(2):117–123

Al-Qarawi AA, Mousa HM, Ali BEH, Abdel-Rahman H, El-Mougy SA (2004) Protective effect of extracts from dates (*Phoenix dactylifera* L.) on carbon tetrachloride-induced hepatotoxicity in rats. Int J Appl Res Vet Med 2(3):176–180

Al-Qarawi AA, Abdel-Rahman H, Ali BH, Mousa HM, El-Mougy SA (2005) The ameliorative effect of dates

(*Phoenix dactylifera* L.) on ethanol-induced gastric ulcer in rats. J Ethnopharmacol 98:313–317

Al-Qarawi AA, Abdel-Rahman H, Mousa HM, Ali BH, El-Mougy SA (2008) Nephroprotective action of *Phoenix dactylifera* in gentamicin-induced nephrotoxicity. Pharm Biol 46(4):227–230

Al-Taher AY (2008) Possible anti-diarrhoeal effect of the date palm (*Phoenix dactylifera* L) spathe aqueous extract in rats. Sci J King Faisal Univ (Basic Appl Sci) 9(1):131–137

Amin ES, Awad O, Abdel Samad M, Iskander MN (1969) Isolation of estrone from Moghat roots and from pollen grains of Egyptian date palm. Phytochem 9:295–297

Barreveld WH (1993) Date Palm Products, FAO Agricultural Services Bulletin No. 101. FAO, Rome, 216 pp

Barrow S (1998) A monograph of *Phoenix* L. (Palame: Coryphoideae). Kew Bull 53:513–575

Besbes S, Blecker C, Deroanne C, Drira N-E, Attia H (2004) Date seeds: chemical composition and characteristic profiles of the lipid fraction. Food Chem 84(4):577–584

Biglari F, AlKarkhi AFM, Easa AM (2008) Antioxidant activity and phenolic content of various date palm (*Phoenix dactylifera*) fruits from Iran. Food Chem 107(4):1636–1641

Biglari F, AlKarkhi AFM, Easa AM (2009) Cluster analysis of antioxidant compounds in dates (*Phoenix dactylifera*): effect of long-term cold storage. Food Chem 112(4):998–1001

Boudries H, Kefalas P, Hornero Méndez D (2007) Carotenoid composition of Algerian date varieties (*Phoenix dactylifera*) at different edible maturation stages. Food Chem 101:1389–1394

Chaira N, Smaali MI, Martinez-Tomé M, Mrabet A, Murcia MA, Ferchichi A (2009) Simple phenolic composition, flavonoid contents and antioxidant capacities in water-methanol extracts of Tunisian common date cultivars (*Phoenix dactylifera* L.). Int J Food Sci Nutr 60(Suppl 7):316–329

Council of Scientific and Industrial Research (1969) The wealth of India. A dictionary of Indian raw materials and industrial products (Raw materials 8). Publications and Information Directorate, New Delhi

Dowson VHW, Aten A (1962) Dates: handling, processing and packing. Food and Agriculture Organisation (FAO), Plant Production and Protection Series No13. FAO. FAO Agricultural Development Paper No. 72, FAO Rome, 392 pp

El-Mougy SA, Abdel-Aziz SA, Al-Shanawany M, Omar A (1991) The gonadotropic activity of Palmae in mature male rats. Alex J Pharm Sci 5:156–159

Foundation for Revitalisation of Local Health Traditions (2008) FRLHT database. htttp://envis.frlht.org.

Hong YJ, Tomas-Barberan FA, Kader AA, Mitchell AE (2006) The flavonoid glycosides and procyanidin composition of Deglet Noor dates (*Phoenix dactylifera*). J Agric Food Chem 54(6):2405–2411

Hussein F, Moustafa S, El-Samiraea FA, El-Zeid A (1976) Studies on physical and chemical characteristics of eighteen date cultivars grown in Saudi Arabia. Indian J Hort 33:107–113

Ishurd O, Kennedy JF (2005) The anti-cancer activity of polysaccharide prepared from Libyan dates (*Phoenix dactylifera* L.). Carbohyd Polym 59(4):531–535

Ishurd O, Ali Y, Wei W, Bashir F, Ali A, Ashour A, Pan Y (2003) An alkali-soluble heteroxylan from seeds of *Phoenix dactylifera* L. Carbohyd Res 338(15): 1609–1612

Morton JF (ed) (1987) Date. In: Fruits of warm climates. Julia F. Morton publishers, Miami, pp 5–11

Nixon RW (1951) The date palm "tree of life" in the subtropical deserts. Econ Bot 5:274–301

Porcher MH et al (1995–2020) Searchable world wide web multilingual multiscript plant name database. Published by The University of Melbourne, Australia. http://www.plantnames.unimelb.edu.au/Sorting/Frontpage.html

Rock W, Rosenblat M, Borochov-Neori H, Volkova N, Judeinstein A, Elias M, Aviram M (2009) Effects of date (*Phoenix dactylifera* L., Medjool or Hallawi variety) consumption by healthy subjects on serum glucose and lipid levels and on serum oxidative status: a pilot study. J Agric Food Chem 57(17):8010–8017

Saafi EB, Trigui M, Thabe R, Hammami M, Achour L (2008) Common date palm in Tunisia: chemical composition of pulp and pits. Int J Food Sci Technol 43(11):2033–2037

Saafi EB, Louedi M, Elfeki A, Zakhama A, Najjar MF, Hammami M, Achour L (2010) Protective effect of date palm fruit extract (*Phoenix dactylifera* L.) on dimethoate induced-oxidative stress in rat liver. Exp Toxicol Pathol, (In Press)

Sahari MA, Barzegar M, Radfar R (2007) Effect of varieties on the composition of dates (*Phoenix dactylifera* L.). Food Sci Technol Int 13(4):269–275

Samarawira I (1983) Date palm, potential source for refined sugar. Econ Bot 37:181–186

U.S. Department of Agriculture, Agricultural Research Service (USDA) (2010) USDA national nutrient database for standard reference, release 23. Nutrient Data Laboratory Home Page. http://www.ars.usda.gov/ba/bhnrc/ndl

Vayalil PK (2002) Antioxidant and antimutagenic properties of aqueous extract of date fruit (*Phoenix dactylifera* L. Arecaceae). J Agric Food Chem 50(3): 610–617

Zaid A, Arias-Jimenez EJ (eds) (2002) Date palm cultivation. Plant Production and Protection Paper 156. Rev. 1. Food and Agricultural Organization of the United Nations, Rome, 110 pp

Zaid A, de Wet PF (2002) Climatic requirements of date palm. In: Zaid A, Arias-Jimenez EJ (eds.) Date palm cultivation. Plant Production and Protection Paper 156. Rev. 1. Food and Agricultural Organization of the United Nations, Rome, pp 57–72

Phoenix reclinata

Scientific Name

Phoenix reclinata Jacq.

Synonyms

Fulchironia senegalensis Lesch., *Phoenix abyssinica* Drude, *Phoenix baoulensis* A. Chev., *Phoenix comorensis* Becc., *Phoenix djalonensis* A. Chev., *Phoenix dybowskii* A. Chev., *Phoenix equinoxialis* Bojer, *Phoenix leonensis* Lodd. ex Kunth, *Phoenix reclinata* var. *comorensis* (Becc.) Jum. & H. Perrier, *Phoenix reclinata* var. *madagascariensis* Becc., *Phoenix reclinata* var. *somalensis* Becc., *Phoenix spinosa* Schumach. & Thonn.

Family

Arecaceae

Common/English Names

Coffee Palm, Dwarf Date-Palm, False Date Palm, Feather Palm, Mukindu Palm, Senegal Date Palm, Swamp Date-Palm, Wild Date Palm

Vernacular Names

Arabic: Wakhale;
Botswana: Kanchindu, Lunchindu (Bemba);
Chinese: Fei Zhou Hai Zao;
Democratic Republic of Congo: Kanchindu, Lunchindu (Bemba), Chisonga (Lunda);
Eritea: Aguseana (Tigrigna);
Ethiopia: Selen, Zembaba (Amharic), Aguseana (Tigrigna);
French: Dattier De Marais, Dattier Du Sénégal, Dattier Sauvage, Dattier Nain Du Sénégal;
German: Senegaldattelpalme;
Ghana: Mia Drum (Ga), Atahe (Wine) (Gbe-Vhe) (Burkill 1998);
Guinea: A-Ngóm (Leaf), Yabà (Cabbage) (Basari) (Burkill 1998);
Guinea-Bissau: Tara (Extracted Fibre (Fula-Pulaar, Manding-Mandinka);
Kenya: Mkindu (Swahili), Gonyoorriya (Boni), Meti (Digo), Gedo (Ilwana), Makindu (Kikuyu), Sosiyot (Kipsigis), Othith (Luo), Ol-Tukai (Maasai), Konchor (Orma), Itikindu (Sanya), Alol (Somali), Mhongana (Taveta), Kigangatehi (Taita), Nakadoki (Turkana), (Beentje 1994), Mangatche (Kilimanjaro District);
Madagascar: Dara, Taratra, Taratsy (Malagasy), Calalou (Morondava);
Malawi: Kanchinda (Nyanja);
Mozambique: Kanchinda (Nyanja);
Namibia: Chisonga, Mukapakapa, Nzalu (Lozi);
Nigeria: Dabino Biri, Kajinjiri, Kabba (Hausa), Deli (Fulani);
Nyanja: Kanchinda
Portuguese: Palmeira-Da-Tara;
Rwanda: Umukindo;

Senegal: Serké (Balanta), Inib, Iñib, I-Nib, I-Ńib (Basari), Gi-Nyɔamèl (Bedik), Tãmbra (Crioulo), Diidioka, Fu Duka, Fudak, Fuduka, Gi Duka, Hu Diak, Hudiak (Diola), Tionkom (The Wine), Tonkom (Wolof);
Sierra Leone: Shaka-Le (Sherbro), Kundi (Mende);
Somalia: Alol (Somali);
South Africa: Datelboom, Kaffer Kofie, Wildedadelboom, Wildedadelpalm (Afrikaans), Makerewa, Shikerewa (Diriko), Dikindu, Makindu (Mbukushu), Kanchinda (Nyanja), Isundu (Zulu);
Tanzania: Kanchindu, Lunchindu (Bemba), Daro, Taratra, Mkindwi (Swahili), Luchingu (Fipa), Kihangaga (Urukindu);
Uganda: Otit (Acholi), Emusogot (Ateso), Musansa (Busoga), Tit (Jonam), Itchi (Madi), Ekingol (Karamojong), Tit (Lango), Wild Date Palm, Enkinu (Luamba), Lukindu, Mukindu, Musansa (Luganda), Makendu (Lugisu), Muyiti (Lugwe), Lukindu, Mukindu (Lunyoro), Kikindu (Lunyuli), Otit, Tit (Luo), L Lusansa (Lusoga), ukindu, Mukindu (Lutoro), Itchi (Madi);
Zambia: Kanchindu, Lunchindu (Bemba), Chisonga, Mukapakapa, Nzalu (Lozi), Kanchinda (Nyanja);
Zimbabwe: Kanchinda (Nyanja).

Origin/Distribution

A native of tropical and southern Africa – Chad, Cote d'Ivoire, Democratic Republic of Congo, Egypt, Eritrea, Ethiopia, Gabon, Gambia, Ghana, Guinea, Kenya, Liberia, Mali, Mauritania, Mozambique, Namibia, Niger, Nigeria, Senegal, Sierra Leone, South Africa, Tanzania, Togo, Uganda, Zambia, Zimbabwe.

Agroecology

Phoenix reclinata is a widely distributed species growing in a diverse range of habitats, from seasonally water-logged or inundated areas along watercourses, in high rainfall areas, in riverine forest, woodlands, grasslands, and even in rainforest areas (restricted to areas of sparse canopy). It occurs also in drier conditions on rocky hillsides, cliffs and grasslands to 3,000 m altitude. It can withstand temperatures down to −3.9°C and is drought resistant. It thrives in full sun but tolerates light shade. The fruits are dispersed by various animals and humans.

Edible Plant Parts and Uses

The fruit is edible raw or cooked and is smaller and inferior to the domesticated date. The roasted seed is used as a coffee substitute. Both the inflorescence and stem are tapped for sap to process palm wine. Palm wine is a popular alcoholic beverage relished by the rural people in Maputaland, Natal, South Africa. Flower buds may be eaten raw or cooked as a vegetable. The heart of the crown (palm cabbage) is eaten as vegetable. The seeds can be dried and ground into flour.

Botany

A dense, clumping, dioecious palm (Plate 1). Stem erect or oblique, 8–10 m high with trunk diameter of 15–25 cm, bark smooth, black ringed by distinct leaf scars and with a dense mass of roots arising from the lower part. Crown consists of 15–30 dark green to yellow green pinnate leaves (Plate 2), leaf sheath fibrous, reddish-brown; pseudopetiole smooth, often channelled adaxially, to 50 cm long. Leaves pinnate, arcuate, 2.5–4.5 m long, leaflets sessile, regularly arranged distally in one plane of orientation but median and proximal leaflets in fascicles of 3–5 and often fanned, 80–130 on each side of rachis, 28–45 × 2.2–3.6 cm; leaflet glossy, dark green, long and narrow, margin minutely crenulate. Inflorescence axillary and branched. Staminate inflorescence erect, prophyll green-yellow in bud, strongly 2-keeled, coriaceous. Staminate flowers creamy-white; calyx cupule 1 mm high; petals with apex acute-acuminate in shape and with jagged margins, 3 (rarely 4), 6–7 × 2–3 mm Pistillate inflorescence erect, arching with weight of fruits; prophyll as for staminate inflorescence; Pistillate flowers usually only one carpel reaching maturity.

Plate 1 Plant with clumping habit

Plate 2 Crown with sparse foliage

Fruit ovoid-ellipsoid or almost obovoid, green ripening yellow to bright orange, 13–20 × 7–13 mm; mesocarp sweet, scarcely fleshy, 1–2 mm thick. Seed obovoid, 12–14 by 5–6 mm; endosperm homogeneous.

Nutritive/Medicinal Properties

Nutrient composition of palm wine from sap of *Phoenix reclinata* (per 100 g) was reported as: moisture 98.3%, protein, 0.2 g, carbohydrate 1.1 g, ash 0.4 g, calcium 0.45 mg, magnesium 5.12 mg, iron 0.07 mg, sodium 5.85 mg, potassium 157 mg, copper 0.05 mg, zinc 0.02 mg, manganese trace, phosphorus 1.74 mg, thiamin 0.01 mg, riboflavin 0.01 mg, niacin 0.5 mg, vitamin C 6.5 mg and alcohol 3.6% v/v (Cunningham and Wehmeyer 1988). The data showed wild date palm wine to be an important source of nicotinic acid and vitamin C. The alcoholic beverage has been recorded as a remedy against urinary infections and parts of the tree used as a remedy for pleurisy.

The following steroid glycosides: methyl proto-Pb; methyl proto-rupicolaside (26-O, β-D-glucopyranosyl (25R)-22-O-methy-furost-5-en-3β, 26-diol 3-O-[β-D-glucopyranosyl-(1→2)-α-L-rhamnopyranosyl-(1→4)-α-L-rhamnopyranosyl-(1→4)][α-L-rhamnopyranosyl-(1→2)]-β-D-glucopyranoside and methyl proto-reclinatoside (26-O-β-D-glucopyranosyl (25R)-22-O-methyl-furost-5-en-3β, 26-diol 3-O-[α-L-rhamnopyranosyl-(1→5)-α-L-arabinofuranosyl-(1→4)] [α-L-rhamnopyranosyl(1→2)]-β-D-glucopyranoside) were isolated from *P. reclinata* leaves (Idaka et al. 1991).

Other Uses

The palm is also grown as avenue planting and as an ornamental specimen plant for large yards, parks, campuses and other spacious areas. The tree can be utilized in the conservation of soil. Trunk wood, which is resistant to white ants and fungi, is used for hut building, beams, poles, making doors, windows and fence posts. Fibres from the stem are made into brushes and brooms, and in the Cape region of South Africa, leaves are used in making the kilts of Xhosa boys taking part in initiation ceremonies. Whole leaves are used as door entrances and covers, or fans for stoking fires and repelling insects. The leaf rachis is used for making thatch, floor mats and fish traps. It also forms a component of wattle for the construction of mud houses. Leaflets from sucker shoots are harvested for making baskets, hats, brushes, building ties, woven dolls and ornaments. Leaves yield a useful dye, and roots yields tannin. Dried inflorescences are used as brooms to sweep areas around dwellings. Roots yield a gum.

Comments

Phoenix reclinata is a protected species in South Africa.

Selected References

Barrow SC (1988) A monograph of *Phoenix* L. (Palmae: Coryphoideae). Kew Bull 53:513–575

Beentje HJ (1994) Kenya trees, shrubs and lianas. National Museums of Kenya, Nairobi

Bekele-Tesemma A, Birnie A, Tengnas B (1993) Useful trees and shrubs for Ethiopia, Regional Soil Conservation Unit (RSCU), Swedish International Development Authority (SIDA)

Burkill HM (1998) Useful plants of West Tropical Africa, Vol 4. Families M-R. Royal Botanic Gardens, Kew, 969 pp

Cunningham AB, Wehmeyer AS (1988) Nutritional value of palm wine from *Hyphaene coriacea* and *Phoenix reclinata* Arecaceae. Econ Bot 42:301–306

Dransfield J, Uhl N, Asmussen C, Baker WJ, Harley M, Lewis C (2008) *Genera Palmarum*. The evolution and classification of palms. Kew Publishing, Royal Botanic Gardens, Kew

Eggeling WJ (1940) The indigenous trees of the Uganda protectorate. Government Printer, Entebbe, 296 pp

Ellison D, Ellison A (2001) Cultivated palms of the world. Briza Publications, Pretoria

Govaerts R, Dransfield J, Zona SF, Hodel, DR, Henderson A (2006) World checklist of Arecaceae. The Board of Trustees of the Royal Botanic Gardens, Kew. Published on the Internet. http://www.kew.org/wcsp/

Hamilton AC (1981) A field guide to Uganda forest trees. Makerere University, Kampala, p 279

Idaka K, Hirai Y, Shoji J (1991) Studies on the constituents of Palmae plants V. steroid saponins and flavonoids of leaves of *Phoenix rupicola* T. Anderson, *Phoenix loureirii* Kunth, *Phoenix reclinata* N. J. Jacquin and *Arecastrum romanzoffianum* Beccari. Chem Pharm Bull (Tokyo) 39:1455–1461

Johnson D (ed) (1991) Palms for human needs in Asia: palm utilization and conservation in India, Indonesia, Malaysia and the Philippines. A.A. Balkema, Rotterdam, 258 pp

Johnson DV (1998) Tropical palms. FAO Regional Office for Asia and the Pacific, Food and Agriculture Organization of the United Nations, Rome, 166 pp

Salacca affinis

Scientific Name

Salacca affinis Griff.

Synonyms

Salacca affinis var. *borneensis* (Becc.), *Salacca borneensis* Becc., *Salacca dubia* Becc.

Family

Arecaceae

Common/English Names

Edible-Fruited Salak Palm, Jungle Salak, Palm Heart (Vegetable), Red Salak, Salak, Snakefruit

Vernacular Names

Borneo: Salak, Ridan, Redan (Iban), Tuncum (Dusun), Birai (Kayan);
French: Coeur De Palmier (Vegetable);
Indonesia: Linsum (Sumatra);
Malaysia: Salak Hutan, Salak Betul (Peninsular).

Origin/Distribution

The species is indigenous to Peninsular Malaysia, Sumatra and Borneo.

Agroecology

The species is a shade-loving palm found scattered in tropical rain forests at wet places or in swamps or near running water.

Edible Plant Parts and Uses

The fruit are sweet to subacid and are eaten by peeling off the scaly, leathery skin. They are sold in rural markets during the harvest season. The palm heart is edible.

Botany

An acaulescant, clumping, armed, dioecious palm, 3–5 m high with long, erect pinnate leaves up to 3 m long which are produced in rosette from short stem. Leaf petioles have sharp long spines. Leaflet is broad, 32 cm long by 7 cm wide, flat in one plan at least towards the leaf tip, evenly spaced along rachis with irregular gaps between.

Inflorescences are erect, spikes hidden in subtending spathes. Male spikes are thinly woolly, sessile, 1.5–2.5 cm long, solitary or in groups of 2–3, spathes 9–16 cm long, bracts connate at base. Female inflorescences have secondary branches, 5–7.5 cm long, bearing tiny alternate short spikes each with three solitary flowers. Fruit is drupe-like, cylindrical, but conical at apex, about 7.5 cm long, dark brown, with flat scales resembling snake-skin and deeply grooved (Plates 1–2). Seeds are usually three, flattened, embedded in sweetish cream-coloured pulp (sarcotesta).

Nutritive/Medicinal Properties

No published information is available on its nutritive and medicinal properties.

Other Uses

The palm can be planted as a hedge or used as fence plant. The petioles can be used for matting and the fronds for thatching.

Comments

The fruits are frequently mistaken for the common salak, *Salacca zalacca*.

Plate 1 Fruit with scaly snake-like skin that peels off easily

Plate 2 Close-up of fruit

Selected References

Burkill IH (1966) A dictionary of the economic products of the Malay Peninsula. Revised reprint. 2 vols. Ministry of Agriculture and Co-operatives, Kuala Lumpur, vol 1 (A–H), pp 1–1240, vol 2 (I–Z), pp 1241–2444

Jansen PCM, Jukema J, Oyen LPA, van Lingen TG (1992) *Salacca affinis* Griffith. In: Verheij EWM, Coronel RE (eds.) Plant resources of South-East Asia No. 2: Edible fruits and nuts. Prosea Foundation, Bogor, pp 355–356

Ridley HN (1922–1925) The flora of the Malay Peninsula, 5 vols. Government of the Straits Settlements and Federated Malay States. L. Reeve & Co., London

Whitmore TC (1973) Palms of Malaya. Oxford University Press, Singapore, 129 pp

Salacca glabrescens

Scientific Name

Salacca glabrescens Griff.

Synonyms

None

Family

Arecaceae

Common/English Names

Jungle Salak

Vernacular Names

Malaysia: Salak Hutan, Poko Rengam;
Thai: Sala, Sala Thai.

Origin/Distribution

The species is indigenous to Thailand and Peninsular Malaysia.

Agroecology

Its native habitat is found near streams or swarms in the warm, wet tropical, lowland rainforest in Thailand and Peninsular Malaysia. It shares similar agroecological requirements with other *Salacca* species. Refer to notes under *Salacca zalacca*.

Edible Plant Parts and Uses

The sarcotesta of ripe fruit is edible – the sarcotesta is sweet but less favoured when compared to fruits of *Salacca zalacca*.

Botany

An acaulescent, clumping, spiny, usually dioecious palm with a very short stem and subterranean stolon (Plates 1–3). Sheaths and petioles bear short rows of sharp spines (Plate 2). Leaf rachis 2–4 m long with 23–30 sigmoid leaflets per side, irregularly arranged in clusters, spreading in different planes, apical leaflet forming a broad, compound terminal leaflet. Male inflorescence to 0.6 m long with to 25 non-hairy flowering branches (Plate 3). Female inflorescence to 0.25 m long, fruits pear shaped, to 5 cm long and 4 cm across, reddish-brown, densely covered in flattened spine-like scales.

Plate 1 Clustering palm habit

Plate 3 Remnants of male inflorescence

Other Uses

No-edible uses are similar to the common salak.

Comments

Selected References

Burkill IH (1966) A dictionary of the economic products of the Malay Peninsula. Revised reprint, 2 vols. Ministry of Agriculture and Co-operatives, Kuala Lumpur, vol 1 (A–H), pp 1–1240. vol 2 (I–Z), pp 1241–2444

Haynes J, McLaughlin J (2000) Edible palms and their uses. University Florida/Miami-Dade County Extension Office Fact Sheet MDCE-00-50

Henderson A (2009) Palms of Southern Asia. (Princeton field guides). Princeton University Press, Princeton, 264 pp

Ridley HN (1922–1925) The flora of the Malay Peninsula, 5 Vols. Government of the Straits Settlements and Federated Malay States. L. Reeve & Co., London

Plate 2 Spiny petioles

Nutritive/Medicinal Properties

No information is available on its nutritive or medicinal properties.

Salacca magnifica

Scientific Name

Salacca magnifica J. P. Mogea

Synonyms

None

Family

Arecaceae

Common/English Names

Greater Salak, Remayong

Vernacular Names

Sarawak: Remayong (Iban), Salak;

Origin/Distribution

Remayong is indigenous to Borneo, in Sarawak it is found naturally in Limbang and Kapit Division.

Agroecology

Being a warm climate species it prefers a tropical climate, but the palm can be grown quite successfully in the warm subtropics. In its native range, it occurs in damp lowland mixed dipterocarp forest at the base of cliffs with yellow clay soil from 160 m to 300 m above sea level. The palm prefers a warm, sheltered, lightly shaded, very moist locality.

Edible Plant Parts and Uses

The acid sweet sarcotesta is eaten in the same way as salak but the quality is inferior to salak.

Botany

A dioecious, clustering, armed palm with male and female flowers on separate plants, acaulescent, almost stemless, usually with a very short, stubby trunk with very long, large undivided, green leaves up to 5 m long. The midrib and frond rachis bear long, sharp whitish spines right to the base. Fruits developed from axillary inflorescences towards the base of the female palm in tight clusters. Each cluster has 30–40 ellipsoid to obovate, 4–6 cm by 4–5 cm fruit with thin, dark

Plate 1 Remayong fruit with long, slender bristles

Plate 2 Remayong fruit on sale in a local market in Kuching

brown skin of united, imbricate scales, each scale ending in a long, slender bristle (Plates 1 and 2). Each fruit thus is covered with long, slender bristles which makes it difficult to remove the skin. Each fruit has 3-whitish segments (sarcotesta) which is dry, crisp and acid-sweet, each segment has one hard, brown seed.

Nutritive/Medicinal Properties

The nutrient composition of the raw ripe fruit per 100 g edible portion was reported as: energy 78 kcal, moisture 79.8%, protein 0.6 g %, fat 0.2%, carbohydrate 18.4%, fibre 0.4%, ash 0.4%, P 12 mg, K 276 mg, Ca 30 mg, Mg 4 mg, Fe 13 μg, Mn 3 μg, Cu 1 μg, Zn 20.4 μg, vitamin C 2 mg (Voon and Kueh 1999).

Other Uses

Non-edible uses are similar to the common salak.

Comments

Fruits are harvested from wild palms.

Selected References

Mogea JP (1980) *Salacca magnifica* Mogea. Reinwardtia 9:468

Voon BH, Kueh HS (1999) The nutritional value of indigenous fruits and vegetables in Sarawak. Asia Pac J Clin Nutr 8(1):24–31

Voon BH, Chin TH, Sim CY, Sabariah P (1988) Wild fruits and vegetables in Sarawak. Sarawak Department of Agriculture, Sarawak, 114 pp

Salacca wallichiana

Scientific Name

Salacca wallichiana C. Mart.

Synonyms

Calamus zalacca Roxb., *Salacca beccarii* Hook.f., *Salacca macrostachya* Griff.

Family

Arecaceae

Common/English Names

Edible-Fruited Rakum Palm, Edible-Fruited Salak Palm, Rakum Palm.

Vernacular Names

Burmese: Yengam
Danish: Salak, Slangeskindsfrugt;
French: Rakam;
Malaysia: Rekam, Asam Kumbang, Kumbar, Salak Kumbar, Kumbak, Salak Rengam (Peninsular);
Philippines: Sala;
Thailand: Sala, Rakam (Ragahm), Cho-La-Ka;
Vietnam: Hình Bộ Luật.

Origin/Distribution

Wild distribution of this species is found in the lowlands of southern Burma, south of 19° North latitude, in the coastal provinces from Bangkok eastwards in Thailand to Indo-China through Peninsular Malaysia and into Sumatra. Both wild (in the forest) and cultivated, the palm is most abundantly found in the hot and humid areas of Thailand with the main producing provinces in Chanthaburi, Trat, Chumphon and Rayong.

Agroecology

Rakum palm thrives best in the hot, humid lowlands with an annual rainfall of 2,500–3,000 mm and optimum temperature range of 22–32°C. It prefers areas with 5–6 dry months for fruit set and development as rain is detrimental to the emerging spadices causing rots and reducing the viability of the pollens. The palm does well in soils ranging from sandy loam to heavy clay in swampy localities. On low-laying wetlands it can tolerate drought very well, but in the dry season irrigation is needed for a good yield. In its natural habitat and in mixed fruit orchards, rakum is commonly found under shade.

Edible Plant Parts and Uses

The sour unripe fruit and fruit of sour wild palms are used as a substitute for lime in cooking and are used in curries. The ripe fruit is sweet and consumed raw. Rakum fruit is sold whole by street vendors or peeled, shrink-wrapped and chilled in packets in the supermarkets in Thailand.

Plate 1 Cluster of developing rakum fruit

Botany

A tillering, creeping, profusely armed, dioecious palm occurring in clumps. In very old palm, the creeping stem can reach 3–4 m, with an erect terminal leaf-bearing part 1 m tall and with many adventitious roots. Roots can reach a length of 2 m but do not extend to a great depth. Leaves are 3–7 m long, pinnate and very spiny and borne on spiny petioles. Spines are flat, linear-triangular and sharp. Leaflets are medium-green, 60–75 cm long by × 6–8 wide. Male inflorescences are about 1 m long, branching into many reddish spadices; flowers have three sepals, three petals and five stamens bearing anthers with functional yellowish pollen. Female inflorescences are 1–2 m long, with 3–8 reddish spadices, each with staminate and hermaphrodite flowers in 1:1 ratio, but both without functional pollen; staminate flowers with a reddish, tubular corolla and five staminodes borne on the corolla throat; hermaphrodite flowers with three pink sepals, fused at the base, ovary trilocular, hairy, with a short, trifid dark red stigma, staminodes 5–6, borne on the corolla throat. Fruit occurs in dense clusters (Plates 1–4). Fruit is a broadly ellipsoid to obovoid drupe, 2.5 cm long, the skin (epicarp) consisting of pale yellow, pinkish or orange-brown scales with sharp, reflexed brittle points (Plates 1–4). Seeds 1–3, flattened and covered with a fleshy, edible sarcotesta.

Plate 2 Close up of immature rakum fruit with the sharp, short bristles

Plate 3 Clusters of ripe spiny rakum fruit

Plate 4 Clusters of spiny, ellipsoid rakum fruits

Nutritive/Medicinal Properties

The fruit is reported to be rich in vitamin B.

A paste of the fruit mixed with other materials is applied to swollen knee joints.

Other Uses

Rakum palm provides a major source of cork, thatch and other construction materials in rural areas. The wild form is grown as a fruit-bearing fence. The petioles of the leaves are used as fishing-rods and the pith of the palm has been reported to be used for gunwales, etc. of boats.

Comments

The fruit has a short post harvest shelf life of 3–4 days at room temperature (25–30°C).

Selected References

Burkill IH (1966) A dictionary of the economic products of the Malay Peninsula. Revised reprint. 2 vols. Ministry of Agriculture and Co-operatives, Kuala Lumpur, vol 1 (A–H). pp 1–1240, vol 2 (I–Z). pp 1241–2444

Morgea JP (1981) Notes on *Salacca wallichaina*. Principes 25(3):120–123

Polprasid P (1992) *Salacca wallichiana* C. Martius. In: Verheij EWM, Coronel RE (eds.) Plant resources of South-East Asia. No. 2. Edible fruits and nuts. Prosea Foundation, Bogor, pp 278–281

Ridley HN (1922–1925) The flora of the Malay Peninsula. 5 vols. Government of the Straits Settlements and Federated Malay States. L. Reeve & Co., London

Salacca zalacca

Scientific Name

Salacca zalacca (Gaertner) Voss

Synonyms

Calamus salakka Willd. ex Steud., *Calamus zalacca* Gaertn. basionym, *Salacca blumeana* Mart., *Salacca edulis* Reinw., *Salacca edulis* var. *amboinensis* Becc., *Salacca rumphii* Wall., nom. illeg., *Salacca zalacca* var. *amboinensis* (Becc.) Mogea, *Salakka edulis* Reinw. ex Blume.

Family

Arecaceae

Common/English Names

Edible-Fruited Salak Palm, Edible Salacca, Salak, Salak Palm, Snake Fruit, Snake Palm, Snake-Skinned Fruit.

Vernacular Names

Brazil: Fruta Cobra (Portuguese);
Burmese: Yingan;
Chinese: Ke Shi Sa La Ka Zong, She Pi Guo;
Danish: Salak, Slangeskindsfrugt;
Dutch: Salak;
French: Palmier Salak, Fruit À Peau De Serpent, Fruit De Palmier À Peau De Serpent, Salacca Aux Fruits À Peau De Serpent;
German: Salakpalme, Salak, Schlagenfrucht, Zalak;
Indonesia: Salak, Salak Pasir (Java), Salubi (Batak), Sala (Buginese, Sulawesi), Hakam, Tusum (Dyak,Kalimantan), Sekumai (Jambi, Sumatra), Salak (Madurese), Sala (Makassar, Sulawesi), Salak (Malay), Sala (Minangkabau, Sumatra);
Japanese: Sarakka Yashi;
Malaysia: Salak;
Spanish: Salaca;
Thai: Sala.

Origin/Distribution

Salak is native to South Sumatra and southwest Java. It was introduced and is now widely cultivated in Thailand, throughout Malaysia and Indonesia as far as the Moluccas (Maluku). It has also been introduced into New Guinea, the Philippines, Queensland and the northern Territory in Australia, the Ponape Island (Caroline Archipelago), China, Surinam, Spain and the Fiji Islands.

Agroecology

Salak thrives under humid tropical lowland conditions with mean annual precipitation of 1,700–3,000 mm. It occurs wild as an understorey tree in the lowland swampy rainforest below 500 m elevation. Fruit yield and quality in Java diminish above 500 m altitude. Salak is usually grown under partial shade as it grows and performs better than in full sun. When grown as the sole crop it is grown under *Gliricidia* shade trees, when grown as mixed crop it is intercropped with coconut, rubber or other perennial fruit trees which provides shade. Young palms require heavy shade which may be reduced after about 1–2 years. Because of its shallow rooting system, salak requires a high water table, rain or irrigation during most of the year, but it is intolerant of flooding. It grows best on podzolic soils (any of a group of acidic, zonal soils having a leached, light-coloured surface) and regosols (well-drained, azonal, mineral soils). In Malaysia, salak is usually cultivated on mineral soils such as well-drained clayey loams, sandy loams and lateritic soils with pH from 4.7 to 5.5.

Edible Plant Parts and Uses

The sarcotesta of ripe fruits are eaten fresh or cooked and they can be canned or processed into fruit juice. Salak has an unique taste that is likened to a combination of apple, pineapple and banana. The sarcotesta are also candied (*manisan salak*). Fresh unripe fruit sarcotesta are used in a spicy fruit *rujak*. Unripe, young fruit sarcotesta are consumed as pickles (*asinan salak*). Sarcotesta from fruits of inferior quality are sometimes cooked with salt. Salak fruit is now being developed into fruit juice, pickle and other food products. The seed kernels of the young fruits of the Javanese cultivar 'Pondoh' are edible. The terminal bud is eaten as palm cabbage. Indonesia exports fresh salak fruit to Singapore, the middle east countries, Holland, Hong Kong and China. Indonesia also exports canned salak to Singapore and Europe.

Botany

An evergreen, acaulescent, very spiny, tillering, usually dioecious palm growing in clumps formed by successive branching at the stem base (Plate 1). The stem is mostly a subterranean stolon, with a short, 1–2 m high, 10–15 cm diameter, erect aerial part bearing the leaves. Roots are superficial, not deep and new roots emerge from the stem immediately under the crown, the internodes are short and crowded. Leaves are pinnatipartite, 3–6 m long. Leaves, leaf-sheaths, petioles and leaflets have numerous long, thin, blackish spines. Petioles are very spiny and 2 m long. Leaflet segments are unequal, linear-lanceolate, with narrowed base, concave, apex caudate and acute, 20–70 cm by 2–7.5 cm. Inflorescence is an axillary compound spadix, stalked, at first covered by spathes; staminate inflorescence 50–100 cm long, composed of 4–12 spadices, each 7–15 cm × 0.7–2 cm, pistillate inflorescence 20–30 cm long, consisting of 1–3 spadices, 7–10 cm long (Plate 2). Flowers are paired in axils of scales; male flowers with reddish, tubular corolla and 6 stamens borne on the corolla throat and a tiny pistillode; female flowers with tubular corolla, yellow-green outside and dark red inside, a trilocular ovary with short 3-fid, red style and 6 staminodes borne on the corolla throat. Fruit is a subglobose to ellipsoid drupe, 15–40 per spadix, measuring 5–7 cm by 5 cm, tapering towards base and rounded at the top; epicarp (skin) comprised of numerous brown to orangey-brown (Plates 3–6), yellow (Plate 7), united, imbricate scales, each scale ending in a fragile prickle. Seeds usually 3 per fruit, each enclosed within 2–8 mm thick, fleshy, firm, mildly aromatic, sweet to subacid, cream-coloured sarcotesta (Plate 8), with a smooth, stony inner part, 23–29 mm × 15–27 mm, which is blackish-brown and trigonous with 2 flat surfaces and a curved one; endosperm white and homogeneous.

Plate 1 Salak tree habit

Plate 2 Assisted pollination, female inflorescence (*center*) being assisted pollinated by placing two mature male inflorescences close to it

Plate 3 Spiny leaf petioles and immature fruit clusters

Plate 4 Ripe orangey-brown salak fruit

Plate 5 Close up of snakeskin-like scales of a dark brown-skinned salak cultivar

Plate 6 Dark brown salak cultivar from Sabah

Nutritive/Medicinal Properties

Proximate nutrient composition of snake fruit per 100 g edible portion based on analyses conducted in Thailand was reported as: energy 51 Cal, moisture 86.4%, protein 0.5 g, fat 0.1 g, carbohydrate 12.1 g, fibre 0.3 g, Ca 8 mg, P 18 mg, Fe 0.4 mg, vitamin B-2 0.2 mg, vitamin C 6 mg (Ministry of Public Health Thailand 1970).

Forty-six compounds were identified in the volatile component of snake fruit, among which methyl esters of branched-chain alkenoic and

Plate 7 A yellow-skinned Gading salak cultivar

Plate 8 Salak fruit with skin removed to show the creamy-white flesh (sarcotesta)

β-hydroxy acids predominated and terpenoids were completely absent. The most abundant components were methyl 3-hydroxy-3-methylpentanoate (25.0%) and methyl (*E*)-3-methylpent-2-enoate (23.4%) (Wong and Tie 1993).

Separate and recent studies reported that a total of 24 odour-active compounds were associated with the aroma of snake fruit (Wijaya et al. 2005). Methyl 3-methylpentanoate was regarded as the character impact odorant of typical snake fruit aroma. 2-Methylbutanoic acid, 3-methylpentanoic acid, and an unknown odorant with very high intensity were found to be responsible for the snake fruit's sweaty odour. Other odorants including methyl 3-methyl-2-butenoate (overripe fruity, ethereal), methyl 3-methyl-2-pentenoate (ethereal, strong green, woody), and 2, 5-dimethyl-4-hydroxy-3[2]-furanone (caramel, sweet, cotton candy-like) were found to contribute to the overall aroma of snake fruit. Methyl dihydrojasmonate and isoeugenol, which also have odour impact, were also identified as snake fruit volatiles. The main differences between the aroma of Pondoh and Gading cultivars could be attributed to the olfactory attributes (metallic, chemical, rubbery, strong green, and woody), which were perceived by most of the panelists in the Pondoh samples but were not detected in the Gading samples.

During the maturation of snake fruit cultivar Pondoh, the contents of sucrose, glucose, fructose, and volatile compounds changed drastically (Supriyadi et al. 2002). The glucose, fructose, and volatile compounds contents showed their maximum levels at the end of maturation; however, the sucrose content fell. The major olfactory volatile compounds responsible for the unique and sweet aroma in salak fruit were identified to be methyl esters of butanoic acids, 2-methylbutanoic acids, hexanoic acids, pentanoic acids, and the corresponding carboxylic acids. Furaneol (4-hydroxy-2,5-dimethyl-3(2 H)-furanone) was also identified as a minor aroma constituent. The methyl esters of carboxylic acids were the characteristic olfactory volatile compounds responsible for the sweet aroma of snake fruit, (Salacca edulis, Reinw) cv. Pondoh (Supriyadi et al. 2003). Pectin methyl transferase was also identified in the flesh. This pectin methyl transferase activity increased during fruit maturation, concomitant with the level of methanol originating from hand-squeezed juice and with the methyl esters extracted from the fruit flesh.

Antioxidant Activity

Salak fruit is a very good source of natural antioxidants. The antioxidants in salak identified included chlorogenic acid, (−)-epicatechin, and singly linked proanthocyanidins that mainly existed as dimers through hexamers of catechin or epicatechin (Shui and Leong 2005). In salak, chlorogenic acid was identified to be an antioxidant of the slow reaction type as it reacted with free radicals much more slowly than either (−)-epicatechin or proanthocyanidins. Total polyphenols (mg of GAE (Gallic acid equivalent)/

100 g of fresh weight) in snake fruit was found to be 217.1 (Haruenkit et al. 2007). Antioxidant activity (µM TE/100 g of FW) of snake fruit as measured by DPPH (2, 2-diphenyl-1-picrylhydrazyl hydrate) and ABTS (2,2'-azinobis-3-ethylbenzothiazoline-6-sulfonic acid) assays was 110.4 and 1507.5 respectively. Separate studies conducted on four Sabah salak varieties reported that the total phenolic and flavonoid contents of the samples were in the range of 12.6–15.0 mg gallic acid equivalent (GAE)/g and 4.9–7.1 mg catechin equivalent/g of dry sample, respectively (Aralas et al. 2009). The antioxidant activities of the extracts (using DPPH assay) were highly correlated with total phenolic and moderately correlated with flavonoid content. The reducing capabilities of the extracts using ferric reducing/antioxidant power (FRAP) assay were moderately correlated with all phytochemicals tested. The results suggested that the phytochemicals and antioxidant activity of salak was slightly affected by variety. The rich phytochemicals and high antioxidant properties of *S. zalacca* indicated that the fruit possessed potential health benefits.

Snake fruit and mangosteen were characterized by polyphenols, proteins and antioxidant potentials and by their influence on plasma lipids and antioxidant activity in wistar rats fed cholesterol were compared (Leontowicz et al. 2006). The content of polyphenols (14.9 and 9.2 mg GAE/g) and antioxidant potential (46.7 and 72.9 µmol Trolox Equivalent/g) in snake fruit were significantly higher than in mangosteen. After 4 weeks diets supplemented with snake fruit and to a lesser degree with mangosteen significantly inhibited the rise in plasma lipids and prevented a decrease in antioxidant activity. Studies showed that, snake fruit and mangosteen contained high levels of bioactive compounds, therefore positively affecting plasma lipid profile and antioxidant activity in rats fed cholesterol-containing diets. Such positive influence was higher in rats fed diet with added snake fruit.

Anticancer Activity

Studies found that snake fruit Thai cultivar Sumalee exhibited antioxidant and antiproliferative activities comparable with kiwi fruit cultivar Hayward (Gorinstein et al. 2009). Both fruits were recommended as supplements to the normal diet. Researchers found similarity between snake fruit (cultivar Sumalee) and kiwi fruit (cultivar Hayward) in the contents of polyphenols (8.15–7.91, mg GAE/g DW), antioxidant values by DPPH (11.28–10.24, µM Trolox Equivalent/g DW), and antiproliferative activities on both human cancer cell lines, Calu-6 for human pulmonary carcinoma (90.5–87.6 cell survival), and SMU-601 for human gastric carcinoma (89.3–87.1% cell survival).

Antihyperuricemic Activity

Recent studies also identified a bioactive compound in snake fruit which could help in the management of gout albeit more studies were needed (Priyatno et al. 2007). Of two compounds isolated from the ethyl acetate extract of snake fruit (*Salacca edulis*) cv. Bongkok, compound 1, 3ß-hydroxy-sitosterol was found inactive in inhibition of the xanthine oxidase enzyme while compound 2 identified as 2-methylester-1-H-pyrrole-4-carboxilyc acid was found to be active with an IC_{50} value of 48.86 ug/ml. Xanthine oxidase catalyzes the oxidation of hypoxanthine and xanthine to uric acid. Gout is caused by high uric acid level in blood and can enhance cardiovascular disorder.

Other Uses

A closely-planted row of salak palms forms an impregnable hedge and the very spiny leaves are also used to construct fences. The bark of the petioles may be used for matting. The leaflets are used for thatching.

Comments

Being dioecious, Salak need to be assisted pollinated for good fruit set and yield. This is usually done by brushing or placing a mature male inflorescences on the mature female inflorescences (Plate 2).

The salak palm grown in northern Sumatra is ascribed to a distinct species, *Salacca sumatrana* Becc. *Salacca zalacca* cultivated elsewhere in Indonesia, is subdivided into two botanic varieties, var. *zalacca* from Java and var. *amboinensis* (Becc.) Mogea from Bali and Ambon. In Indonesia at least 20 intraspecific taxa are distinguished according to place of origin and cultivation, e.g. 'Salak Condet', 'Salak Pondoh', 'Salak Bali', 'Salak Suwaru'; these may obtain cultivar status as vegetative propagation gains importance. Salak 'Bali' is monoecious; the inflorescences bear both hermaphrodite and staminate flowers; the latter produce functional pollen.

Salak pondoh has three more superior variants, namely *pondoh super*, *pondoh hitam* (black pondoh), and *pondoh gading* (ivory-English term for gading/ yellowish-skinned pondoh see image above). Salak Bali has one variant called *'Salak Gula Pasir'* which is very sweet, much sought after and the most expensive.

Selected References

Aralas S, Mohamed M, Abu Bakar MF (2009) Antioxidant properties of selected salak (*Salacca zalacca*) varieties in Sabah, Malaysia. Nutr Food Sci 39(3):243–250

Backer CA, Bakhuizen van den Brink RC Jr (1968) Flora of Java (spermatophytes only), vol 3. Wolters-Noordhoff, Groningen, 761 pp

Burkill IH (1966) A dictionary of the economic products of the Malay Peninsula. Revised reprint. 2 vols, Ministry of Agriculture and Co-operatives, Kuala Lumpur, vol 1 (A–H), pp 1–1240, vol 2 (I–Z), pp 1241–2444

Gorinstein S, Haruenkit R, Poovarodom S, Park YS, Vearasilp S, Suhaj M, Ham KS, Heo BG, Cho JY, Jang HG (2009) The comparative characteristics of snake and kiwi fruits. Food Chem Toxicol 47(8):1884–1891

Haruenkit R, Poovarodom S, Leontowicz H, Leontowicz M, Sajewicz M, Kowalska T, Delgado-Licon E, Rocha-Guzmán NE, Gallegos-Infante JA, Trakhtenberg S, Gorinstein S (2007) Comparative study of health properties and nutritional value of durian, mangosteen, and snake fruit: experiments in vitro and in vivo. J Agric Food Chem 55(14):5842–5849

Leontowicz H, Leontowicz M, Drzewiecki J, Haruenkit R, Poovarodom S, Park Y-S, Jung S-T, Kang S-G, Trakhtenberg S, Gorinstein S (2006) Bioactive properties of snake fruit (*Salacca edulis* Reinw.) and mangosteen (*Garcinia mangostana*) and their influence on plasma lipid profile and antioxidant activity in rats fed cholesterol. Eur Food Res Technol 22(5):697–703

Ministry of Public Health Thailand (1970) Tables of nutrition values in Thai food per 100 gm of edible portion. Office of the Prime Minister, Royal Thai Government, Bangkok

Mogea JP (1982) *Salacca zalacca*, the correct name for the salak palm. Principes 26(2):70–72

Ochse JJ, Bakhuizen van den Brink RC (1980) Vegetables of the Dutch Indies, 3rd edn. Ascher & Co., Amsterdam, 1016 pp

Ochse JJ, Bakhuizen van den Brink RC (1931) Fruits and fruitculture in the Dutch East Indies. G. Kolff & Co., Batavia-C, 180 pp

Ochse JJ, Soule MJ Jr, Dijkman MJ, Wehlburg C (1961) Tropical and subtropical agriculture, 2 vols. Macmillan, New York, 1446 pp

Priyatno LHA, Sukandar EY, Ibrahim S, Adyana IK (2007) Xanthine oxidase inhibitor activity of terpenoid and pyrrole compounds isolated from snake fruit (*Salacca edulis* Reinw.) cv. Bongkok. J Appl Sci 7(20):3127–3130

Ridley HN (1922–1925) The flora of the Malay Peninsula, 5 vols. Government of the Straits Settlements and Federated Malay States, L. Reeve & Co., London

Schuiling DL, Mogea JP (1992) *Salacca zalacca* (Gaertner) Voss. In: Coronel RE, Verheij EWM (eds) Plant resources of South-East Asia. No. 2. Edible fruits and nuts. Prosea Foundation, Bogor, pp 281–284

Shui G, Leong LP (2005) Screening and identification of antioxidants in biological samples using high-performance liquid chromatography-mass spectrometry and its application on *Salacca edulis* Reinw. J Agric Food Chem 53(4):880–886

Soepadmo E (1979) Genetic resources of Malaysian fruit trees. Malays Appl Biol 8(1):33–42

Supriyadi S, Suzuki M, Yoshida K, Muto T, Fujita A, Watanabe N (2002) Changes in the volatile compounds and in the chemical and physical properties of snake fruit (*Salacca edulis* Reinw) cv. Pondoh during maturation. J Agric Food Chem 50(26):7627–7633

Supriyadi, Suzuki M, Wu S, Tomita N, Fujita A, Watanabe N (2003) Biogenesis of volatile methyl esters in snake fruit (*Salacca edulis*, Reinw) cv. Pondoh. Biosci Biotechnol Biochem 67(6):1267–1271

Whitmore TC (1973) Palms of Malaya. Oxford University Press, Singapore, 129 pp

Wijaya H, Ulrich D, Lestari R, Schippel K, Ebert G (2005) Identification of potent odorants in different cultivars of snake fruit [*Salacca zalacca* (Gaert.) voss] using gas chromatography-olfactometry. J Agric Food Chem 53(5):1637–1641

Wong KC, Tie DY (1993) Volatile constituents of salak (*Salacca edulis* Reinw.) fruit. Flavor Fragr J 8:321–324

Serenoa repens

Scientific Name

Serenoa repens (W.Bartram) Small

Synonyms

Brahea serrulata (Michx.) H.Wendl., *Chamaerops serrulata* Michx., *Corypha obliqua* W.Bartram, *Corypha repens* W.Bartram, *Diglossophyllum serrulatum* (Michx.) H.Wendl. ex Salomon, *Sabal serrulata* (Michx.) Schult.f., *Serenoa repens* f. *glauca* Moldenke, *Serenoa serrulata* (Michx.) Hook.f. ex B.D.Jacks.

Family

Arecaceae

Common/English Names

American Dwarf Palm Tree, Cabbage Palm, Saw Palmetto, Scrub Palmetto, Serenoa.

Vernacular Names

Brazil: Saw Palmetto;
Chinese: Ju Chi Zong;
Danish: Dværgpalme, Sabalpalme, Savpalme, Serenoapalme;
Eastonian: Serenoapalm ;
Finnish: Kääpiöpalmu, Sahapalmu;
French: Palmier De L'amerique Du Nord;
German: Sabal, Sägepalme, Zwergpalme;
Hungary: Fűrészpálma;
Icelandic: Freyspálmi;
Italian: Serenoa Repens;
Norwegian: Dvergpalme, Palmetto, Sabelfrukt, Sagpalme, Serenoapalme;
Spanish: Serenoa;
Swedish: Dvergpalm, Dvärgpalm, Sågpalm, Sågpalmetto.

Origin/Distribution

Serenoa repens is endemic to the southeastern United States from South Carolina, Alabama, Georgia, Mississippi to southeastern Louisiana including the Florida peninsula.

Agroecology

Saw palmetto is common in pinelands, prairies, scrublands, maritime forests, and coastal sand dunes, often forming dense swards (Tanner et al. 1996; Dransfield et al. 2008). Saw palmetto grows best in warm-temperate, or humid-subtropical climates. In its natural habitats, the average annual temperature ranges from −4°C to 36°C, average rainfall is approximately 1,000–1,630 mm (Wade and Langdon 1990). This species grows well in either shade or in sun and is especially resistant to fire even though its foliage is highly flammable. It is a frequent invader to sites defoliated by previous fires (Abrahamson 1984). Saw palmetto performs best in dry, well-drained soils rather than in swampy areas (Monk 1965; Abrahamson 1984). Typical soils are high in quartz and fine-grained, but this species is also known to grow on poorly drained sites high in peat (Breininger and Schmalzer 1990).

Edible Plant Parts and Uses

Saw palmetto's fruit is edible raw or cooked. Regular consumption of the fruit is supposed to be very beneficial to the health, improving the digestion and helping to increase weight and strength (Bown 1995). The edible fruits were a staple in the diet of Florida's pre-contact inhabitants. The seeds is also eaten raw or cooked (Bown 1995) and can be ground into flour. The palm heart can be used as a vegetable; honey from bees that visit the flowers is much prized.

Botany

Serenoa repens is a moderate, clustered, shrubby, armed, pleonanthic, hermaphroditic fan palm. Stems are subterranean or prostrate and surface creeping, or rarely erect, covered with persistent leaf sheaths, or becoming striate or smooth with age (Plate 1). Leaves palmate, marcescent, induplicate, 1–2 m long; sheath expanding into a tattered mat of dark brown soft fibres; petiole is flat to slightly rounded and armed with fine teeth; abaxial hastula not well-developed, obtuse; adaxial hastula usually well-developed, obtuse. Leaf blade nearly orbicular, regularly divided to below the middle into narrow stiff, shortly bifid, single-fold glabrous lanceolate segments, basally connate, apices acute or 2-cleft. Inflorescences axillary, within crown of leaves, paniculate up to 60 cm long, ascending, about as long as leaves, with 3–4 orders of branching main axis bearing 2 peduncular bracts above prophyll and with pubescent rachillae. Flowers are bisexual, 5 mm across, with tubular calyx of 3 triangular, slightly imbricate lobes; 3, imbricate, elliptic, reflexed petals, alternating with outer whorl of stamens; stamens 6 in 2 whorls with narrowly triangular, basally connate filaments and dorsifixed, versatile anthers; pistils 3, distinct basally, glabrous with 3 ovules, only 1 developing into fruit; styles are connate, elongate, glabrous; nectaries 3, septal; stigma is minutely 3-lobed, dry. Fruit formed in clusters (Plate 2 & 3). Fruit is an ellipsoidal to subglobose drupe, dark blue to black at maturity; epicarp

Plate 1 Habit of young fruiting palm with subterranean stem

Plate 2 Clusters of developing fruits

Plate 3 Close-up of developing fruits

smooth, mesocarp fleshy without fibres, endocarp thin but somewhat cartilaginous. Seeds brown, ellipsoid, with conspicuous elonagted raphe and the endosperm is bony and homogeneous.

Nutritive/Medicinal Properties

Analysis of the 95% ethanol extract of the saw palmetto berries for the known lipid content showed fatty acids, fatty acid esters, and phytosterols (Anonymous 1997). Other chemical constituents identified in the berries included aliphatic alcohols (C26-30), polyprenic compounds, flavonoids, glucose, galactose, arabinose, uronic acid and other polysaccharides. Major constituents of saw palmetto extract were reported as lauric acid, oleic acid, myristic acid, palmitic acid and linoleic acid (Abe et al. 2009). The fruit also contained myristoleic acid (Iguchi et al. 2001) and monoacylglycerides, such as 1-monolaurin and 1-monomyristin (Shimada et al. 1997). Saw palmetto fruits were found to contain the following phytosterols: campesterol, cycloartenol, lupenone, lupeol, β-sitosterol, and stigmasterol (Sorenson and Sullivan 2007; Bedner et al. 2008). The berries were found to have contain 1.5% volatile oil, comprised of 63% free fatty acids and 37% ethyl esters of those fatty acids. The fatty acids included caproic, caprylic, capric, lauric, palmitic and oleic acids and their ester. The berries also contained β-sitosterol and its glucoside, sitosterol D-glucoside as well as ferulic acid (Duke 1985). 144 steam-volatile components were identified as constituents of the fruits from *S. repens* (Rösler et al. 2009). The main component detected was lauric acid (40.4%).

A comparative study of 14 brands of *Serenoa repens* extract revealed that many of the brands exhibited a significantly different proportional content of free fatty acids (FFAs), methyl and ethyl esters, long-chain esters and glycerides that may have an impact on their clinical efficacy and safety (Habib and Wyllie 2004). Free fatty acids (caproic, caprylic, capric, lauric, myristic, palmitic, palmitoleic, stearic, oleic, linoleic, linolenic, arachidic and eicosenoic acids) in particular, based on previous research were suggested as comprising the active agents of *Serenoa repens* extract for the treatment of lower urinary tract symptoms (LUTS). The mean proportion of FFAs varied from 40.7% to 80.7%. Methyl and ethyl ester content ranged from 1.5% to 16.7% and long-chain ester content varied from 0.7% to 1.36%. Glyceride content ranged from 6.8% to 52.15% while unsaponifiable matter concentrations varied from 1.6% to 2.6%.

Antiprostatic Activity

Saw palmetto (*Serenoa repens*) extract has been reported as the most popular herbal preparation taken for benign prostatic hyperplasia (BPH) (Plosker and Brogden 1996; Markowitz et al. 2003), a common condition in elderly men. Early research indicated that the extract was well-tolerated, safe and afforded mild to moderate improvement in urinary symptoms and flow measures. *Serenoa repens* generated similar improvement in urinary symptoms and flow compared to finasteride but was associated with fewer adverse treatment events (Braeckman et al. 1997; Plosker and Brogden 1996; Di Silverio et al. 1998; Boyle et al. 2000; Marks et al. 2000; Markowitz et al. 2003; Wilt et al. 2002). The extract of *Serenoa repens* in its two dosage forms (160 mg b.i.d. and 320 mg o.d.) was reported to be a safe and effective treatment for the mictional problems associated with BPH (Braeckman et al. 1997). Both dosage forms elicited significant amelioration in efficacy variables: international prostate symptom score (IPSS) (60% after 1 year), quality of life score (85% of patients were satisfied after 1 year of treatment), prostatic volume (12% after 1 year), maximum flow rate (22% after 1 year), mean flow rate (17% after 1 year) and residual urinary volume (16% after 1 year). The extract appeared to offer a potential pharmacologic option for ameliorating BPH symptoms in patients with mild-to-moderate BPH disease.

In an earlier review, Plosker and Brogden (1996), reported that in one large trial, in which >1,000 men with moderate BPH were randomised to receive *Serenoa repens* 160 mg twice daily or finasteride 5 mg once daily for 6 months, similar efficacy was found between the two drugs. *Serenoa repens* was well-tolerated and had higher efficacy than placebo and shared similar efficacy to finasteride in improving symptoms in men with BPH. Studies by Markowitz et al. (2003) indicated that extracts of saw palmetto at generally recommended doses were unlikely to change the disposition of co-administered medications that were primarily dependent on the CYP2D6 or CYP3A4 pathways for suppression. Marks et al. (2000) reported that saw palmetto herbal blend appeared to be a safe, highly desirable alternative for men with moderately symptomatic BPH. The secondary outcome measures of clinical effect in their study were only slightly better (not statistically significant) for saw palmetto herbal blend than placebo. However, saw palmetto herbal blend therapy was associated with epithelial contraction, especially in the transition zone. Separate studies in 70 patients showed that when taken for 3 months, a combination of natural products (cernitin, saw palmetto, β-sitosterol, vitamin E) significantly reduced nocturia and daytime frequency and lessen overall symptomatology of BPH as indicated by an improvement in the total American Urological Association (AUA) Symptom Index score compared to placebo (57 patients) (Preuss et al. 2001). The combination of natural products caused no significant adverse side effects.

However, subsequent studies of higher methodological quality indicated no difference from placebo (Willetts et al. 2003; Bent et al. 2006; Dedhia and McVary 2008). During a double-blind placebo-controlled randomized trial, all 100 participants had some improvement in their symptoms of BPH but there was no significant beneficial effect of *S. repens* extract over placebo in the 12-week trial (Willetts et al. 2003). The latest review of benign prostatic hyperplasia (BPH) conducted by Tacklind et al. (2009) covered nine new trials involving 2,053 additional men (a 64.8% increase). For the main comparison – *Serenoa repens* versus placebo – three trials were added with 419 subjects and 3 endpoints (IPSS, peak urine flow, prostate size). Overall, 5,222 subjects from 30 randomized trials lasting from 4 to 60 weeks were assessed. Twenty-six trials were double blinded and treatment allocation concealment was found to be adequate in 18 studies. *Serenoa repens* was not superior to placebo in improving International Prostate Symptom Score (IPSS) and urinary symptom scores. For nocturia, *Serenoa repens* was significantly better than placebo. A sensitivity analysis, utilizing higher quality, larger trials (40 subjects or more), found no significant difference. *Serenoa repens* was not superior to finasteride or to tamsulosin. Comparing peak urine flow, *Serenoa repens* was not superior to placebo at trial termination. Comparing prostate size at trial termination, there was no significant difference between *Serenoa repens* and placebo. The conclusion was that *Serenoa repens* was not more effective than placebo for treatment of urinary symptoms consistent with benign prostatic hyperplasia.

The extract of *Serenoa repens* (SABA®) administered at the dose of 320 mg/day to 70 patients with benign prostatic hypertrophy, in two studies involving was found to be effective in terms of improvement of the functional symptomatologic profile and favourable tolerability profile (Mantovani 2010). Both studies showed an improvement over the baseline of about 50% of dysuria and pollakisuria, an about 50% rise in micturition rate with positive effects also in terms of reduction of the micturition rate and of prostate size.

Tuncel et al. (2009) found short-term dutasteride and *S. repens* therapies were not superior to control in terms of the decrease in total blood loss during transurethral resection of the prostate (TURP) surgery. No significant statistical differences were found between the treatment and control groups in the total amount of intraoperative blood loss, total blood loss/time, total blood loss/weight of resected tissue, total blood loss/weight/time, serum haemoglobin level change, prostatic microvessel density (MVD) and suburethral MVD. More recently, the data from studies by Anceschi et al. (2010) suggested that pre-treatment of BPH patients with *Serenoa repens* (Permixon®) before transurethral resection of the prostate (TUP) or open prostatectomy surgery for benign prostatic hyperplasia was effective in reducing intra- and post-operative complications. There was also a significant change in hemodynamic parameters and the length of hospitalization was significantly shorter in the pretreated group.

Salmonella enteritidis lipopolysaccharide (LPS) stimulation produced a proinflammatory phenotype in rat peritoneal macrophages (Bonvissuto et al. 2011). *Serenoa repens* (SeR), lycopene (Ly), Se (Selenium) were found to inhibit the inflammatory cascade, but the Ly-Se-SeR association caused a greater inhibitory effect on the expression of COX-2, 5-LOX, and iNOS. This association showed a greater efficacy in decreasing the loss of IκB-α, the enhanced NF-κB binding activity, the elevated mRNA levels of TNF-α, the increased MDA, and nitrite content. The LY-Se-SeR association in-vivo caused a larger inhibitory effect on prostate inflammation induced in rats by partial bladder outlet blockage and might be useful in the treatment of benign prostatic hyperplasia.

Two monoacylglycerides, 1-monolaurin (1) and 1-monomyristin (2) were isolated from 95% ethanol extract of the powdered, dried berries of *Serenoa repens* (Shimada et al. 1997). Compounds 1 and 2 showed moderate biological activities in the brine shrimp lethality test and against renal (A-498) and pancreatic (PACA-2) human tumour cells. Marginal cytotoxicity was exhibited against human prostatic (PC-3) cells. Separate studies demonstrated that the extract from *S. repens* and myristoleic acid a component of the extract stimulated mixed cell death of apoptosis and necrosis in human prostatic LNCaP cells (Iguchi et al. 2001). These results suggested that the extract and myristoleic acid may have potential to be developed as attractive new tools for the treatment of prostate cancer.

Studies showed that the carotenoid, astaxanthin, demonstrated 98% inhibition of 5α-reductase at 300 μg/mL in-vitro (Anderson 2005). Inhibition of 5α-reductase had been reported to decrease the symptoms of benign prostate hyperplasia (BPH) and possibly inhibit or help treat prostate cancer. Alphastat, the combination of astaxanthin and saw palmetto berry lipid extract (SPLE), exhibited a 20% greater inhibition of 5α-reductase than SPLE alone in-vitro. Prostatic carcinoma cells treated for 9 days with astaxanthin in-vitro produced a 24% reduction in growth at 0.1 μg/mL and a 38% reduction at 0.01 μg/mL. SPLE showed a 34% decrease at 0.1 μg/mL. The results indicated that low levels of carotenoid, astaxanthin could inhibit 5α-reductase and decrease the growth of human prostatic cancer cells in-vitro. Astaxanthin added to SPLE showed higher inhibition of 5α-reductase than SPLE alone in-vitro.

Dietary supplementation with the liposterolic extract of saw palmetto extract (SPE) was found effective in controlling prostate cancer (CaP) tumorigenesis (Wadsworth et al. 2007). Treatment with 300 mg/kg/day SPE from 4 to 24 weeks of age significantly lowered the concentration of 5α-dihydrotestosterone (DHT) in the prostate and resulted in a significant enhancement in apoptosis and significant fall in pathological tumour grade and frank tumour incidence. SPE inhibition of prostatic DHT levels supported the hypothesis that suppression of the enzyme 5α-reductase to be a mechanism of action of SPE. Scholtysek et al. (2009) showed that saw palmetto berry extract and its sterol components, β-sitosterol and stigmasterol to be potential anti-tumour agents. The extract and its bioactive components inhibited prostate cancer growth by increasing p53 protein expression and also inhibited carcinoma development by decreasing p21 and p27 protein expression.

Mechanism of Antiprostatic Action

A number of pharmacodynamic effects have been demonstrated with the lipidosterolic extract of *Serenoa repens* (LSESR) in prostate disorders, suggesting multiple mechanisms of action. These included in-vitro inhibition of both type 1 and type 2 isoenzymes of 5 α-reductase; interference with binding of dihydrotestosterone (DHT) to cytosolic androgen receptors in prostate cells; deleterious effects on gene transcription and the regulation of biological responses in these cells (Plosker and Brogden 1996; Di Silverio et al. 1998). Di Silverio et al. (1998) reported a decline of DHT and elevation of testerone in BPH tissue of patients treated with Permixon (n-hexane lipido-sterol extract of *Serenoa repens*) confirming the capacity of this drug to suppress in-vivo 5 α-reductase in human pathological prostate. A marked decrease of epidermal growth factor (EGF), associated with DHT reduction was also noted. Habib et al. (2005) reported that despite *Serenoa repens* effective inhibition of 5α-reductase activity in the prostate, it did not suppress PSA (prostate specific antigen) secretion, confirming the therapeutic advantage of *Serenoa repens* over other 5α-reductase inhibitors as treatment. *Serenoa repens*, unlike other 5α-reductase inhibitors, did not inhibit binding between activated androgen receptor and the steroid receptor-binding consensus in the promoter region of the PSA (prostate specific antigen) gene. Scaglione et al. (2008) reported that all *S. repens* extracts tested, were able to inhibit both isoforms of 5 α-reductase. However, the potency of the extracts appeared to be very different. This was probably due to qualitative and quantitative differences in the active ingredients. Thus, they advocated that the product of each company must be tested to evaluate the clinical efficacy and bioactivity.

Marks et al. (2000) reported that saw palmetto herbal blend (SPHB) therapy was associated with epithelial contraction, especially in the transition zone, indicating a possible mechanism of action underlying the clinical significance detected in other studies. Subsequently the researchers (Veltri et al. 2002) reported that 6 months of SPHB treatment appeared to alter the DNA chromatin structure and organization in prostate epithelial cells. Thus, suggesting a possible molecular basis for tissue changes and therapeutic effect of the compound.

Recently, Abe et al. (2009) demonstrated the binding affinities of fatty acids – lauric acid, oleic acid, myristic acid, and linoleic acid, major constituents of saw palmetto extract (SPE) for pharmacologically relevant (α(1)-adrenergic, muscarinic and 1,4-DHP) calcium channel antagonist receptors. SPE was shown to inhibit 5α-reductase activity in rat liver with an IC_{50} of 101 μg/ml. Similarly, all the fatty acids except palmitic acid inhibited 5α-reductase activity, with IC_{50} values of 42.1–67.6 μg/ml. Recently, Petrangeli et al. (2009) showed that administration of lipido-sterolic extract of *Serenoa repens* (LSESr) (12.5 and 25 μg/ml) resulted in complex changes in cell membrane organization and fluidity of prostate cancer cells that had advanced to hormone-independent status. LSESr administration exerted a biphasic action by both inhibiting proliferation and inducing apoptosis of androgen-independent PC3 prostate cancer cells.

Category III Prostatitis/Chronic Pelvic Pain Syndrome (CP/CPPS)

Studies showed that men with category III prostatitis/chronic pelvic pain syndrome (CP/CPPS) treated with saw palmetto had no appreciable long-term improvement. In contrast, patients treated with finasteride had significant and durable improvement in all various parameters except voiding (Kaplan et al. 2004).

Profluss (made-up of *Serenoa repens* + selenium and lycopene) was found to be a safe and well-tolerated triple therapy for chronic prostatitis/chronic pelvic pain syndrome (CP/CPPS) category IIIA (Morgia et al. 2010). Profluss ameliorated symptoms associated with CP/CPPS IIIa. In the study 102 patients with IIIa CP/CPPS were randomised into two separate groups, each received Profluss or *S. repens* alone for 8 weeks. The mean National Institutes of Health-Chronic Prostatitis Symptom Index (NIH-CPSI) score decreased significantly in both groups. International Prostate Symptoms Score (IPSS) improved significantly in both arms, but more in the Profluss group. Prostate- Specific Antigen (PSA) and white blood cell count decreased significantly only in the Profluss group. The maximum peak flow rate improved more in the Profluss group.

Urinary Tract Improvement

In urinary tract disorders, treatment of men with *S. repens* compounds was found to decrease urinary frequency, increase urinary flow, and ameliorate symptoms of dysuria (painful urination). Studies showed that saw palmetto was well-tolerated and significantly improved lower urinary tract symptoms in men with BPH (Gerber et al. 1998). Significant improvement in the symptom score was noted after 2 months treatment with saw palmetto. An improvement in symptom score of 50% or higher after saw palmetto treatment for 2, 4, and 6 months was observed in 21% (10 of 48), 30% (14 of 47), and 46% (21 of 46) of patients, respectively. There was no significant change in maximum urinary flow rate, postvoid residual urine volume, or detrusor pressure at peak flow among patients. No significant change in mean serum prostate-specific antigen level was noted and major improvement in objective measures of bladder outlet obstruction was not demonstrated. In subsequent double-blind, randomized, placebo-controlled trial, saw palmetto produced a statistically significant improvement in urinary symptoms in men with lower urinary tract symptoms compared with placebo but had no measurable effect on the urinary flow rates (Gerber et al. 2001). A meta-analysis of all available published trials (11 randomized clinical trials and 2 open label trials), involving 2,859 patients of Permixon (*Serenoa repens* extract) in the treatment of men with benign prostatic hyperplasia revealed a significant improvement in peak flow rate and reduction in nocturia that was higher than with placebo (Boyle et al. 2000).

Androgenetic Alopecia (Baldness) Efficacy

The fruit extract of saw palmetto, *Serenoa repens* is one among the many naturally occurring 5 α reductase (5αR) inhibitors which has gained popularity as a magical remedy for androgenetic alopecia or baldness (Murugusundram 2009). Androgenetic alopecia is characterized by the structural miniaturization of androgen-sensitive hair follicles in susceptible individuals and is anatomically defined within a given pattern of the scalp (Prager et al. 2002). Biochemically, one contributing factor of this disorder was found to be the conversion of testosterone (T) to dihydrotestosterone (DHT) via the enzyme 5-α reductase (5αR). This metabolism was also key to the onset and progression of benign prostatic hyperplasia (BPH). Additionally, androgenetic alopecia was shown to be responsive to drugs and agents used to treat BPH. Prager et al. (2002) in a pilot, randomized, double-blind, placebo-controlled trial involving ten males between the ages of 23 and 64 years of age, in good health, with mild to moderate androgenetic alopecia responded highly positively (60%) to treatment with liposterolic extract of *Serenoa repens* (LSESr) and its glycoside, β-sitosterol. In subsequent study, the scientists (Chittur et al. 2009) reported that chronic inflammation of the hair follicle (HF) could be a contributing factor in the pathogenesis of androgenetic alopecia. They found that a blockade of inflammation using a composition containing LSESr as well as two anti-inflammatory agents (carnitine and thioctic acid) could change the expression of molecular markers of inflammation in a well-established in-vitro system. They found that the composition effectively suppressed lipopolysaccharide (LPS)-activated gene expression of chemokines, including CCL17, CXCL6 and LTB(4) associated with pathways involved in inflammation and apoptosis. Their findings suggested that 5-α reductase inhibitors combined with blockade of inflammatory processes afforded a novel two-pronged alternative in the treatment of androgenetic alopecia with improved efficacy over current modalities. Sinclair et al. (2002) reported a case of a women who became sensitized to topical minoxidil following use of an extemporaneous preparation of minoxidil 4% with retinoic acid in a propylene glycol base. She subsequently also became sensitized to saw palmetto (*Serenoa repens*), a topical herbal extract commonly promoted for the treatment of hair loss.

Adverse Effects

Villanueva and González (2009) reported a case of hematuria and coagulopathy in a patient who was using saw palmetto. Recently Lapi et al. (2010) reported a case report of acute liver damage due to *Serenoa repens*. Wargo et al. (2010) reported a case of a 65-year-old male with a medical profile of diabetes, hypertension, hyperlipidemia, gout, Barrett esophagitis, and chronic gastritis who developed acute pancreatitis after taking 1 week of, saw palmetto, for symptoms related to benign prostatic hyperplasia (BPH). He had normal levels of triglycerides, no recent infection or trauma and had no recent increase in alcohol consumption. Ultrasound and computed tomography precluded cholelithiasis and obstruction. The most likely cause of pancreatitis in this case was saw palmetto. The scientists postulated that saw palmetto directly stimulated estrogenic receptors and inhibited progesterone receptors in the prostate tissue. Stimulation of the estrogenic receptors may lead to increased triglyceride levels or induction of a hypercoagulable state that leads to pancreatic necrosis. Additionally, inhibition of cyclooxygenase, a property of saw palmetto, may be linked to acute pancreatitis.

Traditional Medicinal Uses

The use of saw palmetto can be traced back to the Mayan culture that used extracts of the plant as a tonic and an antiseptic. Even the berries of the plant were widely consumed to relieve cough and fever. Traditional indications for the use of saw palmetto include: cystitis, chronic bronchitis, asthma, diabetes, dysentery, indigestion, and for treatment of underdeveloped breasts (Grieve 1971; Foster and Duke 1998; Duke 1985; Bown 1995; Chevallier 1996). The partially dried ripe fruit has been described as aphrodisiac, urinary antiseptic, diuretic, expectorant, sedative and tonic. The fruit has a probable oestrogenic action, it is prescribed in the treatment of impotence, reduced or lack of libido and testicular atrophy in men and to stimulate breast enlargement in women. It is taken internally in the treatment of impotence, debility in elderly men, prostate enlargement and inflammation, bronchial complaints associated with coldness, and wasting diseases. The fruit is also used in the treatment of colds, coughs, irritated mucous membranes, asthma etc. The fruit pulp, or a tincture, is administered to those suffering from wasting disease, general debility and failure to thrive. Saw palmetto is one of the Western herbs that are considered to be anabolic (strengthening and building body tissue and encouraging weight gain). The fruit also has a beneficial effect on the urinary system, helping to reduce the size of an enlarged prostate gland and strengthening the neck of the bladder. A suppository of the powdered fruits, in cocoa butter, has been used as a uterine and vaginal tonic.

Other Uses

S. repens has previously been used to produce pulp for paper products, but it is considered to be of poorer quality compared with other tropical species (Olson and Barnes 1974). Vegetative parts of the plant supply fibre to make brushes, cordage, wax, and roof thatch. The leaf stems have been used in making baskets. The flowers are a favorite nectar source for honey bees, and the sprawling, shrubby palm provides excellent cover for birds, reptiles, and small animals.

Comments

The palm is propagated solely by seeds.

Selected References

Abe M, Ito Y, Oyunzul L, Oki-Fujino T, Yamada S (2009) Pharmacologically relevant receptor binding characteristics and 5 alpha-reductase inhibitory activity of free fatty acids contained in saw palmetto extract. Biol Pharm Bull 32(4):646–50

Abrahamson WG (1984) Species response to fire on the Florida Lake Wales Ridge. Am J Bot 71(1):35–43

Anceschi R, Bisi M, Ghidini N, Ferrari G, Ferrari P (2010) *Serenoa repens* (Permixon(R)) reduces intra- and postoperative complications of surgical treatments of benign prostatic hyperplasia. Minerva Urol Nefrol 62(3):219–23

Anderson ML (2005) A preliminary investigation of the enzymatic inhibition of 5alpha-reduction and growth of prostatic carcinoma cell line LNCap-FGC by natural astaxanthin and Saw Palmetto lipid extract in vitro. J Herb Pharmacother 5(1):17–26

Anonymous (1997) Analysis of 95% alcoholic extract lipid composition. Saw-Palmetto Trading Company. http://www.sawpalmetto.com

Bedner M, Schantz MM, Sander LC, Sharpless KE (2008) Development of liquid chromatographic methods for the determination of phytosterols in standard reference materials containing saw palmetto. J Chromatogr A 1192(1):74–80

Bennett BC, Hicklin JR (1998) Uses of saw palmetto (*Serenoa repens*, Arecaceae) in Florida. Econ Bot 52(4):381–393

Bent S, Kane C, Shinohara K, Neuhaus J, Hudes ES, Goldberg H, Avins AL (2006) Saw palmetto for benign prostatic hyperplasia. N Engl J Med 354(6):557–66

Bonvissuto G, Minutoli L, Morgia G, Bitto A, Polito F, Irrera N, Marini H, Squadrito F, Altavilla D (2011) Effect of *Serenoa repens*, lyocopene, and selenium on proinflammatory iκb-α phenotype activation: an in vitro and in vivo comparison study. Urology 77(1):248.e9–248.e16

Bown D (1995) Encyclopaedia of herbs and their uses. Dorling Kindersley, London, 424 pp

Boyle P, Robertson C, Lowe F, Roehrborn C (2000) Meta-analysis of clinical trials of permixon in the treatment of symptomatic benign prostatic hyperplasia. Urology 55(4):533–539

Braeckman J, Bruhwyler J, Vanderkerckhove K, Géczy J (1997) Efficacy and safety of the extract of *Serenoa repens* in the treatment of benign prostatic hyperplasia: therapeutic equivalence between twice and once daily dosage forms. Phytother Res 11(8):558–563

Breininger DR, Schmalzer PA (1990) Effects of fire and disturbance on plants and birds in Florida oak/palmetto scrub community. Am Midl Nat 123(1):64–74

Chevallier A (1996) The encyclopedia of medicinal plants. Dorling Kindersley, London, 336 pp

Chittur S, Parr B, Marcovici G (2009) Inhibition of inflammatory gene expression in keratinocytes using a composition containing carnitine, thioctic acid and saw palmetto extract. Evid Based Complement Alternat Med. doi:10.1093/ecam/nep102

Dedhia RC, McVary KT (2008) Phytotherapy for lower urinary tract symptoms secondary to benign prostatic hyperplasia. J Urol 179(6):2119–25

Di Silverio F, Monti S, Sciarra A, Varasano PA, Martini C, Lanzara S, D'Eramo G, Di Nicola S, Toscano V (1998) Effects of long-term treatment with *Serenoa repens* (Permixon) on the concentrations and regional distribution of androgens and epidermal growth factor in benign prostatic hyperplasia. Prostate 37(2):77–83

Dransfield J, Uhl N, Asmussen C, Baker WJ, Harley M, Lewis C (2008) Genera Palmarum. The evolution and classification of palms. Kew Publishing, Royal Botanic Gardens, Kew

Duke JA (1985) Handbook of medicinal herbs. CRC, Boca Raton, 443 pp

Foster S, Duke JA (1998) A field guide to medicinal plants in Eastern and Central N. America. Houghton Mifflin Co, New York, 366 pp

Gerber GS, Zagaja GP, Bales GT, Chodak GW, Contreras BA (1998) Saw palmetto (*Serenoa repens*) in men with lower urinary tract symptoms: effects on urodynamic parameters and voiding symptoms. Urology 51(6):1003–1007

Gerber GS, Kuznetsov D, Johnson BC, Burstein JD (2001) Randomized, double-blind, placebo-controlled trial of saw palmetto in men with lower urinary tract symptoms. Urology 58(6):960–964

Grieve M (1971) A modern herbal, 2 vols. Dover publications, New York, 919 pp

Habib FK, Wyllie MG (2004) Not all brands are created equal: a comparison of selected components of different brands of *Serenoa repens* extract. Prost Cancer Prost Dis 7:195–200

Habib FK, Ross M, Ho CK, Lyons V, Chapman K (2005) *Serenoa repens* (Permixon) inhibits the 5alpha-reductase activity of human prostate cancer cell lines without interfering with PSA expression. Int J Cancer 114(2):190–4

Haynes J, McLaughlin J (2000) Edible palms and their uses. University Florida/Miami-Dade County Extension Office Fact Sheet MDCE-00-50

Iguchi K, Okumura N, Usui S, Sajiki H, Hirota K, Hirano K (2001) Myristoleic acid, a cytotoxic component in the extract from *Serenoa repens*, induces apoptosis and necrosis in human prostatic LNCaP cells. Prostate 47(1):59–65

Kaplan SA, Volpe MA, Te AE (2004) A prospective, 1-year trial using saw palmetto versus finasteride in the treatment of category III prostatitis/chronic pelvic pain syndrome. J Urol 171(1):284–288

Lapi F, Gallo E, Giocaliere E, Vietri M, Baronti R, Pieraccini G, Tafi A, Menniti-Ippolito F, Mugelli A, Firenzuoli F, Vannacci A (2010) Acute liver damage due to *Serenoa repens*: a case report. Br J Clin Pharmacol 69(5):558–560

Mantovani F (2010) Serenoa repens in benign prostatic hypertrophy: analysis of 2 Italian studies. Minerva Urol Nefrol 62(4):335–40

Markowitz JS, Donovan JL, Devane CL, Taylor RM, Ruan Y, Wang JS, Chavin KD (2003) Multiple doses of saw palmetto (*Serenoa repens*) did not alter cytochrome P450 2D6 and 3A4 activity in normal volunteers. Clin Pharmacol Ther 74(6):536–42

Marks LS, Partin AW, Epstein JI, Tyler VE, Simon I, Macairan ML, Chan TL, Dorey FJ, Garris JB, Veltri RW, Santos PB, Stonebrook KA, deKernion JB (2000) Effects of a saw palmetto herbal blend in men with symptomatic benign prostatic hyperplasia. J Urol 163(5):1451–6

Moerman D (1998) Native American ethnobotany. Timber Press, Oregon, 927 pp

Monk CD (1965) Southern mixed hardwood forest of north central Florida. Ecol Monogr 35:335–354

Morgia G, Mucciardi G, Galì A, Madonia M, Marchese F, Di Benedetto A, Romano G, Bonvissuto G, Castelli T, Macchione L, Magno C (2010) Treatment of chronic prostatitis/chronic pelvic pain syndrome category IIIA with *Serenoa repens* plus selenium and lycopene (Profluss) versus *S. repens* alone: an Italian randomized multicenter-controlled study. Urol Int 84(4):400–6

Murugusundram S (2009) *Serenoa repens*: does it have any role in the management of androgenetic alopecia? J Cutan Aesthet Surg 2:31–2

Olson DF Jr, Barnes RL (1974) *Serenoa repens* (Bartr.) Small saw-palmetto. In: Schopmeyer CS (ed.) Seeds of woody plants in the United States, vol 450, Agriculture Handbook. U.S. Department of Agriculture, Forest Service, Washington, D.C., pp 769–770

Petrangeli E, Lenti L, Buchetti B, Chinzari P, Sale P, Salvatori L, Ravenna L, Lococo E, Morgante E, Russo A, Frati L, Di Silverio F, Russo MA (2009) Lipido-sterolic extract of *Serenoa repens* (LSESr, Permixon) treatment affects human prostate cancer cell membrane organization. J Cell Physiol 219(1):69–76

Plosker GL, Brogden RN (1996) *Serenoa repens* (Permixon®): a review of its pharmacology and therapeutic efficacy in benign prostatic hyperplasia. Drug Aging 9(5):379–395

Prager N, Bickett K, French N, Marcovici G (2002) A randomized, double-blind, placebo-controlled trial to determine the effectiveness of botanically derived inhibitors of 5-alpha-reductase in the treatment of androgenetic alopecia. J Altern Complement Med 8(2):143–52

Preuss HG, Marcusen C, Regan J, Klimberg IW, Welebir TA, Jones WA (2001) Randomized trial of a combination of natural products (cernitin, saw palmetto, B-sitosterol, vitamin E) on symptoms of benign prostatic hyperplasia (BPH). Int Urol Nephrol 33(2):217–225

Rösler TW, Matusch R, Weber B, Schwarze B (2009) Analysis of the hydrodistillate from the fruits of *Serenoa repens*. Planta Med 75(2):184–6

Scaglione F, Lucini V, Pannacci M, Caronno A, Leone C (2008) Comparison of the potency of different brands of *Serenoa repens* extract on 5alpha-reductase types I and II in prostatic co-cultured epithelial and fibroblast cells. Pharmacol 82(4):270–5

Scholtysek C, Krukiewicz AA, Alonso JL, Sharma KP, Sharma PC, Goldmann WH (2009) Characterizing components of the Saw Palmetto Berry Extract (SPBE) on prostate cancer cell growth and traction. Biochem Biophys Res Commun 379(3):795–8

Shimada H, Tyler VE, McLaughlin JL (1997) Biologically active acylglycerides from the berries of saw-palmetto (*Serenoa repens*). J Nat Prod 60(4):417–8

Sinclair RD, Mallari RS, Tate B (2002) Sensitization to saw palmetto and minoxidil in separate topical extemporaneous treatments for androgenetic alopecia. Aust J Dermatol 43(4):311–2

Sorenson WR, Sullivan D (2007) Determination of campesterol, stigmasterol, and beta-sitosterol in saw palmetto raw materials and dietary supplements by gas chromatography: collaborative study. J AOAC Int 90(3):670–8

Tacklind J, MacDonald R, Rutks I, Wilt TJ (2009) Serenoa repens for benign prostatic hyperplasia. Cochrane Database Syst Rev 2009(2):CD001423

Tanner GW, Mullahey JJ, Maehr D (1996) Saw palmetto: an ecologically and economically important native palm. Wildlife Ecology and Conservation Department, Florida Cooperative Extension Service, Institute of Food and Agricultural Sciences, University of Florida. Circular WEC-109

Tuncel A, Ener K, Han O, Nalcacioglu V, Aydin O, Seckin S, Atan A (2009) Effects of short-term dutasteride and *Serenoa repens* on perioperative bleeding and microvessel density in patients undergoing transurethral resection of the prostate. Scand J Urol Nephrol 43(5):377–82

Veltri RW, Marks LS, Miller MC, Bales WD, Fan J, Macairan ML, Epstein JI, Partin AW (2002) Saw palmetto alters nuclear measurements reflecting DNA content in men with symptomatic BPH: evidence for a possible molecular mechanism. J Urol 60(4):617–22

Villanueva S, González J (2009) Coagulopathy induced by saw palmetto: a case report. Bol Asoc Med P R 101(3):48–50

Wade DD, Langdon OG (1990) Sabal palmetto (Walt.) Lodd. ex. In: Burns RM, Honkala BH (Technical coordinators). Silvics of North America, vol 2 Hardwoods. Agriculture Handbook 654. U.S. Department of Agriculture, Forest Service, Washington, D.C., pp 762–767

Wadsworth TL, Worstell TR, Greenberg NM, Roselli CE (2007) Effects of dietary saw palmetto on the prostate of transgenic adenocarcinoma of the mouse prostate model (TRAMP). Prostate 67(6):661–73

Wargo KA, Allman E, Ibrahim F (2010) A possible case of saw palmetto-induced pancreatitis. South Med J 103(7):683–5

Willetts KE, Clements MS, Champion S, Ehsman S, Eden JA (2003) *Serenoa repens* extract for benign prostate hyperplasia: a randomized controlled trial. BJU Int 92(3):267–70

Wilt T, Ishani A, MacDonald R (2002) Serenoa repens for benign prostatic hyperplasia. Cochrane Database Syst Rev, (3):CD001423

Averrhoa bilimbi

Scientific Name

Averrhoa bilimbi L.

Synonyms

Averrhoa obtusangula Stokes

Family

Averrhoaceae, also in Oxalidaceae

Common/English Names

Bilimbi, Bimbling Plum, Cucumber Tree, Tree Sorrel

Vernacular Names

Argentina: Pepino De Indias;
Brazil: Azedinha, Bilimbi, Bilimbim, Biri-Biri, Limão-De-Caiena, Limão-Japonês;
Catalan: Bilimbi;
Chamorro: Pikue, Pikue Pickle;
Chinese: San Lian, Huang Gua Shu, Mu Hu Gua, Xiang Yang Tao;
Cuba: Grosella China;
Costa Rica: Mimbro, Tiriguro;
Danish: Bilimbi, Bilimbibusk;
Dutch: Blimbing;
El Salvador: Mimbro;
Fiji: Kamkrakh, Belimbines; Belimbi;
French: Blimblim, Blinblin, Carambolier Bilimbi, Cornichomier, Cornichon Des Indes, Zibeline, Zibeline Blonde;
French Guiana: Bilimbi;
German: Bilimbinaum, Gurkenbaum;
Guyana: Birambi;
Haiti: Blimblin;
India: Belambu, Bilimbi, Kamaranga (Hindu), Belambu, Bilimbi, Bimbli, Bimbuli (Kannada), Bimbul (Konkani), Bilimbi, Belimbi, Bilimpi, Elumbipuli, Ilimbai, Irumbanpuli, Irumpuli, Kariccakka, Orakkapuli, Pulinchi, Seemapuli, Vilimbi, Vilumbi, Vilumpi, Wilamju, Wilumpi (Malayalam), Heinahom (Manipuri), Bilamba, Bilambi (Marathi), Koch-Chit-Tamarttai, Pulich-Chakkay, Vilumbi, Pilimpi, Puli-C-Cankay, Pilimbi, Bilimbi, Pulicha, Pulichai, Pilichi, Koccit Tamarattai, Puliccakkay, Puliccakkaymaram, Pulima (Tamil), Bilibili, Bilimbi, Bilumbi, Gommareku, Pulusukaya, Pulasukaya (Telugu);
Indonesia: Bilimbing, Bilimbing Wuluh, Balingbing Bulu, Belimbing Besu, Tjalingting;
Jamaica: Bimbling Plum;
Japanese: Birinbi;
Khmer: Tralong Tong;
Malaysia: Belimbing, Blimbing Assam, Bilimbing Wuluh, Bilimbing Buluk, B'ling, Billing-Billing;
Nicaragua: Mimbro;
Palauan: Imekure, Imekurs, Oterebekii;
Philippines: Kalingiwa (Bisaya), Kiling-Iba (Bikol), Iba (Cebu Bisaya), Pias (Iloko), Puis

(Igorot), Ibag (Manobo), Iba (Panay Bisaya), Iba (Sulu), Iba, Kamias, Kolonanas, Kolonauas, Kalamias (Tagalog), Ibe (Yakan);
Spanish: Bilimbí, Grosella China, Mimbro, Pepino De Indias, Tiriguro, Vinagrillo;
Sri Lanka: Bilin, Bimbiri (Sinhalese);
Surinam: Birambi;
Thailand: Taling Pling, Kaling Pring;
Venezuela: Vinagrillo;
Vietnam: Khe Tau;
Yapese: Kumim.

Origin/Distribution

Bilimbi is native to southeast Asia but now found growing throughout the tropics. The tree is cultivated or found semi-wild throughout Indonesia, Thailand, Malaysia, the Philippines, Sri Lanka, Bangladesh and Myanmar.

Agroecology

Bilimbi is a warm, tropical species, rather cold intolerant compared to carambola and does best in areas with evenly distributed rainfall throughout the year, but with 2–3 months dry season. It grows on a wide range of soil from sandy to limestone soil but does best on rich, well-drained, sandy-loam or clayey loam with plenty of organic matter and with a pH of 5.5–6.5. It grows from near sea level to 750 m altitude.

Edible Plant Parts And Uses

The fruit is extremely sour, high in vitamin C and is eaten raw in salad, *ulam*, *rojak*, *jeruk*, *acar*, or in cooking especially in curries in place of tamarind or tomato as is done in Malaysia. The fruit is commonly used in the popular Malay dishes such *Masak Lemak Ikan* (Fish in coconut milk) and *Asam Pedas* (Hot, spicy, sour dish). It is also processed into jams, jellies, cooling drinks, wine, pickles and preserves. It can replace mango in making chutney. In the Philippines, bilimbi dipped in rock salt, is consumed as a snack. The fruit is also curried or added as flavouring for the common Filipino dish called *sinigang*. In Costa Rica, the raw fruit is prepared as relish and served with rice and beans. In Acheh, Sumatra, it is preserved by sun-drying; the sun-dried bilimbi is called *asam sunti*. Bilimbi and *asam sunti* are popular in Acehnese culinary. A Malaysian nyonya dish also uses sundried salted bilimbi fruit in a duck stew or slices of fresh fruit is used in stir fry with pork pieces and dark soy sauce. In Peninsular Malaysia, the fruits are processed into dried sliced snacks or processed into jam. In Kerala, India, the fruit is used for making pickles, and in Goa and Maharashtra it is consumed raw with salt and spice. The flowers are also sometimes preserved in sugar and consumed.

Botany

A small, perennial tree with a short trunk and sparse thick branches, 4–10 m high. Leaves are alternate, densely crowded at the top of branches, imparipinnate; 30–60 cm long, with 11–37 alternate or subopposite leaflets, ovate or oblong to lanceolate, with obliquely rounded base and acuminate tip, 2–10 cm long by 1.25–1.5 cm wide, downy; medium-green on the upper surface, pale on the underside. Flowers are borne in axillary ramiflorous (on branches) or cauliflorous panicles (on the trunk) (Plate 2). Flower is

Plate 1 Green ellipsoid Bilimbi fruit subtended by a red, star-shaped calyx at the stem-end and crowned with stigma remnants at the apex

Plate 2 Cauliflorous fruit of Bilimbi

Plate 3 Yellow fruited variant of *Averrhoa bilimbi*

bisexual, pentamerous and 10–20 mm long. Sepals are ovate-lanceolate, 4 mm and pubescent. Petals are reddish purple, 13–18 × 3 mm. Stamens are all fertile. Fruit is an ellipsoid-obovoid berry, 4–10 cm long, faintly 5-angular, crowned by a thin, red, star-shaped calyx at the stem-end and tipped with 5 hair-like floral remnants at the apex (Plate 1). The fruit is glossy with a thin skin, green turning to yellowish green or white when ripe, juicy and very acid. Seeds are flattened disc-shaped, small (6 mm) and brown. There is also a yellow fruited variant (Plate 3).

Nutritive/Medicinal Properties

Ripe bilimbi fruits are high in vitamin C and are rich in antioxidant compounds. The proximate food composition per 100 g of edible portion (Morton 1987) was reported as: moisture 94.2–94.7 g, protein 0.61 g, ash 0.31–0.40 g, fibre 0.6 g, phosphorus 11.1 mg, calcium 3.4 mg, iron 1.01 mg, thiamine 0.010 mg, riboflavin 0.026 mg, carotene 0.035 mg, ascorbic acid 15.5 mg and niacin 0.302 mg. Another analysis conducted in Malaysia, reported the food nutrient composition of raw bilimbi fruit per 100 g edible portion as: energy 26 kcal, moisture 94.2 g, protein 0.7 g, fat 1.3 g, carbohydrate 2.8 g, fibre 0.7 g, ash 0.3 g, Ca 12 mg, P 1 mg, Fe 0.4 mg, Na 4 mg, K 80 mg, carotenes 229 μg, vitamin a 38 μg TE, vitamin B1 0.01 mg, vitamin B2 0.08 mg, niacin 1.2 mg and vitamin C 15.7 mg (Tee et al. 1997).

In an analysis conducted in Malaysia, 53 components were identified in the volatile constituents of bilimbi fruit (Wong and Wong 1995). Aliphatic acids accounted for 47.8% of the total volatiles. The main constituents were hexadecanoic acid [palmitic acid] (20.4%), 2-furaldehyde (19.1%) and (Z)-9-octadecenoic acid (10.2%). Among the 12 esters found butyl nicotinate (1.6%) and hexyl nicotinate (1.7%) were dominant.

In another analysis conducted in Cuba, 62 compounds were identified in the volatile constituents of *A. bilimbi* fruit (Pino et al. 2004). Important constituents were a series of C-9 compounds: nonanal, nonanoic acid and (E)-2-nonenal, and (Z)-3-hexenol was also dominant. The exotic aroma character of *A. bilimbi* fruit was attributable to the interaction of fatty-green (C-9 compounds: nonanal, nonanoic acid, (E)-2-nonenal) and green [(Z)-3-hexenol] notes contributing to the complexity of the fruit flavor. The presence of some saturated higher alkanes and its unsaturated homologues (tricosane, 9-tricosene, pentacosane and 9-pentacosene) were also detected.

Bilimbi fruit was found to have higher levels of oxalic acid in half-ripe fruit than in ripe fruit irrespective of seasons (Lima et al. 2001). The oxalic acid levels in half–ripe bilimbi fruit ranged between 9.33 g/100 g (dry season) to 10.32 g/100 g (rainy season) and in ripe fruit 8.57 g/100 g (dry season) to 9.82 g/100 g (rainy season). Ripe fruit has higher levels of vitamin C and total soluble solids (TSS). The level of vitamin C in half-ripe fruit ranged from 20–82 mg/100 g (rainy season) to 32.23 mg/100 g (dry season) and in ripe fruit 36.18 mg/100 g (rainy season) to 60.95 mg/100 g (dry season). The level of TSS in half-ripe fruit ranged from 3.94 °Brix (raining season) to 4.34 °Brix (dry season) and in ripe fruit 4.64 °Brix (rainy season) to 5.06 °Brix (dry season). These high levels of oxalic acid found in bilimbi were probably responsible for its extremely low pH value (0.9–1.5 in both maturity stages). The levels of TSS in carambola were very much higher than in bilimbi especially in the sweet carambola (*Averrhoa carambola*) cultivars like B2, B10 and B17.

Some of the pharmacological properties of the fruit and plant parts are discussed below.

Antimicrobial Activity

Bilimbi was also found to have antibacterial activity. Compared to distilled water, bilimbi and tamarind juice significantly decreased aerobic bacteria (APC), *Listeria monocytogenes* and *Salmonella typhimurium* populations in raw shrimps immediately after washing (0 day) (Wan Norhana et al. 2009). There was a significant difference in bacterial reduction between the dipping (0.40–0.41 log for APC; 0.84 for *L. monocytogenes* and 1.03–1.09 log for *S. typhimurium*) and rubbing (0.68–0.70 log for APC; 1.34–1.58 for *L. monocytogenes* and 1.67–2.00 log for *S. typhimurium*) methods. Regardless of washing treatments or methods, population of *S. typhimurium* declined slightly (5.10–6.29 log cfu/g on day 7 of storage) while populations of *L. monocytogenes* (8.74–9.20 log cfu/g) and APC (8.68–8.92 log cfu/g) increased significantly during refrigerated storage. The pH of experimental shrimps was significantly lowered by 0.15–0.22 pH units after washing with bilimbi and tamarind juice.

Bilimbi fruit was found to contain flavonoids, saponins and triterpenoids but no alkaloids. Studies showed that the chloroform and methanol fruit extracts were active against *Aeromonas hydrophila, Escherichia coli, Klebsiella pneumoniae, Saccharimyces cerevisiae, Staphylococcus aureus, Streptococcus agalactiae* and *Bacillus subtilis* (Nurul Huda et al. 2009). The fruit extracts showed good inhibitory activity against the tested pathogens compared with the standard antibiotic, streptomycin.

Antihypercholesterolemic/ Hypoglycemic Activities

Bilimbi fruit was found to have potent antihypercholesterolemic property. The fruit and its water extract, but not alcohol and hexane extracts, showed notable antihypercholesterolemic activity (Ambili et al. 2009). An active component was isolated from the active fraction, which showed optimum activity at a dose of 0.3 mg/kg. The fruit (125 mg/kg) and its aqueous extract (50 mg/kg) were found to be effective in reducing lipids in the high-fat diet fed rats. Oral administration of the fruit homogenate daily for 15 days did not result in any toxic symptoms up to a dose of 1 g/kg studied suggesting that this fruit could be used as a dietary ingredient to prevent and treat hyperlipidemia.

Recent studies using animal models reported that ethanol leaf extract had hypoglycemic, hypotriglyceridemic, anti-lipid peroxidative and anti-atherogenic properties in streptozotocin-diabetic rats (Pushparaj et al. 2000, 2001; Tan et al. 2005). Ethanol extract of *A. bilimbi* leaves was found to significantly elevate the HDL-cholesterol concentrations by 60% compared with the distilled water control. The extract significantly enhanced the anti-atherogenic index and HDL-cholesterol/total cholesterol ratio but did not affect LDL-cholesterol and total cholesterol concentrations. It also significantly lowered kidney lipid peroxidation level. The studies also reported that the water soluble fraction (AF) was more potent than a butanol soluble fraction (BuF) in amelioration of hyperglycemia and hyperlipidemia in the streptozotocin diabetic rats fed high fat diet. The long term administration of both extracts at a dose of 125 mg/kg significantly decreased triglyceride concentrations and blood glucose when compared to the vehicle. The hepatic glycogen content was significantly elevated in AF–treated rats when compared to diabetic control, however no change was found in the BuF–treated rats. The results indicated that water soluble fractions of the leaf extract may be a potential source for the isolation of active principle(s) for oral anti-diabetic therapy.

Oxalic Toxicity

Owing to it high level of oxalic acid content in bilimbi fruit (Joseph and Mendonca 1989), caution should be exercised in its consumption especially among people with uraemic complications as severe intoxication can arise as in the case of *Averrhoa carambola* (see notes and references in *A. carambola.*).

Traditional Medicinal Uses

In tropical Asia, bilimbi has been widely used in traditional medicine for cough, cold, itches, acne, boils, rheumatism, scurvy, syphilis, diabetes, whooping cough and hypertension.

In Thailand, the leaves are used externally to prevent itching. They can also be used to cure syphilis when taken internally fresh or fermented. It is also used to cure beriberi, biliousness and coughs. In Malaysia, the fresh or fermented leaves of bilimbi are used as a treatment for venereal disease. A leaf infusion is a remedy for coughs and is taken after childbirth as a tonic. A leaf decoction is taken as a medicine to relieve rectal inflammation. It seems to be effective against coughs and thrush. In Sarawak, bilimbi flowers are mixed with bird's nest, milk, steam and taken to mitigate cough in children. The Bidayuhs pound the leaves with bird's nest, fennel seeds, shallot, rock sugar and water and the mixture steam and consumed while still warm to remove wind in children. In the Philippines, the leaves are applied as a warm paste for pruritus or poulticed on itches, swellings of mumps and rheumatism, and on skin eruptions. In Java, a decoction of the leaves is used to treat inflammation of the rectum. The Javanese also apply a paste of them for mumps, rheumatism, and pimples. Elsewhere, they are applied on bites from poisonous creatures. The flower infusion is used for thrush, cold, and cough.

In Java, the fruits combined with pepper in a preparation called "*rujak mircha*" are eaten to induce sweating when people are feeling "under the weather". A paste of pickled bilimbi is smeared all over the body to hasten recovery after a fever. The fruit conserve is administered as a treatment for coughs, beri-beri and biliousness. Syrup prepared from the fruit is taken as a cure for fever and inflammation and to stop rectal bleeding and alleviate internal haemorrhoids.

In India, the fruit is an astringent stomachic and refrigerant and its juice is made into syrup as a cooling drink for reducing fever. It is antiscorbutic and the syrup is used in some minor cases of haemorrhage from the bowels as well as the stomach and internal haemorrhoids. A conserve of the fruit is used in Java for beriberi, biliousness, and coughs.

Other Uses

Bilimbi juice has high levels of oxalic acid, and therefore may be used to remove iron-rust stains from clothes and to impart shine to brassware, and is used to clean the blade of kris (Malay dagger). They also serve as mordants in the preparation of an orange dye for silk fabrics. The wood is white, soft but tough, even-grained but is seldom available for carpentry.

Comments

Bilimbi is a popular back yard tree in villages in southeast Asia.

Selected References

Abas F, Lajis NH, Israf DA, Khozirah S, Umi Kalsom Y (2006) Antioxidant and nitric oxide inhibition activities of selected Malay traditional vegetables. Food Chem 95(4):566–573

Ambili S, Subramoniam A, Nagarajan NS (2009) Studies on the antihyperlipidemic properties of *Averrhoa bilimbi* fruit in rats. Planta Med 75(1):55–58

Backer CA, Bakhuizen van den Brink RC Jr (1963) Flora of java. (Spermatophytes only), vol 1. Noordhoff, Groningen, 648 pp

Burkill IH (1966) A dictionary of the economic products of the Malay Peninsula. Revised reprint. 2 vols. Ministry of Agriculture and Co-operatives, Kuala Lumpur, vol 1 (A–H), pp 1–1240. vol 2 (I–Z) pp 1241–2444

Chai PPK (2006) Medicinal plants of Sarawak. Lee Ming Press, Kuching, 212 pp

Foundation for Revitalisation of Local Health Traditions (2008) FRLHT database. http://envis.frlht.org

Goh SH, Chuah CH, Mok JSL, Soepadmo E (1995) Malaysian medicinal plants for the treatment of cardiovascular diseases. Pelanduk, Malaysia

Joseph J, Mendonca G (1989) Oxalic acid content of carambola (*Averrhoa carambola* L.) and bilimbi (*Averroha bilimbi* L.). Proc Interam Soc Trop Hort 33:117–120

Lennox A, Ragoonath J (1990) Carambola and bilimbi. Fruits 45(5):497–501

Lima VLAGD, Melo EDA, Santos Lima LD (2001) Physicochemical characteristics of bilimbi (*Averrhoa bilimbi* L.). Rev Bras Frutic 23(2):421–423

Morton J (1987) Bilimbi. In: Fruits of warm climates. Julia F. Morton, Miami, pp 128–129

Nurul Huda BAW, Mohd Effendy BAW, Mariam BT, Wan Zuraida BWMZ, Sarah Aqilah BA (2009) Phytochemical screening and antimicrobial efficacy of extracts from *Averrhoa bilimbi* (Oxalidaceace) fruits against human pathogenic bacteria. Pharmacogn J 1(1):62–64

Ochse JJ, Bakhuizen van den Brink RC (1980) Vegetables of the Dutch Indies, 3rd edn. Ascher & Co., Amsterdam, 1016 pp

Pacific Island Ecosystems at Risk (PIER) (1999) Averrhoa bilimbi L., Oxalidaceae. http://www.hear.org/Pier/species/averrhoa_bilimbi.htm

Pino JA, Marbot R, Bello A (2004) Volatile components of *Averrhoa bilimbi* L. fruit grown in Cuba. J Essent Oil Res 16:241–242

Pushparaj P, Tan CH, Tan BKH (2000) Effects of *Averrhoa bilimbi* leaf extract on blood glucose and lipids in streptozotocin-diabetic. J Ethnopharmacol 72(1):69–76

Pushparaj PN, Tan BKH, Tan CH (2001) The mechanism of hypoglycemic action of the semi-purified fractions of *Averrhoa bilimbi* in streptozotocin-diabetic rats. Life Sci 70(5):535–547

Saidin I (2000) Sayuran Tradisional Ulam dan Penyedap Rasa. Penerbit Universiti Kebangsaan Malaysia, Bangi, p 228 (in Malay)

Samson JA (1992) *Averrhoa* L. In: Verheij EWM, Coronel RE (eds) Plant resources of South-East Asia No. 2. Edible fruits and nuts. Prosea Foundation, Bogor

Tan KHB, Tan CH, Pushparaj PN (2005) Anti-diabetic activity of the semi-purified fractions of *Averrhoa bilimbi* in high fat diet fed-streptozotocin-induced diabetic rats. Life Sci 76(24):2827–2839

Tee ES, Noor MI, Azudin MN, Idris K (1997) Nutrient composition of Malaysian foods, 4th edn. Institute for Medical Research, Kuala Lumpur, p 299

Wan Norhana MN, Azman MN, Poole SE, Deeth HC, Dykes GA (2009) Effects of bilimbi (*Averrhoa bilimbi* L.) and tamarind (*Tamarindus indica* L.) juice on *Listeria monocytogenes* Scott A and *Salmonella Typhimurium* ATCC 14028 and the sensory properties of raw shrimps. Int J Food Microbiol 136(1):88–94

Wong KC, Wong SN (1995) Volatile constituents of *Averrhoa bilimbi* L. fruit. J Essen Oil Res 7(6):691–693

Averrhoa carambola

Scientific Name

Averrhoa carambola L.

Synonyms

Averrhoa acutangula Stokes, *Averrhoa pentandra* Blanco, *Connaropsis philippica* Villar.

Family

Averrhoaceae, also in Oxalidaceae

Common/English Names

Carambola, Coolie Tamarind (Trinidad), Coromandel Gooseberry, Country Gooseberry, Five Corner (Australian), Five Fingers (Guyana), Starfruit, Star Pickle

Vernacular Names

Brazil: Camerunga, Carambola, Caramboleiro, Limas De Cayena (Portuguese);
Burmese: Zaung-Ya;
Chinese: Yang-Tao, Ma Fen, Wu Lian Zi;
Costa Rica: Tiriguro;
Danish: Karambol;
Dominican Republic: Vinagrillo;
Dutch: Blimbing, Carambola, Damaksche Blimbing, Stervrucht, Zoeta Vijjhoek;
El Salvador: Pepino De La India;
Fijian: Kamrakh, Wi Ni Idia, Wi Ni Jaina;
French: Carambolier, Carambolier Doux, Pomier De Goa;
French Antilles: Cornichon;
French Guiana: Carambol;
German: Baumstachelbeere, Blimbingbaum, Karambole, Sternfrucht, Gestirnte Pflaume;
Guam: Bilimbines;
Haitian: Zibline;
India: Kardoi, Kordoi, Rohdoi (Assamese), Kamranga (Bengali), Amrenga (Garos, Meghalaya), Kamarakh, Kamaramga, Kamaranga, Kamrak, Karmal, Khamrak (Hindu), Darehuli, Kamarak, Kamaranga, Kirinulli, Daare Huli, Kamaraka Mara, Kamarakshi, Komarakmara, Kamarakshi Mara, Darchuli, Darepuli, Dharehuli, Kamrac, Karmaranga, Kiranelli, Kirinelli, Komaree, Kamaraak Mara, Kamaraakshi, Kamaraakshi Hannu, Kamarakha, Kamraak, Karamaadalu, Kobari Kaayi (Kannada), Dieng Sohtreng (Khasis, Meghalaya), Carambola, Caturappuli, Chaturapuli, Irumpanpuli, Kamarangam, Pulinji, Tamarat-Tuka, Tamaratta, Saturappuli, Catarapuli, Chatarapali, Chaturappuli, Kamaranga, Pulachi, Pulichi, Pulinci, Tamara, Tamaratonga, Thamarathamu (Malayalam), Heinoujom, Seizrak (Manipuri), Kamarakha, Kumrak, Karmare, Karamala, Karmar, Karmal (Marathi), Theiher-Awt, Theiherawt (Mizoram), Koromonga (Oriya),

Brihaddala, Dantasatha, Dharaphala, Dharaphalah, Karmar, Karmara, Karmaranga, Karmarangah, Karuka, Karukah, Mudgara, Pitaphala, Pitaphalah, Rujakara, Shiral, Shukapriya (Sanskrit), Kandasagadam, Tamarttam-Kay, Tamarattai, Tamaraththam, Thamaratham, Sagadam, Sigam, Sisam, Tamaratti, Tamarathai, Cakatam, Cakattai, Caturappuli, Ci, Cicakari, Cicam, Cikam, Cukkirakam, Cukkirakamaram, Kacerukam, Kandasagadam, Kantacatakamaram, Kantacatcam, Kantacatkam, Pukamatitam, Pukamatitamaram, Pulipputtamarattai, Putakecam, Putakecaramaram, Tamarakam, Tamarakamaram, Tamarattai, Tamarattaimaram, Tamarattam, Tamarattankay, Tantacatam, Nattuttamarattai (Tamil), Karamonga, Karomonga, Tamaratamu, Tamarta, Tamarta-Kaya, Karamanooga, Thamaratha (Telugu), Kamrakh, Kamarakha (Urdu);
Indonesia: Belimbing Manis, Belimbing, Belimbing Legee, Blimbing Wana (Java), Bhalingbhing Manes (Madurese),Balingbing, Tjalingtjing Amis (Sundanese); *Japanese*: Gorenshi, Karanbora;
Khmer: Spu;
Laotian: Nak Fuang;
Malaysia: Belimbing Segi, Belimbing Manis, Belimbing, Belimbing Batu, Belimbing Besi, Belimbing Pessegi, Belimbing Sayur, Belimbing Saji, Caramba Carambola, Kambola;
Mexico: Carambolera, Caramboler, Árbol De Pepino;
Pakistan: Kamrak, Kamranga;
Papua New Guinea: Faiv Kona;
Philippines: Balingbing, Balimbing (Bikol), Garahan (Bisaya), Balimbing Balingbing (Cebu Bisaya), Daligan (Iloko), Dalihan, Galuran, Garulan (Ibanag), Malimbin (Samar-Leyte Bisaya), Galañgan (Panay Bisaya), Balimbing (Sulu), Balimbin, Blingbing, Balimbing (Tagalog), Sirinate (Tinggian);
Polish: Karambola;
Slovaščina: Karambola;
Spanish: Arbol De Pepino, Carambola, Carambolera, Carambolero, Tamarindo Chino, Pepino De La India, Tamarindo Dulce;
Sri Lanka: Kamaranga, Kamruk;
Surinam: Blimbing Legi, Fransman-Birambi;
Thailand: Ma Fuang, Fuang, Khe, Sa Bue;
Tonga: Tapanima;
Venzuela: Tamarindo Chino, Tamarindo Dulce;
Vietnamese: Khe, Khe Ta.

Origin/Distribution

The origin of carambola is in West Malesia, the fruit has been cultivated in southeast Asia and Malaysia for centuries. It has naturalised in northern south America. The fruit is commonly cultivated in the provinces of Fujian, Guangdong, Guangxi, Guizhou, Hainan, Sichuan in southern China, in Taiwan and India. It is rather popular in the Philippines and Queensland, Australia, Madagascar and moderately so in some of the South Pacific islands, particularly Tahiti, New Caledonia, Fiji, New Guinea, Guam and Hawaii.

Agroecology

It grows in the hot humid tropics to subtropics as mature trees can tolerate freezing temperatures for short periods and sustain little damage at −2.78°C as in Florida, USA. It grows from near sea level to 1,000 m altitude. It is adaptable to a wide range of soil types from very poor sands such as tin tailings in West Malaysia to clayey soil or limestone soils as in Florida. On the poor sandy soils including tin tailings, it requires abundant organic matter and frequent light irrigation to produce a good crop. On limestone its leaves becomes chlorotic. It thrives best in fertile, rich, well-drained, loamy soils. It needs good drainage and cannot withstand flooding. It grows best in full sun.

Edible Plant Parts and Uses

The sweet star fruit varieties can be eaten out of hand or sliced and used in salads or as garnish in cocktail drinks and beverages. Star fruit is also used as garnish of chicken, pork or fish dishes. Ripe or half-ripe fruit eaten as rojak in Java. Sour fruit is eaten with sugar. The star fruit juice could be used in tropical drinks and smoothies. In

Brazil, the star fruit is served as a fresh beverage, in natura, or as an industrialized juice, as it is also served throughout the world. People living in Brazil sometimes drink up to 500 ml of star fruit juice in a day. The immature or half-ripe fruit is used as vegetables for cooking and prepared into pickles and sweets. In Peninsular Malaysia, unripe fruit is salted as *achar* and also boiled in syrup as *manisan*. Star fruit is also used in chutney, curries, jams, preserves, stewed fruit and tarts. Fruits can be poached in a syrup and enlivened with lime juice. Wine can be produced from fermentation of the ripened and unripened carambola fruit juice using Brewer's yeast starter culture and sugar (Anim and Tano-Debrah 2004; Napahde et al. 2010).

The flowers have been reported to be used in salads in Java.

Plate 1 Different maturity stages of carambola fruit cv B 10

Botany

A low, short-trunked perennial tree with a much-branched, bushy, irregular crown growing to 4–9 m high. Leaves are spirally arranged, alternate, imparipinnate, 15–20 cm long, with 5–13 opposite to sub-opposite leaflets. Each leaflet is subsessile, ovate or ovate-oblong, 2–8.5 cm long, 3–4.5 cm broad, the terminal one being the largest, acuminate, upper surface glabrous, lower sparsely pubescent. Flowers are numerous, small in axillary, tomentose panicles (Plate 7). Pedicel is 2 mm long, terete, glabrous and dark red. Sepals are 3.5 mm long, ovate, imbricate and persistent. Petals are 8–9 mm long, elliptic oblong, lilac to purple and slightly connate above the base. Stamens 10, often 5 antheriferous, alternating with 5 staminodes; filament 2 mm long, curved, persistent; anthers obovate to ovate and basifixed. Ovary, greenish-white, 1.8 mm long, pubescent, with 5 acute lobes; styles 5, glandular, persistent with capitate stigma. Berry is yellow to bright orangey-yellow, waxy and glossy, oblong, 7–13 cm × 5–8 cm, deeply 5–6-ribbed (Plates 1–6), stellate in cross section, very fleshy, sweet to acid sweet, crisp and juicy. Seeds numerous, blackish brown, ovoid, much compressed, 6–12.5 mm long enclosed by slimy aril.

Plate 2 Two major Malaysian commercial cultivars B10 (*left*), B 2 (*right*)

Plate 3 Cv B17 sweetest Malaysian carambola

Plate 4 Cv B6 another Malaysian carambola cultivar. cultivar with a Brix of 18%

Plate 7 *Averrhoa carambola* flowers

Plate 5 Cv Fwang Tung with wavy ribs, grown in Darwin, Australia

Plate 6 Cv. Jungle Gold, another carambola cultivar grown in Darwin, Australia

Nutritive/Medicinal Properties

Nutrients and Other Phytochemicals in the Fruit

Analyses carried out in the United States reported raw carambola fruit (−3% seeds and stem end) to have the following nutrient composition (per 100 g edible portion): water 91.38 g, energy 31 kcal (128 kJ), protein 1.04 g, total lipid 0.33 g, ash 0.52 g, carbohydrates 6.73 g, total dietary fibre 2.8 g, total sugars 3.98 g, Ca 3 mg, Fe 0.08 mg, Mg 10 mg, P 12 mg, K 133 mg, Na 2 mg, Zn 0.12 mg, Cu 0.137 mg, Mn 0.037 mg, Se 0.6 μg, vitamin C 34.4 mg, thiamin 0.014 mg, riboflavin 0.016 mg, niacin 0.367 mg, pantothenic acid 0.391 mg, vitamin B-6 0.017 mg, total folate 12 μg, choline 7.6 mg, vitamin A 61 IU, vitamin E (α-tocopherol) 0.15 mg, total saturated fatty acids 0.019 g, total monounsaturated fatty acids 0.030 g, total polyunsaturated fatty acids 0.184 g, tryptophan 0.008 g, threonine 0.044 g, isoleucine 0.044 g, leucine 0.077 g, lysine 0.077 g, methionine 0.021 g, phenylalanine 0.037 g, tyrosine 0.044 g, valine 0.050 g, arginine 0.021 g, histidine 0.008 g, alanine 0.071 g, aspartic acid 0.098 g, glutamic acid 0.148 g, glycine 0.050 g, proline 0.050 g, serine 0.083 g, β-carotene 25 μg, α-carotene 24 μg, lutein + zeaxanthin 66 μg (USDA 2010).

The fruit is rich in vitamin C and also has vitamin A, vitamin B1 and 2, vitamin B-6, pantothenic acid, α and β carotenes, lutein and zeaxanthin and is low in sodium and fatty acids which is composed primarily of monounsaturated and polyunsaturated fatty acids. It is poor in calcium, magnesium and phosphorus. The fruit contains oxalic acid, and potassium oxalate. Fruits of *Averrhoa carambola* were found to have C13- and C15-norisoprenoid volatile aroma substances (Winterhalter and Schreier 1995). The majority of norisoprenoid volatiles in starfruit were derived from glycosidically bound progenitors from the central portion of the carotenoid chain in the fruit.

Analysis of the fruit juice carried out in Ghana reported carambola fruit to be a suitable raw material for wine production (Anim and Tano-Debrah 2004). Some physico-chemical characteristics of the fruit juice were: pH 2.2, total solids 5.9%, soluble solids 5.0%, titratable acidity 0.66, fat 0.25% and total sugars 2.55%. The analysis of the sugars showed glucose, fructose and sucrose as the predominant sugars present. Vitamin C content was substantially high (35 mg/100 g), but fat content was low (0.25%). The total sugars content and pH were very low. Most brewer's yeasts require a pH range of 4–6; and for adequate alcohol production to form a wine, the sugar content should be about 10%. The physico-chemical characteristics of the juice suggested the feasibility of modifying the juice into a suitable must that could yield adequate amount of alcohol to form a wine. The 2-0-β-D-glucopyranoside of 4-(1′,4′-dihydroxy-2′,2′,6′-trimethylcyclohexyl)-but-3-en-2-ol 1a was isolated from the juice of *Averrhoa carambola* (Lutz et al. 1993). After a pre-separation of the crude glycosidic extract with multi-layer coil countercurrent chromatography the new natural product was peracetylated, purified by HPLC and characterized as pentaacetate 1b.

Phytochemicals in Other Plant Parts

Phytochemicals isolated from flowers included two *O*-glycosyl flavonoid components: quercetin-3-*O*-β-D-glucoside and rutin (Tiwari et al. 1979) and anthocyanins cyanidin-3-*O*-β-D-glucoside, cyanidin-3,5-*O*-β-D-diglucoside (Gunasegaran 1992). GC/MS analysis of the chloroform-soluble fraction of carambola leaves revealed the presence of high percentage of esters (45.06%) dominated by vanillin acetate (7.63%), in addition to nor-pentacosane (11.12%) as aliphatic hydrocarbon (Tadros and Sleem 2004). Galactose and arabinose were identified both in free and combined form in the stems and leaves. GLC of the saponifiable and unsaponifiable fractions of the petroleum ether extracts of the stems and leaves showed that leaves extract contained higher percentage of unsaturated fatty acids (31.87%) compared to the stems (22.17%) where oleic is predominating. On the other hand, the total saturated fatty acids in stems were more than those present in the leaves and palmitic acid was the major one. β-amyrin, stigmasterol and β-sitosterol (13.12%, 12.49% and 11.08% respectively) represented the major constituents of the unsaponifiable fraction and a series of hydrocarbons ranging from n C9–C30 were detected in the stems while stigmasterol and β-sitosterol (14.34% for each) were the majors in case of the leaves and the hydrocarbon series from n C15-C30. From the leaves C-glycoside flavones, such as apigenin-6-C-β-L-fucopyranoside and apigenin-6-C-(2″-O-α-L-rhamnopyranosyl)-β-L-fucopyranoside were isolated (Cabrini et al. 2011). This latter compound, also known as carambola flavone, was isolated along with a known flavone C-glycoside (isovitexin) from the leaves (Araho et al. 2005). From the bark, β-sitosterol, lupeol and anthraquinone glucoside were isolated (Ranganayaki et al. 1980). Two alkyl phenols, namely, 2,5-dimethoxy-3-undecylphenol and 5-methoxy-3-undecylphenol, were isolated together with two known benzoquinones, 5-O-methylembelin and 2-dehydroxy-5-O-methylembelin from the wood of *Averrhoa carambola* (Chakthong et al. 2010).

The extracts of carambola fruit, stems and leaves have been reported to exhibit various pharmacological properties that include: antioxidant, anti-inflammatory, analgesic, antipyretic, anticonvulsant, anti-hyperglycemic (Tadros and Sleem 2004) and other activities.

Antioxidant Activity

Of 17 exotic fruit analysed in Mauritius, the highest antioxidant capacities were observed in red and yellow *Psidium cattleianum*, sweet and acid *Averrhoa carambola*, *Syzygium cumini* and white *Psidium guajava* (Luximon-Ramma et al. 2003). The antioxidant activities of the fruits ranged from 1 to 47 μmol Trolox equivalent antioxidant capacity (TEAC)/g fresh weight and from 0.3 to 34 μmol/g fresh weight FRAP (ferric reducing ability of plasma)/g fresh weight. Total phenolics in the fruits ranged from 118 to 5,638 μg/g fresh weight, proanthocyanidins from 7 to 2,561 μg/g fresh weight, flavonoids from 21 to 712 μg/g fresh weight and vitamin C content from 8 to 1,426 μg/g fresh weight. There were strong correlations between antioxidant activity (assessed by both TEAC and FRAP) and total phenolics and proanthocyanidins. Flavonoids appeared to contribute less to the antioxidant potential of the fruits, while very poor correlations were observed between ascorbate content and antioxidant activity. These fruits were also characterised by high levels of total phenolics and may have potential health benefits.

Of seven local Bangladeshi fruit tested, *A. carambola* exhibited the highest antioxidant activities with IC_{50} of 30 μg/ml in scavenging 1,1-diphenyl-2-picrylhydrazyl (DPPH) free radical (Soubir 2007). *Averrhoa carambola* was found to be a good source of natural antioxidants with polyphenolics as its major antioxidants (Shui and Leong 2004; 2006). Main antioxidants were attributed to L-ascorbic acid and phenolic compounds, (−)epicatechin and gallic acid in gallotannin forms (Shui and Leong 2004). The major antioxidants were initially attributed to singly-linked proanthocyanidins that existed as dimers, trimers, tetramers and pentamers of catechin or epicatechin. The scientists (Shui and Leong 2006), further reported that the residue of star fruit, which was normally discarded during juice drink processing, was found to contain much higher antioxidant activity than the extracted juice. Under optimized extraction conditions, the residue accounted for around 70% of total antioxidant activity (TAA) and total polyphenolic contents. Freeze-dried residue powder, which accounted for around 5% of total weight, had total polyphenolic content of 33.2 mg gallic acid equivalent (GAE)/g sample and total antioxidant activity of 3,490 and 3,412 mg L-ascorbic acid equivalent antioxidant capacity (AEAC) or 5,270 and 5,152 mg trolox equivalent antioxidant capacity (TEAC) per 100 g sample obtained by 2,2′-azino-bis-(3-ethylbenzthiazo-line-6-sulfonic acid) free radical (ABTS+.) and 1,1-diphenyl-2-picryl-hydrazyl (DPPH) scavenging assays, respectively. It was also found to have 510.3 mol ferric reducing/antioxidant power (FRAP) per gram sample. The residue extract also showed strong antioxidant activity in hindering oxidative rancidity of soya bean oil at 110°C. Antioxidant activity and polyphenolic profile of residue extracts were compared with extracts of standardized pyconogenol. High performance liquid chromatography coupled with mass spectrometry (HPLC/MS) showed that major proanthocyanidins in star fruit were different from their isomers in pyconogenol. The high content of phenolics and strong antioxidant activity of residue extracts indicated that starfruit residue powder may impart health benefits when used in functional food products and that residue extracts should also be regarded as potential nutraceutical resources.

Hypoglycaemic Activity

A lyophilised herbal medicine Glico-vitae which contained leaf extract of carambola when administered intragastrically showed absence of acute toxicity and exerted an antihyperglycaemic effect from 30 mg/kg (Provasi et al. 2001). The herbal medicine Glico-vitae was recommended for non-insulin dependent diabetes mellitus (NIDMM). Taiwanese scientists reported that the fruit of *Averrhoa carambola* was found to possess a high level of insoluble fibre-rich fractions (FRFs) including insoluble dietary fibre, alcohol-insoluble solid and water-insoluble solid (46.0–58.2 g/100 g of pomace) (Chau et al. 2004a). These FRFs were mainly composed of pectic substances and hemicellulose. Their physico-chemical properties (e.g. water-holding capacities, swelling properties, and cation-exchange capacities) were significantly higher than those of cellulose. The apparent abilities of these FRFs to adsorb glucose and reduce amylase activity implied that they might help control postprandial serum glucose. Studies showed that the hypoglycemic effects of these insoluble fibre-rich fractions were significantly stronger than that of cellulose (Chau et al. 2004b). Therefore, it was suggested that they could be incorporated as low-calorie bulk ingredients in high-fibre foods to reduce calorie level and help control blood glucose concentration. Further studies demonstrated the pronounced cholesterol-lowering and lipid-lowering effects of water-insoluble fibre-rich fraction in carambola fruit might be attributed to its ability to enhance the excretion of cholesterol and bile acids via faeces (Chau et al. 2004c). The findings suggested that carambola water-insoluble fibre-rich fraction may be used as a promising cholesterol-lowering ingredient in human diets or new formulations of fibre-rich functional foods.

Apigenin-6-C-β-L-fucopyranoside isolated from *Averrhoa carambola* exhibited an acute effect on blood glucose lowering in hyperglycemic rats and promoted glucose-induced insulin secretion (Cazarolli et al. 2009). Its stimulatory effect on glycogen synthesis was noted when muscles were incubated with this flavonoid and also its effect was completely nullified by pre-treatment with insulin signal transduction inhibitors. The apigenin-6-C-β-L-fucopyranoside-induced increase in glycogen synthesis in the rat soleus muscle was found to involve the MAPK-PP1 and PI3K-GSK3 pathways. The study provided evidence for dual effects of apigenin-6-C-β-L-fucopyranoside as an antihyperglycemic (insulin secretion) as well as an insulinomimetic (glycogen synthesis) agent.

Anti-inflammatory Activity

Topically applied ethanolic extract of starfruit leaves was found to decrease edema in a dose-dependent manner, resulting in a maximum inhibition of 73% and an ID_{50} value of 0.05 (range: 0.02–0.13) mg/ear in mice (Cabrini et al. 2011). Myeloperoxidase (MPO) activity was also inhibited by the extract, resulting in a maximum inhibition of 60% (0.6 mg/ear). All of the fractions (hexane, ethyl acetate, and butanol) tested caused inhibition of edema formation and of MPO activity. Treatment with the ethyl acetate fraction was the most effective, resulting in inhibition levels of 75% and 54% for edema formation and MPO activity, respectively. However, treatment of mice with isolated compounds [apigenin-6-C-β-L-fucopyranoside and apigenin-6-C-(2″-O-α-L-rhamnopyranosyl)-β-L-fucopyranoside] did not produce successful results. Apigenin-6-C-(2″-O-α-L-rhamnopyranosyl)-β-L-fucopyranoside caused only a mild reduction in edema formation (28%). Taken together, these preliminary results supported the popular traditional use of *A. carambola* as an anti-inflammatory agent and opened up new possibilities for its use in skin disorders.

Cytochrome Inhibitory Activity

Japanese scientists found that the filtered extracts of the star fruit were capable of altering pharmacokinetics of therapeutic drugs co-administered via CYP3A inhibition, similar to the case with grapefruit (Hidaka et al. 2004; Hosoi et al. 2008). The juice of star fruit showed the most potent inhibition of human cytochrome P450 3A (CYP3A). The stereoisomer of procyanidin B1 and B2 and/or the trimer consisting of catechin and/or epicatechin in carambola fruit were suggested to be potent CYPA3 inhibitory components. Cytochrome P450 3A (CYP3A) is a gene involved in the synthesis of cholesterol, steroids and other lipids. Recent studies reported that star fruit juice exhibited varying degree of inhibitory effect towards different CYP (cytochrome P450) isoforms in the following order: CYP2A6 (0.9) > CYP1A2 (1.4) > CYP2D6 (1.6) > CYP2E1 (2.0) > CYP2C8 (2.2) > CYP2C9 (3.0) > CYP3A4 (3.2) (Zhang et al. 2007). The strongest inhibitory effect was against CYP2A6 and the weakest towards CYP3A4. Additionally, star fruit juice was found not to inhibit the activity of CPR (NADPH-cytochrome P450 reductase). CPR is the electron donor protein for several oxygenase enzymes found on the endoplasmic reticulum of most eukaryotic cells.

Antiulcerogenic Activity

Preliminary studies revealed that the carambola leaf extracts exhibited anti-ulcerogenic activity (Gonçalves et al. 2006). The water-alcohol extract of carambola leaves at doses of 800 and 1,200 mg/kg, p.o., only exhibited significant anti-ulcer activity in the acidified-ethanol-induced ulcer model in rats. ACE (Angiotensin-Converting Enzyme) tested by indomethacin and acute stress ulcerogenic models did not show this activity.

Hypotensive Activity

An aqueous extract of *Averrhoa carambola* leaves exhibited hypotensive effects in rats (Soncini et al. 2011). In normotensive rats, the extract (12.5–50.0 mg/kg, i.v.) induced dose-dependent hypotension. In-vitro, the extract elicited a depression in the E(max) response to phenylephrine without an aletration in sensibility. Also, in a depolarized Ca^{2+}-free medium, extract inhibited $CaCl_2$-induced contractions and caused a concentration-dependent rightward shift of the response curves, indicating that the extract inhibited the contractile mechanisms involving extracellular Ca^{2+} influx. The results supported the Brazilian traditional use for treating hypertension.

Inotropic and Chronotropic Activity

Brazilian scientists reported on the negative inotropic and chronotropic effects on the guinea pig atrium of extracts obtained from *Averrhoa carambola* leaves (Vasconcelos et al. 2005). Caramola leaf extracts (1.5 mg/ml) eliminated the contractile force in guinea pig atria in a concentration-dependent fashion. Among the crude, methanolic, ethanolic, aqueous and acetic extracts, the aqueous one was the most potent ($EC_{50} = 520$ µg/ml). Flavonoids and tannins were found to be the main constituents; Na^+ and K^+ contents in 1.0 mg/ml of aqueous extract were 0.12 and 1.19 mM, respectively. The aqueous extract eliminated the positive Bowditch staircase phenomenon and reduced the inotropic response to $CaCl_2$ (0.17–8.22 mM). In spontaneously beating atria, the aqueous extract stimulated a negative chronotropic effect that was antagonized by 0.1 µM isoproterenol bitartrate. The data relating to the effect of *A. carambola* leaf extract on guineapig atrial contractility and automaticity indicated an L-type Ca^{2+} channel blockade. Further studies (Vasconcelos et al. 2006) found that carambola leaf extract decreased the conduction velocity of atrial impulse and the intraventricular pressure (86%), and increased the conduction time between the right atrium and the His bundle (27%). Based on these results, the researchers recommended that the popular use of such extracts should be avoided because it can promote electrical and mechanical changes in the normal heart.

Toxicity Effects

On the downside, it has been reported that star fruit can lead to a fatal outcome in uraemic patients (Chang et al. 2000; Neto et al. 1998, 2003; Signaté et al. 2009; Tse et al. 2003). Ingestion of star fruit (*Averrhoa carambola*) could induce severe intoxication in subjects with chronic renal failure. Oxalate was found to play a key role in the neurotoxicity of star fruit. The intoxication syndrome consisted of dizziness, hiccup, mental confusion, vomiting, impaired consciousness, muscle twitching and hyperkalaemia shortly after ingestion of star fruit, and leading to mortality in extreme cases. Patients who were promptly treated with haemodialysis, including those with severe intoxication, recovered without sequelae. Patients with severe intoxication who were not treated or treated with peritoneal dialysis did not survive. An excitatory neurotoxin from star fruit was implicated. Recent studies in Brazil reported a neurotoxic fraction (AcTx) from star fruit (Carolino et al. 2005). AcTx is a non-proteic molecule with a molecular weight less than 500, differing from oxalic acid which had been implicated previously. AcTx had no effect on GABA/glutamate uptake or release, or on glutamate binding. However, it specifically inhibited GABA binding in a concentration-dependent manner ($IC_{50}=0.89$ μM). Following cortical administration of AcTx, animals showed behavioral changes, including tonic-clonic seizures, progressing into status epilepticus, accompanied by cortical epileptiform activity. Patients with renal failure on conservative or dialysis treatment should be dissuaded from eating the fruit.

A case of nausea, vomiting, intractable hiccups, and severe encephalopathy accompanied by mental confusion, agitation, disorientation, and seizures in a 53-year-old Brazilian woman with chronic renal failure was reported after ingestion of star fruit juice for 2 days (Auxiliadora-Martins et al. 2010). The patient's ventilatory pattern deteriorated, with development of dyspnea and tachypnea, which resulted in her transfer to an intensive care unit. Despite hemodialysis and adequate septic shock treatment, the patient died on the fifth day after hospital admission. This case demonstrated that star fruit consumption should be considered as a cause of rapid worsening in the renal function of patients with underlying chronic renal failure, potentially resulting in a fatal outcome.

Khoo et al. (2010) found that *A. carambola* juice stored for 0 and 1 hour and given to rats for 14 days were safe to be consumed. However, juice stored for 3 hours exerted toxic effect on rat liver at hepatocellular level. No lethality was found but increment of alanine aminotransferase (ALT) levels was reported in those rats treated with *A. carambola* juice stored for 3 hours.

Traditional Medicinal Uses

Star fruit has been featured in traditional folkloric medicine. The fruit is regarded as laxative, a refrigerant, and an antiscorbutic that excites appetite, a febrifuge and antidysenteric, and a sialagogue and antiphlogistic In Ayurveda, the ripe fruit is considered as digestive, tonic and causes biliousness. The dried fruit, ripe fruit and juice is also used in fever; it is cooling and possesses antiscorbutic properties. A salve made of the fruit is employed to relieve eye afflictions. It is considered as one of the best Indian cooling medicine. The ripe fruit is administered to stop hemorrhages and to relieve bleeding haemorrhoids. In Brazil, carambola is recommended as a diuretic in kidney and bladder complaints, and is believed to be beneficial in the treatment of eczema. In Chinese Materia Medica star fruit is used to quench thirst, increase the salivary secretion, and in fever. Fruits and its fruit juice are used as antioxidant and astringent. The fruit will also benefit hematemesis, melaema, and some other forms of haemorrhage. The Chinese and Annamites employ the fruit in the form of eye-salve against affections of the eyes. It is given, in syrup, as a cooling drink in fevers in the Philippines.

In Indo-China, the flowers are considered to have a vermifuge action and used by the Chinese and Annamites against cutaneous affections. In Malaysia, the crushed leaves or shoots are used by the Malays as an application for chicken-pox, ringworm, and headache. A decoction of the leaves and fruit is given to arrest vomiting. The leaves are applied in fevers. A decoction of the leaves is good for aphthous stomatitis and angina. The seeds may be regarded as a narcotic, anodyne, emetic, and emmenagogue. The powdered seeds are used for asthma and colic. In Sarawak, the Malays and Bidayuhs pound the leaves into a paste and use the warm paste for itchy skin.

Other Uses

The acidic fruit juice may be used for cleaning metal surfaces. The wood is white, turning reddish, moderately hard and close-grained.

Comments

Malaysia is the leading producer of carambola in southeast Asia.

Selected References

Anim G, Tano-Debrah K (2004) Suitability of carambola (*Averrhoa carambola*) fruit juice as a substrate for wine fermentation. Afr J Food Agri Nutr Dev 4(2):Article 10

Araho D, Miyakoshi M, Chou W-H, Kambara T, Mizutani K, Ikeda T (2005) A new flavone C-glycoside from the leaves of *Averrhoa carambola*. Nat Med (Tokyo) 59:113–116

Auxiliadora-Martins M, Alkmin Teixeira GC, da Silva GS, Viana JM, Nicolini EA, Martins-Filho OA, Basile-Filho A (2010) Severe encephalopathy after ingestion of star fruit juice in a patient with chronic renal failure admitted to the intensive care unit. Heart Lung 39(5):448–452

Backer CA, BakhuizenvandenBrink RC Jr (1963) Flora of Java. (Spermatophytes only), vol 1. Noordhoff, Groningen, 648 pp

Burkill IH (1966) A dictionary of the economic products of the Malay Peninsula. Revised reprint, 2 vols, Ministry of Agriculture and Co-operatives, Kuala Lumpur, vol 1 (A–H), pp 1–1240, vol 2 (I–Z), pp 1241–2444

Cabrini DA, Moresco HH, Imazu P, da Silva CD, Pietrovski EF, Mendes DAGB, Prudente AdS, Pizzolatti MG, Brighente IMC, Otuki MF (2011) Analysis of the potential topical anti-inflammatory activity of *Averrhoa carambola* L. in mice. Evid Based Complement Alternat Med, Vol. 2011, Article ID 908059

Campbell CW, Knight RJ Jr, Olszack R (1985) Carambola production in Florida. Proc Fla State Hort Soc 98:145–149

Carolino RO, Beleboni RO, Pizzo AB, Vecchio FD, Garcia-Cairasco N, Moyses-Neto M, Santos WF, Coutinho-Netto J (2005) Convulsant activity and neurochemical alterations induced by a fraction obtained from fruit *Averrhoa carambola* (Oxalidaceae: Geraniales). Neurochem Int 46(7):523–531

Cazarolli LH, Folador P, Moresco HH, Brighente IM, Pizzolatti MG, Silva FR (2009) Stimulatory effect of apigenin-6-C-beta-L-fucopyranoside on insulin secretion and glycogen synthesis. Eur J Med Chem 44(11):4668–4673

Chai PPK (2006) Medicinal plants of Sarawak. Lee Ming Press, Kuching, 212 pp

Chakthong S, Chiraphan C, Jundee C, Chaowalit P, Voravuthikunchai SP (2010) Alkyl phenols from the wood of *Averrhoa carambola*. Chin Chem Lett 21(9):1094–1096

Chang JM, Hwang SJ, Kuo HT, Tsai JC, Guh JY, Chen HC, Tsai JH, Lai YH (2000) Fatal outcome after ingestion of star fruit (*Averrhoa carambola*) in uremic patients. Am J Kidney Dis 35(2):189–193

Chau CF, Chen CH, Lee MH (2004a) Characterization and physicochemical properties of some potential fibres derived from *Averrhoa carambola*. Nahrung 48(1):43–46

Chau CF, Chen CH, Lin CY (2004b) Insoluble fiber-rich fractions derived from *Averrhoa carambola*: hypoglycemic effects determined by in vitro methods. Lebensm Wiss Technol 37(3):331–335

Chau CF, Chen CH, Wang YT (2004c) Effects of a novel pomace fiber on lipid and cholesterol metabolism in the hamster. Nutr Res 24(5):337–345

Foundation for Revitalisation of Local Health Traditions (2008) FRLHT database. http://envis.frlht.org

Gonçalves ST, Baroni S, Bersani-Amado FA, Melo GAN, Cortez DAG, Bersani-Amado CA, Cuman RKN (2006) Preliminary studies on gastric anti-ulcerogenic effects of *Averrhoa carambola* in rats. Acta Farm Bonaerense 25(2):245–247

Gunasegaran R (1992) Flavonoids and anthocyanins of three Oxalidaceae. Fitoterapia 63:89–90

Hidaka M, Fujita K, Ogikubo T, Yamasaki K, Iwakiri T, Okumura M, Kodama H, Arimori K (2004) Potent inhibition by star fruit of human cytochrome P450 3A (CYP3A) activity. Drug Metab Dispos 32(6):581–583

Hosoi S, Shimizu E, Arimori K, Okumura M, Hidaka M, Yamada M, Sakushima A (2008) Analysis of CYP3A inhibitory components of star fruit (*Averrhoa carambola* L.) using liquid chromatography–mass spectrometry. Nat Med (Tokyo) 62(3):345–348

Khoo ZY, Teh CC, Rao NK, Chin JH (2010) Evaluation of the toxic effect of star fruit on serum biochemical parameters in rats. Pharmacogn Mag 6:120–124

Kirtikar KR, Basu BD (1989) Indian medicinal plants, vol 1, 2nd edn. Lalit Mohan Basu, Allahabad

Lutz A, Winterhalter P, Schreier P (1993) The structure of a new ionone glucoside from starfruit (*Averrhoa carambola* L.). Nat Prod Res 3(2):95–99

Luximon-Ramma A, Bahorun T, Crozier A (2003) Antioxidant actions and phenolic and vitamin C contents of common Mauritian exotic fruits. J Sci Food Agri 83(5):496–502

Morton J (1987) Carambola. In: Fruits of warm climates. Julia F. Morton, Miami, pp 125–128

Napahde S, Durve A, Bharati D, Chandra N (2010) Wine production from carambola (*Averrhoa carambola*) juice using *Saccharomyces cerevisiae*. Asian J Exp Biol Sci, SPL 2010:20–23

Neto M, Rob F, Netto J (1998) Intoxication by star fruit (*Averrhoa carambola*) in six dialysis patients? (Preliminary report). Nephrol Dial Transpl 13(3):570–572

Neto MM, da Costa JAC, Garcia-Cairasco N, Netto JC, Nakagawa B, Dantas M (2003) Intoxication by star fruit (*Averrhoa carambola*) in 32 uraemic patients: treatment and outcome. Nephrol Dial Transpl 18:120–125

Ochse JJ, Bakhuizen van den Brink RC (1931) Fruits and fruitculture in the Dutch East Indies. G. Kolff & Co, Batavia-C, 180 pp

Provasi M, Oliveira CE, Martino MC, Pessini LG, Bazotte RB, Cortez DAG (2001) Avaliação da toxicidade e do potencial antihiperglicemiante da *Averrhoa carambola* L. Oxalidaceae Acta Sci 23:665–669 (In Spanish)

Ranganayaki S, Singh R, Singh AK (1980) The chemical examination of the bark of *Averrhoa carambola*. Proc Natl Acad Sci India Sect A 50:61–63

Samson JA (1992) Averrhoa L. In: Verheij EWM, Coronel RE (eds) Plant resources of South-East Asia No. 2. Edible fruits and nuts. Prosea Foundation, Bogor, pp 96–98

Shui G, Leong LP (2004) Analysis of polyphenolic antioxidants in star fruit using liquid chromatography and mass spectrometry. J Chromatogr A 1022(1–2):67–75

Shui G, Leong LP (2006) Residue from star fruit as valuable source for functional food ingredients and antioxidant nutraceuticals. Food Chem 97:277–284

Signaté A, Olindo S, Chausson N, Cassinoto C, Edimo Nana M, Saint Vil M, Cabre P, Smadja D (2009) Star fruit (*Averrhoa carambola*) toxic encephalopathy. Rev Neurol Paris 165(3):268–272, In French

Soncini R, Santiago MB, Orlandi L, Moraes GO, Peloso AL, Dos Santos MH, Alves-da-Silva G, Paffaro VA Jr, Bento AC, Giusti-Paiva A (2011) Hypotensive effect of aqueous extract of Averrhoa carambola L. (Oxalidaceae) in rats: an in vivo and in vitro approach. J Ethnopharmacol 133(2):353–357

Soubir T (2007) Antioxidant activities of some local Bangladeshi fruits (*Artocarpus heterophyllus, Annona squamosa, Terminalia bellirica, Syzygium samarangense, Averrhoa carambola* and *Olea europa*). Sheng Wu Gong Cheng Xue Bao 23(2):257–261

Tadros SH, Sleem AA (2004) Pharmacognostical and biological study of the stem and leaf of *Avehrroa carambola* L. grown in Egypt. Bull Fac Pharm 42:225–246

Thomas S, Patil DA, Patil AG, Naresh C (2008) Pharmacognostic evaluation and physicochemical analysis of *Averrhoa carambola* L. fruit. J Herb Med Toxicol 2(2):51–54

Tiwari KP, Masood M, Minocha PK (1979) Chemical constituents of *Gemlina phillipinensis, Adenocalymna nitida, Allamanda cathartica, Averrhoa carambola* and *Maba buxifolia*. J Indian Chem Soc 56:944

Tse KC, Yip PS, Lam MF, Choy BY, Li FK, Lui SL, Lo WK, Chan TM, Lai KN (2003) Star fruit intoxication in uraemic patients: case series and review of the literature. Int Med J 33(7):314–316

U.S. Department of Agriculture, Agricultural Research Service (USDA) (2010) USDA national nutrient database for standard reference, release 23. Nutrient data laboratory home page. http://www.ars.usda.gov/ba/bhnrc/ndl

Vasconcelos CML, Araujo MS, Silva BA, Conde-Carcia EA (2005) Negative inotropic and chronotropic effects on the guinea pig atrium of extracts obtained from *Averrhoa carambola* L. leaves. Braz J Med Biol Res 38(7):1113–1122

Vasconcelos CM, Araújo MS, Conde-Garcia EA (2006) Electrophysiological effects of the aqueous extract of *Averrhoa carambola* L. leaves on the guinea pig heart. Phytomed 13(7):501–508

Winterhalter P, Schreier P (1995) The generation of norisoprenoid volatiles in starfruit (*Averrhoa carambola* L.): A review. Food Rev Int 11:237–254

Zhang JW, Liu Y, Cheng J, Li W, Ma H, Liu HT, Sun J, Wang LM, He YQ, Wang Y, Wang ZT, Yang L (2007) Inhibition of human liver cytochrome P450 by star fruit juice. J Pharm Sci 10(4):496–503

Averrhoa dolichocarpa

Scientific Name

Averrhoa dolichocarpa Rugayah & Sunarti

Synonyms

None

Family

Averrhoaceae also in Oxalidaceae

Common/English Names

Forest Bilimbi

Vernacular Names

Indonesia: Bilimbing Hutan

Origin/Distribution

The species is indigenous to New Guinea. Wild distribution is found in Papua – Cycloops Nature, Sepik River, Kaiser Willhelmsland (North Papua New Guinea), Yapen Island.

Agroecology

Tropical species in the hot, humid tropical rainforest of New Guinea, agroecological requirements would be similar to that for *Averrhoa bilimbi*. An undercanopy tree that thrives in partial shade and warm, humid conditions.

Edible Plant Parts and Uses

Fruit is edible but sour. Its food uses would be similar to that of *Averrhoa bilimbi*.

Botany

Small, erect, evergreen tree to 8 m high with rough grey brown, lenticellate stem. Leaves are clustered towards the top of the tree or branches (Plate 4). Leaves are alternate, with a pubescent petiole, 4–16 cm long, imparipinnate with 15–60 cm long rachis and 7–13 pairs of opposite to sub-opposite leaflets and a terminal leaflet (Plate 5). Leaflets decreasing in size from top to bottom, narrowly oblong (5.5–11 cm by 3–4.5 cm) to ovate 2–4.5 cm by 1.3–2.8 cm, acute and acuminate at the apex, and unequal and truncate at the base, margin wavy undulating, green, glabrous above, pubescent and punctuate below, midrib densely hairy and prominent on lower

Plate 1 Cauliflorous, immature fruit of *Averrhoa dolichocarpa*

Plate 2 Close up of angled fruit showing the remnants of the style at the apex

Plate 3 Close up of the fruit showing the persistent green calyx at the pedicel end

Plate 4 Alternate, imparipinnate compound leaves clustered towards the apex

Plate 5 Large, sub-opposite, oblong-ovate, dark green leaflets

surface, lateral nerves 5–10 pairs, petiolules pubescent, lateral ones 2 mm long, terminal one longer, 2–3 mm. Inflorescence is cauliflorous, forming crowded clusters of 10–30 or more flowers on the stem (Plate 6); rachis 1–3 cm long, pedicels 3–5 mm long, pubescent. Flowers are small, 1–2.5 cm across, pentamerous and borne on 3–5 mm long, pubescent pedicels. Sepals five, ovate oblong, 6–8 mm long by 3–4 mm wide, recurved at apex, green turning yellowish-brown, glabrous above, pubescent below; petals five,

Plate 6 Cauliflorous flowers arising from the stem

campanulate, coherent at the middle portion, ovate-oblong, recurved, white outside, pinkish purple blotches and white stripes on inner part of reflexed lobes, stamens ten, connate at base, five filaments opposite sepals, 5 mm long, five filaments opposite petals shorter, 3 mm long, anthers sub-rotundate, pistil longer than stamen, ovary 5-angled, five locules, sparsely pubescent, style free and glabrous. Fruit fusiform, 9–12 cm by 2.4–4.5 cm wide, with the persistent calyx and crowned by the persistent remnants of the styles, angled, with five prominent ridges, dark green (Plates 1–3) turning yellow when ripe, star-shaped in cross-section. Seeds small, 7–13 mm by 5 mm covered by a thin transparent aril.

Nutritive/Medicinal Properties

No published information is available on its nutritive and medicinal attributes. It is likely that its nutritive attributes would be similar to that of *Averrhoa bilimbi*.

Other Uses

No published information is available on its miscellaneous uses as its commercial uses and potential are yet to be exploited. It provides a new source of genetic material for breeding and the development and selection of improved cultivars of *Averrhoa carambola* and *A. bilimbi*.

Comments

Averrhoa dolichocarpa is a new and rare *Averrhoa* species.

Selected References

Rugayah, Sunarti S (2008) Two new wild species of *Averrhoa* (Oxalidaceae) from Indonesia. Reiwardtia 12(4):325–334

Samson JA (1992) *Averrhoa* L. In: Verheij EWM, Coronel RE (eds) Plant resources of South-East Asia No. 2. Edible fruits and nuts. Prosea Foundation, Bogor

Averrhoa leucopetala

Scientific Name

Averrhoa leucopetala Rugayah & Sunarti

Synonyms

None

Family

Averrhoaceae also in Oxalidaceae

Common/English Names

Forest Bilimbi

Vernacular Names

Indonesia: Bilimbing Hutan

Origin/Distribution

The species was first discovered in North Sulawesi – Gorontalo (Tangle and Panua Nature Reserve).

Agroecology

Tropical species that occurs wild in the hot, humid, tropical rainforest of Northern Sulawesi. It is an under-canopy tree that thrives in partial shade and humid conditions. Its agroecological requirements would be similar to that for *Averrhoa bilimbi*.

Edible Plant Parts and Uses

Fruit is edible but sour. Culinary uses would be similar to that of *Averrhoa bilimbi*.

Botany

Woody perennial, evergreen shrub or small tree up to 2 m high with sparse branching and an irregular sparse crown. Old stem is rough, lenticellate and grey-brown while juvenile stem is smooth and brown (Plates 2 and 4). Leaves are crowded towards the top, compound, imparipinnate with 4–7 pairs of leaflets and a terminal leaflet and 3–13 cm long, pubescent rachis (Plate 4). Petioles are pubescent and 1.5–6 cm long. Leaflets decrease in size from top down to bottom. Leaflets are dull-green, limp and soft not coriaceous, oblong-lanceolate (7.3 × 2.9 cm) to

Plate 1 Cauliflorous, ridged, fusiform fruit

Plate 3 New growth at the stem apex

Plate 2 Developing fruit with persistent green calyx arising from the rough, lenticellate stem

Plate 4 Compound, imparipinnate leaves crowded towards the stem top

ovate (0.8 by 0.5 cm), with acuminate apex, subcordate to truncate base (Plates 2 and 3), glabrous except near midrib, pubescent and glaucous adaxially. Leaflet has 4–12 pairs secondary nerves and entire, undulating margins with pubescent petiolules, lateral ones 1–2 mm long, terminal 2.5 mm long. Inflorescence is axillary and cauliflorous, comprising small clusters of several flowers, peduncle and rachis green less than 2 cm long, bracts subulate, 1–3 mm long, sparsely hairy and caducous. Flowers are bisexual, pentamerous, white on 3–4 mm long, pubescent pedicels. Sepals five ovate-oblong, 6–7 mm long by 2.5–3 mm broad, apex obtuse, pale green, glabrous above, pubescent below; petals white campanulate, coherent at lower part, lanceolate, rotundate apex, reflexed at the top, stamens ten connate at base, glabrous, antisepala filaments 5, 7 mm long, antipetala filaments five, shorter, 3.5 mm long, anther sub-rotundate, ovary 5-angled, 1 mm long, densely pubescent, styles five, free densely hairy. Fruit is fusiform, 5-ridged, wings rounded at ridge, 5–12 cm by 2.4–3.7 cm diameter, with persistent calyx (Plates 1 and 2), light green tuning to yellow and acid. Seeds are 9 mm by 5 mm with thin transparent aril.

Nutritive/Medicinal Properties

No published information available on its nutritive and medicinal characteristics. It is likely that its nutritive attributes would be similar to that of *Averrhoa bilimbi*.

Other Uses

No published information is available on its non-edible uses as its commercial uses and potential are yet to be exploited. Like *Averrhoa dolichocarpa*, it represents a new source of genetic material for breeding and the development and selection of improved cultivars of *Averrhoa carambola* and *A. bilimbi*.

Comments

Averrhoa leucopetala is a new and rare *Averrhoa* species.

Selected References

Rugayah, Sunarti S (2008) Two new wild species of *Averrhoa* (Oxalidaceae) from Indonesia. Reiwardtia 12(4):325–334

Samson JA (1992) *Averrhoa* L. In: Verheij EWM, Coronel RE (eds) Plant resources of South-East Asia No. 2. Edible fruits and nuts. Prosea Foundation, Bogor

Corylus avellana

Scientific Name

Corylus avellana L.

Synonyms

Corylus avellana f. *aurea* (G. Kirchn.) C. K. Schneid., *Corylus avellana* var. *aurea* G. Kirchn., *Corylus avellana* f. *contorta* (Bean) Rehder, *Corylus avellana* var. *contorta* Bean, *Corylus avellana* f. *fuscorubra* Dippel, *Corylus avellana* var. *fusco-rubra* ined., *Corylus avellana* f. *heterophylla* (Lodd. ex Loudon) Rehder, *Corylus avellana* var. *heterophylla* (Lodd. ex Loudon) Loudon, *Corylus avellana* f. *pendula* (H. Jaeger) Dippel, *Corylus avellana* var. *pendula* H. Jaeger, *Corylus imeretica* Kem.-Nath., *Corylus maxima* Mill., *Corylus pontica* K. Koch, *Corylus tubulosa* Willd.

Family

Betulaceae, also in Corylaceae

Common/English Names

Beaded Hazel, Chinese Filbert, Chinese Hazel, Chinese Hazelnut, Cob, Cobnut, Common Filbert, Common Hazelnut, Curri, European Filbert, European Hazel, Giant Filbert, Hazel Filbert, Hazelnut, Himalayan Hazel, Lambert's Filbert, Siberian Hazel, Tibetan Filbert, Tibetan Hazelnut, Turkish Filbert, Turkish Hazel

Vernacular Names

Afrikaans: Haselaar;
Albanian: Lajthia;
Aragonese: Abellanera;
Arabic: Ailawish, Bamdaq, Jellawz;
Azeri: Findiq;
Basque: Hurritz;
Bosnian: Lijeska;
Brazil: Avelã;
Catalan: Avellaner;
Chinese: Ou Zhen, Ou Zhou Zhen;
Croatian: Obična Lijeska;
Czech: Líska Obecná;
Danish: Almindelig Hassel, Hassel, Hasselnød, Hasselbusk;
Dutch: Haozenoeteboom, Hazelaar;
Eastonian: Harilik Sarapuu;
Esperanto: Avelo;
Finnish: Euroopan Pähkinäpensas, Euroopanpähkinäpensas, Pähkinä, Pähkinäpensas, Pähkinäpensas Tavallinen;
French: Avelinier, Coudrier, Coudrier Noisetier, Noisetier, Noisetier Commun, Noisetier Franc, Noisetier Tubuleux, Noisette;
Friulian: Noglâr;
Galician: Abeleira;

German: Gemeine Hasel, Gewöhnliche Hasel, Hasel, Haselnuß, Haselnuss, Haselstrauch, Lambertnuß, Lambertnuss, Lambertshasel, Zellernuß;
Greek: Fountoukia;
Hungarian: Európai Mogyoró, (Közönséges) Mogyoró, Közönséges Mogyoró;
Icelandic: Hesli, Heslividur;
India: Bindak, Findah (Hindu), Bhangi, Bhotia Badam, Urni (Punjabi), Bindaq, Findaq, Jaluz, Urni (Urdu);
Italian: Avenalla, Cliperia, Nissola, Nizola, Nocchia, Nocchio, Lopima, Nocchiola, Nocciolo, Nocciuolo, Nocciuolo Nucidda, Nucella, Ollana, Vellana;
Japanese: Heezeru Nattsu, Komon Heezeru, Seiyou Hashibami;
Lithuanian: Paprastasis Lazdynas;
Luxembourgish: Hieselter;
Norwegian: Hassel, Hahl, Halt, Hasal, Hati, Hatl, Hessel, Håssel, Nøttebusk;
Philippines: Abelyano (Cebuano), Abelyana (Tagalog);
Polish: Leszczyna, Leszczyna Pospolita, Orzech Laskowy;
Portuguese: Avelã, Aveleira, Aveleira-Comum;
Quecha: Iwrupa Awillanu;
Russian: Leshchina Obyknovennaia, Leščina Obyknovennaja, Oreshnik Obyknovennaia;
Sardinian: Rintzola;
Serbian: Leska;
Slovaščina: Leska Navadna, Navadna Leska;
Slovencina: Lieska Obyčajná;
Spanish: Ablano, Avelaneira, Avellanero, Avellano, Avellano De Lambert, Nochizo;
Swedish: Abellinisk Nöt, Filberthassel, Hassel, Hasselnöt, Hazelträd, Turkisk Hassel, Vanlig Hassel;
Turkish: Adi Fındık;
Upper Sorbian: Wšědna Lěšćina;
Walloon: Côrî;
Welsh: Collen (Coeden);
West Flemish: Oazeloare.

Origin/Distribution

Corylus avellana is indigenous to Europe and western Asia. Recently, using chloroplast microsatellites, hazelnut appeared to have been domesticated independently in three areas: the Mediterranean, Turkey, and Iran (Boccacci and Botta 2009).

Agroecology

Hazel nut thrives in the sub-temperate and Mediterranean climatic zones with a maritime influence where the summers are cool and the winters mild at altitudes of 0–700 m. It is frost hardy and can withstand temperatures down to −8°C. It has a total chill requirement of 1,000–1,500 chill hours (based on 0–7°C). Annual rainfall in key centres for hazelnut production in Europe and USA is generally in the range of 800–1200 mm. Areas that have winter–spring rainfall dominance appear to be more suitable than areas with summer–autumn rainfall dominance, as late summer rains can hamper harvest and may have an adverse effect on nut quality, causing moulds to develop. Hazelnut trees are adversely affected by strong and persistent winds, particularly in the spring. Hot dry winds can cause leaf scorch in summer.

Hazelnut is adaptable to a wide range of soils but deep, well-drained soils of medium texture are preferred. Clay loams and loamy clays that are well structured and well drained are the most suitable soils but it will perform well in deep alluvial loams. A neutral to slightly acidic pH around pH 6 is best. Adequate moisture must be available for good growth and yield.

Edible Plant Parts and Uses

Raw kernels whole or pieces and diced are used as snack food, dessert, in muesli and in a wide range of food products. Blanched kernels with skins removed by heating are used as ingredient in many foods, for example gelati requires varieties that blanch well. Roasted kernels are used in confectionery and bakery products. Hazel nut is widely used in breads, cakes, biscuits, sweets, chocolates, ice-cream, etc. Hazel nut meal made from raw or roasted hazelnuts that have been finely chopped or ground is used as food ingredient in bakery products. Hazelnut meal is also used to make a paste or powder called praline, a flavouring product which is used in a wide range of products such as confectionery. One popular product is *praline belge* or *Belgian chocolates*. Hazel nut is also used in combination with chocolate for chocolate truffles and products such as Nutella. In Australia, hazelnut is used in the manufacture of Cadbury eponymous milk chocolate bar which is the third most popular brand in Australia. In Austria, hazelnut paste is an important ingredient in the world famous *torts* (such as Viennese hazelnut tort). Hazelnut oil obtained from cold pressed raw kernels is a strongly flavoured, clear, yellow cooking oil used in salad dressing and baking; it is rich in vitamin E and monounsaturated fatty acids. Hazelnut butter is considered as a pleasant and nutritious spread. In the United States, hazelnut butter is being promoted as a more nutritious and healthier spread than peanut butter. Hazelnut can also be liquidized and used as a plant milk and is processed into vodka-based Hazelnut liqueurs, such as *Frangelico* which is becoming popular in the United State and Eastern Europe. Hazelnut is popular as a coffee flavouring, especially in the form of hazelnut latte. The flour obtained from residual meal after oil extraction is used as gluten free flour substitute.

Plate 1 Hazel nuts

Plate 2 Hazel nuts – whole and shell removed

Botany

A deciduous shrub or small tree reaching 3–8 m high with a trunk diameter of 50 cm with ascending branches and pubescent twigs. The bark is coppery brown and usually smooth, occasionally exfoliating in thin papery strips. Leaves are borne on densely pubescent petioles; lamina broadly ovate to broadly elliptic, 5–12 by 4–12 cm, moderately thin, base narrowly cordate to narrowly rounded, margins coarsely and doubly serrate, apex abruptly acuminate, sparsely to moderately pubescent on both surface (Plates 3–5). Male catkins, pale yellow, lateral along branchlets on relatively long short shoots, usually in clusters of 2–4, 3–8 by 0.7–1 cm. Female catkins are very small and largely concealed in the buds with only

Plate 3 Emerging spring buds

the bright red 1–3 mm long styles visible. Nuts are found in clusters of 2–4. Each nut is enclosed by much enlarged involucre bracts, distinct nearly to base, expanded, shorter than or only slightly longer than the nuts, with apex deeply lobed;

Plate 4 Juvenile spring flush

Plate 5 Young leaves

bract surfaces pubescent. The nut is roughly sub-spherical to subconical, usually 1.5–2.0 cm long and 1.2–2.0 cm broad, yellowish-brown with a pale scar at the base (Plates 1–2).

Nutritive/Medicinal Properties

Food value of raw, unroasted hazelnut (filbert) per 100 g edible portion exclude 54% shell is as follows (USDA 2010): water 5.31 g, energy 628 kcal (2629 kJ), protein 14.95 g, total lipid (fat) 60.75 g, ash 2.29 g, carbohydrate 16.70 g; total sugars 4.34 g, sucrose 4.20 g, glucose 0.07 g, fructose 0.07 g, starch 0.48 g, minerals – calcium 114 mg, iron, 4.70 mg, magnesium 163 mg, phosphorus 290 mg, potassium 680 mg, sodium 0 mg, zinc 2.45 mg, copper 1.725 mg, manganese 6.175 mg, Se 2.4 μg; vitamins – vitamin C (total ascorbic acid) 6.3 mg, thiamine 0.643 mg, riboflavin 0.113 mg, niacin 1.80 mg, pantothenic acid 0.918 mg, vitamin B-6 0.563 mg, folate (total) 113 μg, choline 45.6 mg, betaine 0.4 mg, vitamin A 20 IU, β-carotene 11 μg, alpha-carotene 3 μg, lutein + zeaxanthin 92 μg, vitamin E (α tocopherol) 15.03 mg, β-tocopherol 0.33 mg, vitamin K (phylloquinone) 14.2 μg; lipids – fatty acids (total saturated) 4.464 g, 16:0 (palmitic acid) 3.097 g, 18:0 (stearic acid) 1.265 g, 20:0 (arachidic acid) 0.102 g; fatty acids (total monounsaturated) 45.652 g, 16:1 undifferentiated (palmitoleic acid) 0.116 g, 18:1 undifferentiated (oleic acid) 45.405 g, 20:1 (gadoleic acid) 0.131 g; fatty acids (total polyunsaturated) 7.920 g, 18:2 undifferentiated (linoleic acid) 7.833 g, 18:3 undifferentiated (linolenic acid) 0.087 g; phytosterols 96 mg, stigmasterol 1 mg, campesterol 6 mg, β-sitosterol 89 mg, amino acids – tryptophan 0.193 g, threonine 0.497 g, isoleucine 0.545 g, leucine 1.063 g, lysine 0.420 g, methionine 0.221 g, cystine 0.277 g, phenylalanine 0.663 g, tyrosine 0.362 g, valine 0.701 g, arginine 2.211 g, histidine 0.432 g, alanine 0.730 g, aspartic acid 1.679 g, glutamic acid 3.710 g, glycine 0.724 g, proline 0.561 g and serine 0.725 g.

Hazelnuts are rich in protein, amino acids (17), complex carbohydrates, dietary fibre, vitamin E, zinc, iron, calcium, potassium, manganese, magnesium, selenium, copper and monounsaturated and polyunsaturated fat. Moreover, they contain significant amounts of thiamine, vitamin Es and vitamin B6, as well as smaller amounts of other B vitamins. The fats are mainly monounsaturated oleic acid. Additionally, one cup (237 ml) of hazelnut flour has 20 g of carbohydrates, 12 g of which are fibre. Hazelnut also contains phytosterol (β-cytosterol), stigmasterol, campesterol, β-sitosterol, vitamin K, choline, betaine, β-carotene, α-carotene, lutein + zeaxanthin and antioxidants.

Analyses carried out in Turkisk laboratories (Alphan et al. 1997) reported hazelnut to be one of the best phyto-sources for iron (5.8 mg/100 g), calcium (160.0 mg/100 g), and zinc (2.2 mg/100 g), the most important minerals for growth and development. Hazelnut was also found to be rich in potassium (655 mg/100 g), which is necessary for nerve stimulation and functioning of muscle tissue. The scientists also found hazelnut to be rich sources for vitamins B_1 (0.33 mg/100 g), and B_2 (0.12 mg/100 g), and very excellent sources for vitamin B_6 (0.24 mg/100 g) and vitamin E (31.4 mg/100 g). Vitamins B_2 and B_6 are especially important nutrients for the school-age children. The hazelnut is also the second best source of vitamin E after plant oils. Vitamin E is essential for the normal functioning of muscle tissue and the reproduction system. It protects the organism against cancer. Vitamin E also prevents the haemolysis of erythrocytes, and thus protects the body against anaemia. The average fat content of hazelnuts was found to be high, 62.7% and 82%, and was dominated by oleic acid. This monounsaturated fatty acid was shown by many workers to increase the level of high density lipoprotein (HDL), in blood. HDL, in turns, lowers blood cholesterol and thus protects against atherosclerosis. According to a long term survey conducted in the United States over 20,000 subjects (Loma Linda University, 1990), the risk of death from coronary heart disease was reduced by half in people consuming nuts at least once a day as compared with those who do not.

Hazelnut was also found to serve as an excellent source of vitamin E (24 mg/100 g) and a good source of water soluble (B complex) vitamins and dietary fibre (Alasalvar et al. 2003a). The major minerals in Tombul hazelnut were potassium, phosphorus, calcium, magnesium, and selenium. The major amino acids were glutamic acid, arginine, and aspartic acid. The three non-essential amino acids and the essential amino acids contributed 44.9% and 30.9% to the total amino acids present, respectively, while lysine and tryptophan were the limiting amino acids in Tombul hazelnut. Twenty-one free amino acids, six sugars, and six organic acids were identified and arginine, sucrose, and malic acid predominated.

The average microelement concentrations in hazelnut varieties were found to vary in the following ranges: B 13.63–23.87 mg/kg; Co 0.47–0.82 mg/kg; Cr 0.22–0.52 mg/kg; Cu 16.23–32.23 mg/kg; Fe 31.60–51.60 mg/kg; Li, 0.035–0.042 mg/kg; Ni 0.58–2.58 mg/kg; Se 0.96–1.39 mg/kg; and Zn 22.03–44.03 mg/kg (Simsek and Aykut 2007). Hazelnuts can be an important microelement source for human nutrition and health. According to the trace element data, a daily consumption of 50 g hazelnut can supply easily about 6% for B, 9% for Co, 19% for Fe, 9% for Ni and 16% for Zn of the recommended daily allowance. On the other hand, Se, Cu and Cr levels of 50 g hazelnuts are higher than the respective daily requirements, but slight overdoses of these elements were found to be non-toxic for human health.

The total phytosterol content in 17 Turkish hazelnut cultivars were found to vary from 1180.4 (cv. Uzunmusa-Ordu) to 2239.4 mg/kg (cv. Cavcava), and the mean was 1581.6 mg/kg (Yorulmaz et al. 2009). One of the most significant commercial Turkish cultivars, Tombul, contained quite low total phytosterols (1297.7 mg/kg). Total and individual phytosterol contents of hazelnut cultivars were significantly different, except for phytostanol and campestanol. The main component was β-sitosterol which ranged from 82.8% to 86.7% in all cultivars. This was followed by campesterol, Δ5-avenasterol, sitostanol and stigmasterol.

The total lipid content of Tombul round hazelnut was 61.2%, of which 98.8% were nonpolar and 1.2% polar constituents (Alasalvar et al. 2003b; 2006a, b). Triacylglycerols were the major nonpolar lipid class and contributed nearly 100% to the total amount. Phosphatidylcholine, phosphatidylethanolamine, and phosphatidylinositol were the most abundant polar lipids, respectively. Sixteen fatty acids were identified, among which oleic acid contributed 82.7% to the total, followed by linoleic, palmitic, and stearic acids. Unsaturated fatty acids accounted for 92.2% of the total fatty acids present. Seven tocol isoforms (four tocopherols and three tocotrienols) and eight phytosterols as well as cholesterol were positively identified and quantified; among these oil soluble bioactives, α-tocopherol (40.40 mg/100 g) and β-sitosterol (134.05 mg/100 g) were predominant in hazelnut oil and contributed 78.74% and 81.28% to the total tocols and phytosterols present, respectively. Tocotrienols were detected in small amounts (1.02% to the total tocols). The data showed crude hazelnut oil extracted from Turkish Tombul hazelnut to be a good source of nutrients, bioactive phytochemicals, and health-promoting components.

Trioleylglycerol, linoleyl-dioleylglycerol, and palmitoyl-dioleylglycerol were the most predominant triacylglycerols throughout hazelnut fruit development (Parcerisa et al. 1999). Phosphatidylcholine, phosphatidylethanolamine, and phosphatidylinositol were the most abundant phospholipids. Phospholipid contents exhibited a steep decline during hazelnut development. Triacylglycerols had a high percentage of monounsaturated fatty acid moieties, whereas phosphatidylcholine had the highest percentage of saturated and monounsaturated fatty acid moieties.

In another study, the main constituent of hazelnut was found to be fats ranging from 56% to 61%, being the nutritional value around 650 kcal per 100 g of fruits (Oliveira et al. 2008). Oleic was the major fatty acid varying between 80.67% in cultivar F. Coutard and 82.63% in cultivar Daviana, followed by linoleic, palmitic, and stearic acids. Aqueous hazelnut extract presented antioxidant activity in a concentration-dependent way, in general with similar behaviour for all cultivars.

Eight phenolic compounds (3-caffeoylquinic acid, 5-caffeoylquinic acid, caffeoyltartaric acid, p-coumaroyltartaric acid, myricetin 3-rhamnoside, quercetin 3-glycoside, quercetin 3-rhamnoside and kaempferol 3-rhamnoside) were identified and quantified in hazelnut leaves (Amaral et al. 2005). Myricetin 3-rhamnoside and quercetin 3-rhamnoside were the major compounds and caffeoyltartaric and p-coumaroyltartaric acids were present in trace amounts.

Antioxidant Activity

Defatted raw hazelnut kernel and hazelnut byproducts (skin, hard shell, green leafy cover, and tree leaf) were found to contain varying amounts of phenolic compounds and exhibited varying antioxidant efficacies (Alasalvar et al. 2003b, 2006a,b). Generally, ethanol extracts of hazelnut by-products (skin, hard shell, green leafy cover, and tree leaf) exhibited stronger activities than hazelnut kernel at all concentrations tested. Among samples, extracts of hazelnut skin, in general, showed excellent antioxidative efficacy and higher phenolic content as compared to other extracts. Five phenolic acids (gallic acid, caffeic acid, p-coumaric acid, ferulic acid, and sinapic acid) were identified and quantified (both free and esterified forms). These results suggested that hazelnut byproducts could potentially be considered as a superior and readily available source of natural antioxidants.

Hazelnuts were found to have one of the highest ORAC values among superfoods (foods with high ORAC values) (Wu et al. 2004). ORAC or Oxygen Radical Absorbance Capacity is the widely used test tube measure of the antioxidant ability to quench (destroy) free radicals or oxidants. Free radicals are bad and cause cancers and are generated during normal activities such as eating, breathing and exercising. Within the category of nuts, hazelnuts (9.649/100 g) scored higher than almonds (4.494/100 g) and pistachios (7.963/100 g), and only pecans (17.940/100 g) and walnuts (13.541/100 g) had higher ORAC scores (Wu et al. 2004).

Hazelnut was found to possess one of the highest PAC (proanthocyanidin) content among superfood. It scored the highest (501/100 g) in the category of nuts tested (Gu et al. 2004). Pecan nut had a PAC level of 494/100 g, cranberries 419/100 g, wild blueberries 329/100 g, pistachios 237/100 g, almond 184/100 g, walnut

87g/100 g. Studies showed that the antioxidant capabilities of PACs were 20 times more potent than vitamin C and 50 times more powerful than vitamin E. The antioxidant and other actions of PACs may help to strengthen blood vessels, suppress platelet aggregation in arteries, reduce risk of cardiovascular disease, lower blood pressure and delay onset of dementia.

Shahidi et al. (2007) similarly found that extracts of hazelnut by-products (skin, hard shell, green leafy cover, and tree leaf) exhibited stronger activities than hazelnut kernel at all concentrations tested. Hazelnut extracts examined exhibited different antioxidative efficacies related to the presence of phenolic compounds. Among samples, extracts of hazelnut skin, generally showed excellent antioxidative efficacy and higher phenolic content as compared to other extracts. These results suggested that hazelnut byproducts could potentially be considered as an excellent and readily available source of natural antioxidants. Hazelnut green leafy cover acetone (HGLCa) extract had the highest content of total phenolics (201 mg of catechin equivalents/g of extract), condensed tannins (542 mg of catechin equivalents/g of extract), and Total Antioxidant Activity (1.29 mmol of Trolox equivalents/g of extract) followed by Hazelnut green leafy cover ethanol (HGLCe) extract, hazelnut kernel acetone (HKa) extract, and hazelnut kernel ethanol (HKe) extract (Alasalvar et al. 2003b; 2006b). Five phenolic acids (gallic acid, caffeic acid, p-coumaric acid, ferulic acid, and sinapic acid) were identified and quantified, among which gallic acid was the most abundant in both free and esterified forms. The order of antioxidant activity in a β-carotene-linoleate model system, the scavenging effect on 2,2-diphenyl-1-picrylhydrazyl (DPPH) radical, and the reducing power in all extracts ranked in the following order: HGLCa > HGLCe > HKa > HKe. These results suggested that both 80% ethanol and acetone were capable of extracting phenolics, but 80% acetone was a more effective extracting solvent. Hazelnut green leafy cover exhibited stronger antioxidant and antiradical activities than hazelnut kernel itself in both extracts and could potentially be considered as an inexpensive source of natural antioxidants.

Schmitzer et al. (2011) confirmed the presence of seven flavan-3-ols (catechin, epicatechin, two procyanidin dimers, and three procyanidin trimers), three flavonols (quercetin pentoside, quercetin-3-O-rhamnoside, and myricetin-3-O-rhamnoside), two hydrobenzoic acids (gallic acid, protocatechulic acid), and one dihydrochalcone (phloretin-2′-O-glucoside) in hazelnuts. Flavonols were only found in whole hazelnut kernels. Individual phenolics content, with the exception of gallic acid, was found to be highest in whole unroasted hazelnuts and was significantly lowered after skin removal. Similarly, total phenolic content and antioxidative potential was reduced when skin was removed. Roasting had a significant negative effect on individual phenolics but did not impact on the total phenolic content and antioxidative potential of kernels. The scientists asserted that from a health promoting phytochemical composition of hazelnuts, the consumption of whole unroasted kernels with skins should be preferential to peeled kernels either roasted or unroasted.

Antibacterial Activity

Tannins fractionated from acetonic extract preparations of phenolic compounds of *Corylus avellana* and other plants were found to exhibit antibacterial activities (Amarowicz et al. 2008). Noteworthy was a relatively high level of activity (62.5–125 µg/ml) of the extract against *Listeria monocytogenes*. Hazelnut extracts exhibited high antimicrobial activity against Gram positive bacteria – *Bacillus cereus, B. subtilis* and *Staphylococcus aureus* (MIC 0.1 mg/ml) (Oliveira et al. 2008).

Anticancer Activity

Both shell and leaf extracts of *Corylus avellana* contained taxanes (Ottaggio et al. 2008). Of these, paclitaxel, 10-deacetylbaccatin III, baccatin III, paclitaxel C, and 7-epipaclitaxel were identified and quantified. The content of total taxanes in leaves was higher than in shell. Hazel extracts also exhibited biological activity, inhibiting metaphase to anaphase transition in a human tumour cell line. Paclitaxel is an effective antineoplastic agent originally extracted in low yield from the bark of *Taxus brevifolia*. Large quantities of hazelnut shell are found as discarded material from food industries and provide good sources of taxane compounds of paclitaxel and other antineoplastic compounds.

Platelet Inhibition Activity

Corylus avellana among other plants used in traditional medicine was also reported to possess inhibitory activity on prostaglandin biosynthesis and platelet activating factor (PAF)-induced exocytosis in-vitro in studies conducted in Sweden (Tunón et al. 1995). *Corylus avellana* extract had some beneficial effects in carbon tetrachloride toxicosis: it reduced hepatocytolysis as well as histological lesions and restored the activity of serum transaminases (GPT and GOT), to normal values in young rats (Rusu et al. 1999).

Allergy Problems

In Europe, hazelnuts (*Corylus avellana*) are a frequent cause of food allergies especially among susceptible consumers. Allergy to hazelnut can be severe and occur at young age. Several important hazelnut allergens had been identified and characterized (Lauer et al. 2004). Specific N-glycans were found to induce strong IgE responses of uncertain clinical relevance, and also glycoproteins such as glycosylated vicilin Cor a 1. Recent studies by Verweij et al. (2011) found that young infants with atopic dermatitis sensitized to hazelnut could display IgE reactivity to Cor a 9, a potentially dangerous hazelnut component.

Traditional Medicinal Uses

The bark, leaves, catkins and fruits have been used in traditional medicine. They have astringent, diaphoretic, febrifuge, nutritive and odontalgic properties. The seed is tonic and stomachic. Hazelnut oil has been reported to have a gentle but constant and effective action in cases of infection with threadworm or pinworm in babies and young children.

Other Uses

Hazelnut can also be planted as a hedge plant. Hazel nut seed oil is also used in paints, cosmetics (as an ingredient of nourishing and moisturising body and hand creams, lotion, soaps, face mask, ointment), wood polish etc. The finely ground seeds are used as an ingredient of face masks in cosmetics. The bark and leaves are a source of tannin. The wood is used for inlay work, small items of furniture, hurdles, wattles, basketry, pea sticks, bows and arrows. The wood also provides a good quality charcoal, used by artists.

Comments

Corylus avellana is readily distinguished from the closely related Filbert (*Corylus maxima*) by the short involucre; in the latter the nut is fully enclosed by a beak-like involucre longer than the nut.

The world's leading producer of hazelnut is Turkey, specifically in the Ordu province. Hazelnuts are also grown extensively in Australia, New Zealand and Chile.

Selected References

Alasalvar C, Shahidi F, Liyanapathirana CM, Ohshima T (2003a) Turkish Tombul hazelnut (*Corylus avellana* L.). 1. Compositional characteristics. J Agric Food Chem 51(13):3790–3796

Alasalvar C, Shahidi F, Ohshima T, Wanasundara U, Yurttas HC, Liyanapathirana CM, Rodrigues FB (2003b) Turkish Tombul hazelnut (*Corylus avellana* L.). 2. Lipid characteristics and oxidative stability. J Agric Food Chem 51(13):3797–3805

Alasalvar C, Amaral JS, Shahidi F (2006a) Functional lipid characteristics of Turkish Tombul hazelnut (*Corylus avellana* L.). J Agric Food Chem 54(26): 10177–10183

Alasalvar C, Karamać M, Amarowicz R, Shahidi F (2006b) Antioxidant and antiradical activities in extracts of hazelnut kernel (*Corylus avellana* L.) and hazelnut green leafy cover. J Agric Food Chem 54(13):4826–4832

Alphan E, Pala M, Açkurt F, Yilmaz T (1997) Nutritional composition of hazelnuts and its effects on glucose and lipid metabolism. Acta Hort (ISHS) 445:305–310

Amaral JS, Ferreres F, Andrade PB, Valentão P, Pinheiro C, Santos A, Seabra R (2005) Phenolic profile of hazelnut (*Corylus avellana* L.) leaves cultivars grown in Portugal. Nat Prod Res 19(2):157–163

Amarowicz R, Dykes GA, Pegg RB (2008) Antibacterial activity of tannin constituents from *Phaseolus vulgaris, Fagoypyrum esculentum, Corylus avellana* and *Juglans nigra*. Fitoterapia 79(3):217–219

Baldwin B (1997) A review of Australian hazelnut production. Acta Hort (ISHS) 445:359–368

Baldwin B (1998) Hazelnuts. In: Hyde KW (ed) The new Rural Industries: a handbook for farmers and investors. Rural Industries and Development Corporation, Canberra, pp 428–435

Baldwin B (2010) Hazelnuts: variety assessment for South-eastern Australia. Rural Industries and Development Corporation, Canberra, p 79 pp, Publication No. 09/0178

Boccacci P, Botta R (2009) Investigating the origin of hazelnut (*Corylus avellana* L.) cultivars using chloroplast microsatellites. Genet Resour Crop Evol 56(6):851–859

Chiej R (1984) Encyclopaedia of medicinal plants. MacDonald & Co., London, 447 pp

Chopra RN, Nayar SL, Chopra IC (1986) Glossary of indian medicinal plants (Including the supplement). Council Scientific Industrial Research, New Delhi, 330 pp

Facciola S (1990) *Cornucopia*. A source book of edible plants. Kampong, Vista, 677 pp

Gu L, Kelm M, Hammerstone JF, Beecher G, Holden J, Haytowitz D, Gebhardt S, Prior R (2004) Concentrations of proanthocyanidins in common foods and estimations of normal consumption. J Nutr 134:613–617

Huxley AJ, Griffiths M, Levy M (eds) (1992) The new RHS dictionary of gardening. Macmillan, London, 4 vols

Lauer I, Foetisch K, Kolarich D, Ballmer-Weber BK, Conti A, Altmann F, Vieths S, Scheurer S (2004) Hazelnut (*Corylus avellana*) vicilin Cor a 11: molecular characterization of a glycoprotein and its allergenic activity. Biochem J 383(2):327–334

Oliveira I, Sousa A, Morais JS, Ferreira IC, Bento A, Estevinho L, Pereira JA (2008) Chemical composition, and antioxidant and antimicrobial activities of three hazelnut (*Corylus avellana* L) cultivars. Food Chem Toxicol 46(5):1801–1807

Ottaggio L, Bestoso F, Armirotti A, Balbi A, Damonte G, Mazzei M, Sancandi M, Miele M (2008) Taxanes from shells and leaves of *Corylus avellana*. J Nat Prod 71(1):58–60

Parcerisa J, Codony R, Boatella J, Rafecas M (1999) Triacylglycerol and phospholipid composition of hazelnut (*Corylus avellana* L.) lipid fraction during fruit development. J Agric Food Chem 47:1410–1415

Rusu MA, Bucur N, Puică C, Tămaş M (1999) Effects of *Corylus avellana* in acetaminophen and CCl4 induced toxicosis. Phytother Res 13(2):120–123

Schmitzer V, Slatnar A, Veberic R, Stampar F, Solar A (2011) Roasting affects phenolic composition and antioxidative activity of hazelnuts (*Corylus avellana* L.). J Food Sci 76(1):S14-19

Shahidi F, Alasalvar C, Liyana-Pathirana CM (2007) Antioxidant phytochemicals in hazelnut kernel (*Corylus avellana* L.) and hazelnut byproducts. J Agric Food Chem 55(4):1212–1220

Simsek A, Aykut O (2007) Evaluation of the microelement profile of Turkish hazelnut (*Corylus avellana* L.) varieties for human nutrition and health. Int J Food Sci Nutr 58(8):677–688

Tunón H, Olavsdotter C, Bohlin L (1995) Evaluation of anti-inflammatory activity of some Swedish medicinal plants. Inhibition of prostaglandin biosynthesis and PAF-induced exocytosis. J Ethnopharmacol 48(2): 61–76

Uphof JCTh (1968) Dictionary of economic plants, 2nd edn. Cramer, Lehre, p 591 pp (1st ed. 1959)

U.S. Department of Agriculture, Agricultural Research Service (USDA) (2010) USDA national nutrient database for standard reference, release 23. Nutrient Data Laboratory Home Page. http://www.ars.usda.gov/ba/bhnrc/ndl

Verweij MM, Hagendorens MM, De Knop KJ, Bridts CH, De Clerck LS, Stevens WJ, Ebo DG (2011) Young infants with atopic dermatitis can display sensitization to Cor a 9, an 11S legumin-like seed-storage protein from hazelnut (*Corylus avellana*). Pediatr Allergy Immunol 22(2):196–201

Wu X, Beecher GR, Holden JM, Haytowitz DB, Gebhardt SE, Prior RL (2004) Lipophilic and hydrophilic antioxidant capacities of common foods in the United States. J Agric Food Chem 52:4026–4037

Yorulmaz A, Velioglu YS, Tekin A, Simsek A, Drover JCG, Ates J (2009) Phytosterols in 17 Turkish hazelnut (*Corylus avellana* L.) cultivars. Eur J Lipid Sci Technol 11(4):402–408

Crescentia cujete

Scientific Name

Crescentia cujete L.

Synonyms

Crescentia acuminata Kunth,, *Crescentia angustifolia* Willd. ex Seem., *Crescentia arborea* Raf., *Crescentia cujete* var. *puberula* Bureau & K. Schum., *Crescentia cuneifolia* Gardner, *Crescentia fasciculata* Miers. *Crescentia ovata* Burm.f., *Crescentia plectantha* Miers, *Crescentia spathulata* Miers.

Family

Bignoniaceae

Common/English Names

Calabash, Calabash Tree, Gourd Tree, Hue Tree

Vernacular Names

Argentina: Japacary;
Brazil: Cabaceira, Coité, Cuia, Cuieira, Cuité, Cujeté;
Columbia: Totumo;
Cook Islands: Kumete, 'Ue (Maori);
Costa Rica: Calabacero;
Danish: Kalabastræ;
Dominican Republic: Higüero;
Dutch: Kalebasboom;
Eastonian: Harilik Kõrvitsapuu;
El Salvador: Guacal;
French: Arbre À Calebasses, Calebassier, Coui, Crescentie;
German: Kalebassenbaum;
Guatemala: Morro;
Hawaiian: La'Amia;
Honduras: Jícaro;
Indonesia: Berenuk (Sundanese);
Italian: Calebassa Guiana, Icara;
Malaysia: Tabu Kayu:
Mexico: Cirian Tecomate, Cujete;
Panama: Jícaro, Totumo;
Philippines: Cujete (Spanish),Cujete (Tagalog);
Papiamento: Kalbas Di Mondi;
Portuguese: Cabaco, Coite, Cuia, Cuieira, Cuite;
Puerto Rico: Higüero;
Samoan: Fagu;
Spanish: Arbol De Las Calabazas, Botote, Botote Pilche, Calabacero, Calabasa, Crescencia, Guacal, Güira, Guiro, Higüero, Jicara, Jícaro, Mate, Morro;
Suriname: Krabasi (Sranan Tongo);
Tahitian: Hue Tumu Ra'Au;
Venezuela: Camasa, Taparo;
Vietnamese: Dào Tiên.

Origin/Distribution

The species is probably native from Mexico, Central America and northern South America, however widely introduced and naturalized in tropical Americas and other continents.

Agroecology

In its native range, the tree is commonly found in thickets and forest margins, on roadsides and in pastures from sea level to 800 m elevation, in areas with a mean annual precipitation between 1,500 and 1,300 mm and mean annual temperature of 26°C. The tree grows in clayey soils with poor drainage subject to frequent floods.

Edible Plant Parts and Uses

Young fruits are pickled or candied, seeds are edible and cooked. In Nicaragua, fresh seeds are ground and mixed with water to make a refreshing, sweet and pleasant drink.

Botany

A small, evergreen perennial tree reaching height of 6–10 m high with pale tan bark, lightly fissured, flaking off in long irregular patches and an open, broad, irregular crown made up of long, arching branches with close-set clusters of leaves (Plates 1–2). Leaves are alternate, often fascicled at the nodes, obovate to oblanceolate, simple, 5–17 cm long, chartaceous, glossy at the upper surface, obtuse or acute at the tip and narrowed at the base (Plate 3). Inflorescence one or two cauliflorous flowers borne on larger branches or trunk, the pedicels lepidote, 1.5 cm long. Flowers with a musty odour; calyx 2 cm long bilabiately split to the base; corolla 4–6 cm long, off-white to yellowish with purplish venation on the lobes and purplish lines on the tube outside, broadly campanulate, the lobes triangular with the apex extended as a narrow point; stamens subexserted;

Plate 1 Irregular, open canopy laden with fruit on branches

Plate 2 Shallowly furrowed corky trunk

Plate 3 Leaves forming in fascicles on the branches

Plate 4 Calabash fruit arising from the main thick branches

Plate 5 Close-up of immature calabash fruit showing the tough, woody and pellucid dotted rind

the ovary rounded conical. Fruit a pepo or calabash, spherical to ovoid-ellipsoid, 15–25 cm in diameter, indehiscent, unilocular, the thin hard shell smooth, lepidote-punctate, green (Plates 3–5) or purplish; seeds small, thin, wingless, 7–8 mm long and 4–6 mm wide, scattered through the pulp of the fruit.

Nutritive/Medicinal Properties

The pulp of *C. cujete* fruit was reported to have the following nutrient composition in both wet and dry samples (Ogbuagu 2008): ash 3.74% (dry), 4.38% (wet); crude protein 7.67% (wet) and 10.01% (dry); crude fibre 4.88% (dry); carbohydrate; 15.65% (wet); 68.13% (dry); dry matter; 31.32% (wet), 87.48% (dry); Na 3.20% (wet) and 0.32% (dry); Ca 0.12% (wet), 0.06% (dry), thiamin 1.50 μg/g (wet), 0.93 μg/g (dry). Both wet and dry samples were free HCN toxicity.

The fruits of *Crescentia cujete* afforded 15 new compounds, three iridoid glucosides (6,8,9), five iridoids (7,11–14), two 3-hydroxyoctanol glycosides (16,17), three 2,4-pentanediol glycosides (18–20) and two 4-hydoxy-2-pentanone glycosides (21,22), along with known compounds, ajugol (1), 6-O-p-hydroxybenzoylajugol (2), aucubin (3), 6-O-p-hydroxybenzoyl-6-epi-aucubin (4), 1-dehydroxy-3, 4-dihydroaucubigenin (5),(10), acanthoside D (15), benzoic acid glucosyl ester (23) and 5-hydroxymethylfurfural (24) (Kaneko et al. 1996). Sixteen iridoids and iridoid glucosides were determined in fruits of *Crescentia cujete* (Kaneko et al. 1997). Eight new compounds, named crescentins I–V and crescentosides A, B and C, and another eight known compounds were identified as ajugol, 6-O-p-hydroxybenzoylajugol, aucubin, 6-O-p-hydroxybenzoyl-6-epiaucubin, agnuside, ningpogenin, 5,7-bisdeoxycynanchoside and a degradation product of glutinoside. The fruits of *Crescentia cujete* afforded eight new compounds, along with four known compounds, acanthoside D, β-D-glucopyransoyl benzoate, (*R*)-1-*O*-β-D-glucopyranosyl-1, 3-octanediol and β-D-fructofuranosyl 6-*O*-(ρ-hydroxybenozoyl)-α-D-glucopyranoside (Kaneko et al. 1998). The structures of the new glycosides were established as three glycosides of (2R,4S)-2,4-pentanediol, two glycosides of (R)-4-hydroxy-2-pentanone, two glycosides of (*R*)-1,3-octanediol and 6-*O*-(ρ-hydroxybenzoyl)-D-glucose. Four new 11-nor-iridoids, 6-O-p-hydroxybenzoyl-10-deoxyeucommiol (1), 6-O-benzoyl-10-deoxyeucommiol (2), 6-O-benzoyl-dihydrocatalpolgenin (a mixture of 3 and 4), as well as two known iridoids, ningpogenin (5) and 6-O-p-hydroxybenzoylaucubin (6), were isolated from the fruits of *Crescentia cujete* (Wang et al. 2010).

Various plant parts of *C. cujete* have been reported to possess pharmacological properties based on laboratory and pre-clinical studies.

Anticancer Activity

Bioassay-directed fractionation of the methyl ethyl ketone extract of *Crescentia cujete* yielded the naphthoquinone 1 (3-hydroxymethylfuro [3,2-b]naphtho[2,3-d]furan-5,10-dione) and naphthoquinone 2 (9-hydroxy-3-hydroxymethylfuro [3,2-b]naphtho[2,3-d]furan-5,10-dione)(Heltzel et al. 1993). Both compounds were cytotoxic, and naphthoquinone 1 showed selective DNA-damaging activity against yeast. Hetzel et al. (1993) further isolated (2S,3S)-3-hydroxy-5, 6-dimethoxydehydroiso-α-lapachone [1], (2R)-5, 6-dimethoxydehydroiso-α-lapachone [2], (2R)-5-methoxydehydroiso-α-lapachone [3], 2-(1-hydroxyethyl) naphtho[2,3-b]furan-4,9-dione [4], 5-hydroxy-2-(1-hydroxyethyl)naphtho[2,3-b] furan-4,9-dione [5], 2-isopropenylnaphtho[2,3-b]furan-4,9-dione [6], and 5-hydroxydehydroiso-α-lapachone [7] (Hetzel et al. 1993). Compounds 1–3 were found to be bioactive, showing selective activity towards DNA-repair-deficient yeast mutants. DNA-repair-deficient yeast mutants is used as a tool to search for DNA-damaging compounds with potential anticancer activity.

Antiinflammatory Activity

A hydroalcoholic extract (80%) of *Crescentia cujete* leaves at an oral dosage of 1,200 mg/kg was found to inhibit paw inflammation in rat induced by formaldehyde injection. The extract was active in a dose-dependent manner, equivalent or higher than sodium diclofenac at 100 mg/kg (Gupta and Esposito Avella 1988).

Antimicrobial Activity

The ethanol extract of leaf and stem of *C. cujete* exhibited in-vitro antibacterial activity against *Bacillus subtilis*, *Pseudomonas aeruginosa*, *Escherichia coli* and *Staphylococcus aureus* (Contreras and Zolla 1982). A hydroalcoholic extract (95%) of *C. cujete* leaf (5 mg/mL) exhibited in-vitro antibacterial activity against *Bacillus subtilis* and *Staphylococcus aureus* (Verpoorte and Dihal 1987). The hydroalcoholic extract of *C. Cujete* leaf inhibited in-vitro the growth of the enteropathogen, *Salmonella typhi* (Cáceres et al. 1990). Fruit pulp of *C. cujete* exhibited in vitro inhibition against strains of *Streptococcus pneumoniae* (Cáceres et al. 1993). The methanolic extracts of the leaves and stem bark *Crescentia cujete* exhibited antimicrobial activity using a range of Gram-positive and Gram-negative bacteria and fungi (Binutu and Lajubutu 1994). Crude extract of *C. cujete* at a concentrations of 5 mg/mL exhibited potent in-vitro antibacterial activity against *Staphylococcus aureus*, *Enterococcus faecalis*, *Streptococcus pneumoniae*, *Streptococcus pyogenes*, *Escherichia coli* and *Candida albicans* (Rojas et al. 2001). The methanolic extract of *C. cujete* leaves exhibited antibacterial activity against Gram negative bacteria - *Staphylococcus aureus*, *Alcaligenes faecalis*, *Pseudomonas fluorescens*, Gram positive bacteria – *Arthrobacter globiformis*, *Enterobacter aerogenes*, *Bacillus cereus*, *Bacillus coagulans*, *Micrococcus roseus*, Gram variable bacteria – *Mycobacterium phlei*, *Mycobacterium rodochrus*, and *Mycobacterium smegmatis* (Meléndez and Capriles 2006).

Antivenom Activity

The unripe fruits of *Crescentia cujete* independently administered by oral, i.p. or i.v. route either before or after an intra-dermal venom injection (10 μg), exhibited moderate neutralizing ability against the hemorrhagic effect of *Bothrops atrox* venom (Otero et al. 2000).

Traditional Medicinal Uses

The fruit of the calabash tree is used by herbalists in Bombay as a pectoral in the form of a poultice of the pulp applied to the chest (Dymock 1883) as cited Bureau of Plant Industry, Philippines (2005). In the West Indies, syrup made from the pulp is much used in dysentery and as a pectoral. An alcoholic extract of near ripe fruit, in doses of 0.10 g, acts as a mild aperient, and that 0.5 g prove strongly drastic. The fresh pulp boiled with water until it forms a black paste, to which vinegar is added has been applied for erysipelas. The bark is employed for diarrhea with mucous (Bocquillon-Limousin (1910) as cited by BPI, (2005)). The pulp of the fruit, in an alcoholic extract, is reported laxative and expectorant. In Western Africa and the Antilles the pulp of the fruit, macerated in water, is considered depurative, cooling, and febrifuge; it is applied to the head in headaches caused by isolation and to burns (Corre and Lejeanne (1887) as cited by BPI, (2005)). De Grosourdy (1864) as cited by BPI, (2005) states that in the Antilles the fresh leaves and tops are ground and use as topicals for wounds as a cicatrizant and in Venezuela a decoction of the bark is given for diarrhoea. In Western Africa, the pulp roasted in ashes is mildly purgative and diuretic and is used for dropsies in the Antilles. Watt and Breyer-Brandwijk (1962) state that the leaves are diuretic in action when taken internally. The flesh is diuretic and a preparation of the seed is administered in Transvaal internally for snake bites and applied to the wound (Watt and Breyer-Brandwijk 1962). In Sumatra a decoction of the bark is used to clean wounds, and the pounded leaves used as a poultice for headache (Heyne 1950).

The fruit of *Crescentia cujete* is used as a folk medicine (local name, Dao Tien) in Vietnam as for expectorant, antitussive, laxative and stomachic (Kaneko et al. 1996). In a non-experimental validation for antidiabetic activity, a leaf decoction of *C. cujete* used as a drink has been reported to yield cyanhidric acid and is believed to stimulate insulin release in Côte-d'Ivoire (N'guessan et al. 2009a). In Côte-d'Ivoire , the leaf decoction is also drunk for hypertension because of its diuretic effect (N'guessan et al. 2009b). The fruit of *Crescentia cujete* as a main ingredient in herbal mixtures is highly regarded by Haitians and is considered as a traditional panacea for several ailments (Volpato et al. 2009). Five formulas have been reported as *miel de güira* (*siwò kalbaz* in Creole), whose main ingredient is the fruit of *Crescentia cujete*. The juice from the fruit pulp mixed with sugar and/or honey and sometimes a small amount of rum, and drunk or eaten for problems of the respiratory system (asthma, catarrh, bronchitis), of the digestive system (stomach pains, intestinal parasites), and of the female reproductive apparatus (infertility).

In traditional Dominican medicine, the fruit of *higüero* is reputed to be a particularly cooling plant (*bien fresco*) that is used for treating all types of infections in the body but is particularly used in preparations for women's health (Yukes and Balick 2010). For gynecological conditions, specifically for "cleansing the reproductive system". In the Dominican Republic, the fresh leaf is heated slightly and the freshly squeezed juice of the leaf is applied to the ear to treat ear infections (Germosén-Robineau 2005). Good results have been seen in the treatment of otalgia (earaches).

Other Uses

Sapwood is pinkish to reddish brown and the heartwood light brown, moderately hard, heavy, strong, flexible and elastic. The wood is used for construction in rural areas, for cattle yokes, agricultural tool handles, wooden wheels, ribs in boat and for firewood. Bent wood is used for baskets and hampers. The bark is stripped for fibre. The ripe fruits, once dry and clean inside, are used for cups, bowls, hand bags, water containers, containers for salt and tortillas, fruit containers, handicrafts richly decorated with paintings or carvings. The dried fruits are also made into musical instruments like maracas or musical rattle. In Brazil, the fibrous lining of the fruit is sometimes used as a substitute for cigarette paper. The tree is also planted as hedge or live fences and as ornamentals.

Comments

The flowers are pollinated by bats at night.

Selected References

Backer CA, van den Brink RCB Jr (1965) Flora of Java (spermatophytes only), 2. Wolters-Noordhoff, Groningen, p 641 pp

Binutu OA, Lajubutu BA (1994) Antimicrobial potentials of some plant species of the Bignoniaceae family. Afr J Med Med Sci 23(3):269–73

Bocquillon-Limousin H (1910) Formulaire des Médicaments Nouveaux, 11e Edition

Bureau of Plant Industry (BPI) (2005) Medicinal plants of the Philippines. Department of Agriculture Republic of Philippines. http://www.bpi.da.gov.ph/Publications/mp/mplants.html

Burkill IH (1966) A dictionary of the economic products of the Malay Peninsula. Revised reprint, 2 volumes, vol 1 (A–H) pp 1–1240, vol 2 (I–Z) pp 1241–2444. Ministry of Agriculture and Co-operatives, Kuala Lumpur

Cáceres A, Cano O, Samayoa B, Aguilar L (1990) Plants used in Guatemala for the treatment of gastrointestinal disorders. 1. Screening of 84 plants against enterobacteria. J Ethnopharmacol 30(1):55–73

Cáceres A, Figueroa L, Taracena AM, Samayoa B (1993) Plants used in Guatemala for the treatment of respiratory diseases. 2: Evaluation of activity of 16 plants against gram-positive bacteria. J Ethnopharmacol 39(1):77–82

Chopra RN, Nayar SL, Chopra IC (1986) Glossary of Indian medicinal plants (including the supplement). Council Scientific Industrial Research, New Delhi, 330 pp

Contreras A, Zolla C (1982) Plantas Tóxicas de México. Instituto Mexicano del Seguro Social, México DF, 271 pp

Corre A, Lejeanne M (1887) Résumé de la Matière Médicale et toxicologique colonial. Doin, Paris, 184 pp

de Grosourdy DR (1864) El Medico Botanico Criollo, vol I. Libreria de Francisco Bracet, Paris

Dymock W (1883) The vegetable materia medica of Western India. Education Society Press, Bombay

Gentry AH (1977) Bignoniaceae [178]. Flora of Ecuador 7, 173 pp

Gentry AH (1980) Bignoniaceae. Part 1. Flora Neotropica, monograph 25 (1). New York Botanical Garden, New York, 130 pp

Germosén-Robineau L (ed) (2005) Farmacopea Vegetal Caribeña, Segunda Edición. Santo Domingo, República Dominicana: enda-caribe,487 pp

Gupta M, Esposito Avella M (1988) Evaluación química y farmacológica de algunas plantas medicinales de TRAMIL. TRAMIL 3, La Havane, Cuba, enda-caribe/MINSAP

Heltzel CE, Gunatilaka AAL, Glass TE, Kingston DGI (1993) Furofuranonaphthoquinones: bioactive compounds with a novel fused ring system from *Crescentia cujete*. Tetrahedron 49(31):6757–6762

Hetzel CE, Gunatilaka AA, Glass TE, Kingston DG, Hoffmann G, Johnson RK (1993) Bioactive furanonaphthoquinones from *Crescentia cujete*. J Nat Prod 56(9):1500–5

Heyne K (1950) De Nuttige Planten van Indonesië [The useful plants of Indonesia], 3rd Edn, 2 volumes. W. van Hoeve, 's-Gravenhage/Bandung, 1660 + CCXLI pp

Kaneko T, Ohtani K, Kasai R, Yamasaki K, Nguyen TN (1996) Iridoids and other glycosides from Vietnamese *Crescentia cujete*. Symp Chem Nat Prod 38:331–336

Kaneko T, Ohtani K, Kasai R, Yamasaki K, Nguyen MD (1997) Iridoids and iridoid glucosides from fruits of *Crescentia cujete*. Phytochem 46(5):907–910

Kaneko T, Ohtani K, Kasai R, Yamasaki K, Nguyen MD (1998) n-Alkyl glycosides and p-hydroxybenzoyloxy glucose from fruits of *Crescentia cujete*. Phytochem 47(2):259–263

Kunkel G (1984) Plants for human consumption: an annotated checklist of the edible phanerogams and ferns. Koeltz Scientific Books, Koenigstein

Meléndez PA, Capriles VA (2006) Antibacterial properties of tropical plants from Puerto Rico. Phytomed 13(4):272–6

Morton JF (1968) The calabash (*Crescentia cujete*) in folk medicine. Econ Bot 22:273–280

N'guessan K, Kouassi KE, Kouadio K (2009a) Ethnobotanical study of plants used to treat diabetes in traditional medicine, by Abbey and Krobou populations of Agboville (Côte-d'Ivoire). Am J Sci Res 4:45–58

N'guessan K, Tiebre M-S, Aké-Assi E, Zirihi GN (2009b) Ethnobotanical study of plants used to treat arterial hypertension, in traditional medicine, by Abbey and Krobou populations of Agboville (Côte-d'Ivoire). Eur J Sci Res 35(1):85–98

Ogbuagu MN (2008) The nutritive and anti-nutritive compositions of calabash (*Crescentia cujete*) fruit pulp. J Anim Vet Adv 7:1069–1072

Otero R, Núñez V, Barona J, Fonnegra R, Jiménez SL, Osorio RG, Saldarriaga M, Díaz A (2000) Snakebites and ethnobotany in the northwest region of Colombia. Part III: Neutralization of the haemorrhagic effect of Bothrops atrox venom. J Ethnopharmacol 73(1–2):233–241

Rojas G, Levaro J, Tortoriello J, Navarro V (2001) Antimicrobial evaluation of certain plants used in Mexican traditional medicine for the treatment of respiratory diseases. J Ethnopharmacol 74(1):97–101

Verpoorte R, Dihal PP (1987) Medicinal plants of Surinam. IV. Antimicrobial activity of some medicinal plants. J Ethnopharmacol 21(3):315–318

Volpato G, Godinez D, Beyra A, Barreto A (2009) Uses of medicinal plants by Haitian immigrants and their descendants in the Province of Camagüey. Cuba J Ethnobiol Ethnomed 5:16

Wang G, Yin W, Zhou ZY, Hsieh KL, Liu JK (2010) New iridoids from the fruits of *Crescentia cujete*. J Asian Nat Prod Res 12(9):770–5

Watt JM, Breyer-Brandwijk MG (1962) The medicinal and poisonous plants of Southern and Eastern Africa, 2nd edn. E. and S. Livingstone, Edinburgh/London, 1457 pp

Yukes JE, Balick MJ (2010) Dominican medicinal plants: a guide for health care providers, 2nd Edition. The New York Botanical Garden. Web text version http://www.nybg.org/files/scientists/mbalick/Dominican%20Medicinal%20Plants_2nd%20Edition%20Manuscript%202010.pdf

Kigelia africana

Scientific Name

Kigelia Africana (Lam.) Benth.

Synonyms

Bignonia africana Lam., *Crescentia pinnata* Jacq., *Kigelia abyssinica* A.Rich., *Kigelia acutifolia* Engl. ex Sprague, *Kigelia aethiopica* Decne, *Kigelia aethopica* Decne. var. *abyssinica* (A. Rich.) Sprague, *Kigelia aethopica* Decne. var. *bornuensis* Sprague, *Kigelia aethiopica* Decne. var. brachycarpa Chiov., *Kigelia aethiopica* Decne. var. stenocarpa Chiov., *Kigelia aethopica* Decne. var. *usambarica* Sprague, *Kigelia aethiopium* (Fenzl) Dandy, *Kigelia africana* (Lam.) Benth. var. *aethiopica* (Decne.) Aubrév. ex Sillans, *Kigelia africana* (Lam.) Benth. var. *elliptica* (Sprague) Sillans, *Kigelia angolensis* Welw. ex Sprague, *Kigelia elliottii* Sprague, *Kigelia elliptica* Sprague, *Kigelia ikbaliae* De Wild., *Kigelia impressa* Sprague, Kigelia lanceolata Sprague, *Kigelia moosa* Sprague, *Kigelia pinnata* (Jacq.) DC., Kigelia pinnata (Jacq.) DC. var. tomentella Sprague, *Kigelia somalensis* Mattei, *Kigelia spragueana* Wernham, *Kigelia talbotii* Hutch. & Dalz., *Kigelia talbotii* Hutch. & Dalz., *Kigelia tristis* A.Chev. Sotor aethiopium Fenzl, *Tanaecium pinnatum* Willd., *Tecoma africana* (Lam.) G.Don, *Tripinnaria africana* Spreng.

Family

Bignoniaceae

Common/English Names

African Sausage Tree, Cucumber Tree, Liver Sausage Tree, Sausage Tree

Vernacular Names

Angola: Mpolata, Mufungufungu, Muzungula (Lozi). Ifungufungu, Mufunofuno (Lunda);
Arabic: Abu Shutor, Abu Sidra, Um Mashatur, Um Shutur;
Botswana: Mpolata, Mufungufungu, Muzungula (Lozi), Muveve, Muzungula (Tongan);
Czech: Kigélie Africká;
Dahomey: Gwam Blipo, Niapopo, Nio (Gbe-Fon);
Danish: Pølsetræ;
Democratic Republic of Congo: Ifungufungu, Mufunofuno (Lunda);
Eastonian: Vorstipuu;
Eritea: Mederba (Tigrina);
French: Arbre-À-Saucisses, Faux Baobab, Saucissonnier;
Fula: Jilahi;
German: Leberwurstbaum;
Ghana: Lele (Adangme), Nana Beretee, O-Nufuten (Akan-Akuapem), Anyafotene (Anyi-Aowin),

Nana Beretee, Nufutene (Asante), Blimmo (Baule), E-Nufutsen, Nufutsen (Fante), Fufɨ́-Akplele, TʃOtotʃO (Ga), Nyakpẽ, Nyãkpekpe (Gbe-Vhe), Anyafotenẽ (Nzema), Anyafσotene (Sehwi), Nufoten, Nufuten, Nufutsen (Twi), Nufuten (Wasa);
Guinea: Dindon, Limbi, Limbi Lamban, Tuda (Manding-Maninka), Tusa (Susu);
India: Marachurai;
Ivory Coast: Tombo (Abe), Assongui (Akan-Brong), Mia Lébé (Akye), Brimau (Anyi), Blima, Bliomo, Brimbo (Baule), Blumo (Kru-Guere), Kuruko (Kulango), Non (Kweni), Findia(N), Findiam, Siai, Sidia, Sidia Findia (Manding-Dyula), Lamban, Lemba, Limbi, Tuda (Maninka);
Mali: Sidiamba (Manding-Bambara), Dindon, Tuda (Maninka), Kombolgna (Songhai);
Mozambique: Kigeli-Keia (Bantu);
Namibia: Mpolata, Mufungufungu, Muzungula (Lozi);
Niger: Kuk, Mechtur (Arabic), Rahunia (Hausa), Bùlóngó (Kanuri);
Nigeria: Um Shutur (Arabic) Ràhéyná (Berom), Úgbòn-Gbón, Ùsuọ́nọ́n (Edo), Ebe-Njok (Ejagham-Etung), Jirlaare, Jirlahi (Fula-Fulfulde), Hantsar Giíwaá, Kiciiciyaa, Noónòn Giíwaá, Ràháínáá, Rawuya (Hausa), Alambọrọgoda, Uturubein, Uturukpa (Igbo), Itemi, Iteni (Igbo, Amankalu), Amọ-Ibi, Uturu-Bein (Igbo, Arochukwu), Izhi (Igbo, Nsokpo), Izhi (Igbo, Nsukka), Alambọrọgọda, Uturu-Bein (Igbo, Onitsha), Óké ogírìsị (Igbo, Owerri), Umu-Aji (Igbo, Umuakpo), Ògírízì (Ijo-Izon, Kolokuma), Bùlóngó (Kanuri), Ubung (Nkem), Béci (Nupe), Itiwa-Enyi (Olulumo-Okuni), Tyembegh (Tiv), Orora, Pandòrò, Uyan (Yoruba);
Russian: Kolbasnoe Derevo;
Senegal: Sidiamba (Manding-Bambara), Dindon, Lambâ, Limbi, Tuda (Maninka), Sayo (Serer), Humbul (Serer-Non), D'abal, Diambal, Dabolé, Dombalé (Wolof);
Sierra Leone: An-Gbonthi, An-Tua (Temne);
South Africa: Worsboom (Afrikaans), Chizutu, Mvula (Nyanja), Umvunguta, Umfongothi (Zulu); Modukguhlu (North Sotho); Muvevha (Venda);
Spanish: Árbol De Las Salchichas;
Sudan: Kombolgna;
Swahili: Mvungavunga, Mvungunya, Mvungwa, Mwegea, Mwicha;
Swedish: Korvträd;
The Gambia: Jirlahi (Fula-Pulaar), Limbi (Manding-Mandinka), Jambal (Wolof);
Togo: Nyakpokpo (Gbe-Gen), Abilu (Tem), Niagpé, Nyakpekpẽ (Vhe);
Uganda: Mussa (Luganda),
Upper Volta: Findia(N), Findiam, Siai, Sidia Findia (Manding-Dyula);
West Cameroons: Èsembea, Èwùŋgè (Duala), Sosong (Koosi), Bulue, Wulule (Kpe);
Zambia: Mufungufungu (Bemba), Mpolata, Mufungufungu, Muzungula (Lozi), Ifungufungu, Mufunofuno (Lunda), Muveve, Muzungula (Tongan);
Zimbabwe: Mubveve, Musonya, Muvhumati, Umvebe.

Origin/Distribution

This species is indigenous to Africa and occurs in:

East Tropical Africa: Kenya, Tanzania. Uganda;

Northeast Tropical Africa: Chad, Eritrea, Ethiopia, Sudan;

Southern Africa: Botswana, Namibia, South Africa – Natal, Transvaal, Swaziland;

West Tropical Africa: Benin, Cote D'Ivoire, Gambia, Ghana, Guinea, Liberia, Mali, Nigeria, Senegal, Sierra Leone, Togo;

West-Central Tropical Africa: Burundi, Cameroon, Equatorial Guinea – Bioko, Rwanda, Zaire.

Agroecology

Kigelia *africana* is a tropical species. It is found on riverbanks, along streams, on floodplains, in open woodland, wet savannah and coastal areas from sea level to 1,200 m altitude. The flowers are nocturnal, foul smelling and bat pollinated.

Edible Plant Parts and Uses

Fresh fruit cannot be eaten as it is said to be a strong purgative and they cause blisters in the mouth and skin. Fruit are prepared for consumption by drying, roasting or fermentation. Roasted fruits are used to flavour beer and aid fermentation. Roasted seeds are used as famine food.

Botany

A wide-spreading, deciduous tree growing to 10–20 m high, with a short, squat light brown trunk and bark that becomes flaky with age. The leaves are alternate and odd pinnate with 5–9 leaflets (Plates 1–2). The leaflets are opposite, glossy, ovate to elliptic-ovate, 8–16 cm long by 3–6 cm wide, coriaceous, scabrid, entire, with acute or retuse or mucronate tip, often oblique base, prominent midrib and veins on under-surface, and strigose. The flowers are maroon red, large, up to 9 cm wide, velvety, nocturnal, with fetid odour and borne in 4–12 flowered panicles on very long, pendulous pedicels (Plates 2–3). Calyx is 2.5–3 cm long, campanulate, usually 5-toothed, or lobed; corolla tube is rather slender, broadly bell-shaped, somewhat curved, and 5-lobed. The fruit is hard, greyish-brown, scurfy, huge, oblong or oblong-cylindric, 20–100 cm in length, 8–18 cm wide, indehiscent, weighing 5–10 kg and pendant on very long peduncles (Plates 1, 2 and 4). Pulp is firm and fibrous containing numerous small seeds.

Plate 1 Sausage-like fruit and odd pinnate leaves

Plate 3 Young flower panicle with 4–12 flowers

Plate 2 Fruit, inflorescence and leaves

Plate 4 Long, sausage-like fruit son long pendulous pedicels

Nutritive/Medicinal Properties

Food value of raw *Kigelia africana* fruit per 100 g edible portion (Leung et al. 1968) was reported as: energy 54 Cal, moisture 85.4%, protein 0.8 g, fat 0.9 g, total carbohydrate 12.2 g, fibre 4.3 g, ash 0.7 g.

The leaf oil of *Kigelia africana* was found to contain 25 volatile components, while the flower oil contained nine (Asekun et al. 2007). Both oils were rich in non-terpenoids; hexadecanoic acid (21.91%, leaf oil; 57.00%, flower oil) was the most abundant in both oils. The other major components were ethyl linoleate (21.73%) and α-pinene (12.28%) in leaf oil and terpenolene (8.26%), myristic acid (7.95%) and linalool (6.71%) in the flower oil.

Various parts of *Kigelia pinnata* were found to contain many classes of bioactive compounds especially iridoids (e.g. catapol), iridoid derivatives (norviburtinal); napthaquinones (including kigelinone) and monoterpenoid-napthaquinones (pinnatal). Other compounds included isocourmarins (including kigelin), furanones, phenylpropanoids, lignans (kigeliol), sterols (including β-sitosterol and stigmasterol) and flavonoids (including quercetin, luteolin) (Houghton 2002).

Fruits of *Kigelia pinnata* yielded a new phenylpropanoid derivative identified as 6-p-coumaroyl-sucrose together with ten known phenylpropanoid and phenylethanoid derivatives and a flavonoid glycoside (Gouda et al. 2006). From the fruits of *Kigelia pinnata*, a furanone derivative 3-(2'-hydroxyethyl)-5-(2"-hydroxypropyl)-dihydrofuran-2(3 H)-one, and four new iridoids named: 7-hydroxy viteoid II, 7-hydroxy eucommic acid, 7-hydroxy-10-deoxyeucommiol and 10-deoxyeucommiol were also isolated together with seven known iridoids, jiofuran, jioglutolide, 1-dehydroxy-3,4-dihydroaucubigenin, des-p-hydroxybenzoyl kisasagenol B, ajugol, verminoside and 6-trans-caffeoyl ajugol (Gouda et al. 2003). The flavonol quercetin and four flavonones luteolin, its 6-OH analogue and corresponding 7-O-glucosides were isolated from the leaves and fruits (El-Sayyad 1982). Sitosterol was isolated from *K. pinnata* fruit (Khan 1998). β-sitosterol-3-0-α-L-rhamnopyranoside was isolated from flowers of *Kigelia pinnata* (Sushil and Dhindsa 2007).

The major iridoids found in the root-bark and stem-bark of *Kigelia pinnata* were the catalpol derivatives esterified with phenylpropanoic acid derivatives at C-6 and these were indentified as specioside 2, verminoside 3 and minecoside 4 (Houghton and Akunyili 1993). Minor constituents which consisted of two known naphthaquinones and three new aromatic monoterpenes were isolated from the root bark (Akunyili and Houghton 1993). The three new compounds were named isopinnatal, kigelinol and isokigelinol. From the root bark of *Kigelia pinnata*, two aldehydes, norviburtinal (6-formylcyclo-penta[c]pyran) and pinnatal, 11-formyl-2, 3, 3a, 5, 10, 10a, 11, 11a-octahydro-8-hydroxy-3, 11-dimethyl-3, 10a-epoxy-1 H -cyclopent[b]anthracene-5, 10-dione, were isolated (Joshi et al. 1982). Kigelin (8-hydroxy-6, 7-dimethoxy-3-methyl-3,4-dihydroisocoumarin) was the major constituent of the root heartwood (Govindachari et al. 1971). They also isolated 6-methoxymellein and stigmasterol and lapachol from the roots. Kigelinone, 5- or 8-hydroxy-2-(1-hydroxyethyl) naphtho [2,3-b] furan-4, 9-dione, a phytochemical analog of naphtho [2,3-b]furan-4,9-dione (furanonaphtho-quinone FNQ]) compounds, was isolated from the inner bark of *Kigelia pinnata*, which was known to have antitumor activity (Inoue et al. 1981). Two pairs of monoterpenoid-naphthoquinone compounds: pinnatal and isopinnatal and kigelinol and isokigelinol, were also isolated from the roots and fruit (Akunyili and Houghton, 1993). O-glucoside were isolated from the leaves and fruits, three isocoumarins 6-methoxymellein, kigelin, 6-deme-thylkigelin from the roots, lignan kigeliol from wood and neoligan balanophonin was isolated from the stem bark (Houghton 2002). Other bioactive compounds isolated from the various plant parts are discussed below.

Kigelia africana is well known for its anticancer, anti-inflammatory, antimicrobial and anti-amoebic activities besides other pharmacological activities.

Anti-inflammatory and Analgesic Activities

K. africana fruits were found to contain verminoside, an iridoid, as a major constituent, and a series of polyphenols such as verbascoside. In-vitro assays showed that verminoside had significant anti-inflammatory effects, inhibiting both iNOS expression and NO release in the LPS-induced J774.A1 macrophage cell line (Picerno et al. 2005). Cytotoxicity and cutaneous irritation of the extract and both compounds revealed that the crude extract and verminoside did not affect cell viability in-vitro either in cells grown in monolayers (ML) or in the reconstituted human epidermis (RHE, 3D) model; neither caused release of pro-inflammatory mediators or histo-morphological modification of RHE.

Methanol extract of *K. pinnata* fruit (100, 200 and 400 mg/kg, p.o) exhibited a dose dependent and significant inhibition in mice and rats in all the experimental models such as formaldehyde-induced paw edema, acetic acid–induced vascular permeability, cotton pellet-induced granuloma, estimation of plasma MDA levels and carrageenan induced peritonitis models (Carey et al. 2008). The anti-inflammatory activity observed was comparable to the respective standard drugs. Preliminary phytochemical screening revealed the presence of flavonoids, carbohydrates, glycosides, proteins and alkaloids. Acute toxicity studies were performed and produced no mortality in dose up to 2,000 mg/kg, p.o. The results gave a scientific basis to the traditional uses of *K. pinnata* particularly involved in inflammation. Methanolic extract of *Kigelia pinnata* flower at doses of 100, 200 and 400 mg/kg b.w. was found to exhibit significant anti-inflammatory in carrageenan-induced paw edema model in rats (Carey et al. 2010) It also exhibited significant analgesic activity in acetic acid-induced writhing, hot plate and formalin-induced paw licking models in mice. The results of the experimental study strongly supported the traditional use of this plant for inflammatory and pain disorders.

The ethanolic extract of *Kigelia africana*, was evaluated for analgesic property using acetic acid induced mouse writhing and hot plate reaction time and anti-inflammatory property using the carrageenan induced paw edema (Owolabi and Omogbai 2007). The extract showed significant dose dependent reduction of the number of writhes with 500 mg/kg body weight dose giving the highest reduction. The extract showed an insignificant elongation of the hot plate reaction time. In the carrageen induced paw edema, a dose dependent significant inhibition was observed between the 2nd and 5th hour. The data indicated that the ethanolic stem bark extract of *K. africana* had significant analgesic and anti inflammatory activity. Inhibition of the synthesis of prostaglandins and other inflammatory mediators probably accounted for the analgesic and anti-inflammatory properties.

Antimicrobial Activity

Fractionation of the methanolic extracts of the root and fruits of *Kigelia pinnata* led to the isolation of the naphthoquinones kigelinone, isopinnatal, dehydro-α-lapachone, and lapachol and the phenylpropanoids p-coumaric acid and ferulic acid. The compounds were found responsible for the observed antibacterial and antifungal activity of the root and kigelinone and caffeic acid from the fruits (Binutu et al. 1996). Antimicrobial activity was exhibited against *Staphylococcus aureus, Bacillus subtilise, Corynebacterium diptheriae, Aspergillus flavus, Aspergillus niger, Candida albicans* and *Pullularia pullularis (Aureobasium* sp.).

Stem bark and fruit extracts showed similar antibacterial activity against Gram-negative and Gram-positive bacteria using the microtitre plate bioassay (Grace et al. 2002). A mixture of three fatty acids exhibiting antibacterial effects was isolated from the ethyl acetate extract of the fruits. Palmitic acid, already known to possess antibacterial activity, was the major compound in this mixture. These results confirmed the antibacterial activity of *K. africana* fruits and stem bark, and supported the traditional use of the plant in therapy of bacterial infections. Separate studies showed that the crude aqueous bark extracts showed significant antimicrobial activity which could be partially explained by the activity of the iridoids present (Akunyili et al. 1991). The aqueous or dilute alcohol extract of *Kigelia* pinnata roots were used as a treatment for STDs in many parts of Africa. Extracts of the roots equivalent to those obtained using the traditional methods were found to contain the iridoids specioside 2 and minecoside 4 as major constituents. The extracts, as well as two of the isolated iridoids, were tested and also their 1/10 and 1/100 dilutions, against four representative species of bacteria viz. *Staphylococcus aureus* (causes impetigo and skin abscesses), *Bacillus subtilis*, *Escherichia coli* (responsible for abscesses), and *Pseudomonas aeruginosa* (causes skin sepsis and infections), and the yeast *Candida albicans* (a fungus that causes thrush) in the absence and presence of the enzyme emulsin. This enzyme converted catalpol-type iridoids to their more anti-microbially active non-sugar containing aglycones. The results showed that the aqueous extract had strong activity, even in the absence of emulsion, against all the bacteria tested but especially against the yeast *C. albicans*.

The methanol extracts of the bark presented a higher antimicrobial activity than the aqueous and chloroform extracts (Jeyachandran and Mahesh 2007). It exhibited the greatest activity against *Salmonella typhi* and *Proteus vulgaris*, moderate activity against *E. coli*, *Staphylococcus aureus* and *Bacillus cereus*. The remaining strains viz., *Enterobacter aerogens*, *Klebsiella pneumonia* and *Pseudomonas aeruginosa* were less affected. Results supported the traditional use of *Kigelia africana* bark as a good source of antimicrobial agent.

Anticancer/Antineoplastic Activity

Serial dilutions of standardised water, ethanol, and dichloromethane extracts of the stem and fruits of *Kigelia pinnata*, lapachol, a possible constituent of these extracts, together with known therapeutic anti-neoplastic agents were tested for their growth inhibitory effects against four melanoma cell lines and a renal cell carcinoma line (Caki-2) using two different (MTT and SRB) assays (Houghton et al. 1994; Jackson et al. 1995, 1996, 2000). Significant inhibitory activity was shown by the dichloromethane extract of the stem-bark and lapachol (continuous exposure). Moreover, activity was dose-dependent, the extract being less active after 1 hour exposure. In another study, crude dichloromethane extracts of *Kigelia pinnata* stem bark and fruit showed cytotoxic activity in-vitro against cultured melanoma and other cancer cell lines using the sulphorhodamine B assay (Jackson et al. 2000). From the active fractions, major compound isolated included norviburtinal and β-sitosterol. Noviburtinal was the most active compound but had little selectivity for melanoma cell lines while isopinnatal showed some cytotoxic activity. β-sitosterol was comparatively inactive. Chemosensitivity of the melanoma cell lines to the stem bark extract was greater than that observed for the renal adenocarcinoma line. However, sensitivity to lapachol was similar amongst the five cell lines. Other fractions which did not contain norviburtinal also displayed appreciable activity, thus indicating that other cytotoxic compounds were present.

Studies demonstrated that the ethanolic extract of *Kigelia africana* fruits possessed significant anti-neoplastic and anti-inflammatory activities, thus giving credence to the traditional use of these plant extract in the treatment of cancer and oedema (Kolodziej 1997). Ethanolic extract of *Kigelia africana* fruits showed moderate cytotoxic activity in the brine shrimp (*Artemia salina*) nauplii bioassay with an LC50 of 7,500 μg/ml. It did not induce any DNA damage in the *Salmonella typhimurium* strains TA 98 and TA 100 disc spot mutagenicity assay. The acute toxicity test showed an LD 50 of 1.3 g/kg i.p. in female Swiss mice. Oral administration of extract to mice resulted in a significant inhibition in the tumour incidence and burden by 67% and 76% respectively in the benzo(a)pyrene-induced fore-stomach tumourigenesis model. The extract evinced marked anti-inflammatory effects in female Wistar rats as reflected by a significant inhibition of the increase in rat paw circumference of 72% (100 mg/kg) and 54% (200 mg/kg) which was caused by injection of fresh egg albumen.

Crude fractions of *K. pinnata* fruits demonstrated in-vitro cytotoxicity against human melanoma cells (Higgins et al. 2010). Compounds isolated and identified included the isocoumarins, demethylkigelin and kigelin; fatty acids, oleic and heneicosanoic acids; the furonaphthoquinone, 2-(1-hydroxyethyl)-naphtho[2,3-b]furan-4,9-dione and ferulic acid. The most potent series of these compounds, furonaphthoquinones, also demonstrated a cytotoxic effect in two human breast cancer cell lines tested.

Antioxidant Activity

Kigelia africana plant extract significantly reduced the production of thiobarbituric acid reactive substances (TBARS) in a concentration-dependent manner in all the tested pro-oxidant-induced oxidative stresses (Olalye and Rocha 2007). The results of the study suggested that the use of the plant in the treatment of various diseases, especially liver disease, was probably due to their ability to act as antioxidants.

Antidiarrhoeal Activity

Research revealed that *Kigelia pinnata* leaf had antidiarrheal property (Akah 1996). Evidence for antidiarrheal activity was provided by the reduced faecal output and protection from castor oil-induced diarrhoea in animals treated with aqueous leaf extract. The extract remarkably decreased the propulsive movement of the gastrointestinal contents. On the isolated guineapig ileum, the extract did not appreciably affect acetylcholine and histamine induced contractions, but significantly reduced nicotine evoked contractions. The i.p. LD_{50} of the extract in mice was estimated to be 785.65 mg/kg.

Central Nervous System (CNS) Activity

CNS stimulant effect of the ethanolic stem bark extract of *Kigelia africana* was studied in mice using the barbiturate induced sleeping time and the Rota rod bar to check the extract's effect on muscle coordination (Owolabi et al. 2008). The results showed that the extract at all doses tested decreased the duration of sleeping time when compared to the control group that received distilled water. This difference in sleeping time was significant and to be dose dependent. The extract gave a shorter duration of sleeping time compared to caffeine, (at 400 mg/kg dose) indicating better stimulant properties. In comparison with diazepam the extract at all doses tested, also gave a shorter duration of sleep. On the rota rod, the animals maintained their balance indicting that the extract had no sedative effect.

Fertility Promoting Activity

The fruit of *Kigelia africana* is used in south eastern Nigeria by local traditional healers for treatment of fertility abnormalities especially in male and female adults. Studies by Ogbeche et al. (2002) revealed that male rats treated with the methanol extract of the fruit showed greater sperm motility 95.14, when compared to that of male rats treated with testosterone and the untreated control rats which had 89.34 and 82.81 respectively. The result also showed that 77.8% of mated pro-estrus female rats were pregnant after a four day circle. There was also a 66.7% pregnancy rate in females mated with testosterone induced male rats and 44.4% for the control, showing that the methanol extract had greater fertility effects on male rats. Research also demonstrated that *Kigelia africana* fruit extract if given within 4 weeks of treatment with *Carica papaya* reversed the deleterious effects on semen parameters, whereas if given after 10 weeks, the damage remains unreversed (Air et al. 2003). When both extracts were given from the beginning, semen parameters remained very low suggesting that *Kigelia africana* was able to reverse papaya induced testicular damage if administered within a certain window period.

Kigelia africana fruit extract (KAFE) treatment was found to reverse the deleterious effects of cisplatin in the testis of male rats (Azu et al. 2010a, b). Cisplatin treatment caused over 37.5%

mortality of SD rats. Cisplatin-treatment negatively affected the histoarchitecture of the seminiferous tubules with massive loss of spermatogenic cells and caused a significant reduction in testicular weight/volume, seminiferous tubules diameter and cross-sectional areas. KAFE post-treatment resulted in focal vacuolar changes in the seminiferous tubules of the SD rats and positively improved the other parameters. KAFE alone and as prophylaxis significantly increased body weight, serum testosterone and follicle-stimulating hormone. It exhibited a significant rise in catalase activity, a fall in malondialdehyde and enhanced glutathione levels. These parameters were negatively affected by cisplatin treatment The cytoprotection against cisplatin-induced testicular damage by KAFE was deemed to be via an antioxidant modulatory pathway as male infertility was frequently accompanied by increased testicular or seminal fluid oxidative stress.

Antifertility Activity

Ethanol extract of *Kigelia pinnata* was reported to show 60–70% anti-implantation activity in female albino rats (Prakash et al. 1985).

Renal Protective Activity

Azu et al. (2010c) found that the methanol extract of *Kigelia africana* fruit extract (KAFE) had a protective against cisplatin induced renal toxicity in male rats. Over 28 days, cisplatin treated rats suffered significant loss in body weight, augmentation in blood urea nitrogen and serum creatinine levels as well as tubular necrosis. However, KAFE alone and as a prophylaxis significantly improved these parameters. Though post treatment of animals with KAFE after cisplatin did not restore serum catalase activity it was lower than those receiving cisplatin alone. The results suggested that KAFE may protect against cisplatin induced renal toxicity hence might serve as a novel combination agent with cisplatin to limit renal injury.

Antiamoebic, Antiplasmodial Activity

Butanol extract from *Kigelia pinnata* stem barks exhibited in-vitro antiamoebic activity against *Entamoeba histolytica* (Bharti et al. 2006). Three known iridoids specioside, verminoside and minecoside in the extract were tested against HK-9 strain of *Entamoeba histolytica* for their in-vitro antiamoebic evaluation and metronidazole was used as reference drug. It was found that verminoside was more active (two-fold) as compared to the standard drug while specioside showed comparable antiamoebic activity with metronidazole.

Four naphthoquinoids from *Kigelia pinnata* root bark were assessed in-vitro against chloroquine-sensitive (T9-96) and -resistant (K1) *Plasmodium falciparum* strains and for cytotoxicity using KB cells (Weiss et al. 2000). 2-(1-Hydroxyethyl)naphtho[2,3-b]furan-4,9-dione possessed good activity against both strains [IC(50) values 627 nM (K1), 718 nM (T9-96)]. Isopinnatal, kigelinol, and isokigelinol exhibited lower activity against both strains.

Antitrypanosomal Activity

Dichloromethane extracts of the root bark and stem bark of *Kigelia* exhibited antitrypanosomal activity against *Trypanosoma brucei brucei* in-vitro (Moideen et al. 1999). Activity-guided fractionation led to the isolation of four naphthoquinones: 2-(1-hydroxyethyl)-naphtho[2,3-b]furan-4,9-quinone (1), isopinnatal (2), kigelinol (3), and isokigelinol (4) from both the root and stem bark of the plant. Compound 1 with a furanonaphthoquinone structure was found to possess pronounced activity against both parasites, *T. brucei brucei* and *T. brucei rhodesiense* with IC_{50} values of 0.12 and 0.045 μM, respectively, although it was less active than the standard drug pentamidine. Compounds 2, 3, and 4 also exhibited activity against the parasites, although to a smaller extent. The activities of the compounds were further assessed by comparison with the cytotoxic activities obtained against KB cell lines. The root bark extract exhibited anti-leishmanial activity against *Leishmania major* promastigotes (Moideen et al. 1997).

Molluscicidal Activity

Methanol and crude extracts of *K. Africana* were also reported to have molluscicidal activity (Kela et al. 1989).

Toxicity Studies

The subchronic toxicity of the aqueous antidiabetic herbal extract ADD-199, prepared from *Maytenus senegalensis*, *Annona senegalensis*, *Kigelia africana* and *Lannea welwitschii*, and administered at a daily dose of 100 or 500 mg/kg body weight over 30 days, was investigated in male Wistar albino rats (Nyarko et al. 2005). It was found that ADD-199 did not affect plasma aspartate aminotransferase (AST), alanine aminotransferase (ALT), alkaline phosphatase (ALP) and albumin or creatinine kinase (CK) levels. It also did not affect plasma creatinine and urea levels. Further, ADD-199 neither affected PCV (Packed cell volume) nor blood haemoglobin, RBC, reticulocytes, platelets, lymphocytes and granulocyte levels. It, however, caused significant dose-dependent reductions in WBC counts at day 15 with varying degrees of recovery by day 30. It also reduced the rate of body weight increases after week 3. However, no changes were observed in organ weights at termination. ADD-199 did not significantly affect zoxazolamine-induced paralysis and pentobarbital-induced sleeping times as well as certain CYP isozyme activities in rats. These findings suggested that ADD-199 had no overt organ specific toxicity and did not demonstrate a potential for drug interactions via CYP-mediated metabolism in the rat on subchronic administration.

Traditional Medicinal Uses

Kigelia africana has a long history of use in African traditional herbal medicine. Traditional healers have used kigelia to treat a wide range of skin ailments, from fungal infections, ringworm, boils, psoriasis and eczema, through to the more serious diseases, such as leprosy, syphilis and skin cancer. It has also been employed as remedy for internal ailments such as dysentery, tapeworm, post-partum haemorrhaging, malaria, diabetes, pneumonia, piles and toothache and used a purgative. Aqueous extracts of the stem bark of this plant is used for the treatment of gynaecological conditions, venereal diseases, dysentery, epilepsy, wounds, sores, abscesses, diarrhoea and oedema. *K. africana* is widely used to treat gynecological disorders. Aqueous preparation of the roots, fruits and flowers are administered orally or as a virginal pessary while the fruits and bark are used to promote breast development in young women or in contrast to reduce swelling and mastitis of the breasts (Gouda et al. 2003).

The bark contains only a bitter principle and tannic acid, and is used in Northern Nigeria both for venereal diseases such as syphilitic conditions and for gonorrhoea and the bark; and fruits are used in the treatment of malignant tumours of the breast. In Southern Nigeria, it is similarly used, and also is given along with the fruit as a wash and drink for young children. In Malawi, the dried root bark is used to treat cancer of the uterus and alimentary tract. People in the Gold Coast use the bark as a cure for rheumatism and dysentery. In Sierra Leone, heated bark is applied to women's breast to hasten their return to normal after a suckling child has been weaned, while the Sambaa in Tanzania use the bark for the treatment of swelling of the breasts. In Cote D'Ivoire, the fruit decoction is used as a remedy for elephantiasis and oedema of the legs. The unripe fruit is used in Central Africa as a dressing in cases of syphilis and rheumatism. Tongas of the Zambezi valley apply the powdered fruit as a dressing to ulcers. The Tongan women regularly apply cosmetic preparations of *Kigelia* fruit to their faces to ensure a blemish-free complexion. Today, a number of beauty products and skin ointments are prepared from *Kigelia* fruit extracts. Rich in flavonoids and saponins this extract from *Kigelia* is clinically reported to improve firmness and elasticity of the skin between 50 and 70% by acting on the muscles. The fruit has also found traditional use as an aphrodisiac.

Other Uses

The tree is widely grown as an ornamental tree in tropical and subtropical regions for its decorative flowers and unusual fruit and also as a shade tree. Kigelia leaves are an important livestock fodder, and the fruits are much prized by monkeys and elephants.

The tough wood is used for shelving and fruit boxes, makoros (boat), yokes, oars and dugout canoes are made from the tree. Roots are said to yield a bright yellow dye.

Comments

The genus *Kigelia* is a monotypic genus comprising only one species, *Kigelia Africana*.

Selected References

Air A, Fio D, Noronha CC, Okanlawon AO (2003) Aqueous extract of the bark of *Kigelia africana* reverses early testicular damage induced by methanol extract of *Carica papaya*. Nig J Health Biomed Sci 2(2):87–89

Akah PA (1996) Antidiarrheal activity of *Kigelia africana* in experimental animals. J Herbs Spices Med Plants 4(2):31–36

Akah PA, Nwambie AI (1994) Evaluation of Nigerian traditional medicines 1: plants used for rheumatic (inflammatory) disorders. J Ethnopharmacol 42:179–182

Akunyili DN, Houghton PJ (1993) Meroterpenoids and naphthaquinones from *Kigelia pinnata* bark. Phytochem 32:1015–1018

Akunyili DN, Houghton PJ, Raman A (1991) Antimicrobial activities of the stembark of *Kigelia pinnata*. J Ethnopharmacol 35:173–177

Asekun OT, Olusegun E, Adebola O (2007) The volatile constituents of the leaves and flowers of *Kigelia africana* Benth. Flav Fragr J 22(1):21–23

Azu OO, Duru FIO, Osinubi AA, Noronha CC, Elesha SO, Okanlawon AO (2010a) Preliminary study on the antioxidant effect of *Kigelia africana* fruit extract (Bignoniacieae) in male Sprague-Dawley rats. Afr J Biotechnol 9(9):1374–1381

Azu OO, Duru FIO, Osinubi AA, Oremosu AA, Noronha CC, Elesha SO, Okanlawon AO (2010b) Histomorphometric effects of *Kigelia africana* (Bignoniaceae) fruit extract on the testis following short-term treatment with cisplatin in male Sprague–Dawley rats. Middle East Fertil Soc J 15(3):200–8

Azu OO, Duru FIO, Osinubi AA, Oremosu AA, Noronha CC, Elesha SO, Okanlawon AO (2010c) Protective agent, *Kigelia africana* fruit extract, against cisplatin induced kidney oxidant injury in Sprague Dawley rats. Asian J Pharm Clin Res 3(2):84–88

Bharti N, Singh S, Naqvi F, Azam A (2006) Isolation and in vitro antiamoebic activity of iridoids isolated from *Kigelia pinnata*. ARKIVOC X:69–76

Binutu OA, Adesogan KE, Okogun JI (1996) Antibacterial and antifungal compounds from *Kigelia pinnata*. Planta Med 62(4):352–3

Burkill HM (1985) The useful plants of West Tropical Africa, vol 1. Royal Botanical Gardens, Kew, pp 960

Carey MW, Babu MJD, Rao VN, Mohan KG (2008) Anti-inflammatory activity of the fruit of *Kigelia pinnata* DC. Pharmacologyonline 2:234–245

Carey MW, Rao NV, Kumar BR, Mohan GK (2010) Anti-inflammatory and analgesic activities of methanolic extract of *Kigelia pinnata* DC flower. J Ethnopharmacol 130(1):179–82

El-Sayyad SM (1982) Flavonoids of the leaves and fruits of *Kigelia pinnata*. Fitoterapia 52:189–91

Gabriel OA, Olubunmi A (2009) Comprehensive scientific demystification of *Kigelia africana*: a review. Afr J Pure Appl Chem 3(9):158–164

Gouda YG, Abdel-baky AM, Darwish FM, Mohamed KM, Kasai R, Yamasaki K (2003) Iridoids from *Kigelia pinnata* DC. fruits. Phytochem 63:887–892

Gouda YG, Abdel-Baky AM, Mohamed KM, Darwish FM, Kasai R, Yamasaki K (2006) Phenylpropanoid and phenylethanoid derivatives from *Kigelia pinnata* DC. fruits. Nat Prod Res 20(10):935–9

Govindachari TR, Patankar SJ, Viswanathan N (1971) Isolation and structure of 2 new dihydroisocoumarins from *Kigelia pinnata*. Phytochem 10:1603–06

Grace OM, Light ME, Lindsey K, Mulholland DA, Van Staden J, Jäger AK (2002) Antibacterial activity and isolation of active compounds from fruit of the traditional African medicinal tree *Kigelia africana*. S Afr J Bot 68(2):220–222

Heine H (1963) Bignoniaceae. In: Hepper FN (ed.) Flora of West Tropical Africa, vol 2, 2nd edn. Crown Agents for Overseas Governments and Administrations, London, pp 383–388

Higgins CA, Bell T, Delbederi Z, Feutren-Burton S, McClean B, O'Dowd C, Watters W, Armstrong P, Waugh D, vanden Berg H (2010) Growth inhibitory activity of extracted material and isolated compounds from the fruits of *Kigelia pinnata*. Planta Med 76(16):1840–1846

Houghton PJ (2002) The sausage tree (*Kigelia pinnata*): ethnobotany and recent scientific work. S Afr J Bot 68:14–20

Houghton PJ, Akunyili DN (1993a) Iridoids from *Kigelia pinnata* bark. Fitoterapia 64:473–474

Houghton PJ, Akunyili DN (1993b) Meroterpenoids and naphthaquinones from *Kigelia pinnata* bark. Phytochem 32:1015–1018

Houghton PJ, Jackson SJ, Browning M, Photiou A, Retsas S (1994) Activity of extracts of *Kigelia pinnata* against melanoma and renal carcinoma cells. Planta Med 60:430–433

Houghton PJ, Jackson S, Photiou A, Retsas S (1995) Cytotoxic activity of extracts of the fruit *Kigelia pinnata*. J Pharm Pharmacol 47:1086

Huxley AJ, Griffiths M, Levy M (eds) (1992) The new RHS dictionary of gardening. Macmillan, New York, 4 Vols

Inoue K, Inoue H, Chen C-C (1981) Quinones and related compounds in higher plants. 17. A naphthoquinone and a lignan from the wood of *Kigelia pinnata*. Phytochem 20:2271–2276

Jackson, S.J., Houghton, P.J., Photiou, A. and Retsas, S. 1995. Effects of fruit extracts and naphthaquinones isolated from *Kigelia pinnata* on melanoma cell lines. *Br. J. Cancer*, 71 (Supplement) 24:62.

Jackson S, Houghton PJ, Photiou A, Retsas S (1996) The isolation of a novel antineoplastic compound from a bioassay-guided fractionation of stem bark extracts of *Kigelia pinnata*. Br J Cancer 73(suppl XXVI):68

Jackson SJ, Houghton PJ, Retsas S, Photiou A (2000) In vitro cytotoxicity of norviburtinal and isopinnatal from *Kigelia pinnata* against cancer cell lines. Planta Med 66(8):758–61

Jeyachandran R, Mahesh A (2007) Antimicrobial evaluation of *Kigelia africana* (Lam). Res J Microbiol 2(8):645–649

Joshi KC, Singh P, Taneja S, Cox PJ, Howie RA, Thomson RH (1982) New terpenoid aldehydes from *Kigelia pinnata*: crystal structure of pinnatal. Tetrahedron 38:2703–8

Kela SL, Ogunsusi RA, Ogbogu N, Nwude VC (1989) Screening of some Nigerian plants for molluscidal activity. Rev Elev Med Vet Pays Trop 42(2):195–202

Khan MR (1998) Cytotoxicity assay of some Bignoniaceae. Fitoterapia 69(6):538–40

Kolodziej H (1997) Protective role of *Kigelia africana* fruits against benzo[a]pyrene-induced forestomach tumourgenesis in mice and against albumen induced inflammation in rats. Pharm Pharmacol Lett 7:67–70

Leung W-TW, Busson F, Jardin C (1968) Food composition table for use in Africa. FAO, Rome, 306 pp

Moideen SVK, Houghton PJ, Croft SL (1997) Antileishmanial activity of *Kigelia pinnata* root bark constituents. J Pharm Pharmacol 49:114

Moideen SVK, Houghton PJ, Rock P, Croft SL, Aboagye-Nyame F (1999) Activity of extracts and naphthoquinones from *Kigelia pinnata* against *Trypanosoma bruceibrucei* and *Trypanosoma brucei rhodiesiense*. Planta Med 65:493–588

Nyarko AK, Okine LKN, Wedzi RK, Addo PA, Ofosuhene M (2005) Subchronic toxicity studies of the antidiabetic herbal preparation ADD-199 in the rat: absence of organ toxicity and modulation of cytochrome P450. J Ethnopharmacol 97(2):319–325

Ogbeche KA, Ogunbiyi YO, Duru FIO (2002) Effect of methanol extract of *Kigelia africana* on sperm motility and fertility in rats. Niger J Health Biomed Sci 2:113–116

Olalye MT, Rocha JB (2007) Commonly used tropical medicinal plants exhibit distinct in vitro antioxidant activities against hepatotoxins in rat liver. Exp Toxicol Pathol 58(6):433–8

Owolabi OJ, Omogbai EKI (2007) Analgesic and anti-inflammatory activities of the ethanolic stem bark extract of *Kigelia africana* (Bignoniaceae). Afr J Biotechnol 6(5):582–585

Owolabi OJ, Omogbai EKI, Obasuyi O (2007) Antifungal and antibacterial activities of the ethanolic and aqueous extract of *Kigelia africana* (Bignoniaceae) stem bark. Afr J Biotechnol 6(15):1677–1680

Owolabi OJ, Amaechina FC, Eledan AB (2008) Central nervous system stimulant effect of the ethanolic extract of *Kigelia africana*. J Med Plants Res 2(2):20–23

Palmer E, Pitman N (1972) Trees of Southern Africa. Balkema, Cape Town

Picerno P, Autore G, Marzocco S, Meloni M, Sanogo R, Aquino RP (2005) Anti-inflammatory activity of verminoside from *Kigelia africana* and evaluation of cutaneous irritation in cell cultures and reconstituted human epidermis. J Nat Prod 68(11):1610–1614

Prakash AO, Saxena V, Shukla S, Tewari RK, Mathur S, Gupta A, Sharma S, Mathur R (1985) Anti-implantation activity of some indigenous plants in rats. Acta Eur Fertil 16(6):441–8

Sharma PK, Kaul MK (1993) Specific ethnomedicinal significance of *Kigelia africana* in India. Fitoterapia 65(5):467–468

Sushil DMK, Dhindsa KS (2007) Beta-sitosterol-3-0-alpha-L-rhamnopyranoside from *Kigelia pinnata* DC flowers. Int J Agric Biol Res 23:51–6

van Steenis CGGJ (1977) Bignoniaceae. In: van Steenis CGGJ (ed.) Flora Malesiana, Ser. I, 8 Sijthoff & Noordhoff Alphenaan den Rijn, 577 pp, pp 114–186

Weiss CR, Moideen SVK, Croft SL, Houghton PJ (2000) Activity of extracts and isolated naphthoquinones from *Kigelia pinnata* against *Plasmodium falciparum*. J Nat Prod 63:1306–1309

Oroxylum indicum

Scientific Name

Oroxylum indicum (Linnaeus) Bentham ex Kurz

Synonyms

Bignonia indica Linnaeus; *Bignonia indica pentandra* Loureiro; *Bignonia quadripinnata* Blanco, *Calosanthes indica* (Linnaeus) Blume, *Hippoxylon indicum* Raf., *Spathodea indica* Pers.

Family

Bignoniaceae

Common/English Names

Broken Bones Plant, Indian Trumpet Flower, Midday Marvel, Tree of Damocles.

Vernacular Names

Chinese: Mu Hu Die;
French: Calosanthe, Oroxyle, Oroxylon;
India: Kotodu (Andhra Pradesh), Bhatghila, Bhat Ghilla, Dingari, Toguna, (Assamese), Banahata, Bhaluksukti, Shone, Sona (Bengali), Khiring (Garo), Syonakh (Gujarat), Arlu, Assarsauna, Bhut-Vrisha, Dirghavrinta, Fari, Kharkath, Kutanat, Manduk (the flower), Patroma, Pharkhat, Pharrai, Pharri, Putivriksha, Sauma, Sauna, Shallaka, Shuran, Shyona, Snapatha, Son, Sonapatha, Syona, Ullu, Urru, Tetu, Sora, Sauno, Tantia, Tantia Pharhar, Tatplanga, Tat-Palanga, Tat Palanga, Tentu, Vatuk (Hindu), Dieng Kawait Blai, Naora (Jaintia), Alangi, Aane Mungu, Aanemungu, Alanda, Anemungu, Arantala, Baagi, Baagimara, Bagi, Buna Paale, Bunepale, Dinchalamara, Dundu Karra, Dudunmkara Mara, Dundukara, Heggo Gate, Heggudatimara, Hegguggate, Hemmara, Hoare Mara, Hone, Hore, Konanakombina Mara, Marasouthe Mara, Mokka, Mokkaa Mara, Mokkada, Nata, Nala, Patagani Mara, Salaa Mara, Shonatha, Shukanaasa, Sonepatte, Sonepatte Mara, Tattuna, Teta, Tetu, Tetutigadi, Thigudi Mara, Tigadu, Tigda, (Kannada), Dieng Titkongling, Ja Ranghon (Khasi), Davamadak (Konkani), Aralu, Arantal, Chorikonnam, Corikkonna, Palagapayani, Palagapaiyani, Palagappayani, Palakapaiyana, Palakappayana, Palakapayani, Palakappayyana, Palakappayyani, Palega-Pajaneli, Palegapajoneli, Payyalanta, Peiam, Valpathiri, Vashrppathiri, Vellapadiri, Veluttapatiri, (Malayalam), Shamba (Manipuri), Davamdak, Dhavandaka, Dinda, Kkharasinga, Taitu, Tayitu, Tetu, Thaentu, Thaitu, Thetu, Titu, Ulu, (Marathi), Archangkawm (Mizoram), Archangkawm, Phanphania, Tatelo (Oriya), Adhvantashatrava, Arala, Aralu, Araluka, Aratu, Bhalluka, Bhallukah, Bhantuka, Bhutapushpa, Bhutataka, Dirghavrintaka, Dirghavrnta, Dirghavrukah, Dundukah, Kandarpa, Katambhara, Katanga, Katvanga, Kurkata,

Kutannata, Mandukaparna, Mandukaparnah, Mayurajangha, Munduka-Purna, Nata, Nihsarah, Ntakah, Padavriksha, Paripadapa, Patrorna, Pitangah, Prathusimbhi, Prithushimba, Priyajiva, Priyajivi, Prthusimba, Prthusimbi, Putivriksha, Ruksha, Shallaka, Shona, Shoshana, Shukanasa, Shyonaka, Sonaka, Sukanasa, Svarnavalkala, Syonaka, Syonakah, Tintuka, Tintukah, Tuntuka, Vatu, Vishanut, (Sanskrit), Achi, Arandai, Aratu, Arulandai, Arulantai, Conam, Conam, Cori-Konnai, Corikkonrai, Corikonnai, Paiyaralantai, Palagaipayani, Palagai-Payani, Palaiyudaichi, Palaiyudaycci, Palai-y-utaicci, Pana, Panamaram, Peiarlankei, Perum Vakai, Peruvaagai, Peyarulandai, Peyarulantai, Pudabudham, Puta-puspam (the flower), Tirkkaviruttam, Tuvakkucaram, Tuvantacattiravam Vangam, Vangamaram, Veluttapatiri, (Tamil), Aaku Maanu, Dondlup, Dundilamu, Dundilum, Dundillamu, Dundillum, Kaligottu Mandookaparnamu,, Manduka-parnamu, Mokka Vepa, Mokkavaepa, Mokkavepa, Nemali Chettu, Pamania, Pampana, Pampena, Pamini,Pampina, Pampini, Peddakatti, Sukanaasamu, Suka-nasamu, Tundilamu (Telugu), Sona Patha (Urdu);
Indonesia: Bungli (Malay), Bungli, Kajeng Jaler, Kayu Lanang, Mungli, Wungli (Java), Dhang-Pedhangan (Madurese), Kapung-Kapung (Sumatra), Pong-Porang (Sundanese);
Malaysia: Beka, Beka Kampong, Berak, Bikir, Bikir Hangkap, Biji Lunang Bonglai Kayu, Bolai Kayu, Boli, Boloi, Bongloi, Kankatang, Kulai, Merlai, Merulai;
Nepal: Tatola;
Philippines: Pinka-Pinkahan;
Russian: Oroksilum Indijskij;
Thailand: Lin Fa, Malit Mai, Phaeka;
Tibetan: So Na Ka;
Vietnam: Nuc Nac, Nguc Ngac, Nam Hoang Ba, Moc Ho Diep, May Ca, Phac Ca (Tay), Coca Lien (Thai), Ngong Pang Daing (Dao), P So Lung (K'ho).

Origin/Distribution

The tree is indigenous to south Asia, China and South-east Asia – in Fujian, Guangdong, Guangxi, Guizhou, Sichuan, Yunnan (Southern China), Taiwan, Bhutan, Cambodia, India, Indonesia (Java, Sumatra), Laos, Malaysia, Myanmar, Nepal, Philippines, Thailand and Vietnam.

Agroecology

The species occurs in tropical and subtropical areas, usually at low altitudes of 500–900 m in open forests, river banks, roadsides, slopes, old clearings, by rice fields or in cultivated grounds. It grows well in full sun and moist soil and is partially shade tolerant when young.

Edible Plant Parts and Uses

Young fruits (Plates 1–2) are bitter but edible and taste better after cooking and is commonly consumed in Thailand and Malaysia as a traditional vegetable. Young leaves and fruit are boiled and eaten with salted fish in Vietnam. The fruit is grilled and the outer skin remove and the inner parts are sliced and eaten with salted fish. The young tender leafy shoots and leaves (Plates 3–4) are eaten after cooking in Malaysia, Myanmar, Laos, Kampuchea, Vietnam, Java and the Philippines. In Malaysia, the young shoots and leaves are steep in hot water and made into *kerabu* (mixed vegetable salad) or used as ingredient in *nasi ulam* (herb rice). In the Philippines, young shoots, young pods and flowers are cooked as vegetables and may be eaten raw as salad. The young pods are similarly used in *kerabu*. In Andhra Pradesh, the flowers are cooked and eaten as vegetables. In Java, the young leaves, flowers and the bark of the trunk are eaten uncooked with rice usually prepared as *sambal pepeh*. This is a hotchpotch of the leaves and flowers pounded with red onions, *muncang* (twin nut), *sereh* (piper betle leaves), chilli, *terasi* (pounded salted dried shrimps or small fish), *lempuyang pait* (bitter *Zingiber* sp. root), and *humbut sohun* (mungbean vermicelli). Young fruits with seeds removed are cut into pieces and boiled and eaten with rice. The bark is bitter but is chewed as a delicacy.

Botany

An evergreen or partly deciduous, sparingly branched, medium-sized tree 6–10 m tall (Plate 5). Trunk is glabrous, stout 15–20 cm across, slightly fissured with grey-brown bark. Twigs are grey, stout with large scars. Leaves are opposite, 2(–4)-pinnately compound, 60–130 cm long, with terminal leaflet. Leaflets are triangular-ovate, shortly petiolate, 5–13 × 3–9 cm, glabrous, with cordate or oblique base, entire margin and acuminate apex (Plates 3–4). Inflorescences are terminal, erect, 40–180 cm long. Flowers are large, with a fetid odour, fleshy, tubular, with small bracts and borne on 3–7 cm long pedicel. Calyx is purple, campanulate, 2.0–4.5 × 2–3 cm, glabrous and membranous. Corolla is purple-red, fleshy, tubular, 3–9 × 1–1.5 cm; with upper lip 2-lobed, lower lip 3-lobed. Stamens are inserted at the middle of the corolla tube. Ovary is oblong and style with 2 parted stigma. Fruit is a woody capsule when mature, flattened dorisi-ventrally, long and broad 40–120 cm × 5–9 cm and 1 cm thick, sabre-like (Plates 1–2), bivalved, yellowish

Plate 1 Hard, dorsi-ventrally flattened, sword-like fruit

Plate 2 Beka fruit in Kota Bahru market, Kelantan

Plate 3 Young leaves (Daun beka) sold as vegetable in Kota Bahru market, Kelantan

Plate 4 Young leaves and shoot eaten as vegetable

Plate 5 Foliage – pinnate compound leaves with 2–4 pairs of leaflets

green strongly tinged with violate later becoming grey. Seeds numerous, rounded, yellowish white, with transparent membranous wing.

Nutritive/Medicinal Properties

The young shoots were reported to be nutritious and to have the following food nutrient composition: dry matter 20 g, 1.29 g fibre, 0.92 g sugar, 5.92 g protein, 3.48 mg vitamin C, 97 mg B-carotene, 26 mg Ca and 1.26 mg Fe (Kuo 2002).

The chemical constituents of *O. indicum* was found to consist mainly of flavonoids, glycoside and volatile oil (Yin et al. 2007). It also contained pterocarpan and rhodioside with p-hydroxyphenylethanols, cyclohexanols and also baicalein −7-O-diglucoside. From the ethanol extract of the seeds of *Oroxylum indicum* a flavone glucuronide named oroxindin was isolated (Nair and Joshi 1979). Its structure was established as 5-hydroxy-8-methoxy-7-0-β-D-glucopyranuronosyl flavone (wogonin-7-0-β-D-glucuronide). Four flavonoids, chrysin, baicalein, baicalein-7-O-glucoside, baicalein-7-O-diglucoside (Oroxylin B) and one unknown flavonoid were isolated and purified from the seeds of *Oroxylum indicum* (Chen et al. 2003; Chen et al. 2005). Chrysin (160.9 mg, 97.3% in purity), baicalein (130.4 mg, 97.6% in purity), baicalein-7-O-glucoside (314.0 mg, 98.3% in purity), baicalein-7-O-diglucoside (179.1 mg, 99.2% in purity), and a new chrysin-diglucoside (21.7 mg, 98.8% in purity) were obtained from 911.6 mg ethyl acetate extract of *Oroxylum indicum* (Yuan et al. 2008). Liu et al. (2010) in separate studies also isolated and purified baicalein-7-O-glucoside and baicalein-7-O-diglucoside. Stigmast-5-en-3-ol was isolated from the roots (Ali et al. 1998). Four bioactive phytoconstituents namely chrysin, baicalein, biochanin-A and ellagic acid were identified and quantified from the roots of *O. indicum* (Zaveri et al. 2008). Pterocarpans, derivatives of benzo-pyrano-furano-benzenes such as hexyl oroxylopterocarpan (6H-Benzofuro[3,2-c]furo[3,2-g][1]benzopyran-8-ol,2-hexyl-1,2,6a,11a-tetrahydro-, (6aR,11aR)-) was isolated from *O. indicum* stem bark (Ali et al. 1999). Other pterocarpans isolated included methyl oroxylopterocarpan, heptyl oroxylopterocarpan and dodecanyl oroxylopterocarpan.

Scientific laboratory and pre-clinical studies have confirmed that *Oroxylum indicum* plant extracts have antioxidant, anti-cancerous, anti-inflammatory, antimutagenic, immunomodulatory, gastroprotective and antimicrobial properties.

Antioxidant Activity

Leaves of *Oroxylum indicum* together with leaves of three other species out of 127 species were classified in the very high class group for both TEAC (Trolox equivalent antioxidant capacity) and SOS (superoxide scavenging) antioxidant activity. TEAC values on a dry weight basis ranged from 0 to 2,105 μmol TE/g and SOS values ranged from 0 to 6,206 μmol ascorbate equivalent (AE) /g. The majority of vegetables hovered between 200 μmol TE/g and 400 μmol AE/ g (Yang et al 2006). Different extracts of *O. indicum* leaves showed concentration dependent free radical scavenging activity (Raghbir et al. 2008). The scavenging effect of plant extracts and standard (L-ascorbic acid) on the diphenyl-picryl-hydrazyl (DPPH) radical was reported to decrease in the following order: L-ascorbic acid > ethyl acetate > methanol > water and was found to be 97.4%, 61.4%, 40.8% and 29.2% at concentration of 100 μg/ml, respectively. The results were expressed as IC_{50} values. Ascorbic acid which was used as a standard showed an IC_{50} of 24.0 μg/ml, whereas, the crude ethyl acetate, methanolic and water extracts of leaves of *O. indicum* showed antioxidant activity with IC_{50} values of 49.0, 55.0 and 42.5 respectively at 100 μg/ml concentration.

Oroxylum indicum was found to have antioxidant activity using Heinz body inhibition assay and ABTS radical cation decolorization assay (Palasuwan et al. 2005). In a recent study in Assam, *Oroxylum indicum*, *Ipomoea aquatica* and *Moringa oleifera* exhibited strong antioxidant activity as compared to other plants (Vimal et al. 2009). *Oroxylum indicum* showed the highest antioxidant activity using 2, 2-diphenyl-picryl-hydrazyl (DPPH) free radical assay. The n-butanol extracts of stem and root bark of *Oroxylum indicum* exhibited very significant scavenging activity when tested using ABTS radical cation decolorization assay, scavenging of nitric oxide radical, reduction of ferric ions by ortho- phenanthroline colour method and inhibition of lipid peroxide formation as compared to standard ascorbic acid (Zaveri and Dhru 2010). *O. indicum* leaf and root bark methanol extracts exhibited highest free radical scavenging activity compared to stem bark, stem and fruit extracts (Mishra et al. 2010). Leaf extract showed maximum reductive ability and was found to contain maximum amount of polyphenolic compounds followed by root bark and stem bark extracts. The highest free radical activity was attributed to the presence of polyphenolic compounds. Total phenolic content and inhibition of lipid peroxidation in both menthol and aqueous bark extracts of *O. indicum* were found to be dosage-dependent (Ashok Kumar et al. 2010). Both extracts exhibited appreciable free radical scavenging and ferric reducing abilities. Both extracts showed moderate levels of DNA protection against oxidative stress. Total phenolic content was found to have significant positive correlations with free radical scavenging and reducing power, as well as lipid peroxidation inhibition.

Anticancer Activity

Five plants out of 118 Thai medicinal plants, namely *Micromelum minutum, Oroxylum indicum, Cuscuta chinensis, Azadirachta indica*, and *Litsea petiolata*, exhibited significant activity with antimutagenic ED_{90} values lower than 5 µl/plate (0.1 mg of dry plant material equivalent) as assayed by the Ames test (Nakahara et al. 2002). Another research reported that extracts from *O. indicum* displayed anti-proliferative activity on MCF7 and MDA-MB-231 breast cancer cell lines in tests conducted on medicinal plants in Bangladesh (Lambertini et al. 2004). *O. indicum* extract also displayed inhibition of Heinz body induction in an in-vitro model and also possessed antioxidant activity as assayed by the ABTS radical cation decolourization assay indicating its potential anticarcinogenic activity.

Of 11 plants used in Bangladeshi folk medicine tested using the brine shrimp lethality assay, sea urchin eggs assay, hemolysis assay and MTT assay using tumour cell lines, the extract of *Oroxylum indicum* showed the highest toxicity on all tumour cell lines tested, with an IC_{50} of 19.6 µg/ml for CEM (lymphoblastic leukaemia cell line), 14.2 µg/ml for HL-60 (human leukaemia-60 cell line), 17.2 µg/ml for B-16 (mouse melanoma cell line) and 32.5 µg/ml for HCT-8 (intestinal adenocarcinoma cell line) (Costa-Lotufo et al. 2005). On the sea urchin eggs assay, it inhibited the progression of cell cycle since the first cleavage (IC_{50} =13.5 µg/ml). Only the extracts of *Oroxylum indicum, Moringa oleifera* and *Aegles marmelos* could be considered as potential sources of anticancer compounds.

A methanolic extract of the fruits of *Oroxylum indicum*, which is widely used in traditional Chinese herbal medicine for its anti-inflammatory, anti-pyretic and anti-hypersensitivity effects, inhibited in-vitro proliferation of HL-60 cells (human leukaemia-60 cell line) (Roy et al. 2007). The flavonoid baicalein was found as an active component in the extract. Baicalein caused a 50% inhibition of proliferation HL-60 cells at concentrations of 25–30 µM. The inhibition of proliferation of HL-60 cells due to 36–48 hours exposure to 10 or 20 µM baicalein was associated with the accumulation of cells at S or G2M phases. However, proliferation inhibition at a higher dose may be associated with induction by apoptosis, as evidenced by the typical nuclear fragmentation using DNA fragmentation assay and terminal deoxynucleotidyl transferase-mediated dUTP nick end labelling (TUNEL). The results indicated that baicalein had anti-tumour effects on human cancer cells, and *Oroxylum indicum* extract could be used in supplementary cancer therapy.

From the fruit pods of *Oroxylum indicum*, two flavones, oroxylin A and chrysin, and a triterpene carboxylic acid, ursolic acid (UA), were identified as inhibitors of O^{2-} generation in xanthine (XA)-xanthine oxidase (XOD) assay XA/AOD system (Jiwajinda et al. 2002). These compounds also exhibited marked inhibitory effects on the tumour promoter 12-O-tetradecanoylphorbol-13-acetate (TPA)-induced O^{2-} generation in dimethylsulfoxide (DMSO)-differentiated HL-60 cells. The results indicated *O. indicum* to be a promising source of antioxidants with chemopreventive potential.

The root bark of *O. indicum* was found to contain flavonoids that showed cytotoxicity activity. The cytotoxicity of two compounds 2,5-dihydroxy-6,7-dimethoxy flavone (1) and 3,7,3′,5′-tetramethoxy-4-hydroxy flavone (2) and crude petroleum ether, ethyl acetate and methanol extracts was demonstrated by brine shrimp lethality bioassay (Uddin et al. 2003a). The LC_{50} values of compound 1 and 2 and petroleum ether, ethyl acetate and methanol extracts were found to be 14.12, 3.80, 3.97, 6.02 and 8.70 µg/ml, respectively.

Oroxylum indicum was one of 29 positive plants out of 93 plant species that exhibited in-vitro photo-cytotoxic activity when screened by means of a cell viability test using a human leukaemia cell-line HL60 (Ong et al. 2009). The plant extract was able to reduce the in-vitro cell viability by more than 50% when exposed to 9.6 J/cm(2) of a broad spectrum light when tested at a concentration of 20 µg/mL.

Several natural flavonoid compounds isolated from *Oroxylum indicum* exhibited in-vitro proprotein convertase inhibitory property when tested against the fluorogenic peptide substrate, Boc-RVRR-MCA (Boc = tert-butyloxy carbonyl, MCA = 4-methyl coumarin7-amide) (Majumdar et al. 2010). These flavonoids also obstructed efficiently the proportein convertase (PC) 4-mediated processing of a fluorogenic peptide derived from the processing site of its substrate, pro-Insulin Growth Factor-1 (proIGF-1). Proprotein Convertases (PCs) belong to a family of calcium-dependent endoproteases that play major roles in the processing of inactive precursor proteins producing their bioactive mature forms that are implicated in a wide variety of diseases including cancer, viral and bacterial infections. This anti-protease activity may provide a rationale for the observed anticancer and anti-HIV properties of some of these flavonoid compounds. In another study, the menthol and aqueous bark extracts of *O. indicum* demonstrated extensive cytotoxicity in both in MDA-MB-435S (melanoma cell line) and Hep3B (human hepatoma) cell lines, with the methanolic extract showing higher cytotoxic potential (Ashok Kumar et al. 2010).

Antimutagenic Activity

The methanol extracts of five Thai plants, *Micromelum minutum, Oroxylum indicum, Cuscuta chinensis, Azadirachta indica,* and *Litsea petiolata,* were found to exhibit significant activity with antimutagenic ED_{90} values lower than 5 µl/plate using the Ames Test (Nakahara et al. 2002). In another paper, Nakahara et al. (2001), reported that a methanolic extract of *Oroxylum indicum* strongly inhibited the mutagenicity of Trp-P-1 in an Ames test. The major antimutagenic constituent was identified as baicalein with an IC_{50} value of 2.78 µM. The potent antimutagenicity of the extract was correlated with the high content (3.95%, dry weight) of baicalein. Baicalein acted as a desmutagen since it inhibited the N-hydroxylation of Trp-P-2.

Gastroprotective/Antiulcerogenic Activity

The petroleum ether (96%) and n-butanol (99%) fractions of *O. indicum* root bark showed maximum inhibition of gastric lesions against ethanol-induced gastric mucosal damage (Khandhar et al. 2006). The results were comparable with the reference standard, omeprazole. In the ethanol-induced gastric ulcer model, treatment with both the active fractions and omeprazole showed significant antioxidant activity as evident from the reduction in the extent of lipid peroxidation that was measured in terms of malondialdehyde (MDA), along with significant increases in the superoxide dismutase (SOD), catalase (CAT), and reduced glutathione levels (GSH), when compared with the control group. Baicalein was found to be a major flavonoid present in both petroleum ether and n-butanol hydrosylate. The mechanism of its anti-ulcer activity was attributed to a reduction in gastric acid secretion and antioxidant activities leading to gastric cytoprotection. This activity was postulated to be linked to the presence of baicalein in the root bark. Two new flavonoid glycosides (1, 2), along with seven known compounds (3–9) were isolated from the stem bark of *Oroxylum indicum* (Hari Babu et al. 2010). Anti-ulcer activity of the compounds were studied on various models in rats.

Anti-inflammatory Activity

The aqueous extracts of stem bark *O. indicum* and *Derris scandens* significantly reduced myeloperoxide release effects associated with anti-inflammatory activity partially supporting the use of the plant for arthritis in Thailand (Laupattarakasem et al. 2003). Research in Vietnam reported that *Oroxylum indicum*'s flavonoids had anti-inflammatory effects in-vivo in dextran-induced oedema of mouse paw (Lê and Nguyễn 2005). When *Oroxylum indicum*'s flavonoids were combined with α-chymotrypsin, their anti-inflammatory activity was augmented. Siriwatanametanon et al. (2010) provided in-vitro evidence for the use of the Thai plants, most importantly *Gynura pseudochina* var. *hispida, Oroxylum indicum* and *Muehlenbeckia platyclada* as Thai anti-inflammatory remedies. Among all the nine species tested, *Gynura pseudochina* var. *hispida* and *Oroxylum indicum* showed the most promising NF-kappaB inhibitory effects with the lowest IC_{50} values (41.96 and 47.45 μg/ml, respectively). Additionally, *Oroxylum indicum* exhibited a high level of antioxidant activity by inhibiting lipid-peroxidation (IC_{50} 0.08 μg/ml).

Antimicrobial Activity

Ali et al. (1998) reported that chrysin (5,7-dihydroxyflavone) from *O. indicum*, at 5 μg concentration inhibited the growth of the Gram negative bacteria *Escherichia coli* and *Pseudomonas aeruginosa* at a rate comparable to that of streptomycin. Baicalein, in contrast, with an additional hydroxyl group at C-6, did not inhibit the growth of Gram-negative bacteria and showed a weak effect against the Gram-positive, spore-forming *Bacillus subtilis* and the Gram-positive coccus *Staphylococcus aureus*.

The crude petroleum ether and ethyl acetate extracts of *O. indicum* and the two flavonoids 2,5-dihydroxy-6,7-dimethoxy flavone (Compound 1) and 3, 7, 3′, 5′-tetramethoxy-4′-hydroxy flavone (Compound 2) showed mild to moderate activity against all bacteria (five Gram-positive and nine Gram-negative) and seven pathogenic fungi, whereas the methanol extract showed little activity against bacteria but moderate activity against fungi (Sayeed et al. 2003; Uddin et al. 2003b) The minimum inhibitory concentration (MIC) of the compound 1 and 2 against *Bacillus subtilis, Staphylococcus aureus, Escherichia* coli and *Shigella dysenteriae* were found to vary between 64 and 128 μg/ml.

Chrysin (5,7-dihydroxy flavone), obtained from *Oroxylum indicum*, was reported to exhibit numerous biological activities including anticancer, anti-inflammatory, and antiallergic activities (Suresh Babu et al. 2006). Most of the derivatives displayed significant activity against a panel of susceptible and resistant Gram-positive and Gram-negative organisms as compared to their parent compound (chrysin). The hexane bark extract of *O. indicum* showed significant antibacterial activity based on inhibition zones against *Bacillus megaterium* (12 mm), *Salmonella paratyphi* (12 mm), *Vibrio mimicus* (13 mm), *Vibrio parahemolyticus* (11 mm), *Pseudomonas aeruginosa* (11 mm) (Islam et al. 2010). The chloroform bark extract showed prominent activity against *Vibrio mimicus* (18 mm); significant activity against *Bacillus megaterium* (12 mm), *Salmonella paratyphi* (11 mm), *Vibrio parahemolyticus* (12 mm), *Salmonella typhi* (12 mm), *Shigella boydii* (11 mm), *Bacillus cereus* (11 mm), *Escherichia coli* (11 mm), *Bacillus subtilis* (11 mm). The carbon tetrachloride bark extract demonstrated significant zones of inhibition >15 mm, thus demonstrating potent activity against almost all bacterial strains except *Shigella dysenteriae* (10 mm), *Shigella boydii* (11 mm), which showed lower susceptibility to the crude extract. The extracts had inhibitory effects against all three fungi (*Saccharomyces cerevisiae, Candida albicans* and *Aspergillus niger*) tested with the carbon tetrachloride fraction being most active against *Candida albicans*, while the chloroform and hexane fraction were most active against *Aspergillus niger*.

From *Oroxylum indicum* stem bark, three flavones viz., baicalein, oroxylin and pinostrobin together with one sterol, stigmast-7-en-3-ol were isolated (Luitel, et al. 2010). Baicalein and oroxylin were found to be active against brine shrimp with LC_{50} values 10.0 and 36.0 µg/ml and also exhibited antimicrobial activity on both Gram-positive and Gram-negative bacteria with MIC values 4.0 and 8.0 mg/ml respectively.

Immunomodulatory Activity

The root bark of *O. indicum* was found to have immuno-stimulating/immunomodulatory activity (Zaveri et al. 2006). Research showed that treatment with the n-butanol fraction of *O. indicum* root bark evoked a significant elevation in circulating haemagglutinating antibody titers during secondary antibody responses, indicating a potentiation of certain aspects of the humoral response. The treatment also resulted in a prominent elevation in paw edema formation, indicating augmented host delayed-type hypersensitivity (DTH) response. Additionally, the antioxidant potential of the drug was exhibited by large reductions in whole blood malondialdehyde (MDA) content along with an elevation in the activities/levels of superoxide dismutase (SOD), catalase (CAT) and reduced glutathione (GSH). Furthermore, histopathologic analysis of lymphoid tissues exhibited an augmentation in cellularity, e.g., T-lymphocytes and sinusoids, in the treatment group. The reported immunomodulatory activity of active *O. indicum* fraction may be attributed to its ability to enhance specific immune responses (both humoral and cell-mediated) as well as its antioxidant potential.

Genotoxic and Mutagenic Activity

A study reported that oral administration of a nitrosated *Oroxylum indicum* fraction to rats, was mutagenic without S9 mix to *Salmonella typhimurium* TA98 and TA100 (Tepsuwan et al. 1992). Administration of the nitrosated fraction at doses of 0.7–2.8 g/kg body weight also induced dose-dependent increases, up to 11-fold, in replicative DNA synthesis in the stomach pyloric mucosa 16 h after its administration. Further, administration of the nitrosated fraction at doses of 0.25–2.0 g/kg body weight elicited dose-dependent rises, up to 100-fold, in ornithine decarboxylase activity in the stomach pyloric mucosa with a peak 4h after its administration. These results demonstrated that the nitrosated *O. indicum* fraction had genotoxic and cell proliferative activity in the pyloric mucosa of rat stomach in-vivo.

Traditional Medicinal Uses

Oroxylum indicum is widely regarded as medicinal from India to Indonesia and use in folk medicine. The bitter bark is employed for intestinal complaints. It is astringent and tonic and used mainly for diarrhoea and dysentery. A bark decoction has been used for dysentery in Malaysia and for constipation in the Philippines, where the bark was also deemed antidysenteric, diaphoretic and antirheumatic. The leaves have wider application. A leaf decoction is drunk for stomach-ache; it causes eructations and so brings relief. It is also taken for loss of appetite and rheumatism. The leaf decoction is also used externally as a hot fomentation in cholera, fever, childbirth and rheumatic swellings. Boiled leaves are employed as application in labour, after labour and for dysentery and hot leaves applied upon the enlarged spleen. Leaves are applied to the cheek for toothache and as poultices for headache. Any part of the tree may be used for making a decoction used externally in childbirth. The bitter seed are used in bolus for rheumatism.

In Vietnam, the bark is used in treating jaundice, urticaria, angina, cough, gastralgia, inflammation of urinary tract, urodynia, hematuria, infantile rash and measles. Seeds are used for treating cough, bronchitis, gastralgia, abdominal pains and slow healing ulcers.

In India, *Oroxylum indicum* is commonly used in folk medicine as a cure of various diseases. The root bark is used in fever, bronchitis, intestinal worms, leukoderma, asthma, inflammation, anal troubles as tonic, astringent, antidiarrheal, antidysenteric, diaphoretic and antirheumatic drugs. The decoction of the bark is taken for curing gastric ulcer and the paste of the bark is applied to mouth for cancer, scabies, tonsil pain and other diseases. The paste is also used for the animal wounds and the decoction for de-worming. The fruit and seeds are used as expectorant, purgative and bitter tonic. In Hindu medicine, the root, bark, stem and leaf are prescribed for snake- bites, diarrhoea and dysentery.

The decoction of the bark is taken for curing gastric ulcer and a paste made of the bark powder is applied for mouth cancer, scabies and other skin diseases. The seed is ground with fire-soot and the paste applied to the neck for quick relief of tonsil pain. Also, a paste made of the bark is applied to the wounds of animals to kill maggots. Decoction of the bark is given to animals for de-worming. The seeds and bark are used medicinally for alleviating pain and as an antiphlogistic medicine.

Other Uses

The tree is sometimes grown as an ornamental for its strange appearance. The soft wood is used for matchwood in the Philippines. Tanning and dyeing can be done with the pods and bark. Dye obtained from the bark is used to decorate rattan baskets. The seeds are used as cattle medicine in India. The sword-like fruit or a branch of the tree is used by the farmers to kill crabs in wet paddy fields.

Comments

Flowers are bat-pollinated and blooms at night emitting a strong, fetid odour which attracts bats.

Selected References

Ali RM, Houghton PJ, Raman A, Hoult JRS (1998) Antimicrobial and antiinflammatory activities of extracts and constituents of *Oroxylum indicum*. Phytomed 5:375–381

Ali M, Chaudhary A, Ramachandram R (1999) New pterocarpans from *Oroxylum indicum* stem bark. Indian J Chem (Sect B) 38:950–952

Ashok Kumar R, Rajkumar V, Guha G, Mathew L (2010) Therapeutic potentials of *Oroxylum indicum* bark extracts. Chin J Nat Med 8(2):121–126

Burkill IH (1966) A dictionary of the economic products of the Malay Peninsula. Revised reprint, 2 volumes, vol 1 (A–H) pp 1–1240, vol 2 (I–Z) pp 1241–2444. Ministry of Agriculture and Co-operatives, Kuala Lumpur

Chen LJ, Games DE, Jones J (2003) Isolation and identification of four flavonoid constituents from the seeds of *Oroxylum indicum* by high-speed counter-current chromatography. J Chromatogr A 988(1):95–105

Chen LJ, Song H, Lan XQ, Games DE, Sutherland IA (2005) Comparison of high-speed counter-current chromatography instruments for the separation of the extracts of the seeds of *Oroxylum indicum*. J Chromatogr A 1063(1–2):241–245

Costa-Lotufo LV, Khan MT, Ather A, Wilke DV, Jimenez PC, Pessoa C, de Moraes ME, de Moraes MO (2005) Studies of the anticancer potential of plants used in Bangladeshi folk medicine. J Ethnopharmacol 99(1):21–30

Council of Scientific and Industrial Research (1966) The wealth of India. A dictionary of Indian raw materials and industrial products (Raw materials 7). Publications and Information Directorate, New Delhi

Foundation for Revitalisation of Local Health Traditions (2008) FRLHT Database. http://envis.frlht.org

Hari Babu T, Manjulatha K, Suresh Kumar G, Hymavathi A, Tiwari AK, Purohit M, Madhusudana Rao J, Suresh Babu K (2010) Gastroprotective flavonoid constituents from *Oroxylum indicum* Vent. Bioorg Med Chem Lett 20(1):117–120

Islam MK, Eti IZ, Chowdhury JA (2010) Phytochemical and antimicrobial analysis on the extracts of *Oroxylum indicum* Linn. stem-bark. Iran J Pharm Ther 9:25–28

Jiwajinda S, Santisopasri V, Murakami A, Kim OK, Kim HW, Ohigashi H (2002) Suppressive effects of edible Thai plants on superoxide and nitric oxide generation. Asian Pac J Cancer Prev 3(3):215–223

Khandhar M, Sha M, Santani D, Jain S (2006) Antiulcer activity of the root bark of *Oroxylum indicum* against experimental gastric ulcers. Pharm Biol 44(5):363–370

Kuo, G. 2002. Project 9. Collaborative research and networks for vegetable production. AVRDC progress report 2002, pp 116–122

Lambertini E, Piva R, Khan MT, Lampronti I, Bianchi N, Borgatti M, Gambari R (2004) Effects of extracts from Bangladeshi medicinal plants on in vitro proliferation of human breast cancer cell lines and expression of estrogen receptor alpha gene. Int J Oncol 24(2):419–423

Laupattarakasem P, Houghton PJ, Hoult JR, Itharat A (2003) An evaluation of the activity related to inflammation of four plants used in Thailand to treat arthritis. J Ethnopharmacol 85(2–3):207–215

Lê TDH, Nguyễn XT (2005) Study on influence of flavonoids from *Oroxylum indicum* Vent. towards alpha-chymotrypsine in relationship with the process of inflammation. TC Dược Học 8:23–26 (In Vietnamese)

Liu R, Xu L, Li A, Sun A (2010) Preparative isolation of flavonoid compounds from *Oroxylum indicum* by high-speed counter-current chromatography by using ionic liquids as the modifier of two-phase solvent system. J Sep Sci 33(8):1058–1063

Luitel HN, Rajbhandari M, Kalauni SK, Awale S, Kazuo Masuda K, Gewali MB (2010) Chemical constituents from *Oroxylum indicum* (L) Kurz of Nepalese Origin. Sci World 8(8):66–68

Majumdar S, Mohanta BC, Chowdhury DR, Banik R, Dinda B, Basak A (2010) Proprotein convertase inhibitory activities of flavonoids isolated from *Oroxylum indicum*. Curr Med Chem 17(19):2049–2058

Mao AA (2002) *Oroxylum indicum* Vent. – a potential anticancer medicinal plant. Indian J Tradit Knowl 1(1):17–21

Mishra SL, Sinhamahapatra PK, Nayak A, Das R, Sannigrahi S (2010) *In vitro* antioxidant potential of different parts of *Oroxylum indicum*: a comparative study. Indian J Pharm Sci 72:267–269

Nair AGR, Joshi BS (1979) Oroxindin – A new flavone glucuronide from *Oroxylum indicum* Vent. J Chem Sci 88(5):323–327

Nakahara K, Onishi KM, Oro H, Yoshida M, Trakoontivakorn G (2001) Antimutagenic activity against Trp-p-1 of the edible Thai plant, *Oroxylum indicum* Vent. Biosci Biotechnol Biochem 65:2358–2360

Nakahara K, Trakoontivakorn G, Alzoreky NS, Ono H, Onishi-Kameyama M, Yoshida M (2002) Antimutagenicity of some edible Thai plants, and a bioactive carbazole alkaloid, mahanine, isolated from *Micromelum minutum*. J Agric Food Chem 50(17):4796–4802

National Institute of Materia Medica (1999) Selected medicinal plants in Vietnam, vol 2. Science and Technology Publishing House, Hanoi, 460 pp

Nguyen, VD and Doan TN (1989) *Medicinal Plants in Vietnam* World Health Organization (WHO), Regional Publications, Western Pacific Series No 3. WHO, Regional Office for the Western Pacific, Manila, the Philippines and Institute of Materia Medica, Hanoi, Vietnam.

Ochse JJ, Bakhuizen van den Brink RC (1980) Vegetables of the Dutch Indies, 3rd edn. Ascher & Co., Amsterdam, 1016 pp

Ong CY, Ling SK, Ali RM, Chee CF, Samah ZA, Ho AS, Teo SH, Lee HB (2009) Systematic analysis of in vitro photo-cytotoxic activity in extracts from terrestrial plants in Peninsula Malaysia for photodynamic therapy. J Photochem Photobiol B 96(3):216–222

Palasuwan A, Soogarun S, Lertlum T, Pradniwat P, Wiwanitkit V (2005) Inhibition of Heinz body induction in an in vitro model and total antioxidant activity of medicinal Thai plants. Asian Pac J Cancer Prev 6(4):458–463

Pongpangan S, Poobrasert S (1991) Edible and poisonous plants in Thai forests. O.S. Printing House, Bangkok, 176 pp

Raghbir C, Gupta VS, Sharma N, Kumar N, Singh B (2008) In vitro antioxidant activity from leaves of *Oroxylum indicum* (L.) Vent. – a north Indian highly threatened and vulnerable medicinal plant. J Pharm Res 1(1):65–71

Reddy KN, Pattanaik C, Reddy CS, Raju VS (2007) Traditional knowledge on wild food plants in Andhra Pradesh, India. Indian J Tradit Knowl 6(1):223–229

Roy MK, Nakahara K, Na TV, Trakoontivakorn G, Takenaka M, Isobe S, Tsushida T (2007) Baicalein, a flavonoid extracted from a methanolic extract of *Oroxylum indicum* inhibits proliferation of a cancer cell line in vitro via induction of apoptosis. Pharmazie 62(2):149–153

Saidin I (2000) Sayuran Tradisional Ulam dan Penyedap Rasa. Penerbit Universiti Kebangsaan Malaysia, Bangi, p 228 (In Malay)

Sayeed A, Islam A, Rahman AA, Mohal Khan GRMA, Uddin K, Sadik MG, Khatun S (2003) Biological activities of extracts and two flavonoids from *Oroxylum indicum* Vent. (Bignoniaceae). J Biol Sci 3(3):371–375

Siriwatanametanon N, Fiebich BL, Efferth T, Prieto JM, Heinrich M (2010) Traditionally used Thai medicinal plants: in vitro anti-inflammatory, anticancer and antioxidant activities. J Ethnopharmacol 130(2):196–207

Suresh Babu K, Hari Babu T, Srinivas PV, Hara Kishore K, Murthy US, Rao JM (2006) Synthesis and biological evaluation of novel C (7) modified chrysin analogues as antibacterial agents. Bioorg Med Chem Lett 16(1):221–224

Tanaka Y, Nguyen VK (2007) Edible wild plants of Vietnam: the Bountiful Garden. Orchid Press, Bangkok, 175 pp

Tepsuwan A, Furihata C, Rojanapo W, Matsushima T (1992) Genotoxicity and cell proliferative activity of a nitrosated *Oroxylum indicum* Vent fraction in the pyloric mucosa of rat stomach. Mutat Res 281(1):55–61

Uddin K, Sayeed A, Islam A, Rahman AA, Ali A, Astaq Mohal Khan GRM, Sadik MG (2003a) Purification, characterization and cytotoxic activity of two flavonoids from *Oroxylum indicum* Vent. (Bignoniaceae). Asian J Plant Sci 2(6):515–518

Uddin K, Sayeed A, Islam A, Rahman AA, Khatun S, Astaq Mohal Khan GRM, Sadik MG (2003b) Biological activities of extracts and two flavonoids from *Oroxylum indicum* Vent. (Bignoniaceae). J Biol Sci 3(3):371–375

Vimal K, Gogoi BJ, Meghvansi MK, Singh L, Srivastava RB, Deka DC (2009) Determining the antioxidant activity of certain medicinal plants of Sonitpur, (Assam), India using DPPH assay. J Phytol 1(1):49–56

Yang RY, Tsou SCS, Lee TC, Wu WJ, Hanson PM, GKuo G, Engle LM, Lai PY (2006) Distribution of 127 edible plant species for antioxidant activities by two assays. J Sci Food Agric 86(14):2395–2403

Yin WG, Li ML, Kang C (2007) Advances in the studies of *Oroxylum indicum*. Zhongguo Zhong Yao Za Zhi 32(19):1965–1970 (In Chinese)

Yuan Y, Luo H, Chen L (2008) Linear scale-up of the separation of active components from *Oroxylum indicum* using high-speed counter-current chromatography. Se Pu 26(4):489–493 (In Chinese)

Zaveri M, Dhru B (2010) In-vitro antioxidant potential of stem and root bark of *Oroxylum indicum*. J Global Pharma Technol 2(8):41–48

Zaveri M, Gohil P, Jain S (2006) Immunostimulant activity of n-butanol fraction of root bark of *Oroxylum indicum* Vent. J Immunotoxicol 3(2):83–99

Zaveri M, Khandhar A, Jain S (2008) Quantification of baicalein, chrysin, biochanin-a and ellagic acid in root bark of *Oroxylum indicum* by RP-HPLC with UV detection. Eur J Anal Chem 3:245–257

Parmentiera aculeata

Scientific Name

Parmentiera aculeata (Kunth) Seem.

Synonyms

Crescentia aculeata Kunth, *Crescentia edulis* Moc. ex A. DC., *Crescentia musaecarpa* Zaldivar ex C. Heller, *Parmentiera aculeata* (Kunth) L.O. Williams, *Parmentiera edulis* DC., *Parmentiera edulis* Raf., *Parmentiera foliolosa* Miers, *Parmentiera lanceolata* Miers.

Family

Bignoniaceae

Common/English Names

Candle Tree, Cucumber Tree, Cow Okra, Food Candle Tree, Yellowtaper Candletree

Vernacular Names

Belize: K'at (Maya), Cow Okra, Kat, Wild Okra;
Costa Rica: Cuajilote (Spanish);
Dominican Republic: Guineito (Spanish);
El Salvador: Cuajilote (Spanish);
French: Parmentiéra Épineux;
Guatemala: Cuajilote, Guajilote (Spanish);
Mexico: Auve-Quec (Chontal), Tzote (Hausteco), Chote, Chote Cuáhuitl, Coxilotl, Cuaxilot "Palo De Jilote", Cuaxilotl (Nahuatl), Tyacuanajun (Mixteco), Camburito, Chachi, Chayote, Chote, Chotes, Cuachilote, Cuajiote, Cuajilote, Guachilote, Guajilote, Peino De Árbol, Turi (Spanish), Puch (Tepehua), Puxni (Totonaco), Guetoxiga (Zapoteco), Tzutzu (Zoque), Cuajxilutl.

Origin/Distribution

The species is native to Central America – Mexico, Belize, Costa Rica, El Salvador; Guatemala and Honduras. It has naturalized in northern Australia.

Agroecology

The species occurs in moist or dry thickets or lowland tropical forest, often along rocky watercourses, mainly at 1,200 m elevation or less in its native range. It is hardy, but needs a frost–free, well-drained, moderately fertile site and prefers full sun. It is cultivated commonly around dwellings in the drier regions.

Edible Plant Parts and Uses

Its short, plump, banana-shaped fruit is edible and sweet and has a flavour similar to sugar cane. It is eaten raw (fresh), cooked or roasted, or made into pickles and preserves.

Botany

Small or medium-sized, evergreen tree to 10 m high with a short thick trunk and pale bark, spreading crown, the branches subterete, usually armed with a short, stout spine at each node. Leaves are opposite or sub-opposite, often alternate on young branchlets, often fasciculate in the axils of the 2 spines, 3–5 leaflets (usually 5), the petioles long usually narrowly winged (Plates 1–3). Leaflets are entire, elliptic to obovate, each 4–6 cm long. Flowers are large or small, campanulate, greenish to greenish-white with brown purple lines, solitary or fasciculate from nodes on old wood along branches and trunk, or terminating the branchlets. Calyx is spathaceous, 2.5–3.5 cm. long, closed in bud, cleft on one side in anthesis and soon deciduous, glandular-lepidote. Corolla is campanulate (trumpet-shaped) and somewhat curved, 5–6.5 cm long, the limb somewhat two-lipped. Stamens are slightly exserted with glabrous anthers. Disc large and pulvinate. Ovary is oblong, glandular-lepidote and 2-celled. Fruit is subcylindric, candle-shaped, short and plump, 10–12 cm long, 7 cm wide, indehiscent, costate (ridged) and often slightly curved, pointed, yellow-green to dull yellow and waxy

Plate 1 Fruits ripe and immature and leaves with leaflets

Plate 2 Pendulous immature fruits

(Plates 1–4), with fleshy, edible pericarp. Seeds are small, numerous, not winged, imbedded in the sweetish pulp.

Nutritive/Medicinal Properties

P. aculeata is one of many plants that are used in Mexican folk medicine as herbal extracts for the treatment of diabetes mellitus (Andrade-Cetto and Heinrich 2005). Studies confirmed that the plant possessed hypoglycaemic activity. Studies showed that the administration of 300 mg/kg of chloroform extracts from *Parmentiera edulis* and *Bouvardia terniflora* and hexane from *Brickellia veronicaefolia* to alloxan-induced diabetic mice decreased the blood glucose levels in 43.75%, 58.56% and 72.13%, respectively (Pérez-Gutiérrez et al. 1998). These extracts administered to normal mice reduced blood glucose levels in 29.61%, 33.42% and 39.84%, respectively. The results

Plate 3 Ripe fruit among the dense foliage

Plate 4 Close-up of ripe fruits

supported the use of these of plant extracts in traditional medicine for diabetes treatment. In another study, a guaianolide compound, lactucin-8-O-methylacrylate, was isolated from the chloroform extract of the dried fruits of *Parmentiera edulis* (Perez et al. 2000). The compound lowered blood sugar levels after administration to alloxan-diabetic mice.

Studies in Guatemala reported that fruits of *Parmentiera edulis* exhibited significant in-vitro inhibitory activity against *Neisseria gonorrhoeae* strains isolated from symptomatic patients (Cáceres et al. 1995). *N. gonorrhoeae* is a gram negative bacterium responsible for the sexually transmitted disease, gonorrhoea.

Traditional Medicinal Uses

All parts of the plant, roots, branches, leaves, bark, flowers and fruits are used in traditional medicine of indigenous Mexican communities since ancient days. The Aztecs used it for renal disorders, dropsy, gastric indigestion, colds and ear infections (otitis). They drank daily a tea made from the leaves for ear infections they soaked a cotton ball in this mixture and inserted in the ear. The plant is used in herbal remedies for diabetes. The fruit is reported to be a good remedy for colds, and the roots are used as a diuretic in traditional medicine.

Other Uses

The species is frequently cultivated as a shade tree and ornamental in the tropics. It is especially cultivated in Mexico for its many uses - food, fodder, production of domestic tools, handicrafts and medicine.

Comments

See also notes on *Parmentiera cereifera*.

Selected References

Andrade-Cetto A, Heinrich M (2005) Mexican plants with hypoglycaemic effect used in the treatment of diabetes. J Ethnopharmacol 99(3):325–348
Cáceres A, Menéndez H, Méndez E, Cohobón E, Samayoa BE, Jauregui E, Peralta E, Carrillo G (1995) Antigonorrhoeal activity of plants used in Guatemala for the treatment of sexually transmitted diseases J. Ethnopharmacology 48(2):85–8
Gentry AH (1980) Bignoniaceae part I (Crescentieae and Tourrettieae). In: Organization for Flora Neotropica, ed. Flora Neotrop Monogr 25(1): 100–102
Gentry AH (1992) A synopsis of Bignoniaceae ethnobotany and economic botany. Ann Mo Bot Gard 79(1): 53–64
Maisch JM (ed) (1885) Materia Medica of the New Mexican Pharmacopoeia Part 3. Am J Pharm 57(9). http://www.swsbm.com/AJP/AJP_1885_No_9.pdf
Perez RM, Perez C, Zavala MA, Perez S, Hernandez H, Lagunes F (2000) Hypoglycemic effects of lactucin-8-O-methylacrylate of *Parmentiera edulis* fruit. J Ethnopharmacol 71(3):391–4
Pérez-Gutiérrez RM, Pérez-González C, Zavala-Sánchez MA, Pérez-Gutiérrez S (1998) Hypoglycemic activity of *Bouvardia terniflora*, *Brickellia veronicaefolia*, and *Parmentiera edulis*. Salud Pública Méx 40(4): 354–8
Rehm S, Espig G (1991) The cultivated plants of the tropics and subtropics: cultivation, economic value, utilization. Margraf, Weikersheim, 552 pp
van Steenis CGGJ (1977) Bignoniaceae. In: van Steenis CGGJ (ed) Flora Malesiana, Ser. I, 8. Sijthoff & Noordhoff Alphen aan den Rijn, 577 pp

Parmentiera cereifera

Scientific Name

Parmentiera cereifera Seem.

Synonyms

None

Family

Bignoniaceae

Common/English Names

Candle Tree, Candlestick Tree, Panama Candle Tree

Vernacular Names

Brazil: Árvore-De-Velas (Portuguese)
Costa Rica: Palo De Velas;
Dominican Republic: Arbol De Vela, Palo De Cera, Palo De Vela;
Dutch: Kaarsenboom;
French: Parmentiéra Du Panama;
German: Kerzenbaum;
Indonesia: Pohon lilin;
India: Mom Brikshya (Bengali), Mom Batti (Hindu);
Panama: Velo;
Puerto Rico: Palo De Vela;
Venezuela: Palo De Vela.

Origin/Distribution

The species is native to Central America – Panama and dispersed to many neighbouring countries and warm tropical countries as an ornamental.

Agroecology

The species grows well in tropical regions but will survive light frosts. Typically found in thicket vegetation at an altitude of sea level to 881 m. The Candlestick Tree is fairly adaptable to a range of soils and climatic conditions.

Edible Plant Parts and Uses

Fruits and seeds are edible in the same way as *P. aculeata*.

Botany

A small, evergreen, erect tree to 6 m high. Leaves opposite on young branchlets, trifoliate, on winged, 1.5–5 cm long petioles (Plate 4). Leaflets sub-sessile, glabrous, elliptic or elliptic obovate,

Plate 1 Trunk and cauliflorous, immature fruit

Plate 2 Green, immature fruit and greenish – white flowers

Plate 3 Cluster of ridged, elongate, candle-like fruit

Plate 4 Trifoliolate leaves with entire, simple leaflets

acuminate, simple, entire, 1.5–7 cm by 0.75–3.5 cm. Flowers solitary or in 2–4 flowered fascicles on old wood, pedicels 1.75–3 cm, calyx 3–4 cm long, corolla greenish-white, campanulate, recurved 5 broadly rounded lobes, stamens 4 inserted at base of widened part of the corolla tube (Plates 1–2). Fruits cauliflorous, pendulous, elongate, tapering pointed, candle-shaped berry 40–60 cm long by 1.5–2.5 cm, green turning waxy, yellow when ripe (Plates 1–3), seeds ellipsoid, compressed, 2.5–3.5 mm long embedded in fragrant edible, fleshy pulp.

Nutritive/Medicinal Properties

The plant is grown as an ornamental and medicinal plant. The leaves contain an unidentified glycoside, and traces of hydrocyanic acid have been detected in the fruits, leaves and roots (Deshaprabhu 1966; Quisumbing 1978).

Other Uses

The tree is also commonly grown as an ornamental for its strange appearance when fruits are ripening. Fruits and leaves have been fed to cattle in Panama.

Comments

See also notes on the other closely related edible species, *Parmentieria aculeata*.

Selected References

Backer CA, Bakhuizen van den Brink RC Jr (1965) Flora of Java, (spermatophytes only), vol 2. Wolters-Noordhoff, Groningen, 641 pp

Burkill HM (1985) The useful plants of West Tropical Africa, vol 1, Families A–D. Royal Botanical Gardens, Kew, 960 pp

Deshaprabhu SB (ed) (1966) The wealth of India, raw materials, vol 7, N–Pe. C.S.I.R, New Delhi, 330 pp

Gentry AH (1980) Bignoniaceae part I (Crescentieae and Tourrettieae). In: Organization for Flora Neotropica (ed), Flora Neotrop Monogr 25(1): 100–102

Gentry AH (1992) A synopsis of Bignoniaceae ethnobotany and economic botany. Ann Mo Bot Gard 79(1):53–64

Quisumbing E (1978) Medicinal plants of the Philippines. Katha Publishing Co, Quezon City, 1262 pp

Rehm S, Espig G (1991) The cultivated plants of the tropics and subtropics: cultivation, economic value, utilization. Margraf, Weikersheim, 552 pp

van Steenis CGGJ (1977) Bignoniaceae. In: van Steenis CGGJ (ed) Flora Malesiana, Ser. I, 8. Sijthoff & Noordhoff Alphen aan den Rijn, 577 pp, pp 114–186

Bixa orellana

Scientific Name

Bixa orellana L.

Synonyms

Bixa acuminata Bojer, *Bixa americana* Poir., *Bixa katagensis* Delpierre, *Bixa odorata* Ruiz & Pav. ex G. Don, *Bixa orellana* f. *leiocarpa* (Kuntze) J.F. Macbr., *Bixa orellana* var. *leiocarpa* (Kuntze) Standl. & L.O. Williams, *Bixa orleana* Noronha, *Bixa purpurea* Sweet, *Bixa tinctaria* Salisb., *Bixa upatensis* Ram. Goyena, *Orellana americana* (Poir.) Kuntze, *Orellana americana* var. *leiocarpa* Kuntze, *Orellana americana* var. *normalis* Kuntze, *Orellana orellana* (L.) Kuntze.

Family

Bixaceae

Common/English Names

Annato, Annatto, Annatto Plant, Annatto Seeds, Annatto Tree, Arnatto, Arnotto Dye Plant, Lipstick Plant, Lipstick Tree, Lipsticktree, Yellow Dye

Vernacular Names

Arabic: Galuga;
Bangladesh: Latkan, Utkana;
Brazil: Açafroa, Açafrão Do Brasil, Açafroeira Da Terra, Anato, Annato, Anoto, Colorau, Urucú, Urucú, Urucú Bravo, Urucú Da Mata, Urucú-Bravo, Urucú, Urucum (Portuguese);
Bulgarian: Achiote, Ačiote, Ачиоте;
Chamorro: Achiote, Achoti;
Chinese: Hong Mu, Yan Zhi Mu, Yan Zhi Shu;
Columbia: Achihuite;
Cook Islands: 'Uaeva (Maori);
Côte d'Ivoire: Kuiguéhé (Kru-Guere);
Creole: Chiót, Woukou Rocou;
Czech: Annata;
Danish: Annatobusken, Orleantræ, Smørfarvetræ;
Dutch: Achiote, Anatto, Annotto, Orleaan, Rocou;
Eastonian: Annatopõõsa, Värvibiksa;
Fijian: Nggesa, Nggisa, Qesa;
Finnish: Annaatto;
French: Achiote, Annato, Atole, Rocou, Rocouyer, Roucou, Roucou (West Indies), Roucouyer;
German: Anatto, Anattosamen, Anattostrauch, Annatto, Annattosamen, Orleansamen, Orleansbaum, Orleansstrauch, Orleanstrauch, Orleanstrauch Orlean-Strauch;
Ghana: Daagyene, Konin (Akan-Asante), Dagyiri (Brong), Á'jama (Ga), Bernitiku (Gbe-Vhe), Brɔfo Agyama (Twi), Daagyeni (Wasa);
Guatemala: Achiote (Mayan);

Guinea-Bissau: Djanfaraná (Crioulo), Djambaraná (Fula-Pulaar);
Hawaiian: 'Alaea, 'Alaea La'Au, Kūmauna;
Hungarian: Bjoul, Orleánfa, Orleánfa, Ruku;
India: Jolandhar, Jorot Goch (Assamese), Kong, Kuombi, Latkan, Lotkan, Sindure (Bengali), Sinduri (Gujarati), Gowpurgee, Latkan, Latkhan, Senduria, Sinduria, Sinduriya, Vatkana (Hindu), Aarnatu, Arnattu, Bangaara Kaayi Bangarakayi, Bhangarakai, Bhangarakayi, Chaayulitha, Chayalitha, Jaaphra, Jaaphredu, Japhredu, Japrero, Kappu Mankaali, Kappumankala, Kesari, Kesari Rangu, Kuppamanhale, Rangamalar, Rangamali, Ranggida, Rangoomalai, Rangu Maale, Rangu Malle, Rangumale, Rungamali, Sannajabbale, Sannajapali (Kannada), Korangumunga, Korungoomungal, Kurangamanchil, Kurannamannal, Kurannumannal, Kuppamanjal, Kuppamannal (Malayalam), Ureirom (Manipuri), Kaesari, Kesari, Keshri, Kesui, Kesuri, Kisri, Sendri, Shendri, (Marathi), Rawngsen (Mizoram), Jampra, Lotkons, Rawngsen (Oriya), Karachhada, Raktabija, Raktapushpa, Shonapushpi, Sindurapuspi, Sinduri, Sindurpushpi, Sunomala, Trivapushpi, Virpushpa (Sanskrit), Amudadaram, Amutataram, Amuttaram, Amuttiram, Aruna, Avam, Camankalikam, Cappira, Cappiran, Carani, Caruni, Curanacanimaram, Curattunacani, Irakumancal, Iram, Jaffra-Maram, Japhara, Kantukam, Karankumankal, Konkamaram, Konkaram, Konkaramam, Konkaramaram, Konkarayam, Konkari, Kunkumam, Kunguman, Kuppaimancal, Kurakumancal, Kurankumancal, Mancatti, Mancitti, Manjitti, Nakamucikai, Nakamucikaimaram, Naravam, Naravu, Naravucaram, Punkavi, Sabara, Sappiravirai, Sappiravirai, Tumpacitamaram, Turumam, Uragumanjai, Uragumanjal, Varagumanjalmantiravanci, Vennaivirai (Tamil), Jaabara, Jaabura, Jaapharaa Chettu, Jabaru Kaya, Jabura, Jaffra, Jaffrachettu, Jafra (Telugu);
Indonesian: Galuga, Galinggem (Sundanese), Galuga, Kasumba Kling, Pachar Kling, Somba Kling;
Italian: Annatto, Anotto;
Japanese: Achiote, Anatoo, Anatto, Beni No Ki, Beninoki;
Khmer: Châm´ Puu, Châm´ Puu Chrâluëk;
Korean: A Chi O Te, A Na To, A-Ci-O-Te, A-Na-To, Achiote, Anato, Bik Sa Sok;

Laotian: Kh´Am, Sômz Phuu, Satii;
Malaysia: Jarak Belanda, Kesum, Kesumba, Kesumba Kling, Kunyit Jawa, Sumpeh, Suntak;
Nahuatl: Achiote, Achiote Caspi, Achiotl;
Nepalese: Sindur, Sindure;
Nigeria: Úhíé, Úhíé Arọ̀ (Igbo), Mkpụlụ Ọ́fị́á (Igbo-Awka), Úfíé, Uhia Nkum (Igbo-Ibusa), Ula, Ula Machuku (Igbo-Onitsha), Úhíé Ṇkū, Úhíé (Igbo-Umuahia), Oçùn Búkẹ (Yoruba);
Palauan: Búrek, Burk;
Panama: Achote;
Papua New Guinea: Pen;
Philippines: Achuete (Bikol), Sotis (Cebu Bisaya), Asoti (Ibanag), Achiti, Achuete, Asuite, Atsuite (Iloko), Apatut (Gaddang), Achuete (Panay Bisaya), Chotes (Samar-Leyte Bisaya), Achuete, Atsiute (Sambali), Chanang, Janang (Sulu), Achote, Achoete, Achuete, Asuti, Atseuete (Tagalog);
Polish: Arnota, Arnota Właściwa;
Portuguese: Açafroa, Açafroa Do Brasil, Açafroeira-Da-Terra, Açafroa-Do-Brasil;
Puerto Rico: Anatta, Annato, Bija, Bija, Bijol, Chacuanguarica, Pumacua;
Russian: Achiote, Annato, Biksa, Biksa Orel´Ina, Pomadnoe Derevo;
Samoan: Loa, Loa;
Serbian: Orlean-Drvo;
Sierra Leone: Yellow Dye (English), Kamgo-Poto (Bulom-Kim), Bundu (Kono), Rɛd-Rɛd (Krio), Kudonia (Limba-Tonko), Mbundona (Loko), Mbundσ, Pu-Bundσ (Mende), Kamonyi (Susu), Lugbagbel-La (Susu-Dyalonke) A-Kam-A-Loli (Temne);
Slovak: Anatto;
Spanish: Achioytello, Roucou;
Swahili: Mzingefuri;
Swedish: Annattobuske;
Tahitian: 'Uaefa;
Thai: Kam Sêt, Kam Saed, Kam Tai, Kham Faet, Kham Ngae, Kham Ngo, Kham Saet (Bangkok), Kham Thai, Sati;
Togo: Berniticu (Gbe-Vhe), Kirane (Tem-Tshaudjo);
Turkish: Arnatto;
Venezuela: Onoto;
Vietnamese: Diêù Nhuôm, Ht Iu Màu, Siêm Phung;
Yapese: Rang.

Origin/Distribution

Annatto is native to continental tropical America. It has been introduced elsewhere in the tropics and sub-tropics becoming naturalized in the margins of forest, in thickets, and in waste places.

Agroecology

Annatto requires a frost-free, warm, humid climate with a mean annual temperature: 20–26°C and a sunny location up to an elevation of 2,000 m. It can grow in tropical to subtropical climates where rainfall is evenly distributed throughout the year with mean annual rainfall of 1,250–2,000 mm. It can be grown on almost all soil types, with a preference for well-drained, neutral and slightly alkaline soils. It grows into a larger tree when planted in deeper and more fertile soil, rich in organic matter. It is intolerant of shade.

Edible Plant Parts And Uses

In Mexico, Spain and the Philippines, the thin pigmented pulp covering the seeds is used as a spice/condiment. Annato has been used as a natural dye in a variety of food products, drugs and cosmetics, and also in Brazilian cuisine as a condiment (*colorau*). Annatto was introduced into the Philippines by the Spaniards and is widely used in the Philippines as a spice. The seeds are ground and used as a subtly flavoured and colourful additive in Latin American, Jamaican and Filipino cuisine. The seeds yield a non-carcinogenic dye, bixin which is extracted and used as colouring for food, salad oil, rice and dairy products butter, cheese (Edam, Munster, Red Leicester and Red Cheshire cheeses), margarine, ice cream and bakery products. The famous Jamaican dish of akee and slated cod fish is served in a sauce coloured with annatto. The seeds are sold as paste or powder. Annato is a main ingredient in the Yucatecan spice mixture *recado rojo*, or *achiote paste*. Annato is also a vital ingredient of *cochinita pibil*, the spicy pork dish popular in Mexico.

The red dye from achiote was used by the Mayan culture in Guatemala to colour and flavour cacao beverages. The roots has also been reported to impart the taste and the colour of saffron to meats.

Botany

An evergreen perennial arborescent shrub or small tree, 2–10 m high; trunk up to 10 cm in diameter; bark light to dark brown, tough, smooth, sometimes fissured. Leaves spirally arranged, simple, stipulate, ovate, 8–24 × 5–15 cm, shallowly cordate at base, acuminate at apex, green or dark green above, grey or brownish-green beneath on long, terete petiole, 2.5–12 cm long (Plate 1). Flowers in terminal branched panicles, 8–50 flowered, fragrant, 4–6 cm across; sepals 4–5, free, obovate, 1–1.2 cm long, covered with reddish-brown scales; petals 4–7, obovate, 2–3 × 1–2 cm, pinkish or whitish; stamens numerous, 1.6 cm long; anthers violet; pistil 1.6 cm long, composed of hairy 1-celled, superior ovary; style thickened upwards, with a short, bilobed stigma. Fruit broadly elongated ovoid capsule, 2–5 × 2–4.0 cm, flattened, 2 valved, densely covered with long soft bristles, green turning greenish-brown, yellowish-red or bright red (Plates 2–3) when mature; seeds numerous, obovoid and angular, 4–5 mm, small with bright orange-red fleshy coats (arils) (Plate 4).

Plate 1 Fruit (red variety) and foliage of annatto

Plate 2 Close-up of annatto fruit

Plate 3 Yellowish-red variety of annatto

Plate 4 Annatto seeds with bright orange-red fleshy coats

Nutritive/Medicinal Properties

Annatto seed was found to contain a relatively high amount of protein, which fluctuated between 13% and 17% (Bressani et al. 1983). Crude fibre levels were also high, about 16%, phosphorus content was also high but calcium was low. Its protein contained adequate levels of tryptophan and lysine, but was found to be low in methionine, isoleucine, leucine, phenylalanine and threonine. Thirty-five components were identified in the seed oil of *Bixa orellana,* of which (Z,E)-farnesyl acetate (11.6%Io), occidentalol acetate (9.7%), spathulenol (9.6%) and ishwarane (9.1%) were the major components (Pino and Correa 2003). Galindo-Cuspinera et al. (2002) reported the presence of 107 volatile compounds, in oil- soluble and water-soluble annatto extracts. They identified 51 aromatic compounds and tentatively identified 56 compounds responsible for the characteristic fragrance. The following compounds were identified in the aqueous and oily fractions: alcohols, aldehydes, alkanes, alkenes, ketones, esters and acids heterocyclic compounds, monoterpenes (geraniol, linalool), sesquiterpenes. Samples of commercial annatto food colouring formulations were found to contain the aromatic hydrocarbons toluene which ranged from <5 to 12 mg/kg toluene; and m-xylene from <5 to 200 mg/kg (Scotter et al. 2000). Bixin-in-oil formulations contained the highest m-xylene concentrations. The results provided evidence for the thermal degradation of annatto during source extraction and processing, resulting in contamination by internal generation of both bixin and norbixin types with aromatic hydrocarbons.

The seeds were found to contain carotenoid pigments and geranylgeraniol; the two major carotenoids were bixin and norbixin (Chao et al. 1991). Other carotenoids identified included β-carotene, cryptoxanthin, lutein, zeaxanthin and methyl bixin (Tirimanna 1981). The structure of an apocarotenoid (1% of total carotenoid) isolated from the seed coat of *Bixa Orellana* fruit was confirmed as methyl 9'Z-apo-6'-lycopenoate(methyl 9'Z-6'-apo-ψ-caroten-6'-oate) (Mercadante et al. 1996). The well-known C40-carotenes phytoene, phytofluene, ζ-carotene and neurosporene were also identified. The seeds also contained minor carotenoids: 6-geranylgeranyl 8'-methyl-6,8'-diapocaroten-6,8'-dioate; 6-geranylgeranyl 6'-methyl (9'Z)-6,6'-diapocaroten-6,6'-dioate and 6-geranylgeranyl 6'-methyl-6,6'-diapocaroten-6,6'-dioate (Mercadante et al. 1999). Carotenoids are regarded as effective antioxidants, antimutagenic and anticarcinogenic agents.

Yellow to orange color of annatto was reported to be derived from the outer layer of seeds of *Bixa orellana* (Chattopadhyay et al. 2008). The carotenoids, bixin, and norbixin, were found to be responsible for appearance of yellow to orange color. The pH and solubility was found to affect the color hue; the greater the solubility in oil, the brighter was the color. Water soluble, oil soluble, and oil/water dispersible forms of annatto are available.

Phytochemical evaluation revealed the presence of carbohydrates, steroids, alkaloids, proteins, flavonoids, terpenoids, phenolics, tannins and glycosides in annatto leaves (Radhika and Nasreen Begum 2010).

Annato extracts have different pharmacological properties including the antitumor, anti-inflammatory, astringent, emollient, antiseptic, antibacterial, antioxidant, cicatrizant among others and has been prescribed in wounds and burns cure (de la Lourido Perez and Martinez Sanchez 2010). Other pharmacological properties are also discussed below.

Hypoglycaemic and Hyperglycaemic Activities

Annatto seed extract was found to have both hypoglycaemic and hyperglycaemic activities. Studies showed that a crude annatto seed extract exhibited either glucose lowering or hyperglycaemia-inducing activity depending on how it was further manipulated (Russell et al. 2008). The annatto extract was found to reduce blood glucose levels in fasting normoglycaemic and streptozotocin-induced diabetic dogs. Additionally, in normal dogs, it suppressed the postprandial rise in blood glucose after an oral glucose load. The extract also caused an increase in insulin-to-glucose ratio in normal dogs. It was concluded that *Bixa orellana* lowered blood glucose by stimulating peripheral utilization of glucose. In earlier studies the scientists found that annatto's hypoglycaemic episodes seen in the dogs was mediated by an increase in plasma insulin concentration as well as an increase in insulin binding on the insulin receptor due to elevated affinity of the ligand for the receptor (Russell et al. 2005). In another earlier study, this purified substance trans-bixin from annatto seed was demonstrated, in anaesthetised mongrel dogs, to cause hyperglycaemia (Morrison et al. 1991). Concomitant electron microscopy of tissue biopsies, revealed damage to mitochondria and endoplasmic reticulum mainly in the liver and pancreas. When dogs were fed on a diet fortified with riboflavin, there was neither demonstrable tissue damage nor associated hyperglycaemia.

Studies showed that ingestion of norbixin (another major carotenoid constituent of annatto seed) did not induce any detectable DNA breakage in liver and kidney but caused a considerable impairment in plasma glucose levels of rats and mice (Fernandes et al. 2002). In rats exposed to doses of 0.8, 7.5 and 68 mg/kg (annatto extract) and 0.8, 8.5 and 74 mg/kg (norbixin) no toxicity was detected by plasma chemistry. In mice exposed to doses of 56 and 351 mg/kg (annatto extract) and 0.8, 7.6, 66 and 274 mg/kg, norbixin caused an increase in plasma alanine aminotransferase (ALT) activity while both norbixin and annatto extract caused a decrease in plasma total protein and globulins. However, no signs of toxicity were detected in the liver by histopathological analysis. No enhancement in DNA breakage was detected in liver or kidney from mice treated with annatto pigments, as evaluated by the comet assay. Nevertheless, there was a notable effect of norbixin on the glycemia of both rodent species. In rats, norbixin induced hyperglycemia that ranged from 26.9% (8.5 mg/kg norbixin) to 52.6% (74 mg/kg norbixin) above control levels. In mice, norbixin induced hypoglycemia that ranged from 14.4% (0.8 mg/kg norbixin) to 21.5% (66 mg/kg norbixin) below control levels. Rats and mice treated with annatto pigments showed hyperinsulinemia and hypoinsulinemia, respectively indicating that pancreatic β-cells were functional. Additional studies showed that under low serum conditions (2% fetal bovine serum (FBS)), a protective effect of norbixin against H_2O_2-induced DNA breakage was inversely related to its concentration, a protection ranging from 41% to 21% (Kovary et al. 2001). At higher concentrations of norbixin, oxidative DNA breakage was still augmented, even in the presence of a high serum concentration (10% FBS). Under normal conditions, norbixin per se had no detectable genotoxic or cytotoxic effects on murine fibroblasts. The antimutagenic potential of norbixin against oxidative mutagens was also evaluated by the *Salmonella typhimurium* assay, with a maximum inhibition of 87% against the mutagenicity induced by H_2O_2.

Neuropharmacological Activity

In the pentobarbitone-induced hypnosis test, the methanol extract of *Bixa orellana* leaves statistically reduced the time for the onset of sleep at 500 mg/kg dose and (dose-dependently) increased the total sleeping time at 250 and 500 mg/kg dose (Shilpi et al. 2006). A statistically significant decrease in locomotor activity was observed at all doses in the open-field and hole-cross tests.

Anticonvulsant and Analgesic Activity

In the strychnine-induced anticonvulsant test, the methanol extract of *Bixa orellana* leaves increased the average survival time of the test animals (statistically significant at 250 and 500 mg/kg) (Shilpi et al. 2006). The extract significantly and dose-dependently reduced the writhing reflex in the acetic acid-induced writhing test.

Antidiarrhoeal Activity

Anti-diarrhoeal activity was supported by a statistically significant decrease in the total number of stools (including wet stools) in castor oil-induced diarrhoea model (Shilpi et al. 2006). A statistically significant delay in the passage of charcoal meal was observed at 500 mg/kg in the gastrointestinal motility test.

Antioxidant Activity

The natural food color annatto and γ-tocopherol were found to effectively inhibited hydroperoxide formation (Haila et al. 1996). Annatto had been used for over two centuries as a food color especially in cheese and in various other food products.

Kiokias and Gordon (2003) evaluated the antioxidant effects of β-carotene, oil-soluble (bixin) and water-soluble (norbixin) annatto preparations and mixtures of these carotenoids with virgin olive oil polar extract in bulk olive oil and oil-in-water emulsions stored at 60°C. Norbixin was the only carotenoid that inhibited the oxidative deterioration of lipids in both systems. Though bixin and β-carotene did not prevent autoxidation, their mixtures with the polar extract from virgin olive oil augmented the antioxidant effect of the olive oil extract. Norbixin (2 mM) was of similar activity to δ-tocopherol (0.1 mM) in stored oil. The combination of norbixin with ascorbic acid or ascorbyl palmitate in oil exhibited a decrease in formation of volatile oxidation products but not in peroxide value, compared with the analogous sample lacking norbixin. In olive oil-in-water emulsions, norbixin (2 mM) reduced hydroperoxide formation to a similar extent as δ-tocopherol (0.1 mM), which in turn was a better antioxidant than α-tocopherol. A synergistic effect between norbixin and ascorbic acid or ascorbyl palmitate was observed in the emulsion systems. In another study, norbixin was found to protect against DNA damage induced by agents such as UV radiation, hydrogen peroxide (H_2O_2) and superoxide anions (O_2) using *Escherichia coli* cells

(Júnior et al. 2005). Norbixin increased survival at least ten times. The SOS induction by UVC radiation was inhibited 2.3 times more when cells were grown in the presence of norbixin. The results of the study by Oboh et al. (2010) showed that intraperitoneal administration of cyclophosphamide (75 mg/kg of body weight) caused a significant increase in the malondialdehyde (MDA) content of the brain; however, dietary inclusion of annatto seed extracts (0.1% and 0.2%) caused dose-dependent significant reduction in the MDA content of the brain. Likewise, the extracts also caused dose-dependent inhibition of the elevated serum glutamate oxaloacetate transaminase (SGOT), glutamate pyruvate transaminase (SGPT), alkaline phosphatase and total bilirubin. However, the non-polar extract had significantly higher inhibitory effects on the elevated MDA production in brain and serum liver function markers. This higher protective effect of the non-polar extract could be attributed to its higher antioxidant properties as typified by its significantly greater reducing power, free-radical scavenging and Fe (II) chelating ability. They concluded that dietary inclusion of annato seed extracts as food colourant could prevent oxidative stress occasioned by cyclophosphamide administration, but the non-polar extract was a better protectant. The methanol extract of *Bixa orellana* leaves exhibited radical scavenging properties in the DPPH assay ($IC_{50} = 22.36$ µg/ml) (Shilpi et al. 2006).

Anticancer/Antimutagenic Activity

Júnior et al. (2005) found that norbixin had antimutagenic properties, with a maximum inhibition of H_2O_2-induced mutagenic activity of 87%, based on the *Salmonella* mutagenicity test. Bixin isolated from *B. orellana* exhibited COX-1 and COX-2 inhibition and showed a dose-dependent growth inhibition against breast, colon, stomach, central nervous system, and lung tumour cells, respectively (Reddy et al. 2005).

The constituent cis-bixin from annatto seeds was found to induce cytotoxicity in a wide variety of tumour cell lines (IC_{50} values from 10 to 50 µM, 24-hours exposures) and, importantly, also selectively killed freshly collected patient multiple myeloma cells and highly drug-resistant multiple myeloma cell lines (Tibodeau et al. 2010). Mechanistic studies indicated that cis-bixin-induced cytotoxicity was significantly diminished by co-treatment with glutathione or N-acetylcysteine (NAC); whereas fluorescence-activated cell sorting (FACS) assays demonstrated that cis-bixin rapidly promoted cellular reactive oxygen species (ROS) in dose- and time-dependent manner, collectively implicating ROS as contributory to cis-bixin-induced cytotoxicity. Further the scientists found that cis-bixin inhibited both thioredoxin (Trx) and thioredoxin reductase (TrxR1) activities at concentrations comparable to those required for cytotoxicity, implicating the inhibition of these redox enzymes as potentially contributing to its ability to impose cellular ROS and to kill cancer cells. Their studies indicated that the annatto constituent cis-bixin had unique selective antimyeloma activity that appeared to be mediated through effects on redox signaling.

Anticlastogenic Activity

Studies showed that bixin was not cytotoxic or clastogenic, when compared to untreated control. A marked decrease in the MI (mitotic index) values compared to the untreated control and an increased percentage of aberrant metaphases were seen in all cultures treated with clastogen cisplatin (cDDP) (Antunes et al. 2005). The carotenoid efficiency in reducing the inhibitory effect of cDDP on lymphocyte MI was concentration-dependent. Cultures simultaneously treated with bixin and cDDP showed a statistically significant reduction in total chromosomal aberrations and aberrant metaphases. It was postulated that bixin may have acted as an antioxidant by intercepting free radicals generated by cDDP. The data obtained in the study suggested that dietary carotenoids may act as protective agents against clastogenic effects of antitumor agents.

Diuretic Activity

The methanolic extract of dried annatto leaf powder exhibited significant diuretic activity at a dose of 500 mg/kg body weight by augmenting the total volume of urine and contents of sodium, potassium and chloride in urine when compared to standard drug, Furosemide and peanut oil as control and vehicle for the extracts (Radhika et al. 2010).

Radioprotective Activity

Bixa orellana seed extract was found to have radioprotective activity. Recent studies found that pre-treatment by intraperitoneal injection of hydroalcoholic extract of seeds to healthy adult Swiss mice prior to whole body irradiation of 2.0 Gy gamma radiation resulted in a significant reduction in the percentage of aberrant metaphases and in the different types of aberration scored (Karchuli and Ganesh 2009). Radiation (4.0 Gy) increased the number of aberrant cells from less than 1% in controls to almost 20%. The extract was not toxic at 1,500 mg/kg body weight. Being non toxic and easily available natural source, *Bixa orellana* extract was suggested to have potential for use as a radioprotective for human beings.

Antimicrobial Activity

Commercial water-soluble annatto extracts afforded two fractions with antimicrobial activity against *Staphylococcus aureus* (Galindo-Cuspinera and Rankin 2005). Carotenoids 9′-cis-norbixin (UV(max) 460 and 489 nm) and all-trans-norbixin (UV(max) 287, 470, and 494 nm) were found as the major components in both fractions. Results indicated that 9′-cis-norbixin and all-trans-norbixin were responsible for the antimicrobial properties of annatto. Previous reports indicated that commercial annatto extracts had biological activities against microorganisms of significance to food fermentation, preservation, and safety.

B. orellana was also found to possess antimicrobial activity. Ethanol extract of *Bixa orellana* was inhibitory to *Escherichia coli* with a minimum inhibitory concentration (MIC) of 0.8 μg/ml compared to gentamycin sulfate (0.9 8 μg/ml) and MIC of 0.2 μg/ml against *Bacillus cereus* compared to 0.5 μg/ml for gentamycin sulphate (Rojas et al. 2006). Another study reported that ethanol extracts from the different parts of *B. orellana* showed differential antimicrobial activity. It was found that the extracts of leaves showed maximum in-vitro activity against *Bacillus pumilus* followed by the extracts from the roots and hypocotyls (Castello et al. 2002). The callus derived from different explants also showed antimicrobial activity. The leaf callus showed maximum activity. The zone of inhibition for the diluted extracts of in-vitro hypocotyls and roots and their corresponding calli showed minimum zone of inhibition at concentration 24 mg/ml, whereas the diluted extract of in-vitro leaves and leaf derived callus showed minimum zone of inhibition at 16 mg/ml. Methanol extract of *Bixa orellana* leaves exhibited antibacterial activity against selected causative agents of diarrhoea and dysentery, including *Shigella dysenteriae* (Shilpi et al. 2006). Bark extract of *B. orellana* showed in-vitro antimicrobial activity against strains of the gonorrhoea pathogen, *Neisseria gonorrhoeae* (Cáceres et al. 1995).

B. orellana and Prostate Cancer

In a 12-month, double-blind, placebo-controlled study involving 136 men with benign prostatic hyperplasia (BPH) presenting moderate lower urinary tract symptoms (LUTS), use of *Bixa orellana* (annatto) at a dose of 250 mg three times daily failed to improve symptoms (Zegarra et al. 2007). Patients with BPH that presented moderate LUTS did not show any benefit receiving annatto when compared to placebo.

Toxicity/Safety Studies

Scientific toxicity studies reported that low consumption rates of annatto is safe. The No-Observed-Adverse-Effect-Level (NOAEL) was judged to be a dietary level of 0.1% (69 mg/kg body weight/

day for males, 76 mg/kg body weight/day for females) of annatto extract (norbixin) under experimental conditions (Hagiwara et al. 2003). Marked elevation in absolute and relative liver weights was also found in both sexes of the 0.9% and 0.3% groups, but not the 0.1% group. Hepatocyte hypertrophy was evident and an additional electron microscopic examination demonstrated this to be linked to abundant mitochondria after exposure to a dietary level of 0.9% annatto extract for 2 weeks. Another toxicity study in rats reported male body weight gain, but had no effect on either food intake or food conversion efficiency as revealed by haematological and plasma biochemical examination (Bautista et al. 2004). Kidney apoptosis occurred in 20% treated female rats in restricted areas without proliferation or tubular segments modification. The scientists suggested that annatto was not toxic to rats.

Studies also indicated that annatto, was neither mutagenic nor an inhibitor of induced mutations in bone marrow cells from Swiss male mice treated with 1,330 and 5330ppm annatto incorporated into the diet (Alves de Lima et al. 2003). However, the scientists cautioned that high doses (10,760ppm) may increase the effect of a mutagen.

In another study, annatto seed, containing 5% bixin, administered in the animal diet at concentrations of 20, 200, and 1,000 ppm (0.07; 0.80 and 4.23 bixin/kg body weight/day, respectively), continuously during 2 weeks before, or 8 weeks after diethylnitrosamine (DEN) treatment showed that annatto was neither genotoxic nor carcinogenic at the highest concentration tested (1,000 ppm) (Agner et al. 2004). No protective effects were observed in both glutathione S-transferase (GST)-P foci development and comet assays. The studies concluded that under the experimental conditions, annatto showed no hepatocarcinogenic effect or modifying potential against DEN-induced DNA damage and preneoplastic foci in the rat liver. In a subsequent paper, the scientists found that dietary administration of 1,000 ppm annatto neither induce DNA damage in blood and colon cells nor aberrant crypt foci in rat distal colon (Agner et al. 2005). Conversely, annatto was successful in inhibiting the number of crypts/colon (animal), but not in the incidence of dimethylhydrazine (DMH)-induced ACF, mainly when administered after DMH. However, no antigenotoxic effect was observed in colon cells. The results suggested possible chemopreventive effects of annatto through their modulation of the cryptal cell proliferation but not at the initiation stage of colon carcinogenesis. A separate study also reported that bixin played an important role in the induction of cytochromes P450 (CYP1A and CYP2B) in the liver. Annatto especially at 250 mg/kg containing 70 mg bixin/kg induced the activities of the liver monooxygenases (De-Oliveira et al. 2003).

Traditional Medicinal Uses

The achiote has been traced back to the ancient Peru Indians who employed it as a principal coloring agent in foods, for body paints and as a coloring for arts, crafts and murals. Parts of the plant have been used to make folkloric medicinal remedies for such conditions as sunstroke, tonsilitis, burns, leprosy, pleurisy, apnoea, rectal discomfort, and headaches. Today in Peru traditional medicine, achiote used to treat heartburn and stomach distress caused by spicy foods, and as a mild diuretic and mild purgative while traditional medicine recommends it as a vaginal antispetic and cicatrizant, as a wash for skin infections, and for liver and stomach disorders. The Indians of the West Indies used the yellow pigment as body paint for war and purposes of magic, especially for the lips, which is the origin of the plant's nickname, lipstick tree. Applied as body paint it absorbs ultraviolet light and repels mosquitoes and flies. Annatto paste filters out the ultraviolet rays of sunlight, thereby protecting the skin from excessive sunburn. The use of the dye in the hair by men of the Tsáchila of Ecuador is the origin of their usual Spanish name, the *Colorados*.

Leaves and shoots have been used as antiseptic, astringent, febrifugal and aphrodisiac in traditional medicine. When soaked or boiled in water, compounds from the anatto's leaves were used to treat a variety of ailments, such as diarrhea, dysentery, sores, rashes, and swellings. Leaves were applied to the head and to sprains to relieve aches; a decoction was gargled as a cure for mouth and throat infections. Leaves were also be used in baths to relieve colic or to get rid of worms in children and for snakebite. They were applied as poultices as topicals to relive headache. The leaves, pounded and macerated in water, were diuretic and a remedy for gonorrhea.

The sap from fruits was also used to treat type 2 diabetes, and fungal infections.

A macerated seed decoction was taken orally for relief of fever, and the pulp surrounding the seed made into an astringent drink used to treat dysentery and kidney infection. A waxy substance that has paralytic action on mammalian intestinal parasites is present in the seed coat. Seeds were used as expectorant.

In the Philippines, annatto dye was much used with lime as an external application in erysipelas. The red, resinous substance of the seeds was considered an efficient remedy for certain skin diseases. The fine powder which covers the seeds was used as haemostatic, and internally as a stomachic. The seeds being slightly astringent and in decoction were very good remedy for gonorrhea. The seeds also reported to posses antiperiodic and antipyretic properties, but to a lesser extent. The pulp covering the seeds being astringent and slightly purgative was given in dysentery and kidney ailments. In Uruguay the ground pulp of the seeds, if applied immediately to burns, was said to prevent the formation of blisters or scars. The pulp was also prescribed for stomachache. The seeds were said to be an antidote to cassava and *Jatropha curcas* poisoning. The oil of the seeds has been reported to be effective in leprosy.

The root-bark is antiperiodic and antipyretic. In the Philippines, a decoction of the bark was employed in febrile catarrhs. In French Guiana, an infusion was prescribed as a purgative in dysentery. A root decoction was taken orally to control asthma. Oliguria and jaundice were treated using root teas; infusions of root in water and rum were used to treat venereal diseases.

Other Uses

Annatto is planted as wind shelter and for hedges as goats and cattle do not eat the leaves. It is also planted as shade trees, and as live support trees in vanilla plantations. The flowers are a source of nectar for honey. The dye from the seed is used for cosmetics in nail gloss polish, hair oil, lipstick, soap and floor wax, wood polish, shoe polish, brass lacquer, and wood stain. This dye was also used by Amerindians as war paint and was used in dying wool, cotton and silk. Bixin extracted from the seed coat is used in India as an insect repellent. Bixa meal, which remains after extraction of the pigment from the seed, is a useful additive to poultry feed and can replace 30% of the maize in the food. Fibre for cordage has been obtained from the bark. Bark from the branches of the trees yields a water-soluble gum that is similar to gum Arabic. Bixa wood is also used as fuel wood.

Comments

The species is very polymorphous and is sometimes split up into several taxa.

Selected References

Agner AR, Barbisan LF, Scolastici C, Salvadori DM (2004) Absence of carcinogenic and anticarcinogenic effects of annatto in the rat liver medium-term assay. Food Chem Toxicol 42(10):1687–1693

Agner AR, Bazo AP, Ribeiro LR, Salvadori DM (2005) DNA damage and aberrant crypt foci as putative biomarkers to evaluate the chemopreventive effect of annatto (*Bixa orellana* L.) in rat colon carcinogenesis. Mutat Res 582(1–2):146–154

Alves de Lima RO, Azevedo L, Ribeiro LR, Salvadori DM (2003) Study on the mutagenicity and antimutagenicity of a natural food colour (annatto) in mouse bone marrow cells. Food Chem Toxicol 41(2):189–192

Antunes LM, Pascoal LM, Bianchi MdeL, Dias FL (2005) Evaluation of the clastogenicity and anticlastogenicity of the carotenoid bixin in human lymphocyte cultures. Mutat Res 585(1–2):113–119

Aparnathi K, Lata DR, Sharma RS (1990) Annatto (*Bixa orellana* L.) – its cultivation, preparation and usage. Int J Trop Agric 8(1):80–88

Bautista AR, Moreira EL, Batista MS, Miranda MS, Gomes IC (2004) Subacute toxicity assessment of annatto in rat. Food Chem Toxicol 42(4):625–629

Bressani R, Porta-España de Barneón F, Braham JE, Elías LG, Gómez-Brenes R (1983) Chemical composition, amino acid content and nutritive value of the protein of the annatto seed (*Bixa orellana*, L.). Arch Latinoam Nutr 3(2):356–376 [Article in Spanish]

Burkill IH (1966) A dictionary of the economic products of the Malay Peninsula. Revised reprint, 2 volumes, vol 1 (A–H) pp 1–1240, vol 2 (I–Z) pp 1241—2444. Ministry of Agriculture and Co-operatives, Kuala Lumpur

Burkill HM (1985) The useful plants of West Tropical Africa, vol 1, 2nd edn, Families A-D. Royal Botanical Gardens, Kew, 960 pp

Cáceres A, Menéndez H, Méndez E, Cohobón E, Samayoa BE, Jauregui E, Peralta E, Carrillo G (1995) Antigonorrhoeal activity of plants used in Guatemala for the treatment of sexually transmitted diseases. J Ethnopharmacol 48(2):85–88

Castello MC, Phatak A, Chandra N, Sharon M (2002) Antimicrobial activity of crude extracts from plant parts and corresponding calli of *Bixa orellana* L. Indian J Exp Biol 40(12):1378–1381

Chao RC, Mulvaney SJ, Sanson DR, Hsieh F-H, Tempesta MS (1991) Supercritical CO2 extraction of annatto (*Bixa orellana*) pigments and some characteristics of the color extracts. J Food Sci 56(1):80–83

Chattopadhyay P, Chatterjee S, Sen SK (2008) Biotechnological potential of natural food grade biocolorants. Afr J Biotechnol 7:2972–2985

De-Oliveira AC, Silva IB, Manhaes-Rocha DA, Paumgartten FJ (2003) Induction of liver monooxygenases by annatto and bixin in female rats. Braz J Med Biol Res 36(1):113–118

Fernandes AC, Almeida CA, Albano F, Laranja GA, Felzenszwalb I, Lage CL, de Sa CC, Moura AS, Kovary K (2002) Norbixin ingestion did not induce any detectable DNA breakage in liver and kidney but caused a considerable impairment in plasma glucose levels of rats and mice. J Nutr Biochem 13(7):411–420

Foundation for Revitalisation of Local Health Traditions (2008) FRLHT database. htttp://envis.frlht.org

Galindo-Cuspinera V, Rankin SA (2005) Bioautography and chemical characterization of antimicrobial compound(s) in commercial water-soluble annatto extracts. J Agric Food Chem 53(7):2524–2529

Galindo-Cuspinera V, Lubran MB, Rankin SA (2002) Comparison of volatile compounds in water- and oil-soluble annatto (*Bixa orellana* L.) extracts. J Agric Food Chem 50(7):2010–2015

Hagiwara A, Imai N, Ichihara T, Sano M, Tamano S, Aoki H, Yasuhara K, Koda T, Nakamura M, Shirai T (2003) A thirteen-week oral toxicity study of annatto extract (norbixin), a natural food color extracted from the seed coat of annatto (*Bixa orellana* L.), in Sprague-Dawley rats. Food Chem Toxicol 41(8):1157–1164

Haila KM, Lievonen SM, Heinonen MI (1996) Effects of lutein, lycopene, annatto and γ-tocopherol on autoxidation of triglycerides. J Agric Food Chem 44:2096–2100

Júnior AC, Asad LM, Oliveira EB, Kovary K, Asad NR, Felzenszwal I (2005) Antigenotoxic and antimutagenic potential of an annatto pigment (norbixin) against oxidative stress. Genet Mol Res 4(1):94–99

Karchuli MS, Ganesh N (2009) Protective effect of *Bixa orellana* L. against radiation induced chromosomal aberration in Swiss albino mice. Int J Phytomed 1:18–21

Kiokias S, Gordon MH (2003) Antioxidant properties of annatto carotenoids. Food Chem 83(4):523–529

Kovary K, Louvain TS, Costa e Silva MC, Albano F, Pires BB, Laranja GA, Lage CL, Felzenszwalb I (2001) Biochemical behaviour of norbixin during in vitro DNA damage induced by reactive oxygen species. Br J Nutr 85(4):431–440

de la Lourido Perez HC, Martinez Sanchez G (2010) The *Bixa orellana* L. in treatment of stomatology affections: a subject that hasn't studied yet. Rev Cubana Farm 44:2 [Online]

Mercadante AZ, Steck A, Pfander H (1999) Three minor carotenoids from annatto (*Bixa orellana*) seeds. Phytochem 52(1):135–139

Mercadante Z, Steck A, Rodriguez-Amaya D, Pfander H, Britton G (1996) Isolation of methyl 9′Z-apo-6′-lycopenoate from *Bixa orellana*. Phytochem 41(4):1201–1203

Morrison EY, Thompson H, Pascoe K, West M, Fletcher C (1991) Extraction of an hyperglycaemic principle from the annatto (*Bixa orellana*), a medicinal plant in the West Indies. Trop Geogr Med 43(1–2):184–188

Oboh G, Akomolafe TL, Adefegha SA, Adetuyi AO (2010) Inhibition of cyclophosphamide-induced oxidative stress in rat brain by polar and non-polar extracts of Annatto (*Bixa orellana*) seeds. Exp Toxicol Pathol 63(3):257–262

Pacific Island Ecosystems at Risk (PIER) (1999) *Bixa orellana* L. Bixaceae. http://www.hear.org/Pier/species/Bixa_orellana.htm

Perry LM (1980) Medicinal plants of East and Southeast Asia: attributed properties and uses. MIT Press, Cambridge/London, 620 pp

Pino JA, Correa MT (2003) Chemical composition of the essential oil from annatto (*Bixa orellana* L.) seeds. J Essent Oil Res 15(2):66–67

Rajendran R (1992) *Bixa orellana* L. In: Lemmens RHMJ, Wulijarni-Soetjipto N (eds) Plant resources of South East Asia No 3. Dye and tannin-producing plants. Prosea Foundation, Bogor, pp 50–53

Radhika B, Nasreen Begum SK (2010) Pharmacognostic and preliminary phytochemical evaluation of the leaves of *Bixa orellana*. Pharmacogn J 2(7):132–136

Radhika B, Nasreen Begum SK, Reddy VM (2010) Diuretic activity of *Bixa orellana* Linn. leaf extracts. Indian J Nat Prod Res 1(3):353–355

Reddy MK, Alexander-Lindo RL, Nair MG (2005) Relative inhibition of lipid peroxidation, cyclooxygenase enzymes, and human tumor cell proliferation by natural food colors. J Agric Food Chem 53(23):9268–9273

Rojas JJ, Ochoa VJ, Ocampo SA, Muñoz JF (2006) Screening for antimicrobial activity of ten medicinal plants used in Colombian folkloric medicine: a possible alternative in the treatment of non-nosocomial infections. BMC Complement Altern Med 6:2

Russell KR, Morrison EY, Ragoobirsingh D (2005) The effect of annatto on insulin binding properties in the dog. Phytother Res 19(5):433–436

Russell KR, Omoruyi FO, Pascoe KO, Morrison EY (2008) Hypoglycaemic activity of *Bixa orellana* extract in the dog. Methods Find Exp Clin Pharmacol 30(4):301–305

Scotter MJ, Wilson LA, Appleton GP, Castle L (2000) Analysis of annatto (*Bixa orellana*) food coloring formulations. 2. Determination of aromatic hydrocarbon thermal degradation products by gas chromatography. J Agric Food Chem 48(2):484–488

Shilpi JA, Taufiq-Ur-Rahman M, Uddin SJ, Alam MS, Sadhu SK, Seidel V (2006) Preliminary pharmacological screening of *Bixa orellana* L. leaves. J Ethnopharmacol 108(2):264–271

Tibodeau JD, Isham CR, Bible KC (2010) Annatto constituent cis-bixin has selective antimyeloma effects mediated by oxidative stress and associated with inhibition of thioredoxin and thioredoxin reductase. Antioxid Redox Signal 13(7):987–997

Tirimanna ASL (1981) Study of the carotenoid pigments of *Bixa orellana* L. Seeds by thin layer chromatography. Microchim Acta 76(1–2):11–16

Yang Q, Gilbert MG (2007) Bixaceae. In: Wu ZY, Raven PH, Hong DY (eds). Flora of China, vol 13 (Clusiaceae through Araliaceae). Science Press, Beijing, and Missouri Botanical Garden Press, St. Louis

Zegarra L, Vaisberg A, Loza C, Aguirre RL, Campos M, Fernandez I, Talla O, Villegas L (2007) Double-blind randomized placebo-controlled study of *Bixa orellana* in patients with lower urinary tract symptoms associated to benign prostatic hyperplasia. Int Br J Urol 33(4):493–500

Adansonia digitata

Scientific Name

Adansonia digitata L.

Synonyms

Adansonia baobab Gaertn., *Adansonia baobab* L., *Adansonia digitata* var. *congolensis* A. Chev., *Adansonia sphaerocarpa* A.Chev., *Adansonia sulcata* A. Chev.

Family

Bombacaceae, also placed in Malvaceae

Common/English Names

African Baobab, Baobab, Bottle Tree, Cream Of Tartar Tree, Dead-Rat Tree, Ethiopian Sour Tree, Judas Fruit, Monkey-Bread Tree, Senegal Calabash, Sour Gourd, Upside Tree

Vernacular Names

Arabic: Buhibab, Hama-Hamaraya, Hijid, Gangeleis (Fruit);
Burkina Faso: Trega, Twega, Toayga (More), Ngigne (Senufo);
Chad: Hamar, Hamaraya (Arabic);
Cote d'Ivoire: Ngigne (Senufo), Fromdo (Baule);
Danish: Abebrødstræ, Baobab;
Dutch: Apenbroodboom, Kremetartboom;
Egypt: Habhab (Arabic);
French: Arbre Aux Calebasses, Arbre De Mille Ans, Baobab Africain, Baobab De Mozambique, Calebassier Du Sengal, Pain De Singe (Fruit);
Ethiopia: Bamba, Fertata, Baobub (Amargna), Hemmer, Dumma (Tigre), Bamba (Yao);
German: Affenbrotbaum;
India: Gadhagachh (Bengali), Bukha, Sumpura (Gujerati), Gorakh-Imli, Hathi-Khatiyan (Hindu), Gorakh Chinch (Marathi), Sarpadandi (Sanskrit), Papparappuli, Anaipluiya-Marum (Tamil), Brahma-Mlinka, Brahmaaamlika, Seemasinta (Telugu);
Kenya: Olimesera, Ol-Unisera (Masai), Muramba (Meru), Mwambo (Kamba);
Malawi: Mlonje (Yao), Mnambe, Mlambe (Chichewa), Mbuye (Nkonde);
Mali: Ngigne (Senufo), Oro (Dogon), Konian, Ko (Sonrai), Konian (Dierma), Sira (Bambara), Babbe, Boki, Olohi (Peulh), Sira, Sito (Mandinke);
Mauritania: Teidoum (Arabic);
Mozambique: Mbuyu, Majoni Ya Mbuyu (Swahili);
Niger: Kouka, Kuka (Hausa);
Nigeria: Kouka, Kuka (Hausa), Boki, Bokki (Fulani), Oshe (Yoruba);
Polish: Baobab Wlasciwy;
Portuguese: Baobá, Cabacerve;
Senegal: Goui, Gouis, Lalo, Boui (Wolof), Bak (Serer), Boubakakou (Dirla Foghy);
Somalia: Yag (Somali), Mbuyu (Swahili);

South Africa: (Kremetart, Kremetartboom, Mubuyu, Muyu, Mbuyu, Mkulukumba, Mlambe Afrikaans and Others), Isimuhu, Umdhimulu (Zulu);
Sudan: Gunguleiz (Fruit), Tebeldi, Mumr, Homeira, Dungwol (Arabic);
Tanzania: Olimesera, Ol-Unisera (Masai), Majoni Ya Mbuyu (Swahili);
West Indies: Mapou Zombi (Creole);
Zimbabwe: Umnkhomo (Ndebele).

Origin/Distribution

Very recent studies using chloroplast DNA phylogeography has reported that *A. digitata* probably originated in West Africa and migrated subsequently throughout the tropical parts of Africa, and beyond, by natural and human-mediated terrestrial and overseas dispersal. The natural geographical distribution of the African baobab comprises most of tropical Africa, but also small patches of southern Arabia, and several Atlantic and Indian Ocean islands surrounding the African continent, notably including Madagascar. In Africa, *A. digitata* is found in Angola, Benin, Botswana, Burkina Faso, Cameroon Cape Verde, Chad, Congo, Cote d'Ivoire, Ethiopia, Eritrea, Gambia, Ghana, Guinea, Kenya, Malawi, Mali, Mauritania, Mozambique, Namibia, Niger, Nigeria, Senegal, Sierra Leone, Somalia, South Africa, Sudan, Tanzania, Togo, Zambia, and Zimbabwe. It was also introduced to Central African Republic, Comoros, Democratic Congo, Egypt, Gabon, Madagascar, Sao Tome, Zaire and introduced as park/garden trees elsewhere in the tropics.

Agroecology

Baobab has such a wide distribution in the continent of Africa commensurate with its wide ecological tolerance. It appears that its extant distribution could be determined by minimum requirement for a certain annual precipitation and the need for a dry season. It is adapted to zones with 100–1,000 mm annual precipitation, but trees are often stunted in lower rainfall areas. Typically it occurs as a scattered tree in the savannah and along tracks and associated with habituation, commonly from 450 to 600 m altitude although in Ethiopia it can be found at 1,500 m elevation. Its mean annual temperature range is 20–30°C, but it can tolerate well high temperatures up to 40–42°C (in West Africa), resistant to fire, and survive low temperature as long as there is no frost. It is drought tolerant and frost sensitive. Baobab thrives on a wide range of soils which include stony, non-agricultural soil, rocky and lateritic soils, calcareous soils, clays, sands, alluvial silts and various kinds of loams. Its is found in poorly drained soils in Zimbabwe, poorly drained plains in the Zambezi delta, sandy soils overlying compacted slit and liable to flooding in the heavy rainy seasons in Nigeria. It grows in full sun.

Edible Plant Parts and Uses

Baobab (*Adansonia digitata*) is a multi-purpose tree species. Tender root, tubers, twigs, fruit, seeds, leaves and flowers are edible and are common ingredients in traditional dishes in rural areas in Africa. Common food products of *A. digitata* among the ethnic communities in Benin include dough, gruel, drink (from pulp sauces from leaves, seeds kernel) flavouring agents (from kernels).

Its fruit pulp has very high vitamin C content (~10 times that of orange), and can be used in seasoning, as an appetizer and to make juices (De Caluwé et al. 2010). Seeds contain appreciable quantities of crude protein, digestible carbohydrates and oil, whereas they have high levels of lysine, thiamine, Ca and Fe. They can be eaten fresh or dried, ground into flour and thus added to soups and stews. Processing eliminates a number of anti-nutritional factors present in the seed. Baobab leaves are superior in nutritional quality to fruit pulp, and contain significant levels of vitamin A. The leaves are a staple for many populations in Africa, and are eaten fresh or dried.

The uncooked fruit pulp is used in preparing refreshing drinks with a pleasant wine-gum flavor in rural areas and also has become a popular ingredient in ice products in urban areas. This drink is also used to treat fevers and other complaints. In some areas baobab 'milk' is very popular. It is very nutritious and is made from dried pulp mixed into a solution. In northern Nigerian, the fruit pulp emulsion is mixed with milk as a drink. The pulp is used in food preparations and also processed into sweets. Fresh pulp is also eaten. Whole fruit or pulp can be stored for several months under dry conditions. In Tanzania, the fruit pulp is added to aid fermentation of sugar cane for beer making. The acid pulp is also used as a substitute for cream of tartar in baking. The powdered fruit husk and peduncle may be smoked as a substitute for tobacco.

Seeds are used as thickening agent in soups. Fermented seeds are used in flavouring soups. In Sudan the seeds are pounded whole into a coarse meal dried and added to soups and other dishes like '*Burma*'. Baobab seeds are ground with peanuts and added to water and sugar to make a sauce eaten with porridge. Roasted seeds is used as a side dish substituting peanut and used as snacks. It is also used as coffee substitute. Seeds are pressed for oil as an oilseed meal for cooking. The shoots and roots of germinating seeds and seedlings are also eaten.

Leaves are a staple food source for rural populations in many parts off Africa particularly in the central region of the continent. Among the people who comprise the Hausa ethnic group in Northern Nigeria in particular, ground leaves serves as the main ingredient of a soup called '*miyar kuka*'. Young leaves are chopped and cooked in a sauce or eaten as spinach. In parts of northern Nigeria, especially Borno and Yobe states, where the leave is eaten as soup condiment. Dried green leaves are used throughout the year. The dried leaves are powdered and used for cooking. The powder is called '*lalo*' in Mali and sold in many local markets. In Mali use of the leaves in sauce is usually in combination with *Parkia biglobosa* seeds, onion, okra, pepper, ginger sometimes meat but more often fish. The sauce is used with a rich gruel made form millet, sorghum or maize but also from couscous rice.

Flowers can be eaten raw or used to flavour drinks. In West Africa, the tender roots are reputed to be cooked and eaten presumably in times of famine. The Temne of Sierra Leone believe that a root decoction taken with food causes stoutness.

Botany

The tree is not particularly tall growing up to 18–25 m tall but the trunk is short but massive, as much as 10–14 m in girth, often deeply fluted and typically cylindrical or bottle-shaped (Plate 1). The trunk is in fact used to store water during dry periods. It's not uncommon for old trees to have several huge trunks branching off near the ground, and a tree (18.3 m) tall can have a spread of more than 30.5 m. The bark is smooth, reddish-brown or greyish with a purplish tinge or rough and wrinkled. It has a shallow (<2m deep) but

Plate 1 Tree habit

Plate 2 Digitate leaves

extensive lateral root system, greater than the height of the tree which accounts for it high water holding capacity. The roots end in tubers. The leaves are clustered at the ends of short, stocky branches, alternate, simple or digitate. Young trees usually have simple leaves (Plate 2). Leaves are 2–3 foliolate at the commencement of the season and are early deciduous, later maturing leaves 5–7–(9)-foliate. Mature leaf may attain 20 cm across. Leaflets are elliptic to obovate-elliptic, 5–15 cm by 2–7 cm, sessile to sub-sessile, with acuminate tip, decurrent base, entire margin stellate pubescent below becoming glabrescent or glabrous. Stipule is early caducous, subulate or narrowly deltoid, 2–5 mm long, glabrous except for the ciliate margin. Flowers are pendulous, solitary or paired, axillary, large and showy. Pedicel variable length with two small caducous bracteoles. Calyx 3–5 lobed, fused into a disc below but dived half or more above, lobes triangular or oblong-triangular, green and tomentose outside, cream and villous inside, lobes reflexed. Corolla 5 white, overlapping, obovate petals base shortly clawed, apex rounded, sparely hairy to glabrous. Stamens numerous, 720–1,600, forming a cylindrical staminal tube petals, 2–6 cm long, upper part of filaments free and reflexed to form a ring, anthers reniform. Ovary 5–10 locular, conical to globose, silky tomentose. Style exserted, reflexed or erect, villous below and glabrous above. Stigma white with 5–10 fimbriate-papillose lobes. Fruit variable, globose to ovoid to oblong-cylindrical, 7.5–54 cm long by 7.5–20 wide, apex pointed or obtuse, covered by velvety yellowish hairs. Pericarp 8–10 mm thick, woody enclosing dry mealy pulp. Seeds reniform, dark brown to reddish black 10–13 × 8–10 × 4–5 mm, embedded in pulp.

Nutritive/Medicinal Properties

Food nutrient composition of the baobab fruit pulp per 100 g had been reported by Osman (2004) as: moisture 10.4%, protein 3.2%, fat 0.3%, ash 4.5%, crude fibre 5.4%, carbohydrate 76.2%; energy 320.3 kcal/100 g, potassium 1240 mg, sodium 27.9 mg, calcium 295 mg, magnesium 90 mg, iron 9.3 mg, copper 1.6 mg, Zn 1.8 mg; amino acids (g/100 g protein) – aspartic acid 6.4 g, glutamic acid 6.5 g, serine 3.2 g, histidine 1.2 g, glycine 2.9 g, threonine 2.8 g, arginine 7.6 g, alanine 3.3 g, prolamine 2.2 g, tyrosine 20.6 g, methionine 0.2 g, valine 4.8 g, phenylalanine 4.4 g, isoleucine 4.3 g, leucine 4.3 g, lysine 1.7 g and cysteine 1 g.

Nour et al. (1980) reported that the fruit pulp contained on a dry weight basis, 79.3% soluble solids, alcohol insoluble solids 57.3%, total sugars 23.2%, reducing sugars 19.9%, total pectin (galacturonic acid) 56.2%, ascorbic acid 300 mg/100 g and pH 3.3.

Food nutrient composition of the baobab seed per 100 g (Leung 1968) had been reported as: moisture 4.3%, protein 18.4%, fat 12.2%, ash 3.8%, crude fibre 16.2%, carbohydrate 45.1%; energy 363.8 kcal/100 g, potassium 910 mg, sodium 28.3 mg, calcium 410 mg, magnesium 270 mg, iron 6.4 mg, copper 2.6 mg, zinc 5.2 mg; amino acids (g/100 g protein) – aspartic acid 10.3 g, glutamic acid 23.7 g, serine 6.1 g, histidine 2.2 g, glycine 8.6 g, threonine 3.8 g, arginine 8.0 g, alanine 7.1 g, prolamine 6.9 g, tyrosine 1.5 g, methionine 1.0 g, valine 5.9 g, phenylalanine 4.0 g, isoleucine 3.6 g, leucine 7.0 g, lysine 5.0 g, cysteine 1.5 g.

Fatty acid profile of baobab seed oil had been reported by Osman (2004) as: saturated fatty acids 31.7%, C14:0 (myristic acid) 0.2%, C16:0 (palmitic acid) 24.2%, C18:0 (stearic acid) 4.6%, C20:0 (arachidic acid) 1.3%, C22:0 (behenic acid) 0.7%, C24:0 (lignoceric acid) 0.2%;

monounsaturated fatty acids 37%, C17:1 (heptadecenoic acid) 0.3%, C18:1(oleic acid) 35.8%, C20:1 (gadoleic acid) 0.9%; polyunsaturated fatty acids 31.7%, C18:2 (linoleic acid) 30.7%, C18:3 (linolenic acid) 1.0%.

The above data indicate that the seed is a good source of fat, protein and crude fibre and the fruit pulp is a rich source of carbohydrate and has low fat. The sugars accounted for 43.62 of the carbohydrates in the fruit pulp (Odetokun 1996). The five types of sugar identified in the pulp were glucose, fructose, sucrose, maltose, and raffinose. Both the seed and fruit pulp were excellent sources of potassium, calcium and magnesium and poor sources of iron, zinc and copper. Both seed and fruit pulp contained high amounts of glutamic acid, aspartic acid and arginine and low amounts of the sulfur containing amino acids. Proline and valine were the limiting amino acids in the pulp and seed. The seeds were found to have a relatively high amount of essential amino acids. Baobab fruit pulp was deemed a valuable and important source of ascorbic acid (Vitamin C) with a very high vitamin C content ranging from 1623 to 4991 mg/kg, almost 10 times that of orange (Sidibé et al. 1998).

Baobab seed oil had been reported to be an excellent source of mono- and polyunsaturated fatty acids. The major fatty acid was oleic (35.8%) followed by linoleic (30.7%) and palmitic 24.2%.

The baobab fruit as is common with all plant fruits, was found to contain naturally occurring anti-nutritional substances such as trypsin inhibitors, phytate and tannin that could interfere with utilization of the baobab fruit pulp and seed. The seed and pulp were reported to have similar low levels and the seed had a higher level of phytic acid and tannin compared to the fruit pulp (Osman 2004).

Nutrient composition of raw baobab leaves per 100g edible portion had been reported as: energy 69 kcal, moisture 77%, protein 3.8 g, fat 0.3 g, carbohydrates 16.1 g, dietary fibre 2.8 g, ash 2.8 g, Ca 402 mg, P 65 mg, ascorbic acid 52 mg (Leung et al. 1968).

Nutrient composition of dried baobab leaves per 100 g edible portion (Leung et al. 1968) was reported as follows: energy 282 kcal, moisture 11.8%, protein 12.3 g, fat 3.1 g, carbohydrates 63.2 g, dietary fibre 9.7 g, ash 9.6 g, Ca 2241 mg, P 275 mg, Fe 24 mg, K 9170 mg, thiamin 0.63 mg, riboflavin 0.82 mg, niacin 0.44 mg ascorbic acid traces.

Baobab leaf was reported to contain 10.6% (dry weight) protein and an amino acid composition which compared favourably to that of an "ideal" protein: valine (5.9%), phenylalanine+tyrosine (9.6%), isoleucine (6.3%), lysine (5.7%), arginine (8.5%), threonine (3.9%), cysteine+methionine (4.8%), tryptophan (1.5%). In terms of mineral content, baobab leaf was found to be an excellent source of calcium, iron, potassium, magnesium, manganese, molybdenum, phosphorus, and zinc (Yazzie et al. 1994). The highest level of pro-vitamin A was detected in young leaves especially when used as dried material (Sidibé et al. 1996), expressed in retinol equivalent it was between 9 and 27 mg/kg. The data indicated that in terms of both quality and quantity, baobab leaf could serve as a significant protein and mineral source in the staple food of the local population.

Several plant parts of *A. digitata* have interesting anti-oxidant, anti-inflammatory antimicrobial, anti-viral, hepatoprotective, anti-diarrhoeal, and trypanocidal properties, and baobab has been used extensively since ancient times in traditional medicine.

Antioxidant Activity

Baobab fruit pulp and leaves had been reported as excellent sources of antioxidant in comparison to orange, kiwi fruit, apple, strawberry and an OPC (oligomeric proantocyanidins) rich vegetal extract (Vertuani et al. 2002). In terms of water soluble antioxidant capacity, highest value was observed in baobab fruit pulp and dry leaves 6–7 mmol/g of Trolox, followed by glycolic extract from leaves (4 mmol/g), strawberry (0.9 mmol/g), kiwi (0.34 mmol/g), apple (0.16 mmol/g) and orange (0.1 mmol/g). Pulp and leaf extracts had slightly higher trolox values than the OPC rich extract (6.16 mmol/g). On the basis of lipid-soluble antioxidant capacity, baobab fruit again ranked first at 4.4148 mmol/g followed by the 90% OPC rich vegetal extract (4.093 mmol/g) and dry leaves at 2.35 mmol/g. The other plant products had very low lipid-soluble antioxidants. When compared together using the Integral Antioxidant Capacity (IAC) value, the ranking was as follows: Baobab fruit ≥ OPC rich vegetal extract > baobab dry leaves >> baobab leaves glycolic extract >>> strawberry fresh fruit pulp > kiwi fresh fruit pulp > orange fresh fruit pulp > apple fresh fruit pulp.

In a study of 14 wild edible fruits in Burkina Faso, *Detarium microcarpum* fruit had been reported to have the highest phenolic and the highest flavonoid content, followed by that of *Adansonia digitata, Ziziphus mauritiana, Ximenia americana* and *Lannea macrocarpa* (Lamien-Meda et al. 2008). There was a strong correlation between total phenolic and flavonoid levels and antioxidant activities. Studies showed that the beverages prepared from *Parinari curatelifolia, Strychnos spinosa* and *Adansonia digitata* were capable of acting as antioxidant sources as they displayed significant radical scavenging properties (Nhukarume et al. 2010). *Adansonia digitata* had the highest and comparable antioxidant activities to *Citrus sinensis* (orange). The total phenolic, ascorbic acid, proantocyanidin and flavonoid contents ranged between 12 and 58 mg GAE/100 ml, 0.00–51.26 mg/100 ml, 0.35–1.071% and 18.3–124 mg/100 ml, respectively. There was a positive correlation between antioxidant activities and phenolic compounds content but there was no clear relationship between proanthocyanidin content and antioxidant activity.

Anti-inflammatory Activity

Aqueous extract of the baobab fruit pulp produced a marked anti-inflammatory activity (Ramadan et al. 1994). It reduced the size of pedal swelling induced by formalin, the effect being comparable to that produced by standard phenylbutazone (15 mg/kg). This anti-inflammatory activity was related to the presence of sterols, saponins and triterpenes in the aqueous extract. The extract also generated a marked analgesic activity in mice at a dose of 800 mg/kg reaction time was significantly longer than the control and resembled that induced by acetylsalicylic acid (ASA) or Aspirin (50 mg/kg). The extract also displayed marked antipyretic activity as the rectal temperature was significantly decreased compared with control values of the hyperthermic rats, equivalent to that generated by a standard dose of ASA. These findings supported the use of the plant as antipyretic and febrifuge in folk medicine.

Hepatoprotective Activity

The aqueous extract of the *Adansonia digitata* fruit pulp exhibited hepatoprotective activity against chemical toxicity induced by CCL4 in rats (Al-Qarawi et al. 2003). The aqueous extract exhibited significant hepatoprotective activity and consumption of Adansonia digitata fruit may play an important part in human resistance to liver damage in areas in which the plant is consumed. The mechanism of liver protection was unknown, but could possibly result from triterpenoids, β-sitosterol, β-amyrin palmitate, or/and α-amyrin, and ursolic acid in the fruit.

Antiviral Activity

Root bark extract displayed strong activity against three viruses tested, Herpes Simplex, polio and Sindbis viruses, the leaf extract was active against the former two viruses particularly against Herpes simples virus but was not active against the polio virus (Ananil et al. 2000). The bark and leaf extract also showed cytotoxicity activity. Also aqueous extracts of Baobab bark were found to posses hypocholesterolemic and haemolytic activities (Sodipo and Mohamd 1990).

Antidiarrhoeal Activity

Studies using baobab fruit pulp solution to treat young children with diarrhoea found that WHO standard solution for diarrhoea was superior, but not statistically significant in term of duration of diarrhoea and weight gain from baobab fruit pulp solution treatment (Tal-Dia et al. 1997). Baobab fruit based solution presented additional advantages: nutritional, economic and cultural.

Antimicrobial Activity

Results of antibiotic studies showed that methanolic extract of *A. digitata* root bark or leaves had antibacterial activity against *Staphylococcus aureus*, *Streptococcus faecalis*, *Bacillus subtilis*, *Escherichia coli* and *Mycobacterium phlei* (Ananil et al. 2000). Ethanol extracts of *Adansonia digitata* at 75% and 100% concentrations gave greatest inhibitory activity against *Escherichia coli* isolated from urine and water sources compared with test antibiotics (Yagoub 2008). Aqueous extracts of the stem bark (ASBE) and root bark (ARBE) showed significant antimicrobial activity against *Bacillus anthracis, Bacillus subtilis* and *Staphylococcus aureus* (Masola et al. 2009). ASBE was also active against *E. coli* and *Salmonella* sp. which were resistant to ARBE. Both ERBE (ethanol root bark extract) and ESBE (ethanol stem bark extract) showed activity against *B. subtilis*. ESBE showed activity against *E. coli* and *S. aureus* which were resistant to ERBE. Both aqueous and ethanolic stem and root bark extracts did not show any antimicrobial activity against *Psuedomonas aeruginosa* and *Candida albicans*. The antimicrobial activity of crude stem bark extract against bacteria was attributed to the presence of tannins, phlobatannins, terpenoids and saponins in the stem bark. The attributing factor for antibacterial activity of crude root bark extract could be due to the presence of terpenoids in the root bark.

Trypanocidal Activity

Studies showed that the aqueous and methanolic root extracts of baobab possessed reversal antisickling properties but not inhibitory antisickling activity (Adesanya et al. 1988). The absence of inhibitory activity in-vitro did not justify the use of *A. digitata* for the prevention of sickling crises. Methanolic extract from root of *A. digitata* had also been reported to have in-vitro trypanocidal activity against *Trypanosoma congolense* and *T. brucei brucei* (Atawodi et al. 2003).

Adverse Hypoalbuminemia Effect

The methanolic leaf extract of *Adansonia digitata* had no significant effect on serum electrolyte levels except serum albumin in normal and alcohol fed rats (Geidam et al. 2004). The serum albumin levels of normal rats and alcohol fed rats was significantly reduced when fed with the leaf extract. The hypoalbuminemia that resulted on the administration of the extract was postulated to be due to the possible action of *Adansonia digitata* on liver function that consequently affected the production of albumin. Physiological effects of this were decreases in osmotic pressure and in transportation of some organic molecules carried by albumin. No significant effect was observed in the level of total protein.

Traditional Medicinal Uses

The bark, roots, leaves, fruits and seeds are widely used by indigenous peoples for human and animal medicines (Wickens 1982; Gebauer et al. 2002;Wickens and Lowe 2008; Masola et al. 2009). In African traditional folk medicine the fruit pulp and leaves have been reported to be used as antipyretic or febrifuge to overcome fever. The fruit pulp is also therapeutically employed as analgesic, anti diarrhoea, anti-dysentery and for treatment of haemoptysis, small-pox and measles. Powdered seeds are used for dysentery and to promote perspiration (a diaphoretic) and hiccough. Oil extracted from seeds is used for inflamed gums and to ease infected teeth. Leaves are used medicinally as a diaphoretic, an expectorant and as a prophylactic against fever and as an astringent. The leaves in infusion also have hyposensitive and antihistamine properties being employed to treat kidneys and bladder diseases, asthma, blood clearing, general fatigue, ophthalmia, otitis, diarrhoea, internal pains, dysentery, inflammation, insect bites, guinea worms etc. The bark is used as a substitute for quinine as a prophylactic against fever especially that caused by malaria. An aqueous extract of baobab bark is drunk in Nigerian traditional medicine as a treatment for sickle cell anaemia.

In Indian traditional medicine, a decoction of the bark is used internally as a refrigerant, antipyretic and antiperiodic. Powdered leaves are used as to check excessive perspiration. Young leaves are crushed into a poultice for painful swellings. Pulp is used internally with buttermilk in cases of diarrhoea and dysentery.

Other Uses

Hollow trees provide reservoirs for fresh water particularly in western Sudan. They are also used as shelter and for storage and in West Africa and are used as tombs. The wood is spongy and light and is used mainly for fuel but is also used for making wide light canoes, wooden platters, trays, floats for fishing nets. The green bark furnishes a dye and is used for decoration. The bark yields a strong fibre which is used in making ropes, cordage, harness straps, string for musical instrument, baskets, fibre cloth, mats, fishing nets bags, and hats. In East Africa, a soluble red dye is obtained from the roots. The bark is also used for tanning. The root bark is also used as string or rope for making fishing nets, socks, mats, etc. The pollens of flowers are mixed with water to form glue. The husks of the fruit may be used as dishes or fashioned into containers or snuff boxes and fishing floats. In Tanzania, they are used as water dippers. They can be used as fuel and provide a potash rich ash suitable for soap making. The fruit and leaves are also nutritious and palatable to animals and constitute an important feed for cattle in the savannah areas especially in the arid zones. In ethnoveterinary medicine, it has been reported that bark of *A. digitata* was used for treatment of diarrhoea in poultry (Guèye 1999) and fruits for treatment of Newcastle disease in poultry (Guèye 1999).

During the past decade there has been an increased interest by several European companies in baobab pharmaceutical and toiletry products, such as powdered and tableted pulp, dried leaf powders, leaf bath 'salts' and foams, seed bath oils, lubrications and anti-aging skin preparations (Wickens and Lowe 2008). The disintegration of thermolabile active ingredients, such as essential oils, vitamins, etc. and the caramelisation of the sugars during manufacture are prevented by the use of ultrasound extraction at ambient temperatures. This ensures a better qualitative/quantitative ratio without any modification of the active ingredients.

Comments

See also notes on the other closely related edible species, *Adansonia gregorii*.

Selected References

Adesanya SA, Idowu TB, Elujoba AA (1988) Antisickling activity of Adansonia digitata. Planta Med 54(4): 374–375

Al-Qarawi AA, Al-Damegh MA, El-Mougy SA (2003) Hepatoprotective influence of Adansonia digitata pulp. J Herbs Spices Med Plants 10(3):1–6

Ananil K, Hudson JB, de Souza C, Akpaganal K, Tower GHN, Amason JT, Gbeassor M (2000) Investigation of medicinal plants of Togo for antiviral and antimicrobial activities. Pharm Biol 39(1):40–45

Atawodi SE, Bulus T, Ibrahim S, Ameh DA, Nok AJ, Mamman M, Galadima M (2003) In vitro trypanocidal effect of methanolic extract of some Nigerian savannah plants. Afr J Biotechnol 2:317–321

Chadare FJ, Houhouigan JD, Linnemann AR, Nout MJR, van Boekel MAJS (2008) Indigenous knowledge and processing of Adansonia digitata L. food products. Ecol Food Nutr 47(4):338–362

De Caluwé E, Halamová K, Van Damme P (2010) Baobab (Adansonia digitata L.): A review of traditional uses, phytochemistry and pharmacology. In: Juliani HR, Simon J, Ho CT (eds) African natural plant products: new discoveries and challenges in chemistry and quality. American Chemical Society Symposium Series, vol 1021, Chapter 4, pp 51–84

Gebauer J, El-Siddig K, Ebert G (2002) Baobab (Adansonia digitata L.): a review on a multipurpose tree with promising future in Sudan. Gartenbauwissenschaft 67(4):155–160

Geidam MA, Oyesola SA, Kokori AM (2004) Effects of methanolic leaf extract of Adansonia digitata (Linn) on serum electrolyte levels in normal and alcohol fed rats. Pak J Biol Sci 7(8):1404–1406

Guèye EE (1999) Ethnoveterinary medicine against poultry diseases in African villages. World Poult Sci J 55:187–198

Lamien-Meda A, Lamien CE, Compaoré MM, Meda RN, Kiendrebeogo M, Zeba B, Millogo JF, Nacoulma OG (2008) Polyphenol content and antioxidant activity of fourteen wild edible fruits from Burkina Faso. Molecules 13(3):581–594

Leung W-TW, Busson F, Jardin C (1968) Food composition table for use in Africa. FAO and US Department of Health, Education, and Welfare, Rome/Washington. http://www.fao.org/docrep/003/X6877E/X6877E00.htm#TOC

Masola SN, Mosha RD, Wambura PN (2009) Assessment of antimicrobial activity of crude extracts of stem and root barks from *Adansonia digitata* (Bombacaceae) (African baobab). Afr J Biotechnol 8(19): 5076–5083

Nhukarume L, Chikwambi Z, Muchuweti M, Chipurura B (2010) Phenolic content and antioxidant capacities of *Parinari curatelifolia*, *Strychnos spinosa* and *Adansonia digitata*. J Food Biochem 34:207–221

Nour AA, Magboul BI, Kheiri NH (1980) Chemical composition of baobab fruit (*Adansonia digitata* L.). Trop Sci 22:383–388

Odetokun SM (1996) The nutritive value of Baobab fruit (*Andasonia digitata*). Riv Ital Sostanze Gr 73(8): 371–373

Osman MA (2004) Chemical and nutrient analysis of Baobab (*Adansonia digitata*) fruit and seed protein solubility. Plant Foods Hum Nutr 59:29–33

Pock Tsy JML, Lumaret R, Mayne D, Vall AO, Abutaba YI, Sagna M, Raoseta SOR, Danthu P (2009) Chloroplast DNA phylogeography suggests a West African centre of origin for the baobab, *Adansonia digitata* L. (Bombacoideae, Malvaceae). Mol Ecol 18(8):1707–1715

Ramadan A, Harraz FM, El-Mougy SA (1994) Antiinflammatory, analgesic and antipyretic effects of the fruit pulp of *Adansonia digitata*. Fitoterapia 65(5):418–422

Sidibe M, Williams JT (2002) Baobab. *Adansonia digitata*. International Centre for Underutilised Crops, Southampton, 96 pp

Sidibé M, Scheuring JF, Tembely D, Sidibé MM, Hofman P, Frigg M (1996) Baobab – homegrown vitamin C for Africa. Agroforest Today 8(2):13–15

Sidibé M, Scheuring JF, Kone M, Hofman P, Frigg M (1998) More on baobab – home grown vitamin C: some trees have more than others consistently. Agroforest Today 10:10

Sodipo OA, Mohamd SL (1990) Foam forming and haemolytic activities of the extract of the Baobab tree. Nig J Basic Appl Sci 4(182):41–52

Tal-Dia A, Toure K, Sarr O, Sarr M, Cisse MF, Garnier P, Wone I (1997) A baobab solution for the prevention and treatment of acute dehydration in infantile diarrhoea. Dakar Méd 42(1):68–73 (In French)

Vertuani S, Braccioli E, Buzzoni V, Manfredini S (2002) Antioxidant capacity of *Adansonia digitata* fruit pulp and leaves. Acta Phytother 5(2):2–7

Wickens GE (1982) The baobab – Africa's upside-down tree. Kew Bull 37:173–209

Wickens GE, Lowe P (2008) Human and veterinary medicine. In: The baobabs: Pachycauls of Africa, Madagascar and Australia. Springer, pp 81–100, 500 pp

Yagoub SO (2008) Anti-microbial activity of *Tamarindus indica* and *Adansonia digitata* extracts against *E. coli* isolated from urine and water specimens. Res J Microbiol 3(3):193–197

Yazzie D, Vanderjagt DJ, Pastuszyn A, Okolo A, Glew RH (1994) The amino acid and mineral content of baobab (*Adansonia digitata* L.) leaves. J Food Compos Anal 7(3):189–193

Adansonia gregorii

Scientific Name

Adansonia gregorii F. Muell.

Synonyms

Adansonia gibbosa (A.Cunn.) Guymer ex D. Baum

Family

Bombacaceae, also placed in Malvaceae

Common/English Names

Australian Baobab, Boab, Cream Of Tartar Tree, Bottle Tree, Dead Rat Tree, Gouty Stem Tree, Sour Gourd.

Vernacular Names

Australia: Gadawon, Larrgadi, Larrgadiy (Nyulnyulan) Kimberly region, North Western Australia.

Origin/Distribution

Australian boab is indigenous to the Kimberley region of north Western Australia. Its range extends from the sandy plains on the Logue River between Broome and Derby, to the Victoria River Basin in the Northern Territory.

Agroecology

Australian baobab occurs on the plains of the Ord and Fitzroy valleys on rocky crops and on the limestone hills of the Oscar and Napier Ranges from near sea level to 400 m elevation within the latitudinal range of 14–18°S. The tree is also found on cockatoo sands around Kununurra and also on soils derived from basalt and sandstone. In its native habitat, the temperature ranges from 13°C to 18°C during the winter months to 35–39°C during the summer months with mean annual rainfall of 500–1500 mm.

Edible Plant Parts and Uses

Various parts of the tree are edible. Aborigines in north Western Australia eats the fruit pulp, fresh young leaves, seeds and tuberous roots. In times of food scarcity, the sap from the trunk and branches is eaten and the fibrous pith from the trunk and branches can be eaten after boiling. The leaves have a delicate peppery flavour and nutty taste and may have great potential for salads and as a garnish. The boab leaves are quite high in vitamins C and A (Johnson et al. 2006). The tuberous root can be cut into straws for salads, dips and stir-fries and can be used in soups. The taste and texture is similar to water chestnuts or radish. The boab tubers are high in iron and potassium, with a high level of protein (for a vegetable) and fibre and a relatively low fat content (Johnson et al. 2006).

The seeds are eaten roasted. The pulp of the fruit has an agreeable, acid taste like cream of tartar and is very high in vitamin C.

In a recent study, the tuberous roots of seedlings and leaves have been shown to have great prospects for commercialising the crop with very promising positive market response for the roots and leaves as vegetables particularly for the hospitality industry (Johnson et al. 2002).

Botany

A deciduous, small to large tree, 9–12 m high with a short massive swollen cylindrical to bottle shaped trunk with a girth reaching 15.7 m and a smooth to pock-marked greyish bark (Plates 1–2).

The tap roots of seedlings are fusiform and tuberous (Plates 4–6) and the roots of older trees are relatively shallow but extensive and end in tubers. The leaves are clustered at the ends of short, stocky branches, alternate, simple or digitate (Plates 4, 5, 7). Seedlings usually have simple lanceolate leaves at the early growth stage subsequent leaves are digitate, 3–5 foliolate. Leaves on

Plate 2 Massive, bottle-shaped trunk

Plate 1 Tree habit

Plate 3 Close-up of flower

Plate 4 Seedlings grown for its tuberous roots (P. Johnson)

Plate 5 Seedling with tuberous roots (P. Johnson)

Plate 6 Paring of the tuber skin (P. Johnson)

Plate 7 Fruits, leaves and fruit ornaments (P. Johnson)

adult trees are alternate, digitate 5–7 foliolate, on terete, 7–11 cm long glabrous petiole, with a pair of subulate deciduous stipules at the base of leaf. Leaflets dark green above, dull green beneath, glabrous, broadly lanceolate to ovate, usually 9–11 by 3–4 cm, margin entire, with prominent mid vein below, and borne on short petiolules. Flowers solitary, pedicellate, 9×3 cm, white and showy (Plate 3). Sepals 5 fused with apex free, pubescent, brown and triangular and reflex at flower anthesis. Petals oblanceolate, 11 by 2 cm, white, waxy. Stamens numerous >100, 5–6 cm long, exserted, bases fused to form short tube, outer surface of tube pubescent. Ovary 5–10 locules, conical to globose; style longer than stamens crowed by spherical stigma. Fruit variable, globose to ovoid to oblong-cylindrical, 15–25 cm by 10–20 cm wide, apex pointed or obtuse, covered by velvety yellowish hairs (Plate 7). Pericarp 8–10mm thick, woody enclosing dry mealy pulp. Seeds reniform, dark brown to black 10–15×5–10 mm, embedded in powdery pulp.

Nutritive/Medicinal Properties

Nutrient composition for boab fruit pulp/100 g (Brand Miller et al. 1993) was reported as follows: energy 354 kJ, moisture 61.8%, ash 3 g, protein 4.6 g, fat 1.9 g, total dietary fibre 25.7 g, nitrogen 0.74 g, potassium 1088 mg, sodium 54 mg calcium 0.68 mg, magnesium 46 mg, zinc 0.6 mg and niacin equivalent 0.8 mg.

Nutrient composition for boab seed/100 g (Brand Miller et al. 1993) was reported as follows: energy 626 kJ, moisture 9.2%, ash 5.3 g, protein 13.9 g, fat 6.1 g, total dietary fibre 20.5 g, nitrogen 2.22 g, potassium 1476 mg, sodium 21 mg calcium 184 mg, phosphorus 396 mg, magnesium 316 mg, zinc 3.6 mg, lead 0.2 ug, niacin equivalent 2.3 mg, thiamine 0.39 mg, riboflavin 0.04 mg and vitamin C 6 mg.

Nutrient composition for boab root samples (Johnson et al. 2002) was reported as follows: moisture 6.7%, ash 7.9%, protein (crude) 0.26%, fat 3.6%, starch 16.7%, total carbohydrates 16%, crude fibre 21%, sugars 19%, potassium 2.73%, sodium 0.05%, calcium 0.68%, magnesium 0.36%, sulphur 0.06%, chlorine 0.61%, boron 13 mg/kg, copper 4.6 mg/kg, iron 630 mg/kg, manganese 28 mg/kg, zinc 20 mg/kg, total tannin (catechin) 2.8% and phytate 1.02%.

Boab seeds are rich in proteins, energy, fats and mineral like potassium, calcium, magnesium, and zinc than the fruit pulp and also has thiamin, niacin, riboflavin and small amount of vitamin C. The boab fruit pulp is high in potassium, protein, fibre and sodium but has no vitamin C.

Boab roots were found to be a moderate source of starch and sugars. Potassium levels were also reasonably high. No toxic compounds were identified in the samples (Johnson et al. 2002).

Adansonia gregorii flower and leaf extract exhibited in-vitro antibacterial activity against *Pseudomonas fluorescens* and *Bacillus cereus* respectively (Cock 2008).

The aborigines in Australia have used the fruit for medicine.

Other Uses

For thousands of years, the Australian Aborigines found the boab tree to be of economic importance as various parts of the tree provided food, medicine, emergency supplies of water, fibre, glue and shelter. Almost every part of the tree was found to have some use. The fruit is used to carve decorative ornaments depicting aboriginal cultural scenes. The porous trunk has been used for moisture and sap. The bark of the tree is used for making twine and the gum as glue. Cattle will eat most parts of a fallen tree. Early European settlers to the region soon found uses for the tree in the form of shelter, stock-food in times of drought and even turning the hollowed centres of the trees into the occasional prison.

Comments

See also notes on the other closely related edible species, *Adansonia digitata*.

Selected References

Baum DA (1995) A systematic revision of *Adansonia* (Bombacaceae). Ann Mo Bot Gard 82:440–470

Boland DJ, Brooker MIH, Chippendale GM, Hall N, Hyland BPM, Johmston RD, Kleinig DA, McDonald MW, Turner JD (2006) Forest trees of Australia, 5th edn. CSIRO, Collingwood, 768 pp

Bowman DMJS (1997) Observations on the demography of the Australian Boab (*Adansonia gibbosa*) in the North-west of the Northern Territory, Australia. Aust J Bot 45(5):893–904

Brand Miller J, James KW, Maggiore P (1993) Tables of composition of Australian Aboriginal foods. Aboriginal Studies Press, Canberra

Brock J (1988) Top end native plants. John Brock, Darwin, 354 pp

Cock IE (2008) Antibacterial activity of selected Australian native plant extracts. Internet J Microbiol 4(2):1–8

FAO (1988) Traditional food plants: a resource book for promoting the exploitation and consumption of food plants in arid, semi-arid and sub-humid lands of East Africa. FAO Food and Nutrition Paper 42. FAO, Rome

Johnson PR, Robinson C, Green E (2002) The prospects of commercialising boab roots as a vegetable. Rural Industries Research and Development Corporation (RIRDC) Publication No 02/020, Canberra, 21 pp

Johnson PR, Green EJ, Crowhurst M, Robinson CJ (2006) Commercialisation of boab tubers. Rural Industries Research and Development Corporation (RIRDC) Publication No 06/022, Canberra

Sidibe M, Williams JT (2002) Baobab. *Adansonia digitata*. International Centre for Underutilised Crops, Southampton, 96 pp

Wickens GE, Lowe P (2008) Human and veterinary medicine. In: The baobabs: Pachycauls of Africa, Madagascar and Australia. Springer, Berlin, pp 81–100, 500 pp

Ceiba pentandra

Scientific Name

Ceiba pentandra (L.) Gaertn.

Synonyms

Bombax cumanense Kunth, *Bombax guineense* Schum. & Thonn., *Bombax mompoxense* Kunth, *Bombax occidentale* Spreng., *Bombax orientale* Spreng., *Bombax pentandrum* L. (Basionym), *Ceiba anfractuosa* (DC.) M. Gómez, *Ceiba caribaea* (DC.) A. Chev., *Ceiba casearia* Medik., *Ceiba guineensis* (Schum. & Thonn.) A. Chev., *Ceiba guineensis* var. *ampla* A. Chev., *Ceiba occidentalis* (Spreng.) Burkill, *Ceiba pentandra* fo. *albolana* Ulbr., *Ceiba pentandra* fo. *grisea* Ulbr., *Ceiba pentandra* var. *caribaea* Bakh., *Ceiba pentandra* var. *clausa* Ulbr., *Ceiba pentandra* var. *dehiscens* Ulbr., *Ceiba pentandra* var. *indica* Bakhuisen, *Ceiba thonnerii* A. Chev., *Ceiba thonningii* A. Chev., *Eriodendron anfractuosum* DC., *Eriodendron anfractuosum* α *indicum* DC., *Eriodendron anfractuosum* var. *africanum* DC., *Eriodendron anfractuosum* var. *caribaeum* DC., *Eriodendron anfractuosum* var. *indicum* DC., *Eriodendron caribaeum* G. Don ex Loud., *Eriodendron guineense* G. Don ex Loud., *Eriodendron occidentale* (Spreng.) G. Don, *Eriodendron orientale* (Spreng.) Kostel., *Eriodendron pentandrum* Kurz, *Gossampinus alba* Hamilt., *Xylon pentandrum* Kuntze.

Family

Bombacaceae, also in Malvaceae

Common/English Names

Ceiba, Corkwood, Cotton Silk Tree, Cotton Tree, Cottonwood Tree, Java Cotton, Java Kapok, Kapok, Kapok Tree, Kapoktree, Silk Cotton Tree, Silk-Cottontree, White Silk Cotton Tree, White Silk-Cottontree, White-Flowered Silk-Cotton Tree

Vernacular Names

Arabic: Rum (Chad), Shajaret Al Kutun;
Australia: Paina;
Bambara: Bàna, Bànan (Mali);
Benin: Gounma (Bariba) Araba;
Bolivia: Hoja De Yuca, Toborochi (Spanish),
Brazil: Arvore-Da-Lã, Arvore-Da-Seda, Barriguda De Espinho, Mai-Das-Arvores, Paina-Lisa, Paineira, Samaúma-Cabeluda, Samaúna Da Várzea, Samaúma-Lisa, Sumaúma, Sumaúma-Barriguda, Sumaúma-Branca, Sumaúma-Da-Várzea, Sumaúma-De-Macaco, Sumaúma-Rosada, Sumaúma-Verdadeira (Portuguese); **British West Indies**: Corkwood;
Burmese: Thinbawle;
Cameroon: Douma (Sangmelima, Ebolowa), Dum (Yaounde), Bouma (Douala), Djam (Bangangte), Nfuma, Nkouma, Ogouma;

Caribbean: Koemaka, Makau;
Central African Republic: Bouma (Lissongo), Fromager (Local French);
Chamorro: Algidon, Algodon De Manila, Algodon Di Manila, Atgidon Di Manila, Atgodon Di Manila;
Chinese: Ji Bei, Ji Bei Mian, Ji Bei Mu Mian Zhua Wa Mu Mian;
Colombia: Bonga, Ceiba Blanca, Ceiba De Lana (Spanish);
Congo: Bokuma, Bosengo, Mofumo;
Cook Islands: Mama'U, Mama'U, Mama'U, Tumu Mama'U, Vavai (Maori);
Cote I'voire: **Cote D'ivoire**: Gbi (Abe), Enivé (Abure), Akuondi, Gna (Akan-Asante), Muong, Nguéhié, Won (Akye), Egniè, Egnien, Enia, Enya (Anyi), Egna, Etchui (Avikam), Angbo, Gna, Gniè, Gnien, Nyé (Baule), Guima (Brong), Guê (Dan), Banatan, Bantan, Bantignei, Bantiguehi (Fula-Fulfulde), Gué, Molongué (Gagu), Diô, Tiô, Do, Tshyo, Go, Djô (Guere), Go (Kru-Bete), Ton'go, Ton'ko, Toonko (Kulango), Dangué, N-Gué, Gwɛ̃ɛ̃, Tyɛ̀nɛ̃ (Kweni), Ague, Allotegue, Anié (Kyama), Bana, Bana-Bandan (Manding-Maninka), Ghê (Mano),Tshyo (Ngere), Eguina, Eniémé, Enyam'gua, Enyan'gua (Nzema), Sérigné (Senufo-Tagwana),Tiu ('Onele');
Chuukese: Kkaton, Kkóton, Koton;
Curaçao: Ceiba;
Dahomey: Guénesso (Baseda), Guma (Batonnun), Gbê (Busa), Bantan (Dendi), Bentan Habu, Dehon, Gué Dehunsu, Patin Dehun (Fon), Linihi, Rinihi (Fula-Fulfulde), Adjoro Hun, Gpati Dêkrun, Hun-Ti (Gbe-Fon), Hunsufu (Hweda), Hunti (Vhe), Igi Èégun, Igi Àràbà, Ogufé (Yoruba-Nago);
Danish: Kapoktræ, Silkebomuldstræ;
Democratic Republic of Congo: Ntunturu (Yanzi);
Dutch: Kapokboom;
Fijian: Semar, Vauvau Ni Vavalangi;
Finnish: Capoc, Kapokkipuu, Seiba;
French: Bois Coton, Capoc, Faux Contonni, Coton, Capoquier, Cotonnier De l'Inde, Faux Cotonnier, Fromager, Fromager Commun, Fromager Des Antilles, Fromager D'indo-Malaisie, Fromager Inerme Du Golf De Guinée, Kapokier, Faux Kapokier, Kapokier A Fleurs Blandes, Kapokier, Kapokier De Java, Kapokier Du Togo;
French Guiana: Cotonnier Grand Bois;
French West Indies: Arbre À Cotton, Fromagier;
Gabon: Nfuma, Nkouma, Ogouma;
Gambia: Bosanobo (Diola-Flup), Bantehi (Fula-Pulaar), Bantalŋforo, Bantaŋ (Manding-Mandinka), Bentenki, Betenbi (Wolof);
German: Baumwollbaum, Fuma, Kapokbaum, Wollbaum;
Ghana: Lɛno, Sokpe (Adangme), Onyima (Akan-Asante), Enyaa (Anyi), Enya (Anyi-Aowin), Onyima (Asante),Nye (Baule),Gbang (Bimoba), Danta, Ekile, Σdanta-Pu (Brong), Gongu, Goni (Dagaari), Gumbihi, Guna, Gunga, Gunguma-Gumdi Gungumli (Dagbani), Onyāā, Onyina, Ayigbe Ogbedei, Ayigbe, Onyāá, Onyãí (Fante), Atepre, Euti, Ofu, Wudese (Gbe-Vhe), Gung (Grusi), Kàkèlθ, Kàkĩ́líyà, Kakre (Guang-Gonja), Rimi (Huasa), Bufo-Sōgbum, Kpugbum (Konkomba); Kekyafu (Krachi), Gunga (Mampruli), Gonga (Nankanni), Enyenna, Enyεnσa (Nzema), Enyaa (Sehwi), Kuŋkomo, Kuŋ-Kumuŋ (Sasaala), Onyã-Hene, Onyina (Twi), Euti, Ofua, Ofwho, Vule, Atepre, Be (Vhe), Enyena, Onyina (Vulgar), Onya, Onyina (Wasa);
Greek: Καπόκ;
Guinea: Kö-Porō Kö (Baga), Bantignei, Bantignei Bentégniévi (Fula-Pulaar), Banda (Kissi), Bara (Kono), Uyé (Kpelle), Banan (Manding-Bambara), Bana, Ban-Bandan, Bandan (Manika), Bantan (Pular), Kondé (Susu), Am-Polon (Temne); *Guinea-Bissau*: Psáhè, Rumbum (Balanta), Brêgue (Biafada), Cob-Be (Bidyogo), Poilão, Polom (Crioulo), Bantanhe (Fula-Pulaar), Bantaŋ(-Ô) Bantango, Bintaforo (Manding-Mandinka), Péntia (Mandyak), Pèntè, Pentene (Mankanya) Metchene, N'teme, N'tene (Pepel);
Guyana: Kumae, Kumak, Kumaka (Macushi), Silk Cotton Tree (Creole), Wiring (Wapisiana);
India: Schwet Simul, Setsimul (Bengali), Hattian, Katan, Safedsemal, Saphed Simal, Safed-Semal (Hindu), Apurani, Biliburaga, Buraga, Dudi, Elava, Apoorni, Bilibaralu, Bilibooruga, Bili Booruga, Booruga, Doodi, Doodihatthi, Dudi, Marali (Kannada), Ilavu, Mullillapappula,

Nakuli, Panja, Panni, Pannimaram, Panniyala, Pula (Malayalam), Moreh Tera (Manipuri), Pandhari, Salmali, Safetasavara, Samali, Shamieula, Saalmali, Safedsavara (Marathi), Japanpang (Mizoram), Chirayu, Kutasalmali, Kutashalmali, Kutsitashalmali, Moch, Rochana, Salmali, Shvetasalmali, Sthirayu, Svetashalmali (Sanskrit), Ilavam, Karukkanam, Pancu, Panji, Ulagamaram, Ilavu, Panjimaram, Ilavam Pinchu, Ilavumaram, Acikai, Akikai, Apurani, Ayika, Ayikam, Calamali, Calamirittiran, Callaki, Callakimaram, Camani, Camani, Canamali, Canamali, Canamali, Canamalikam, Canamalikamaram, Caranamalam, Cilesmavarttani, Cilettumavarttani, Ciracivi, Cirancivi, Citakapi, Citalicimitta, Cittanmuki, Cittanmukimaram, Cukumaranul, Cukumaratuli, Cukumaratuvi, Emanarvalli, Icanam, Ilakumaram, Ilavam, Ilavamaram, Irakaputpam, Iyamanarvalli, Kantapalam, Kantapalamaram, Kariyacu, Katicakali, Katicakalimaram, Kukkati, Matacaram, Pancu, Piccila, Purani, Putpapani, Terru, Terrukam, Tolam, Tolaya, Tulakikam, Tulakikamaram, Tulam, Tulani, Tulavirukkam, Tuli, Tulini, Tulipalai, Turanimoca, Mocacani, Mocacanimaram, Mocai, Mocani, Mocaniriyacam, Nariyilavamaram, Nariyilavu, Nattilavu, Tuyarkantam, Ulakumaram, Ulavamaram (Tamil), Kadami, Tellaburaga, Tella Buruga, Booruga, Thellabooraga (Telugu), Smabal (Urdu);
Indonesia: Randu (Java), Randu (Sundanese), Kapuek, Kapeh Panji, Panji (Sumatra);
Italian: Albero Del Kapok, Pianta Del Kapok;
Japanese: Kapokku, Kiwata Kapokku;
Khmer: Koo, Kor;
Kosraean: Cutin, Kuhtin;
Laotian: Kokuiyu, Nguiz Baanz;
Liberia: Gwe, Gwèh (Dan), Dju (Kru-Guere), Geh, Guéh (Mano), Nguwa (Mende);
Mali: Dámu, Jiǔ, J Ǔ (Dogon), Bantignei, Bantiguehi (Fula-Pulaar), Bana(n) (Manding-Bambara), Bana, Bana-Bandan (Maninka);
Malaysia: Kabu-Kabu, Kekabu, Kabu Kabes, Kapok, Kapok-Kapok, Mengapas, Randu;
Marquesan: Puatio, Uruuru, Uruuruvaikirita;
Marshallese: Bulik, Kotin;
Mexico: Pochota, Pochote, Yaxché (Spanish);
Nauruan: Duwoduwo;

Niger: Bantiguehi (Fula-Fulfulde), Forgo (Songhai-Zarma);
Nigeria: Ukem Akabi, Ù-Mùùm, U-Muum, Àrù-Mùùm (Abua), Úkúm (Anaang), Rum (Arabic-Shuwa), Bokum (Bokyi), Gbée, Gbiê-Li (Busa), Okha (Edo), Úkím (Efik), Bantahi, (Fulfulde), Gehi, Gyehi (Gwari), Rimi, Rimin, Rini, Masar (Hausa), Úkím (Ibibio), Agwu, Agwugu (Igala), Ákpū, Ákpū-Ugu, Ákpū Ógwū, Ákpū Ùdèlè, Mbom (Igbo), Afalafase (Ijo), Àsìsàghà, Ogungbologhá (Ijo-Izon), ēgungun (Isekiri), Ahe (Isoko), ìsághái̯ (Izon), Torn (Kanuri), Kúci (Nupe), Gbê-Siê, Konngô, Vàmbè (Shanga), Óháhèn (Urhob), Araba, Okha (Vulgar), Okho (Yekhee), Àràbà, ẹ̀ẹ̀gun, Ogungun (Yoruba);
Niuean: Vavae;
Palauan: Kalngebárd, Kalngebard, Kerrekar Ngebard;
Peru: Ceibo, Huimba (Spanish);
Philippines: Sanglai (Bontak), Boboi, Kayo (Bikol), Boi-Boi, Daldol, Doldol, Kapoc, Kayo (Bisaya), Bulak-Dondol, Dogdol, Dondol, Gapas (Cebu Bisaya), Gataova (Ifugao), Basanglai, Dondol, Kapas-Sanglai, Kulak (Iloko), Bulak, Bulak-Kastila, Kasanglai (Pampangan), Kapas (Pangasinan), Kapak, Kapas (Sambali), Kapoc, Kapuk (Sulu), Balios, Boboi, Buboi, Bulak, Bulak-Kahoi, Bulak-Sina (Tagalog), Sanglai (Tinggian);
Pohnpeian: Cottin, Koatoa, Koatun;
Portuguese: Barriguda, Mafumeira, Paina, Poilão, Polão, Sumauma Da Mata, Sumauma De Terra Firme, Sumaumeira;
Samoan: Vavae, Vavae, Vavae Papalagi;
Senegal: Kidem (Banyun), A-Ndín (Basari), Gi-Ndii (Bedik), Busana, Busanay, Étufay (Diola), Bâtigéhi, Bâtinévi (Fula-Pulaar), Bamâri, Batân (Manding-Bambara), Kantaŋ (Mandinka), Bana, Bana-Bâdâ, Bana-Bandan, Bâtân, Busana (Maninka), Buday, Budey (Non), M-Buday (Serer), Len (Serer-Non), Bêtéñé (Wolof), Bêtanô (Soce);
Sierra Leone: Poloŋɛ, Polon-Dɛ (Bulom), Banta (Fula-Pulaar), Sona (Gola), G-Banda (Kissi), G-Banda (Kono), Kɔtin-Tri (Krio), Kutɛnɛ (Limba), N-Gukhσ(I) (Loko), Gbandaŋ (Manding-Mandinka), Banda (Maninka), Nguwa (Mende), Konde (Susu), Konde-Na Kundi-Na, (Susu-Dyalonke,) Am-Poloŋ, G-Banda (Vai);

Spanish: Árbol Capoc, Ceiba, Ceibo, Pochote Sumauma Kapok, Arbol De Seda, Arbol De La Seda, Capoquero, Ceiba Juca, Mosmote, Peem, Yuca;
Sri Lanka: Elavam, Imbul, Kottapulung, Pulung, Pulunimbal (Sinhalese);
Surinam: Kumaka (Arawak), Jumbi Tree, Kankantrie, Silk (Creole), Kankantrie, Katoenboom (Dutch), Fuma, Koddobakkoe;
Swahili: Mbuyu, Msufi;
Swedish: Kapok;
Tahitian: Komiro, Pacae, Vavai;
Thai: Ngio Sai (Northern Thailand), Nun, Ngui Noi, Ngio Soi, Ngao, Nun Tale;
Togo: Bubumbu, Bufu (Bassari), Aloe, Eloe (Gbe), Huti, Vuti, Wuti (Gbe-Fon), Lovi (Gen), Botu, Botu-Kisemto, Botu-Kocholemotu Botu, Kolombolu Komu, Kpong (Kabre), Igboa, Ju, Juna (Kposo), Bahun, Gomu-Schiere, Ubombĕ (Moore-Nawdam), Agú, Oguvé (Nago), Bagbasse, Komu (Tem), Aloe, Eloe, Evu, Ewu, Vu, Wu, Wudese, Wuti (Vhe), Huti, Vuti, Wuti (Yoruba-Ife);
Upper Volta: Pi (Bobo), Banda(Dyula), Bantan, Bantignei (Fula-Fulfulde),Rimi(Hausa), Belon (Kirma), Banan (Manding-Bambara), Gunga (Moore), Bantan, Bonetan (Songhai-Zarma), Blo (Turuka);
Venezuela: Ceiba Yuca, Ceibo Jabillo (Spanish);
Vietnamese: Bông Gòn;
West Cameroons: Bŭma, Kabò (Duala), Buma, Wuma (Kpe), Bum (Lundu);
Tongan: Vavae;
Yapese: Batte Ni Gan' Ken.

Origin/Distribution

Kapok is native to tropical America and Africa – Mexico, Central America, northern South America and the Caribbean, and tropical west Africa. Now introduced elsewhere in the tropics, in Asia and the South Pacific.

Agroecology

Kapok thrives in various environments from savannah to forest, in backyards, along roadsides in open cultivated land from sea level to 1,000 m. It thrives on rich volcanic soils but also tolerates poor soils. It requires a marked dry season and sheds its leaves during drought.

Edible Plant Parts And Uses

Young leaves, immature fruits and petals are eaten as vegetables. Young capsules are eaten fresh or cooked or in soup; seeds eaten pounded in soup or roasted. In Thailand, the inner part of young capsules are eaten as vegetable. The young leaves are eaten as a leafy vegetable. The young leaves are reported to be very good sources of calcium and iron. The young leaves are sometimes cooked and eaten in West Africa as a soup herb. The sprouts are also edible. In south east Asia – Peninsular Malaysia, Java and Celebes, the seeds are toasted and eaten. The Hausa of Northern Nigeria roast the seed to prepare a foodstuff called *gandido*, and in Ghana the seeds are prepared into a paste called *kantong* which is taken as a food or as a seasoning. In West Africa, the seeds are pounded and ground to a meal, and cooked in soup. In the Amazon basin, the seeds are used as a source of edible oil. Kapok oil is used as a substitute or adulterant for cotton-seed oil and olive oil. In West Cameroons, the whole flower, or more usually just the calyx, is eaten.

Botany

This is an erect, deciduous, buttressed tree grows up to 15 m high. The trunk is cylindric with grey bark, usually bearing scattered, large spines. The branches in distinct horizontal whorls. The leaves are palmately compound, with 5–8 lanceolate leaflets, 6–18 cm long, acuminate, glaucous, entire on 7–20 cm long petioles. The flowers are numerous, creamy white, 5-merous and about 3 cm long and clustered on the branchlets. The capsules are pendulous, dehiscent, oblong-ellipsoid, 7–15 cm in length, and 5 cm wide, green turning brown when mature (Plates 1–2). They contain numerous brown seeds, which are compressed-globose, smooth, and embedded in fine, silky hairs.

Plate 1 Kapok fruit clusters

Plate 2 Close-up of immature kapok fruit

Nutritive/Medicinal Properties

Dried kapok seeds were reported to have the following food composition per 100 g edible portion: energy 494 kcal, moisture 6.8%, protein 30.9 g, fat 36.5 g, total carbohydrate 20.1 g, fibre 1.5 g, ash 5.7 g, Ca 214 mg (Leung et al. 1968).

Kapok oil, was found to be similar to groundnut oil with potential applications in industrial manufacturing as well as for food. Its composition was variable according to region but was of the sequence: oleic acid 50–53%, linoleic acid 26–29%, palmitic acid 10–16%, stearic acid 2–5%, traces of arachidic and myristic acids, and phytosterol (Kerharo and Adam 1974; Quisumbing 1978).

Two new isoflavones, pentandrin and pentandrin glucoside, were isolated from the stem barks of *Ceiba pentandra* along with β-sitosterol and its 3-O-β-D-glucopyranoside (Ngounou et al. 2000). From the 80% ethanol extract of the bark of *Ceiba pentandra* a new isoflavone glycoside, 5-hydroxy-7,4′, 5′-trimethoxyisoflavone 3′-O-α-L-arabinofuranosyl(1→6)-β-D-glucopyranoside was isolated along with known isoflavones, vavain and vavain glucoside. A naphthoquinone, 2,7-dihydroxy-8-formyl-5-isopropyl-3-methyl-1,4-naphthoquinone together with a known naphthoquinone,8-formyl-7-hydroxy-5-isopropyl-2-methoxy-3-methyl-1,4-naphthoquinone were isolated from the heartwood (Kishore et al. 2003).

Anti-inflammatory Activity

Three isoflavone glucoside: vavain 3′-O-β-d-glucoside (1) and its aglycon, vavain (2), and flavan-3-ol (+)-catechin, isolated from the bark of *Ceiba pentandra*, exhibited IC_{50} values of 381, 97, and 80 μM on inhibition of cyclooxygenase-1-catalyzed prostaglandin biosynthesis (Noreen et al. 1998). When further tested for their inhibitory effects on cyclooxygenase-2-catalyzed prostaglandin biosynthesis, compounds 1 and 2 were found to be inactive ($IC_{50} > 1,200$ and >900 μM, respectively). From the 80% EtOH extract of the bark of *Ceiba pentandra*, a new isoflavone glycoside, 5-hydroxy-7,4′,5′-trimethoxyisoflavone 3′-O-α-L-arabinofuranosyl (1.RAR.6-β-D-glucopyranoside) was isolated along with known isoflavones, vavain and vavain glucoside (Ueda et al. 2002).

Hypoglycaemic Activity

Ceiba pentandra was found to exhibit hypoglycaemic and antidiabetic activities. Studies showed that oral administration of root bark methylene chloride/methanol extract of *Ceiba pentandra* was capable of ameliorating hyperglycaemia in streptozotocin-induced type-2 diabetic rats (Dzeufiet et al. 2006, 2007). The methylene chloride/methanol extract of *Ceiba pentandra* treatment significantly decreased the intake of both food and water as well as the levels of blood glucose, serum cholesterol, triglyceride, creatinine and urea, in comparison with diabetic controls. The treatment also improved impaired glucose tolerance but no effect was observed in the level of hepatic glycogen. The effect of *Ceiba pentandra* (40 mg/kg) was more outstanding when compared to glibenclamide in lowering blood glucose, with the added benefit of appreciably decreasing serum cholesterol and triglyceride concentrations. Results indicated it to be a potential source for isolation of new orally active agent(s) for anti-diabetic therapy. Separate studies also confirmed that the stem bark of *Ceiba pentandra* had hypoglycemic properties (Ladeji et al. 2003). Administration of the aqueous bark extract caused a statistically significant reduction in plasma glucose level in streptozotocin induced diabetic rats. The extract appeared non-toxic as evidenced by normal serum levels of aspartate aminotransferase (AST), ALT, ALP and bilirubin.

Studies demonstrates that *C. pentandra* leaves at moderate concentrations, exhibited both hypoglycaemic and hypolipidaemic effects in alloxan induced diabetic rats in a concentration dependent manner (Aloke et al. 2010). Plasma glucose was found to be significantly decreased in the treated rats when compared with the controls, with feed containing 20% of the leaves exerting the greatest effect. All the groups exhibited increases in body weight, which was least in group that received 20% of the experimental feed. However while high density lipoprotein was significantly raised in the treated animals, low density lipoprotein and very low density lipoproteins and triglycerides showed significant decreases. Also total protein and albumin were raised in the treated groups compared with the controls. The results indicated that *C. pentandra* leaves could be of importance in the treatment of diabetes and its associated complications such as coronary artery disease.

Antidrepanocytary Activity

Ceiba pentandra was one of 12 plants screened, which exhibited significant anti-drepanocytary (anti-sickle cell anaemia) activity, supporting the claims of the traditional healers (Mpiana et al. 2007). The stem bark also had trypanocidal activity. The stem bark was able to reduce the parasitaemia caused by *Trypanosoma brucei brucei* in animals treated at the dose of 100 mg/kg body weight (intraperitoneally, twice daily for 3 days) and of 150 mg/kg body weight (per orally, twice daily for 3 days), respectively.

Antiangiogenic Activity

Methanol extract of *Ceiba pentandra* stem exhibited the strongest inhibitory activity on the tube-like formation induced by human umbilical venous endothelial (HUVE) cells in the in-vitro angiogenesis assay with inhibition percentages of 89.3% and 87.5% at 30 and 100 µgram/ml, respectively (Nam et al. 2003).

Antimicrobial Activity

Two sesquiterpene lactones, 2,7-dimethoxy-5-isopropyl-3-methyl-8,1-naphthalene carbolactone and 2-hydroxy-5-isopropyl-7-methoxy-3-methyl-8,1-naphthalene carbolactone from the root bark of *Ceiba pentandra* showed moderate antimicrobial activity (Rao et al. 1993).

Antiulcerogenic Activity

Aqueous extracts of *Ceiba pentandra* showed antiulcerogenic effects (Ibara et al. 2007). Studies in rats showed that aqueous extract *of C. pentandra* at a dose of 400 mg/kg per os decreased significantly the decrease of pH and the formation of lesions induced by indometacin, compared to control group. The aqueous extract was well tolerated by animals. Until the dose of 3,200 mg/kg per os, no mortality was observed.

Hepatoprotective Activity

Studies by Bairwa et al. (2010) found that the ethyl acetate fraction of methanolic extract of

Ceiba pentandra possessed hepatoprotective potential against paracetamol-induced hepatotoxicity in rats. The ethyl acetate fraction (400 mg/kg) of *C. pentandra* stem bark administered orally to the rats with hepatotoxicity induced by paracetamol (3 g/kg) caused a significant reduction in serum enzymes glutamic-oxaloacetic transaminase (GOT), alanine aminotransferase (ALT), aspartate aminotransferase (AST), alkaline phosphatase (ALP), total bilirubin content. This was also evident from histopathological screening.

Antivenom Activity

Aqueous methanol of *Ceiba pentandra* leaves extract was found to have potent antivenom activity against *Echis ocellatus* snake venom (Sarkiyayi et al. 2010). Haemolysis in mice due to venom was significantly reduced by the extract from 66% to 27.4%. The in-vivo studies demonstrated that there were significant reduction in packed cell volume, total protein and haemoglobin concentrations for the venom group and there was only slight changes in the venom extract and control groups suggesting that the leaf extract had some inhibitory effect on the venom activity. The purified phospholipase incubated with the extract revealed neutralization effect against the phospholipase A2 activity. The results suggested that *Ceiba pentandra* leaf extract possessed potent snake venom-neutralizing capacity and could be use as an antidote for snakebite envenomation.

Toxicity Studies

Studies showed that the aqueous methanol extract of *Ceiba pentandra* leaves had very low toxicity profile in all the tested animals and the plant was relatively safe for herbal oral medication (Sarkiyayi et al. 2009). For acute toxicity test, the LD_{50} was found to be over 5,000 mg/kg. For chronic toxicity test there was no significant change in body weight and haematological and serum biochemical parameters . apart from urea, all other parameters were within physiological acceptable range. The results further revealed that there was a reduction in creatinine (88 µMol/L) and total protein (57.4 g/L), while aspartate aminotransferase (AST) (19 IU/L), alanine aminotransferase (ALT) (10 IU/L) and alkaline phosphatase (ALP) (204 IU/L) concentrations increased significantly. There was no significant change in packed cell volume and bilirubin concentrations.

Traditional Medicinal Uses

Various parts of the tree are used in traditional folk medicine (Kerharo and Bouquet 1950; Burkill 1966; Adegoke et al. 1968; Bouquet 1969; Burkill 1985; Dalziel 1955; Irvine 1961; Bouquet and Debray 1974; Kerharo and Adam 1974; Quisumbing 1978; Watt and Breyer-Brandwijk 1962; Oliver-Bever 1986; Noumi and Tchakonang 2001; Tapsoba and Deschamps 2006; Magassouba, et al. 2007).

Ceiba pentandra is used as an additive to some versions of the psychoactive, hallucinogenic drink *Ayahuasca* by the Amerindians of Amazonian Columbian. The plant has been used for relieving cough and as hair shampoo. In French Guiana, a decoction of the flowers is used for constipation. The ripe fruit is regarded as demulcent and astringent. The latex is used as styptic for treating diarrhoea, dysentery, menorrhagia and given with milk as a cooling laxative for incontinence in children.

The leaves are deemed medicinal among the Malays in Malaysia, being used in the same way as those of the cotton-plant. In Singapore, an infusion, prepared by pounding the leaves with an onion and a little turmeric, and adding water, is administered for coughs. In Java, infusion of the leaves is used for coughs, hoarseness, intestinal catarrh, and urethritis. In India, the tender leaves are administered for gonorrhoea. In Java, the leaves are used for cleaning the hair. In Senegal, West Africa, a decoction of the leaves is dropped into eyes to treat inflammation of the conjunctivitis, freshly pounded leaves are steeped in water and drunk for general fatigue, the leaves are also used to counter fatigue and lumbago. The leaves are used in Nigeria as an alterative and laxative, and an infusion is given as a cure for colic in man and in stock, when pounded to a fine state, the leaves are applied as a curative dressing on sores. In Guinea, a wet poultice of pulped leaves is used to maturate tumours.

In the Philippines, the bark is employed as a vomitive and as an aphrodisiac. The bark is brewed into a decoction regarded as specific in febrile catarrh and is used for type 2 diabetes and for headaches. The bark is diuretic and, in sufficient quantity, produces vomiting and is used in Cambodia, Laos, Vietnam for fevers, gonorrhoea and diarrhoea. In Malaya, the Malays used it for treating asthma and colds in children. In Java, the bark with areca nuts, nutmegs, and sugar-candy, is taken as a diuretic, and is thought good for stone in the bladder. In Mexico, a decoction of the bark taken internally has emetic, diuretic, and antispasmodic properties. The bark, which has tannin, is pounded and macerated in cold water and applied to swollen fingers.

In Congo, a bark-decoction is taken by mouth to relieve stomach complaints, diarrhoea, hernia, blennorrhoea, heart-trouble and asthma, and in mouth-washes and gargles for gingivitis, aphthae and sometimes toothache. In Nigeria, a bark-infusion is employed as a febrifuge. Homeopathically a bark-decoction is given to ricketty children, and bark-sap is administered to sterile women to promote conception in Ivory Coast-Upper Volta and in Congo. In the Sangmelima region of southern Cameroon, the stem bark or petiole of *C. pentadra* is used as a mechanical abortifacient. In Guinea Conakry, West Africa, a decoction of the stem bark is used as antiseptic against infection. Kapok oil is said to be used in Nigeria for rheumatism. In Cote D'Ivoire, a tisane is taken for diarrhoea and localized oedemas and a decoction is used to wash sores, furuncles and leprous macules. In Senegal, a decoction of the stem, branch bark is used as gargle for inflamed gums and gingivitis and also in Burkina Faso.

In India, the root is medicinal, a decoction being given for chronic dysentery, diarrhoea, ascites, diabetes and anasarca. The young roots, dried in the shade and powdered, form a chief ingredient in aphrodisiac medicines. The tap-root of the young plant is useful in gonorrhoea and dysentery. In Myanmar, the root is regarded as tonic. In Ivory Coast-Upper Volta, the root is used in various remedies for leprosy.

Other Uses

Today kapok tree is widely cultivated in tropics, especially in South America, Thailand, Cambodia, Indonesia, Kenya, Tanzania, Indonesia, Philippines and India, mainly as a crop, also as fetish or medicinal plant. The tree is also used as hedge, shade tree or as live support for growing pepper, *Piper nigrum*. The leaves are used as goat-fodder. The fibres from the pod are light, very buoyant, resilient, highly flammable and resistant to water. The fibre which is highly flammable is used as a fuel in fire pistons, in Thailand called *taban fai*, and in fireworks. The fibres have been extensively used for padding and stuffing pillows, cushions, mattresses, upholstery, *zafus*, teddy bears and insulation. They are also employed in making life-jackets. The fibres are fibre can be used for surgical preparations, replacing cottonwool and also make good acoustic insulant. The light wood is used for the production of canoes, boats, toys, paper, dishes, boxes, drums, carved figures, idols, stamps and dies, for modelling, musical instruments, boxes, rice mortar, wooden clogs and sandals, furniture, matches and wood-wool. Plane articles such as doors, table-tops, plates, trays, etc., can be made from the buttresses. The seeds contain up to 25% oil and this kapok oil, extracted from the seeds, is used for fuel and for the manufacture of the soap, paints, ointments, and as a substitute or adulterant for cotton-seed oil and olive oil. The oil is also used as lubricant and illuminant. The fresh cake is valuable as stock feed and kapok seed meal has potential as feed for broiler chicken and as fertilisers. The ashes of the fruit were used by dyers in Malaya. Wood-ash is widely used in Africa as a kitchen salt and in soap-making. In Casamance, Senegal, ash from the pods is used in making snuff and in Guinea at Fouta Djalon in the indigo industry. The seed cake is a good cattle-feed and also a good agricultural fertilizer.

Malaysian kapok has shown great potential as an effective natural oil sorbent, owing to high sorption and retention capacity, structural stability and high reusability (Abdullah et al. 2010). Recent studies indicated that Kapok fibre has potential industrial application in the cleaning up of oil spills. Kapok is a natural hollow hydrophobic-oleophilic fibrous sorbent for oil spill cleanup (Lim and Huang 2007). Studies showed that kapok's sorption capacities for the three oils were significantly higher than those of polypropylene (PP), a widely used commercial oil sorbent for oil spill cleanup. The hydrophobic-oleophilic characteristics of the kapok fibre could be attributed to its waxy surface, while its large lumen contributed to its excellent oil absorbency and retention capacity. A deep-bed kapok filtration column was found to be successful in achieving oily water separation under different packing densities and flow rates (Rahmah and Abdullah 2010). The oil and water front movements were shown to be influenced by the affinity of liquid to kapok fibers. Results indicated the excellent physicochemical property of Malaysian kapok for oil removal from water.

Activated carbon prepared from *Ceiba pentandra* hulls, an agricultural solid waste by-product has potential industrial application as adsorbent in the removal of metals like, copper, cadmium, lead and zinc from aqueous solutions (Rao et al. 2006). The adsorbent exhibited good sorption potential for copper and cadmium at pH 6.0. Maximum desorption of 90% for copper and 88% for cadmium occurred with 0.2 M HCl.

Comments

This tree is the official national tree of Puerto Rico and Guatemala.

Selected References

Abdullah MA, Rahmah AU, Man Z (2010) Physicochemical and sorption characteristics of Malaysian *Ceiba pentandra* (L.) Gaertn. as a natural oil sorbent. J Hazard Mater 177(1–3):683–91

Adegoke EA, Akinsaya A, Naqui SHZ (1968) Studies of Nigerian medicinal plants I. A preliminary survey of plant alkaloids. J West Afr Sci Assoc 13:13–33

Aloke C, Nwachukwu N, Idenyi JN, Ugwuja EI, Nwachi EU, Edeogu CO, Ogah O (2010) Hypoglycaemic and hypolipidaemic effects of feed formulated with *Ceiba pentandra* leaves in alloxan induced diabetic rats. Aust J Basic Appl Sci 4(9):4473–4477

Bairwa NK, Sethiya NK, Mishra SH (2010) Protective effect of stem bark of *Ceiba pentandra* Linn. against paracetamol-induced hepatotoxicity in rats. Pharmacogn Res 2:26–30

Bouquet A (1969) *Féticheurs et Médecines Traditionnelles du Congo (Brazzaville)*. Mémoires. O.R.S.T.O.M., 36, Paris, 282 pp

Bouquet A, Debray M (1974) *Plantes Médicinales de la Côte d'Ivoire*. Mémoires O.R.S.T.O.M., 32, Paris, 231 pp

Burkill IH (1966) A dictionary of the economic products of the Malay Peninsula. Revised reprint, 2 volumes, vol 1 (A–H) pp 1–1240, vol 2 (I–Z), pp 1241–2444. Ministry of Agriculture and Co-operatives, Kuala Lumpur

Burkill HM (1985) The useful plants of West Tropical Africa, vol 1, Families A–D. Royal Botanical Gardens, Kew, 960 pp

Dalziel JM (1955) The useful plants of West Tropical Africa (Reprint of 1937 edn) Crown Agents for Overseas Governments and Administrations, London, 612 pp

Dzeufiet PDD, Tedong L, Asongalem EA, Dimo T, Sokeng SD, Kamtchouing P (2006) Hypoglycaemic effect of methylene chloride/methanol root extract of *Ceiba pentandra* in normal and diabetic rats. Indian J Pharmacol 38:194–7

Dzeufiet PD, Ohandja DY, Tédong L, Asongalem EA, Dimo T, Sokeng SD, Kamtchouing P (2007) Antidiabetic effect of *Ceiba pentandra* extract on streptozotocin-induced non-insulin-dependent diabetic (NIDDM) rats. Afr J Tradit Complement Altern Med 4(1):47–54

Foundation for Revitalisation of Local Health Traditions (2008) FRLHT database. htttp://envis.frlht.org

Ibara JR, Elion Itou RDG, Ouamba JM, Diatewa M, Gbeassor M, Abena AA (2007) Preliminary evaluation of antiulcerogenic activity of *Ceiba pentandra* Gaertn and *Helicrysum mechowianum* Klatt in rats. J Med Sci 7:485–488

Irvine FR (1961) Woody plants of Ghana: with special reference to their uses. Oxford University Press, London, p 868

Kerharo J, Adam JG (1974) La Pharmacopée Sénégalaise Traditionnelle. Plantes Médicinales et Toxiques. Editions Vigot Frères, Paris, 1011 pp (In French)

Kerharo J, Bouquet A (1950) Plantes médicinales & Toxiques de la Côte-d'Ivoire-Haute-Volta. Vigot Freres, Paris

Kishore PH, Reddy MV, Gunasekar D, Caux C, Bodo B (2003) A new naphthoquinone from *Ceiba pentandra*. J Asian Nat Prod Res 5(3):227–30

Ladeji O, Omekarah I, Solomon M (2003) Hypoglycemic properties of aqueous bark extract of *Ceiba pentandra* in streptozotocin-induced diabetic rats. J Ethnopharmacol 84(2–3):139–42

Leung W-TW, Busson F, Jardin C (1968) Food composition table for use in Africa. FAO, Rome, 306 pp

Lim TT, Huang X (2007) Evaluation of kapok (*Ceiba pentandra* (L.) Gaertn.) as a natural hollow hydrophobic-oleophilic fibrous sorbent for oil spill cleanup. Chemosphere 66(5):955–63

Magassouba FB, Dialloa A, Kouyaté M, Mara F, Bangoura O, Camara A, Traoré S, Diall AK, Camara G, Traoré S, Keita A, Camara MK, Barry R, Keita S, Oularé K, Barry MS, Donzo M, Camara K, Toté K, Vanden Berghe D, Totté J, Pieters L (2007) Ethnobotanical survey and antibacterial activity of some plants used in Guinean traditional medicine. J Ethnopharmacol 114:44–53

Mpiana PT, Tshibangu DS, Shetonde OM, Ngbolua KN (2007) In vitro antidrepanocytary activity (anti-sickle cell anemia) of some congolese plants. Phytomedicine 14(2–3):192–5

Nam NH, Kim HM, Bae KH, Ahn BZ (2003) Inhibitory effects of Vietnamese medicinal plants on tube-like formation of human umbilical venous cells. Phytother Res 17(2):107–11

Ngounou FN, Meli AL, Lontsi D, Sondengam BL, Atta-Ur-Rahman, Choudhary MI, Malik S, Akhtar F (2000) New isoflavones from *Ceiba pentandra*. Phytochem 54(1):107–110

Noreen Y, el-Seedi H, Perera P, Bohlin L (1998) Two new isoflavones from *Ceiba pentandra* and their effect on cyclooxygenase-catalyzed prostaglandin biosynthesis. J Nat Prod 61(1):8–12

Noumi E, Tchakonang NYC (2001) Plants used as abortifacients in the Sangmelima region of Southern Cameroon. J Ethnopharmacol 76:263–268

Ochse JJ, Bakhuizen van den Brink RC (1980) Vegetables of the Dutch East Indies, 3rd edn. Asher & Co., Amsterdam, 1016 pp

Oliver-Bever B (1986) Medicinal plants in Tropical West Africa. Cambridge University Press, Cambridge, 375 pp

Pacific Island Ecosystems at Risk (PIER) (1999) *Ceiba pentandra* (L.) Gaertn., Bombacaceae. http://www.hear.org/Pier/species/ceiba_pentandra.htm

Pongapangan S, Poobrasert S (1991) Edible and poisonous plants in Thai forests. O.S. Printing House, Bangkok, 176 pp

Quisumbing E (1978) Medicinal plants of the Philippines. Katha Publishing Co., Quezon City, 1262 pp

Rahmah AU, Abdullah MA (2010) Evaluation of Malaysian *Ceiba pentandra* (L.) Gaertn. for oily water filtration using factorial design. Desalination 266(1–3):51–55

Rao KV, Sreeramulu K, Gunasekar D, Ramesh D (1993) Two new sesquiterpene lactones from *Ceiba pentandra*. J Nat Prod 56(12):2041–5

Rao MM, Ramesh A, Rao GP, Seshaiah K (2006) Removal of copper and cadmium from the aqueous solutions by activated carbon derived from *Ceiba pentandra* hulls. J Hazard Mater 129(1–3):123–9

Sarkiyayi S, Ibrahim S, Abubakar MS (2009) Toxicological studies of *Ceiba pentandra* Linn. Afr J Biochem Res 3(7):279–281

Sarkiyayi S, Ibrahim S, Abubakar MS (2010) Studies on antivenom activity of *Ceiba pentandra* leaves' aqueous methanol extract against *Echis ocellatus*' snake venom. Res J Appl Sci Eng Technol 2(7):687–694

Tapsoba H, Deschamps JP (2006) Use of medicinal plants for the treatment of oral diseases in Burkina Faso. J Ethnopharmacol 104:68–78

Tropicos Org. Missouri Botanical Garden (Jan 2009). <http://www.tropicos.org

Ueda H, Kaneda N, Kawanishi K, Alves SM, Moriyasu M (2002) A new isoflavone glycoside from *Ceiba pentandra* (L.) Gaertner. Chem Pharm Bull 50(3):403–404

Watt JM, Breyer-Brandwijk MG (1962) The Medicinal and poisonous plants of Southern and Eastern Africa, 2nd edn. E. and S. Livingstone, Edinburgh/London, 1457 pp

Yang Q, Gilbert MG (2007) *Ceiba* Miller. In: Wu ZY, Raven PH, Hong DY (eds) Flora of China, vol 12, Hippocastanaceae through Theaceae. Science Press/Missouri Botanical Garden Press, Beijing/St. Louis

Durio dulcis

Scientific Name

Durio dulcis Becc.

Synonyms

Durio conicus Becc.

Family

Bombacaceae, also placed in Malvaceae, Durionaceae.

Common/English Names

Red Durian, Lahong

Vernacular Names

Borneo: Dianjau, Durian Bala (Kenya-Dyak), Durian Merah (Malay), Lahong Lajang Lajung (Dyak), Layung, Pesasang (Tidung), Tutong (Iban), Durian Marangang, Durian Tinggang, Duyen, Kusi, Lahit Manuk.

Origin/Distribution

D. dulcis is native to Sumatra and Borneo (Sarawak, Brunei, Sabah, West-, Central-, South- and East-Kalimantan). Occasionally also cultivated for the fruits.

Agroecology

A strictly tropical species, it occurs wild in lowland tropical rainforests. It is found scattered in mixed dipterocarp forest on sandy clay soils and friable clay loams, commonly in the lowlands but also on hillsides and ridges with sandy soils, sometimes on limestone from 20–800 m altitude. It is also cultivated.

Edible Plant Parts and Uses

The deep yellow pulp is edible and sweet and eaten raw or cooked. In Sabah, red durian pulp is fried with onions and chilli and served as a side dish.

Botany

A fairly large, evergreen, under-canopy, perennial tree up to 49 m tall, with bole up to 85 cm in diameter and large buttresses. The bark surface is rough, shallowly fissured or irregularly flaky and reddish-brown. Stipules are present, but deciduous, falling off early. Leaves are alternate, green above, silvery brown below, simple, penniveined, elliptical or obovate-elliptical, 7–14 cm × 3.5–6 cm, densely scaly below with entire margin (Plate 2). Flowers occur in short inflorescence clustered on older branches, with 2-lobed epicalyx, pink petals up to 45 mm long and stamens in bundles. Fruit is a globose capsule, up to 15 cm across, dark red to dark brown-red with slender 15–20 mm long, thin spines (Plate 1) and very odorous. Seeds are dark brown, each completely covered by a deep yellow to reddish, sweet aril.

Nutritive/Medicinal Properties

No information on its nutritive properties or medicinal properties has been published.

Other Uses

D. dulcis is one of the principal sources of durian timber in Sarawak.

Comments

The species has been deemed as a vulnerable species.

Plate 1 Spiny, globose red Lahong fruit with yellow arils

Plate 2 Elliptic, simple leaves and rough, flaky bark of Lahong

Selected References

Jansen PCM, Jukema J, Oyen LPA, van Lingen TG (1992) *Durio dulcis* Becc. In: Coronel RE, Verheij EWM (eds.) Plant Resources of South-East Asia No 2. Edible Fruits and Nuts. Prosea Foundation, Bogor, p 330

Kochummen KM (1972) Bombacaceae. In: Whitmore TC (ed.) Tree flora of Malaya, vol 1. Longman, Kuala Lumpur, pp 100–120

Kostermans AJGH (1958) The genus *Durio* Adans. (Bombac.). Reinwardtia 4(3):357–460

Slik JWF (2006) Trees of Sungai Wain. Nationaal Herbarium Nederland. http://www.nationaalherbarium.nl/sungaiwain/

Soegeng-Reksodihardjo W (1962) The species of *Durio* with edible fruits. Econ Bot 16:270–282

Voon BH, Chin TH, Sim CY, Sabariah P (1988) Wild fruits and vegetables in Sarawak. Sarawak Department of Agriculture, Kuching, 114 pp

Yap SK, Martawijaya A, Miller RB, Lemmens RHMJ (1995) *Durio* Adans. In: Lemmens RHMJ, Soerianegara I, Wong WC (eds.) Plant resources of South-East Asia No 5(2). Timber trees: minor commercial timbers. Prosea Foundation, Bogor, pp 215–225

Durio graveolens

Scientific Name

Durio graveolens Becc.

Synonyms

None

Family

Bombacaceae, also placed in Malvaceae, Durionaceae.

Common/English Names

Orange-Fleshed Durian, Red-Fleshed Durian, Yellow Durian

Vernacular Names

Borneo – Indonesia (Kalimantan), Malaysia (Sarawak, Sabah): Durian Burong, Durian Kuning, Durian Merah (Malay), Durian Isu (Iban), Durian Umot (Bidayuh), Durian Anggang Durian Ajan, Tabela, Tabelak, Tuala, Tuwala, (Dyak- Kenya), Alau, Dujen, Durian, Durian Alau, Durian Daun Dungoh, Durian Hutan, Durian Pipit, Lai Bengang, Merang Kunyit, Pasang, Tongkai;
Indonesia: Tinambela (Batak, Sumatra);
Malaysia: Durian Rimba, Durian Burung, Durian Kuning, Durian Merah, Durian Otak Udang Galah (Peninsular), Durian Burong, Durian Kuning, Durian Merah (Malay, Sarawak), Durian Isu (Iban, Sarawak), Durian umot (Bidayuh, Sarawak), Durian Merah (Malay, Sabah);
Thailand: Thurian-Rakka (Peninsular).

Origin/Distribution

Wild distributions occur in Peninsular Malaysia, Sumatra and Borneo (Sarawak, Brunei, Sabah, West-, Central- and East-Kalimantan); cultivated in Sabah, Sarawak and Brunei.

Agroecology

D. graveolens is strictly a tropical species that thrives in a wet and hot climate. It is commonly found in wet lowland forests on clay-rich soils in mixed dipterocarp forest as well as on shale ridges up to 1,000 m elevation. In secondary forests, it is usually present as a pre-disturbance remnant tree. It also occurs often along swamps and riversides, but also on hillsides and ridges.

The tree can be grown on well-drained, peat soil but the fertile alluvial plains are preferred.

Edible Plant Parts and Uses

The fruit has sweet deep yellow to orange or crimson-coloured pulp which is eaten raw. The pulp has a fragrance of roasted almonds. The deep crimson coloured aril of the fruit is highly esteemed and are given names like *durian otak udang galah* to signify the red colour of the 'brain of a crayfish'. Red-fleshed durian is traditionally added to *soup sayur*, an Indonesian soup made with fresh water fish.

Botany

A large, upper canopy tree up to 50 m tall, with straight, cylindrical 85 cm bole branchless for up to 25 m and up to 100 cm in diameter having steep buttresses, bark reddish-brown or greyish-mauve, smooth, to flaky. Stipules are present but dropped early. Leaves are simple, entire, elliptical to oblong, 10–26 cm × 4–10 cm, abruptly pointed apex, rounded base, coriaceous, glabrous above, densely copper-brown scaly below (Plate 1). Flowers occurs in short cymes on branches; calyx thin, 3–5 lobed sac-like at base; petals spatulate, 25–35 mm long, white; stamens and staminodes in 5 free bundles; ovary globose to ovoid, with greenish-white style and yellow capitate stigma. Fruit is a globose to ovoid capsule, about 10–15 cm in diameter, greenish-yellow, dirty yellow to orange-yellow, with sharp pyramidal 1 cm long spines (Plates 1–10), dehis-

Plate 2 Sub-globose, solitary spiny immature fruit, on a branch

Plate 3 Globose green fruit with long and sharp spines

Plate 1 Spiny, globose fruit and leaves with densely copper-brown scaly lower surfaces

Plate 4 Golden-orange arils of ripe fruit

Plate 5 White locule wall and sharp spines on the rind

Plate 8 Yellow fruit with yellow aril

Plate 6 Green fruit with red arils

Plate 9 Orangey-yellow fruit

Plate 7 Yellow globose fruit

Plate 10 Orangey-yellow fruit with red aril

cent into 5 valves while still attached to the branch. Seeds are ellipsoid, 4 cm × 2 cm, glossy brown, completely enclosed by a fleshy aril with a diversity of colour ranging from pale yellow, deep yellow, orange to crimson red (Plates 4–6, 8, 9–11).

Plate 11 Red and orange arils removed from fruit

Nutritive/Medicinal Properties

The proximate nutrient composition of the fruit of *Durio graveolens* per 100 g edible portion (pulp) based on analyses made in Sarawak (Voon and Kueh 1999) was reported as: water 66.7%, energy 152 kcal, protein 2.6%, fat 6.2%, carbohydrates 21.5%, crude fibre 2.0%, ash 1.0%, P 43 mg, K 529 mg, Ca 10 mg, Mg 27 mg, Fe 0.6 mg, Mn 4 ppm, Cu 7 ppm, Zn 5.9 ppm, vitamin C 10.4 mg.

The Iban in Sarawak would boil the bark from a mature tree and use the solution to bath babies 1 day after birth as they believed that this would strengthen the skin of prematurely-borne babies.

Other Uses

Durio graveolens is one of the important sources of durian timber in Sarawak.

Comments

The fruit of *D. graveolens* opens while still on the tree whereas the fruit of *D. dulcis* remains unopened and drops to the ground.

Durian suluk, also known as durian siunggong, is a natural hybrid between *Durio zibethinus* and *Durio graveolens*, and retains the flavour and texture of *D. zibethinus* with subtle, burnt caramel overtones of *D. graveolens*. Durian simpor is a mild-flavoured, yellow-fleshed variant of *D. graveolens*.

Selected References

Chai PPK (2006) Medicinal plants of Sarawak. Lee Ming Press, Kuching, pp 212

Jansen PCM, Jukema J, Oyen LPA, van Lingen TG (1992) *Durio graveolens* Becc. In: Coronel RE, Verheij EWM (eds.) Plant resources of South-East Asia No 2. Edible fruits and nuts. Prosea Foundation, Bogor, p 330

Kochummen KM (1972) Bombacaceae. In: Whitmore TC (ed.) Tree flora of Malaya, vol 1. Longman, Kaula Lumpur, pp 100–120

Kostermans AJGH (1958) The genus *Durio* Adans. (Bombac.). Reinwardtia 4(3):57–460

Slik JWF (2006) Trees of Sungai Wain. Nationaal Herbarium Nederland. http://www.nationaalherbarium.nl/sungaiwain/

Soegeng-Reksodihardjo W (1962) The species of *Durio* with edible fruits. Econ Bot 16:270–282

Voon BH, Kueh HS (1999) The nutritional value of indigenous fruits and vegetables in Sarawak. Asia Pac J Clin Nutr 8(1):24–31

Voon BH, Chin TH, Sim CY, Sabariah P (1988) Wild fruits and vegetables in Sarawak. Sarawak Department of Agriculture, Kuching, 114 pp

Wong WWW, Chong TC, Tanank J, Ramba H, Kalitu N (2007) Fruits of Sabah, vol 1. Department of Agriculture, Sabah, 126 pp

Yap SK, Martawijaya A, Miller RB, Lemmens RHMJ (1995) *Durio* Adans. In: Lemmens RHMJ, Soerianegara I, Wong WC (eds.) Plant resources of South-East Asia No 5(2). Timber trees: minor commercial timbers. Prosea Foundation, Bogor, pp 215–225

Durio kinabaluensis

Scientific Name

Durio kinabaluensis Kosterm. & Soegeng

Synonyms

Durio kutejensis (Hassk.) Becc. forma *kinabaluensis* Bakh.

Family

Bombacaceae, also placed in Malvaceae, Durionaceae.

Common/English Names

Durian Kinabalu

Vernacular Names

Sabah: Durian Kinabalu.

Origin/Distribution

The species is a native of Sabah, East Malaysia (Kostermans 1958). A wild durian known only from the foothills of the Crocker Range and Mount Kinabalu, mainly outside Park boundaries.

Agroecology

Durio kinabaluensis is adapted to high altitude, on hill slopes above 1,000 m above sea level in the tropics (Salma 2000).

Edible Plant Parts and Uses

The aril of ripe fruit is sweet and edible.

Botany

Large tree with smooth grey or brown branches, juvenile parts covered with coppery brown scales. Leaves alternate, chartaceous, elliptic or oblong, 10–18 cm by 4–7 cm, apex acuminate, acumen slender, upper surface glossy, reticulation, densely prominulose, lower surface with a dense layer of coppery scales, midrib prominent, reticulation obscure, lateral nerves 9–14 pairs, petiole 2 cm long. Inflorescence ramiflorous, unbranched, few-flowered on thick pedicels, 3–4 cm long, epicalyx irregularly splitting into two, deciduous, inside glabrous, outside with dense brown scales. Calyx grey-cream urceolate with 5 deltoid lobes, 0.75 cm long, tube 2–2.5 cm, inside glabrous outside densely covered with brown scales. Petals 5 reddish-pink, 5–5.5 cm long sparsely covered with scales and stellate hairs outside inside glabrous, stamens reddish–pink, phalanges connate at base up to 10 cm long, slender each phalange consist of 3–4 stamens, anther globose; ovary cream ovoid

Plate 1 Ripe Kinbalu durian with sharp conical spines

densely covered with scales and stellate hairs, 0.8 cm long, style densely covered with greyish stellate hairs, stigma inconspicuous, yellow. Fruit is subglobose, ellipsoid to ovoid, 14–18 cm long by 10–14 cm across, pale-greenish yellow to dirty yellow and yellowish–brown covered with densely packed pyramidal spines 1–1.5 cm slightly curved at apex (Plates 1–4), 5 valved, seeds dark brown, 2–4 cm shrunken (Plate 4). The aril is thick, pulpy creamy-white to pale yellow, sweet. Fruits dropped unopened recognised by distinct long remnants strand of the red stamen phalange at the stalk end (Plates 1–3).

Plate 2 Fruit with creamy-white pulp and the characteristic reddish stamen strands

Nutritive/Medicinal Properties

No published information is available on its nutritive and medicinal properties.

Other Uses

Refer to notes on durian, *Durio zibethinus*.

Plate 3 Close-up of aril, thorn and long stamen strands

Comments

D. kinabaleunsis can be distinguished from *Durio kutejensis* by its smaller flowers, longer filaments and stamens in 5 phalanges and scaleless upper part of the petals, epicalyx is deciduous as opposed to sub-persistent in *D. kutenjensis*.

Based on PCR-RFLP (polymerase chain reaction – restriction fragment length polymorphism) on the *ndh*C-*trn*V gene, Santoso et al. (2005) grouped 10 *Durio* species into five distinct clusters. The dendrogram revealed the formation of five clusters. Three of those, the first, fourth and

Plate 4 Shrivelled, brown-coated seeds

fifth clusters, were scattered off and occupied by single species of *D. kinabaluensis*, *D. grandiflorus* and *D. testudinarum* in each cluster, respectively. They had low genetic similarities to other species, indicating that independent evolutionary processes involving *ndh*C-*trn*V region might have occurred on those three species. The second cluster consists of four accessions belonging to three species, i.e. *D. oxleyanus*, two accesions of *D. graveolens* (orange fruit and green fruit), and *D. dulcis*. A similar trend was observed in the third cluster, which consisted of four species, i.e. *D. kotejensis*, *D. sabahensis*, *D. purureus* and *D. zibethinus*. This cluster had high genetic similarities amongst the four species, indicating the presence of some associations in the evolutionary processes involving *ndh*C-*trn*V region in the four species. In contrast, based on PCR-RFLP on the *rbc*L gene, the accessions were grouped into three distinct clusters, and generally showed low variations. The dendrogram revealed the formation of three clusters. The first cluster consisted of three accessions belonging to *D. kinabaluensis* and the two accessions of *D. graveolens*. The second cluster consisted of six species, i.e. *D. oxleyanus*, *D. dulcis*, *D. purureus*, *D. testudinarum*, *D. kotejensis* and *D. grandiflorus*, and the third consisted of two species, *D. sabahensis* and *D. zibethinus*.

According to Santoso et al. (2005) *Ndh*C-*trn*V gene is more reliable for phylogeny analysis in lower taxonomic level of *Durio* or to be used for diversity analysis, while *rbc*L gene is a reliable marker to be used for phylogeny analysis at higher taxonomic level.

Selected References

Kostermans AJGH (1958) The genus *Durio* Adans. (Bombac.). Reinwardtia 4(3):357–460

Salma I (2000) The significance of pollen morphology in the taxonomy of the genus *Durio* (Bombaceae). Gard Bull (Singapore) 52:261–271

Santoso PJ, Saleh GB, Saleh NM, Napis S (2005) Phylogenetic relationships amongst 10 *Durio* species based on PCR-RFLP analysis of two chloroplast genes. Indonesian J Agric Sci 6(1):20–27

Durio kutejensis

Scientific Name

Durio kutejensis (Hassk.) Becc.

Synonyms

Lahia kutejensis Hassk.

Family

Bombacaceae, also placed in Malvaceae, Durionaceae.

Common/English Names

Borneo Durian, Durian Nyekek

Vernacular Names

Borneo: Paken (Punan Malinau, East Kalimantan), Ruwat, Ruat (Lundanye, East Kalimantan), Pekeing (Merap, East Kalimantan), Lezing Bala (Kenyah Uma', East Kalimantan), Dian Bala, Durian, Durian Isu, Durian Kuning, Durian Tinggang, Durian Utan, Kawai, Lai, Layuk, Lembunyu, Luas, Paku, Pekawai, Putuk, Ruas, Se;
Brunei: Durian Pulu;
Malaysia: Durian Kuning (Peninsular), Nyekek, Ukak Pakan (Iban, Sarawk), Durian Kuning, Puloh (Malay, Sarawak), Durian Lai, Durian Luas (Sabah);

Origin/Distribution

The species is indigenous to Java and Borneo (Sarawak, Brunei, Sabah, West-, Central-, South- and East-Kalimantan).

Agroecology

D. kutejensis occurs wild in foothills of central Borneo; in Sarawak, it occurs locally on fertile clay-rich soils on undulating land in undisturbed mixed dipterocarp forest up to 800 m altitude, but usually lower. In secondary forests, it is usually present as a pre-disturbance remnant tree, or planted, mostly on hillsides and ridges on sandy to clayey soils and also on limestone. It is also found growing in well-drained, alluvial plains and river basin from 20–200 m elevation and appears to tolerate fluctuating water table.

Edible Plant Parts and Uses

The thick pulp (aril) is sweet, pleasantly fragrant and mild which has appeal to those who may otherwise find the fragrance of durian too strong.

Botany

An upper canopy tree up to 40 m tall, with bole of 68 cm diameter, older tree is branchless up to 12 m with low, rounded buttresses and a rough and flaky, grey to reddish-brown bark (Plates 1, 3). Stipules are present, but shed early. Leaves are alternate, distichous, simple, penniveined, elliptical-oblong, 10–33 cm× 3–12 cm, glossy green, glabrous above, densely pale golden-brown scaly below with rounded bases and entire margin (Plates 2–3). Flowers are foetid, in 3–20 flowered cymes on older branches, with 2-lobed epicalyx and petals up to 90 mm long, dark red, lanceolate-obovate and revolute; ovary ovoid, 5 locular; stamens 45–55 free, opening by a slit (Plate 4). Fruit is an ovoid to ellipsoid, pentangular capsule, up to 20 cm× 12 cm, green maturing to orange yellow, covered with slightly curved soft, short spines 1–1.5 cm long and with a persistent style (Plates 5–9). Seeds are ellipsoid, up to 4 cm long, glossy brown, completely enclosed by a fleshy, orangey yellow to orange, mildly fragrant, sweet aril (Plates 8–10).

Plate 3 Scaly rough trunk and large, oblong leaves

Plate 1 Tree habit of young *Durio kutejensis* tree

Plate 4 Bright red flowers of *Durio kutenjensis*

Plate 2 Leaves of *Durio kutenjensis*

Plate 5 Ovoid fruit of *Durio kutenjensis*

Plate 6 Pentangular sub-ovoid fruit of *Durio kutenjensis*

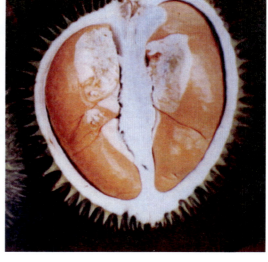

Plate 9 Ovoid fruit with orange aril

Plate 7 Ellipsoid fruit of *Durio kutenjensis*

Plate 10 Fruit with golden yellow arils and brown seeds

Plate 8 Ellipsoid fruit with deep orange aril

Nutritive/Medicinal Properties

The proximate nutrient composition of the fruit of *Durio kutejensis* per 100 g edible portion (aril) based on analyses made in Sarawak (Voon and Kueh 1999) was reported as: water 61.5%, energy 149 kcal, protein 2.6%, fat 1.7%, carbohydrates 30.9%, crude fibre 1.9%, ash 1.5%, P 25 mg, K 362 mg, Ca 19 mg, Mg 19 mg, Fe 0.7 mg, Mn 5 ppm, Cu 3.2 ppm, Zn 7.3 ppm, vitamin C 15.9 mg.

In a study conducted in Malaysia, the total carotene content (mg/100 g) of selected underutilized tropical fruit (Khoo et al. 2008) in decreasing sequence was: jentik-jentik (*Baccaurea polyneura*) 19.83 mg > cerapu 2 (*Garcinia prainiana*) 15.81 mg > durian nyekak 2 (*Durio kutejensis*) 14.97 mg > tampoi kuning (*Baccaurea reticulata*) 13.71 mg > durian nyekak (1) 11.16 mg > cerapu 1 6.89 mg > bacang 1 (*Mangifera foetida*) 4.81 mg > kuini (*Mangifera odorata*) 3.95 mg > jambu susu (*Syzygium malaccense*) 3.35 mg > bacang (2) 3.25 mg > durian daun (*Durio lowianus*) 3.04 mg > bacang (3) 2.58 mg > tampoi putih (*Baccaurea macrocarpa*) 1.47 mg > jambu mawar (*Syzygium jambos*) 1.41 mg. β-carotene content was determined by HPLC to be the highest in jentik-jentik 17,46 mg followed by cerapu (2) 14.59 mg, durian nyekak (2) 10.99 mg, tampoi kuning 10.72 mg, durian nyekak (1) 7.57 mg and cerapu (1) 5.58 mg. These underutilized fruits were found to have acceptable amounts of carotenoids and to be potential antioxidant fruits.

Wood bark extract from *Durio kutejensis* yielded new triterpenes 3β-O-trans-caffeoyl-2α-hydroxyolean-12-en-28-oic acid and 3β-O-trans-caffeoyl-2α-hydroxytaraxest-12-en-28-oic acid together with four known compounds, maslinic acid, arjunolic acid, 2,6-dimethoxy-*p*-benzoquinone, and fraxidin (Rudiyansyah and Garson 2006).

Other Uses

Its timber is sold as durian wood.

Comments

An important characteristic of *Durio kutejensis* is that the calyx and its filaments are firmly attached to the base of the fruit.

Selected References

Backer CA, Bakhuizen van den Brink RC Jr (1963) Flora of Java, vol 1. Noordhoff, Groningen, 648 pp

Khoo HE, Ismail A, Mohd.-Esa N, Idris S (2008) Carotenoid content of underutilized tropical fruits. Plant Foods Hum Nutr 63:170–175

Kochummen KM (1972) Bombacaceae. In: Whitmore TC (ed.) Tree flora of Malaya, vol 1. Longman Malaysia, Kuala Lumpur, pp 100–120

Kostermans AJGH (1958) The genus *Durio* Adans. (Bombac.). Reinwardtia 4(3):357–460

Munawaroh E, Purwanto Y (2009) Studi hasil hutan non kayu di Kabupaten Malinau, Kalimantan Timur (In Indonesian). Paper presented at the 6th basic science national seminar, Universitas Brawijaya, Indonesia, 21 Feb 2009. http://fisika.brawijaya.ac.id/bss-ub/proceeding/PDF%20FILES/BSS_146_2.pdf

Rudiyansyah, Garson MJ (2006) Secondary metabolites from the wood bark of *Durio zibethinus* and *Durio kutejensis*. J Nat Prod 69(8):1218–1221

Slik JWF (2006) Trees of Sungai Wain. Nationaal Herbarium Nederland. http://www.nationaalherbarium.nl/sungaiwain/

Soegeng-Reksodihardjo W (1962) The species of *Durio* with edible fruits. Econ Bot 16:270–282

Voon BH, Kueh HS (1999) The nutritional value of indigenous fruits and vegetables in Sarawak. Asia Pac J Clin Nutr 8(1):24–31

Voon BH, Chin TH, Sim CY, Sabariah P (1988) Wild fruits and vegetables in Sarawak. Sarawak Department of Agriculture, Kuching, 114 pp

Wong WWW, Chong TC, Tanank J, Ramba H, Kalitu N (2007) Fruits of Sabah, vol 1. Department of Agriculture, Sabah, 126 pp

Yap SK, Martawijaya A, Miller RB, Lemmens RHMJ (1995) *Durio* Adans. In: Lemmens RHMJ, Soerianegara I, Wong WC (eds.) Plant resources of South-East Asia No 5(2). Timber trees: minor commercial timbers. Prosea Foundation, Bogor, pp 215–225

Durio oxleyanus

Scientific Name

Durio oxleyanus Griffith

Synonyms

None

Family

Bombacaceae, also placed in Malvaceae, Durionaceae.

Common/English Names

Oxyleyanus Durian, Durian Lai, Isu

Vernacular Names

Borneo: Dian, Durian Lai, Lai Bengang, Kartungan, Kerantongan, Kerantungan, Ketungan, Kutongan, Sukang;
Indonesia: Kerantongan, Kerantungan, Ladyin Tedak (Kalimantan), Durian Daun (Sumatra), Tungen (Punan Malinau, East Kalimantan), Kayu Keritungen (Lundanye, East Kalimantan), Yang Luang (Merap, East Kalimantan), Lezing Da'Eng (Kenyah Uma', East Kalimantan), Durian Beludu, Koroyot, Tarutung, Durian Rimba;
Malaysia: Durian Daun, Durian Hutan (Peninsular), Durian Isu (Iban, Sarawak), Durian Isu, Durian Beludu (Malay, Sarawak), Durian Sukang (Dusun, Sabah).

Origin/Distribution

The species is indigenous in Peninsular Malaysia, Sumatra and Borneo (Sarawak, Sabah, West-, Central- and East-Kalimantan).

Agroecology

D. oxleyanus usually occurs in moist locations in lowland mixed dipterocarp primary forests. In secondary forests, it is usually present as a pre-disturbance remnant tree. It is sometimes cultivated for the fruits. It occurs usually on frequently flooded clay-rich alluvium on hillsides and ridges with sandy soils up to altitudes of 400 m in Malaysia, and up to 690 m in Kalimantan.

Edible Plant Parts and Uses

The pulp of ripe fruit is edible. It has a smooth creamy texture, sweet with an excellent unique durian flavour that includes undertones of banana and grapes.

Botany

An upper canopy tree up to 50 m tall and 90 cm bole with bole branchless for up to 30 m and with buttresses up to 3 m high. The bark is very rough and deeply fissured, dark brown to dark rusty peeling off in long pieces. Stipules present but soon falling. Twigs are brown with brown scales. Leaves are alternate, simple, penni-veined, broadly elliptical to oblong (Plate 4), 7–20 cm × 3–7.5 cm, dark green, densely covered with greyish stellate hairs adaxially with 15–20 pairs of lateral veins. Flowers in irregular cymes fascicled on twigs or on older branches. Flowers with 4 toothed sepals, 4 white to cream, 15 mm long petals, stamens in bundles alternating with 4 free filaments. Fruit

Plate 3 Brown, plump, ellipsoid seed

Plate 1 Ripe Isu fruit with sharp, slightly curved, pyramidal spines

Plate 4 Large, simple, elliptic-oblong leaves of Isu

globose, up to 20 cm in diameter, outsidegreyish-green with large, stiff, broadly pyramidal, slightly curved spines (Plates 1–2). The flesh is yellow, smooth-textured and sweet enclosing a glossy, brown, ellipsoid seed (Plate 3).

Nutritive/Medicinal Properties

Nutritive value of the edible aril of *D. oxyleyanus* has not been published.

In local traditional medicine, an extract from the bark is taken against malaria and grated seed is applied to ulcerations and wounds. The bark of *Durio oxleyanus* yielded two new lignan ethers, threo-carolignan Y and erythro-carolignan Y, together with lignin boehmenan X, erythro-carolignan X, and boehmenan (Lambert and Garson 2010).

Plate 2 Isu fruit with bright yellow aril

Other Uses

The timber is used for planks in house building.

Comments

Genetic erosion of wild *Durio oxleyanus* has been reported in Indonesia and Malaysia.

Selected References

Burkill IH (1966) A dictionary of the economic products of the Malay Peninsula. Revised reprint, 2 volumes, vol 1 (A–H) pp 1–1240, vol 2 (I–Z) pp 1241–2444. Ministry of Agriculture and Co-operatives, Kuala Lumpur

Kochummen KM (1972) Bombacaceae. In: Whitmore TC (ed.) Tree flora of Malaya, vol 1. Longman Malaysia, Kuala Lumpur, pp 100–120

Kostermans AJGH (1958) The genus *Durio* Adans. (Bombac.). Reinwardtia 4(3):357–460

Lambert LK, Garson MJ (2010) Lignans and triterpenes from the bark of *Durio carinatus* and *Durio oxleyanus*. J Nat Prod 73(10):1649–1654

Munawaroh E, Purwanto Y (2009) Studi hasil hutan non kayu di Kabupaten Malinau, Kalimantan Timur (In Indonesian). Paper presented at the 6th basic science national seminar, Universitas Brawijaya, Indonesia, 21 Feb 2009. http://fisika.brawijaya.ac.id/bss-ub/proceeding/PDF%20FILES/BSS_146_2.pdf

Slik JWF (2006) Trees of Sungai Wain. Nationaal Herbarium Nederland. http://www.nationaalherbarium.nl/sungaiwain/

Soegeng-Reksodihardjo W (1962) The species of *Durio* with edible fruits. Econ Bot 16:270–282

Voon BH, Chin TH, Sim CY, Sabariah P (1988) Wild fruits and vegetables in Sarawak. Sarawak Department of Agriculture, Kuching, 114 pp

Wong WWW, Chong TC, Tanank J, Ramba H, Kalitu N (2007) Fruits of Sabah, vol 1. Department of Agriculture, Sabah, 126 pp

Yap SK, Martawijaya A, Miller RB, Lemmens RHMJ (1995) *Durio* Adans. In: Lemmens RHMJ, Soerianegara I, Wong WC (eds.) Plant resources of South-East Asia No 5(2). Timber trees: minor commercial timbers. Prosea Foundation, Bogor, pp 215–225

Durio testudinarum

Scientific Name

Durio testudinarum Becc.

Synonyms

None

Family

Bombacaceae, also placed in Malvaceae, Durionaceae.

Common/English Names

Tortoise Durian, Turtle Durian, Durian Kura-kura

Vernacular Names

East Malaysia: Durian Kakura, Durian Kura-Kura, Duria Terkura, Lujian Beramatai

Origin/Distribution

The species is native to Brunei Darussalam, Indonesia (Kalimantan) and Malaysia (Sabah, Sarawak)

Agroecology

A species restricted to the tropical lowland mixed dipterocarp forests. It thrives in a humid, equatorial climate with short dry season with annual rainfall of 2,000–4,000 mm but heavy rains prior to flower initiation affect normal flowering, and fruit formation. Like durian, it grows on different types of soils with 5–6.5 pH, provided they are moist, well-drained and rich in organic matter. Deep silt or loams with good drainage and high level of fertility are ideal for its cultivation.

Edible Plant Parts and Uses

Arils are yellow, fleshy and sweet with a strong aroma and are edible.

Botany

A medium-sized, evergreen tree up to 25 m with a trunk diameter of 35 cm, low-rooted buttress root circumference. Trunk is straight, grey-brown, rough and shallowly fissured (Plate 1). Leaf is ellipsoidal, 19–26 cm by 6–9 cm with acuminate tip, shiny green above, and lower surface covered with tightly adpressed, short, fimbriate scales (Plate 2). Flowers are white in fascicles on gnarls on the trunk usually close to the buttressed base. The flowers are generally pentamerous with a

Plate 1 Fruit ramiflorous formed mainly at the base of the shallowly buttressed trunk (L Eng)

Plate 2 Leaves – upper and lower surfaces

Plate 3 Ellipsoid, sub-globose fruit with sharp pyramidal spines (L Eng)

Plate 4 Close-up of ripe fruit

fused calyx usually consisting of five lobes stamina phalanges. The pollen grains are not smooth but bluntly macropositively sculptured. Fruit is ellipsoid, ovoid to sub-globose capsule, greenish brown turning to yellow, with broad-based, conical-pyramidal spines 7–10 mm long and borne on 8–10 cm long peduncle (Plate 2) and formed close to the base of the trunk (Plates 3–4). The aril is yellow, fleshy sweet, strongly aromatic and completely envelopes the seed.

Nutritive/Medicinal Properties

Nothing has been published on its nutritive attributes and medicinal value and uses.

Other Uses

The fruit is occasionally cultivated for its edible fruit and the wood is reported to be used as durian timber.

Comments

Some botanists believed that the species is conspecific with *Durio affinis*.

Selected References

Brown MJ (1997) Durio – a bibliographic review. In: Arora RK, Rao RV, Rao AN (eds.). IPGRI Office for South Asia, New Delhi

Jansen PCM, Jukema J, Oyen LPA, van Lingen TG (1992) *Durio testudinarum* Becc. In: Coronel RE, Verheij EWM (eds.) Plant resources of South-East Asia No 2. Edible fruits and nuts. Prosea Foundation, Bogor, p 331

Kochummen KM (1972) Bombacaceae. In: Whitmore TC (ed.) Tree flora of Malaya, vol 1. Longman Malaysia, Kuala Lumpur, pp 100–120

Kostermans AJGH (1958) The genus *Durio* Adans. (Bombac.). Reinwardtia 4(3):357–460

Kostermans AJGH, Soegeng-Reksodihardjo W (1958) A monograph of the genus Durio Adans. Part I, Bornean Species. Communication 61, Forest Research Institute, Bogor, 80 pp

Whitmore TC, Tantra IGM, Sutisna U (eds) (1989) Tree flora of Indonesia. Forest Research and Development Centre, Bogor, 181 pp

Wong WWW, Chong TC, Tanank J, Ramba H, Kalitu N (2007) Fruits of Sabah, vol 1. Department of Agriculture, Sabah, 126 pp

Yap SK, Martawijaya A, Miller RB, Lemmens RHMJ (1995) *Durio* Adans. In: Lemmens RHMJ, Soerianegara I, Wong WC (eds.) Plant resources of South-East Asia No 5(2). Timber trees: minor commercial timbers. Prosea Foundation, Bogor, pp 215–225

Durio zibethinus

Scientific Name

Durio zibethinus Murray

Synonyms

Durio foetida Thunb., *Durio acuminatissima* Merr.

Family

Bombacaceae, also placed in Malvaceae, Durionaceae.

Common/English Names

Civet-Cat Fruit Tree, Civet Fruit, Common Durian, Durian, Kampong Durian.

Vernacular Names

Borneo (Kalimantan, Sarawak, Sabah): Catu, Dian, Duhuian, Durian, Durian Puteh, Jatu, Kalang, Lampun, Lujian;
Burmese: Du-Yin;
Chinese: Liu Lian, Lau Lin;
Czech: Durian Cibetkový;
Danish: Durian;
Dutch: Doerian, Stinkvrucht;
Eastonian: Harilik Durianipuu;
Finnish: Durio, Durian;
German: Durianbaum, Stinkfrucht, Zibetbaum;
Indonesia: Dereyan (Aceh, Sumatra), Tarutung (Alas, Sumatra), Durene (Alfoersch, Ambon), Duren (Bali), Durian, Durin, Drotong, Tarutung (Batak, Sumatra), Aduria, Duria (Bima, Timor), Duriang (Boeginesch, Sulawesi), Duliango (Boeol, Manado, Sulawesi), Durian, Ehriane, Urian, Warian (Boeroe, Maluku), Dulen, Duran, Rulen (East Ceram, Maluku), Tulene, Turen, Turiane (West Ceram, Maluku), Tuleno, Tureno (S. Ceram, Maluku), Doso (Punan Malinau, East Kalimantan), Derian (Lundanye, East Kalimantan), Nyan Kale (Merap, East Kalimantan), Lezing Talang (Kenyah Uma', East Kalimantan), Dahian, Dahujan, Dahuian, Dian, Duhuian, Duian, Lampun, Ruian, Jatu (Dyak, Kalimantan), Duren (Gajo, Sumatra), Juranaa, Juria (Goram, Maluku), Julia, Duria (Gorontalo, Sulawesi), Durian, Duriang (N. Halmaheira), Durian Gaja (S. Halmaheira), Duran (Kai, Maluku), Duren (Javanese), Durecan (Kambang), Lezing Talang (Kenya Uma' Lung, Kalimantan), Derian, Durian Gedecan (Lampong, Sumatra), Derian (Lun Daye, Mentarang, Kalimantan), Ambetan, Dhurin (Madurese), Duriang (Makassar, Sulawesi), Dian, Duren Borneo, Durian (Malay), Durian, Duriang (Mandar, Sulawesi), Duriat (Mentawi, Sumatra), Nyan Kale (Merap, Malinau, Kalimantan), Duian, Durian (Minangkabau, Sumatra), Dureo (Nias, Sumatra), Dureno, Turenjo, Tureno (Oelias, Maluku), Doso (Punam, Malinau, Kalimantan), Duren (Sasak, Lombok), Wo Kerara (Sawoe, Timor), Tulian, Turian

(Simaloer, Sumatra), Dahia, Dahiang, N'dalia, 'Ndia, 'Ndihain (Soela, Ternate), Wu Karara (Soemba, Timor), Duwoyan, Duwuan, Roya, Rupyan (Sulawesi Utara), Kadu (Sundanese), Duria (Tidore, Ternate), Dulian (Toli-Toli, Sulawesi), Duhian, Duria, Luria, Made, Oria, Tamadue (Toraja, Sulawesi);
Japanese: Durian;
Khmer: Thureen;
Korean: Du Ri An;
Laotian: Thourien;
Malaysia: Durian (General), Hampak, Pendok, Penak, Penek Daun, Pendok Sengkuit (Semang), Shempa, Sempa (Sakai), Pele Diyan (Besisi), Tuang (Jakun);
Norwegian: Durian;
Portuguese: Durião;
Philippines: Durio, Dulian (Bagobo), Duryan (Iloko), Durian, Dulian (Lanao), Dulian (Maguindanao), Duyan, Dulian (Sulu), Duryan (Tagalog);
Slovenian: Durian;
Spanish: Durión;
Swedish: Durio, Durian;
Thai: Thurian, Rian;
Vietnamese: Sau Rieng.

Origin/Distribution

Durian is native to southeast Asia. It is found wild or semi-wild in South Tenasserim, Lower Burma, and around villages in peninsular Malaya. Wild durian trees are still found in Borneo and Sumatra. Borneo is the centre of diversity for *Durio* species. Durian is commonly cultivated along roads or in commercial orchards in Thailand, Laos, Kampuchea, Vietnam Malaysia, Brunei, Indonesia and the Philippines. Durian is also cultivated from south-eastern India and Ceylon to New Guinea, and the Pacific Islands, Hawaii, Honduras, the Antilles and Zanzibar in Africa. In Australia, it is cultivated northern Australia around Darwin in the Northern Territory and from Tully to Cairns in Queensland.

Agroecology

Durian thrives in a hot, humid and wet climate with average annual temperatures of 24–30°C, relative humidity of 75–90% and annual precipitation of 1,600–4,000 mm. It needs well-distributed rainfall, but a relatively dry spell stimulates and synchronizes flowering. It thrives best in the lowlands, it does not grow well and fruit above 800 m altitude. Durian prefers a well-drained, fertile loamy soil rich in organic matter with pH of 5–6.5. It is intolerant of water logging which will cause destructive fungal root and trunk rot diseases. Durian cannot withstand more than 0.02% salinity in the soil.

Edible Plant Parts and Uses

Durian is often dubbed the 'King of Fruit' and its fruit is much prized and relished for its sweet highly nutritious, edible pulpy aril which is eaten fresh (Plates 5–7). The pulp can be kept frozen for several months by quick freezing to sub-zero temperatures then thawed and eaten. Australia currently only allows frozen whole durian or frozen durian segments to be imported. Some even eat the raw pulp as an accompaniment with rice. Durian is also processed into a wide assortment of edible delectables which include ice-cream, smoothies, milk shakes, sherbets, cappuccinos, ice kacang, cakes, yule logs, jam or *sri kaya* (durian pulp, eggs and coconut), *agar-agar durian, kueh dodol, pengat durian, bubuh* durian (porridge with mungbean), *pulut durian* (glutinous rice with durian and coconut milk), *tempoyak* (a concoction with salt and prawn paste), lempuk (sausage shaped durian cake) and other nyonya preparations like *apong balik* (folded, crispy pancake coated with butter, ground peanuts and sugar on the inside.) *Tempoya*k can be eaten either cooked or uncooked, is normally eaten with rice, and can also be used for making curry. The pulp is also used for moon-cake fillings, pastries, buns, traditional Malay candy, sweets and biscuits.

Plate 1 Large fruit of Monthong cultivar thinned to 2–3 fruit per fruiting branch

Plate 2 Unthinned cluster of small fruit

Plate 3 Close-up of the sharp pyramidal spines on the fruit

Malaysians make both sugared and salted preserves from durian. When durian is minced with salt, onions and vinegar, it is called *boder*.

Popular durian recipes in Indonesia include *dodol durian, sri kaya durian, wajik durian, es santan durian, es telur durian, kolak durian* and *pepes bunga durian kadupandak*. In Sumatra, a traditional dish is *ikan brengkes* which is fish cooked in a durian-based sauce. Another Sumatran specialty is *sambal tempoyak*, made from the fermented durian pulp, coconut milk, and a collection of spicy *sambal* (chilli, prawn paste condiment). Javanese prepare the pulp as a sauce to be served with rice; they also blend the minced pulp with minced onion, salt and diluted vinegar as a kind of relish; and they add half-ripe arils to certain dishes. In Palembang, Sumatra and Java, the pulp is fermented in earthen jars, sometimes smoked, and eaten as a special side dish. In Thailand, blocks of durian paste are sold in the markets. Durian paste is made from arils of over-ripe fruit. Thai people also use the pulp, glutinous rice and coconut cream in making a popular dessert. The arils are also made into a preserved confection called *thurian kuan* and durian chips which are on sale in supermarkets throughout the year. Durian chips are made from immature,

Plate 4 Ripe Monthong durian on sale in a Bangkok market

Plate 5 A prized Indonesian cultivar with deep golden yellow arils

Plate 7 Durian arils sold fresh or frozen in styrofoam plates

Plate 6 Well-developed locule with 3 arils (left) and malformed fruit with one large aril per locule right

unripe aril. Durian powder is made from fully mature but unripe pulp and is used as essence in coconut milk, ice-cream, cookies, biscuits, candy beverages and other products requiring durian flavour including tooth-paste for durian fanciers. Durian arils are canned in syrup for export.

The pulp of unripe fruit is boiled whole and eaten as a vegetable. The petals of durian flowers are eaten in the Batak provinces of Indonesia and in East Malaysia where it is used in curries.

The durian seeds, which are the size of chestnuts, can be eaten after they are boiled, roasted or fried in coconut oil. In Java, the seeds are sliced thin and cooked with sugar as a confectionery or the slice seeds are dried or baked in coconut oil with spices and eaten with *rijsttafel*. *Rijsttafel* is an elaborate Javanese meal that consists of rice and various food dishes: curried meats, fish, chicken, vegetables, fruits, relishes, pickles, sauces, condiments, nuts, eggs, and etc. Young tender leaves and shoots are occasionally cooked as greens.

Plate 8 Corymb fascicles of ramiflorous durian flowers

Plate 9 Durian flowers sold as vegetables in Kuching market

Botany

A large, evergreen, perennial tree growing to 25–40 m tall with a bole diameter of 50–120 cm, a short straight rough, dark brown peeling bark and irregular dense or open canopy (Plate 1). Leaves are alternate, petiolate (1.5–3 cm long) simple, oblong-lanceolate, or elliptic-obovate, 15–25 cm by 5–9 cm wide, entire margin, obtuse or rounded base and acuminate apex, adaxial surface is dark green, glossy, glabrous and abaxial surface with coppery brown hairy scales. Flowers are borne in corymbose fascicles of 5–30 flowers on branches, i.e. the durian is ramiflorous (Plate 8). Flowers are creamy-white, perfect, odorous, borne on 4–7 cm long pedicels with an outer epicalyx splitting into 4 segments and a saccate calyx with 3–5 pointed, concave lobes 2–3 cm high, pubescent and golden bronze; 5 nectaries situated at the base of the calyx; petals are spatulate, 3.5–5 cm long, becoming reflexed; stamens are white, joined at the lower half forming 5 distinct phalanges at the upper half, each phalange with 10–12 stamens bearing reniform anthers which dehisces by a slit; ovary is ovoid to ellipsoid, five ribbed and covered with fimbriate scales and superior with a long slender, pinkish coloured style pubescent at the top and crowned by a yellow papillose and capitellate stigma (Plate 9). Flowers are protogynous with the stigma maturing first before the anther dehisces releasing their pollens thus favouring outcrossing. Fruit is a 5-loculicidal capsule, varies greatly in size and shape, generally ovoid or ovoid-oblong to nearly round 15–30 cm long and 10–15 cm wide, green to yellowish-brown and covered with sharp pointed, thick pyramidal spines that are variable in length and which develop from the peltate scales of the ovary (Plates 1–5). Seeds are chestnut-brown, plump and the number of seeds per locule or segment varies from 0- (2–4) – 6 and the seeds are enclosed by the thick, sweet, creamy, pulpy aril which varies in colour from white, cream, yellow to golden yellow.

Nutritive/Medicinal Properties

Analyses carried out in the United States reported raw durian fruit, *Durio zibethinus* (minus 68% seeds, and shell) to have the following proximate composition (per 100 g edible portion): water 64.99 g, energy 147 kcal (615 kJ), protein 1.47 g, total lipid 5.33 g, ash 1.12 g, carbohydrates 27.09 g, total dietary fibre 3.8 g, total sugars 1.36 g, Ca 6 mg, Fe 0.43 mg, Mg 30 mg, P 39 mg, K 436 mg, Na 2 mg, Zn 0.28 mg, Cu 0.207 mg, Mn 0.325 mg, vitamin C 19.7 mg, thiamin 0.374 mg, riboflavin 0.200 mg, niacin 1.074 mg, pantothenic acid 0.230 mg, vitamin B-6 0.316 mg, total folate 36 µg, vitamin A 44 IU, β-carotene 23 µg and α-carotene 6 µg (USDA 2010).

Another analysis provided the following food nutrient composition of durian per 100 g edible portion (Leung *et al.* 1972): energy 124 kcal, moisture 66.8 g, protein 2.5 g, fat 1.6 g, total carbohydrates 28.3 g, fibre 1.4 g, ash 0.8 g, Ca 20 mg, P 63 mg, Fe 0.9 mg, K 601 mg, Na 1 mg, β-carotene equivalent 10 µg, vitamin B1 (thiamin) 0.27 mg, vitamin B2 (riboflavin) 0.29 mg, niacin 1.2 mg and vitamin C 37 mg.

Fresh durian arils were found to contain about 1.47–2.5 g of protein and all the essential amino acid which was much higher than found in fruits like dates. Studies reported that the following amino acid composition of durian arils (Zanariah and Noor Rehan 1987) in terms of mg/100 g fresh weight: isoleucine 85.8 g, isoleucine 85.8 mg, leucine 143 mg, lysine 124.8 mg, methionine 44.2 mg, histidine 52 mg, cystine 78 mg, phenylalanine 78 mg, tyrosine 57.2 mg, threonine 67.6 mg, valine 122.2 mg and the essential amino acids (g/16 gN): lysine 4.8 g, histidine 2.0 g, arginine 2.1 g, aspartic acid 9.3 g, threonine 2.6 g, serine 3.9 g, glutamic acid 11.9 g, proline 3.8 g, glycine 4.1 g, alanine 8.4 g, cystine 3.0 g, valine 4.7 g, methionine 1.7 g, leucine 5.5 g, isoleucine 3.3 g, tyrosine 2.2 g and phenylalanine 3.0 g.

Durian pulp of the five different Malaysian durian cultivars (D2, D24, MDUR78, D101 and Chuk) were found to contain 30 esters, 5 ketones and 16 sulphurous compounds in its aroma composition (Voon et al. 2007). Diethyl disulphide, ethyl-n-propyl disulphide, diethyl trisulphide and ethanethiol were the predominant sulphur-containing compounds in all the cultivars. The major esters present in durian were either ethyl propanoate, ethyl-2-methyl butanoate, or propyl-2-methylbutanoate and their levels varied within cultivars. A strong correlation was observed between sensory properties with flavour compound and physicochemical characteristics of the fruit. A separate study reported a total of 63 constituents were identified in the volatile component of 3 different clones of durian, comprising 30 esters, 16 sulphur-containing compounds, 5 ketones, 8 alcohols and 4 miscellaneous compounds (Wong and Tie 1995). All three clones produced approximately the same proportions of esters and of ketones. There was much variation in the content of sulphur volatiles among the three clones. Identified among this chemical class were some uncommon plant volatiles such as cis- and trans3,5-dimethyl-1,2,4-trithiolane and S-alkyl esters of alkylthiocarboxylic acids. Another study reported that one of the three strongest durian odorants was identified as 3,5-dimethyl-1,2,4-trithiolane among the sulphurous odorants (Weenen et al. 1996). Ethyl 2-methylbutanoate was found to have the highest odor impact among the non-sulfurous odorants in durian. A much earlier studies (Moser et al 1980) ascribed the characteristic foetid odour to hydrogen sulphide, hydrodisulphides, dialkyl polysulphides, ethyl esters and 1.1-diethoxyethane.

Durian is an extremely rich fruit with very high food value and contains phytonutrient antioxidants beneficial to health. Durian fruit was found to contain a high amount of sugar, vitamin C, potassium, and the serotoninergic amino acid tryptophan, and to be a fair source of carbohydrates, proteins, and vitamin Bs.

Nutritive composition of durian seed was reported as follows (Leung et al. 1972) with 2 sets of values raw seeds (minus seed coat) and the second for cooked seeds (minus seed coat): moisture (51.5, 51.1%), fat (0.4, 0.2%), protein (2.6, 1.5%), total carbohydrates (43.6, 46.2%), crude fibre (–, 0.7%), ash (1.9, 1.0%), calcium (17, 39 mg/100 g), phosphorus (68, 87 mg/100 g), iron (1.0, 0.6 mg/100 g), sodium (3, – mg/100 g), potassium (962, – mg/100 g), b-carotene equivalents (250, – µg/100 g), riboflavin (0.05, 0.05 mg/100 g), thiamine (–, 0.03 mg/100 g) and niacin (0.9, 0.9 mg/100 g).

On the lipid components of both the arils and seeds of durian, it was reported that cyclopropene fatty acids such as sterculic, dihydrosterculic and malvalic acids were present in the uncooked seeds but not in the aril (Berry 1980a, b). Due to the toxic and perhaps carcinogenic nature of these substances, it would be unwise to ingest uncooked durian seeds although raw durian seeds are never consumed.

Two carbohydrate fractions from durian fruit hulls, crude (DFI) and purified (DFII), useful pharmaceutical excipient in tablet and liquid preparations, were isolated and characterised (Pongsamart and Panmuang 1998). Powders of DFI and DFII was light-brown and of creamy-white in color, respectively. The powders swelled and formed a viscous mass when suspended in water. DFI and DFII comprised 54.8% and 41.5% ash, 9.1% and 12.0% water, respectively. Crude fibre was absent in the fractions. DFI had 19.33% carbon and 2.72% hydrogen while DFII had 22.89% carbon and 3.24% hydrogen. Carbohydrate contents in DFI and DFII were 9.2 and 9.5% expressed as glucose respectively. DFI had 4 sugar components identical to glucose, rhamnose, arabinose and fructose in a ratio of 18:2:2:2:1. DFII showed three components identical to glucose, rhamnose and arabinose in a ratio of 3:1:1. Mineral analysis indicated the presence of Ca, Mg, K and Na as major components and Fe, Al, Mn, Zn, Si and Cu as minor components. 0.5% DFII powder provided satisfactory food formulation to produce stable jelly products. The principal components of the pharmaceutically useful water-soluble polysaccharides isolated from durian rinds were found to be pectic polysaccharides (Hokputsa et al. 2004).

Wood bark extract from *Durio zibethinus* afforded two new triterpenoids, namely, methyl 27-O-trans-caffeoylcylicodiscate and methyl 27-O-cis-caffeoylcylicodiscate, a new phenolic, 1,2-diarylpropane-3-ol, and seven known compounds, fraxidin, eucryphin, boehmenan, threo-carolignan E, (−)-(3R,4 S)-4-hydroxymellein, methyl protocatechuate, and (+)-(R)-de-O-methyllasiodiplodin (Rudiyansyah and Garson 2006).

Antioxidant Activity

Recent studies reported that total polyphenols, flavonoids, anthocyanins and flavanols in ripe durian were significantly higher than in mature and overripe fruits in the durian cv Mon Thong (Arancibia-Avila et al. 2008). Free polyphenols and flavonoids were at lower levels than hydrolyzed ones. Caffeic acid and quercetin were the dominant antioxidant substances in ripe durian. In these fruits, methanol extracts contained a relatively high capacity of 74.9% inhibition using β-carotene–linoleic acid assay. Ferric-reducing/antioxidant power (FRAP) and cupric-reducing antioxidant capacity (CUPRAC) assays supported this finding. The correlation coefficients between polyphenols and antioxidant capacities of durian samples with all applied assays were about 0.98. In conclusion, the bioactivity of ripe durian was high and the total polyphenols were the main contributors to the overall antioxidant capacity. The scientist further reported that the total polyphenols, flavonoids, flavanols, ascorbic acid, tannins and the antioxidant activity determined by four assays (CUPRAC, DPPH, ABTS and FRAP) differed in immature, mature, ripe and overripe samples (Haruenkit et al. 2010). The content of polyphenols and antioxidant activity were the highest in over-ripe durian, flavonoids were the highest in ripe durian, and flavanols and antiproliferative activity were the highest in mature durian.

The antioxidant activities of different durian cultivars at the same stage of ripening (Mon Thong, Chani, Kan Yao, Pung Manee and Kradum) were compared in order to choose the best as a supplement in the human diet (Toledo et al. 2008). Total polyphenols (mg gallic acid equivalent/100 g fresh weight (FW)) and flavonoids (mg catechin equivalent (CE)/100 g FW) in Mon Thong (361.4 and 93.9) were significantly higher than in Kradum (271.5 and 69.2) and Kan

Yao (283.2 and 72.1). The free polyphenols and flavonoids exhibited lower results than the hydrolyzed ones. Anthocyanins (μg cyanidin-3-glucoside equivalent/100 g FW) and flavanols (μg CE/100 g FW) were significantly higher in Mon Thong (427.3 and 171.4) than in Kradum (320.2 and 128.6) and Kan Yao (335.3 and 134.4). Caffeic acid and quercetin were the dominant bioactive substances in Mon Thong cultivar. The antioxidant activity (μM Trolox equivalent/100 g FW) of Mon Thong cultivar (260.8, 1,075.6 and 2,352.7) determined by ferric-reducing/antioxidant power (FRAP), cupric reducing antioxidant capacity (CUPRAC) and 2,2′-azinobis(3-ethylbenzothiazoline-6-sulfonic acid) with Trolox equivalent antioxidant capacity (TEAC) assays was significantly higher than in Kradum (197.4, 806.5 and 1,773.2) and in Kan Yao (204.7, 845.5 and 1,843.6). The correlation coefficients between polyphenols, flavonoids, flavanols and FRAP, CUPRAC and TEAC capacities were between 0.89 and 0.98. The studies found that the bioactivity of durian cultivars Mon Thong, Chani and Pung Manee was high and the total polyphenols were the main contributors to the overall antioxidant capacity. The results in-vitro are comparable with other fruits that widely used in human diets. Therefore, the researchers suggested that durian could be used as a supplement for nutritional and healthy purposes, especially Durian Mon Thong, Chani and Pung Manee. Diets supplemented with Mon Thong and to a lesser degree with Chani and Kan Yao significantly hindered the rise in the plasma lipids and improved antioxidant activity, protein and metabolic status.

In another study comparing three different fruit: durian, snake fruit (*Zalacca edulis*) and mangosteen (*Garcinia mangostana*), total polyphenols (mg GAE/100 g of FW) and flavonoids (mg CE/100 g FW) and antioxidant activity were found to be significantly higher in durian and snake fruit than inmangosteen respectively: durian (309.7 and 61.2), snake fruit (217.1 and 85.1), mangosteen (190.3 and 54.1) (Haruenkit et al. 2007). Antioxidant activity (μM TE/100 g of FW) of durian measured by DPPH and ABTS assays (228.2 and 2016.3) was significantly higher than in snake fruit (110.4 and 1507.5) and mangosteen (79.1 and 1268.6). HPLC/DAD analysis of durian (μg/100 g of FW) showed that quercetin (1214.23) was present at levels three times that of caffeic acid, and twice as high as p-coumaric and cinnamic acids. Dietary fibres, minerals, and trace elements were comparable. The studies indicated that durian could be used as a supplement for nutritional and health purposes.

Nutritional and bioactive values of durian were found comparable with indices in mango and avocado (Gorinstein et al. 2010). These fruits were found to contain high, comparable quantities of basic nutritional and antioxidant compounds, and to possess high antioxidant potentials. All fruits exhibited a high level of correlation between the contents of phenolic compounds (flavonoids and phenolic acids) and the antioxidant potential. The methods used (three-dimensional fluorescence, FTIR spectroscopy, radical scavenging assays) were suitable for bioactivity determination of these fruits. In order to receive best results, a combination of these fruits was suggested to be included in the diet. The contents of total fibre, proteins and fats were significantly higher in avocado, and carbohydrates were significantly lower in avocado than in the two other fruits. Similarity was found between durian, mango and avocado in polyphenols (9.88, 12.06 and 10.69, mg gallic acid equivalents/g dry weight), and in antioxidant assays such as CUPRAC (27.46, 40.45 and 36.29, μM Trolox equivalent (TE)/g d.w.) and FRAP (23.22, 34.62 and 18.47, μoM TE/g dry weight), respectively.

Antihypercholesterolemic/ Hypoglycemic Activities

Further studies showed that the bioactive compound in ripe durian pulp positively affects the plasma lipid profile, the plasma glucose and the antioxidant activity in rats fed cholesterol enriched diets (Leontowicz et al. 2007). Studies showed that total polyphenols (mgGAE/100 g FW) and flavonoids (mg CE/100 g FW) in ripe durian (358.8 and 95.4) were significantly higher than in mature (216.1 and 39.9) and over-ripe

(283.3 and 53.5) in Monthong cultivar. Antioxidant capacity (μMTE/100 g FW) in total polyphenol extracts of ripe durian measured by 1,1-diphenyl-2-picrylhydrazyl radical (DPPH) and [2,2′-azinobis (3-ethylbenzothiazoline-6-sulfonic acid)] (ABTS) assays (259.4 and 2341.8) were significantly higher than that of mature (151.6 and 1394.6) and overripe (201.7 and 1812.2) samples. The correlation coefficients between the bioactive compounds in different stages of ripening and their antioxidant capacities were high (R2 = 0.99). Dietary studies using male Wistar rats found that diets containing ripe and to a lesser degree mature and overripe durian significantly hindered the rise in plasma lipids and also hindered a decrease in plasma antioxidant activity. The nitrogen retention in rats of the cholesterol /ripe fruit group was significantly higher (63.6%) than in other diet groups and the level of the plasma glucose remained normal. A decrease in fibrinogen fraction with ripe durian included in rat's diets was shown by electrophoretic separation. The study showed that ripe durian contained higher quantity of bioactive compounds, had higher antioxidant capacity and nutritional value. It positively affected the plasma lipid profile, the plasma glucose and the antioxidant activity in rats fed cholesterol enriched diets. Therefore, the ripe durian supplemented diet could be beneficial for patients suffering from hypercholesterolemia and diabetes. These findings dispelled the folkloric belief that durian is a high-glycemic or high-fat food that would enhance high blood pressure and that its consumption should be minimised.

Durian is recommended as a good source of raw fats by several raw food advocates, while others classify it as a high-glycemic or high-fat food, recommending that its consumption be minimised. Contrariwise, recent studies showed durian to be a low G.I. fruit. Among several Malaysian fruits tested, the glycemic index (G.I.) of pineapple, 82, was significantly greater than those of papaya, 58, watermelon, 55, and durian, 49 (Robert et al. 2008). The studies found that pineapple had a high glycemic index, whereas papaya was intermediate and watermelon and durian were low glycemic index foods.

Antiatherosclerotic Activity

In another study, Leontowicz et al. (2008) investigated the properties of Mon Thong, Chani and Kan Yao durian (*Durio zibethinus* Murr.) cultivars to find the best one as a supplement to antiatherosclerotic diet. Total polyphenols (361.4 mg GAE/100 g FW), flavonoids (93.9 mg CE/100 g FW) and total antioxidant capacity determined by DPPH and β-carotene-linoleic acid assays (261.3 μMTE/100 g F.W. and 77.8% of inhibition) were maximal in Mon Thong in comparison with Chani and Kan Yao and showed a good correlation between these three variables (R2 = 0.9859). The properties of Mon Thong, Chani and Kan Yao durian cultivars were compared in-vitro and in-vivo studies in order to find the best one as a supplement to antiatherosclerotic diet. Total polyphenols (361.4 mg GAE/100 g F.W.), flavonoids (93.9 mg CE/100 g F.W.) and total antioxidant capacity determined by DPPH and β-carotene-linoleic acid assays (261.3 μM TE/100 g F.W. and 77.8% of inhibition) were maximal in Mon Thong in comparison with Chani and Kan Yao and showed a good correlation between these three variables (R2 = 0.9859). Diets supplemented with cholesterol and Mon Thong and to a lesser extent with Chani and Kan Yao significantly hindered the rise in the plasma lipids (TC − 8.7%, 16.1% and 10.3% and (b) LDL-C − 20.1%, 31.3% and 23.5% for the Chol/Kan Yao, Chol/Mon Thong and Chol/Chani, respectively) and the decrease in plasma antioxidant activity. Nitrogen retention remained significantly higher in Chol/Mon Thong than in other diet groups. Diet supplemented with Mon Thong affected the composition of plasma fibrinogen in rats. No lesions were found in the examined tissue of heart and brains. Mon Thong cultivar was found to be preferable for the supplementation of the diet as it positively influenced the lipid, antioxidant, protein and metabolic status. Based on the results of this study the scientist recommended that durian cultivars could be used as a relatively new source of antioxidants.

Another research reported that the polysaccharide gel (PG) from fruit-hulls of durian gave satisfactory results in being useful as an excipient in pharmaceutical and food preparation (Tippayakul et al. 2005). The polysaccharide gel exhibited lipid entrapment property. A significant relationship was found between the decreases of absorbed cholesterol into reverted rat jejunum with respect to increasing concentration of the polysaccharide gel. These results suggested that durian polysaccharide gel was able to entrap lipids and may have potential use as medicinal dietary food for lipid management in patients.

Antimicrobial Activity

The polysaccharide gel (PG) extracted from durian fruit-hulls exhibited in-vitro inhibitory activities against two bacterial species, *Staphylococcus aureus* and *Escherichia coli* but were inactive against two yeast species, *Candida albicans* and *Saccharomyces cerevisiae* (Lipipun et al. 2002). Polysaccharide gel (PG) extract of durian fruit-hulls exhibited potent antibacterial activity against *Staphylococcus aureus*, *Staphylococcus epidermidis*, *Micrococcus luteus*, *Bacillus subtilis*, *Escherichia coli* and *Lactobacillus pentosus* (Pongsamart et al. 2005). Minimal inhibitory concentrations (MICs) of PG were 6.4 mg/mL against *B. subtilis*, *E. coli*, *S. epidermidis* and *M. luteus*; 12.8 mg/mL against *S. aureus* and 25.6 mg/mL against *L. pentosus*. Minimal bactericidal concentrations (MBCs) of PG were 25.6 mg/mL against *B. subtilis*, *E. coli*, *S. aureus*, *S. epidermidis* and *M. luteus*; and 51.2 mg/mL against *L. pentosus*. PG film dressing was successfully prepared using a plasticizer, 10% propylene glycol or 15% glycerin. Soft fibre of PG dressing patch was also prepared by freeze-drying a solution of 0.5% PG in water. PG film showed antibacterial effect against the susceptible bacteria tested. The results suggested PG dressing to have potential benefit as a new water soluble antibacterial dressing for open wounds. Alcohol-free hand sanitizer preparation consisting of polysaccharide gel from durian fruit hulls, tea tree oil, betel vine oil, menthol, castor oil derivative, lanolin derivative, propylene glycol, triethanolamine was found to have good antimicrobial activity against *Staphylococcus aureus*, *Staphylococcus epidermidis*, *Bacillus subtilis*, *Micrococcus luteus*, *Escherichia coli*, *Proteus vulgaris*, *Salmonella typhimurium*, *Krebsiella pneumoniae*, *Pseudomonas aeruginosa* and *Candida albicans* (Pongsamart et al. 2006).

The polysaccharide gel (PG) extract from durian fruit-rinds was found to have antibacterial activity against bovine mastitis bacterial isolates (Maktrirat et al. 2008). Studies showed that PG had poetical to be used as an antibacterial postmilking teat dip. In-vitro time-kill studies demonstrated that bacterial counts of field isolates of *Streptococcus agalactiae*, *Streptococcus dysagalactiae*, *Streptococcus uberis*, *Streptococcus bovis* and *Streptococcus acidominimus* were reduced by more than 90% within 1 minute. In-vivo studies by experimental challenge using *S. agalactiae* revealed that PG teat dip completely inhibited (100%) new intramammary infections (IMI). Significant difference in total new IMI number was observed between dairy cow quarters in PG-treated 0 (0%) and control group 5 (6.17%).

Musikapong et al. (2005) found that the minimum bactericidal concentration of durian rind polysaccharide gel (PG) extract against oral pathogenic bacteria *Streptococcus mutans* and *Actinobacillus actinomycetemcomitans* was 35 mg/ml within 24 hours. Recently, the scientists (Thunyakipisal et al. 2010) reported that the pectic polysaccharide gel (PG) at 150 mg/ml and 1-minute exposure exhibited bactericidal activity against *Streptococcus mutans* and *Aggregatibacter actinomycetemcomitans*. Bacteria treated with either 0.1% chlorhexidine or 50–150 mg/ml PG produced blebs, irregular-shaped cells, and disrupted cells as observed under scanning electron microscopy. The results suggested the potential of the durian polysacahride gel to be used as a natural antibacterial ingredient in oral hygiene products.

Oral administration of durian rind pectic polysaccharide gel (PG) in shrimp diets revealed immunostimulating potential and disease resistance in *Penaeus monodon* (black tiger shrimp) (Pholdaeng and Pongsamart 2010). PG inhibited growth of the shrimp bacterial pathogen, *Vibrio harveyi*. The immune response was evaluated by analysis of prophenoloxidase activity and total haemocyte count in the shrimp fed PG-supplemented diets for 12 weeks. Prophenoloxidase activity in shrimp fed the 1%, 2% and 3% PG-supplemented diet and total hemocyte count in shrimp fed the 1% and 2% PG-supplemented diet were greater than those of the control group. The percent survival was higher in groups fed the 1–3% PG-supplemented diets in challenge tests with either white spot syndrome virus (WSSV) or the bacterium *V. harveyi* than that of the control group. The group fed 2% PG-supplemented diet exhibited highest RSP (relative percent survival) value for disease resistance of 100% at day 6 and 36% at day 4 in treated shrimp against viral and bacterial infection, respectively. Mortality of shrimps fed PG-supplemented diets against WSSV infection was also found to be much lower than that of the control group.

Cosmetic Use and Wound Healing Activity

Polysaccharide gel (PG) from durian's fruit hulls can be used for cosmetic formulations (Futrakul et al. 2010). The PG formulation showed significant effect on skin capacitance after 28 and 56 days and significantly enhanced skin firmness after 56 days of treatment with non-allergic effect as there were no positive patch tests. Polysaccharide gel (PG) from durian's fruit hulls has application for use in the preparations of food and pharmaceutical products.

Studies showed that durian fruit rind polysaccharide gel (PG) prepared as a film dressing hastened wound healing in swines (Chansiripornchai et al. 2005). The results demonstrated that PG dressing film treated wounds showed more rapid wound closure, slightly and less tissue reactions than those in the control group (treated with 1% Lugol's solution) and the treatment 2 group (applied with 1% Lugol's solution and covered with PG dressing film) at the point-end (18 days) of the study. Chansiripornchai and Pongsamart (2008) treated two female mongrel dogs with infected open wounds and skin necrosis with durian fruit hull polysaccharide gel (PG) film dressing. Complete wound healing in both dogs occurred within 14 weeks of treatment. The results showed the benefit of using PG film dressings in pharmaceutical and veterinary applications.

Toxicity Studies

Animals studies showed durian fruit hull polysaccharide gel extracts (PG) to be safe and with no toxic effect when used at high oral doses (2 g/kg) (Pongsamart et al. 2001) or in long term consumption at oral doses of 0.25 and 0.5 g/kg/day for a long period of 60–100 days (Pongsamart et al. 2002). Both studies showed no liver injuries in PG treated animals. Relative body weight gain profile in PG treated mice was not different from control. No significant elevation of serum enzymes, like alkaline phosphatase (ALP), aspartate aminotransferase (AST; SGOT) and alanine aminotransferase (ALT; SGPT) was found in PG treated animals. The clinical data of glucose, cholesterol, creatinine and blood urea nitrogen of PG-treated groups were not significantly different from control groups of mice and rats.

Durian a Hyperthermic Fruit?

Some folkloric beliefs regard durian to be a "heaty" fruit, i.e., a fruit with warming properties and its consumption should be accompanied by eating mangosteen fruit which has cooling properties. Another assertion is that durian has special body-warming (hyperthermic) properties and should not be consumed with paracetamol due to a risk of toxic effects (Chua et al. 2008). Contrary to what some believe, even though durian was found to increase body temperature in some rats, this increment was not significant. No consistent pattern of blood pressure change was observed and serum chemical analysis for alanine aminotransferase (ALT) also did not show any significant change. However, rats receiving the durian-paracetamol combination showed a significant drop in body temperature, which may explain the belief that the two mixtures are incompatible and toxic.

Durian and Alcohol Consumption

Another belief is that durian should not be taken with alcohol beverages (beer, brandy, whisky or wine). Recent studies demonstrated that durian fruit extract rich in sulphur compounds exhibited a dose-dependent inhibition of yeast aldehyde dehydrogenase (Maninang et al. 2009). Symptoms were reminiscent of the disulfiram–ethanol reaction (DER) arising from the inhibition of aldehyde dehydrogenase. Sulphur-rich TLC (thin layer chromatography) fruit extract fractions that eluted farthest from the origin effected the greatest inhibitory action. The studies provided a scientific basis of the adverse, or at times lethal, effect of ingesting durian while imbibing alcohol. Other folkloric beliefs that need substantiation are that: (a) pregnant women or people with high blood pressure are traditionally advised not to consume durian and (b) durian is also considered to have aphrodisiac properties.

Traditional Medicinal Uses

Durian also has some medicinal uses in traditional medicine in southeast Asia. Durian is regarded as a general tonic. The fruit is also believed to be an aphrodisiac, an abortifacient and to improve menstruation. In Peninsular Malaysia, a decoction of the leaves and roots were prescribed as an antipyretic. The juice of fresh leaves had been used as an ingredient in a lotion for fevers and the roots were used for treating fever, either pulverised and rubbed on the body or as a decoction. The leaf juice was applied on the head of a fever patient. A traditional Malay prescription for fever, reported drinking the boiled decoction of the roots of *Hibiscus rosa-sinensis, Nephelium longan, Durio zibethinus, Nephelium mutabile* and *Artocarpus integrifolia*, and boiling the leaves of all of these species and using as a poultice. Another traditional treatment for fever was to bath the patient's head for several days with water containing crushed leaf extracts of *Curculigo latifolia, Gleichenia linearis, Nephelium lappaceum* and *Durio zibethinus*. The leaves were employed in medicinal baths for people with jaundice. Decoctions of the leaves and fruits were applied to swellings and skin diseases. The ash of the burned hull was taken after childbirth. The pulpy flesh was said to be antihelminthic and to serve as a vermifuge. In Java, the fruit hulls had been used externally for treating skin ailments.

Other Uses

The sapwood is white, the heartwood light red-brown, soft, coarse, not durable nor termite-resistant. It is used for masts and interiors of huts in Peninsular Malaysia. Durian rind extracts have been found to be suitable for the preparation of granules and to have desirous binding properties for the manufacture of tablets. The empty rinds are also used in Indonesia as a source of fuel; in the Moluccas islands (Maluku), the husk of the durian fruit is used as fuel to smoke fish.

Ashes of the fruit hull have been used in Patani in Thailand and in Pekan, Pahang for making silk white and as a lye with other dyes. Studies showed that activated carbon synthesised under vacuum pyrolysis from durian peel exhibited better properties and adsorption capacities for Basic Green 4 dye than that under nitrogen atmospheric pyrolysis (Nuithitikul et al. 2010). The HCl treatment of the synthesised activated carbon improved properties and adsorption capacities of activated carbons.

Studies by Chansiripornchai et al. (2008) suggested the potential of using durian fruit rind polysaccharide gel (PG) as a poultry feed supplement for induction of an immune response to Newcastle disease in chicken and for lowering their cholesterol levels, thus providing a source of healthy meat for humans. The results showed hemagglutination-inhibition (HI) titers against Newcastle disease in the PG-fed chicken and lower antibody levels in those that did not receive PG in the feed. Cholesterol in the pectoral muscle of the PG-fed chicken was lower than those that received no PG in the feed.

Comments

Each during growing country in southeast Asia has a preponderance of cultivars which differ in fruit shape, size, eating quality and harvesting period.

Selected References

Arancibia-Avila P, Toledo F, ParkYS JST, Kang SG, Heo BG, Lee SH, Sajewicz M, Kowalska T, Gorinstein S (2008) Antioxidant properties of durian fruit as influenced by ripening. LWT Food Sci Technol 41(10): 118–2125

Backer CA, Bakhuizen van den Brink RC Jr (1963) Flora of Java, vol 1. Noordhoff, Groningen, 648 pp

Berry SK (1980a) Cyclopropenoid fatty acids in some Malaysian edible seeds and nuts. J Food Sci Technol India 17(5):224–227

Berry SK (1980b) Cyclopropene fatty acids in some Malaysian edible seeds and nuts: I. Durian *Durio zibethinus* (Murr.). Lipids 15(6):452–455

Berry SK (1980c) Fatty acid composition and organoleptic quality of four clones of durian (*Durio zibethinus* Murr.). Seminar Nasional Buah-Buahan Malaysia, Serdang, Selangor, 5–7 Nov 1980. Paper No. 9, 8 pp

Brown MJ (1997) Durio – a bibliographic review. In: Arora RK, Rao RV, Rao AN (eds). IPGRI Office for South Asia, New Delhi

Burkill IH (1966) A dictionary of the economic products of the Malay Peninsula. Revised reprint, 2 volumes, vol 1 (A–H) pp 1–1240, vol 2 (I–Z) pp 1241–2444. Ministry of Agriculture and Co-operatives, Kuala Lumpur

Burkill IH, Haniff M (1930) Malay village medicine, prescriptions collected. Gard Bull 6:165–321

Chansiripornchai P, Pongsamart S (2008) Treatment of infected open wounds on two dogs using a film dressing of polysaccharide gel extracted from the hulls of durian (*Durio zibethinus* Murr.): case report. Thai J Vet Med 38:55–61

Chansiripornchai P, Pongsamart S, Nakchat O, Pramatwinai C, Ransipipat A (2005) The efficacy of polysaccharide gel extracted from fruit-hulls of durian (*Durio zibethinus* L.) for wound healing in pig skin. Acta Hort 679:37–43

Chansiripornchai N, Chansiripornchai P, Pongsamart S (2008) A preliminary study of polysaccharide gel extracted from the fruit hulls of durian (*Durio zibethinus* Murr.) on immune responses and cholesterol reduction in chicken. Acta Hort 786:57–60

Chua YA, Nurhaslina H, Gan SH (2008) Hyperthermic effects of *Durio zibethinus* and its interaction with paracetamol. Meth Find Exp Clin Pharmacol 30(10): 739–743

De Padua LS, Lugod GC, Pancho JV (1977–1983) Handbook on Philippine medicinal plants, 4 volumes. Documentation and Information Section, Office of the Director of Research, University of the Philippines at Los Banos, The Philippines.

Futrakul B, Kanlayavattanakul M, Krisdaphong P (2010) Biophysic evaluation of polysaccharide gel from durian's fruit hulls for skin moisturizer. Int J Cosmet Sci 32(3):211–215

Gorinstein S, Haruenkit R, Poovarodom S, Vearasilp S, Ruamsuke P, Namiesnik J, Leontowicz M, Leontowicz H, Suhaj M, Sheng GP (2010) Some analytical assays for the determination of bioactivity of exotic fruits. Phytochem Anal 21(4):355–362

Haruenkit R, Poovarodom S, Leontowicz H, Leontowicz M, Sajewicz M, Kowalska T, Delgado-Licon E, Rocha-Guzmán NE, Gallegos-Infante JA, Trakhtenberg S, Gorinstein S (2007) Comparative study of health properties and nutritional value of durian, mangosteen, and snake fruit: experiments in vitro and in vivo. J Agric Food Chem 55(14):5842–5849

Haruenkit R, Poovarodom S, Vearasilp S, Namiesnik J, Sliwka-Kaszynska M, Park YS, Heo SG, Cho JY, Jang HG, Gorinstein S (2010) Comparison of bioactive compounds, antioxidant and antiproliferative activities of Mon Thong durian during ripening. Food Chem 118(3):540–547

Hokputsa S, Gerddit W, Pongsamart S, Inngjerdingen K, Heinze T, Koschella A, Harding SE, Paulsen BS (2004) Water-soluble polysaccharides with pharmaceutical importance from durian rinds (*Durio zibethinus* Murr.): isolation, fractionation, characterisation and bioactivity. Carbohydr Polym 56(4):471–481

Kochummen KM (1972) Bombacaceae. In: Whitmore TC (ed.) Tree flora of Malaysia, vol 1. Longman Malaysia, Kuala Lumpur, pp 100–120

Kostermans AJGH (1958) The genus *Durio* Adans. (Bombac.). Reinwardtia 4(3):357–460

Leontowicz M, Leontowicz H, Jastrzebski Z, Jesion I, Haruenkit R, Poovarodom S, Katrich E, Tashma Z, Drzewiecki J, Trakhtenberg S, Gorinstein S (2007) The nutritional and metabolic indices in rats fed cholesterol-containing diets supplemented with durian at different stages of ripening. Biofactors 29(2–3):123–136

Leontowicz H, Leontowicz M, Haruenkit R, Poovarodom S, Jastrzebski Z, Drzewiecki J, Ayala AL, Jesion I, Trakhtenberg S, Gorinstein S (2008) Durian (*Durio zibethinus* Murr.) cultivars as nutritional supplementation to rat's diets. Food Chem Toxicol 46(2):581–589

Leung W-TW, Butrum RR, Huang Chang F, Narayana Rao M, Polacchi W (1972) Food composition table for use in East Asia. FAO, Rome, 347 pp

Lim TK (1990) Durian diseases and disorders. Tropical Press, Kuala Lumpur, 95 pp

Lim TK (1995) Durian. In: Coombs B (ed) Horticulture Australia—the complete reference of the Australian horticultural industry. Morescope Publishing, Victoria

Lim TK, Luders L (1998) Durian flowering, pollination and incompatibility studies. Ann Appl Biol 132:151–165

Lipipun V, Nantawanit N, Pongsamart S (2002) Antimicrobial activity (in vitro) of polysaccharide gel from durian fruit-hulls. Songklanakarin J Sci Technol 24(1):31–38

Maktrirat R, Pongsamart S, Ajariyakhajorn K, Chansiripornchai P (2008) Bactericidal effect of post-milking teat dip prepared from polysaccharide gel from durian rinds on streptococci causing clinical bovine mastitis. Acta Hort 796:33–39

Maninang JS, Lizada MCC, Gemma H (2009) Inhibition of aldehyde dehydrogenase enzyme by Durian (*Durio zibethinus* Murray) fruit extract. Food Chem 117(2):352–355

Morton JF (1987) Durian. In: Fruits of warm climates. Julia F. Morton, Miami, pp 287–291

Moser R, Duvel D, Greve R (1980) Volatile constituents and fatty acid composition of lipids in *Durio zibethinus*. Phytochemistry 19(1):79–81

Munawaroh E, Purwanto Y (2009) Studi hasil hutan non kayu di Kabupaten Malinau, Kalimantan Timur (In Indonesian). Paper presented at the 6th basic science national seminar, 21 Feb 2009, Universitas Brawijaya, Indonesia. http://fisika.brawijaya.ac.id/bss-ub/proceeding/PDF%20FILES/BSS_146_2.pdf

Musikapong P, Thunyakitpisal P, Pongsamart S (2005) Antimicrobial activity of polysaccharide gel from durian fruit-halls against *Streptococcus mutans* and *Actinobacillus actinomycetemcomitans*. CU Dent J 28:137–144

Nanthachai S (ed.) (1994) Durian fruit development, postharvest physiology, handling and marketing in ASEAN. ASEAN Food Handling Bureau, Kuala Lumpur, 156 pp

Nuithitikul K, Srikhun S, Hirunpraditkoon S (2010) Influences of pyrolysis condition and acid treatment on properties of durian peel-based activated carbon. Bioresour Technol 101(1):426–429

Ochse JJ, Bakhuizen van den Brink RC (1931) Fruits and fruitculture in the Dutch East Indies. G. Kolff & Co, Batavia-C, 180 pp

Ochse JJ, Bakhuizen van den Brink RC (1980) Vegetables of the Dutch Indies, 3rd edn. Ascher & Co., Amsterdam, 1016 pp

Pholdaeng K, Pongsamart S (2010) Studies on the immunomodulatory effect of polysaccharide gel extracted from *Durio zibethinus* in *Penaeus monodon* shrimp against *Vibrio harveyi* and WSSV. Fish Shellfish Immunol 28(4):555–561

Pongsamart S, Panmuang T (1998) Isolation of polysaccharides from fruit-hulls of durian (*Durio zibethinus* L.). Songklanakarin J Sci Technol 20:323–332

Pongsamart S, Sukrong S, Tawatsin A (2001) The determination of toxic effects at a high oral dose of polysaccharide gel extracts from fruit-hulls of durian (*Durio zibethinus* L.) in mice and rats. Songklanakarin J Sci Technol 23:56–62

Pongsamart S, Tawatsin A, Sukrong S (2002) Long-term consumption of polysaccharide gel from durian fruit-hull in mice. Songklanakarin J Sci Technol 24: 649–661

Pongsamart S, Nanatawanit N, Lertchaipon J, Lipipun V (2005) Novel water soluble antibacterial dressing of durian polysaccharide gel. Acta Hort 678:65–73

Pongsamart S, Lipipun V, Jesadanont S, Pongwiwatana U (2006) Antimicrobial polysaccharide of durian-rinds and natural essential oils combination in an effective non-alcoholic antiseptic lotion for hands. Planta Med 72:996–997

Robert SD, Ismail AA, Winn T, Wolever TM (2008) Glycemic index of common Malaysian fruits. Asia Pac J Clin Nutr 17(1):35–39

Rudiyansyah, Garson MJ (2006) Secondary metabolites from the wood bark of *Durio zibethinus* and *Durio kutejensis*. J Nat Prod 69(8):1218–1221

Soegeng-Reksodihardjo W (1962) The species of *Durio* with edible fruits. Econ Bot 16:270–282

Subhadrabandhu S, Ketsa S (2001) Durian king of tropical fruit. Daphne Brassell Assoc Ltd and CABI Publishing, 175 pp

Subhadrabandhu S, Schneemann JMP, Verheij EWM (1992) *Durio zibethinis* Murray. In: Verheij EWM, Coronel RE (eds.) Plant resources of South-East Asia. No. 2: edible fruits and nuts. Prosea Foundation, Bogor, pp 157–161

Thunyakipisal P, Saladyanant T, Hongprasong N, Pongsamart S, Apinhasmit W (2010) Antibacterial activity of polysaccharide gel extract from fruit rinds of *Durio zibethinus* Murr. against oral pathogenic bacteria. J Invest Clin Dent 1(2):12–125

Tippayakul C, Pongsamart S, Suksomtip M (2005) Lipid entrapment property of polysaccharide gel (PG) extracted from fruit-hulls of durian (*Durio zibethinus* Murr. Cv. Mon-Thong). Songklanakarin J Sci Technol 27(2):291–300

Toledo F, Arancibia-Avila P, Park YS, Jung ST, Kang SG, Heo BG, Drzewiecki J, Zachwieja Z, Zagrodzki P, Pasko P, Gorinstein S (2008) Screening of the antioxidant and nutritional properties, phenolic contents and proteins of five durian cultivars. Int J Food Sci Nutr 7:1–13

Umprayn K, Chanpaparp K, Pongsamart S (1990a) The studies of durian rind extracts as an aqueous binder I: evaluation of granule properties. Thai J Pharm Sci 15(2):95–115

Umprayn K, Chanpaparp K, Pongsamart S (1990b) The studies of durian rind extracts as an aqueous binder II: evaluation of tablets properties. Thai J Pharm Sci 15(3):173–186

U.S. Department of Agriculture, Agricultural Research Service (USDA) (2010) USDA national nutrient database for standard reference, release 23. Nutrient Data Laboratory Home Page. http://www.ars.usda.gov/ba/bhnrc/ndl

Voon YY, Abdul Hamid NS, Rusul G, Osman A, Quek SY (2007) Characterisation of Malaysian durian (*Durio zibethinus* Murr.) cultivars: relationship of physiochemical and flavour properties with sensorial properties. Food Chem 103:1217–1227

Weenen H, Koolhaas WE, Apriyantono A (1996) Sulfur-containing volatiles of durian fruits (*Durio zibethinus* Murr.). J Agric Food Chem 44(10):3291–3293

Wong KC, Tie DY (1995) Volatile constituents of durian (*Durio zibethinus* Murr.). Flav Fragr J 10(2):79–83

Yap SK, Martawijaya A, Miller RB, Lemmens RHMJ (1995) *Durio* Adans. In: Lemmens RHMJ, Soerianegara I, Wong WC (eds.) Plant resources of South-East Asia No 5(2). Timber trees: minor commercial timbers. Prosea Foundation, Bogor, pp 215–225

Zanariah J, Noor Rehan J (1987) Protein and amino acid profiles of some Malaysian fruits. MARDI Res Bull Malaysia 15(1):1–7

Pachira aquatica

Scientific Name

Pachira aquatica Aubl.

Synonyms

Bombax aquaticum (Aubl.) K. Schum. (Basionym), *Bombax insigne* Wall., *Bombax macrocarpum* (Schltdl. & Cham.) K. Schum., *Bombax rigidifolium* Ducke, *Carolinea macrocarpa* Schltdl. & Cham., *Carolinea princeps* L. f. nom. illeg., *Pachira aquatica* var. *occidentalis* Cuatrec., *Pachira aquatica* var. *surinamensis* Decne., *Pachira grandiflora* Tussac, *Pachira longiflora* (Mart. & Zucc.) Decne., *Pachira macrocarpa* (Schltdl. & Cham.) Walp., *Pachira pustulifera* Pittier, *Pachira spruceana* Decne., *Pachira villosula* Pittier, *Pachira villulosa* Pittier, *Sophia carolina* L.

Family

Bombacaceae, also placed in Malvaceae

Common/English Names

American Chestnut, Fortune Tree, Guiana Chestnut, Guyana Chestnut, Malabar Chestnut, Money Plant, Money Tree, Provision Tree, Saba Nut, Shaving Brush Tree, Wild Cocoa

Vernacular Names

Brazil: Cacau-selvagem, Carolina, Castanha das Guianas, Castanheiro-do-maranhão, Embiruçu, Mamorana, Paineira De Cuba, Mulungu;
Chinese: Gua Li;
Columbia: Zapotolongo, Zapoton;
Czech: Bombakopsis vodní;
Eastonian: Vesipahhiira;
French: Cacaoyer-Rivière, Châtaigner Sauvage, Châtaignier De La Guyane, Noisetier De La Guyane, Pachirier Aquatique;
German: Wilder Cacobaum, Wilder Kakaobaum;
Guyana: Wild Cocoa;
Mexico: Apompo, Zapote bobo;
Portuguese: Astanheiro Do Maranhão, Cacau Selvagem, Castanheira Da água, Castanheirio Do Maranhão, Castanheiro Da Guyana, Castanheiro De Guiana, Mamorana, Monguba, Mungaba;
Spanish: Apombo, Apompo, Cacao Cimarron, Cacao De Monte, Cacao De Playa, Castanho De Agua, Castanho Silvestre, Castanho, Castaño De Agua, Castaño De La Guayana, Ceibo De Argua, Ceibo De Arroya, Chila Blanca, Jelinjoche, Palo De Boya Teton, Pumpunjuche, Quirillo, Sapotolon, Sunzapote, Tsine, Zapote Bobo, Zapote De Argua, Zapotolongo, Zapoton;
Venezuela: Castanho

Origin/Distribution

The species is indigenous to the southern parts of Mexico to Northern Peru and Northern Brazil. It has been introduced and cultivated throughout the tropics and sub-tropics.

Agroecology

Pachira aquatica is a tropical wetland tree that grows alongside rivers and estuaries and tropical swamps in its native range. It grows from sea level to 1,300 m in frost free areas with average temperatures of 24°C or higher and annual precipitation between 1,000 and 2,000 mm. It does best in sites with well-drained fertile sandy loam or sandy clay soil in full sun or partial shade and needs protection from drying winds. It is a hardy plant and will adapt very well to different conditions. This species is very tolerant of drought and shade and are sold in many countries as ornamental houseplants.

Plate 1 Palmate leaves of Malabar chestnut

Plate 2 Flower bud of Malabar chestnut

Edible Plant Parts And Uses

The seed have a delicious chestnut/peanut flavour. The seeds are roasted or fried in oil and eaten. The seeds after roasting are sometimes used for the preparation of beverages. They also can be ground into a flour for baking bread. The seeds also yield 58% of a white, inodorous fat which, when refined, is suitable for cooking. The young leaves and flowers are cooked and eaten as vegetable.

Plate 3 Opened flower of Malabar chestnut

Botany

A small to medium-sized, much-branched deciduous tree that can reach 17 m in height with a stout trunk of 90 cm diameter and buttress base. The bark is greenish when young, grey when mature. Petiole 11–15 cm long, tomentose. Leaves are alternate, palmately compound, with 5–11 leaflets (Plates 1–3) and on 11–15 cm long, tomentose petioles. Leaflets are ovate-elliptical or elliptic oblong, 13–28 cm long by 5–8 cm wide, lustrous green, margin entire, base cuneate, apex acuminate, abaxially tomentose, adaxially glabrous, subsessile to slightly petiolulate. Flowers are solitary, large, up to 31 cm long, showy, olive green, axillary to leaf near the shoot

Plate 4 Stems braided together and sold as ornamental trees

Plate 5 Young fruit of Malabar chestnut

Plate 6 Old fruit and mature seeds of Malabar chestnut

apex on 2 cm long pedicel (Plates 2–3). Flower – calyx is campanulate tubular obscurely 3–6 toothed; petals 5, 15–20 cm long reflexed (curled back to the base), greenish-yellow, sparsely velvety; stamens 200–250, staminal tube rather short divided into many groups of 8–10 filaments, yellowish-white below and red distally, anther oblong-linear, ovary cylindrical; style longer than stamen, red with 5-lobed stigma (Plate 3). Fruit is a pyriform, ellipsoid or subglobose dehiscent capsule up to 30 cm long by 12 cm thick, hard, olive-green when immature turning to dark brown, woody and tomentose when mature (Plates 5–6). Fruit has five valves densely packed with numerous seeds in the non-fibrous pulp. Seed is subglobose, 2–2.5 cm by 1.5–2 cm, dark brown with white spiral markings (Plate 6).

Nutritive/Medicinal Properties

P. aquatica seeds were found to contain 9% water, 10% starch, 16% protein and 40–50% fat; the yellow fat having physical and chemical characteristics resembling those of palm oil but containing toxic and possibly carcinogenic cyclopropenic fatty acids (Burkill 1985; Arkoll and Clement 1989). The fat obtained from *P. aquatica* seeds contained a high proportion (about 60%) of palmitic acid and about 25% of acids of the sterculic acid type which may contain hydroxyl groups. The increase in viscosity which occurred when the fat was heated to 250° was attributed to the presence of these acids (De Bruin et al. 1963).

P. aquatica seeds were found to be rich in amino acids in particular tryptophan, threonine, phenylalanine and tyrosine (Oliveira et al. 2000).

The contents of the latter two were reported to be higher than those reported for human milk, chicken egg and cow's milk. Haemagglutinating and trypsin inhibitor activities were also found to be present in the seeds of *P. aquatic* (Oliveira et al. 2000). Raw seeds of *P. aquatica* were found to be highly toxic when fed to rats. The seed also contained a lipase that showed preference for esters of long-chain fatty acids, but exhibited significant activity against a wide range of substrates (Polizelli et al. 2008).

The volatiles in the flowers of *P. aquatica* were found to be (E,E)-α-farnesene (19.2%), β-caryophyllene (11.5%), trans-linalool oxide (pyranoid) (7.2%), elemol (5.6%), phenylacetaldehyde (5.3), cis-and trans-linalool oxide (furanoids) (5.2% and 4.2%), and palmitic acid (4.3%) (Zoghbi et al. 2003).

Toasted seeds can also be ground and prepared as a chocolate and is used as a tonic in folkloric medicine. The young leaves are soaked in water to produce a liquid used for protection against poisoning and as an antidote for the bites of poisonous animals. In Nicaragua, the bark is used medicinally to treat stomach complaints and headaches while a tisane from the boiled bark is used for blood tonic.

Other Uses

Pachira aquatica is used as hedges, live fence posts and avenue trees. It is an excellent ornamental species that flowers, even as a shrub and has become popular in Europe. In Asia, the stems of young seedlings are braided (Plate 4) to make braided bonsai money plant tree which has become highly prized and in great demand in East Asia and south east Asia. They are symbolically associated with good financial fortune and are typically seen in businesses, sometimes with red ribbons or other auspicious ornamentation attached. The trees play an important role in Taiwan's agricultural export economy. It also provides a white and soft wood which is suitable for manufacturing paper, yielding 36% cellulose paste. Bark is used for caulking boats and cordage and yields a dark red dye. Seed oil has been reported to have industrial potential for manufacturing soap.

Comments

In ornamental nurseries, the plant is usually propagated from seeds and also from cuttings.

Selected References

Arkoll DB, Clement CR (1989) Potential new food crops from the Amazon. In: Wickens GE, Haq N, Day P (eds.) New crops for food and industry. Chapman and Hall, London, pp 150–165

Barrett B (1994) Medicinal plants of Nicaragua's Atlantic Coast. Econ Bot 48:8–20

Burkill HM (1985) The useful plants of West Tropical Africa, vol 1, Families A-D. Royal Botanical Gardens, Kew, 960 pp

De Bruin A, Heesterman JE, Mills MR (1963) A preliminary examination of the fat from *Pachira aquatica*. J Sci Food Agric 14(10):758–760

Duke JA (1989) Handbook of nuts. CRC, Boca Raton

FAO (1986) Food and fruit-bearing forest species, 3: examples from Latin America. FAO forestry paper no. 44/3. FAO, Rome, 327 pp

Hedrick UP (1972) Sturtevant's edible plants of the world. Dover Publications, Mineola, 686 pp

Instituto Nacional de Pesquisas da Amazonia (INPA) (1986) *Food and Fruit Bearing Forest Species 3: Examples From Latin America*. Forestry Paper 44-3, Food and Agriculture Organization of The United Nations, Rome. 332 pp.

Menninger EA (1977) Edible nuts of the world. Horticultural Books, Stuart, 175 pp

Oliveira JTA, Vasconcelos IM, Bezerra LCNM, Silveira SB, Monteiro ACO, Moreira RA (2000) Composition and nutritional properties of seeds from *Pachira aquatica* Aubl., *Sterculia striata* St Hil. et Naud and *Terminalia catappa* Linn. Food Chem 70(2):185–191

Polizelli PP, Facchini FD, Cabral H, Bonilla-Rodriguez GO (2008) A new lipase isolated from oleaginous seeds from *Pachira aquatica* (Bombacaceae). Appl Biochem Biotechnol 150(3):233–242

Standley PC, Steyermark JA (1949) Bombacaceae. In: Standley PC, Steyermark JA (eds.) Flora of Guatemala part VI. Fieldiana Bot 24(6): 386–403

Ya T, Gilbert MG, Dorr LJ (2007) Bombacaceae. In: Wu ZY, Raven PH, Hong DY (eds.) Flora of China, vol 12, Hippocastanaceae through Theaceae. Science Press/Missouri Botanical Garden Press, Beijing/St. Louis

Zoghbi MDGB, Andrade EHA, Maia JGS (2003) Volatiles from flowers of *Pachira aquatica* Aubl. J Essent Oil Bear Plants 6(2):116–119

Pachira insignis

Scientific Name

Pachira insignis (Sw.) Sw. Ex Savigny

Synonyms

Bombax spectabile Ulbr., *Bombax spruceanum* Ducke, *Carolinea insignis* Sw., *Pachira spruceana* Decne.

Family

Bombacaceae, also placed in Malvaceae

Common/English Names

Guiana Chestnut, Malabar Chestnut, Trinidad Pachira, Wild Breadnut, Wild Chataigne, Wild Chestnut

Vernacular Names

Brazil: Cacao-Selvagem, Carolina, Castanheiro-Do-Maranhão, Mamorana (Portuguese);
Eastonian: Tore Pahhiira;
French: Châtaigner De La Côte D'espagne, Châtaignier D'amérique, Marronier, Pachirier Remarquable;
Spanish: Castaño, Estrella, Mamorana Grande.

Origin/Distribution

The species is native to Central America and South America – French Guiana, Guyana, Suriname, Venezuela, Brazil, Bolivia, Columbia, Ecuador and Peru.

Agroecology

In its native tropical range, it occurs from 0 to 4,000 m altitude in full sun to partial shade. Optimum mean annual temperature is around 26–29°C. It has moderate water requirement.

Edible Plant Parts and Uses

The seeds, young leaves and flowers serve as food. The seeds taste like peanuts and are eaten raw, roasted or fried.

Botany

A large, branched, evergreen tree growing to 20–30 m high with grey –green trunk. Leaves are compound, palmate on 5–10 mm petiole with 5–9 leaflets (Plates 1–2). Leaflets obovate, rounded at tips, 35 cm long by 15 cm wide, sometimes with brown scale beneath, glossy-green. Flower whitish, terminal or in pairs in leaf axils, on 7 cm pedicel, calyx dark red, petals

Plate 1 Fruits and leaves of Guiana chestnut

Plate 2 Close-up of fruit and palmate leaf

5 white, narrowly strap-shaped with hundreds of reddish and white stamens, style up to 30 cm long. Fruit oblong-ovoid shaped, 14 cm long by 9 cm across with a brown, scurfy hard, tough rind (Plates 1–2) and 1 cm thick valves and 4 large, irregularly angled and edible seeds.

Nutritive/Medicinal Properties

No information has been published on its nutritive and medicinal attributes.

Other Uses

The tree is occasionally grown as an ornamental both for its striking fruits and flowers.

Comments

See also notes on the other closely related edible species, *Pachira aquatica*.

Selected References

Brako L, Zarucchi JL (1993) Catalogue of the flowering plants and gymnosperms of Peru, vol 45, Monographs in systematic botany from the Missouri Botanical Garden, St. Louis Mo, pp i–xl, 1–1286

d'Eeckenbrugge GC, Ferla DL (2000) Fruits from America an ethnobotanical inventory. IPGRI/CIRAD-FLHOR. http://www.cefe.cnrs.fr/ibc/pdf/coppens/Fruits%20from%20America/fruits_from_america.htm

Funk V, Hollowell T, Berry P, Kelloff C, Alexander SN (2007) Checklist of the plants of the Guiana shield (Venezuela: Amazonas, Bolivar, Delta Amacuro; Guyana, Surinam, French Guiana). Contrib US Natl Herb 55:1–584

Hedrick UP (1972) Sturtevant's edible plants of the world. Dover Publications, New York, 686 pp

Quararibea cordata

Scientific Name

Quararibea cordata (Humb. & Bonpl.) Vischer

Synonyms

Matisia cordata Humb. & Bonpl.

Family

Bombacaceae, also placed in Malvaceae.

Common/English Names

Chupa-Chupa, Matisia, Sapote Columbiana, South American Sapote

Vernacular Names

Brazil: Sapota, Sapote-Do-Peru, Sapota-Do-Solimóes (Portuguese);
Columbia: Chupa-chupa, Zapote Amarillo, Zapote, Zapote Chuchupa, Zapote Chupachupa, Zapote Chupa, Sapote De Monte, Sapotillo;
Ecuador: Sapote, Zapote;
French: Sapote Du Pérou;
Peru: Sapote De Monte, Sapotillo, Zapote De Monte, Zapote Chupachupa, Zapote Chupa;
Spanish: Firolisto, Chupa-Chupa, Mamey Colorado, Zapote De Monte, Sapote, Sapote De Monte, Sapotillo, Zapote, Zapote De Monte.

Origin/Distribution

Matisia occurs wild in the humid lowland rainforests of Peru, Ecuador and adjacent areas of Brazil, especially around the mouth of the Javari River. It is common in the western part of Amazonas, southwestern Venezuela, and in the Cauca and Magdalena Valleys of Colombia. It thrives especially well near the sea at Tumaco, Colombia. The fruits are commonly sold in the markets of Antioquia, Buenaventura and Bogotá, Colombia; Puerto Viejo, Ecuador; the Brazilian towns of Tefé, Esperanca, Sao Paulo de Olivenca, Tabetinga, Benjamin Constant and Atalaia do Norte.

Matisia was introduced and cultivated elsewhere in the tropics e.g. in south Florida and North Queensland, Australia.

Agroecology

Matisia is a tropical to subtropical species that flourishes in the warm, humid lowland rainforest. In Ecuador, it can be found from near sea-level to 1,200–2,000 m. The tree grows well in full sun and needs abundant moisture. It does best in deep, well-drained, soil high in organic matter. It needs protection from strong winds and cold. It is frost intolerant and cannot withstand flooding.

Plate 1 Matisia fruit with thick, tomentose, brownish rind

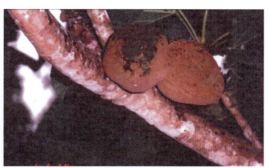

Plate 2 Two fruit arising from one fascicle on the branch

Edible Plant Parts and Uses

The fruit is highly esteemed in South America. The fruits are eaten fresh or processed into juice, drinks and nectar. The fruits are also added to desserts.

Botany

A semi-deciduous, much branched, erect, large tree, 15–40 m high with light brown bark and copious, gummy yellow sap. It is sometimes buttressed and has stiff branches in tiered whorls of five above a straight cylindrical rugose trunk and green branchlets. Leaves are alternate, long-petioled, clustered in whorls near the ends of the branches, cordate to suborbicular, 15–30 cm long and nearly as wide, membranous, glabrous, seven-nerves arising from the base. Flowers occur in fascicles of three, pale rose coloured, short pedicelled, 2.5 cm wide, with two or three bracts near base; calyx is two to five-lobed and tomentose; corolla is sub-bilabiate, with two petals a little smaller than the other three, all obovate; stamen tube with five linear lobes with about 12 anthers on each lobe; style shorter than stamens, puberulent; ovary five-angled and stigma five-sulcate and capitate. Fruit is oval, ovoid to elliptic with a prominent, rounded knob at the apex and is capped with a two to five-lobed, tomentose, leathery, strongly persistent calyx at the base, 10–15 cm long by 8–10 cm wide. The rind is thick, leathery, greenish-brown and tomentose (Plates 1–3). The pulp is orange-yellow, soft, juicy, sweet and contain two to five hard, 4 cm × 2.5 cm seed from which long fibres extend through the pulp.

Plate 3 Matisia fruit with orangey-yellow, fleshy pulp and a thick rind

Nutritive/Medicinal Properties

Analyses made in Ecuador (Morton 1987) reported that nutrient composition of chupa-chupa fruit per 100 g edible portion was: moisture 85.3 g; protein, 0.129 g; fat, 0.10 g; fibre, 0.5 g, ash, 0.38 g, calcium 18.4 mg, phosphorus, 28.5 mg; iron, 0.44 mg; thiamine, 0.031 mg; riboflavin, 0.023 mg; niacin, 0.33 mg; ascorbic acid, 9.7 mg.

The antioxidant activity of the fruit was determined by free radical 1,1-diphenyl-2-picrylhydrazyl (DPPH) method to be $IC_{50} = 21.4$ mg/ml and phenolic compound content of 102.9 mg/100 g, which was lower than that reported for orange

(*Citrus sinensis*) IC_{50} = 18.3 8 mg/ml and phenolic compound content of 64.2 mg/ml (Franco 2006).

The bark of *Odontocarya tripetala* is boiled with *Matisia cordata* and hot peppers to expel worms in folkloric medicine in Amazonia.

Other Uses

Matisia wood is used for construction, furniture, and as fuel-wood.

Comments

The plant is propagated by seeds, stem cuttings and by grafting.

Selected References

Duke JA, Martinez RV (1994) Amazonian ethnobotanical dictionary. CRC, Boca Raton, 224 pp

Franco EM (2006) Actividad antioxidante in vitro de las bebidas de frutas. Bebidas -Alfa Editores Técnicos, Junio/Julio:20–27 (In Spanish)

Hodge WH (1960) The South American "Sapote". Econ Bot 14:203–206

Instituto Nacional de Pesquisas da Amazonia (INPA) (1986) *Food and Fruit Bearing Forest Species 3: Examples From Latin America*. Forestry Paper 44-3, Food and Agriculture Organization of The United Nations, Rome. 332 pp.

Macbride JF (1936) Flora of Peru. Chicago: USA Field museum of natural history, botanical series, vol XIII. pp 1936–1971, 6 parts

Morton JF (1987) Chupa-Chupa. In: Fruits of warm climates. Julia F. Morton, Miami, pp 291–292

Whitman WF (1976) South American sapote. Proc Fla State Hort Soc 89:227–229

Ananas comosus

Scientific Name

Ananas comosus L.

Synonyms

Ananas ananas (L.) Voss, *Ananas bracteatus* var. *hondurensis* Bertoni, *Ananas domestica* Rumph., *Ananas parguazensis* L.A. Camargo & L.B. Sm., *Ananas sativus* (Lindl.) Schult.f., *Ananas sativus* Schult. & Schult. f., *Ananassa sativa* Lindl., *Ananas ananas* (L.) Voss, *Ananas duckei* hort., nom. inval., *Ananas sativus* var. *duckei* Camargo, nom. nud., *Bromelia ananas* L., *Bromelia ananas* Willd., *Bromelia comosa* L. (basionym).

Family

Bromeliaceae

Common/English Names

Pineapple

Vernacular Names

Angola: Ananas, Abacasi (Portuguese);
Arabic: Anânâs;
Argentina: Ananá;
Brazil: Abacaxi, Abacaxí-Do-Mato, Abacaxizeiro, Ananás-Selvagem, Gravatá, Nana (Portuguese);
Burmese: Na Naq Thì;
Catalan: Ananas;
Chinese: Bo Luo, Feng Li;
Czech: Ananasovník Setý;
Danish: Almindelig Ananas, Ananas;
Democratic Republic of Congo: Langa (Ngbandi);
Dutch: Ananas;
Eastonian: Ananass;
Finnish: Ananas, Hedelmaeananas;
French: Ananas, Ananas Commun, Pain De Sucre;
Gabon: Ngubi (Apindji), Dilanga (Bahumbu), Difubu (Balumbu), Iba (Banzabi), Difubu (Eshira), Nkoc-Nsèc, Nkuba (Fang), Ananas (French), Iguwu (Galoa), Iguwu (Mpongwé), Ikoko-Ny'atanga (Nkomi);
German: Ananas;
Greek: Ananas;
Guinea Conakry: Kwito;
Hungarian: Ananász;
India: Aanaaras, Aanaaros, Aanarasa, Ananas, Anarosh (Bengali), Ananas, Anannaasa (Hindu), Anasahannu, Ananasu Hannu (Kannada), Ananas (Konkani), Ananas, Kaitha Chakka, Kaitachchakka (Malayalam), Keehom (Manipuri), Ananus (Marathi), Anamnasam, Bahunetraphalam (Sanskrit), Anachi Pazham, Anasippazham (Tamil), Anaasa, Annasapandu (Telugu), Ananas, Annaanas, Anannaas (Urdu);
Indonesia: Anes (Acheh), Arnasinu, Kanasi, Kurnasin, Mangala, Nanasi (Ambon), Manas

(Bali), Hanas, Henas, Honas, Kenas, Onas (Batak), Aruna (Bima), Lalato (Boeol, Manado), Kalnasi (Boeroe, Ambon), Pandang, Panrang (Bugis), Ai Nasi, Kai Nasi, Thuan Bababa (E. Ceram), Bankalo, Kampora, Kampala (W. Ceram), Anasu, Bangala, Bangkala, Kai Nsu, Kambala, Kampala (S. Ceram), Belasan, Kayu Ujan, Kanas, Malaka, Urousan, Samblaka (Dyak), Ekaha-uku (Enggano, Sumatra), Ananas, Peda, Pedang (Flores), Nas (Gajo), Nanati (Gorontalo), Nanhi, Nanasi (N Halmaheira), Nanas (Java), Esne (Kisar), Kanas, Kanjas, Nas (Lampong), Lanas, Nanas (Madurese), Piamber (N. New Guinea), Manilmap, Haramina, Miniap (W. New guinea), Pandang (Makassar), Enas, Kanas, Nanas, Nenas (Malay, Lingga, Manado), Onas (Malay, Mandailing), Asit, Masit (Mentawai), Aneh, Naneh (Minangkabau), Gona (Nias), Ranasi, (Nufur, New Guinea), Anasui, Nanasu (Oelias), Nanasi (Sangir), Nanas (Sasak), Nana, Wo Nama (Sawoe), Panda Jawa (Soemba), Pedang (Solor), Busa, Naasi, Nanasi, Pinang, Tuis, Tuis In Balanda, Tui Mongondow, Tuis Ne Walanda, Wusa (N. Sulawesi), Danas, Ganas (Sundanese), Nanasi, Parangena (Talaud), Ngewu (Tanimbar), Nanas (Ternate), Nanasi (Tidore), Edan, Ekam, Hedam (Timor), Nanasi (Toraja);
Italian: Ananasso, Ananasso Ordinario;
Khmer: Maneas, Moneah, M'noah;
Korean: P'a In Ae P'ul;
Laotian: Ananas, Màak Nat;
Lithuanian: Ananas;
Madagascar: Mananasy;
Malaysia: Berunai (Iban), Kabar (Kelabit), Nenas, Nanas (Malay);
Maldives: Alanaasi (Dhivehi);
Nepalese: Bhui Katar;
Nigeria: Ogede-Oyinbo (Lagos State), Ehin-Ahun, Ekukkun, Ope-Oyinbo, (Okeigbo);
Norwegian: Ananas;
Papiamento: Anasa;
Philippines: Apagdan, Pangdan (Bontok), Piña (Cebu Bisaya), Piña (Spanish), Piña, Piya (Tagalog);
Polish: Ananas, Ananas Jadalny;
Polynesia: Napolo;
Portuguese: Abacaxi, Ananás;
Romanian: Ananas;
Russian: Ananas;
Serbian: Ananas;
Sierra Leone: Lime;
Slovaščina: Ananas;
Spanish: Ananá, Piña, Piña De América, Piña Tropical;
Swedish: Ananas;
Thai: Sàp Pàrót, Bonat, Yaannat;
Tonga: Fain;
Turkish: Ananas, Festa;
Uganda: Nanasi (Bulamogi), Enanasi (Rukonjo), Enanasi (Runyankole), Enanasi (Runyaruguru);
Vietnamese: Trái Qua, Thom Dua.

Origin/Distribution

Pineapple is native to the South American Tropics and was widely introduced elsewhere during the sixteenth and seventeenth centuries. The crop is now widely grown throughout the tropics and subtropics. The international pineapple canning industry is based on plantations in Thailand, the Philippines, Malaysia and north Sumatra as well as in Hawaii, Brazil, Taiwan, South Africa, Kenya, Ivory Coast, Mexico and Puerto Rico.

Agroecology

Pineapples are grown mainly between latitudes 24°N and 25°S in the tropics and subtropics, principally at lower altitudes. The optimum temperature range for pineapple is 23–32°C, although it can be grown in areas where temperature drops as low as 10°C. However, the plant does not tolerate frost and the fruit is sensitive to sunburn. Pineapple is tolerant to drought and a wide range of rainfall; 1,000–1,500 mm per annum is considered optimal. A well-drained, sandy loam is preferred, with a high organic matter content and pH 4.5–6.5. However, plants can be grown over a wide range of soil types, such as the acid peats (pH 3–4.5) in Malaysia. Drainage should be perfect, because waterlogged plants quickly succumb to root rot. Crop duration increases substantially further away from the equator and at higher elevations. In Kenya, pineapple is grown at elevations of 1,800 m where fruits develop a sugar:acid ratio of 16:1 which is ideal for canning. At higher elevations fruits become too acidic.

Edible Plant Parts And Uses

Pineapple is best eaten fresh and raw as dessert fruit or in fruit salads. It can be chopped into various size pieces and cooked in a number of dishes in soups, sauces, casserole, stews, curries with meat, fish and prawns, stir-fry dishes with meat and other vegetables. It also can be frozen for later use. Freezing is another way of preserving pineapple. The fruit is peeled in the usual way and cut into cubes, put it into plastic bags, seal and freeze. When it is thawed and eaten, it will not be as crisp as fresh pineapple.

Dried pineapple slices from firm-ripe fruit not over-ripe fruit makes delicious and nutritious snacks for children. It can be sold in stores to take the place of less nutritious snacks. Dried pineapple can also be used in meat and fish dishes or in desserts.

The fruit is also amenable to canning as slices, spirals, chunks, spears, titbits and cubes. Pieces may be canned mixed with other fruits, e.g. rambutan in Thailand. The fruit can be made into pickles, chutneys, crystallized and glacé fruit, candies, sweets, jams, juices, syrups cakes, pastries, ice-cream, sherbets, drinks, punch, beverages and wine. Pineapple juice can also be frozen by pouring the juice into ice-cube trays and freeze. When frozen remove the cubes from the trays and store frozen in sealed plastic bags.

The flesh adhering to the shell after peeling is scraped and made into pineapple crush or juice. The latter is also made by crushing the shell, core and bits and pieces of flesh which could not be used as specified cuts. The by-products of canning can be used as cattle feed or to produce pineapple wine or vinegar.

Meat can be softened by soaking it in a pineapple marinade. To make this marinade, cooking oil and lemon juice are mixed together in the ratio 2:1. Slices of fresh pineapple are added to the mixture and the meat is immersed in this marinade for 2 h before cooking it.

Young leaves in bud stage (the "heart" or cabbage) and the inflorescences or young fruit are also eaten. In Java, the cabbage is eaten raw or cooked as *lalab* or mixed in *sayur*. The flower spike and young spurious fruit are first peeled, sliced, steamed or cooked and eaten as *lalab* or mixed in *sayur (lodeh)* with rice. Half-ripe spurious fruits are peeled, sliced and eaten uncooked with a *sambal* paste like *petis*.

Botany

A perennial or biennial, herbaceous, plant sometimes spinescent, succulent, up to 1 m tall (Plate 4). Leaves are long, up to 1 m or more, 5–8 cm wide, sword-like, arranged in a tight spiral around a short stem, margins coarsely and laxly spinose serrate, green, often variegated, or red or brown streaked (Plate 5). Inflorescence is compact with numerous (up to 200) reddish-purple, sessile, trimerous flowers, each subtended by a pointed bract. Flower with three short, fleshy, free sepals; three violet or reddish petals forming a tube enclosing six stamens and a narrow style with trifid stigma. Fruit is a coenocarpium (syncarp) formed by an extensive thickening of the axis of the inflorescence and by the fusion of the small berry-like individual fruits; the hard rind of the fruit is formed by the fusion of the persistent sepals and floral bracts. The fruit is oval to cylindrical, yellowish to orange, often greenish; about 15–25 cm long and 14–17 cm in diameter, weighing 1–2.5 kg; the fruit is surmounted by a rosette of short, stiff spirally arranged leaves, called the 'crown' (Plates 1–3, 4); the flesh is pale to golden yellow, juicy and usually seedless due to self incompatibility and the use of triploid cultivars.

Nutritive/Medicinal Properties

Analyses carried out in the United States reported that raw, pineapple (excludes 49% refuse of the core, crown and parings) had the following nutrient composition (per 100 g edible portion): water

Plate 1 (**a** & **b**) Morris Pineapple variety

Plate 2 Sarawak pineapple variety

Plate 3 Close-up of pineapple fruit

Plate 4 Pineapple plantation

Plate 5 Developing, immature pineapple fruit and leaves

86.00 g, energy 50 kcal (209 kJ), protein 0.54 g, total lipid 0.12 g, ash 0.22 g, carbohydrates 13.12 g, total dietary fibre 1.4 g, total sugars 9.85 g, sucrose 5.99 g, glucose 1.73 g, fructose 2.12 g, Ca 13 mg, Fe 0.29 mg, Mg 12 mg, P 8 mg, K 109 mg, Na 1 mg, Zn 0.12 mg, Cu 0.110 mg, Mn 0.927 mg, Se 0.1 μg, vitamin C 47.8 mg, thiamine 0.079 mg, riboflavin 0.032 mg, niacin 0.500 mg, pantothenic acid 0.213 mg, vitamin B-6 0.112 mg, total folate 18 µg, choline 5.5 mg, betaine 0.1 mg, vitamin A 58 IU, vitamin E (α-tocopherol) 0.02 mg, vitamin K (phylloquinone) 0.7 µg, total saturated fatty acids 0.009 g, 16:0 (palmitic acid) 0.005 g, 18:0 (stearic acid) 0.003 g; total monounsaturated fatty acids 0.013 g, 16:1 undifferentiated (palmitoleic acid) 0.001 g, 18:1 undifferentiated (oleic acid) 0.012; total polyunsaturated fatty acids 0.040 g, 18:2 undifferentiated (linoleic acid) 0.023 g, 18:3 undifferentiated (linolenic acid) 0.017 g; phytosterols 6 mg, tryptophan 0.005 g, threonine 0.019 g, isoleucine 0.019 g, leucine 0.024 g, lysine 0.026 g, methionine 0.012 g, cystine 0.014 g, phenylalanine 0.021 g, tyrosine 0.019 g, valine 0.024 g, arginine 0.019 g, histidine 0.010 g, alanine 0.033 g, aspartic acid 0.121 g, glutamic acid 0.079 g, glycine 0.024 g, proline 0.017 g, serine 0.035 g and β-carotene 35 µg (USDA 2010).

Pineapple is a rich source of vitamin C, which keeps body tissues strong, helps the body use iron, and helps chemical actions in the body. It is also a good source of vitamin A and β carotene and phytosterols. It is a fair source of vitamin B_1 (thiamin), which helps the body to convert carbohydrates into energy and heat. It is also a good source of manganese.

One hundred and eighteen volatile compounds were identified in fresh pineapple pulp (Teai et al. 2001). These included seven hydrocarbons (3.3%), nine sulfur compounds (10.3%), 42 esters (44.9%), 10 lactones (11.5%), 11 carbonyl compounds (4.7%), 14 acids (7.3%), 11 alcohols and phenols (3.8%), and 14 miscellaneous compounds (14.3%). Four compounds were found at a level higher than 1 mg/kg, methyl octanoate (1.49 ppm), methyl 3-(methylthio)propanoate (1.14 ppm), methyl hexanoate (1.1 ppm) and 3-methyl-2,5-furanedione (1.07 ppm), three others were detected at a level between 0.5 ppm and 1 ppm and 25 at a level between 0.1 ppm and 0.5 ppm. Of these 118 compounds, 71 have already been described in the literature to be present in pineapple aroma. The study showed that among the 47 newly detected volatile constituents, five free phenolic compounds, four sulfur compounds and a family of furanoid compounds were observed for the first time in the composition of the pineapple aroma. In another study, the following compounds were identified as key odorants in fresh pineapple flavor: 4-hydroxy-2,5-dimethyl-3(2H)-furanone (HDF; sweet, pineapple-like, caramel-like), ethyl 2-methylpropanoate (fruity), ethyl 2-methylbutanoate (fruity) followed by methyl 2-methylbutanoate (fruity, apple-like) and 1-(E,Z)-3,5-undecatriene (fresh, pineapple-like) (Tokitomo et al. 2005). A mixture of these 12 odorants in concentrations equal to those in the fresh pineapple resulted in an odor profile similar to that of the fresh juice. Additionally, the results of omission tests showed that HDF and ethyl 2-methylbutanoate were character impact odorants in fresh pineapple.

Analyses of pineapple volatiles showed that pineapple crown contained C6 aldehydes and alcohols while the pulp and intact fruit were characterized by a diverse array of esters, hydrocarbons, alcohols and carbonyl compounds (Takeoka and Butter 1989). The following compounds were found to be important contributors to fresh pineapple aroma: 2,5-dimethyl-4-hydroxy-3(2H)-furanone, methyl 2-methylbutanoate, ethyl 2-methylbutanoate, ethyl acetate, ethyl hexanoate, ethyl butanoate, ethyl 2-methylpropanoate, methyl hexanoate and methyl butanoate. A total of 183 volatile compounds were identified in ripe pineapple fruit (Takashi and Katsumi 2002). Among these, 11 sulfur-containing compounds were found in the aroma extract, including eight components newly identified such as methyl S-(methylthio) malonate and ethyl 5-(methylthio)pentanote. Furaneol, mesifurane, γ-octalactone, methional, and methyl S-(methylthio) malonate were identified as the major odor-active components of ripened pineapple aroma.

Analysis of juices made from fresh-cut pineapple fruit revealed the known predominance of esters, with methyl 2-methylbutanoate, methyl 3-(methylthio)-propanoate, methyl butanoate, methyl hexanoate, ethyl hexanoate and ethyl 3-(methylthio)-propanoate, as well as 2,5-dimethyl-4-methoxy-3(2H)-furanone (mesifurane) and 2,5-dimethyl-4-hydroxy-3(2H)-furanone (furaneol) as major constituents (Elss et al. 2005). A similar flavour profile was rarely found in water phases/recovery aromas. A few of the commercial single strength pineapple juices had fruit-related flavour profiles, juices produced from concentrates mostly exhibited a flavour composition similar to that of concentrates, i.e. they were predominantly governed by their contents of furaneol and did not exhibit the fruit-related ester distribution. Correspondingly, the pineapple jams under study were poor in typical pineapple constituents.

Fruit bromelain FA2, was determined to be the main proteinase component of the juice of pineapple fruit (Yamada et al. 1976). The amino acid composition of FA2 was not significantly different from that of stem bromelain, except for a much lower lysine content and a lower alanine content relative to glycine in FA2. FA2 being a non-glycoprotein contained neither amino sugars nor neutral carbohydrates. Pineapple stem was also found to be a rich source of bromelain, the thiohydryl protease (Heinicke and Gortner 1957). Bromelain also contained a peroxidase, acid phosphatase, several protease inhibitors and organically bound calcium (Kelly 1996).Two forms of an acidic bromelain proteinase called SAB/a and SBA/b were isolated from crude bromelain, an extract from pineapple stem (Harrach et al. 1998). Partial N-terminal amino acid sequences (11 residues) were identical to SBA/a and SBA/b and similar with those of stem bromelain, the basic main proteinase of the pineapple stem, and fruit bromelain, the acidic main proteinase of the pineapple fruit. Both components were highly glycosylated. The comparison of the catalytic properties of SBA/a with those of SBA/b revealed no major differences in the hydrolysis of three peptidyl-NH-Mec substrates and in the inhibition profiles using chicken cystatin and E-64, indicating that these components could be regarded as two forms of a single enzyme. Both forms were scarcely inhibited by chicken cystatin and slowly inactivated by E-64 (cysteine protease inhibitor), hence were not normal cysteine proteinases of the papain superfamily.

A mixture of ananain and comosain purified from crude pineapple stem extract was found to contain numerous closely related enzyme forms (Napper et al. 1994). Structural and kinetic analyses revealed comosain to be closely related to stem bromelain, whereas ananain differed significantly from both comosain and stem bromelain. Some differences were found between comosain and stem bromelain in amino acid composition and kinetic specificity towards the epoxide inhibitor E-64. Ananain was found to have five isolatable alternative forms of ananain. Three of the enzyme forms displayed ananain-like amidolytic activity, whereas the other two forms were inactive. Thiol-stoichiometry determinations revealed that the active enzyme forms contained one free thiol, while the inactive forms lacked the reactive thiol needed for enzyme activity.

Many workers have reported bromelain to possess the following pharmacological properties: proteolytic activity; anticancer activity; inhibition of platelet aggregation; fibrinolytic activity; anti-inflammatory action; skin debridement properties. These biological functions of bromelain, a non-toxic compound, have therapeutic values in (a) immunomodulating; (b) tumour growth; (c) blood coagulation; (d) inflammatory changes; (e) debridement of third degree burns; (f) enhancement of absorption of drugs especially antibiotics; and also in angina pectoris, bronchitis, sinusitis, surgical traumas, thrombophlebitis, pyelonephritis (Taussig and Batkin 1988; Maurer 2001; Tochi et al. 2008; Chobotova et al. 2009). Pineapple plant especially the fruit and leaves have been reported to have anti-diarrhoeal, cardioprotective, abortifacient, anti-diabetic, anti-dyslipidemic, anti-oxidative and *Pityriasis lichenoides* chronic activities as elaborated below. Research on animals also reported that fresh pineapple had a reconstructive effect on the hepatic metabolism.

Antioxidant Activity

Pineapple has several beneficial, nutritive properties including antioxidant activity. Studies by Mhatre et al. (2009), showed that pineapple fruit extract from non-transformed and transformed pineapple (with the magainin gene construct, for disease resistance) both exhibited high antioxidant activity with oxygen radical absorbance capacity (ORAC) values varying from 9.5 to 26.4, similar to or greater than those for some fruits and vegetables. The main antioxidant compounds present were ascorbic acid, quercetin, flavone-3-ols, flavones, cinnamic acids. The transformed pineapple core extract exhibited slightly higher level of ascorbic acid and much higher content of flavon-3-ols. The study indicated that the transformation event had caused only marginal difference in antioxidant activity.

Studies in Thailand reported that pineapple cultivars differed in antioxidant activity and that other compounds besides vitamin C, such as phenolic compounds and β-carotene were also accountable for the antioxidant activity (Kongsuwan et al. 2009). Vitamin C contents were 18.88 and 6.45 mg/100 g FW in "Phulae" and "Nanglae" cultivars respectively. The vitamin C content was 3-fold higher in "Phulae" than "Nanglae" cultivars. The β-carotene contents were relatively low in both cultivars but they were significantly higher in "Phulae" (3.35 μg/100 g FW) than "Nanglae" cultivar (1.41 μg/100 g FW). Further, total phenolic contents were 26.20 and 20.28 mg GAE/100 FW in "Phulae" and "Nanglae" respectively. It was found that "Nanglae" had significantly better scavenging ability than "Phulae"cultivar. The DPPH radical scavenging capacities in aqueous extract of pineapple pulp were 152.93 and 118.18 μmol TE/100 g FW for "Nanglae" and "Phulae" cultivars respectively. Both cultivars exhibited reducing power. The FRAP value of "Nanglae" cultivar was 205.73 μmol AAE/100 g FW indicating significantly greater reducing power than in "Phulae" (165.28 μmol AAE/100 g FW).

Haripyaree et al. (2010) reported that pineapple fruit phenolics scavenged DPPH, superoxide anion radicals and hydroxyl radicals in a dose dependent manner. In DPPH assay, the IC_{50} values of pineapple phenolics, ascorbic acid and pyragallol were 12.2, 17.82 and 15.92 μg/ml respectively. In superoxide anion and hydroxyl radical scavenging activities, the IC_{50} values of pineapple phenolics were 11.42 and 55.292 μg/ml; for ascorbic acid 49.62, 48.52 μg/ml and that of pyragallol was 15.672 and 60.62 μg/ml. The IC_{50} value was lowest in pineapple phenolics than ascorbic acid and pyragallol in DPPH and superoxide anion assays, but it was higher than ascorbic acid and lower than pyragallol in the hydroxyl radical assay. The lower the IC_{50} values, the higher the antioxidant activities. The phenolics extracted from this variety of pineapple exhibited outstanding free radical scavenging activity. The result showed that pineapple and its active constituents may be employed in further antioxidative therapy.

Anti-inflammatory Activity

Bromelain has been used clinically as anti-inflammatory agents in arthritis, rheumatoid arthritis, soft tissues injuries, colonic inflammation, chronic pain, edema, ankle injury, skeletal muscle injury, sinusitis, rhinosinusitis, airway allergic disease and asthma (Blonstein 1960; Taub 1966; Seltzer 1967; Izaka et al., 1972; Taussig and Batkin 1988; Kelly1996; Maurer 2001; Desser et al. 2001; Manhart et al. 2002; Walker et al. 2002; Kerkhoffs et al. 2004; Hale et al. 2005; Onken et al. 2008; Tochi et al. 2008).

In one controlled study, 74 boxers with bruises on their faces and upper bodies were administered bromelain until all signs of bruising had disappeared; another 72 boxers were given placebo. All symptoms of bruising disappeared in 58 of the group taking bromelain within 4 days, compared to only 10 taking placebo (Blonstein 1960; Blonstein 1969). In another study involving 73 patients with acute thrombophlebitis, bromelain, in addition to analgesics, was shown to decrease all symptoms of inflammation, including, pain, edema, tenderness, skin temperature, and disability (Seligman 1962, 1969). Unfortunately, this study was not double-blind. A double-blind, placebo-controlled study evaluated 160 women who received episiotomies (surgical cuts in the perineum) during childbirth (Zatuchni and Colombi 1967). Participants administered 40 mg of bromelain four times daily for 3 days, beginning 4 h after delivery, showed a statistically significant reduction in swelling, inflammation, and pain. Ninety percent of patients taking bromelain demonstrated excellent or good responses compared to 44% in the placebo group. However, another double-blind study of 158 women who received episiotomies failed to find significant benefit (Howat and Lewis 1972). Enteral application (intraduodenally) of bromelain in rats resulted in a significant decrease of the edema induced in the hindleg but the parenteral application only resulted in a minimal therapeutic effect (Uhlig and Seifert 1981). Although enterally applied enzymes are thought to be degraded in the gut, the better results were obtained after enteral administration of bromelain. This supported the observation that enzymes could be absorbed by the gut without losing their biological properties.

Studies demonstrated that Phlogenzym (a preparation consisting of the proteases bromelain and trypsin and the antioxidant rutosid) was effective as an adjuvant with antibiotics and supportive treatment for early improvement of pediatric patients with sepsis in a double-blind, randomised, controlled phase III study at a tertiary care centre in Mumbai involving 60 children (Shahid et al. 2002.) Median time taken for fever to subside was 3 days in the phlogenzym group (n=30; 17 boys) compared to 4 days in the placebo group (n=30; 22 boys). Further, haemodynamic support was needed for 2 days in the phlogenzym group but 3 days in the placebo group. The modified Glasgow coma scale score was normalized in 3 days in the phlogenzym group versus 5.5 days in the placebo group. Oral feeds could be started in 4 days in the phlogenzym group versus 5 days in the placebo group and two patients died in the placebo group.

Anti-inflammatory Activity – Sinusitis/ AAD (Allergic Airway Disease) Therapy

Bromelain was found to confer both anti-inflammatory and mucolytic properties thereby decreasing congestion and irritation of the mucous membranes bombarded during allergy season (Rimoldi et al. 1978). The mucolytic properties are particularly important, since the addition of copious or thick mucous creates a superb breeding ground for bacteria leading frequently to a secondary opportunistic infection moving into the sinuses or bronchial airways, causing one of the inflammatory "itis" conditions: sinusitis, conjunctivitis, rhinitis, pharyngitis and potential bronchitis. A 2005 German clinical study found 116 children under age 11 years with acute sinusitis exhibited statistically significant faster symptom recovery with bromelain compared with standard treatment (Braun et al. 2005). The shortest period of symptoms were observed in patients treated with bromelain as an isolated therapy confirming clinical findings from the 1960s. Once 1967 study reported 85% of sinusitis patients receiving bromelain obtained complete recovery from inflammation of the nasal mucosa compared to 40% in the placebo group (Taub 1967). In another 1967 study involving a double-blind trial, 48 patients with moderately severe to severe sinusitis were administered bromelain or placebo for 6 days (Ryan 1967). All patients were placed on standard therapy for sinusitis, which included antihistamines, analgesics, and antibiotics. Upon termination of the study, inflammation was reduced in 83% of those taking bromelain compared to 52% of the placebo group. Breathing difficulty was relieved in 78% of the bromelain group and 68% of the placebo group. In summary, good to excellent results were observed in 87% of patients treated with bromelain compared to 68% on placebo. Benefits were also seen in two other studies involving a total of more than 100 individuals with sinusitis (Taub 1966; Seltzer 1967). According to the scientists, patients of the bromelain monotherapy group showed a statistically significant speedier recovery from symptoms compared to the other treatment groups (Kelly 1996; Braun et al. 2005).

A recent review reported that three randomised controlled trials (RCTs) tested bromelain in either acute sinusitis (2 RCTs) or patients of mixed diagnosis (chronic and acute sinusitis), generated some positive findings (Guo et al. 2006). Meta-analysis of the two RCTs in acute sinusitis suggested that adjunctive use of bromelain significantly improved some symptoms of acute rhinosinusitis. Bromelain is approved by the German Commission for the treatment of sinus and nasal swelling following ear, nose, and throat surgery or trauma (Blumenthal et al. 2000). Bromelain may help reduce cough and nasal mucus associated with sinusitis, and relieve the swelling and inflammation caused by hay fever. Due to its efficacy after oral administration, its safety and lack of undesired side effects, bromelain has become an internationally revered nutraceutical that helps ameliorate sinusitis and bronchitis. It has also been shown to lessen blood stickiness, angina pectoris, post-surgical traumas, thrombophlebitis, pyelonephritis and can actually enhance absorption of drugs, particularly various antibiotics (Maurer 2001).

Anti-inflammatory Activity – Colonic Inflammation

Bromelain may be a novel therapy for inflammatory bowel disease. Hale and co-worker and other researchers had previously shown that proteolytically active bromelain removed certain cell surface molecules and modified leukocyte migration, activation, and production of cytokines and inflammatory mediators in-vitro. Studies demonstrated that bromelain enzymes could remain intact and proteolytically active within the murine gastrointestinal tract, further supporting the hypothesis that oral bromelain may potentially modify inflammation within the gastrointestinal tract through local proteolytic activity within the colonic microenvironment (Hale 2004). In subsequent studies Hale et al. (2005) reported that daily treatment with oral bromelain starting at age 5 weeks decreased the incidence and severity of spontaneous colitis in C57BL/6 IL-10-/- mice. Bromelain also significantly decreased the clinical and histologic severity of colonic inflammation when administered to piroxicam-exposed IL-10-/- mice with established colitis. Proteolytically active bromelain was found to be required for anti-inflammatory effects in-vivo.

Anti-inflammatory Activity – Arthritis, Osteoarthritis, Rheumatoid Arthritis

Bromelain had been demonstrated to show anti-inflammatory and analgesic properties and may provide a safer alternative or adjunctive treatment for osteoarthritis (Brien et al. 2004). All previous trials, which had been uncontrolled or comparative studies, indicate its potential use for the treatment of osteoarthritis. In a randomized, double-blind study, the oral enzyme ERC containing rutosid and the enzymes bromelain and trypsin was found to be as an effective and safe alternative nonsteroidal anti-inflammatory drugs (NSAID) such as diclofenac in the therapy of painful episodes of osteoarthritis of the knee (Klein and Kullich 2000). Both treatments resulted in marked improvements. Within the 6-week observation period, the mean value of the Lequesne's Algofunctional Index (LFI) decreased from 13.0 to 9.4 in the ERC group (52 patients) and from 12.5 to 9.4 in the diclofenac group (51 pateints). Non-inferiority of ERC was demonstrated by both primary criteria, LFI and complaint index. Considerable improvements were also seen in secondary efficacy criteria, with a slight tendency towards superiority of ERC. In a subsequent double-blind, randomised study, Klein et al. (2006) reported significant non-inferiority from 6 weeks treatment with phlogenzym in patients with osteoarthritis of the hip. However, in this study there was no real difference between Phlogenzym and diclofenac 100 mg/day (with 52 patients per group), implying an equal benefit-risk relation between the substances. Phlogenzym may well be recommended for the treatment of patients with osteoarthritis of the hip with signs of inflammation as indicated by a high pain level. In a randomized placebo-controlled pilot study, the researchers found no statistically significant differences were observed between groups (14 bromelain, 17 placebo) for the primary outcome, nor the WOMAC (Western Ontario MacMaster) subscales or SF36 (Brien et al. 2006). Both treatment groups showed clinically relevant improvement in the WOMAC disability subscale only. Adverse events were generally mild in nature. The result suggested that bromelain was not efficacious as an adjunctive treatment of moderate to severe osteoarthritis.

In a separate prospective, randomized, controlled, single-blind study of 7 weeks duration involving 50 patients aged 40–75 years, with activated osteoarthrosis of knee joint, phlogenzym which consisted of bromelain, trypsin and the anti-oxidant rutosid and an NSAID (diclofenac) separately elicited a reduction in pain and joint tenderness and swelling, and slight improvement in the range of movement in the study group (Tilwe et al. 2001). The reduction in joint tenderness was greater in the study group receiving phlogenzym. The data showed that phlogenzym was as efficacious and well tolerated as diclofenac sodium in the management of active osteoarthrosis over 3 weeks of treatment.

Desser et al. (2001) demonstrated a high correlation between active and latent transforming growth factor-β TGF-β in serum (r=0.8021) in healthy volunteers (n=78) and in patients with rheumatoid arthritis (RA) (n=38), osteomyelofibrosis (OMF) (n=7) and herpes zoster (HZ) (n=7). Treatment with oral proteolytic enzymes (OET) containing papain, bromelain, trypsin, and chymotrypsin had no significant effect on TGF-β1 concentration in healthy volunteers or patients with a normal level of TGF-β1. In patients with elevated TGF-β1 concentration (>50 μg/ml serum), OET reduced TGF-β1 in patients with RA, in OMF and in HZ. These results supported the concept that OET was beneficial in diseases characterized in part by TGF-β1 overproduction.

In another study, bromelain at 200 or 400 mg per day for 3 months was found effective in improving physical symptoms and general well-being in otherwise healthy adults (77) suffering from mild knee pain in a dose-dependent fashion (Walker et al. 2002). In both treatment groups, all WOMAC symptom knee health Index and the Psychological Well-Being Index dimension scores were significantly reduced compared with baseline, with reductions in total symptom score of 41% and 59% in the low and high dose groups respectively. Further, ameriolations in total symptom score and the stiffness and physical function dimensions were significantly greater in the high-dose (400 mg per day) compared with the low-dose group. Compared to baseline, overall psychological well-being was significantly improved in both groups after treatment in the low and high dose groups respectively. Earlier Walker et al. (1992) found that bromelain attenuated skeletal muscle injury induced by lengthening contractions in animals. Maximum isometric tetanic force of injured muscles from animals treated with bromelain were higher than that in untreated animals at 3 and 14 days but not 7 days. The maximum isometric tetanic force of untreated injured muscles were significantly lower than that of contralateral control muscles at all time periods. The number of intact fibres of 3-day untreated injured muscles was lower than the number of intact fibres in control muscles. Masson (1995) found that treatment with bromelain resulted in a marked reduction in all four parameters: swelling, pain at rest and during movement, and tenderness in 59 patients with blunt injuries to the musculoskeletal system. Both swelling and the symptoms of pain were appreciably ameliorated at all evaluation time points as compared with baseline. The tolerability of the preparation was very good, and patient compliance was correspondingly high.

Phlogenzym (rutoside, bromelain, and trypsin) was compared with double combinations, the single substances, and placebo in a: multinational, multicentre, double blind, randomised, parallel group design with eight groups structured according to a factorial design involving a total of 721 patients aged 16–53 years with acute unilateral sprain of the lateral ankle joint (Kerkhoffs et al. 2004). Phlogenzym was not found to be superior to the three two-drug combinations, the three single substances, or placebo for treatment of patients with acute unilateral sprain of the lateral ankle joint. At the primary end point at 7 days, the greatest reduction in pain was found in the bromelain/trypsin group (73.7%). The Phlogenzym group showed a mean reduction of 60.3%, and the placebo group showed a mean reduction of 73.3%. The largest increase in range of motion (median) was in the placebo group (60% change from baseline). The Phlogenzym group showed an average increase of 42.9%. The biggest reduction in swelling was in the trypsin group (3.9% change from baseline). The Phlogenzym group showed a −2.30% change from baseline and the placebo group a −2.90% change. In the small subgroup analysis of patients treated without the support of a Caligamed brace, Phlogenzym was found superior over placebo.

Mechanism of Anti-inflammatory Action

Bromelain was found to exert anti-inflammatory effects in a number of inflammatory models. Bromelain was found to downregulate the production of kinin, a group of pain-inducing polypeptides, by reducing the levels of plasma kininogen. Bromelain was reported also to reduce edematous swelling of airway tissue, such as nasal congestion via plasmin activation and thus help quench the pervasive inflammation (Taussig 1980; Kelly 1996). Bromelain's ability to thin nasal secretions had been shown to be an effective mucolytic agent in other respiratory tract diseases (Rimoldi et al. 1978). Through its activity on the blood-clotting-related substances, fibrinogen and fibrin, bromelain was reported to induce the production and release of anti-inflammatory prostaglandins, while concurrently reducing the production and release of pro-inflammatory prostaglandins and the initiation of prostaglandin El accumulation (inhibiting the release of polymorphonuclear leukocyte lysosomal enzymes) (Felton 1977; Taussig 1980; Ako et al. 1981; Kelly 1996).

The results of studies by Manhart et al. (2002) revealed that phlogenzym (a preparation consisting of the proteases bromelain and trypsin and the antioxidant rutosid) may ameliorate inflammatory process by reducing the number of CD4+ cells and by decreasing INF-γ mRNA levels. Orally administered phlogenzym significantly reduced lymphocyte subpopulations in Peyer's patches (PPs) of healthy and endotoxemic mice. Similarly, the number of splenic lymphocytes in endotoxin-boostered mice was significantly lowered by phlogenzym. The effect of phlogenzym was more marked on T cells than on B cells leading especially to a diminution of CD4+ cells.

Bromelain was found to attenuate development of allergic airway disease (AAD) (Secor et al. 2005). Bromelain treatment of ADD mice, reduced total BAL leukocytes, eosinophils, CD4+ and CD8+ T lymphocytes, CD4+/CD8+ T cell ratio, and interleukin (IL-13). The reduction in AAD outcomes suggested that bromelain may have similar effects in the therapy of human asthma and hypersensitivity disorders. Additional studies showed that bromelain treatment reduced CD25 expression on activated CD4+ T cells in-vitro. The concentration of soluble fraction of CD24, sCD25 was increased in supernatants of bromelain treated activated CD4(+) T cells as compared to control cells, suggesting that bromelain proteolytically cleaved cell-surface CD25 (Secor et al. 2009). This novel mechanism of action identified how bromelain may exert its therapeutic benefits in inflammatory conditions. The reduction in AAD outcomes suggested that bromelain may have similar effects in the therapy of human asthma and hypersensitivity disorders. Hou et al. (2006) found that the cross-linked (CL)-bromelain preparation (cross-linked with organic acids and polysaccharides) exhibited in-vivo and in-vitro anti-inflammatory effects and may have therapeutic potential. The anti-inflammatory effect of CL-bromelain was correlated with reduced lipopolysaccharide (LPS)-induced NF-κB activity and cyclooxygenase 2 (COX-2) mRNA expression in rat livers. Further, CL-bromelain dose-dependently inhibited LPS-induced COX-2 mRNA and prostaglandin E2(PGE2) in BV-2 microglial cells. CL-Bromelain also inhibited the LPS-activated extracellular signal-regulated kinase (ERK), c-Jun N-terminal kinase (JNK), and p38 mitogen-activated protein kinase (MAPK). In subsequent studies, the researchers (Huang et al. 2008) reported that anti-inflammatory activity of bromelain may be attributed to inhibition of signaling pathways by bromelain's proteolytic activity. The results showed that bromelain (50–100 μg/ml) significantly and reversibly reduced tumour necrosis factor (TNF)-α interleukin- (IL)-1β and IL-6 from LPS-induced PBMC and THP-1 cells. This effect was correlated with reduced LPS-induced TNF-α mRNA and NF-κB activity in THP-1 cells. Additionally, bromelain dose-dependently suppressed LPS-induced prostaglandin E(2), thromboxane B(2) and COX-2 mRNA but not COX-1 mRNA. Importantly, bromelain degraded TNF-α and IL-1β molecules, and reduced the expression of surface marker CD14 but not Toll-like receptor 4 from THP-1 cells.

The mechanisms by which exogenous proteinases affect bowel inflammation was elucidated in subsequent studies by Fitzhugh et al. (2008).

Bromelain was found to effectively decrease neutrophil migration to sites of acute inflammation and to support the specific removal of the CD128 chemokine receptor as a potential mechanism of action. Further, bromelain treatment was found to decrease secretion of pro-inflammatory cytokines and chemokines in-vitro (Onken et al. 2008). Significant increases in granulocyte colony-stimulating factor (G-CSF), interferon (IFN)-γ, interleukin (IL)-1β, IL-6, and tumour necrosis factor (TNF) were detected in the media from actively inflamed areas in ulcerative colitis and Crohn's disease as compared with non-inflamed inflammatory bowel disease tissue and non- inflammatory bowel disease controls. In-vitro bromelain treatment was found to reduced secretion of granulocyte colony-stimulating factor, granulocyte-macrophage colony-stimulating factor, interferon IFN-γ, CCL4/macrophage inhibitory protein (MIP)-1β, and tumour necrosis factor by inflamed tissue in inflammatory bowel disease.

Anticancer and Antimetastatic Activity

In the past decades, studies have shown that bromelain has the capacity to modulate key pathways that support malignancy, inhibit the growth of cancerous tumours and enhance the effectiveness of some chemotherapies. Chobotova et al. (2009) suggested that the anti-cancer activity of bromelain resulted from its direct impact on cancer cells and their micro-environment, as well as in the modulation of immune, inflammatory and haemostatic systems.

Bromelain displayed in-vivo anti-tumour activity in animals against the following cancers: P-388 leukemia, sarcoma (S-37), Ehrlich ascitic tumour (EAT), Lewis lung carcinoma (LLC) and ADC-755 mammary adenocarcinoma (Báez et al. 2007). With the exception of MB-F10 melanoma, all other tumour-bearing animals had a significantly increased survival index after bromelain treatment. The largest increase (approximately 318%) was attained in mice bearing EAT ascites and receiving 12.5 mg/kg of bromelain. This antitumoral effect was superior to that of 5-FU, whose survival index was approximately 263%, relative to the untreated control. Bromelain significantly reduced the number of lung metastasis induced by Lewis lung carcinoma transplantation, as observed with 5-FU. The antitumoral activity of bromelain against S-37 and EAT, tumour models sensitive to immune system mediators, and the unchanged tumour progression in the metastatic model suggested that the antimetastatic action resulted from a mechanism independent of the primary antitumoral effect.

In another study, bromelain, added to mice laboratory chow decreased lung metastasis of Lewis lung cancer cells implanted sub-cutaneously (s.c.) (Batkin et al. 1988). This antimetastatic potential was demonstrated by both the active and inactive bromelain with or without proteolytic, anticoagulant properties. Intraperitoneal (i.p.) or s.c. administration of bromelain significantly reduced local tumour weight, however, lung colonization was non-significantly reduced (Beuth and Braun 2005). Bromelain incubation of sarcoma L-1 cells significantly reduced their tumorigenic/metastatic capacities. Bromelain treatment after tumour cell inoculation significantly reduced local tumour growth, experimental lung metastasis, however, to a lesser, non-significant degree. Bromelain had also been used to enhance the effectiveness of some chemotherapies (Gerard 1972).

Studies suggested that orally applied bromelain stimulated the deficient monocytic cytotoxicity of mammary tumor patients, which may partially explain its proposed antitumour activity (Eckert et al. 1999). About 40% of the patients responded to bromelain with an enhancement of cytotoxicity from 7.8% to 54% in bMAK-cell activity and from 16% to 47% in MAK-cell (macrophage-activated killer cell) activity. Bromelain was less effective on the higher cytotoxicity of monocytes from healthy donors, but stimulated the secretion of IL-1β from monocytes. Bromelain had no effects on the impaired patients NK (natural killer)- and LAK (lymphokine-activated killer) -cell activity, but reduced the LAK-cell activity of healthy donors. Bromelain reduced the expression of CD44, but weakly increased CD11a and CD62L expression on patient lymphocytes, whereas CD16 remained unchanged. In-vitro bromelain application to lymphocytes had similar effects, with higher reduction rates of CD44 and CD16 expression. As to coagulation parameters in plasma of healthy donors, the activated partial thromboplastin time was increased from 38 to 46 seconds, leaving prothrombin time and plasminogen unchanged.

Tysnes et al. (2001) reported that several complementary assays demonstrated that bromelain significantly and reversibly reduced glioma cell adhesion, migration, and invasion without affecting cell viability, even after treatment periods extending over several months. They found that cyclic AMP response element (CRE)-mediated signaling processes were suppressed. These results indicated that bromelain exerted its antiinvasive effects in gliomas by proteolysis, signaling cascades, and translational attenuation.

Kalra et al. (2008) recorded antitumorigenic activity of bromelain in 7,12-dimethylbenz(a)anthracene (DMBA)-initiated and 12-O-tetradecanoylphorbol-13-acetate(TPA)-promoted 2-stage mouse skin model. Results showed that bromelain application impeded the onset of tumorigenesis and reduced the cumulative number of tumours, tumour volume and the average number of tumours/mouse. The scientists concluded that bromelain induced apoptosis-related proteins along with inhibition of NF-κB-driven Cox-2 expression by blocking the MAPK and Akt/protein kinase B signaling in DMBA-TPA-induced mouse skin tumours, which may account for its anti-tumorigenic effects.

Antiplatelet Aggregation Activity

Bromelain was found to facilitate healthy blood flow and circulatory health by decreasing the bloods ability to clot and reducing risk factors for recurrent heart attack and/or stroke. In a study of 47 patients with various disorders having edema and inflammation placed on intensive bromelain therapy for 1 week were found not to be adversely affected by clotting of blood at the therapeutic dosages (Cirelli and Smyth 1963). Changes in bleeding, coagulation, and prothrombin times were very slight and without practical significance. Oral administration of bromelian significantly reduced platelet sensitivity to adenosine phosphate induced aggregation (Heinecke et al. 1972). When administered orally daily to a group of 20 volunteers with a history of heart attack or stroke, bromelain decreased blood platelets aggregation in 17 of them and normalise conditions in eight of the nine volunteers with high aggregation values. The combination of fibrinolytic and antithrombic properties appeared to be effective and two large scale tests on heart patients elicited a practically complete elimination of thrombosis (Felton 1980). It was postulated that the body could maintain thrombin at a level too low to cause platelet aggregation but adequate to stimulate release of prostaglandins and enzymes for more than 24 h from a single dose of the pineapple enzymes. Since bromelain therapy led to formation of platelets with increased resistance to aggregation, it followed that the dominant endogenous prostaglandins being produced must be from the group that increased platelet cyclicAMP levels (prostacyclin, PGE1, etc.). A fibrinolysis enzyme activator was isolated from commercial bromelain (Ako et al. 1981). The active substance had no intrinsic fibrinolytic activity but appeared to enhance the activity of one of the fibrinolytic enzymes. The presence of the fibrinolysis enzyme activator in commercial bromelain may explain some of the physiological and clinical effects observed following the oral administration of the enzyme preparation.

Bromelain was found to have antithrombotic and anticoagulant activities in-vivo (Metzig et al. 1999). Preincubation of platelets with bromelain (10 μg/mL) completely prevented the thrombin (0.2 U/mL) induced platelet aggregation. Papain was less active in preventing platelet aggregation. In-vitro, bromelain (0.1 μg/mL) reduced the adhesion of bound, thrombin stimulated, fluorescent labelled platelets to bovine aorta endothelial cells. Further, preincubation of platelets with bromelain, prior to thrombin activation, reduced the platelet adhesion to the endothelial cells to the low binding value of unstimulated platelets. Bromelain, orally applied at 60 mg/kg body weight, suppressed the thrombus formation in a time dependent fashion, the maximum being after 2 h in 11% of arterioles and 6% of venules. Intravenous application at 30 mg/kg bromelain was slightly more active in reducing thrombus formation in arterioles (13%) and venules (5%), suggesting that orally applied bromelain was biologically active. These results may help to explain some of the clinical effects observed after bromelain treatment in patients with thrombosis and related diseases.

Morita et al. (1979) showed in-vitro that bromelain inhibited platelet aggregation in a dose dependent manner. Platelet aggregation inhibitory factors were also isolated and characterised; fraction II and III showed both proteolytic activity and inhibition of platelet aggregation, but no peroxidatic activity and demonstrated γ mobility. It was suggested that the proteolytic activity was associated with the inhibition of platelet aggregation, since oxidation of fractions II and III with sodium tetrathionate abolished both activities. The mechanism of inhibition of platelet aggregation by bromelain was postulated to involve its influence on the prostaglandin synthetic pathway of platelets.

Bromelain was found to decrease platelet count, platelet aggregation and platelet activity in-vitro in blood samples taken from the antecubital vein of 10 healthy male non-smokers (Gläser and Hilberg 2006). Platelet count decreased after incubation with 2.5 and 5 mg bromelain/ml from 277 platelets/nl before to 256 and 247 platelets/nl after the treatment. The ADP (adenosine diphosphate) and TRAP-6 (thrombin receptor activator for peptide 6) induced platelet aggregation led to a significant decrease after the incubation with 2.5 mg bromelain/ml (ADP: 48.6%; TRAP-6: 49.6%) or 5 mg bromelain/ml (ADP: 5.0%; TRAP-6: 9.0%) in comparison to control (ADP: 81.4%; TRAP-6: 77.4%). The percentage of unstimulated CD62P positive platelets was minimally higher after incubation with 5 mg bromelain/ml (0.57% PC) in comparison to control (0.22% PC), but after TRAP-6 stimulation the incubation with 5 mg bromelain/ml led to a notable decrease in comparison to the untreated control (50.4–0.9% PC). The changes of CD62P (TRAP-stimulated) and the results of platelet aggregation after incubation with bromelain in-vitro demonstrated the potential of bromelain as a substance for platelet inhibition.

Cardioprotective Activity

Bromelain may also reduce the severity of angina and support the rhythm and health of the heart muscle and may reduce pain and inflammation associated with blood clots (Taussig and Nieper 1979). Administration of bromelain to patients with angina pectoris resulted in the disappearance of symptoms in all patients within 4–90 days. In a another study, it was found that after discontinuing bromelain, angina attacks reemerged after a variable period of time, often triggered by stressful experiences (Nieper 1978). A drastic reduction in the incidence of coronary infarct after administration of potassium and magnesium orotate along with 120–400 mg of bromelain per day was also demonstrated (Nieper 1977).

Bromelain exhibited cardioprotective effects against ischemia-reperfusion injury through Akt/FOXO pathway in rat myocardium (Juhasz et al. 2008). Bromelain treatment exhibited higher left ventricular functional recovery throughout reperfusion compared with the controls. Aortic flow was also elevated in bromelain treatment when compared with that in untreated rats (11 vs. 1 ml). Furthermore, bromelain treatment reduced both the infarct size (34% vs. 43%) and the degree of apoptosis (28% vs. 37%) compared with the control animals. Western blot analysis showed an increased phosphorylation of both Akt and FOXO3A in the treatment group compared with the control. These results demonstrated that bromelain triggered an Akt-dependent survival pathway in the heart, revealing a novel mechanism of cardioprotective action and a potential therapeutic target against ischemia-reperfusion injury.

Enzymatic Debridement Activity

In a prospective, non-comparative study, a bromelain-derived debriding agent, Debridase, was found to be efficacious and safe in 130 patients with 332 deep second degree and third degree burns treated between 1984 and 1999 (Rosenberg et al. 2004). The percentage debridement by number of applications was 89% for a single application, 77% for two, and 62% for three Debridase applications, respectively. There were no significant adverse events. The availability of such a fast acting, reliable and complication-free enzymatic debriding agent may open new horizons and provide a new treatment modality for burns. the use of a proteolytic enzyme complex derived, isolated and purified from pineapple stems in Enzymatic debridement proved to be an innovative, rapid, effective, selective and safe method of postburn necrotic skin removal (Koller et al. 2008). The major advantages of the procedure included minimal invasivity, rapidity, effectiveness, ease of performing debridement at the bedside, minimal or no loss of blood and minimal interference with natural wound healing processes. Preliminary experience with this treatment method showed that in most of the cases treated the debridement was excellent, safe and rapid. The average duration of the debridement was less than 4 hours. The debridement was accompanied by minor to moderate pain which could be treated by analgesic medications. No serious adverse events or reactions have been observed during the study. The time for healing was comparable with the standard of care methods.

Immunomodulatory Activity

Several studies suggested that bromelain may enhance the body's natural immune defenses.

Bromelain, a mixture of cysteine proteases, was reported to modulate immunological responses and had been proposed to be of clinical use. Studies showed that bromelain increased T cell receptor (TCR) and anti-CD28-mediated T cell proliferation in splenocyte cultures by enhancing the costimulatory activity of accessory cell populations (Engwerda et al. 2001a). In-vivo, bromelain enhanced T-cell-dependent, Ag-specific, B cell antibody responses. Further, bromelain induced a concomitant decrease in splenic IL-2 mRNA accumulation in immunized mice. Collectively, these data showed that bromelain could simultaneously enhance and inhibit T cell responses in-vitro and in-vivo through a stimulatory effect on accessory cells and a direct inhibitory action on T cells. The results suggested that bromelain may have important implications for the use of exogenous cysteine proteases as vaccine adjuvants or immunomodulatory agents.

Studies suggested that bromelain possessed immuno-stimulatory properties (Brakebusch et al. 2001). Phagocytes of patients with humoral immune system X-linked agammaglobulinaemia (XLA) appeared to be particularly susceptible to this stimulation. After incubation with bromelain and trypsin (10 µg/ml each) rate of phagocytosis was nearly doubled (phagocyte activity), T2 of respiratory burst was reduced by 65% and killing was accelerated by 27%. Incubation of phagocytes from healthy controls with bromelain and trypsin accelerated phagocytosis to a level comparable to that of untreated phagocytes from patients with XLA and also accelerated reactive oxygen species (ROS) production. In contrast to phagocytes from patients with XLA, phagocytes of patients with common variable immuno deficiency showed a similar stimulation by B/T like healthy controls.

(Hale et al. 2006) demonstrated that proteolytically active antigens such as bromelain could induce both systemic and mucosal immune responses following repeated oral exposure. Significantly higher anti-bromelain antibody titers were seen in interleukin IL-10-deficient versus wild-type mice, suggesting that simultaneous treatments that decrease IL-10 activity may further enhance systemic antibody responses following oral exposure. The antibodies generated did not affect the proteolytic activity of bromelain.

Engwerda et al. (2001b) showed that bromelain, a mixture of cysteine proteases, enhanced interferon-γ (IFN-γ)-mediated nitric oxide and tumour necrosis factor a (TNFα) production by macrophages (Engwerda et al. 2001b). Bromelain enhanced or acted synergistically with IFN-γ receptor-mediated signals. These effects were seen in both RAW 264.7, a macrophage cell line, and primary macrophage populations. Bromelain also increased IL-2- and IL-12-mediated IFN-γ production. These results indicated a potential role for bromelain in the activation of inflammatory responses in situations where they may be deficient, such as may occur in immunocompromised individuals. Earlier studies indicated that the biological effects observed after oral administration of polyenzyme preparations (bromelain, papain or amylase) were related to their ability to induce cytokine production (Desser et al. 1993). Interferon IFN-α and IFN-γ which had no effect alone, synergistically increased TNF (tumour necrosis factor) production when applied together with the enzymes. Further, it was demonstrated that after ingestion of milligram doses of the polyenzyme preparation, peripheral blood mononuclear cells of healthy donors acquired the ability to produce TNF-α, interleukin IL-1 β and IL-6 when incubated ex-vivo with IFN-γ. This may explain the antitumor effects of such enzymes. The results also suggested that polyenzyme preparations may have a stronger immunomodulatory effect when used in combination with IFN-γ.

Drug Potentiating Activity

Bromelain has been used to enhance the effectiveness of certain antibiotic therapies. It may boost the bacteria-fighting properties of antibiotics used to fight pneumonia, bronchitis, sinusitis, and more. In treating a variety of infections, bromelain was found to potentiate the action of antibiotics such as penicillin, chloramphenicol, and erythromycin (Neubauer 1961). In one study, 22 out of 23 people who had previously not responded to these antibiotics did so after supplementing bromelain, taken four times per day. When taken with amoxicillin, bromelain was shown to increase absorption of amoxicillin in humans (Tinozzi and Venegoni 1978). When 80 mg of bromelain was taken together with amoxicillin and tetracycline, blood levels of both drugs increased (Luerti and Vignali 1978). The determination of levels of amoxicillin and tetracycline in serum, uterus, ovarian tube and ovary in patients treated with these antibiotics and bromelain proved to be higher than in patients treated with the same antibiotics and placebo or indomethacin. It was concluded that bromelain favoured the absorption and tissue penetration of antibiotics by mechanisms that were not conferred by other anti-inflammatory agents.

Hepatoprotective Activity

Because of its immunomodulatory action, the protease bromelain afforded a novel alternative for the treatment of hepatic ischemia/reperfusion (I/R) injury (Bahde et al. 2007). In sham-operated animals, treatment with 10 mg/kg b.w. bromelain evoked a disturbed microcirculation with increased leukocyte adherence, apoptosis rate, Kupffer cell activation, and endothelial cell damage. Six hours after central venous catheter (CVC) placement and administration of 10 mg/kg body weight of bromelain, aspartate transferase and alanine transferase levels were significantly elevated. After ischemia/reperfusion, rats treated with 0.1 mg/kg b.w. bromelain exhibited an improved microcirculation, reduction in leukocyte adhesion, apoptosis rates, Kupffer cell activation and endothelial cell damage enhanced eNOS expression, and significantly decreased AST levels compared with untreated animals.

Another study reported that overdosing of Doliprane® (2 g/kg of live weight) affected the biochemical level and the normal functioning of the liver of the wistar rats intoxicated per os; fresh pineapple treatment had a restorative effect on the hepatic metabolism of the intoxicated rats (Jacques et al. 2008). Doliprane intoxicated animals without pineapple treatment had highly significant increase in the rates of alkaline phosphatase (ALP), glutamate oxaloacetic transaminase (GOT) and glutamate pyruvate transaminase (GPT) enzymes. These increases were attenuated by pineapple treatment.

Hypolipidemic Activity

Studies showed that pineapple leaves had potential to be a new natural product for the treatment of hyperlipidemia (Xie et al. 2007).

The ethanolic extract of pineapple leaves (0.40 g/kg) significantly inhibited the increase in serum triglycerides by 40% in fructose-fed mice. In mice induced by alloxan and high-fat diets, serum total cholesterol remained at a high level (180–220 mg/daily) within 7 days of removing high-fat diets but attained normal level (120–140 mg/daily) after leaf extract treatment. Also, the extract (0.40 and 0.80 g/kg) significantly inhibited serum lipids enhancement in Triton WR-1,339-induced hyperlipidemic mice. The results suggested that the leaf extract exerted its actions through mechanisms of inhibiting HMGCoA reductase and stimulating lipoprotein lipase activities. Its active mechanisms was different from those with fibrates but may be partly similar to those with statins.

Antidiabetic Activity

Studies showed that ethanolic extract of *Ananas comosus* leaves had anti-diabetic, anti-dyslipidemic and anti-oxidative activities (Xie et al. 2005). The extract at the dose of 0.40 g/kg significantly suppressed elevation in blood glucose in diabetic rats in oral glucose tolerance test, but did not cause any hypoglycaemic activity in normal rats. It also significantly inhibited the elevation in postprandial triglycerides (TG) concentrations in both normal and diabetic rats in olive oil load test. After 15 days of treatment of diabetic dyslipidemic rats, the extract significantly lowered blood glucose (−51.0%), TG (−50.1%,), TC (−23.3%,), LDL-c (−47.9%) and glycated albumin (−25.4%) levels, significantly elevated serum high-density lipoprotein cholesterol concentrations (66.2%) and prevented lower body weight of diabetes (11.8%), significantly reduced lipid peroxidation productions of blood (−27.8%), brain (−31.6%), liver (−44.5%) and kidneys (−72.2%) compared with those in untreated diabetic dyslipidemic rats. The scientists also reported that the ethanolic extracts of pineapple leaves at an oral dose of 0.40 g/kg could significantly ameliorate sensitivity to exogenous insulin (Xie et al. 2006). After a subacute treatment, the extract also could inhibit the development of insulin resistance in high-fat diet-fed and low-dose streptozotozin-treated diabetic rats following the test of loss of tolbutamide-induced blood glucose lowering action. For intravenous insulin/glucose infusion test, high-fat diet-fed and low-dose alloxan-treated Wistar rats were associated with insulin resistance, which was improved after pineapple leaf extract or fenofibrate treatment. Pineapple leaf extract application inhibited the development of insulin resistance in HepG2 cells. Collectively, the results suggested that pineapple leaf extract may improve insulin sensitivity in type 2 diabetes and could be developed into a new potential natural product for handling of insulin resistance in diabetic patients. Bioactive phenolic compounds detected in the ethanolic extract of pineapple leaves included eight phenylpropane diglycerides, together with two hydroxycinnamic acids, three hydroxycinnamoyl quinic acids, four phenylpropane monoglycerides, three flavones and six phenylpropanoid glycosides (Ma et al. 2007).Other phenolic compounds isolated from pineapple leaves included ananasate, 1-O-caffeoylglycerol, 1-O-p-coumaroylglycerol, caffeic acid, p-coumaric acid, β-sitosterol and daucosterol (Wang et al. 2006).

Antidiarrhoeal Activity

Studies showed that bromelain was 62% effective in inhibiting *Escherichia coli* heat-labile enterotoxin-induced secretion, 51% efficacious against *Vibrio cholerae* toxin, and 35% efficacious against heat-stable enterotoxin (Mynott et al. 1997). Bromelain also prevented secretory changes caused by prostaglandin E2, theophylline, calcium-ionophore A23187, 8-bromoadenosine 3′:5′-cyclic monophosphate, and 8-bromoguanosine 3′:5′-cyclic monophosphate, well-known intracellular mediators of ion secretion. The results indicated that bromelain prevented intestinal fluid secretion mediated by secretagogues that act via adenosine 3′:5′-cyclic monophosphate, guanosine 3′:5′-cyclic monophosphate, and calcium-dependent signaling cascades. In separate studies bromelain (12.5 mg, 125 mg) orally administered for 10 days was successful in reducing the incidence of enterotoxigenic *Escherichia coli* (K88+ ETEC) diarrhoea and protected piglets from life threatening disease (Chandler and Mynott 1998). Bromelain treated pigs also had significantly enhanced weight gain compared with untreated pigs. Bromelain only temporarily inhibited K88+ ETEC receptor activity in-vivo and protected against ETEC induced diarrhoea. Bromelain may have potential as an effective prophylaxis against enterotoxigenic *Escherichia coli* infection.

Post-operative Ileus Activity

Studies showed that bromelain enhanced the wet weight, dry weight, water content and number of fecal pellets in laparotomized, mechanically manipulated rats, suggesting improvement of postoperative ileus (Wen et al. 2006). Additionally, bromelain treatment impeded over-expressed iNOS mRNA and revitalised down-regulated inhibitor κB-α mRNA in the colon of the postoperative rats. Bromelain also impeded lipopolysaccharide (LPS)-induced nitrite overproduction in macrophage cell lines and LPS-induced NF-κB luciferase reporter gene expression in RAW264.7 macrophages transfected with NF-κB luciferase reporter gene. The findings suggested that bromelain facilitated defecation in postoperative rats, partially by inhibiting colonic iNOS overexpression via NF-κB pathway. The data indicated that bromelain may benefit patients with postoperative ileus. Ileus continues to be a common consequence of abdominal surgery, causing significant patient discomfort and often leading to more serious problems.

Antifertility Activity

The juice of the unripe fruits of *A. comosus* administered at 50 ml/day for 7 days demonstrated encouraging anti-implantation activity showing 40% of implants only in rats (Garg et al. 1970). All the steroid compounds in pineapple: ergosterol peroxide; β-sitosterol; 5-stigmastene-3β, 7 α-diol; 5-stigmastene-3β-7β-diol dibenzoate; 7-oxo-5-stigmasten-3β-ol benzoate, and 5α-stigmastene-3β,5,6β-triol 3-monobenzoate were reported to exert some degree of abortifacient activity (Pakrashi and Basak 1976).

Pityriasis Lichenoides Chronica Efficacy

Bromelain was found to be an effective therapeutic option for Pityriasis lichenoides chronica (PLC) a skin disease of unknown etiology; its efficacy could be related to its anti-inflammatory, immunomodulatory and/or antiviral properties (Massimiliano et al. 2007). Eight PLC patients (three males and five females) showed full clinical recovery after bromelain treatment. In 12 months of follow up, two patients experienced relapse 5–6 months after suspension of bromelaintherapy but responded to another brief cycle of therapy. No side effects were encountered during therapy.

Diuretic Activity

Extracts of five indigenous Thai medicinal having ethnomedical application in the treatment of dysuria were investigated for their diuretic activity (Sripanidkulchai et al. 2001). Root extracts of *Ananas comosus* and *Carica papaya*, administered orally to rats at a dose of 10 mg/kg, significantly enhanced urine output which was 79% and 74%, respectively, similar to the effect of an equivalent dose of hydrochlorothiazide. Both plant extracts and hydrochlorothiazide produced similar profiles of urinary electrolyte excretion. The analyses of the urinary osmolality and electrolyte excretion per unit time suggested the discerned effect of *A. comosus* was inherent, whereas that of *Carica papaya* may have resulted from a high salt content of this extract.

Antidepressant Activity

The anti-depressant effect of pineapple juice was found to be comparable to standard depressants, fluoxetine (20 mg/kg) and imipramine (15 mg/kg) in mice (Milind and Pooja 2010). It was found that pineapple juice (5–20%) significantly reduced immobility time in both the forced swim test and tail suspension test. It also reversed hypothermia induced by reserpine. It also inhibited the monoamine oxidase MAO-A and MOA-B activity and significantly reduced malondialdeyhde levels.

Allergy Problem

Bromelains are commonly used in pharmaceutical industries, food production and in diagnostic laboratories. Bromelains were known to cause IgE-mediated reactions of both the immediate type and the 'late phase reaction of immediate type reaction' with predominantly respiratory symptoms (Gailhofer et al. 1988). Findings indicated bromelain to be a strong sensitizer; sensitization typically occurring through inhalation and not ingestion, and bromelain allergy to be occupationally acquired.

Hypertension and Bromelain

Despites bromelain's low toxicity, caution should be exercised when administering bromelain to individuals with hypertension, since one report indicated that individuals with pre-existing hypertension might experience tachycardia following high doses of bromelain (Gutfreund et al. 1978).

Traditional Medicinal Uses

Traditional ethno-medicinal uses of the pineapple fruit include its use as a vermifuge, diuretic, abortifacient, indigestion, and for gonorrhoea; and the leaves as a purgative, anthelmintic and for venereal diseases. Unripe pineapple fruit is used as abortifacient by women and as vermifuge for children in Moluccas. In Java, as an emmenagogue. In Malacca, unripe fruit is used as a diuretic and in gonorrhoea. In India, ripe fruit juice is regarded as a diuretic and taken in large quantities causes strong uterine contractions. In Brazil, the acid of ripe fruit is deemed to be good for dyspepsia, the tryptic ferment aiding indigestion. In the Philippines, juice of ripe fruit is regarded as diuretic, gently laxative, cooling and digestive.

In Peninsular Malaysia, a sweetened decoction of the leaves was drunk for venereal diseases. Sap from the leaves was used as a purgative, and anthleminthic, and for hiccough in India. Both the root and fruit are sometimes eaten or applied topically as an anti-inflammatory and as a proteolytic agent.

Other Uses

A proteolytic enzyme bromelain extracted from stem and fruit juice are used in meat tenderizers, in chill-proofing beer, bating of hides (Heinicke and Gortner 1957), manufacturing precooked cereals, and in certain cosmetics. Only fresh, uncooked pineapple has this special ingredient. Once pineapple is cooked or canned, the special substance is destroyed. Bromelain is also nematicidal. Also the juice from canning can be is fermented for alcohol. Pineapple bran, the residue after juicing, is high in vitamin A, and is used in livestock feed. From the juice may be extracted organic acids such as citric acid, malic and ascorbic or on fermentation, alcohol.

Fine but strong fibres (bromelia fibre, crowa fibre) can be extracted from the leaves and are used for ropes, baskets, and fine textiles. In the Philippines and Taiwan, the fibres from the leaves are woven into a fine pina cloth. Pineapple is occasionally planted as a hedge.

Comments

Raw pineapple will also destroy gelatin. Use only cooked pineapple in any recipe that contains gelatin.

Selected References

Ako H, Cheung AH, Matsuura PK (1981) Isolation of a fibrinolysis enzyme activator from commercial bromelain. Arch Int Pharmacodyn Thér 254:157–167

Báez R, Lopes MTP, Salas CE, Hernández M (2007) In vivo antitumoral activity of stem pineapple (*Ananas comosus*) bromelain. Planta Med 73:1377–1383

Bahde R, Palmes D, Minin E, Stratmann U, Diller R, Haier J, Spiegel HU (2007) Bromelain ameliorates hepatic microcirculation after warm ischemia. J Surg Res 139(1):88–96

Batkin S, Taussig SJ, Szekerezes J (1988) Antimetastatic effect of bromelain with or without its proteolytic and anticoagulant activity. J Cancer Res Clin Oncol 114:507–508

Beuth J, Braun JM (2005) Modulation of murine tumor growth and colonization by bromelaine, an extract of the pineapple plant (*Ananas comosum* L.). In Vivo 19(2):483–5

Blonstein JL (1960) Control of swelling in boxing injuries. Practitioner 185:78

Blonstein JL (1969) Control of swelling in boxing injuries. Practitioner 203:206

Blumenthal M, Goldberg A, Brinkman J (eds.) (2000) Herbal medicine: expanded commission e monographs. Integrative Medicine Communications, Boston, pp 33–35

Brakebusch M, Wintergerst U, Petropoulou T, Notheis G, Husfeld L, Belohradsky BH, Adam D (2001) Bromelain is an accelerator of phagocytosis, respiratory burst and killing of *Candida albicans* by human granulocytes and monocytes. Eur J Med Res 6(5):193–200

Braun JM, Schneider B, Beuth HJ (2005) Therapeutic use, efficiency and safety of the proteolytic pineapple enzyme Bromelain-POS in children with acute sinusitis in Germany. In Vivo 19(2):417–421

Brien S, Lewith G, Walker A, Hicks SM, Middleton D (2004) Bromelain as a treatment for osteoarthritis: a review of clinical studies. Evid Based Complement Alternat Med 1(3):251–257

Brien S, Lewith G, Walker AF, Middleton R, Prescott P, Bundy R (2006) Bromelain as an adjunctive treatment for moderate-to-severe osteoarthritis of the knee: a randomized placebo-controlled pilot study. QJM 99(12):841–850

Burkill IH (1966) A dictionary of the economic products of the Malay Peninsula. Revised reprint. 2 vols. Ministry of Agriculture and Co-operatives, Kuala Lumpur, vol 1 (A–H), pp 1–1240, vol 2 (I–Z), pp 1241–2444

Chandler DS, Mynott TL (1998) Bromelain protects piglets from diarrhea caused by oral challenge with K88 positive enterotoxigenic *Escherichia coli*. Gut 43:196–202

Chobotova K, Vernallis AB, Abdul Majid FA (2009) Bromelain's activity and potential as an anti-cancer agent: current evidence and perspectives. Cancer Lett 290(2):148–156

Cirelli MG, Smyth RD (1963) Effects of bromelain anti-edema therapy on coagulation, bleeding, and prothrombin times. J New Drugs 3:37–39

Collins JL (1968) The pineapple: botany, cultivation and utilization. Hill, London, 294 pp

Desser L, Rehberger A, Kokron E, Paukovits W (1993) Cytokine synthesis in human peripheral blood mononuclear cells after oral administration of polyenzyme preparations. Oncology 50(6):403–407

Desser L, Holomanova D, Zavadova E, Pavelka K, Mohr T, Herbacek I (2001) Oral therapy with proteolytic enzymes decreases excessive TGF-beta levels in human blood. Cancer Chemother Pharmacol 47(Suppl):S10–S15

Eckert K, Grabowska E, Stange R, Schneider U, Eschmann K, Maurer HR (1999) Effects of oral bromelain administration on the impaired immunocytotoxicity of mononuclear cells from mammary tumor patients. Oncol Rep 6(6):1191–1199

Elss S, Preston C, Hertzig C, Heckel F, Richling E, Schreier P (2005) Aroma profiles of pineapple fruit (*Ananas comosus* [L.]Merr.) and pineapple products. LWT – Food Sci Technol 38(3):263–274

Engwerda CR, Andrew D, Ladhams A, Mynott TL (2001a) Bromelain modulates T cell and B cell immune responses in vitro and in vivo. Cell Immunol 210(1):66–75

Engwerda CR, Andrew D, Murphy M, Mynott TL (2001b) Bromelain activates murine macrophages and natural killer cells in vitro. Cell Immunol 210(1):5–10

Felton GE (1977) Does kinin released by pineapple stem bromelain stimulate production of prostaglandin El-like compounds? Hawaii Med J 36:39–47

Felton GE (1980) Fibrinolytic and antithrombotic action of bromelain may eliminate thrombosis in heart patients. Med Hypotheses 6(11):1123–1133

Fitzhugh DJ, Shan S, Dewhirst MW, Hale LP (2008) Bromelain treatment decreases neutrophil migration to sites of inflammation. Clin Immunol 128(1):66–74

Gailhofer G, Wilders-Truschnig M, Smolle J, Ludvan M (1988) Asthma caused by bromelain: an occupational allergy. Clin Allergy 18(5):445–450

Garg SK, Saksena SK, Chaudhury RR (1970) Antifertility screening of plants. VI. Effect five indigenous plants early pregnancy albino rats. Indian J Med Res 58(9):1285–1289

Gerard G (1972) Anti-cancer therapy with bromelain. Agressologie 13:261–274

Gläser D, Hilberg T (2006) The influence of bromelain on platelet count and platelet activity in vitro. Platelets 17(1):37–41

Guo R, Canter PH, Ernst E (2006) Herbal medicines for the treatment of rhinosinusitis: a systematic review. Otolaryngol Head Neck Surg 135(4):496–506

Gutfreund AE, Taussig SJ, Morris AK (1978) Effect of oral bromelain on blood pressure and heart rate of hypertensive patients. Hawaii Med J 37(5):143–146

Hale LP (2004) Proteolytic activity and immunogenicity of oral bromelain within the gastrointestinal tract of mice. Int Immunopharmacol 4(2):255–264

Hale LP, Greer PK, Trinh CT, Gottfried MR (2005) Treatment with oral bromelain decreases colonic inflammation in the IL-10-deficient murine model of inflammatory bowel disease. Clin Immunol 116(2):135–142

Hale LP, Fitzhugh DJ, Staats HF (2006) Oral immunogenicity of the plant proteinase bromelain. Int Immunopharmacol 6(13–14):2038–2046

Haripyaree A, Guneshwor K, Damayanti M (2010) Evaluation of antioxidant properties of phenolics extracted from *Ananas comosus* L. Not Sci Biol 2(2):68–71

Harrach T, Eckert K, Maurer HR, Machleidt I, Machleidt W, Nuck R (1998) Isolation and characterization of two forms of an acidic bromelain stem proteinase. J Protein Chem 17(4):351–361

Heinicke RM, Gortner WA (1957) Stem bromelain-a new protease preparation from pineapple plants. Econ Bot 11(3):225–234

Heinicke RM, van der Wal L, Yokohama M (1972) Effect of bromelain (Ananase) on human platelet aggregation. Experientia 28(7):844–845

Hou RC-W, Chen YS, Huang JR, Jeng KC (2006) Cross-linked bromelain inhibits -induced cytokine production involving cellular signaling suppression in rats. Lipopolysaccharide. J Agric Food Chem 54(6): 2193–2198

Howat RC, Lewis GD (1972) The effect of bromelain therapy on episiotomy wounds—a double-blind controlled clinical trial. J Obstet Gynaecol 79:951–953

Huang JR, Wu CC, Hou RC-W, Jeng KC (2008) Bromelain inhibits lipopolysaccharide-induced cytokine production in human THP-1 monocytes via the removal of CD14. Immunol Invest 37(4):263–277

Hunter RG, Henry GW, Heinicke RM (1957) The action of papain and bromelain on the uterus. Am J Obstet Gynecol 73:867–873

Izaka KI, Yamada M, Kawano T, Suyama T (1972) Gastrointestinal absorption and antiinflammatory effect of bromelain. Jpn J Pharmacol 22(4): 519–534

Jacques DT, Marc KT, Marius A (2008) Biochemical effectiveness in liver detoxication of fresh pineapple (*Ananas comosus*) with the wistar rats, previously intoxicated by Doliprane®. J Cell Anim Biol 2(2):031–035

Juhasz B, Thirunavukkarasu M, Pant R, Zhan L, Penumathsa SV, Secor ER Jr, Srivastava S, Raychaudhuri U, Menon VP, Otani H, Thrall RS, Maulik N (2008) Bromelain induces cardioprotection against ischemia-reperfusion injury through Akt/FOXO pathway in rat myocardium. Am J Physiol Heart Circ Physiol 294(3):H1365–H1370

Kalra N, Bhui K, Roy P, Srivastava S, George J, Prasad S, Shukla Y (2008) Regulation of p53, nuclear factor kappaB and cyclooxygenase-2 expression by bromelain through targeting mitogen-activated protein kinase pathway in mouse skin. Toxicol Appl Pharmacol 226(1):30–37

Kelly GS (1996) Bromelain: a literature review and discussion of its therapeutic applications. Altern Med Rev 1:243–257

Kerkhoffs GM, Struijs PA, de Wit C, Rahlfs VW, Zwipp H, van Dijk CN (2004) A double blind, randomised, parallel group study on the efficacy and safety of treating acute lateral ankle sprain with oral hydrolytic enzymes. Br J Sports Med 38(4):431–435

Klein G, Kullich W (2000) Short-term treatment of painful osteoarthritis of the knee with oral enzymes. A randomized, double-blind study versus diclofenac. Clin Drug Invest 19(1):15–23

Klein G, Kullich W, Schnitker J, Schwann H (2006) Efficacy and tolerance of an oral enzyme combination in painful osteoarthritis of the hip. A double-blind, randomised study comparing oral enzymes with non-steroidal anti-inflammatory drugs. Clin Exp Rheumatol 24(1):25–30

Knill-Jones RP, Pearce H, Batten J, Williams R (1970) Comparative trial of Nutrizym in chronic pancreatic insufficiency. Br Med J 4:21–24

Koller J, Bukovcan P, Orság M, Kvalténi R, Gräffinger I (2008) Enzymatic necrolysis of acute deep burns–report of preliminary results with 22 patients. Acta Chir Plast 50(4):109–114

Kongsuwan A, Suthiluk P, Theppakorn T, Srilaong V, Setha S (2009) Bioactive compounds and antioxidant capacities of phulae and nanglae pineapple. Asian J Food Agro-Ind 2(Special Issue):S44–S50

Luerti M, Vignali M (1978) Influence of bromelain on penetration of antibiotics in uterus, salpinx and ovary. Drugs Exp Clin Res 4(1):45–48

Ma C, Xiao SY, Li ZG, Wang W, Du LJ (2007) Characterization of active phenolic components in the ethanolic extract of *Ananas comosus* L. leaves using high-performance liquid chromatography with diode array detection and tandem mass spectrometry. J Chromatogr A 1165(1–2):39–44

Manhart N, Akomeah R, Bergmeister H, Spittler A, Ploner M, Roth E (2002) Administration of proteolytic enzymes bromelain and trypsin diminish the number of CD4+ cells and the interferon-gamma response in Peyer's patches and spleen in endotoxemic balb/c mice. Cell Immunol 215(2):113–119

Massimiliano R, Pietro R, Paolo S, Sara P, Michele F (2007) Role of bromelain in the treatment of patients with pityriasis lichenoides chronica. J Dermatol Treat 18(4):219–222

Masson M (1995) Bromelain in blunt injuries of the locomotor system. A study of observed applications in general practice. Fortschr Med 113:303–306 (In German)

Maurer HR (2001) Bromelain: biochemistry, pharmacology and medical use. Cell Mol Life Sci 58(9): 1234–1245

Metzig C, Grabowska E, Eckert K, Rehse K, Maurer HR (1999) Bromelain proteases reduce human platelet aggregation in vitro, adhesion to bovine endothelial cells and thrombus formation in rat vessels in vivo. In Vivo 13:7–12

Mhatre M, Tilak-Jain J, De S, Devasagayam TPA (2009) Evaluation of the antioxidant activity of non-transformed and transformed pineapple: a comparative study. Food Chem Toxicol 47(11):2696–2702

Milind P, Pooja G (2010) Eat pineapple a day to keep depression at bay. Int J Res Ayurveda Pharm 1(2):439–448

Morita AH, Uchida DA, Taussig SJ, Chou SC, Hokama Y (1979) Chromatographic fractionation and characterization of the active platelet aggregation inhibitory factor from bromelain. Arch Int Pharmacodyn Thér 239:340–350

Mynott TL, Guandalini S, Raimondi F, Fasano A (1997) Bromelain prevents secretion caused by *Vibrio cholerae* and *Escherichia coli* enterotoxins in rabbit ileum in vitro. Gastroenterol 113(1):175–184

Napper AD, Bennett SP, Borowski M, Holdridge MB, Leonard MJ, Rogers EE, Duan Y, Laursen RA, Reinhold B, Shames SL (1994) Purification and characterization of multiple forms of the pineapple-stem-derived cysteine proteinases ananain and comosain. Biochem J 30(3):727–735

Neubauer RA (1961) A plant protease for potentiation of and possible replacement of antibiotics. Exp Med Surg 19:143–160

Nieper HA (1977) Decrease of the incidence of coronary heart infarct by Mg- and K-orotate and bromelain. Acta Med Empirica 12:614–618

Nieper HA (1978) Effect of bromelain on coronary heart disease and angina pectoris. Acta Med Empirica 5:274–278

Ochse JJ, Bakhuizen van den Brink RC (1931) Fruits and fruitculture in the Dutch East Indies. G. Kolff & Co, Batavia-C, 180 pp

Ochse JJ, Bakhuizen van den Brink RC (1980) Vegetables of the Dutch Indies, 3rd edn. Ascher & Co., Amsterdam, 1016 pp

Onken JE, Greer PK, Calingaert B, Hale LP (2008) Bromelain treatment decreases secretion of pro-inflammatory cytokines and chemokines by colon biopsies in vitro. Clin Immunol 126(3):345–352

Pakrashi A, Basak B (1976) Abortifacient effect of steroids from *Ananas comosus* and their analogues on mice. J Reprod Fertil 46(2):461–462

Porcher MH et al (1995 – 2020) Searchable world wide web multilingual multiscript plant name database. Published by The University of Melbourne, Australia http://www.plantnames.unimelb.edu.au/Sorting/Frontpage.html

Purseglove JW (1972) Tropical crops. Monocotyledons. 1 & 2. Longman, London, 607 pp

Rimoldi R, Ginesu F, Giura R (1978) The use of bromelain in pneumological therapy. Drugs Exp Clin Res 4:55–66

Rosenberg L, Lapid O, Bogdanov-Berezovsky A, Glesinger R, Krieger Y, Silberstein E, Sagi A, Judkins K, Singer AJ (2004) Safety and efficacy of a proteolytic enzyme for enzymatic burn debridement: a preliminary report. Burns 30(8):843–850

Rovenska E, Svik K, Stancikova M, Rovensky J (2001) Inhibitory effect of enzyme therapy and combination therapy with cyclosporin A on collagen-induced arthritis. Clin Exp Rheumatol 19(3):303–309

Ryan RE (1967) A double-blind clinical evaluation of bromelains in the treatment of acute sinusitis. Headache 7:13–17

Schafer A, Adelman B (1985) Plasmin inhibition of platelet function and of arachidonic acid metabolism. J Clin Invest 75:456–461

Secor ER Jr, Carson WF 4th, Cloutier MM, Guernsey LA, Schramm CM, Wu CA, Thrall RS (2005) Bromelain exerts anti-inflammatory effects in an ovalbumin-induced murine model of allergic airway disease. Cell Immunol 237(1):68–75

Secor ER Jr, Singh A, Guernsey LA, McNamara JT, Zhan L, Maulik N, Thrall RS (2009) Bromelain treatment reduces CD25 expression on activated CD4+ T cells in vitro. Int Immunopharmacol 9(3):340–346

Seligman B (1962) Bromelain: an anti-inflammatory agent. Angiology 13:508–510

Seligman B (1969) Oral bromelains as adjuncts in the treatment of acute thrombophlebitis. Angiology 20(1):22–26

Seltzer AP (1967) Adjunctive use of bromelains in sinusitis: a controlled study. Eye Ear Nose Throat Mon 46:1281–1288

Shahid SK, Turakhia NH, Kundra M, Shanbag P, Daftary GV, Schiess W (2002) Efficacy and safety of phlogenzym–a protease formulation, in sepsis in children. J Assoc Physicians India 50:527–531

Sripanidkulchai B, Wongpanich V, Laupattarakasem P, Suwansaksri J, Jirakulsomchok D (2001) Diuretic effects of selected Thai indigenous medicinal plants in rats. J Ethnopharmacol 75(2–3):185–190

Takashi A, Katsumi U (2002) Volatile components of ripened pineapple (*Ananas comosus* [L.] Merr). Koryo, Terupen oyobi Seiyu Kagaku ni kansuru Toronkai Koen Yoshishu 46:31–33 (In Japanese)

Takeoka G, Butter RG (1989) Volatile constituents of pineapple (*Ananas Comosus* [L.] Merr.). In: Teranishi R, Buttery RG, Shahidi F (eds.) Flavor chemistry trends and developments. American Chemical Society, Washington, pp 223–237

Tassman GC, Zafran JN, Zayon GM (1964) Evaluation of a plant proteolytic enzyme for the control of inflammation and pain. J Dent Med 19:73–77

Tassman GC, Zafran JN, Zayon GM (1965) A double-blind crossover study of a plant proteolytic enzyme in oral surgery. J Dent Med 20:51–54

Taub SJ (1966) The use of ananase in sinusitis: a study of 60 patients. Eye Ear Nose Throat Mon 45:96–98

Taub SJ (1967) The use of bromelains in sinusitis: a double-blind clinical evaluation. Eye Ear Nose Throat Mon 46:361–362

Taussig S (1980) The mechanism of the physiological action of bromelain. Med Hypotheses 6:99–104

Taussig SJ, Batkin S (1988) Bromelain, the enzyme complex of pineapple (*Ananas comosus*) and its clinical application. An update. J Ethnopharmacol 22(2):191–203

Taussig SJ, Nieper HA (1979) Bromelain: its use in prevention and treatment of cardiovascular disease, present status. J Int Assoc Prev Med 6:139–151

Teai T, Claude-Lafontaine A, Schippa C, Cozzolino F (2001) Volatile compounds in fresh pulp of pineapple (*Ananas comosus* [L.] Merr.) from French Polynesia. J Essent Oil Res 13(5):314–318

Tilwe GH, Beria S, Turakhia NH, Daftary GV, Schiess W (2001) Efficacy and tolerability of oral enzyme therapy as compared to diclofenac in active osteoarthrosis of knee joint: an open randomized controlled clinical trial. J Assoc Physicians India 49:617–621

Tinozzi S, Venegoni A (1978) Effect of bromelain on serum and tissue levels of amoxicillin. Drugs Exp Clin Res 4(1):39–44

Tochi BN, Wang Z, Xu SY, Zhang W (2008) Therapeutic application of pineapple protease (bromelain): a review. Pak J Nutr 7(4):513–520

Tokitomo Y, Steinhaus M, Büttner A, Schieberle P (2005) Odor-active constituents in fresh pineapple (*Ananas comosus* [L.] Merr.) by quantitative and sensory evaluation. Biosci Biotechnol Biochem 69(7):1323–1330

Tysnes BB, Maurer HR, Porwol T, Probst B, Bjerkvig R, Hoover F (2001) Bromelain reversibly inhibits invasive properties of glioma cells. Neoplasia 3(6):469–479

U.S. Department of Agriculture, Agricultural Research Service (2010) USDA National Nutrient Database for Standard Reference, Release 23. Nutrient Data Laboratory Home Page, http://www.ars.usda.gov/ba/bhnrc/ndl

Uhlig G, Seifert J (1981) The effect of proteolytic enzymes (traumanase) on posttraumatic edema. Fortschr Med 99:554–556 (In German)

Walker JA, Cerny FJ, Cotter JR, Burton HW (1992) Attenuation of contraction-induced skeletal muscle injury by bromelain. Med Sci Sports Exerc 24:20–25

Walker AF, Bundy R, Hicks SM, Middleton RW (2002) Bromelain reduces mild acute knee pain and improves well-being in a dose-dependent fashion in an open study of otherwise healthy adults. Phytomed 9(8):681–686

Wang W, Ding Y, Xing DM, Wang JP, Du LJ (2006) Studies on phenolic constituents from leaves of pineapple (*Ananas comosus*). Zhongguo Zhong Yao Za Zhi 31(15):1242–1244 (In Chinese)

Wee YC, Thongtham MLC (1991) *Ananas comosus* (L.) Merr. In: Verheij EWM, Coronel RE (eds.) Plant resources of South-East Asia No. 2: edible fruits and nuts. Pudoc, Wageningen, pp 66–71

Wen S, Huang TH, Li GQ, Yamahara J, Roufogalis BD, Li Y (2006) Bromelain improves decrease in defecation in postoperative rats: modulation of colonic gene expression of inducible nitric oxide synthase. Life Sci 78(9):995–1002

Xie WD, Xing DM, Sun H, Wang W, Ding Y, Du LJ (2005) The effects of *Ananas comosus* L. leaves on diabetic-dyslipidemic rats induced by alloxan and a high-fat/high-cholesterol diet. Am J Chin Med 33(1):95–105

Xie W, Wang W, Su H, Xing D, Pan Y, Du L (2006) Effect of ethanolic extracts of *Ananas comosus* L. leaves on insulin sensitivity in rats and HepG2. Comp Biochem Physiol C Toxicol Pharmacol 143(4):429–435

Xie W, Wang W, Su H, Xing D, Cai G, Du L (2007) Hypolipidemic mechanisms of *Ananas comosus* L. leaves in mice: different from fibrates but similar to statins. J Pharmacol Sci 103(3):267–274

Yamada F, Takahashi N, Murachi T (1976) Purification and characterization of a proteinase from pineapple fruit, fruit bromelain FA2. J Biochem 79(6):1223–1234

Zatuchni GI, Colombi DJ (1967) Bromelains therapy for the prevention of episiotomy pain. Obstet Gynecol 29:275–278

Canarium decumanum

Scientific Name

Canarium decumanum Gaertn.

Synonyms

Canariopsis decumana Bl. ex Miq., *Pimela decumana* Blume

Family

Burseraceae

Common/English Names

Pig Canary

Vernacular Names

Indonesia: Minang (Papua), Kenari Sabrang, Kenari Besar, Jilapat, Kedungdong, Jelemuq, Pamatodon;
Malaysia: Pomotodon, Jelamu (Sabah), Pomotodon (Sarawak).

Origin/Distribution

The species is indigenous to Malesia: East Malaysia (Sabah, Sarawak), Indonesia – (Kalimantan, Moluccas, Papua) and Papua New Guinea.

Agroecology

Strictly a warm tropical species, *C. decumanum* occurs naturally in the hot and wet tropical rainforest of Malesia.

Edible Plant Parts and Uses

The large kernels are edible raw or roasted.

Botany

A large canopy, evergreen, tall, perennial tree with a cylindrical bole 20–25 m long and 50–150 cm diameter and large thick buttresses and reaching heights greater than 50 m (Plates 1–2). The bark is grey, rough and flaky and the bark with resinous ducts with exudates that turns red on exposure to air. Leaves spirally arranged, imparipinnate with 5–11 opposite leaflets, ovate-oblong, base rounded apex long acuminate, margin obscurely dentate-serrate, shortly pubescent to glabrous, 8–26 cm by 4–8 cm, venation pinnate, upper surface green lower surface greenish-brown because of the dense reddish hairs. Flowers bisexual in much branched, pubescent, odoriferous. Axillary panicle near the top of branchlets. Flower has three sepals, three free, yellowish-green petals, six free stamens, ovary inferior, carpels joined, three-loculed and a solitary style. Fruit is a large grey, indehiscent, ovoid

Plate 1 Tree habit

Plate 2 Strong and well developed buttresses

Plate 3 Tree label at the Bogor Botanic Gardens, Java

drupe, 7–8.5 cm by 4.5–6 cm, seed (pyrene) ribbed, 1 cm across.

Nutritive/Medicinal Properties

No published data is available on its nutritive value or medicinal uses.

Other Uses

The species is an occasional timber species; it provides a good and durable timber that is used for making boats, knives and other implements (Plate 3). The resin is used to glue metals, wood and when diluted in coconut oil is used for caulking of boats.

Comments

See also notes on other related, edible *Canarium* species.

Selected References

Backer CA, Bakhuizen van den Brink Jr RC (1965) Flora of Java (spermatophytes only), vol 2. Wolters-Noordhoff, Groningen, 641 pp

Conn BJ, Damas KQ (2006+) Guide to trees of Papua New Guinea (http://www.pngplants.org/PNGtrees)

Lanoeroe S, Kesaulija EM, Rahawarin YY (2005) Pemanfaatan jenis tumbuhan berkayu sebagai bahan baku perahu tradisional oleh Suku Yachai di Kabupaten Mappi. Biodiversitas 6(3):212–216

Leenhouts PW (1956) Burseraceae. In: van Steenis CGGJ (ed.) Flora malesiana, Series I. 5(2) (pp. 209–296), Noordhoff-Kolff Djakarta, 595 pp

Slik JWF (2006) *Trees of Sungai Wain*. Nationaal Herbarium Nederland. http://www.nationaalherbarium.nl/sungaiwain/

Canarium indicum

Scientific Name

Canarium indicum L.

Synonyms

Canarium amboinense Hochreut., *Canarium commune* L., *Canarium grandistipulatum* Lauterb., *Canarium mehenbethene* Gaertn., *Canarium moluccanum* Blume, *Canarium nungi* Guill., *Canarium shortlandicum* Reching., *Canarium subtruncatum* Engl., *Canarium zephyrinum* Rumphius.

Family

Burseraceae

Common/English Names

Blume Galip, Canarium Almond, Canarium Nut, Galip, Galip Nut, Java Almond, Java Olive, Kanari, Nangai Nut, Ngali Nut

Vernacular Names

French: Noix De Nangaille, Ngoli, Galap, La Nangaille, Noix De Kanari;
German: Indisches Kanaribaum, Kanarinuß, Galipnuß;
India: Agarbati, Dhup (Hindi);
Indonesia: Jal, Jar (Ambon), Kanari Bagéa (Maluku), Kenari Ambon (Sundanese), Buah Kenari;
Malaysia: Canari, Ngali, Nangai, Kenari, Pokok Kenari;
Papua New Guinea: Baga, Galip, Galip Nut (General), Hinuei (New Ireland), a ngallip Lawele (New Britain);
Solomon Islands: Eghe (Savo Island), Nina, Ninge, Voia (Santa Cruz), Angari (Santa Ana), Ngari (Kausage/Simbo and Varisi), Ngoeta (Marovo), Nolepo (Garciosa Bay), Nyia Nyinge (Ayiwo), Okete (Roviana), Sela (Guadalcanal), Voi'A (Vaiakau) Ngali, Ngali Nut (Kwara'Ae), Aneri, Kwakora, Olaipa;
Vanuatu: Nangrau (Aneityum Island/Anelghowat Village), Nangai (Bislama), Bunnige, Punnige, Nige Kava, Nige Karia (Epi Island/Moriu), Ngapor, Ngaqov (Gaua, Banks Group/Lambot, Namasari Villages), Negerdove (Loh Island/Lungharagi Village), Ngaetua (Maewo Island/Naone Village), Vungaingai, Vungigae (Malo Island/Naviaru Village)Nanae, Vanae (Santo Island/Sarete Village), Nangae (Santo Island/Narango Village), Ngna, Nangan, Nanga (Santo Island/Hog Harbour), Ngeta, Ngev Tentel (Vanua Lava, Banks Group/Mosina Village).

Origin/Distribution

Galip is native to the humid, lowland zones of eastern Indonesia (Maluku, Ambon, West Papua), Papua New Guinea, the Solomon

Islands, and Vanuatu. It is frequently cultivated in Melanesia. Also cultivated in Australia, Taiwan, Fiji, Hawaii, Honduras and Trinidad.

Agroecology

In its native area, it occurs in the humid, wet lowland rainforest, secondary forest, old garden areas, and it is widely planted around villages and settlements at altitudes of 0–1,000 m. Galip is well-adapted to the hot tropical conditions with mean annual temperatures of 25–32°C, with summer or uniform annual rainfall of 1,800–4,000 mm with optimum range usually between 2,500 and 3,500 mm. It is tolerant of annual precipitation of 6,000 mm. It also grows in sub-tropical conditions with temperatures from 17°C to 25°C and is sensitive to low cold temperatures and frost. It prefers medium to heavy-textured soils such as loams, sandy clay loams, clays, clay loams, and sandy clays, of moderate to high fertility, well-drained to slightly impeded drainage, with good organic matter content and a pH of 4.5–6.5, but will tolerate pH levels up to pH 7.4. Galip is intolerant of shallow, infertile, or saline soils. It performs best in full sun but can tolerate partial shade of 20–70%; juvenile trees need to be partially shaded. It can withstand short periods of drought, strong winds and salt sprays.

Edible Plant Parts and Uses

In Melanesia except in Fiji, oil-rich seed kernels are highly esteemed as a food in the local diet. The kernels after removal of the seed shell is eaten raw, baked or roasted. It is often eaten as snack food, or incorporated in other food with some staple root crops, added to soups, or eaten with megapode eggs in the Solomon Islands. In Vanuatu, they are mixed with tuber puddings. The kernels can be dried by smoking and be kept for many months for future use. Crushed kernels are eaten as toppings with ice-cream. Oil from the seeds is sometimes used as a substitute for coconut oil for cooking food. The oil could potentially be blended with other oils.

Botany

An evergreen, large, monoecious, or dioecious tree, up to 40 m tall and with a fluted, buttressed, 1m diameter trunk and heavy lateral branches and a dense canopy. Leaves are imparipinnate, three to seven jugate, bright green. Leaflet is oblong-obovate to oblong-lanceolate, large, 7–28 cm by 3.5–11 cm, obtusely acuminate with sub-undulating, entire margin. Pseudostipules are persistent, large, leafy, ovate and serrate-dentate with fringed margins. Flowers arise in terminal panicles, with deciduous stipule and bract at the base of flower. Flowers are small, 1 cm across, yellowish-white, perianth trimerous with pubescent outer surface; stamens six, joined and free of the perianth; ovary is superior, three-loculed with one style. Infructescences are large with up to 30 fruits borne on pendulous pedicels (Plate 1). Fruit is an ovoid to elliptic-oblong drupe, 3–6 cm × 2–3 cm, generally green when unripe (Plate 2),

Plate 1 Terminal infructescence and foliage of Galip nut

Plate 2 Infructescence of unripe galip fruit

Plate 3 Ripe black, elliptic-oblong galip fruit with intact flesh and with flesh removed (*left*)

Plate 4 Stony, hard nut-in – shell of Galip with flesh removed

Plate 5 Galip kernels trigonous and with brown testa

turning deep dark green to black or blue-black when ripe (Plate 3). The nut in shell is stony hard, rounded or 3–6 sided in cross section (Plate 4). Seed (kernel) is usually trigonous, 1 cm across, with brown testa (Plate 5).

Nutritive/Medicinal Properties

The nutrient composition of raw galip nut per 100 g edible portion (exclude hard shell) was reported as: energy 439 kcal/1,838 kJ, water 35.4 g, protein 8.2 g, fat 45.9 g, sugar 0.2 g, starch 0.3 g, ash 2.6 g, fibre 10.6 g, β-carotene equivalent 165 μg, riboflavin 0.06 mg, thiamin 0.13 mg, niacin 1.7 mg, vitamin C 8 mg, Na 18 mg, K 627 mg, Ca 44 mg, Fe 3.5 mg, Mg 284 mg, Zn 2.4 mg, Cu 1.6 mg, and Mn 1.1 mg (English et al. 1996).

Galip nut was found to contain 70–80% oil, 13% protein, 7% starch (Howes 1948; Macrae et al. 1993). Kernel oil of *C. indicum* was found to contain about 50% saturated fat (34% palmitic and 13% stearic), 38% monosaturated (oleic) and 14% polyunsaturated (linoleic) (Nevenimo et al. 2007). Galip kernels were found to be rich in oils (67–75.4%) and to contain almost equal proportion of saturated and unsaturated fatty acid (Leakey et al. 2008). Strong positive relationships were found between anti-oxidant activity measured in mg ascorbate equivalents and phenolic content expressed in mg catechin equivalents with r^2 of 0.843, fat content versus energy r^2 of 0.0993, fat content versus carbohydrate content r^2 of 0.823. There was little variation in protein content but large variation in vitamin E contents especially in β-tocopherol among the kernels of trees sampled from various areas in Papua New Guinea. Besides β-tocopherol which predominated in kernels, α-, δ- and γ- tocopherols were also present in kernels. Galip kernels also possessed anti-inflammatory activity in-vitro as demonstrated by the inhibition of prostaglandin (PGE_2) production in 3T3 Swiss Albino fibroblast cells by the kernel oil (Leakey et al. 2008). The kernel oil inhibited PGE_2 production comparable to aspirin.

Galip kernel oil is commonly used for cooking but have application as a medicinal product particularly in cosmetic and skin care product. *Canarium* kernel oil has been included in a product marketed as an anti-inflammatory while patents on a product for the prevention and treatment of arthritis have been registered in a number of countries.

Oily extracts of *Canarium* are already known as anti-inflammatory agents for combating wrinkles and loss of skin firmness. Maestro et al. (2009) reported that a cosmetic skincare Canarium product, intended to prevent and/or treat at least one cutaneous sign of aging was under patent consideration then. The treatment comprised the topical application to the skin of a composition containing at least one active agent capable of stimulating tensin 1 expression. The active agents include extracts (in particular of resin or gum) of elemi, i.e. of *Canarium indicum* (or *Canarium commune*) or of *Canarium luzonicum*.

In the western Solomon Islands, the bark has been reported to be used in traditional medicine for chest pains.

Other Uses

Galip is occasionally used as a shade tree in plantations, avenue tree and as windbreak. Its timber is suitable for light construction, canoes, boats, tools and crafts. Some trees have a decorative wood with veneer potential. These timbers are suitable for: general construction, mouldings, interior finish, veneers, boxes, bowls, utility furniture, flooring, lining, joinery, doors and window frames, cabinet work and chip board, but is not suitable for exterior uses unless treated with preservatives. The wood is soft and also used as firewood. The rotten logs provide a source of edible insect larvae. In the past, oil from the seed kernel was used for lighting. The resinous exudate from the trunk is used in caulking of canoes. The hard shells of the nut are used for fuel and for carving into jewellery. Also it can be burned in kilns to produce clean, dense and high-grade charcoal fuel, which can be refined to 'activated carbon' for pharmaceutical uses. Shells can also be used as bedding for horticultural crops. In Papua New Guinea, the shell is used to make pipes for tobacco smoking. The testa can be used as an ingredient of animal foods.

C. commune is one of 11 seed oil species that have great potential for biodiesel as its fatty acid methyl esters in its oil meet all of the major biodiesel requirements in the USA (ASTM D 6751–02, ASTM PS 121–99), Germany (DIN V 51606) and the European Union (EN 14214) (Azam et al. 2005).

Comments

Canarium indicum has two recognised botanical varieties: the type variety *indicum* and another variety *platycerioideum*. The latter, uncommon, variety occurs in West Papua (Indonesia) and has larger leaves and fruits.

Selected References

Azam MM, Waris A, Nahar NM (2005) Prospects and potential of fatty acid methyl esters of some non-traditional seed oils for use as biodiesel in India. Biomass Bioenergy 29(4):293–302

Backer CA, Bakhuizen van den Brink Jr RC (1965) Flora of Java (spermatophytes only), vol 2. Wolters-Noordhoff, Groningen, 641 pp

Bourke M (1996) Edible indigenous nuts in Papua New Guinea: their potential for commercial development. Aust New Crops Newsl 5:3–4

Burkill IH (1966) A dictionary of the economic products of the Malay Peninsula. Revised reprint. 2 vols. Ministry of Agriculture and Co-operatives, Kuala Lumpur, vol 1 (A–H), pp 1–1240, vol 2 (I–Z), pp 1241–2444

Conn BJ, Damas KQ (2006+) Guide to trees of Papua New Guinea (http://www.pngplants.org/PNGtrees)

English RM, Aalbersberg W, Scheelings P (1996) Pacific Island foods – description and nutrient composition of 78 local foods. IAS Technical Report 96/02, ACIAR Report 9306. Institute of Applied Science, University of the South Pacific, Suva, Fiji, 94 pp

Evans BR (1991) The Agronomy on Ngali nut (*Canarium spp.*) in Solomon Islands. Research Bulletin No. 9. Division of Research, Dodo Creek Research Station, Honiara, Solomon Islands, 27 pp

Evans BR (1991) The production, processing and marketing of Ngali nut (*Canarium spp.*) in Solomon Islands. Report to UK Overseas Development Administration, London. Dodo Creek Research Station, Honiara, Solomon Islands, 37 pp

Howes FN (1948) Nuts. Their production and everyday use. Faber and Faber, London, 264 pp

Jansen PCM, Jukema J, Oyen LPA, van Lingen TG (1991) *Canarium indicum* L. In: Verheij EWM, Coronel RE (eds.) Plant resources of South-East Asia, No. 2. Edible fruits and nuts. Prosea Foundation, Bogor, p 322

Leakey R, Fuller S, Treloar T, Stevenson L, Hunter D, Nevenimo T, Binifa J, Moxon J (2008) Characterization of tree-to-tree variation in morphological, nutritional and medicinal properties of *Canarium indicum* nuts. Agrofor Syst 73(1):77–87

Leenhouts PW (1956) Burseraceae. In: van Steenis CGGJ (ed.) Flora malesiana. Series I. 5(2) (pp 209–296), Noordhoff-Kolff Djakarta, 595 pp

Leenhouts PW (1965) Notes on *Canarium* (Burseraceae) in the Solomon Islands. Blumea 13(1):163–166

Lemmens RHMJ, Soerianegara I, Wong WC (eds.) (1995) Plant resources of South-East Asia No 5(2). Timber trees: minor commercial timbers. Backhuys Publishers, Leiden

Macrae R, Robinson RK, Sadler MJ (1993) Encyclopaedia of food science. Food technology and nutrition, vol 5, Malt-pesticides and herbicides. Academic, London

Maestro Y, Saintigny G, Bernard FX (2009) Cosmetic use of an active agent capable of stimulating tensin 1 expression. www.faqs.org/patents/app/20090028897

Nevenimo T, Moxon J, Wemin J, Johnston M, Bunt C, Leakey R (2007) Domestication potential and marketing of *Canarium indicum* nuts in the Pacific: 1. A literature review. Agrofor Syst 69(2):117–134

Thomson LAJ, Evans B (2006) *Canarium indicum* var. *indicum* and *C. harveyi* (canarium nut), ver. 2.1. In: Elevitch CR (ed.) Species profiles for Pacific Island Agroforestry. Permanent Agriculture Resources (PAR), Holualoa, Hawai'i. http://www.traditionaltree.org

Walter A, Sam C (1996) Indigenous nut trees in Vanuatu: ethnobotany and variability. In: Stevens ML, Bourke RM, Evans BR (eds) South Pacific indigenous nuts. ACIAR Proceedings No.69, pp 56–66

Yen DE (1993) The origins of subsistence agriculture in Oceania and the potentials for future tropical food crops. Econ Bot 47:3–14

Yen DE (1994) Melanesian arboriculture: historical perspectives with emphasis on the genus *Canarium*. In: Stevens ML, Bourke RM, Evans BR (eds) South Pacific indigenous nuts. ACIAR Proceedings No.69, pp 36–44

Canarium odontophyllum

Scientific Name

Canarium odontophyllum Miq.

Synonyms

Canarium beccarii Engl. in DC, *Canarium multifidum* H.J. Lam, *Canarium palawanense* Elm.

Family

Burseraceae

Common/English Names

Dabai, Sibu Olive

Vernacular Names

Borneo: Bundui-Bundui, Dabai (Sarawak), Dabang, Dabu, Danau Majang, Dawai, Kembayau (Brunei), Kembayau, Kembayu (Sabah), Defai (Punan Malinau, East Kalimantan), Bua Pasan (Lundaye, East Kalimantan), Bua Labai (Merap, East Kalimantan), Keravei (Kenyah Uma', East Kalimantan), Kedongdong, Kumbayan, Kurihang, Saluan.

Origin/Distribution

Dabai is indigenous to Borneo – the area of greatest diversity is located in Brunei Darussalam and the Brunei Bay region. In Sarawak dabai can be found along the riverbansk in Sibu, Kapit and Sarikei divisions. It is also found in Pahlawan and Sumatra.

Agroecology

Dabai occurs in undisturbed lowland forests up to 700 m altitude; mostly on hillsides and ridges, but sometimes also in swamps or along rivers in its native tropical range. Quite often, dabai is found on limestone. In secondary forests, it is usually present as a pre-disturbance remnant tree.

Edible Plant Parts and Uses

The fruit is edible. The whole fruit is soaked 5–10 minutes in hot water to soften the mesocarp and eaten with sugar or with a little soy sauce or salt, with a meal or as a savoury snack. The fruit represents a rich energy source with good amounts of oils and proteins. The kernel or nut containing the whitish compressed cotyledon is also edible. It has a fine flavour, crunchy with a rich mealy, oily taste.

Dabai's popularity as an exotic and health fruit among the local populace in Sarawak has increased over the years with the Agriculture Department producing recipes for its use in pizza, fried rice, mixed vegetables, maki (dried seaweed roll), pickles as well as desserts, mayonnaise, chips, ice-cream and salad sauce based on the fruit. The Department has also developed a freezing technique for the highly perishable fruit and is now able to retain its freshness up to 7 days and dabai fruit could last up to 1 year when kept in cold storage (Lau and Fatimah 2007). In Sarawak, the fruit has been used by local restaurants as ingredient in their dishes.

Botany

A very handsome, medium-sized, upright tree up to 36 m high with 86 cm trunk diameter and branched canopy at the top of the trunk (Plates 1–2). Leaves are large, thin and finely pubescent, pinnate leaves (Plate 3), shallowly odd pinnate with serrated margin and penniveined; lowest pair of leaflets often reduced to pseudo-stipules with fimbriate margin (Plate 4). New foliage emerges in flushes of green or red velvet. Branchlets and petioles have resinous ducts. Panicles are terminal, female panicles are short and with fewer flowers. Flower is white-yellow, 10 mm across, with dentate-lobed, campanulate calyx and corolla pubescent outside consisting of five petals. Male flower has six stamens; female flower has six staminodes, a three-loculed ovary, short style with three-lobed stigma. Fruit is a fleshy oblong to ellipsoidal drupe, (35–40 mm long by 20–25 mm wide), white when immature, turning purplish-black when ripe with a thin, edible skin (Plates 5–8). The white or yellow flesh inside is 4–7 mm thick (Plates 5–6) and covers a single large, three-angled pointed seed. Terminal clusters of olive-like fruits are held above the dark green foliage (Plate 5).

Plate 2 Crown of dabai tree

Plate 1 A cultivated dabai tree

Plate 3 Odd-pinnate leaf of dabai

Plate 4 Distinctive, fimbriate stipules

Plate 6 Olive-like, developing dabai fruit

Plate 7 Ripe, purplish-black dabai fruit

Plate 5 Terminal panicles of dabai fruit amongst the large, pinnate leaves

Plate 8 Ripe dabai fruit on sale in the Sibu market in Sarawak

Nutritive/Medicinal Properties

The fruit is delicious and nutritious and contains protein, fat and carbohydrate. The proximate composition per 100 g edible portion of the fruit pulp was reported as : energy 339 kcal, moisture 41.3%, protein 3.8%, fat 26.2%, carbohydrate 22.1%, crude fibre 4.3%, ash 2.3%, P 65 mg, K 810 mg, Ca 200 mg, Mg 106 mg, Fe 1.3 mg, Mn 8ug, Cu 7.0ug, Zn 4.7ug (Voon and Kueh 1999).

The physical properties of skin, flesh, and kernel of *Canarium odontophyllum* fruit were studied by Azrina et al. (2009). Averaged values of big and small fruits were determined for length (4.10; 3.74 cm), width (2.79; 2.40 cm), thickness (0.50; 0.40 cm), sphericity index (67.37; 41.28%), aspect ratio (43.62; 65.79%), whole fruit mass (18.28; 12.73 g), skin mass (1.02; 0.86 g), flesh mass (11.22; 7.81 g) and kernel mass (6.79; 5.84 g).

The chemical characteristics of the oil extracted from dabai flesh were 7.99 mEq peroxide/kg fat for peroxide value, 8.4 mg KOH for acid value, 52.8 g iodine/100 g fat for iodine value and 181.62 mg KOH/g fat for saponification value (Alina and Azrina 2008). Whereas the characteristics for the oil extracted from kernel were 16.0 mEq peroxide/kg fat for peroxide value, 5.55 mg KOH for acid value, 37.7 g iodine/100 g fat for iodine value and 171.33 mg KOH/g fat for saponification value. For colour determination, the L, a, b values for dabai flesh oil were 75.5, 8.98, 49.33, respectively while the values for dabai kernel oil were 64.12, −3.05, 32.91for L, a, b, respectively. The vitamin C content in the flesh and skin were 2.59 mg vitamin C/100 g and 5.85 mg vitamin C/100 g, respectively. The chemical results obtained indicated that both oils had similar values to palm oil. The vitamin C content of dabai fruit was shown to be similar to that of the olive fruit. Hence, the researchers asserted that these extracted oils could be recommended for use as margarine and as edible oil for commercial use.

The content of total saturated fatty acid was found to be 44.4% in dabai (*Canarium odontophyllum*) pulp, kernel (60.8%), palm oil (47.9%) and olive (25.5%) oils (Azrina et al. 2010b). Palmitic, myristic, oleic and linoleic acids were detected in dabai pulp oil (36.1, 5.8, 41.5, 11.8%) dabai kernel oil (46.4, 9.3, 35.1, 2.8%), palm oil (33.8, 9.2, 39.7, 10.9%) and olive oil (9.9, 12.9, 64.4, 5.1%) respectively. Vitamin E was not detected in the dabai pulp oil, while palm oil had the highest vitamin E content, followed by dabai kernel oil and olive oil. Further, the physicochemical characteristics of dabai oils exhibited better quality than the studied commercial oils. The results also showed that dabai pulp and kernel oils had good fatty acid composition and a high potential to be developed into healthy cooking oils.

Carotenoids were isolated and identified from peel, pulp and seed fractions of *Canarium odontophyllum*, and their antioxidant capacities were evaluated (Prasad et al. 2011). All -trans-β-carotene was present in a large amount in peel (69.5 mg/kg), followed by pulp (31.1 mg/kg) and seed (15.1 mg/kg). Additionally, 15-cis-β-carotene, 9-cis-β-carotene and 13-cis-β-carotenes were also major contributors to carotenoid contents in peel, pulp and seed fractions. Pulp exhibited excellent β-carotene bleaching activity, significantly greater than peel and seed; high 1,1-diphenyl-2-picrylhydrazyl (DPPH) radical-scavenging activity, whereas peel showed significantly higher scavenging activity of 2,2'-azino-bis (3-ethylbenzthiazoline-6-sulphonic acid) (ABTS) radicals. All the extracts exhibited good inhibitory effect against hydrogen peroxide-induced haemoglobin oxidation, ranging from 45.3% to 59.7%. The results suggested that dabai could be explored and be promoted as a potential source of natural antioxidants.

Average antioxidant values of fermented dabai flesh (FCO) (Nur Naazira Iman and Norhaizan 2008) were reported in [mM Trolox equivalent (TE)/g] activity of FCO as 23.59, 14.75 and 11.37 as evalauted by the DPPH (2, 2-diphenyl-1-picryl-hydrazyl), BCB (β-carotene bleaching), and FRAP (ferric reducing/antioxidant power) assays respectively. Total phenolic content (TPC) and amount of phytic acid for FCO was 2.39 mg GAE/g and 0.04 g/100 g respectively. The value of phytic acid in FCO was lower when compared with *Glycine max* (Soya bean) (1.31 g/100 g) and *Arachis hypogea* (groundnut) (1.42 g/100 g). Antioxidant activity of FCO was significantly lower than BHT (butylated hydroxytolune) in DPPH assay. There was no significant difference between antioxidant activity of FCO and BHT in BCB and FRAP assay. Findings obtained showed that the antioxidant activity of FCO was high and comparable to BHT. Total phenolic content and amount of phytic acid for fermented dabai flesh was 2.39 mg GAE/g and 0.04 g g/100 g respectively. The value of phytic acid in fermented dabai was lower compared with *Glycine max* (soya bean) (1.31 g/100 g) and *Arachis hypogea* (groundnut) (1.42 g/100 g). Besides, the low amount of phytic acid in fermented dabai might be due to fermentation process that was known to destroy the phytic acid.

The average antioxidant properties (mM TE/g FM) in skin, flesh, and kernel of *Canarium odontophyllum* were 16.46, 20.54 and 8.89, respectively by DPPH assay; 151.24, 70.58 and 5.65, respectively by FRAP assay; and 47.9, 11.61 and 3.00, respectively by β-Carotene bleaching method (Azrina et al. 2010a). The averaged OH scavenging activity (mg DMSOE/mg FM) in skin, flesh, and kernel of *Canarium odontophyllum* were 43.33, 7.81 and 3.31, respectively. The averaged total phenolic content (mg GAE/100 g FM) were 387.5, 267.0 and 51.0 for skin, flesh, and kernel respectively. Antioxidant activities were positively correlated with the total phenolic content.

In another study, antioxidant capacities of ethyl-acetate, butanol, and water fractions of peel, pulp, and seeds of *Canarium odontophyllum* were determined using various in-vitro antioxidant models. Ethyl-acetate fraction of dabai fruit peel exhibited the highest total phenolic, total flavonoid content, and antioxidant activities compared to pulp, seeds, and other solvent fractions (Prasad et al. 2010). Great variation in phenolic content was observed in peel, pulp, and seed fractions of *Canarium odontophyllum* ranging from 68–10 mg GAE/g in ethyl acetate fractions, 35–4 mg GAE/g in butanol fractions, 18–3 mg GAE/g in water fractions. Ethyl acetate fraction of peel (EAFPE) exhibited the highest scavenging activity (95.5%) compared to other fractions at a concentration of 40 μg/ml. All the fractions showed a good total antioxidant capacity, which was concentration dependent. The total antioxidant activity of EAFPE at 20 μg/ml was 0.17 significantly higher than butanol and water fractions. *Canarium odontophyllum* peel exhibited the highest reducing power Ferric Reducing Antioxidant Power (FRAP). Followed by seed and pulp. Among dabai peel, the ethyl acetate fraction exhibited a strong reducing power and was higher than BHT. The result showed that all the fractions had good protective effect against hydrogen peroxide induced hemoglobin oxidation. The percentage inhibition of hemoglobin oxidation varied significantly in all the tested fractions with the greatest activity (50%) observed in ethyl acetate seed fraction, while the least (33%) was in water pulp fraction. A positive correlation ($R^2=0.775$) was found between total phenolic contents and DPPH radical scavenging activity. Similar observations ($R^2=0.804$) was also observed for FRAP and total phenolic content. The results suggested that the peel extract of *C. odontophyllum* could be potentially used as a source of natural antioxidant agent.

Other Uses

Soap has been developed from Dabai fruit. The tree is a minor timber species and the wood is used as kedondong.

Comments

Dabai fruit is also sought after by wild animals like the lemurs.

Selected References

Alina M, Azrina A (2008) Determination of chemical characteristic of oils extracted from flesh and kernel of *Canarium odontophyllum* and vitamin C content in flesh and skin of *Canarium odontophyllum*. p S56. In: Abstracts of the 23rd scientific conference of the nutrition society of Malaysia, Kuala Lumpur, 27–28 March 2008. Mal J Nutr 14(2):S1–S87

Azrina A, Nurul Nadiah MN, Zulkhairi A, Amin I (2009) Physical properties of skin, flesh, and kernel of *Canarium odontophyllum* fruit. J Food Agri Environ 7(3&4):55–57

Azrina A, Nurul Nadiah MN, Amin I (2010a) Antioxidant properties of methanolic extract of *Canarium odontophyllum* fruit. Int Food Res J 17:319–326

Azrina A, Prasad KN, Khoo HE, Nurnadia AA, Alina M, Amin I, Zulkhairi A (2010b) Comparison of fatty acids, vitamin E and physicochemical properties of *Canarium odontophyllum* Miq. (dabai), olive and palm oils: Nutritional and physicochemical properties of dabai oils. J Food Compos Anal 23(8):2–6

Lau CY, Fatimah O (2007) Cold storage of dabai fruit (*Canarium odontophyllum* Miq.): effect of packaging methods and storage duration on quality. In: Proceedings of the Sarawak fruit seminar 2007. Kingwood Hotel, Sibu, Sarawak, 8–9 August 2007

Lemmens RHMJ, Soerianegara I, Wong WC (eds.) (1995) Plant esources of South-East Asia No. 5(2) timber trees: Minor commercial timbers. Backhuys Publishers, Leiden, the Netherlands. p. 103.

Munawaroh E, Purwanto Y (2009) Studi hasil hutan non kayu di Kabupaten Malinau, Kalimantan Timur. In: Paper presented at the 6th basic science national seminar, Universitas Brawijaya, Indonesia (In Indonesian), 21 February 2009, http://fisika.brawijaya.ac.id/bss-ub/proceeding/PDF%20FILES/BSS_146_2.pdf

Nur Naazira Iman A, Norhaizan ME (2008) Polyphenol, antioxidant activity and phytic acid of fermented *Canarium odonthophyllum*. p S65. In: Abstracts of the 23 rd scientific conference of the nutrition society of Malaysia, Kuala Lumpur, 27–28 March 2008. Mal J Nutr 14(2):S1–S87

Prasad KN, Chew LY, Khoo HE, Kong KW, Azrina A, Amin I (2010) Antioxidant capacities of peel, pulp, and seed fractions of *Canarium odontophyllum* Miq. fruit. J Biomed Biotechnol, Article ID 871379, 8 p

Prasad KN, Chew LY, Khoo HE, Yang B, Azrina A, Amin I (2011) Carotenoids and antioxidant capacities from *Canarium odontophyllum* Miq. Fruit Food Chem 124(4):1549–1555

Slik JWF (2006) Trees of Sungai Wain. Nationaal Herbarium Nederland http://www.nationaalherbarium.nl/sungaiwain/

Voon BH, Kueh HS (1999) The nutritional value of indigenous fruits and vegetables in Sarawak. Asia Pac J Clin Nutr 8(1):24–31

Voon BH, Chin TH, Sim CY, Sabariah P (1988) Wild fruits and vegetables in Sarawak. Sarawak Department of Agriculture, Kuching, 114 pp

Canarium vulgare

Scientific Name

Canarium vulgare Leenh.

Synonyms

Canarium commune auct. non. L.

Family

Burseraceae

Common/English Names

Chinese Olive, Java Almond, Kenari Nut, Kenari Nut Tree, Wild Almond.

Vernacular Names

Danish: Kanarietræ;
Estonian: Harilik Kanaarium;
French: Amande De Java;
German: Kanarinuß;
Indonesia: Jal, Jar (Ambon), Kanak, Kanali, Kenari, Kituwa, (Java), Haki, Hika, Jal, Jar, Njiha Ambon, Njiha Furu (Maluku), Kodja (Nusa Tenggara), Kanari Panjang (Sumatra), Dedi, Ipalo, Reri (Sulawesi), Knarje (Timor), Ki Kanari (Sundanese);
Japanese: Kanari Noki, Kanariya Noki;
Malaysia: Pokok Kenari, Rata Kukana;
Spanish: Almendra De Java, Canari.

Origin/Distribution

Canarium vulgare is native to Eastern Malaysia, Papua, Papua New Guinea (Morobe) and Indonesia – Alor, Nusa Tenggara Timor (Eastern Lesser Sunda Islands) and Maluku (Moluccas). It is especially endemic in Ambon. It has been introduced to India, Sri Lanka and elsewhere in the tropics.

Agroecology

Kenari occurs in tropical primary or secondary rainforest at low and medium altitudes. Its agroecological requirements are similar to that describe for *Canarium indicum*.

Edible Plant Parts and Uses

The kernels are edible after removal from the hard shell and eaten raw, roasted, baked or smoked. The kernels are eaten in Sri Lanka as a dessert nut, and are made into bread in the Celebes (Sulawesi). In Malaysia, crushed nuts are used as sprinkle over cakes, and in the past used as cooking oil in areas where coconut oil was scare. The kernel oil can be used in an emulsion as an artificial milk for feeding infants in the tropics.

Plate 1 Dense foliage and terminal racemose inflorescence of kenari

Plate 2 Clusters of terminal infructescences of kenari and imparipinnate leaves with large, entire, elliptic-oblong leaflets

Plate 3 Close-up of unripe, green ovoid-oblong kenari fruit

Plate 4 Unripe, green fruit (*left*) and ripe, purplish-black kenari fruit (*right*)

Botany

A large, dense canopy, evergreen, dioecious, perennial tree, growing to 45 m high with fluted, buttressed trunk, greyish-brown rough, lenticelled bark and stilt (prop) roots. Leaves are spirally arranged, imparipinnate, dark green, 5–15 leaflets, pseudostiplues present at base of petioles, subulate and uncinated (Plates 1–2). Leaflets are entire, glabrous, elliptic-oblong, 2–5 cm long, with pinnate venation. Flowers in dense, pubescent terminal panicles. Flowers, small, creamy white, actinomorphic, unisexual with male and female flowers on different trees. Male flowers with campanulate calyx three-lobed, 3–4 mm long; six free stamens and corolla pale yellow turning brown, 5–6 mm long. Female flowers with urceolate calyx, shortly three-dentate and 5.5–6 mm long; corolla pale yellow turning brown, ovary superior, carpels joined (when more than one), locules three and style glabrous with three-lobed stigma. Fruit is an ovoid-oblong, indehiscent,

Plate 5 Sign in Indonesian describing the origin and uses of kenari wood

non-fleshy drupe, 3.5–5 cm long and 1.5–3 cm wide, green when unripe turning to purplish black when ripe (Plates 2–4). Seeds 1–3, 1–3 cm long by 1–1.5 cm across.

Nutritive/Medicinal Properties

Chinese olive has been referred to as *Canarium vulgare* by the USDA. Chinese olive has been regarded to be demulcent, expectorant, immunostimulant, irritant, rubefacient, stimulant and vulnerary. It has been used for treating cold, gastrosis, gonorrhea, rheumatism, immunodepression, ulcer and sores.

Other Uses

Canarium vulgare is occasionally used as a shade tree in plantations and for its timber under the trade name Kedondong. Its wood is used for house construction and building boats (Plate 5).

Comments

See also notes on the other closely related edible Canarium species.

Selected References

Backer CA, Bakhuizen van den Brink RC Jr (1965) Flora of Java (Spermatophytes only). Vol. 2. Wolters-Noordhoff, Groningen, 641 pp

Conn BJ, Damas KQ (2006+) Guide to trees of Papua New Guinea. http://www.pngplants.org/PNGtrees

Duke JA, Bogenschutz-Godwin MJ, DuCellier J, Duke PA (2002) CRC handbook of medicinal plants, 2nd edn. CRC, Boca Raton, 936 pp

Hedrick UP (1972) Sturtevant's edible plants of the world. Dover Publications, New York, 686 pp

Leenhouts PW (1956) Burseraceae. In: van Steenis CGGJ (ed.) Flora malesiana. Series I. 5(2) (pp 209-296), Noordhoff-Kolff Djakarta, 595 pp

Menninger EA (1977) Edible nuts of the world. Horticultural Books, Stuart, 175 pp

Yen DE (1993) The origins of subsistence agriculture in Oceania and the potentials for future tropical food crops. Econ Bot 47:3–14

Dacryodes rostrata

Scientific Name

Dacryodes rostrata (Blume) H.J. Lam.

Synonyms

Canarium articulatum Engl. ex Koord., *Canarium caudatifolium* Merr., *Canarium crassifolium* Merr., *Canarium cuspidatum* Merr., *Canarium gilvescens* Miq., *Canarium kadondon* Benn. in Hook.f., *Canarium minahassae* Koord., *Canarium montanum* Korth. ex Miq., *Canarium reticulatum* Ridl., *Canarium rostriferum* Miq., *Canarium rostriferum* var. *cuspidatum* Miq., *Dracontomelon cuspidatum* Bl., *Hemisantiria rostrata* H.J. Lam, *Santiria montana* Bl., *Santiria rostrata* Bl., *Santiria samarensis* Merr

Family

Burseraceae

Common/English Names

Kembayau

Vernacular Names

Borneo: Ampadu Kalui, Kalasu, Kambayan, Karamu, Kedongdong, Kelamok Maruk, Kembayan Aie, Kembayan Burong, Kembayau, Kembayau Lamak, Kembayau Teta, Kembayan Utan, Keramuh, Kumabang, Langsat-Langsat, Masam, Merading, Merasam, Njihah, Peninasan, Piramuh, Tindau, Ungit;
Brunei: Ungit;
Indonesia: Asem Begomdang (Sumatra), Rengas Burung (Bangka), Kembajau (Kalimantan), Kayu Minyak, Kedondong (General), Keramuh (Punan Malinau, East Kalimantan), Kedamu (Lundanye, East Kalimantan), Buwa Keramau (Merap, East Kalimantan), Keramu (Kenyah Uma', East Kalimantan);
Malaysia: Kedondong Kerut, Kembayau (Peninsular), Kelamok Maruk, Kembayau, Keramoh (Sarawak), Kamubang (Sabah);
Philippines: Palaspas (Bikol), Pili-Hanai (Bisaya), Lunai (Lanao);
Vietnamese: Trám Xuyên.

Origin/Distribution

Wild distribution of the species is found in Indo-China, Thailand, Peninsular Malaysia, Sumatra, Sulawesi, Borneo (Sarawak, Brunei, Sabah, West-, Central- and East-Kalimantan) and the Philippines.

Agroecology

The species occurs wild in tropical lowland undisturbed mixed dipterocarp forests and on hills up to 700 m altitude and occasionally in swamps on most soil types, including limestone. In secondary forests, it is usually present as a pre-disturbance remnant tree.

Edible Plant Parts and Uses

The pulp of the fruits is edible. Fruits are immersed in luke-warm water for 10–15 minutes, removed and sprinkle with salt and eaten. The fruit pulp can be preserved in salt and fried as a dish. The fleshy seed cotyledons are also edible.

Botany

An evergreen, perennial tree up to 40 m tall with 1 m trunk diameter and low buttresses (Plate 1). Leaves are alternate, pinnate, 2–10-jugate. Leaflets are ovate-oblong, papery, with asymmetrical base, leaf apex elongated and widening at the tip (Plate 3). Leaf rachis is strongly swollen at junction with leaf stalk and hairy, petiole with or without resinous ducts. Panicles are axillary often combined into a terminal inflorescence (Plate 2). Flowers three to-numerous, small, 3 mm across, yellowish–white, calyx copular and shortly dentate. Male flowers have six stamens connate to periphery of disk; female flowers have six staminodes, three-loculed ovary with sessile three-lobed stigma. Fruit is an ovoid to oblongoid, fleshy drupe, 2–4 cm × 1–2 cm, yellow-brown to purplish black (Plate 4), containing 1 hard seed.

Plate 1 Tree habit

Plate 2 Young fruits

Plate 3 Fruit and pinnate leaves

Plate 4 Ripe fruit of *Dacryodes rostrata*

Nutritive/Medicinal Properties

Kembayau fruit is pleasant, nutritious and has commercial potential. The proximate composition per 100 g edible portion of the fruit pulp was reported as: energy 241 kcal, moisture 56.2%, protein 3.4%, fat 16.1%, carbohydrate 20.7%, crude fibre 0.2%, ash 1.3%, p 35 mg, K 399 mg, Ca 83 mg, Mg 83 mg, Fe 1.1 mg, Mn 84 μg, Cu 7.5 μg, Zn 10 (Voon and Kueh 1999).

Kembayau seeds and pulp were found to be rich in fat, while peels had the highest ash contents (Kong et al. 2011). Potassium was the most abundant mineral in peels (380.72–1112.00 mg/100 g). In kembayau fruits, total flavonoid content (1012.74–28,022.28 mg rutin equivalent/100 g) was higher than total phenolic and total monomeric anthocyanin contents. Kembayau seeds exhibited high flavonoid and phenolic contents compared to the contents in peels and pulp. Antioxidant capacities were also higher in seeds as reflected by Trolox equivalent antioxidant capacity assay (51.39–74.59 mmol TE/100 g), ferric reducing antioxidant power assay (530.05–556.98 mmol Fe^{2+}/100 g) and by 1,1-diphenyl-2-picryl hydrazyl radical scavenging activity (92.18–92.19%) when compared to peels and pulp. Findings suggested that pulp and peels of kembayau fruit may be an important source of energy and minerals for human consumption, while seeds have a good potential as antioxidants.

Other Uses

The wood is used as timber for planks and paddy-pounders. The resin is used to make torches.

Comments

This species has two forms: forma *cuspidata* and forma *rostrata*, both are very common and widespread.

Selected References

Burkill IH (1966). A dictionary of the economic products of the Malay Peninsula. Revised reprint. 2 vols. Ministry of Agriculture and Co-operatives, Kuala Lumpur, vol 1 (A–H), pp 1–1240, vol 2 (I–Z), pp 1241–2444

Jansen PCM, Jukema J, Oyen LPA, van Lingen TG (1991) *Dacryodes rostrata* (Blume) H.J. Lam. In: Verheij EWM, Coronel RE (eds.) Plant resources of South-East Asia no. 2. Edible fruits and nuts. Prosea Foundation, Bogor, p 327

Kong KW, Chew LY, Prasad KN, Lau CY, Ismail A, Sun J Hosseinpoursarmadi B (2011) Nutritional constituents and antioxidant properties of indigenous kembayau (*Dacryodes rostrata* (Blume) H. J. Lam) fruits. Food Res. Int. (In press)

Munawaroh E, Purwanto Y (2009) Studi hasil hutan non kayu di Kabupaten Malinau, Kalimantan Timur. Paper presented at the 6th basic science national seminar, Universitas Brawijaya, Indonesia, (In Indonesian) 21 February 2009. http://fisika.brawijaya.ac.id/bss-ub/proceeding/PDF%20FILES/BSS_146_2.pdf

Slik JWF (2006) Trees of Sungai Wain. Nationaal Herbarium Nederland. http://www.nationaalherbarium.nl/sungaiwain/

Voon BH, Kueh HS (1999) The nutritional value of indigenous fruits and vegetables in Sarawak. Asia Pac J Clin Nutr 8(1):24–31

Voon BH, Chin TH, Sim CY, Sabariah P (1988) Wild fruits and vegetables in Sarawak. Sarawak Department of Agriculture, Sarawak, 114 pp

Cereus repandus

Scientific Name

Cereus repandus (L.) Mill.

Synonyms

Cactus peruvianus L., *Cactus repandus* L. basionym, *Cephalocereus remolinensis* (Backeb.) Borg, *Cereus albispinus* Salm-Dyck, *Cereus atroviridis* Backeb., *Cereus grenadensis* Britt. & Rose, *Cereus margaritensis* J. R. Johnston, *Cereus margaritensis* (Johnst.) var. *micracanthus* Hummel., *Cereus repandus* Haw., *Cereus peruvianus* L., *Cereus peruvianus* (L.) Mill., *Cereus peruvianus* Britton and Rose, *Cereus remolinensis* Backeb., *Pilocereus albispinus* Salm-Dyck ex Foerst.-Rümpler, *Pilocereus remolinensis* (Backeb.) Backeb., *Pilocereus repandus* (L.) Schum., *Piptanthocereus peruvianus* (L.) Riccobono, *Stenocereus peruvianus* R. Kiesling, *Subpilocereus atroviridis* (Backeb.) Backeb., *Subpilocereus grenadensis* (Britt. & Rose) Backeb., *Subpilocereus margaritensis* (Johnst.) Backeb., *Subpilocereus margaritensis* (Johnst.) Backeb. var. *micracanthus* (Hummel.) Trujillo & Ponce, *Subpilocereus remolinensis* (Backeb.) Backeb., *Subpilocereus repandus* (L.) Backeb.

Family

Cactaceae

Common/English Names

Apple Cactus, Giant Club Cactus, Hedge Cactus, Night Blooming Cereus, Peruvian Apple, Peruvian Apple Cactus, Peruvian Tree, Peruvian Tree Cactus.

Vernacular Names

Curaçao: Breebee, Kadoesji, Cadushi, Cadusji, Kadushi;
Danish: Søjlekaktus;
French: Ciege Du Pérou;
German: Apfelkaktus;
Israel: Koubo (Trade Name);
Papiamento: Kadushi;
Spanish: Pitahaya, Pitaya;
Venezuela: Cardon Lefaria, Cardon Blanco, Dato Blanco;
Vietnamese: Long Cot, Noc Tru, Xuong Rong Khe.

Origin/Distribution

Peruvian apple is native to Central America and south America – Venezuela, Dutch Antilles, Paraguay, Uruguay, Brazil and Argentina. Endemic to the Dutch Antilles on the islands Aruba, Bonaire, and Curaçao, where it often grows in thickets and the arid plains of north western Venezuela. It is also found on the dry coast of northern Venezuela and on the small offshore islands. It is cultivated as a commercial crop in Israel.

Agroecology

Peruvian apple grows well in full sun in well-drained sandy soils. It can thrive in marginal, infertile, semi-arid- arid conditions. Low winter temperatures, saline water or waterlogged conditions adversely impacts on its growth.

Edible Plant Parts and Uses

The fruit is sweet and edible. The fluted stems (cladodes) after cutting away the spines and peeling off the skin are eaten fresh with other vegetables by the Khmer people in Tra Vinh province in Vietnam. In Aruba and Curaçao, the green flesh is either dried and powdered, or freshly eaten in a mucilaginous soup known as "*cadada*" in Aruba. It is usually cooked with bits of salt, pork, goat meat, fish, sweet red peppers and tomatoes.

Botany

A large erect, branched, thorny columnar cactus with distinct cylindrical columnar, segmented, grey-green to dull green, 10 cm diameter stems, and ascending branches. It grows to 10 m high (Plate 1). Stems succulent, ribbed six to nine vertical ribs 15 cm thick, Areoles small widely separated. And borne atop branches surrounded by clumps of grey spines. Clusters of glochidiate spines emerge at regular points along the ridges.

Flowers very large, 25–35 cm by 12–15 cm wide, white, corolla outer green inner whorl white, stamen numerous, stigma 12–15, ovary inferior, flower nocturnal. Fruit berry pale green to red to violet red elongate, oblong to egg-shaped, smooth, 5 cm by 3 cm (Plates 1–3) containing many small crunchy black seeds embedded in the white pulp.

Nutritive/Medicinal Properties

Soluble sugars content in the fruit was reported to increase fivefold during ripening (Ninio et al. 2003). Glucose and fructose were the main sugars in the fruit pulp, and each increased from 0.5 to 5.5 g/100 g fresh weight during ripening. The polysaccharides content decreased during ripening from 1.4 to 0.4 g/100 g fresh weight.

Plate 2 Ripe and unripe ovoid apple cactus fruits

Plate 1 Columnar stems with flowers and fruits

Plate 3 Ripe oblongish apple cactus fruits

The titratable acidity decreased and the pH increased during ripening. The major organic acid in the fruit was malic acid, which decreased from 0.75 g/100 g fresh weight at the mature green stage to 0.355 g/100 g fresh weight in ripe fruit. Citric, succinic, and oxalic acids occurred in concentrations lower than 0.07 g/100 g fresh weight. Prominent accumulation of aroma volatiles occurred toward the end of the ripening process. The main volatile found in the ripe fruit was linalool, attaining concentrations of 1.5–3.5 µg/g fresh weight.

Alkyl esters were found in the wax from *Cereus peruvianus* (Dembitsky and Rezanka 1996). The wax esters were composed of fatty acids up to 30:0 and alcohols up to iso-34:0. The major fatty acids were 16:0, 16:1, 18:0 and 18:1. The major fatty alcohols were 16:0, 18:0 and 18:1, together with 26:0 (9.8%) and 26:1 (12%). Iso- and anteiso-fatty acids (5.5%) and fatty alcohols (11.3%) were also detected. More than 600 isomers of the wax esters were identified. Very-long-chain alkyl esters were detected in the wax from *Cereus peruvianus* (Rezanka and Dembitsky 1998). More than 80 isomeric alkyl wax esters were identified.

Agricultural wastes from s *Cereus peruvianus* and *Opuntia ficus indica* after extraction of natural polyelectrolytes could be utilised for protein production by solid substrate fermentation with *Aspergillus niger* (Oliveira et al. 2001). After 120 h of fermentation the protein increased by 12.8%. Aspartic acid (1.27%), threonine (0.97%), glutamic acid (0.88%), valine (0.70%), serine (0.68%), arginine (0.82%), and phenylalanine (0.51%) were the principal amino acids produced.

The waxy pecto-cellulosic cuticle of cladodes of *Cereus peruvianus* was found to be a source of an α-D-polygalacturonic or pectic acid (35–40% yield, on a dry wt based on the wax-free pectocellulose layer) (Alvarez et al. 1995). Cereus cuticle pectate (sodium salt) tended to gel above a concentration of 1%, a useful property that could be more easily obtained by the inclusion of sucrose, light addition of calcium salt, and/or mild acidification.

Cut stems have cooling property. In Curaçao, the green layer of flesh, found between the cuticle and the yellow-coloured, firm layer of the shoot, is sliced off, sun or fire dried, and consumed to combat diarrhoea.

Other Uses

The plant is use as ornamental or landscape plant in rockeries and as living hedge fence. Stems when dry serve as torches and fuel. The thick trunks are used for carpentry and furniture. Sliced stems have been used as soap substitute to either wash dishes or one's body.

Various interesting compounds were found in *Cactus peruvianus* (Alvarez et al. 1992). Included were acidic gum and cellulose as the highly polymerized carbohydrate components, and a complex waxy lipid fraction. The major gum fraction was found to be an uronylated rhamnoarabinogalactan. A pigment-free powdered gum was obtained by precipitating and washing the fresh mucilage with 2–3 vol of ethanol. The almost protein-free polysaccharide formed viscous solution upon redissolution. The possible uses for the pretreated cactus gum was investigated as an adjuvant in the flocculation of water impurities and in formulation of cosmetics.

Comments

This cactus blooms at night.

Selected References

Alvarez M, Costa SC, Utumi H, Huber A, Beck R, Fontana JD (1992) The anionic glycan from the cactus *Cereus peruvianus*: structural features and potential uses. Appl Biochem Biotechnol 34–35(1):283–295

Alvarez M, Costa SC, Huber A, Baron M, Fontana JD (1995) The cuticle of the cactus *Cereus peruvianus* as a source of a homo-α-d –galacturonan. Appl Biochem Biotechnol 51–52(1):367–377

Anderson EF (2001) The cactus family. Timber Press, Portland, 776 pp

Dembitsky VM, Rezanka T (1996) Molecular species of wax esters in *Cereus peruvianus*. Phytochem 42(4):1075–1080

Lohmüller FA (2005) The botanical system of the plants. http://www.f-lohmueller.de/botany/gen/c/Cereus.htm

Mizrahi Y, Nerd A, Sitrit Y (2002) New fruits for arid climates. In: Janick J, Whipkey A (eds.) Trends in new crops and new uses. ASHS Press, Alexandria, pp 378–384

Morton JF (1967) Cadushi (*Cereus repandus* Mill.), a useful cactus of Curacao. Econ Bot 21(2):185–191

Morton JF (1987) Cactaceae, strawberry pear. In: Fruits of warm climates. Julia F. Morton, Miami, pp 347–348

Nerd A, Aronson JA, Mizrahi Y (1990) Introduction and domestication of rare and wild fruit and nut trees for desert areas. In: Janick J, Simon JE (eds) Advances in new crops. Timber Press, Portland, pp 353–363

Ninio R, Lewinsohn E, Mizrahi Y, Sitrit Y (2003) Changes in sugars, acids, and volatiles during ripening of koubo [*Cereus peruvianus* (L.) Miller] fruits. J Agric Food Chem 51(3):797–801

Oliveira MA, Rodrigues C, dos Reis EM, Nozaki J (2001) Production of fungal protein by solid substrate fermentation of cactus *Cereus peruvianus* and *Opuntia ficus indica*. Quim Nova 24(3):307–310

Rezanka T, Dembitsky VM (1998) Very-long-chain alkyl esters in *Cereus peruvianus* wax. Phytochem 47(6): 1145–1148

Tanaka Y, Nguyen VK (2007) Edible wild plants of Vietnam: the bountiful garden. Orchid Press, Bangkok, 175 pp

Hylocereus megalanthus

Scientific Name

Hylocereus megalanthus (K. Schumann ex Vaupel) Ralf Bauer (2003)

Synonyms

Cereus megalanthus K. Schum. ex Vaupel (Basionym), *Mediocactus megalanthus* (K. Schum. ex Vaupel) Britton & Rose, *Selenicereus megalanthus* (Schum) Britton & Rose.

Family

Cactaceae

Common/English Names

Pitahaya, Pitaya, Yellow Dragonfruit, Yellow Pitahaya

Vernacular Names

German: Gelbe Pitaya, Gelbe Pitahaya;
Portuguese: Pitaya amarela;
Russian: Drakonov frukt;
Spanish: Pitahaya amarilla, Pitajaya amarilla, Pitaya amarilla;
Swedish: Gul Pitahaya.

Origin/Distribution

The yellow pitaya is native to Colombia, Peru, Bolivia, Ecuador and Venezuela. It is commercially grown to a limited extent in Columbia. Fruits are occasionally exported to Europe and Canada.

Agroecology

In its native range, yellow pitaya occurs in riverine forests. It grows well in tropical and subtropical climates, free of frosts and freezes. Sub-zero temperatures are detrimental but it will recover from brief duration of low temperatures. Yellow pitaya tolerates cool or warm climates, provided temperatures do not exceed 38°C. Extreme high temperature results in extensive stem damage. Initial estimates from native areas suggest that optimum temperatures for growth are 18–25°C. It tolerates some shade and may be injured by extreme sunlight although it thrives in sunny position in their native countries. Like other pitayas, it is adapted to a wide range of soils provided they are well-drained including calcareous and moderately saline soils. As with other fruit crops, they may show minor element deficiencies in the poor, high-pH soils. It thrives in soils high in organic matter or where this is wanting compost or manure needs to be added. Yellow pitaya appears to tolerate windy conditions, however, very strong winds or hurricanes may cause considerable damage to trellises or supports and consequently to the plants.

Edible Plant Parts and Uses

Like other pitayas, the fruit is popularly eaten chilled, out of hand. It is also used to flavor drinks and pastries. The fruit may be converted into juice or wine; the flowers can be eaten or steeped as tea.

Botany

A terrestrial or epiphytic cactus vine, with fleshy, procumbent, scandent (climbing) or pendant stems producing aerial roots. Stems are robust, green, three-ribbed, 1.5 cm thick, with slightly undulating margins, white areoles bearing 1–3 yellowish spines, 2–3 mm long. Flowers are nocturnal, large, white and funnel-shaped, 32–38 cm long; pericarpel (hypanthium) ovoid to globose with large flattened tubercles and felt-like, spiny areoles subtended by small bracteoles. Outer tepals are long, green, triangular-acute; inner tepals 100 cm long, 3.5 cm wide, white; stamens numerous inserted in two zones, yellow; style yellow, stigma lobes numerous, green. Fruit: ovoid, tuberculate, spiny, yellow (Plate 1) with numerous black seeds embedded in a sweet, juicy white pulp and much smaller than the red pitayas.

Nutritive/Medicinal Properties

The nutrient value of fresh yellow pitaya fruit per 100 g edible portion was reported as: water 85%, energy 50 cal, carbohydrate 13.2 g, protein 0.4 g, fat 0.1 g, fibre 0.5 g, ash 0.4 g, Ca 10 mg, P 16 mg, Fe 0.3 mg, thiamine 0 mg, riboflavin 0 mg, niacin 0.2 mg, ascorbic acid 4 mg (ICBF 1992).

Other Uses

Yellow pitahaya is an impressive ornamental plant.

Comments

Recent research suggest that this species an allotetraploid ($2n = 4x = 44$) derived from natural hybridization between two closely related diploid taxa, *Hylocereus* and *Selenicereus* (Tel-zur et al. 2004). The two species possibly involved, as being native in the same area, are *Hylocereus costaricensis* and *Selenicereus inermis*.

Note: Differential characters between:

Hylocereus megalanthus – floral tube or pericarpel 32–38 cm long with large flattened tubercles.

Hylocereus setaceus – floral tube or pericarpel 19–22 cm with small tubercles.

Plate 1 Yellow pitahaya fruits

Selected References

Anderson EF (2001) The cactus family. Timber Press, Portland, 776 pp

ICBF (1992) Tabla de composición de alimentos. ICBF. Sexta edición, INCAP y FAO (Cited in *El Cultivo de Pitaya y su Posicionamiento en el Mercado*. http://www.angelfire.com/ia2/ingenieriaagricola/ pitaya.htm)

Mizrahi Y, Nerd A (1996) New crops as a possible solution to the troubled Israeli export market. In: Janick J (ed.) Progress in new crops. ASHS Press, Alexandria, pp 56–64

Mizrahi Y, Nerd A (1999) Climbing and columnar cacti: new arid land fruit crops. In: Janick J (ed.) Perspectives on new crops and new uses. ASHS Press, Alexandria, pp 358–366

Mizrahi Y, Nerd A, Sitrit Y (2002) New fruits for arid climates. In: Janick J, Whipkey A (eds.) Trends in new crops and new uses. ASHS Press, Alexandria, pp 378–384

Nerd A, Mizrahi Y (1998) Fruit development and ripening in yellow pitaya. J Am Soc Hort Sci 123: 560–562

Tel-Zur N, Abbo S, Bar-Zvi D, Mizrahi Y (2004) Genetic relationships among *Hylocereus* and *Selenicereus* vine cacti (Cactaceae): evidence from hybridization and cytological studies. Ann Bot (Lond) 94(4):527–534

Hylocereus polyrhizus

Scientific Name

Hylocereus polyrhizus (F. A. C. Weber) Britton & Rose

Synonyms

Cereus polyrhizus F.A.C. Weber.

Family

Cacataceae

Common/English Names

Belle Of The Night, Common Night Blooming Cereus, Conderella Plant, Dragon Fruit, Nanettikafruit, Night Blooming Cereus, Pitahaya, Red-fleshed Dragon Fruit, Red-fleshed Pitahaya, Red Pitahaya, Queen Of The Night, Strawberry Pear.

Vernacular Names

Brazil: Cardo-Ananás, Jaramacarú, Rei-Da-Noite, Pitaya Vermelha De Polpa Branca, Pitaya Vermelha Depolpa Vermelha (Portuguese);
Chinese: Huǒ Lóng Guǒ, Lóng Zhū Guǒ;
Columbia: Pitahaya Blanca Pitahaya Roja;
Czech: Kaktus Zvlněný;
Danish: Pitahaya;
Eastonian: Maasik-Metskaktus;
French: Cierge-Lezard, Fruit Du Dragon, Belle De Nuit, Pitahya Dulce, Pitahaya Rouge, Pitaya, Poire De Chardon, Raquette Tortue;
German: Distelbirne, Drachenfrucht, Echte Stachelbirne, Königin Der Nacht, Rotepitahaya, Waldkaktus;
Hawaiian: Pāniniokapunahou, Pāpipi Pua;
Indonesia: Buag naga, buah naga merah;
Malaysia: Dragonfruit, Buah Naga;
Mauritius: Débousse-To-Fesse;
Mexico: Junco Tapatío Pitahaya De Cardó, Pitahaya Dulce, Pitahaya Orejona, Pitahaya Roja, Pitahaya Blanca, Reina De La Noche, Tasajo (Spanish);
Philippines: Caliz;
Puerto Rico: Flor De Caliz, Pitajava;
Portguese: Cato-Barse, Cardo-Ananaz, Pitaya Vermelha De Polpa Branca , Rainha Da Noite;
Slovenian: Pitaja;
Spanish: Chacam, Chak-Wob, Pitahaya, Flor De Caliz, Pitaja, Junco, Junco Tapatio, Pithaya Orejona, Reina De La Noche, Tasajo, Zacamb;
Swedish: Skogskaktus, Röd Pitahaya;
Thai: Geow Mangon;
Venezuela: Pitahaya Roja, Pitahaya Blanca;
Vietnamese: Thanh Long.

Origin/Distribution

This species is reported to be indigenous to Nicaragua, Costa Rica and Panama (Anderson 2001).

Agroecology

As described for *Hylocereus undatus*.

Edible Plant Parts and Uses

Similar edible uses as described for *Hylocereus undatus*.

Botany

H. polyrhizus (Web.) Britton & Rose has a sprawling to climbing habit, profusely branched, epiphytic, lithophytic, and terrestrial, to several meters long. Stems three- to four-angled, usually three-angled, dark green, becoming whitish with age, internode 0.5–2 m long, 3–10 cm wide. Ribs smooth to wavy, often horny, with concave faces. Areoles on prominences bearing spines. Spines 2–6, bulbous basally, reddish, becoming gray, needle-like, straight, 2–10 mm long. Flowers are borne singly or in small clusters, creamy white, 25–31 cm long, 25–30 cm in diameter, outer reddish perianth segments especially at the tips (Plates 1–3); floral tubes straight to strongly curved; pericarpels indistinct, stigma lobes short and yellowish not bifid. Fruits ellipsoidal or oblong, scarlet, 7–12 cm long, covered with scales that vary in size; flesh red to purplish-red with many tiny, black seeds (Plates 4–5).

Plate 2 Young flowers and fruit

Plate 3 Close-up of opened Pitahaya flower

Plate 1 Flower of red-fleshed pitaya

Plate 4 Whole and halved fruit

Plate 5 Red-flesh and tiny black seeds

Nutritive/Medicinal Properties

Nutrient composition of the edible fruit flesh or red fleshed pitaya was reported as: moisture 82.5–83 g, protein 0.159–0.229 g, fat 0.21–0.61 g, crude fibre 0.7–0.9 g, vitamin C 8–9 mg/l (Jaafar et al. 2009). The premature stem was found to have higher nutrient values than the mature stem or fruit flesh. The stem had 96–98 g moisture, 0.12–0.27 g protein, 0.09–0.23 g fat, 0.02–0.05 g crude fibre, vitamin C 63.71–132,95 mg/l, ash 0.03–0.09 g, glucose 0.236–0.552 g/l. The nutrient-rich stem could be processed into freeze-dried powder which could be rehydrated and be used as food components for health and therapeutic benefits. In another analysis, Pitahaya fruits were reported to have a low vitamin C content (116–171 µg/g of fresh pulp without seeds), high contents of betacyanins (0.32–0.41 mg/g) and phenolic compounds (5.6–6.8 µmol Eq gallic acid/g), and high antioxidant ORAC value of (8.8–11.3 µmol Eq Trolox/g) (Vaillant et al. 2005).

Essential fatty acids, namely, linoleic acid and linolenic acid was found to comprise a significant percentage of the unsaturated fatty acids of the seed oil extract of both pitaya species *Hylocereus undatus* and *Hylocereus polyrhizus* (Azis et al. 2009). Both pitaya species exhibited two oleic acid isomers. Both pitaya species contained about 50% essential fatty acids C18:2 (48%) and C18:3 (1.5%). Studies showed that pitaya (*Hylocereus undatus* and *Hylocereus polyrhizus*) seed oil contained a high amount of oil (18.33–28.37%) (Lim et al. 2010). The three major fatty acids in the *H. undatus* seed oil (WFSO) and *H. polyrhizus* seed oil (RFSO) were linoleic, oleic, and palmitic acids. The total tocopherol contents in the WFSO and RFSO were 36.70 and 43.50 mg/100 g, respectively. The phytosterol compounds identified in the WFSO and RFSO were cholesterol, campesterol, stigmasterol, and β-sitosterol. Seven phenolic acid compounds were identified in the WFSO and RFSO, namely, gallic, vanillic, syringic, protocatechuic, p-hydroxybenzoic, p-coumaric, and caffeic acids. This study revealed that pitaya seed oil has a high level of functional lipids and could be used as a new source of essential oil.

Betalains were found to responsible for the main antioxidant capacity of purple pitaya juices while non-betalainic phenolic compounds contributed only to a minor extent (Esquivel et al. 2007). Additionally, several biosynthetic precursors were detected. Notably, decarboxylated and dehydrogenated betalains were identified as genuine compounds of the juices. Gallic acid was also identified for the first time in pitaya fruits. While the phenolic profiles generally differed between genotypes, phenolic compound composition of 'Rosa' resembled that of *H. polyrhizus* with respect to total contents of betacyanins, betalainic precursors, phyllocactin and cyclo-Dopa malonyl-glucosides. Studies revealed light as the major factor of betalain pigment degradation in red pitaya fruit (Woo et al. 2010b). Refrigeration storage (4°C) condition without light exposure managed to preserve the colour of fruit juice for up to 3 weeks.

The dark red flesh was found to be a rich source of nutrients and minerals such as vitamin B1, vitamin B2, vitamin B3 and vitamin C, protein, fat, carbohydrate, crude fibre, flavonoid, thiamin, niacin, pyridoxine, kobalamin, glucose, phenolic, betacyanins, polyphenol, carotene, phosphorus, iron and phytoalbumin (Le Bellec et al. 2006). *Hylocereus polyrhizus* was found to be rich in phytoalbumin antioxidants, fibre, vitamin C and minerals (Mahattanatawee et al. 2006). It was also found to be rich in potassium, protein, fibre, sodium and calcium (Khalili et al. 2006).

Fruit growth of red-fleshed dragon fruit (*Hylocereus polyrhizus*) was found to follow a sigmoid growth pattern (Jamaludin et al. 2011). Significant changes in colour were obtained in both peel and pulp from green to red-violet colour at ripening. Red-violet betacyanin was manifested earlier in pulp at 25 days after pollination (DAP), followed by peel 4–5 days later, and finally both peel and pulp turned full red-violet by 30 DAP. There was a significant increase in soluble solids concentration and titratable acidity with the continuous increase in betacyanin content as DAP progressed.

In general, white fleshed pitaya fruits had higher total soluble solid (TSS) than red fleshed pitaya (Wu and Chen 1997). TSS varied in the different fruit parts in decreasing order of fruit core, stylar-end, stem-end and peripheral parts. Glucose, fructose and sucrose were the main soluble sugars and their contents followed a similar trend except for sucrose. In white fleshed fruit, glucose varied from 104.3 mg/g in the core to 64.3 mg/g in the peripheral part, fructose 64.9 mg/g (core) to 40.1 mg/g (peripheral), sucrose 5.4 mg/g (core) to 7.5 mg/g (peripheral). For red fleshed fruit, glucose varied from 68.1 mg/g in the core to 41.9 mg/g in the (peripheral part, fructose 49.9 mg/g (core) to 32.3 mg/g peripheral), sucrose 4.2 mg/g (core) to 6.8 mg/g (peripheral). Distribution of sugars in various parts of the fruit was closely related to the different enzyme activities. In both red and white-fleshed fruits, core had the highest value of acid, invertase, total invertase and amylase activities followed by the stylar-end, stem-end and peripheral parts.

The deep purple or deep red colour of the pulp was found to be contributed by a set of pigments known as the betalains, nitrogen-containing pigments (Harivaindaran et al. 2008), made up of the red-violet betacyanins and yellow betaxanthins with maximum absorptivity at 535 and 480 nm, respectively (Herbach et al. 2006). Betalains are of commercial interest not only for food coloring, but also for their anti-oxidant properties for protection against certain oxidative stress-related disorders. *Hylocereus polyrhizus* was found to be rich in betalains, containing 525.3 mg/l of betacyanins in the juice but with only 5.3 mg/l of betaxanthins (Stintzing et al. 2003).

The betalain pattern consisted of 10 betacyanin (Stintzing et al. 2002). Five of them were indisputably assigned to bougainvillein-r-I, betanin, isobetanin, phyllocactin, and iso-phyllocactin. The remaining betacyanins were tentatively identified as (6′-O-3-hydroxy-3-methyl-glutaryl)-betanin, its C15-stereoisomer, and (6′-O-3-hydroxy-3–butyryl)-betanin, respectively. In subsequent studies 4 betacyanin pigments: betanin, isobetanin, phyllocactin and hylocerenin were isolated (Stintzing et al. 2004). Fruits of *H. polyrhizus* exhibited the highest relative concentration of hylocerenin, a recently discovered pigment, and a high relative concentration of phyllocactin (Wybraniec and Mizrahi 2002) in addition to betanin (Wybraniec et al. 2001). Hylocerenin and isohylocerenin, present in fruits at relative concentrations of 11.7% and 5.8%, respectively, were probably responsible for the fluorescent color of the fruit pulp. Hitherto unknown pigments: betanidin 5-O-(2′-O-β-d-apiofuranosyl)-β-d-glucopyranoside, betanidin 5-O-(4′-O-malonyl)-β-d-glucopyranoside and betanidin 5-O-[(5″-O-E-sinapoyl)-2′-O-β-d-apiofuranosyl]-β-d-glucopyranoside were subsequently isolated from the peel and flash (Wybraniec et al. 2007). The total concentration of betacyanins in the pitahaya fruit peel of 25, 30 and 35 DAA (days after flower anthesis) was 0.24, 3.99 and 8.72 mg/ml, respectively (Phebe et al. 2009). The fruit flesh contained 2.40, 7.93 and 11.70 mg/ml betacyanins at 25, 30 and 35 DAA, respectively, which was higher than peel. Betanin and isobetanin were identified.

Red pitaya peel (*Hylocereus polyrhizus*), discarded during processing was found to comprise about 22% of the whole fruit weight and to have potential for recovery of value-added materials (Jamilah et al. 2010). Physico-chemical analysis of the discarded pitaya peel revealed that the moisture content of the peel was approximately 92.7% and it was low in total soluble solids, protein, ash and fat content. However, the betacyanin pigment (150.46 mg/100 g) and pectin (10.8%) were high in the peel. Glucose, maltose and fructose were detected in the peel but not sucrose and galactose. The peel also had very high insoluble and soluble dietary fibre which had exhibited a good ratio of insoluble dietary fibre to soluble dietary fibre (3.8:1.0). Woo et al. (2010a) extracted pectin from dried fruit peels of red pitahya. Highest yield of 14.86% was obtained in 60 min extraction at pH 3.5. Highest glucose content was observed in sample extracted for 60 min, pH 4.0, whilst sample with the highest degree of esterification was obtained from treatment at pH 4.0 for 120 minutes.

Acidity of the fruit flesh was found to be generally low, between 2.4 and 3.4, resulting in a high sugar-to-acid ratio, giving a poor sensory quality when juice was consumed alone (Stintzing et al. 2003). The main organic acids present in pitahaya juice were citric acid and L-lactic acid. Protein content varied considerably from 0.3% to 1.5%; these differences may be due to the methodologies applied, or because of possible interference from betalain, the nitrogen-containing pigment responsible for the red color. The main amino acid present in pitahaya juice appeared to be proline with a notable high content of (1.1–1.6) g/l of juice (Stintzing et al. 2003). Pitahaya juice also had a high antiradical activity (around 10 µmol Trolox equivalent) assessed by the ORAC method with fluorescein, a value very similar to beetroot. Pitahaya juice was found to have the functional properties of a natural food colorant with high antioxidant potency (Vaillant et al. 2005).

Antioxidant Activity

The total phenolic contents of flesh (42.4 mg of gallic acid equivalents (GAE)/100 g of flesh fresh weight) and peel (39.7 mg of GAE/100 g of peel fresh weight) of red pitahaya were similar (Wu et al. 2006). The flavonoid contents of flesh and peel did not vary much (7.21 mg vs. 8.33 mg of catechin equivalents/100 g of flesh and peel matters). The concentrations of betacyanins expressed as betanin equivalents per 100 g of fresh flesh and peel were 10.3 mg and 13.8 mg, respectively. The antioxidant activity, measured by the DPPH method at EC_{50}, was 22.4 and 118 µmol vitamin C equivalents/g of flesh and peel dried extract. The values of EC_{50}, determined by the $ABTS^+$ approach, were 28.3 and 175 µmol of trolox equivalents antioxidant capacity (TEAC)/g of flesh and peel dried extract, respectively. The results indicated that the flesh and peel were both rich in polyphenols and were good sources of antioxidants. Rebecca et al. (2010) reported 86.10 mg of total polyphenolic compound in 0.50 g of dried dragon fruit extract expressed in gallic acid equivalent, and 2.30 mg catechin/g. The reducing power assay further confirmed the antioxidant activity present in dragon fruit where the reducing capability increased from 0.18 to 2.37 with the increase of dry weight sample from 0.03 to 0.5 g. The DPPH radical scavenging activity determination showed that the effective concentration (EC_{50}) for dragon fruit was 2.90 mM vitamin C equivalents/g dried extract.

Studies showed that the addition of white and red dragon fruit into yogurt increased the milk fermentation rate, lactic acid content, syneresis percentage, antioxidant activity, and total phenolics content in yogurt (Zainoldin and Baba 2010). Milk fermentation rate was augmented in red dragon fruit yogurt for all doses (−0.3606 to −0.4126 pH/h) while only white dragon fruit yogurt with 20% and 30% (w/w) composition showed increment in fermentation rate (−0.3471 to −0.3609 pH/h) compared to plain yogurt (−0.3369 pH/h). All dragon fruit enriched yogurts generally showed lower pH readings of pH 3.95–4.03 compared to plain yogurt, pH 4.–5. Both fruit yogurts showed a higher lactic acid percentage (1.14–1.23%) compared to plain yogurt (1.08%). Significantly higher syneresis percentage (57.19–70.32%) compared to plain yogurt (52.93%) were observed in all pitahaya fruit enriched yogurts. The antioxidant activity of plain yogurt (19.16%) was augmented by the presence of white and red dragon fruit (24.97–45.74%). All fruit enriched yogurt exhibited an increment in total phenolic content (36.44–64.43 mg/ml) compared to plain yogurt (20.25 mg/ml). However, the addition of white and red dragon fruit did not increase the proteolysis of milk during fermentation.

Hypocholesterolemic Activity

Khalili et al. (2009) reported the total phenolic content in red pitaya to be 46.06 mg GAE/100 g fresh weight and antioxidant activity as 76.10% using FTC (ferric thiocyanate) method. TBA (thiobarbituric acid) analysis also showed red pitaya extract had high antioxidant effect (72.90%). An in-vivo study also showed red pitaya had hypocholesterolemic effect on induced hypercholesterolemia rats. All groups that received red pitaya supplementation had high antioxidant properties and showed good results in managing the lipid profile. It was suggested that the consumption of red pitaya demonstrated the potential to reduce dyslipidemia and may play a role in the prevention of cardiovascular disease.

Anticancer Activity

The antiproliferative study on B16F10 melanoma cells revealed that pitahaya peel (EC_{50} 25.0 μg of peel matter) component was a stronger inhibitor of the growth of B16F10 melanoma cancer cells than the flesh (Wu et al. 2006).

Other Pharmacological Properties

Red pulp of pitaya fruit had generated a lot of interest as a source of natural red colour for food colouring (Harivaindaran et al. 2008), cosmetic industry and health potential for improving eyesight, preventing hypertension and combating anaemia (Raveh et al. 1998; Lee 2002; Stintzing et al. 2002). The medicinal qualities of red pitaya also included alleviating common stomach ailment and it had been recommend for hypercholesterolemia, diabetic and anaemic persons (Raveh et al. 1998; Lee 2002).

Ingestion of excessive amounts of red-fleshed dragon fruit was found to result in pseudohaematuria, a harmless reddish discoloration of the urine and faeces (MMR 2008).

Other Uses

Pitahaya is also planted as an ornamental and occasionally planted as flowering hedge plant.

Comments

In southeast Asia, Pitahaya is usually cultivated on trellises or on live tree supports.

Selected References

Anderson EF (2001) The cactus family. Timber Press, Portland, 776 pp

Azis A, Jamilah B, Chin PT, Russly R, Roselina K, Chia CL (2009) Essential fatty acids of pitaya (dragon fruit) seed oil. Food Chem 114:561–564

Esquivel P, Stintzing FC, Carle R (2007) Phenolic compound profiles and their corresponding antioxidant capacity of purple pitaya (*Hylocereus* sp.) genotypes. Z Naturforsch C 62(9–10):636–644

Harivaindaran KV, Rebecca OPS, Chandran S (2008) Study of optimal temperature, pH and stability of dragon fruit (*Hylocereus polyrhizus*) peel for use as potential natural colorant. Pak J Biol Sci 15(11):2259–2263

Herbach KM, Stintzing FC, Carle R (2006) Betalain stability and degradation-structural and chromatic aspects. J Food Sci 71:R41–R50

Jaafar RA, Abdul Rahman ARB, Che Mahmod NZ, Vasudevan R (2009) Proximate analysis of dragon fruit (*Hylocereus polyhizus*). Am J Appl Sci 6(7):1341–1346

Jamaludin NA, Ding P, Hamid AA (2011) Physicochemical and structural changes of red-fleshed dragon fruit (*Hylocereus polyrhizus*) during fruit development. J Sci Food Agric 91(2):278–285

Jamilah B, Shu CE, Kharidah M, Dzulkifly MA, Noranizan A (2010) Physico-chemical characteristics of red pitaya (*Hylocereus polyrhizus*) peel. Int Food Res J 17:59–65

Khalili RMA, Norhayati AH, Rokiah MY, Asmah R, Mohd Nasir MT, Siti Muskinah M (2006) Proximate composition and selected mineral determination in organically grown red pitaya (*Hylocereus sp.*). J Trop Agri Food Sci 34(2):269–276

Khalili RMA, Norhayati AH, Rokiah MY, Asmah R, Siti Muskinah M, Abdul Manaf A (2009) Hypocholesterolemic effect of red pitaya (*Hylocereus* sp.) on hypercholesterolemia induced rats. Int Food Res J 16:431–440

Le Bellec F, Vaillant F, Imbert E (2006) Pitahaya (*Hylocereus* spp.): a new fruit crop, a market with a future. Fruits 61(4):237–250

Lee VT (2002) Current status of the Vietnamase rural economy and measures for its vitalation and improving farmer's income. 9th JIRCAS international symposium 2002 – 'Value-addition to agriculture products'. pp 42–48

Lim HK, Chin Ping Tan CP, Roselina Karim R, Abdul Azis Ariffin AA, Jamilah Bakar J (2010) Chemical composition and DSC thermal properties of two species of *Hylocereus* cacti seed oil: *Hylocereus undatus* and *Hylocereus polyrhizus*. Food Chem 119(4):1326–1331

Luders L, McMahon G (2006) The pitaya or dragon fruit. Department of Primary Industries and Fisheries. Northern Territory, Australia, *Agnote D 42*

Mahattanatawee K, Manthey JA, Luzio G, Talcott ST, Goodner K, Baldwin EA (2006) Total antioxidant activity and fiber content of select Florida-grown tropical fruits. J Agric Food Chem 54:7355–7363

Mizrahi Y, Nerd A, Nobel PS (1996) Cacti as crops. Hortic Review 18:291–320

MMR (Malaysian Medical Resources) (2008) Pseudohaematuria due to dragonfruit ingestion. http://www.medcine.com.my/wp/?p=3152.

Phebe D, Chew MK, Suraini AA, Lai OM, Janna OA (2009) Red-fleshed pitaya (*Hylocereus polyrhizus*) fruit colour and betacyanin content depend on maturity. Int Food Res J 16:233–242

Raveh E, Nerd A, Mizrahi Y (1998) Responses of two hemi-epiphytic fruit crop cacti to different degrees of shade. Sci Hortic 73:151–164

Rebecca OPS, Boyce AN, Chandran S (2010) Pigment identification and antioxidant properties of red dragon fruit *(Hylocereus polyrhizus)*. Afr J Biotechnol 9(10):1450–1454

Stintzing FC, Schieber A, Carle R (2002) Betacyanins in fruits from red-purple pitaya, *Hylocereus polyrhizus* (Weber) Britton & Rose. Food Chem 77(1):101–106

Stintzing FC, Schieber A, Carle R (2003) Evaluation of colour properties and chemical quality parameters of cactus juices. Eur Food Res Technol 216:303–311

Stintzing FC, Klaiber I, Beifuss U, Carle R (2004) Structural investigations on betacyanin pigments by LC NMR and 2D NMR spectroscopy. Phytochem 65:415–422

Vaillant F, Perez A, Davila I, Dornier M, Reynes M (2005) Colorant and antioxidant properties of red pitahaya (*Hylocereus* sp.). Fruits 60:1–7

Woo KK, Chong YY, Li Hiong SK, Tang PY (2010a) Pectin extraction and characterization from red dragon fruit (*Hylocereus polyrhizus*): a preliminary study. J Biol Sci 10:631–636

Woo KK, Ngou FH, Ngo LS, Soong WK, Tang PY (2010b) Stability of betalain pigment from red dragon fruit (*Hylocereus polyrhizus*). Am J Food Technol 6:140–148

Wu MC, Chen CS (1997) Variation of sugar content in various parts of pitahaya fruit. Proc Fla State Hortic Soc 110:225–227

Wu LC, Hsu HW, Chen YC, Chiu CC, Lin YI, Annie Ho JA (2006) Antioxidant and antiproliferative activities of red pitaya. Food Chem 95:319–327

Wybraniec S, Mizrahi Y (2002) Fruit flesh betacyanin pigments in *Hylocereus* cacti. J Agri Food Chem 50:6086–6089

Wybraniec S, Platzner I, Geresh S, Gottlieb HE, Haimberg M, Mogilnitzki M (2001) Betacyanins from vine cactus *Hylocereus polyrhizus*. Phytochemistry 58:1209–1212

Wybraniec S, Nowak-Wydra B, Mitka K, Kowalski P, Mizrahi Y (2007) Minor betalains in fruits of *Hylocereus* species. Phytochem 68:251–259

Zainoldin KH, Baba AS (2010) The Effect of *Hylocereus polyrhizus* and *Hylocereus undatus* on physicochemical, proteolysis, and antioxidant activity in yogurt. Int J Med Sci 1(2):93–98

Hylocereus undatus

Scientific Name

*Hylocereus undatu*s (Haw.) Britt. et Rose

Synonyms

Cereus triangularis Haw., *Cereus triangularis* var. *aphyllus* (Haw.) Jacq., *Cereus triangularis* var. *major* (Haw.) DC., *Cereus tricostatus* Gosselin, *Cereus trigonus* Plum; *Cereus trigonus* var. *guatemalensis* (Haw.) Eichlam; *Cereus undatus* Haw., *Hylocereus guatemalensis* (Eichlam) Britton & Rose, *Hylocereus tricostatus* (Rol.-Goss) Britton et Rose.

Family

Cactaceae

Common/English Names

Belle Of The Night, Common Night Blooming Cereus, Conderella Plant, Dragon Fruit, Nanettika Fruit, Night Blooming Cereus, Pitahaya, Red Pitahaya, Queen Of The Night, Strawberry Pear, White-fleshed Dragon Fruit, White-fleshed Pitahaya.

Vernacular Names

Brazil: Cardo-Ananás, Jaramacarú, Rei-Da-Noite, Pitaya Vermelha De Polpa Branca (Portuguese);
Chinese: Huǒ Lóng Guǒ, Liang Tian Chi, Lóng Zhū Guǒ;
Columbia: Flor De Calis, Pitahaya Blanca Pitahaya Roja, Pitjaya;
Czech: Kaktus Zvlněný;
Danish: Pitahaya;
Eastonian: Maasik-Metskaktus;
French: Cierge-Lezard, Belle De Nuit, Fruit Du Dragon, Pitahya Dulce, Pitahaya Rouge, Pitaya, Poire De Chardon, Raquette Tortue;
German: Distelbirne, Drachenfrucht, Echte Stachelbirne, Königin Der Nacht, Rotepitahaya, Waldkaktus;
Hawaiian: Pāniniokapunahou, Panani O'Ka, Pāpipi Pua;
Hungarian: Pithayakaktusz;
Indonesia: Buah Naga, Buah Naga Merah;
Israel: Pitaya;
Malaysia: Dragonfruit, Buah Naga;
Mauritius: Débousse-To-Fesse;
Mexico: Junco Tapatío Pitahaya De Cardó, Pitahaya Dulce, Pitahaya Orejona, Pitahaya Roja, Pitahaya Blanca, Reina De La Noche, Tasajo (Spanish);
Philippines: Caliz;

Puerto Rico: Flor De Caliz, Junco, Junco Tapatio, Orijona, Pitajava, Reina De La Noche;
Portguese: Cato-Barse, Cardo-Ananaz, Pitaya Vermelha De Polpa Branca , Rainha Da Noite;
Slovaščina: Pitaja;
Spanish: Chacam, Chak-Wob, Pitahaya, Flor De Caliz, Pitaja, Junco, Junco Tapatio, Pithaya Orejona, Reina De La Noche, Tasajo, Zacamb;
Sri Lanka: Dragon fruit;
Swedish: Distelbirn, Echtestachelbrin, Röd Pitahaya, Skogskaktus;
Thai: Geow Mangon;
Venezuela: Flor De Calis, Pitahaya Blanca, Pitahaya Roja;
Vietnamese: Thanh Long.

Origin/Distribution

*Hylocereu*s species is endemic to Latin America. Exact origin of dragon fruit is not known but is believed to be in the area from southern Mexico, the Pacific side of Guatemala and Costa Rica, and El Salvador. It is commonly cultivated and naturalized throughout tropical American lowlands, the West Indies, the Bahamas, Bermuda, southern Florida and the tropics of the Old World. It has been widely introduced and escaped in tropical Asia, Australia, and South America. It is cultivated commercially in Israel, Thailand, Philippines, Okinawa (Japan), Taiwan, Sri Lanka, southern China, Malaysia, Vietnam, Indonesia and northern Australia.

Agroecology

Dragon fruit is a shade tolerant vine cactus from the tropical forest of Central America. It has a wide range of agroecological adaptability. This cactus is able to tolerate drought, heat, poor soil, and cold. The modification of the stem for water storage, the reduction or absence of leaves, the waxy surfaces, and night-time opening of the tissues for carbon dioxide uptake (the CAM process), enable the plants to tolerate harsh conditions The roots are non-succulent and require small amounts of water and cooler temperatures. It can be cultivated in the tropics or dry subtropics in areas with mean temperatures of 20–30°C. It can tolerate maximum temperature of 38–40°C and low temperature of near zero. Stem damage occurs at 45°C and cell mortality occurs at −1.5°C. Dragon fruit is not fastidious of soil type but is intolerant of saline soils or waterlogged conditions. It grows under full sun or partial shade. It grows best on well-drained red-yellow podzolic, reddish brown earth or lateritic soils especially when supplemented with organic matter at pH of 5.5–6.5. Its moisture requirement is modest, 600–1,500 mm mean annual rainfall, excessive rain can lead to flower drop and fruit rot. In its natural habitat it climbs up trees or rock. For commercial cultivation artificial supports such as concrete or wooden post, fences and trees has to be provided for this epiphyte.

Edible Plant Parts and Uses

Dragon fruit is highly relished especially if chilled and cut in half so that the flesh can be eaten with a spoon or eaten in fruit salad. The flesh is firm and crisp, with a delicately sweet and lingering flavour. The juice is also enjoyed as a cool, refreshing drink and used as a base for beverages. The juicy flesh can also be mixed with milk or sugar, used in marmalades, jellies, ices and soft drinks. A syrup made of the whole fruit is used to colour pastries and candy.

The unopened flower buds are cooked and eaten as a vegetable. The flowers are also harvested before anthesis and dried for subsequent use as vegetables in soups. Vietnam exports dried dragon fruit blooms which can be obtained from Asian stores in Australia.

Botany

A climbing or scrambling succulent hemi-epiphytic, cactus reaching heights of 6–10 m with spreading, drooping branches (Plate 1). Its stems are three-winged, the wings 2–3 cm wide, crenate

Plate 1 Plant habit on a trellis

Plate 2 Large, white flowers of white-fleshed dragon fruit

Plate 3 Open pitahaya flower

with green, fleshy, much-branched stems with scalloped, horny margins. They scramble over rocks or bushes, climb and form dense masses in trees, and cling to walls, by means of numerous, strong aerial roots. Areoles about 4–6 cm apart, 2–5 mm diameter, with 1–4 conical spines 1–3 mm long. Flowers nocturnal, fragrant, 25–30 cm long, 15–25 cm by 15–34 cm; sepaloid perianth parts greenish white, linear to linear-lanceolate, 10–15 cm long, 1–1.5 cm wide, inner perianth parts white, oblanceolate, 10–15 cm long, ca 2.5 cm wide (Plates 2–3); stamens numerous with long cream-colored staminal filaments; style cream-colored, 17.5–20 cm long; stigma lobes up to 24, cream-colored. Fruit large, red, non-spiny berry, subglobose, ellipsoid, oblong-oval, 5–12.5 cm long by 4–9 cm across, coated with the bright-red, fleshy, ovate bases of scales red and greenish at the tips (Plates 4–7). Within is white or deep red, juicy, sweet pulp containing innumerable tiny black, partly hollow seeds (Plates 6–7).

Nutritive/Medicinal Properties

The nutrient composition of the fruit per 100 g edible portion was reported as: energy 67.70 kcal, water 85.30%, protein 1.10 g, fat 0.57 g, fibre 11.34 g, Ca 10.20 mg, P 27.50 mg, Mg 38.90 mg,

Plate 4 Developing pitahaya fruits

Plate 6 L. S. of white-fleshed pitahaya fruit

Plate 5 Pitahaya fruit on sale in the market

Plate 7 Pitahaya fruit sliced for consumption

K 3.37 mg, Fe 0.70 mg, Na 8.90 mg, Zn 0.35 mg, sorbitol 32.70 mg, fructose 3.20 mg, vitamin C 3 mg, niacin 2.8 mg (FAMA 2006).

In general, white fleshed pitaya fruits were found to have higher total soluble solid (TSS) than red fleshed pitaya (Wu and Chen 1997). TSS varied in the different fruit parts in decreasing sequence of fruit core, stylar-end, stem-end and peripheral parts. Glucose, fructose and sucrose were the main soluble sugars and their concentrations followed a similar trend except for sucrose. In white fleshed fruit, glucose varied from 104.3 mg/g in the core to 64.3 mg/g in the peripheral part, fructose 64.9 mg/g in the core to 40.1 mg/g in the peripheral, sucrose 5.4 mg/g in the core to 7.5 mg/g in the peripheral. For red fleshed fruit, glucose varied from 68.1 mg/g in the core to 41.9 mg/g in the peripheral part, fructose 49.9 mg/g in the core to 32.3 mg/g in the peripheral, sucrose 4.2 mg/g in the core to 6.8 mg/g in the peripheral. Distribution of sugars in various parts of the fruit was closely associated with the different enzyme activities. In both red and white-fleshed fruits, the core had the highest content of acid, invertase, total invertase and amylase activities followed by the stylar-end, stem-end and peripheral parts.

Studies in Japan showed that mature pithaya fruit had higher total soluble solids (TSS) contents in the centre of the fruit compared with other sections (Nomura et al. 2005). Malic acid was the main acid in the mature fruit, with no clear seasonal pattern. Changes in glucose and fructose contents exhibited a similar pattern to that of total soluble solids, but the values were higher in mature Autumn fruit than in mature Summer fruit, and differed among fruit sections. Some studies reported that *Hylocerus undatus* with white flesh had higher soluble solids than *Hylocerus* fruit with red flesh and the core of the fruit exhibited higher levels than the peripheral parts. The soluble solids comprised mainly of reducing sugars such as glucose 46 g/L, fructose 18 g/L (Stintzing et al. 2003) and sucrose 2.8–7.5% of total sugars (Wu and Chen 1997).

The deep purple or deep red colour of the pulp was found to be due to a set of nitrogen-containing pigments known as the betalains (Harivaindaran et al. 2008), made up of the red-violet betacyanins and yellow betaxanthins with maximum absorptivity at 535 and 480 nm, respectively (Herbach et al. 2006). According to Stintzing et al. (2002) and Wybraniec and Mizrahi (2002), at least seven identified betacyanins in the *Hylocereus* genus namely: betanin, isobetanin, lhyllocactin, isophyllocactin, betanidin, isobetanidin and bougainvillein-R-I all had identical absorption spectra that contributed to the deep purple coloured pulp.

Studies showed that the addition of white and red dragon fruit into yogurt augmented the milk fermentation rate, lactic acid content, syneresis percentage, antioxidant activity, and total phenolics content in yogurt (Zainoldin and Baba 2010). Milk fermentation rate was increased in red dragon fruit yogurt for all doses (−0.3606 to −0.4126 pH/h) while only white dragon fruit yogurt with 20% and 30% (w/w) composition showed increase in fermentation rate (−0.3471 to −0.3609 pH/hour) compared to plain yogurt (−0.3369 pH/hour). All dragon fruit enriched yogurts generally exhibited lower pH readings (pH 3.95–4.03) compared to plain yogurt (pH 4.05). Both fruit yogurts showed a higher lactic acid percentage (1.14–1.23%) compared to plain yogurt (1.08%). Significantly higher syneresis percentage (57.19–70.32%) compared to plain yogurt (52.93%) were noted in all fruit enriched yogurts. The antioxidant activity of plain yogurt (19.16%) was increased by the presence of white and red dragon fruit (24.97–45.74%). All fruit enriched yogurt exhibited an increase in total phenolic content (36.44–64.43 mg/ml) compared to plain yogurt (20.25 mg/ml). However, the addition of white and red dragon fruit did not increase the proteolysis of milk during fermentation.

The seed oil was found to contain about 50% essential fatty acids C18:2 (48%) and C18:3 (1.5%). The fatty acid composition of *H. undatus* seed oil was determined as follows: 14:0 myristic acid 0.3%; 16:0 palmitic acid 17.1%; 18:0 stearic acid 4.37%; palmitoleic acid 0.61%; 16:1 undifferentiated oleic acid 23.8%; cis-vaccenic acid (11-cis-octadecenoic acid) 2.81%, 18:2 undifferentiated linoleic acid 50.1%; 18:3 linolenic acid 0.98% (Azis et al. 2009). Another study showed that pitaya (*Hylocereus undatus* and *Hylocereus polyrhizus*) seed oil contained a high amount of oil (18.33–28.37%) (Lim et al. 2010). The three major fatty acids in the *H. undatus* seed oil (WFSO) and *H. polyrhizus* seed oil (RFSO) were linoleic, oleic, and palmitic acids. The total tocopherol contents in the WFSO and RFSO were 36.70 and 43.50 mg/100 g, respectively. The phytosterol compounds identified in the WFSO and RFSO were cholesterol, β-sitosterol, campesterol and stigmasterol. Seven phenolic acid compounds were identified in the WFSO and RFSO, namely, caffeic, gallic, p-coumaric, p-hydroxybenzoic, protocatechuic, syringic and vanillic acids. This study revealed that pitaya seed oil possessed a high level of functional lipids and could be used as a new source of essential oil.

Research in Mexico reported that two new pentacyclic triterpene taraxast-20-ene-3α-ol and taraxast-12, 20(30)-dien-3α-ol isolated from chloroform extract of leaves of *Hylocereus undatus* exhibited protective activity against the elevation of skin vascular permeability in rabbits (Gutiérrez et al. 2007). The scientists also reported that in streptozotocin-diabetic rats, where healing was delayed, topical applications of *H. undatus* (aqueous extracts of leaves, rind, fruit pulp and flowers) enhanced hydroxyproline level, tensile strength, total proteins, DNA collagen content and better epithelization thereby facilitating healing (Gutiérrez et al. 2005). *H. undatus* had no hypoglycaemic activity.

The sap of the stems of *H. undatus* has been employed as a vermifuge but it is said to be caustic and hazardous (Morton 1987). The air-dried, powdered stems contain *B*-sitosterol.

Other Uses

Pitahaya is also planted as an ornamental and occasionally planted as flowering hedge plant.

Comments

H. undatus can be differentiated from *H. polyrhizus* on the basis of their outer perianth segments which are greenish white in *H. undatus* and yellowish with red tip and edges in *H. polyrhizus*.

Selected References

Azis A, Jamilah B, Chin PT, Russly R, Roselina K, Chia CL (2009) Essential fatty acids of pitaya (dragon fruit) seed oil. Food Chem 114:561–564

FAMA (2006) Dragon fruit (*Hylocereus undatus*). http://eshoppe.famaxchange.org/index.php?ac=147&ch=ctlg_fspd&pg=ctlg_fspd_fruit&tpt=eshoppe

Gunasena HPM, Pushpakumara DKNG, Kariyawasam M (2007) Dragon fruit (*Hylocerus undatus* (Haw.) Britton and Rose), pp 110–141. In: Pushpakumara DKNG, Gunasena HPM, Singh VP (eds) Underutilized fruit trees in Sri Lanka, vol 1. World Agroforestry Centre (ICRAF), South Asia Regional Office, New Delhi, 504 pp

Gutiérrez RMP, Vargas Solis P, Ortiz H YD (2005) Wound healing properties of *Hylocereus undatus* on diabetic rats. Phytother Res 19:665–8

Gutiérrez RMP, Vargas Solis R, Baez EG, Flores JMM (2007) Microvascular protective activity in rabbits of triterpenes from *Hylocereus undatus*. J Nat Med 61(3):296–301

Harivaindaran KV, Rebecca OPS, Chandran S (2008) Study of optimal temperature, pH and stability of dragon fruit (*Hylocereus polyrhizus*) peel for use as potential natural colorant. Pak J Biol Sci 15(11):2259–63

Herbach KM, Stintzing FC, Carle R (2006) Betalain stability and degradation – structural and chromatic aspects. J Food Sci 71:R41–R50

Le Bellec F, Vaillant F, Imbert E (2006) Pitahaya (*Hylocereus* spp.): a new fruit crop, a market with a future. Fruits 61(4):237–250

Li ZY, Taylor NP (2007) Cacataceae. In: Wu ZY, Raven PH, Hong DY (eds) Flora of China. Vol. 13 (Clusiaceae through Araliaceae). Science Press/Missouri Botanical Garden Press, Beijing/St. Louis

Lim HK, Chin Ping Tan CP, Roselina Karim R, Abdul Azis Ariffin AA, Jamilah Bakar J (2010) Chemical composition and DSC thermal properties of two species of *Hylocereus* cacti seed oil: *Hylocereus undatus* and *Hylocereus polyrhizus*. Food Chem 119(4):1326–1331

Luders L, McMahon G (2006) The pitaya or dragon fruit. Department of Primary Industries & Fisheries. Northern Territory, Australia, Agnote D 42

Mizrahi Y, Nerd A, Nobel PS (1996) Cacti as crops. Hort Reviews 18:291–320

Morton J (1987) Strawberry pear. In: Fruits of warm climates. Julia F. Morton, Miami, pp 347–348

Nerd A, Mizrahi Y (1996) Reproductive biology of cacti fruit crops. Hort Review 18:321–346

Nomura K, Ide M, Yonemoto Y (2005) Changes in sugars and acids in pitaya (*Hylocereus undatus*) fruit during development. J Hort Sci Biotechnol 80(6):711–715

Stintzing FC, Schieber A, Carle R (2002) Betacyanins in fruits from red-purple pitaya, *Hylocereus polyrhizus* (Weber) Britton & Rose. Food Chem 77(1):101–106

Stintzing FC, Schieber A, Carle R (2003) Evaluation of colour properties and chemical quality parameters of cactus juices. Eur Food Res Technol 216:303–311

Wu MC, Chen CS (1997) Variation of sugar content in various parts of pitahaya fruit. Proc Fla State Hort Soc 110:225–227

Wybraniec S, Mizrahi Y (2002) Fruit flesh betacyanin pigments in *Hylocereus* cacti. J Agri Food Chem 50:6086–6089

Zainoldin KH, Baba AS (2010) The Effect of *Hylocereus polyrhizus* and *Hylocereus undatus* on physicochemical, proteolysis, and antioxidant activity in yogurt. Int J Med Sci 1(2):93–98

Nopalea cochenillifera

Scientific Name

Nopalea cochenillifera (L.) Salm-Dyck

Synonyms

Cactus campechianus Thierry ex Steud., *Cactus cochenillifer* L. (basionym), *Opuntia cochenillifera* (L.) Mill.[1]

Family

Cacataceae

Common/English Names

Cochineal Cactus, Cochineal Nopal Cactus, Cochineal-Plant, Nopal Cactus, Nopalea Grande, Nopales Opuntia, Prickly Pear, Tunita, Velvet Opuntia, Warm Hand, Wooly Joint Prickly Pear.

Vernacular Names

Anguilla: Cochineal Cactus, French Prickle;
Aztec: Nopal Nochetzli;
Bermuda: Cochineal Cactus;
Chinese: Yan Zhi Zhang;
French: Cochenillier, Nopal A Cochenille, Raquette Espagnole;
India: Puchikallai (Tamil);
Mexico: Nopal, Tuna;
Norwegian: Cochenillekaktus;
Philippines: Arapal (Bis-Cebuano);
Portuguese: Cacto-De-Cochonilha, Palma, Palma-De-Engorda, Palma-Doce, Palma-Miuda, Palmatória;
Saint Helena: English Tungy, Opuntia, Prickly Pear, White Tungi, White Tungy;
Sierra Leone: Tandaletando (Kisse), Ngele-Gonu (Mende);
Spanish: Nopal Chamacuero, Nopal De Cochinilla, Nopal De La Cochinilla;
Venezeula: Tuna Real;

Origin/Distribution

This cactus is native to Mexico. Wild distribution is found in tropical central America, Mexico and Jamaica. This cactus is frequently cultivated in tropical and subtropical America, also in the Mediterranean, Canary island, tropical Africa, India and Southeast Asia.

---

[1] Some authorities now recognize this species in the genus *Nopalea*, as distinct from *Opuntia*, based on pollen-morphological differences (Li and Taylor 2007).

Agroecology

Nopalea cochenillifera is a shrub-like xerophytic plant with good adaptation to semiarid climatic regimes. Studies showed that it tolerated high-temperatures, as LT_{50} (the temperature killing 50% of the cells, based on uptake of the vital stain neutral red) averaged 57°C for stem chlorenchyma cells and root cortical cells for plants at day/night air temperatures of 25/20°C; the plants were relatively susceptible to low temperatures, for which LT_{50} averaged −7°C (Nobel and Zutta 2008) As air temperatures were reduced by 15°C (10/5°C), LT_{50} for low temperatures decreased 1.1°C for stems and roots; as they were raised by 20°C (45/40°C), LT_{50} for high temperatures increased 4.2°C for stems and 3.4°C for roots. This acclimation leads to advantageous seasonal adjustments to ambient temperatures. The increased low-temperature tolerance during drought was accompanied by a decreased stem water potential and water content.

It is found in hammocks, fields, sandy soils at low altitudes and on mountain slopes in its native range.

Plate 1 Plant habit – woody trunk and cladodes

Plate 2 Flowers and young fruit

Edible Plant Parts and Uses

Fruit and cladodes are edible. The fruits are eaten fresh or processed into drinks, fruit salads etc as with *Opuntia ficus–indica*. Nopalea tender young pads or cladodes, are known as "nopalitos verdes". They are commonly consumed as a fresh or cooked green vegetable in Mexico and in some parts of the United States. The plants have been used as host plants for the cochineal insects (*Dactylopius coccus*) used for the commercial production of a dye. The dye can be used as a natural colouring of foods, for soft drinks and many cosmetics (including lipsticks).

Botany

Shrubs or small trees, 2–4 m tall. Trunk when present is terete and woody. Stem segments (cladodes) are green, elliptic to narrowly obovate, 8–40(–50) × 5–7.5(–15) cm, thick, margin entire, base and apex rounded (Plate 1). Areoles 2 mm in diameter, 2–3 cm apart. Spines usually absent, when present: 1–3 per areole, spreading, greyish tan, acicular, 3–9 mm; glochids inconspicuous and early deciduous. Leaves conic, 3–4 mm, early deciduous. Flowers 1.2–1.5 cm in diam., erect (Plate 2). Sepaloid tepals with brilliant red or green midrib, spatulate, largest ones ovate-deltoid, 5–12 × 6–9 mm, margin entire, apex acute. Petaloid tepals bright red, ovate to obovate, 1.3–1.5 × 0.6–1 cm, margin entire or undulate, apex rounded or acute. Filaments pink, 3–4 cm; anthers pink, 1.5 mm. Style pink, 4–4.5 cm; stigmas 6–8, greenish, 3 mm, nectar chamber elliptic to obconic. Fruit red, ellipsoid, 3–5 × 2.5–3 cm, umbilicus developed but not conspicuous. Seeds gray or tan, thickened discoid, 3 mm in diameter, slightly pubescent.

Nutritive/Medicinal Properties

The cladodes are rich in minerals, vitamins and amino acids and are low in lipids.

The nutrient value per 100 g edible portion of nopales (cladodes) of *Nopalea cochenillifera* minus 4% refuse of spines and dark spots (USDA 2010) was reported as follows: water 94.12 g, energy 16 kcal (66 kJ), protein 1.32 g, total lipid 0.09 g, ash 1.14 g, carbohydrate 3.33 g, total dietary fibre 2.2 g, total sugars 1.15 g, calcium 164 mg, iron 0.59 mg, magnesium 52 mg, phosphorus 16 mg, potassium 257 mg, sodium 21 mg, zinc 0.25 mg, copper 0.052 mg, manganese 0.457 mg, selenium 0.7 µg, vitamin C 9.3 mg, thiamine 0.012 mg, riboflavin 0.041 mg, niacin 0.410 mg, pantothenic acid 0.167 mg, vitamin B-6 0.070 mg, total folate 3 µg, total choline 7.3 mg, vitamin A RAE 23 µg, β-carotene 250 µg, α-carotene 48 µg, vitamin A 457 IU, vitamin K (phylloquinone) 5.3 µg, total saturated fatty acids 0.016 g, 16:0 0.012 g, 18:0 0.002 g; total monounsaturated fatty acids 0.018 g, 16:1 undifferentiated 0.002 g, 18:1 undifferentiated 0.017 g, total polyunsaturated fatty acids 0.050 g, 18:2 undifferentiated 0.044 g, 18:3 undifferentiated 0.005 g; tryptophan 0.014 g, threonine 0.040 g, isoleucine 0.049 g, leucine 0.077 g, lysine 0.059 g, methionine 0.015 g, cystine 0.008 g, phenylalanine 0.049 g, tyrosine 0.029 g, valine 0.059 g, arginine 0.052 g, histidine 0.025 g, alanine 0.050 g, aspartic acid 0.086 g, glutamic acid 0.145 g, glycine 0.046 g, proline 0.043 g and serine 0.043 g.

Nopalea cochenillifera was found to be most suitable for harvest at the end of the rapid growth phase, at which time they were 11–13 cm long, weighed about 40 g, and had a chemical composition similar to that of certain *Opuntia* species used as a vegetable (Nerd et al. 1997). The cladodes exhibited typical acidity changes of CAM plants, with titratable acidity measured at harvest fluctuating from a morning high level of 620 mmol H^+/m^2 to an evening low level of 69 mmol H^+/m^2 (155 and 17 mmol H^+ kg/FW, respectively). Acidity of fresh morning-harvested cladodes stored in an open place or a room declined to an acceptable level of 400 mmol H^+/m^2 (100 mmol H^+ kg/FW) by the second day. Dark storage studies indicated that cladodes had to be protected from low temperature (4°C) because of chilling injuries and maintenance of acidity. With respect to both acidity and weight loss, the most favourable treatment for long-term storage was wrapping the cladodes in PVC film and storing them at either 12°C or 20°C.

Nopalea cochenillifera (NC) extracts of both fresh and dried pads exhibited antibacterial and antifungal activities (Gomez-Flores et al. 2006). MICs of fresh NC hexanic, chloroformic and ethanolic fractions against *Candida albicans* were 250, 250 and 3.9 µg/ml respectively; MICs of fresh NC against *Salmonella enterica* var. *thyphimurium* were 15.6, 62.5 and 3.9 µg/ml respectively; and MICs of fresh NC against *Escherichia coli* were 500, NA (no activity) and 3.9 µg/ml respectively. In addition, MICs of dried NC hexanic, chloroformic and ethanolic fractions against *C. albicans* were 31.2, 31.2 and 3.9 µg/ml respectively. MICs of dried NC against *Salmonella enterica* var. *thyphimurium* were 15.6, NA and 3.9 µg/ml respectively; and MICs of dried NC against *Escherichia coli* were NA, 62.5 µg/ml and NA respectively.

Nopalea cochenillifera was also shown to inhibit herpes simplex virus type 1 infection (Szuchman et al. 1999).

On the downside, *Nopalea cochinellifera* was reported to increase blood glucose levels in treated mice (Villaseñor and Lamadrid 2006).

Nopalea cochinellifera has been reported to be used for hypertension in Trinidad and Tobago (Lans 2006). In Tamil Nadu, India, the cladodes were used for toothache (Ganesan 2008). In Senegal, the cladodes are used as poultices for bruises.

Other Uses

The stem segments, (pads or cladodes), of *Nopalea cochenillifera* are used as food, fodder, and poultices, and for rearing cochineal insects to obtain a red dye (once a major industry). This species may have been selected for spinelessness in Mexico, much like *Opuntia ficus-indica*, to ease the culturing and collection of cochineal scale insects for red dye. With the invention of synthetic dyes the cultivation area decreased since the end of the eighteenth century. But recent demands for natural dyes, (synthetic ones have been linked to cancer) resulted in new interest in the cochineal dye. The cactus is also planted as an ornamental plant.

Comments

In Mexico, annual production of this vegetal is about 600,000 tons, with the area of Milpa Alta, D.F. as the most important producer. In the United States, *Nopalea cochenillifera* clone 1,308 is produced in the States of California and Texas, with an annual production of 5,000 tons (Felker et al. 2006).

Selected References

Anderson EF (2001) The cactus family. Timber Press, Portland, 776 pp

Felker P, Paterson A, Jenderek MM (2006) Forage potential of *Opuntia* clones maintained by the USDA, National Plant Germplasm System (NPGS). Crop Sci 46:2161–2168

Ganesan S (2008) Traditional oral care medicinal plant survey of Tamil Nadu. Nat Prod Rad 7(2):166–172

Gomez-Flores R, Tamez-Guerra P, Tamez-Guerra R, Rodriguez-Padilla C, Monreal-Cuevas E, Hauad-Marroquin LA, Cordova-Puente C, Rangel-Llanas A (2006) In vitro antibacterial and antifungal activities of *Nopalea cochenillifera* pad extracts. Am J Infect Dis 2(1):1–8

Lans CA (2006) Ethnomedicines used in Trinidad and Tobago for urinary problems and diabetes mellitus. J Ethnobiol Ethnomed 2:45

Li ZY, Taylor NP (2007) Cacataceae. In: Wu ZY, Raven PH, Hong DY (eds.) Flora of China. Vol. 13 (Clusiaceae through Araliaceae). Science Press/Missouri Botanical Garden Press, Beijing/St. Louis

Myre M (1974) Los cactos forrajeiros. I. Cultura e adaptaçao de *Nopalea cochenillifera* (L.) Salm-Dyck em Regioes de Moçambique. Agron Moçamb 8:19–30

Nerd A, Dumoutier M, Mizrahi Y (1997) Properties and postharvest behavior of the vegetable cactus *Nopalea cochenillifera*. Posthar Biol Technol 10(2):135–143

Nobel PS, Zutta BR (2008) Temperature tolerances for stems and roots of two cultivated cacti, *Nopalea cochenillifera* and *Opuntia robusta*: acclimation, light, and drought. J Arid Environ 72(5):633–642

Pinkava DJ (2003a) Cactaceae cactus family: Part 6. *Opuntia* P. Miller Prickly-pears. J Arizona-Nevada Acad Sci 35:137–150

Pinkava DJ (2003b) *Nopalea* Salm-Dyck. In: Flora of North America Editorial Committee (ed.) Flora of North America North of Mexico, vol 4. Oxford University Press, New York/Oxford

Szuchman A, Tal J, Mizrahi Y (1999) Antiviral properties in cladodes of the cactus *Nopalea cochenillifera* (L.). In: XVI International botanical congress, Israel. Abstract 5409, Poster 2429

U.S. Department of Agriculture, Agricultural Research Service (USDA) (2010) USDA National Nutrient Database for Standard Reference, Release 23. Nutrient Data Laboratory Home Page. http://www.ars.usda.gov/ba/bhnrc/ndl

Villaseñor IM, Lamadrid MR (2006) Comparative antihyperglycemic potentials of medicinal plants. J Ethnopharmacol 104(1–2):129–31

Opuntia ficus-indica

Scientific Name

Opuntia ficus-indica (L.) Miller

Synonyms

Cactus chinensis Roxb., *Cactus decumanus* Willd.; *Cactus elongatus* Willd., *Cactus ficus-indica* L(basionym)., *Cactus lanceolatus* Haw., *Cactus opuntia* Gussone, *Cactus opuntia* Linnaeus, *Opuntia amyclaea* Ten., *Opuntia chinensis* (Roxb.) K. Koch, *Opuntia compressa* J. F. Macbride; *Opuntia cordobensis* Speg., *Opuntia crassa* Haw., *Opuntia decumana* (Willd.) Haw., *Opuntia elongata* (Willd.) Haw., *Opuntia ficus-barbarica* Berger, *Opuntia ficus-indica* var. *gymnocarpa* (F. A. C. Weber) Speg., *Opuntia gymnocarpa* F. A. C. Weber, *Opuntia hispanica* Griffiths, *Opuntia lanceolata* (Haw.) Haw., *Opuntia maxima* Mill.; *Opuntia megacantha* Salm-Dyck; *Opuntia occidentalis* Auct., *Opuntia parvula* Salm-Dyck, *Opuntia paraguayensis* K. Schum., *Opuntia tuna-blanca* Speg., *Opuntia undulata* Griff., *Opuntia undosa* Griff., *Opuntia vulgaris* Miller, *Platyopuntia cordobensis* (Speg.) F. Ritter, *Platyopuntia ficus-indica* (Mill.) F. Ritter.

Family

Cactaceae

Common/English Names

Barbary-Fig, Indian Fig, Indian Fig Opuntia, Indian-Fig Prickly-Pear, Mission Cactus, Mission Prickly-Pear, Nopal, Nopalito, Prickly Pear, Smooth Mountain Prickly-Pear, Smooth Prickly-Pear, Spineless Cactus, Tuberous Prickly-Pear, Tuna Cactus.

Vernacular Names

***Afrikaan**s*: Boereturksvy, Doringblaar, Grootdoringturksvy, Kaalblaar;
Algeria: Kermus Ennsara, Kerma (Arabic);
Arabic: Beles;
Algeria: Seruit (Spineless Form), Hendi, Karmouz En-Nsara, Tin Shawki (Arabic);
Argentina: Higuera, Higuera Chumba, Higuera De Pala, Nopal, Nopal De Castilla, Nopal Pelón, Tuna;
Bolivia: Higuera, Higuera Chumba, Higuera De Pala, Nopal, Nopal De Castilla, Nopal Pelón, Tuna;
Brazil: Figo-Da-Índia, Figueira-Da-Barbária, Figueira-Da-Índia, Figueira-Do-Inferno;
Catalan: Chumbo, Chumbua, Fifuera De Moro;
Chile: Higuera, Higuera Chumba, Higuera De Pala, Nopal, Nopal De Castilla, Nopal Pelón, Tuna;
Chinese: Ci Li, Xian Tao, Li Guo Xian Ren Zhang, Feng Lon Guo, Feng Lung Kuo, Huo Long Guo, Hua Lung Kuo;

Czech: Opuncie Smokvoňovitá;
Danish: Ægte Figenkaktus, Figen-Kaktus, Indisk Figen, Kaktusfigen;
Dutch: Barbarijse Vijgen, Bloedvijg, Cactusvijg, Schijfcactus, Woestijnvijg;
Eastonian: Suureviljaline Viigikaktus;
Egypt: Seruit (Spineless Form), Hendi, Karmouz En-Nsara, Tin Shawki (Arabic);
Eritrea: Beles;
Ethiopia: Beles, Qulqual-Aabashaa;
Finnish: Viikunaopuntia;
French: Chardon D Inde, Figue De Barbarie, Figuier À Raquettes, Figuier D'inde, Opunce, Raquette; Hawaiian: Pānini, Pāpipi;
German: Echter Feigenkaktus, Feigendistel, Feigenkaktus, Feigenopuntie, Frucht Des Feigenkactus, Kaktusfeige, Kaktusfeigen, Stachelfeigen, Indianische Feige, Indische Feige, Nopal, Nopalitos;
Greek: Fragkosykia;
Hungarian: Közönséges Fügekaktusz;
Italian: Fico D'india, Opunzia;
Japanese: Opunchia;
Libya: Seruit (Spineless Form), Hendi, Karmouz En-Nsara, Tin Shawki (Arabic);
Madagascar: Raketa;
Maltese: Bajtra;
Morocco: Âknâri Lemselmîn, El-Kerm, Handiyya, Hindîr, Hindiya, Kermous(Arabic), Âknari, Taknârit (Berbers), Figuier De Barbarie, Figuier D'inde, Cactus Raquette (Local French);
Norwegian: Fikenkaktus;
Peru: Higuera, Higuera Chumba, Higuera De Pala, Nopal, Nopal De Castilla, Nopal Pelón, Tuna;
Philippines: Nopal, Palad (Bikol), Dapal (Bisaya), Abacus (Bontok), Dila-Dila (Iloko), Dilang-Baka (Tagalog);
Polish: Opuncja Figa Indyjska;
Portuguese: Figo-Da-Índia, Figo-Da-Espanha, Figueira-Da-Barbária, Jamaracá, Jurumbeba, Orelha-De-Onça, Palma-De-Gado, Palma-Gigante, Figo De Pitoira, Figueira Da India, Palmatoria Sem Espinhos, Tabaibo;
Russian: Opuntsiia Indiiskaia, Samyi, Obyknovennyi;
Sicily: Ficodinnia;
South Africa: Itolofiya;
Spain: Cardón De México, Chumba, Chumbera, Higo Chimbo, Higo Chumbo, Higo De Cactus, Higo De Pala, Higo México, Higos Chumbos, Higos De La India, Tuna De España, Tuna De Castilla, Tuna Española, Tuna Mansa, Tuna Real;
Swedish: Fikonkaktus, Fikonopuntia;
Tunisia: Seruit (Spineless Form), Hendi, Karmouz En-Nsara, Tin Shawki (Arabic).

Origin/Distribution

The biogeographic and evolutionary origins of *Opuntia ficus-indica* have been obscured by ancient and widespread cultivation and naturalization. The origin of this species has been investigated through the use of Bayesian phylogenetic analyses of nrITS DNA sequences. These analyses has supported the following hypotheses: that *O. ficus-indica* is a close relative of a group of arborescent, opuntioid cactus from southern and central Mexico; that the nucleus of domestication for this species is in central Mexico; and that the taxonomic concept of *O. ficus-indica* may include clones derived from multiple lineages and therefore be polyphyletic (Griffith 2004).

Opuntia ficus-indica is a long-domesticated cactus crop that plays an important role in agricultural economies throughout arid and semi-arid areas of the world. Evidence exists relating to the use of *Opuntia* as human food at least 9,000–1,200 years ago. The leading *Opuntia* tuna producing countries are Mexico, Spain, Sicily, and the coasts of Southern Italy, Morocco, Algeria, Egypt, Saudi Arabia, Israel, Palestine, Chile, Brazil, Turkey and northern Africa, as well as in Eritrea and Ethiopia. Mexico exports to the US and Japan, while Italy and North Africa produce good quality fruit. There is also appreciable plantings in Jeju Island in south Korea with an area of over 2,000 ha of commercial production and an annual production of approximately 2,500 tons. *Opuntia ficus-indica* is grown for commercial purposes as food and medicinal plant in Jeju and has been made the official Natural Monument Number 429 by Bukjeju-gun Province (Lee et al. 2005).

Agroecology

Opuntioid cacti including *Opuntia ficus-indica* are adaptable to a range of climatic conditions from tropical to subtropical, semi-arid to arid from 0–600 to 2,600 m altitude. They are recognized as ideal crops for semi-arid and arid climatic regimes as they are hardy, extremely drought and heat tolerant. Owing to crassulacean acid metabolism, these plants exhibit a high water-use efficiency with a relatively high potential of biomass production (Lopez 1995; Russell and Felker 1987). They have been reported to be four- to five-times more efficient in converting water to dry matter than the most efficient grasses. Some *Opuntia* species including *O. ficus indica* grow rapidly with fresh-fruit yields of 8,000–12,000 kg/ha/year or more and dry-matter vegetative production of 20,000–50,000 kg/ha/year (Russell and Felker 1987).

Opuntia ficus-indica is found in coastal chaparral, sage scrub, arid uplands, washes, canyons, disturbed sites, rocky slopes and river banks. It prefers a mean annual temperature of 18–26°C but can survive brief periods of frost down to −2°C to −3°C, and mean annual rainfall of 150–600 mm. It adapts well to poor soils but thrives best on deep sandy soils or slightly acid, free-draining soils. The plant is adversely sensitive to prolonged water-logging as such poorly drained clayey soils should be avoided. Excessive moisture is detrimental and can lead to disease and death, particularly in cool conditions. If the climate is very dry, clayey soils are tolerated. In wetter areas, higher pH is preferable (close to neutral or slightly above) whereas in dry areas lower pH may be tolerated. Very acid soils may not be suitable for nopalito. Cactus pear grows in full sun and requires no shelter from wind. The only requirement is that it be planted and left to grow on its own, without fertilizer or watering.

Edible Plant Parts and Uses

Opuntia ficus-indica is widely eaten throughout Latin America, the Mediterranean and the Middle East. It is also eaten in Europe, Australia, New Zealand, and South Korea. The ripe pulp ranges in flavour from acid-sweet to very sweet and taste like a sweet, juicy watermelon and is usually eaten fresh or chilled. The fleshy, white/yellowish or reddish/purple pulp contains many tiny hard seeds that are usually swallowed, but should be avoided by those who have problems with seeds. The pulp can be dried, made into compotes, jellies, jams, candy, preserves, "tuna cheese", alcoholic and non-alcoholic beverages. Currently, it is cultivated on Cheju Island for use in the manufacture of health foods such as a tea, jam and juice. Mexicans have used Opuntia since ancient times to make an alcoholic drink called *colonche*. In the Province of Enna, Sicily, a cactus pear-flavored liqueur called *Ficodi* is produced. In Malta, a liqueur called *Bajtra* is made from the fruit. On Saint Helena Island, the prickly pear also gives its name to locally distilled liqueur, *Tungi Spirit*. In Chile, wine, alcohol and vinegar can be developed from the fruits by various processes of fermentation (Bustos 1981).

The large flattened jointed stems (platyclades, pads or cladodes) of *Opuntia* species are known in Mexico as nopales or nopalitos. The young green pads and fruits have been a source of nourishment for the Indians of the Americas well before the colonial period, deep within Pre-Columbian times (Gutierrez 1998). They are usually harvested before the spines hardened and used as vegetable or pickles. Nopales are generally sold fresh, bottled, or canned, less often dried, to prepare nopalitos. Nopales are also featured in New Mexican cuisine, and are gaining popularity elsewhere in the United States. In Mexico, the pads are sliced into strips skinned or unskinned, and fried with eggs and jalapeños chillies served as a breakfast treat. A refreshing salad is commonly made from nopales, tomatoes, and onions topped with a vinegar and mustard dressing. Nopales can be flavored with green chilies and a touch of vinegar, rolled in a flour tortilla. They are used in Mexican dishes such as *huevos con nopales* (eggs with nopal), *carne con nopales* (meat with nopal) or *tacos de nopales* (tacos with nopal). Other recipes include *nopalitos con queso* (a mixture of nopales, onions, and chilies, stir fried and sprinkled with cheese) and *revoltijos* (potatoes, shrimp, and nopal strips cooked in mole rojo, a red chili sauce) in many parts of Mexico especially during Lent. The flowers are good foraging sources for bees to produce reasonable honey.

Opuntia ficus-indica produces edible tender stems (cladodes) and fruits with a high nutritional value in terms of minerals, protein, dietary fibre and phytochemicals; however, around 20% of fresh weight of cladodes and 45% of fresh weight of fruits are by-products that include dietary fibre and natural antioxidant compounds (Bensadón et al. 2010). Mucilage from *Opuntia ficus-indica* can be processed into edible films (Espino-Díaz et al. 2010). Mucilage from nopal could represent a good option for the development of edible films in countries where nopal is highly produced at low cost, constituting a processing alternative for nopal.

Botany

Erect spreading shrub to 3 m high sometimes to 5 m, with a short, sturdy trunk. 30–45 cm diameter. Stems dull gray-green, glabrous consisting of a series of jointed fleshy, elliptic, ovate to broadly obovate segments (platyclades) 25–60 cm long, 20–40 cm wide, margins more or less entire to low tuberculate (Plates 1–3). Areoles 7–11 per diagonal row across midstem segment, rhombic to subcircular, 2–4(–5) mm across, with tufts of short, <2 mm, fine, barbed bristles (glochids) and occasionally with 1–6 white or yellowish spines 1–3 cm long. Leaves small, scale-like produced beneath the areole. Flowers, actinomorphic, bisexual, large, 6–7 cm long, 5–7 cm in diameter; outer perianth parts(sepaloid) broadly ovate or obovate, to 2 cm, yellow with a green or reddish centres, 10–20 mm long, 15–20 mm wide, inner perianth parts (petaloid), obovate to oblong-obovate, 2.5–3.5 × 1.5–2 cm, yellow to orangey yellow, rotate, 25–30 mm long, 15–20 mm wide; anthers and filaments yellow; style reddish, 15 mm long; stigma lobes 8–10, yellow (Plate 2). Fruit greenish white to yellow, yellowish brown, or reddish purple, depending on the variety, fleshy, barrel-shaped, 5–10 cm long, 4–9 cm in diameter, areoles 40–60 bearing tufts of fine barbed bristles with cavity at he stylar end (umbilicus) low and concave (Plates 1, 3–4). Seeds rounded, 5 mm diameter, pale brown embedded in red to yellowish-white fleshy pulp.

Plate 2 Yellow, large flower

Plate 1 Ripe and immature fruits, and flower arising from the cladode

Plate 3 Closer view of immature and ripe fruits

Plate 4 Cactus pear fruit on sale in the market

Nutritive/Medicinal Properties

The nutrient value of raw prickly pear fruit (greenish variety) per 100 g edible portion excluding refuse of 255 of seeds, skin and bud end (USDA 2010) was reported as follows: water 87.55 g, energy 41 kcal (172 kJ), protein 0.73 g, total lipid (fat) 0.51 g, ash 1.64 g, carbohydrates 9.57 g, dietary fibre 3.6 g, calcium 56 mg, iron 0.30 mg, magnesium 85 mg, phosphorus 24 mg, potassium 220 mg, sodium 5 mg, zinc 0.12 mg, copper 0.080 mg, selenium 0.6 μg, vitamin C 14 mg, thaimin 0.014 mg, riboflavin 0.060 mg, niacin 0.460 mg, vitamin B-6 0.060 mg, total folate 6 μg, vitamin A IU 43 IU (2 μg RAE), total suttrated fatty acids 0.067 g, 16:0 0.052 g, 18:0 0.010 g; total monounsaturated fatty accids 0.075 g, 16:1 undiferentiated 0.002 g, 18:1 undiferentiated 0,072 g, 20:1 0.001 g; total polyunsaturated fatty acids 0.213 g, 18:2 undifferentiated 0.186 g, 18:3 undifferentiated 0.023 g, β-carotene 25 μg and β-cryptoxanthin 3 μg.

The most abundant component of the pulp and skin of *Opuntia ficus indica* was found to be ethanol-soluble carbohydrates on a dry weight basis (El Kossori et al. 1998). Pulp contained glucose (35%) and fructose (29%) while the skin contained essentially glucose (21%). Protein content was 5.1% (pulp), 8.3% (skin) and 11.8% (seeds). Starch was found in each of the three parts of the fruit. Pulp fibres were rich in pectin (14.4%), while the skin and seeds were rich in cellulose (29.1% and 45.1%, respectively). Skin was also rich in calcium (2.09%) and potassium (3.4%). Another report stated that the fruits contained more than 8.85% w/w of sugars composed of pentoses, hexoses, hexuronic acids and polysaccharides. In addition, organic acids were identified accompanied by ascorbic acid which is present in a 0.094% (El-Moghazy et al. 1982).

The edible pulp of cactus pear (*Opuntia ficus-indica*) cultivars Sicilian red (Sanguigna), yellow (Surfarina), and white (Muscaredda) were found to contain biothiols, taurine, and lipid-soluble antioxidants such as tocopherols and carotenoids (Tesoriere et al. 2005). The yellow cultivar had the highest level of reduced glutathione (GSH, 8.1 mg/100 g pulp), whereas the white cultivar showed the highest amount of cysteine (1.21 mg/100 g pulp). Taurine amounted to 11.7 mg/100 g in the yellow pulp, while lower levels were found in the others. With the exception of kaempferol in the yellow cultivar (2.7 μg/100 g pulp), the edible pulp of cactus pear was not a good source of flavonols. Very low amounts of lipid-soluble antioxidant vitamins such as vitamin E and carotenoids were determined in all cultivars. As a consequence of industrial processing, a total loss of GSH and β-carotene and a net decrease of vitamin C and cysteine were found in the fruit juice, whereas betalains, taurine, and vitamin E appeared to be less susceptible to degradation.

In *Opuntia undulata* and *O. ficus-indica* fruits, both betacyanins and betaxanthins were identified (Castellar et al. 2003). Prickly pear fruits were found to contain betalain pigments, with indicaxanthin and betanin as the major ones (Fernández-López and Almela 2001). Five betaxanthins (yellow-orange water-soluble pigments) were found in cactus pear (*Opuntia ficus-indica* cv. Gialla) fruit and characterized as the immonium adducts of betalamic acid with serine, gamma-aminobutyric acid, valine, isoleucine, and phenylalanine (Stintzing et al. 2002). *Opuntia ficus-indica* proved to be a rich source of dietary taurine (2-aminoethanesulfonic acid) (Stintzing et al. 1999). Using L-taurine as the amino compound, a new betaxanthin was synthesized by partial synthesis and determined to be the taurine-immonium-conjugate of betalamic acid. Also, betalamic acid could be detected in yellow and orange coloured cultivars of *Opuntia ficus-indica*.

Fruit peel of *Opuntia ficus indica* was found to contain a series of soluble pectic polysaccahrides in the peel (Habibi et al. 2005c) and arabinogalactan polysaccharide (Habibi et al. 2004). The seed pericarp was found to contain water soluble d-xylans (4-O-methyl-D-glucurono)-D-xylans, with 4-O-α-D-glucopyranosyluronic acid groups linked at C-2 of a $(1 \rightarrow 4)$-β-D-xylan (Habibi et al. 2002), and seed endospermum to contain d-xylans (Habibi et al. 2005b) and arabinan-rich polysaccharides (Habibi et al. 2005a).

Opuntia ficus-indica seeds were found to contain on average 71.5 g/kg dry matter, 61.9 g/kg crude oil, 9.4 g/kg protein, 507.4 g/kg crude fibre, 12.3 g/kg ash and 409.0 g/kg carbohydrate (Coskuner and Tekin 2003). The fatty acid composition of prickly pear seed oil consisted of 1.3–1.9 g/kg myristic (14:0), 132.1–156.0 g/kg palmitic (16:0), 14.4–18.5 g/kg palmitoleic (16:1), 33.1–47.9 g/kg stearic (18:0), 210.5–256.0 g/kg oleic (18:1), 522.5–577.6 g/kg linoleic (18:2), 2.9–9.7 g/kg linolenic (18:3), 4.2–6.6 g/kg arachidic (20:0) and 2.1–3.0 g/kg behenic (22:0) acids, which was comparable with that of corn oil. Another paper reported that *Opuntia ficus-indica* seeds contained 16.6% protein, 17.2% fat, 49.6% fibre and 3.0% ash (Sawaya et al. 1983). The meal showed a high amount of iron (9.45 mg%). The contents of Mg, P, K, Zn and Cu were nutritionally significant contributing approximately 10–20% of the Recommended Dietary Allowances (RDA) of these elements per 100 g of dry weight. The amount of Ca represented less than 10% of the RDA for that element. Aspartic acid, glutamic acid, arginine and glycine were the most abundant amino acids comprising about half of the total amino acids content. The seeds were rich in sulfur amino acids (methionine+cystine). Lysine was the first limiting amino acid resulting in a chemical score of 62 for the protein. The in-vitro protein digestibility and the calculated protein efficiency ratio were 77% and 1.82 compared to 90% and 2.50, respectively, for ANRC casein. A major albumin was isolated and characterised from the seeds of *Opuntia ficus-indica* (Uchoa et al. 1998). The amino acid composition of this protein shared similarities with the amino acid composition of several proteins from the 2S albumin storage protein family. The N-terminal amino acid sequence of this protein was elucidated as Asp-Pro-Tyr-Trp-Glu-Gln-Arg.

Total lipids (TL) in lyophilised seeds and pulp of *Opuntia ficus-indica* were 98.8 (dry weight) and 8.70 g/kg, respectively (Ramadan and Mörsel 2003). High amounts of neutral lipids were found (87.0% of TL) in seed oil, while glycolipids and phospholipids occurred at high levels in pulp oil (52.9% of TL). In both oils, linoleic acid was the predominant fatty acid, followed by palmitic and oleic acids. Pulp oil had higher amounts of trienes, γ- and α-linolenic acids, while α-linolenic acid was only detected at low levels in seed oil. The sterol marker, β-sitosterol, comprised 72% and 49% of the total sterol content in seed and pulp oils, respectively. Vitamin E level was higher in the pulp oil than in the seed oil, whereas γ-tocopherol was the major component in seed oil and δ-tocopherol was the main constituent in pulp oil. β-Carotene was also higher in pulp oil than in seed oil. Both oils had similar level of vitamin K_1 (0.05% of TL).

Opuntia ficus-indica seed oil was found to have the following quality parameters: fatty acid content (as oleic acid): 2.5%, iodine number: Wijs (105.5), peroxide number (10 meq O2/Kg), UV absorption: K232=3.15; K268=0.22; ΔK=+0.01, all in accordance with the parameters of other common vegetable oils (Salvo et al. 2002). The acidic fraction revealed a high degree of unsaturation (82.3%) with monounsaturated and polyunsaturated contents of 21.0% and 61.3%, respectively. The sterolic fraction was composed mainly of β-sitosterol (61.4%), campesterol (16.5%), stigmasterol (4.2%) and other more unsaturated sterols (6.2%). Alcoholic fraction contained mainly hexacosanol (73.7%), along with low quantities of docosanol (10.7%) and octacosanol (8.1%) and smaller amounts of other components. Among triterpenic alcohols, the most abundant component was found to be 24-methylcycloarthenol (34.2%), followed by cycloarthenol (29.3%), and β-amyrin (2.2%). Based on these results, *Opuntia ficus-indica* seed oil appeared to be a good potential source of cooking oil.

Cactus pear syrup contained 52.38% reducing sugar and had an ash content of 1.4%, a viscosity of 27.05 cps (counts per second), water activity (Aw) of 0.83, a pH of 4.31, a density of 1.2900 g/ml, and an acidity (as citric acid) of 0.74% (Sáenz et al. 1998). Compared with traditional sweeteners such as fructose and glucose syrup, the acidity was greater than that of HFCS (high fructose corn syrup) of 0.035%, and the ash values were considered a little high compared to glucose syrup which was 1.0%. Sensory evaluation showed that cactus pear syrup and glucose had the same relative sweetness, but lower than fructose; cactus pear syrup had a relative sweetness value of 67 with respect to sucrose, 100.

The acidity of the prickly pear juice was found too be very low (0.02%) and the pH very high (6.4–6.5) when compared with levels found in other common fruit juices (Gurrieri et al. 2000). The sugar content (mainly glucose and fructose) was found to be very high (11–12%), and also L-ascorbic acid occurred in appreciable amount (31–38 mg/100 g). Among the minerals, a high level of manganese (1.7–2.9 ppm) and good contents of iron (0.6–1.2 ppm) and zinc (0.3–0.4 ppm) were found. In particular, such ions appear to be present mainly in the thick skin of the fruit and the pulp. Pectin methylesterase (PME) was found in traces and was not highly active. The results obtained was deemed to be very satisfactory, and the juice had been widely appreciated when compared with other products commonly available on the market such as pear and peach juices.

The proximate compositions of the crushed pads and fruits were reported by Lee et al. (2005). The major minerals were Ca (4391.2–2086.9 mg%), K (1932.1–2608.7 mg%), and Mg (800.6–1984.8 mg%). The main amino acid was glutamic acid, comprising 16.3% of total amino acids in fruit and 25.2% in pad. Dihydroflavonols were identified as (+)-trans-dihydrokaempferol and (+)-trans-dihydroquercetin. Citric acid methyl esters extracted from fruits showed inhibitory activities against monoamine oxidase-B. The presence of trimethyl citrate had been reported in other plants, but 1,3-dimethyl citrate and 1-monomethyl citrate had not been previously reported. The results of pharmacological efficacy tests, including serum biochemical and hematological parameters, autonomic nervous system, anti-inflammatory, analgestic activity, anti-diabetic activity, antithrombotic, anticoagulant, dopamine beta-hydroxylase, monoamine oxidase activity, hyperlipidemia, the respiratory system, antigastric, and anti-ulcerative actions indicated that the fruit and pad of *Opuntia* had functional food benefits.

The computed biological value of prickly pear protein in the pads was 72.6, relative to egg protein (Teles et al. 1984). Trace amounts of malonic, malic and citric acids in pad material collected at 1,800 h were: traces 0.95 and 0.31 mg/g, respectively. In similar material analysed at 0600 h, the concentrations of those acids were 0.36, 9.85, and 1.78 mg/g, respectively. Since there was a significant accumulation of the acids during the evening, crassulacean-type metabolism was suggested. The crude protein content of *Opuntia ficus indica* pads on a dry weight basis was 11.03%, oven dried whole plant 4.82% (Teles et al. 1997). The amino acids contents for the pads (nopal) as % protein (g/16 g N) were found as follows: lysine 5.446%, histidine 2.205%, arginine 4.493%, aspartic acid 7.539%, threonine 3.487%, serine 3.472%, glutamic acid 14.478%, proline 3.621%, glycine 4.301%, alanine 6.512%, cysteine traces, valine 4.972%, methionine 1.714%, isoleucine 3.755%, leucine 6.381%, tyrosine 2.917%, phenylalanine 3.854%. The limiting amino acids were arginine, cystein and methionine.

From the stems (pads) and fruits of *Opuntia ficus-indica* var. *saboten,* eight flavonoids, kaempferol (1), quercetin (2), kaempferol 3-methyl ether (3), quercetin 3-methyl ether (4), narcissin (5), (+)-dihydrokaempferol (aromadendrin, 6), (+)-dihydroquercetin (taxifolin, 7), eriodictyol (8), and two terpenoids, (6S, 9S)-3-oxo-α-ionol β-d-glucopyranoside (9) and corchoionoside C (10) were isolated and identified (Lee et al. 2003). Among these isolates, compounds 3–5 and 8–10 were reported for the first time from the stems and fruits of *O. ficus indica* var. *saboten*. The isolation of two alkaloids (indicaxanthin and neobetanin) (Impellizzeri and Piattelli 1972; Strack et al. 1987) and five flavonoids (isorhamnetin, quercetin, kaempferol, dihydrokaempferol, and dihydroquercetin) from the fruits of this species (Jeong et al. 1999) were also reported. A butanol fraction, from the methanolic extract of *Opuntia ficus-indica* var. *saboten*, on purification yielded three chemical components: isorhamnetin 3-O-(6''-O-E-feruloyl)neohesperidoside; (6R)-9,10-dihydroxy-4,7-megastigmadien-3-one-9-O-β-D-glucopyranoside and (6S)-9,10-dihydroxy-4,7-megastigmadien-3-one-9-O-β-D-glucopyranoside along with 15 known compounds (Saleem et al. 2006).

Mescaline, tyramine and N-methyltyramine were isolated and identified from the cladodes of *Opuntia ficus-indica* and from the flower petals, the flavonoids penduletin, kaempferol, luteolin, quercitrin and rutin were isolated (El-Moghazy et al. 1982). Mescaline is a naturally-occurring psychedelic alkaloid of the phenethylamine class and produces hallucinations. It is mainly used as a recreational drug, an entheogen, and a tool to supplement various practices for transcendence, including in meditation, psychonautics, art projects, and psychedelic psychotherapy.

The pads (nopales) of the prickly pear cactus were reported to be a good source of minerals and other nutrients on the basis of compositional analysis (McConn and Nakata 2004). Studies showed that they were rich in minerals. Their tissue calcium was not freely available due to sequestration in the form of calcium oxalate crystals. Nopales had also been reported to be rich in insoluble and especially soluble dietary fibre, vitamins (especially vitamin A, vitamin C, and vitamin K, but also riboflavin and vitamin B6) and minerals (especially magnesium, potassium, and manganese, but also iron and copper). Folate contents in the cladodes of varying sizes was found to range from 5.5 to 5.62 ng/g (Ortiz-Escobar et al. 2010).

The fruit and cladodes of *Opuntia ficus-indica* have many pharmacological attributes and traditional medicinal uses.

Antioxidant Activity

The fruits and cladodes of *Opuntia ficus-indica* were found to be rich in antioxidants phytochemicals. *Opuntia ficus-indica* fruit was found to contain vitamin C and betalain pigments which exhibited in-vitro radical-scavenging properties and antioxidant (Tesoriere et al. 2004). Consumption of cactus pear fruit positively affected the body's redox balance, decreased oxidative damage to lipids, and improved antioxidant status in healthy humans. Supplementation with vitamin C at a similar dosage enhanced overall antioxidant defense but did not significantly affect body oxidative stress. Components of cactus pear fruit other than antioxidant vitamins were postulated to play a role in the observed effects.

Sicilian cultivars of prickly pear *(Opuntia ficus indica)* produce yellow, red, and white fruits, due to the combination of two betalain pigments, the purple-red betanin and the yellow-orange indicaxanthin. According to a spectrophotometric analysis, the yellow cultivar exhibited the highest amount of betalains, followed by the red and white cultivars (Butera et al. 2002). Indicaxanthin accounted for about 99% of betalains in the white fruit, while the ratio of betanin to indicaxanthin varied from 1:8 (w:w) in the yellow fruit to 2:1 (w:w) in the red one. Polyphenol pigments were negligible components only in the red fruit. When measured as 6-hydroxy-2,5,7,8-tetramethylchroman-2-carboxylic acid (Trolox) equivalents per gram of pulp, the methanolic fruit extracts showed distinct antioxidant activity. Vitamin C did not account for more than 40% of the measured activity. Further, the extracts dose-dependently inhibited the organic hydroperoxide-stimulated red cell membrane lipid oxidation, as well as the metal-dependent and -independent low-density lipoprotein oxidation. The extract from the white fruit showed the highest protection in all models of lipid oxidation. Purified betanin and indicaxanthin were more effective than Trolox at scavenging the [2,2'-azinobis(3-ethylbenzothiazoline-6-sulfonic acid)] diammonium salt cation radical. Betanin underwent complex formation through chelation with Cu^{2+}, whereas indicaxanthin was not altered. These findings suggested that the above betalains contributed to the antioxidant activity of prickly pear fruits. In another study, of three species of Spanish red-skinned cactus pear fruits (*Opuntia ficus-indica, Opuntia undulata* and *Opuntia stricta*), *Opuntia ficus-indica* fruit extract exhibited the strongest antioxidant capacity as evaluated by the 2,2′-azinobis(3-ethylbenzothiazoline-6-sulfonic acid) (Trolox equivalent antioxidant capacity) method and the 2,2-diphenyl-1-picrylhydrazyl radical (DDPH) method (Fernández-López et al. 2010). It also had the highest taurine content. *O. stricta* fruits were the richest in ascorbic acid and total phenolics, whereas *O. undulata* fruits showed the highest carotenoid content. Quercetin and isorhamnetin were the main flavonoids detected.

Catus pear (*Opuntia ficus-indica*) fruit extracts (CPFE) successfully controlled lipid peroxidation in fish oils and fish oil-in-water emulsion at different stages of the pathway in a dose-dependent manner (Siriwardhana and Jeon 2004). Inhibition of the lipid peroxidation in fish oil and fish oil-in-water emulsion was successfully improved by increasing the level of CPFE from 0.01% to 0.1%. CPFE regulated conjugated diene formation from lipid radicals, demonstrating lower conjugated diene hydroperoxide values than its control counterpart. Further, it regulated the addition of oxygen to conjugated dienes to form lipid peroxyl radicals, resulting in lower weight gain. CPFE also recorded a lower peroxide value than its control counterpart, indicating its inhibitory effect on peroxyl radical formation. Such multiple and integrated effects controlled the overall lipid peroxidation, resulting in lower TBARS values than the control. Characterization of the cactus pear antioxidants proved that those antioxidants were heat-resistant, although the color of the cactus pear pigments disappeared rapidly.

A butanol fraction, from the methanolic extract of *Opuntia ficus-indica* var. *saboten*, on purification yielded three chemical components: isorhamnetin 3-O-(6″-O-E-feruloyl) neohesperidoside (1), (6R)-9,10-dihydroxy-4,7-megastigmadien-3-one-9-O-beta-D-glucopyranoside (2) and (6S)-9,10-dihydroxy-4,7-megastigmadien-3-one-9-O-beta-D-glucopyranoside (3) along with 15 known compounds (Saleem et al. 2006). In a DPPH radical scavenging assay, compound 1 showed moderate inhibitory activity ($IC_{50} = 45.58$ μg/ml).

Cactus pear fruit (*Opuntia ficus-indica*) extract (CPFE) exhibited scavenging activity of 1,1-diphenyl-2-picrylhydrazyl adducts at 0.125, 0.25 and 0.5 mg/ml with values of 32.5, 56.7 and 71.6%, respectively (Siriwardhana et al. 2006). CPFE at 0.25 and 0.5 mg/mL scavenged 19.2 and 63.6% of 5,5-dimethyl-1-pyrroline N-oxide-OH adducts. Alkyl radical scavenging effect of CPFE at 0.0625, 0.125, 0.25 and 0.5 mg/ml was 38.0%, 42.4%, 67.5% and 78.8%, respectively. CPFE reduced the H_2O_2-induced DNA damage in lymphocytes (CPFE at 0.1 mg/ml totally inhibited the damage). The data demonstrated that both antioxidative and DNA damage-reduction activities were increased with increasing CPFE dosage, suggesting cactus pear constituents as potential sources of raw material for pharmaceutical and functional food industries.

The juice of whole fruits of Sicilian cultivars of prickly pear (*Opuntia ficus-indica*) showed antioxidant activity in the DPPH test, attributable to the phenolic compounds that were effective radical scavengers (Galati et al. 2003). In the juice, ferulic acid was the chief derivative of hydroxycinnamic acid and the mean concentration of total phenolic compounds was 746 μg/mL. The flavonoid fraction, consisted of rutin and isorhamnetin derivatives. Administration of the juice was found to inhibit the ulcerogenic activity of ethanol in rat. Light microscopy observations revealed an increase in mucus production and the restoration of the normal mucosal architecture.

Studies showed the ability of betalains from cactus pear fruit to protect endothelium from cytokine-induced redox state alteration, through ICAM-1 inhibition (Gentile et al. 2004). These antixodant pigments could contribute to reduce the risk of oxidative stress-correlated diseases and should be incorporated into dietary constituents.

Trolox-equivalent antioxidant capacity (TEAC) and oxygen radical absorbance capacity (ORAC) values were found to be highly correlated with each other and to values for total phenolics, betalain contents, and ascorbic acid concentrations in differently colored cactus pear clones (nine *Opuntia ficus-indica* clones and one

O. robusta clone) (Stintzing et al. 2005). Total phenolics had the greatest contribution to ORAC and TEAC values. The red and yellow betalains were absent in lime green colored cactus fruits. The ratio and concentration of these pigments were responsible for the yellow, orange, red, and purple colors in the other clones. Progeny of purple and lime green colored parents were characterized by 12% and 88% of plants bearing lime green and purple fruit, respectively. Besides known pigments typical of Cactaceae, two unexpected betalains were identified. Whereas gomphrenin I was found for the first time in tissues of cactus plants, methionine-betaxanthin had never been described before as a genuine betalain. In addition to their alleged health-promoting properties, various combinations of yellow betaxanthins and red-purple betacyanins may allow the development of new food products without using artificial colorants.

The ethanol extract of the stem (cladode) of *Opuntia ficus-indica* var. *saboten* (OFS) exhibited a concentration-dependent inhibition of linoleic acid oxidation in a thiocyanate assay system (Lee et al. 2002a). Further, the OFS extract showed dose-dependent free-radical scavenging activity, including DPPH radicals, superoxide anions (O^{2-}), and hydroxyl radicals (OH^-), using different assay systems. The OFS ethanol extract was also found to be effective in protecting plasmid DNA against strand breakage induced by hydroxyl radicals in a Fenton's reaction mixture. Moreover, the extract showed significant dose-dependent protection of mouse splenocytes against glucose oxidase-mediated cytotoxicity. The high amount of phenolics (180.3 mg/g) in the OFS extract was postulated to be the active compounds responsible for the antioxidant properties.

The flavonoids quercetin, (+)-dihydroquercetin, and quercetin 3-methyl ether, isolated from the ethyl acetate fractions of the fruits and stems of *Opuntia ficus-indica* var. *saboten* exhibited protective effects against oxidative neuronal injuries induced in primary cultured rat cortical cells (Dok-Go et al. 2003). Their antioxidant activities were evaluated by using three different cell-free bioassays. Quercetin was found to inhibit H_2O_2- or xanthine (X)/xanthine oxidase (XO)-induced oxidative neuronal cell injury, with an estimated IC_{50} of 4–5 µg/ml, but was not protective at concentrations of 30 µg/ml and above. (+)-dihydroquercetin dose-dependently inhibited oxidative neuronal injuries, but it was less potent than quercetin. In contrast, quercetin 3-methyl ether strongly inhibited H_2O_2-and X/XO-induced neuronal injuries, with IC_{50} values of 0.6 and 0.7 µg/ml, respectively. All three compounds distinctly inhibited lipid peroxidation and scavenged 1,1-diphenyl-2-picrylhydrazyl free radicals. Further, quercetin and quercetin 3-methyl ether were shown to inhibit XO activity in-vitro, with respective IC_{50} values of 10.67 and 42.01 µg/ml. These results indicated that quercetin, (+)-dihydroquercetin, and quercetin 3-methyl ether were the active antioxidant principles in the fruits and stems of *Opuntia ficus-indica* var. *saboten* exhibiting neuroprotective actions against the oxidative injuries induced in cortical cell cultures. Further, quercetin 3-methyl ether appeared to be the most potent neuroprotectant of the three flavonoids isolated from this plant.

In a study on the effect of air-drying flow rates on the amount and antioxidant capacity of extracts of *Opuntia ficus-indica* cladodes, nopal drying at an air flow rate of 3 m/s at 45°C exhibited higher values of phenols, flavonoids and flavonols (Gallegos-Infante et al. 2009). The best value of low-density lipoprotein inhibition and deoxyribose was found at 1,000 µg/ml. The air flow rate affected the amount of polyphenols and the OH^- radical scavenging, but did not alter the chain-breaking and the low-density lipoprotein inhibition activities.

Anti-inflammatory Activity

Ethanol extracts of both *Opuntia ficus-indica* fruits and stem extracts inhibited the writhing syndrome induced by acetic acid, indicating that they possessed analgesic effect (Park et al. 1998). The oral administrations of both extracts suppressed carrageenan-induced rat paw edema and also exhibited potent inhibition in the leukocyte migration of CMC-pouch model in rats. Further, the extracts curbed the release of β-glucuronidase, a lysosomal enzyme in rat neutrophils. The extracts also exhibited the protective effect on gastric mucosal layers. From the results it was suggested that the cactus extracts contained anti-inflammatory action with protective effect against gastric lesions. The active anti-inflammatory principle was isolated and identified as β-sitosterol (Park et al. 2001).

O. ficus-indica (OFI) can be used as a hangover cure. An extract of the OFI plant exhibited moderate effect on reducing hangover symptoms, apparently by inhibiting the production of inflammatory mediators (Wiese et al. 2004). The symptoms of the alcohol hangover were largely due to the activation of inflammation. The severity of the alcohol hangover may be related to inflammation induced by impurities in the alcohol beverage and byproducts of alcohol metabolism. The extract of the OFI was found to diminish the inflammatory response to stressful stimuli.

In-vitro studies indicated that the extracts of *Opuntia ficus-indica* cladodes were able to neutralise the harmful effects of proinflammatory cytokine interleukin-1 beta (IL-1β) in human chondrocyte culture (Panico et al. 2007). IL-1beta stimulated the production of key molecules released during chronic inflammatory events such as nitric oxide (NO), glycosaminoglycans (GAGs), prostaglandins (PGE (2)) and reactive oxygen species (ROS) in human chondrocyte culture. The data showed the protective effect of the extracts of *Opuntia ficus-indica* cladodes in cartilage alteration, which appeared greater than that induced by hyaluronic acid (HA) commonly employed as visco-supplementation in the treatment of joint diseases.

Anticancer Activity

A major betacyanin pigment, isolated from the fruits of *Opuntia ficus-indica*, showed antiproliferative activity on human chronic myeloid leukemia cell line (K562) (Sreekanth et al. 2007). The results showed dose and time dependent reduction in the proliferation of K562 cells treated with betanin with an IC_{50} of 40 μM. Further studies involving scanning and transmission electron microscopy revealed the apoptotic characteristics such as chromatin condensation, cell shrinkage and membrane blebbing. Agarose electrophoresis of genomic DNA of cells treated with betanin showed fragmentation pattern typical for apoptotic cells. Flow cytometric analysis of cells treated with 40 μM betanin exhibited 28.4% of cells in sub G0/G1 phase. Further studies demonstrated that betanin induced apoptosis in K562 cells through the intrinsic pathway mediated by the release of cytochrome C from mitochondria into the cytosol, and PARP (poly ADP ribose) cleavage.

Opuntia ficus indica cladodes was found to possess a cytoprotective action against ethanol-induced ulcer in the rat, and was suggested to be attributed to mucilages and not significantly to pectin, two major carbohydrate polymers components of the cladodes (Galati et al. 2007). Clinical studies indicated that a dry flower preparation of the cactus improved subjectively the discomforts associated with prostatic hypertrophy (Palevitch et al. 1993).

Antigenotoxic Activity

The simultaneous administration of *Opuntia ficus-indica* cladodes extract with zearalenone (ZEN), a potent estrogenic metabolite, resulted in an efficient inhibition of micronuclei aberration (Zourgui et al. 2009). The number of micronucleated polychromatic erythrocyte (MNPCE) decreased from 71.3 for animals treated with Zen to 32.6 for animals treated with cactus cladodes, chromosomal aberrations frequency decreased from 38.3% to 18.6% in bone marrow cells and of DNA fragmentation compared to the group treated with ZEN alone. ZEN was genotoxic to Balb/c mice, inducing DNA damage as indicated by DNA fragmentation, micronuclei and chromosomal aberrations in bone marrow cells. It was also observed that cactus cladodes extract assayed alone at high dose (100 mg/kg b.w.) was found completely safe and did not induce any genotoxic effects. It could be concluded that cactus cladodes extract was effective in the protection against ZEN genotoxicity. This could be relevant, particularly with the emergent demand for natural products which may neutralize the genotoxic effects of the multiple food contaminants and therefore prevent multiple human diseases.

Separate studies clearly showed that zearalenone (Zen) induced significant alterations in all tested oxidative stress markers the MDA level, the protein carbonyls generation, the catalase activity and the expression of the heat shock proteins (Hsp) (Zourgui et al. 2008). Oxidative damage appeared to be a key determinant of ZEN induced toxicity in both liver and kidney of Balb/c mice. The combined treatment of ZEN with the lowest tested dose of cactus extracts (25 mg/kg b.w.) showed a total reduction of ZEN induced oxidative damage for all tested markers. The scientists concluded that cactus cladodes extract was effective in the protection against ZEN hazards. Zearalenone is one of the most widely distributed fusarial mycotoxins encountered at high incidence in many foodstuffs. It is associated with different reproductive disorders in animals.

Antihypercholesterolemic, Antihyperlipidemic Activities

Rats fed 12% nopal for a month were found to have lower weight gains when compared with counterparts fed 6% nopal or the control diet (Medellín et al. 1998). Consumption of nopal did not affect glucose, total cholesterol and HDL cholesterol levels. However, rats fed raw nopal at the 12% concentration level had a 34% reduction in LDL cholesterol levels. It was thus concluded that raw nopal had a potentially beneficial effect for hypercholesterolemic individuals.

In another study, diets enriched with cactus pear oil (CPO) (2.5%, wt/wt) or seeds (CPS) (33%, wt/wt), for 9 weeks significantly decreased the atherogenic index compared to the control diet fed with standard diet, but serum cholesterol level was only reduced by the supplementation with CPO diet (Ennouri et al. 2007). No differences in pancreas, kidney or liver weights were noted in the CPS diet whereas the CPO diet elicited a significant increase in liver and pancreas weights. No differences in serum lipids were observed among the groups, whereas liver lipids showed slight variations. The results thus, indicated that the supplementation with CPO or CPS could be effective in decreasing the atherogenic risk factors in rats.

NeOpuntia, patented, dehydrated, *Opuntia ficus-indica* cladode was found to exhibit improvement of blood lipid parameters in a monocentric, randomized, placebo controlled, double-blind, 6-week study involving 68 women, aged 20–55 years, with metabolic syndrome and a body mass index between 25 and 40 (Linarès et al. 2007). All volunteers followed well balanced diets with controlled lipid input. NeOpuntia or placebo capsules were taken at a dosage of 1.6 g per meal. For the 42 females above 45 years of age, there was a significant increase in HDL-C levels with NeOpuntia and a tendency toward decreased triglyceride levels. Simultaneously, there was a decrease in HDL-C levels with placebo. Overall, for the entire study population, similar but less pronounced tendencies were demonstrated. At the study end, 39% of the NeOpuntia group, but only 8% of the placebo group, were no longer diagnosed with metabolic syndrome. The results indicated an advantage of using NeOpuntia in dietary supplements and functional foods because of improvement of blood lipid parameters associated with cardiovascular risks.

A study on the effects of a diet supplement of *Opuntia ficus-indica* (OFI) on heart-rate variability involving 10 high level athletes using a randomised, placebo, cross over trial found that the OFI diet supplement increased high frequency and low frequency activities and decreased heart rate (Schmitt et al. 2008).

Prickly pear cactus has hypocholesterolemic activity. Studies demonstrated that the intake of prickly pear pectin lowered plasma LDL concentrations by raising the hepatic apolipoprotein B/E receptor expression and escalating receptor mediated LDL turnover in guinea pigs that were fed a high lard and cholesterol (LC) diet (Fernandez et al. 1992). The guinea pigs on the prickly pear pectin (LC-P) diet demonstrated a reduction in plasma LDL cholesterol concentrations by 33%, whereas plasma VLDL and HDL cholesterol levels were unaltered. This resulted in an overall decrease of 28% in total cholesterol for the LC-P diet group. Hepatic apolipoprotein B/E receptor expression was 60% higher in guinea pigs on the LC-P diet. In a subsequent study, the scientists (Fernandez et al. 1994) reported that total plasma cholesterol was significantly lower in the LC-P than the LC group. The cholesterol reduction was specific to LDL because VLDL and HDL cholesterol plasma concentrations were unaffected. They determined that this hypocholesterolemic effect of prickly pear pectin on total plasma cholesterol and LDL levels did not result from a reduction in cholesterol absorption, but may be due to its effects on LDL receptor expression and LDL turnover. This reduced hepatic acyl CoA: cholesterol acyltransferase (ACAT) activity while not affecting hepatic HMG-CoA reductase activity, may have a major effect on hepatic cholesterol homeostasis.

The glycoprotein (90 kDa) (OFI) isolated from *Opuntia ficus-indica* var. *saboten* was found to decrease plasma lipid level through scavenging of intracellular radicals in Triton WR-1339-induced mice (Oh and Lim 2006). Oral administration of OFI glycoprotein [50 mg/kg body weight (BW)] for 2 weeks resulted in a significant decrease of plasma lipid levels such as total cholesterol (TC), triglyceride (TG), and low-density lipoprotein (LDL)in Triton WR-1339-treated mice. Mice induced by Triton WR-1339 exhibited significantly elevated levels of TC, TG and LDL, whereas the high-density lipoprotein (HDL) level was lowered. However, the values were reversed in Triton WR-1339-treated mice with pre-treatment with OFI glycoprotein. The data also showed that pre-treatment with OFI glycoprotein resulted in reduction of thiobarbituric acid-reactive substances (TBARS) level and a rise of nitric oxide (NO) level in Triton WR-1339-treated mice, while the activities of antioxidant enzyme [superoxide dismutase (SOD), catalase (CAT) and glutathione peroxidase (GPx)] were augmented. Therefore, it was speculated that the OFI glycoprotein would be effective in reduction of plasma lipid levels.

Cactus pear (*Opuntia ficus-indica*) seed oil was found to be rich in polyunsaturated fatty acids (Ennouri et al. 2006a, b). The main fatty acids were C16:0, C18:0, C18:1, C18:2 with an exceptional level of linoleic acid, up to 700 g/kg, and a total content of unsaturated fatty acids of 884.8 g/kg. The results indicated a significant decrease in serum glucose concentration (22%) over the control group. However, an increase in the concentration of glycogen was observed in liver and muscle. Blood cholesterol and low density lipoprotein (LDL)-cholesterol decreased in the treated group. High density lipoprotein (HDL)-cholesterol concentration remained unchanged during the treatment. These findings supported the nutritional value of cactus pear as a natural source of edible oil containing essential fatty acids. Additional studies indicated a significant decrease in body weight of rats receiving a diet partially substituted with *O. ficus indica* powder seeds, probably due to a significant decrease in serum-free thyroxin (FT(4)) compared to the control group (Ennouri et al. 2006b). In the treated group, a reduction of glucose concentration in blood and an increase of glycogen in liver and skeletal muscle were observed. A significant elevation in HDL-cholesterol was found in the group receiving the supplemented diet with *O. ficus indica* powder seeds. These results suggested that *O. ficus-indica* seeds can be used as a health food.

Gastroprotective/Antiulcerogenic Activity

Pre-treatment of rats with lyophilized cladodes revealed a protective action against ethanol-induced ulcer which were confirmed by ultrastructural changes observed by transmission electronic microscopy (TEM) (Galati et al. 2001). It was postulated that mucilage of *Opuntia ficus-indica* was involved. This supported the use of *Opuntia ficus-indica* cladodes for the treatment of gastric ulcer in Sicilian folk medicine. In another study, mucilage obtained from cladodes of *Opuntia ficus-indica* reversed gastric mucosal alterations in rats caused by chronic gastric mucosa injury induced by ethanol (Vázquez-Ramírez et al. 2006). Ethanol produced the histological profile of gastritis characterized by loss of the surface epithelium and infiltration of polymorphonuclear leukocytes with concomitant decrease of phosphatidylcholine and increased cholesterol in the plasma membranes of the gastric mucosa. Further, cytosolic activity increased while the activity of alcohol dehydrogenases decreased. The administration of cladode mucilage immediately rectified these enzymatic changes. The cladode mucilage promptly accelerated restoration of the ethanol-induced histological alterations and the disturbances in plasma membranes of gastric mucosa, showing a univocal anti-inflammatory effect. The activity of 5'-nucleotidase correlated with the changes in lipid composition and the fluidity of gastric mucosal plasma membranes. The beneficial action of mucilage appeared correlated with stabilization of plasma membranes of damaged gastric mucosa.

Opuntia ficus-indica cladodes exhibited cytoprotective effects against experimental ethanol-induced ulcer in rats (Galati et al. 2002a). The major components of cladodes were found to be carbohydrate-containing polymers, which consisted of a mixture of mucilage and pectin. When cladodes were administered as a preventive therapy, the gastric mucosa was kept under normal condition by preventing mucus dissolution caused by ethanol and favouring mucus production. An increase of mucus production was also observed during the course of the curative treatment. The treatment with *O. ficus-indica* cladodes provoked an increase in the number of secretory cells. The gastric fibroblasts were postulated to be involved in the antiulcer activity.

Dried stem powder of *Opuntia ficus-indica* var. *saboten* (OF-s) showed significant inhibition in HCl/ethanol-induced gastric lesion at the doses of 200 and 600 mg/kg p.o. and in HCl/aspirin-induced gastric lesion at 600 mg/kg p.o (Lee et al. 2002a, b). OF-s also showed significant inhibition in indomethacin-induced gastric lesion at the doses of 200 and 600 mg/kg, p.o. However, it did not affect both the aspirin-induced and Shay ulcers in rats. It also did not affect gastric juice secretion, acid output and pH. These data indicated that OF-s possessed pronounced inhibitory action on gastric lesion without antiulcer activity in rats.

Alimi et al. (2010) reported that a methanolic root extract of *Opuntia ficus indica* f. *inermis* (ORE) demonstrated DPPH radical scavenging activity and reducing power with EC_{50} of 118.65 and 300 µg/ml respectively. In-vivo the pre-treatment of rats with ranitidine (50 mg/kg) and 200, 400, and 800 mg/kg doses of ORE significantly reduced the 80% ethanol induced-ulcer lesion, with a rate of 82.68%, 49.21%, 83.13%, and 92.59% respectively. Pre-treatment also thwarted the depletion of antioxidant enzymes, superoxide dismutase (SOD), catalase (CAT), glutathione peroxidase (GPx), total glutathione (GSH), and obstructed the increase of myeloperoxidase (MPO) and malondialdehyde (MDA) in rat stomach tissues when compared with the ethanol group. Also pre-treatment with ORE showed a dose-dependent attenuation of histopathology changes induced by ethanol. Phenolic and flavonoids contents, radical scavenging activity, and reducing power, were implicated for the antiulcer property of ORE.

Hypoglycaemic/Antidiabetic Activity

Studies on human patients showed that nopal extract did not reduce fasting glycemia in diabetic subjects (Frati-Munari et al. 1998). Nevertheless, the extract diminished the increase of serum glucose following a dextrose load. Peak serum glucose was 20.3 mg/dl (lower in the test with nopal than in the control one). Dehydrated extract of nopal did not show acute hypoglycemic effect, although it reduced postprandial hyperglycemia.

Polysaccharides isolated from *Opuntia ficus-indica* (POLOF) caused significant anti-hyperglycemic effect in intragastric-induced mice with temporary hyperglycemia, the results obtained was elucidated by a reduction in the intestinal absorption of glucose (Alarcon-Aguilar et al. 2003). POLOF did not show significant effect with sub-cutaneous induced temporary hyperglycemia.

Aguilar et al. (1996) conducted a preliminary meta-analysis of the six studies and found that, ingestion of prickly pear cactus reduced serum glucose levels among persons with diabetes on the order of 10–30 mg/dL at 30–180 minutes post-ingestion when measured as the difference from baseline. These results represented intervention group data only as the control group data reported in the published studies were of insufficient detail to use in the preliminary meta-analysis. Definitive conclusions were not made at this time. The preliminary findings however, did suggest a strong possibility of a true metabolic effect for persons with diabetes and ingestion of prickly pear cactus but a clinical trial incorporating tight metabolic control design parameters was warranted.

A recent study showed that the proprietary product OpunDia (*Opuntia ficus-indica*) had acute blood glucose lowering effects in obese pre-diabetic men and women (Godard et al. 2010). In the double-blind placebo controlled study (16 weeks), participants (age range of 20–50 years) were randomly assigned to one of the two groups and given a 16-week supply of either the 200 mg OpunDia (n = 15), or placebo (n = 14). There was a statistically significant reduction in the blood glucose concentrations at the 60, 90 and 120 minutes time points with the pre-oral glucose tolerance test (OGTT) compared to the OpunDia bolus trial. There were no between-group differences found with the OGTT time points, area under the curve, blood chemistry variables (insulin, hsCRP, adiponectin, proinsulin, Hb1Ac), diet analysis variables (carbohydrates, fat, protein and total kcals), body composition variables (fat mass, fat free mass, percent body fat and total body weight), or blood chemistry safety parameters (comprehensive metabolic panel) pre-to-post 16-week intervention. The results supported the traditional use of *Opuntia ficus-indica* for blood glucose management.

A recent study conducted by Butterweck et al. (2011) with the traditional cladode *Opuntia ficus-indica* extract revealed that maximum effects on blood glucose and insulin were observed after oral administration in a dose range of 6–176 mg/kg. The proprietary *Opuntia ficus-indica* blend (stem/fruit skin ratio 75/25) significantly lowered blood glucose levels in the glucose tolerance test to a similar extent as the traditional aqueous cladode extract when administered in a dose of 6 mg/kg. In contrast to the aqueous extract, the proprietary blend highly significantly elevated basal plasma insulin levels indicating a direct action on pancreatic beta cells. The results suggested that both *Opuntia ficus-indica* extracts exerted hypoglycemic activities in rats in doses as low as 6 mg/kg but that the effects of the proprietary stem/fruit blend were more pronounced. The results also supported the traditional use of the cladodes to treat diabetes mellitus in Mexico.

Neuroprotective Activity

Opuntia ficus-indica (MEOF) exhibited neuroprotective action against N-methyl-d-aspartate (NMDA)-, kainate (KA)-, and oxygen-glucose deprivation (OGD)-induced neuronal injury in cultured mouse cortical cells (Kim et al. 2006). Treatment of neuronal cultures with MEOF (30, 300, and 1,000 µg/ml) dose-dependently inhibited NMDA (25 µM)-, KA (30 µM)-, and OGD (50 minutes)-induced neurotoxicity. The butanol fraction of *Opuntia ficus-indica* (300 µg/ml) significantly reduced NMDA (20 µM)-induced delayed neurotoxicity by 27%. When gerbils were given doses of 4.0 g/kg (3 days) and 1.0 g/kg (4 weeks), the neuronal damage in the hippocampal region was reduced by 32% and 36%, respectively. These results suggested that the preventive administration of *Opuntia ficus-indica* extracts may be helpful in alleviating the excitotoxic neuronal damage induced by global ischemia.

A butanol fraction obtained from 50% ethanol extracts of *Opuntia ficus-indica* var. *saboten* stem and its hydrolysis product inhibited the production of nitric oxide (NO) in lipopolysaccharide (LPS)-activated microglia in a dose dependent fashion (IC$_{50}$ 15.9, 4.2 µg/mL, respectively) (Lee et al. 2006). They also suppressed the expression of protein and mRNA of by inducible nitric oxide synthase (iNOS) in LPS-activated microglial cells at higher than 30 µg/mL. They also inhibited the degradation of

I-kappaB-alpha in activated microglia. Further, they exhibited strong activity of peroxynitrite scavenging in a cell free bioassay system. These results implied that *Opuntia ficus- indica* may have neuroprotective activity through the inhibition of NO production by activated microglial cells and peroxynitrite scavenging activity.

Memory Enhancement Activity

Recent studies suggested that the subchronic administration of n-butanolic extract of *O. ficus-indica* var. *saboten* (BOF) enhanced long-term memory, and that this effect was partially mediated by ERK-CREB-BDNF signaling and the survival of immature neurons (Kim et al. 2010). After the oral administration of BOF for 7 days, the latency time in the passive avoidance task was significantly increased relative to vehicle-treated controls. Western blotting showed that the expression levels of brain-derived neurotrophic factor (BDNF), phosphorylated cAMP response element binding-protein (pCREB), and phosphorylated extracellular signal-regulated kinase (pERK) 1/2 were significantly elevated in hippocampal tissue after 7 days of BOF administration. Doublecortin and 5-bromo-2-deoxyuridine immunostaining also showed that BOF significantly enhanced the survival of immature neurons, but did not affect neuronal cell proliferation in the subgranular zone of the hippocampal dentate gyrus.

Hepatoprotective Activity

The juice of *Opuntia ficus-indica* fruit (prickly pear) displayed protective effects against carbon tetrachloride (CCl_4)-induced hepatotoxicity in rats (Galati et al. 2005). The fruit juice contained many phenol compounds, ascorbic acid, betalains, betacyanins, and a flavonoid fraction, which consisted mainly of rutin and isorhamnetin derivatives. Hepato-protection may be related to the flavonoid fraction of the juice, but other compounds, such as vitamin C and betalains could, synergistically, negate many degenerative processes by means of their antioxidant activity.

Opuntia ficus-indica cladode extract was found to protect against liver damage induced in male Swiss mice by an organophosphorous insecticide, the chlorpyrifos (CPF) (Ncibi et al. 2008). CPF affected significantly all parameters (liver weight, alanine amino transferase (ALAT), aspartate amino transferase (ASAT), phosphatase alkaline (PAL), lactate dehydrogenase (LDH), cholesterol and albumin in serum) studied. However, when this pesticide was administrated together with the cladode extract, a recovery of all their levels was observed. In contrast, cladode extract alone did not affect the studied parameters. These results indicated firstly that CPF was hepatotoxic and secondly that *Opuntia ficus indica* stem extract protected the liver and decreased the toxicity induced by this organophosphorous pesticide.

Wound Healing Activity

The methanolic extract of *Opuntia ficus-indica* stems and its hexane, ethyl acetate, *n*-butanol and aqueous fractions were evaluated for their wound healing activity in rats (Park and Chun 2001). The extract and less polar fractions showed significant effects.

Studies showed that lyophilized polysaccharide extracts obtained from *O. ficus-indica* cladodes when topically applied for 6 days on large, full-thickness wounds in rats induced a beneficial effect on cutaneous repair (Trombetta et al. 2006). The extract accelerated the re-epithelization and remodelling phases of skin lesions also by affecting cell-matrix interactions and by modulating laminin deposition. Furthermore, the wound-healing effect was more marked for polysaccharides with a molecular weight (MW) ranging 104–106 Da than for those with MW greater than 106 Da.

Diuretic Activity

Studies showed that *O. ficus-indica* cladode, fruit and flower infusions significantly increased diueresis (Galati et al. 2002b). This effect was more evident with the fruit infusion and was particularly significant during the chronic treatment. The fruit infusion showed also antiuric effect. In all experiments, cladode, flower and fruit infusions showed a modest but not significant increase in natriuresis and kaliuresis.

Dehydrated extract of the prickly pear fruit *Opuntia ficus indica*, Cacti-Nea (R), exhibited chronic diuretic and antioxidant effects in Wistar rats (Bisson et al. 2010). Cacti-Nea (R) orally administered daily for 7 days at the dose of 240 mg/kg/day significantly increased the urine volumes excreted by rats in comparison with the control group. It also exhibited a trend to reduce significantly the body weight gain of rats. No significant differences were observed in the urine concentration of sodium, potassium and uric acid in rats treated with hydrochlorothiazide at the dose of 10 mg/kg/day in comparison with the control group. The chronic diuretic effects of Cacti-Nea (R) were comparable with that of the standard drug hydrochlorothiazide. Chronic oral administration of Cacti-Nea (R) significantly increased the blood globular levels of glutathione peroxidase in comparison with control and hydrochlorothiazide groups.

Antidepressant Activity

Three derivatives of methyl citrate and 1-methyl malate were isolated from the fruits of *Opuntia ficus-indica* var. *saboten* Makino through in-vitro bioassay-guided isolation for the inhibition on monoamine oxidase (MAO) (Han et al. 2001). The IC_{50} values for MAO-B of 1-monomethyl citrate, 1,3-dimethyl citrate, trimethyl citrate and 1-methyl malate were 0.19, 0.23, 0.61 and 0.25 mM, respectively. Unusually high or low levels of MAOs in the body were found to be associated with depression. However, on MAO-A, their inhibitions showed only marginal activity.

Antispasmodic Activity

Opuntia ficus indica fruit pulp extract, (CFE) (10–320 mg of fresh fruit pulp equivalents/mL of organ bath) was found to reduce dose-dependently the spontaneous mouse ileum contractions (Baldassano et al. 2010). This effect was unaffected by tetrodotoxin, a neuronal blocker, N(omega)-nitro-l-arginine methyl ester, a nitric oxide synthase blocker, tetraethylammonium, a potassium channel blocker, or atropine, a muscarinic receptor antagonist. CFE also decreased the contractions elicited by carbachol, without affecting the contractions evoked by high extracellular potassium. Indicaxanthin, but not ascorbic acid, assayed at concentrations comparable with their content in CFE, simulated the CFE effects. The data showed that CFE was able to exert direct antispasmodic effects on the intestinal motility. The fruit pigment indicaxanthin appeared to be the main constituent responsible for the fruit pulp extract-induced effects.

Antiallergenic Activity

Lim (2010) found that the glycoprotein (90 kDa) isolated from *Opuntia ficus-indica* var. *saboten* (OFI glycoprotein) exhibited in-vivo and in-vitro anti-allergic activity in ICR mice and RBL-2 H3 cells respectively. The OFI glycoprotein (5 mg/kg) inhibited histamine and beta-hexosaminidase release, lactate dehydrogenase (LDH), and interleukin 4 (IL-4) in mice serum. Also, OFI glycoprotein (25 μg/ml) had suppressive effects on the expression of MAPK (ERK1/2), and on protein expression of anti-allergic proteins (iNOS and COX-2). Thus, it was speculated that the OFI glycoprotein obstructed anti-allergic signal transduction pathways.

Antimicrobial Activity

Cladodes of *Opuntia ficus-indica* were found to have lectins (Santana et al. 2009). Lectins are hemagglutinating proteins with carbohydrate binding sites; these proteins have biological properties including antimicrobial activity. *Opuntia ficus-indica* lectin (OfiL) and cladode crude extract (CE20%) demonstrated antifungal activity against *Colletotrichum gloeosporioides*,

Candida albicans, *Fusarium descencelulare*, *Fusarium lateritium*, *Fusarium moniliforme*, *Fusarium oxysporum* and *Fusarium solani*. The lectin was mainly active on *C. albicans*. CE20% hemagglutinating activity was detected with rabbit, chicken and human erythrocytes, A and O types. The hemagglutinating activity was high at pH 5.0, thermostable, inhibited by glycoproteins and stimulated with Ca^{2+} or Mg^{2+}.

Protective Effects Against Nickel-Induced Toxicity

Opuntia ficus-indica cladode extract exhibited a protective effect against nickel-induced toxicity in Wistar rats (Hfaiedh et al. 2008). Significant increases of lactate dehydrogenase, aspartate aminotransferase, alanine aminotransferase activities and of cholesterol, triglycerides and glucose levels were observed in blood of nickel-treated rats injected daily, for 10 days, with either nickel chloride solution (4 mg (30 μmol)/kg body weight). In the liver, nickel chloride was found to induce an oxidative stress evidenced by an increase in lipid peroxidation and changes in antioxidant enzymes activities. Superoxide-dismutase activity was found to be increased whereas glutathione peroxidase and catalase activities were decreased. These changes did not occur in animals previously given cactus juice, demonstrating a protective effect of this vegetal extract.

Traditional Medicinal Uses

The cladodes, flowers and fruit of *Opuntia ficus-indica* have been used in traditional medicine since pre-Hispanic times and some of the traditional uses have been scientifically supported by studies as discussed above. The plant has diuretic, analgesic, cardiotonic, laxative and antiparasitic properties. The plant has been used since to treat kidney ailments and burns, to induce childbirth and as an antidiabetic (Lopez 1995). The juice of nopalito was used to treat nausea, fever, and ulcers, while the roasted fruit used to cure coughs and the rind to cure kidney diseases. *Opuntia ficus-indica* cladodes are used in traditional medicine of many countries for their cicatrisant activity (Galati et al. 2002a). In traditional medicine extracts of polysaccharide-containing plants are widely employed for the treatment of skin and epithelium wounds and of mucous membrane irritation. The extracts of *Opuntia ficus-indica* cladodes are used in folk medicine for their anti-ulcer and wound-healing activities (Trombetta et al. 2006). In Sicily folk medicine, *Opuntia ficus indica* cladodes are used for the treatment of gastric ulcer (Galati et al. 2001). Its fruits and stems have been used in Korean folk medicine to treat diabetes, hypertension, asthma, dyspepsia, burns, edema and indigestion (Ahn 1998; Lopez 1995; Kim et al. 2010). *Opuntia ficus-indica* is employed for the treatments of inflammation, burns, and edema as a folk medicinal plant in Jeju Island (Lee et al. 2005). Ground or pureed young pads can be used for first aid treatment like the *Aloe vera* plant.

Other Uses

The most extensive use of *Opuntia ficus-indica* occurs in Brazil where it has been grown as a fodder for cattle, sheep and goat for about 100 years. Local dairymen assert that nopalito pads are important for good lactation, imparts a better flavour and quality to the milk and enhances better quality for butter. Recently, the cattle industry of the Southwest United States has begun to cultivate *Opuntia ficus-indica* as a fresh source of feed for cattle. During the frequent, unpredictable droughts, propane torches known as "pear burners" are used to singe the spines off cactus pads so that they can be eaten by livestock. The seeds of prickly pear are also suitable as animal feed (Coskuner and Tekin 2003).

Opuntia ficus-indica and other *Opuntia* spp have been cultivated from pre-Columbian times in noplaries to serve as a host plant for cochineal insects, which produce useful, valuable red and purple dyes (Donkin 1977).

Opuntia hedges play a major role in erosion control and land-slope partitioning particularly when established along contours (Le Houérou 2002). The cactus is also used for land reclamation where degraded landscape can be rehabilitated. In Tunisia and Algeria for instance, stony and rocky slope have been rehabilitated by planting cacti on contours. The cactus also helps in soil improvement, maintaining soil fertility via their geobiogene and trace element cycling activities, enriching the top soil in organic matter and improving its structure and the stability of its aggregates, hence permeability and water uptake balance. It is often used as defensive hedges for the protection of gardens, orchards and olive groves through out North America and in parts of Italy and Spain. These hedges demarcate boundaries as well.

In Mexico, nopal juice is used in the preparation of lime mortar (Cárdenas et al. 1998). The lime mortar used in this way has been used for centuries for the restoration and protection of historical buildings because of an enhanced improvement against water penetration and cracking. Studies showed that this was due to the formation of an interpenetrated network of nopal polysaccharide mucilage into the hydroxide base. Nopal mucilage used as a binding and waterproofing agent in adobe. Adobe is a natural building material made from sand, clay, and water, with some kind of fibrous or organic material (nopals) which is shaped into bricks using frames and dried in the sun.

Other uses of the cactus include industrial products such as alcohol, soap, pigments, pectins, and oils. Studies are being conducted to evaluate its potential for bioethanol production. A team of experts from the Cajamar Foundation and Almeria Albaida Recursos Naturales y Medioambiente (Spain) are conducting feasibility studies on the production of bioethanol in semi-arid lands using prickly-pear (*Opuntia ficus indica*) and the tobacco tree (*Nicotiana glauca*) as feedstocks (Anonymous 2009). These two plant species have been found to be ideally adapted to conditions of extreme water shortage and at the same time these plants have high energy biomass due to the fermentation process of their organic matter. The use of semi-arid lands are appropriate plantation sites for bioenergy crop plantations because these lands are not usually used for food-crop cultivations. Experimental plantations of the crops have been initiated for the eventual bioethanol production. The project covers the "complete cycle of biofuels", from the bioenergy crop production, to biomass-feedstock processing (ethanol production), and eventually the application of the biofuel-ethanol in the motor industry.

Comments

Opuntia-ficus-indica is regraded as a noxious invasive weed in many countries.

Selected References

Aguilar C, Ramirez C, Castededa-Andrade I, Frati-Munari AC, Medina R, Mulrow C, Pugh J (1996) Opuntia (prickly pear cactus) and metabolic control among patients with diabetes mellitus (abstract). Annu Meeting Int Soc Technol Assess Health Care 12:14

Ahn DK (1998) Illustrated book of Korean medicinal herbs. Kyohaksa, Seoul, p 497

Alarcon-Aguilar FJ, Valdes-Arzate A, Xolalpa-Molina S, Banderas-Dorantes T, Jimenez-Estrada M, Hernandez-Galicia E, Roman-Ramos R (2003) Hypoglycemic activity of two polysaccharides isolated from *Opuntia ficus-indica* and *O. streptacantha*. Proc West Pharmacol Soc 46:139–142

Alimi H, Hfaiedh N, Bouoni Z, Hfaiedh M, Sakly M, Zourgui L, Rhouma KB (2010) Antioxidant and antiulcerogenic activities of *Opuntia ficus indica* f. *inermis* root extract in rats. Phytomedicine 17(14): 1120–1126

Anderson EF (2001) The cactus family. Timber Press, Portland, 776 pp

Anonymous (2009) Prickly pears and tobacco farmed for bioethanol. http://www.thebioenergysite.com/articles/478/prickly-pears-and-tobacco-farmed-for-bioethanol

Bacardi-Gascon M, Duenas-Mena D, Jimenez-Cruz A (2007) Lowering effect on postprandial glycemic response of nopales added to Mexican breakfasts. Diabetes Care 30(5):1264–1265

Baldassano S, Tesoriere L, Rotondo A, Serio R, Livrea MA, Mulè F (2010) Inhibition of the mechanical activity of mouse ileum by cactus pear (*Opuntia ficus indica*, L, Mill.) fruit extract and its pigment indicaxanthin. J Agric Food Chem 58(13):7565–7571

Barbera G, Carimi F, Inglese P (1992) Past and present role of the Indian-fig prickly-pear (*Opuntia ficus-indica* (L.) Miller, Cactaceae) in the agriculture of Sicily. Econ Bot 46:10–20

Bensadón S, Hervert-Hernández D, Sáyago-Ayerdi SG, Goñi I (2010) By-products of Opuntia ficus-indica as a source of antioxidant dietary fiber. Plant Foods Hum Nutr 65(3):210–216

Bisson JF, Daubié S, Hidalgo S, Guillemet D, Linarés E (2010) Diuretic and antioxidant effects of Cacti-Nea ((R)), a dehydrated water extract from prickly pear fruit, in rats. Phytother Res 24(4):587–594

Borrego EF, Murillo M, Flores V, Parga V, Guzman E (1990) Industrial potential of fifteen varieties of nopal (prickly pear, Opuntia spp.). In: Naqui H, Estilai A, Ting IP (eds.) New industrial crops and products. Proceedings of The First International Conference on New Industrial Crops and Products, October 8-12, 1990. California, USA: Association of The Industrial Crops and Products and The University of Arizona, pp 215–218

Brutsch MO, Helmuth GZ (1990) The pricky pear (Opuntia ficus-indica [Cactaceae]) in South Africa: utilization of the naturalized weed and of the cultivated plants. Econ Bot 47(2):154–162

Bustos OE (1981) Alcoholic beverage from Chilean Opuntia ficus indica. Am J Enol Vitic 32(3):228–229

Butera D, Tesoriere L, Gaudio FD, Bongiorno A, Allegra M, Pintaudi AM, Kohen R, Livrea MA (2002) Antioxidant activities of Sicilian prickly pear (Opuntia ficus indica) fruit extracts and reducing properties of its betalains: betanin and indicaxanthin. J Agric Food Chem 50(23):6895–6901

Butterweck V, Semlin L, Feistel B, Pischel I, Bauer K, Verspohl EJ (2011) Comparative evaluation of two different Opuntia ficus-indica extracts for blood sugar lowering effects in rats. Phytother Res 25(3)370–375

Cárdenas A, Arguelles WM, Goycoolea FM (1998) On the possible role of Opuntia ficus-indica mucilage in lime mortar performance in the protection of historical buildings. J Profess Assoc Cactus Dev 3:64–71

Castellar R, Obón JM, Alacid M, Fernández-López JA (2003) Color properties and stability of betacyanins from Opuntia fruits. J Agric Food Chem 51(9):2772–2776

Coskuner Y, Tekin A (2003) Monitoring of seed composition of prickly pear (Opuntia ficus-indica L.) fruits during maturation period. J Sci Food Agric 83(8):846–849

Dok-Go H, Lee KH, Kim HJ, Lee EH, Lee J, Song YS, Lee YH, Jin C, Lee YS, Cho J (2003) Neuroprotective effects of antioxidative flavonoids, quercetin, (+)-dihydroquercetin and quercetin 3-methyl ether, isolated from Opuntia ficus-indica var. saboten. Brain Res 965(1–2):130–136

Donkin R (1977) Spanish red: an ethnogeographical study of cochineal and the Opuntia cactus. T Am Philos Soc 67:1–77

El Kossori RL, Villaume C, El Boustani E, Sauvaire Y, Méjean L (1998) Composition of pulp, skin and seeds of prickly pears fruit (Opuntia ficus indica sp.). Plant Foods Hum Nutr 52(3):263–270

El-Moghazy AM, El-Sayyad SM, Abdel-Baky AM, Bechalt EY (1982) A phytochemical study of Opuntia ficus indica (L.) Mill cultivated in Egypt. Egypt J Pharm Sci 23(1–4):247–254

Ennouri M, Fetoui H, Bourret E, Zeghal N, Attia H (2006a) Evaluation of some biological parameters of Opuntia ficus indica. 1. Influence of a seed oil supplemented diet on rats. Bioresour Technol 97(12):1382–1386

Ennouri M, Fetoui H, Bourret E, Zeghal N, Guermazi F, Attia H (2006b) Evaluation of some biological parameters of Opuntia ficus indica. 2. Influence of seed supplemented diet on rats. Bioresour Technol 97(16):2136–2140

Ennouri M, Fetoui H, Hammami M, Bourret E, Attia H, Zeghal N (2007) Effects of diet supplementation with cactus pear seeds and oil on serum and liver lipid parameters in rats. Food Chem 101(1):248–253

Espino-Díaz M, de Jesús Ornelas-Paz J, Martínez-Téllez MA, Santillán C, Barbosa-Cánovas GV, Zamudio-Flores PB, Olivas GI (2010) Development and characterization of edible films based on mucilage of Opuntia ficus-indica (L.). J Food Sci 75(6):E347–E352

FAO (1988) Traditional food plants: a resource book for promoting the exploitation and consumption of food plants in arid, semi-arid and sub-humid lands of East Africa. FAO food and nutrition paper 42. FAO, Rome

Fernandez ML, Lin ECK, Trejo A, McNamara DJ (1992) Prickly pear (Opuntia sp.) pectin reverses low density lipoprotein receptor suppression induced by a hypercholesterolemic diet in guinea pigs. J Nutr 22(12):2230–2240

Fernandez ML, Lin ECK, Trejo A, McNamara DJ (1994) Prickly Pear (Opuntia sp.) pectin alters hepatic cholesterol metabolism without affecting cholesterol absorption in guinea pigs fed a hypercholesterolemic diet (Biochemical and Molecular Roles of Nutrients). J Nutr 24(6):817–823

Fernández-López JA, Almela L (2001) Application of high-performance liquid chromatography to the characterization of the betalain pigments in prickly pear fruits. J Chromatogr A 913(1–2):415–420

Fernández-López JA, Almela L, Obón JM, Castellar R (2010) Determination of antioxidant constituents in cactus pear fruits. Plant Foods Hum Nutr 65(3):253–259

Frati-Munari AC, de León C, Ariza-Andraca R, Bañales-Ham MB, López-Ledesma R, Lozoya X (1998) Effect of a dehydrated extract of nopal (Opuntia ficus indica Mill.) on blood glucose. Arch Invest Med (Mex) 20(3):211–216 (In Spanish)

Galati EM, Monforte MT, Tripodo MM, d'Aquino A, Mondello MR (2001) Antiulcer activity of Opuntia ficus indica (L.) Mill. (Cactaceae): ultrastructural study. J Ethnopharmacol 76(1):1–9

Galati EM, Pergolizzi S, Miceli N, Monforte MT, Tripodo MM (2002a) Study on the increment of the production of gastric mucus in rats treated with Opuntia ficus indica (L.) Mill. cladodes. J Ethnopharmacol 83(3):229–233

Galati EM, Tripodo MM, Trovato A, Miceli N, Monforte MT (2002b) Biological effect of Opuntia ficus indica

(L.) Mill. (Cactaceae) waste matter. Note I: diuretic activity. J Ethnopharmacol 79(1):17–21

Galati EM, Mondello MR, Giuffrida D, Dugo G, Miceli N, Pergolizzi S, Taviano MF (2003) Chemical characterization and biological effects of Sicilian *Opuntia ficus indica* (L.) Mill. fruit juice: antioxidant and antiulcerogenic activity. J Agric Food Chem 51(17):4903–4908

Galati EM, Mondello MR, Lauriano ER, Taviano MF, Galluzzo M, Miceli N (2005) *Opuntia ficus indica* (L.) Mill. fruit juice protects liver from carbon tetrachloride-induced injury. Phytother Res 19(9):796–800

Galati EM, Monforte MT, Miceli N, Mondello MR, Taviano MF, Galluzzo M, Tripodo MM (2007) *Opuntia ficus indica* (L.) Mill. mucilages show cytoprotective effect on gastric mucosa in rat. Phytother Res 21(4):344–346

Gallegos-Infante JA, Rocha-Guzman NE, González-Laredo RF, Reynoso-Camacho R, Medina-Torres L, Cervantes-Cardozo V (2009) Effect of air flow rate on the polyphenols content and antioxidant capacity of convective dried cactus pear cladodes (*Opuntia ficus indica*). Int J Food Sci Nutr 60(Suppl 2):80–87

Gentile C, Tesoriere L, Allegra M, Livrea MA, D'Alessio P (2004) Antioxidant betalains from cactus pear (*Opuntia ficus-indica*) inhibit endothelial ICAM-1 expression. Ann NY Acad Sci 1028:481–486

Godard MP, Ewing BA, Pischel I, Ziegler A, Benedek B, Feistel B (2010) Acute blood glucose lowering effects and long-term safety of OpunDia supplementation in pre-diabetic males and females. J Ethnopharmacol 130(3):631–634

Griffith MP (2004) The origins of an important cactus crop, *Opuntia ficus-indica* (Cactaceae): new molecular evidence. Am J Bot 91:1915–1921

Gurrieri S, Miceli L, Lanza CM, Tomaselli F, Bonomo RP, Rizzarelli E (2000) Chemical characterization of Sicilian pickly pear (*Opuntia ficus indica*) and perspectives for the storage of its juice. J Agric Food Chem 48(11):5424–5431

Gutierrez MA (1998) Medicinal use of the Latin food staple nopales: the prickly pear cactus. Nutrition Bytes 4(2), Article 3. http://repositories.cdlib.org/uclabiolchem/nutritionbytes/vol4/iss2/art3.

Habibi Y, Mahrouz M, Vignon MR (2002) Isolation and structure of -xylans from pericarp seeds of *Opuntia ficus-indica* prickly pear fruits. Carbohydr Res 337(17):1593–1598

Habibi Y, Mahrouz M, Marais M-F, Vignon MR (2004) An arabinogalactan from the skin of *Opuntia ficus-indica* prickly pear fruits. Carbohydr Res 339(6):1201–1205

Habibi Y, Mahrouz M, Vignon MR (2005a) Arabinan-rich polysaccharides isolated and characterized from the endosperm of the seed of *Opuntia ficus-indica* prickly pear fruits. Carbohydr Polymer 60(3):319–329

Habibi Y, Mahrouz M, Vignon MR (2005b) d-Xylans from seed endosperm of *Opuntia ficus-indica* prickly pear fruits. CR Chim 8(6–7):1123–1128

Habibi Y, Mahrouz M, Vignon MR (2005c) Isolation and structural characterization of protopectin from the skin of *Opuntia ficus-indica* prickly pear fruits. Carbohydr Polymer 60(2):205–213

Halloy S (2002) *Nopalito (Opuntia ficus indica)*. New Zealand Crop & Food Research [Broadsheet] Number 137. http://www.crop.cri.nz/home/products-services/publications/broadsheets/137Nopalito.pdf

Han YN, Choo Y, Lee YC, Moon YI, Kim SD, Choi JW (2001) Monoamine oxidase B inhibitors from the fruits of *Opuntia ficus-indica* var. *saboten*. Arch Pharm Res 24(1):51–54

Hfaiedh N, Allagui MS, Hfaiedh M, Feki AE, Zourgui L, Croute F (2008) Protective effect of cactus (*Opuntia ficus indica*) cladode extract upon nickel-induced toxicity in rats. Food Chem Toxicol 46(12):3759–3763

Hoskins JR, Mcfayden RE, Murray ND (1988) Distribution and biological control of cactus pears in eastern Australia. Plant Protect Q 3:115–123

Hu SY (2005) Food plants of China. The Chinese University Press, Hong Kong, 844 pp

Impellizzeri G, Piattelli M (1972) Biosynthesis of indicaxanthin in *Opuntia ficus-indica* fruits. Phytochem 11:2499–2502

Jeong SJ, Jun KY, Kang TH, Ko EB, Kim YC (1999) Flavonoids from the fruits of *Opuntia ficus-indica* var. *saboten*. Saengyak Hakhoechi 30:84–86

Kim J-H, Park S-H, Ha H-J, Moon C-J, Shin T-K, Kim J-M, Lee N-H, Kim H-C, Jang K-J, Myung-Bok Wie M-B (2006) *Opuntia ficus-indica* attenuates neuronal injury in in vitro and in vivo models of cerebral ischemia. J Ethnopharmacol 104(1–2):257–262

Kim JM, Kim DH, Park SJ, Park DH, Jung SY, Kim HJ, Lee YS, Jin C, Ryu JH (2010) The n-butanolic extract of *Opuntia ficus-indica* var. *saboten* enhances long-term memory in the passive avoidance task in mice. Prog Neuropsychopharmacol Biol Psychiatry 34(6):1011–1017

Le Houérou HN (2002) Cacti (*Opuntia* spp.) as a fodder crop for marginal lands in the Mediterranean basin. Acta Hort (ISHS) 581:21–46

Lee EB, Hyun JE, Li DW, Moon YI (2002a) Effects of *Opuntia ficus-indica* var. Saboten stem on gastric damages in rats. Arch Pharm Res 25(1):67–70

Lee JC, Kim HR, Kim J, Jang YS (2002b) Antioxidant property of an ethanol extract of the stem of *Opuntia ficus-indica* var. *saboten*. J Agric Food Chem 50(22):6490–6496

Lee EH, Kim HJ, Song YS, Jin CB, Lee KT, Cho JS, Lee YS (2003) Constituents of the stems and fruits of *Opuntia ficus-indica* var. *saboten*. Arch Pharm Res 26(12):1018–1023

Lee YC, Pyo Y-H, Ahn C-K, Kim S-H (2005) Food functionality of *Opuntia ficus-indica* var. cultivated in Jeju Island. J Food Sci Nutr 10:103–110

Lee MH, Kim JY, Yoon JH, Lim HJ, Kim TH, Jin C, Kwak WJ, Han CK, Ryu JH (2006) Inhibition of nitric oxide synthase expression in activated microglia and peroxynitrite scavenging activity by *Opuntia ficus indica* var. *saboten*. Phytother Res 20(9):742–747

Li ZY, Taylor NP (2007) Cacataceae. In: Wu ZY, Raven PH, Hong DY (eds.) Flora of China. Vol. 13 (Clusiaceae through Araliaceae). Science Press/Missouri Botanical Garden Press, Beijing/St. Louis

Lim KT (2010) Inhibitory effect of glycoprotein isolated from *Opuntia ficus-indica* var. *saboten* Makino on activities of allergy-mediators in compound 48/80-stimulated mast cells. Cell Immunol 264(1):78–85

Linarès E. Thimonier C, Degre M (2007) The effect of NeOpuntia on blood lipid parameters–risk factors for the metabolic syndrome (syndrome X). Adv Ther 24(5):1115–1125

Lopez D (1995) A review: use of the fruits and stem of prickly pear cactus (*Opuntia* spp) into human food. Food Sci Technol Int 1:65–74

Mann J (1970) Cacti naturalised in Australia and their control. Department of Lands, Brisbane, 128 pp

McConn M, Nakata P (2004) Oxalate reduces calcium availability in the pads of the prickly pear cactus through formation of calcium oxalate crystals. J Agric Food Chem 52(5):1371–1374

Medellín MLC, Saldívar SOS, de la Garza JV (1998) Effect of raw and cooked nopal (*Opuntia ficus indica*) ingestion on growth and profile of total cholesterol, lipoproteins, and blood glucose in rats. Arch Latinoam Nutr 48(4):316–323, In Spanish

Ncibi S, Ben Othman M, Akacha A, Krifi MN, Zourgui L (2008) *Opuntia ficus indica* extract protects against chlorpyrifos-induced damage on mice liver. Food Chem Toxicol 46(2):797–802

Oh P-S, Lim K-T (2006) Glycoprotein (90 kDa) isolated from *Opuntia ficus-indica* var. *saboten* Makino lowers plasma lipid level through scavenging of intracellular radicals in Triton WR-1339-induced mice. Biol Pharm Bull 29:1391–1396

Ortiz-Escobar TB, Valverde-González ME, Paredes-López O (2010) Determination of the folate content in cladodes of nopal (*Opuntia ficus indica*) by microbiological assay utilizing *Lactobacillus casei* (ATCC 7469) and enzyme-linked immunosorbent assay. J Agric Food Chem 58(10):6472–6475

Palevitch D, Earon G, Levin I (1993) Treatment of benign prostatic hypertrophy with *Opuntia ficus-indica* (L.) Miller. J Herbs Spices Med Plants 2(1):45–49

Panico AM, Cardile V, Garufi F, Puglia C, Bonina F, Ronsisvalle S (2007) Effect of hyaluronic acid and polysaccharides from *Opuntia ficus indica* (L.) cladodes on the metabolism of human chondrocyte cultures. J Ethnopharmacol 111(2):315–321

Park E-H. Chun M-J (2001) Wound healing activity of *Opuntia ficus-indica*. Fitoterapia 72(2):165–167

Park EH, Kahng JH, Paek EA (1998) Studies on the pharmacological action of cactus: identification of its antiinflammatory effect. Arch Pharm Res 21(1):30–34

Park EH, Kahng JH, Lee SH, Shin KH (2001) An antiinflammatory principle from cactus. Fitoterapia 72(3):288–290

Parsons WT, Cuthbertson EG (2001) Noxious weeds of Australia, 2nd edn. CSIRO, Collingwood, p 712

Pinkava DJ (2003) *Opuntia* Miller. In: Flora of North America Editorial Committee (ed.) Flora of North America North of Mexico. Oxford University Press, New York/Oxford

Porcher MH et al (1995–2020) Searchable world wide web multilingual multiscript plant name database. Published by The University of Melbourne, Australia. http://www.plantnames.unimelb.edu.au/Sorting/Frontpage.html

Ramadan MF, Mörsel J-T (2003) Oil cactus pear (*Opuntia ficus-indica* L.). Food Chem 82(3):339–345

Russell CE, Felker P (1987) The prickly pears (*Opuntia* spp., Cactaceae): a source of human and animal food in semiarid regions. Econ Bot 41:433–445

Sáenz C, Estévez AM, Sepúlveda E, Mecklenburg P (1998) Cactus pear fruit: a new source for a natural sweetener. Plant Foods Hum Nutr 52(2):141–149

Sáenz-Hernandez C, Corrales-Garcia J, Aquino-Pérez G (2002) Nopalitos, mucilage, fiber, and cochineal. In: Nobel PS (ed.) Cacti: biology and uses. University of California, Berkeley, pp 211–234

Saleem M, Kim HJ, Han CK, Jin C, Lee YS (2006) Secondary metabolites from *Opuntia ficus-indica* var. *saboten*. Phytochemistry 67(13):1390–1394

Salvo F, Galati EM, Lo Curto S, Tripodo MM (2002) Chemical characterization of *Opuntia ficus-indica* seed oil. Acta Hort (ISHS) 581:283–289

Santana GMS, Albuquerque LP, Simões DA, Coelho LCBB, Paiva PMG, Gusmão NB (2009) Isolation of lectin from *Opuntia ficus-indica* cladodes. Acta Hort (ISHS) 811:281–286

Sawaya WN, Khalil JK, Al-Mohammad MM (1983) Nutritive value of prickly pear seeds, *Opuntia ficus-indica*. Plant Foods Hum Nutr 33(1):91–97

Schmitt L, Fouillot JP, Nicolet G, Midol A (2008) *Opuntia ficus indica*'s effect on heart-rate variability in high-level athletes. Int J Sport Nutr Exerc Metab 18(2):169–178

Siriwardhana N, Jeon Y-J (2004) Antioxidative effect of cactus pear fruit *(Opuntia ficus-indica)* extract on lipid peroxidation inhibition in oils and emulsion model systems. Eur Food Res Technol 219(4):369–376

Siriwardhana N, Shahidi F, Jeon Y-J (2006) Potential antioxidative effects of cactus pear fruit (*Opuntia ficus-indica*) extract on radical scavenging and DNA damage reduction in human peripheral lymphocytes. J Food Lipids 13(4):445–458

Sreekanth D, Arunasree MK, Roy KR, Chandramohan Reddy T, Reddy GV, Reddanna P (2007) Betanin a betacyanin pigment purified from fruits of *Opuntia ficus-indica* induces apoptosis in human chronic myeloid leukemia cell line-K562. Phytomed 14(11):739–746

Stintzing FC, Schieber A, Carle R (1999) Amino acid composition and betaxanthin formation in fruits from *Opuntia ficus-indica*. Planta Med 65(7):632–635

Stintzing FC, Schieber A, Carle R (2002) Identification of betalains from yellow beet (*Beta vulgaris* L.) and cactus pear [*Opuntia ficus-indica* (L.) Mill.] by high-performance liquid chromatography-electrospray ionization mass spectrometry. J Agric Food Chem 50(8):2302–2307

Stintzing FC, Herbach KM, Mosshammer MR, Carle R, Yi W, Sellappan S, Akoh CC, Bunch R, Felker P (2005) Color, betalain pattern, and antioxidant properties of cactus pear (*Opuntia* spp.) clones. J Agric Food Chem 53(2):442–451

Strack D, Engel U, Wray V (1987) Neobetanin: a new natural plant constituent. Phytochemistry 26:2399–2400

Teles FFF, Stull JW, Brown WH, Whiting FM (1984) Amino and organic acids of the prickly pear cactus (*Opuntia ficus indica* L.). J Sci Food Agric 35(4):421–425

Teles FFF, Whiting FM, Price RL, Borges VEL (1997) Protein and amino acids of nopal (*Opuntia ficus indica* L.). Revista Ceres 44(252):205–214

Tesoriere L, Butera D, Pintaudi AM, Allegra M, Livrea MA (2004) Supplementation with cactus pear (*Opuntia ficus-indica*) fruit decreases oxidative stress in healthy humans: a comparative study with vitamin C. Am J Clin Nutr 80(2):391–395

Tesoriere L, Fazzari M, Allegra M, Livrea MA (2005) Biothiols, taurine, and lipid-soluble antioxidants in the edible pulp of Sicilian cactus pear (*Opuntia ficus-indica*) fruits and changes of bioactive juice components upon industrial processing. J Agric Food Chem 53(20):7851–7855

Trejo-Gonzalez A, Gabriel-Ortiz G, Puebla-Perez AM, Huizar-Contrera MD, Munguia-Mazariegos M, Mejia-Arreguin S, Calva E (1996) A purified extract from prickly pear cactus (*Opuntia fuliginosa*) controls experimentally induced diabetes in rats. J Ethnopharmacol 55:27–33

Trombetta D, Puglia C, Perri D, Licata A, Pergolizzi S, Lauriano ER, De Pasquale A, Saija A, Bonina FB (2006) Effect of polysaccharides from *Opuntia ficus-indica* (L.) cladodes on the healing of dermal wounds in the rat. Phytomed 13(5):352–358

U.S. Department of Agriculture, Agricultural Research Service (2009) USDA National Nutrient Database for Standard Reference, Release 22. Nutrient Data Laboratory Home Page. http://www.ars.usda.gov/ba/bhnrc/ndl

Uchoa AF, Souza PA, Zarate RM, Gomes-Filho E, Campos FA (1998) Isolation and characterization of a reserve protein from the seeds of *Opuntia ficus-indica* (Cactaceae). Braz J Med Biol Res 31(6):757–761

Vázquez-Ramírez R, Olguín-Martínez M, Kubli-Garfias C, Hernández-Muñoz R (2006) Reversing gastric mucosal alterations during ethanol-induced chronic gastritis in rats by oral administration of *Opuntia ficus-indica* mucilage. World J Gastroenterol 12(27):4318–4324

Wagner WL, Herbst DR, Sohmer SH (1999) Manual of the flowering plants of Hawaii. Revised edition. Bernice P. Bishop Museum special publication. University of Hawai'i Press/Bishop Museum Press, Honolulu, 1919 pp (two vols)

Wiese J, McPherson S, Odden MC, Shlipak MG (2004) Effect of *Opuntia ficus indica* on symptoms of the alcohol hangover. Arch Intern Med 164:1334–1340

Zourgui L, Golli EE, Bouaziz C, Bacha H, Hassen W (2008) Cactus (*Opuntia ficus-indica*) cladodes prevent oxidative damage induced by the mycotoxin zearalenone in Balb/C mice. Food Chem Toxicol 46(5):1817–1824

Zourgui L, Ayed-Boussema I, Ayed Y, Bacha H, Hassen W (2009) The antigenotoxic activities of cactus (*Opuntia ficus-indica*) cladodes against the mycotoxin zearalenone in Balb/c mice: prevention of micronuclei, chromosome aberrations and DNA fragmentation. Food Chem Toxicol 47(3):662–667

Opuntia monacantha

Scientific Name

Opuntia monacantha Haw.

Synonyms

Cactus indicus Roxb., *Cactus monacanthos* Willd., *Opuntia lemaireana* Console ex F. A. C. Weber, *Opuntia vulgaris* auct. mult. Non P.Mill., *Opuntia vulgaris* var. *lemaireana* (Console ex F. A. C. Weber) Backeb.

An earlier name formerly and widely applied to *Opuntia monacantha* is *Opuntia vulgaris* Miller. This confused name has now been typified to become a synonym of *Opuntia ficus-indica* (Linnaeus) Miller (Leuenberger 1993).

Family

Cactaceae

Common/English Names

Barberry Fig, Barbary Fig, Cochineal Fig, Cochineal Prickly-Pear, Common Prickly Pear, Drooping Prickly Pear, Indian Fig, Prickly Pear

Vernacular Names

Afrikaans: Luisiesturksvy, Suurturksvy;
Ascension, Saint Helena: Opuntia, Prickly Pear, Red Tungi, Red Tungy, Round Red Prickly Pear;
Brazil: Cardo-Palmático Palmatória;
Chinese: Dan Ci Xian Ren Zhang;
India: Nag Phena (Hindu), Naga Kulli (Kannada);
Portuguese: Arumbeva, Monducuru, Urumbeba;
Samoan: Lauaufai Va;
Somalia: Quecis, Tiin, Tiin-Hindi.

Origin/Distribution

This species is indigenous to South America – Argentina, Paraguay, Uruguay and Brazil. It has been widely introduced and naturalized in tropical and subtropical regions around the world.

Agroecology

This *Opuntia* species is found in Semiarid, warm temperate to subtropical areas. It occurs on most soil types including sand and calcareous soils. It grows in dry sites in full sun in flats, slopes, agricultural areas or wastelands from near sea level to 2,000 m.

Edible Plant Parts and Uses

Fruits are edible raw or made into jams and jellies, beverages and alcohol. A very strong spirit which tastes of whiskey may be distilled from the fruits. The young cladodes or pads (stem segments) called nopals are also eaten and have been candied.

Botany

An erect succulent shrub with shallow fibrous roots, commonly 2–3 m but can reach 4m, stems grey green to light green. Stem one main, woody at base, with many side branches made up of fleshy, jointed, glossy green, obovate, narrowly so, obovate-oblong, oblong, or oblanceolate segments (platyclades), up to 45 cm long by 15 cm wide and 1.5 cm thick, glossy, upper segments drooping, each areole bears a few short very fine barded bristles and some areoles have 1–4 sharp spines about 5 cm long (Plates 1–2). Leaves small, scale-like, produced beneath the areoles. Flowers 6 cm diameter borne on the margins of segments; outer perianth parts yellow with a reddish median stripe or yellow with red markings on the back; inner perianth parts yellow to orange, rotate, 25–40 mm long, 12–40 mm wide; staminal filaments green to white; style green, 12–20 mm long; stigma lobes 8–10, cream yellow (Plates 1–2). Fruit green turning yellow to reddish purple, obovoid, 5–7.5 × 4–5 cm, umbilicus slightly depressed bear-

Plate 1 Cladodes with sharp long spines and flowers

Plate 2 Flower and developing fruits

Plate 3 Ripe and near ripe fruits

ing tufts of fine barbed bristles in the areoles (Plates 2–3). Seeds smooth, pale brown, irregularly elliptic, 4 × 3 mm, embedded in the reddish, pulpy flesh.

Nutritive/Medicinal Properties

The nutritive value of this species would be similar to that described for *Opuntia ficus-indica* and *Opuntia stricta*. This cactus has a few pharmacological properties such as antioxidant and antidiabetic activities.

Antioxidant Activity

Flavonoids, kaempferol and isorhamnetin were isolated from the methanol extract of the cladodes of *O. monacantha* (Valente et al. 2010).

The water (91.1%), ash, protein, fibre and lipid contents (15.0, 5.4, 18.5 and 1.4 g/100 g, respectively) were shown to be quite similar to the mean values of other *Opuntia* spp., some widely used as food and forage. The well-known free-radical DPPH scavenging activity of the isolated flavonoids reinforced the contribution of these compounds to the antioxidant activity of the *O. monacantha* cladodes.

Antidiabetic Activity

The polysaccharides of *Opuntia monacantha* cladode (POMC) was found to decrease the daily water consumption in streptozotocin-induced diabetic rats comparable to dimethylbiguanide, a commercial anti-diabetic drug (Yang et al. 2008). An increase to food intake was also shown for streptozotocin-induced diabetic rats administered by POMC. POMC had beneficial effects on the improvement in the control of blood glucose and serum lipid level. Daily treatment with 100–300 mg/kg body weight of POMC for 4 weeks produced a significant decrease on blood glucose level in streptozotocin-induced diabetic rats and also enhanced the cardioprotective lipid HDL level. The insulin level in streptozotocin-induced diabetic rats was not significantly affected by POMC and dimethylbiguanide treatment. Separate studies indicated that these polysaccharides could improve the control in blood glucose and serum lipid levels of streptozotocin-induced diabetic rats. POMC had two major fractions, POMC V and VI (Zhao et al. 2007). POMC V, which had a molecular weight of 28.7 kDa, was comprised mainly of rhamnose, arabinose and glucose in the molar ratio of 9.15:1.00:6.84, with 3.07% (w/w) of glucuronic acid, while POMC VI, which had a molecular weight of 10.8 kDa, was comprised mainly of rhamnose, mannose and glucose in the molar ratio of 8.72:1.00:6.19, with 4.68% (w/w) of glucuronic acid. The evaluation of anti-glycation activity suggested that POMC had good potential for inhibiting the formation of advanced glycation end-products. Time- and dose-dependent effects were also observed for all POMC samples. In another paper, POMC was collected into three fractions, POMC I, II, III. The results showed that inhibition of the formation of glycosylation end products (AGEs) process, at the fourth week, POMC II showed more potent inhibition than the same dose of aminoguanidine (Yang et al. 2007). With respect to inhibition of aldose reductase (AR) activity, POMC II exhibited the strongest inhibitory effect, but significantly lower than the positive control.

Traditional Medicinal Uses

Opuntia cladode has been used traditionally as an herbal medicine for treating diabetes, burns, bronchial asthma and indigestion in many countries over the world (Yan et al. 2008). This cactus is one of the many ethnomedicinal plants used by villagers from the Dharapuram Taluk, Tamil Nadu state, India. It was employed as a contraceptive. One teaspoon of the plant powder mixed with sugar was taken on empty stomach, from the first day of menstrual cycle up to 20 days (Balakrishnan et al. 2009).

Other Uses

The cactus was introduced to Ascension for the purpose of enriching the soil and preventing the evaporation of moisture. Before synthetic dyes were produced *O. monacantha* plants were cultivated for the purpose of supporting populations of *Dactylopius coccus*. When crushed the bodies of this Mexican scale insect produce a carmine-coloured dye.

Comments

Opuntia monacantha like most *Opuntia* species is deemed a noxious, invasive weed.

Selected References

Anderson EF (2001) The cactus family. Timber Press, Portland, 776 pp

Balakrishnan V, Prema P, Ravindran KC, Robinson JP (2009) Ethnobotanical studies among villagers from Dharapuram Taluk, Tamil Nadu, India. Global J Pharmacol 3(1):8–14

Hoskins JR, Mcfayden RE, Murray ND (1988) Distribution and biological control of cactus pears in eastern Australia. Plant Protect Q 3:115–23

Leuenberger BE (1993) Interpretation and typification of *Cactus opuntia* L., *Opuntia vulgaris* Mill., and *O. humifusa* (Rafin.) Rafin. (Cactaceae). Taxon 42:419–429

Li ZY, Taylor NP (2007) Cacataceae. In: Wu ZY, Raven PH, Hong DY (eds) Flora of China. Vol. 13 (Clusiaceae through Araliaceae). Science Press/Missouri Botanical Garden Press, Beijing/St. Louis

Mann J (1970) Cacti naturalised in Australia and their control. Department of Lands, Brisbane, 128 pp

Parsons WT, Cuthbertson EG (2001) Noxious weeds of Australia, 2nd edn. CSIRO, Collingwood, p 712

Pinkava DJ (2003) *Opuntia* Miller. In: Flora of North America Editorial Committee (ed) Flora of North America North of Mexico, vol 4. Oxford University Press, New York/Oxford

Valente LMM, da Paixão D, do Nascimento AC, dos Santos PFP, Scheinvar LA, Moura MRL, Tinoco LW, Gomes LNF, da Silva JFM (2010) Antiradical activity, nutritional potential and flavonoids of the cladodes of *Opuntia monacantha* (Cactaceae). Food Chem 123(4):1127–1131

Yang N, Zhao M, Yang B, Zhu B, Fang F (2007) Inhibition of polysaccharides from *Opuntia monacantha* on development of advanced glycated end products and activity of aldose reductase. Nat Prod Res Dev 19(4):568–571

Yang N, Zhao M, Zhu B, Yang B, Chen C, Cui C, Jiang Y (2008) Anti-diabetic effects of polysaccharides from *Opuntia monacantha* cladode in normal and streptozotocin-induced diabetic rats. Innovat Food Sci Emerg Technol 9(4):570–574

Zhao M, Yang N, Yang B, Jiang YA, Zhang G (2007) Structural characterization of water-soluble polysaccharides from *Opuntia monacantha* cladodes in relation to their anti-glycated activities. Food Chem 105(4):1480–1486

Opuntia stricta

Scientific Name

Opuntia stricta (Haw.) Haw.

Synonyms

Cactus strictus Haw. (basionym), *Cactus dillenii* Ker Gawl., *Cactus opuntia* var. *inermis* DC., *Opuntia anahuacensis* Griffiths, *Opuntia dillenii* (Ker Gawl.) Haw., *Opuntia dillenii* var. *tehuantepecana*, *Opuntia horrida* Salm-Dyck ex DC., *Opuntia inermis* (DC.) DC., *Opuntia maritima* Raf, *Opuntia stricta* var. *dillenii* (Ker Gawl.) L.D. Benson, *Opuntia vulgaris* var. *balearica* F. A. C. Weber.

Family

Cataceae

Common/English Names

Araluen Pear, Australian Pest Pear, Coastal Prickly Pear, Coastal Prickly-Pear, Common Pest Pear, Common Prickly Pear, Dildo, Eltham Indian Fig, Eltham Indian-Fig, Erect Prickly Pear, Erect Prickly-Pear, Erect Pricklypear, Erect Pricklypear Cactus, Gayndah Pear, Pakan, Pest Pear Of Australia, Pest Prickly-Pear, Pest Pricklypear, Sour Prickly Pear, Southern Spineless Cactus, Spineless Prickly Pear, Spiny Pest Pear, Sweet Prickly Pear, Sweet Prickly-Pear.

Vernacular Names

Afrikaan: Suurturksvy;
Chinese: Xiang-Ren-Zhang, Hsian-Jen-Chang, Xian-Tao, Hsian-T'ao;
German: Feigenkaktus;
India: Nagajemudu, Nagadali (Andhra Pradesh), Nagphana (Bengal), Chorhathalo (Gujrat), Nagphan, Chhittarthor (Himachal Pradesh), Hathhathoria, Naghhana (Hindi), Papaskalli, Chappatigalli (Karnatka), Palakakkali, Nagamullu (Kerala), Chapal (Maharashtra), Nagophenia (Orissa), Chittarthohar (Punjab), Mahavriksha, Vajrakantaka (Sanskrit), Nagathali, Sappathikalli (Tamilnadu);
Mexico: Yaaxpakan (Spanish);
Portuguese: Opúntia, Palma-De-Espinho, Palmatória;
Spanish: Chumbera, Nopal Estricto.

Origin/Distribution

This species is native to the Caribbean region, tropical and subtropical coast of eastern North America, and adjacent South America.

Agroecology

O. stricta is a tropical – subtropical species. It occurs in agricultural areas, ruderal, disturbed areas, scrub, shrublands, tundra, urban areas, lowland grassland and woody grassland, dry sclerophyll forest and woodland, rocky slopes and riparian situations near water courses.

Edible Plant Parts and Uses

The fruits are edible raw or made into jams and jellies. Flowers produce reasonable honey. The segments (platyclads) have been candied. *Opuntia stricta* fruit juice is a potential source of betacyanin pigments which can be used as a natural red-purple food colorant. In India, a very good jam was prepared from this fruit as well as an alcoholic drink.

Plate 2 Close-up of fruits

Botany

A spreading to erect, succulent shrub, 1–2 m high with shallow fibrous roots. Stems dull green or bluish green, glabrous consisting of a series of flattened, jointed fleshy, obovate segments (platyclades), 30 cm long by 15 cm wide by 1–2 cm thick, tuberculate, making margins appear scalloped between raised areoles, areoles three to five per diagonal row across midstem segment (Plates 1–2). Each areole with tufts of short fine, barbed bristles (glochids) and occasionally one to two straight or curve, yellow aging brown, stout spines 2–4 cm long. Leaves small, conical l scale-like produced beneath the areoles (eyes) on young segments and are shed as the segment matures.

Flowers, hermaphrodite, cyclic, epigynous, actinomorphic, 6–8 cm across, sessile with fleshy base borne on segment margins with inner tepals lemon yellow throughout, 25–30 mm; filaments yellow; anthers yellow; style and stigma lobes yellowish. Fruit skin red to purplish red, stipitate, with a depressed cavity in one end, pear-shaped, ellipsoid or barrel-shaped, 4–6 cm long, 2.5–4 cm diameter, areoles bearing tufts of fine barbed bristles (Plates 1–2), flesh reddish, juicy, pulpy. Seeds rounded, 4–5 mm diameter, pale brown, with slightly irregular surface and embedded in the pulp.

Nutritive/Medicinal Properties

Nutrient value per 100 g edible portion of *O. stricta* fruit was reported by Brand Miller et al. (1993) as: energy 17 g, moisture 82.9 g, nitrogen 0.13 g, protein 0.8 g, fat 0.1 g, calcium 230 mg, copper 0.1 mg, iron 0.4 mg, magnesium 43 mg, phosphorus 15 mg, sodium 700 mg, zinc 0.7 mg, niacin (from tryptophan or protein) 0.1 mg, niacin equivalents 0.1 mg, vitamin C 1 mg.

Opuntia stricta fruit juice was found to be a potential source of betacyanin pigments and could be used as a natural red-purple food colorant. *O. stricta* fruits had only betacyanins (betanin and isobetanin) and not betaxantins (Castellar et al. 2003). *O. stricta* fruits showed the highest betacyanin content (80 mg/100 g

Plate 1 Cladodes and fruits

fresh fruit). The red food colorant betanin was obtained from *O. stricta* fruit by fermentation of juice and homogenized fruits (Castellar et al. 2008). Liquid concentrated betanin was obtained with low viscosity and sugar free. Besides, bioethanol was obtained as byproduct. Studies showed that an attractive, vivid red-purple powder food colorant could be obtained by co-current spray drying of *O. stricta* fruit juices with a bench-scale two fluid nozzle spray dryer (Obón et al. 2009). This colorant was successfully applied in two food model systems: a yogurt and a soft-drink. The rind of the fruit yielded an arabinogalactan consisting of arabinose and galactose in the ratio 1:3 (Srivastava and Pande 1974).

In another study, of three species of Spanish red-skinned cactus pear fruits (*Opuntia ficus-indica, Opuntia undulata* and *Opuntia stricta*), *Opuntia ficus-indica* fruit extract had the strongest antioxidant capacity as evaluated by the 2,2′-azinobis(3-ethylbenzothiazoline-6-sulfonic acid) (Trolox equivalent antioxidant capacity) method and the 2, 2-diphenyl-1-picrylhydrazyl radical method (Fernández-López et al. 2010). It also had the highest taurine content. *O. stricta* fruits were the richest in ascorbic acid and total phenolics, whereas *O. undulata* fruits showed the highest carotenoid content. Quercetin and isorhamnetin were the main flavonoids detected.

From the 80% ethanolic extract of the cladodes, 4-ethoxyl-6-hydroxymethyl-α-pyrone, 3-O-methyl isorhamnein, 1-heptanecanol, vanillic acid, isorhamnetin-3-O-β-D-rutinoside and rutin were isolated (Qiu et al. 2003). In subsequent studies, from the 80% ethanolic extract of the cladode, the following compounds were isolated and identified: daucosterol, p-hydroxybenzoicacid, L-(−)-malic acid, (E)-ferulic acid, opuntiol and opuntioside (Qiu et al. 2005) and from the aqueous-ethanol stem extract, two new α-pyrones, named opuntioside II and opuntioside III (Qiu et al. 2007). Two novel C29-5β-sterols, opuntisterol [(24R)-24-ethyl-5β-cholest-9-ene-6β,12α-diol] and opuntisteroside [(24R)-24-ethyl-6β-[(β-d-glucopyranosyl) oxy]-5β-cholest-9-ene-12α-ol], together with nine known compounds, β-sitosterol, taraxerol, friedelin, methyl linoleate, 7-oxositosterol, 6β-hydroxystigmast-4-ene-3-one, daucosterol, methyl eucomate, and eucomic acid were isolated from the stem cladodes (Jiang et al. 2006).

Wang et al. (2009) identified 32 constituents in the volatile oil from *Opuntia dillenii*. The main compounds were: isobutyl phthalate 27.492%, palmitic acid 16.716%, butyl phthalate 11.257%, menthol 6.722%, linoleic acid 5.995%, nonanal 4.594%, hexanal 3.614% and dodecanic acid 3.244%. Thirty-five components were in the identified in volatile oil extracted from *Opuntia dillenii* by steam distillation (Jin and Ma 2010). The main components were: trans-phytol (36.57%), himachalene (10.89%), spathulenol (8.35%), aromadendrene (7.24%), caryophyllene (5.90%), hexadecanoic acid (5.45%), guaiene (3.05%), phenylacetaldehyde (2.74%) and hexanal (2.36%).

Some of the pharmacological properties of the fruit and stem cladodes are discussed below.

Antioxidant Activity

The aqueous ethanolic extract from the fresh cladodes of *Opuntia dillenii* showed potent radical scanvenging activity (Qiu et al. 2002). Three new compounds, opuntioside I, 4-ethoxyl-6-hydroxymethyl-α-pyrone, and kaempferol 7-O-β-D-glucopyranosyl-(1→4)-β-D-glucopyranoside, were isolated from the extract and the radical scavenging activity of these principal constituents were also determined.

Supercritical carbon dioxide extraction of seed oil from Opuntia dillenii gave a maximum extraction yield of 6.65% (Liu et al. 2009). The main fatty acids were linolenic acid (66.56%), palmitic acid (19.78%), stearic acid (9.01%) and linoleic acid (2.65%). The antioxidant activity of seed oil was found to be concentration-dependent and nearly comparable to the references ascorbic acid and butylated hydroxytoluene (BHT) as demonstrated by means of 2,2-diphenyl-1-picrylhydrazyl (DPPH) radical-scavenging assay and β-carotene bleaching test.

Antihyperglycaemic Activity

The fruit was found to have antihyperglycaemic activity. A single oral dose of the fruit juice (5 ml/kg) was found to reduce significantly the percent increase of glucose-induced hyperglycemia in both normoglycemic and alloxan-induced diabetic rabbits (Perfumi and Tacconi 1996). The hypoglycemic effect of the juice was comparable to that of an oral dose of tolbutamide (100 mg/kg) under the same oral glucose-tolerance conditions in normoglycemic rabbits; however, the juice did not increase glucose-induced plasma insulin levels. Further, the antihyperglycemic effect of the drug was not detected when the glucose load was administered intravenously. The data suggested that this cactus produced hypoglycemia mainly by lowering intestinal absorption of glucose. Other different mechanisms of action such as the presence of an orally active insulin-like compound could also be involved. During the oral toxicity study of the crude drug, rats given doses up to 50 ml/kg exhibited no symptoms of toxicity.

Anti-inflammatory Activity

This cactus was found to possess anti-inflammatory activity. Intra-peritoneal administration of the fruit extract (100–400 mg/kg) inhibited, in a dose-related fashion, carrageenan-induced paw edema in rats (Loro et al. 1999). A dose-dependent effect was obtained against chemical (writhing test) and thermic (hot plate test) stimuli, respectively, with doses of 50 and 100 mg/kg. In another study, the alcohol extract of the flowers elicited the most potent anti-inflammatory effect compared with the fruit and cladode using the carrageenan-induced rat paw oedema model and a pronounced analgesic action at a dose of 200 mg/kg using electric current as a noxious stimulus method (Ahmed et al. 2005). Three flavonoid glycosides, namely, kaempferol 3-O-arabinoside, isorhamnetin-3-O-glucoside and isorhamnetin-3-O-rutinoside were isolated from the extract. The aqueous ethanolic extract from the fresh cladodes was found to show anti-inflammatory activity (Qiu et al. 2007). Two new α-pyrones, designated opuntioside II and opuntioside III, were isolated from the extract together with six known compounds.

Hypotensive Activity

Methanolic extract of the cladodes and its pure compound α-pyrone glycoside, opuntioside-I exhibited potent hypotensive activity in normotensive rats. Both the extract and opuntioside-I exhibited comparable effect of 44–54% fall in mean arterial blood pressure (MABP) at the dose of 10 mg/kg (Saleem et al. 2005). Histopathology revealed adverse effects of high doses on liver and spleen of the experimental rats. However, no mortality occurred in rats even at the high doses of 1,000 mg/kg/d and 900 mg/kg/d of the extract and opuntioside-I respectively.

Neuroprotective Activity

Cactus polysaccharides (CP), demonstrated neuroprotective effects in rat brain slices (Huang et al. 2009). Pre-treatment with CP prior to hydrogen peroxide exposure significantly raised cell viability, decreased hydrogen peroxide-induced apoptosis, and reduced both intracellular and total accumulation of reactive oxygen species (ROS) production. Additionally, CP also reversed the augmentation of Bax/Bcl-2 mRNA ratio, the downstream cascade following ROS. The results suggested that CP may be a potential compound for the therapy of ischemia and oxidative stress-induced neurodegenerative disease.

Antispermatogenic Activity

The cactus was found to have anti-spermatogenic activity. Oral administration of the phylloclade extract to male rats caused a significant reduction in the weights of testes, epididymides, seminal vesicle and ventral prostate (Gupta et al. 2002). The production of spermatid was decreased by

88.06% in *Opuntia* treated rats. The populations of preleptotene spermatocytes and spermatogonia were lowered by 59.7% and 61.65% and secondary spermatocytes by 63.32%, respectively. The cross sectional surface area of Sertoli cells was reduced significantly. The seminiferous tubule and Leydig cell nuclear area were also decreased significantly when compared to controls. Motility of the spermatozoa was diminished significantly. *Opuntia* lowered the fertility of male rats by 100%.

Traditional Medicinal Uses

This cactus is used in folk medicine as an antidiabetic and anti-inflammatory (Ahmed et al. 2005). Stem segments have been used medicinally for treatment of whooping coughs and diabetes. The fresh fruit is used as an antidiabetic agent in Canary Islands folk medicine (Perfumi and Tacconi 1996). In India, the plant is bitter, hot, laxative; stomachic, carminative and antipyretic use to treat biliousness, burning, leucoderma, urinary complaints, tumours, loss of consciousness, piles, inflammations, anaemia, ulcers and the enlargement of the spleen. The heated juice of the plant has been employed for tumours and leucoderma. The fruit is refrigerant and the ripe fruits, when eaten, dye the urine red. They are also said to be useful in gonorrhoea. The leaves mashed up and applied as a poultice are said to allay heat and inflammation. The leaf made into a pulp is applied to the eyes in the case of ophthalmia. The flowers are employed for bronchitis and asthma. The underground roots are used for inducing quick vomiting in the case of persons bitten by poisonous snakes.

Other Uses

Worldwide, the cactus is cultivated as hedges and living fences, also for sand dune fixation. The segments have been used after removal of the spines as a fodder crop.

Comments

A noxious, invasive weed. Cactus glochids easily detach from the plant and become lodged in the skin, causing irritation upon contact with the tufts that cover the plant, each tuft containing hundreds of tiny barbs.

Selected References

Ahmed MS, El Tanbouly ND, Islam WT, Sleem AA, El Senousy AS (2005) Antiinflammatory flavonoids from *Opuntia dillenii* (Ker-Gawl) Haw. flowers growing in Egypt. Phytother Res 19(9):807–809

Brand Miller J, James KW, Maggiore P (1993) Tables of composition of Australian aboriginal foods. Aboriginal Studies Press, Canberra

Castellar R, Obón JM, Alacid M, Fernández-López JA (2003) Color properties and stability of betacyanins from *Opuntia* fruits. J Agric Food Chem 51(9):2772–2776

Castellar MR, Obon JM, Alacid M, Fernandez-Lopez JA (2008) Fermentation of *Opuntia stricta* (Haw.) fruits for betalains concentration. J Agric Food Chem 56(11):4253–4257

Fernández-López JA, Almela L, Obón JM, Castellar R (2010) Determination of antioxidant constituents in cactus pear fruits. Plant Foods Hum Nutr 65(3): 253–259

George AS (ed.) (1984) Phytolaccaceae to Chenopodiaceae. Flora of Australia, vol 4. AGPS, Canberra, p 71

Gupta RS, Sharma R, Sharma A, Chaudhudery R, Bhatnager AK, Dobhal MP, Joshi YC, Sharma MC (2002) Antispermatogenic effect and chemical investigation of *Opuntia dillenii*. Pharm Biol 40(6):411–415

Hoskins JR, Mcfayden RE, Murray ND (1988) Distribution and biological control of cactus pears in eastern Australia. Plant Protect Q 3:115–123

Hu SY (2005) Food plants of China. The Chinese University Press, Hong Kong, 844 pp

Huang X, Li Q, Li H, Guo L (2009) Neuroprotective and antioxidative effect of cactus polysaccharides in vivo and in vitro. Cell Mol Neurobiol 29(8):1211–1221

Jiang J, Li Y, Chen Z, Min Z, Lou F (2006) Two novel C29-5β-sterols from the stems of *Opuntia dillenii*. Steroids 71(13–14):1073–1077

Jin H, Ma C (2010) Analysis of chemical constituents of volatile oil in *Opuntia dillenii* by GC-MS. J Anhui Agric Sci 24:13060–13061 (In Chinese)

Kirtikar KR, Basu BD (1989) Indian medicinal plants, vol I, 2nd edn. Lalit Mohan Basu, Allahabad

Li ZY, Taylor NP (2007) Cacataceae. In: Wu ZY, Raven PH, Hong DY (eds) Flora of China. Vol. 13 (Clusiaceae through Araliaceae). Science Press/Missouri Botanical Garden Press, Beijing/St. Louis

Liu W, Fu YF, Zu YG, Mei-Hong Tong MH, Wu N, Liu XL, Zhang S (2009) Supercritical carbon dioxide extraction of seed oil from *Opuntia dillenii* Haw. and its antioxidant activity. Food Chem 114(1):334–339

Loro JF, del Rio I, Pérez-Santana L (1999) Preliminary studies of analgesic and anti-inflammatory properties of *Opuntia dillenii* aqueous extract. J Ethnopharmacol 67(2):213–218

Mann J (1970) Cacti naturalised in Australia and their control. Department of Lands, Brisbane, 128 pp

Obón JM, Castellar MR, Alacid M, Fernández-López JA (2009) Production of a red–purple food colorant from *Opuntia stricta* fruits by spray drying and its application in food model systems. J Food Eng 90(4):471–479

Parmar C, Kaushal MK (1982) *Opuntia dillenii*, pp 54–57. In: Wild fruits of the Sub-Himalayan region. Kalyani Publishers, New Delhi. 136 pp

Parsons WT, Cuthbertson EG (2001) Noxious weeds of Australia, 2nd edn. CSIRO, Collingwood, p 712

Perfumi M, Tacconi R (1996) Antihyperglycemic effect of fresh *Opuntia dillenii* fruit from Tenerife (Canary Islands). Pharm Biol 34(1):41–47

Pinkava DJ (2003) *Opuntia* Miller. In: Flora of North America Editorial Committee (ed) Flora of North America North of Mexico, vol 4. Oxford University Press, New York and Oxford

Qiu Y, Chen Y, Pei Y, Matsuda H, Yoshikawa M (2002) Constituents with radical scavenging effect from *Opuntia dillenii*: structures of new alpha-pyrones and flavonol glycoside. Chem Pharm Bull (Tokyo) 50(11):1507–1510

Qiu YK, Dou DQ, Pei YP, Yoshikawa M, Matsuda H, Chen YJ (2003) The isolation and identification of a new alpha-pyrone from *Opuntia dillenii*. Yao Xue Xue Bao 38(7):523–525 (In Chinese)

Qiu YK, Dou DQ, Pei YP, Yoshikawa M, Matsuda H, Chen YJ (2005) Study on chemical constituents from *Opuntia dillenii*. Zhongguo Zhong Yao Za Zhi 30(23):1824–1826 (In Chinese)

Qiu YK, Zhao YY, Dou DQ, Xu BX, Liu K (2007) Two new alpha-pyrones and other components from the cladodes of *Opuntia dillenii*. Arch Pharm Res 30(6):665–669

Saleem R, Ahmad M, Azmat A, Ahmad SI, Faizi Z, Abidi L, Faizi S (2005) Hypotensive activity, toxicology and histopathology of opuntioside-I and methanolic extract of *Opuntia dillenii*. Biol Pharm Bull 28(10):1844–1851

Srivastava BK, Pande CS (1974) Arabinogalactan from the pods of *Opuntia dillenii*. Planta Med 25(1):92–97

Wang K, Ding L, Liu J, Gao Y, An L, Liao C (2009) Determination of chemical constituents of the volatile oil from *Opuntia dillenii* (Ker-Gaw.) Haw. Biotechnology 5:54–55

Carica papaya

Scientific Name

Carica papaya L.

Synonyms

Carica bourgeaei Solms, *Carica citriformis* Jacq., *Carica cubensis* Solms, *Carica hermaphrodita* Blanco, *Carica jamaicensis* Urb., *Carica jimenezii* (Bertoni in J. B. Jimenez) Bertoni, *Carica mamaya* Vell., *Carica papaya* fo. *mamaya* Stellfeld, *Carica papaya* fo. *portoricensis* Solms, *Carica papaya* var. *bady* Aké Assi, *Carica papaya* var. *jimenezii* Bertoni in J. B. Jimenez, *Carica peltata* Hook. & Arn., *Carica pinnatifida* Heilborn, *Carica portorricensis* (Solms) Urb., *Carica posopora* L., *Carica rochefortii* Solms, *Carica sativa* Tussac, *Papaya bourgeaei* (Solms) Kuntze, *Papaya carica* Gaertn., *Papaya cimarrona* Sint. ex Kuntze, *Papaya citriformis* (Jacq.) A. DC., *Papaya communis* Noronha, *Papaya cubensis* (Solms) Kuntze, *Papaya cucumerina* Noronha, *Papaya edulis* Bojer, *Papaya edulis* var. *macrocarpa* Bojer, *Papaya edulis* var. *pyriformis* Bojer, *Papaya hermaphrodita* Blanco, *Papaya papaya* (L.) H. Karst., *Papaya peltata* (Hook. & Arn.) Kuntze, *Papaya rochefortii* (Solms) Kuntze, *Papaya vulgaris* A. DC., *Vasconcellea peltata* (Hook. & Arn.) A. DC.

Family

Caricaceae

Common/English Names

Papaya, Papaw, Mummy Apple, Melon Tree

Vernacular Names

Angola: (I)Mama (Umbundu), (Otchi)Mama (Umubumbu), Mamoeiro, Papais (Portuguese);
Arabic: Aanabahe-Hindi, Amba-Hindi, Arandkharbuza;
Benin: Adopuba, Douba, Kpinma (Adja), Carabossi, Krakrambossi (Bariba), Aguipa (Dassa), Carabossi (Dendi), Kpentin, Kpin (Fon), Papayer (French), Adubati (Gèn), Gbekpetin (Goun), Adouba (Mina), Gbékpé (Nagot), Igbékpé (Pédah), Papayi (Peuhl), Korokotoré (Somba), Igi Bekpe (Yoruba), Ibepe, Ibepe Dudu;
Brazzaville: Moloolo (Beembé), Papayer (French), Maloolo (Laadi), Mouloolo (Lari), Nlolo (Yoombe);
Brazil: Chamburé, Mamão, Mamoeiro, Papaia;
Burkina Faso: Bofré, Boflé (Abron), Bofré, Boflé (Ashanti), Baké (Gagou), Vatré, Vatou, Fakwaou (Guéré), Bofré, Boflé (Koulango), Papaya (Moore), Badié (Shien);

Burmese: Thimbaw;
Burundi: Ipapayi, Umupapayi (Kirundi);
Cameroon: Pahpah (Bamileke), Papaya;
Carolinian: Bweibwayúl Mwel, Bweibwayúl Wal;
Central African Republic: Nkovo (Banda);
Chinese: Fan Mu Gua, Mukua-Wan-Shou-Kuo;
Chuukese: Baibai, Bwebwao, Bwebwao Kipwae, Kipau, Kippwau, Kipwai, Kipwpwaaw, Momiap, Momwiyáp, Pwaipwai, Pwáyipwáy, Pwipwai;
Comoros: Pwapari (Anjouan Island);
Cook Islands: Ninitā, Nītā, Vī Angai Puaka, Vī Puaka, Vī Puaka, Vī Puaka, Vī Puaka, Vīnītā (Maori);
Cuba: Fruta Bomba, Lechosa;
Czech: Papája Melounová, Papája Obecná;
Danish: Almindelig Papaya, Melontræ, Papaya;
Democratic Republic of Congo: Papayi (Kikongo), Papai (Kinande), Dipaya Paya (Kitandu), Paya Paya (Kiyaka), Mutie Papayi (Kiyanzi), Dilolo (Kiyombe), Mouloolo (Lari), Papai (Lega), Pai-Pai (Lingala), Papai (Mashi), Ipapayi (Shi), Mpapai, Mpapayu, Pai-Pai, Payipayi (Swahili), Payipayi (Topoke), Mutshi Wa Tshipayi - Payi (Tshiluba), Pay-Pay (Yanzi);
Dominican Republic: Lechosa;
Dutch: Meloenboom, Meloenboom Soort, Papaya, Papaya Soort;
Eastonian: Harilik Papaya, Vili: Papaya;
Ethiopia: Papaya (Amhara), Papaya (Gedeoffa), Paappa (Konta), Papaya (Omotic), Papaya (Afaan Oromo);
Fijian: Maoli, Oleti, Papita, Seaki, Weleti, Wi;
Finland: Papaija;
French: Arbre De Melon, Papago, Papaye, Papayier;
French Guiana: Papaye, Papayer (French), Mau (Wayapi);
Gabon: Lolo (Apindji), Mulolè, Mulolo (Baduma), Délolo (Bakèlè), Géroro (Baléngi), Mulolu (Balumbu), Lolo (Banzabi), Mulolu (Bapunu), Mulolu (Bavarama), Mulolu (Bavili), Élolo (Bavové), Mulolu (Bavungu), Ilolo (Benga), Ilolo (Béséki), Mulolu (Eshira), Alola (Fang), Ololo (Galoa), Ololo (Ivéa), Papayer, Arbre À Melons (Local French), Mulolu (Masangu), Oti A Papayi (Mindumu), Édodo (Mitsogo), Ololo (Mpongwè), Ololo (Nkomi), Mulola (Ngowé), Ololo (Orungu);
German: Baummelone, Melonenbaum, Papajabaum, Papajapflanze, Papaya, Papayabaum;
Ghana: Bofere (Asante-Twi), Góndílí (Dagomba), Kwalentia (Lobi), Brofe (Southern);
Guinea: Yeletiga (Guerze), Iritike (Manon);
Guinea Conakry: Boudi Baga (Foula Du Fouta-Djalon), Yiridye;
Guyana: Ma PaYa Yik, Map, Pa Ya Yik (Patamona), Papaw, Papaya;
Hawaiian: Hē'Ī, Mīkana, Milikana, Papaia;
Hungarian: Dinnyefa, Papája;
I-Kiribati: Te Babaia, Te Mwemweara, Te Papaw;
India: Amita (Assamese); Pappaiya, Papeya, Penpe, Pepiya, (Bengali), Pappayu (Gujerati), Andakharbuja, Arandkharbuza, Papal, Papaya, Papeeta, Papiitaa, Papita, Papiya, Pappita, Pepiya, Popaiya, Popaiyah, Popaiya Pappita, Pepiya, Popaiya, Popaiyah, Popaiya (Hindu), Akka Thangi Hannu, Bappangaayi, Bappangayi, Barangi, Boppayi-Hannu, Boppe, Goppe, Goppe Hannu, Goppen, Nda Karbhooja Pangi, Papaya, Pappaya, Pappangaye, Pappangayi, Pappayi, Parangi, Parangi Hannu Mara, Parangi Mara, Peragi, Perinji, Pharangi-Hannu, Piranji, Poppaya (Kannada), Poppayi (Konkani), Apappaya-Pazham, Kaplam, Kappalam, Karmmatti, Karmmos, Karmmosu, Karmosu, Karmmusu, Karumusa, Karutha, Omakai, Ommai, Papajamaram, Pappali, Pappaya, Pappaya-Pazham, Pappayam, Pappayambalam, Pappayampazham, Poppoia, Umbalay (Malayalam), Awathabi (Manipuri), Papaaya, Papai, Papaya, Papay, Pappayi, Popai, Popay (Marathi), Thingfanghma, Thingfanghana (Mizoram), Thingfanghma, Amritobonda, Amrut Bhanda (Oriya), Brahmairandah, Chirbhita, Erand Karkati, Erandachirbhata, Erandachirbhita, Erandakarakati, Erandakarkati, Madhukarkati, Nalikadala (Sanskrit), Cittamukkicam, Cittamukkikamaram, Conkarikam, Conkarikamaram, Kaniyamanakku, Kariyamanakku, Karpakkini, Karpakkinimaram, Maniyamanakku, Pacalai, Pappai, Pappali, Pappali-Pazham, Pappay, Pappayi, Pappayi-Pazham, Parangi, Parangiyamanakku, Pasalai, Poppayi, Puppali Pullum, Pappali, Pappalikaimaram, Pappalippal, Parankiyamanakku, Parankiyamanakku (Tamil), Bapaipundoo,

Bappayi, Bappayya, Bobbasi, Booppamkaya, Boppai, Boppaayi, Boppayi, Boppasa, Boppayi-Pandu, Boppayya, Chettu, Madana Anapa, Madana-Anapakaya, Madanaanapa, Madananaba, Madhurnakam, Madhurnakamu, Paringi, Poppaya, Parindhi (Telugu), Arand Kharbuza, Papiitaa, Papitha, Papita Desi (Urdu);
Indonesia: Gandul Bali, Gandul Kates, Kamplong, Tela Gantung, Gedang, Kates, Ketalah (Javanese), Kates (Madurese), Gedang (Sundanese), Papaya;
Italian: Papaia;
Ivory Coast: Bofré, Boflé (Abron), Eplé (Aboure), Bofré, Boflé (Ashanti), Boflè (Baoule), Kpakpa (Ebrie), Baké (Gagou), Vatré, Vatou, Fakwaou (Guéré), Vadien (Gouro), Bofré, Boflé (Koulango), Badien (Malinke), Badié (Shien);
Japanese: Papaiya, Popoo;
Kapingamarangi: Memeapu;
Kenya: Poi Ppoi (Luo), Mûbabaî (Kikuyu), Mubabai (Mbeere);
Khmer: Ihong, Doeum Lahong;
Korean: Pa Pa Ya;
Kosraean: Es;
Kwara'Ae: Takafo;
Laotian: Houng;
Madagascar: Mapaza, Papay;
Malaysia: Rungan (Iban), Betik, Betek (Malay), Majan (Kelabit), Kuntaia Kepaya (Jakun), Papaya;
Mali: Manje (Bambara), Maye (Bwa), Papayé (Malinke), Manayi (Minyanka), Manayi (Senoufo);
Marshallese: Kehnap, Keinabbu, Keinabu, Keinapu, Kenabu, Kinabu, Kinapau;
Mauritius & Rodrigues: Papayer;
Mokilese: Mamiyap;
Naruan: Dababaia;
Nepali: Mewa;
Nigeria: Uhro (Bini), Nukuhi (Fulfude), Ibepe, Gwanda (Hausa), Udia Edi, Popo, Ukpod (Ibibio), Ojo-Mgbimgbi (Ibo), Okpurukwa, Okwulu Oyibo (Igbo), Ugboja (Igede), Gwada (Koma), Ibepe (Okeigbo), Ibepe (Ondo), Ibepe (Owomode), Ibepe (Yoruba), Ekebo;
Niuean: Loku, Loku Fua Ku, Loku Fua Leleva, Loku Fua Magaia;
Nikuoro: Mami;
Norway: Papaya;
Pakistan: Papeeta;
Paluan: Babai, Bóbai, Bobai, Ebingel;
Papiamento: Papaya;
Persian: Aanabahe-Hindi, Amba-Hindi;
Philippines: Tapayas (Bikol), Lapaya (Bontok), Kapayas (Samar-Leyte Bisaya), Papyas (Subanum), Kapayas, Papaye (Sulu), Kapaya, Lapaya, Papaya (Tagalog);
Pingelapan: Kaineap;
Pohnpeian: Memiap, Mohmiyap, Momiyap, Mommyapple;
Polish: Melonowiec Wlasciwy, Papaja;
Puerto Rico: Lechosa;
Republic of Guinea: Yiridyi (Manika), Buudhi (Pular), Fofia (Soso);
Reunion: Papaye, Papayer;
Russian: Papaia;
Rwanda: Ipapayi;
Samoan: Esi, Esi, Esi Loa;
Senegal: Mandé, Mandu, Papia, Papiu (Bambara), Bum Papa, Bupapay (Diola), Papayer (Local French), Impapakèye (Niominka), Boodie Baga, Papayi, Papayo (Peul), Papayo (Sérère), Papayi (Tocolor), Papayo (Wolof);
Seychelles: Papayer;
Sierra Leone: Fakalii (Cibemba), Fakai Laa (Mende), An Papai (Temne), Fakai, Fakalj;
Slovaščina: Papaja;
Songsorol: Babaia;
South Africa: Mupapawe (Luvenda);
Spanish: Mamón, Melón Zapote, Papaya, Papayero, Papayo;
Sri Lanka: Papol', Guslabu (Sinhalese), Pawpaw;
Surinam: Papaja, Papaya;
Surinam Javan: Kates;
Swahili: Papayu, Papai;
Swedish: Melonträd, Papaya;
Tahitian: I'Ita;
Tanzania: Papai Dume (Swahili);
Thai: Loko, Malako, Malakor, Ma Kuai Thet, Sa Kui Se;
Togo: Brofudé (Akassélem), Adibati (Evé), Aghidi (Fé), Adibati (Ewé), Gbekpe (Fon), Somolu (Kagyé), Adubati (Mina), Debleti (Ouatchi), Gui - Bekpe (Yoruba);

Tongan: Lesi;
Tongarevan: Nīnītā;
Tuvaluan: Esi;
Uilithian: Bwebwae;
Venezuela: Lechosa;
Vietnamese: Dudu;
Uganda: Epaipai (Ngakarimojong), Papaali (Luganda), Mapapari (Runyankole), Mapapari (Runyaruguru), Amapapaali, Apapalo, Mupapali Omuisaiza;
Woleaian: Beibaay, Bweibwae, Bweibwai, Pai Wai;
Yapese: Babae, Babay, Baiwai, Waiwai;
Zambia: Paw-Paw.

Origin/Distribution

Papaya is indigenous to tropical Mexico, Central America and northern South America. Papaya has been distributed throughout the tropics and subtropics, where it is extensively cultivated, and as far north and south as 32°latitude.

Agroecology

Papaya grows in the tropics to subtropics in areas ranging from warm dry to moist temperate through to very dry to wet tropical zones from near sea level to 1,500 m. It grows in areas with annual temperature of 16–33°C but thrives best in areas with temperatures of 16.2–26.6°C and evenly distributed annual rainfall from 600–4,000 mm. It is adaptable to a wide range of soil types from calcareous soils, rocky volcanic soils, acid sulphate soils to peat soils but grows best in well-drained, moist, friable, organic matter rich, light soil – loams and sandy loams. The plant can grow at soil pH 5–7 but pH 6–6.5 is most desirable. The plant is not frost tolerant and abhors water-logged soil which leads to root rots as well as salty or saline soils. It is killed by frost and does not tolerate shade, flooding, strong winds and heavy clayey soils.

Edible Plant Parts And Uses

Ripe papaya fruits are eaten as dessert fruit often with lemon or lime juice squeezed over the flesh. The fruit of papaya is widely used in the food industry for the production of juices, tonic drinks, syrup, canned fruit pieces/slices, jellies, preserves, marmalade and jams. In the Philippines, the jam is often flavoured with calamansi juice. The partially ripe fruits are also cut into pieces and steeped in vinegar and sugar and the pieces are also dehydrated and candied and made into crystallised fruit slices. Unripe to partial ripe fruit are used as salad and as vegetable for cooking in stews, curries and soups. In Malaysia, the unripe fruit is shredded and the shredded pieces are used to stuff deseeded green chillies which are then pickled in vinegar called *Acha*r. The green fruit is also extensively used as a pickle called *hachara* in the Philippines. Green papaya is widely used in Thai cuisine raw as in a special papaya salad or cooked in soup and curries. The juice and flesh from green fruit (which contains the enzyme papain) are used to tenderize pork, beef, and fish. Fermented papaya preparation (FPP) is a natural health food that is commercially sold in Japan and the Philippines and is widely consumed by elderly people. This nutriceutical, bio-normalizer product has antioxidant action, inhibitory effect on oxidative DNA damage and tissue injury, being a potent OH scavenger.

Papaya flowers are eaten cooked as vegetables in south east Asia. Young leaves are also used as vegetables, steamed or cooked and eaten as spinach or in curries . In the Pacific islands the leaves are also used as tenderiser – by wrapping meat and octopus in the leaves. Papain, mainly from the latex of unripe fruit and trunk is widely used as a meat tenderiser, in the beverage and chilling beer industries and for making chewing gum. Crushed papaya seeds can be added to minced meat for *koftas* (spicy meatballs) or use as marinade for meat. The pounded fruit pulp can also be similarly used. Crushed or ground papaya seeds can be used as a substitute for pepper and is added to salad dressings or sauces to serve with fish. They also add texture and flavour to a fruit salad.

Botany

Small, soft-wooded, herbaceous tree with white milky sap growing to 4 m high (Plates 1–2). Trunk thick straight, hollow-centred with scars of fallen leaves and tapering from base to apex and terminating in a cluster of leaves at the top (Plate 2). Leaves are large, alternate, close together, 30–60 cm long, palmately divided into 5 to –7 irregularly cut lobes, and are borne on leaf-stalks 40–100 cm long, 1–3 cm in diameter. Plants mostly dioecious rarely monoecious with fragrant and nocturnal white or cream-coloured flowers. The male flowers are borne on long peduncled, pendulous racemes, 30–100 cm long. Flower in clusters, sessile and consists of small 5-lobed calyx, 5-lobed creamy yellow corolla tube, 10 stamens in two whorls, outer whorl of the stamens shortly stalked, filaments inner most sessile, basifixed 2-celled anthers which dehisces longitudinally (Plates 3–4). In female plant, the female flowers are solitary and axillary on short stalks, and consist of a small, 5-lobed calyx, 5 twisted, pale-yellow lanceolate, obtuse petals and a large ovary, bearing 5 dilated, subsessile stigmas. The

Plate 2 Tree habit of a yellow-skinned papaya variety

Plate 1 Pyriform papaya fruits

Plate 3 Male inflorescences of a male tree

Plate 4 Papaya flowers on sale in a local market

Plate 7 Globose- subglobose papaya fruits

Plate 5 Papaya fruits at various stages of ripening

Plate 8 A large obovate papaya fruit

Plate 6 An elongated, large papaya variety

Plate 9 Seeds and flesh of sliced papaya

ovary is globular, 1-celled, and contains numerous ovules attached to 5 parietal placentae. Fruit various shaped, large elongated-oblong, spherical, ovoid, obovate, or pyriform berry (Plates, 1, 2, 5–8), large usually 20–30 × 8–15 cm, turning yellow or orange with yellow or orange flesh. Seeds black, wrinkled, each enclosed in mucilaginous membrane, oval in shape, 2 mm in diameter enclosed in the hollow centre of the fruit (Plate 9) arranged in five rows.

Nutritive/Medicinal Properties

The nutrient value per 100 g edible portion of ripe orange-fleshed papaya was reported as: water 88.83 g, energy 39 kcal (163 kJ), protein 0.61 g, total lipid 0.14 g, ash 0.61 g, carbohydrate 9.81 g, total dietary fibre 1.8 g, total sugars 5.90 g, minerals – Ca 24 mg, Fe, 0.10 mg, Mg 10 mg, P 5 mg, K 257 mg, Na 3 g, Zn 0.07 mg, Cu 0.016 mg, Mn 0.011 mg, Se 0.6 μg, vitamins – vitamin C 61.8 mg, thiamin, 0.027 mg, riboflavin 0.032 mg, niacin 0.338 mg, pantothenic acid 0.218 mg, vitamin B-6 0.019 mg, total folate 38 μg, total choline 6.1 mg, vitamin A 1,094 IU, vitamin E (α-tocopherol) 0.73 mg, vitamin K (phylloquinone) 2.6 μg; total saturated fatty acids 0.043 g, 12:0 (lauric acid) 0.001 g, 14:0 (myristic acid) 0.007 g, 16:0 (palmitic acid) 0.032 g, 18:0 (stearic acid) 0.002 g; total monounsaturated fatty acids 0.038 g, 16:1 undifferentiated (palmitoleic acid) 0.020 g, 18:1 undifferentiated (oleic acid) 0.018 g; total polyunsaturated fatty acids 0.031 g, 18:2 undifferentiated (linoleic acid) 0.006 g, 18:3 undifferentiated (linolenic acid) 0.025 g; amino acids – tryptophan 0.008 g, threonine 0.011 g, isoleucine 0.008 g, leucine 0.016 g, methionine 0.002 g, phenylalanine 0.009 g, valine 0.010 g, arginine 0.010 g, histidine 0.005 g, alanine 0.014 g, aspartic acid 0.049 g, glutamic acid 0.033 g, glycine 0.018 g, proline 0.010 g, serine 0.015 g; β-carotene 276 μg; β-cryptoxanthin 761 μg, lutein + zeaxanthin 75 μg (USDA 2010).

Papaya fruit is a very rich source of vitamin A, β- carotene, β-cryptoxanthin, vitamin C, Lutein and zeaxanthin and is low in lipids, calories and sodium. Papaya is also a fair source of folate, dietary fibre, potassium and vitamin E. It contains small amounts of iron, thiamine, riboflavin, niacin pantothenic acid, vitamin B6, calcium, phosphorus, It is also rich in antioxidant nutrients such as carotenes and flavonoids. Papaya is one of the main fruits which have been recommended for prevention of vitamin A deficiency in Sri Lanka. Papaya is rich in vitamin C. Indeed, one piece of fruit contains a full day's supply of vitamin C. The health benefits of vitamin C are long and varied. Researchers are currently studying vitamin C as an alternative cancer treatment. It is a powerful antioxidant and must be obtained from food because the human body cannot create its own vitamin C. Vitamin E is another well known antioxidant found in the papaya, well known for skin health and as a blood thinner. β-carotene, which the body converts to vitamin A, is another nutrient provided by the papaya. The whole papaya fruit is an excellent source of dietary fibre, which is also necessary for digestive health.

Yellow-fleshed and red fleshed papayas were found to differ in the carotenoid profile and organisation of carotenoids in the cell (Chandrika et al. 2003). Yellow-fleshed papaya showed small carotenoid globules dispersed all over the cell, whereas in red-fleshed papaya the carotenoids were accumulated in one large globule. The major carotenoids of yellow-fleshed papaya were the provitamin A carotenoids β-carotene (1.4 μg/g dry weight DW) and β-cryptoxanthin (15.4 μg/g DW) and the non-provitamin A carotenoid ζ-carotene (15.1 μg/g DW), corresponding theoretically to (1,516) μg/kg DW mean retinol equivalent (RE). Red-fleshed papaya contained the provitamin A carotenoids β1-carotene (7.0 μg/g DW), β-cryptoxanthin (16.9 μg/g DW) and β-carotene-5,6-epoxide (2.9 μg/g DW), and the non-provitamin A carotenoids lycopene (11.5 μg/g DW) and ζ-carotene (9.9 μg/g DW), corresponding theoretically to 2,815 μg/kg DW mean RE.

Papaya was found to be a rich fruit source of carotenoids. The following carotenoids were found in the pulp of several papaya varieties (Rodriguez-Amaya 1999): common papaya – β-carotene 1.2 μg/g, ζ-carotene 0.8 μg/g, β-zeacarotene 0.1 μg/g, β-cryptoxanthin-5,6-epoxide 2 μg/g, β-cryptoxanthin 8.1 μg/g, cryptoflavin 0.8 μg/g; in the Solo variety – β-carotene 2.5 μg/g, ζ-carotene 1.4 μg/g, γ-carotene 0.2 μg/g, β-cryptoxanthin-5,6-epoxide traces, β-cryptoxanthin 9.1 μg/g, antheraxanthin traces, lycopene 21 μg/g; in the Fomosa variety – β-carotene 1.4 μg/g, ζ-carotene 1.7 μg/g, β-cryptoxanthin-5,6-epoxide (3.8), β-cryptoxanthin 5.3 μg/g, antheraxanthin 1.8, lycopene 19 μg/g.

One hundred and sixty-six compounds were identified in the aroma concentrate of papaya fruit, of which 77 were identified for the first time as papaya volatiles (Pino et al. 2003). Among the identified compounds, methyl butanoate, ethyl butanoate, 3-methyl-1-butanol and 1-butanol were found to be the major components. The esters of lower fatty acids were considered to contribute much to the typical papaya flavour. Fermentation with brewer's yeast and distillation yielded 4% alcohol, of which 91.8% was ethanol, 4.8%.methanol, 2.2% N-propanol, and 1.2% unknown (non-alcohol) (Sharma and Ogbeide 1982).

The latex from unripe fruit and stem of *Carica papaya* was found to be a rich source of the endopeptidases papain, chymopapain, protease omega (Brocklehurst and Salih 1985), glycyl endopeptidase and caricain (Dubois et al. 1988; Azarkan et al. 2003). Chymopapain was found to exist in its native state, two thiol forms. These enzymes were present in the laticifers at a concentration greater than 1 mM. Papaya latex also contained other enzymes as minor constituents. Several of these enzymes, namely a class-II and a class-III chitinase, an inhibitor of serine proteinases and the presence of a glutaminyl cyclotransferase and also β-1,3-glucanase and of a cystatin were also speculated (Azarkan et al. 2003). Papain has many pharmacological properties and medicinal, food and industrial uses.

Tang (1971) found that the concentration of benzyl isothiocyanate decreased in the fruit pulp but increased in the seeds with maturity of the papaya fruit. Free benzyl isothiocyanate was detected in both the wax layer and the vapor which emanated from intact green papayas. In a subsequent study, Tang (1973) reported that macerated papaya seeds and pulp contained benzyl isothiocyanate, produced by the enzymatic hydrolysis of benzyl glucosinolate by thioglucosidase. In mature papaya seeds, the enzyme, thioglucosidase was found in sarcotesta but not in endosperms, while the reverse was true for benzyl glucosinolate, which comprised more than 6% (w/w) of the endosperms. Both the enzyme and substrate were present in embryos and the amount of the benzyl glucosinolate was 3.9% (w/w). In immature papaya pulp, benzyl glucosinolate was localized principally in the latex, ranging from 7.3% to 11.6% of the dry wt of latex fluid. No thioglucosidase activity was found in papaya latex. Papaya fruit was found to be a good dietary source for glucosinolate and isothiocyanates (Rossetto et al. 2008). The pulp, peel and seeds were found to contain benzylglucosinolates and benzylisothiocyanates. Volatile benzylisothiocyanate was also confirmed in the internal cavity of the fruit during ripening. The highest benzylglucosinolate levels were detected in seeds, followed by the peel and pulp and the content decreased in all tissues during fruit development. Similarly, the levels of benzylisothiocyanates were much higher in the seeds than the peel and pulp. *Carica papaya* seed was found to contain a non-mitogenic lectin caricin (Togun et al. 2005). Papaya seed extract also showed a very high activity of myrosinase and, without myrosinase inactivation, produced 460 μmol of benzyl isothiocyanate in 100 g of seed. The seeds also contained glucosinolate (benzyl glucosinolate, glucotropaeolin) and the enzyme myrosinase, a glycoside, sinigrin, and carpasemine (Panse and Patanjpe 1943; Akah et al. 1997; Eno et al. 2000; Seigler et al. 2002; Wilson et al. 2002; Nakamura et al. 2007).

Papaya seed was found to be a rich source of proteins (27.8% undefatted, 44.4% defatted), lipids (28.3% undefatted) and crude fibre (22.6% undefatted, 31.8% defatted) (Marfo et al. 1986a). Of the toxicants estimated, glucosinolates was found in the highest proportion. The seed was low in free monosaccharides with sucrose predominating (75.0% of total sugars). Mineral content was generally low. Ca and P occurred in considerable amounts (17.340 μg/g and 10.250 μ/g, respectively). The seed oil was low in iodine value (74.8), free fatty acids (0.94%) and carotene (0.02 μg/g). The major fatty acid was oleic acid, C18:1 (79.1%). Globulins were found to constitute the bulk of the seed protein (53.9%) (Marfo et al. 1986b). The amino acid profile was not too different from other plant protein sources. However, papaya seed appeared deficient in many amino acids. Compared to soya bean meal and protein concentrate, the papaya products were inferior in terms of functional properties.

An earlier study found that the oil from papaya seeds contained the following acids: lauric, 0.4%; myristic, 0.4%; palmitic, 16.2%; stearic, 5.0%; arachidic, 0.9%; behenic, 1.6%; hexadecenoic, 0.8%; oleic, 74.3%; and linoleic, 0.4% (Badami and Daulatabad 1967). Another paper reported that the fatty oil of the seeds contained 16.97% saturated acids (11.38% palmitic, 5.25% stearic, and 0.31% arachidic) and 78.63% unsaturated acids (76.5% oleic and 2.13% linoleic) (Duke 1993). Crude protein and crude fibre contents of the defatted papaya seed meal were found to be especially high, 40.0% and 48.9% (Chan et al. 1978). Benzylisothiocyanate content in the seed oil was 0.56% (w/w), and the benzylglucosinolate content in the seed meal was 1.0 (w/w).

Sarcotesta of papaya seed had higher percentages of ash, crude protein, and crude fibre than did endosperm, but was lacking in fat (Passera and Spettoli 1981); while the endosperm contained 60% fat. The oil extract exhibited very high contents of oleic and palmitic acids. The essential amino acid profiles of endosperm and sarcotesta protein were compared. The results indicated that the endosperm protein was a good potential source of supplemental protein. The predominant fatty acid in the papaya seed oil was oleic acid (72–78%), with some palmitic (12–14%), stearic (4–5%) and linoleic (2.5–3.5%) acid (Puangsri et al. 2005). The main triacylglycerols (TAGs) were sn-glycerol-oleate-oleate-oleate (OOO) (43.5–45.5%) and 1-palmitoyl-dioleoyl glycerol (POO)+stearoyl-oleoyl-linoleoyl glycerol (SOL) (29.5–30.5%). Generally, the color of the oil was reddish yellow. The iodine and the saponification values of the petroleum ether-extracted oil were found to be 66.0 and 154.7, respectively, while those of the enzyme extracted oil (a-amylase, neutral protease, cellulose and pectinase) were 66.2–69.3 and 154.2–161.7, respectively. The unsaponifiable matter of the oil extracted using different enzymes varied between 2.07% and 2.90% and were significantly different from that of the solvent-extracted oil (1.39%).

Papaya leaves were reported to contain per 100 g: 74 cal, 77.5 g water, 7.0 g protein, 2.0 g fat, 11.3 g total carbohydrate, 1.8 g fibre, 2.2 g ash, 344 mg Ca, 142 mg P, 0.8 mg Fe, 16 mg Na, 652 mg K, 11,565 μg β-carotene equivalent, 0.09 mg thiamine, 0.48 mg riboflavin, 2.1 mg niacin, and 140 mg ascorbic acid, as well 136 mg vitamin E (Duke 1993). Leaves were found to contain the glycoside, carposide, and the alkaloids carpaine, pseudocarpaine, dehydrocarpaine I and II (Hornick et al. 1978; Khuzhaev and Aripova 2000). *Carica papaya* leaves contained phenolic acids as the main compound, while chlorogenic acid was found in trace amounts, compared to the flavonoids and coumarin compounds (Canini et al. 2007). Some of the compound included – protocatechuic acid, p-coumaric acid, caffeic acid, chlorogenic acid, quercetin 5,7-dimethoxycoumarin and kaempferol. The majority of the free phenolics were flavonols, such as kaempferol and quercetin, while the bound phenolics were mainly phenolic acids. Phenolic compounds are important components in vegetable foods, infusions and teas for their beneficial effects on human health (Croft 1998). The presence of such phenolic and coumarin compounds *in Carica papaya* leaves, could partially explain the pharmacological properties of this plant and demonstrated its importance in nutrition and daily intake (Canini et al. 2007).

Functional Nutraceutical Food Supplement

Fermented papaya preparation (FPP) (a product of yeast fermentation of *Carica papaya* Linn) as a food supplement was found to have many functional nutraceutical benefits (Aruoma et al. 2010). Studies in chronic and degenerative disease conditions (such as thalassemia, cirrhosis, diabetes and aging) and performance sports showed that FPP favorably modulated immunological, hematological, inflammatory, vascular and oxidative stress damage parameters. Neuroprotective potential evaluated in an Alzheimer's disease cell model showed that the toxicity of the β-amyloid could be significantly regulated by FPP. FPP regulated the H_2O_2-induced ERK, Akt and p38 mitogen activation with the reduction of p38 phosphorylation induced by H_2O_2. FPP was also found to reduce the extent of the H_2O_2-induced DNA damage, an outcome corroborated by similar effects obtained in the benzo[a]pyrene treated cells. No genotoxic effect was observed in experiments with FPP exposed to HepG2 cells nor was FPP toxic to the PC12 cells. Oxidative stress-induced cell damage and inflammation are implicated in a variety of cancers, diabetes, arthritis, cardiovascular dysfunctions, neurodegenerative disorders (such as stroke, Alzheimer's disease, and Parkinson's disease), exercise physiology (including performance sports) and aging. These conditions could potentially benefit from functional nutraceutical/food supplements like fermented papaya preparation exhibiting anti-inflammatory, antioxidant, immunostimulatory (at the level of the mucus membrane) and induction of antioxidant enzymes.

Antioxidant Activity

Papaya juice was found to be safe and to have antioxidative stress potential comparable to the standard antioxidant compound α-tocopherol. In-vitro evaluation of the antioxidant effects of papaya showed that the highest antioxidant activity (80%) was observed with a concentration of 17.6 mg/ml (Mehdipour et al. 2006). Blood lipid peroxidation levels reduced significantly after administration of all doses of papaya juice (100, 200, 400 mg/kg/day) to 35.5%, 39.5% and 40.86% of the control, respectively, compared with a value of 28.8% for vitamin E. The blood total antioxidant power was increased significantly by all doses of papaya juice (100, 200, 400 mg/kg/day) to 11.11%, 23.58% and 23.14% of the control, respectively. The value for vitamin E was 18.44%.

Fermented papaya preparation (FPP), a natural health food product obtained by biofermentation of *Carica papaya*, was shown to limit oxidative stress both in-vitro and in-vivo (Fibach et al. 2010). In β-hemoglobinopathies, such as β-thalassemia (thal) and sickle cell anemia, the primary defects were mutations in the β-globin gene and many aspects of the pathophysiology were facilitated by oxidative stress. Their studies indicated that in both groups, βthal patients: β-thal, major and intermedia (in Israel) and E-β-thal (in Singapore), FPP treatment enhanced the content of reduced glutathione (GSH) in red blood cells (RBC), and diminished their reactive oxygen species (ROS) generation, membrane lipid peroxidation, and externalization of phosphatidylserine (PS), indicating amelioration of their oxidative status, without a significant change in the hematological parameters.

Anticancer Activity

Glucosinolates and isothiocyanates from papaya were found to induce phase two detoxication enzymes, to boost antioxidant status, and to protect animals against chemically induced cancer (Shapiro et al. 2001). Glucosinolates were hydrolyzed by myrosinase (an enzyme found in plants and bowel microflora) to form isothiocyanates. These bioactive compounds were found to be chemoprotective and to protect against cancers of the lungs and alimentary tract (Johnson 2002). Isothiocyanates also inhibited mitosis and stimulated apoptosis in human tumour cells, in-vitro and in-vivo. Researchers separated and quantified the contents of benzyl isothiocyanate and the corresponding glucosinolate (benzyl glucosinolate, glucotropaeolin) in papaya pulp and seed (Nakamura et al. 2007). The papaya seed with myrosinase inactivation contained >1 mmol of benzyl glucosinolate in 100 g of fresh seed. This content was equivalent to that of Karami daikon (the hottest Japanese white radish) or that of cress. The papaya seed extract also showed a very high activity of myrosinase and, without myrosinase inactivation, produced 460 μmol of benzyl isothiocyanate per 100 g of seed. In contrast, papaya pulp contained an undetectable amount of benzyl glucosinolate and showed no significant myrosinase activity. The n-hexane extract of the papaya seed homogenate was highly effective in inhibiting superoxide generation and apoptosis induction in HL-60 (human promyelocytic leukemia) cells, the activities of which are comparable to those of authentic benzyl isothiocyanate.

Otsuki et al. (2010) found significant growth inhibitory activity of the *Carica papaya* leaf (CP) extract on tumour cell lines. In human peripheral blood mononuclear cells (PBMC), the production of IL-2 and IL-4 was reduced following the addition of CP extract, whereas production of IL-12p40, IL-12p70, IFN-γ and TNF-α was augmented without growth inhibition. Additionally, cytotoxicity of activated PBMC against K562 (human erythromyeloblastoid leukemia) was enhanced by the addition of CP extract. Moreover, microarray analyses showed that the expression of 23 immunomodulatory genes, classified by gene ontology analysis, was elevated by the addition of CP extract. The active component of CP extract, which inhibited tumour cell growth and stimulated anti-tumour effects, was found to be the fraction with molecular weight <1,000. The results suggested that the CP leaf extract may potentially provide the means for the treatment and prevention of selected human diseases such as cancer, various allergic disorders, and may also serve as immunoadjuvant for vaccine therapy.

Antiviral Activity

Research conducted reported that patients with Hepatitis C virus (HCV)-related cirrhosis showed a significant imbalance of redox status (low antioxidants/high oxidative stress markers), recorded a significant improvement of redox status when administered a fermented papaya preparation (FPP) during bedtime for 6 months (Marotta et al. 2006, 2007). FPP significantly lowered 8-hydroxydeoxy-guanidine (8-OHdG) level in circulating leukocyte DNA and the improvement of cytokine balance with FPP was significantly better than with vitamin E treatment. Findings suggested a potential supportive role of antioxidants/immunomodulators as FPP in hepatitis C virus patients. The fermented papaya preparation (FPP)-supplemented group showed a significant enhancement of the antioxidant protection with glutathune-S transferase M1 (GSTM1) (−) and of plasma DNA adduct, irrespective of the GSTM1 genotype. Only the GSTM1 (−) subgroup was the one that, under FPP treatment, increased lymphocyte 8-OHdG. Data showed that FPP could be a promising nutraceutical for improving antioxidant-defense in elderly patients even without any overt antioxidant-deficiency state.

Neuroprotective Activity

Scientists found that the fermented papaya preparation (FPP) made with yeast had antioxidant actions and may be a promising prophylactic food against the age related and neurological diseases associated with free radicals such as cancer, diabetes and especially in neurological disorders, for example, Parkinson's disease or Alzheimer's disease (Imao et al. 1998). Oral administration of the fermented papaya preparation for 4 weeks decreased the elevated lipid peroxide levels in the ipsilateral 30 minutes after injection of iron solution by iron into the left cortex of rats. The fermented papaya preparation also increased superoxide dismutase activity in the cortex and hippocampus.

Studies found that fermented papaya preparation (FPP) was able to attenuate β-amyloid precursor protein (Zhang et al. 2006); the deposition of β-amyloid (Abeta) is related in the pathogenesis of Alzheimer's disease (AD). It was found that the abnormal interactions of Abeta with metal ions such as copper were implicated in the process of Abeta deposition and oxidative stress in AD brains. Copper was found to trigger the Abeta neurotoxicity in SH-SY5Y cells overexpressing the Swedish mutant form of human APP (APPsw) in a concentration dependent manner. Fermented papaya preparation (FPP) showed high free radical scavenging ability in-vivo and in-vitro. FPP post-treatment increased cell viability and decreased the intracellular $[Ca^{2+}]i$, reactive oxygen species (ROS) generation such as hydroxyl free radical and superoxide anion and nitric oxide (NO) accumulation in the cell. The results also showed that FPP prevented the cell apoptosis through bax/bcl-2 sensitive pathway.

Separate studies found that fermented papaya preparation (FPP) derived from *Carica papaya* possessed the ability to modulate oxidative DNA damage due to H_2O_2 in rat pheochromocytoma

(PC12) cells and protection of brain oxidative damage in hypertensive rats (Aruoma et al. 2006). Cells pre-treated with FPP (50 μg/ml) prior to incubation with H_2O_2 had significantly increased viability and maintenance of morphology and shape. A significant reduction of DNA damage was observed at concentrations >or=10 μg/ml FPP, with 50 μg/ml FPP reducing the genotoxic effect of H_2O_2 by about 1.5-fold compared to only H_2O_2 exposed cells.

Anti-Inflammatory Activity

Ethanolic extract of *Carica papaya* leaves was found to significantly reduce paw oedema in the carrageenan test (Owoyele et al. 2008). Likewise the extract produced significant reduction in the amount of granuloma. In the formaldehyde arthritis model, the extracts significantly reduced the persistent oedema from the 4th day to the 10th day of the investigation. The study demonstrated the anti-inflammatory activity of *Carica papaya* leaves.

Immunomodulatory Activity

Different bioactive fractions of papaya seed extract possessed immunomodulatory activity (Mojica-Henshaw et al. 2003) Three major observations were made in the study: (1) the crude *Carica papaya* seed extract and two other bioactive fractions significantly promoted the phytohemagglutinin responsiveness of lymphocytes; (2) none of the *Carica papaya* seed extract (at the concentrations used in the study) was able to protect the lymphocytes from the toxic effects of chromium; and (3) some of the bioactive fractions of *Carica papaya* seed extract were able to significantly inhibit the classical complement-mediated hemolytic pathway. These findings provided evidence for immunostimulatory and anti-inflammatory actions of *Carica papaya* seed extract.

Hepatoprotective Activity

A study reported that the ethanol and aqueous extracts of *Carica papaya* showed remarkable hepatoprotective activity against carbon tetrachloride induced hepatotoxicity (Rajkapoor et al. 2002). The activity was evaluated by using biochemical parameters such as serum aspartate amino transferase (AST), alanine amino transferase (ALT), alkaline phosphatase, total bilirubin and gamma glutamate transpeptidase (GGTP).

Antidiabetic Activity

Research confirmed the empirical experience that fermented papaya preparation (FPP) use could induce a significant decrease in plasma sugar levels in both healthy subjects and type 2 diabetic patients (Danese et al. 2006). This hypoglycaemic effect, associated with clinical signs, induced the diabetic patients to reduce the dosage of their antidiabetic oral therapy. Thus, the FPP administration was suggested as an adjuvant drug to be use in the oral antidiabetic therapy in type 2 diabetes mellitus.

Purgative/Antidysenteric Activity

Aqueous extract of *Carica papaya* roots was found to increase the propulsive movement of the intestinal contents in rats, and increases the number of wet faeces (Akah et al. 1997). On the guinea pig isolated ileum, the extract elicited sustained contractions blocked by atropine. The results suggested a purgative effect mediated via the cholinergic system.

Wound Healing Activity

Carica papaya was reported to be used in The Gambia at the Royal Victoria Hospital, Banjul in the Paediatric Unit as the major component of burns dressings, where it was found to be well tolerated by the children (Starley et al. 1999). Cheap and widely available, the pulp of the papaya fruit was mashed and applied daily to full thickness and infected burns. It appeared to be effective in desloughing necrotic tissue, preventing burn wound infection, and providing a granulating wound suitable for the application of a split thickness skin graft. Possible mechanisms of action postulated included the activity of proteolytic enzymes chymopapain and papain, as well as an antimicrobial activity. Papaya was also reported to be widely used by nurses in Faculty of Medical Sciences, University of the West Indies, Jamaica as a form of dressing for chronic ulcers and there was a need for standardisation of its preparation and application (Hewitt et al. 2000). Topical application of the unripe fruit promoted desloughing, granulation and healing and reduced odour in chronic skin ulcers. Papaya was considered to be more cost effective and effective than other topical applications in the treatment of chronic ulcers. There was some difficulty in preparation of the fruit and occasionally a sensation of burning was reported by the patients. There was concern about the use of a non-sterile, non-standardised procedure but there were no reports of wound infection from its use.

Studies reported that *Carica papaya* promoted significant wound healing in diabetic rats. Aqueous extract of *C. papaya* fruit produced 77% reduction in the wound area in streptozotocin-induced diabetic rats when compared to controls which was 59% (Nayak et al. 2007). The extract treated wounds were found to epithelize faster as compared to controls. The wet and dry granulation tissue weight and hydroxyproline content increased significantly when compared to controls. The extract exhibited antimicrobial activity against the five organisms tested. Separate studies found that treatment with the phytopreparation from papaya accelerated wound healing and reduced the severity of local inflammation in rats with burn wounds (Mikhal'chik et al. 2004). The effect of this phytopreparation was postulated to be related to an increase in the effectiveness of intracellular bacterial killing by tissue phagocytes due to the inhibition of bacterial catalase. Antioxidant activity of the preparation decreased the risk of oxidative damage to tissues.

Another research found that aqueous extracts of green papaya epicarp (GPE) applied on induced wounds on mice stimulated and completed healing in shorter periods (13 days) than that required while using aqueous extracts of ripe papaya epicarp (RPE) (17 days), sterile water (18 days) and Solcoseryl ointment (21 days) (Anuar et al. 2008). Mice given RPE and misoprostol, an abortive drug, respectively experienced embryonic resorption while this effect was observed in none of the mice given GPE and water. The average body weight of live pups delivered by mice given GPE was significantly lower than those delivered by mice given water. Differences in composition may have contributed to the different wound healing and abortive effects of green and ripe papaya.

A standardized fermented papaya preparation (FPP) negated the gain in blood glucose and improved the lipid profile in adult obese diabetic (db/db) mice after 8 weeks of oral supplementation (Collard and Roy 2010). However, FPP did not influence weight gain during the supplementation period. FPP (0.2 g/kg body weight) supplementation for 8 weeks before wounding was effective in giving wound closure. Studies on viable macrophages isolated from the wound site demonstrated that FPP supplementation improved respiratory-burst function as well as inducible NO production. Diabetic mice supplemented with FPP showed a higher abundance of CD68 as well as CD31 antibodies at the wound site, suggesting effective recruitment of monocytes and an improved proangiogenic response. This work provided evidence that diabetic-wound outcomes may benefit from FPP supplementation by specifically influencing the response of wound-site macrophages and the subsequent angiogenic response.

Antimicrobial Activity

Carica papaya was found to have antibacterial effects that could be useful in treating chronic skin ulcers to promote healing. *Carica papaya* seeds were reported to contain anti-bacterial activity that inhibited growth of Gram-positive and Gram-negative organisms (Dawkins et al. 2003). Seed extracts from the fruit showed inhibition of the following bacteria in the following ranking: *Bacillus cereus* > *Escherichia coli* > *Streptococcus faecalis*

> *Staphylococcus* > *Proteus vulgaris* > *Shigella flexneri*. No significant difference was found in bacterial sensitivity between immature, mature and ripe fruits. No inhibition zone was produced by epicarp and endocarp extracts.

The meat, seed and pulp of *Carica papaya* were reported to be bacteriostatic against several enteropathogens such as *Bacillus subtilis, Enterobacter cloacae, Escherichia coli, Salmonella typhi, Staphylococcus aureus, Proteus vulgaris, Pseudomonas aeruginosa*, and *Klebsiella pneumoniae* (Osato et al. 1993). The same parts of papaya were indisputably demonstrated by electron spin resonance spectrometry to scavenge 1,1-diphenyl-2-picrylhydrazyl, hydroxyl) and superoxide radicals with the seed giving the highest activity at concentrations (IC_{50}) of 2.1, 10.0 and 8.7 mg/ml, respectively. The superoxide dismutase (SOD)-like activity in the meat, seed and pulp amounts to about 32, 98 and 33 units/ml; comparable to those of soybean paste miso, rice bran and baker's yeast. Vitamin C, malic acid, citric acid and glucose were found to be some of the possible antioxidative components in papaya. The study correlated the bacteriostatic activity of papaya with its scavenging action on superoxide and hydroxyl radicals which could be part of the cellular metabolism of such enteropathogens. The findings were indicative of the patho-physiological role of these reactive oxygen species in gastrointestinal diseases and papaya's ability to counteract the oxidative stress. Another research reported that ripe and unripe *Carica papaya* fruits (epicarp, endocarp, seeds) extracts produced very significant antibacterial activity against *Staphylococcus aureus, Bacillus cereus, Escherichia coli, Pseudomonas aeruginosa* and *Shigella flexneri* (Emeruwa 1982). Extracts of leaves showed no activity.

Studies showed also that papaya seed could be used as an effective antibacterial agent for selected bacteria. The minimum inhibitory concentration of papaya seed extract for 50% of the test bacteria – *Escherichia coli* and *Staphylococcus aureus* was 18.38 mg/ml and for *Salmonella typhi* the MIC was at 11.8 mg/ml of extract (Yismaw et al. 2008). However, the growth inhibitory effect of papaya seed extract was not observed for *Pseudomonas aeruginosa* up to 26.25 mg/ml of extract.

Carica papaya latex was found to inhibit the growth of *Candida albicans* (Giordani et al. 1991, 1996, 1997). Latex proteins appeared to be responsible for this antifungal effect. The minimum protein concentration for producing a complete inhibition was estimated to be about 138 µg/ml. Only α-D-mannosidase and N-acetyl-β-D-glucosaminidase were present in latex in significant amounts. The two enzymes showed a limited inhibitory effect on yeast growth, α-D-mannosidase being more efficient than N-acetyl-β-D-glucosaminidase. A mixture of the two enzymes showed a synergistic action on the inhibition of the yeast growth. When *C. albicans* was cultured in medium supplemented with N-acetyl-β-D-glucosaminidase a lack of polysaccharides was observed in the outermost and innermost layers of fungal cell wall. When *C. albicans* was cultured on medium supplemented with D(+)-glucosamine, an inhibitor of N-acetyl-β-D-glucosaminidase, growth was inhibited (34%) in a similar manner. Addition of D(+)-glucosamine during the exponential growth phase also had a fungistatic effect (26%). A mixture of *Carica papaya* latex (0.41 mg protein/ml) and the fungicide fluconazole (2 µg ml/) showed a synergistic action on the inhibition of *Candida albicans* growth. This synergistic effect resulted in partial cell wall degradation as indicated by transmission electron microscopy observations.

Antihelminthic and Antiamoebic Activities

Carica papaya seeds were found to have antihelminthic and anti-amoebic activities. Research reported that significantly more children given elixir prepared from air-dried *C. papaya* seeds and honey (CPH) than those given honey had their stools cleared of parasites [23 of 30 (76.7%) vs. 5 of 30 (16.7%)] (Okeniyi et al. 2007). There were no harmful effects. The stool clearance rate for the various types of parasites encountered was between 71.4% and 100% following CPH elixir treatment compared with 0–15.4% with honey. The results suggested that air-dried *C. papaya* seeds were efficacious in treating human intestinal parasites and without significant side effects. Thus, the researchers asserted that its consumption offered a cheap, natural, harmless, readily available monotherapy and preventive strategy against intestinal parasitosis, especially in tropical communities.

Results from the in-vitro antiamoebic activity of some Congolese plant extracts used as antidiarrhoeic in traditional medicine reported that mature papaya seeds exhibited antiamoebic activity (Tona et al. 1998). However, metronidazole used as reference product showed a more pronounced activity. It was also reported that extract of unripe pawpaw had antisickling agent and that this antisickling agent resided in the ethyl acetate fraction of the extract (Oduola et al. 2006). This fraction was found to prevent sickling of haemoglobin SS, (Hb SS) red cells and reversed sickled Hb SS red cells in 2% sodium metabisulphite whereas the butanol and aqueous fractions were inactive.

Papaya seeds were also reported to have anthelmintic properties, the active compound being benzyl isothiocyanate (Kermanshai et al. 2001). A positive correlation between anthelmintic activity and benzyl isothiocyanate content of papaya seeds was found. A 10 hour incubation of crude seed extracts at room temperature led to a decrease in anthelmintic activity. An ethanol extract of papaya seeds exhibited inhibitory effect on jejunal contractions that was significantly irreversible (Adebiyi and Adaikan 2005). Previous studies indicated that benzyl isothiocyanate (BITC) was the main bioactive compound responsible for the anthelmintic activity of papaya seeds. In the present study, standard BITC (0.01–0.64 mmol/L) also caused significant irreversible inhibition of jejunal contractions. It was thus observed that at the toxic level required to kill and expel intestinal worms in-vivo, BITC may also cause impairment of intestinal functions. *Carica papaya* seed preparations are used in traditional medicine to expel intestinal worms in human and ruminants.

Studies reported that papaya latex also showed antiparasitic activity and had potential role as an anthelmintic against patent intestinal nematodes of mammalian hosts (Satrija et al. 1994, 1995). *Heligmosomoides polygyrus* infected mice were given papaya latex suspended in water at dose levels of 0 g (group A), 2 g (group B), 4 g (group C), 6 g (group D) and 8 g (group E) of papaya latex/kg body weight. Antiparasitic efficacy based on post-mortem worm counts were 55.5%, 60.3%, 67.9% and 84.5% in groups B, C, D and E, respectively (Satrija et al. 1995). They also reported separately that papaya latex was also effective against *Ascaris suum* in pigs and had potential as a herbal medicine for livestock and humans (Satrija et al. 1994). Results of post mortem counts on day 7 post treatment revealed worm count reductions of 39.5%, 80.1% and 100% in groups B (2 g latex/kg body weight), C (4 g latex/kg body weight), and D (8 g latex/kg body weight), respectively. Some of the pigs receiving the highest dose of the latex showed mild diarrhoea on the day following treatment. Separate studies reported that cysteine proteinases in the crude papaya latex, papain and chymopapain had potent effect against the rodent gastrointestinal nematode, *Heligmosomoides polygyrus* (as test animal) (Stepek et al. 2007). The mechanism of action of these plant enzymes (i.e. an attack on the protective cuticle of the worm) suggests that resistance would be slow to develop in the field. The efficacy and mode of action make plant cysteine proteinases potential candidates for a novel class of anthelmintics urgently required for the treatment of humans and domestic livestock.

Galactopoietic Activity

Studies found that rats fed with *Carica papaya* roots (CP) for 15 days, had significantly higher average weights of mammary glands than those of the control groups (Tossawanchuntra and Aritajat 2005). The level of alkaline phosphatase in rats fed with CP at 500 mg/kg/day was highest while the level of prolactin in rats fed with CP at 400 mg/kg/day was significantly higher than other groups. Rats fed with CP at 400 and 1,000 mg/kg/day contained higher protein content than the control groups. Rats fed with CP at

400 mg/kg/day for 21 days had higher level of alkaline phosphatase and the average weight of their mammary glands was significant higher than the control groups. The levels of prolactin in rats fed with CP at 400 and 500 mg/kg/day were significant higher than the control ones. It was concluded that feeding lactating rats with papaya roots for 15 and 21 days could stimulate higher lactating ability by increasing the weight of mammary gland, alkaline phosphatase, prolactin and protein content.

Cardiovascular Protective Activities

An alcoholic extract of *Carica papaya* was reported to induce central muscle relaxation in male rats (Gupta et al. 1990). The behavioral effects of the extract were associated with an initial desynchronization of the electroencephalogram (EEG) and an increased activity of the electromyogram (EMG). The extract at doses ≥50 mg/kg (i.p.) completely protected the rats against pentylenetetrazol-induced seizures, while doses of 5 mg/kg (i.p.) gave 50% protection. The extract at doses of 100 and 200 mg/kg (i.p.) also gave 100% protection against maximal electroshock-induced convulsions.

In-vitro studies using isolated rabbit arterial (aorta, renal and vertebral) strips showed that the crude ethanol extract prepared from the unripened fruit of *Carica papaya* (10 µg/mL) produced relaxation of vascular muscle tone which was, however, diminished by phentolamine (0.5–1.5 µg/mL) (Eno et al. 2000). It was concluded that the fruit juice of *C. papaya* probably contained antihypertensive agent(s) which exhibited mainly α-adrenoceptor activity. Another research reported that papaya leaf, and *Mentha arvensis* (mint leaf) extracts exhibited more than 50% relaxing effect on aortic ring preparations, while *Piper betle* and *Cymbopogon citratus* (lemongrass stalk) showed comparable vasorelaxation on isolated perfused mesenteric artery preparation (Runnie et al. 2004). The study demonstrated that many edible plants common in Asian diets possess potential health benefits, affording protection at the vascular endothelium level.

Carpaine one of the major alkaloid components of papaya leaves was found to have an effect on the circulatory system (Hornick et al. 1978). Using Wistar male rats under pentabarbital (30 mg/kg) anaesthesia, increasing dosages of carpaine from 0.5 to 2.0 mg/kg resulted in gradual reductions in systolic, diastolic, and mean arterial blood pressure. Carpaine, 2 mg/kg, decreased cardiac output, stroke volume, stroke work, and cardiac power, but the calculated total peripheral resistance remained unaltered. It was concluded from these results that carpaine directly affected the myocardium.

To investigate their potentially toxic effects on mammalian vascular smooth muscle, pentane extracts of papaya seeds and the chief active ingredient in the extracts, benzyl isothiocyanate (BITC), were tested for their effects on the contraction of strips of dog carotid artery (Wilson et al. 2002). BITC and the papaya seed extract caused relaxation when added to tissue strips that had been pre-contracted with phenylephrine (PE). The scientists concluded that these extracts, when present in high concentration, were cytotoxic by increasing the membrane permeability to Ca^{2+}, and that the vascular effects of papaya seed extracts were consistent with the notion that BITC was the chief bio-active ingredient.

Gastroprotective/Antiulcerogenic Activity

Research showed that gastric ulcer index was significantly reduced in rats pretreated with *Carica papaya* leaf (CPL) extract as compared with alcohol treated controls (Indran et al. 2008). The in-vitro studies using 2,2-Diphenyl-1-Picryl-Hydrazyl (DPPH) assay showed strong antioxidant nature of CPL extract. Biochemical analysis indicated that the acute alcohol induced damage was manifested in the changes of blood oxidative indices, and CPL extract offered some protection with decrease in plasma lipid peroxidation level and increased erythrocyte glutathione peroxidase activity. *Carica papaya* leaf may potentially serve as a good therapeutic agent for protection against gastric ulcer and oxidative stress.

Studies showed that *Carica papaya* extract significantly reduced the ulcer index, lipid peroxide levels and alkaline phosphatase activity in the rats with aspirin induced ulcers (Ologundudu et al. 2008). It also maintained the activity of the antioxidant enzyme, catalase in the gastric mucosa, to near normalcy when compared with water controls. The results indicated that *Carica papaya* may exert its gastroprotective effect by a free radical scavenging action. This study suggested that *Carica papaya* may have considerable therapeutic potential in the treatment of gastric diseases.

The aqueous (AE) and methanol (ME) extracts of whole unripe *Carica papaya* fruit significantly reduced the ulcer index in rats in two experimental models compared to the control group (Ezike et al. 2009). ME showed a better protection against indomethacin-induced ulcers, whereas AE was more effective against ethanol-induced gastric ulcers. The extracts also significantly inhibited intestinal motility, with ME showing higher activity. Oral administration of AE and ME up to 5,000 mg/kg did not produce lethality or signs of acute toxicity in mice after 24 hours. The extracts of unripe *C. papaya* was found to contain terpenoids, alkaloids, flavonoids, carbohydrates, glycosides, saponins, and steroids. They postulated that the cytoprotective and antimotility properties of the extracts may account for the anti-ulcer property of the unripe fruit.

Nephroprotective Activity

Studies showed that intraperitoneal injection of CCl_4 caused a significant increase in the serum levels of uric acid, urea and creatinine and induced histological features of severe tubule-interstitial necrosis in rats (Olagunju et al. 2009). However, increases in the measured biochemical parameters were significantly diminished in rats pre-treated with the graded oral doses of the aqueous seed extract of *Carica papaya* (CPE), in dose related manner. Maximum nephroprotection was offered by the extract at 400 mg/kg/day CPE which lasted up to 3 hours post-CCl_4 exposure.

The study concluded that CPE had nephroprotective effect on CCl_4 renal injured rats, an effect which could be mediated by any of the phytocomponents present in it via either antioxidant and/or free radical scavenging mechanism(s).

Tocolytic Effect

Studies reported that *Carica papaya* seeds extracted with 80% ethanol (EEPS) caused concentration-dependent tocolysis of uterine strips isolated from gravid and non-gravid rats (Adebiyi et al. 2003). Prostaglandin F2α and oxytocin-induced contractions of the isolated rat uterus were also inhibited in a concentration-dependent manner by EEPS. Recoveries of the uterine activity after EEPS-induced uterine quiescence were very weak. Higher concentration of EEPS caused prompt uterine inactivity which was also significantly irreversible. Earlier studies had demonstrated BITC-induced functional and morphological derangement of isolated uterus. The scientists concluded that at high concentration, EEPS was capable of causing irreversible uterine tocolysis probably due to the deleterious effect of BITC (its chief phytochemical) on the myometrium.

Arbotifacient, Antifertility Activities

Research indicated that normal consumption of ripe papaya during pregnancy may not pose any significant abortifacient danger; however, the unripe or semi-ripe papaya (containing high concentration of the latex that produces marked uterine contractions) could be unsafe in pregnancy (Adebiyi et al. 2002). There was no significant difference in the number of implantation sites and viable foetuses in the rats given ripe papaya relative to the control. No sign of foetal or maternal toxicity was observed in all the groups. In the in-vitro study, ripe papaya juice did not show any significant contractile effect on uterine smooth muscles isolated from pregnant and non-pregnant rats. Conversely, crude papaya latex induced spasmodic contraction of the uterine muscles similar to oxytocin and prostaglandin F2α.

In another study, female Sprague-Dawley rats given aqueous extract of *Carica papaya* seeds at various doses showed no significant differences in total body weight in foetuses exposed to these regimes (Oderinde et al. 2002). However, in the group treated with 100 mg/kg body weight, there was a significant increase in the implantation sites and foetal weight was significantly decreased compared to the controls. No dead or malformed foetuses were found. However, in the group treated with 800 mg/kg body weight, there was obvious vaginal bleeding but no treatment related increase in implantation sites compared with control. There was however, complete resorption of about 30% of the foetuses. The surviving foetuses were stunted when compared with the control but were without any external malformations. The results indicated that low dose aqueous crude extract of *Carica papaya* seeds did not adversely affect prenatal development. The altered toxicological profile indicated that the abortifacient property was a high dose side effect.

Another research reported that crude aqueous extract of *Carica papaya* seeds had contraceptive efficacy. Cauda epididymal sperm motility and count was reduced significantly at low and high dose regimens both in the oral as well as the intramuscular groups (Lohiya et al. 1994). The reduced sperm motility was associated with morphological defects. Testicular sperm counts and testicular weight were also reduced in all the treatment groups except the low dose intramuscular group. Body weight and toxicological observations did not show any untoward response. Fertility and all other associated changes returned to normal within 45 and 30 days of treatment cessation in the oral and intramuscular groups, respectively. The data revealed that reversible sterility could be induced in male rats by papaya seeds aqueous extract treatment without adverse effects on libido and toxicological profile. Separate research found that *Carica papaya* seed extract exhibited a significant dose dependent suppression of cauda epididymal sperm motility in male mice coinciding with a decrease in sperm count and viability (Verma et al. 2006). When tested 45 days after the withdrawal of treatment, complete normalcy was restored, proving that the induced effects were transient. Another study showed that oral administration of *C. papaya* seed extract to rats prevented ovum fertilization, lowered sperm cell counts, caused sperm cell degeneration, and induced testicular cell lesion (Udoh et al. 2005). These observations led to the conclusion that *C. papaya* seed extract oral administration could induce reversible male infertility and therefore could be used for pharmaceutical development of a male contraceptive.

The aqueous extract of *Carica papaya* seeds did not manifest any estrogenic effects in male mice, and LD_{50} studies indicated its nontoxic nature (Chinoy et al. 1994). The body weight and weights of reproductive organs, kidney, and adrenal were not affected, indicating that the extract did not promote body weight gain through obesity or water retention. The serum SGOT, SGPT, protein, and cholesterol levels were also within the normal range in the extract-treated mice, suggesting that the extract did not affect liver function or cholesterol and protein metabolism. The data suggested that the aqueous extracts of papaya seeds was safe and could serve as an effective male contraceptive in rodents.

Research also showed that aqueous crude extract of the bark of *Carica papaya* had contraceptive potential (Kusemiju et al. 2002). The contraceptive potential of an aqueous crude extract of the bark administered orally was reported at the dose regime of 5 and 10 ml/animal/day for 4 weeks. Although the body weight or the weights of reproductive organs, kidneys, adrenals remained unchanged during the course of treatment, there was significant alteration in the histology of the testis and semen analysis when compared to the intact control group. The seminiferous tubules of rats treated with low doses of *Carica papaya* for 4 weeks showed no significant histological changes compared with the control. At the high dose concentration there was disorganisation in some of the seminiferous tubules with arrest of spermatogenesis beyond the level of spermatocytes. There was also widening of the lumen of the tubules. Seminal analysis showed some of the motile spermatozoa with abnormal feature in all the experimental groups. There was a dose-dependent suppression of sperm progressive forward motility which coincided with a decrease in sperm count, viability and a dose-dependent increase in percentage abnormal spermatozoa during the 14–28 days experimental period. Taken together, the study showed that complete loss of fertility was attributed to decline in sperm motility and alteration in their morphology and suggested that the aqueous extracts of *Carica papaya* bark was safe and could serve as an effective male contraceptive in animals.

The long term daily oral administration of methanol sub-fraction (MSF) of the seeds of *Carica papaya* was found to affect sperm parameters without adverse side effects and was clinically safe as a male contraceptive (Goyal et al. 2010). MSF treatment at various dose regimens, daily for 52 weeks did not show significant changes in body weight, organs weight, food and water intake and pre-terminal deaths compared to those of control animals. Sperm count and viability in 50 mg/kg body weight treated animals and the weight of epididymis, seminal vesicle and prostate of all the treated animals showed significant reduction compared to control. Cauda epididymal spermatozoa of 50 mg/kg body weight treated animals were immotile. Azoospermia was observed in 100, 250 and 500 mg/kg body weight treated animals. Serum clinical parameters, serum testosterone and histopathology of vital organs were comparable to those of control animals. Histology of testis revealed adverse effects on the process of spermatogenesis, while the histology of epididymis, seminal vesicles and ventral prostate showed no changes compared to control.

Gliadin Detoxifying Activity

Papaya latex was found to have gluten-detoxifying enzymes (Cornell et al. 2010). Using chromatographic techniques, coupled with different enzymes assays, the activity in papaya latex eas found to be largely due to caricain and to a lesser extent, chymopapain and glutamine cyclotransferase. Fractions rich in caricain would be suitable for enzyme therapy in gluten intolerance and appeared to have synergistic action with porcine intestinal extracts.

Thrombocyte Stimulation Activity

Papaya leaf preparation was found to stimulate production of thrombocytes (Sathasivam et al. 2009). Significantly higher mean counts of thrombocytes in mice were observed at 1, 2, 4, 8, 10 and 12 hours after dosing with the *C. papaya* leaf formulation as compared to the mean count at hour 0. There was only a non-significant rise of thrombocyte counts in the group having received saline solution, possibly the expression of a normal circadian rhythm in mice. The group having received palm oil only showed a protracted increase of platelet counts that was significant at hours 8 and 48 hours and obviously the result of a hitherto unknown stimulation of thrombocyte release.

Genotoxic, Cytotoxic and Toxicity Studies

In one study, different doses, 50, 100, 150, 200 and 250 mg/kg of the aqueous extract of unripe *Carica papaya* were administered daily to the rat for 6 weeks and the oral acute toxicity of (LD_{50}) was determined (Oduola et al. 2007). Liver, kidney and bone marrow function tests were assessed using standard techniques. The results obtained for biochemical and haematological parameters for the control and experimental groups showed no statistical significant different. This result showed that the intake of unripe *C. papaya* extract had no adverse effects on the functions of these organs in rats.

The results of recent studies by da Silva et al. (2010) indicated that papain was not cytotoxic to *E. coli* strains AB1157 (wild type), PQ65 (*rfa*), BH20 (*fpg*), and BW9091 (*xth*A) at the tested concentrations. The results obtained with WP2 Mutoxitest demonstrated that all the tested papain concentrations did not produce mutagenic activity with *E. coli* IC203 (WP2 *uvr*A *oxy*R pKM101) and IC204 (WP2 *uvr*A del *umu*DC). In the circular plasmid DNA treatment, papain did not induce single or double strand breaks in DNA in-vitro. The results reinforced the idea that the papain was neither a cytotoxicity nor genotoxic agent. The study indicated that papain was not toxic and/or mutagenic in bacterial systems. Further, the data showed papain to be an antioxidant agent against H_2O_2-induced damage.

Traditional Medicinal Uses

Different parts of *Carica papaya* tree is used in traditional medicine in many countries in different ways (Burkill 1966; Morton 1977; Quisumbing 1978; Chopra et al. 1986; Morton 1987; Gill 1992; WHO 1998; Ross 2003).

Fruit

The ripe fruit is stomachic, laxative, digestive and carminative. Various preparations of *Carica papaya* containing papain has a wide range of reputed medicinal application in traditional medicine abortifacient, galactagogue, analgesic, amebicide, antibiotic, antibacterial, cardiotonic, cholagogue, digestive, emmenagogue, febrifuge, hypotensive, laxative, pectoral, stomachic, vermifuge. Papaya juice is used in the treatment of warts, corns, sinuses, and chronic forms of skin indurations such as scaly eczema, cutaneous tubercles, and other hardness of the skin, produced by irritation, etc. and is used as a cosmetic to remove freckles. The use of papaya for medicinal purposes is traced back to the early Caribbean, who have long employed the ripe fruit as a cosmetic. The notable complexion of these people is attributed to the use of the pulp as a skin soap. Mashed ripe papaya mixed with kalamansi juice is use as a treatment for acne in the Philippines. Papaya juice is also employed in the treatment of epithelioma, glossal fissures and ulcerations, and particularly syphilitic ulcerations of the throat, mouth and tongue; chronic suppurative inflammation of the middle ear and for ringworm. The green fruit is used laxative and diuretic. Unripe papaya has been used for contraceptive purposes by traditional healers in Pakistan, India and Sri Lanka. Dietary papaya is reported to reduce urine acidity in humans. The ripe fruit is alternative, and, if eaten regularly, corrects habitual constipation and is also laxative. It is useful also for bleeding piles and for dyspepsia. Green fruit is also said to be ecbolic. In Asia, the latex is smeared on the mouth of the uterus as ecbolic. Javanese believe that eating papaya prevents rheumatism. The unripe fruit is used traditionally among the Yoruba tribe of Nigeria for the treatment of various human and veterinary diseases including malaria, hypertension, diabetes mellitus, hypercholesterolaemia, jaundice, intestinal helminthiasis.

Papain/Latex

Two important compounds in papaya – chymopapain and papain, are believed to aid in digestion. Papain has been used in a tonic and as a remedy for fermentative dyspepsia or gastric affections with painful acid eructations, flatulence, deficiency of the gastric juice, excess of the unhealthy mucous in the stomach, intestinal irritation, and constipation. It is used in solution to dissolve the fibrinous membrane in croup or diphtheria, a solution in glycerin being painted on the pharynx every few minutes. It has been applied with good results to ulcers and fissures of the tongue, and, in the form of a pigment prepared with borax and water, removes warts, corns epithelioma, tubercles and other horny excrescences of the skin. Papain is used effectively as an anthelmintic and to treat arthritis. Papaya pills are promoted for use as natural antacids, for ulcer relief and to relieve constipation.

The milky juice (latex) which contains papain is used in psoriasis, ringworm, and prescribed for the removal of cancerous growths in Cuba. It is applied in India to the oesoteri for inducing abortion. It is similarly used by the Malays. The latex is also used as a styptic and as a vermifuge. Latex from unripe fruit or trunk is used as debridement (removal of purulent exudate and blood clots from wound and ulcer) applied as external applications to burns and scalds. Latex is used locally as antiseptic.

Seeds

Seeds are considered alexeritic, abortifacient, counter-irritant, emmenagogue, and anthelmintic.

Papaya seeds are chewed to freshen breath in India and widely used as pessaries and as medicine for piles and flatulence. The seeds are said to quench thirst and are also used as vermifuge. Dried pulverized seed is taken for worm infestation. In the West Indies, the powder of the seeds is used as a vermifuge. In India, it is used as a powerful abortifacient and given as an emmenagogue.

Leaves

The leaves have been used as a heart tonic, febrifuge, vermifuge, and in the treatment of gastric problems, fever and amoebic dysentery. Young leaves, and to lesser degree, other parts, contains carpain, an active bitter alkaloid, which has a depressing action on heart. The fresh leaves have been used as a dressing for foul wounds. Poultice of crushed papaya leaves has been used for rheumatic complaints in the Philippines and to reduce elephantoid growths. The leaves dipped in hot water or warmed over a fire are applied locally for nervous pains. A decoction of the leaves is employed as a remedy for asthma. Leaf is smoked for asthma relief in various remote areas. In the Gold Coast, an infusion of the dried leaves is drunk to cure stomach troubles. Tea decoction of dried leaves is employed for a variety of stomach troubles and cystitis.

Flowers

An infusion of the flowers is reported to have emmenagogue, febrifuge, and pectoral properties. Decoction of boiled flowers or powdered seeds is reported to promote menstruation. Infusion of male flowers with honey is used for cough, hoarseness, bronchitis, laryngitis and tracheitis. Flowers have also been used as therapy for jaundice.

Bark/Roots

A decoction of the roots is digestive and tonic and is much used in the therapy of dyspepsia. The roots are used for yaws and for piles. The roots are also used in the Gold Coast as an abortifacient. The fresh roots are rubefacient. Inner bark is used for sore teeth. A poultice of roots is used for centipede bites. Sinapisms prepared from the root are also said to help tumours of the uterus. The root infusion is used for syphilis in Africa. Infusion of roots is said to remove urinal concretions.

Other Uses

Fragrant papaya flowers are used in garlands. Papain, the proteolytic enzyme has a wealth of industrial uses. Papain is employed in the tanning industry, for bating skins and hides, in the pharmaceutical industries and for degumming natural silk and softening wool. It is used for meat-tenderizers and in chewing gums. Nearly 80% of American beer is treated with papain, which digests the precipitable protein fragments and then the beer remains clear on cooling. Cosmetically, it is used in some dentifrices, shampoos, and face-lifting preparations. Papain is also used in the manufacture of rubber from *Hevea*.

In recent years, *Carica papaya* lipase (CPL) is attracting more and more interest. This hydrolase, being tightly bonded to the water-insoluble fraction of crude papain, is considered an active "naturally immobilized" biocatalyst (Domínguez de María et al. 2006). To date, several CPL applications have already been described: (i) fats and oils modification, derived from the sn-3 selectivity of CPL as well as from its preference for short-chain fatty acids; (ii) esterification and inter-esterification reactions in organic media, accepting a wide range of acids and alcohols as substrates; (iii) more recently, the asymmetric resolution of different non-steroidal anti-inflammatory drugs (NSAIDs), 2-(chlorophenoxy) propionic acids, and non-natural amino acids.

A novel α-amylase inhibitor with molecular mass of 4,562 Da, was isolated from papaya seeds (*Carica papaya*) with deleterious activity against cowpea weevil, *Callosobruchus maculates* enzymes (Farias et al. 2007). The α-amylase inhibitors were able to increase larval mortality (50%) and also decrease insect fecundity and adult longevity. The results indicated that this inhibitor probably could be utilised, through genetic engineering, in the development of transgenic plants with enhanced resistance toward cowpea weevil.

Fermentation with brewer's yeast and distillation yielded 4% alcohol, of which 91.8% was ethanol, 4.8%.methanol, 2.2% N-propanol, and 1.2% unknown (non-alcohol) (Sharma and Ogbeide 1982). This raised the possibility of using papaya as a renewable energy resource for the production of alcohol fuels.

Comments

Brazil is the leading producer of papayas followed by Nigeria, India and Mexico. The leading exporting countries of papayas include Mexico, Malaysia, Brazil and USA(Hawaii).

Selected References

Adebiyi A, Adaikan PG (2005) Modulation of jejunal contractions by extract of Carica papaya L. seeds. Phytother Res 19(7):628–632

Adebiyi A, Adaikan PG, Prasad RNV (2002) Papaya (Carica papaya) consumption is unsafe in pregnancy: fact or fable? Scientific evaluation of a common belief in some parts of Asia using a rat model. Br J Nutr 88:199–203

Adebiyi A, Adaikan PG, Prasad RN (2003) Tocolytic and toxic activity of papaya seed extract on isolated rat uterus. Life Sci 74(5):581–592

Akah PA, Oli AN, Enwerem NM, Gamaniel K (1997) Preliminary studies on purgative effect of Carica papaya root extract. Fitoterapia 68(4):327–331

Anuar NS, Zahari SS, Taib IA, Rahman MT (2008) Effect of green and ripe Carica papaya epicarp extracts on wound healing and during pregnancy. Food Chem Toxicol 46(7): 2384–2389

Aruoma OI, Colognato R, Fontana I, Gartlon J, Migliore L, Koike K, Coecke S, Lamy E, Mersch-Sundermann V, Laurenza I, Benzi L, Yoshino F, Kobayashi K, Lee MC (2006) Molecular effects of fermented papaya preparation on oxidative damage, MAP Kinase activation and modulation of the benzo[a]pyrene mediated genotoxicity. Biofactors 26(2):147–159

Aruoma OI, Hayashi Y, Marotta F, Mantello P, Rachmilewitz E, Montagnier L (2010) Applications and bioefficacy of the functional food supplement fermented papaya preparation. Toxicol 278(1):6–16

Azarkan M, El Moussaoui A, van Wuytswinkel D, Dehon G, Looze Y (2003) Fractionation and purification of the enzymes stored in the latex of Carica papaya. J Chromatogr B Anal Technol Biomed Life Sci 79(1–2):229–238

Backer CA, van den Bakhuizen Brink Jr RC (1963) Flora of Java (spermatophytes only), vol 1. Noordhoff, Groningen, 648 pp

Badami C, Daulatabad CD (1967) The component acids of Carica papaya (caricaceae) seed oil. J Sci Food Agri 18(8):360–361

Brocklehurst K, Salih E (1985) Fresh non-fruit latex of Carica papaya contains papain, multiple forms of chymopapain A and papaya proteinase OMEGA. Biochem J 228(2):525–527

Burkill I H (1966) A dictionary of the economic products of the Malay Peninsula. Revised reprint. 2 vols. Ministry of Agriculture and Co-operatives, Kuala Lumpur, vol 1 (A–H) pp 1–1240, vol 2 (I–Z) pp. 1241–2444

Canini A, Alesiani D, D'Arcangelo G, Tagliatesta P (2007) Gas chromatography–mass spectrometry analysis of phenolic compounds from Carica papaya L. leaf. J Food Compos Anal 20(7):584–590

Chan HT, Heu RA, Tang C-S, Okazaki EN, Ishizaki SM (1978) Composition of papaya seeds. J Food Sci 43:255–256

Chandrika UG, Jansz ER, Wickramasinghe SMDN, Warnasuriya ND (2003) Carotenoids in yellow- and red-fleshed papaya (Carica papaya L). J Sci Food Agric 83(12):1279–1282

Chinoy NJ, D'Souza JM, Padman P (1994) Effects of crude aqueous extract of Carica papaya seeds in male albino mice. Reprod Toxicol 8(1):75–79

Chopra RN, Nayar SL, Chopra IC (1986) Glossary of Indian medicinal plants (including the supplement). Council Scientific Industrial Research, New Delhi, 330 pp

Collard E, Roy S (2010) Improved function of diabetic wound-site macrophages and accelerated wound closure in response to oral supplementation of a fermented papaya preparation. Antioxid Redox Signal 13(5):599–606

Cornell HJ, Doherty W, Stelmasiak T (2010) Papaya latex enzymes capable of detoxification of gliadin. Amino Acids 38(1):155–165

Croft KD (1998) The chemistry and biological effects of flavonoids and phenolic acids. Ann NY Acad Sci 854:435–442

da Silva CR, Oliveira MB, Motta ES, de Almeida GS, Varanda LL, de Pádula M, Leitão AC, Caldeira-de-Araújo A (2010) Genotoxic and cytotoxic safety evaluation of papain (Carica papaya L.) using in vitro assays. J Biomed Biotechnol 2010: article ID197898

Danese C, Esposito D, D'Alfonso V, Cirene M, Ambrosino M, Colotto M (2006) Plasma glucose level decreases as collateral effect of fermented papaya preparation use. Clin Ter 157(3):195–198

Dawkins G, Hewitt H, Wint Y, Obiefuna PCM, Wint B (2003) Antibacterial effects of *Carica papaya* fruit on common wound organisms. West Indian Med J 52(4):290–292

Domínguez de María P, Sinisterra JV, Tsai SW, Alcántara AR (2006) *Carica papaya* lipase (CPL): an emerging and versatile biocatalyst. Biotechnol Adv 24(5): 493–499

Dubois T, Jacquet A, Schnek AG, Looze Y (1988) The thiol proteinases from the latex of *Carica papaya* L. I. Fractionation, purification and preliminary characterization. Biol Chem Hoppe Seyler 369(8):733–740

Duke JA (1993) CRC handbook of alternative cash crops. CRC, London, 544 pp

Emeruwa AC (1982) Antibacterial substance from *Carica papaya* fruit extract. J Nat Prod 45(2):123–127

Eno AE, Owo OI, Itam EH, Konya RS (2000) Blood pressure depression by the fruit juice of *Carica papaya* (L.) in renal and DOCA-induced hypertension in the rat. Phytother Res 14(4):235–239

Ezike AC, Akah PA, Okoli CO, Ezeuchenne NA, Ezeugwu S (2009) *Carica papaya* (Paw-Paw) unripe fruit may be beneficial in ulcer. J Med Food 12(6):1268–1273

Farias LA, Pelegrini PB, Grossi-de-Sá MF, Neto SM, Bloch C Jr, Laumann RA, Noronha EF, Franco OL (2007) Isolation of a novel *Carica papaya* α-amylase inhibitor with deleterious activity toward *Callosobruchus maculatus*. Pest Biochem Physiol 87(3):255–260

Fibach E, Tan ES, Jamuar S, Ng I, Amer J, Rachmilewitz EA (2010) Amelioration of oxidative stress in red blood cells from patients with beta-thalassemia major and intermedia and E-beta-thalassemia following administration of a fermented papaya preparation. Phytother Res 24(9):1334–1338

Foundation for Revitalisation of Local Health Traditions (2008) FRLHT Database. htttp://envis.frlht.org

Gill LS (1992) *Carica papaya* L. In: ethnomedicinal uses of plants in Nigeria. UNIBEN Press: Benin City, pp 57–58

Giordani R, Cardenas ML, Moulin-Traffort J, Regli P (1996) Fungicidal activity of latex sap from *Carica papaya* and antifungal effect of D(+)-glucosamine on *Candida albicans* growth. Mycoses 39(3–4):103–110

Giordani R, Gachon C, Moulin-Traffort J, Regli P (1997) A synergistic effect of *Carica papaya* latex sap and fluconazole on *Candida albicans* growth. Mycoses 40(11–12):429–437

Giordani R, Siepaio M, Moulin-Traffort J, Régli P (1991) Antifungal action of *Carica papaya* latex: isolation of fungal cell wall hydrolysing enzymes. Mycoses 34(11–12):469–477

Goyal S, Manivannan B, Ansari AS, Jain SC, Lohiya NK (2010) Safety evaluation of long term oral treatment of methanol sub-fraction of the seeds of *Carica papaya* as a male contraceptive in albino rats. J Ethnopharmacol 127(2):286–291

Gupta A, Wambebe CO, Parsons DL (1990) Central and cardiovascular effects of the alcoholic extract of the leaves of *Carica papaya*. Int J Crude Drug Res 28(4):257–266

Harsha J, Chinoy NJ (1996) Reversible antifertility effects of benzene extract of papaya seed on female rats. Phytother Res 10(4):327–328

Hewitt H, Whittle S, Lopez S, Bailey E, Weaver S (2000) Topical use of papaya in chronic skin ulcer therapy in Jamaica. West Indian Med J 49(1):32–33

Hornick CA, Sanders LI, Lin YC (1978) Effect of carpaine, a papaya alkaloid, on the circulatory function in the rat. Res Commun Chem Pathol Pharmacol 22(2):277–289

Imao K, Wang H, Komatsu M, Hiramatsu M (1998) Free radical scavenging activity of fermented papaya preparation and its effect on lipid peroxide level and superoxide dismutase activity in iron-induced epileptic foci of rats. Biochem Mol Biol Int 45(1):11–23

Indran M, Mahmood AA, Kuppusamy UR (2008) Protective effect of *Carica papaya* L leaf extract against alcohol induced acute gastric damage and blood oxidative stress in rats. West Indian Med J 57(4):323–326

Johnson IT (2002) Glucosinolates: bioavailability and importance to health. Int J Vitam Nutr Res 72(1):26–31

Kermanshai R, McCarry BE, Rosenfeld J, Summers PS, Weretilnyk EA, Sorger GJ (2001) Benzyl isothiocyanate is the chief or sole anthelmintic in papaya seed extracts. Phytochem 57:427–435

Khuzhaev VU, Aripova SF (2000) Pseudocarpaine from *Carica papaya*. Chem Nat Comp 36(4):418–420

Kusemiju O, Noronha C, Okanlawon A (2002) The effect of crude extract of the bark of *Carica papaya* on the seminiferous tubules of male Sprague-Dawley rats. Niger Postgrad Med J 9(4):205–209

Lohiya NK, Goyal RB, Jayaprakash D, Ansari AS, Sharma S (1994) Antifertility effects of aqueous extract of *Carica papaya* seeds in male rats. Planta Med 60(5):400–404

Lohiya NK, Mishra PK, Pathak N, Manivannan B, Bhande SS, Panneerdoss S, Sriram S (2005) Efficacy trial on the purified compounds of the seeds of *Carica papaya* for male contraception in albino rat. Reprod Toxicol 20(1):135–148

Marfo EK, Oke OL, Afolabi OA (1986a) Chemical composition of papaya (*Carica papaya*) seeds. Food Chem 22(4): 259–266

Marfo EK, Oke OL, Afolabi OA (1986b) Some studies on the proteins of *Carica papaya* seeds. Food Chem 22(4): 267–277

Marotta F, Weksler M, Naito Y, Yoshida C, Yoshioka M, Marandola P (2006) Nutraceutical supplementation: effect of a fermented papaya preparation on redox status and DNA damage in healthy elderly individuals and relationship with GSTM1 genotype: a randomized, placebo-controlled, cross-over study. Ann NY Acad Sci 1067:400–407

Marotta F, Yoshida C, Barreto R, Naito Y, Packer LJ (2007) Oxidative-inflammatory damage in cirrhosis: effect of vitamin E and a fermented papaya preparation. Gastroenterol Hepatol 22(5):697–703

Mehdipour S, Yasa N, Dehghan G, Khorasani R, Mohammadirad A, Rahimi R, Abdollahi M (2006)

Antioxidant potentials of Iranian *Carica papaya* juice in vitro and in vivo are comparable to α-tocopherol. Phytother Res 20(7):591–594

Mikhal'chik EV, Ivanova AV, Anurov MV, Titkova SM, Pen'kov LY, Kharaeva ZF, Korkina LG (2004) Wound-healing effect of papaya-based preparation in experimental thermal trauma. Bull Exp Biol Med 137(6):560–562

Mojica-Henshaw MP, Francisco AD, De Guzman F, Tigno XT (2003) Possible immunomodulatory actions of *Carica papaya* seed extract. Clin Hemorheol Microcirc 29(3–4):219–229

Morton JF (1977) Major medicinal plants: botany, culture, and uses. Charles Thomas, Springfield, 431 pp

Morton JF (1987) The papaya. Fruits of warm climates. Creative Resources Systems, Inc, Winterville, pp 336–346

Nakamura Y, Yoshimoto M, Murata Y, Shimoishi Y, Asai Y, Park EY, Sato K, Nakamura Y (2007) Papaya seed represents a rich source of biologically active isothiocyanate. J Agric Food Chem 55(11):4407–4413

Nayak SB, Pinto PL, Maharaj D (2007) Wound healing activity of *Carica papaya* L. in experimentally induced diabetic rats. Indian J Exp Biol 45(8):739–743

Ochse JJ, Bakhuizen van den Brink RC (1980) Vegetables of the Dutch Indies, 3rd edn. Ascher & Co., Amsterdam, 1016 pp

Oderinde O, Noronha C, Oremosu A, Kusemiju T, Okanlawon OA (2002) Abortifacient properties of aqueous extract of *Carica papaya* (Linn) seeds on female Sprague-Dawley rats. Niger Postgrad Med J 9(2):95–98

Oduola T, Adeniyi FAA, Ogunyemi EO, Bello IS, Idowu TO (2006) Antisickling agent in an extract of unripe pawpaw [*Carica papaya*]: is it real? Afr J Biotechnol 5:1947–1949

Oduola T, Adeniyi FAA, Ogunyemi EO, Bello IS, Idowu TO, Subair HG (2007) Toxicity studies on an unripe *Carica papaya* aqueous extract: biochemical and haematological effects in wistar albino rats. J Med Plants Res 1(1):1–4

Okeniyi JAO, Ogunlesi TA, Oyelami OA, Adeyemi LA (2007) Effectiveness of dried *Carica papaya* seeds against human intestinal parasitosis: a pilot study. J Med Food 10(1):194–196

Olagunju JA, Adeneye AA, Fagbohunka BS, Bisuga NA, Ketiku AO, Benebo AS, Olufowobi OM, Adeoye AG, Alimi MA, Adeleke AG (2009) Nephroprotective activities of the aqueous seed extract of *Carica papaya* Linn. in carbon tetrachloride induced renal injured Wistar rats: a dose- and time-dependent study. Biol Méd 1(1):11–19

Ologundudu A, Lawal AO, Ololade IA, Omonkhua AA, Obi FO (2008) The anti-ulcerogenic activity of aqueous extract of *Carica papaya* fruit on aspirin – induced ulcer in rats. Internet J Toxicol 5(2)

Osato JA, Santiago LA, Remo GM, Cuadra MS, Mori A (1993) Antimicrobial and antioxidant activities of unripe papaya. Life Sci 53(17):1383–1389

Otsuki N, Dang NH, Kumagai E, Kondo A, Iwata S, Morimoto C (2010) Aqueous extract of *Carica papaya* leaves exhibits anti-tumor activity and immunomodulatory effects. J Ethnopharmacol 127(3):760–767

Owoyele BV, Adebukola OM, Funmilayo AA, Soladoye AO (2008) Anti-inflammatory activities of ethanolic extract of *Carica papaya* leaves. Inflammopharmacol 16(4):168–173

Pacific Island Ecosystems at Risk (PIER) (2004) *Carica papaya* L. Caricaceae. http://www.hear.org/Pier/species/carica_papaya.htm

Panse TB, Patanjpe AS (1943) A study of carpasemine isolated form *Carica papaya* seeds. Proc Indian Acad Sci 18:140–142

Passera C, Spettoli P (1981) Chemical composition of papaya seeds. Plant Foods Hum Nutr 31(1):77–83

Pino JA, Almora K, Marbot R (2003) Volatile components of papaya (*Carica papaya* L., Maradol variety) fruit. Flav Fragr J 18(6):492–496

Porcher M H et al. (1995–2020) Searchable world wide web multilingual multiscript plant name database.The University of Melbourne, Australia. http://www.plantnames.unimelb.edu.au/Sorting/Frontpage.html

Pousset JL, Boum B, Cave A (1981) Antihaemolytic action of xylitol isolated from *Carica papaya* bark. Planta Med 41(1):40–47

Puangsri T, Abdulkarim SM, Ghazali HM (2005) Properties of *Carica papaya* L (papaya) seed oil following extractions using solvent and aqueous enzymatic methods. J Food Lipids 12:62–76

Quisumbing E (1978) Medicinal plants of the Philippines. Katha Publishing Co, Quezon City, 1262 pp

Rajkapoor B, Jayakar B, Kavimani S, Murugesh N (2002) Effect of dried fruits of *Carica papaya* Linn on hepatotoxicity. Biol Pharm Bull 25(12):1645–1646

Rodriguez-Amaya DB (1999) Latin American food sources of carotenoids. Arch Latinoam Nutri 49(1–5):74S–84S

Ross IA (2003) Medicinal plants of the world. Volume I: chemical constituents, traditional and modern uses, 2nd edn. Humana, Totawa, 489 pp

Rossetto MRM, Oliveira do Nascimento JR, Purgatto E, Fabi JP, Lajolo FM, Cordenunsi BR (2008) Benzylglucosinolate, benzylisothiocyanate, and myrosinase activity in papaya fruit during development and ripening. J Agric Food Chem 56(20):9592–9599

Runnie I, Salleh MN, Mohamed S, Head RJ, Abeywardena MY (2004) Vasorelaxation induced by common edible tropical plant extracts in isolated rat aorta and mesenteric vascular bed. J Ethnopharmacol 92(2–3):311–316

Sathasivam K, Ramanathan S, Mansor SM, Haris MR, Wernsdorfer WH (2009) Thrombocyte counts in mice after the administration of papaya leaf suspension. Wien Klin Wochenschr 121(Suppl 3):19–22

Satrija F, Nansen P, Bjørn H, Murtini S, He S (1994) Effect of papaya latex against *Ascaris suum* in naturally infected pigs. J Helminthol 68(4):343–346

Satrija F, Nansen P, Murtini S, He S (1995) Anthelmintic activity of papaya latex against patent *Heligmosomoides polygyrus* infections in mice. J Ethnopharmacol 48(3):161–164

Schwab W, Schreier P (1988) Aryl beta -D-glucosides from *Carica papaya* fruit. Phytochem 27(6):1813–1816

Seigler DS, Pauli GF, Nahrstedt A, Leen R (2002) Cyanogenic allosides and glucosides from *Passiflora edulis* and *Carica papaya*. Phytochemistry 60:873–882

Shapiro TA, Fahey JW, Wade KL, Stephenson KK, Talalay P (2001) Chemoprotective glucosinolates and isothiocyanates of broccoli sprouts: metabolism and excretion in humans. Cancer Epidemiol Biomarkers Prev 10(5):501–508

Sharma VC, Ogbeide ON (1982) Pawpaw as a renewable energy resource for the production of alcohol fuels. Energy 7(10):871–873

Sheu F, Shyu YT (1996) Determination of benzyl isothiocyanate in papaya fruit by solid phase extraction and gas chromatography. J Food Drug Anal 4(4):327–334

Starley IF, Mohammed P, Schneider G, Bickler SW (1999) The treatment of paediatric burns using topical papaya. Burns 25(7):636–639

Stepek G, Buttle DJ, Duce IR, Lowe A, Behnke JM (2007) Assessment of the anthelmintic effect of natural plant cysteine proteinases against the gastrointestinal nematode, *Heligmosomoides polygyrus*, in vitro. Parasitology 134(10): 1409–1419

Tang CS (1971) Benzyl isothiocyanate of papaya fruit. Phytochemistry 10(1):117–121

Tang CS (1973) Localization of benzyl glucosinolate and thioglucosidase in *Carica papaya* fruit. Phytochem 12(4):769–773

Thomas KD, Ajani B (1987) Antisickling agent in an extract of unripe pawpaw fruit (*Carica papaya*). Trans R Soc Trop Med Hyg 81:510–511

Togun RA, Ekerete MO, Emma-Okon BO, Binutu OO, Aboderin A (2005) Haemagglutinins from the seeds of *Carica papaya* (Linn). Nig J Biochem Mol Biol 20:67–74

Tona L, Kambu K, Ngimbi N, Cimanga K, Vlietinck AJ (1998) Antiamoebic and phytochemical screening of some Congolese medicinal plants. J Ethnopharmacol 61(1):57–65

Tossawanchuntra G, Aritajat S (2005) Effect of aqueous extract of *Carica papaya* dry root powder on lactation of albino rats. Acta Hort ISHS 678:85–90

Tropicos Org. Nomenclatural and specimen database of the Missouri Botanical Garden. http://www.tropicos.org/Home.aspx

Udoh FV, Udoh PB, Umoh EE (2005) Activity of alkaloid extract of *Carica papaya* seeds on reproductive functions in male wistar rats. Pharm Biol 43(6):563–567

U.S. Department of Agriculture, Agricultural Research Service (2009) USDA national nutrient database for standard reference, Release 22. Nutrient Data Laboratory Home Page. http://www.ars.usda.gov/ba/bhnrc/ndl

Verma RJ, Nambiar D, Chinoy NJ (2006) Toxicological effects of *Carica papaya* seed extract on spermatozoa of mice. J Appl Toxicol 26(6):533–535

Villegas VN (1992) *Carica papaya* L. In: Verheij EWM, Coronel RE (eds.) Plant resources of south East Asia No 2. Edible fruits and nuts. Prosea Foundation, Bogor, pp 108–112

Wilson RK, Kwan TK, Kwan CY, Sorger GJ (2002) Effects of papaya seed extract and benzyl isothiocyanate on vascular contraction. Life Sci 71(5):497–507

World Health Organisation (WHO) (1998) Medicinal plant in the South Pacific. WHO Regional Publications Western Pacific Series No 19. 151 pp

Yismaw G, Tessema B, Mulu A, Tiruneh M (2008) The invitro assessment of antibacterial effect of papaya seed extract against bacterial pathogens isolated from urine, wound and stool. Ethiop Med J 46(1):71–77

Zhang J, Mori A, Chen Q, Zhao B (2006) Fermented papaya preparation attenuates beta-amyloid precursor protein: beta-amyloid-mediated copper neurotoxicity in beta-amyloid precursor protein and beta-amyloid precursor protein Swedish mutation over-expressing SH-SY5Y cells. Neurosci 143(1):63–72

Vasconcellea × heilbornii

Scientific Name

Vasconcellea × heilbornii (V. M. Badillo) V. M. Badillo

Synonyms

Carica × heilbornii V.M. Badillo (basionym), *Carica × heilbornii* nothovar. *pentagona* (Heilborn) V.M. Badillo, *Carica × heilbornii* nothovar. *chrysopetala* (Heilborn) V.M. Badillo, *Carica chrysopetala* Heilborn, *Carica pentagona* Heilborn, *Vasconcellea × heilbornii* nothovar. *chrysopetala*.

Family

Caricaceae

Common/English Names

Babaco, Champagne Fruit

Vernacular Names

Ecuador: Toronchi, Toronche;
German: Babaco;
Spanish: Babaco, Chamburo, Chamburo De Castilla, Papayo Calentano, Poronchi, Toronche De Castilla.

Origin/Distribution

Vasconcellea × heilbornii is a hybrid taxon. (See comments below on its genetical origin). Its area of origin is in the Andes of Southern Colombia and Ecuador – Azuay, Canar, Chimborazo, Loja, Pichinch and Tungurahua. It is also being grown successfully in New Zealand, Australia, Israel, Italy, Guernsey, and California. New Zealand growers are already shipping babaco to Japan and the United States.

Agroecology

In its natural range, babaco is found at high altitudes of 1,500–3,000 m; it is the most cold-tolerant plant in the genus *Vasconcellea*. Babaco is a cool, sub-temperate crop, it will withstand temperatures down to −2°C, although it may lose most of its leaves. Its optimum temperature is 15–20°C and annual rainfall requirement of 600–1,400 mm for optimum growth and development. It thrives best in full sun, requiring at least 4.5 hours of sun exposure but intense sun exposure can burn its foliage. It prefers a fertile, well-drained, sandy clayey soil, rich in organic matter and with pH of 6.5–7. It needs to be planted in localities free from strong winds as it has a shallow root system. Babaco is intolerant of saline soil or saline water. The babaco is also suited to container culture and is excellent for green-houses. Babaco fruits after 12 months of planting and have a life span of about 8 years. It is more prolific in bearing than papaya.

Edible Plant Parts and Uses

The fruit skin and whitish/creamy pulp is eaten, fresh, stewed, processed into juice, concentrate, dehydrated fruit powder, sherbets, marmalade, dried preserves, fruit pieces in syrup and candies. Cooked, they are used in sauces, jams, pie-fillings and pickles. Additionally, they are added to cheese cakes and dairy products like yogurt. The flesh of the babaco is very juicy, slightly acidic and low in sugar. The unique flavour has been described as having overtones of lemon, pineapple and papaya.

Botany

Babaco is a small herbaceous shrub, 1.5–3 m high (Plate 1) but can grow taller with an erect softwood, semi-lignous trunk lined with leaf scars like papaya. It has a shallow root system with flashy tuberous roots of 40 cm diameter and numerous lateral roots. It occasionally branched around the base. Leaves are alternate, large, up to 40 cm across, coriaceous, green, palmate with 5–7 lobes with distinct mid rib and lateral veins like papaya and borne on long hollow petioles that radiate from the trunk (Plate 2). The flowers are all female and are solitary on the end of a long pendulous stalk, arising from the leaf axil.

Flowers are 3.5 cm across, campanulate, with yellow petals and dark green sepals. Fruits are formed parthenocarpically, an aspermous berry with no ovarian cavity and seeds, pentagonal (five-sided), with five longitudinal ridges, rounded at the stem end and pointed at the apex like a torpedo, 20–30 cm long and 12–18 cm wide, green (Plates 2–3) turning yellow to golden yellow on ripening. The flesh is thick, whitish to cream-coloured, juicy, acid-sweet with the core filled with spongy tissues.

Nutritive/Medicinal Properties

Nutrient value of babaco fruit per 100 g edible portion (98%) was reported as follows (Wills et al. 1985, 1986): energy 89 kJ, moisture 93.9 g, nitrogen 0.21 g, protein 1.3 g, fat 0.1 g, ash 0.2 g, total sugars 3.1 g, fructose 1.1 g, glucose 1.3 g, sucrose 0.7 g, available carbohydrate 3.1 g, total

Plate 1 Tree habit of babaco

Plate 2 Large, palmate babaco leaves

Plate 3 Close-up of torpedo-shaped babaco fruits

dietary fibre 0.9 g, calcium 11 mg, iron 0.4 mg, magnesium 6 mg, potassium 140 mg, sodium 2 mg, zinc 0.1 mg, thiamine 0.03 mg, riboflavin 0.02 mg, niacin 0.5 mg, niacin derived from tryptophan or protein 0.2 mg, niacin equivalents 0.7 mg, vitamin C 23 mg, α-carotene 10 μg, β-carotene 160 μg, cryptoxanthin 20 μg, β-carotene equivalents 175 μg, retinol equivalents 29 μg, citric acid, 0.2 g and malic acid 0.5 g.

Among the 32 components of the volatile flavour of babaco, ethyl butanote and ethyl hexanoate were the major contributors to its volatile flavour (Shaw et al. 1985). Barbeni et al. (1990) identified 119 aromatic compounds in babaco volatiles that included 9 hydrocarbons, 20 alcohols, 9 carbonyls, 9 acids, 60 esters, 3 lactones, 2 sulphur and 7 miscellaneous components with the major components of butan-1-ol and hexan-1-ol. The aroma of babaco appeared to be due to the collective effect of a number of components.

The crude babaco latex exhibited equivalent proteolytic activities (5.73 units/mg) and lipolytic activities (1.01 units/mg) compared to the well-known *Carica papaya*, the commercially source for papain (4.57 units/mg and 0.90 units/mg respectively) (Dhuique-Mayer et al. 2001). Therefore, in interesterification reactions, *Carica pentagona* latex showed interesting lipase properties (0.77 units/mg) higher than commercial *Carica papaya* latex (0.28 units/mg). Like the latex from papaya, the crude latex of babaco could be used as biocatalyst for fats and oils and proteins as demonstrated in its good lipase and papain activities for proteolysis, lipolysis, and interesterification reactions (Dhuique-Mayer et al. 2001). Further studies (Dhuique-Mayer et al. 2003) showed that babaco possessed also triacylglycerols alcoholysis reaction and esterification reactions performed in solvent free system. The use of this new biocatalyst can be successfully exploited for lipid biotransformations in interesterification reactions but also via triacylglycerol alcoholysis and in esterification reactions. The crude lipase prepared from *Carica pentagona* latex was found to be as an effective enantioselective biocatalyst for the hydrolytic resolution of (R,S)-naproxen 2,2,2-trifluoroethyl ester in water-saturated organic solvents (Chen et al. 2005). Babaco latex displayed a higher activity on triacylglycerols with short chain and unsaturated fatty acids (Cambon et al. 2008). In contrast, papaya latex had a slight sn-3 stereopreference. Both biocatalysts are 1,3-regioselective with ratios for 1,2-2,3-diacylglycerols/1,3-diacylglycerol of 6.5 and 21 for babaco and papaya, respectively.

Other Uses

Babaco is a rich source of proteolytic enzymes like papain which is used as a meat tenderiser. Babaco is a natural biocatalyst that can be successfully exploited for lipid biotransformations in interesterification reactions but also via triacylglycerol alcoholysis and in esterification reactions.

Comments

The hybrid taxon, *Vasconcellea × heilbornii* has long been considered as natural sterile hybrid originating from hybridisation between *Vasconcellea stipulata* and *Vasconcellea cundinamarcensis* (Horovitz and Jimemez 1967; Badillo 1971, 1993) based on morphological characters. However with the introduction of molecular techniques into *Vasconcellea* research its origin was shown to be far more complex than a single hybridisation event (Van Droogenbroeck et al. 2006; Kyndt et al. 2006). Three different scenarios for its origin was proposed by Droogenbroeck et al. (2006) including (i) a common maternal progenitor for *Vasconcellea × heilbornii* and *V. weberbaueri*, (ii) a common ancestral polymorphism in *V. stipulata* and *V. weberbaueri* or (iii) a triple hybridization event involving *V. stipulata*, *V. weberbaueri* and *V. cundinamarcensis*. Kyndt et al. (2006) presented results from microsatellite amplification of nuclear and chloroplast simple sequence repeat (SSR) primers, showing that only the first scenario was possible. They maintained that *V. × heilbornii* and *V. weberbaueri* most likely shared a common maternal ancestor from which they had independently diverged. *V. × heilbornii* individuals had been crossing since then with *V. stipulata* and though less frequently with *V. cundinamarcensis*, both of which had acted as pollen donors. Additionally, the fact that single individual of *V. × heilbornii* displayed halotype W and also some distant nuclear relationship with *V. weberbaueri* suggested that there was occasional hybridization between these taxa.

Selected References

Badillo VM (1971) Monographia de la Familia Caricaceae. Universidad Central de Venezuela, Maracay

Badillo VM (1993) Caricaceae Segundo Esquema. Revista de la Facultad de Agronomía de la Universidad Central, Maracay, 111 pp

Badillo VM (2000) *Carica* L. vs. *Vasconcellea* St.-Hil. (Caricaceae) con la rehabilitacion de este ultimo. Ernstia 10(2):74–79

Barbeni M, Guarda PA, Villa M, Cabella P, Pivetti F, Ciaccio F (1990) Identification and sensory analysis of volatile constituents of babaco fruit (*Carica pentagona* Heilborn). Flav Fragr J 5(1):27–33

Cambon E, Rodriguez JA, Pina M, Arondel V, Carriere F, Turon F, Ruales J, Villeneuve P (2008) Characterization of typo-, regio-, and stereo-selectivities of babaco latex lipase in aqueous and organic media. Biotechnol Lett 30(4): 769–774

Chen CC, Tsai SW, Villeneuve P (2005) Enantioselective hydrolysis of (R, S)-naproxen 2,2,2-trifluoroethyl ester in water-saturated solvents via lipases from *Carica pentagona* Heilborn and *Carica papaya*. J Mol Catal B Enzym 34(1–6):51–57

Dhuique-Mayer C, Caro Y, Pina M, Ruales J, Dornier M, Graille J (2001) Biocatalytic properties of lipase in crude latex from babaco fruit (*Carica pentagona*). Biotechnol Lett 23(13):1021–1024

Dhuique-Mayer C, Villarreal L, Caro Y, Ruales J, Villeneuve P, Pina M (2003) Lipase activity in alcoholysis and esterification reactions of crude latex from babaco fruit (*Carica pentagona*). Ol Corps Gras, Lipides 10(3):232–234

Horovitz S, Jimemez H (1967) Cruziamentos interespecificos e intergenericos en Caricaceas y sus implicaciones fitotecnicas. Agron Trop 17:323–344

Kempler C, Kabaluk T (1996) Babaco (*Carica pentagona* Heilb.): a possible crop for the greenhouse. Hortscience 31(5): 785–788

Kyndt T, Van Droogenbroeck B, Haegeman A, Roldan-Ruiz I, Gheysen G (2006) Cross-species microsatellite amplification in *Vasconcellea* and related genera and their use in germplam classification. Genome 49:786–798

National Research Council (1989) Lost crops of the incas: little-known plants of the Andes with promise for worldwide cultivation. BOSTID, National Research Council. National Academy Press, Washington, D.C, 428 pp.

Scheldeman X, Romero Motoche JP, Van Damme P (2001) Highland papayas in southern Ecuador: need for conservation actions. Acta Hortic ISHS 575:199–205

Shaw GJ, Allen JM, Visser FR (1985) Volatile flavor components of babaco fruit (*Carica pentagona*, Heilborn). J Agric Food Chem 33(5):795–797

Tankard G (1987) Tropical fruit an Australian guide to growing and using exotic fruits. Viking O'Neil/Thomas Nelson, Australia, 152 pp

USDA, ARS, National Genetic Resources Program. Germplasm Resources Information Network – (GRIN) [Online Database]. National Germplasm Resources Laboratory, Beltsville, Maryland. URL: http://www.ars-grin.gov/cgi-bin/npgs/html/index.pl

Van Droogenbroeck B, Romeijn-Peeters E, Kyndt T, Van Thuyne W, Goetghebeur P, Romero-Motochi JP, Gheysen G (2006) Evidence of natural hybridization and introgression between *Vasconcellea* spp. (Caricaceae) from southern Ecuador revealed by morphological and chloroplast, mitochondrial and DNA markers. Ann Bot (London) 97(5):793–805

Villareal L, Dhuique-Mayer C, Dornier M, Ruales J, Reynes M (2003) Evaluation de l'interet du babaco (*Carica pentagona* Heilb.). Fruits 58:39–52

Viteri P (1992) El cultivo del babaco en el Ecuador. Manual N19, Programa de Frutales. Instituto Nacional de Investigaciones Agrocuarius (INIAP), Ecuador

Wills RBH, Lim JSK, Greenfield H (1985) Nutrient composition of babaco fruit. J Plant Foods 6:165–166

Wills RBH, Lim JSK, Greenfield H (1986) Composition of Australian foods. 31. Tropical and sub-tropical fruit. Food Technol Aust 38(3):118–123

Juniperus communis

Scientific Name

Juniperus communis L.

Synonyms

Juniperus canadensis Lodd. ex Burgsd., *Juniperus communis* subsp. *nana* (Willd.) Syme, *Juniperus communis* L. var. *communis, Juniperus communis* var. *depressa* Pursh, *Juniperus communis* var. *montana* Aiton, *Juniperus communis* var. *montana* Neilr., *Juniperus communis* var. *saxatilis* Pall., *Juniperus densa* Gord., *Juniperus nana* Willd., *Juniperus sibirica* Burgsd.

Family

Cupressaceae

Common/English Names

Common Juniper, Fairy Circle, Dwarf Juniper, Ground Juniper, Hackmatack, Horse Savin, Juniper, Juniper Berry, Malchangel, Mountain Common Juniper, Mountain Juniper, Old Field Common Juniper, Prostrate Juniper, Siberian Juniper

Vernacular Names

Albanian: Dellinjë E Rëndomtë, Dëllinja, Dullinjë;
Arabic: Abhal, Arar, Habbul Arar, Habbul-Aaraar, Shajratulla;
Armenian: Ardoog, Artuch;
Brazil: Zimbro;
Bulgarian: Khvojna;
Burmese: Kyauk-Hkak-Pan;
Catalan: Ginebre, Ginebre Comú, Ginebre Mascle, Ginebre Negral, Ginebre Ver, Ginebrer, Ginebrera, Ginebró;
Chinese: Du Song, Kuli;
Croatian: Borovica;
Czech: Jalovec Obecný, Jalovec Obecný Pravý;
French: Baies De Genièvre (Berries), Genévrier, Genévrier Commun, Genibre, Genièvre, Thériaque Des Paysans;
Danish: Almindelig Ene, Ene, Enebær, Enebærtræ, Enebærbusk, Junipero;
Dutch: Jeneverbes, Jeneverboom, Jeneverboom Soort;
Eastonian: Harilik Kadakas, Kadakamarjad;
Finnish: Kataja, Katajanmarja, Koti Kataja, Metsäkataja;
French: Genévrier, Genévrier Commun;
Galician: Xenebro Común;
German: Dürrenstaude, Feuerbaum, Gemeine Wacholder, Gemeiner Wachholder, Gewöhnlicher

Wacholder, Heide-Wacholder, Krammetsbeerstrauch, Krammetsbeerbaum, Kranawettstrauch, Kranawitt, Kraunbaum, Machandel, Machandelbaum, Macholder, Sadebaum, Wachholder, Wacholderbeeren (Berries), Weidewacholder;
Greek: Agriokedros, Arkeuthos, I Koeni, Arkeuthos Koenos, Gioniperos Koinos, Mnesitheos, Thalassokedro, Thamnokedro;
Hungarian: Boróka, Borókabogyó, Közönséges Boróka;
Icelandic: Einiber, Einir;
India: Hayusha (Bengali), Palash (Gujarati), Abhal, Aaraar, Hauber, Haubera, Jhora, Padmak, Bither, Guggal (Hindu), Padma Beeja (Kannada), Hosh, Hosha (Marathi), Halvulba (Punjabi), Aparajita, Ashvathaphala, Dhmankshanashini, Habusha, Hapusa, Hapusha, Havusa, Kachhughna, Kaphaghni, Kapotapanka, Matsyagandha, Plihahantri, Plihashatru, Svalpaphala, Vapusha, Vigandhika, Vishaghni, Visra, Visraganga (Sanskrit), Abhal (Urdu);
Irish: Aiteal;
Italian: Bacca Di Ginepro, Coccola Di Ginepro, Ginepro, Ginepro Commune;
Japanese: Junipaa, Juniperusu Kommunisu, Nezu, Seiyou Byakushin, Seiyou Nezu, Seiyou Suzu, Seiyou Toshou, Toshou;
Korean: Du Song, Chunipo (Junipeo), Chyunipo (Jyunipeo);
Latvian: Kadiķa Ogas (Berries), Paegļi, Zviedrijas Kadiķis;
Lithuanian: Paprastasis Kadagys;
Macedonian: Borovinka, Klekovača, Smreka;
Nepalese: Dhupi, Dhupisin, Pamo, Phar;
Norwegian: Brakje, Bresk, Brisk, Bruse, Eine, Einer, Ener, Sprakje;
Pakistan: Bhentri (Hindko), Ghojar (Pushto), Bantha (Kohistani), Pama, Petthri (Punjabi), Abhal (Urdu);
Persian: Sarv Kohi, Sarv Kuhi, Tukhme Rahal;
Polish: Jalowiec Pospolity;
Portuguese: Fruto-De-Genebra, Genebreiro, Junípero, Junípero-Comum, Junipo, Zimbrão, Zimbreiro, Zimbro, Zimbro-Anão, Zimbro-Comum, Zimbro-Rasteiro, Zimbro-Vulgar;
Serbian: Kleka;
Slovaščina: Brin, Brin Navadni, Brinove Jagode, Navadni Brin;
Slovencina: Borievka, Borievka Obyčajná;
Spanish: Baya De Enebro (Berries), Cedro, Enebrina, Enebriza, Enebro, Enebro Albar, Enebro Común, Enebro Espinoso, Enebro Morisquillo, Enebro Real, Enebrosa, Ginebro, Ginebro Real, Grojo, Junipero, Junípro, Nebrina, Nebrini, Nebro, Sabino;
Swahili: Mreteni;
Swedish: En, Enbär, Enbuske, Eneträ, Vanlig En;
Switzerland: Gemeiner Wachholder, Reckholder;

Tibetan: Spa Ma I Bras Bu, Spa Mai Bras Bu;
Turkish: Adi Ardiç, Ardiç Meyvasi (Berries), Ardıç Yemişi, Ephel;
Ukrainian: Yalivets Zvychajnyj, Yalovets Zvychajnyj;
Vietnamese: Cây Bách Xù;
Zulu: Uhlobo Lwesihlahlana.

Origin/Distribution

This is the most widespread conifer in the world, native to temperate Eurasia, and North America north of Mexico, from 30°N latitude, occupying an extraordinary range of habitats up in the northern hemisphere to the margin of the Arctic.

Agroecology

The species is adapted to temperate climatic conditions. It is found in low and high altitudes up to 3,000 m in the alpine region, throughout the northern hemisphere from cool temperate to polar regions, in heaths, moorlands, wooded hillsides, chalk down, dry open woods, gravelly ridges, outcrops, sand terraces, maritime escarpments and open rocky slopes. It prefers full sun as it is intolerant of shade and is drought tolerant. The species is adaptable to a wide range of soils and soil pH, from poor rocky soils to acidic and calcareous sands, loams, or marls.

Edible Plant Parts and Uses

Juniper is an important spice in many European cuisines, especially in the Alpine regions. The blue-black Juniper berries, are too bitter and astringent to eat raw and are usually sold dried and used to flavour meats, game birds, venison, duck, rabbit, pork, ham, lamb, sauces and stuffings. Dried berries are added to vegetable pates, game, venison and marinades. It adds flavour to potatoes, sauerkraut sausages and casseroles. Juniper berry sauce is often a popular flavoring choice for quail, pheasant, veal, rabbit, venison and other meat dishes. Juniper berries should be used sparingly as they impart a strong taste. They are generally crushed before use to release their flavour. Fresh berries are used to make a conserve to accompany cold meats. To make a conserve, the fresh ripe berries are cooked in water till soft and the pulp is crushed. Three times the berries weight equivalent of sugar is added to the berry puree and stirred vigorously and left to cool. The berries are dried and ground and used in the preparation of a mush or cake.

The berries are used as a spice to flavour gin, liquers, bitters, Swedish beer and other beverages (tea) and cordials. In Sweden, a beer is made that is regarded as a healthy drink. Incidentally, the word 'gin' is derived from the French word for juniper berry, *genièvre*, which is also the French name for gin. Distilled gin is made by redistilling neutral grain spirit and raw cane sugar which has been flavoured with juniper berries. Juniper berries are also used as the primary flavor in the liquor *jenever*. *Jenever*, is the juniper-flavored and strongly alcoholic traditional liquor of the Netherlands, Belgium and Northern France. *Borovick*a, the Slovak national alcoholic beverage is also flavoured with juniper berry extract. Juniper is the key ingredient of *sahti*, a traditional Finnish ale made from a variety of cereals, malted and unmalted, including barley, rye, wheat, and oats. The roasted seed can be used as a coffee substitute. Juniper leaves can be used fresh or dried to flavour grilled fish. The wood and leaves can be used on a barbeque to give a subtle flavour to meat.

Botany

An evergreen, spreading multi-stemmed or columnar, dioecious, coniferous shrub to 1 m tall, or erect small tree to 3 m high with a usually depressed crown. Bark is very thin, reddish brown, scaly and fibrous. Leaves are persistent, needle-like to narrowly lance-shaped, 5–12 mm long (sometimes to 15 mm), jointed at base, stiff, very prickly; whitish-silver above, dark green below; arranged in whorls of 3s, sessile. Fruit developed from female cones made up of green, ovate or acuminate scales. Fruit is berry-like, globose to sub-globose, 6–12 mm in diameter, green when young then red and bluish black when mature, glaucous, always covered with white bloom, maturing in the second season (Plate 2). Each berry-cone usually has three fused scales,

Plate 1 Leaves and male cones

Plate 2 Ripe and unripe juniper berries

each scale with a single seed. Each cone berry commonly has two to three seeds. Seeds are ~5 mm. Male cones are sessile or stalked, yellowish (Plate 1), 2–3 mm long, catkin-like on separate plants and fall soon after shedding their pollens.

Nutritive/Medicinal Properties

Juniper berries were found to contain volatile oil (0.2–3.4%), sugars, resin, vitamin C flavonoids, proanthocyanidines, lignan desoxypodophyllotoxin and its isomer desoxypicropodophyllotoxin, diterpene acids and sesquiterpenes, (Ladner 2005). Five diterpenes were isolated, from *J. communis* berries one of which was a new labdane diterpene 15,16-epoxy-12-hydroxy-8(17),13(16),14-labdatrien-19-oic acid (Martin et al. 2006). More than 200 constituents were detected in the supercritical CO_2 extracts of ground *J. communis* berries, and the contents of 50 compounds were reported (Barjaktarović, et al. 2005). The percentage yields of monoterpene, sesquiterpene, oxygenated monoterpene, and oxygenated sesquiterpene hydrocarbon groups extracted were dependent on the extraction time and conditions. At all pressures, monoterpene hydrocarbons were almost completely extracted from the berries in the first 0.6 h while oxygenated monoterpenes were completely extracted at 100 bar in 0.5 h and at 90 bar in 1.2 h. Oxygenated sesquiterpenes were extracted at constant rates at 100 and 90 bar in 1.2 and 3 h, respectively.

The essential oil from unripe and ripe berries was predominated by α-pinene (39.7–64.9%) chemotype (Butkienë et al. 2004). Myrcene was the second major constituent (4.8–19.6%) followed by α-Cadinol (2.7–7.1%). The amounts of α-pinene, sabinene, β-pinene and bornyl acetate decreased during ripening of berries. An inverse correlation was determined for myrcene, terpinen-4-ol and α-terpineol. Monoterpene hydrocarbons (61.4–80.8%) prevailed in the essential oils. Hydrocarbons decreased markedly in berries during ripening. The major part (81.5–88.6%) of essential oils comprised compounds (including esters) with four carbon skeletons: 2.6-dimethyloctane, menthane, pinane and cadinane. The identified compounds made up 94.8–99.4% of the essential oils. Juniper essential oil (Juniperi aetheroleum) obtained from the juniper berry, was found to have α-pinene (29.17%) and β-pinene (17.84%), sabinene (13.55%), limonene (5.52%), and myrcene (0.33%) as the main compounds (Pepeljnjak et al. 2005).

Another analysis reported that juniper oil contained 41 and 27 components in the needles and berries, respectively (Shahmir et al. 2003). The needle oil consisted mainly of sabinene (40.7%), α-pinene (12.5%) and terpinen-4-ol (12.3%). The berry oil contained sabinene (36.8%), α-pinene (20%), limonene (10.6%), germacrene D (8.2%) and myrcene (4.8%) as the main components.

The essential oil yields (w/w) were 0.78%, 0.70% and 0.12% for leaves, berries and wood, respectively (Marongiu et al. 2006). The leaf essential oil was made up of limonene (36.2%), β-selinene (15.2%) and α-terpinyl acetate (5.3%). The berry oil was composed mainly of: limonene (40.1%), germacrene D (17.2%) and α-pinene (4.7%). The oil derived from the wood mainly consisted of: limonene (8.9%), α-terpinyl acetate (9.7%) and germacrene D (8.6%). Recently, Adams et al. (2010) reported on the variation in composition of the leaf essential oils of *J. communis* varieties in the Western Hemisphere. The oil of *Juniperus communis* var. *jackii* was found to have the most distinct oil and contained moderate amounts of α-pinene (16.1–18.9%), d-3-carene (17.9–28.4%) and β-phellandrene (9.2–13.4%) along with several diterpenes including isoabienol. *Juniperus communis* var. *charlottensis* oil had a very high concentration of α-pinene (59.3%), with moderate amounts of β-pinene (5.9%), δ-3-carene (3.6%) and β-phellandrene (2.9%). The oil of *J. communis* var. *depressa* was dominated by α-pinene (53.9%), δ-3-carene (9.3%) and β-pinene (5.5%) and was unusual in also containing citronellol, citronellyl acetate, neryl acetate and geranyl acetate with no diterpenes. The oil of *J. communis* var. *megistocarpa* possessed large amounts of α-pinene (58.5%), limonene (20.4%) and β-pinene (5.0%) and, like *J. communis* var. *depressa*, its oil also contained citronellol, citronellyl acetate, neryl acetate and geranyl acetate. Putative *J. communis* var. *saxatilis* from Idaho, USA, had a high concentration of a-pinene (56.5%) with moderate amounts of δ-3-carene (11.5%), β-pinene (5.4%), myrcene (4.5%) and β-phellandrene (3.1%). The oils of these North American varieties were compared to *J. communis* var. *communis* (Sweden) and *J. communis* var. *saxatilis* (Switzerland). A survey of approximately 65 species of *Juniperus* revealed that isoabienol was present in eight species.

Two neolignan glycosides (junipercomnosides A and B) were isolated from aerial parts of *Juniperus communis* var. *depressa* along with two known neolignan glycosides and seven flavonoid glycosides (Nakanishi et al. 2004). A new monoterpene glucoside and three new natural megastigmane glycosides were isolated along with a known megastigmane glucoside from leafy twigs of *Juniperus communis* var. *depressa* collected in Oregon, U.S.A. (Nakanishi et al. 2005). A new flavone xyloside, 1, and 2 new flavan-3-ol glucosides, 3 and 4, were isolated together with three known flavones, 2, 11, and 12, five known flavans, 5–9, and a known dihydrochalcone, 10, from the stems and leaves of *Juniperus communis* var. *depressa* collected in Oregon, U.S.A. (Iida et al. 2007). Two new phenylpropanoid glycosides were isolated from the leaves and stems of *Juniperus communis* var. *depressa* along with 14 known compounds (Iida et al. 2010).

Juniper plant parts have been reported to have various pharmacological attributes as discussed below:

Antioxidant Activity

Juniper berries were found to contain 16 different compounds comprising flavonoids, such as isoscutellarein and 8-hydroxyluteolin or hypolaetin glycosides, and six biflavonoids, among them amentoflavone, hynokiflavone, cupressoflavone, and methyl-biflavones (Innocenti et al. 2007). The flavonoid concentration in the berries ranged between 1.46 and 3.79 mg/g of fresh pulp, while the content of the biflavonoids was always lower, varying between 0.14 and 1.38 mg/g of fresh weight. These flavonoids were reported to contribute to the berry's strong antioxidant activity.

Both the water and the ethanol extracts of Juniper fruit exhibited strong total antioxidant activity (Elmastas et al. 2006). The concentrations of 20, 40, and 60 µg/mL of water and ethanol extracts of juniper fruit showed 75%, 88%, 93%, 73%, 84%, and 92% inhibition on peroxidation of linoleic acid emulsion, respectively. In contrast, 60 µg/ml of standard antioxidant such as butylated hydroxyanisole (BHA), butylated hydroxytoluene (BHT), and α-tocopherol exhibited 96, 96, and 61% inhibition on peroxidation of linoleic acid emulsion, respectively. However, both extracts of juniper had effective reducing power, free radical scavenging, superoxide anion radical scavenging, hydrogen peroxide scavenging, and metal chelating activities at similar concentrations (20, 40, and 60 µg/ml). Additionally, total phenolic compounds in both aqueous and ethanolic juniper extracts were determined as gallic acid equivalents. Thus, the results indicated that juniper had in-vitro antioxidant properties and these may be major reasons for the inhibition of lipid peroxidation properties.

Miceli et al. (2009) found that total polyphenols (Folin–Ciocalteau method) were three-fold higher in *Juniperus communis* L. var. *communis* (Jcc) (59.17 mg GAE/g extract) than in *Juniperus communis* L. var. *saxatilis* (Jcs) (17.64 mg GAE/g extract). Flavonoid and biflavonoid content, were higher in Jcc (25,947 and 4,346 µg/g extract) than in Jcs (5,387 and 1,944 µg/g extract). The HPLC analysis of Jcc afforded 16 flavonoids; hypolaetin-7-pentoside and quercetin-hexoside were the main compounds. Moreover, gossypetin-hexoside-pentoside and gossypetin-hexoside were identified for the first time in Jcc berries. In Jcs eight flavonoids were identified: quercetin-hexoside and isoscutellarein-8-O-hexoside were the most abundant compounds. The in-vitro antioxidant activity was determined using different methods; Jcc was found to be more active than Jcs in the DPPH test (IC_{50} of 0.63 mg/ml and 1.84 mg/ml), in reducing power assay (12.82 ASE/ml and 64.14 ASE/ml), and in TBA assay

(IC_{50} of 4.44 µg/ml and 120.07 µg/ml). By contrast, Jcs exhibited greater Fe^{2+} chelating ability than Jcc. The extracts displayed antimicrobial capacity only against Gram-positive bacteria.

Antimicrobial/Antiparasitic Activity

Juniper essential oil was found to have antimicrobial activity (Pepeljnjak et al. 2005). Juniper essential oil showed bactericidal activities against Gram-positive and Gram negative bacteria with MIC values between 8% and 70% (v/v). All Gram positive bacteria (*Bacillus cereus*, *Bacillus subtilis*, *Micrococcus flavus*, *Micrococcus luteus*, *Staphylococcus aureus*, *Staphylococcus epidermidis* and *Enterococcus faecalis*) were sensitive to the oil. Among the Gram negative bacteria, *Citrobacter freundii* and *Escherichia coli* were resistant while *Serratia* sp., *Salmonella enteritidis*, *Proteus mirabilis*, *Shigella sonnei*, *Klebsiella oxytoca* and *Yersinia enterocolitica* were susceptible to the oil. The oil had stronger fungicidal activity with MIC values below 10% (v/v) than bactericidal activity. All the yeast and yeast-like fungi (*Candida albicans*, *Candida krusei*, *Candida tropicalis*, *Candida parapsilosis*, *Candida glabrata*, *Candida kefyr*, and *Candida lusitaniae*) were sensitive to the juniper essential oil. Among the dermatophyte species tested, only *Microsporum canis* was resistant, while *Microsporum gypseum*, *Trichophyton mentagrophytes* and *Trichophyton rubrum* were susceptible. The strongest fungicidal activity was recorded against *Candida* spp. (MIC from 0.78% to 2%, V/V) and dermatophytes (from 0.39% to 2%, V/V). The lowest values of *MIC* of the essential oil against fungal strains indicated that the main compounds present in the oil-terpene hydrocarbons (pinenes, sabinene, mircene and limonene) had a stronger antifungal than antibacterial activity. Juniper needles are known to be rich in terpenoids and phenolics. Martz et al. (2009) found that the concentration of these compounds increased with latitude and altitude with, however, a stronger latitudinal effect (a higher content of monoterpenoids, proanthocyanidins, and flavonols in northern latitudes). Analysis of methanolic extracts showed quite good activity against both antibiotic-sensitive and antibiotic-resistant *Staphylococcus aureus* strains and suggested an important role of the soluble phenolic fraction. They also found that the relative lack of toxicity of juniper extracts on keratinocytes and fibroblastic cells, raised the possibility of their use in preventing bacterial skin infection.

The antimycobacterial activity of *Juniperus communis* was attributed to a sesquiterpene identified as longifolene and two diterpenes, totarol and trans-communic acid (Gordien et al. 2009). Totarol showed the highest activity against *Mycobacterium tuberculosis* H(37)Rv (MIC of 73.7 µM) and against the isoniazid-, streptomycin-, and moxifloxacin-resistant variants (MIC of 38.4, 83.4 and 60 µM, respectively). Longifolene and totarol were most active against the rifampicin-resistant variant (MICs of 24 and 20.2 µoM, respectively). Totarol showed the greatest activity in the low oxygen recovery assay (LORA) (MIC of 81.3 µM) and against all non-tuberculous mycobacteria species (*Mycobacterium aurum*, *Mycobacterium phlei*, *Mycobacterium fortuitum* and *Mycobacterium smegmatis*), (MICs in the range of 7–14 µM). Trans-communic acid showed good activity against *Mycobacterium aurum* (MIC of 13.2 µM). The low selectivity indices (SI) obtained following cytotoxicity studies indicated that the isolated terpenoids were relatively toxic towards mammalian cells. The presence of antimycobacterial terpenoids in *Juniperus communis* aerial parts and roots justified, to some extent, the ethnomedicinal use of this species as a traditional anti-tuberculosis (TB) and other respiratory diseases remedy. In another study, totarol was found to inhibit *Leishmania donovani* promastigotes with IC_{50} values of 3.5–4.6 µg/ml (Samoylenko et al. 2008). Additionally, totarol demonstrated nematicidal and antifouling activities against *Caenorhabditis elegans* and *Artemia salina* at a concentration of 80 and 1 µg/ml, respectively.

All the crude leaf extracts (aqueous, methanol, ethanol, chloroform and hexane) of *Juniperus communis* were found to be effective in inhibiting the growth of five pathogenic multidrug resistant bacteria (*Bacillus subtilis*, *Erwinia chrysanthemi*, *Escherichia coli*, *Agrobacterium tumefaciens* and *Xanthomonas phaseoli*) (Sati and Joshi 2010). The hexane extract showed best inhibition against the test microorganisms followed by ethanol, methanol and chloroform extract. The inhibitory activity of these extracts was found to be very effective as compared to ampicillin (10 μg) and erythromycin (15 μg) standard antibiotics which were used as positive control against these test microorganisms. In another study, the following extracts, methanolic, ethanolic, chloroformic and petroleum ether of *J. communis* leaves were found to be effective against *Staphylococcus aureus, Micrococcus luteus, Pseudomonas aeruginosa, Salmonella typhimurium, Klebsiella oxytoca, Listeria monocytogenes,* and *Escherichia coli* (Kumar et al. 2010). The chloroform extract was most effective against all the bacterial strains.

Hypoglycemic Activity

Studies showed that juniper berry had hypoglycemic activity (De Medina et al. 1994). Juniper berry decoction lowered glycemic levels in normoglycemic rats at a dose of 250 mg/kg. This effect was postulated to be achieved through: (a) an increase of peripheral glucose consumption; (b) a potentiation of glucose-induced insulin secretion. The administration of the decoction (125 mg total berries/kg) to streptozotocin-diabetic rats for 24 days resulted in a significant reduction both in blood glucose levels and in the mortality index, as well as the prevention of body weight loss. Ten aqueous plant extracts including *Juniperus communis* with proven antihyperglycemic properties were examined at a concentration of 50 g plant extract/l using an in-vitro method to assess their possible effects on glucose diffusion across the gastrointestinal tract (Gallagher et al. 2003). *J. communis* extract was found to reduce significantly glucose movement but were less effective than agrimony (*Agrimony eupatoria*) and avocado.

Anticancer Activity

Juniper leaf extract was found to have anti-cancerous activity (Marongiu et al. 2006). Juniper leaf extract inhibited the proliferation of cell derived from haematological and solid human tumours. It showed the same potency and selectivity as etoposide (an anti-cancer chemotherapy drug) and proved to be 8–100-fold more potent than 6-MP6 (Mercaptopurine, an anti-cancer chemotherapy drug). *Juniperus communis* berry extracts was demonstrated in-vitro to kill a large amount of the liver and colon carcinomas and myosarcoma cells in G2, M and G0 phases of the mitotic cell cycle.

Diuretic Activity

Juniper berries was found to have diuretic activity (Stanic et al. 1998). An infusion and the essential oil of juniper berries (*Juniperus communis*) as well as terpinen-4-ol elicited diuresis response in rats. The effect of the smallest dose of vasopressin (ADH) (0.004 IU/100 g b.w.) intraperitoneally administered was comparable to the effect of a 10% drug infusion and 0.1% essential oil solution, while the effect of 0.4 IU/100 g b.w. of ADH was comparable to that of a 0.01% solution of terpinen-4-ol. Among tested preparations a 10% drug infusion exerted the most considerable diuretic activity. The results suggested that the diuretic activity of juniper berries cannot be attributed only to the essential oil but also to hydrophilic drug constituents.

Platelet Inhibition Activity

Extracts of *Juniperus communis* exhibited inhibitory activity on human platelet-type 12(S)-lipoxygenase [12(S)-LOX] (Schneider et al. 2004). The methylene chloride extracts of Juniperi lignum (wood), Juniperi pseudo-fructus (false fruit) and the ethyl acetate extract of Juniperi pseudo-fructus showed a significant inhibition on the production of 12(S)-HETE [12(S)-hydroxy-5,8,10,14-eicosatetraenoic acid] at 100 μg/ml (54.0%, 66.2% and 76.2%, respectively). From the methylene chloride extract of the wood, cryptojaponol and β-sitosterol were isolated as active inhibitory compounds (inhibition activity at 100 μg/ml=55.4% [IC_{50}= 257.5 μM] and 25.0%, respectively). Further, a lipid fraction containing unsaturated fatty acids contributed to the in-vitro activity of the crude extract.

Traditional Medicinal Uses

Juniper berries have long been used in traditional medicine by many cultures. Its medicinal attributes include: diuretic, antiseptic, aromatic, rubefacient, stomachic and antirheumatic. Traditional uses include cystitis, flatulence and colic. Native Americans also used juniper berries as a female contraceptive. Juniper has been used in dyspepsia (upset stomach) as a berry tea, in eczema and other skin diseases as cade oil or juniper oil. Juniper berries act as a strong urinary tract disinfectant if consumed and were used by American Indians as an herbal remedy for urinary tract infections. It also has been used for treatments of kidney ailments, gout, rheumatism and other arthritic conditions by indigenous peoples from Eurasia. Western tribes combined the berries with *Berberis* root bark in an herbal tea to treat diabetes. Juniper was used by Great Basin Indians as a blood tonic. Native Americans from the Pacific Northwest used tonics made from the branches to treat colds, flu, arthritis, muscle aches, and kidney problems. Juniper tincture is applied topically for some skin conditions and baldness.

Other Uses

Juniperus communis is widely used as an ornamental shrub, it provides good ground cover on stony or sandy sites. The wood is fine grained, durable, and reddish with white sapwood but posses no commercial value. It has been reported to be used for fuel (especially for pellet-stoves), fire-wood, fence-posts, cement and particle boards, wall board, cordwood housing, parquet, paper, chemical derivatives, activated carbon and small wood items. In Scandinavia, juniper wood is used for making containers for storing small quantities of dairy products such as butter and cheese and also for making wooden butter knives. Juniper wood is resistant to fungal decay and insects even when exposed to the soil. This resinous wood yields much more tar than drier ones. Therefore, juniper wood should make an excellent wood preservative. Juniper is a powerful antiseptic and used in insecticides and perfumes. Juniper is safely employed as a fragrance in soaps, shampoos, cosmetics, sachets and other products.

Comments

Juniperus communis is listed as threatened by the Conifer Specialist Group (1998).

Selected References

Adams RP (1993) *Juniperus* Linnaeus. In: Flora of North America Editorial Committee (ed.) Flora of North America North of Mexico, vol 2. Oxford University Press, Oxford

Adams RP, Beauchamp PS, Dev V, Bathala RM (2010) The leaf essential oils of *Juniperus communis* L. varieties in North America and the NMR and MS data for isoabienol. J Essent Oil Res 22(1):23–28

Barjaktarović B, Sovilj M, Knez Z (2005) Chemical composition of *Juniperus communis* L. fruits supercritical CO_2 extracts: dependence on pressure and extraction time. J Agric Food Chem 53(7):2630–2636

Butkienė R, Nivinskienė O, Mockutė D (2004) Chemical composition of unripe and ripe berry essential oils of *Juniperus communis* L. growing wild in Vilnius district. Chemija 15(4):57–63

Conifer Specialist Group (1998) *Juniperus communis*. In: IUCN 2009. IUCN Red List of Threatened Species. Version 2009.2. <www.iucnredlist.org>

De Medina FS, Gamez MJ, Jimenez I, Jimenez J, Osuna JI, Zarzuelo A (1994) Hypoglycemic activity of juniper berries. Planta Med 60(3):197–200

Elmastas M, Gülcin I, Beydemir S, Küfrevioglu ÖI, Aboul-Enein HY (2006) A study on the in vitro antioxidant activity of juniper (*Juniperus communis* L.) fruit extracts. Anal Lett 39(1):47–65

Foundation for Revitalisation of Local Health Traditions (2008) FRLHT database. htttp://envis.frlht.org.

Gallagher AM, Flatt PR, Duff G, Abdel-Wahab YHA (2003) The effects of traditional antidiabetic plants on in vitro glucose diffusion. Nutr Res 23(3):413–424

Gordien AY, Gray AI, Franzblau SG, Seidel V (2009) Antimycobacterial terpenoids from *Juniperus communis* L. (Cuppressaceae). J Ethnopharmacol 126(3):500–505

Iida N, Inatomi Y, Murata H, Inada A, Murata J, Lang FA, Matsuura N, Nakanishi T (2007) A new flavone xyloside and two new flavan-3-ol glucosides from *Juniperus communis* var. *depressa*. Chem Biodivers 4(1):32–42

Iida N, Inatomi Y, Murata H, Murata J, Lang FA, Tanaka T, Nakanishi T, Inada A (2010) New phenylpropanoid glycosides from *Juniperus communis* var. *depressa*. Chem Pharm Bull (Tokyo) 58(5):742–746

Innocenti M, Michelozzi M, Giaccherini C, Ieri F, Vincieri FF, Mulinacci N (2007) Flavonoids and biflavonoids in Tuscan berries of *Juniperus communis* L.: detection and quantitation by HPLC/DAD/ESI/MS. J Agric Food Chem 55(16):6596–6602

Kumar P, Bhatt RP, Singh L, Shailja, Chandra H, Prasad R (2010) Identification of phytochemical content and antibacterial activity of *Juniperus communis* leaves. Int J Biotechnol Biochem 6(1):87–91

Ladner J (2005) *Juniperus communis* L. In: Cook BG, Pengelly BC, Brown SD, Donnelly JL, Eagles DA, Franco MA, Hanson J, Mullen BF, Partridge IJ, Peters M, Schultze-Kraft R (eds.) Tropical forages: an interactive selection tool. CSIRO/DPI&F (Qld)/CIAT/ILRI, Brisbane [CD-ROM]

Marongiu B, Porcedda S, Piras A, Sanna G, Murreddu M, Loddo R (2006) Extraction of *Juniperus communis* L. ssp. nana Willd. essential oil by supercritical carbon dioxide. Flav Fragr J 21(1):148–154

Martin AM, Queiroz EF, Marston A, Hostettmann K (2006) Labdane diterpenes from *Juniperus communis* L. berries. Phytochem Anal 17(1):32–35

Martz F, Peltola R, Fontanay S, Duval RE, Julkunen-Tiitto R, Stark S (2009) Effect of latitude and altitude on the terpenoid and soluble phenolic composition of juniper (*Juniperus communis*) needles and evaluation of their antibacterial activity in the boreal zone. J Agric Food Chem 57(20):9575–9584

Miceli N, Trovato A, Dugo P, Cacciola F, Donato P, Marino A, Bellinghieri V, La Barbera TM, Güvenç A, Taviano MF (2009) Comparative analysis of flavonoid profile, antioxidant and antimicrobial activity of the berries of *Juniperus communis* L. var. *communis* and *Juniperus communis* L. var. *saxatilis* Pall. from Turkey. J Agric Food Chem 57(15):6570–6577

Nakanishi T, Iida N, Inatomi Y, Murata H, Inada A, Murata J, Lang FA, Iinuma M, Tanaka T (2004) Neolignan and flavonoid glycosides in *Juniperus communis* var. *depressa*. Phytochem 65(2):207–213

Nakanishi T, Iida N, Inatomi Y, Murata H, Inada A, Murata J, Lang FA, Iinuma M, Tanaka T, Sakagami Y (2005) A monoterpene glucoside and three megastigmane glycosides from *Juniperus communis* var. *depressa*. Chem Pharm Bull (Tokyo) 53(7):783–787

Pepeljnjak S, Kosalec I, Kalodera Z, Blazević N (2005) Antimicrobial activity of juniper berry essential oil (*Juniperus communis* L., Cupressaceae). Acta Pharm 55:417–422

Porcher MH et al. (1995–2020) Searchable world wide web multilingual multiscript plant name database. The University of Melbourne, Australia. http://www.plant-names.unimelb.edu.au/Sorting/Frontpage.html

Samoylenko V, Dunbar DC, Gafur MA, Khan SI, Ross SA, Mossa JS, El-Feraly FS, Tekwani BL, Bosselaers J, Muhammad I (2008) Antiparasitic, nematicidal and antifouling constituents from *Juniperus* berries. Phytother Res 22(12):1570–1576

Sati SC, Joshi S (2010) Antibacterial potential of leaf extracts of *Juniperus communis* L. from Kumaun Himalaya. Afr J Microbiol Res 4(12):1291–1294

Schneider I, Gibbons S, Bucar F (2004) Inhibitory activity of *Juniperus communis* on 12(S)-HETE production in human platelets. Planta Med 70(5):471–474

Shahmir F, Ahmadi L, Mirza M, Korori SAA (2003) Secretory elements of needles and berries of *Juniperus communis* L. ssp. *communis* and its volatile constituents. Flav Fragr J 18(5):425–428

Stanic G, Samarzija I, Blazevic N (1998) Time-dependent diuretic response in rats treated with Juniper berry preparations. Phytother Res 12(7):494–497

Cycas revoluta

Scientific Name

Cycas revoluta Thunb.

Synonyms

Cycas inermis C.A.J. Oudemans, *Cycas miquelii* Warb., *Cycas revoluta* β *inermis* Miq., *Cycas revoluta* var. *brevifrons* Miq., *Cycas revoluta* var. *planifolia* Miq., *Cycas revoluta* var. *prolifera* Siebold & Zucc., *Cycas revoluta* var. *robusta* Messeri.

Family

Cycadaceae

Common/English Names

False Sago, Sago Palm, Sago Palm of Japan, King Sago Palm, Sago Cycas, Sotetsu Nut

Vernacular Names

Burmese: Monmg-Tain;
Chinese: Feng-Wei-Jiao-Yr, Su Tie, Tie Shu [Bonta and Osborne 2007];
Czech: Cykas Japonský, Cykas Zavinutý;
Danish: Japansk Cykas, Sagopalme, Tilbagerullet Sagotræ;
Eastonian: Rahu-Palmlehik;
Finnish: Saagopalmu;
French: Cycas Du Japon, Cycas Sagoutier;
German: Eingerollter Palmfarn, Japanischer Sagobaum, Japanischer Sagopalmfarn, Sagopalme;
India: Madanagameswari (Tamil);
Indonesia: Pakis Haji;
Italian: Japan Sikasi, Palma A Sagù;
Japan (Main Islands): Sotetsu (Preferred), Ban Shou, Hou Bi, Hou Bi Shou, Sha Ka Shou, Tesshou, Tessio, Tosso [Bonta and Osborne 2007];
Japan (Ryuku Islands): Hichichi, Hitichi, Satetsu, Shichichi, Sichi, Sichidzi, Sidzichi, Sidzidzui, Sihittu, Sirichi, Sitechi, Sitichi, Sitidzi, Sitochi, Sitsuchi, Sitsudzu, Situchi, Suidzu, Susitykuki, Sutachi, Suticha, Sutichi, Sutta, Sutuku, Suutichi, Syutta, Syutto, Tsudzu (Plant), Kyungama, Mii, Nadzu, Nari, Sutitsi-Nari, Yanabu (Seed) [Bonta and Osborne 2007];
Korea: So Ch'eol;
Malaysia: Paku Laut;
Philippines: Oliva (Spanish), Bait, Pitago Oliba (Tagalog);
Russian: Sagovnik Ponikaiushchii, Sagovnik Ponikaiushchii Drevovidnyi (Trunk), Tsikas Ponikaiushchii;
Spanish: Cicas Revoluta, Palma Sagú, Sagutero;
Swedish: Kottepalm;
Thai: Pron-Tha-Le;
Turkey: Sago Hurma Ag;
Vietnamese: Thuen Tú Ê.

Origin/Distribution

The species is endemic to the southern Japanese islands of Kyushu, Ryukyu, Mitsuhama and Satsuma. It is found widely distributed in colonies on steep to precipitous stony sites on hillsides.

Reports of natural occurrences in coastal Fukien Province of China (Lianjiang Xian, Ningde Xian, and some islands) have not been substantiated, although circumstantial support for these claims is strong (Ye 1999).

Agroecology

This species is adapted to the temperate and subtropical areas but can also be grown in the tropics. It is a hardy cycad, frost and drought tolerant and is sensitive to water-logging. Frond damage can occur at temperatures below −10°C. It prefers a sunny, well drained spot, with deep soil and some organic matter, but will still thrive in less than ideal conditions. It will tolerate light shading and can be grown in glasshouses, as an indoor plant, it is also excellent as a potted or container plant and is a prized bonsai specimen.

Edible Plant Parts and Uses

The seeds are used as food source in its native areas. The principal use of sotetsu in Amami islands, Japan has been for food especially during times of famine. Currently the local product using soetsu is *nari-miso* or *sotetsu miso* made from fermentation of pulverised sotetsu seed/flour and brown rice. *Nari miso* has been getting high consumer scores for its unique taste and low salt content. In addition, tests have been conducted to produce noodles and confectionery from the starch in the seeds and stem (Kira and Miyoshi 2000).

The pith is very rich in edible starch, and is used for making sago which must be carefully washed to leach out toxins contained in the pith.

Botany

A medium-sized cycad with arborescent trunk, 1–8 m high and 35–95 cm diameter (Plate 1) with numerous suckers arising from the base and occasional offsets produced on the trunk, apex tomentose; bark gray-black, scaly. Leaves deep green, semi glossy, 40–100 or more, 1-pinnate, 0.7–1.4(−1.8) m × 20–25(−28) cm petiole subtetragonal in cross section, 10–20 cm, with 6–18 spines along each side, leaf blade oblong- or elliptic-lanceolate, strongly "V"-shaped in cross section, recurved, brown tomentose when young. Leaflets in 60–150 pairs, horizontally inserted at approximately 45° above rachis (Plates 2–3), not glaucous

Plate 1 Plant habit of *Cycas revoluta*

Plate 2 Crown foliage

Plate 3 Female cone early stage

Plate 4 Close-up of more advanced female cone

when mature, straight to subfalcate, 10–20 cm × 4–7 mm, leathery, sparsely pubescent abaxially, base decurrent, margin strongly recurved, apex acuminate. Male cones 10–40 cm by 4–6 cm across, narrowly cylindrical to ovoid, hairy, brown, sporophylls, narrowly wedge shaped with a short upcurved point. Female cone loose and open, brown and hairy (Plates 3–4), sporophylls densely covered with brown hairs, apical lobe ovate, 40–60 mm by 20 mm, margins deeply laciniate with 12–18 tapered lobes, ovules 4–6 on each sporophyll, densely hairy. Seeds 2(–5), orange to red, slightly pruinose, obovoid or ellipsoid, somewhat compressed, (3–)4–5 × 2.5–3.5 cm, sparsely hairy; sclerotesta smooth, or longitudinally grooved.

Nutritive/Medicinal Properties

Two non-protein amino acids, cycasindene (3-[3'-amino-indenyl-2']-alanine) and cycasthioamide (N-[glycinyl-alaninyl-11-thio]-5-one-pipecolic acid), along with eight known non-protein amino acids, were isolated from the seeds of *Cycas revoluta* (Pan et al. 1997a, b). The seeds also contained ß-D-Glucosidase (Yagi et al. 1985a). This enzyme catalyzes the hydrolysis of terminal non-reducing residues in β-D-glucosides with the release of glucose.

The following flavonoids were found in *C. revoluta* leaves: 2,3-Dihydroamentoflavone, 2,3-dihydrohinokiflavone, amentoflavone, Hinokiflavone and sotetsuflavone (Geiger and Pfleiderer 1971). Leaves of *Cycas revoluta* were found to contain inhibitors of the human enzyme, cytochrome P-450 aromatase and thus may be efficacious in treating estrogen-dependent tumour (Kowalska et al. 1995). A novel lectin belonging to the jacalin-related lectin family was isolated from leaves of *Cycas revoluta* (Yagi et al. 2002). The lectin has a carbohydrate-binding specificity similar to those of mannose recognizing, jacalin-related lectins.

Volatiles emitted from the male and female cones of *C. revoluta* were found to be dominated by estragole (4-allylanisole) (67.0–92.7%), with small amounts of other benzenoids, e.g., anethole, methyl salicylate, methyl eugenol, and ethyl benzoate (Azuma and Kono 2006). Several fatty acid esters were also detected.

Antibacterial Activity

The chloroform extract of *C. revoluta* leaflets afforded 12 compounds (Moawad et al. 2010), seven of which are reported for the first time from this species. The isolated compounds included 14 biflavonoids, three lignans, three flavan-3-ols, two flavone-C-glucosides, two NOR-isoprenoids, and one flavanone. Several compounds 2, 6, and 18 displayed moderate antibacterial activity against *Staphylococcus aureus* (IC_{50} values of 3.9, 9.7, and 8.2 µM, respectively) and methicillin-resistant *S. aureus* (MRSA; IC_{50} values of 5.9, 12.5, and 11.5 µoM, respectively).

Anti-Nutritional Factors

Cycasin, a toxin and an azoxyglycoside was isolated from *Cycas revoluta* seeds and its structure determined by Nishida et al. (1955) and from the stem (Nishida et al. 1956). The crystal and molecular structure of Cycasin, (Z)-β-D-glucopyranosyloxy-NNO-azoxymethane was characterised by Tate et al. (1995). Cycasin is chemically closely related to macrozamin except for its sugar moiety which is D-glucose in cycasin. Other azoyglycosides isolated include neocycasin A (Nishida et al. 1959), neocycasin B and macrozamin from seeds (Nagahama et al. 1959). Other azoxyglycosides isolated from *C. revoluta* included neocycasin C formed by transglycosylation with cycad emulsion (Nagahama et al. 1960); and neocycasin E, β-cellobiosyloxy-azoxymethane (Nagahama et al. 1961). Numata et al. (1960) isolated several azoxyglycosides cyasin and its tetra-acetyl derivative and neocyasin A and B and also traces of macrozamin were from male strobili and seeds of *C. revoluta*.

The seeds also contained neurotoxins: neurotoxin ß-N-methylamino-L-alanine (BMAA) (Pan et al. 1997b; Krüger et al. 2010) and δ-N-oxalyl-ornithine (Pan et al. 1997a). Seeds of *Cycas revoluta* were found to contain 6.µg/g of free BMAA (Krüger et al. 2010). A new azoxyglycoside containing isomaltose: neocycasin Bα was isolated from a crude fraction (Yagi et al. 1985b). Yagi et al. (1985c) also reported on the formation of three new trisaccharide-azoxyglycosides by transglucosylation by a β-D-glucosidase from cycad seeds. These azoxyglycosides, named neocycasin H, I, and J, were identified as O-β-D-glucopyranosyl- (1→4)-O-β-D-glucopyranosyl-(1→3)-O-β-D-glucopyranoside of methylazoxymethanol (MAM), O-β-D-glucopyranosyl-(1→3)-[O-β-D-glucopyranosyl-(1→6)]-O-β-D-glucopyranoside of MAM, and O-β-D-glucopyranosyl-(1→3)-[O-β-D-xylopyranosyl-(1→6)]-O-β-D-glucopyranoside of MAM, respectively. Cycasin concentration was greatest in leaves and similarly for seeds and pith of *C. revoluta* while macrozamin was absent from seeds and present in low concentration in leaves and pith tissues (Yagi and Tadera 1987).

Cycasin and its aglycone methylazoxymethanol conjugates were reported to have toxic properties as alkylating agent such as neurotoxicity, hepatotoxicity and carcinogenicity in various experimental animals rats, mice, cattle, goats, pigs and monkeys (Laqueur 1964; Laqueur and Matsumoto 1966; Laqueur et al. 1967; Hirono and Shibuya 1967; Laqueur and Spatz. 1968; Hirono et al. 1968a, b; Sieber et al. 1980; Shimizu et al. 1986). Goats fed with cycasin (4–6 mg/kg daily) exhibited weakness and paretic signs in the hindquarters (Shimizu et al. 1986). They showed anorexia, weight loss, anaemia and decrease in serum γ glutamyl transpeptidase levels. Some also had hepatic lesions and leukomyelopathy consisting of demyelination and axonal swelling or disappearance in the spinal cord similar to that observed in cycad poisoning in grazing cattle ingesting the leaves (Kobayahsi et al. 1984). Kobayashi et al. (1984) and Yasuda et al. (1984) reported that cattle that ingested leaves exhibited ataxia and paralysis in hindquarters. Yasuda and Shimizu (1998) reported that ingestion of the leaves cause neurotoxicity to the CNS in cattle. Sieber et al. (1980) found that long-term administration of cycasin and/or methylazoxymethanol (MAM) acetate by oral and intra perintoneal routes was hepatotoxic (produced toxic hepatitis and cirrhosis) and carcinogenic, producing hepatocellular carcinomas, renal carcinomas, squamous cell carcinomas of the esophagus, and adenocarcinomas of the small intestine in old-world monkeys. Kono et al. (1985) reported that toxicity of cycasin in guinea pigs in the liver and other organs was very similar to those reported in other experimental animals. MAM and cycasin were also reported to be mutagenic using a modified Ames *Salmonella typhimurium* test (Matsushima et al. 1979).

Studies in Taiwan reported on the cyanogenic poisoning potential of *Cycas* seed ingestion in 21 human cases (Chang et al. 2004). Severe vomiting was the most striking symptom. All patients except one presented with gastrointestinal disturbance, and 90% sought medical care at the emergency department. Within 24 h, all patients had recovered. The levels of blood cyanide or thiocyanate were higher than normal in some cases but did not reach the toxic range. Spencer et al. (1987) reported on the association of cycad use and motor neurone disease (MND) in Kii peninsula of Japan. Subsequent studies by Iwami et al. (1993, 1994) did not support the hypothesis that cycad used as food, medicine, toys, and garden trees was an etiological factor of MND in Kii peninsula, that was common to the western Pacific foci. In the western Pacific, results of studies by Khabazian et al. (2002) supported the hypothesis that cycad consumption may be an important factor in the etiology of amyotrophic lateral sclerosis (ALS) -parkinsonism dementia complex (ALS-PDC), and further suggested that some sterol β-d-glucosides may be involved as potential neurotoxin in other neurodegenerative disorders. Studies by Kuzuhara (2007) found that there had been no differences in drinking water and food consumption between the residents in the high amyotrophic lateral sclerosis (ALS) incidence area and those in the neighbouring low incidence areas, and none of the patients had habits of eating the cycad, flying fox or any other odd materials. These findings suggested that genetic factors may be etiologically primary and environmental factors may modify the clinical phenotypes. The result of a very recent follow-up study reported by Kihira et al. (2010) indicated that the recent incidence of ALS in the southern part of the Kii Peninsula, especially in the Koza, Kozagawa, and Kushimoto area (K area) was high, the age of onset had recently become higher and the number of bulbar-onset patients had increased. All the above-mentioned findings were postulated to be attributed to an increase in the senility rate in the population. Between 2000 and 2008, the age-adjusted incidence in ALS for women in K area, especially in the Kozagawa district, was high, indicating an increase in that the incidence of ALS among women in this area after 2000. The factors responsible for the high incidence of ALS in this area remain to be clarified.

Traditional Medicinal Uses

According to Duke and Ayensu (1985), the terminal shoot is astringent and diuretic; the seed is emmenagogue, expectorant and tonic. It has been used in the treatment of rheumatism. Substances extracted from the seeds had been used to inhibit the growth of malignant tumours and the leaves were used in the treatment of cancer and hepatoma.

In China, the fruit had been used as expectorant and tonic. In Japan's Kii Peninsula, a "tonic" is made from the dried seeds *of C. revoluta*.

Other Uses

Cycas revoluta is the most popular ornamental and widely cultivated of the cycads. It makes an excellent landscape plant, as well being very well suited to pot culture, and even bonsai.

Sosotesu has also been used as manure, a pesticide and in floral arrangement and as substitute for medicine (Kira and Miyoshi 2000). Leaves are used in floral arrangement and for growing mushrooms. Seeds are sold as such and are used for planting out seedlings for sale as ornamentals.

Comments

Cycas revoluta is extremely poisonous to animals including domestic pets when ingested.

Selected References

Azuma H, Kono M (2006) Estragole (4-allylanisole) is the primary compound in volatiles emitted from the male and female cones of *Cycas revoluta*. J Plant Res 119:671–676

Bonta M, Osborne R (2007) Cycads in the vernacular: a compendium of local names. In: Proceedings of the 7th International Conference on Cycad Biology, Xalapa, Mexico, Jan 2005. Mem. New York Bot. Gard. 97:143–175

Chang SS, Chan YL, Wu ML, Deng JF, Chiu TF, Chen JC, Wang FL, Tseng CP (2004) Acute *Cycas* seed poisoning in Taiwan. Clin Toxicol 42(1):49–54

Chen J, Stevenson DW (1999) Cycadaceae persoon. In: Wu ZY, Raven PH (eds.) Flora of China, vol 4, Cycadaceae through Fagaceae. Science Press/Missouri Botanical Garden Press, Beijing/St. Louis

Council of Scientific and Industrial Research (1950) The wealth of India. A dictionary of Indian raw materials and industrial products (raw materials 2). Publications and Information Directorate, New Delhi

Duke JA, Ayensu ES (1985) Medicinal plants of China, vols 1 and 2. Reference Publications, Inc., Algonac 705 pp

Geiger H, de Pfleiderer WG (1971) Über 2,3-dihydrobiflavone in *Cycas revoluta*. Phytochem 10(8): 1936–1938

Hirono I, Shibuya C (1967) Induction of a neurological disorder by cycasin in mice. Nature 216:1311–1312

Hirono I, Laquer GL, Spatz M (1968a) Transplantability of cycasin-induced tmrours in rats with emphasis on nephroblastomas. J Natl Cancer Inst 40:1011–1025

Hirono I, Laquer GL, Spatz M (1968b) Tumour induction in Fischer and Osborne-Mendel rats by a single administration of cycasin. J Natl Cancer Inst 40:1003–1010

Iwami O, Niki Y, Watanabe T, Ikeda M (1993) Motor neuron disease on the Kii Peninsula of Japan: cycad exposure. Neuroepidemiol 12(6):307–312

Iwami O, Watanabe T, Moon CS, Nakatsuka H, Ikeda M (1994) Motor neuron disease on the Kii Peninsula of Japan: excess manganese intake from food coupled with low magnesium in drinking water as a risk factor. Sci Total Environ 149(1–2): 121–135

Jones DL (1993) Cycads of the world. New Holland, Australia, 312 pp

Khabazian I, Bains JS, Williams DE, Cheung J, Wilson JM, Pasqualotto BA, Pelech SL, Andersen RJ, Wang YT, Liu L, Nagai A, Kim SU, Craig UK, Shaw CA (2002) Isolation of various forms of sterol beta-D-glucosides from the seed of *Cycas circinalis*: neurotoxicity and implication for ALS-parkisonism dementia complex. J Neurochem 82(3):516–528

Kihira T, Yoshida S, Murata K, Ishiguti H, Kondo T, Kohmoto J, Okamoto K, Kokubo Y, Kuzuhara S (2010) Changes in the incidence and clinical features of ALS in the Koza, Kozagawa, and Kushimoto area of the Kii Peninsula–from the 1960s to the 2000s (follow-up study). Brain Nerve 62(1):72–80 (In Japanese)

Kira K, Miyoshi A (2000) Utilization of sotetsu (*Cycas revoluta*) in the Amami Islands. Res Bull Kagoshima Univ For 28:31–37 (In Japanese)

Kobayashi A, Tadera K, Yagi F, Kono I, Sakamoto T, Yasuda N (1984) Studies on poisoning of grazing cattle due to ingestion of *Cycas revoluta* Thunb. Bull Fac Agric Kagoshima Univ 34:119–129

Kono I, Takehara E, Shimizu T, Yasuda N, Kobayashi A, Tadera K, Yagi F (1985) Experimental studies on effects of cycasin on guinea pigs. Bull Fac Agric Kagoshima Univ 35:159–169

Kowalska MY, Itzhak Y, Puett D (1995) Presence of aromatase inhibitors in cycads. J Ethnopharmacol 47(3):113–116

Krüger T, Mönch B, Oppenhäuser S, Luckas B (2010) LC-MS/MS determination of the isomeric neurotoxins BMAA (beta-N-methylamino-L-alanine) and DAB (2,4-diaminobutyric acid) in cyanobacteria and seeds of *Cycas revoluta* and *Lathyrus latifolius*. Toxicon 55(2–3):547–557

Kuzuhara S (2007) Revisit to Kii ALS–the innovated concept of ALS-Parkinsonism-dementia complex, clinicopathological features, epidemiology and etiology. Brain Nerve 59(10): 1065–1074 (In Japanese)

Laqueur GL (1964) Carcinogenic effects of cycad meal and cycasin, methylazoxymenthanol glycoside in rats and effects of cycasin in germ free rats. Fed Proc 23:1386–1387

Laqueur GL, Matsumoto H (1966) Neoplasms in female Fischer rats following intraperitoneal injection with methylazoxymethanol (MAM) and synthetic MAM acetate. J Natl Cancer Inst 37:217–232

Laqueur GL, Spatz M (1968)Toxicology of cycasin. Cancer Res 28:2262–2267

Laqueur GL, McDaniel EG, Matsumoto H (1967) Tumour induction in germfree rats with injection with methylazoxymethanol (MAM) and synthetic MAM acetate. J Natl Cancer Inst 39:355–371

Matsushima T, Matsumoto H, Shirai A, Sawamura M, Sugimura T (1979) Mutagenicity of the naturally occurring carcinogen cycasin and synthetic methyl azoxy methanol conjugates in *Salmonella typhimurium*. Cancer Res 39:3780–3782

Moawad A, Hetta M, Zjawiony JK, Jacob MR, Hifnawy M, Marais JP, Ferreira D (2010) Phytochemical investigation of *Cycas circinalis* and *Cycas revoluta* leaflets: moderately active antibacterial biflavonoids. Planta Med 76(8):796–802

Nagahama T, Numata T, Nishida K (1959) Studies on some new azoxy glycosides of *Cycas revoluta* Thunb. II. Neocycasin B and macrozamin. Bull Agric Chem Soc Jpn 23:556–557

Nagahama T, Nishida K, Numata T (1960) Studies on some new azoxyglycosides of *Cycas revoluta* Thunb. VI. The structure of neocycasin C formed by transglycosylation with cycad emulsin. Bull Agric Chem Soc Jpn 24:536–537

Nagahama T, Nishida K, Numata T (1961) Neocycasin E, B-cellobiosyloxy-azoxymethane. Studies on some new azoxylglycosides of *Cycas revoluta* Thunb. VII. Agric Biol Chem 25:937–938

Nishida K, Kobayashi A, Nagahama T (1955) Cycasin, a new toxic glycoside of *Cycas revoluta* Thunb. I. Isolation and structure of cycasin. Bull Agric Chem Soc Jpn 19:77–84

Nishida K, Kobayashi A, Nagahama T (1956) Cycasin, a new toxic glycoside of *Cycas revoluta* Thunb. III. Isolation of cycasin from the stem of Japanese cycad. J Jpn Biochem Soc Seigaku 28:70–74

Nishida K, Kobayashi A, Nagahama T, Numata T (1959) Studies on some new azoxy glycosides of *Cycas revoluta* Thunb. I. On neocycasin A, beta-laminaribiosyloxyazoxymethane. Bull Agric Chem Soc Jpn 23:460–464

Numata T, Nagahama T, Nishida K (1960) Some new azoxyglycosides of *Cycas revoluta*. IV azoxyglycosides found in male strobil. Mem Fac Agric Kagoshima Univ 4:5–8

Pan M, Mabry TJ, Beale JM, Mamiya BM (1997a) Nonprotein amino acids from *Cycas revoluta*. Phytochem 45(3):517–519

Pan M, Mabry TJ, Cao P, Moini M (1997b) Identification of nonprotein amino acids from cycad seeds as N-ethoxycarbonyl ethyl ester derivatives by positive chemical-ionization gas chromatography-mass spectrometry. J Chromatogr A 787(1–2):288–294

Shimizu T, Yasuda N, Kono I, Yagi F, Tadera K, Kobayashi A (1986) Hepatic and spinal lesions in goats chronically intoxicated with cycasin. Jpn J Vet Sci 48(6):1291–1295

Sieber SM, Correa H, Dalgard DW, McIntyre ICR, Adamson RH (1980) Carcinogenicity and hepatotoxicity of cycasin and its aglycone methylazoxymethanol acetate in nonhuman primates. J Natl Cancer Inst 65:177–189

Spencer PS, Ohta M, Palmer VS (1987) Cycad use and motor neurone disease in Kii Peninsula of Japan. Lancet 330(8573):1462–1463

Tate ME, Delaere IM, Jones GP, Tiekink ERT (1995) Crystal and molecular structure of cycasin, (Z)-β-D-Glucopyranosyloxy-NNO-azoxymethane. Aust J Chem 48(5):1059–1063

Thieret JW (1958) Economic botany of the cycads. Econ Bot 12:3–4

Usher G (1974) A dictionary of plants used by man. Constable, London, 619 pp

Yagi F, Tadera K (1987) Azoxyglycoside contents in seeds of several species and various parts of Japanese cycad. Agric Biol Chem 51:1719–1722

Yagi F, Hatanaka M, Tadera K, Kobayashi A (1985a) β-D-glucosidase from seeds of Japanese cycad, *Cycas revoluta* Thunb.: properties and substrate specificity. J Biochem 97(1):119–126

Yagi F, Tadera K, Kobayashi A (1985b) A new azoxyglycoside containing isomaltose: neocycasin Bα. Agric Biol Chem 49(5):1531–1532

Yagi F, Tadera K, Kobayashi A (1985c) Formation of three new trisaccharide-azoxyglycosides by transglucosylation by cycad β-glucosidase. Agric Biol Chem 49(10):2985–2990

Yagi F, Iwaya T, Haraguchi T, Goldstein IJ (2002) The lectin from leaves of Japanese cycad, *Cycas revoluta* Thunb. (Gymnosperm) is a member of the jacalin-related family. FEBS J 269(17):4335–4341

Yasuda N, Shimizu T (1998) Cycad poisoning in cattle in Japan. Studies on spontaneous and experimental cases. J Toxicol Sci 23:126–128

Yasuda N, Kono I, Shimizu T, Kobayashi A, Tadera K, Yagi F (1984) Pathological studies on poisoning of grazing cattle due to ingestion of cycad, *Cycas revoluta* Thunb.: the lesions and their distribution in the spinal cord. Bull Fac Agric Kagoshima Univ 34:131–137

Ye ZY (1999) The investigation into the past and the current state of *Cycas revoluta* in Fujian. In: Chen CG (ed.). Biology and conservation of cycads. Proceedings of the 4th International Conference on Cycad Biology. International Academic Publishers, Beijing, pp 73–74

Medical Glossary

AAD allergic airway disease, an inflammatory disorder of the airways caused by allergens.

AAPH 2,2′-azobis(2-amidinopropane) dihydrochloride, a water-soluble azo compound used extensively as a free radical generator, often in the study of lipid peroxidation and the characterization of antioxidants.

Abeta aggregation amyloid beta protein (Abeta) aggregation is associated with Alzheimer's disease (AD), it is a major component of the extracellular plaque found in AD brains.

Abdominal distension referring to generalised distension of most or all of the abdomen. Also referred to as stomach bloating often caused by a sudden increase in fibre from consumption of vegetables, fruits and beans.

Ablation therapy the destruction of small areas of myocardial tissue, usually by application of electrical or chemical energy, in the treatment of some tachyarrhythmias.

Abortifacient a substance that causes or induces abortion.

Abortivum a substance inducing abortion.

Abscess a swollen infected, inflamed area filled with pus in body tissues.

ABTS 2.2 azinobis-3-ethylhenthiazoline-6-sulfonic acid, a type of mediator in chemical reaction kinetics of specific enzymes.

ACAT acyl CoA: cholesterol acyltransferase.

ACE see angiotensin-converting enzyme.

Acetogenins natural products from the plants of the family Annonaceae; are very potent inhibitors of the NADH-ubiquinone reductase (Complex I) activity of mammalian mitochondria.

Acetylcholinesterase (AChE) is an enzyme that degrades (through its hydrolytic activity) the neurotransmitter acetylcholine, producing choline.

Acne vulga´ris also known as chronic acne, usually occurring in adolescence, with comedones (blackheads), papules (red pimples), nodules (inflamed acne spots), and pustules (small inflamed pus-filled lesions) on the face, neck, and upper part of the trunk.

Acidosis increased acidity.

Acquired immunodeficiency syndrome (AIDS) an epidemic disease caused by an infection by human immunodeficiency virus (HIV-1, HIV-2), retrovirus that causes immune system failure and debilitation and is often accompanied by infections such as tuberculosis.

Acridone an organic compound based on the acridine skeleton, with a carbonyl group at the 9 position.

ACTH adrenocorticotropic hormone (or corticotropin), a polypeptide tropic hormone produced and secreted by the anterior pituitary gland.

Activating transcription factor (ATF) a protein (gene) that binds to specific DNA sequences regulating the transfer or transcription of information from DNA to mRNA.

Activator protein-1 (AP-1) a heterodimeric protein transcription factor that regulates gene expression in response to a variety of stimuli, including cytokines, growth factors, stress, and bacterial and viral infections. AP-1 in turn regulates a number of cellular processes including differentiation, proliferation, and apoptosis.

Acyl-CoA dehydrogenases a group of enzymes that catalyzes the initial step in each cycle of fatty acid β-oxidation in the mitochondria of cells.

Adaptogen a term used by herbalists to refer to a natural herb product that increases the body's resistance to stresses such as trauma, stress and fatigue.

Adaptogenic increasing the resistance of the body to stress.

Addison's disease is a rare endocrine disorder. It occurs when the adrenal glands cannot produce sufficient hormones (corticosteroids). It is also known as chronic adrenal insufficiency, hypocortisolism or hypocorticism.

Adenocarcinoma a cancer originating in glandular tissue.

Adenoma a benign tumour from a glandular origin.

Adenosine receptors a class of purinergic, G-protein coupled receptors with adenosine as endogenous ligand. In humans, there are four adenosine receptors. A_1 receptors and A_{2A} play roles in the heart, regulating myocardial oxygen consumption and coronary blood flow, while the A_{2A} receptor also has broader anti-inflammatory effects throughout the body. These two receptors also have important roles in the brain, regulating the release of other neurotransmitters such as dopamine and glutamate, while the A_{2B} and A_3 receptors are located mainly peripherally and are involved in inflammation and immune responses.

ADH see alcohol dehydrogenase.

Adipocyte a fat cell involved in the synthesis and storage of fats.

Adiponectin a protein in humans that modulates several physiological processes, such as metabolism of glucose and fatty acids, and immune responses.

Adipose tissues body fat, loose connective tissue composed of adipocytes (fat cells).

Adoptogen containing smooth pro-stressors which reduce reactivity of host defense systems and decrease damaging effects of various stressors due to increased basal level of mediators involved in the stress response.

Adrenal glands star-shaped endocrine glands that sit on top of the kidneys.

Adrenalectomized having had the adrenal glands surgically removed.

Adrenergic having to de with adrenaline (epinephrine) and/or noradrenaline (norepinephrine).

Adrenergic receptors a class of G protein-coupled receptors that are targets of the noradrenaline (norepinephrine) and adrenaline (epinephrine).

Adulterant an impure ingredient added into a preparation.

Advanced Glycation End products (AGEs) resultant products of a chain of chemical reactions after an initial glycation reaction. AGEs may play an important adverse role in process of atherosclerosis, diabetes, aging and chronic renal failure.

Aegilops an ulcer or fistula in the inner corner of the eye.

Afferent something that so conducts or carries towards, such as a blood vessel, fibre, or nerve.

Agammaglobulinaemia an inherited disorder in which there are very low levels of protective immune proteins called immunoglobulins. Cf. x-linked agammaglobulinaemia.

Agalactia lack of milk after parturition (birth).

Agglutinin a protein substance, such as an antibody, that is capable of causing agglutination (clumping) of a particular antigen.

Agglutination clumping of particles.

Agonist a drug that binds to a receptor of a cell and triggers a response by the cell.

Ague a fever (such as from malaria) that is marked by paroxysms of chills, fever, and sweating that recurs with regular intervals.

AHR AhR, aryl hydrocarbon receptor, a cytosolic protein transcription factor.

AIDS see Acquired Immunodeficiency Syndrome.

Akathisia a movement disorder often caused by long-term use of antipsychotic medications.

Akt signaling pathway Akt are protein kinases involved in mammalian cellular signaling, inhibits apoptotic processes.

Akt/FoxO pathway Cellular processes involving Akt and FoxO transcription factors that play a role in angiogenesis and vasculogenesis.

Alanine transaminase (ALT) also called Serum Glutamic Pyruvate Transaminase (SGPT) or Alanine aminotransferase (ALAT), an enzyme present in hepatocytes (liver cells). When a cell is damaged, it leaks this enzyme into the blood.

ALAT, (Alanine aminotransferase) see Alanine transaminase.

Albumin water soluble proteins found in egg white, blood serum, milk, various animal tissues and plant juices and tissues.

Albuminaria excessive amount of albumin in the urine, a symptom of severe kidney disease.

Aldose reductase, aldehyde reductase an enzyme in carbohydrate metabolism that converts glucose to sorbitol.

Alexipharmic an antidote, remedy for poison.

Alexiteric a preservative against contagious and infectious diseases, and the effects of poisons.

Alcohol dehydrogenase (ADH) an enzyme involved in the break-down of alcohol.

Alkaline phosphatase (ALP) an enzyme in the cells lining the biliary ducts of the liver. ALP levels in plasma will rise with large bile duct obstruction, intrahepatic cholestasis or infiltrative diseases of the liver. ALP is also present in bone and placental tissues.

Allergenic having the properties of an antigen (allergen), immunogenic.

Allergic pertaining to, caused, affected with, or the nature of the allergy.

Allergy a hypersensitivity state induced by exposure to a particular antigen (allergen) resulting in harmful immunologic reactions on subsequent exposures. The term is usually used to refer to hypersensitivity to an environmental antigen (atopic allergy or contact dermatitis) or to drug allergy.

Allogeneic cells or tissues which are genetically different because they are derived from separate individuals of the same species. Also refers to a type of immunological reaction that occurs when cells are transplanted into a genetically different recipient.

Allostasis the process of achieving stability, or homeostasis, through physiological or behavioral change.

Alopecia is the loss of hair on the body.

Alopecia areata is a particular disorder affecting hair growth (loss of hair) in the scalp and elsewhere.

ALP see Alkaline phosphatase.

Alpha-adrenoceptor receptors postulated to exist on nerve cell membranes of the sympathetic nervous system in order to explain the specificity of certain agents that affect only some sympathetic activities (such as vasoconstriction and relaxation of intestinal muscles and contraction of smooth muscles).

Alpha amylase α-amylase a major form of amylase found in humans and other mammals that cleaves alpha-bonds of large sugar molecules.

ALT see Alanine transaminase.

Alterative a medication or treatment which gradually induces a change, and restores healthy functions without sensible evacuations.

Alveolar macrophage a vigorously phagocytic macrophage on the epithelial surface of lung alveoli that ingests carbon and other inhaled particulate matter. Also called coniophage or dust cell.

Alzheimer's disease a degenerative, organic, mental disease characterized by progressive brain deterioration and dementia, usually occurring after the age of 50.

Amastigote refers to a cell that does not have any flagella, used mainly to describe a certain phase in the life-cycle of trypanosome protozoans.

Amenorrhea the condition when a woman fails to have menstrual periods.

Amidolytic cleavage of the amide structure.

Amoebiasis state of being infected by amoeba such as *Entamoeba histolytica*.

Amoebicidal lethal to amoeba.

Amyloid beta (Aβ or Abeta) a peptide of 39–43 amino acids that appear to be the main constituent of amyloid plaques in the brains of Alzheimer's disease patients.

Amyotrophy progressive wasting of muscle tissues. *adj.* amyotrophic.

Anaemia a blood disorder in which the blood is deficient in red blood cells and in haemoglobin.

Anaesthesia condition of having sensation temporarily suppressed.

Anaesthetic a substance that decreases partially or totally nerve the sense of pain.

Analeptic a central nervous system (CNS) stimulant medication.

Analgesia term describing relief, reduction or suppression of pain. *adj.* analgetic.

Analgesic a substance that relieves or reduces pain.

Anaphoretic an antiperspirant.

Anaphylaxis a severe, life-threatening allergic response that may be characterized by symptoms such as reduced blood pressure, wheezing, vomiting or diarrhea.

Anaphylactic *adj.* see anaphylaxis.

Anaphylotoxins are fragments (C3a, C4a or C5a) that are produced during the pathways of the complement system. They can trigger release of substances of endothelial cells, mast cells or phagocytes, which produce a local inflammatory response.

Anaplasia a reversion of differentiation in cells and is characteristic of malignant neoplasms (tumours).

Anaplastic *adj.* see anaplasia.

Anasarca accumulation of great quantity of fluid in body tissues.

Androgen male sex hormone in vertebrates. Androgens may be used in patients with breast cancer to treat recurrence of the disease.

Angina pectoris, Angina chest pain or chest discomfort that occurs when the heart muscle does not get enough blood.

Angiogenesis a physiological process involving the growth of new blood vessels from pre-existing vessels.

Angiogenic *adj.* see angiogenesis.

Angiotensin an oligopeptide hormone in the blood that causes blood vessels to constrict, and drives blood pressure up. It is part of the renin-angiotensin system.

Angiotensin-converting enzyme (ACE) an exopeptidase, a circulating enzyme that participates in the body's renin-angiotensin system (RAS) which mediates extracellular volume (i.e. that of the blood plasma, lymph and interstitial fluid), and arterial vasoconstriction.

Ankylosing spondylitis (AS) is a type of inflammatory arthritis that targets the joints of the spine.

Annexitis also called adnexitis is a pelvic inflammatory disease involving the inflammation of the ovaries or fallopian tubes.

Anodyne a substance that relieves or soothes pain by lessening the sensitivity of the brain or nervous system. Also called an analgesic.

Anorectal relating to the rectum and anus.

Anorectics appetite suppressants, substances which reduce the desire to eat. Used on a short term basis clinically to treat obesity. Also called anorexigenics.

Anorexia lack or loss of desire to eat.

Anorexic having no appetite to eat.

Anorexigenics see anorectics.

Antagonist a substance that acts against and blocks an action.

Antalgic a substance used to relive a painful condition.

Antecubital vein This vein is located in the antecubital fossa -the area of the arm in front of the elbow.

Anterior uveitis is the most common form of ocular inflammation that often causes a painful red eye.

Anthelmintic an agent or substance that is destructive to worms and used for expulsion of internal parasitic worms in animals and humans.

Anthocyanins a subgroup of antioxidant flavonoids, are glucosides of anthocyanidins. Which are beneficial to health. They occur as water-soluble vacuolar pigments that may appear red, purple, or blue according to pH in plants.

Anthrax a bacterial disease of cattle and ship that can be transmitted to man though unprocessed wool.

Anthropometric pertaining to the study of human body measurements.

Antiamoebic a substance that destroys or suppresses parasitic amoebae.

Antiamyloidogenic compounds that inhibit the formation of Alzheimer's β-amyloid fibrils (fAβ) from amyloid β-peptide (Aβ) and destabilize fAβ.

Antianaphylactic agent that can prevent the occurrence of anaphylaxis (life threatening allergic response).

Antiangiogenic a drug or substance used to stop the growth of tumours and progression of cancers by limiting the pathologic formation of new blood vessels (angiogenesis).

Antiarrhythmic a substance to correct irregular heartbeats and restore the normal rhythm.

Antiasmathic drug that treats or ameliorates asthma.

Antiatherogenic that protects against atherogenesis, the formation of atheromas (plaques) in arteries.

Antibacterial substance that kills or inhibits bacteria.

Antibilious an agent or substance which helps remove excess bile from the body.

Antibiotic a chemical substance produced by a microorganism which has the capacity to inhibit the growth of or to kill other microorganisms.

Antiblennorrhagic a substance that treats blenorrhagia a conjunctival inflammation resulting in mucus discharge.

Antibody a gamma globulin protein produced by a kind of white blood cell called the plasma cell in the blood used by the immune system to identify and neutralize foreign objects (antigen).

Anticarcinomic a substance that kills or inhibits carcinomas (any cancer that arises in epithelium/tissue cells).

Anticephalalgic headache-relieving or preventing.

Anticestodal a chemical destructive to tapeworms.

Anticholesterolemic a substance that can prevent the build up of cholesterol.

Anticlastogenic having a suppressing effect of chromosomal aberrations.

Anticoagulant a substance that thins the blood and acts to inhibit blood platelets from sticking together.

Antidepressant a substance that suppresses depression or sadness.

Antidiabetic a substance that prevents or alleviates diabetes. Also called antidiabetogenic.

Antidiarrhoeal having the property of stopping or correcting diarrhoea, an agent having such action.

Antidote a remedy for counteracting a poison.

Antidopaminergic a term for a chemical that prevents or counteracts the effects of dopamine.

Antidrepanocytary anti-sickle cell anaemia.

Antidysenteric an agent used to reduce or treat dysentery and diarrhea.

Antidyslipidemic agent that will reduce the abnormal amount of lipids and lipoproteins in the blood.

Anti-edematous reduces or suppresses edema.

Anti-emetic an agent that stops vomiting.

Antifebrile a substance that reduces fever, also called antipyretic.

Antifeedant preventing something from being eaten.

Antifertility agent that inhibits formation of ova and sperm and disrupts the process of fertilization (antizygotic).

Antifilarial effective against human filarial worms.

Antifungal an agent that kills or inhibits the growth of fungi.

Antigen a substance that prompts the production of antibodies and can cause an immune response. *adj.* antigenic.

Antigenotoxic an agent that inhibits DNA adduct formation, stimulates DNA repair mechanisms, and possesses antioxidant functions.

Antiganacratia anti- menstruation.

Antigastralgic preventing or alleviating gastric colic.

Antihematic agent that stops vomiting.

Antihemorrhagic an agent which stops or prevents bleeding.

Antihepatotoxic counteracting injuries to the liver.

Antiherpetic having activity against Herpes Simplex Virus (HSV).

Antihistamine an agent used to counteract the effects of histamine production in allergic reactions.

Antihyperalgesia the ability to block enhanced sensitivity to pain, usually produced by nerve injury or inflammation, to nociceptive stimuli. *adj.* antihyperalgesic.

Antihypercholesterolemia term to describe lowering of cholesterol level in the blood or blood serum.

Antihypercholesterolemic agent that lowers chlosterol level in the blood or blood serum.

Antihyperlidemic promoting a reduction of lipid levels in the blood, or an agent that has this action.

Antihypersensitive a substance used to treat excessive reactivity to any stimuli.

Antihypertensive a drug used in medicine and pharmacology to treat hypertension (high blood pressure).

Antiinflammatory a substance used to reduce or prevent inflammation.

Antileishmanial inhibiting the growth and proliferation of *Leishmania* a genus of flagellate protozoans that are parasitic in the tissues of vertebrates.

Antileprotic therapeutically effective against leprosy.

Antilithiatic an agent that reduces or suppresses urinary calculi (stones) and acts to dissolve those already present.

Antileukaemic anticancer drugs that are used to treat leukemia.

Antimalarial an agent used to treat malaria and/or kill the malaria-causing organism, *Plasmodium* spp.

Antimelanogenesis obstruct production of melanin.

Antimicrobial a substance that destroys or inhibits growth of disease-causing bacteria, viruses, fungi and other microorganisms.

Antimitotic inhibiting or preventing mitosis.

Antimutagenic an agent that inhibits mutations.

Antimycotic antifungal.

Antineoplastic said of a drug intended to inhibit or prevent the maturation and proliferation of neoplasms that may become malignant, by targeting the DNA.

Antineuralgic a substance that stops intense intermittent pain, usually of the head or face, caused by neuralgia.

Antinociception reduction in pain: a reduction in pain sensitivity produced within neurons when an endorphin or similar opium-containing substance opioid combines with a receptor.

Antinociceptive having an analgesic effect.

Antinutrient are natural or synthetic compounds that interfere with the absorption of nutrients and are commonly found in food sources and beverages.

Antioestrogen a substance that inhibits the biological effects of female sex hormones.

Antiophidian anti venoms of snake.

Antiosteoporotic substance that can prevent osteoporosis.

Antiovulatory substance suppressing ovulation.

Antioxidant a chemical compound or substance that inhibits oxidation and protects against free radical activity and lipid oxidation such as vitamin E, vitamin C, or beta-carotene (converted to vitamin B), carotenoids and flavonoids which are thought to protect body cells from the damaging effects of oxidation. Many foods including fruit and vegetables contain compounds with antioxidant properties. Antioxidants may also reduce the risks of cancer and age-related macular degeneration(AMD).

Antipaludic antimalarial.

Antiperiodic substance that prevents the recurrence of symptoms of a disease e.g. malaria.

Antiperspirant a substance that inhibits sweating. Also called antisudorific, anaphoretic.

Antiphlogistic a traditional term for a substance used against inflammation, an anti-inflammatory.

Antiplatelet agent drug that decreases platelet aggregation and inhibits thrombus formation.

Antiplasmodial suppressing or destroying plasmodia.

Antiproliferative preventing or inhibiting the reproduction of similar cells.

Antiprostatic drug to treat the prostate.

Antiprotozoal suppressing the growth or reproduction of protozoa.

Antipruritic alleviating or preventing itching.

Antipyretic a substance that reduces fever or quells it. Also known as antithermic.

Antirheumatic relieving or preventing rheumatism.

Antiscorbutic a substance or plant rich in vitamin C that is used to counteract scurvy.
Antisecretory inhibiting or diminishing secretion.
Antisense refers to antisense RNA strand because its sequence of nucleotides is the complement of message sense. When mRNA forms a duplex with a complementary antisense RNA sequence, translation of the mRNA into the protein is blocked. This may slow or halt the growth of cancer cells.
Antiseptic preventing decay or putrefaction, a substance inhibiting the growth and development of microorganisms.
Anti-sickling agent an agent used to prevent or reverse the pathological events leading to sickling of erythrocytes in sickle cell conditions.
Antispasmodic a substance that relieves spasms or inhibits the contraction of smooth muscles; smooth muscle relaxant, muscle-relaxer.
Antispermatogenic preventing or suppressing the production of semen or spermatozoa.
Antisudorific see antiperspirant.
Antisyphilitic a drug (or other chemical agent) that is effective against syphilis.
Antithermic a substance that reduces fever and temperature. Also known as antipyretic.
Antithrombotic preventing or interfering with the formation of thrombi.
Antitoxin an antibody with the ability to neutralize a specific toxin.
Antitumoral substance that acts against the growth, development or spread of a tumour.
Antitussive a substance that depresses coughing.
Antiulcerogenic an agent used to protect against the formation of ulcers, or is used for the treatment of ulcers.
Antivenin an agent used against the venom of a snake, spider, or other venomous animal or insect.
Antivinous an agent or substance that treats addiction to alcohol.
Antiviral substance that destroys or inhibits the growth and viability of infectious viruses.
Antivomitive a substance that reduces or suppresses vomiting.
Antizygotic see antifertility.
Anuria absence of urine production.
Anxiolytic a drug prescribed for the treatment of symptoms of anxiety.
Aperient a substance that acts as a mild laxative by increasing fluids in the bowel.
Aperitif an appetite stimulant.
Aphonia loss of the voice resulting from disease, injury to the vocal cords, or various psychological causes, such as hysteria.
Aphrodisiac an agent that increases sexual activity and libido and/or improves sexual performance.
Apnoea suspension of external breathing.
Apolipoprotein B (APOB) primary apolipoprotein of low-density lipoproteins which is responsible for carrying cholesterol to tissues.
Apoplexy a condition in which the brain's function stops with loss of voluntary motion and sense.
Appendicitis is a condition characterized by inflammation of the appendix. Also called epityphlitis.
Appetite stimulant a substance to increase or stimulate the appetite. Also called aperitif.
Aphthae white, painful oral ulcer of unknown cause.
Apthous ulcer canker sore in the lining of the mouth.
Aphthous stomatitis a canker sore, a type of painful oral ulcer or sore inside the mouth or upper throat, caused by a break in the mucous membrane. Also called aphthous ulcer.
Apolipoprotein A-I (APOA1) a major protein component of high density lipoprotein (HDL) in plasma. The protein promotes cholesterol efflux from tissues to the liver for excretion.
Apolipoprotein B (APOB) is the primary apolipoprotein of low-density lipoproteins (LDL or "bad cholesterol"), which is responsible for carrying cholesterol to tissues.
Apolipoprotein E (APOE) the apolipoprotein found on intermediate density lipoprotein and chylomicron that binds to a specific receptor on liver and peripheral cells.
Apoptogenic ability to cause death of cells.
Apoptosis death of cells.
Aphthous ulcer also known as a canker sore, is a type of oral ulcer, which presents as a

painful open sore inside the mouth or upper throat.

Apurinic lyase a DNA enzyme that catalyses a chemical reaction.

Arachidonate cascade includes the cyclooxygenase (COX) pathway to form prostanoids and the lipoxygenase (LOX) pathway to generate several oxygenated fatty acids, collectively called eicosanoids.

Ariboflavinosis a condition caused by the dietary deficiency of riboflavin that is characterized by mouth lesions, seborrhea, and vascularization.

Aromatase an enzyme involved in the production of estrogen that acts by catalyzing the conversion of testosterone (an androgen) to estradiol (an estrogen). Aromatase is located in estrogen-producing cells in the adrenal glands, ovaries, placenta, testicles, adipose (fat) tissue, and brain.

Aromatic having a pleasant, fragrant odour.

Aromatherapy a form of alternative medicine that uses volatile liquid plant materials, such as essential oils and other scented compounds from plants for the purpose of affecting a person's mood or health.

Arrhythmias abnormal heart rhythms that can cause the heart to pump less effectively. Also called dysrhythmias.

Arsenicosis see arsenism.

Arsenism an incommunicable disease resulting from the ingestion of ground water containing unsafe levels of arsenic, also known as arsenicosis.

Arteriosclerosis imprecise term for various disorders of arteries, particularly hardening due to fibrosis or calcium deposition, often used as a synonym for atherosclerosis.

Arthralgia is pain in the joints from many possible causes.

Arthritis inflammation of the joints of the body.

ASAT or AST aspartate aminotransferase, see aspartate transaminase.

Ascaris a genus of parasitic intestinal round worms.

Ascites abnormal accumulation of fluid within the abdominal or peritoneal cavity.

Ascorbic acid See vitamin C.

Aspartate transaminase (AST) also called Serum Glutamic Oxaloacetic Transaminase (SGOT) or aspartate aminotransferase (ASAT) is similar to ALT in that it is another enzyme associated with liver parenchymal cells. It is increased in acute liver damage, but is also present in red blood cells, and cardiac and skeletal muscle and is therefore not specific to the liver.

Asphyxia failure or suppression of the respiratory process due to obstruction of air flow to the lungs or to the lack of oxygen in inspired air.

Asphyxiation the process of undergoing asphyxia.

Asthenia a nonspecific symptom characterized by loss of energy, strength and feeling of weakness.

Asthenopia weakness or fatigue of the eyes, usually accompanied by headache and dimming of vision. *adj.* asthenopic.

Asthma a chronic illness involving the respiratory system in which the airway occasionally constricts, becomes inflamed, and is lined with excessive amounts of mucus, often in response to one or more triggers.

Astringent a substance that contracts blood vessels and certain body tissues (such as mucous membranes) with the effect of reducing secretion and excretion of fluids and/or has a drying effect.

Ataxia (loss of co-ordination) results from the degeneration of nerve tissue in the spinal cord and of nerves that control muscle movement in the arms and legs.

ATF-2 activating transcription factor 2.

Athlete's foot a contagious skin disease caused by parasitic fungi affecting the foot, hands, causing itching, blisters and cracking. Also called dermatophytosis.

Atherogenic having the capacity to start or accelerate the process of atherogenesis.

Atherogenesis the formation of lipid deposits in the arteries.

Atheroma a deposit or degenerative accumulation of lipid-containing plaques on the innermost layer of the wall of an artery.

Atherosclerosis the condition in which an artery wall thickens as the result of a build-up of fatty materials such as cholesterol.

Atherothrombosis medical condition characterized by an unpredictable, sudden disruption (rupture or erosion/fissure) of an atherosclerotic plaque, which leads to platelet activation and thrombus formation.

Atonic lacking normal tone or strength.

Atony insufficient muscular tone.

Atresia a congenital medical condition in which a body orifice or passage in the body is abnormally closed or absent.

Atretic ovarian follicles an involuted or closed ovarian follicle.

Atrial fibrillation is the most common cardiac arrhythmia (abnormal heart rhythm) and involves the two upper chambers (atria) of the heart.

Attention-deficit hyperactivity disorder (ADHD, ADD or AD/HD) is a neurobehavioral developmental disorder, primarily characterized by "the co-existence of attentional problems and hyperactivity.

Augmerosen a drug that may kill cancer cells by blocking the production of a protein that makes cancer cells live longer. Also called bcl-2 antisense oligonucleotide.

Auricular of or relating to the auricle or the ear in general.

Aurones [2-benzylidenebenzofuran-3(2H)-ones] are the secondary plant metabolites and is a subgroup of flavonoids. See flavonoids.

Autolysin an enzyme that hydrolyzes and destroys the components of a biological cell or a tissue in which it is produced.

Autophagy digestion of the cell contents by enzymes in the same cell.

Autopsy examination of a cadaver to determine or confirm the cause of death.

Avidity Index describes the collective interactions between antibodies and a multivalent antigen.

Avulsed teeth is tooth that has been knocked out.

Ayurvedic traditional Hindu system of medicine based largely on homeopathy and naturopathy.

Azoospermia is the medical condition of a male not having any measurable level of sperm in his semen.

Azotaemia a higher than normal blood level of urea or other nitrogen containing compounds in the blood.

Babesia a protozoan parasite (malaria–like) of the blood that causes a hemolytic disease known as Babesiosis.

Babesiosis malaria-like parasitic disease caused by Babesia, a genus of protozoal piroplasms.

Bactericidal lethal to bacteria.

Balanitis is an inflammation of the glans (head) of the penis.

BALB/c mice Balb/c mouse was developed in 1923 by McDowell. It is a popular strain and is used in many different research disciplines, but most often in the production of monoclonal antibodies.

Balm aromatic oily resin from certain trees and shrubs used in medicine.

Baroreceptor a type of interoceptor that is stimulated by pressure changes, as those in blood vessel wall.

Barrett's esophagus (Barrett esophagitis) a disorder in which the lining of the esophagus is damaged by stomach acid.

Basophil a type of white blood cell with coarse granules within the cytoplasm and a bilobate (two-lobed) nucleus.

BCL-2 a family of apoptosis rtegulator proteins in humans encoded by the B-cell lymphoma 2 (BCL-2) gene.

BCL-2 antisense oligonucleotide see augmerosen.

BCR/ABL a chimeric oncogene, from fusion of BCR and ABL cancer genes associated with chronic myelogenous leukaemia.

Bechic a remedy or treatment of cough.

Belching, or burping refers to the noisy release of air or gas from the stomach through the mouth.

Beri-beri is a disease caused by a deficiency of thiamine (vitamin B_1) that affects many systems of the body, including the muscles, heart, nerves, and digestive system.

Beta-carotene naturally-occurring retinol (vitamin A) precursor obtained from certain fruits and vegetables with potential antineoplastic and chemopreventive activities. As an antioxidant, beta carotene inhibits free-radical damage to DNA. This agent also induces cell differentiation and apoptosis of some tumour cell types, particularly in early stages of tumorigenesis, and enhances immune system activity by stimulating the release of natural killer cells, lymphocytes, and monocytes.

Beta-thalassemia an inherited blood disorder that reduces the production of hemoglobin.

Beta-lactamase enzymes produced by some bacteria that are responsible for their resistance to beta-lactam antibiotics like penicillins.

BHT butylated hydroxytoluene (phenolic compound), an antioxidant used in foods, cosmetics, pharmaceuticals, and petroleum products.

Bifidobacterium is a genus of Gram-positive, non-motile, often branched anaerobic bacteria. Bifidobacteria are one of the major genera of bacteria that make up the gut flora. Bifidobacteria aid in digestion, are associated with a lower incidence of allergies and also prevent some forms of tumour growth. Some bifidobacteria are being used as probiotics.

Bile fluid secreted by the liver and discharged into the duodenum where it is integral in the digestion and absorption of fats.

Bilharzia, bilharziosis see Schistosomiasis.

Biliary relating to the bile or the organs in which the bile is contained or transported.

Biliary infections infection of organ(s) associated with bile, comprise: (a) acute cholecystitis: an acute inflammation of the gallbladder wall; (b) cholangitis: inflammation of the bile ducts.

Biliousness old term used in the eighteenth and nineteenth centuries pertaining to bad digestion, stomach pains, constipation, and excessive flatulence.

Bilirubin a breakdown product of heme (a part of haemoglobin in red blood cells) produced by the liver that is excreted in bile which causes a yellow discoloration of the skin and eyes when it accumulates in those organs.

Biotin also known as vitamin B7. See vitamin B7.

Bitter a medicinal agent with a bitter taste and used as a tonic, alterative or appetizer.

Blackhead see comedone.

Blain see chilblain.

Blastocyst blastocyst is an embryonic structure formed in the early embryogenesis of mammals, after the formation of the morula, but before implantation.

Blastocystotoxic agent that suppresses further development of the blastocyst through to the ovum stage.

Blebbing Bulging e.g. membrane blebbing also called membrane bulging or ballooning.

Bleeding diathesis is an unusual susceptibility to bleeding (hemorrhage) due to a defect in the system of coagulation.

Blennorrhagia gonorrhea.

Blennorrhea inordinate discharge of mucus, especially a gonorrheal discharge from the urethra or vagina.

Blepharitis inflammation of the eyelids.

Blister thin vesicle on the skin containing serum and caused by rubbing, friction or burn.

Blood brain barrier (BBB) is a separation of circulating blood and cerebrospinal fluid (CSF) in the central nervous system (CNS).

Boil localized pyrogenic, painful infection, originating in a hair follicle.

Borborygmus rumbling noise caused by the muscular contractions of peristalsis, the process that moves the contents of the stomach and intestines downward.

Bowman Birk inhibitors type of serine proteinase inhibitor.

Bouillon a broth in French cuisine.

Bradicardia as applied to adult medicine, is defined as a resting heart rate of under 60 beats per minute.

Bradyphrenia referring to the slowness of thought common to many disorders of the brain.

Brain derived neutrophic factor (BDNF) a protein member of the neutrophin family that plays an important role in the growth, maintenance, function and survival of neurons. The protein molecule is involved in the modulation of cognitive and emotional functions and in the treatment of a variety of mental disorders.

Bright's disease chronic nephritis.

Bronchial inflammation see bronchitis.

Bronchiectasis a condition in which the airways within the lungs (bronchial tubes) become damaged and widened.

Bronchitis is an inflammation of the main air passages (bronchi) to your lungs.

Bronchoalveolar lavage (BAL) a medical procedure in which a bronchoscope is passed through the mouth or nose into the lungs and fluid is squirted into a small part of the lung and then recollected for examination.

Bronchopneumonia or bronchial pneumonia; inflammation of the lungs beginning in the terminal bronchioles.

Broncho-pulmonary relating to the bronchi and lungs.

Bronchospasm is a difficulty in breathing caused by a sudden constriction of the muscles in the walls of the bronchioles as occurs in asthma.

Brown fat brown adipose tissue (BAT) in mammals, its primary function is to generate body heat in animals or newborns that do not shiver.

Bubo inflamed, swollen lymph node in the neck or groin.

Buccal of or relating to the cheeks or the mouth cavity.

Bullae blisters; circumscribed, fluid-containing, elevated lesions of the skin, usually more than 5 mm in diameter.

Bursitis condition characterized by inflammation of one or more bursae (small sacs) of synovial fluid in the body.

C-jun NH(2)-terminal kinase enzymes that belong to the family of the MAPK superfamily of protein kinases. These kinases mediate a plethora of cellular responses to such stressful stimuli, including apoptosis and production of inflammatory and immunoregulatory cytokines in diverse cell systems. *cf:* MAPK.

c-Src a cellular non-receptor tyrosine kinase.

Cachexia physical wasting with loss of weight, muscle atrophy, fatigue, weakness caused by disease.

Caco-2 cell line a continuous line of heterogeneous human epithelial colorectal adenocarcinoma cells.

Cadaver a dead body, corpse.

Ca^{2+} ATPase (PMCA) is a transport protein in the plasma membrane of cells that serves to remove calcium (Ca^{2+}) from the cell.

Calcium (Ca) is the most abundant mineral in the body found mainly in bones and teeth. It is required for muscle contraction, blood vessel expansion and contraction, secretion of hormones and enzymes, and transmitting impulses throughout the nervous system. Dietary sources include milk, yoghurt, cheese, Chinese cabbage, kale, broccoli, some green leafy vegetables, fortified cereals, beverages and soybean products.

Calcium channel blockers (CCBs) a class of drugs and natural substances that disrupt the calcium ($Ca2+$) conduction of calcium channels.

Calculus (calculi) hardened, mineral deposits that can form a blockage in the urinary system.

Calculi infection most calculi arise in the kidney when urine becomes supersaturated with a salt that is capable of forming solid crystals. Symptoms arise as these calculi become impacted within the ureter as they pass toward the urinary bladder.

Caligo dimness or obscurity of sight, dependent upon a speck on the cornea.

Calmodulin is a CALcium MODULated proteIn that can bind to and regulate a multitude of different protein targets, thereby affecting many different cellular functions.

cAMP dependent pathway cyclic adenosine monophosphate is a G protein-coupled receptor triggered signaling cascade used in cell communication in living organisms.

Cancer a malignant neoplasm or tumour in nay part of the body.

Candidiasis infections caused by members of the fungus genus *Candida* that range from superficial, such as oral thrush and vaginitis, to systemic and potentially life-threatening diseases.

Canker see chancre.

Carboxypeptidase an enzyme that hydrolyzes the carboxy-terminal (C-terminal) end of a peptide bond. It is synthesized in the pancreas and secreted into the small intestine.

Carbuncle is an abscess larger than a boil, usually with one or more openings draining pus onto the skin.

Carcinogenesis production of carcinomas. *adj* carcinogenic.

Carcinoma any malignant cancer that arises from epithelial cells.

Carcinosarcoma a rare tumour containing carcinomatous and sarcomatous components.

Cardiac relating to, situated near or affecting the heart.

Cardiac asthma acute attack of dyspnoea with wheezing resulting from a cardiac disorder.

Cardialgia heartburn.

Cardinolides cardiac glycosides with a 5-membered lactone ring in the side chain of the steroid aglycone.

Cardinolide glycoside cardenolides that contain structural groups derived from sugars.

Cardioactive having an effect on the heart.

Cardiogenic shock is characterized by a decreased pumping ability of the heart that causes a shock-like state associated with an inadequate circulation of blood due to primary failure of the ventricles of the heart to function effectively.

Cardiomyocytes cardiac muscle cells.

Cardiotonic something which strengthens, tones, or regulates heart functions without overt stimulation or depression.

Cardiovascular pertaining to the heart and blood vessels.

Caries tooth decay, commonly called cavities.

Cariogenic leading to the production of caries.

Carminative substance that stops the formation of intestinal gas and helps expel gas that has already formed, relieving flatulence: relieving flatulence or colic by expelling gas.

Carnitine palmitoyltransferase I (CPT1) also known as carnitine acyltransferase I or CAT1 is a mitochondrial enzyme, involved in converting long chain fatty acid into energy.

Carotenes are a large group of intense red and yellow pigments found in all plants ; these are hydrocarbon carotenoids (subclass of tetraterpenes) and the principal carotene is beta-carotene which is a precursor of vitamin A.

Carotenoids a class of natural fat-soluble pigments found principally in plants, belonging to a subgroup of terpenoids containing eight isoprene units forming a C40 polyene chain. Carotenoids play an important potential role in human health by acting as biological antioxidants. See also carotenes.

Carotenodermia yellow skin discoloration caused by excess blood carotene.

Carpopedal spasm spasm of the hand or foot, or of the thumbs and great toes.

Capases cysteine-aspartic acid proteases, are a family of cysteine proteases, which play essential roles in apoptosis (programmed cell death).

Catalase (CAT) enzyme in living organisms that catalyses the decomposition of hydrogen peroxide to water and oxygen.

Catalepsy indefinitely prolonged maintenance of a fixed body posture; seen in severe cases of catatonic schizophrenia.

Catamenia menstruation.

Cataplasia Degenerative reversion of cells or tissue to a less differentiated form.

Cataplasm a medicated poultice or plaster. A soft moist mass, often warm and medicated, that is spread over the skin to treat an inflamed, aching or painful area, to improve the circulation.

Cataractogenesis formation of cataracts.

Catarrh, Catarrhal inflammation of the mucous membranes especially of the nose and throat.

Catechins are polyphenolic antioxidant plant metabolites. They belong to the family of flavonoids; tea is a rich source of catechins. See flavonoids.

Catecholamines hormones that are released by the adrenal glands in response to stress.

Cathartic is a substance which accelerates defecation.

Caustic having a corrosive or burning effect.

Cauterization a medical term describing the burning of the body to remove or close a part of it.

CD 28 is one of the molecules expressed on T cells that provide co-stimulatory signals, which are required for T cell (lymphocytes) activation.

CD31 also known as PECAM-1 (Platelet Endothelial Cell Adhesion Molecule-1), a member of the immunoglobulin superfamily, that mediates cell-to-cell adhesion.

CD68 a glycoprotein expressed on monocytes/macrophages which binds to low density lipoprotein.

Cecal ligation tying up the cecam.

Cell adhesion molecules (CAM) glycoproteins located on the surface of cell membranes involved with binding of other cells or with the extra-cellular matrix.

Cellular respiration is the set of the metabolic reactions and processes that take place in organisms' cells to convert biochemical energy from nutrients into adenosine triphosphate (ATP), and then release waste products. The reactions involved in respiration are catabolic reactions that involve the oxidation of one molecule and the reduction of another.

Cellulitis a bacterial infection of the skin that tends to occur in areas that have been damaged or inflamed.

Central nervous system part of the vertebrate nervous system comprising the brain and spinal cord.

Central venous catheter a catheter placed into the large vein in the neck, chest or groin.

Cephalagia pain in the head, a headache.

Cephalic relating to the head.

Ceramide oligosides oligosides with an N-acetyl-sphingosine moiety.

Cerebral embolism a blockage of blood flow through a vessel in the brain by a blood clot that formed elsewhere in the body and traveled to the brain.

Cerebral ischemia is the localized reduction of blood flow to the brain or parts of the brain due to arterial obstruction or systematic hyperfusion.

Cerebral infarction is the ischemic kind of stroke due to a disturbance in the blood vessels supplying blood to the brain.

Cerebral tonic substance that can alleviate poor concentration and memory, restlessness, uneasiness, and insomnia.

Cerebrosides are glycosphingolipids which are important components in animal muscle and nerve cell membranes.

Cerebrovascular disease is a group of brain dysfunctions related to disease of the blood vessels supplying the brain.

cGMP cyclic guanosine monophosphate is a cyclic nucleotide derived from guanosine triphosphate (GTP). cGMP is a common regulator of ion channel conductance, glycogenolysis, and cellular apoptosis. It also relaxes smooth muscle tissues.

Chalcones a subgroup of flavonoids.

Chancre a painless lesion formed during the primary stage of syphilis.

Chemoembolization a procedure in which the blood supply to the tumour is blocked surgically or mechanically and anticancer drugs are administered directly into the tumour.

Chemokines are chemotactic cytokines, which stimulate migration of inflammatory cells towards tissue sites of inflammation.

Chemosensitizer a drug that makes tumour cells more sensitive to the effects of chemotherapy.

Chickenpox is also known as varicella, is a highly contagious illness caused by primary infection with varicella zoster virus (VZV). The virus causes red, itchy bumps on the body.

Chilblains small, itchy, painful lumps that develop on the skin. They develop as an abnormal response to cold. Also called perniosis or blain.

Chlorosis Iron deficiency anemia characterized by greenish yellow colour.

Cholagogue is a medicinal agent which promotes the discharge of bile from the system.

Cholecalcifereol a form of vitamin D, also called vitamin D3. See vitamin D.

Cholecyst gall bladder.

Cholecystitis inflammation of the gall bladder.

Cholecystokinin a peptide hormone that plays a key role in facilitating digestion in the small intestine.

Cholethiasis presence of gall stones (calculi) in the gall bladder.

Cholera an infectious gastroenteritis caused by enterotoxin-producing strains of the bacterium *Vibrio cholera* and characterized by severe, watery diarrhea.

Choleretic stimulation of the production of bile by the liver.

Cholesterol a soft, waxy, steroid substance found among the lipids (fats) in the bloodstream and in all our body's cells.

Choline a water soluble, organic compound, usually grouped within the Vitamin B complex. It is an essential nutrient and is needed for physiological functions such as structural integrity and signaling roles for cell membranes, cholinergic neuro-transmission (acetylcholine synthesis).

Cholinergic activated by or capable of liberating acetylcholine, especially in the parasympathetic nervous system.

Cholinergic system a system of nerve cells that uses acetylcholine in transmitting nerve impulses.

Cholinomimetic having an action similar to that of acetylcholine; called also parasympathomimetic.

Chonotropic affecting the time or rate, as the rate of contraction of the heart.

Choriocarcinoma a quick-growing malignant, trophoblastic, aggressive cancer that occurs in a woman's uterus (womb).

Chromium (Cr) is required in trace amounts in humans for sugar and lipid metabolism. Its deficiency may cause a disease called chromium deficiency. It is found in cereals, legumes, nuts and animal sources.

Chromosome long pieces of DNA found in the center (nucleus) of cells.

Chronic persisting over extended periods.

Chyle a milky bodily fluid consisting of lymph and emulsified fats, or free fatty acids.

Chylomicrons are large lipoprotein particles that transport dietary lipids from the intestines to other locations in the body. Chylomicrons are one of the five major groups of lipoproteins (chylomicrons, VLDL, IDL, LDL, HDL) that enable fats and cholesterol to move within the water-based solution of the bloodstream.

Chylorus milky (having fat emulsion).

Chyluria also called chylous urine, is a medical condition involving the presence of chyle (emulsified fat) in the urine stream, which results in urine appearing milky.

Chymopapain an enzyme derived from papaya, used in medicine and to tenderize meat.

Cicatrizant the term used to describe a product that promotes healing through the formation of scar tissue.

Cirrhosis chronic liver disease characterized by replacement of liver tissue by fibrous scar tissue and regenerative nodules/lumps leading progressively to loss of liver function.

C-Kit Receptor a protein-tyrosine kinase receptor that is specific for stem cell factor. this interaction is crucial for the development of hematopoietic, gonadal, and pigment stem cells.

Clastogen is an agent that can cause one of two types of structural changes, breaks in chromosomes that result in the gain, loss, or rearrangements of chromosomal segments. *Adj.* clastogenic.

Claudication limping, impairment in walking.

Clyster enema.

C-myc codes for a protein that binds to the DNA of other genes and is therefore a transcription factor.

CNS Depressant anything that depresses, or slows, the sympathetic impulses of the central nervous system (i.e., respiratory rate, heart rate).

Coagulopathy a defect in the body's mechanism for blood clotting, causing susceptibility to bleeding.

Cobalamin vitamin B12. See vitamin B12.

Co-carcinogen a chemical that promotes the effects of a carcinogen in the production of cancer.

Cold an acute inflammation of the mucous membrane of the respiratory tract especially of the nose and throat caused by a virus and accompanied by sneezing and coughing.

Collagen protein that is the major constituent of cartilage and other connective tissue; comprises the amino acids hydroxyproline, proline, glycine, and hydroxylysine.

Collagenases enzymes that break the peptide bonds in collagen.

Colic a broad term which refers to episodes of uncontrollable, extended crying in a baby who is otherwise healthy and well fed.

Colitis inflammatory bowel disease affecting the tissue that lines the gastrointestinal system.

Collyrium a lotion or liquid wash used as a cleanser for the eyes, particularly in diseases of the eye.

Colorectal relating to the colon or rectum.

Coma a state of unconsciousness from which a patient cannot be aroused.

Comedone a blocked, open sebaceous gland where the secretions oxidize, turning black. Also called blackhead.

Comitogen agent that is considered not to induce cell growth alone but to promote the effect of the mitogen.

Concoction a combination of crude ingredients that are prepared or cooked together.

Condyloma, Condylomata acuminata genital warts, venereal warts, anal wart or anogenital wart, a highly contagious sexually transmitted infection caused by epidermotropic human papillomavirus (HPV).

Conjunctivitis sore, red and sticky eyes caused by eye infection.

Constipation a very common gastrointestinal disorder characterised by the passing of hard, dry bowel motions (stools) and difficulty of bowel motion.

Consumption term used to describe wasting of tissues including but not limited to tuberculosis.

Consumptive afflicted with or associated with pulmonary tuberculosis.

Contraceptive an agent that reduces the likelihood of or prevents conception.

Contraindication a condition which makes a particular treatment or procedure inadvisable.

Contralateral muscle muscle of opposite limb (leg or arm).

Contusion another term for a bruise. A bruise, or contusion, is caused when blood vessels are damaged or broken as the result of a blow to the skin.

Convulsant a drug or physical disturbance that induces convulsion.

Convulsion rapid and uncontrollable shaking of the body.

Coolant that which reduces body temperature.

Copper (Cu) is essential in all plants and animals. It is found in a variety of enzymes, including the copper centers of cytochrome c oxidase and the enzyme superoxide dismutase (containing copper and zinc). In addition to its enzymatic roles, copper is used for biological electron transport. Because of its role in facilitating iron uptake, copper deficiency can often produce anemia-like symptoms. Dietary sources include curry powder, mushroom, nuts, seeds, wheat germ, whole grains and animal meat.

Copulation to engage in coitus or sexual intercourse. *adj.* copulatory.

Cordial a preparation that is stimulating to the heart.

Corn or callus is a patch of hard, thickened skin on the foot that is formed in response to pressure or friction.

Corticosteroids a class of steroid hormones that are produced in the adrenal cortex, used clinically for hormone replacement therapy, for suppressing ACTH secretion, for suppression of immune response and as antineoplastic, anti-allergic and anti-inflammatory agents.

Corticosterone a 21-carbon steroid hormone of the corticosteroid type produced in the cortex of the adrenal glands.

Cortisol is a corticosteroid hormone made by the adrenal glands.

Cornification is the process of forming an epidermal barrier in stratified squamous epithelial tissue.

Coryza a word describing the symptoms of a head cold. It describes the inflammation of the mucus membranes lining the nasal cavity which usually gives rise to the symptoms of nasal congestion and loss of smell, among other symptoms.

COX-1 see cyclooxygenase -1.

COX-2 see cyclooxygenase-2.

CPY1B1, CPY1A1 a member of the cytochrome P450 superfamily of heme-thiolate monooxygenase enzymes.

Creatin a nitrogenous organic acid that occurs naturally in vertebrates and helps to supply energy to muscle.

Creatine phosphokinase (CPK, CK) enzyme that catalyses the conversion of creatine and consumes adenosine triphosphate (ATP) to create phosphocreatine and adenosine diphosphate (ADP).

CREB cAMP response element-binding, a protein that is a transcription factor that binds to certain DNA sequences called cAMP response elements.

Crohn Disease an inflammatory disease of the intestines that affect any part of the gastrointestinal tract.

Crossover study a longitudinal, balance study in which participants receive a sequence of different treatments or exposures.

Croup is an infection of the throat (larynx) and windpipe (trachea) that is caused by a virus (Also called laryngotracheobronchitis).

Curretage surgical procedure in which a body cavity or tissue is scraped with a sharp instrument or aspirated with a cannula.

Cutaneous pertaining to the skin.

Cyanogenesis generation of cyanide. *adj.* cyanogenetic.

Cyclooxygenase (COX) an enzyme that is responsible for the formation of prostanoids – prostaglandins, prostacyclins, and thromboxanes that are each involved in the inflammatory response. Two different COX enzymes existed, now known as COX-1 and COX-2.

Cyclooxygenase-1 (COX-1) is known to be present in most tissues. In the gastrointestinal tract, COX-1 maintains the normal lining of the stomach. The enzyme is also involved in kidney and platelet function.

Cyclooxygenase-2 (COX-2) is primarily present at sites of inflammation.

Cysteine proteases are enzymes that that degrade polypeptides possessing a common catalytic mechanism that involves a nucleophilic cysteine thiol in a catalytic triad. They are found in fruits like papaya, pineapple, and kiwifruit.

Cystitis a common urinary tract infection that occurs when bacteria travel up the urethra, infect the urine and inflame the bladder lining.

Cystorrhea discharge of mucus from the bladder.

Cytochrome P450 3A CYP3A a very large and diverse superfamily of heme-thiolate proteins found in all domains of life. This group of enzymes catalyze many reactions involved in drug metabolism and synthesis of cholesterol, steroids and other lipids.

Cytokine nonantibody proteins secreted by certain cells of the immune system which carry signals locally between cells. They are a category of signaling molecules that are used extensively in cellular communication.

Cytopathic any detectable, degenerative changes in the host cell due to infection.

Cytoprotective protecting cells from noxious chemicals or other stimuli.

Cytosolic relates to the fluid of the cytoplasm in cells.

Cytostatic preventing the growth and proliferation of cells.

Cytotoxic of or relating to substances that are toxic to cells; cell-killing.

D-galactosamine an amino sugar with unique hepatotoxic properties in animals.

Dandruff scurf, dead, scaly skin among the hair.

Dartre condition of dry, scaly skin.

Debility weakness, relaxation of muscular fibre.

Debridement is the process of removing non-living tissue from pressure ulcers, burns, and other wounds.

Debriding agent substance that cleans and treat certain types of wounds, burns, ulcers.

Deciduogenic relating to the uterus lining that is shed off at childbirth.

Decoction a medical preparation made by boiling the ingredients.

Decongestant a substance that relieves or reduces nasal or bronchial congestion.

Defibrinated plasma blood whose plasma component has had fibrinogen and fibrin removed.

Degranulation cellular process that releases antimicrobial cytotoxic molecules from secretory vesicles called granules found inside some cells.

Demulcent an agent that soothes internal membranes. Also called emollient.

Dendritic cells are immune cells and form part of the mammalian immune system, functioning as antigen presenting cells.

Dentition a term that describes all of the upper and lower teeth collectively.

Deobstruent a medicine which removes obstructions; also called an aperient.

Deoxypyridinoline (Dpd) a crosslink product of collagen molecules found in bone and excreted in urine during bone degradation.

Depilatory an agent for removing or destroying hair.

Depressant a substance that diminish functional activity, usually by depressing the nervous system.

Depurative an agent used to cleanse or purify the blood, it eliminates toxins and purifies the system.

Dermatitis inflammation of the skin causing discomfort such as eczema.

Dermatophyte a fungus parasitic on the skin.

Dermatosis is a broad term that refers to any disease of the skin, especially one that is not accompanied by inflammation.

Dermonecrotic pertaining to or causing necrosis of the skin.

Desquamation the shedding of the outer layers of the skin.

Detoxifier a substance that promotes the removal of toxins from a system or organ.

Diabetes a metabolic disorder associated with inadequate secretion or utilization of insulin and characterized by frequent urination and persistent thirst. See diabetes mellitus.

Diabetes mellitus (DM) (sometimes called "sugar diabetes") is a set of chronic, metabolic disease conditions characterized by high blood sugar (glucose) levels that result from defects in insulin secretion, or action, or both. Diabetes mellitus appears in two forms.

Diabetes mellitus type I (formerly known as juvenile onset diabetes), caused by deficiency of the pancreatic hormone insulin as a result of destruction of insulin-producing beta cells of the pancreas. Lack of insulin causes an increase of fasting blood glucose that begins to appear in the urine above the renal threshold.

Diabetes mellitus type II (formerly called non-insulin-dependent diabetes mellitus or adult-onset diabetes), the disorder is characterized by high blood glucose in the context of insulin resistance and relative insulin deficiency in which insulin is available but cannot be properly utilized.

Diads two adjacent structural units in a polymer molecule.

Dialysis is a method of removing toxic substances (impurities or wastes) from the blood when the kidneys are unable to do so.

Diaphoresis is profuse sweating commonly associated with shock and other medical emergency conditions.

Diaphoretic a substance that induces perspiration. Also called sudorific.

Diaphyseal pertaining to or affecting the shaft of a long bone (diaphysis).

Diaphysis the main or mid section (shaft) of a long bone.

Diarrhoea a profuse, frequent and loose discharge from the bowels.

Diastolic referring to the time when the heart is in a period of relaxation and dilatation (expansion). *cf.* systolic.

Dieresis surgical separation of parts.

Dietary fibre is a term that refers to a group of food components that pass through the stomach and small intestine undigested and reach the large intestine virtually unchanged. Scientific evidence suggest that a diet high in dietary fibre can be of value for treating or preventing such disorders as constipation, irritable bowel syndrome, diverticular disease, hiatus hernia and haemorrhoids. Some components of dietary fibre may also be of value in reducing the level of cholesterol in blood and thereby decreasing a risk factor for coronary heart disease and the development of gallstones. Dietary fibre is beneficial in the treatment of some diabetics.

Digalactosyl diglycerides are the major lipid components of chloroplasts.

Diosgenin a steroid-like substance that is involved in the production of the hormone progesterone, extracted from roots of *Dioscorea* yam.

Dipsomania pathological use of alcohol.

Discutient an agent (as a medicinal application) which serves to disperse morbid matter.

Disinfectant an agent that prevents the spread of infection, bacteria or communicable disease.

Diuresis increased urination.

Diuretic a substance that increases urination (diuresis).

Diverticular disease is a condition affecting the large bowel or colon and is thought to be caused by eating too little fibre.

DMBA 7,12-Dimethylbenzanthracene. A polycyclic aromatic hydrocarbon found in tobacco smoke that is a potent carcinogen.

DNA deoxyribonucleic acid, a nucleic acid that contains the genetic instructions used in the development and functioning of all known living organisms.

DOCA desoxycorticosterone acetate – a steroid chemical used as replacement therapy in Addison's disease.

Dopamine a catecholamine neurotransmitter that occurs in a wide variety of animals, including both vertebrates and invertebrates.

Dopaminergic relating to, or activated by the neurotransmitter, dopamine.

Double blind refer to a clinical trial or experiment in which neither the subject nor the researcher knows which treatment any particular subject is receiving.

Douche a localised spray of liquid directed into a body cavity or onto a part.

DPPH 2,2 diphenyl -1- picryl-hydrazyl – a crystalline, stable free radical used as an inhibitor of free radical reactions.

Dracunculiasis also called guinea worm disease (GWD), is a parasitic infection caused by the nematode, *Dracunculus medinensis*.

Dropsy an old term for the swelling of soft tissues due to the accumulation of excess water. *adj.* dropsical.

Dysentery (formerly known as flux or the bloody flux) is a disorder of the digestive system that results in severe diarrhea containing mucus and blood in the feces. It is caused usually by a bacterium called *Shigella*.

Dysesthesia an unpleasant abnormal sensation produced by normal stimuli.

Dyskinesia the impairment of the power of voluntary movement, resulting in fragmentary or incomplete movements. *adj.* dyskinetic.

Dyslipidemia abnormality in or abnormal amount of, lipids and lipoproteins in the blood.

Dysmenorrhea is a menstrual condition characterized by severe and frequent menstrual cramps and pain associated with menstruation.

Dysmotility syndrome a vague, descriptive term used to describe diseases of the muscles of the gastrointestinal tract (esophagus, stomach, small and large intestines).

Dyspedia indigestion followed by nausea.

Dyspepsia refers to a symptom complex of epigastric pain or discomfort. It is often defined as chronic or recurrent discomfort centered in the upper abdomen and can be caused by a variety of conditions.

Dysphagia swallowing disorder.

Dysphonia a voice disorder, an impairment in the ability to produce voice sounds using the vocal organs.

Dysplasia refers to abnormality in development.

Dyspnoea shortness of breath, difficulty in breathing.

Dysrhythmias see arrhythmias.

Dystocia abnormal or difficult child birth or labour.

Dystonia a neurological movement disorder characterized by prolonged, repetitive muscle contractions that may cause twisting or jerking movements of muscles.

Dysuria refers to difficult and painful urination.

EC 50 median effective concentration that produces desired effects in 50% of the test population.

Ecbolic a drug (as an ergot alkaloid) that tends to increase uterine contractions and that is used especially to facilitate delivery.

Ecchymosis skin discoloration caused by the escape of blood into the tissues from ruptured blood vessels.

ECG see electrocardiography.

EC–SOD extracellular superoxide dismutase, a tissue enzyme mainly found in the extracellular matrix of tissues. It participates in the detoxification of reactive oxygen species by catalyzing the dismutation of superoxide radicals.

Eczema is broadly applied to a range of persistent skin conditions. These include dryness and recurring skin rashes which are characterized by one or more of these symptoms: redness, skin edema, itching and dryness, crusting, flaking, blistering, cracking, oozing, or bleeding.

Eczematous rash dry, scaly, itchy rash.

ED 50 is defined as the dose producing a response that is 50% of the maximum obtainable.

Edema formerly known as dropsy or hydropsy, is characterized swelling caused by abnormal accumulation of fluid beneath the skin, or in one or more cavities of the body. It usually occurs in the feet, ankles and legs, but it can involve the entire body.

Edematogenic producing or causing edema.

EGFR proteins epidermal growth factor receptor (EGFR) proteins – Protein kinases are enzymes that transfer a phosphate group from a phosphate donor onto an acceptor amino acid in a substrate protein.

EGR-1 early growth response 1, a human gene.

Eicosanoids are signaling molecules made by oxygenation of arachidonic acid, a 20-carbon essential fatty acid, includes prostaglandins and related compounds.

Elastase a serine protease that also hydrolyses amides and esters.

Electrocardiography or ECG, is a transthoracic interpretation of the electrical activity of the heart over time captured and externally recorded by skin electrodes.

Electromyogram (EMG) a test used to record the electrical activity of muscles. An electromyogram (EMG) is also called a myogram.

Electuary a medicinal paste composed of powders, or other medical ingredients, incorporated with sweeteners to hide the taste, suitable for oral administration.

Elephantiasis a disorder characterized by chronic thickened and edematous tissue on the genitals and legs due to various causes.

Embrocation lotion or liniment that relieves muscle or joint pains.

Embryotoxic term that describes any chemical which is harmful to an embryo.

Emesis vomiting, throwing up.

Emetic an agent that induces vomiting, *cf*: antiemetic.

Emetocathartic causing vomiting and purging.

Emmenagogue a substance that stimulates, initiates, and/or promotes menstrual flow. Emmenagogues are used in herbal medicine to balance and restore the normal function of the female reproductive system.

Emollient an agent that has a protective and soothing action on the surfaces of the skin and membranes.

Emulsion a preparation formed by the suspension of very finely divided oily or resinous liquid in another liquid.

Encephalitis inflammation of the brain.

Encephalopathy a disorder or disease of the brain.

Endocytosis is the process by which cells absorb material (molecules such as proteins) from outside the cell by engulfing it with their cell membrane.

Endometriosis is a common and often painful disorder of the female reproductive system. The two most common symptoms of endometriosis are pain and infertility.

Endometritis refers to inflammation of the endometrium, the inner lining of the uterus.

Endometrium the inner lining of the uterus.

Endoplasmic reticulum is a network of tubules, vesicles and sacs around the nucleus that are interconnected.

Endosteum the thin layer of cells lining the medullary cavity of a bone.

Endosteul pertaining to the endosteum.

Endotoxemia the presence of endotoxins in the blood, which may result in shock. *adj.* endotoxemic.

Endotoxin toxins associated with certain bacteria, unlike an 'exotoxin' is not secreted in soluble form by live bacteria, but is a structural component in the bacteria which is released mainly when bacteria are lysed.

Enema liquid injected into the rectum either as a purgative or medicine, Also called clyster.

Enteral term used to describe the intestines or other parts of the digestive tract.

Enteral administration involves the esophagus, stomach, and small and large intestines (i.e., the gastrointestinal tract).

Enteritis refers to inflammation of the small intestine.

Enterocolic disorder inflamed bowel disease.

Enterocytes tall columnar cells in the small intestinal mucosa that are responsible for the final digestion and absorption of nutrients.

Enterohemorrhagic causing bloody diarrhea and colitis, said of pathogenic microorganisms.

Enteropooling increased fluids and electrolytes within the lumen of the intestines due to increased levels of prostaglandins.

Enterotoxin is a protein toxin released by a microorganism in the intestine.

Enterotoxigenic of or being an organism containing or producing an enterotoxin.

Entheogen a substance taken to induce a spiritual experience.

Enuresis bed-wetting, a disorder of elimination that involves the voluntary or involuntary release of urine into bedding, clothing, or other inappropriate places.

Enophthalmos a condition in which the eye falls back into the socket and inhibits proper eyelid function.

Envenomation is the entry of venom into a person's body, and it may cause localised or systemic poisoning.

Eosinophilia the state of having a high concentration of eosinophils (eosinophil granulocytes) in the blood.

Eosinophils (or, less commonly, acidophils), are white blood cells that are one of the immune system components.

Epididymis a structure within the scrotum attached to the backside of the testis and whose coiled duct provides storage, transit and maturation of spermatozoa.

Epididymitis a medical condition in which there is inflammation of the epididymis.

Epigastralgia pain in the epigastric region.

Epilepsy a common chronic neurological disorder that is characterized by recurrent unprovoked seizures.

Epileptiform resembling epilepsy or its manifestations.

Epileptogenesis a process by which a normal brain develops epilepsy, a chronic condition in which seizures occur. *adj.* epileptogenic.

Episiotomy a surgical incision through the perineum made to enlarge the vagina and assist childbirth.

Epistaxis acute hemorrhage from the nostril, nasal cavity, or nasopharynx (nose-bleed).

Epstein Barr Virus herpesvirus that is the causative agent of infectious mononucleosis. It is also associated with various types of human cancers.

Epithelioma a usually benign skin disease most commonly occurring on the face, around the eyelids and on the scalp.

Ergocalciferol a form of vitamin D, also called vitamin D2. See vitamin D.

Ergonic increasing capacity for bodily or mental labor especially by eliminating fatigue symptoms.

ERK (extracellular signal regulated kinases) widely expressed protein kinase intracellular signalling molecules which are involved in functions including the regulation of meiosis, mitosis, and postmitotic functions in differentiated cells.

Eructation the act of belching or of casting up wind from the stomach through the mouth.

Eruption a visible rash or cutaneous disruption.

Erysipelas is an intensely red *Streptococcus* bacterial infection that occurs on the face and lower extremities.

Erythema abnormal redness and inflammation of the skin, due to vasodilation.

Erythematous characterized by erythema.

Erythroleukoplakia an abnormal patch of red and white tissue that forms on mucous membranes in the mouth and may become cancer. Tobacco (smoking and chewing) and alcohol may increase the risk of erythroleukoplakia.

Erythropoietin (EPO) a hormone produced by the kidney that promotes the formation of red blood cells (erythrocytes) in the bone marrow.

Eschar a slough or piece of dead tissue that is cast off from the surface of the skin.

Escharotic capable of producing an eschar; a caustic or corrosive agent.

Estradiol is the predominant sex hormone present in females, also called oestradiol.

Estrogen female hormone produced by the ovaries that play an important role in the estrous cycle in women.

Estrogen receptor (ER) is a protein found in high concentrations in the cytoplasm of breast, uterus, hypothalamus, and anterior hypophysis cells; ER levels are measured to determine a breast CA's potential for response to hormonal manipulation.

Estrogen receptor positive (ER+) means that estrogen is causing the tumour to grow, and that the breast cancer should respond well to hormone suppression treatments.

Estrogen receptor negative (ER-) tumour is not driven by estrogen and need another test to determine the most effective treatment.

Estrogenic relating to estrogen or producing estrus.

Estrus sexual excitement or heat of female; or period of this characterized by changes in the sex organs.

Euglycaemia normal blood glucose concentration.

Exanthematous characterized by or of the nature of an eruption or rash.

Excitotoxicity is the pathological process by which neurons are damaged and killed by glutamate and similar substances.

Excipient a pharmacologically inert substance used as a diluent or vehicle for the active ingredients of a medication.

Exocytosis the cellular process by which cells excrete waste products or chemical transmitters.

Exophthalmos or exophthalmia or proptosis is a bulging of the eye anteriorly out of the orbit.

Expectorant an agent that increases bronchial mucous secretion by promoting liquefaction of the sticky mucous and expelling it from the body.

Exteroceptive responsiveness to stimuli that are external to an organism.

Extrapyramidal side effects are a group of symptoms (tremour, slurred speech, akathisia, dystonia, anxiety, paranoia and bradyphrenia) that can occur in persons taking antipsychotic medications.

Extravasation discharge or escape, as of blood from the vein into the surrounding tissues.

Familial amyloid polyneuropathy (FAP) also called Corino de Andrade's disease, a neurodegenerative autosomal dominant genetically transmitted, fatal, incurable disease.

Familial adenomatous polyposis (FAP) is an inherited condition in which numerous polyps form mainly in the epithelium of the large intestine.

Familial dysautonomia a genetic disorder that affects the development and survival of autonomic and sensory nerve cells.

FasL or CD95L Fas ligand is a type-II transmembrane protein that belongs to the tumour necrosis factor (TNF) family.

Fauces the passage leading from the back of the mouth into the pharynx.

Favus a chronic skin infection, usually of the scalp, caused by the fungus, *Trichophyton schoenleinii* and characterized by the development of thick, yellow crusts over the hair follicles. Also termed tinea favosa.

Febrifuge an agent that reduces fever. Also called an antipyretic.

Febrile pertaining to or characterized by fever.

Fetotoxic toxic to the fetus.

Fibrates hypolipidemic agents primarily used for decreasing serum triglycerides, while increasing High density lipoprotein (HDL).

Fibril a small slender fibre or filament.

Fibrin insoluble protein that forms the essential portion of the blood clot.

Fibrinolysis a normal ongoing process that dissolves fibrin and results in the removal of small blood clots.

Fribinolytic causing the dissolution of fibrin by enzymatic action.

Fibroblast type of cell that synthesizes the extracellular matrix and collagen, the structural framework (stroma) for animal tissues, and play a critical role in wound healing.

Fibromyalgia a common and complex chronic pain disorder that affects people physically, mentally and socially. Symptoms include debilitating fatigue, sleep disturbance, and joint stiffness. Also referred to as FM or FMS.

Fibrosarcoma a malignant tumour derived from fibrous connective tissue and characterized by immature proliferating fibroblasts or undifferentiated anaplastic spindle cells.

Fibrosis the formation of fibrous tissue as a reparative or reactive process.

Filarial pertaining to a thread-like nematode worm.

Filariasis a parasitic and infectious tropical disease, that is caused by thread-like filarial nematode worms in the superfamily Filarioidea.

Fistula an abnormal connection between two parts inside of the body.

5'-Nucleotidase (5'-ribonucleotide phosphohydrolase), here before Fistula-in-ano

Fistula-in-ano a track connecting the internal anal canal to the skin surrounding the anal orifice.

5'-Nucleotidase (5'-ribonucleotide phosphohydrolase), an intrinsic membrane glycoprotein present as an ectoenzyme in a wide variety of mammalian cells, hydrolyzes 5'-nucleotides to their corresponding nucleosides.

Flatulence is the presence of a mixture of gases known as flatus in the digestive tract of mammals expelled from the rectum. Excessive flatulence can be caused by lactose intolerance, certain foods or a sudden switch to a high fibre.

Flavans a subgroup of flavonoids. See flavonoids.

Flavanols a subgroup of flavonoids, are a class of flavonoids that use the 2-phenyl-3,4-dihydro-2H-chromen-3-ol skeleton. These compounds include the catechins and the catechin gallates. They are found in chocolate, fruits and vegetables. See flavonoids.

Flavanones a subgroup of flavonoids, constitute >90% of total flavonoids in citrus. The major dietary flavanones are hesperetin, naringenin and eriodictyol.

Flavivirus A family of viruses transmitted by mosquitoes and ticks that cause some important diseases, including dengue, yellow fever, tick-borne encephalitis and West Nile fever.

Flavones a subgroup of flavonoids based on the backbone of 2-phenylchromen-4-one (2-phenyl-1-benzopyran-4-one). Flavones are mainly found in cereals and herbs.

Flavonoids (or bioflavonoids) are a group of polyphenolic antioxidant compounds in that are occur in plant as secondary metabolites. They are responsible for the colour of fruit and vegetables. Twelve basic classes (chemical types) of flavonoids have been recognized: flavones, isoflavones, flavans, flavanones, flavanols, flavanolols, anthocyanidins, catechins (including proanthocyanidins), leukoanthocyanidins, chalcones, dihydrochalcones, and aurones. Apart from their antioxidant activity, flavonoids are known for their ability to strengthen capillary walls, thus assisting circulation and helping to prevent and treat bruising, varicose veins, bleeding gums and nosebleeds, heavy menstrual bleeding and are also anti-inflammatory.

Flourine F is an essential chemical element that is required for maintenance of healthy bones and teeth and to reduce tooth decay. It is found in sea weeds, tea, water, seafood and dairy products.

Fluorosis a dental health condition caused by a child receiving too much fluoride during tooth development.

Flux an excessive discharge of fluid.

FMD (Flow Mediated Dilation) a measure of endothelial dysfunction which is used to evaluate cardiovascular risk.

Follicular atresia the break-down of the ovarian follicles.

Fomentation treatment by the application of war, moist substance.

Fontanelle soft spot on an infant's skull.

Framboesia see yaws.

FRAP ferric reducing ability of plasma, an assay used to assess antioxidant property.

Friedreich's ataxia is a genetic inherited disorder that causes progressive damage to the nervous system resulting in symptoms ranging from muscle weakness and speech problems to heart disease. *cf.* ataxia.

Functional food is any fresh or processed food claimed to have a health-promoting or disease-preventing property beyond the basic function of supplying nutrients. Also called medicinal food.

Furuncle is a skin disease caused by the infection of hair follicles usually caused by *Staphylococcus aureus,* resulting in the localized accumulation of pus and dead tissue.

Furunculosis skin condition characterized by persistent, recurring boils.

GABA gamma aminobutyric acid, required as an inhibitory neurotransmitter to block the transmission of an impulse from one cell to another in the central nervous system, which prevents over-firing of the nerve cells. It is used to treat both epilepsy and hypertension.

GADD 152 a pro-apoptotic gene.

Galactogogue a substance that promotes the flow of milk.

Galactophoritis inflammation of the milk ducts.

Galactopoietic increasing the flow of milk; milk-producing.

Gall bladder a small, pear-shaped muscular sac, located under the right lobe of the liver, in which bile secreted by the liver is stored until needed by the body for digestion. Also called cholecyst, cholecystis.

Gallic acid equivalent (GAE) measures the total phenol content in terms of the standard Gallic acid by the Folin-Ciocalteau assay.

Gamma GT (GGT) Gamma-glutamyl transpeptidase, aliver enzyme.

Gastralgia (heart burn) – pain in the stomach or abdominal region. It is caused by excess of acid, or an accumulation of gas, in the stomach.

Gastric pertaining to or affecting the stomach.

Gastric emptying refers to the speed at which food and drink leave the stomach.

Gastritis inflammation of the stomach.

Gastrocnemius muscle the big calf muscle at the rear of the lower leg.

Gastrotonic (Gastroprotective) substance that strengthens, tones, or regulates gastric functions (or protects from injury) without overt stimulation or depression.

Gavage forced feeding.

Gene silencing suppression of the expression of a gene.

Genotoxin a chemical or other agent that damages cellular DNA, resulting in mutations or cancer.

Genotoxic describes a poisonous substance which harms an organism by damaging its DNA thereby capable of causing mutations or cancer.

Ghrelin a gastrointestinal peptide hormone secreted by epithelial cells in the stomach lining.

Gingival Index an index describing the clinical severity of gingival inflammation as well as its location.

Gingivitis refers to gingival inflammation induced by bacterial biofilms (also called plaque) adherent to tooth surfaces.

Glaucoma a group of eye diseases in which the optic nerve at the back of the eye is slowly destroyed, leading to impaired vision and blindness.

Gleet a chronic inflammation (as gonorrhea) of a bodily orifice usually accompanied by an abnormal discharge.

Glioma is a type of tumour that starts in the brain or spine. It is called a glioma because it arises from glial cells.

Glomerulonephritis (GN) a renal disease characterized by inflammation of the glomeruli, or small blood vessels in the kidneys. Also known as glomerular nephritis. *Adj.* glomerulonephritic.

Glomerulosclerosis a hardening of the glomerulus in the kidney.

Glossal pertaining to the tongue.

GLP-1 glucagon-like peptide-1 is derived from the transcription product of the proglucagon gene, associate with type 2-diabetes therapy.

Gluconeogenesis a metabolic pathway that results in the generation of glucose from non-carbohydrate carbon substrates such as lactate. *adj.* gluconeogenic.

Glucose transporters (GLUT or SLC2A family) are a family of membrane proteins found in most mammalian cells.

Glucosyltranferase an enzyme that enable the transfer of glucose.

Glucuronidation a phase II detoxification pathway occurring in the liver in which glucuronic acid is conjugated with toxins.

Glutamic oxaloacetate transaminase (GOT) catalyzes the transfer of an amino group from an amino acid (Glu) to a 2-keto-acid to generate a new amino acid and the residual 2-keto-acid of the donor amino acid.

Glutamic pyruvate transaminase (GPT) see Alanine aminotransferase.

Glutathione (GSH) a tripeptide produced in the human liver and plays a key role in intermediary metabolism, immune response and health. It plays an important role in scavenging free radicals and protects cells against several toxic oxygen-derived chemical species.

Glutathione peroxidase (GPX) the general name of an enzyme family with peroxidase activity whose main biological role is to protect the organism from oxidative damage.

Glutathione S-transferase (GST) a major group of detoxification enzymes that participate in the detoxification of reactive electrophilic compounds by catalysing their conjugation to glutathione.

Glycation or glycosylation a chemical reaction in which glycosyl groups are added to a protein to produce a glycoprotein.

Glycogenolysis is the catabolism of glycogen by removal of a glucose monomer through cleavage with inorganic phosphate to produce glucose-1-phosphate.

Glycometabolism metabolism (oxidation) of glucose to produce energy.

Glycosuria or glucosuria is an abnormal condition of osmotic diuresis due to excretion of glucose by the kidneys into the urine.

Glycosylases a family of enzymes involved in base excision repair.

Goitre an enlargement of the thyroid gland leading to swelling of the neck or larynx.

Gonadotroph a basophilic cell of the anterior pituitary specialized to secrete follicle-stimulating hormone or luteinizing hormone.

Gonatropins protein hormones secreted by gonadotrope cells of the pituitary gland of vertebrates.

Gonorrhoea a common sexually transmitted bacterial infection caused by the bacterium *Neisseria gonorrhoeae*.

Gout a disorder caused by a build-up of a waste product, uric acid, in the bloodstream. Excess uric acid settles in joints causing inflammation, pain and swelling.

G-protein-coupled receptors (GPCRs) comprise a large and diverse family of proteins whose primary function is to transduce extracellular stimuli into cells.

Granulation the condition or appearance of being granulated (becoming grain-like).

Gravel sand-like concretions of uric acid, calcium oxalate, and mineral salts formed in the passages of the biliary and urinary tracts.

Gripe water is a home remedy for babies with colic, gas, teething pain or other stomach ailments. Its ingredients vary, and may include alcohol, bicarbonate, ginger, dill, fennel and chamomile.

Grippe an epidemic catarrh; older term for influenza.

GSH see Glutathione.

GSH-Px Glutathione peroxidase, general name of an enzyme family with peroxidase activity whose main biological role is to protect the organism from oxidative damage.

GSSG glutathione disulfides are biologically important intracellular thiols, and alterations in the GSH/GSSG ratio are often used to assess exposure of cells to oxidative stress.

GSTM glutathione S transferase M1, a major group of detoxification enzymes.

GSTM 2 glutathione S transferase M2, a major group of detoxification enzymes.

Gynecopathy any or various diseases specific to women.

Haemagogic promoting a flow of blood.

Haematemesis, Hematemesis is the vomiting of blood.

Haematinic improving the quality of the blood, its haemoglobin level and the number of erythrocytes.

Haematochezia passage of stools containing blood.

Haematochyluria, hematochyluria the discharge of blood and chyle (emulsified fat) in the urine, see also chyluria.

Haematoma, hematoma a localized accumulation of blood in a tissue or spece composed of clotted blood.

Haematometra, hematometra a medical condition involving bleeding of or near the uterus.

Haematopoiesis, hematopoiesis formation of blood cellular components from the haematopoietic stem cells.

Haematopoietic *(adj)* relating to the formation and development of blood cells.

Haematuria, Hematuria is the presence of blood in the urine. Hematuria is a sign that something is causing abnormal bleeding in a person's genitourinary tract.

Haeme oxygenase (HO-1, encoded by Hmox1) is an inducible protein activated in systemic inflammatory conditions by oxidant stress, an enzyme that catalyzes degradation of heme.

Haemochromatosis is a condition in which the body takes in too much iron.

Haemodialysis, Hemodialysis a method for removing waste products such as potassium and urea, as well as free water from the blood when the kidneys are in renal failure.

Haemolyis lysis of red blood cells and the release of haemoglobin into the surrounding fluid (plasma). *adj.* haemolytic.

Haemoptysis, hemoptysis is the coughing up of blood from the respiratory tract. The blood can come from the nose, mouth, throat, the airway passages leading to the lungs.

Haemorrhage, hemaorrhage bleeding, discharge of blood from blood vessels.

Haemorrhoids, Hemorrhoids a painful condition in which the veins around the anus or lower rectum are enlarged, swollen and inflamed. Also called piles.

Haemostasis, hemostasis a complex process which causes the bleeding process to stop.

Haemostatic, hemostatic something that stops bleeding.

Halitosis (bad breath) a common condition caused by sulfur-producing bacteria that live within the surface of the tongue and in the throat.

Hallucinogen drug that produces hallucinogen.

Hallucinogenic inducing hallucinations.

Hapten a small molecule that can elicit an immune response only when attached to a large carrier such as a protein.

HBeAg hepatitis B e antigen.

HBsAg hepatitis B s antigen.

Heartburn burning sensation in the stomach and esophagus caused by excessive acidity of the stomach fluids.

Heat rash any condition aggravated by heat or hot weather such as intertrigo.

Helminthiasis a disease in which a part of the body is infested with worms such as pinworm, roundworm or tapeworm.

Hemagglutination a specific form of agglutination that involves red blood cells.

Hemagglutination–inhibition test measures of the ability of soluble antigen to inhibit the agglutination of antigen-coated red blood cells by antibodies.

Hemagglutinin refers to a substance that causes red blood cells to agglutinate.

Hemangioma blood vessel.

Hematocrit is a blood test that measures the percentage of the volume of whole blood that is made up of red blood cells.

Hematopoietic pertaining to the formation of blood or blood cells.

Hematopoietic stem cell is a cell isolated from the blood or bone marrow that can renew itself, and can differentiate to a variety of specialized cells.

Heme oxygenase-1 (HO-1) an enzyme that catalyses the degrdation of heme; an inducible stress protein, confers cytoprotection against oxidative stress in-vitro and in-vivo.

Hemoglobinopathies genetic defects that produce abnormal hemoglobins and anemia.

Hemolytic anemia anemia due to hemolysis, the breakdown of red blood cells in the blood vessels or elsewhere in the body.

Hemorrhagic colitis an acute gateroenteritis characterized by overtly bloody diarrhea that is caused by *Escherichia coli* infection.

Hemolytic-uremic syndrome is a disease characterized by hemolytic anemia, acute renal failure (uremia) and a low platelet count.

Hepa-1c1c7 a type of hepatoma cells.

Hepatalgia pain or discomfort in the liver area.

Heptalgia pain in the liver and spleen.

Hepatectomy the surgical removal of part or all of the liver.

Hepatic relating to the liver.

Hepatic cirrhosis affecting the liver, characterize by hepatic fibrosis and regenerative nodules.

Hepatitis inflammation of the liver.

Hepatitis A (formerly known as infectious hepatitis) is an acute infectious disease of the liver caused by the hepatovirus hepatitis A virus.

Hepatocarcinogenesis represents a linear and progressive cancerous process in the liver in which successively more aberrant monoclonal populations of hepatocytes evolve.

Hepatocellular carcinoma (HCC) also called malignant hepatoma, is a primary malignancy (cancer) of the liver.

Hepatocytolysis cytotoxicity (dissolution) of liver cells.

Hepatoma cancer of the liver.

Hepatopathy a disease or disorder of the liver.

Hepatoprotective (liver protector) a substance that helps protect the liver from damage by toxins, chemicals or other disease processes.

Hepatoregenerative a compound that promotes hepatocellular regeneration, repairs and restores liver function to optimum performance.

Hepatotonic (liver tonic) a substance that is tonic to the liver – usually employed to normalize liver enzymes and function.

Hernia occurs when part of an internal organ bulges through a weak area of muscle.

HER- 2 human epidermal growth factor receptor 2, a protein giving higher aggressiveness in breast cancer, also known as ErbB-2, ERBB2.

Herpes a chronic inflammation of the skin or mucous membrane characterized by the development of vesicles on an inflammatory base.

Herpes simplex virus 1 and 2 – (HSV-1 and HSV-2) are two species of the herpes virus family which cause a variety of illnesses/infections in humans such cold sores, chickenpox or varicella, shingles or herpes zoster (VZV), cytomegalovirus (CMV), and various cancers, and can cause brain inflammation (encephalitis). HSV-1 is commonly associated with herpes outbreaks of the face known as cold sores or fever blisters, whereas HSV-2 is more often associated with genital herpes. They are also called Human Herpes Virus 1 and 2 (HHV-1 and HHV-2) and are neurotropic and neuroinvasive viruses; they enter and hide in the human nervous system, accounting for their durability in the human body.

Herpes zoster or simply zoster, commonly known as shingles and also known as zona, is a viral disease characterized by a painful skin rash with blisters.

Heterophobia term used to describe irrational fear of, aversion to, or discrimination against heterosexuals.

HDL-C (HDL Cholesterol) high density lipoprotein-cholesterol, also called "good cholesterol". See also high-density lipoprotein.

Hiatus hernia occurs when the upper part of the stomach pushes its way through a tear in the diaphragm.

High-density lipoprotein (HDL) is one of the five major groups of lipoproteins which enable cholesterol and triglycerides to be transported within the water based blood stream. HDL can remove cholesterol from atheroma within arteries and transport it back to the liver for excretion or re-utilization—which is the main reason why HDL-bound cholesterol is sometimes called "good cholesterol", or HDL-C. A high level of HDL-C seems to protect against cardiovascular diseases. cf. LDL.

HGPRT, HPRT (hypoxanthine-guanine phosphoribosyl transferase) an enzyme that catalyzes the conversion of 5-phosphoribosyl-1-pyrophosphate and hypoxanthine, guanine, or 6-mercaptopurine to the corresponding 5'-mononucleotides and pyrophosphate. The enzyme is important in purine biosynthesis as well as central nervous system functions.

Hippocampus a ridge in the floor of each lateral ventricle of the brain that consists mainly of gray matter.

Hippocampal pertaining to the hippocampus.

Histaminergic liberated or activated by histamine, relating to the effects of histamine at histamine receptors of target tissues.

Histaminergic receptors are types of G-protein coupled receptors with histamine as their endogenous ligand.

HIV see Human immunodeficiency virus.

Hives (urticaria) is a skin rash characterised by circular wheals of reddened and itching skin.

Hodgkin's disease disease characterized by enlargement of the lymph glands, spleen and anemia.

Homeodomain transcription factor a protein domain encoded by a homeobox. Homeobox genes encode transcription factors which typically switch on cascades of other genes.

Homeostasis describes the physical and chemical parameters that an organism must maintain to allow proper functioning of its component cells.

Homeotherapy treatment or prevention of disease with a substance similar but not identical to the causative agent of the disease.

Hormonal (female) substance that has a hormone-like effect similar to that of estrogen and/or a substance used to normalize female hormone levels.

Hormonal (male) substance that has a hormone-like effect similar to that of testosterone and/or a substance used to normalize male hormone levels.

HRT hormone replacement therapy, the administration of the female hormones, oestrogen and progesterone, and sometimes testosterone.

HSP27 is an ATP-independent, 27 kDa heat shock protein chaperone that confers protection against apoptosis.

HSP90 a 90 kDa heat shock protein chaperone that has the ability to regulate a specific subset of cellular signalling proteins that have been implicated in disease processes.

hTERT – (TERT) telomerase reverse transcriptase is a catalytic subunit of the enzyme telomerase in humans. It exerts a novel protective function by binding to mitochondrial DNA, increasing respiratory chain activity and protecting against oxidative stress–induced damage.

HT29 cells are human intestinal epithelial cells which produce the secretory component of Immunoglobulin A (IgA), and carcinoembryonic antigen (CEA).

Human cytomegalovirus (HCMV) a DNA herpes virus which is the leading cause of congenital viral infection and mental retardation.

Human factor X a coagulation factor also known by the eponym Stuart-Prower factor or as thrombokinase, is an enzyme involved in blood coagulation. It synthesized in the liver and requires vitamin K for its synthesis.

Human immunodeficiency virus (HIV) a retrovirus that can lead to acquired immunodeficiency syndrome (AIDS), a condition in humans in which the immune system begins to fail, leading to life-threatening opportunistic infections.

Humoral immune response (HIR) is the aspect of immunity that is mediated by secreted antibodies (as opposed to cell-mediated immunity, which involves T lymphocytes) produced in the cells of the B lymphocyte lineage (B cell).

HUVEC human umbilical vien endothelial cells.

Hyaluronidase enzymes that catalyse the hydrolysis of certain complex carbohydrates like hyaluronic acid and chondroitin sulfates.

Hydatidiform a rare mass or growth that forms inside the uterus at the beginning of a pregnancy.

Hydrocholeretic an agent that stimulates an increased output of bile of low specific gravity.

Hydrogogue a purgative that causes an abundant watery discharge from the bowel.

Hydronephrosis is distension and dilation of the renal pelvis and calyces, usually caused by obstruction of the free flow of urine from the kidney.

Hydrophobia a viral neuroinvasive disease that causes acute encephalitis (inflammation of the brain) in warm-blooded animals. Also called rabies.

Hydropsy see dropsy.

Hyperalgesia an increased sensitivity to pain, which may be caused by damage to nociceptors or peripheral nerves.

Hyperammonemia, hyperammonaemia a metabolic disturbance characterised by an excess of ammonia in the blood.

Hypercholesterolemia high levels of cholesterol in the blood that increase a person's risk for cardiovascular disease leading to stroke or heart attack.

Hyperemia is the increased blood flow that occurs when tissue is active.

Hyperemesis severe and persistent nausea and vomiting (morning sickness) during pregnancy.

Hyperglycemic, hyperglycaemia high blood sugar; is a condition in which an excessive amount of glucose circulates in the blood plasma.

Hyperglycemic a substance that raises blood sugar levels.

Hyperinsulinemia a condition in which there are excess levels of circulating insulin in the blood; also known as pre-diabetes.

Hyperkalemia is an elevated blood level of the electrolyte potassium.

Hyperleptinemia Increased serum leptin level.

Hyperpiesia persistent and pathological high blood pressure for which no specific cause can be found.

Hypermethylation an increase in the inherited methylation of cytosine and adenosine residues in DNA.

Hyperpropulsion using water pressure as a force to move objects; used to dislodge calculi in the urethra.

Hyperpyrexia is an abnormally high fever.

Hypertension commonly referred to as "high blood pressure" or HTN, is a medical condition in which the arterial blood pressure is chronically elevated.

Hypertensive characterized or caused by increased tension or pressure as abnormally high blood pressure.

Hypertriglyceridaemia or hypertriglycemia a disorder that causes high triglycerides in the blood.

Hypertrophy enlargement or overgrowth of an organ.

Hyperuricemia is a condition characterized by abnormally high level of uric acid in the blood.

Hypoadiponectinemia low plasma adiponectin concentrations associated with obesity and type 2 diabetes; that is closely related to the degree of insulin resistance and hyperinsulinemia than to the degree of adiposity and glucose tolerance.

Hypoalbuminemia a medical condition where levels of albumin in blood serum are abnormally low.

Hypocalcemic tetany a disease caused by an abnormally low level of calcium in the blood and characterized by hyperexcitability of the neuromuscular system and results in carpopedal spasms.

Hypocholesterolemic (cholesterol-reducer), a substance that lowers blood cholesterol levels.

Hypocorticism see Addison's disease.

Hypocortisolism see Addison's disease.

Hypoglycemic - an agent that lowers the concentration of glucose (sugar) in the blood.

Hypoperfusion decreased blood flow through an organ, characterised by an imbalance of oxygen demand and oxygen delivery to tissues.

Hypophagic under-eating.

Hypotensive characterised by or causing diminished tension or pressure, as abnormally low blood pressure.

Hypothermia a condition in which an organism's temperature drops below that required for normal metabolism and body functions.

Hypothermic relating to hypothermia, with subnormal body temperature.

Hypoxaemia is the reduction of oxygen specifically in the blood.

Hypoxia a shortage of oxygen in the body.

IC 50 the median maximal inhibitory concentration; a measure of the effectiveness of a compound in inhibiting biological or biochemical function.

Iceterus jaundice, yellowish pigmentation of the skin.

Ichthyotoxic a substance which is poisonous to fish.

Icteric hepatitis an infectious syndrome of hepatitis characterized by jaundice, nausea, fever, right-upper quadrant pain, enlarged liver and transaminitis (increase in alanine aminotransferase (ALT) and/or aspartate aminotransferase (AST)).

Icterus neonatorum jaundice in newborn infants.

Idiopathic of no apparent physical cause.

IgE Immunoglobin E – a class of antibody that plays a role in allergy.

IGFs insulin-like growth factors, polypeptides with high sequence similarity to insulin.

IgG Immunoglobin G – the most abundant immunoglobin (antibody) and is one of the major activators of the complement pathway.

IgM Immunoglobin M – primary antibody against A and B antigens on red blood cells.

IKAP is a scaffold protein of the IvarKappa-Beta kinase complex and a regulator for kinases involved in pro-inflammatory cytokine signaling.

IKappa B or IkB-beta, a protein of the NF-Kappa-B inhibitor family.

Ileus a temporary disruption of intestinal peristalsis due to non-mechanical causes.

Immune modulator a substance that affects or modulates the functioning of the immune system.

Immunodeficiency a state in which the immune system's ability to fight infectious disease is compromised or entirely absent.

Immunogenicity the property enabling a substance to provoke an immune response.

Immunomodulatory capable of modifying or regulating one or more immune functions.

Immunoreactive reacting to particular antigens or haptens.

Immunostimulant agent that stimulates an immune response.

Immunosuppression involves a process that reduces the activation or efficacy of the immune system.

Immunotoxin a man-made protein that consists of a targeting portion linked to a toxin.

Impetigo a contagious, bacterial skin infection characterized by blisters that may itch, caused by a *Streptoccocus* bacterium or *Staphylococcus aureus* and mostly seen in children.

Impotence a sexual dysfunction characterized by the inability to develop or maintain an erection of the penis.

Incontinence (fecal) the inability to control bowel's movement.

Incontinence (urine) the inability to control urine excretion.

Index of structural atypia (ISA) index of structural abnormality.

Induration hardened, as a soft tissue that becomes extremely firm.

Infarct an area of living tissue that undergoes necrosis as a result of obstruction of local blood supply.

Infarction is the process of tissue death (necrosis) caused by blockage of the tissue's blood supply.

Inflammation a protective response of the body to infection, irritation or other injury, aimed at destroying or isolating the injuries and characterized by redness, pain, warmth and swelling.

Influenza a viral infection that affects mainly the nose, throat, bronchi and occasionally, lungs.

Infusion a liquid extract obtained by steeping something (e.g. herbs) that are more volatile or dissolve readily in water, to release their active ingredients without boiling.

Inguinal hernia a henia into the inguinal canal of the groin.

Inhalant a medicinal substance that is administered as a vapor into the upper respiratory passages.

iNOS, inducible nitric oxide synthases through its product, nitric oxide (NO), may contribute to the induction of germ cell apoptosis. It plays a crucial role in early sepsis-related microcirculatory dysfunction.

Inotropic affecting the force of muscle contraction.

Insecticide an agent that destroys insects. *adj.* insecticidal.

Insomnia a sleeping disorder characterized by the inability to fall asleep and/or the inability to remain asleep for a reasonable amount of time.

Insulin a peptide hormone composed of 51 amino acids produced in the islets of Langerhans in the pancreas causes cells in the liver, muscle, and fat tissue to take up glucose from the blood, storing it as glycogen in the liver and muscle. Insulin deficiency is often the cause of diabetes and exogenous insulin is used to control diabetes.

Insulinotropic changing the action of insulin.

Integrase an enzyme produced by a retrovirus (such as HIV) that enables its genetic material to be integrated into the DNA of the infected cell.

Interferons (IFNs) are natural cell-signaling glycoproteins known as cytokines produced by the cells of the immune system of most vertebrates in response to challenges such as viruses, parasites and tumour cells.

Interleukins a group of naturally occurring proteins and is a subset of a larger group of cellular messenger molecules called cytokines, which are modulators of cellular behavior.

Interleukin-1 (IL-1) a cytokine that could induce fever, control lymphocytes, increase the number of bone marrow cells and cause degeneration of bone joints. Also called endogenous pyrogen, lymphocyte activating factor, haemopoetin-1 and mononuclear cell factor, amongst others that IL-1 is composed of two distinct proteins, now called IL-1α and IL-1β.

Interleukin 1 Beta (IL-1β) a cytokine protein produced by activated macrophages. cytokine is an important mediator of the inflammatory response, and is involved in a variety of cellular activities, including cell proliferation, differentiation, and apoptosis.

Interleukin 2 (IL-2) a type of cytokine immune system signaling molecule that is instrumental in the body's natural response to microbial infection.

Interleukin-6 (IL-6) an interleukin that acts as both a pro-inflammatory and anti-inflammatory cytokine.

Interleukin 8 (IL-8) a cytokine produced by macrophages and other cell types such as epithelial cells and is one of the major mediators of the inflammatory response.

Intermediate-density lipoproteins (IDL) is one of the five major groups of lipoproteins (chylomicrons, VLDL, IDL, LDL, HDL) that enable fats and cholesterol to move within the water-based solution of the bloodstream. IDL is further degraded to form LDL particles and, like LDL, can also promote the growth of atheroma and increase cardiovascular diseases.

Interstitial pertaining to the interstitium.

Interstitium the space between cells in a tissue.

Intertrigo an inflammation (rash) caused by microbial infection in skin folds.

Intima innermost layer of an artery or vein.

Intoxicant substance that produce drunkenness or intoxication.

Intraperitoneal (IP) the term used when a chemical is contained within or administered through the peritoneum (the thin, transparent membrane that lines the walls of the abdomen).

Intromission the act of putting one thing into another.

Intubation refers to the placement of a tube into an external or internal orifice of the body.

Iodine (I) is an essential chemical element that is important for hormone development in the human body. Lack of iodine can lead to an enlarged thyroid gland (goitre) or other iodine deficiency disorders including mental retardation and stunted growth in babies and children. Iodine is found in dairy products, seafood, kelp, seaweeds, eggs, some vegetables and iodised salt.

IP see Intraperitoneal.

Iron (Fe) is essential to most life forms and to normal human physiology. In humans, iron is an essential component of proteins involved in oxygen transport and for haemoglobin. It is also essential for the regulation of cell growth and differentiation. A deficiency of iron limits oxygen delivery to cells, resulting in fatigue, poor work performance, and decreased immunity. Conversely, excess amounts of iron can result in toxicity and even death. Dietary sources include, certain cereals, dark green leafy vegetables, dried fruit, legumes, seafood, poultry and meat.

Ischemia an insufficient supply of blood to an organ, usually due to a blocked artery.

Ischuria retention or suppression of urine.

Isoflavones a subgroup of flavonoids in which the basic structure is a 3-phenyl chromane skeleton. They act as phytoestrogens in mammals. See flavonoids.

Isomers substances that are composed of the same elements in the same proportions and hence have the same molecular formula but differ in properties because of differences in the arrangement of atoms.

Jamu traditional Indonesian herbal medicine.

Jaundice refers to the yellow color of the skin and whites of the eyes caused by excess bilirubin in the blood.

JNK (Jun N-terminal Kinase), also known as Stress Activated Protein Kinase (SAPK), belongs to the family of MAP kinases.

Jurkat cells a line of T lymphocyte cells that are used to study acute T cell leukemia.

KB cell a cell line derived from a human carcinoma of the nasopharynx, used as an assay for antineoplastic (anti-tumour) agents.

Kaposi sarcoma herpes virus (KSHV) also known as human herpesvirus-8, is a gamma 2 herpesvirus or rhadinovirus. It plays an important role in the pathogenesis of Kaposi sarcoma (KS), multicentric Castleman disease (MCD) of the plasma cell type, and primary effusion lymphoma and occurs in HIV patients.

Keratin a sulphur-containing protein which is a major component in skin, hair, nails, hooves, horns, and teeth.

Keratinocyte is the major constituent of the epidermis, constituting 95% of the cells found there.

Keratinophilic having an affinity for keratin.

Keratomalacia an eye disorder that leads to a dry cornea.

Kidney stones (calculi) are hardened mineral deposits that form in the kidney.

Kinin is any of various structurally related polypeptides, such as bradykinin, that act locally to induce vasodilation and contraction of smooth muscle.

Kininogen either of two plasma α2-globulins that are kinin precursors.

Knockout gene knockout is a genetic technique in which an organism is engineered to carry genes that have been made inoperative.

Kunitz protease inhibitors a type of protein contained in legume seeds which functions as a protease inhibitor.

Kupffer cells are resident macrophages of the liver and play an important role in its normal physiology and homeostasis as well as participating in the acute and chronic responses of the liver to toxic compounds.

L-Dopa (L-3,4-dihydroxyphenylalanine) is an amino acid that is formed in the liver and converted into dopamine in the brain.

Labour process of childbirth involving muscular contractions.

Lactagogue an agent that increases or stimulates milk flow or production. Also called a galactagogue.

Lactate dehydrogenase (LDH) enzyme that catalyzes the conversion of lactate to pyruvate.

Lactation secretion and production of milk.

Lactic acidosis is a condition caused by the buildup of lactic acid in the body. It leads to acidification of the blood (acidosis), and is considered a distinct form of metabolic acidosis.

LAK cell a lymphokine-activated killer cell i.e. a white blood cell that has been stimulated to kill tumour cells.

Laparotomy a surgical procedure involving an incision through the abdominal wall to gain access into the abdominal cavity. *adj.* laparotomized.

Larvacidal an agent which kills insect or parasite larva.

Laryngitis is an inflammation of the larynx.

Laxation bowel movement.

Laxatives substances that are used to promote bowel movement.

LC 50 median lethal concentration, see LD 50.

LD 50 median lethal dose – the dose required to kill half the members of a tested population. Also called LC 50 (median lethal concentration).

LDL see low-density lipoprotein.

LDL Cholesterol see low-density lipoprotein.

LDL receptor (LDLr) a low-density lipoprotein receptor gene.

Lectins are sugar-binding proteins that are highly specific for their sugar moieties. They play a role in biological recognition phenomena involving cells and proteins.

Leishmaniasis a disease caused by protozoan parasites that belong to the genus *Leishmania* and is transmitted by the bite of certain species of sand fly.

Lenticular opacity also known as or related to cataract.

Leprosy a chronic bacterial disease of the skin and nerves in the hands and feet and, in some cases, the lining of the nose. It is caused by the *Mycobacterium leprae*. Also called Hansen's disease.

Leptin is a 16 kDa protein hormone with important effects in regulating body weight, metabolism and reproductive function.

Lequesne Algofunctional Index is a widespread international instrument (10 questions survey) and recommended by the World Health Organization (WHO) for outcome measurement in hip and knee diseases such as osteoarthritis.

Leucocyte white blood corpuscles, colourless, without haemoglobin that help to combat infection.

Leucoderma a skin abnormality characterized by white spots, bands and patches on the skin; they can also be caused by fungus and tinea. Also see vitiligo.

Leucorrhoea commonly known as whites, refers to a whitish discharge from the female genitals.

Leukemia, leukaemia a cancer of the blood or bone marrow and is characterized by an abnormal proliferation (production by multiplication) of blood cells, usually white blood cells (leukocytes).

Leukocytopenia abnormal decrease in the number of leukocytes (white blood cells) in the blood.

Leukomyelopathy any diseases involving the white matter of the spinal cord.

Leukopenia a decrease in the number of circulating white blood cells.

Leukoplakia condition characterized by white spots or patches on mucous membranes, especially of the mouth and vulva.

Leukotriene a group of hormones that cause the inflammatory symptoms of hay-fever and asthma.

Levarterenol see Norepinephrine.

LexA repressor or Repressor LexA is repressor enzyme that represses SOS response genes

coding for DNA polymerases required for repairing DNA damage.
Libido sexual urge.
Lichen planus a chronic mucocutaneous disease that affects the skin, tongue, and oral mucosa.
Ligroin a volatile,, inflammable fraction of petroleum, obtained by distillation and used as a solvent.
Liniment liquid preparation rubbed on skin, used to relieve muscular aches and pains.
Lipodiatic having lipid and lipoprotein lowering property.
Lipodystrophy a medical condition characterized by abnormal or degenerative conditions of the body's adipose tissue.
Lipolysis is the breakdown of fat stored in fat cells in the body.
Lipoxygenase a family of iron-containing enzymes that catalyse the dioxygenation of polyunsaturated fatty acids in lipids containing a cis,cis-1,4- pentadiene structure.
Lithiasis formation of urinary calculi (stones).
Lithontripic removes stones from kidney, gall bladder.
Lotion a liquids suspension or dispersion of chemicals for external application to the body.
Lovo cells colon cancer cells.
Low-density lipoprotein (LDL) is a type of lipoprotein that transports cholesterol and triglycerides from the liver to peripheral tissues. High levels of LDL cholesterol can signal medical problems like cardiovascular disease, and it is sometimes called "bad cholesterol".
LTB4 a type of leukotriene, a major metabolite in neutrophil polymorphonuclear leukocytes. It stimulates polymorphonuclear cell function (degranulation, formation of oxygen-centered free radicals, arachidonic acid release, and metabolism). It induces skin inflammation.
Luciferase is a generic name for enzymes commonly used in nature for bioluminescence.
Lumbago is the term used to describe general lower back pain.
Lung abscess necrosis of the pulmonary tissue and formation of cavities containing necrotic debris or fluid caused by microbial infections.
Lusitropic an agent that affects diastolic relaxation.

Lutein a carotenoid, occurs naturally as yellow or orange pigment in some fruits and leafy vegetables. It is one of the two carotenoids contained within the retina of the eye. Within the central macula, zeaxanthin predominates, whereas in the peripheral retina, lutein predominates. Lutein is necessary for good vision and may also help prevent or slow down atherosclerosis, the thickening of arteries, which is a major risk for cardiovascular disease.
Luteolysis is the structural and functional degradation of the corpus luteum (CL) that occurs at the end of the luteal phase of both the estrous and menstrual cycles in the absence of pregnancy.
Lymphadenitis-cervical inflammation of the lymph nodes in the neck, usually caused by an infection.
Lymphatitis inflammation of lymph vessels and nodes.
Lymphadenopathy a term meaning "disease of the lymph nodes – lymph node enlargement".
Lymphoblastic pertaining to the production of lymphocytes.
Lymphocyte a small white blood cell (leucocyte) that plays a large role in defending the body against disease. Lymphocytes are responsible for immune responses. There are two main types of lymphocytes: B cells and T cells. Lymphocytes secrete products (lymphokines) that modulate the functional activities of many other types of cells and are often present at sites of chronic inflammation.
Lymphocyte B cells the B cells make antibodies that attack bacteria and toxins.
Lymphocyte T cells T cells attack body cells themselves when they have been taken over by viruses or have become cancerous.
Lymphoma a type of cancer involving cells of the immune system, called lymphocytes.
Lymphopenia abnormally low number of lymphocytes in the blood.
Lysosomes are small, spherical organelles containing digestive enzymes (acid hydrolases and other proteases (cathepsins)).
Maceration softening or separating of parts by soaking in a liquid.

Macrophage a type of large leukocyte that travels in the blood but can leave the bloodstream and enter tissue; like other leukocytes it protects the body by digesting debris and foreign cells.

Macular degeneration a disease that gradually destroys the macula, the central portion of the retina, reducing central vision.

Macules small circumscribed changes in the color of skin that are neither raised (elevated) nor depressed.

Maculopapular describes a rash characterized by raised, spotted lesions.

Magnesium (M g) is the fourth most abundant mineral in the body and is essential to good health. It is important for normal muscle and nerve function, steady heart rhythm, immune system, and strong bones. Magnesium also helps regulate blood sugar levels, promotes normal blood pressure, and is known to be involved in energy metabolism and protein synthesis and plays a role in preventing and managing disorders such as hypertension, cardiovascular disease, and diabetes. Dietary sources include legumes (e.g. soya bean and by-products), nuts, whole unrefined grains, fruit (e.g. banana, apricots), okra and green leafy vegetables.

MAK cell macrophage-activated killer cell, activated nacrophage that much more phagocytic than monocytes.

Malaise a feeling of weakness, lethargy or discomfort as of impending illness.

Malaria is an infection of the blood by *Plasmodium* parasite that is carried from person to person by mosquitoes. There are four species of malaria parasites that infect man: *Plasmodium falciparum*, so called 'malignant tertian fever', is the most serious disease, *Plasmodium vivax*, causing a relapsing form of the disease, *Plasmodium malariae*, and *Plasmodium ovale*.

Malassezia a fungal genus (previously known as *Pityrosporum*) classified as yeasts, naturally found on the skin surfaces of many animals including humans. It can cause hypopigmentation on the chest or back if it becomes an opportunistic infection.

Mammalian target of rapamycin (mTOR) pathway that regulates mitochondrial oxygen consumption and oxidative capacity.

Mammogram an x-ray of the breast to detect tumours.

Mandibular relating to the mandible, the human jaw bone.

Manganese is an essential element for heath. It is an important constituent of some enzymes and an activator of other enzymes in physiologic processes. Manganese superoxide dismutase (MnSOD) is the principal antioxidant enzyme in the mitochondria. Manganese-activated enzymes play important roles in the metabolism of carbohydrates, amino acids, and cholesterol. Manganese is the preferred cofactor of enzymes called glycosyltransferases which are required for the synthesis of proteoglycans that are needed for the formation of healthy cartilage and bone. Dietary source include whole grains, fruit, legumes (soybean and by-products), green leafy vegetables, beetroot and tea.

MAO activity monoamine oxidase activity.

MAPK (Mitogen-activated protein kinase) these kinases are strongly activated in cells subjected to osmotic stress, UV radiation, disregulated K^+ currents, RNA-damaging agents, and a multitude of other stresses, as well as inflammatory cytokines, endotoxin, and withdrawal of a trophic factor. The stress-responsive MAPKs mediate a plethora of cellular responses to such stressful stimuli, including apoptosis and production of inflammatory and immunoregulatory cytokines in diverse cell systems.

Marasmus is one of the three forms of serious protein-energy malnutrition.

Mastectomy surgery to remove a breast.

Masticatory a substance chewed to increase salivation. Also called sialogue.

Mastitis a bacterial infection of the breast which usually occurs in breastfeeding mothers.

Matrix metalloproteinases (MMP) a member of a group of enzymes that can break down proteins, such as collagen, that are normally found in the spaces between cells in tissues (i.e., extracellular matrix proteins). Matrix metalloproteinases are involved in wound healing, angiogenesis, and tumour cell metastasis. See also metalloproteinase.

MBC minimum bacterial concentration – the lowest concentration of antibiotic required to kill an organism.

MDA Malondialdehyde is one of the most frequently used indicators of lipid peroxidation.

Measles an acute, highly communicable rash illness due to a virus transmitted by direct contact with infectious droplets or, less commonly, by airborne spread.

Megaloblastic anemia an anemia that results from inhibition of DNA synthesis in red blood cell production, often due to a deficiency of vitamin B12 or folate and is characterized by many large immature and dysfunctional red blood cells (megaloblasts) in the bone marrow.

Melaene (melena) refers to the black, "tarry" feces that are associated with gastrointestinal hemorrhage.

Melanogenesis production of melanin by living cells.

Melanoma malignant tumour of melanocytes which are found predominantly in skin but also in the bowel and the eye and appear as pigmented lesions.

Melatonin a hormone produced in the brain by the pineal gland, it is important in the regulation of the circadian rhythms of several biological functions.

Menarche the first menstrual cycle, or first menstrual bleeding, in female human beings.

Menorrhagia heavy or too-frequent menstrual periods.

Menopausal refer to permanent cessation of menstruation.

Menses see menstruation.

Menstruation the approximately monthly discharge of blood from the womb in women of childbearing age who are not pregnant. Also called menses. *Adj.* menstrual.

Mesangial cells are specialized cells around blood vessels in the kidneys, at the mesangium.

Metalloproteinase enzymes that breakdown proteins and requiring zinc or calcium atoms for proper function.

Metaphysis is the portion of a long bone between the epiphyses and the diaphysis of the femur.

Metaphyseal pertaining to the metaphysis.

Metaplasia transformation of one type of one mature differentiated cell type into another mature differentiated cell type.

Metastasis is the movement or spreading of cancer cells from one organ or tissue to another.

Metetrus the quiescent period of sexual inactivity between oestrus cycle.

Metroptosis the slipping or falling out of place of an organ (as the uterus).

Metrorrhagia uterine bleeding at irregular intervals, particularly between the expected menstrual periods.

MIC minimum inhibitory concentration – lowest concentration of an antimicrobial that will inhibit the visible growth of a microorganism.

Micelle a submicroscopic aggregation of molecules.

Micellization formation process of micelles.

Microangiopathy (or microvascular disease) is an angiopathy affecting small blood vessels in the body.

Microfilaria a pre-laval parasitic worm of the family Onchocercidae, found in the vector and in the blood or tissue fluid of human host.

Micronuclei small particles consisting of acentric fragments of chromosomes or entire chromosomes, which lag behind at anaphase of cell division.

Microsomal PGE2 synthase is the enzyme that catalyses the final step in prostaglandin E2 (PGE2) biosynthesis.

Microvasculature the finer vessels of the body, as the arterioles, capillaries, and venules.

Micturition urination, act of urinating.

Migraine a neurological syndrome characterized by altered bodily perceptions, severe, painful headaches, and nausea.

Mimosine is an alkaloid, β-3-hydroxy-4 pyridone amino acid, it is a toxic non-protein free amino acid and is an antinutrient.

Mineral apposition rate MAR, rate of addition of new layers of mineral on the trabecular surfaces of bones.

Miscarriage spontaneous abortion.

Mitochondrial complex I the largest enzyme in the mitochondrial respiratory oxidative phosphorylation system.

Mitochondrial permeability transition (MPT) is an increase in the permeability of the mitochondrial membranes to molecules of less than 1,500 Da in molecular weight. MPT is one of the major causes of cell death in a variety of conditions.

Mitogen an agent that triggers mitosis, elicit all the signals necessary to induce cell proliferation.

Mitogenic able to induce mitosis or transformation.

Mitogenicity process of induction of mitosis.

Mitomycin a chemotherapy drug that is given as a treatment for several different types of cancer, including breast, stomach, oesophagus and bladder cancers.

Mitosis cell division in which the nucleus divides into nuclei containing the same number of chromosomes.

MMP matrix metalloproteinases, a group of peptidases involved in degradation of the extracellular matrix (ECM).

Molecular docking is a key tool in structural molecular biology and computer-assisted drug design.

Molluscidal destroying molluscs like snails.

Molt 4 cells MOLT4 cells are lymphoblast-like in morphology and are used for studies of apoptosis, tumour cytotoxicity, tumorigenicity, as well as for antitumour testing.

Molybdenum (Mo) is an essential element that forms part of several enzymes such as xanthine oxidase involved in the oxidation of xanthine to uric acid and use of iron. Molybdenum concentrations also affect protein synthesis, metabolism, and growth. Dietary sources include meat, green beans, eggs, sunflower seeds, wheat flour, lentils, and cereal grain.

Monoamine oxidase A (MAOA) is an isozyme of monoamine oxidase. It preferentially deaminates norepinephrine (noradrenaline), epinephrine (adrenaline), serotonin, and dopamine.

Monoaminergic of or pertaining to neurons that secrete monoamine neurotransmitters (e.g., dopamine, serotonin).

Monoclonal antibodies are produced by fusing single antibody-forming cells to tumour cells grown in culture.

Monocyte large white blood cell that ingest microbes, other cells and foreign matter.

Monogalactosyl diglyceride are the major lipid components of chloroplasts.

Monorrhagia is heavy bleeding and that's usually defined as periods lasting longer than 7 days or excessive bleeding.

Morbidity a diseased state or symptom or can refer either to the incidence rate or to the prevalence rate of a disease.

Morelloflavone a biflavonoid extracted from *Garcinia dulcis*, has shown antioxidative, antiviral, and anti-inflammatory properties.

Morphine the major alkaloid of opium and a potent narcotic analgesic.

MTTP microsomal triglyceride transfer protein that is required for the assembly and secretion of triglyceride-rich lipoproteins from both enterocytes and hepatocytes.

Mucositis painful inflammation and ulceration of the mucous membranes lining the digestive tract.

Mucous relating to mucus.

Mucolytic capable of reducing the viscosity of mucus, or an agent that so acts.

Mucus viscid secretion of the mucous membrane.

Multidrug resistance (MDR) ability of a living cell to show resistance to a wide variety of structurally and functionally unrelated compounds.

Muscarinic receptors are G protein-coupled acetylcholine receptors found in the plasma membranes of certain neurons and other cells.

Mutagen an agent that induces genetic mutation by causing changes in the DNA.

Mutagenic capable of inducing mutation (used mainly for extracellular factors such as X-rays or chemical pollution).

Mycosis an infection or disease caused by a fungus.

Myelocyte is a young cell of the granulocytic series, occurring normally in bone marrow, but not in circulating blood.

Myeloid leukaemia (Chronic) a type of cancer that affects the blood and bone marrow, characterized by excessive number of white blood cells.

Myeloma cancer that arise in the plasma cells a type of white blood cells.

Myeloperoxidase (MPO) is a peroxidase enzyme most abundantly present in neutrophil granulocytes (a subtype of white blood cells). It is an inflammatory enzyme produced by activated leukocytes that predicts risk of coronary heart disease.

Myeloproliferative disorder disease of the bone marrow in which excess cells are produced.

Myocardial relating to heart muscles tissues.

Myocardial infarction (MI) is the rapid development of myocardial necrosis caused by a critical imbalance between oxygen supply and demand of the myocardium.

Myocardial ischemia an intermediate condition in coronary artery disease during which the heart tissue is slowly or suddenly starved of oxygen and other nutrients.

Myopia near – or short-sightedness.

Myosarcoma a malignant muscle tumour.

Myotonia dystrophica an inherited disorder of the muscles and other body systems characterized by progressive muscle weakness, prolonged muscle contractions (myotonia), clouding of the lens of the eye (cataracts), cardiac abnormalities, balding, and infertility.

Myringosclerosis also known as tympanosclerosis or intratympanic tympanosclerosis, is a condition caused by calcification of collagen tissues in the tympanic membrane of the middle ear.

Mytonia a symptom of certain neuromuscular disorders characterized by the slow relaxation of the muscles after voluntary contraction or electrical stimulation.

Myotube a developing skeletal muscle fibre with a tubular appearance.

N-nitrosmorpholine a human carcinogen.

N-nitrosoproline an indicator for N-nitrosation of amines.

NADPH The reduced form of nicotinamide adenine dinucleotide phosphate that serves as an electron carrier.

NAFLD Non-alcoholic fatty liver disease.

Narcotic an agent that produces narcosis, in moderate doses it dulls the senses, relieves pain and induces sleep; in excessive dose it cause stupor, coma, convulsions and death.

Nasopharynx upper part of the alimentary continuous with the nasal passages.

Natriuresis the discharge of excessive large amount of sodium through urine. *adj.* natriuretic.

Nausea sensation of unease and discomfort in the stomach with an urge to vomit.

Necropsy see autopsy.

Necrosis morphological changes that follow cell death, usually involving nuclear and cytoplasmic changes.

Neoplasia abnormal growth of cells, which may lead to a neoplasm, or tumour.

Neoplasm tumour; any new and abnormal growth, specifically one in which cell multiplication is uncontrolled and progressive. Neoplasms may be benign or malignant.

Neoplastic transformation conversion of a tissue with a normal growth pattern into a malignant tumour.

Nephrectomised kidneys surgically removed.

Nephrectomy surgical removal of the kidney.

Nephric relating to or connected with a kidney.

Nephrin is a protein necessary for the proper functioning of the renal filtration barrier.

Nephritis is inflammation of the kidney.

Nephrolithiasis process of forming a kidney stone in the kidney or lower urinary tract.

Nephropathy a disorder of the kidney.

Nephrotic syndrome nonspecific disorder in which the kidney glomeruli are damaged, causing them to leak large amounts of protein.

Nephrotoxicity poisonous effect of some substances, both toxic chemicals and medication, on the kidney.

Nerve growth factor (NGF) a small protein that induces the differentiation and survival of particular target neurons (nerve cells).

Nervine a nerve tonic that acts therapeutically upon the nerves, particularly in the sense of a sedative that serves to calm ruffled nerves.

Neuralgia is a sudden, severe painful disorder of the nerves.

Neuraminidase inhibitors a class of antiviral drugs targeted at the influenza viruses whose mode of action consists of blocking the function of the viral neuraminidase protein, thus preventing the virus from reproducing.

Neurasthenia a condition with symptoms of fatigue, anxiety, headache, impotence, neuralgia and impotence.

Neurasthenic a substance used to treat nerve pain and/or weakness (i.e. neuralgia, sciatica, etc).

Neurite refers to any projection from the cell body of a neuron.

Neuritis an inflammation of the nerve characterized by pain, sensory disturbances and impairment of reflexes. *adj.* neuritic.

Neuritogenesis the first step of neuronal differentiation, takes place as nascent neurites bud from the immediate postmitotic neuronal soma.

Neuroblastoma a common extracranial cancer that forms in nerve tissues, common in infancy.

Neuroleptic refers to the effects on cognition and behavior of antipsychotic drugs that reduce confusion, delusions, hallucinations, and psychomotor agitation in patients with psychoses.

Neuropharmacological relating the effects of drugs on the neurosystem.

Neuroradiology is a subspecialty of radiology focusing on the diagnosis and characterization of abnormalities of the central and peripheral nervous system. *adj.* neuroradiologic.

Neurotrophic relating to neutrophy i.e. the nutrition and maintenance of nervous tissue.

Neutropenia a disorder of the blood, characterized by abnormally low levels of neutrophils.

Neutrophil a type of white blood cell, specifically a form of granulocyte.

Neutrophin protein that induce the survival, development and function of neurons.

NF-kappa B (NF-kB) nuclear factor kappa B, is an ubiquitous rapid response transcription factor in cells involved in immune and inflammatory reactions.

Niacin Vitamin B3. See vitamin B3.

Niacinamide an amide of niacin, also known as nicotinamide. See vitamin B3.

NIH3T3 cells a mouse embryonic fibroblast cell line used in the cultivation of keratinocytes.

Nitrogen (N) is an essential building block of amino and nucleic acids and proteins and is essential to all living organisms. Protein rich vegetables like legumes are rich food sources of nitrogen.

NK cells Natural killer cells, a type of cytotoxic lymphocyte that constitute a major component of the innate immune system.

NMDA receptor N-methyl-D-aspartate receptor. A brain receptor activated by the amino acid glutamate, which when excessively stimulated may cause cognitive defects in Alzheimer's disease.

Nootropics are substances which are claimed to boost human cognitive abilities (the functions and capacities of the brain). Also popularly referred to as "smart drugs", "smart nutrients", "cognitive enhancers" and "brain enhancers".

Noradrenalin see Norepinephrine.

Norepinephrine a substance, both a hormone and neurotransmitter, secreted by the adrenal medulla and the nerve endings of the sympathetic nervous system to cause vasoconstriction and increases in heart rate, blood pressure, and the sugar level of the blood. Also called levarterenol, noradrenalin.

Normoglycaemic having the normal amount of glucose in the blood.

Nosocomial infections infections which are a result of treatment in a hospital or a healthcare service unit, but secondary to the patient's original condition.

Nulliparous term used to describe a woman who has never given birth.

Nyctalopia night blindness, impaired vision in dim light and in the dark, due to impaired function of certain specialized vision cells.

Nycturia excessive urination at night; especially common in older men.

Occulsion closure or blockage (as of a blood vessel).

Odontalgia toothache. *adj.* odontalgic.

Odontopathy any disease of the teeth.

Oedema see edema.

Oligonucleosome type of nucleosome that form the fundamental repeating units of eukaryotic chromatin.

Oligoarthritis an inflammation of two, three or four joints.

Oligospermia or oligozoospermia refers to semen with a low concentration of sperm, commonly associated with male infertility.

Oliguria decreased production of urine.

Omega 3 fatty acids are essential polyunsaturated fatty acids that have in common a final carbon–carbon double bond in the n−3 position. Dietary sources of omega-3 fatty acids include fish oil and certain plant/nut oils. The three most nutritionally important omega 3 fatty acids are alpha-linolenic acid, eicosapentaenoic acid (EPA) and docosahexaenoic acid (DHA). Research indicates that omega 3 fatty acids are important in health promotion and disease and can help prevent a wide range of medical problems, including cardiovascular disease, depression, asthma, and rheumatoid arthritis.

Omega 6 fatty acids are essential polyunsaturated fatty acids that have in common a final carbon–carbon double bond in the n−6 position. Omega-6 fatty acids are considered essential fatty acids (EFAs) found in vegetable oils, nuts and seeds. They are essential to human health but cannot be made in the body. Omega-6 fatty acids – found in vegetable oils, nuts and seeds – are a beneficial part of a heart-healthy eating. Omega-6 and omega-3 PUFA play a crucial role in heart and brain function and in normal growth and development. Linoleic acid (LA) is the main omega-6 fatty acid in foods, accounting for 85–90% of the dietary omega-6 PUFA. Other omega 6 acids include gamma-linolenic acid or GLA, sometimes called gamoleic acid, eicosadienoic acid, arachidonic acid and docosadienoic acid.

Omega 9 fatty acids are not essential polyunsaturated fatty acids that have in common a final carbon–carbon double bond in the n−9 position. Some n−9s are common components of animal fat and vegetable oil. Two n−9 fatty acids important in industry are: oleic acid (18:1, n−9), which is a main component of olive oil and erucic acid (22:1, n−9), which is found in rapeseed, wallflower seed, and mustard seed.

Oncogenes genes carried by tumour viruses that are directly and solely responsible for the neoplastic (tumorous) transformation of host cells.

Ophthalmia severe inflammation of eye, or the conjunctiva or deeper structures of the eye. Also called ophthalmitis.

Ophthalmia (Sympathetic) inflammation of both eyes following trauma to one eye.

Opiate drug derived from the opium plant.

Opioid receptors a group of G-protein coupled receptors located in the brain and various organs that bind opiates or opioid substances.

Optic placode an ectodermal placode from which the lens of the embryonic eye develops; also called lens placode.

ORAC (Oxygen radical absorbance capacity) a method of measuring antioxidant capacities in biological samples.

Oral submucous fibrosis a chronic debilitating disease of the oral cavity characterized by inflammation and progressive fibrosis of the submucosa tissues.

Oral thrush an infection of yeast fungus, *Candida albicans*, in the mucous membranes of the mouth.

Orchidectomy surgery to remove one or both testicles.

Orchidectomised with testis removed.

Orchitis an acute painful inflammatory reaction of the testis secondary to infection by different bacteria and viruses.

Orofacial dyskinesia abnormal involuntary movements involving muscles of the face, mouth, tongue, eyes, and occasionally, the neck – may be unilateral or bilateral, and constant or intermittent.

Ostalgia, Ostealgia pain in the bones, also called osteodynia.

Osteoarthritis is the deterioration of the joints that becomes more common with age.

Osteoarthrosis chronic noninflammatory bone disease.

Osteoblast a mononucleate cell that is responsible for bone formation.

Osteoblastic relating to osteoblasts.

Osteocalcin a noncollagenous protein found in bone and dentin, also refer to as bone gamma-carboxyglutamic acid-containing protein.

Osteoclasts a kind of bone cell that removes bone tissue by removing its mineralized matrix.
Osteoclastogenesis the production of osteoclasts.
Osteodynia pain in the bone.
Osteomalacia refers to the softening of the bones due to defective bone mineralization.
Osteomyelofibrosis a myeloproliferative disorder in which fibrosis and sclerosis finally lead to bone marrow obliteration.
Osteopenia reduction in bone mass, usually caused by a lowered rate of formation of new bone that is insufficient to keep up with the rate of bone destruction.
Osteoporosis a disease of bone that leads to an increased risk of fracture.
Osteoprotegerin also called osteoclastogenesis inhibitory factor (OCIF), a cytokine, which can inhibit the production of osteoclasts.
Osteosacrcoma a malignant bone tumour. Also called osteogenic sarcoma.
Otalgia earache, pain in the ear.
Otic placode a thickening of the ectoderm on the outer surface of a developing embryo from which the ear develops.
Otitis inflammation of the inner or outer parts of the ear.
Otorrhea running drainage (discharge) exiting the ear.
Ovariectomised with one or two ovaries removed.
Ovariectomy surgical removal of one or both ovaries.
Oxidation the process of adding oxygen to a compound, dehydrogenation or increasing the electro-negative charge.
Oxidoreductase activity catalysis of an oxidation-reduction (redox) reaction, a reversible chemical reaction. One substrate acts as a hydrogen or electron donor and becomes oxidized, while the other acts as hydrogen or electron acceptor and becomes reduced.
Oxytocic *adj.* hastening or facilitating childbirth, especially by stimulating contractions of the uterus.
Oxytocin is a mammalian hormone that also acts as a neurotransmitter in the brain. It is best known for its roles in female reproduction: it is released in large amounts after distension of the cervix and vagina during labor, and after stimulation of the nipples, facilitating birth and breastfeeding, respectively.
Oxygen radical absorbance capacity (ORAC) a method of measuring antioxidant capacities in biological samples.
Oxyuriasis infestation by pinworms.
Ozoena discharge of the nostrils caused by chronic inflammation of the nostrils.
P53 (also known as protein 53 or tumour protein 53), is a tumour suppressor protein that in humans is encoded by the TP53 gene.
P-glycoprotein (P-gp, ABCB1, MDR1) a cell membrane-associated drug-exporting protein that transports a variety of drug substrates from cancer cells.
Palpitation rapid pulsation or throbbing of the heart.
Paludism state of having symptoms of malaria characterized by high fever and chills.
Pancreatitis inflammation of the pancreas.
Pantothenic acid vitamin B5. See vitamin B5.
Papain a protein degrading enzyme used medicinally and to tenderize meat.
Papilloma a benign epithelial tumour growing outwardly like in finger-like fronds.
Papule a small, solid, usually inflammatory elevation of the skin that does not contain pus.
Paradontosis is the inflammation of gums and other deeper structures, including the bone.
Paralytic person affected with paralysis, pertaining to paralysis.
Parasympathetic nervous system subsystem of the nervous systems that slows the heart rate and increases intestinal and gland activity and relaxes the sphincter muscles.
Parasympathomimetic having an action resembling that caused by stimulation of the parasympathetic nervous system.
Parenteral administration administration by intravenous, subcutaneous or intramuscular routes.
Paresis a condition characterised by partial loss of movement, or impaired movement.
Paresthesia is an abnormal sensation of the skin, such as burning, numbness, itching, hyperesthesia (increased sensitivity) or tingling, with no apparent physical cause.

Parenteral is a route of administration via the veins that involves piercing the skin or mucous membrane.

Parotitis inflammation of salivary glands.

Paroxysm a sudden outburst of emotion or action, a sudden attack, recurrence or intensification of a disease.

Paroxystic relating to an abnormal event of the body with an abrupt onset and an equally sudden return to normal.

PARP see poly (ADP-ribose) polymerase.

Parturition act of child birth.

PCE/PCN ratio polychromatic erythrocyte/normochromatic erythrocyte ratio use as a measure of cytotoxic effects.

pCREB phosphorylated cAMP (adenosine 3'5' cyclic monophosphate)-response element binding protein.

PDEF acronym for prostate-derived ETS factor, an ETS (epithelial-specific E26 transforming sequence) family member that has been identified as a potential tumour suppressor.

Pectoral pertaining to or used for the chest and respiratory tract.

pERK phosphorylated extracellular signal-regulated kinase, protein kinases involved in many cell functions.

Peliosis see purpura.

Pellagra is a systemic nutritional wasting disease caused by a deficiency of vitamin B3 (niacin).

Pemphigus neonatorum Staphylococcal scalded skin syndrome, a bacterial disease of infants, characterized by elevated vesicles or blebs on a normal or reddened skin.

Peptic ulcer a sore in the lining of the stomach or duodenum, the first part of the small intestine.

(Per os by way of the mouth) Percutanous pertains to a medical procedure where access to inner organs or tissues is done via needle puncture of the skin.

Periapical periodontitis is the inflammation of the tissue adjacent to the tip of the tooth's root.

Perifuse To flush a fresh supply of bathing fluid around all of the outside surfaces of a small piece of tissue immersed in it.

Periodontal ligament (PDL) is a group of specialized connective tissue fibres that essentially attach a tooth to the bony socket.

Periodontitis is a severe form of gingivitis in which the inflammation of the gums extends to the supporting structures of the tooth. Also called pyorrhea.

Peripheral neuropathy refers to damage to nerves of the peripheral nervous system.

Peristalsis a series of organized, wave-like muscle contractions that occur throughout the digestive tract.

Perlingual through or by way of the tongue.

Perniosis an abnormal reaction to cold that occurs most frequently in women, children, and the elderly. Also called chilblains.

Peroxisome proliferator-activated receptors (PPARs) a family of nuclear receptors that are involved in lipid metabolism, differentiation, proliferation, cell death, and inflammation.

Pertussis whooping cough, sever cough.

Peyers Patches patches of lymphoid tissue or lymphoid nodules on the walls of the ileal-small intestine.

PGE-2 Prostaglandin E2, a hormone-like substance that is released by blood vessel walls in response to infection or inflammation that acts on the brain to induce fever.

Phagocytes are the white blood cells that protect the body by ingesting (phagocytosing) harmful foreign particles, bacteria and dead or dying cells. *adj.* phagocytic.

Phagocytosis is process the human body uses to destroy dead or foreign cells.

Pharmacognosis the branch of pharmacology that studies the composition, use, and history of drugs.

Pharmacopoeia authoritative treatise containing directions for the identification of drug samples and the preparation of compound medicines, and published by the authority of a government or a medical or pharmaceutical society and in a broader sense is a general reference work for pharmaceutical drug specifications.

Pharyngitis, Pharyngolaryngitis inflammation of the pharynx and the larynx.

Phenolics class of chemical compounds consisting of a hydroxyl group (-OH) bonded directly to an aromatic hydrocarbon group.

Pheochromocytoma is a tumour that usually originates from the adrenal glands' chromaffin cells, causing overproduction of catecholamines, powerful hormones that induce high blood pressure and other symptoms.

Phlebitis is an inflammation of a vein, usually in the legs.

Phlegm abnormally viscid mucus secreted by the mucosa of the respiratory passages during certain infectious processes.

Phlegmon a spreading, diffuse inflammation of the soft or connective tissue due to infection by Streptococci bacteria.

Phoroglucinol a white, crystalline compound used as an antispasmodic, analytical reagent, and decalcifier of bone specimens for microscopic examination.

Phosphatidylglycerol is a glycerophospholipid found in pulmonary active surface lipoprotein and consists of a L-glycerol 3-phosphate backbone ester-bonded to either saturated or unsaturated fatty acids on carbons 1 and 2.

Phosphatidylinositol 3-kinases (PI 3-kinases or PI3Ks) a group of enzymes involved in cellular functions such as cell growth, proliferation, differentiation, motility, survival and intracellular trafficking, which in turn are involved in cancer.

Phospholipase an enzyme that hydrolyzes phospholipids into fatty acids and other lipophilic substances.

Phospholipase A2 (PLA2) a small lipolytic enzyme that releases fatty acids from the second carbon group of glycerol. Plays an essential role in the synthesis of prostaglandins and leukotrienes.

Phospholipase C enzymes that cleaves phospholipase.

Phospholipase C gamma (PLC gamma) enzymes that cleaves phospholipase in cellular proliferation and differentiation, and its enzymatic activity is upregulated by a variety of growth factors and hormones.

Phosphorus (P) is an essential mineral that makes up 1% of a person's total body weight and is found in the bones and teeth. It plays an important role in the body's utilization of carbohydrates and fats; in the synthesis of protein for the growth, maintenance, and repair of cells and tissues. It is also crucial for the production of ATP, a molecule the body uses to store energy. Main sources are meat and milk; fruits and vegetables provides small amounts.

Photoaging is the term that describes damage to the skin caused by intense and chronic exposure to sunlight resulting in premature aging of the skin.

Photocarcinogenesis represents the sum of a complex of simultaneous and sequential biochemical events that ultimately lead to the occurrence of skin cancer.

Photopsia an affection of the eye, in which the patient perceives luminous rays, flashes, coruscations, etc.

Photosensitivity sensitivity toward light.

Phthisis an archaic name for tuberculosis.

Phytohemagglutinin a lectin found in plant that is involved in the stimulation of lymphocyte proliferation.

Phytonutrients certain organic components of plants, that are thought to promote human health. Fruits, vegetables, grains, legumes, nuts and teas are rich sources of phytonutrients. Phytonutrients are not 'essential' for life. Also called phytochemicals.

Phytosterols a group of steroid alcohols, cholesterol-like phytochemicals naturally occurring in plants like vegetable oils, nuts and legumes.

Piebaldism rare autosomal dominant disorder of melanocyte development characterized by distinct patches of skin and hair that contain no pigment.

Piles see haemorrhoids.

Pityriasis lichenoides is a rare skin disorder of unknown aetiology characterised by multiple papules and plaques.

PKC protein kinase C, a membrane bound enzyme that phosphorylates different intracellular proteins and raised intracellular Ca levels.

PKC Delta inhibitors Protein Kinase C delta inhibitors that induce apoptosis of haematopoietic cell lines.

Placode a platelike epithelial thickening in the embryo where some organ or structure later develops.

Plasma the yellow-colored liquid component of blood, in which blood cells are suspended

Plasma kallikrien a serine protease, synthesized in the liver and circulates in the plasma.

Plasmin a proteinase enzyme that is responsible for digesting fibrin in blood clots.

Plasminogen the proenzyme of plasmin, whose primary role is the degradation of fibrin in the vasculature.

Plaster poultice.

Platelet activating factor (PAF) is an acetylated derivative of glycerophosphorylcholine, released by basophils and mast cells in immediate hypersensitive reactions and macrophages and neutrophils in other inflammatory reactions. One of its main effects is to induce platelet aggregation.

PLC gamma phospholipase C gamma plays a central role in signal transduction.

Pleurisy is an inflammation of the pleura, the lining of the pleural cavity surrounding the lungs, which can cause painful respiration and other symptoms. Also known as pleuritis.

Pneumonia an inflammatory illness of the lung caused by bacteria or viruses.

Pneumotoxicity damage to lung tissues.

Poliomyelitis is a highly infectious viral disease that may attack the central nervous system and is characterized by symptoms that range from a mild non-paralytic infection to total paralysis in a matter of hours; also called polio or infantile paralysis.

Poly (ADP-ribose) polymerase (PARP) a protein involved in a number of cellular processes especially DNA repair and programmed cell death.

Polyarthritis is any type of arthritis which involves five or more joints.

Polychromatic erythrocyte (PCE) an immature red blood cell containing RNA, that can be differentiated by appropriate staining techniques from a normochromatic erythrocyte (NCE), which lacks RNA.

Polycythaemia a type of blood disorder characterised by the production of too many red blood cells.

(Polydypsia excessive thirst) Polymorphnuclear having a lobed nucleus. Used especially of neutrophilic white blood cells.

Polyneuritis widespread inflammation of the nerves.

Polyneuritis gallinarum a nervous disorder in birds and poultry.

Polyp a growth that protrudes from a mucous membrane.

Polyphagia medical term for excessive hunger or eating.

Polyuria a condition characterized by the passage of large volumes of urine with an increase in urinary frequency.

Pomade a thick oily dressing.

Porphyrin any of a class of water-soluble, nitrogenous biological pigments.

Postpartum Depression depression after pregnancy; also called postnatal depression.

Postprandial after mealtime.

Potassium (K) is an element that's essential for the body's growth and maintenance. It's necessary to keep a normal water balance between the cells and body fluids, for cellular enzyme activities and plays an essential role in the response of nerves to stimulation and in the contraction of muscles. Potassium is found in many plant foods and fish (tuna, halibut): chard, mushrooms, spinach, fennel, kale, mustard greens, Brussel sprouts, broccoli, cauliflower, cabbage winter squash, eggplant, cantaloupe, tomatoes, parsley, cucumber, bell pepper, turmeric, ginger root, apricots, strawberries, avocado and banana.

Poultice is a soft moist mass, often heated and medicated, that is spread on cloth over the skin to treat an aching, inflamed, or painful part of the body. Also called cataplasm.

PPARs peroxisome proliferator-activated receptors – a group of nuclear receptor proteins that function as transcription factors regulating the expression of genes.

Prebiotics a category of functional food, defined as non-digestible food ingredients that beneficially affect the host by selectively stimulating the growth and/or activity of one or a limited number of bacteria in the colon, and thus improve host health. *cf.* probiotics.

Pre-eclampsia see toxemia.

Prenidatory referring to the time period between fertilization and implantation.

Prenylated flavones flavones with an isoprenyl group in the 8-position, has been reported to have good anti-inflammatory properties.

Proangiogenic promote angiogensis (formation and development of new blood vessels).

Probiotics are dietary supplements and live microorganisms containing potentially beneficial bacteria or yeasts that are taken into the alimentary system for healthy intestinal functions. *cf.* prebiotics.

Progestational of or relating to the phase of the menstrual cycle immediately following ovulation, characterized by secretion of progesterone.

Prognosis medical term to describe the likely outcome of an illness.

Prolapsus to fall or slip out of place.

Proliferating cell nuclear antigen (PCNA) a new marker to study human colonic cell proliferation.

Proliferative vitreoretinopathy (PVR) a most common cause of failure in retinal reattachment surgery. Characterized by the formation of cellular membrane on both surfaces of the retina and in the vitreous.

Promastigote the flagellate stage in the development of trypanosomatid protozoa, characterized by a free anterior flagellum.

Promyelocytic leukemia a subtype of acute myelogenous leukemia (AML), a cancer of the blood and bone marrow.

Pro-oxidants chemicals that induce oxidative stress, either through creating reactive oxygen species or inhibiting antioxidant systems.

Prophylaxis prevention or protection against disease.

Proptosis see exophthalmos.

Prostacyclin a prostaglandin that is a metabolite of arachidonic acid, inhibits platelet aggregation, and dilates blood vessels.

Prostaglandins a family of C 20 lipid compounds found in various tissues, associated with muscular contraction and the inflammation response such as swelling, pain, stiffness, redness and warmth.

Prostaglandin E2 (PEG -2) one of the prostaglandins, a group of hormone-like substances that participate in a wide range of body functions such as the contraction and relaxation of smooth muscle, the dilation and constriction of blood vessels, control of blood pressure, and modulation of inflammation.

Prostaglandin E synthase an enzyme that in humans is encoded by the glutathione-dependent PTGES gene.

Prostanoids term used to describe a subclass of eicosanoids (products of COX pathway) consisting of: the prostaglandins (mediators of inflammatory and anaphylactic reactions), the thromboxanes (mediators of vasoconstriction) and the prostacyclins (active in the resolution phase of inflammation.)

Prostate a gland that surround the urethra at the bladder in the male.

Prostate cancer a disease in which cancer develops in the prostate, a gland in the male reproductive system. Symptoms include pain, difficulty in urinating, erectile dysfunction and other symptoms.

Protein kinase C (PKC) a family of enzymes involved in controlling the function of other proteins through the phosphorylation of hydroxyl groups of serine and threonine amino acid residues on these proteins. PKC enzymes play important roles in several signal transduction cascades.

Protein tyrosine phosphatase (PTP) a group of enzymes that remove phosphate groups from phosphorylated tyrosine residues on proteins.

Proteinase a protease (enzyme) involved in the hydrolytic breakdown of proteins, usually by splitting them into polypeptide chains.

Proteinuria means the presence of an excess of serum proteins in the urine.

Proteolysis cleavage of the peptide bonds in protein forming smaller polypeptides. *adj.* proteolytic.

Proteomics the large-scale study of proteins, particularly their structures and functions.

Prothrombin blood-clotting protein that is converted to the active form, factor IIa, or thrombin, by cleavage.

Prothyroid good for thyroid function.

Proto-oncogene A normal gene which, when altered by mutation, becomes an oncogene that can contribute to cancer.

Prurigo a general term used to describe itchy eruptions of the skin.

Pruritis defined as an unpleasant sensation on the skin that provokes the desire to rub or scratch the area to obtain relief; itch, itching. adj. pruritic.

PSA Prostate Specific Antigen, a protein which is secreted into ejaculate fluid by the healthy prostate. One of its functions is to aid sperm movement.

Psoriasis a common chronic, non-contagious autoimmune dermatosis that affects the skin and joints.

Psychoactive having effects on the mind or behavior.

Psychonautics exploration of the psyche by means of approaches usch as meditation, prayer, lucid dreaming, brain wave entrainment etc.

Psychotomimetic hallucinogenic.

Psychotropic capable of affecting the mind, emotions, and behavior.

Pthysis silicosis with tuberculosis.

Ptosis drooping of the upper eye lid.

PTP protein tyrosine phosphatase.

PTPIB protein tyrosine phosphatase 1B

Puerperal pertaining to child birth.

Pulmonary embolism a blockage (blood clot) of the main artery of the lung.

Purgative a substance used to cleanse or purge, especially causing the immediate evacuation of the bowel.

Purpura is the appearance of red or purple discolorations on the skin that do not blanch on applying pressure. Also called peliosis.

Purulent containing pus discharge.

Pustule small, inflamed, pus-filled lesions.

Pyelonephritis an ascending urinary tract infection that has reached the pyelum (pelvis) of the kidney.

Pyodermatitis refers to inflammation of the skin.

Pyorrhea see periodontitis.

Pyretic referring to fever.

Pyrexia fever of unknown origin.

Pyridoxal a chemical form of vitamin B6. See vitamin B6.

Pyridoxamine a chemical form of vitamin B6. See vitamin B6.

Pyridoxine a chemical form of vitamin B6. See vitamin B6.

Pyrolysis decomposition or transformation of a compound caused by heat. *adj.* pyrolytic.

PYY Peptide a 36 amino acid peptide secreted by L cells of the distal small intestine and colon that inhibits gastric and pancreatic secretion.

QT interval is a measure of the time between the start of the Q wave and the end of the T wave in the heart's electrical cycle. A prolonged QT interval is a biomarker for ventricular tachyarrhythmias and a risk factor for sudden death.

Quorum sensing (QS) the control of gene expression in response to cell density, is used by both gram-negative and gram-positive bacteria to regulate a variety of physiological functions.

Radioprotective serving to protect or aiding in protecting against the injurious effect of radiations.

RAGE is the receptor for advanced glycation end products, a multiligand receptor that propagates cellular dysfunction in several inflammatory disorders, in tumours and in diabetes.

RAS see renin-angiotensin system or recurrent aphthous stomatitis.

Rash a temporary eruption on the skin, see uticaria.

Reactive oxygen species species such as superoxide, hydrogen peroxide, and hydroxyl radical. At low levels, these species may function in cell signalling processes. At higher levels, these species may damage cellular macromolecules (such as DNA and RNA) and participate in apoptosis (programmed cell death).

Rec A is a 38 kDa Escherichia coli protein essential for the repair and maintenance of DNA.

Recticulocyte non-nucleated stage in the development of the red blood cell.

Recticulocyte lysate cell lysate produced from reticulocytes, used as an in-vitro translation system.

Recticuloendothelial system part of the immune system, consists of the phagocytic cells located in reticular connective tissue, primarily monocytes and macrophages.

Recurrent aphthous stomatitis, or RAS is a common, painful condition in which recurring ovoid or round ulcers affect the oral mucosa.

Redox homeostasis is considered as the cumulative action of all free radical reactions and antioxidant defenses in different tissues.

Refrigerant a medicine or an application for allaying heat, fever or its symptoms.

Renal calculi kidney stones.

Renin-angiotensin system (RAS) also called the renin-angiotensin-aldosterone system (RAAS) is a hormone system that regulates blood pressure and water (fluid) balance.

Reperfusion the restoration of blood flow to an organ or tissue that has had its blood supply cut off, as after a heart attack.

Reporter gene a transfected gene that produces a signal, such as green fluorescence, when it is expressed.

Resistin A cysteine-rich protein secreted by adipose tissue of mice and rats.

Resolutive a substance that induces subsidence of inflammation.

Resolvent reduce inflammation or swelling.

Resorb to absorb or assimilate a product of the body such as an exudates or cellular growth.

Restenosis is the reoccurrence of stenosis, a narrowing of a blood vessel, leading to restricted blood flow.

Resveratrol is a phytoalexin produced naturally by several plants when under attack by pathogens such as bacteria or fungi. It is a potent antioxidant found in red grapes and other plants.

Retinol a form of vitamin A, see vitamin A.

Retinopathy a general term that refers to some form of non-inflammatory damage to the retina of the eye.

Revulsive counterirritant, used for swellings.

Rheumatic pertaining to rheumatism or to abnormalities of the musculoskeletal system.

Rheumatism, Rheumatic disorder, Rheumatic diseases refers to various painful medical conditions which affect bones, joints, muscles, tendons. Rheumatic diseases are characterized by the signs of inflammation – redness, heat, swelling, and pain.

Rheumatoid arthritis (RA) is a chronic, systemic autoimmune disorder that most commonly causes inflammation and tissue damage in joints (arthritis) and tendon sheaths, together with anemia.

Rhinitis irritation and inflammation of some internal areas of the nose and the primary symptom of rhinitis is a runny nose.

Rhinoplasty is surgery to repair or reshape the nose.

Rhinorrhea commonly known as a runny nose, characterized by an unusually significant amount of nasal discharge.

Rhinosinusitis inflammation of the nasal cavity and sinuses.

Rho GTPases Rho-guanosine triphosphate hydrolase enzymes are molecular switches that regulate many essential cellular processes, including actin dynamics, gene transcription, cell-cycle progression and cell adhesion.

Ribosome inactivating proteins protein that are capable of inactivating ribosomes.

Rickets is a softening of the bones in children potentially leading to fractures and deformity.

Ringworm dermatophytosis, a skin infection caused by fungus.

Roborant restoring strength or vigour, a tonic.

Rotavirus the most common cause of infectious diarrhea (gastroenteritis) in young children and infants, one of several viruses that causes infections called stomach flu.

Rubefacient a substance for external application that produces redness of the skin e.g. by causing dilation of the capillaries and an increase in blood.

S.C. abbreviation for sub-cutaneous, beneath the layer of skin.

S-T segment the portion of an electrocardiogram between the end of the QRS complex and the beginning of the T wave. Elevation or depression of the S-T segment is the characteristics of myocardial ischemia or injury and coronary artery disease.

Sapraemia see septicaemia.

Sarcoma cancer of the connective or supportive tissue (bone, cartilage, fat, muscle, blood vessels) and soft tissues.

SARS Severe acute respiratory syndrome, the name of a potentially fatal new respiratory disease in humans which is caused by the SARS coronavirus (SARS-CoV)

Satiety state of feeling satiated, fully satisfied (appetite or desire).

Scabies a transmissible ectoparasite skin infection characterized by superficial burrows, intense pruritus (itching) and secondary infection.

Scarlatina scarlet fever, an acute, contagious disease caused by infection with group A streptococcal bacteria.

Schistosomiasis is a parasitic disease caused by several species of fluke of the genus *Schistosoma*. Also known as bilharzia, bilharziosis or snail fever.

Schizophrenia a psychotic disorder (or a group of disorders) marked by severely impaired thinking, emotions, and behaviors.

Sciatica a condition characterised by pain deep in the buttock often radiating down the back of the leg along the sciatic nerve.

Scleroderma a disease of the body's connective tissue. The most common symptom is a thickening and hardening of the skin, particularly of the hands and face.

Scrofula a tuberculous infection of the skin on the neck caused by the bacterium *Mycobacterium tuberculosis*.

Scrophulosis see scrofula.

Scurf abnormal skin condition in which small flakes or sales become detached.

Scurvy a state of dietary deficiency of vitamin C (ascorbic acid) which is required for the synthesis of collagen in humans.

Secretagogue a substance that causes another substance to be secreted.

Sedative having a soothing, calming, or tranquilizing effect; reducing or relieving stress, irritability, or excitement.

Seizure the physical findings or changes in behavior that occur after an episode of abnormal electrical activity in the brain.

Selenium (Se) a trace mineral that is essential to good health but required only in tiny amounts; it is incorporated into proteins to make selenoproteins, which are important antioxidant enzymes. It is found in avocado, brazil nut, lentils, sunflower seeds, tomato, whole grain cereals, seaweed, seafood and meat.

Sensorineural bradyacuasia hearing impairment of the inner ear resulting from damage to the sensory hair cells or to the nerves that supply the inner ear.

Sepsis a condition in which the body is fighting a severe infection that has spread via the bloodstream.

Sequela an abnormal pathological condition resulting from a disease, injury or trauma.

Serine proteinase peptide hydrolases which have an active centre histidine and serine involved in the catalytic process.

Serotonergic liberating, activated by, or involving serotonin in the transmission of nerve impulses.

Serotonin a monoamine neurotransmitter synthesized in serotonergic neurons in the central nervous system.

Sepsis condition in which the body is fighting a severe infection that has spread via the bloodstream.

Septicaemia a systemic disease associated with the presence and persistence of pathogenic microorganisms or their toxins in the blood.

Sequelae a pathological condition resulting from a prior disease, injury, or attack.

Sexual potentiator increases sexual activity and potency, enhances sexual performance due to increased blood flow and efficient metabolism.

Sexually transmitted diseases (STD) infections that are transmitted through sexual activity.

SGOT, Serum glutamic oxaloacetic transaminase an enzyme that is normally present in liver and heart cells. SGOT is released into blood when the liver or heart is damaged. Also called aspartate transaminase (AST).

SGPT, Serum glutamic pyruvic transaminase an enzyme normally present in serum and body tissues, especially in the liver; it is released into the serum as a result of tissue injury, also called Alanine transaminase (ALT),

Shiga–like toxin a toxin produced by the bacterium *Escherichia coli* which disrupts the function of ribosomes, also known as verotoxin.

Shiga toxigenic *Escherichia coli* **(STEC)** comprises a diverse group of organisms capable of causing severe gastrointestinal disease in humans.

Shiga toxin a toxin produced by the bacterium *Shigella dysenteriae*, which disrupts the function of ribosomes.

Shingles skin rash caused by the Zoster virus (same virus that causes chicken pox) and is medically termed Herpes zoster.

Sialogogue salivation-promoter, a substance used to increase or promote the excretion of saliva.

Sialoproteins glycoproteins that contain sialic acid as one of their carbohydrates.

Sialyation reaction with sialic acid or its derivatives; used especially with oligosaccharides.

Sialyltransferases enzymes that transfer sialic acid to nascent oligosaccharide.

Sickle cell disease is an inherited blood disorder that affects red blood cells. People with sickle cell disease have red blood cells that contain mostly hemoglobin S, an abnormal type of hemoglobin. Sometimes these red blood cells become sickle-shaped (crescent shaped) and have difficulty passing through small blood vessels.

Side stich is an intense stabbing pain under the lower edge of the ribcage that occurs while exercising.

Signal transduction cascade refers to a series of sequential events that transfer a signal through a series of intermediate molecules until final regulatory molecules, such as transcription factors, are modified in response to the signal.

Silicon (Si) is required in minute amounts by the body and is important for the development of healthy hair and the prevention of nervous disorders. Lettuce is the best natural source of Silicon.

Sinapism signifies an external application, in the form of a soft plaster, or poultice.

Sinusitis inflammation of the nasal sinuses.

SIRC cells Statens Seruminstitut Rabbit Cornea (SIRC) cell line.

Skp1 (S-phase kinase-associated protein 1) is a core component of SCF ubiquitin ligases and mediates protein degradation.

Smallpox is an acute, contagious and devastating disease in humans caused by Variola virus and have resulted in high mortality over the centuries.

Snuff powder inhaled through the nose.

SOD superoxide dismutase, is an enzyme that repairs cells and reduces the damage done to them by superoxide, the most common free radical in the body.

Sodium (Na) is an essential nutrient required for health. Sodium cations are important in neuron (brain and nerve) function, and in influencing osmotic balance between cells and the interstitial fluid and in maintenance of total body fluid homeostasis. Extra intake may cause a harmful effect on health. Sodium is naturally supplied by salt intake with food.

Soleus muscle smaller calf muscle lower down the leg and under the gastrocnemius muscle.

Somites mesodermal structures formed during embryonic development that give rise to segmented body parts such as the muscles of the body wall.

Somites Mesodermal structures formed during embryonic development that give rise to segmented body parts such as the muscles of the body wall.

Soporific a sleep inducing drug.

SOS response a global response to DNA damage in which the cell cycle is arrested and DNA repair and mutagenesis are induced.

Spasmolytic checking spasms, see antispasmodic.

Spermatorrhoea medically an involuntary ejaculation/drooling of semen usually nocturnal emissions.

Spermidine an important polyamine in DNA synthesis and gene expression.

Spleen organ that filters blood and prevents infection.

Spleen tyrosine kinase (SYK) is an enigmatic protein tyrosine kinase functional in a number of diverse cellular processes such as the regulation of immune and inflammatory responses.

Splenitis inflammation of the spleen.

Splenocyte is a monocyte, one of the five major types of white blood cell, and is characteristically found in the splenic tissue.

Splenomegaly is an enlargement of the spleen.

Sprain to twist a ligament or muscle of a joint without dislocating the bone.

Sprue is a chronic disorder of the small intestine caused by sensitivity to gluten, a protein found in wheat and rye and to a lesser extent oats and barley. It causes poor absorption by the intestine of fat, protein, carbohydrates, iron, water, and vitamins A, D, E, and K.

Sputum matter coughed up and usually ejected from the mouth, including saliva, foreign material, and substances such as mucus or phlegm, from the respiratory tract.

Stanch to stop or check the flow of a bodily fluid like blood from a wound.

Statin a type of lipid-lowering drug.

Status epilepticus refers to a life-threatening condition in which the brain is in a state of persistent seizure.

STD sexually transmitted disease.

Steatorrhea is the presence of excess fat in feces which appear frothy, foul smelling and floats because of the high fat content.

Steatosis refer to the deposition of fat in the interstitial spaces of an organ like the liver, fatty liver disease.

Sterility inability to produce offspring, also called asepsis.

Steroidogenic relating to steroidogenisis, the production of steroids.

Stimulant a substance that promotes the activity of a body system or function.

Stomachic (digestive stimulant), an agent that stimulates or strengthens the activity of the stomach; used as a tonic to improve the appetite and digestive processes.

Stomatitis oral inflammation and ulcers, may be mild and localized or severe, widespread, and painful.

Stomatology medical study of the mouth and its diseases.

Stool faeces.

Strangury is the painful passage of small quantities of urine which are expelled slowly by straining with severe urgency; it is usually accompanied with the unsatisfying feeling of a remaining volume inside and a desire to pass something that will not pass.

Straub tail condition in which an animal carries its tail in an erect (vertical or nearly vertical) position.

Striae gravidarum a cutaneous condition characterized by stretch marks on the abdomen during and following pregnancy.

Stricture an abnormal constriction of the internal passageway within a tubular structure such as a vessel or duct

Strongyloidiasis an intestinal parasitic infection in humans caused by two species of the parasitic nematode *Strongyloides*. The nematode or round worms are also called thread worms.

Styptic a short stick of medication, usually anhydrous aluminum sulfate (a type of alum) or titanium dioxide, which is used for stanching blood by causing blood vessels to contract at the site of the wound. Also called hemostatic pencil. see antihaemorrhagic.

Subarachnoid hemorrhage is bleeding in the area between the brain and the thin tissues that cover the brain.

Sudatory medicine that causes or increases sweating. Also see sudorific.

Sudorific a substance that causes sweating.

Sulfur Sulfur is an essential component of all living cells. Sulfur is important for the synthesis of sulfur-containing amino acids, all polypeptides, proteins, and enzymes such as glutathione an important sulfur-containing tripeptide which plays a role in cells as a source of chemical reduction potential. Sulfur is also important for hair formation. Good plant sources are garlic, onion, leeks and other Alliaceous vegetables, Brassicaceous vegetables like cauliflower, cabbages, Brussels sprout, Kale; legumes – beans, green and red gram, soybeans; horse radish, water cress, wheat germ.

Superior mesenteric artery (SMA) arises from the anterior surface of the abdominal aorta, just inferior to the origin of the celiac trunk, and supplies the intestine from the lower part of the duodenum to the left colic flexure and the pancreas.

Superoxidae mutase (SOD) antioxidant enzyme.

Suppuration the formation of pus, the act of becoming converted into and discharging pus.

Supraorbital located above the orbit of the eye.

SYK, Spleen tyrosine kinase is a human protein and gene. Syk plays a similar role in transmitting signals from a variety of cell surface receptors including CD74, Fc Receptor, and integrins.

Sympathetic nervous system the part of the autonomic nervous system originating in the thoracic and lumbar regions of the spinal cord that in general inhibits or opposes the physiological effects of the parasympathetic nervous system, as in tending to reduce digestive secretions or speed up the heart.

Syncope fainting, sudden loss of consciousness followed by the return of wakefulness.

Syneresis expulsion of liquid from a gel, as contraction of a blood clot and expulsion of liquid.

Syngeneic genetically identical or closely related, so as to allow tissue transplant; immunologically compatible.

Synovial lubricating fluid secreted by synovial membranes, as those of the joints.

Synoviocyte located in the synovial membrane, there are two types. Type A cells are more numerous, have phagocytic characteristics and produce degradative enzymes. Type B cells produce synovial fluid, which lubricates the joint and nurtures nourishes the articular cartilage.

Syphilis is perhaps the best known of all the STD's. Syphilis is transmitted by direct contact with infection sores, called chancres, syphilitic skin rashes, or mucous patches on the tongue and mouth during kissing, necking, petting, or sexual intercourse. It can also be transmitted from a pregnant woman to a fetus after the fourth month of pregnancy.

Systolic the blood pressure when the heart is contracting. It is specifically the maximum arterial pressure during contraction of the left ventricle of the heart.

Tachyarrhythmia any disturbance of the heart rhythm in which the heart rate is abnormally increased.

Tachycardia a false heart rate applied to adults to rates over 100 beats per minute.

Tachyphylaxia a decreased response to a medicine given over a period of time so that larger doses are required to produce the same response.

Tachypnea abnormally fast breathing.

Taenia a parasitic tapeworm or flatworm of the genus, *Taenia*.

Taeniacide an agent that kills tapeworms.

TBARS see thiobarbituric acid reactive substances.

T-cell a type of white blood cell that attacks virus-infected cells, foreign cells and cancer cells.

TCA cycle see Tricarboxylic acid cycle.

TCID50 median tissue culture infective dose; that amount of a pathogenic agent that will produce pathological change in 50% of cell cultures.

Tendonitis is inflammation of a tendon.

Tenesmus a strong desire to defaecate.

Teratogen is an agent that can cause malformations of an embryo or fetus. *adj.* teratogenic.

Testicular torsion torsion of the spermatic cord, is a surgical emergency because it causes strangulation gonadal blood supply with subsequent testicular necrosis and atrophy.

Tetanus an acute, potentially fatal disease caused by tetanus bacilli multiplying at the site of an injury and producing an exotoxin that reaches the central nervous system producing prolonged contraction of skeletal muscle fibres. Also called lockjaw.

Tete acute dermatitis caused by both bacterial and fungal infection.

Tetter any of a number of skin diseases.

Th cells or T helper cells a subgroup of lymphocytes that helps other white blood cells in immunologic processes.

Thermogenic tending to produce heat, applied to drugs or food (fat burning food).

Thiobarbituric acid reactive substances (TBARS) a well-established method for screening and monitoring lipid peroxidation.

Thrombocythaemia a blood condition characterize by a high number of platelets in the blood.

Thrombocytopenia a condition when the bone marrow does not produce enough platelets (thrombocytes) like in leukaemia.

Thromboembolism formation in a blood vessel of a clot (thrombus) that breaks loose and is carried by the blood stream to plug another vessel.

Thrombogenesis formation of a thrombus or blood clot.

Thrombophlebitis occurs when there is inflammation and clot in a surface vein.

Thromboplastin an enzyme liberated from blood platelets that converts prothrombin into thrombin as blood starts to clot, also called thrombokinase.

Thrombosis the formation or presence of a thrombus (clot).

Thromboxanes any of several compounds, originally derived from prostaglandin precursors in platelets, that stimulate aggregation of platelets and constriction of blood vessels.

Thromboxane B2 the inactive product of thromboxane.

Thrombus a fibrinous clot formed in a blood vessel or in a chamber of the heart.

Thrush a common mycotic infection caused by yeast, *Candida albicans*, in the digestive tract or vagina. In children it is characterized by white spots on the tongue.

Thymocytes are T cell precursors which develop in the thymus.

Thyrotoxicosis or hyperthyroidism – an overactive thyroid gland, producing excessive circulating free thyroxine and free triiodothyronine, or both.

TIMP-3 a human gene belongs to the tissue inhibitor of matrix metalloproteinases (MMP) gene family. see MMP.

Tincture solution of a drug in alcohol.

Tinea ringworm, fungal infection on the skin.

Tinea favosa See favus.

Tinnitus a noise in the ears, as ringing, buzzing, roaring, clicking, etc.

Tisane a herbal infusion used as tea or for medicinal purposes.

TNF alpha cachexin or cachectin and formally known as tumour necrosis factor-alpha, a cytokine involved in systemic inflammation. primary role of TNF is in the regulation of immune cells. TNF is also able to induce apoptotic cell death, to induce inflammation, and to inhibit tumorigenesis and viral replication.

Tocolytics medications used to suppress premature labor.

Tocopherol fat soluble organic compounds belonging to vitamin E group. See vitamin E.

Tocotrienol fat soluble organic compounds belonging to vitamin E group. See vitamin E.

Toll-like receptors (TLRs) are a class of proteins that play a key role in the innate immune system.

Tonic substance that acts to restore, balance, tone, strengthen, or invigorate a body system without overt stimulation or depression.

Tonic clonic seizure a type of generalized seizure that affects the entire brain.

Tonsillitis an inflammatory condition of the tonsils due to bacteria, allergies or respiratory problems.

Topoisomerases a class of enzymes involved in the regulation of DNA supercoiling.

Total parenteral nutrition (TPN) is a method of feeding that bypasses the gastrointestinal tract.

Toxemia is the presence of abnormal substances in the blood, but the term is also used for a serious condition in pregnancy that involves hypertension and proteinuria. Also called preeclampsia.

Tracheitis is a bacterial infection of the trachea; also known as bacterial tracheitis or acute bacterial tracheitis.

Trachoma a contagious disease of the conjunctiva and cornea of the eye, producing painful sensitivity to strong light and excessive tearing.

TRAIL acronym for tumour necrosis factor-related apoptosis-inducing ligand, is a cytokine that preferentially induces apoptosis in tumour cells.

Tranquilizer a substance drug used in calming person suffering from nervous tension or anxiety.

Transaminase also called aminotransferase is an enzyme that catalyzes a type of reaction between an amino acid and an α-keto acid.

Transaminitis increase in alanine aminotransferase (ALT) and/or aspartate aminotransferase (AST) to >5 times the upper limit of normal.

Transcatheter arterial chemoembolization (TACE) is an interventional radiology procedure involving percutaneous access of to the hepatic artery and passing a catheter through the abdominal artery aorta followed by radiology. It is used extensively in the palliative treatment of unresectable hepatocellular carcinoma (HCC)

Transcriptional activators are proteins that bind to DNA and stimulate transcription of nearby genes.

Transcriptional coactivator PGC-1 a potent transcriptional coactivator that regulates oxidative metabolism in a variety of tissues.

Transcriptome profiling to identify genes involved in peroxisome assembly and function.

TRAP 6 thrombin receptor activating peptide with 6 amino acids.

Tremorine a chemical that produces a tremor resembling parkinsonian tremor.

Triacylglycerols or triacylglyceride, is a glyceride in which the glycerol is esterified with three fatty acids.

Tricarboxylic acid cycle (TCA cycle) a series of enzymatic reactions in aerobic organisms involving oxidative metabolism of acetyl units and producing high-energy phosphate compounds, which serve as the main source of cellular energy. Also called citric acid cycle, Krebs cycle.

Trichophytosis infection by fungi of the genus *Trichophyto*n.

Trigeminal neuralgia (TN) is a neuropathic disorder of one or both of the facial trigeminal nerves, also known as prosopalgia.

Triglycerides a type of fat (lipids) found in the blood stream.

Trismus continuous contraction of the muscles of the jaw, specifically as a symptom of tetanus, or lockjaw; inability to open mouth fully.

TrKB receptor also known as TrKB tyrosine kinase , a protein in humans that acts as a catalytic receptor for several neutrophins.

Trolox Equivalent measures the antioxidant capacity of a given substance, as compared to the standard, Trolox also referred to as TEAC (Trolox equivalent antioxidant capacity).

Trypanocidal destructive to trypanosomes.

Trypanosomes protozoan of the genus *Trypanosoma*.

Trypanosomiasis human disease or an infection caused by a trypanosome.

Trypsin an enzyme of pancreatic juice that hydrolyzes proteins into smaller polypeptide units.

Trypsin inhibitor small protein synthesized in the exocrine pancreas which prevents conversion of trypsinogen to trypsin, so protecting itself against trypsin digestion.

Tuberculosis (TB) is a bacterial infection of the lungs caused by a bacterium called *Mycobacterium tuberculosis,* characterized by the formation of lesions (tubercles) and necrosis in the lung tissues and other organs.

Tumorigenesis formation or production of tumours.

Tumour an abnormal swelling of the body other than those caused by direct injury.

Tussis a cough.

Tympanic membrane ear drum.

Tympanitis infection or inflammation of the inner ear.

Tympanophonia increased resonance of one's own voice, breath sounds, arterial murmurs, etc., noted especially in disease of the middle ear.

Tympanosclerosis see myringoslcerosis.

Tyrosinase a copper containing enzyme found in animals and plants that catalyses the oxidation of phenols (such as tyrosine) and the production of melanin and other pigments from tyrosine by oxidation.

UCP1 an uncoupling protein found in the mitochondria of brown adipose tissue used to generate heat by non-shivering thermogenesis.

UCP – 2 enzyme uncoupling protein 2 enzyme, a mitochondrial protein expressed in adipocytes.

Ulcer an open sore on an external or internal body surface usually accompanied by disintegration of tissue and pus.

Ulcerative colitis is one of two types of inflammatory bowel disease – a condition that causes the bowel to become inflamed and red.

Ulemorrhagia bleeding of the gums.
Ulitis inflammation of the gums.
Unguent ointment.
Unilateral ureteral obstruction unilateral blockage of urine flow through the ureter of one kidney, resulting in a backup of urine, distension of the renal pelvis and calyces, and hydronephrosis.
Uraemia an excess in the blood of urea, creatinine and other nitrogenous end products of protein and amino acids metabolism, more correctly referred to as azotaemia.
Uraemic of or involving excess nitrogenous waste products in the urine (usually due to kidney insufficiency).
Urethra tube conveying urine from the bladder to the external urethral orifice.
Urethritis is an inflammation of the urethra caused by infection.
Urinary pertaining to the passage of urine.
Urinogenital relating to the genital and urinary organs or functions.
Urodynia pain on urination.
Urolithiasis formation of stone in the urinary tract (kidney bladder or urethra).
Urticant a substance that causes wheals to form.
Urticaria (or hives) is a skin condition, commonly caused by an allergic reaction, that is characterized by raised red skin welts.
Uterine relating to the uterus.
Uterine relaxant an agent that relaxes the muscles in the uterus.
Uterine stimulant an agent that stimulates the uterus (and often employed during active childbirth).
Uterotonic giving muscular tone to the uterus.
Uterotrophic causing an effect on the uterus.
Uterus womb.
Vagotomy the surgical cutting of the vagus nerve to reduce acid secretion in the stomach.
Vagus nerve a cranial nerve, that is, a nerve connected to the brain. The vagus nerve has branches to most of the major organs in the body, including the larynx, throat, windpipe, lungs, heart, and most of the digestive system.
Variola or smallpox, a contagious disease unique to humans, caused by either of two virus variants, *Variola major* and *Variola minor*. The disease is characterised by fever, weakness and skin eruption with pustules that form scabs that leave scars.
Varicose veins are veins that have become enlarged and twisted.
Vasa vasorum is a network of small blood vessels that supply large blood vessels. *plur.* vasa vasori.
Vascular endothelial growth factor (VEGF) a polypeptide chemical produced by cells that stimulates the growth of new blood vessels.
Vasoconstrictor drug that causes constriction of blood vessels.
Vasodilator drug that causes dilation or relaxation of blood vessels.
Vasodilatory causing the widening of the lumen of blood vessels.
Vasospasm refers to a condition in which blood vessels spasm, leading to vasoconstriction and subsequently to tissue ischemia and death (necrosis).
VEGF Vascular endothelial growth factor.
Venereal disease (VD) term given to the diseases syphilis and gonorrhoea.
Venule a small vein, especially one joining capillaries to larger veins.
Vermifuge a substance used to expel worms from the intestines.
Verotoxin a Shiga-like toxin produced by *Escherichia coli*, which disrupts the function of ribosomes, causing acute renal failure.
Verruca plana is a reddish-brown or flesh-colored, slightly raised, flat-surfaced, well-demarcated papule on the hand and face, also called flat wart.
Vertigo an illusory, sensory perception that the surroundings or one's own body are revolving; dizziness.
Very-low-density lipoprotein (VLDL) a type of lipoprotein made by the liver. VLDL is one of the five major groups of lipoproteins (chylomicrons, VLDL, intermediate-density lipoprotein, low-density lipoprotein, high-density lipoprotein (HDL)) that enable fats and cholesterol to move within the water-based solution of the bloodstream. VLDL is converted in the bloodstream to low-density lipoprotein (LDL).

Vesical calculus calculi (stones) in the urinary bladder

Vesicant a substance that causes tissue blistering.

Viremia a medical condition where viruses enter the bloodstream and hence have access to the rest of the body.

Visceral fat intra-abdominal fat, is located inside the peritoneal cavity, packed in between internal organs and torso.

Vitamin any complex, organic compound, found in various food or sometimes synthesized in the body, required in tiny amounts and are essential for the regulation of metabolism, normal growth and function of the body.

Vitamin A retinol, fat-soluble vitamins that play an important role in vision, bone growth, reproduction, cell division, and cell differentiation, helps regulate the immune system in preventing or fighting off infections. Vitamin A that is found in colorful fruits and vegetables is called provitamin A carotenoid. They can be made into retinol in the body. Deficiency of vitamin A results in night blindness and keratomalacia.

Vitamin B1 also called thiamine, water-soluble vitamins, dissolve easily in water, and in general, are readily excreted from the body they are not readily stored, consistent daily intake is important. It functions as coenzyme in the metabolism of carbohydrates and branched chain amino acids, and other cellular processes. Deficiency results in beri-beri disease.

Vitamin B2 also called riboflavin, an essential water-soluble vitamin that functions as coenzyme in redox reactions. Deficiency causes ariboflavinosis.

Vitamin B3 comprises niacin and niacinamide, water-soluble vitamin that function as coenzyme or co-substrate for many redox reactions and is required for energy metabolism. Deficiency causes pellagra.

Vitamin B5 also called pantothenic acid, a water-soluble vitamin that function as coenzyme in fatty acid metabolism. Deficiency causes paresthesia.

Vitamin B6 water-soluble vitamin, exists in three major chemical forms: pyridoxine, pyridoxal, and pyridoxamine. Vitamin B6 is needed in enzymes involved in protein metabolism, red blood cell metabolism, efficient functioning of nervous and immune systems and hemoglobin formation. Deficiency causes anaemia and peripheral neuropathy.

Vitamin B 7 also called biotin or vitamin H, an essential water-soluble vitamin, is involved in the synthesis of fatty acids amino acids and glucose, in energy metabolism. Biotin promotes normal health of sweat glands, bone marrow, male gonads, blood cells, nerve tissue, skin and hair, Deficiency causes dermatitis and enteritis.

Vitamin B9 also called folic acid, an essential water-soluble vitamin. Folate is especially important during periods of rapid cell division and growth such as infancy and pregnancy. Deficiency during pregnancy is associated with birth defects such as neural tube defects. Folate is also important for production of red blood cells and prevent anemia. Folate is needed to make DNA and RNA, the building blocks of cells. It also helps prevent changes to DNA that may lead to cancer.

Vitamin B12 a water-soluble vitamin, also called cobalamin as it contains the metal cobalt. It helps maintain healthy nerve cells and red blood cells, and DNA production. Vitamin B12 is bound to the protein in food. Deficiency causes megaloblastic anaemia.

Vitamin C also known as ascorbic acid is an essential water-soluble vitamin. It functions as cofactor for reactions requiring reduced copper or iron metallonzyme and as a protective antioxidant. Deficiency of vitamin C causes scurvy.

Vitamin D a group of fat-soluble, prohormone vitamin, the two major forms of which are vitamin D2 (or ergocalciferol) and vitamin D3 (or cholecalciferol). Vitamin D obtained from sun exposure, food, and supplements is biologically inert and must undergo two hydroxylations in the body for activation. Vitamin D is essential for promoting calcium absorption in the gut and maintaining adequate serum calcium and phosphate concentrations to enable normal growth and mineralization of bone and prevent hypocalcemic tetany. Deficiency causes rickets and osteomalacia. Vitamin D has other roles in human health, including modulation of neuromuscular and immune function, reduction of inflammation and modulation of many genes encoding proteins that regulate cell proliferation, differentiation, and apoptosis.

Vitamin E is the collective name for a group of fat-soluble compounds and exists in eight chemical forms (alpha-, beta-, gamma-, and delta-tocopherol and alpha-, beta-, gamma-, and delta-tocotrienol). It has pronounced antioxidant activities stopping the formation of Reactive Oxygen Species when fat undergoes oxidation and help prevent or delay the chronic diseases associated with free radicals. Besides its antioxidant activities, vitamin E is involved in immune function, cell signaling, regulation of gene expression, and other metabolic processes. Deficiency is very rare but can cause mild hemolytic anemia in newborn infants.

Vitamin K a group of fat soluble vitamin and consist of vitamin K_1 which is also known as phylloquinone or phytomenadione (also called phytonadione) and vitamin K_2 (menaquinone, menatetrenone). Vitamin K plays an important role in blood clotting. Deficiency is very rare but can cause bleeding diathesis.

Vitamin P a substance or mixture of substances obtained from various plant sources, identified as citrin or a mixture of bioflavonoids, thought to but not proven to be useful in reducing the extent of hemorrhage.

Vitiligo a chronic skin disease that causes loss of pigment, resulting in irregular pale patches of skin. It occurs when the melanocytes, cells responsible for skin pigmentation, die or are unable to function. Also called leucoderma.

Vitreoretinopathy see proliferative vitreoretinopathy.

VLDL see very low density lipoproteins.

Vomitive substance that causes vomiting.

Vulnerary (wound healer), a substance used to heal wounds and promote tissue formation.

Wart an infectious skin tumour caused by a viral infection.

Welt see wheal.

Wheal a firm, elevated swelling of the skin. Also called a weal or welt.

White fat white adipose tissue (WAT) in mammals, store of energy. cf. brown fat.

Whitlow painful infection of the hand involving one or more fingers that typically affects the terminal phalanx.

Whooping cough acute infectious disease usually in children caused by a *Bacillus* bacterium and accompanied by catarrh of the respiratory passages and repeated bouts of coughing.

X-linked agammaglobulinemia also known as X-linked hypogammaglobulinemia, XLA, Bruton type agammaglobulinemia, Bruton syndrome, or sex-linked agammaglobulinemia; a rare x-linked genetic disorder that affects the body's ability to fight infection.

Xanthine oxidase a flavoprotein enzyme containing a molybdenum cofactor (Moco) and (Fe_2S_2) clusters, involved in purine metabolism. In humans, inhibition of xanthine oxidase reduces the production of uric acid, and prevent hyperuricemia and gout.

Xanthones unique class of biologically active phenol compounds with the molecular formula C13H8O2 possessing antioxidant properties, discovered in the mangosteen fruit.

Xenobiotics a chemical (as a drug, pesticide, or carcinogen) that is foreign to a living organism.

Xenograft a surgical graft of tissue from one species to an unlike species.

Xerophthalmia a medical condition in which the eye fails to produce tears.

Yaws an infectious tropical infection of the skin, bones and joints caused by the spirochete bacterium *Treponema pertenue*, characterized by papules and pappiloma with subsequent deformation of the skins, bone and joints; also called framboesia.

Yellow fever is a viral disease that is transmitted to humans through the bite of infected mosquitoes. Illness ranges in severity from an influenza-like syndrome to severe hepatitis and hemorrhagic fever. Yellow fever virus (YFV) is maintained in nature by mosquito-borne transmission between nonhuman primates.

Zeaxanthin a common carotenoid, found naturally as coloured pigments in many fruit vegetables and leafy vegetables. It is important for good vision and is one of the two carotenoids contained within the retina of the eye. Within the central macula, zeaxanthin predominates, whereas in the peripheral retina, lutein predominates.

Zinc (Zn) is an essential mineral for health. It is involved in numerous aspects of cellular metabolism: catalytic activity of enzymes, immune function, protein synthesis, wound healing, DNA synthesis, and cell division. It also supports normal growth and development during pregnancy, childhood, and adolescence and is required for proper sense of taste and smell. Dietary sources include beans, nuts, pumpkin seeds, sunflower seeds, whole wheat bread and animal sources.

Scientific Glossary

Abaxial facing away from the axis, as of the surface of an organ.
Abscission shedding of leaves, flowers, or fruits following the formation of the abscission zone.
Acaulescent lacking a stem, or stem very much reduced.
Accrescent increasing in size after flowering or with age.
Achene a dry, small, one-seeded, indehiscent one-seeded fruit formed from a superior ovary of one carpel as in sunflower.
Acid soil soil that maintains a pH of less than 7.0.
Acidulous acid or sour in taste.
Actinomorphic having radial symmetry, capable of being divided into symmetrical halves by any plane, refers to a flower, calyx or corolla.
Aculeate having sharp prickles.
Acuminate tapering gradually to a sharp point.
Acute (Botany) tapering at an angle of less than 90° before terminating in a point as of leaf apex and base.
Adaxial side closest to the stem axis.
Aldephous having stamens united together by their filaments.
Adherent touching without organic fusion as of floral parts of different whorls.
Adnate united with another unlike part as of stamens attached to petals.
Adpressed lying close to another organ but not fused to it.
Adventitious arising in abnormal positions, e.g. roots arising from the stem, branches or leaves, buds arising elsewhere than in the axils of leaves
Adventive Not native to and not fully established in a new habitat or environment; locally or temporarily naturalized. e.g. an adventive weed.
Aestivation refers to positional arrangement of the floral parts in the bud before it opens.
Akinete a thick-walled dormant cell derived from the enlargement of a vegetative cell. It serves as a survival structure.
Alfisols soil with a clay-enriched subsoil and relatively high native fertility, having undergone only moderate leaching, containing aluminium, iron and with at least 35% base saturation, meaning that calcium, magnesium, and potassium are relatively abundant.
Alkaline soil soil that maintains a pH above 7.0, usually containing large amounts of calcium, sodium, and magnesium, and is less soluble than acidic soils.
Alkaloids naturally occurring bitter, complex organic-chemical compounds containing basic nitrogen and oxygen atoms and having various pharmacological effects on humans and other animals.
Alternate leaves or buds that are spaced along opposite sides of stem at different levels.
Allomorphic with a shape or form different from the typical.

Alluvial soil a fine-grained fertile soil deposited by water flowing over flood plains or in river beds.

Alluvium soil or sediments deposited by a river or other running water.

Amplexicaul clasping the stem as base of certain leaves.

Anatomizing interconnecting network as applied to leaf veins.

Andisols are soils formed in volcanic ash and containing high proportions of glass and amorphous colloidal materials.

Androdioecious with male flowers and bisexual flowers on separate plants.

Androecium male parts of a flower; comprising the stamens of one flower.

Androgynous with male and female flowers in distinct parts of the same inflorescence.

Andromonoecious having male flowers and bisexual flowers on the same plant.

Angiosperm a division of seed plants with the ovules borne in an ovary.

Annual a plant which completes its life cycle within a year.

Annular shaped like or forming a ring.

Annulus circle or ring-like structure or marking; the portion of the corolla which forms a fleshy, raised ring.

Anthelate an open, paniculate cyme.

Anther the part of the stamen containing pollen sac which produces the pollen.

Antheriferous containing anthers.

Anthesis the period between the opening of the bud and the onset of flower withering.

Anthocarp a false fruit consisting of the true fruit and the base of the perianth.

Anthocyanidins are common plant pigments. They are the sugar-free counterparts of anthocyanins.

Anthocyanins a subgroup of antioxidant flavonoids, are glucosides of anthocyanidins. They occur as water-soluble vacuolar pigments that may appear red, purple, or blue according to pH in plants.

Antipetala situated opposite petals.

Antisepala situated opposite sepals.

Antrorse directed forward upwards.

Apetalous lacking petals as of flowers with no corolla.

Apical meristem active growing point. A zone of cell division at the tip of the stem or the root.

Apically towards the apex or tip of a structure.

Apiculate ending abruptly in a short, sharp, small point.

Apiculum a short, pointed, flexible tip.

Apocarpous carpels separate in single individual pistils.

Apopetalous with separate petals, not united to other petals.

Aposepalous with separate sepals, not united to other sepals.

Appressed pressed closely to another structure but not fused or united.

Aquatic a plant living in or on water for all or a considerable part of its life span.

Arachnoid (Botany) formed of or covered with long, delicate hairs or fibres.

Arborescent resembling a tree; applied to non-woody plants attaining tree height and to shrubs tending to become tree-like in size.

Arbuscular mycorrhiza (AM) a type of mycorrhiza in which the fungus (of the phylum Glomeromycota) penetrates the cortical cells of the roots of a vascular plant and form unique structures such as arbuscules and vesicles. These fungi help plants to capture nutrients such as phosphorus and micronutrients from the soil.

Archegonium a flask-shaped female reproductive organ in mosses, ferns, and other related plants.

Areolate with areolea.

Areole (Botany) a small, specialized, cushion-like area on a cactus from which hairs, glochids, spines, branches, or flowers may arise; an irregular angular specs marked out on a surface e.g. fruit surface. *pl.* areolea.

Aril specialized outgrowth from the funiculus (attachment point of the seed) (or hilum) that encloses or is attached to the seed. *adj.* arillate.

Arillode a false aril; an aril originating from the micropyle instead of from the funicle or chalaza of the ovule, e.g. mace of nutmeg.

Aristate bristle-like part or appendage, e.g. awns of grains and grasses.
Aristulate having a small, stiff, bristle-like part or appendage; a diminutive of aristate
Articulate jointed; usually breaking easily at the nodes or point of articulation into segments.
Ascending arched upwards in the lower part and becoming erect in the upper part.
Ascospore spore produced in the ascus in Ascomycete fungi.
Ascus is the sexual spore-bearing cell produced in Ascomycete fungi. *pl.* asci.
Asperulous refers to a rough surface with short, hard projections.
Attenuate tapered or tapering gradually to a point.
Auricle an ear-like appendage that occurs at the base of some leaves or corolla.
Auriculate having auricles.
Awn a hair-like or bristle-like appendage on a larger structure.
Axil upper angle between a lateral organ, such as a leaf petiole and the stem that bears it.
Axile situated along the central axis of an ovary having two or more locules, as in axile placentation.
Axillary arising or growing in an axil.
Baccate beery-like, pulpy or fleshy.
Barbate bearded, having tufts of hairs.
Barbellae short, stiff, hair-like bristles. *adj.* barbellate.
Bark is the outermost layers of stems and roots of woody plants.
Basal relating to, situated at, arising from or forming the base.
Basaltic soil soil derived from basalt, a common extrusive volcanic rock.
Basidiospore a reproductive spore produced by Basidiomycete fungi.
Basidium a microscopic, spore-producing structure found on the hymenophore of fruiting bodies of Basidiomycete fungi.
Basifixed attached by the base, as certain anthers are to their filaments.
Basionym the synonym of a scientific name that supplies the epithet for the correct name.
Beak a prominent apical projection, especially of a carpel or fruit. *adj.* beaked.
Bearded having a tuft of hairs.
Berry a fleshy or pulpy indehiscent fruit from a single ovary with the seed(s) embedded in the fleshy tissue of the pericarp.
Biconvex convex on both sides.
Biennial completing the full cycle from germination to fruiting in more than one, but not more than 2 years.
Bifid forked, divided into two parts.
Bifoliolate having two leaflets.
Bilabiate having two lips as of a corolla or calyx with segments fused into an upper and lower lip.
Bipinnate twice pinnate; the primary leaflets being again divided into secondary leaflets.
Bipinnatisect refers to a pinnately compound leaf, in which each leaflet is again divided into pinnae.
Biserrate doubly serrate; with smaller regular, asymmetric teeth on the margins of larger teeth.
Bisexual having both sexes, as in a flower bearing both stamens and pistil, hermaphrodite or perfect.
Biternate Twice ternate; with three pinnae each divided into three pinnules.
Blade lamina; part of the leaf above the sheath or petiole.
Blotched see variegated.
Bole main trunk of tree from the base to the first branch.
Brachyblast a short, axillary, densely crowded branchlet or shoot of limited growth, in which the internodes elongate little or not at all.
Bracket fungus shelf fungus.
Bract a leaf-like structure, different in form from the foliage leaves, associated with an inflorescence or flower. *adj.* bracteate.
Bracteate possessing bracts.
Bracteolate having bracteoles.
Bracteole a small, secondary, bract-like structure borne singly or in a pair on the pedicel or calyx of a flower. *adj.* bracteolate.
Bristle a stiff hair.
Bulb a modified underground axis that is short and crowned by a mass of usually fleshy, imbricate scales. *adj.* bulbous.

Bulbil A small bulb or bulb-shaped body, especially one borne in the leaf axil or an inflorescence, and usually produced for asexual reproduction.

Bullate puckered, blistered.

Burr type of seed or fruit with short, stiff bristles or hooks or may refer to a deformed type of wood in which the grain has been misformed.

Bush low, dense shrub without a pronounced trunk.

Buttress supporting, projecting outgrowth from base of a tree trunk as in some Rhizophoraceae and Moraceae.

Caducous shedding or falling early before maturity refers to sepals and petals.

Caespitose growing densely in tufts or clumps; having short, closely packed stems.

Calcareous composed of or containing lime or limestone.

Calcrete a hardpan consisting gravel and sand cemented by calcium.

Callus a condition of thickened raised mass of hardened tissue on leaves or other plant parts often formed after an injury but sometimes a normal feature. A callus also can refer to an undifferentiated plant cell mass grown on a culture medium. *n.* callosity. *pl.* calli, callosities. *adj.* callose.

Calyptra the protective cap or hood covering the spore case of a moss or related plant.

Calyptrate operculate, having a calyptra.

Calyx outer floral whorl usually consisting of free sepals or fused sepals (calyx tube) and calyx lobes. It encloses the flower while it is still a bud. *adj.* calycine.

Calyx lobe one of the free upper parts of the calyx which may be present when the lower part is united into a tube.

Calyx tube the tubular fused part of the calyx, often cup shaped or bell shaped, when it is free from the corolla.

Campanulate shaped like a bell refers to calyx or corolla.

Canaliculate having groove or grooves.

Candelabriform having the shape of a tall branched candle-stick.

Canescent covered with short, fine whitish or grayish hairs or down.

Canopy uppermost leafy stratum of a tree.

Cap see pileus.

Capitate growing together in a head. Also means enlarged and globular at the tip.

Capitulum a flower head or inflorescence having a dense cluster of sessile, or almost sessile, flowers or florets.

Capsule a dry, dehiscent fruit formed from two or more united carpels and dehiscing at maturity by sections called valves to release the seeds. *adj.* capsular.

Carinate keeled.

Carpel a simple pistil consisting of ovary, ovules, style and stigma. *adj.* carpellary.

Carpogonium female reproductive organ in red algae. *pl.* carpogonia.

Carpophore part of the receptacle which is lengthened between the carpels as a central axis; any fruiting body or fruiting structure of a fungus.

Cartilaginous sinewy, having a firm, tough, flexible texture (in respect of leaf margins).

Caryopsis a simple dry, indehiscent fruit formed from a single ovary with the seed coat united with the ovary wall as in grasses and cereals.

Cataphyll a reduced or scarcely developed leaf at the start of a plant's life (i.e., cotyledons) or in the early stages of leaf development.

Catkin a slim, cylindrical, pendulous flower spike usually with unisexual flowers.

Caudate having a narrow, tail-like appendage.

Caudex thickened, usually underground base of the stem.

Caulescent having a well developed aerial stem.

Cauliflory botanical term referring to plants which flower and fruit from their main stems or woody trunks. *adj.* cauliflorus.

Cauline borne on the aerial part of a stem.

Chaffy having thin, membranous scales in the inflorescence as in the flower heads of the sunflower family.

Chalaza the basal region of the ovule where the stalk is attached.

Chartaceous papery, of paper-like texture.

Chasmogamous describing flowers in which pollination takes place while the flower is open.

Chloroplast a chlorophyll-containing organelle (plastid) that gives the green colour to leaves and stems. Plastids harness light energy that is used to fix carbon dioxide in the process called photosynthesis.

Chromoplast plastid containing colored pigments apart from chlorophyll.

Chromosomes thread-shaped structures that occur in pairs in the nucleus of a cell, containing the genetic information of living organisms.

Cilia hairs along the margin of a leaf or corolla lobe.

Ciliate with a fringe of hairs on the margin as of the corolla lobes or leaf.

Ciliolate minutely ciliate.

Cilium a straight, usually erect hair on a margin or ridge. *pl.* cilia.

Cincinnus a monochasial cyme in which the lateral branches arise alternately on opposite sides of the false axis.

Circinnate spirally coiled, with the tip innermost.

Circumscissile opening by a transverse line around the circumference as of a fruit.

Cladode the modified photosynthetic stem of a plant whose foliage leaves are much reduced or absent. *cf.* cladophyll, phyllode.

Cladophyll A photosynthetic branch or portion of a stem that resembles and functions as a leaf, like in asparagus. *cf.* cladode, phyllode.

Clamp connection In the Basidiomycetes fungi, a lateral connection or outgrowth formed between two adjoining cells of a hypha and arching over the septum between them.

Clavate club shaped thickened at one end refer to fruit or other organs.

Claw the conspicuously narrowed basal part of a flat structure.

Clay a naturally occurring material composed primarily of fine-grained minerals like kaolinite, montmorrillonite-smectite or illite which exhibit plasticity through a variable range of water content, and which can be hardened when dried and/or fired.

Clayey resembling or containing a large proportion of clay.

Cleft incised halfway down.

Cleistogamous refers to a flower in which fertilization occurs within the bud i.e. without the flower opening. *cf.* chasmogamous.

Climber growing more or less upwards by leaning or twining around another structure.

Clone all the plants reproduced, vegetatively, from a single parent thus having the same genetic make-up as the parent.

Coccus one of the sections of a distinctly lobed fruit which becomes separate at maturity; sometimes called a mericarp. *pl.* cocci.

Coenocarpium a fleshy, multiple pseudocarp formed from an inflorescence rather than a single flower.

Coherent touching without organic fusion, referring to parts normally together, e.g. floral parts of the same whorl. *cf.* adherent, adnate, connate.

Collar boundary between the above- and below ground parts of the plant axis.

Colliculate having small elevations.

Column a structure formed by the united style, stigma and stamen(s) as in Asclepiadaceae and Orchidaceae.

Comose tufted with hairs at the ends as of seeds.

Composite having two types of florets as of the flowers in the sunflower family, Asteraceae.

Compost organic matter (like leaves, mulch, manure, etc) that breaks down in soil releasing its nutrients.

Compound describe a leaf that is further divided into leaflets or pinnae or flower with more than a single floret.

Compressed flattened in one plane.

Conceptacles specialised cavities of marine algae that contain the reproductive organs.

Concolorous uniformly coloured, as in upper and lower surfaces. *cf.* discolorous

Conduplicate folded together lengthwise.

Cone a reproductive structure composed of an axis (branch) bearing sterile bract-like organs and seed or pollen bearing structures. Applied to Gymnospermae, Lycopodiaceae, Casuarinaceae and also in some members of Proteaceae.

Conic cone shaped, attached at the broader end.

Conic-capitate a cone-shaped head of flowers.

Connate fused to another structure of the same kind. *cf.* adherent, adnate, coherent.

Connective the tissue separating two lobes of an anther.

Connivent converging.

Conspecific within or belonging to the same species.

Contorted twisted.

Convolute refers to an arrangement of petals in a bud where each has one side overlapping the adjacent petal.

Cordate heart-shaped as of leaves.

Core central part.

Coriaceous leathery texture as of leaves.

Corm a short, swollen, fleshy, underground plant stem that serves as a food storage organ used by some plants to survive winter or other adverse conditions

Cormel a miniature, new corm produced on a mature corm.

Corolla the inner floral whorl of a flower, usually consisting of free petals or a petals fused forming a corolla tube and corolla lobes. *adj.* corolline.

Corona a crown-like section of the staminal column, usually with the inner and outer lobes as in the Stapelieae.

Coroniform crown shaped, as in the pappus of Asteraceae.

Cortex the outer of the stem or root of a plant, bounded on the outside by the epidermis and on the inside by the endodermis containing undifferentiated cells.

Corymb a flat-topped, short, broad inflorescence, in which the flowers, through unequal pedicels, are in one horizontal plane and the youngest in the centre. *adj.* corymbose

Costa a thickened, linear ridge or the midrib of the pinna in ferns. *adj.* costate.

Costapalmate having definite costa (midrib) unlike the typical palmate leaf, but the leaflets are arranged radially like in a palmate leaf.

Cotyledon the primary seed leaf within the embryo of a seed.

Cover crop crop grown in between trees or in fields primarily to protect the soil from erosion, to improve soil fertility and to keep off weeds.

Crenate round-toothed or scalloped as of leaf margins.

Crenulate minutely crenate, very strongly scalloped.

Crisped with a curled or twisted edge.

Crozier shaped like a shepherd's crook.

Crustaceous like a crust; having a hard crust or shell.

Cucullate having the shape of a cowl or hood, hooded.

Culm the main aerial stem of the Graminae (grasses, sedges, rushes and other monocots).

Culm sheath the plant casing (similar to a leaf) that protects the young bamboo shoot during growth, attached at each node of culm.

Cultigen plant species or race known only in cultivation.

Cultivar cultivated variety; an assemblage of cultivated individuals distinguished by any characters significant for the purposes of agriculture, forestry or horticulture, and which, when reproduced, retains its distinguishing features.

Cuneate wedge-shaped, obtriangular.

Cupular cup-shaped, having a cupule.

Cupule a small cup-shaped structure or organ, like the cup at the base of an acorn.

Cusp an elongated, usually rigid, acute point. *cf.* mucro.

Cuspidate terminating in or tipped with a sharp firm point or cusp. *cf.* mucronate.

Cuspidulate constricted into a minute cusp. *cf.* cuspidate.

Cyathiform in the form of a cup, a little widened at the top.

Cyathium a specialised type of inflorescence of plants in the genus Euphorbia and Chamaesyce in which the unisexual flowers are clustered together within a bract-like envelope. *pl.* cyathia.

Cylindric tubular or rod shaped.

Cylindric-acuminate elongated and tapering to a point.

Cymbiform boat shaped, elongated and having the upper surface decidedly concave.

Cyme an inflorescence in which the lateral axis grows more strongly than the main axis with the oldest flower in the centre or at the ends. *adj.* cymose

Cymule a small cyme or one or a few flowers.

Cystidium a relatively large cell found on the hymenium of a Basidiomycete, for example, on the surface of a mushroom.

Cystocarp fruitlike structure (sporocarp) developed after fertilization in the red algae.

Deciduous falling off or shedding at maturity or a specific season or stage of growth.

Decorticate to remove the bark, rind or husk from an organ; to strip of its bark; to come off as a skin.

Decompound as of a compound leaf; consisting of divisions that are themselves compound.

Decumbent prostrate, laying or growing on the ground but with ascending tips. *cf.* ascending, procumbent.

Decurrent having the leaf base tapering down to a narrow wing that extends to the stem.

Decussate having paired organs with successive pairs at right angles to give four rows as of leaves.

Deflexed bent downwards.

Dehisce to split open at maturity, as in a capsule.

Dehiscent splitting open at maturity to release the contents. *cf.* indehiscent.

Deltate triangular shape.

Deltoid shaped like an equilateral triangle.

Dendritic branching from a main stem or axis like the branches of a tree.

Dentate with sharp, rather coarse teeth perpendicular to the margin.

Denticulate finely toothed.

Diageotropic the tendency of growing parts, such as roots, to grow at right angle to the line of gravity.

Diadelphous having stamens in two bundles as in Papilionaceae flowers.

Dichasium a cymose inflorescence in which the branches are opposite and approximately equal. *pl.* dichasia. *adj.* dichasial.

Dichotomous divided into two parts.

Dicotyledon angiosperm with two cotyledons.

Didymous arranged or occurring in pairs as of anthers, having two lobes.

Digitate having digits or fingerlike projections.

Dikaryophyses or dendrophydia, irregularly, strongly branched terminal hyphae in the Hymenomycetes (class of Basidiomycetes) fungi.

Dimorphic having or occurring in two forms, as of stamens of two different lengths or a plant having two kinds of leaves.

Dioecious with male and female unisexual flowers on separate plants. *cf.* monoecious.

Diploid a condition in which the chromosomes in the nucleus of a cell exist as pairs, one set being derived from the female parent and the other from the male.

Diplobiontic life cycle life cycle that exhibits alternation of generations, which features of spore-producing multicellular sporophytes and gamete-producing multicellular gametophytes. mitoses occur in both the diploid and haploid phases.

Diplontic life cycle or gametic meiosis, wherein instead of immediately dividing meiotically to produce haploid cells, the zygote divides mitotically to produce a multicellular diploid individual or a group of more diploid cells.

Dipterocarpous trees of the family Dipterocarpaceae, with two-winged fruit found mainly in tropical lowland rainforest.

Disc (Botany) refers to the usually disc shaped receptacle of the flower head in Asteraceae; also the fleshy nectariferous organ usually between the stamens and ovary; also used for the enlarged style-end in Proteaceae.

Disc floret the central, tubular 4 or 5-toothed or lobed floret on the disc of an inflorescence, as of flower head of Asteraceae.

Disciform flat and rounded in shaped. *cf.* discoid, radiate.

Discoid resembling a disc; having a flat, circular form; disk-shaped *cf.* disciform, radiate.

Discolorous having two colours, as of a leaf which has different colors on the two surfaces. *cf.* concolorous.

Dispersal dissemination of seeds.

Distal site of any structure farthest from the point of attachment. *cf.* proximal.

Distichous referring to two rows of upright leaves in the same plane.

Dithecous having two thecae.

Divaricate diverging at a wide angle.

Domatium a part of a plant (e.g., a leaf) that has been modified to provide protection for other organisms. *pl.* domatia.

Dormancy a resting period in the life of a plant during which growth slows or appears to stop.

Dorsal referring to the back surface.

Dorsifixed attached to the back as of anthers.

Drupaceous resembling a drupe.

Drupe a fleshy fruit with a single seed enclosed in a hard shell (endocarp) which is tissue embedded in succulent tissue (mesocarp) surrounded by a thin outer skin (epicarp). *adj.* drupaceous.

Drupelet a small drupe.

Ebracteate without bracts.

Echinate bearing stiff, stout, bristly, prickly hairs.

Edaphic refers to plant communities that are distinguished by soil conditions rather than by the climate.

Eglandular without glands. *cf.* glandular.

Ellipsoid a 3-dimensional shape; elliptic in outline.

Elliptic having a 2-dimensional shape of an ellipse or flattened circle.

Elongate extended, stretched out.

Emarginate refers to leaf with a broad, shallow notch at the apex. *cf.* retuse.

Embryo (Botany) a minute rudimentary plant contained within a seed or an archegonium, composed of the embryonic axis (shoot end and root end).

Endemic prevalent in or peculiar to a particular geographical locality or region.

Endocarp The hard innermost layer of the pericarp of many fruits.

Endosperm tissue that surrounds and nourishes the embryo in the angiosperm seed.

Endospermous refers to seeds having an endosperm.

Endotrophic as of mycorrhiza obtaining nutrients from inside.

Ensilage the process of preserving green food for livestock in an undried condition in airtight conditions. Also called silaging.

Entire having a smooth, continuous margin without any incisions or teeth as of a leaf.

Entisols soils that do not show any profile development other than an A horizon.

Ephemeral transitory, short-lived.

Epicalyx a whorl of bracts, subtending and resembling a calyx.

Epicarp outermost layer of the pericarp of a fruit.

Epicormic attached to the corm.

Epicotyl the upper portion of the embryonic axis, above the cotyledons and below the first true leaves.

Epigeal above grounds with cotyledons raised above ground.

Epiparasite an organism parasitic on another that parasitizes a third.

Epipetalous borne on the petals, as of stamens.

Epiphyte a plant growing on, but not parasitic on, another plant, deriving its moisture and nutrients from the air and rain e.g. some Orchidaceae. *adj.* epiphytic.

Erect upright, vertical.

Essential oils volatile products obtained from a natural source; refers to volatile products obtained by steam or water distillation in a strict sense.

Etiolation to cause (a plant) to develop without chlorophyll by preventing exposure to sunlight.

Eutrophic having waters rich in mineral and organic nutrients that promote a proliferation of plant life, especially algae, which reduces the dissolved oxygen content and often causes the extinction of other organisms.

Excentric off the true centre.

Excrescence abnormal outgrowth.

Excurrent projecting beyond the tip, as the midrib of a leaf or bract.

Exserted sticking out, protruding beyond some enclosing organ, as of stamens which project beyond the corolla or perianth.

Exstipulate without stipules. *cf.* stipulate.

Extra-floral outside the flower.

Extrose turned outwards or away from the axis as of anthers. *cf.* introrse, latrorse.

Falcate sickle shaped, crescent-shaped.

Fascicle a cluster or bundle of stems, flowers, stamens. *adj.* fasciculate.

Fasciclode staminode bundles.

Fastigiate a tree in which the branches grow almost vertically.
Ferrosols soils with an iron oxide content of greater than 5%.
Ferruginous rust coloured, reddish-brown.
Fertile having functional sexual parts which are capable of fertilisation and seed production. *cf.* sterile.
Filament the stalk of a stamen supporting and subtending the anther.
Filiform Having the form of or resembling a thread or filament.
Fimbriate fringed.
Fixed oils non volatile oils, triglycerides of fatty acids.
Flaccid limp and weak.
Flag leaf the uppermost leaf on the stem.
Flaky in the shape of flakes or scales.
Flexuous zig-zagging, sinuous, bending, as of a stem.
Floccose covered with tufts of soft woolly hairs.
Floral tube a flower tube usually formed by the basal fusion of the perianth and stamens.
Floret one of the small individual flowers of sunflower family or the reduced flower of the grasses, including the lemma and palea.
Flower the sexual reproductive organ of flowering plants, typically consisting of gynoecium, androecium and perianth or calyx and/or corolla and the axis bearing these parts.
Fluted as of a trunk with grooves and folds.
Fodder plant material, fresh or dried fed to animals.
Foliaceous leaf-like.
Foliar pertaining to a leaf.
Foliolate pertaining to leaflets, used with a number prefix to denote the number of leaflets.
Foliose leaf-like.
Follicle (Botany) a dry fruit, derived from a single carpel and dehiscing along one suture.
Forb any herb that is not grass or grass-like.
Free central placentation The arrangement of ovules on a central column that is not connected to the ovary wall by partitions, as in the ovaries of the carnation and primrose.
Frond the leaf of a fern or cycad.
Fruit ripened ovary with adnate parts.
Fugacious shedding off early.

Fulvous yellow, tawny.
Funiculus (Botany) short stalk which attaches the ovule to the ovary wall.
Fusiform a 3-dimensional shape; spindle shaped, i.e. broad in the centre and tapering at both ends thick, but tapering at both ends.
Gamete a reproductive cell that fuses with another gamete to form a zygote. Gametes are haploid, (they contain half the normal (diploid) number of chromosomes); thus when two fuse, the diploid number is restored.
Gametophyte The gamete-producing phase in a plant characterized by alternation of generations.
Gamosepalous with sepals united or partially united.
Gall-flower short styled flower that do not develop into a fruit but are adapted for the development of a specific wasp within the fruit e.g. in the fig.
Geniculate bent like a knee, refer to awns and filaments.
Geocarpic where the fruit are pushed into the soil by the gynophore and mature.
Geophyte a plant that stores food in an underground storage organ e.g. a tuber, bulb or rhizome and has subterranean buds which form aerial growth.
Geotextile are permeable fabrics which, when used in association with soil, have the ability to separate, filter, reinforce, protect, or drain.
Glabrescent becoming glabrous.
Glabrous smooth, hairless without pubescence.
Gland a secretory organ, e.g. a nectary, extrafloral nectary or a gland tipped, hair-like or wart-like organ. *adj.* glandular. *cf.* eglandular.
Glaucous pale blue-green in colour, covered with a whitish bloom that rubs off readily.
Gley soils a hydric soil which exhibits a greenish-blue-grey soil color due to wetland conditions.
Globose spherical in shape.
Globular a three-dimensional shape; spherical or orbicular; circular in outline.
Glochids tiny, finely barbed hair-like spines found on the areoles of some cacti and other plants.
Glochidiate having glochids.

Glochidote plant having glochids.

Glume one of the two small, sterile bracts at the base of the grass spikelet, called the lower and upper glumes, due to their position on the rachilla. Also used in Apiaceae, Cyperaceae for the very small bracts on the spikelet in which each flower is subtended by one floral glume. *adj.* glumaceous.

Guttation the appearance of drops of xylem sap on the tips or edges of leaves of some vascular plants, such as grasses and bamboos.

Guttule small droplet.

Gymnosperm a group of spermatophyte seed-bearing plants with ovules on scales, which are usually arranged in cone-like structures and not borne in an ovary. *cf.* angiosperm.

Gynoecium the female organ of a flower; a collective term for the pistil, carpel or carpels.

Gynomonoecious having female flowers and bisexual flowers on the same plant. *cf.* andromonoecious.

Gynophore stalk that bears the pistil/carpel.

Habit the general growth form of a plant, comprising its size, shape, texture and stem orientation, the locality in which the plant grows..

Halophyte a plant adapted to living in highly saline habitats. Also a plant that accumulates high concentrations of salt in its tissues. *adj.* halophytic.

Hapaxanthic refer to palms which flowers only once and then dies. c.f. pleonanthic.

Haploid condition where nucleus or cell has a single set of unpaired chromosomes, the haploid number is designated as n.

Haplontic life cycle or zygotic meiosis wherein meiosis of a zygote immediately after karyogamy, produces haploid cells which produces more or larger haploid cells ending its diploid phase.

Hastate having the shape of an arrowhead but with the basal lobes pointing outward at right angles as of a leaf.

Hastula a piece of plant material at the junction of the petiole and the leaf blade; the hastula can be found on the top of the leaf, adaxial or the bottom, abaxial or both sides.

Heartwood wood from the inner portion of a tree.

Heliophilous sun-loving, tolerates high level of sunlight.

Heliotropic growing towards sunlight.

Herb a plant which is non-woody or woody at the base only, the above ground stems usually being ephemeral. *adj.* herbaceous.

Herbaceous resembling a herb, having a habit of a herb.

Hermaphrodite bisexual, bearing flowers with both androecium and gynoecium in the same flower. *adj.* hermaphroditic.

Heterocyst a differentiated cyanobacterial cell that carries out nitrogen fixation.

Heterogamous bearing separate male and female flowers, or bisexual and female flowers, or florets in an inflorescence or flower head, e.g. some Asteraceae in which the ray florets may be neuter or unisexual and the disk florets may be bisexual. *cf.* homogamous.

Heteromorphous having two or more distinct forms. *cf.* homomorphous.

Heterophyllous having leaves of different form.

Heterosporous producing spores of two sizes, the larger giving rise to megagametophytes (female), the smaller giving rise to microgametophytes (male). Refer to the ferns and fern allies. *cf.* homosporous.

Heterostylous having styles of two different lengths or forms.

Heterostyly the condition in which flowers on polymorphous plants have styles of different lengths, thereby facilitating cross-pollination.

Hilar of or relating to a hilum.

Hilum The scar on a seed, indicating the point of attachment to the funiculus.

Hirsute bearing long coarse hairs.

Hispid bearing stiff, short, rough hairs or bristles.

Hispidulous minutely hispid.

Histosol soil comprising primarily of organic materials, having 40 cm or more of organic soil material in the upper 80 cm.

Hoary covered with a greyish layer of very short, closely interwoven hairs.

Holdfast an organ or structure of attachment, especially the basal, root-like formation by which certain seaweeds or other algae are attached to a substrate.

Holocarpic having the entire thallus developed into a fruiting body or sporangium.

Homochromous having all the florets of the same colour in the same flower head *cf.* heterochromous.

Homogamous bearing flowers or florets that do not differ sexually *cf.* heterogamous.

Homogenous endosperm endosperm with even surface that lacks invaginations or infoldings of the surrounding tissue.

Homogonium a part of a filament of a cyanobacterium that detaches and grows by cell division into a new filament. *pl.* homogonia.

Homomorphous uniform, with only one form. *cf.* heteromorphous.

Homosporous producing one kind of spores. Refer to the ferns and fern allies. *cf.* heterosporous.

Hurd fibre long pith fibre of the stem.

Hyaline colourless, almost transparent.

Hybrid the first generation progeny of the sexual union of plants belonging to different taxa.

Hybridisation the crossing of individuals from different species or taxa.

Hydathode a type of secretory tissue in leaves, usually of Angiosperms, that secretes water through pores in the epidermis or margin of the leaf.

Hydrophilous water loving; requiring water in order to be fertilized, referring to many aquatic plants.

Hygrochastic applied to plants in which the opening of the fruits is caused by the absorption of water.

Hygrophilous living in water or moist places.

Hymenial cystidia the cells of the hymenium develop into basidia or asci, while in others some cells develop into sterile cells called cystidia.

Hymenium spore-bearing layer of cells in certain fungi containing asci (Ascomycetes) or basidia (Basidiomycetes).

Hypanthium cup-like receptacles of some dicotyledonous flowers formed by the fusion of the calyx, corolla, and androecium that surrounds the ovary which bears the sepals, petals and stamens.

Hypha is a long, branching filamentous cell of a fungus, and also of unrelated Actinobacteria. *pl.* hyphae.

Hypocotyl the portion of the stem below the cotyledons.

Hypodermis the cell layer beneath the epidermis of the pericarp.

Hypogeal below ground as of germination of seed.

Imbricate closely packed and overlapping. *cf.* valvate.

Imparipinnate pinnately compound with a single terminal leaflet and hence with an odd number of leaflets. *cf.* paripinnate.

Inceptisols old soils that have no accumulation of clays, iron, aluminium or organic matter.

Incised cut jaggedly with very deep teeth.

Included referring to stamens which do not project beyond the corolla or to valves which do not extend beyond the rim of a capsular fruit. *cf.* exserted.

Incurved curved inwards; curved towards the base or apex.

Indefinite numerous and variable in number.

Indehiscent not opening or splitting to release the contents at maturity as of fruit. *cf.* dehiscent.

Indumentum covering of fine hairs or bristles commonly found on external parts of plants.

Indurate to become hard, often the hardening developed only at maturity.

Indusium an enclosing membrane, covering the sorus of a fern. Also used for the modified style end or pollen-cup of some Goodeniaceae (including Brunoniaceae). *adj.* indusiate.

Inferior said of an ovary or fruit that has sepals, petals and stamens above the ovary. *cf.* superior.

Inflated enlarged and hollow except in the case of a fruit which may contain a seed. *cf.* swollen.

Inflexed Bent or curved inward or downward, as petals or sepals.

Inflorescence a flower cluster or the arrangement of flowers in relation to the axis and to each other on a plant.

Infrafoliar located below the leaves.

Infraspecific referring to any taxon below the species rank.

Infructescence the fruiting stage of an inflorescence.

Inrolled curved inwards.

Integuments two distinct tissue layers that surround the nucellus of the ovule, forming the testa or seed coat when mature.

Intercalary of growth, between the apex and the base; of cells, spores, etc., between two cells.

Interfoliar inter leaf.

Internode portion of the stem, culm, branch, or rhizome between two nodes or points of attachment of the leaves.

Interpetiolar as of stipules positioned between petioles of opposite leaves.

Intrastaminal within the stamens.

Intricate entangled, complex.

Introduced not indigenous; not native to the area in which it now occurs.

Introrse turned inwards or towards the axis or pistil as of anthers. *cf.* extrorse, latrorse.

Involucre a whorl of bracts or leaves that surround one to many flowers or an entire inflorescence.

Involute having the margins rolled inwards, referring to a leaf or other flat organ.

Jugate of a pinnate leaf; having leaflets in pairs.

Juvenile young or immature, used here for leaves formed on a young plant which are different in morphology from those formed on an older plant.

Keel a longitudinal ridge, at the back of the leaf. Also the two lower fused petals of a 'pea' flower in the Papilionaceae, which form a boat-like structure around the stamens and styles, also called carina. *adj.* keeled. *cf.* standard, wing.

Labellum the modified lowest of the three petals forming the corolla of an orchid, usually larger than the other two petals, and often spurred.

Laciniate fringed; having a fringe of slender, narrow, pointed lobes cut into narrow lobes.

Lamella a gill-shaped structure: fine sheets of material held adjacent to one another.

Lamina the blade of the leaf or frond.

Lanate wooly, covered with long hairs which are loosely curled together like wool.

Lanceolate lance-shaped in outline, tapering from a broad base to the apex.

Landrace plants adapted to the natural environment in which they grow, developing naturally with minimal assistance or guidance from humans and usually possess more diverse phenotypes and genotypes.

Laterite reddish–coloured soils rich in iron oxide, formed by weathering of rocks under oxidizing and leaching conditions, commonly found in tropical and subtropical regions. *adj.* lateritic.

Latex a milky, clear or sometimes coloured sap of diverse composition exuded by some plants.

Latrorse turned sideways, i.e. not towards or away from the axis as of anthers dehiscing longitudinally on the side. *cf.* extrorse, introse.

Lax loose or limp, not densely arranged or crowded.

Leaflet one of the ultimate segments of a compound leaf.

Lectotype a specimen chosen after the original description to be the type.

Lemma the lower of two bracts (scales) of a grass floret, usually enclosing the palea, lodicules, stamens and ovary.

Lenticel is a lens shaped opening that allows gases to be exchanged between air and the inner tissues of a plant, commonly found on young bark, or the surface of the fruit.

Lenticellate dotted with lenticels.

Lenticular shaped like a biconvex lens. *cf.* lentiform.

Lentiform shaped like a biconvex lens, *cf.* lenticular.

Leptomorphic temperate, running bamboo rhizome; usually thinner then the culms they support and the internodes are long and hollow.

Liane a woody climbing or twining plant.

Lignotuber a woody, usually underground, tuberous rootstock often giving rise to numerous aerial stems.

Ligulate small and tongue shaped or with a little tongue shaped appendage or ligule, star shaped as of florets of Asteraceae.

Ligule a strap-shaped corolla in the flowers of Asteraceae; also a thin membranous outgrowth from the inner junction of the grass leaf sheath and blade. *cf.* ligulate.

Limb the expanded portion of the calyx tube or the corolla tube, or the large branch of a tree.

Linear a 2-dimensional shape, narrow with nearly parallel sides.

Linguiform tongue shaped *cf.* ligulate.

Lithosol a kind of shallow soils lacking well-defined horizons and composed of imperfectly weathered fragments of rock.

Littoral of or on a shore, especially seashore.

Loam a type of soil mad up of sand, silt, and clay in relative concentration of 40%-40%-20% respectively.

Lobed divided but not to the base.

Loculicidal opening into the cells, when a ripe capsule splits along the back.

Loculus cavity or chamber of an ovary. *pl.* loculi.

Lodicules two small structures below the ovary which, at flowering, swell up and force open the enclosing bracts, exposing the stamens and carpel.

Lyrate pinnately lobed, with a large terminal lobe and smaller laterals ones which become progressively smaller towards the base.

Macronutrients chemical elements which are needed in large quantities for growth and development by plants and include nitrogen, phosphorus, potassium, and magnesium.

Maculate spotted.

Mallee a growth habit in which several to many woody stems arise separately from a lignotuber; usually applied to certain low-growing species of *Eucalyptus*.

Mangrove a distinctive vegetation type of trees and shrubs with modified roots, often viviparous, occupying the saline coastal habitats that are subject to periodic tidal inundation.

Marcescent withering or to decay without falling off.

Margin the edge of the leaf blade.

Medulla the pith in the stems or roots of certain plants; or the central portion of a thallus in certain lichens.

Megasporangium the sporangium containing megaspores in fern and fern allies. *cf.* microsporangium.

Megaspore the large spore which may develop into the female gametophyte in heterosporous ferns and fern allies. *cf.* microspore.

Megasporophyll a leaflike structure that bears megasporangia.

Megastrobilus (female cone, seed cone, or ovulate cone) contains ovules within which, when fertilized by pollen, become seeds. The female cone structure varies more markedly between the different conifer families.

Meiosis the process of cell division that results in the formation of haploid cells from diploid cells to produce gametes.

Mericarp a one-seeded portion of an initially syncarpous fruit (schizocarp) which splits apart at maturity. *Cf.* coccus.

Meristem the region of active cell division in plants, from which permanent tissue is derived. *adj.* meristematic

-merous used with a number prefix to denote the basic number of the three outer floral whorls, e.g. a 5-merous flower may have five sepals, ten petals and 15 stamens.

Mesic moderately wet.

Mesocarp the middle layer of the fruit wall derived from the middle layer of the carpel wall. *cf.* endocarp, exocarp, pericarp.

Mesophytes terrestrial plants which are adapted to neither a particularly dry nor particularly wet environment.

Micropyle the small opening in a plant ovule through which the pollen tube passes in order to effect fertilisation.

Microsporangium the sporangium containing microspores in pteridophytes. *cf.* megasporangium.

Microspore a small spore which gives rise to the male gametophyte in heterosporous pteridophytes. Also for a pollen grain. *cf.* megaspore.

Midvein the main vascular supply of a simple leaf blade or lamina. Also called mid-rib.

Mitosis is a process of cell division which results in the production of two daughter cells from a single parent cell.

Mollisols soils with deep, high organic matter, nutrient-enriched surface soil (A horizon), typically between 60 and 80 cm thick.

Monadelphous applied to stamens united by their filaments into a single bundle.

Monocarpic refer to plants that flower, set seeds and then die.

Monochasial a cyme having a single flower on each axis.

Monocotyledon angiosperm having one cotyledon.

Monoecious having both male and female unisexual flowers on the same individual plant. *cf.* dioecious.

Monoembryonic seed the seed contains only one embryo, a true sexual (zygotic) embryo. polyembryonic seed.

Monolete a spore that has a simple linear scar.

Monopodial with a main terminal growing point producing many lateral branches progressively. *cf.* sympodial.

Monotypic of a genus with one species or a family with one genus; in general, applied to any taxon with only one immediately subordinate taxon.

Montane refers to highland areas located below the subalpine zone.

Mucilage a soft, moist, viscous, sticky secretion. *adj.* mucilaginous.

Mucous (Botany) slimy.

Mucro a sharp, pointed part or organ, especially a sharp terminal point, as of a leaf.

Mucronate ending with a short, sharp tip or mucro, resembling a spine. cf. cuspidate, muticous.

Mucronulate with a very small mucro; a diminutive of mucronate.

Mulch protective cover of plant (organic) or non-plant material placed over the soil, primarily to modify and improve the effects of the local microclimate and to control weeds.

Multiple fruit a fruit that is formed from a cluster of flowers.

Muricate covered with numerous short hard outgrowths. *cf.* papillose.

Muriculate with numerous minute hard outgrowths; a diminutive of muricate.

Muticous blunt, lacking a sharp point. *cf.* mucronate.

Mycorrhiza the mutualistic symbiosis (non-pathogenic association) between soil-borne fungi with the roots of higher plants.

Mycorrhiza (vesicular arbuscular) endomycorrhiza living in the roots of higher plants producing inter-and intracellular fungal growth in root cortex and forming specific fungal structures, referred to as vesicles and arbuscles. *abbrev.* VAM.

Native a plant indigenous to the locality or region.

Naviculate boat-shaped.

Necrotic applied to dead tissue.

Nectariferous having one or more nectaries.

Nectary a nectar secretory gland; commonly in a flower, sometimes on leaves, fronds or stems.

Nervation venation, a pattern of veins or nerves as of leaf.

Node the joint between segments of a culm, stem, branch, or rhizome; the point of the stem that gives rise to the leaf and bud.

Nodule a small knoblike outgrowth, as those found on the roots of many leguminous, that containing *Rhizobium* bacteria which fixes nitrogen in the soil.

Nomen Illegitimum illegitimate taxon deemed as superfluous at its time of publication either because the taxon to which it was applied already has a name, or because the name has already been applied to another plant. *abbrev.* nom. illeg.

Nomen Nudum the name of a taxon which has never been validated by a description. *abbrev.* nom. nud.

Nomen Dubium an invalid proposed taxonomic name because it is not accompanied by a definition or description of the taxon to which it applies. *abbrev.* nom. dub.

Nucellus central portion of an ovule in which the embryo sac develops.

Nucellar embryony a form of seed reproduction in which the nucellar tissue which surrounds the embryo sac can produce additional embryos (polyembryony) which are genetically identical to the parent plant. This is found in many citrus species and in mango.

Nut a dry indehiscent 1-celled fruit with a hard pericarp.

Nutlet a small. 1-seeded, indehiscent lobe of a divided fruit.

Ob- prefix meaning inversely or opposite to.

Obconic a 3-dimensional shape; inversely conic; cone shaped, conic with the vertex pointing downward.

Obcordate inversely cordate, broad and notched at the tip; heart shaped but attached at the pointed end.

Obdeltate inversely deltate; deltate with the broadest part at the apex.

Oblanceolate inversely lanceolate, lance-shaped but broadest above the middle and tapering toward the base as of leaf.

Oblate having the shape of a spheroid with the equatorial diameter greater than the polar diameter; being flattened at the poles.

Oblong longer than broad with sides nearly parallel to each other.

Obovate inversely ovate, broadest above the middle.

Obpyramidal resembling a 4-sided pyramid attached at the apex with the square base facing away from the attachment.

Obpyriform inversely pyriform, resembling a pear which is attached at the narrower end. *cf.* pyriform.

Obspathulate inversely spathulate; resembling a spoon but attached at the broadest end. *cf.* spathulate.

Obtriangular inversely triangular; triangular but attached at the apex. *cf.* triangular.

Obtrullate inversely trullate; resembling a trowel blade with the broadest axis above the middle. *cf.* trullate.

Obtuse with a blunt or rounded tip, the converging edges separated by an angle greater than 90°.

-oid suffix denoting a 3-dimensional shape, e.g. spheroid.

Ochraceous a dull yellow color.

Ocreate having a tube-like covering around some stems, formed of the united stipules; sheathed.

Oleaginous oily.

Oligotrophic lacking in plant nutrients and having a large amount of dissolved oxygen throughout.

Operculum a lid or cover that becomes detached at maturity by abscission, e.g. in *Eucalyptus*, also a cap or lid covering the bud and formed by fusion or cohesion of sepals and/or petals. *adj.* operculate.

Opposite describing leaves or other organs which are borne at the same level but on opposite sides of the stem. *cf.* alternate.

Orbicular of circular outline, disc-like.

Order a taxonomic rank between class and family used in the classification of organisms, i.e. a group of families believed to be closely related.

Orifice an opening or aperture.

Organosols soils not regularly inundated by marine waters and containing a specific thickness of organic materials within the upper part of the profile.

Ovary the female part of the pistil of a flower which contains the ovules (immature seeds).

Ovate egg-shaped, usually with reference to two dimensions.

Ovoid egg-shaped, usually with reference to three dimensions.

Ovule the young, immature seed in the ovary which becomes a seed after fertilisation. *adj.* ovular..

Ovulode a sterile reduced ovule borne on the placenta, commonly occurring in Myrtaceae.

Oxisols refer to ferralsols.

Pachymorphic describes the short, thick, rhizomes of clumping bamboos with short, thick and solid internode (except the bud-bearing internodes, which are more elongated). *cf.* sympodial.

Palate (Botany) a raised appendage on the lower lip of a corolla which partially or completely closes the throat.

Palea the upper of the two membraneous bracts of a grass floret, usually enclosing the lodicules, stamens and ovary. *pl.* paleae. *adj.* paleal. *cf.* lemma.

Paleate having glumes.

Palm heart refers to soft, tender inner core and growing bud of certain palm trees which are eaten as vegetables. Also called heart of palm, palmito, burglar's thigh, chonta or swamp cabbage.

Palmate describing a leaf which is divided into several lobes or leaflets which arise from the same point. *adj.* palmately.

Palmito see palm heart.

Palustrial paludal, swampy, marshy.
Palustrine marshy, swampy.
Palustrine herb vegetation that is rooted below water but grows above the surface in wetland system.
Panduriform fiddle shaped, usually with reference to two dimensions.
Panicle a compound, indeterminate, racemose inflorescence in which the main axis bears lateral racemes or spikes. *adj.* paniculate.
Pantropical distributed through-out the tropics.
Papilionaceous butterfly-like, said of the pea flower or flowers of Papilionaceae, flowers which are zygomorphic with imbricate petals, one broad upper one, two narrower lateral ones and two narrower lower ones.
Papilla a small, superficial protuberance on the surface of an organ being an outgrowth of one epidermal cell. *pl.* papillae. *adj.* papillose.
Papillate having papillae.
Papillose covered with papillae.
Pappus a tuft (or ring) of hairs, bristles or scales borne above the ovary and outside the corolla as in Asteraceae often persisting as a tuft of hairs on a fruit. *adj.* pappose.
Papyraceous resembling parchment of paper.
Parenchyma undifferentiated plant tissue composed of more or less uniform cells.
Parietal describes the attachment of ovules to the outer walls of the ovaries.
Paripinnate pinnate with an even number of leaflets and without a terminal leaflet. *cf.* imparipinnate.
-partite divided almost to the base into segments, the number of segments written as a prefix.
Patelliform shaped like a limpet shell; cap-shaped and without whorls.
Patent diverging from the axis almost at right angles.
Peat is an accumulation of partially decayed vegetation matter.
Pectin a group of water-soluble colloidal carbohydrates of high molecular weight found in certain ripe fruits.
Pectinate pinnatifid with narrow segments resembling the teeth of a comb.
Pedicel the stalk of the flower or stalk of a spikelet in Poaceae. *adj.* pedicellate.
Pedicellate having pedicel.
Peduncle a stalk supporting an inflorescence. *adj.* pedunculate
Pellucid allowing the passage of light; transparent or translucent.
Pellucid-dotted copiously dotted with immersed, pellucid, resinous glands.
Peltate with the petiole attached to the lower surface of the leaf blade.
Pendant hanging down.
Pendulous drooping, as of ovules.
Penniveined or penni-nerved pinnately veined.
Pentamerous in five parts.
Perennial a plant that completes it life cycle or lives for more than 2 years. *cf.* annual, biennial.
Perfoliate a leaf with the basal lobes united around--and apparently pierced by--the stem.
Pergamentaceous parchment-like.
Perianth the two outer floral whorls of the Angiosperm flower; commonly used when the calyx and the corolla are not readily distinguishable (as in monocotyledons).
Pericarp (Botany). The wall of a ripened ovary; fruit wall composed of the exocarp, mesocarp and endocarp.
Persistent remaining attached; not falling off. *cf.* caduceus.
Petal free segment of the corolla. *adj.* petaline. *cf.* lobe.
Petiolar relating to the petiole.
Petiolate having petiole.
Petiole leaf stalk. *adj.* petiolate.
Petiolulate supported by its own petiolule.
Petiolule the stalk of a leaflet in a compound leaf. adj. petiolulate.
pH is a measure of the acidity or basicity of a solution. It is defined as the cologarithm of the activity of dissolved hydrogen ions (H+).
Phenology the study of periodic plant life cycle events as influenced by seasonal and interannual variations in climate.
Phyllary a bract of the involucre of a composite plant, term for one of the scale-like bracts beneath the flower-head in Asteraceae.
Phylloclade a flattened, photosynthetic branch or stem that resembles or performs the func-

tion of a leaf, with the true leaves represented by scales.

Phyllode a petiole that function as a leaf. *adj.* phyllodineous. *cf.* cladode.

Phyllopodia refer to the reduced, scale-like leaves found on the outermost portion of the corm where they seem to persist longer than typical sporophylls as in the fern Isoetes.

Phytoremediation describes the treatment of environmental problems (bioremediation) through the use of plants which mitigate the environmental problem without the need to excavate the contaminant material and dispose of it elsewhere.

Pileus (Botany) cap of mushroom.

Piliferous (Botany) bearing or producing hairs, as of an organ with the apex having long, hair-like extensions.

Pilose covered with fine soft hairs.

Pinna a primary division of the blade of a compound leaf or frond. *pl.* pinnae.

Pinnate bearing leaflets on each side of a central axis of a compound leaf; divided into pinnae.

Pinnatifid, pinnatilobed a pinnate leaf parted approximately halfway to midrib; when divided to almost to the mid rib described as deeply pinnatifid or pinnatisect.

Pinnatisect lobed or divided almost to the midrib.

Pinnule a leaflet of a bipinnate compound leaf.

Pistil female part of the flower comprising the ovary, style, and stigma.

Pistillate having one or more pistils; having pistils but no stamens.

Placenta the region within the ovary to which ovules are attached. *pl.* placentae.

Placentation the arrangement of the placentae and ovules in the ovary.

Plano- a prefix meaning level or flat.

Pleonanthic refer to palms in which the stem does not die after flowering.

Plicate folded like a fan.

Plumose feather-like, with fine hairs arising laterally from a central axis; feathery.

Pneumatophore modified root which allows gaseous exchange in mud-dwelling shrubs, e.g. mangroves.

Pod a dry one to many-seeded dehiscent fruit, as applied to the fruit of Fabaceae i.e. Caesalpiniaceae, Mimosaceae and Papilionaceae.

Podzol, Podsolic soil any of a group of acidic, zonal soils having a leached, light-coloured, gray and ashy appearance. Also called spodosol.

Pollen cone male cone or microstrobilus or pollen cone is structurally similar across all conifers, extending out from a central axis are microsporophylls (modified leaves). Under each microsporophyll is one or several microsporangia (pollen sacs).

Pollinia the paired, waxy pollen masses of flowers of orchids and milkweeds.

Polyandrous (Botany) having an indefinite number of stamens.

Polyembryonic seed seeds contain many embryos, most of which are asexual (nucellar) in origin and genetically identical to the maternal parent.

Polygamous with unisexual and bisexual flowers on the same or on different individuals of the same species.

Polymorphic with different morphological variants.

Polypetalous (Botany) having a corolla composed of distinct, separable petals.

Pome a fleshy fruit where the succulent tissues are developed from the receptacle.

Pore a tiny opening.

Premorse Abruptly truncated, as though bitten or broken off as of a leaf.

Procumbent trailing or spreading along the ground but not rooting at the nodes, referring to stems. *cf.* ascending, decumbent, erect.

Prophyll a plant structure that resembles a leaf.

Prostrate lying flat on the ground.

Protandous relating to a flower in which the anthers release their pollen before the stigma of the same flower becomes receptive.

Proximal end of any structure closest to the point of attachment. *cf.* distal.

Pruinose having a thick, waxy, powdery coating or bloom.

Pseudocarp a false fruit, largely made up of tissue that is not derived from the ovary but from floral parts such as the receptacle and calyx.

Pteridophyte a vascular plant which reproduces by spores; the ferns and fern allies.
Puberulent covered with minute hairs or very fine down; finely pubescent.
Puberulous covered with a minute down.
Pubescent covered with short, soft hairs.
Pulvinate having a swelling, pulvinus at the base as a leaf stalk.
Pulvinus swelling at the base of leaf stalk.
Pulviniform swelling or bulging.
Punctate marked with translucent dots or glands.
Punctiform marked by or composed of points or dots.
Punctulate marked with minute dots; a diminutive of punctate.
Pusticulate characterized by small pustules.
Pyrene the stone or pit of a drupe, consisting of the hardened endocarp and seed.
Pyriform pear-shaped, a 3-dimensional shape; attached at the broader end. *cf.* obpyriform.
Pyxidium seed capsule having a circular lid (operculum) which falls off to release the seed.
Raceme an indeterminate inflorescence with a simple, elongated axis and pedicellate flowers, youngest at the top. *adj.* racemose.
Rachilla the main axis of a grass spikelet.
Rachis the main axis of the spike or other inflorescence of grasses or a compound leaf.
Radiate arranged around a common centre; as of an inflorescence of Asteraceae with marginal, female or neuter, ligulate ray-florets and central, perfect or functionally male, tubular, disc florets. *cf.* disciform, discoid.
Radical arising from the root or its crown, or the part of a plant embryo that develops into a root.
Ray the marginal portion of the inflorescence of Asteraceae and Apiaceae when distinct from the disc. Also, the spreading branches of a compound umbel.
Receptacle the region at the end of a pedicel or on an axis which bears one or more flowers. *adj.* receptacular.
Recurved curved downwards or backwards.
Reflexed bent or turned downward.
Regosol soil that is young and undeveloped, characterized by medium to fine-textured unconsolidated parent material that maybe alluvial in origin and lacks a significant horizon layer formation.
Reniform kidney shaped in outline.
Repand with slightly undulate margin.
Replicate folded back, as in some corolla lobes.
Resinous producing sticky resin.
Resupinate twisted through 180°.
Reticulate having the appearance of a network.
Retrorse bent or directed downwards or backwards. *cf.* antrorse.
Retuse with a very blunt and slightly notched apex. *cf.* emarginated.
Revolute with the margins inrolled on the lower (abaxial) surface.
Rhizine a root-like filament or hair growing from the stems of mosses or on lichens.
Rhizoid root-like filaments in a moss, fern, fungus, etc. that attach the plant to the substratum.
Rhizome a prostrate or underground stem consisting of a series of nodes and internodes with adventitious roots and which generally grows horizontally.
Rhizophore a stilt-like outgrowth of the stem which branches into roots on contact with the substrate.
Rhombic shaped like a rhombus.
Rhomboid shaped like a rhombus.
Rib a distinct vein or linear marking, often raised as a linear ridge.
Riparian along the river margins, interface between land and a stream.
Rosette a tuft of leaves or other organs arranged spirally like petals in a rose, ranging in form from a hemispherical tuft to a flat whorl. *adj.* rosetted, rosulate.
Rostrate beaked; the apex tapered into a slender, usually obtuse point.
Rostrum a beak-like extension.
Rosulate having a rosette.
Rotate wheel shaped; refers to a corolla with a very short tube and a broad upper part which is flared at right angles to the tube. *cf.* salverform.
Rotundate rounded; especially at the end or ends.
Rugae refers to a series of ridges produced by folding of the wall of an organ.

Rugose deeply wrinkled.
Rugulose finely wrinkled.
Ruminate (Animal) chew repeatedly over an extended period.
Ruminate endosperm uneven endosperm surface that is often highly enlarged by ingrowths or infoldings of the surrounding tissue. cf. homogenous endosperm.
Saccate pouched.
Sagittate shaped like an arrow head.
Saline soils soils that contain excessive levels of salts that reduce plant growth and vigor by altering water uptake and causing ion-specific toxicities or imbalances.
Salinity is characterised by high electrical conductivities and low sodium ion concentrations compared to calcium and magnesium
Salverform applies to a gamopetalous corolla having a slender tube and an abruptly expanded limb.
Samara an indehiscent, winged, dry fruit.
Sand a naturally occurring granular material composed of finely divided rock and mineral particles range in diameter from 0.0625 μm to 2 mm. *adj.* sandy
Saponins are plant glycosides with a distinctive foaming characteristic. They are found in many plants, but get their name from the soapwort plant (*Saponaria*).
Saprophytic living on and deriving nourishment from dead organic matter.
Sapwood outer woody layer of the tree just adjacent to and below the bark.
Sarcotesta outermost fleshy covering of Cycad seeds below which is the sclerotesta.
Scabrid scurfy, covered with surface abrasions, irregular projections or delicate scales.
Scabrous rough to the touch.
Scale dry bract or leaf.
Scandent refer to plants, climbing.
Scape erect flowering stem, usually leafless, rising from the crown or roots of a plant. *adj.* scapose.
Scapigerous with a scape.
Scarious dry, thin and membranous.
Schizocarp a dry fruit which splits into longitudinally multiple parts called mericarps or cocci. *adj.* schizocarpous.

Sclerotesta the innermost fleshy coating of cycad seeds, usually located directly below the sarcotesta.
Scorpoid refers to a cymose inflorescence in which the main axis appears to coil.
Scutellum (Botany) any of various parts shaped like a shield.
Secondary venation arrangement of the lateral veins arising from the midrib in the leaf lamina.
Secund with the flowers all turned in the same direction.
Sedge a plant of the family Apiaceae, Cyperaceae.
Segmented constricted into divisions.
Seminal root or seed root originate from the scutellar node located within the seed embryo and are composed of the radicle and lateral seminal roots.
Senescence refers to the biological changes which take place in plants as they age.
Sepal free segment of the calyx. *adj.* sepaline.
Septum a partition or cross wall. *pl.* septa. *adj.* septate.
Seriate arranged in rows.
Sericeous silky; covered with close-pressed, fine, straight silky hairs.
Serrate toothed like a saw; with regular, asymmetric teeth pointing forward.
Serrated toothed margin.
Serratures serrated margin.
Serrulate with minute teeth on the margin.
Sessile without a stalk.
Seta a bristle or stiff hair. *pl.* setae. *adj.* setose, setaceous.
Setaceous bristle-like.
Setate with bristles.
Setiform bristle shaped.
Setulose with minute bristles.
Sheathing clasping or enveloping the stem.
Shrub a woody plant usually less than 5 m high and many-branched without a distinct main stem except at ground level.
Silicula a broad, dry, usually dehiscent fruit derived from two or more carpels which usually dehisce along two sutures. *cf.* siliqua.
Siliqua a silicula which is at least twice as long as broad.

Silt is soil or rock derived granular material of a grain size between sand and clay, grain particles ranging from 0.004 to 0.06 mm in diameter. *adj.* silty.

Simple refer to a leaf or other structure that is not divided into parts. *cf.* compound.

Sinuate with deep wavy margin.

Sinuous wavy.

Sinus an opening or groove, as occurs between the bases of two petals.

Sodicity is characterised by low electrical conductivities and high sodium ion concentrations compared to calcium and magnesium.

Sodic soils contains high levels of sodium salts that affects soil structure, inhibits water movement and causes poor germination and crop establishment and plant toxicity.

Soil pH is a measure of the acidity or basicity of the soil. See pH.

Solitary usually refer to flowers which are borne singly, and not grouped into an inflorescence or clustered.

Sorocarp fruiting body formed by some cellular slime moulds, has both stalk and spore mass.

Sorophore stalk bearing the sorocarp.

Sorus a discrete aggregate of sporangia in ferns. *pl.* sori

Spadix fleshy spike-like inflorescence with an unbranched, usually thickened axis and small embedded flowers often surrounded by a spathe. *pl.* spadices.

Spathe a large bract ensheathing an inflorescence or its peduncle. *adj.* spathaceous.

Spatheate like or with a spathe.

Spathulate spatula or spoon shaped; broad at the tip and narrowed towards the base.

Spicate borne in or forming a spike.

Spiculate spikelet-bearing.

Spike an unbranched, indeterminate inflorescence with sessile flowers or spiklets. *adj.* spicate, spiciform.

Spikelet a small or secondary spike characteristics of the grasses and sedges and, generally composed of two glumes and one or more florets. Also applied to the small spike-like inflorescence or inflorescence units commonly found in Apiaceae.

Spine a stiff, sharp, pointed structure, formed by modification of a plant organ. *adj.* spinose.

Spinescent ending in a spine; modified to form a spine

Spinulate covered with small spines.

Spinulose with small spines over the surface.

Spodosol see podsol.

Sporangium a spore bearing structure found in ferns, fern allies and gymnosperms. *pl.* sporangia. *adj.* sporangial.

Sporocarp a stalked specialized fruiting structure formed from modified sporophylls, containing sporangia or spores as found in ferns and fern allies.

Sporophore a spore-bearing structure, especially in fungi.

Sporophyll a leaf or bract which bears or subtends sporangia in the fern allies, ferns and gymnosperms.

Sporophyte the spore-producing phase in the life cycle of a plant that exhibits alternation of generations.

Spreading bending or spreading outwards and horizontally.

Spur a tubular or saclike extension of the corolla or calyx of a flower.

Squama structure shaped like a fish scale. *pl.* squamae.

Squamous covered in scales.

Squarrose having rough or spreading scale-like processes.

Stamen the male part of a flower, consisting typically of a stalk (filament) and a pollen-bearing portion (anther). *adj.* staminal, staminate.

Staminate unisexual flower bearing stamens but no functional pistils.

Staminode a sterile or abortive stamen, often reduced in size and lacking anther. *adj.* staminodial.

Standard refers to the adaxial petal in the flower of Papilionaceae. *cf.* keel, wing.

Starch a polysaccharide carbohydrate consisting of a large number of glucose units joined together by glycosidic bonds α-1-4 linkages.

Stellate star shaped, applies to hairs.

Stem the main axis of a plant, developed from the plumule of the embryo and typically bearing leaves.

Sterile lacking any functional sexual parts which are capable of fertilisation and seed production.

Stigma the sticky receptive tip of an ovary with or without a style which is receptive to pollen.

Stilt root a supporting root arising from the stem some distance above the ground as in some mangroves, sometimes also known as a prop root.

Stipe a stalk that support some other structure like the frond, ovary or fruit.

Stipel secondary stipule at the base of a leaflet. *pl.* stipellae. *adj.* stipellate.

Stipitate having a stalk or stipe, usually of an ovary or fruit.

Stipulated having stipules.

Stipule small leaf-like, scale-like or bristle-like appendages at the base of the leaf or on the petiole. *adj.* stipulate.

Stolon a horizontal, creeping stem rooting at the nodes and giving rise to another plant at its tip.

Stoloniferous bearing stolon or stolons.

Stoma a pore in the epidermis of the leaf or stem for gaseous exchange. *pl.* stomata.

Stone the hard endocarp of a drupe, containing the seed or seeds.

Stramineous chaffy; straw-liked.

Striae parallel longitudinal lines or ridges. *adj.* striate.

Striate marked with fine longitudinal parallel lines or ridges.

Strigose bearing stiff, straight, closely appressed hair; often the hairs have swollen bases.

Strobilus a cone-like structure formed from sporophylls or sporangiophores. *pl.* strobili

Style the part of the pistil between the stigma and ovary.

Sub- a prefix meaning nearly or almost, as in subglobose or subequal.

Subcarnose nearly fleshy.

Sub-family taxonomic rank between the family and tribe.

Subglobose nearly spherical in shape.

Subretuse faintly notched at the apex.

Subsessile nearly stalkless or sessile.

Subshrub intermediate between a herb and shrub.

Subspecies a taxonomic rank subordinate to species.

Substrate surface on which a plant or organism grows or attached to.

Subtend attached below something.

Subulate narrow and tapering gradually to a fine point, awl-shaped.

Succulent fleshy, juicy, soft in texture and usually thickened.

Sulcate grooved longitudinally with deep furrows.

Sulcus a groove or depression running along the internodes of culms or branches.

Superior refers to the ovary is free and mostly above the level of insertion of the sepals, and petals. *cf.* inferior.

Suture line of dehiscence.

Swidden slash-and-burn or shifting cultivation.

Symbiosis describes close and often long-term mutualistic and benefical interactions between different organisms.

Sympetalous having petals united.

Sympodial refers to a specialized lateral growth pattern in which the apical meristem. *cf* monopodial.

Synangium an organ composed of united sporangia, divided internally into cells, each containing spores. *pl.* synangia.

Syncarp an aggregate or multiple fruit formed from two or more united carpels with a single style. *adj.* syncarpous.

Syncarpous carpels fused forming a compound pistil.

Syconium a type of pseudocarp formed from a hollow receptacle with small flowers attached to the inner wall. After fertilization the ovaries of the female flowers develop into one-seeded achenes, e.g. fig.

Tannins group of plant-derived phenolic compounds.

Taxon the taxonomic group of plants of any rank. e.g. a family, genus, species or any infraspecific category. *pl.* taxa.

Tendril a slender, threadlike organ formed from a modified stem, leaf or leaflet which, by coiling around objects, supports a climbing plant.

Tepal a segment of the perianth in a flower in which all the perianth segments are similar in appearance, and are not differentiated into calyx and corolla; a sepal or petal.

Tetrasporangium a sporangium containing four haploid spores as found in some algae.

Terete having a circular shape when cross-sectioned or a cylindrical shape that tapers at each end.
Terminal at the apex or distal end.
Ternate in threes as of leaf with three leaflets.
Testa a seed coat, outer integument of a seed.
Thallus plant body of algae, fungi, and other lower organisms.
Thyrse a dense, panicle-like inflorescence, as of the lilac, in which the lateral branches terminate in cymes.
Tomentose refers to plant hairs that are bent and matted forming a wooly coating.
Tomentellose mildly tomentose.
Torus receptacle of a flower.
Transpiration evaporation of water from the plant through leaf and stem pores.
Tree that has many secondary branches supported clear of the ground on a single main stem or trunk.
Triangular shaped like a triangle, 3-angled and 3-sided.
Tribe a category intermediate in rank between subfamily and genus.
Trichome a hair-like outgrowth of the epidermis.
Trichotomous divided almost equally into three parts or elements.
Tridentate three toothed or three pronged.
Trifid divided or cleft into three parts or lobes.
Trifoliate having three leaves.
Trifoliolate a leaf having three leaflets.
Trifurcate having three forks or branches.
Trigonous obtusely three-angled; triangular in cross-section with plane faces.
Tripartite consisting of three parts.
Tripinnate relating to leaves, pinnately divided three times with pinnate pinnules.
Tripliveined main laterals arising above base of lamina.
Triploid describing a nucleus or cell that has three times (3n) the haploid number (n) of chromosomes.
Triveined main laterals arising at the base of lamina.
Triquetrous three-edged; acutely 3-angled.
Trullate with the widest axis below the middle and with straight margins; ovate but margins straight and angled below middle, trowel-shaped.
Truncate with an abruptly transverse end as if cut off.
Tuber a stem, usually underground, enlarged as a storage organ and with minute scale-like leaves and buds. *adj.* tuberous.
Tubercle a wart-like protuberance. *adj.* tuberculate.
Tuberculate bearing tubercles; covered with warty lumps.
Tuberization formation of tubers in the soil.
Tuft a densely packed cluster arising from an axis. *adj.* tufted.
Turbinate having the shape of a top; cone-shaped, with the apex downward, inversely conic.
Turgid distended by water or other liquid.
Turion the tender young, scaly shoot such as asparagus, developed from an underground bud without branches or leaves.
Turnery articles made by the process of turning.
Twining winding spirally.
Ultisols mineral soils with no calcareous material, have less than 10% weatherable minerals in the extreme top layer of soil, and with less the 35% base saturation throughout the soil.
Umbel an inflorescence of pedicellate flowers of almost equal length arising from one point on top of the peduncle. *adj.* umbellate.
Umbellet a secondary umbel of a compound umbel. *cf.* umbellule.
Umbellule an, a secondary umbel of a compound umbel. *cf.* umbellet.
Uncinate bent at the end like a hook; unciform.
Undershrub subshrub; a small, usually sparsely branched woody shrub less than 1 m high. *cf.* shrub.
Undulate with an edge/margin or edges wavy in a vertical plane; may vary from weakly to strongly undulate or crisped. *cf.* crisped.
Unifoliolate a compound leaf which has been reduced to a single, usually terminal leaflet.
Uniform with one form, e.g. having stamens of a similar length or having one kind of leaf. *cf.* dimorphic.
Uniseriate arranged in one row or at one level.
Unisexual with one sex only, either bearing the anthers with pollen, or an ovary with ovules, referring to a flower, inflorescence or individual plant. *cf.* bisexual.

Urceolate shaped like a jug, urn or pitcher.

Utricle a small bladdery pericarp.

Valvate meeting without overlapping, as of sepals or petals in bud. *cf.* imbricate.

Valve one of the sections or portions into which a capsule separates when ripe.

Variant any definable individual or group of individuals which may or may not be regarded as representing a formal taxon after examination.

Variegate, variegated diverse in colour or marked with irregular patches of different colours, blotched.

Variety a taxonomic rank below that of subspecies.

Vein (Botany) a strand of vascular bundle tissue.

Velum a flap of tissue covering the sporangium in the fern, Isoetes.

Velutinous having the surface covered with a fine and dense silky pubescence of short fine hairs; velvety. *cf.* sericeous

Venation distribution or arrangement of veins in a leaf.

Veneer thin sheet of wood.

Ventral (Botany) facing the central axis, opposed to dorsal.

Vernation the arrangement of young leaves or fronds in a bud or at a stem apex. *cf.* circinnate

Verrucose warty

Verticil a circular arrangement, as of flowers, leaves, or hairs, growing about a central point; a whorl.

Verticillaster false whorl composed of a pair of opposite cymes as in Lamiaceae.

Verticillate whorled, arranged in one or more whorls.

Vertisol a soil with a high content of expansive montmorillonite clay that forms deep cracks in drier seasons or years.

Vertosols soils that both contain more than 35% clay and possess deep cracks wider than 5 mm during most years.

Vesicle a small bladdery sac or cavity filled with air or fluid. *adj.* vesicular.

Vestigial the remaining trace or remnant of an organ which seemingly lost all or most of its original function in a species through evolution.

Vestiture covering; the type of hairiness, scaliness or other covering commonly found on the external parts of plants. *cf.* indumentums.

Vibratile capable of to and for motion.

Villose covered with long, fine, soft hairs, finer than in pilose.

Villous covered with soft, shaggy unmatted hairs.

Vine a climbing or trailing plant.

Violaxanthin is a natural xanthophyll pigment with an orange color found in a variety of plants like pansies.

Viscid sticky, being of a consistency that resists flow.

Viviparous describes seeds or fruit which sprout before they fall from the parent plant.

Whorl a ring-like arrangement of leaves, sepals, stamens or other organs around an axis.

Winged having a flat, often membranous expansion or flange, e.g. on a seed, stem or one of the two lateral petals of a Papilionaceous flower or one of the petal-like sepals of Polygalaceae. *cf.* keel, standard.

Xanthophylls are yellow, carotenoid pigments found in plants. They are oxidized derivatives of carotenes.

Xeromorphic plant with special modified structure to help the plant to adapt to dry conditions.

Xerophyte a plant which naturally grows in dry regions and is often structurally modified to withstand dry conditions.

Zygomorphic having only one plane of symmetry, usually the vertical plane, referring to a flower, calyx or corolla. *cf.* actinomorphic.

Zygote the fist cell formed by the union of two gametes in sexual reproduction. *adj.* zygotic.

Common Name Index

A
Achiote, 515–517, 523
Adonidia palm, 257
African baobab, 527, 528
African fan palm, 293
African oil palm, 335, 342
African sausage tree, 486
Agrimony, 729
Alligator apple, 180
Almonds, 140, 142, 409, 476, 553, 619, 630
Ambarella, 160–164
American chestnut, 584
American dwarf palm tree, 438
American pepper, 153
Annato, 515–517, 519, 521, 523
Annatto
    plant, 515
    tree, 515
Apple, 48, 50–53, 55–58, 60, 62, 63, 93, 99, 160, 161, 171–173, 176, 180–184, 186, 190, 201, 202, 205, 207, 209, 212, 216, 221, 222, 227, 293, 417, 433, 531, 532, 594, 595, 597, 598, 609, 636, 637, 693
Apple cactus, 636, 637
Apricot, 169
Araluen pear, 687
Arctic kiwi, 5
Areca
    nut, 260–274, 278, 547
    nut/dohra, 269
    nut palm, 260
    quid, 266, 267. 269, 270, 278, 279
    seed, 269
Arenavirus, 313
Arenga fruit, 281
Arenga palm, 280, 282
Areng palm, 280
Argus pheasant-tree, 75
Arnatto, 515, 516
Arnotto dye plant, 515
Asam kumbang, 135, 429
Asam paya, 396, 397
Asian palmyra palm, 293
Atemoya, 171–174
Atlantic salmon, 379
Australian areca palm, 277
Australian baobab, 536
Australian pepper, 153
Australian pest pear, 687
Avocado, 98, 576, 729

B
Babaco, 718–720
Baby kiwi, 5
Bacang, 83–85, 129, 562
Bambangan, 131–134
Banana, 1, 45, 131, 252, 253, 282, 305, 433, 509, 563
Baobab, 486, 527–534, 536
Barbary fig, 660, 683
Barberry fig, 660, 683
Beaded hazel, 471
Beetroot, 647
Belle of the night, 643, 650
Beluno, 79, 121, 122
Bengal currant, 241
Betel
    nut, 258, 260, 262, 266, 267, 272, 273
    nut palm, 260
    palm, 260
    pepper, 262
    quid, 270, 271, 273
    vine, 578
Betel quid, 264–273, 278, 279
Big num num, 237
Bilimbi, 448–452, 465, 467, 468, 470
Bimbling plum, 448
Binjai, 79–81, 121, 123
Biriba, 221–223
Biribah, 221
Blackcurrant, 34
Black elder, 30, 38
Black-fiber palm, 280
Black sugar palm, 280
Black tiger shrimp, 578
Blume galip, 619
Boab, 536–539
Bob wood, 180
Borassus palm, 293
Bore tree, 30
Borneo rubber, 249
Bottle tree, 527, 536
Bour tree, 30
Bower actinidia, 5
Bower vine, 5

Brab tree, 293
Brazilian pawpaw, 190
Brazilian peppertree, 153
Brazilian plum, 166
Breadfruit-vine, 252
Brine, 91, 92, 167, 183, 197, 235, 442, 491, 501, 502, 504
Broken bones plant, 497
Bullock/bull's heart, 201, 203
Bungua, 277
Burahol, 227–229
Burmese plum, 72
Bush banana, 231

C

Cabbage looper, 217
Cabbage palm, 263, 278, 282, 303, 336, 400, 420, 438
Cactus pear, 662, 664, 666–668, 671–674, 678, 689
Calabash, 480, 482, 483, 527
Calabash tree, 480, 483
California pepper tree, 153
Cambodian palm, 293
Canarium almond, 619
Canarium nut, 619
Candlenut, 328
Candlestick tree, 512
Candle tree, 508, 512
Carambola, 449, 451, 452, 454–462, 467, 470
Carissa, 237–245
Cashew, 45–63, 150
Cashew nut, 45, 46, 48, 50, 52–54, 56–59, 61–63
Cassava, 524
Castor oil, 415, 492, 520, 578
Catechu, 260–274, 277, 278
Catfish, 381
Catus pear, 668
Cerapu, 84, 85, 129, 562
Ceriman, 252–254
Champagne fruit, 718
Cherimoya of the lowlands, 176
Cherry, 169
Chilean pepper tree, 153
Chile plum, 166
Chilli, 48, 70, 79, 91, 92, 162, 167, 498, 550, 571
Chinese actinidia, 12
Chinese betel-nut, 257
Chinese filbert, 471
Chinese gooseberry, 20, 28
Chinese hazel, 471
Chinese hazelnut, 471
Chinese olive, 630, 632
Christmas palm, 257
Christ thorns, 241
Ciruela, 166, 167
Civet-cat fruit tree, 569
Civet fruit, 569
Clumping betel nut, 277
Coastal prickly pear, 687
Cob, 471

Cobnut, 471
Cochineal cactus, 656
Cochineal fig, 683
Cochineal nopal cactus, 656
Cochineal-plant, 656
Cochineal prickly-pear, 683
Cockroach, 158, 198
Cocktail kiwi, 5
Cocoa, 274, 303, 323, 337–339, 377, 445, 584
Coco de mer palm, 399
Coconut
    milk, 48, 92, 93, 305, 308–309, 314, 315, 320, 322, 323, 405, 449, 570, 573
    oil, 114, 216, 217, 274, 306, 309–325, 329, 338, 344, 352, 354–356, 379, 573, 617, 620, 630
    palm, 301, 302, 310, 323, 326, 354
    tree, 303, 323, 328
    water, 295, 305–308, 310, 314, 317, 320, 322, 327, 402
Coffee, 48, 291, 303, 305, 306, 338, 410, 419, 420, 473, 529, 725
Coffee palm, 419
Common custard apple, 201
Common durian, 569
Common elder, 30
Common (black) elder, 38
Common filbert, 471
Common hazelnut, 471
Common juniper, 723
Common night blooming cereus, 643, 650
Common pest pear, 687
Common pistache, 142
Common pistachio, 142
Common prickly pear, 683, 687
Conderella plant, 643, 650
Coolie tamarind (Trinidad), 454
Copra, 305, 306, 311, 315, 317, 323–325, 383, 401
Copra oil, 311, 315, 317
Corkwood, 180, 540
Corn, 92, 253, 291, 305, 316, 343, 354, 405, 666
Corn oil, 314, 317, 339, 353, 354, 356, 361, 362, 665
Coromandel gooseberry, 454
Cottonseed oil, 356, 543, 547
Cotton silk tree, 540, 541
Cotton tree, 540, 541
Cottonwood tree, 540
Country gooseberry, 454
Cow apple, 180
Cow okra, 508
Cowpea weevil, 713, 714
CPO. *See* Crude palm oil
Cranberries, 476
Cream of tartar tree, 527, 536
Crude palm oil (CPO), 336, 337, 340, 342–349, 356, 361, 375, 377, 378, 671
Cucumber tree, 448, 486, 508
Custard apple, 171–173, 176, 186, 190, 201, 205, 207, 216, 221
Cut-leaf-philodendron, 252

Common Name Index   819

D
Dabai, 624–628
Dakka, 262
Danewort, 30
Date palm, 407–410, 412–416
Dates, 336, 407–417, 574
Dead Rat tree, 527, 536
Delicious monster, 252
Dessert kiwi, 5
Diamondback moth, 217
Dildo, 687
Doleib, 293
Double coconut, 399
Doub palm, 293
Dragon fruit, 643, 645, 647, 648, 650–652, 654
Drooping prickly pear, 683
Dura, 286, 336, 343, 383, 393–395
Durian, 84, 85, 98, 129, 190, 550–553, 555–557, 559, 562, 563, 566, 568–580
Durian daun, 85, 129, 552, 562, 563
Durian lai, 559, 563
Durian nyekak, 84, 85, 129, 562
Dwarf date-palm, 419
Dwarf elder, 38
Dwarf juniper, 723
Dwarf royal palm, 257

E
Edible-fruited rakum palm, 429
Edible-fruited salak palm, 396, 423, 429, 432
Edible salacca, 432
Egg mango, 124
Elder, 30, 31, 33, 37–41
Elderberry, 30–36, 39, 40
Elder bush, 30
Eltham Indian fig, 687
Erect prickly pear, 687
Erect prickly-pear cactus, 687
Escobilla, 153
Ethiopian sour tree, 527
European alder, 30
European black elder, 30
European black elderberry, 30
European elder, 30
European elderberry, 30, 32
European filbert, 471
European hazel, 471

F
Fairy circle, 723
False date palm, 419
False pepper, 153
False sago, 732
Feather palm, 419
Fennel, 262, 452
Finger root, 231
Five corner (Australian), 454
Five fingers (Guyana), 454

Food candle tree, 508
Fortune tree, 584
Fragrant mango, 93, 127
Fruit salad plant, 252

G
Galip, 619–622
Galip nut, 619–621
Gandaria, 69, 72
Gayndah pear, 687
Giant club cactus, 636
Giant filbert, 471
Ginger, 92, 529
Golden apple, 160, 161
Gold kiwi, 12, 27
Gold kiwifruit, 12, 15–17, 25, 27
Gomuti palm, 280
Gooseberry, 5, 20, 28, 454
Gourd tree, 480
Gouty stem tree, 536
Grapefruit, 56, 209, 460
Grapes, 1, 6, 24, 52, 563
Grass carp, 381
Greater salak, 427
Great fan palm, 293
Green almond, 142
Green-fleshed actinidia, 20
Green gram, 305
Green kiwi, 17, 27
Green kiwifruit, 10, 15–18, 20, 25, 27
Ground juniper, 723
Groundnut, 311, 627
Groundnut oil (GNO), 311, 337, 362, 544
Guanabana, 186, 190, 191
Guar, 321
Guiana chestnut, 584, 588, 589
Gutka, 272
Guyana chestnut, 584

H
Hackmatack, 723
Hardy kiwi, 5
Hardy kiwifruit, 10
Hawthorn, 105
Hazel filbert, 471
Hazelnut, 471–478432–439
Head lice, 198, 216, 322
Hedge cactus, 636
Hepatitis C virus, 272, 703
Herpes simplex virus, 35, 36, 111, 195, 244, 312, 532, 658
Herpes simplex virus type 1, 35, 36, 312, 658
Himalayan hazel, 471
HIV. *See* Human immunodeficiency virus
Hog plum, 160, 166
Horse savin, 723
Hue tree, 480
Human immunodeficiency virus (HIV), 35, 101, 183
Hurricaneplant, 252

I

Ice-apple, 293
Ilama, 176, 177
Indian earthworms, 205
Indian fig, 660, 683, 687
Indian fig opuntia, 660
Indian-fig prickly-pear, 660
Indian mango, 87
Indian mombin, 160
Indian nut, 260
Indian trumpet flower, 497
Isu, 552, 559, 563, 564

J

Jackfruit, 305
Jamaican apple, 201
Jamaica plum, 166
Jambu
    mawar, 85, 129, 562
    susu, 84, 129, 562
Japanese actinidia, 5
Java almond, 619, 630
Java cotton, 540
Java kapok, 540
Java olive, 619, 630
Jentik-jentik, 84, 85, 129, 562
Jew plum, 160, 161
Jocote, 46, 166–168
Jocote fruit, 168
Judas fruit, 527
June plum, 160, 161
Jungle salak, 423, 425
Junin virus, 313
Juniper, 723, 725–730
Juniper berry, 723, 725, 726, 729
Juniperi lignum, 729
Juniperi pseudo-fructus, 729

K

Kalamansi, 712
Kampong durian, 569
Kanari, 619, 630
Kapok tree, 540, 542–544, 547, 548
Karanda, 237, 241, 243
Kemang, 121, 122
Kembayau, 624, 633, 635
Kenari nut tree, 630
Kepel, 227, 229
Kepel apple, 227
Kerpis palm, 257
Kidney-nut, 45
King sago palm, 732
Kiwi
    fruit, 10, 16, 17, 20–25, 27, 436, 531
    gold, 12–15, 17, 18, 27
Kiwiberry, 5–7, 10
Kiwifruit, 10, 12, 15–17, 20–28

Kubal madu, 247, 248
Kubal tusu, 249, 250
Kuini, 84, 85, 127, 129, 569
Kurwini mango, 127
Kwini, 127–129
Kwini mango, 127

L

Lamantan, 138, 139
Lambert's filbert, 471
Lancewood, 221
Large num num, 237
Lemon, 52, 83, 91–93, 99, 162, 163, 176, 278,
    404, 409, 595, 688, 696, 708, 719
Lemongrass, 708
Lettuce, 34
Lice, 198, 205, 216, 217, 322
Lime, 6, 48, 91–93, 128, 172, 176, 245, 253,
    262, 267, 272, 328, 430, 449, 455, 456,
    524, 594, 634, 669, 678, 696
Lipstick plant, 515
Lipstick tree, 515, 523
Liver sausage tree, 486
Locust and wild honey, 252
Lontar palm, 293
Love nuts, 399

M

Macaw fat, 335
Malabar chestnut, 584–586, 588
Malayan pit viper, 113
Malchangel, 723
Maldive coconut, 399
Mango, 1, 48, 52, 69, 72, 73, 87–115, 124–127, 129,
    150, 161, 169, 291, 449
Mangosteen, 436, 576, 579
Mango tree, 87, 97, 99, 125
Mangrove annona, 180
Mangrove palm, 402
Manila palm, 257
Marian plum, 69, 72
Marian tree, 72
Mastic tree, 153
Mauby, 320
Melon tree, 693
Merrill palm, 257
Mexican breadfruit, 252
Midday marvel, 497
Mini kiwi, 5
Mint, 91, 708
Mission cactus, 660
Mission prickly-pear, 660
Molle, molle del peru, 153–158
Money plant, 584, 587
Money tree, 584
Mon fruit, 75, 77
Monkey apple, 180

Monkey-bread tree, 527
Monstera, 252–256
Monster fruit, 252
Mountain common juniper, 723
Mountain juniper, 723
Mountain soursop, 186, 187
Mud mussels, 405
Mukindu palm, 419
Mummy apple, 693

N

Nanettika fruit, 643, 650
Nangai nut, 619
Natal plum, 237, 239
Nematodes, 322, 707
Netted custard apple, 201
New guinea walnut, 75, 77
Ngali nut, 619
Night blooming cereus, 636, 643, 650
Nipa, 402–405
Nipah, 402, 405
Nipa palm, 402–405
Nopal, 656, 660–663, 666, 669, 671, 673, 678, 687
Nopal cactus, 656
Nopalea grande, 656
Nopales opuntia, 656
Nopalito, 660, 662, 677
Nypa palm, 402

O

Oil palm, 335–367, 369–383, 393, 394
Okra, 508, 529
Old field common juniper, 723
Olive, 84, 162, 404, 585, 586, 619, 624
Olive oil, 311, 318, 344, 352, 355, 356, 520, 543, 547, 609, 624–627
Onion, 1, 34, 91, 92, 498, 529, 546, 550, 571, 662
Orange, 6, 13, 14, 22, 48, 51, 52, 70, 73, 93, 98, 114, 128, 164, 168, 169, 172, 181, 194, 239, 247, 248, 250, 264, 278, 288, 336, 341, 349, 397, 404, 409, 421, 430, 452, 517–519, 528, 531, 532, 552–555, 558, 560, 561, 591, 595, 664, 665, 667, 669, 684, 698, 734
Orange-fleshed durian, 552
Otaheite apple, 160
Ovo, 166, 167, 169
Ox-heart, 201
Oxyleyanus durian, 563

P

Pacific walnut, 75
Padi mango, 124–125
Padi mango fruit, 125
Pajang, 85, 129, 131–134
Pakan, 559, 687

Palm
    chestnut, 285, 287, 295, 404, 573, 584–586, 588, 589
    oil, 324, 336, 337, 339, 340, 343–346, 393, 586, 711,
    olein, 337–340, 342, 346–349, 352–355, 369, 372, 377–379
    stearin, 337–339, 345, 354, 378, 379
Palm kernel oil (PKO), 336–338, 342, 346, 375, 376, 378, 379
Palm-pressed mesocarp fiber (PPF), 347, 379
Palmyra, 293–299
Palmyrah, 297
Palmyra palm, 293–299
Panama candle tree, 512
Papaw, 92, 693–694
Papaya, 48, 92, 492, 577, 610, 693–714
Papua New Guinea walnut, 75, 77
Parainfluenza viruses, 150
Pawpaw, 55, 56, 707
Peach, 169, 286, 290, 666
Peach palm, 285
Peanut oil, 354, 356, 522
Peanuts, 91, 144, 337, 354, 356, 473, 522, 529, 570, 585, 588,
Pecans, 50, 146, 476, 687
Pejibaye, 285, 287–290
Pejibaye fruit, 289, 290
Pelajau, 140, 141
Pelong tree, 140
Pepper, 83, 143, 153, 154, 262, 283, 383, 452, 529, 547, 697
Pepper berry tree, 153
Pepperina, 153
Pepper rose, 153
Pepper tree, 153
Peruvian apple, 636, 637
Peruvian apple cactus, 636
Peruvian mastic tree, 153
Peruvian pepper, 153
Peruvian pepper-tree, 153
Peruvian tree, 153, 636
Peruvian tree cactus, 636
Pest pear of Australia, 687
Pest prickly-pear, 687
Pewa nut, 285
Pig canary, 616
Pinang palm, 260
Pine, 50, 146, 165, 252, 577, 598
Pineapple, 1, 48, 52, 169, 171, 253, 433, 577–581, 605, 606, 608–611, 719
Pineapple sweetsop, 171
Pine fruit tree, 252
Pink pepper, 153
Pink pepper corns, 153
Pipe tree, 30
Pine wood nematode, 165
Pisifera, 343, 393–395
    oil palm, 393
    palm, 393, 394
Pistachio nut, 142, 146–150

Pistachios, 143–148, 150, 476
Pistacia nut, 142
Pitabu, 247
Pitahaya, 636, 640–655
Pitaya, 221, 636, 640–648, 650, 653, 654
Plajau, 140
Plum, 69, 72, 160, 161, 166, 169, 237, 239, 241, 448, 650
Plum-mango, 72
Poison ivy, 150
Polio, 532
Poliovirus, 244
Polynesian plum, 160,161
Pond apple, 180–184
Potato, 60, 305, 336, 340
Prickly custard apple, 190
Prickly pear, 656, 660–668, 671, 673–677, 683, 687
Prickly pear cactus, 667, 671, 673
Prostrate juniper, 723
Provision tree, 584
Psitachio, 144
Purple jobo, 166
Purple mombin, 166
Purple pitaya, 645
Purple plum, 166

Q
Queen of the night, 643,650

R
Radish, 34, 537, 702
Rakam, 429
Rakum, 430, 431
Rakum palm, 429
Rape seed, 324, 339
RBD. *See* Refined, bleached and deodorized
RBDO. *See* Refined, bleached, and deodorized palm oil
RBD palm olein, 378
RBDPO. *See* Refined, bleached and deodorized palm olein
Red
  dragon fruit, 647, 654,
  mombin, 166–169
  pitayas,641
Red-fleshed dragon fruit, 645, 646
Red-fleshed durian, 552
Red-fleshed pitaya, 644–646, 653
Red palm oil (RPO), 337, 346, 348–351, 355, 357, 360, 362, 363, 379
Red palm olein (RPOL), 340, 346, 349, 369
Red palm shortening (RPS), 340
Red pitahaya, 643, 647
Refined, bleached and deodorized (RBD), 306, 309, 336, 337, 346, 355, 357, 378, 379
Refined, bleached, and deodorized palm oil (RBDO), 346, 361
Refined, bleached and deodorized palm olein (RBDPO), 340, 346, 354, 361, 362, 378
Remayong, 427, 428

Rollinia, 197, 221–225
Ron palm, 293
RPO. *See* Red palm oil
RPOL. *See* Red palm olein
RPS. *See* Red palm shortening

S
Saba nut, 584
Sago cycas, 732
Sago palm, 732
Sago palm of Japan, 732
Sagwire-palme, 280
Saipan mango, 127
Salad oil, 306, 337, 356, 517
Salak, 396, 423–429, 432–437
Salak palm, 396, 423, 429, 432, 437
Sambu, 30
Sausage tree, 486
Saw palmetto, 438–445
Scarlet plum, 166
Scrub palmetto, 438
Sea apple, 293
Sea coconut, 399
Senegal calabash, 527
Senegal date palm, 419
Serenoa, 438–445
Serapit, 247
Sesame seeds, 144, 373, 409
Seychelles coconut, 399
Seychelles nut, 399
Seychelles nut palm, 399
Seychelles palm nut tree, 399
Shaving brush tree, 584
Shrimp, 6, 21, 48, 70, 91, 162, 183, 197, 328, 337, 442, 491, 501, 502, 504, 578, 579, 662
Siberian gooseberry,5
Siberian hazel, 471
Siberian juniper, 723
Sibu olive, 624
Silk cotton tree, 540, 541
Sindbis virus, 244, 532
Siriguela, 169
Smooth mountain prickly-pear, 660
Smooth prickly-pear, 660
Snail, 58, 197
Snake
  fruit, 432, 434–436, 576
  palm, 432
Snake-skinned fruit, 432
Sotetsu, 732, 733
Sotetsu nut, 732
Sour gourd, 527, 536
Sour prickly pear, 687
Soursop, 169, 186, 187, 190–194, 197, 198, 205
Soursop fruit, 193, 194
Southern spineless cactus, 687
Soya bean, 314, 361, 459, 627, 700
Soybean oil, 337, 340, 353, 354, 359, 364, 369, 379
Spanish plum, 166

Spelt nipa, 402
Spineless cactus, 660, 687
Spineless prickly pear, 687
Spiny pest pear, 687
Split-leaf philodendron, 252
Starfruit, 454, 458–460
Star fruit, 455, 456, 459–462
Star pickle, 454
Strawberry peach, 20
Strawberry pear, 643, 650
Sugar apple, 173, 176, 190, 202, 207, 209, 212, 221, 222
Sugar cane, 296, 325, 405, 509, 529, 725
Sugar palm, 91, 280, 282, 293, 295, 410
Sumac, 150
Sunflower, 146, 320, 356, 358, 359
Sunflower seed oil (SSO), 356, 359, 360
Swamp date-palm, 419
Sweet potato, 305
Sweet prickly pear, 687
Sweet sop, 171, 201, 207–211, 216
Swiss cheese plant, 252

T
Tahitian quince, 160
Tala palm, 293
Tal-palm, 293
Tamarind, 83, 93, 162, 396, 449, 451, 454, 455
Tampoi kuning, 84, 85, 129, 562
Tampoi putih, 85, 129, 562
Tapioca, 282, 305
Taravine, 5
Taro, 305, 328
Tea tree, 578
Tenera, 335, 343, 348, 383, 393–395
Terebinth nut, 142
Thai cobra, 113
Tibetan filbert, 471
Tibetan hazelnut, 471
Tilapia, 379, 381
Tobacco tree, 678
Toddy palm, 280, 293
Tree of damocles, 497
Tree of medicine, 30
Tree of music, 30
Tree sorrel, 448
Triandra palm, 277–279
Trinidad pachira, 588
Tuberous prickly-pear, 660
Tuna cactus, 660
Tunita, 656
Turkish filbert, 471
Turkish hazel, 471, 475
Tussar silkworm, 245

U
Upside tree, 527

V
Veitchia palm, 257
Velvet opuntia, 656
Vesicular stomatitis virus, 313
Vi apple, 160
Vinepear, 5
Virgin coconut oil (VCO), 306, 309–311, 315, 317, 318, 321, 324
Virgin olive oil (VOO), 318, 520

W
Walnuts, 75, 77, 409, 476
Warm hand, 656
Water coconut, 295, 305–308, 310, 314, 317, 320, 322, 327, 402
Watermelon, 577, 662
Weeping pepper, 153
White dragon fruit, 647, 650, 654
White-flowered silk-cotton tree, 540
White pelon tree, 140
White silk cotton tree, 540
White spot syndrome, 579
Wi apple, 160
Wild almond, 630
Wild areca palm, 277
Wild blueberries, 476
Wild breadnut, 588
Wild cachiman, 221
Wild chataigne, 588
Wild chestnut, 588
Wild cocoa, 584
Wild custard apple, 186, 221
Wild date, 297, 407, 419–421
Wild date palm, 407, 419–421
Wild plum, 166
Wild soursop, 186
Wild sugar apple, 221, 222
Window-leaf, 252
Wine palm, 293, 294, 336, 339, 343, 420, 421
Wooly joint prickly pear, 656

Y
Yellow dragonfruit, 640
Yellow durian, 552
Yellow dye, 158, 495, 515, 516
Yellow-fleshed actinidia, 12
Yellow pitahaya, 640, 641
Yellow plum, 160
Yellowtaper candletree, 508

Scientific Name Index

A

Acajuba occidentalis, 45
Acinetobacter calcoacetica, 157
Actinidia arguta, 5–10
Actinidia arguta var. *arguta*, 10
Actinidia arguta var. *cordifolia*, 5
Actinidia arguta var. *curta*, 5
Actinidia arguta var. *dunnii*, 5
Actinidia arguta var. *giraldii*, 7, 10
Actinidia arguta var. *megalocarpa*, 5
Actinidia arguta var. *purpurea*, 5, 10
Actinidia arguta var. *rufa*, 5
Actinidia callosa var. *rufa*, 5
Actinidia chartacea, 5
Actinidia chinensis, 12–18, 20, 23, 25, 27
Actinidia chinensis f. *jinggangshanensis*, 12
Actinidia chinensis var. *chinensis*, 12
Actinidia chinensis var. *deliciosa*, 20
Actinidia chinensis var. *hispida*, 20
Actinidia chinensis var. *jinggangshanensis*, 12
Actinidia chinensis var. *setosa*, 20
Actinidia cordifolia, 5
Actinidia deliciosa, 7, 10, 13–17, 20–28
Actinidia giraldii, 5
Actinidia latifolia var. *deliciosa*, 20
Actinidia megalocarpa, 5
Actinidia melanandra var. *latifolia*, 5
Actinidia multipetaloides, 12
Actinidia platyphylla, 5
Actinidia purpurea, 5
Actinidia rufa, 7
Actinobacillus actinomycetemcomitans, 37, 578
Adansonia baobab, 527, 528
Adansonia digitata, 527–534, 539
Adansonia digitata var. *congolensis*, 527
Adansonia gibbosa, 536
Adansonia gregorii, 534, 536–539
Adansonia sphaerocarpa, 527
Adansonia sulcata, 527
Adonidia merrillii, 257–259
Aedes aegypti, 59
Aegles marmelos, 501
Aeromonas hydrophila, 451
Aggregatibacter actinomycetemcomitans, 578
Agrimony eupatoria, 729
Agrobacterium tumefaciens, 728
Alcaligenes faecalis, 157, 483
Aleurites moluccana, 328
Alpinia officinarum, 264
Alternaria alternata, 157
Anacardium amilcarianum, 45
Anacardium corymbosum, 64
Anacardium curatellifolium, 45
Anacardium excelsum, 64
Anacardium giganteum, 64
Anacardium humile, 64
Anacardium kuhlmannianum, 45
Anacardium mediterraneum, 45
Anacardium microcarpum, 45
Anacardium microsepalum, 64
Anacardium occidentale, 45–64
Anacardium occidentale var. *americanum*, 45
Anacardium occidentale var. *gardneri*, 45
Anacardium occidentale var. *indicum*, 45
Anacardium occidentale var. *longifolium*, 45
Anacardium othonianum, 45
Anacardium rondonianum, 45
Anacardium subcordatum, 45
Ananassa sativa, 593
Ananas ananas, 593, 594
Ananas bracteatus var. *hondurensis*, 593
Ananas comosus, 593–611
Ananas domestica, 593
Ananas duckei, 593
Ananas parguazensis, 593
Ananas sativus, 593
Ananas sativus var. *duckei*, 593
Ancylocladus coriaceus, 249
Ancylocladus firmus, 249
Ancylocladus minutiflorus, 249
Ancylocladus nodosus, 249
Ancylocladus vriesianus, 249
Annona asiatica, 207
Annona atemoya, 171–174
Annona australis, 180
Annona biflora, 207, 221
Annona bonplandiana, 190
Annona cearaensis, 190
Annona cherimolia, 171, 197
Annona chrysocarpa, 180
Annona cinerea, 207
Annona diversifolia, 176–179
Annona forskahlii, 207
Annona glabra, 180–184, 207

Annona humboldtiana, 180, 201
Annona humboldtii, 180, 201
Annona klainii, 180
Annona laevis, 201
Annona laurifolia, 180
Annona longifolia, 201
Annona lutescens, 201
Annona macrocarpa, 190
Annona marcgravii, 186
Annona microcarpa, 221
Annona montana, 186–189
Annona mucosa, 221
Annona muricata, 174, 186, 188, 190–198, 203, 204
Annona muricata var. *borinquensis* Morales, 190
Annona obtusiflora, 221
Annona obtusifolia, 221
Annona palustris, 180
Annona peruviana, 180
Annona pisonis, 180, 186
Annona pterocarpa, 221
Annona pteropetala, 221
Annona reticulata, 194, 201–205, 221
Annona reticulata var. *mucosa*, 221
Annona senegalensis, 235
Annona sericea, 190
Annona sieberi, 221
Annona sphaerocarpa, 186
Annona squamosa, 171, 194, 198, 205, 207–218
Annona uliginosa, 180
Anona bonplandiana, 190
Anona cearaensis, 190
Anona macrocarpa, 190
Anopheles stephensi, 218
Anopheles subpictus, 59
Antura edulis, 240
Antura hadiensis, 240
Areca alicae, 277
Areca borneensis, 277
Areca catechu, 260–274, 277, 278
Areca catechu var. *dulcissima*, 264
Areca faufel, 260
Areca himalayana, 260
Areca hortensis, 260
Areca humilis, 277
Areca laxa, 277
Areca nagensis, 277
Areca nigra, 260
Areca polystachya, 277
Areca triandra, 277–279
Areca triandra var. *bancana*, 277
Arenga gamuto, 280
Arenga pinnata, 280–284
Arenga saccarifera, 280
Arduina brownii, 240
Arduina campenonii, 240
Arduina edulis, 240
Arduina grandiflora, 237
Arduina inermis, 240
Arduina laxiflora, 240
Arduina macrocarpa, 237

Arduina xylopicron, 240
Artemia salina, 197, 491, 728
Arthrobacter globiformis, 483
Artocarpus integrifolia, 580
Artocarpus odoratissimus, 133
Ascaris galli, 59
Ascaris suum, 707
Asparagus officinalis, 40
Aspergillus flavus, 112, 490
Aspergillus fumigatus, 112
Aspergillus niger, 490, 503, 504, 638
Aspergillus ochraceus, 157
Aspergillus parasiticus, 157
Averrhoa acutangula, 454
Averrhoa bilimbi, 448–452, 465, 467, 468, 470
Averrhoa carambola, 451, 452, 545–462, 467, 470
Averrhoa dolichocarpa, 465–467, 470
Averrhoa leucopetala, 468–470
Averrhoa obtusangula, 448
Averrhoa pentandra, 454
Azadirachta indica, 501, 502

B
Baccaurea macrocarpa, 85, 129, 562
Baccaurea polyneura, 84, 129, 562
Baccaurea reticulata, 84, 129, 562
Bacillus cereus, 112, 150, 157, 477, 483, 491, 503, 522, 539, 705, 706, 728
Bacillus coagulans, 483
Bacillus megaterium, 503
Bacillus pumilus, 112, 312, 522
Bacillus subtilis, 40, 112, 157, 189, 216, 233, 312, 451, 483, 490, 491, 503, 533, 578, 706, 728
Bactris ciliata, 285
Bactris gasipaes, 285–292
Bactris insignis, 285
Bactris macana, 285
Bactris speciosa, 285
Bactris speciosa var. *chichagui*, 285
Bactris utilis, 285
Beneckea natriegens, 157
Berberis, 730
Bignonia africana, 486
Bignonia indica, 497
Bignonia indica pentandra, 497
Bignonia quadripinnata, 497
Biomphalaria glabrata, 58, 197
Biomphalaria straminea, 58
Biomphalaria tenagophila, 58
Bixa acuminata, 515
Bixa americana, 515
Bixa katagensis, 515
Bixa odorata, 515
Bixa orleana Noronha, 515
Bixa orellana, 515–524
Bixa orellana f. *leiocarpa*, 515
Bixa orellana var. *leiocarpa*, 515
Bixa purpurea, 515
Bixa tinctaria, 515

Bixa upatensis, 515
Blattella germanica, 158, 198
Bombax aquaticum, 584
Bombax cumanense, 540
Bombax guineense, 540
Bombax insigne, 584
Bombax macrocarpum, 584
Bombax mompoxense, 540
Bombax occidentale, 540
Bombax orientale, 540
Bombax pentandrum, 540
Bombax rigidifolium, 584
Bombax spectabile, 588
Bombax spruceanum, 588
Borassus flabellifer, 293–299
Borassus flabellifer var. *aethiopicum*, 293
Borassus flabelliformis, 293
Borassus gomutus, 280
Borassus sonneratii, 399
Borassus sundaicus, 293
Borassus tunicatus, 293
Bothrops atrox, 483
Bouea angustifolia, 72
Bouea burmanica, 72
Bouea burmanica var. *kurzii*, 72
Bouea burmanica var. *microphylla*, 72
Bouea burmanica var. *roxburghii*, 72
Bouea diversifolia, 72
Bouea gandaria, 69
Bouea macrophylla, 69–71
Bouea microphylla, 72
Bouea myrsinoides, 72
Bouea oppositifolia, 71–74
Bouvardia terniflora, 509
Brahea serrulata, 438
Brevibacterium ammoniagenes, 58
Brickellia veronicaefolia, 509
Brochothrix thermosphacata, 157
Bromelia ananas, 593
Bromelia comosa, 593
Bursaphelenchus xylophilus, 165

C
Cabucala brachyantha, 240
Cactus campechianus, 656
Cactus chinensis, 660
Cactus cochenillifer, 656
Cactus decumanus, 660
Cactus dillenii, 687
Cactus elongatus, 660
Cactus ficus-indica, 660
Cactus indicus, 683
Cactus lanceolatus, 660
Cactus monacanthos, 683
Cactus opuntia, 660, 687
Cactus opuntia var. *inermis*, 687
Cactus peruvianus, 636, 638
Cactus repandus basionym, 636
Cactus strictus, 687

Caenorhabditis elegans, 728
Calamus salakka, 432
Calamus zalacca, 429, 432
Calappa nucifera, 301
Calloselasma rhodostoma, 113, 266
Callosobruchus maculates, 713
Calosanthes indica, 497
Canariopsis decumana, 616
Canarium amboinense, 619
Canarium articulatum, 633
Canarium beccarii, 624
Canarium caudatifolium, 633
Canarium commune, 619, 622, 630
Canarium crassifolium, 633
Canarium cuspidatum, 633
Canarium decumanum, 616
Canarium gilvescens, 633
Canarium grandistipulatum, 619
Canarium indicum, 619–622, 630
Canarium kadondon, 633
Canarium luzonicum, 622
Canarium mehenbethene, 619
Canarium minahassae, 619
Canarium moluccanum, 619
Canarium montanum, 633
Canarium multifidum, 624
Canarium nungi, 619
Canarium odontophyllum, 624–628
Canarium palawanense, 624
Canarium reticulatum, 633
Canarium rostriferum, 633
Canarium rostriferum var.*cuspidatum*, 633
Canarium shortlandicum, 619
Canarium subtruncatum, 619
Canarium vulgare, 630–632
Canarium zephyrinum, 619
Candida albicans, 25, 150, 312, 375, 483, 490, 491, 503, 504, 533, 578, 658, 677, 706, 728, 776, 788
Candida glabrata, 728
Candida kefyr, 728
Candida krusei, 313, 728
Candida lusitaniae, 728
Candida parapsilosis, 150, 728
Candida tropicalis, 728
Carandas edulis, 240
Carica bourgeaei, 693
Carica chrysopetala, 718
Carica citriformis, 693
Carica cubensis, 693
Carica hermaphrodita, 693
Carica jamaicensis, 693
Carica jimenezii, 693
Carica mamaya, 693
Carica papaya, 492, 610, 693–714, 720
Carica papaya fo. *mamaya*, 693
Carica papaya fo. *portoricensis*, 693
Carica papaya var. *bady*, 693
Carica papaya var. *jimenezii*, 693
Carica peltata, 693
Carica pentagona, 718, 720

Carica pinnatifida, 693
Carica portorricensis, 693
Carica posopora, 693
Carica rochefortii, 693
Carica sativa, 693
Carica × heilbornii, 718
Carica × heilbornii nothovar. *Chrysopetala*, 718
Carica × heilbornii nothovar. *pentagona*, 718
Carissa africana, 237
Carissa brownii F. Muell. nom. illeg., 240
Carissa brownii var. *angustifolia*, 240
Carissa brownii var. *ovata*, 240
Carissa campenonii, 240
Carissa candolleana, 240
Carissa carandas, 237, 240, 243–245
Carissa carandas var. *congesta*, 240, 244
Carissa carandas var. *paucinervia*, 240
Carissa cochinchinensis, 240
Carissa comorensis, 240
Carissa congesta, 240
Carissa coriacea, 240
Carissa cornifolia, 240
Carissa dalzellii, 240
Carissa densiflora, 240
Carissa densiflora var. *microphylla*, 240
Carissa diffusa, 240
Carissa dulcis, 240
Carissa edulis, 240
Carissa edulis f.*nummularis*, 240
Carissa edulis f.*pubescens*, 240
Carissa edulis subsp. *madagascariensis*, 240
Carissa edulis var. *ambungana*, 240
Carissa edulis var. *comorensis*, 240
Carissa edulis subsp. *continentalis*, 240
Carissa edulis var. *densiflora*, 240
Carissa edulis var. *horrida*, 240
Carissa edulis var. *lucubea*, 240
Carissa edulis var. *major*, 240
Carissa edulis var. *microphylla*, 240
Carissa edulis var. *nummularis* 240
Carissa edulis var. *revoluta*, 240
Carissa edulis var. *sechellensis*, 240
Carissa edulis var. *septentrionalis*, 240
Carissa edulis var. *subtrinervia*, 240
Carissa edulis var. *tomentosa*, 240
Carissa gangetica, 240
Carissa grandiflora, 237
Carissa hirsuta, 240
Carissa horrida, 240
Carissa inermis, 240
Carissa lanceolata, 240
Carissa laotica, 240
Carissa laotica var. *ferruginea*, 240
Carissa laxiflora, 240
Carissa macrocarpa, 237–239
Carissa macrophylla, 240
Carissa madagascariensis, 240
Carissa mitis, 240
Carissa obovata, 240
Carissa oleoides, 240

Carissa opaca Stapf ex Haines, 240
Carissa ovata, 240
Carissa ovata var. *pubescens*, 240
Carissa ovata var. *stolonifera*, 240
Carissa papuana, 240
Carissa paucinervia, 240
Carissa pilosa, 240
Carissa praetermissa, 237
Carissa pubescens, 240
Carissa revoluta, 240
Carissa richardiana, 240
Carissa scabra, 240
Carissa sechellensis, 240
Carissa septentrionalis, 240
Carissa spinarum, 240–245
Carissa stolonifera, 240
Carissa suavissima, 240
Carissa tomentosa, 240
Carissa velutina, 241
Carissa villosa, 241
Carissa xylopicron, 241
Carissa yunnanensis, 241
Carolinea insignis, 588
Carolinea macrocarpa, 584
Carolinea princeps, 584
Caryota onusta, 280
Cassuvium pomiferum, 45
Cassuvium reniforme, 45
Ceiba anfractuosa, 540
Ceiba caribaea, 540
Ceiba casearia, 540
Ceiba guineensis, 540
Ceiba guineensis var. *ampla*, 540
Ceiba occidentalis, 540
Ceiba pentandra, 540–548
Ceiba pentandra fo. *albolana*, 540
Ceiba pentandra fo. *grisea*, 540
Ceiba pentandra var. *caribaea*, 540
Ceiba pentandra var. *clausa*, 540
Ceiba pentandra var. *dehiscens*, 540
Ceiba pentandra var. *indica*, 540
Ceiba thonnerii, 540
Ceiba thonningii, 540
Cephalocereus remolinensis, 636
Cereus albispinus, 636
Cereus atroviridis, 636
Cereus grenadensis, 636
Cereus margaritensis, 636
Cereus margaritensis var. *micracanthus*, 636
Cereus megalanthus, 640
Cereus peruvianus, 636, 638
Cereus polyrhizus, 643
Cereus remolinensis, 636
Cereus repandus, 636–638
Cereus triangularis, 650
Cereus triangularis var. *aphyllus*, 650
Cereus triangularis var. *major*, 650
Cereus tricostatus, 650
Cereus trigonus, 650
Cereus trigonus var. *guatemalensis*, 650

Cereus undatus, 650
Chamaerops serrulata, 438
Chilocarpus brachyanthus, 247
Cinnamomum cassia, 264
Citrobacter freundii, 157, 728
Citrullus vulgaris, 110
Citrus aurantifolia, 414
Citrus sinensis, 114, 164, 169, 194, 532, 592
Clarias gariepinus, 381
Clerodendron splendens, 233
Clostridium sporogenes, 157
Clostridium tetani, 112
Cocos indica, 301
Cocos maldivica, 399
Cocos nana, 301
Cocos nucifera, 297, 301–329
Cocos nucifera var. *synphyllica*, 301
Cocos nypa, 402
Colletotrichum gloeosporioides, 676
Colubrina arborescens, 320
Comeurya cumingiana, 75
Condonum malaccense, 160
Connaropsis philippica, 454
Corylus avellana f. *aurea*, 471
Corylus avellana f. *contorta*, 471
Corylus avellana f. *fuscorubra*, 471
Corylus avellana f. *heterophylla*, 471
Corylus avellana f. *pendula*, 471
Corylus avellana var. *aurea*, 471
Corylus avellana var. *contorta*, 471
Corylus avellana var. *fusco-rubra*, 471
Corylus avellana var. *heterophylla*, 471
Corylus avellana var. *pendula*, 471
Corylus imeretica, 471
Corylus maxima, 471, 478
Corylus pontica, 471
Corylus tubulosa, 471
Corynebacterium diptheriae, 490
Corynebacterium xerosis, 58
Corypha obliqua, 438
Corypha repens, 438
Crataegus oxyacantha, 105
Crescentia aculeata, 508
Crescentia acuminata, 480
Crescentia angustifolia, 480
Crescentia arborea, 480
Crescentia cujete, 480–484
Crescentia cujete var. *puberula*, 480
Crescentia cuneifolia, 480
Crescentia edulis, 508
Crescentia fasciculata, 480
Crescentia musaecarpa, 508
Crescentia ovata, 480
Crescentia pinnata, 486
Crescentia plectantha, 480
Crescentia spathulata, 480
Crocidolomia binotalis, 184
Cryptococcus neoformans, 312
Cryptosporidium parvum, 112
Cucumis melo, 110

Culex quinquefasciatus, 218
Curculigo latifolia, 580
Curcuma, 128
Cuscuta chinensis, 501, 502
Cycas inermis, 732
Cycas miquelii, 732
Cycas revoluta, 732–736
Cycas revoluta var. *brevifrons*, 732
Cycas revoluta var. *planifolia*, 732
Cycas revoluta var. *prolifera*, 732
Cycas revoluta var. *robusta*, 732
Cymbopogon citratus, 708

D

Dacryodes rostrata, 633–635
Dactylopius coccus, 657, 685
Damnacanthus esquirolii, 241
Derris scandens, 503
Detarium microcarpum, 532
Diglossophyllum serrulatum, 438
Dracontomelon brachyphyllum, 75
Dracontomelon celebicum, 75
Dracontomelon cumingiana, 75
Dracontomelon cuspidatum, 633
Dracontomelon dao, 75–77
Dracontomelon edule, 75
Dracontomelon lamiyo, 75
Dracontomelon laxum, 75
Dracontomelon mangiferum var. *puberulum*, 75
Dracontomelon mangiferum var. *pubescens*, 75
Dracontomelon sylvestre, 75
Dracontomelum mangiferum, 75
Dracontomelum puberulum, 75
Durio, 557, 570
Durio acuminatissima, 569
Durio affinis, 568
Durio conicus, 550
Durio dulcis, 550–551
Durio foetida, 569
Durio grandiflorus, 558
Durio graveolens, 552–555
Durio kinabaluensis, 556–558
Durio kutejensis, 84, 129, 556, 559–562
Durio kutejensis forma *kinabaluensis*, 556
Durio lowianus, 85, 129, 562
Durio oxleyanus, 558, 563–565
Durio purureus, 558
Durio sabahensis, 558
Durio testudinarum, 558, 566–568
Durio zibethinus, 555, 557, 569–581

E

Echis ocellatus, 546
Elaeis dybowskii, 335
Elaeis guineensis, 335–383, 393
Elaeis guineensis f. *androgyna*, 335
Elaeis guineensis f. *caryolitica*, 335
Elaeis guineensis f. *dioica*, 335

Elaeis guineensis f. *dura*, 335
Elaeis guineensis f. *fatua*, 335
Elaeis guineensis f. *ramosa*, 335
Elaeis guineensis f. *semidura*, 335
Elaeis guineensis f. *tenera*, 335
Elaeis guineensis subsp. *nigrescens*, 335
Elaeis guineensis subsp. *virescens*, 335
Elaeis guineensis var. *albescens*, 335
Elaeis guineensis var. *angulosa*, 335
Elaeis guineensis var. *ceredia*, 335
Elaeis guineensis var. *compressa*, 335
Elaeis guineensis var. *dura*, 395
Elaeis guineensis var. *gracilinux*, 335
Elaeis guineensis var. *idolatrica*, 335
Elaeis guineensis var. *intermedia*, 335
Elaeis guineensis var. *leucocarpa*, 335
Elaeis guineensis var. *macrocarpa*, 335
Elaeis guineensis var. *macrocarya*, 335
Elaeis guineensis var. *macrophylla*, 335
Elaeis guineensis var. *macrosperma*, 335
Elaeis guineensis var. *madagascariensis*, 335
Elaeis guineensis var. *microsperma*, 335
Elaeis guineensis var. *pisifera*, 335, 393–395
Elaeis guineensis var. *repanda*, 335
Elaeis guineensis var. *rostrata*, 335
Elaeis guineensis var. *sempernigra*, 335
Elaeis guineensis var. *spectabilis*, 335
Elaeis guineensis var. *tenera*, 348, 395
Elaeis macrophylla, 335
Elaeis madagascariensis, 335
Elaeis melanococca, 335
Elaeis melanococca var. *semicircularis*, 335
Elaeis nigrescens, 335
Elaeis tenera, 343
Elaeis virescens, 335
Eleiodoxa conferta, 396–397
Eleiodoxa microcarpa, 396
Eleiodoxa orthoschista, 396
Eleiodoxa scortechinii, 396
Eleiodoxa xantholepis, 396
Emblica officinalis, 104
Entamoeba histolytica, 197, 493
Enterobacter aerogenes, 157, 483, 491
Enterobacter cloacae, 706
Enterococcus faecalis, 150, 483, 728
Epidermophyton floccosum, 157
Eriodendron anfractuosum, 540
Eriodendron anfractuosuma a indicum, 540
Eriodendron anfractuosum var. *africanum*, 540
Eriodendron anfractuosum var. *caribaeum*, 540
Eriodendron anfractuosum var. *indicum*, 540
Eriodendron caribaeum, 540
Eriodendron guineense, 540
Eriodendron occidentale, 540
Eriodendron orientale, 540
Eriodendron pentandrum, 540
Erwinia chrysanthemi, 728
Escherichia coli, 17, 25, 57, 112, 113, 150, 157, 169, 179, 216, 233, 298, 312, 451, 483, 491, 503, 522, 533, 578, 609, 610, 658, 705, 706, 728, 729

Evia amara var. *tuberculosa*, 160
Evia dulcis, 160

F
Fonsecaea pedrosoi, 312
Fulchironia senegalensis, 419
Fusarium culmorum, 157
Fusarium descencelulare, 677
Fusarium lateritium, 677
Fusarium moniliforme, 677
Fusarium oxysporum, 677
Fusarium solani, 677

G
Garcinia mangostana, 576
Garcinia prainiana, 84, 129, 562
Geloina coaxans, 405
Gleichenia linearis, 580
Glycyrrhiza uralensis, 264
Gomutus saccharifer, 280
Gossampinus alba, 540
Guanabanus muricatus, 190
Guanabanus palustris, 180
Guanabanus squamosus, 207
Guilelma chontaduro, 285
Guilielma ciliata, 285
Guilielma gasipaes, 285
Guilielma gasipaes var. *chichagui*, 285
Guilielma gasipaes var. *chontaduro*, 285
Guilielma gasipaes var. *coccinea*, 285
Guilielma gasipaes var. *flava*, 285
Guilielma gasipaes var. *ochracea*, 285
Guilielma insignis, 285
Guilielma macana, 285
Guilielma microcarpa, 285
Guilielma speciosa, 285
Guilielma speciosa var. *coccinea*, 285
Guilielma speciosa var. *flava*, 285
Guilielma speciosa var. *mitis*, 285
Guilielma speciosa var. *ochracea*, 285
Guilielma utilis, 285
Gynura pseudochina var. *hispida*, 503

H
Haemonchus contortus, 217
Helicobacter pylori, 15, 25, 58, 149
Heligmosomoides polygyrus, 707
Hemisantiria rostrata, 633
Henosepilachna vigintiocto-punctata, 184
Hibiscus rosa-sinensis, 580
Hippoxylon indicum, 497
Histoplasma capsulatum, 157
Holarrhena antidysenterica, 105
Hylocereus, 641, 654
Hylocereus costaricensis, 641
Hylocereus guatemalensis, 650
Hylocereus megalanthus, 640–641

Hylocereus polyrhizus, 643–648, 654
Hylocereus setaceus, 641
Hylocereus tricostatus, 650
Hylocereus undatus, 644, 645, 650–655
Hypericum perforatum, 36

I
Ipomoea aquatica, 500

J
Jasminonerium africanum, 237
Jasminonerium densiflorum, 241
Jasminonerium dulce, 241
Jasminonerium edule, 241
Jasminonerium grandiflorum, 237
Jasminonerium inerme, 241
Jasminonerium laxiflorum, 241
Jasminonerium macrocarpum, 237
Jasminonerium madagascariense, 241
Jasminonerium ovatum, 241
Jasminonerium pubescens, 241
Jasminonerium sechellense, 241
Jasminonerium suavissimum, 241
Jasminonerium tomentosum, 241
Jasminonerium xylopicron 241
Jatropha curcas, 524
Juniperus canadensis, 723
Juniperus communis L. var., 723–730
Juniperus communis subsp. nana, 723
Juniperus communis var. *charlottensis*, 726
Juniperus communis var. *communis*, 723
Juniperus communis var. *depressa*, 723, 727
Juniperus communis var. *jackii*, 726
Juniperus communis var. *megistocarpa*, 726
Juniperus communis var. *montana*, 723
Juniperus communis var. *saxatilis*, 723, 727
Juniperus densa, 723
Juniperus nana, 723
Juniperus sibirica, 723

K
Kigelia, 494, 495
Kigelia abyssinica, 486
Kigelia acutifolia, 486
Kigelia aethiopica, 486
Kigelia aethiopica var. *stenocarpa*, 486
Kigelia aethiopium, 486
Kigelia aethopica var. *abyssinica*, 486
Kigelia aethopica var. *bornuensis*, 486
Kigelia aethopica var. *usambarica*, 486
Kigelia africana, 486–495
Kigelia africana var. *aethiopica*, 486
Kigelia africana var. *elliptica*, 486
Kigelia angolensis, 486
Kigelia elliottii, 486
Kigelia elliptica, 486
Kigelia ikbaliae, 486
Kigelia impressa, 486
Kigelia lanceolata, 486
Kigelia moosa, 486
Kigelia pinnata, 486, 489–493
Kigelia pinnata (*Artemia salina*), 491
Kigelia somalensis, 486
Kigelia spragueana, 486
Kigelia talbotii, 486
Kigelia tristis, 486
Klebsiella, 113
Klebsiella oxytoca, 729
Klebsiella pneumoniae, 112, 157, 451, 706

L
Lactobacillus pentosus, 578
Lactuca sativa, 34
Lahia kutejensis, 559
Lannea macrocarpa, 532
Lannea welwitschii, 494
Lawsonia inermis, 105
Leishmania amazonensis, 321
Leishmania braziliensis, 197
Leishmania donovani, 150, 225, 728
Leishmania panamensis, 197
Leuconostoc cremoris, 157
Listeria monocytogenes, 157, 451, 477, 729
Litsea petiolata, 501–502
Lodoicea callipyge, 399
Lodoicea maldivica, 399–401
Lodoicea seychellarum, 399
Lodoicea sonneratii, 399
Loranthus europeas, 414

M
Maackia amurensis, 38
Mamestra brassica, 184
Mangifera amba, 87
Mangifera anisodora, 87
Mangifera arbor, 87
Mangifera austroindica, 87
Mangifera austro-yunnanensis, 87
Mangifera balba, 87
Mangifera bompardii, 87
Mangifera caesia, 79–81, 121–123
Mangifera caesia var. *kemanga*, 121
Mangifera caesia var. *verticillata*, 79
Mangifera caesia var. *wanji*, 79
Mangifera domestica, 87
Mangifera equina, 87
Mangifera foetida, 82–86, 127, 129, 133, 562
Mangifera foetida var. *bakkill*, 127
Mangifera foetida var. *bombom*, 127
Mangifera foetida var. *kawini*, 127
Mangifera foetida var. *mollis*, 127
Mangifera foetida var. *odorata*, 127
Mangifera fragrans, 87
Mangifera gladiata, 87

Mangifera indica, 87–115, 124, 126
Mangifera integrifolia, 87
Mangifera kemanga, 81, 121–123
Mangifera kukula, 87
Mangifera langong, 135
Mangifera laurina, 124–126
Mangifera linnaei, 87
Mangifera longipes, 124
Mangifera longipetiolata, 135
Mangifera maingayi, 135
Mangifera maritima, 87
Mangifera mekongensis, 87
Mangifera montana, 87
Mangifera oblongifolia, 127
Mangifera odorata, 84, 85, 127–130, 133, 562
Mangifera odorata var. *pubescens*, 127
Mangifera oppositifolia, 72
Mangifera oppositifolia var. *microphylla*, 72
Mangifera oppositifolia var. *roxburghii*, 72
Mangifera orophila, 87
Mangifera oryza, 87
Mangifera pajang, 85, 129, 131–134
Mangifera parih, 124
Mangifera polycarpa, 121
Mangifera quadrifida, 135–137
Mangifera quadrifida var. *spathulaefolia*, 135
Mangifera racemosa, 87
Mangifera rigida, 135
Mangifera rostrata, 87
Mangifera rubra, 87
Mangifera rubropetala, 87
Mangifera sativa, 87
Mangifera siamensis, 87
Mangifera similis, 138–139
Mangifera spathulaefolia, 135
Mangifera sugenda, 87
Mangifera sumatrana, 124
Mangifera sylvatica, 87
Mangifera torquenda, 138
Mangifera verticillata, 79
Mangifera viridis, 87
Martinezia ciliata, 285
Matania laotica, 72
Matisia cordata, 590, 592
Maytenus senegalensis, 494
Mediocactus megalanthus, 640
Mentha arvensis, 708
Micrococcus flavus, 728
Micrococcus luteus, 189, 578, 728, 729
Micrococcus roseus, 483
Micromelum minutum, 501, 502
Microsporum canis, 157, 728
Microsporum ferrugineum, 157
Microsporum gypseum, 157, 728
Molinema desetae, 197
Monstera borsigiana, 252
Monstera deliciosa, 252–256
Monstera deliciosa var. *borsigiana*, 252
Monstera deliciosa var. *sierrana*, 252
Monstera lennea, 252

Monstera tacanaensis, 252
Moringa oleifera, 500, 501
Muehlenbeckia platyclada, 503
Mycobacterium aurum, 728
Mycobacterium fortuitum, 728
Mycobacterium phlei, 483, 533, 728
Mycobacterium rodochrus, 483
Mycobacterium smegmatis, 233, 483, 728
Mycobacterium tuberculosis, 728

N
Naja naja kaouthia, 113, 266
Neisseria gonorrhoeae, 510, 522
Nenga nagensis, 277
Nephelium lappaceum, 580
Nephelium longan, 580
Nephelium mutabile, 580
Nephotettix cincticeps, 184
Nicotiana glauca, 678
Nipa fruticans, 402
Nipa litoralis, 402
Nocardia brasiliensis, 197
Nopalea cochenillifera, 656–659
Normanbya merrillii, 257
Nothoprotium sumatranum, 140
Nypa arborescens, 402
Nypa fruticans, 402–405

O
Opuntia, 656, 658, 660–662, 664, 666, 668, 677
Opuntia amyclaea, 660
Opuntia anahuacensis, 687
Opuntia chinensis, 660
Opuntia cochenillifera, 656
Opuntia compressa, 660
Opuntia cordobensis, 660
Opuntia crassa, 660
Opuntia decumana, 660
Opuntia dillenii, 687, 689
Opuntia dillenii var. *tehuantepecana*, 687
Opuntia elongata, 660
Opuntia ficus-barbarica, 660
Opuntia ficus-indica, 660–678, 683, 684, 689
Opuntia ficus-indica f. *inermis*, 673
Opuntia ficus-indica var. *gymnocarpa*, 660
Opuntia ficus-indica var. *saboten*, 666–669, 672–676
Opuntia gymnocarpa, 660
Opuntia hispanica, 660
Opuntia horrida, 687
Opuntia inermis, 687
Opuntia lanceolata, 660
Opuntia lemaireana, 683
Opuntia maritima, 687
Opuntia maxima, 660
Opuntia megacantha, 660
Opuntia monacantha, 683–685
Opuntia occidentalis, 660
Opuntia paraguayensis, 660

Opuntia parvula, 660
Opuntia robusta, 669
Opuntia stricta, 668, 684, 687–691
Opuntia stricta var. *dillenii*, 687
Opuntia tuna-blanca, 660
Opuntia undosa, 660
Opuntia undulata, 660, 664, 668, 689
Opuntia vulgaris, 660, 683
Opuntia vulgaris var. *balearica*, 687
Opuntia vulgaris var. *lemaireana*, 683
Orellana americana, 515,
Orellana americana var. *leiocarpa*, 515
Orellana americana var. *normalis*, 515
Orellana orellana, 515
Oreochromis niloticus, 379, 381
Oroxylum indicum, 497–505

P
Pachira aquatica, 584–587, 589
Pachira aquatica var. *occidentalis*, 584
Pachira aquatica var. *surinamensis*, 584
Pachira grandiflora, 584
Pachira insignis, 588–589
Pachira longiflora, 584
Pachira macrocarpa, 584
Pachira pustulifera, 584
Pachira spruceana, 584, 588
Pachira villosula, 584
Paeonia suffruticosa, 264
Paliurus dao, 75
Paliurus edulis, 75
Paliurus lamiyo, 75
Palma cocos, 301
Palma oleosa, 335
Papaya bourgeaei, 693
Papaya carica, 492, 610, 693–714, 720
Papaya cimarrona, 693
Papaya citriformis, 693
Papaya communis, 693
Papaya cubensis, 693
Papaya cucumerina, 693
Papaya edulis, 693
Papaya edulis var. *macrocarpa*, 693
Papaya edulis var. *pyriformis*, 693
Papaya hermaphrodita, 693
Papaya papaya, 693
Papaya peltata, 693
Papaya rochefortii, 693
Papaya vulgaris, 693
Parinari curatelifolia, 532
Parkia biglobosa, 529
Parmentiera aculeata, 508–511
Parmentiera cereifera, 512–514
Parmentiera edulis, 508–510
Parmentiera foliolosa, 508
Parmentiera lanceolata, 508
Penaeus monodon, 578
Penicillium chrysogenum, 57, 381
Pentaspadon minutiflora, 140

Pentaspadon moszkowskii, 140
Pentaspadon motleyi, 140–141
Pentaspadon officinalis, 140
Pherentima posthuma, 205
Philodendron anatomicum, 252
Phoenix abyssinica, 419
Phoenix baoulensis, 419
Phoenix comorensis, 419
Phoenix dactylifera, 407–417
Phoenix djalonensis, 419
Phoenix dybowskii, 419
Phoenix equinoxialis, 419
Phoenix leonensis, 419
Phoenix reclinata, 419–422
Phoenix reclinata var. *comorensis*, 419
Phoenix reclinata var. *madagascariensis*, 419
Phoenix reclinata var. *somalensis*, 419
Phoenix spinosa, 419
Phoenix sylvestris, 297
Pholidocarpus tunicatus, 293
Phoneix reclinata, 419–422
Pilocereus albispinus, 636
Pilocereus remolinensis, 636
Pilocereus repandus, 636
Pimela decumana, 616
Piper betle, 262, 708
Piper nigrum, 547
Piptanthocereus peruvianus, 636
Pistachio vera, 148
Pistacia nigricans, 142
Pistacia officinarum, 142
Pistacia reticulata, 142
Pistacia terebinthus, 142
Pistacia trifolia, 142
Pistacia variifolia, 142
Pistacia vera, 142–151
Pityrosporum ovale, 58
Plasmodium berghei berghei, 235
Plasmodium falciparum, 112, 150, 197, 235, 493, 771
Plasmodium yoelii nigeriensis, 112
Platyopuntia cordobensis, 660
Platyopuntia ficus-indica, 660
Plutella xylostella, 184, 217
Plutella xylostella, 184
Pomum draconum, 75
Pomum draconum silvestre, 75
Porphyromonas gingivalis, 37, 57
Poupartia dulcis, 160
Poupartia mangifera, 75
Propionibacterium acnes, 57, 58
Proteus spp., 150
Proteus vulgaris, 113, 157, 491, 578, 706,
Pseudomonas aeruginosa, 111, 112, 150, 157, 179, 216, 233, 312, 483, 491, 503, 578, 706, 729
Pseudomonas fluorescens, 483, 539
Psidium cattleianum, 459
Psidium guajava, 59, 459
Psuedomonas aeruginosa, 533
Ptychosperma polystachyum, 277
Pullularia pullularis, 490

Q
Quararibea cordata, 590–592

R
Raphanus sativus, 34
Rhus novoguineensis, 140
Ricinus communis, 38
Rollinia curvipetala, 221
Rollinia deliciosa, 221
Rollinia exsucca, 197
Rollinia jimenezii, 221
Rollinia jimenezii var. *nelsonii*, 221
Rollinia mucosa, 221–225
Rollinia mucosa subsp. *Aequatorialis*, 221
Rollinia mucosa subsp. *Portoricensis*, 221
Rollinia mucosa var. *macropoda*, 221
Rollinia mucosa var. *neglecta*, 221
Rollinia neglecta, 221
Rollinia obtusiflora, 221
Rollinia orthopetala, 221
Rollinia permensis, 221
Rollinia pittieri, 197
Rollinia pterocarpa, 221
Rollinia pulchrinervia, 221
Rollinnia sieberi, 221

S
Sabal serrulata, 438
Saccharomyces cerevisiae, 53, 57, 63, 112, 578,
Saguerus australasicus, 280
Saguerus pinnatus, 280
Saguerus pinnatus saccharifer, 280
Saguerus rumphii, 280
Saguerus saccharifer, 280
Sagus gomutus, 280
Salacca affinis, 423–424
Salacca affinis var. *borneensis*, 423
Salacca beccarii, 429
Salacca blumeana, 432
Salacca borneensis, 423
Salacca conferta, 396
Salacca dubia, 423
Salacca edulis, 432, 435, 436
Salacca edulis var. *amboinensis*, 432
Salacca glabrescens, 425–426
Salacca macrostachya, 429
Salacca magnifica, 427–428
Salacca rumphii, 432
Salacca scortechinii, 396
Salacca sumatrana, 437
Salacca wallichiana, 429–431
Salacca zalacca, 424, 425, 432–437
Salacca zalacca var. *amboinensis*, 432, 437
Salacca zalacca var. *zalacca*, 432
Salakka edulis, 432
Salmonella, 61, 413, 521, 533
Salmonella agona, 112
Salmonella enterica var. *thyphimurium*, 658
Salmonella enteritidis, 169, 442, 728
Salmonella paratyphi, 503
Salmonella sp., 533
Salmonella typhi, 169, 179, 233, 491, 706
Salmonella typhimurium, 60, 234, 298, 451, 491, 504, 520, 578, 729, 735
Sambucus alba, 30
Sambucus arborescens, 30
Sambucus canadensis, 33
Sambucus ebulus, 38
Sambucus florida, 30
Sambucus laciniata, 30
Sambucus medullosa, 30
Sambucus nigra, 30–41
Sambucus pyramidata, 30
Sambucus virescens, 30
Sambucus vulgaris, 30
Santiria montana, 633
Santiria rostrata, 633
Santiria samarensis, 633
Saponaria officinalis, 36
Schinus angustifolia, 153
Schinus areira, 153
Schinus bituminosus, 153
Schinus huigan, 153
Schinus molle, 153–158
Schinus molle var. *areira*, 153
Schinus molle var. *argentifolius*, 153
Schinus molle var. *huigan*, 153
Schinus molle var. *huyngan*, 153
Schinus occidentalis, 153
Schinus terebinthifolius, 155, 156
Selenicereus inermis, 641
Selenicereus megalanthus, 640
Serenoa repens, 438–445
Serenoa repens f. *glauca*, 438
Serenoa serrulata, 438
Serratia marcescens, 157
Serratia sp., 728
Shigella boydii, 503
Shigella dysenteriae, 169, 503, 522
Shigella flexneri, 169, 706
Shigella sonnei, 728
Sitophilus oryzae, 217
Sophia carolina, 584
Sotor aethiopium, 486
Spathodea indica, 497
Spondias cirouella, 166
Spondias crispula, 166
Spondias cytherea, 160–165
Spondias dulcis, 160, 166
Spondias mangifera, 160
Spondias mangifera var. *tuberculosa*, 160
Spondias mexicana, 166
Spondias mombin, 166
Spondias myrobalanus, 166
Spondias nigrescens, 166
Spondias purpurea, 166–170
Spondias purpurea var. *munita*, 166
Spondias radlkoferi, 166

Staphylococcus, 706
Staphylococcus aureus, 57, 58, 111, 112, 150, 216, 233, 312, 313, 318, 451, 477, 483, 490, 491, 503, 522, 533, 578, 706, 728, 729, 735
Staphylococcus burahol, 208–210
Staphylococcus citreus, 112
Staphylococcus epidermidis, 578, 728
Stelechocarpus burahol, 227–229
Stenocereus peruvianus, 636
Straphylococcus aureus, 233, 451
Streptococcus acidominimus, 578
Streptococcus agalactiae, 451, 578
Streptococcus aureus, 58, 112–113
Streptococcus bovis, 578
Streptococcus dysagalactiae, 578
Streptococcus faecalis, 533, 705
Streptococcus mitis, 112
Streptococcus mutans, 57, 58, 112, 149, 578
Streptococcus pneumoniae, 112, 483
Streptococcus pyogenes, 233, 483
Streptococcus salivarious, 149
Streptococcus sanguis, 149
Streptococcus sobrinus, 149
Streptococcus uberis, 578
Strychnos spinosa, 532
Sublimia areca, 260
Subpilocereus atroviridis, 636
Subpilocereus grenadensis, 636
Subpilocereus margaritensis, 636
Subpilocereus margaritensis var. *micracanthus*, 636
Subpilocereus remolinensis, 636
Subpilocereus repandus, 636
Swertia chirata, 105
Syzygium cumini, 459
Syzygium jambos, 85, 129, 562
Syzygium malaccense, 85, 129, 562

T

Tabernaemontana macrocarpa, 249
Tanaecium pinnatum, 486
Tecoma africana, 486
Terminalia, 165
Terminalia bellerica, 57
Tetranychus urticae, 184
Thermoascus aurantiacus, 112
Tornelia fragrans, 252
Triatoma infestans, 158
Tribolium castaneum, 158
Trichinella spiralis, 113
Trichoderma reesei, 112
Trichoderma viride, 341
Trichophyton equinum, 157
Trichophyton mentagrophytes, 157, 728
Trichophyton rubrum, 157, 728
Trichophyton tonsurans, 157
Trichoplusia ni, 217
Tripinnaria africana, 486
Trochostigma argutum, 5
Trochostigma rufa, 5

Trogoderma granarium, 158
Tropidopetalum javanicum, 69
Trypanosoma brucei, 235, 493, 533, 545
Trypanosoma brucei brucei, 493, 545
Trypanosoma brucei rhodesiense, 493
Trypanosoma congolense, 533
Trypanosoma cruzi, 197

U

Unona macrocarpa, 231
Urnularia rufescens, 247
Uvaria burahol, 227
Uvaria chamae, 231–236
Uvaria cristata, 231
Uvaria cylindrica, 231
Uvaria echinata, 231
Uvaria nigrescens, 231

V

Vahea angustifolia, 247
Vasconcellea cundinamarcensis, 721
Vasconcellea Xheilbornii nothovar. cgrysopetala, 718
Vasconcellea peltata, 693
Vasconcellea stipulata, 721
Vasconcellea weberbaueri, 721
Vasconcellea xheilbornii, 718–721
Veitchia merrillii, 257
Vibrio cholerae, 609
Vibrio harveyi, 578
Vibrio mimicus, 503
Vibrio parahemolyticus, 503
Vibrio spp., 233

W

Warmingia pauciflora, 166
Willughbeia angustifolia var. *gracilior*, 247
Willughbeia angustifolia, 247–248
Willughbeia apiculata, 247
Willughbeia burbidgei, 249
Willughbeia coriacea, 249–250
Willughbeia elmeri, 247
Willughbeia firma, 249
Willughbeia firma var. *oblongifolia*, 249
Willughbeia firma var. *macrophylla*, 249
Willughbeia firma var. *obtusifolia*, 249
Willughbeia firma var. *platyphylla*, 249
Willughbeia minutiflora, 249
Willughbeia nodosa, 249
Willughbeia rufescens, 247
Willughbeia vrieseana, 249
Willughbeiopsis rufescens, 247

X

Xanthomonas phaseoli, 728
Ximenia americana, 532
Xylon pentandrum, 540

Xylopia aromatica, 197
Xylopia frutescens, 207
Xylopiastrum macrocarpum, 231

Y
Yersinia enterocolitica, 728

Z
Zalacca edulis, 576
Zingiber officinalis, 414
Zingiber sp., 498
Ziziphus mauritiana, 532

Printed by Publishers' Graphics LLC
BT20121026.19.32.129